RC
376.5
G46
2003

# GENETICS OF
MOVEMENT DISORDERS

# GENETICS OF MOVEMENT DISORDERS

Edited by
## STEFAN-M. PULST

*Carmen and Louis Warschaw Chair in Neurology*
*Division of Neurology, Cedars-Sinai Medical Center*
*Professor of Medicine and Neurobiology, UCLA School of Medicine*
*Los Angeles, California 90048*

An imprint of Elsevier Science

Amsterdam   Boston   London   New York   Oxford   Paris
San Diego   San Francisco   Singapore   Sydney   Tokyo

This book is printed on acid-free paper. ∞

Copyright © 2003, Elsevier Science (USA).

All Rights Reserved.
No part of this publication may be reproduced or transmitted in any form or by any means, electronic or mechanical, including photocopy, recording, or any information storage and retrieval system, without permission in writing from the publisher.

Requests for permission to make copies of any part of the work should be mailed to:
Permissions Department, Academic Press, 6277 Sea Harbor Drive,
Orlando, Florida 32887-6777

Academic Press
*An imprint of Elsevier Science.*
525 B Street, Suite 1900, San Diego, California 92101-4495, USA
http://www.academicpress.com

Academic Press
84 Theobald's Road, London WC1X 8RR, UK
http://www.academicpress.com

Library of Congress Catalog Card Number: 2002109272

International Standard Book Number: 0-12-566652-7

PRINTED IN THE UNITED STATES OF AMERICA
02  03  04  05  06  07  MM  9  8  7  6  5  4  3  2  1

*To Wolfgang, Johannes, Sebastian, and Tilman with whom I have shared so much more than just the same Y chromosome.*

# Contents

Contributors xv
Foreword xix
Preface xxi

## 1. Introduction to Medical Genetics and Methods of DNA Testing
STEFAN-M. PULST

I. Concepts and Terminology 2
II. Patterns of Inheritance 2
III. Molecular Genetic Tools 11
IV. Molecular Genetic Testing 14
V. Animal Models 15
VI. Web-Based Information for Genetic Diagnosis and Testing 16
References 17

## 2. Inherited Ataxias: An Introduction
STEFAN-M. PULST

I. Classification 19
II. Recently Identified Ataxias 22
III. Worldwide Prevalence 23
IV. Mechanisms of Disease 25
V. Genotype/Phenotype Correlations and Modifying Loci 29
VI. Progression and Treatment 30
References 30

## 3. Spinocerebellar Ataxia 1 (SCA1)
HARRY T. ORR

I. Summary 35
II. SCA1—The Phenotype 35
III. The *SCA1* Gene 36
IV. Models of Disease 38
V. Treatment 41
References 41

## 4. Spinocerebellar Ataxia 2 (SCA2)
STEFAN-M. PULST

I. Summary 45
II. Phenotype 45
III. Normal and Abnormal Gene Function 47
IV. Diagnosis 50
V. Neurophysiology 51
VI. Neuroimaging 51
VII. Neuropathology 51
VIII. Animal Models 51
IX. Genotype/Phenotype/Modifying Alleles 52
X. Treatment 53
References 53

## 5. Spinocerebellar Ataxia 3— Machado-Joseph Disease (SCA3)
HENRY PAULSON AND S. H. SUBRAMONY

I. Historical Introduction 57
II. Prevalence of MJD 58
III. Phenotype 58
IV. Diagnosis 60
V. Neuropathology 61
VI. Molecular Genetics 62
VII. Phenotype-Genotype Correlation 62
VIII. Pathogenic Mechanisms and Models 63
IX. Treatment 65
References 65

## 6. Spinocerebellar Ataxia Type 4 (SCA4)
HIDEHIRO MIZUSAWA

I. Summary   71
II. Phenotype   71
III. Gene Locus   71
IV. Diagnostic and Ancillary Tests   72
V. Neuropathology   72
VI. ADCCA or Pure Cerebellar Ataxia Linked to SCA4 Locus   72
References   73

## 7. Spinocerebellar Ataxia 5 (SCA5)
LAURA P. W. RANUM, KATHERINE A. DICK, AND J. W. DAY

I. Introduction   75
II. Anticipation   77
III. Genetic and Physical Mapping   77
IV. Repeat Expansion Detection and Rapid Cloning   78
V. Clinical Features   78
VI. Neuroimaging and Neuropathology   79
VII. Conclusions   79
References   80

## 8. Spinocerebellar Ataxia 6 (SCA6)
JOANNA C. JEN

I. Introduction   81
II. Clinical Features   81
III. Genetics   81
IV. Diagnosis   82
V. Molecular Pathogenesis   82
VI. Neuropathology   83
VII. Animal Models   83
VIII. Treatment   83
References   83

## 9. Spinocerebellar Ataxia 7 (SCA7)
ANNE-SOPHIE LEBRE, GIOVANNI STEVANIN, AND ALEXIS BRICE

I. Summary   85
II. Phenotype   85
III. Gene   86
IV. Diagnostic and Ancillary Tests   87
V. Neuroimaging   88
VI. Neuropathology   88
VII. Cellular and Animal Models of Disease   89
VIII. Genotype/Phenotype Correlations/Modifying Alleles   90
IX. Treatment   91
X. Conclusion   91
References   92

## 10. Spinocerebellar Ataxia 8 (SCA8)
MICHAEL D. KOOB

I. Summary   95
II. Phenotype   95
III. Gene   96
IV. Diagnostic and Ancillary Tests   97
V. Cellular and Animal Models of Disease   98
VI. Genotype/Phenotype Correlation and Modifying Alleles   98
VII. Treatment   101
References   101

## 11. Spinocerebellar Ataxia 10 (SCA10)
TOHRU MATSUURA AND TETSUO ASHIZAWA

I. Summary   103
II. Phenotype   103
III. The SCA10 Gene   104
IV. Instability of the Expanded ATTCT Repeat   108
V. Diagnosis   108
VI. Genotype-Phenotype Correlation   110
VII. Population Genetics   111
VIII. Models and Disease Mechanism of the ATTCT Expansion   111
IX. Treatment   113
References   113

## 12. Spinocerebellar Ataxia 11 (SCA11)
HEMA VAKHARIA, MIN-KYU OH, AND STEFAN-M. PULST

I. Summary   117
II. Phenotype   117
III. Gene   118
IV. Neuroimaging and Ancillary Tests   119
References   119

## 13. Spinocerebellar Ataxia 12 (SCA12)
SUSAN E. HOLMES, SHWETA CHOUDHRY, CHISTOPHER A. ROSS, ELIZABETH O'HEARN, ACHAL K. SRIVASTAVA, RUSSELL L. MARGOLIS, SAMIR K. BRAHMACHARI, AND SATISH JAIN

I. Introduction   121
II. Phenotype of SCA12   122
III. Normal and Abnormal Gene Function   123
IV. Diagnosis   128

V. Treatment 130
References 130

## 14. Spinocerebellar Ataxia 13, 14, and 16
HIROTO FUJIGASAKI, ALEXANDRA DÜRR, GIOVANNI STEVANIN, AND ALEXIS BRICE

I. Summary 133
II. Phenotype 134
III. Gene 134
IV. Diagnostic and Ancillary Tests 136
V. Neuroimaging 136
VI. Neuropathology 136
VII. Cellular and Animal Models of Disease 137
VIII. Genotype/Phenotype Correlation/Modifying Alleles 137
IX. Treatment 137
References 137

## 15. Spinocerebellar Ataxia 17 (SCA17)
SHOJI TSUJI

I. Summary 139
II. Phenotype 139
III. Gene 140
IV. Diagnosis 140
V. Neuropathology 140
VI. Neuroimaging 140
VII. Cellular and Animal Models of Disease 140
VIII. Treatment 140
References 140

## 16. Dentatorubral-Pallidoluysian Atrophy (DRPLA)
SHOJI TSUJI

I. Phenotype 143
II. Molecular Genetics of DRPLA 144
III. Diagnostic and Ancillary Tests 145
IV. Cellular and Animal Models of Disease 146
V. Treatment 148
References 148

## 17. Ataxia in Prion Diseases
LEV G. GOLDFARB

I. The *PRNP* Gene and Protein Products 151
II. Phenotypes 153
III. Diagnostic and Ancillary Tests 159
IV. Cellular and Animal Models of Disease 160

V. Treatment and Management 161
References 161

## 18. Friedreich Ataxia
MASSIMO PANDOLFO

I. Summary 165
II. Phenotype 165
III. Gene 169
IV. Diagnostic and Ancillary Tests 170
V. Pathology 173
VI. Cellular and Animal Models of Disease 174
VII. Genotype/Phenotype Correlations/Modifying Alleles 174
VIII. Treatment 174
References 175

## 19. Familial Ataxia with Isolated Vitamin E Deficiency (AVED)
FAYCAL HENTATI, RIM AMOURI, MONCEF FEKI, SANA GABSI-GHERAIRI, AND SAMIR BELAL

I. Introduction 179
II. Phenotype 180
III. Gene 181
IV. Diagnostic and Ancillary Tests 183
V. Cellular and Animal Models of the Disease 184
VI. Genotype/Phenotype Correlations—Modifying Alleles 185
VII. Treatment 185
References 185

## 20. Autosomal Recessive Spastic Ataxia of Charlevoix-Saguenay (ARSACS/SACS)—No Longer a Local Disease
ANDREA RICHTER

I. Phenotype 189
II. Gene 191
III. Diagnostic and Ancillary Tests 191
IV. Neuroimaging 192
V. Neuropathology 192
VI. Genotype/Phenotype Correlations 192
References 193

## 21. Ataxia–Telangiectasia
KAI TREUNER AND CARROLEE BARLOW

I. Phenotype 195
II. ATM Gene and Function Based on Human Data 197

III. Biochemical Targets of ATM Kinase Activity 197
IV. Diagnosis 198
V. Cellular and Animal Models of the Disease 198
VI. Animal Models of A–T 200
VII. Genotype/Phenotype Correlations 201
VIII. Treatment 202
IX. Syndromes Related to A–T 202
References 202

## 22. Episodic and Intermittent Ataxias
JOANNA C. JEN AND ROBERT W. BALOH

I. Clinical Features 205
II. Genetics 207
III. Diagnosis 208
IV. *In Vitro* and *In Vivo* Models 209
V. Treatment 210
References 211

## 23. Multiple System Atrophy
CLIFF SHULTS AND SID GILMAN

I. Introduction 213
II. Clinical Features 214
III. Diagnosis of MSA 217
IV. Neuroimaging 220
V. Pathology of MSA 223
VI. Epidemiology and Environmental Risk Factors 224
VII. Genetics 225
VIII. Management and Treatment 226
References 226

## 24. Metabolic and Mitochondrial Ataxias
ENRICO BERTINI, CARLO DIONISI-VICI, AND MASSIMO ZEVIANI

I. Ataxia in Mitochondrial Disorders 231
II. Ataxia in Lipid Disorders 237
III. Ataxia in Lysosomal Disorders 240
IV. Ataxias Associated with Other Metabolic Disorders 244
References 248

## 25. Diagnostic Evaluation of Ataxic Patients
SUSAN L. PERLMAN

I. Defining the Neurologic Phenotype in Patients with Ataxia as the Primary Symptom 254
II. Determining whether the Disease is Genetic 263
References 267

## 26. Parkinson's Disease: Genetic Epidemiology and Overview
CONNIE MARRAS AND CAROLINE TANNER

I. Introduction 273
II. The Clinical and Pathological Features of Parkinson's Disease 274
III. Diagnosis of Parkinson's Disease 274
IV. The Treatment of Parkinson's Disease 275
V. Challenges Investigating the Etiology of Parkinson's Disease 275
VI. Studies of Familial Aggregation 276
VII. Twin Studies 280
VIII. Single Gene Associations 281
IX. Conclusion 283
References 283

## 27. PARK1 and α-Synuclein: A New Era in Parkinson's Research
J. WILLIAM LANGSTON, LAWRENCE I. GOLBE, AND SEUNG-JAE LEE

I. Introduction 287
II. The Contursi Kindred 288
III. The Clinical Phenotype of the A53T Mutation 289
IV. Parkinsonism Due to a A30P Mutation 291
V. Gene Function 292
VI. Aggregation of α-Synuclein 294
VII. Diagnosis 298
VIII. Neuroimaging 298
IX. Animal Models of α-Synucleinopathies 298
X. Is the A53T Phenotype Parkinson's Disease? 299
XI. Treatment 300
XII. Conclusions 300
References 300

## 28. Parkin Mutations (Park2)
YOSHIKUNI MIZUNO, SHIUCHI ASAKAWA, TOSHIAKI SUZUKI, NOBUTAKA HATTORI, SHINSEI MINOSHIMA, TOMOKI CHIBA, HIROYO YOSHINO, NOBUYOSHI SHIMIZU, AND KEIJI TANAKA

I. Introduction 305
II. Gene 306
III. Diagnostic and Ancillary Tests 308
IV. Neuroimaging 311
V. Neuropathology 311
VI. Cellular and Animal Models 312
VII. Genotype and Phenotype Correlations/Modifying Alleles 312
VIII. Treatment 312
References 312

## 29. PARK3, Ubiquitin Hydrolase-L1 and Other PD Loci
REJKO KRÜGER AND OLAF RIESS

I. *PARK3*   315
II. *PARK4*   316
III. *PARK5*   317
IV. *PARK6*   317
V. *PARK7*   319
VI. *PARK8* and *PARK9*   319
VII. Other Linked PD Loci   319
References   321

## 30. *tau* Genetics in Frontotemporal Lobe Dementia, Progressive Supranuclear Palsy, and Corticobasal Degeneration
JOSEPH J. HIGGINS

I. Introduction   325
II. Anatomy of the *tau* Gene   326
III. *tau* Genetics and Molecular Function   329
IV. Clinical Phenotypes Caused by *tau* Gene Mutations   331
V. Clinical Genetics in FTD, PSP, and CBD   333
VI. *tau* Genetics and Transgenic Models of Disease   335
VII. Treatment   335
References   336

## 31. Wilson Disease
JOHN H. MENKES

I. Introduction   341
II. Phenotype   341
III. Normal and Abnormal Gene Function   343
IV. Diagnosis   346
V. Neuroimaging   346
VI. Pathologic Anatomy   348
VII. Animal Models   348
VIII. Treatment   348
References   350

## 32. Essential Tremor
ELAN D. LOUIS AND RUTH OTTMAN

I. Phenotype   353
II. Gene   355
III. Diagnostic and Ancillary Tests   356
IV. Neuroimaging   356
V. Neuropathology   357
VI. Cellular and Animal Models of Disease   357
VII. Genotype/Phenotype Correlations/Modifying Alleles   358
VIII. Treatment   358
References   360

## 33. Molecular Biology of Huntington's Disease (HD) and HD-Like Disorders
DAVID C. RUBINSZTEIN

I. Summary   365
II. Symptomatology of HD   366
III. Neuropathology   366
IV. Neuroimaging   366
V. Genetics of HD   367
VI. Diagnostic and Predictive Testing   368
VII. Gene, Normal Gene, and Abnormal Gene Function   369
VIII. Huntingtin Aggregates   370
IX. Cell Death in HD   372
X. Early Changes in Gene Expression   374
XI. Excitotoxicity and Impaired Energy Production   374
XII. Animal Models   375
XIII. Genotype/Phenotype/Modifying Alleles   375
XIV. Treatment   376
XV. HD-Like Disorders   376
References   377

## 34. Paroxysmal Dyskinesias
KAILASH P. BHATIA

I. Historical Aspects and Classification   385
II. Pathophysiology   390
III. Future Directions   391
IV. Concluding Summary   391
References   392

## 35. Primary Dystonias
ULRICH MÜLLER

I. Autosomal Dominant Dystonias   397
II. Autosomal Recessive Dystonias   400
III. X-Linked Recessive Dystonias   400
IV. Animal Models of Dystonia   402
References   403

## 36. DYT1 Dystonia
LAURIE J. OZELIUS AND SUSAN B. BRESSMAN

I. Summary   407
II. Phenotype   407
III. Gene   409

IV. Diagnostic and Ancillary Tests 413
  V. Neuroimaging 413
  VI. Neuropathology 413
  VII. Cellular and Animal Models of Disease 414
  VIII. Treatment 414
References 415

### 37. Dopa-Responsive Dystonia
HIROSHI ICHINOSE, CHIHO SUMI-ICHINOSE, TOSHIHARU NAGATSU, AND TAKAHIDE NOMURA

  I. Introduction 419
  II. Phenotype and Treatment 420
  III. Causative Gene 420
  IV. Diagnosis 422
  V. Genotype/Phenotype Correlation 423
  VI. Mechanism of Dominant Inheritance 424
  VII. The Mechanism of Neuronal Selectivity— A Study with an Animal Model of Biopterin Deficiency 424
References 426

### 38. Hallervorden-Spatz Syndrom
SUSAN J. HAYFLICK

  I. Introduction 429
  II. Phenotype 429
  III. Gene 433
  IV. Diagnostic and Ancillary Tests 434
  V. Cellular and Animal Models of Disease 437
  VI. Genotype/Phenotype Correlations/Modifying Alleles 437
  VII. Treatment 438
References 439

### 39. Genetics of Familial Idiopathic Basal Ganglia Calcification (FIBGC)
MARIA-JESUS SOBRIDO AND DANIEL H. GESCHWIND

  I. Clinical Phenotype 443
  II. Genetics 444
  III. Diagnostic and Ancillary Tests 444
  IV. Neuroimaging 446
  V. Neuropathology 447
  VI. Pathogenesis and Models of Disease 447
  VII. Genotype-Phenotype Correlations 448
  VIII. Treatment 448
References 448

### 40. Myoclonus and Myoclonus-Dystonias
CHRISTINE KLEIN

  I. Summary 451
  II. Phenotype 452
  III. Gene(s) 460
  IV. Diagnostic and Ancillary Tests 464
  V. Neuroimaging 465
  VI. Neuropathology 465
  VII. Cellular and Animal Models of Disease 465
  VIII. Genotype/Phenotype Correlations 465
  IX. Treatment 469
References 470

### 41. Mitochondrial Mutations in Parkinson's Disease and Dystonias
DAVID K. SIMON

  I. Mitochondrial Genetics 474
  II. Parkinson's Disease 474
  III. Dystonia 480
  IV. Treatment Implications 482
  V. Conclusions 483
References 483

### 42. Genetics of Gilles de la Tourette Syndrome
DAVID L. PAULS, STEFAN-M. PULST, AND MATTHEW W. STATE

  I. The GTS Phenotype 491
  II. GTS is Heritable 492
  III. Segregation Analyses of GTS Family Data 493
  IV. The Search for Genes in GTS 494
  V. Cytogenetic and Molecular Cytogenetic Approaches 496
  VI. Neuroimaging 497
  VII. Treatment 497
  VIII. Summary and Future Prospects 497
References 498

### 43. The Genetics of Restless Legs Syndrome
ANDY PEIFFER

  I. Introduction 503
  II. Clinical Features 503
  III. Prevalence and Progression 504
  IV. Genetic Studies 504
  V. Diagnosis 506

VI. Neuroimaging and Neurophysiological Studies 507
VII. Treatment 507
VIII. Summary 508
References 508

## 44. Other Adult-Onset Movement Disorders with a Genetic Basis
JAMES P. SUTTON

I. Inborn Errors of Metabolism 511
II. Disorders of Heavy Metal Metabolism 518
III. Movement Disorders Associated with Hematological Disease 520
IV. Other Rare Disorders 531
V. Summary 533
References 533

## 45. Ethical Issues in Genetic Testing for Movement Disorders
MARTHA A. NANCE, THOMAS D. BIRD, AND STEFAN-M. PULST

I. Introduction 541
II. Understanding the Role of Molecular Diagnostics in the Management of Neurological Disorders 542
III. Ethical Principles 545
IV. Conclusions 549
References 549

Index 551

# Contributors

*Numbers in parentheses indicate the pages on which the authors' contributions begin.*

**Rim Amouri** (179) Department of Neurology, Institut National de Neurologie, La Rabta, Tunis, Tunisia

**Shiuchi Asakawa** (305) Department of Molecular Biology, Keio University School of Medicine, Tokyo 160-8582, Japan

**Tetsuo Ashizawa** (103) Department of Neurology, The University of Texas Medical Branch, Galveston, Texas 77555

**Robert W. Baloh** (205) Department of Neurology, UCLA School of Medicine, University of California, Los Angeles, California 90095

**Carrolee Barlow** (195) The Laboratory of Genetics, The Salk Institute for Biological Studies, La Jolla, California 92037

**Samir Belal** (179) Department of Neurology, Institut National de Neurologie, La Rabta, Tunis, Tunisia

**Enrico Bertini** (231) Department of Neurosciences, Laboratory of Molecular Medicine, Bambino Gesu' Children Research Hospital, Rome, Italy

**Kailash P. Bhatia** (385) University Department of Clinical Neurology, Institute of Neurology, Queen Square, University College London, London WC1N, United Kingdom

**Thomas D. Bird** (541) Department of Medicine and Neurology, University of Washington, Seattle, Washington 98108

**Samir K. Brahmachari** (121) Functional Genomics Unit, Center for Biochemical Technology, CSIR Delhi, India

**Susan B. Bressman** (407) Department for Neurology, Beth Israel Medical Center, New York, New York 10003

**Alexis Brice** (85, 133) Neurologie et Thérapeutique Expérimentale, INSERM U289, Département Génétique, Cytogénétique et Embryologie, Groupe Hôspitalier Pitié-Salpêtrière, 75651 Paris Cedex, France

**Tomoki Chiba** (305) Department of Molecular Oncology, The Tokyo Metropolitan Institute of Medical Science, Tokyo, Japan

**Shweta Choudhry** (121) Functional Genomics Unit, Center for Biochemical Technology, CSIR Delhi, India

**J.W. Day** (75) Department of Neurology, Institute of Human Genetics, University of Minnesota, Minneapolis, Minnesota 55455

**Katherine A. Dick** (75) Department of Genetics, Cell Biology, and Development, Institute of Human Genetics, University of Minnesota, Minneapolis, Minnesota 55455

**Carlo Dionisi-Vici** (231) Department of Neurosciences, Division of Metabolic Disorders, Bambino Gesu' Children Research Hospital, Rome, Italy

**Alexandra Dürr** (133) INSERM U289 and Département Génétique, Cytogénétique et Embryologie, Groupe Hôspitalier Pitié-Salpêtrière, 75651 Paris Cedex, France

**Moncef Feki** (179) Hospital La Rabta, Servie Biochimie, Tunis, Tunisia

**Hiroto Fujigasaki** (133) INSERM U289, Groupe Hôspitalier Pitié-Salpêtrière, 75651 Paris Cedex, France and Department of Neurology and Neurological Science, Graduate School, Tokyo Medical and Dental University, Tokyo, Japan

**Sana Gabsi-Gherairi** (179) Department of Neurology, Institut National de Neurologie, La Rabta, Tunis, Tunisia

**Daniel H. Geschwind** (443) Department of Neurology, University of California, Los Angeles, California 90095

**Sid Gilman** (213) Department of Neurology, University of Michigan, Ann Arbor, Michigan 48109

**Lawrence I. Golbe** (287) UMDNJ-Robert Wood Johnson Medical School, New Brunswick, New Jersey 08901

**Lev G. Goldfarb** (151) National Institute of Neurological Disorders and Stroke, National Institutes of Health, Bethesda, Maryland 20892

**Nobutaka Hattori** (305) Department of Neurology, Juntendo University School of Medicine, Tokyo 113-8421, Japan

**Susan J. Hayflick** (429) Molecular and Medical Genetics, Pediatrics, and Neurology, Oregon Health & Science University, Portland, Oregon 97201

**Faycal Hentati** (179) Department of Neurology, Institut National de Neurologie, La Rabta, Tunis, Tunisia

**Joseph J. Higgins** (325) Center for Human Genetics and Child Neurology, Mid-Hudson Family Health Institute, New Paltz, New York 12561

**Susan E. Holmes** (121) Department of Psychiatry, Division of Neurobiology, Johns Hopkins University School of Medicine, Baltimore, Maryland 21287

**Hiroshi Ichinose** (419) Institute for Comprehensive Medical Science, Fujita Health University, Toyoake, Japan

**Satish Jain** (121) Department of Neurology, Neurosciences Center, All India Institute of Medical Sciences, New Delhi, India

**Joanna C. Jen** (81, 205) Department of Neurology, UCLA School of Medicine, The University of California, Los Angeles, California 90095

**Christine Klein** (451) Department of Neurology, Medical University of Lübeck, 23538 Lübeck, Germany

**Michael D. Koob** (95) Institute of Human Genetics, University of Minnesota, Minneapolis, Minnesota 55455

**Rejko Krüger** (315) Department of Neurology, University of Tübingen, Tübingen, Germany

**J. William Langston** (287) Parkinson's Institute, Sunnyvale, California 94089

**Anne-Sophie Lebre** (85) Neurologie et Thérapeutique Expérimentale, INSERM U289, Département Génétique, Cytogénétique et Embryologie, Groupe Hôspitalier Pitié-Salpêtrière, 75651 Paris Cedex, France

**Seung-Jae Lee** (287) Parkinson's Institute, Sunnyvale, California 94089

**Elan D. Louis** (353) G.H. Sergievsky Center and Department of Neurology, College of Physicians and Surgeons, Columbia University, New York, New York 10032

**Russell L. Margolis** (121) Department of Psychiatry, Division of Neurobiology, Johns Hopkins University School of Medicine, Baltimore, Maryland 21287

**Connie Marras** (273) The Parkinson's Institute, Sunnyvale, California 94089

**Tohru Matsuura** (103) Department of Neurology, Baylor College of Medicine, Houston, Texas 77030

**John H. Menkes** (341) Division of Pediatric Neurology, Cedars-Sinai Medical Center, Los Angeles, California 90048

**Shinsei Minoshima** (305) Department of Molecular Biology, Keio University School of Medicine, Tokyo 160-8582, Japan

**Yoshikuni Mizuno** (305) Department of Neurology, Juntendo University School of Medicine, Tokyo 113-8421, Japan

**Hidehiro Mizusawa** (71) Department of Neurology and Neurological Science, Graduate School, Tokyo Medical and Dental University, Tokyo, 113-8519 Japan

**Ulrich Müller** (395) Institut für Humangenetik, Justus-Liebig-Universität, D35392 Giessen, Germany

**Toshiharu Nagatsu** (419) Institute for Comprehensive Medical Science, Fujita Health University, Toyoake, Japan

**Martha A. Nance** (541) Park Nicollet Clinic, St. Louis Park, Minnesota 55426

**Takahide Nomura** (419) Department of Pharmacology, School of Medicine, Fujita Health University, Toyoake, Japan

**Min-Kyu Oh** (117) Division of Neurology, Cedars-Sinai Medical Center, Los Angeles, California 90048

**Elizabeth O'Hearn** (121) Departments of Neurology and Neuroscience, Johns Hopkins University School of Medicine, Baltimore, Maryland 21287

**Harry T. Orr** (35) Departments of Laboratory Medicine and Pathology, and Genetics, Cell Biology and Development, Institute of Human Genetics, University of Minnesota, Minneapolis, Minnesota 55455

**Ruth Ottman** (353) G.H. Sergievsky Center, College of Physicians and Surgeons, Columbia University, New York, New York 10032

**Laurie J. Ozelius** (407) Department of Molecular Genetics, Albert Einstein College of Medicine, Bronx, New York 10461

**Massimo Pandolfo** (165) Université Libre de Bruxelles-Hôpital Erasme, B-1070 Brussels, Belgium

**David L. Pauls** (491) Department of Psychiatry, Massachusetts General Hospital, Harvard Medical School, Charlestown, Massachusetts 02129

**Henry Paulson** (57) Department of Neurology, University of Iowa Roy J. and Lucille A. Carver College of Medicine, Iowa City, Iowa 52242

**Andy Peiffer** (503) Department of Pediatrics, Division of Medical Genetics, University of Utah, Salt Lake City, Utah 84112

**Susan L. Perlman** (253) Department of Neurology, UCLA School of Medicine, University of California, Los Angeles, California 90095

**Stefan-M. Pulst** (1, 19, 45, 117, 491, 541) Division of Neurology, Cedars-Sinai Medical Center, Departments of Medicine and Neurobiology, UCLA School of Medicine, Los Angeles, California 90048

**Laura P.W. Ranum** (75) Department of Genetics, Cell Biology, and Development, Institute of Human Genetics, University of Minnesota, Minneapolis, Minnesota 55455

**Andrea Richter** (189) Service de Génétique Médicale, Hôpital Sainte-Justine, Département de Pédiatrie, Université de Montréal, Montréal, Québec, Canada

**Olaf Riess** (315) Department of Neurology, University of Tübingen, Tübingen, Germany

**Christopher A. Ross** (121) Departments of Psychiatry and Neuroscience, Johns Hopkins University School of Medicine, Baltimore, Maryland 21205

**David C. Rubinsztein** (365) Department of Medical Genetics, Cambridge Institute for Medical Research, Addenbrooke's Hospital, Cambridge, CB2 2XY, United Kingdom

**Nobuyoshi Shimizu** (305) Department of Molecular Biology, Keio University School of Medicine, Tokyo 160-8582, Japan

**Cliff Shults** (213) Department of Neurosciences, University of California San Diego and Neurology Service, VA San Diego Healthcare System, La Jolla, California 92093

**David K. Simon** (473) Department of Neurology, Beth Israel Deaconess Medical Center and Harvard Medical School, Boston, Massachusetts 02115

**Maria-Jesus Sobrido** (443) Department of Neurology, Hospital Universitario de Salamanca, 37000 Salamanca, Spain

**Achal K. Srivastava** (121) Department of Neurology, Neurosciences Center, All India Institute of Medical Sciences, New Delhi, India

**Matthew W. State** (491) Child Study Center, Yale University School of Medicine, New Haven Connecticut 06520

**Giovanni Stevanin** (85, 133) Neurologie et Thérapeutique Expérimentale, INSERM U289, Groupe Hôspitalier Pitié-Salpêtrière, 75651 Paris Cedex 13, France

**S. H. Subramony** (57) Department of Neurology, University of Mississippi Medical Center, Jackson, Mississippi 39216

**Chiho Sumi-Ichinose** (419) Department of Pharmacology, School of Medicine, Fujita Health University, Toyoake, Japan

**James P. Sutton** (511) California Neuroscience Institute, Oxnard, California 93030

**Toshiaki Suzuki** (305) Department of Molecular Oncology, The Tokyo Metropolitan Institute of Medical Science, Tokyo, Japan

**Keiji Tanaka** (305) Department of Molecular Oncology, The Tokyo Metropolitan Institute of Medical Science, Tokyo, Japan

**Caroline Tanner** (273) The Parkinson's Institute, Sunnyvale, California 94089

**Kai Treuner** (195) The Laboratory of Genetics, The Salk Institute for Biological Studies, La Jolla, California 92037

**Shoji Tsuji** (139, 143) Department of Neurology, Brain Research Institute, Niigata University, Niigata 951, Japan and Department of Neurology, University of Tokyo, Tokyo 113–8655, Japan

**Hema Vakharia** (117) Division of Neurology, Cedars-Sinai Medical Center, Los Angeles, California 90048

**Hiroyo Yoshino** (305) Department of Neurology, Juntendo University School of Medicine, Tokyo 113-8421, Japan

**Massimo Zeviani** (231) Division of Biochemistry and Genetics, and Division of Child Neurology, Istituto Nazionale Neurologico "C. Besta," Milano, Italy

# Foreword

The field of movement disorders has witnessed an explosion of research activity in the past decade, with better understanding of nosology, physiology, pharmacology, and treatment. But perhaps the most explosive growth has been the identification of genes and the mapping of gene locations for many of these disorders. The availability of simple diagnostic tests for the genes involved saves the physician countless hours and saves patients money. Equally important is our newly gained knowledge of the clinical phenomenology of various movement disorders because we can now identify them using gene tests. For example, the identification of the gene for Machado-Joseph disease and its subsequent labeling as SCA-3 has provided a fuller understanding of how common this disorder, previously thought to be rare, actually is. The research has also revealed the extensive clinical spectrum by which the abnormal gene manifests itself.

The editor of this book, Stefan-M. Pulst, is a neurologist and neuroscientist who is equally at home in the clinical phenomenology of movement disorders and their molecular-genetic analysis. This is reflected in the structure of the chapters that take the reader from phenotype to genotype including descriptions of neuroimaging, neuropathology, and neurophysiology when appropriate. The neuroscientist will find the discussions of normal and abnormal gene function of interest, while clinical neurologists and geneticists will use this volume to clarify issues of genotype/phenotype correlations. The editor has chosen contributors who are both leaders in the field of neurogenetics and experienced physicians who use genetic testing in their clinical practice and who are familiar with the promises and pitfalls of molecular tests.

Movement disorders are neurological syndromes in which there is either an excess of movement or a paucity of voluntary and automatic movements unrelated to weakness or spasticity. Disorders in the excess movement category are commonly referred to as *hyperkinesias* (excessive movements), *dyskinesias* (unnatural movements), and *abnormal involuntary movements*). Even though the terms are interchangeable, dyskinesias is the one most often used. The five major categories of dyskinesias, in alphabetical order, are: chorea, dystonia, myoclonus, tics, and tremor. Within each category, there are many entities and etiologies.

Movement disorders due to a paucity of movement are referred to as *hypokinesia* (decreased amplitude of movement), but *bradykinesia* (slowness of movement) and *akinesia* (loss of movement) could be reasonable alternative names. The parkinsonian syndromes are the most common cause of such paucity of movement; other hypokinetic disorders represent only a small group of patients. Basically, movement disorders can be conveniently divided into parkinsonism and all other types; there is about an equal number of patients in each of these two groups.

Thus, the time is ripe for *Genetics of Movement Disorders*, in which experts gathered by the editor summarize the genetic information that has threatened to overwhelm us. We are just at the beginning of this rich field, with many more genetic discoveries to come. At an exciting time both in movement disorders and neurogenetics, this book superbly brings the two fields together. *Genetics of Movement Disorders* presents a unique new breed of neurology text, a successful synthesis of molecular neuroscience and clinical medicine. Future researchers and current clinicians will value this volume as an ideal entry point.

*Stanley Fahn, M.D.*
Columbia University College of Physicians & Surgeons
New York City, NY

# Preface

Movement disorders have shaped neurologic thinking for more than a century. They provide visible and striking afflictions of the nervous system often combining symptoms attributable to hyper- or hypoexcitability of specific neuronal groups with features of neurodegeneration. Nosologic classification of these disorders has thus relied on the phenotype, in recent years aided by neuroimaging and to a lesser degree by neurophysiologic investigations. Positional or candidate gene-based identification of disease genes has now led to new genotypically based classifications. DNA tests increasingly substitute for clinical, morphological, or biochemical definitions as the gold standard for diagnosis. But this century has already brought with it a classification based on changes at the protein level. Terms such as tauopathy or polyglutamine disease indicate that classifications based on protein abnormalities may unite what the genotype has divided.

This book is written for neurologists, neuroscientists, geneticists, and genetic counselors. For those not familiar with genetic terminology and methods, the first chapter provides an introduction. The following chapters are divided according to classic definitions of movement disorders taking into account that a patient does not present with a genotype written on his or her forehead but with symptoms and signs of disease. All authors are familiar with the differential diagnosis and treatment of the movement disorders covered in their chapters. The use of DNA-based testing in the evaluation of the patient with a movement disorder has received particular emphasis. The genotype is correlated with the phenotype to aid the physician in counseling and to provide the basic scientist with a model for the effect of mutant alleles at the organismal level.

The focus of genetics in the first decade of the new millennium is moving from the identification of alleles causing mendelian diseases to the analysis of common traits. Single gene defects in addition to causing a specific disease with high probability may be understood as identifying genes that may have other alleles that modify more common traits. Reduced penetrance or variable expressivity of specific alleles can be traced back to interaction with other alleles or environmental factors that ameliorate or worsen the phenotype. The chapters on Parkinson disease and Progressive Supranuclear Palsy serve as examples for this group of diseases encompassing both rare single gene defects and examples of more common predisposing alleles in the general population.

Molecular investigations have provided insights into movement disorders that might have seemed audacious or even preposterous just a few years ago. A case in point is the development of disease models in rodents or invertebrates based on the knockout of genes or the expression of mutated transgenes. Although these models may not be perfect replicas of their respective human diseases, they recapitulate salient pathological, biochemical, or functional features. Neuroscientists and geneticists are now using these models to investigate pathways, to identify modifiers of disease pathogenesis, and to test novel therapies. All chapters address animal models and *in vitro* studies which will make this book of interest to the molecular biologist working on fundamental issues of human CNS diseases.

My thanks go to all the contributors who did not tire of up-dating chapters to include recently identified disease genes and disease mechanisms. Without Cindy Minor's excellent editorial assistance and kind insistence the book would have seen a much later publication date. Hilary Rowe served as a knowledgeable Publishing Editor and Paul Gottehrer competently supervised production management. Thanks also go to members of my laboratory, past and present, for daring experiments and stimulating discussions and to Carmen Warschaw and Teddi Winograd for their support of academic neurology. Finally, in the name of all the contributors I wish to acknowledge patients and families whose participation in genetic studies made the discovery of disease genes possible.

*Stefan-M. Pulst, M.D., Dr. med.*
Los Angeles, June 2002

CHAPTER 1

# Introduction to Medical Genetics and Methods of DNA Testing

STEFAN-M. PULST

*Division of Neurology, Cedars-Sinai Medical Center*
*Departments of Medicine and Neurobiology, UCLA School of Medicine*
*Los Angeles, California 90048*

I. Concepts and Terminology
II. Patterns of Inheritance
   A. Phenotype
   B. Dominance and Recessiveness
   C. Inheritance Patterns
   D. The Structure of a Gene
   E. Genetic Linkage Analysis
   F. Types of Mutations
III. Molecular Genetic Tools
   A. Polymerase Chain Reaction
   B. Southern Blot Analysis
   C. DNA Sequencing
   D. Analysis of RNA Transcripts and Proteins
   E. Detection of DNA Polymorphisms
IV. Molecular Genetic Testing
   A. Direct Mutation Detection
   B. Indirect Analysis Using Linked Genetic Markers
V. Animal Models
   A. Expression of Foreign Transgenes
   B. Gene Targeting (Knock-Out, Knock-In)
VI. Web-Based Information for Genetic Diagnosis and Testing
   Acknowledgments
   References

This chapter will provide some of the necessary basics for those clinicians not familiar with medical genetic terminology and molecular-genetic methods. For the neuroscientist it may provide a brief overview of the benefits and problems associated with an approach to the nervous system based on the identification of disease genes. The first part will review concepts and terminology, the second part current methodology. Thus, mutation types are covered in the first part and their detection in the second part of this chapter.

For centuries neurology and the classification of neurologic illnesses have mainly relied on the neurologic phenotype. This was particularly true for the movement disorders leading to keen observations on subtle variations of disorders affecting movement control. Effective symptomatic treatments were developed based on the increasing understanding of physiology and pathology of movement disorders. In the last decades neuroimaging has played an ever-increasing role in the understanding of these disorders.

Grouping disorders based on inheritance patterns did not necessarily result in an improved classification. Diseases like the dominant ataxias shared a similar inheritance pattern, but phenotypes did not "breed true" within families and overlap between families was significant. We now know that mutations in several different genes can cause ataxia with an added twist conferred by the presence of dynamic mutations. A truly novel understanding and grouping of disorders was only possible with the advent of linkage analysis and eventually isolation of the causative disease genes. Recently, new concepts have emerged that are based on protein structure or function such as that of the polyglutamine disorders.

Movement disorders have been at the forefront of neurologic diseases that were investigated using genetic methods to map disease genes and ultimately to identify the genes themselves. A movement disorder was actually the first neurologic illness mapped to a specific chromosome. This disorder, now known as spinocerebellar ataxia type 1, was linked to the HLA region on human chromosome 6 using polymorphic protein markers (Jackson *et al.*, 1977).

This predated by three years the suggestion that restriction fragment length polymorphisms (RFLPs) could be used to establish a genetic map of the human genome (Botstein *et al.*, 1980). Another movement disorder, Huntington's disease (HD), was the first neurologic illness that was genetically mapped using this new class of polymorphic DNA markers (Gusella *et al.*, 1983).

## I. CONCEPTS AND TERMINOLOGY

The following summarizes some of the basic knowledge concerning molecular and medical genetics. For more detailed discussions the reader is referred to other textbooks (Jorde *et al.*, 2000; Pulst, 2000, Strachan and Read, 1999; Watson *et al.*; 1992). Additional information and figures can be found on the Web (Human Genome web site, 2002; Genetics science learning center web site, 2002). Instead of a systematic review, preference was given to terms and methods that are used in subsequent chapters.

## II. PATTERNS OF INHERITANCE

Mendelian or unifactorial inheritance refers to a pattern of inheritance that can be explained on the basis of mutation in a single gene. Thus, the presence or absence of a genetic character depends on the genotype at a single locus. Monogenic traits are also referred to as Mendelian traits, because they follow the well-delineated patterns of inheritance first described by Mendel in 1865. In contrast to previous animal and plant breeders, Mendel did not stop at the descriptive level, but was able to conclude that there must be dominant and recessive traits by analysis of the ratios of observed traits in the offspring. Mendel's original publications can now be viewed on the Web in the original German or translated into English (Blumberg, 2002).

In contrast to monogenic disorders, complex genetic traits cannot be explained on the basis of mutations in a single specific gene, and a phenotype is only observed when mutations in several genes have occurred with or without a significant contribution from environmental factors. After a focus on single gene disorders, movement disorder genetics is now also beginning to analyze complex diseases where multiple genes and environmental factors must act in concert to cause disease (see Chapter 26).

### A. Phenotype

Whereas genotype refers to a person's DNA sequences at a specific chromosomal locus, the term phenotype describes what can be observed clinically. In addition to the "clinical" phenotype, one can also examine cellular or biochemical phenotypes which may not be directly observable by physical examination. The genotype describes the nature of the two alleles (for autosomal genes) for a specific gene. Allele is the term for different forms of a gene based on different DNA sequences at this locus (for more details see Fig. 1.7).

There may not be a defined relationship between genotype at a locus and phenotype. Mutations in different genes may produce a similar phenotype (nonallelic heterogeneity). For example, mutations in different genes on different chromosomes may cause familial Parkinson disease (see Chapter 26). These observations suggest, but do not prove that the proteins encoded by these genes may be involved in the same cellular pathway as has recently been suggested for the interaction of parkin and αsynuclein (Shimura *et al.*, 2001).

On the other hand, different types of mutations in the same gene may result in different phenotypes (allelic heterogeneity). Examples are mutations in the CACNA1A gene that can result in episodic ataxia, familial hemiplegic migraine, or a progressive degenerative ataxia, designated as SCA6 (see Chapters 8 and 22).

Finally, the same mutation can result in very different phenotypes or in significantly different ages of onset or disease progression. Phenotypic variability associated with a specific mutation may be caused by several genetic and nongenetic factors. The phenotype is determined not only by the genotype at the disease locus, but also by genotypes at other loci. These factors are often referred to as modifying loci, modifying genes, or modifying alleles. Stochastic events, such as mitotic loss of alleles or the partition of mitochondria into daughter cells are important in determining phenotype (see Chapters 24 and 41). In addition, the phenotype may be influenced by the interaction of environmental factors with the genotype, a hypothesis recently favored to explain the majority of Parkinson disease cases (see Chapter 26).

### B. Dominance and Recessiveness

Fundamental to the understanding of Mendelian inheritance are the concepts of dominance and recessiveness. Dominance and recessiveness are properties of traits, not of genes, since different mutations in the same gene on the autosomes may show dominant or recessive inheritance.

Dominance is not a property intrinsic to a particular allele, but describes the relationship between it and the corresponding allele on the homologous chromosome with regard to a particular trait. A trait is dominant, if it is manifest in the heterozygote. Dominant alleles exert their phenotypic effect despite the presence of a normal (wild-type) allele on the second homologous chromosome. Thus, if the phenotypes associated with genotypes AA and AB are identical but are different from the BB phenotype, the A allele is dominant to allele B. Conversely, the B allele is recessive to allele A. Recessive mutations lead to phenotypic consequences only when both alleles contain mutations. If the mutations on both alleles are different, this is referred to as compound heterozygosity. An example would be a patient with Friedreich's ataxia whose mutation is an expansion of the intronic GAA repeat on one allele, whereas the mutation in the other allele is a missense mutation (see Chapter 18).

The same mutation can lead to recessive or dominant traits. Sickle cell anemia is a recessive disease. The trait "sickle cell anemia" is manifested in homozygotes with the HemoglobinS (HbS) mutation. However, the trait "sickling," the aggregation of red blood cells at low oxygen tensions, is a dominant trait, since it is seen in heterozygotes that carry one allele with the HbS mutation and one wild-type allele.

Most human dominant syndromes occur only in heterozygotes. Some geneticists refer to dominant mutations that have the same phenotype in heterozygotes and homozygotes as "true" dominant. This is distinguished from semi-dominant when the heterozygote AB has an intermediate phenotype between the phenotypes of AA and BB. This is probably the case for most polyglutamine diseases (and the respective mouse models), where individuals with two mutant alleles have a more severe phenotype. If the phenotypes of AB, AA, and BB are identical, alleles A and B are said to be co-dominant.

## 1. Mechanisms of Dominance

The majority of mutations result in an inactive gene product (see Section II.F). The function of the remaining normal allele, however, is sufficient in most cases to guarantee normal cellular function. Therefore, most mutant alleles are recessive. When the function of the remaining allele is not sufficient to maintain normal function, this is referred to as haploinsufficiency.

*a. Gain of Function*

Most dominant mutations lead to a gain of function. This may be due to the loss of negative regulatory domains, loss of normal protein degradation, or abnormal protein processing or protein-protein interaction. A toxic gain of function has been suggested for mutant proteins containing extended polyglutamine stretches (see Chapter 2). Although mutations that lead to truncation of the protein can result in dominant alleles (see EA-2, Chapter 22), dominant mutations are more commonly missense mutations leading to amino acid substitution.

*b. Loss of Function Mutations Associated with Dominant Traits*

Loss of function mutations may be dominant, when they involve a critical rate-limiting step in a metabolic pathway, or involve regulatory genes or structural genes that are sensitive to gene dosage effects. The gene encoding guanosine triphosphate cyclohydrolase I (GCH) is mutated in Dopa-responsive dystonia (see Chapter 37). Although mutations in genes encoding enzymes are usually recessive, mutations in this enzyme are associated with a phenotype, probably because enzyme levels above 50% of normal are necessary in critical neuronal populations.

Dominant-negative mutations result in a mutant protein that interferes with the action of the normal protein. For example, in a homodimeric protein (a protein complex made up of two identical proteins), the mutation of one allele will result in only 25% of the resulting dimers having a normal composition.

Certain mutations may combine gains and losses of function. Examples are the polyglutamine diseases such as HD. Expansion of the polyglutamine tract results in a gain of toxic function. At the same time studies of transcription factors that interact with huntingtin have suggested that huntingtin with expanded polyQ domains sequesters specific transcription factors thus resulting in a loss of function (see Chapter 33).

### C. Inheritance Patterns

The relationship between members of a family are conveniently indicated by a pedigree notation (Fig. 1.1). In addition to their blood relationship, phenotypic and genotypic information can be included in one graphic display. The first individual ascertained in a pedigree (proband) is indicated by an arrow.

## 1. The Multigeneration Pedigree

One of the most frequent misconceptions encountered in the routine genetic assessment of movement disorder patients is the fact that physicians are content with obtaining a family history restricted to siblings or parents. When determining inheritance patterns it is crucial to obtain a three-generation pedigree including all first and second degree relatives. Without the history and examination of many at-risk individuals, it may be impossible to distinguish autosomal recessive inheritance from dominant inheritance with reduced penetrance, dominant inheritance with anticipation, or from other forms of inheritance. X-linked recessive inheritance requires the absence of transmission through males, but this may only be apparent when a larger number of male transmissions can be studied.

The following may serve as an example. We published a pedigree of Portuguese ancestry with atypical Parkinsonism and an inheritance pattern suggestive of autosomal dominant inheritance. On account of the Portuguese background and phenotype we assumed that this pedigree likely segregated a mutation in the SCA3 gene, although on a

| Symbol | | Comments |
|---|---|---|
| □ | 1. Individual (Male) | Assign gender by phenotype. |
| ● | 2a. Affected individual (Female) | Key/legend used to define shading or other fill. (e.g., hatches, dots, etc.) |
| ▨ | 2b. Affected individual (Male) | With ≥2 conditions, the individual's symbol should be partitioned accordingly, each segment shaded with a different fill and defined in legend. |
| ⑤ | 3. Multiple individuals (Number known for female) | Number of siblings written inside symbol. (Affected individuals should not be grouped). |
| ◇ d. 35y | 4. Deceased individual (Sex Unknown) | If death is known, write "d." with age at death below symbol. |
| ■ | 5a. Proband | First affected family member coming to medical attention. |
| ○ | 5b. Consultand | Individual(s) seeking genetic counseling/testing. |
| △ male | 6. Spontaneous abortion (SAB) | If due to an ectopic pregnancy, write ECT below symbol. Also note sex below symbol. |
| ▲ female | 7. Affected SAB | If gestational age known, write below symbol. Key/legend used to define shading. |
| ⊠ sex unknown | 8. Termination of pregnancy (TOP) | No abbreviations used for sake of consistency. |
| ⊙ | 9. Obligate carrier (will **not** manifest disease) | Normal phenotype and negative test result. (e.g., Woman with normal physical exam and carrier of a mutation in the frataxin gene) |
| ▯ | 10. Asymptomatic/Presymptomatic carrier | Clinically unaffected at this time but could later exhibit symptoms. |
| □—○ | 11. Genetic line of descent (vertical or diagonal) | Biologic parents shown connected by horizontal line. Offspring are connected to parents by a vertical line. |
| Monozygotic / Dizygotic | 12. Twins | A horizontal line between the symbols implies a relationship line. |
| □=○ | 13. Consanguinity | If degree of relationship is not obvious from the pedigree, it should be stated above the relationship line. (e.g., third cousins) |

**FIGURE 1.1** Pedigree symbols according to Bennett et al. (1995). For use in pedigrees see also Figure 1.2.

formal basis maternal inheritance could not be excluded (Sutton and Pulst, 1997). Subsequent molecular analysis determined that the family did not carry expansions in the SCA3 CAG repeat nor was there linkage to the SCA3 locus, but the family segregated a mutation in the mitochondrial genome (see also Fig. 41.2). Maternal inheritance was not striking due to lack of affected males having a significant number of offspring (Simon et al., 1999).

## 2. Autosomal Dominant

In autosomal dominant disorders, a mutation in a single gene on any of the 22 autosomes produces clinical symptoms or signs. The disease or mutant allele is dominant to the normal (wild-type) allele, and the disease phenotype is seen in heterozygotes. Offspring are at 50% risk to inherit the disease. Offspring who inherit the normal allele will not develop the disease nor pass it on to their offspring. The disease appears over multiple generations which appears as "vertical transmission" in the pedigree notation. Males and females are evenly affected, and the disease is passed on from affected fathers or mothers to male and female offspring with equal probability.

Penetrance in an individual who carries a disease allele may not show any manifestations of the disease phenotype. However, this individual may transmit the disease to the next generation. Studies have shown that in pedigrees with familial torsion dystonia approximately 30% of individuals who are known to carry the DYT1 mutation do not show signs of a movement disorder (Pauls and Korczyn, 1990). Thus, the penetrance of mutations in the DYT1 gene is considered to be 30%. Penetrance is also related to the thoroughness of the clinical examination and the use of ancillary studies (see Fig. 1.2).

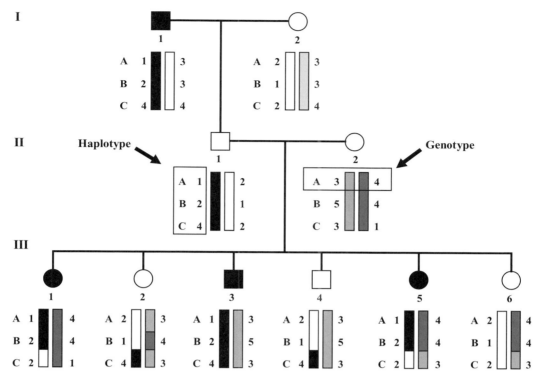

**FIGURE 1.2** Pedigree segregating an autosomal dominant trait; individual II:1 is non-penetrant. The segregation of alleles for three marker loci (A, B, and C) is shown. Vertical boxes provide a depiction of chromosome segments allowing easy visualization of recombination events. A double recombination is shown for the maternally inherited chromosome in individual III:2. The haplotype for markers A,B,C is boxed in individual II:1 (a1-b2-c-4), the genotype for marker A is boxed in individual II:2 (a3/4).

The terms penetrance and expressivity of a mutation need to be clearly differentiated. Penetrance is an all-or-none phenomenon. Signs of a given phenotype (clinical, biochemical, imaging, etc.) are either present or not present. Variable expressivity describes the extent and variability of the expression of the phenotype. Variable expressivity of a mutation can refer to variable severity of disease symptoms or variable age of onset, but also to expression of completely different symptoms in carriers of the mutation. The typical mutation in the DYT1 gene, i.e., deletion of a GAG codon, causes generalized dystonia in most individuals. However, in one family, individuals with this mutation developed writer"s cramp. This is an example of variable expressivity of the mutation (Gasser *et al.*, 1998).

### 3. Autosomal Recessive

When both copies of a gene need to be mutated in order to produce a phenotype, the disorder is inherited as an autosomal recessive trait. Only individuals homozygous for the mutant allele will develop the disease, whereas heterozygous individuals (one normal copy and one mutated copy) are clinically normal. Since heterozygotes may pass on an abnormal allele to their offspring, they are called carriers. Parents of affected individuals are usually carriers of the disease gene, and each parent contributes one abnormal copy to the offspring. Disease risk to siblings is 25%, and 50% of siblings are at risk to be carriers. In contrast to autosomal dominant inheritance, where vertical transmission is observed, horizontal aggregation is typical for recessive disorders, where multiple individuals in one generation are affected (Fig. 1.3).

The likelihood of encountering recessive disorders is increased in specific populations that have a high frequency of mutant alleles. For example, Tay-Sachs disease is common in Ashkenazi Jews with a heterozygote frequency of 1 in 30, whereas the frequency in Caucasians is only 1 in 3000. Consanguinity may be present in pedigrees with autosomal recessive inheritance. The probability that a mating was consanguineous increases when the mutant allele is very rare. For many of the rare neurologic recessive diseases, consanguinity becomes a factor. For example, at the low mutant gene frequency of 0.008 for Wilson disease, half of the cases are the result of consanguineous matings (Saito, 1988). Consanguinity is seen in many cultures of the Mideast.

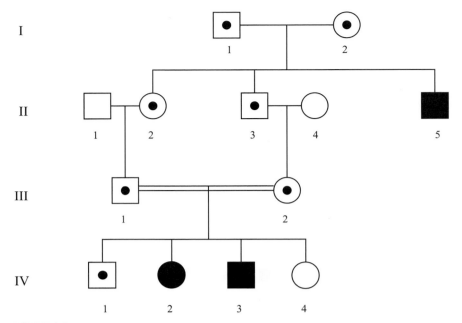

**FIGURE 1.3** Pedigree segregating an autosomal recessive trait. Note consanguineous marriage in generation III. Obligate carrier status is indicated symbol with a dot.

## 4. X-linked Inheritance

X-linked recessive disorders show a phenotype only in affected males, but transmission of the disease occurs through unaffected mothers. Male-to-male transmission excludes X-linked inheritance. Since females have two X chromosomes, females are usually clinically normal in X-linked recessive disorders. Sons of carrier females are at 50% risk to inherit the mutation, and to develop clinical signs (Fig. 1.4). Affected males carry one copy of the mutated allele, but no normal allele. This state is called hemizygosity, and differs from heterozygosity in the lack of a second (normal) allele. Daughters of affected males will be carriers with 100% risk, because all paternal X chromosomes carry the mutated allele.

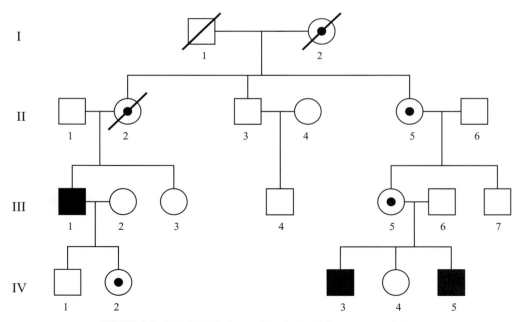

**FIGURE 1.4** X-linked inheritance. Note lack of father-son transmission.

Although carrier females are usually clinically normal, occasional carrier females are mildly symptomatic or show biochemical signs of the disease. These individuals are called manifesting heterozygotes. The basis for this is X-inactivation, a random process that leads to inactivation of one copy of the X chromosome. Inactivation of one X chromosome compensates for the presence of two X chromosomes in females. (Only a small part of the X chromosome escapes inactivation.) Although in most carrier females the normal and abnormal X chromosome are inactivated approximately half of the time, in occasional individuals the majority of normal chromosomes are inactivated by chance resulting in (usually mild) expression of disease manifestations. Examples of manifesting heterozygotes are seen in adrenoleukodystrophy.

### 5. Mitochondrial Inheritance

The mitochondrial genome is distinct from the nuclear genome and contained in a circular DNA molecule with 16,569 bp. A cell contains several hundred mitochondria in its cytoplasm and each single mitochondrion contains up to 10 circular DNA molecules (see Chapters 24 and 41). The mitochondrial DNA of a fetus is solely contributed from the mother; no mitochondria are contributed by the sperm. This explains why mitochondrial disorders are only transmitted through the mother (Fig. 1.5). However, the number of mutant mitochondria in each offspring may vary, resulting in extremely variable phenotypes. All children are at risk to inherit at least some mutated mitochondria.

### 6. Other Genetic Mechanisms Affecting Phenotype

Issues of nonpenetrance, modifying alleles, and interaction of genotypes with environmental factors have been discussed above. Two mechanisms that have now been defined at the molecular level will be briefly reviewed here.

#### a. Anticipation

In several diseases, a phenomenon of earlier disease onset in successive generations has been observed (see also Fig. 2.1). However, this observation was difficult to distinguish from the biased ascertainment of probands or from random variations in the age of onset. With the identification of unstable DNA repeats, this phenomenon now has a molecular basis. Earlier disease onset and increased severity in subsequent generations are correlated with increasing expansion of the DNA repeat. It is likely that this mechanism underlies a significant number of neurodegenerative diseases, although anticipation may at times be difficult to recognize.

#### b. Imprinting

Most genes are expressed in near equal amounts from the maternal and paternal copy of the gene. Imprinted genes, however, are expressed from only one of the two homologous chromosomes. For mutations in imprinted genes the parental origin of the mutation matters. For example, if the maternally inherited allele is silenced by imprinting, only mutations of the paternal allele have phenotypic consequences (Nicholls *et al.*, 1998). A well-

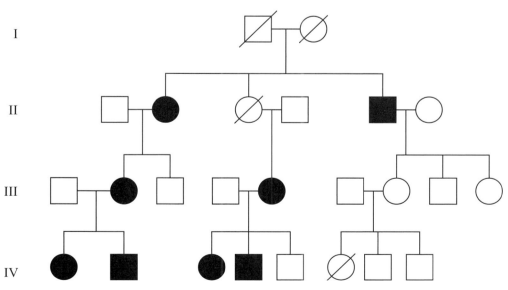

FIGURE 1.5 Mitochondrial inheritance. Note the lack of transmission through affected males.

known example of imprinting is deletion of genes in chromosome 15q12 giving rise to Prader-Willi or Angelman syndrome. At the molecular level, different DNA methylation patterns underlie genomic imprinting. Imprinting should not be confused with sex-specific expression of a phenotype where the sex of the mutation carrier is important, not the sex of the parent from whom the mutation is inherited.

## 7. Complex Inheritance Patterns

Despite the presence of familial aggregation, inspection of pedigrees for many diseases may not disclose a Mendelian inheritance pattern. Familial aggregation may have environmental causes, may be due to the interaction of more than one gene (polygenic trait), or the result of environmental and genetic interactions (complex trait).

Twin studies can be an important tool to estimate the approximate contributions of environmental and genetic factors to disease by comparing monozygotic and dizygotic twin pairs (see Chapters 26 and 42). If a phenotype is caused by the interaction of many genes (such as in autism), twin studies may indicate very high heritability, but most cases will occur as singletons (reviewed in Alarcon et al., 2002).

## 8. Phenotypic Variation in Mendelian Disorders

Subsequent Chapters in this book will bear ample witness to the fact that individuals carrying identical mutations can have different disease severity or a different age of disease onset. Phenotypic variation in Mendelian disorders can be regarded as a special case of complex inheritance, and environmental factors and lifestyle choices can clearly play a role. But genetic factors can be important as well. These include the presence of DNA polymorphisms in the allele bearing the disease-causing mutation or variation in promoter sequences that reduce or enhance expression of this allele. The precise DNA sequence of the second allele can have an effect. For example, if the second allele has normal, but slightly reduced function, this could result in earlier disease onset in that particular individual.

Alleles at other loci (also called modifier alleles) can influence the phenotype. Modifiers of disease severity are being identified in animal models as well as in humans (see particularly Chapters 3–5). Imprinting of disease or modifier alleles can have an effect. But modifiers are not limited to the nuclear genome. The mitochondrial genome may be different between different individuals, and subtle variation in its sequence may only be apparent in the presence of a disease allele.

Stochastic factors describe events that for lack of better explanations appear to occur at random. These include random loss of parts of chromosomes, a mechanism important in the pathogenesis of human tumors. In disorders caused by unstable CAG repeats stochastic variation of DNA repeat length has been observed between different CNS regions (see Chapter 3). This variation cannot be detected when determining repeat length in white blood cells.

## D. The Structure of a Gene

The term gene was coined in 1909 by Johannsen. Genes are the units of heredity and encode information that results in the production of functional end products, either proteins or RNA molecules. Figure 1.6 graphically depicts the structure of a gene, and the processes leading from genomic DNA to transcription into messenger RNA (mRNA).

The part of a gene that gives rise to an mRNA transcript, the expressed or coding region, is not continuously encoded. Instead, a gene contains interspersed non-coding DNA sequences. These DNA sequences are called introns and are present in the primary transcript, but are later spliced out when the mature mRNA is formed. Those parts of the gene that remain in the mature transcript and (mostly) code for protein are called exons. The process of splicing is not invariant. By alternative splicing of exons, transcript containing different exons can be generated from one gene, thus giving rise to more than one protein from a single gene. Mutations do not only occur in the coding regions of genes, but may also lie in introns affecting the process of transcript splicing (see, for example, parkin mutations).

After modifications, the processed mRNAs (without introns) are transported out of the nucleus, hence the name messenger RNA, because this type of RNA transports the DNA encoded information to the cytoplasm (Papavassiliou, 1995). In the cytoplasm, the ribosomal machinery translates this information into an amino acid sequence and synthesizes the protein (von Hippel, 1998).

## 1. DNA Polymorphisms and Alleles

The variant DNA sequences at a particular chromosomal location (locus) are referred to as alleles (Fig. 1.7). When the DNA sequence change causes a disease state, we speak of a mutant allele as compared to the normal or wild-type allele. Thus, a patient with Huntington's disease, an autosomal dominant disease, has one mutant allele (with an expanded CAG repeat) and one normal or wild-type allele. Those genetic variations that are not causative for a phenotype, but are associated with a disease phenotype are referred to as risk or predisposing alleles. For example, certain alleles of the *tau* gene are found at increased frequency in patients with progressive supranuclear palsy (see Chapter 30).

Allelic sequence variation is conventionally referred to as a DNA polymorphism if more than one allele (DNA

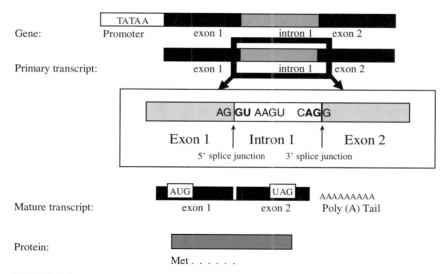

**FIGURE 1.6** Structure of a gene. Simplified representation of the organization of a gene. TATAA box: sequence frequently found in gene promoters. Splicing of the primary transcript results in the formation of the mature transcript. The boxed insert shows the common splice signals (AG-GUA and CAG-G) that result in removal of intronic sequences. The bolded sequences are always present. The mature transcript is transported into cytoplasm and translated into protein. AUG: translation start site. The portion of exon 1 preceding the first AUG codon is not translated (5'-UTR: untranslated region); UAG: stop codon, end of translation. The part of the mRNA transcript beyond the stop codon is designated the 3'-UTR.

**DNA Polymorphisms and Alleles**

- Wildtype allele 1

ATT TCT GCT AAT TAT AAC GAG TTA ATC AGC ACC GAG AAT ATA AAT GCA
Ile  Ser Ala Asn Tyr Asn Glu Leu Ile Ser Thr Glu Asn Ile Asn Glu

- Wildtype allele 2

ATT TCT GCT AAT TAT AAC GAG TTG ATC AGC ACC GAG AAT ATA AAT GCA
Ile  Ser Ala Asn Tyr Asn Glu Leu Ile Ser Thr Glu Asn Ile Asn Glu

- Predisposing allele

ATT TCT GCT AAT TAT AAC GAG ATA ATC AGC ACC GAG AAT ATA AAT GCA
Ile  Ser Ala Asn Tyr Asn Glu **Ile** Ile Ser Thr Glu Asn Ile Asn Glu

- Mutant allele

ATT TCT GCT AAT TAT AAC GAG TTA ACC AGC ACC GAG AAT ATA AAT GCA
Ile  Ser Ala Asn Tyr Asn Glu Leu **Thr** Ser Thr Glu Asn Ile Asn Glu

**FIGURE 1.7** DNA sequence changes give rise to normal (wildtype), predisposition, and pathogenic (mutant) alleles. In wild-type allele 2 a single base pair change in the third position of the TTA codon changes the DNA sequence, but not the amino acid sequence. In a third allele a change in a first position base pair changes the amino acid sequence. This again could be a normal variant, a predisposition allele or a mutant allele. Without established phenotypic association, the severity of the likely effect on protein function can only be estimated. Thus a change of a leucine (Leu) to an isoleucine (Ile) is most likely less severe than a change from an isoleucine to a threonine (Thr) as is seen in the "mutant" allele.

variant) is found with a frequency of at least 1% in human populations. In other words, if a locus has 2 or more alleles that occur with a frequency of >1%, the locus is referred to as polymorphic. The limit of 1% was chosen to exclude chance recurrence of a DNA sequence change. Rare DNA polymorphisms, however, with a frequency of less than 1% exist and allele frequencies may vary greatly between different populations. The distinction between mutant alleles and very rare polymorphisms not associated with disease can be difficult at times.

## 2. Genotypes and Haplotypes

When identical alleles occur on both copies of a chromosome pair, the person is said to be homozygous at this locus. When the alleles differ, the person is heterozygous. If one allele is missing, the person is hemizygous at that locus. For example, all males are hemizygous for alleles on the X chromosome.

The alleles at a specific locus define the genotype (see Fig. 1.2). For example, the genotype of a patient with spinocerebellar ataxia type 2 (SCA2) for the trinucleotide repeat in the *ataxin-2* gene may be designated as $(CAG)_{37}/(CAG)_{22}$. This refers to the fact that this individual has 37 CAG repeats in this gene on one chromosome 12 and 22 repeats on the other chromosome 12 (see also Fig. 2.1).

A *string* of alleles on the *same* chromosome is referred to as a chromosomal haplotype (see Fig. 1.2). Haplotype analysis can be important for gene identification when most cases of a disease derive from a common founder. This phenomenon can be especially prominent in isolated populations as recently exemplified by the identification of the ARSACS gene (see Chapter 20).

## E. Genetic Linkage Analysis

Genetic linkage analysis is one of the most important tools in the process of identifying disease genes. A recent brief review of these concepts is provided elsewhere (Pulst, 1999). Genetic linkage refers to the observation that genes that are physically close on a chromosome are inherited together and thus appear genetically linked. With increasing physical distance between two genes, the probability of their separation during meiotic chiasma formation increases.

By comparing the inheritance pattern of the disease phenotype with that of DNA marker alleles a chromosomal location for the disease gene can be assigned. A marker very close to the disease gene basically shows no recombination with the disease trait, but markers further distant or on other chromosomes show a segregation pattern that is completely different. As Fig. 1.2 illustrates, the marker B is more closely linked to the disease than either marker A or C. The marker allele B-2 segregates with the disease trait, whereas for the other two markers recombination events are observed.

The significance of observed linkage depends on the number of meioses in which the two loci remain linked. It is intuitively obvious that the observation of linkage in 4 meioses is less significant than the observation of linkage in 20 meioses. A measure for the likelihood of linkage is the logarithm of the odds (lod) score. The lod score Z is the logarithm of the odds that the loci are linked divided by the odds that the loci are unlinked (Ott, 1999). An analysis of a pedigree with formal lod score analysis can also be found in Chapter 12 (Table 12.1 and Figure 12.1).

Once the relationship between a disease trait and DNA markers is established, genetic linkage can be used to follow the inheritance of the chromosomal region containing the mutant allele. This is discussed further in Section IV.B.

## 1. Linkage Disequilibrium and Association

Linkage and association should not be confused with one another. Linkage relates to the physical location of genetic loci and refers to the relationship of *loci*. Association describes the concurrence of a specific allele with another trait, and thus refers to the relationship of *alleles* at a frequency greater than predicted by chance. In order to study association, one has to determine allele frequencies in unrelated cases and compare them to the allele frequencies found in controls. For association studies, it is imperative to repeat the analysis in different patient populations to minimize effects attributable to population stratification, particularly when individuals with the disease belong to a genetically distinct subset of the population. This problem can be circumvented by using parents as controls (transmission disequilibrium test, TDT). The transmission of specific parental alleles to affected offspring is scored and statistically analyzed (for recent review see George and Laud, 2002).

Association studies may point to genetic factors involved in the pathogenesis or susceptibility of a disease (see also Fig. 30.8). Association studies are very powerful for the detection of alleles that have a relatively small effect, but also require that the tested polymorphism be relatively close to the genetic variant that is responsible for the phenotypic variation. This often means that the polymorphism itself has a causative effect or is at least located within the gene of interest.

Linkage disequilibrium refers to the occurrence of specific alleles at two loci with a frequency greater than expected by chance. If the alleles at locus A are a1 and a2 with frequencies of 70 and 30%, and alleles at the locus B are b1 and b2 with frequencies of 60% and 40%, the expected frequencies of haplotypes would be a1b1 = 0.42, a1b2 = 0.28, a2b1 = 0.18, and a2b2 = 0.12. Even if the two loci were closely linked, unrestricted recombination would result in allelic combinations that are close to the frequencies given above. When a particular combination occurs at a higher frequency, for example a2b2 at a frequency of 25%, this is called linkage disequilibrium.

At the population level, linkage disequilibrium can be used to investigate processes such as mutation, recombination, admixture, and selection. It is a powerful tool for genetic mapping. When a disease mutation arises on a founder chromosome and not much time has elapsed since the mutational event, the disease mutation will be in linkage disequilibrium with alleles from loci close to the gene.

### F. Types of Mutations

Mutations (heritable changes) in DNA are the basis of genetic variation. Types of mutations range from changes of a single base pair in a gene, to deletions or duplications of entire exons or even genes. Some mutations involve large alterations of parts of a chromosome or involve an entire chromosome such a trisomy for chromosome 21. Chromosome abnormalities cause a significant proportion of genetic disease. They are the leading cause of pregnancy loss and mental retardation. Structural chromosomal abnormalities usually affect several genes and cause phenotypes involving malformations and dysfunction of several organ systems. Therefore, structural chromosome abnormalities would be a very rare cause of a phenotype dominated by a movement disorder without dysmorphologic features or involvement of other organ systems. Smaller deletions, however, can involve a single gene or parts of a gene, for example, deletions involving the parkin gene.

#### 1. Single Gene Mutations

Mutations that affect single genes are most commonly located in the coding region of genes, at intron-exon boundaries, or in regulatory sequences. These changes can be due to changes of single base pairs, or insertions/deletions of one or multiple base pairs.

Single base pair substitutions—also called point mutations—result when a single base pair is replaced by another base pair. Many single base pair mutations, even when located in the coding region of a gene, may not change the amino acid sequence due to the redundancy of the genetic code. These are often located in the third position of a codon and are called silent substitutions. Although a base pair substitution may not change the amino acid sequence, it may nevertheless be disease causing by introducing cryptic splice signals.

Others base pair substitutions may result in a change in the amino acid sequence (missense mutation) or produce a change into one of the three stop codons (nonsense mutation). A third type of mutation may change an existing stop codon into an amino acid coding codon (stop codon mutation), resulting in the translation of an elongated polypeptide.

Not all amino acid changes are pathogenic. Especially when the amino acid change is conservative (substitution with a similar amino acid), the resulting protein may represent a normal variant and have normal function. At times it may be difficult to distinguish a normal variant from a disease-causing sequence change, especially when the function of a protein is poorly understood. Such an example is discussed with regard to a missense change in the D2 dopamine receptor (DRD2) gene (see Chapter 40).

Deletions or insertions of one or several base pairs of DNA represent another mutation type. If the change involves 3 bp or a multiple thereof, amino acids are added or deleted from the protein, but the reading frame and the remainder of the protein remain intact. Deletion of a single GAG codon in torsin is sufficient to cause dystonia type I (DYT1) (see Chapter 36).

Deletions/insertions that are not a multiple of three will alter the codons and the resulting amino acid sequence downstream of the deletion/insertion. This change in the reading frame (frameshift mutations) usually results in a shortened polypeptide, because the frameshift will result in the recognition of a premature stop codon.

#### 2. Splice Site Mutations

Base pair substitutions or insertions/deletions can also interfere with the proper processing of the primary transcript. These splice site mutations alter the GT sequence at the 5′-donor site or the AG sequence at the 3′-acceptor site or alter sequences near these sites (see boxed insert in Fig. 1.6). These mutations may result in the deletion of an entire exon from the mature transcript or may result in the recognition of normally unused or cryptic splice sites. Depending on the number of bases in the exon, exon deletion by splice site mutations can leave the reading frame intact or change it, often resulting in a truncated protein.

Truncated and usually nonfunctional protein products can thus be generated by nonsense mutations, by deletions/insertions resulting in a frame-shift, and by exon deletion due to small interstitial chromosomal deletion. Shortened proteins can also be generated by skipping of exons due to mutations involving exon/intron boundaries.

### III. MOLECULAR GENETIC TOOLS

This section will briefly introduce the reader to some of the methods used in molecular genetic diagnosis. Detailed protocols and more in-depth descriptions of various procedures can be found in laboratory manuals and books about molecular biology (Ausubel *et al.*, 1999, Pulst, S. M.,

## A. Polymerase Chain Reaction

DNA-based diagnosis was revolutionized by the introduction of the polymerase chain reaction (PCR; Saiki *et al.*, 1985; Mullis, 1990). In an ingeniously simple procedure PCR combines the sequence specificity of restriction enzymes with amplification previously only possible by cloning of restriction fragments. Sequence specificity is provided by the annealing of oligonucleotide probes complementary to the DNA sequence of interest, and amplification is achieved by repeated rounds of oligonucleotide-primed DNA synthesis.

The specificity of PCR is primarily determined by the annealing temperature. When the annealing temperature is lowered, the primers will anneal to DNA sequences that are not perfectly matched and other fragments may be amplified. For molecular diagnosis, however, it is of utmost importance that amplification occur in a specific fashion both for normal and mutated alleles (Fig. 1.8).

Almost all DNA-based testing involves PCR. Mutations involving deletions/insertions can be detected as a variation in amplicon length. This approach is typically used for disorders involving DNA repeats such as the polyQ disorders or for Friedreich ataxia. For the detection of base pair substitutions PCR serves as the basis for sequence analysis of mutant alleles. For the detection of single base changes or smaller deletions/insertions that change restriction sites PCR amplification is often followed by digestion of PCR products with restriction endonucleases.

Although PCR is a highly specific and sensitive methodology it has technical limitations that need to be recognized when interpreting DNA test results. Any setting that results in nonamplification of the mutant allele may result in a false-negative DNA test, because amplification from the wild-type allele will proceed normally. Thus,

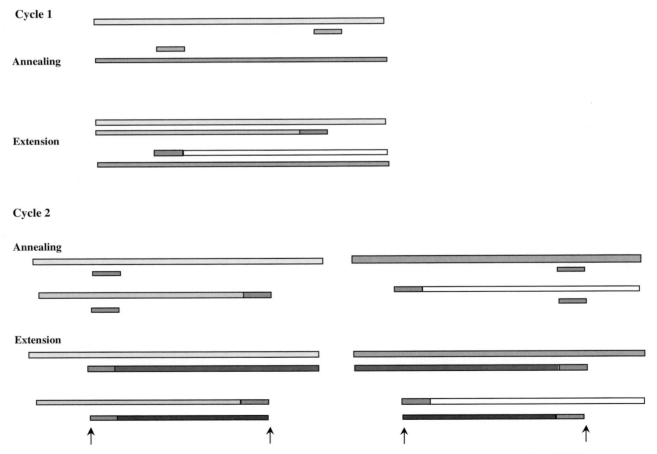

**FIGURE 1.8** Schematic of the polymerase chain reaction (PCR). Annealing of primers and extension of DNA fragments are shown for two cycles. Note that the DNA fragments shown between the arrows have a length defined by the spacing of the oligonucleotide primers.

non-amplification of the mutant allele may appear as homozygosity for the wild-type allele.

For example, elongation of the PCR product may be impeded by secondary structure of the intervening DNA sequence or by large insertions. This is particularly relevant for mutations caused by large DNA repeats. If a mutation disrupts annealing of the oligonucleotide primer, an amplicon from the mutant allele is not generated, and thus will escape detection. Similarly, if a genomic deletion involves the entire length of DNA that is to be amplified (such as an entire exon) it will not be detected, because the homologous DNA on the other chromosome will generate an amplicon.

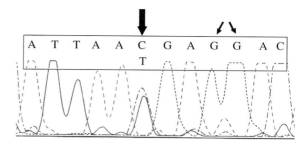

FIGURE 1.9 Automated sequencing; note that peak height varies for DNA bases based on sequence context (short arrows). A DNA variant is detected as a double peak (long arrow). In this case peaks for both bases have approximately the same height, but this is not necessarily the rule, leading to potential errors in automated base calling.

### B. Southern Blot Analysis

The nature of some mutations may require the use of Southern blot analysis. Southern blot analysis refers to a transfer or DNA fragments to a support membrane after separation of the fragments by electrophoresis. DNA fragments are generated in a sequence-specific fashion by digestion with restriction enzymes resulting in DNA fragments cleaved at specific DNA sequences. These are then separated according to size by gel electrophoresis, and DNA fragments transferred to a nylon membrane by capillary action (Southern *et al.*, 1975). The relevant fragment is detected by hybridization with a complementary DNA probe.

Southern blot analysis is technically more demanding and time-consuming then PCR analysis, and its use in DNA diagnosis is limited to settings when PCR analysis is not possible, because Taq polymerase cannot efficiently amplify across very large repeat expansions. An example is the detection of large DNA repeat expansion mutations such as SCA10 (see Chapter 11). Southern blot analysis is also used to detect large genomic rearrangements.

### C. DNA Sequencing

Several sequencing methods exist. Only the chain termination method will be further described, because it is the method currently used for automated sequencing. In this method, a DNA fragment is provided as template for the synthesis of new DNA strands using a DNA polymerase (Sanger *et al.*, 1977). The reaction is primed by a sequencing primer usually 17–22 bp in length which specifically binds to the region being sequenced. In addition to regular deoxynucleotide triphosphates (dNTPs, N stands for any of the four bases), smaller amounts of *dideoxy*nucleotide triphosphates (ddNTPs) are added to the reaction mix. Although ddNTPs are incorporated into the newly synthesized DNA chain, they cause abrupt termination of the chain due to lack of the 3′ hydroxyl group preventing formation of the phosphodiester bond (Fig. 1.9).

Chain termination will occur randomly, whenever a ddNTP is incorporated into the growing chain instead of dNTP. The length of the fragments can be determined by polyacrylamide gel or capillary electrophoresis and the fragments visualized by labeling with radioisotopes or fluorophores.

Although sequence-based mutation detection may appear straightforward, technical difficulties can occur. DNA structure may interfere with the abundance of a particular fragment (seen as a weak band in a radioactive gel or a decreased peak in automated sequencing). Since most sequencing for mutations occurs on genomic DNA (with two alleles present except for X-chromosomal sequences in males), detection of sequence differences occurs in the heterozygote state. A weak peak marking a single base mutation may be masked by the presence of a normal-sized peak from the normal allele. This can be circumvented by adjustments in sequence analysis software, by sequencing in forward and reverse directions, and by use of different sequencing primers.

### D. Analysis of RNA Transcripts and Proteins

Genetic tests are certainly not limited to the analysis of DNA (see Chapter 45). Metabolic disorders have long been diagnosed by determination of enzyme activity (see Chapter 24). RNA molecules can be analyzed directly in a process called Northern blotting in a play of words on Southern blotting. Using reverse transcribed PCR which transcribes RNA into DNA and then amplifies it, RNA molecules can be analyzed for size changes. This strategy usually requires expression of the respective gene in an easily accessible tissue such as blood or skin. Analysis of RNA transcripts is particularly useful, when mutations involve deletion of entire exons (which may be missed by exon-specific genomic amplification) resulting in changes of amplicon length.

In a process called Western blotting proteins can be transferred to a membrane and analyzed using antibodies.

Truncating mutations in genes with long open reading frames can also be detected by *in vitro* translation of the RNA and detection of truncated proteins by gel electrophoresis.

### E. Detection of DNA Polymorphisms

The simplest and most abundant DNA polymorphisms are those that are generated by substitution of a single base pair. These polymorphisms are designated SNPs for single nucleotide polymorphisms. SNPs are almost exclusively bi-allelic and thus not highly polymorphic, but this limitation is overcome by their frequency of approximately one polymorphism per 1000 bp. The ability to probe for SNPs using DNAs arrayed on a chip or glass slide provides the prospect of automated genotyping at extremely high resolution. The first genetic maps based on SNPs are being assembled (Wang *et al.*, 1998). SNPs are increasingly used for the analysis of polygenic or complex traits, for example, SNPs in the tau gene in Parkinson's disease (Martin *et al.*, 2001).

DNA variants are not restricted to SNPs. Length variations in short tandem repeats such as bi- or trinucleotide repeats have commonly been used for linkage analysis (Weber and May, 1989). At this point in time, short tandem repeats (STRs) are the most commonly used DNA polymorphisms for genetic testing by segregation analysis (see Section IV.B).

## IV. MOLECULAR GENETIC TESTING

Despite the fact that all mutation detection is in the end sequence-based, various technologies and strategies are employed depending on the nature and frequency of mutations and the size of the gene to be analyzed. DNA-based diagnosis falls into two broad categories: Direct mutation detection and indirect detection of a mutated gene using segregation (linkage) analysis. When the disease gene has been identified and the types of mutations are limited, direct mutational analysis is possible. With ever improving sequencing technology even genes with multiple non-recurring mutations distributed over many exons such as the *CACNA1A* gene are now amenable to direct mutation analysis.

For some diseases, however, a large number of different mutations exist without mutational "hot spots," so that indirect testing is technically more feasible than direct mutational analysis. A similar situation arises when the chromosomal location of a disease gene is known, but the gene itself has not been identified. In the above two scenarios, DNA diagnosis can be performed using polymorphic DNA markers, which are used to track the disease chromosome in a given family. As discussed below, there are many important differences between direct and indirect testing with regard to requirements for DNAs from affected family members, and the sensitivity and specificity of the information obtained.

### A. Direct Mutation Detection

For diseases caused by a specific mutation, direct detection of the mutation is simple and inexpensive. The direct genetic test more closely resembles a conventional laboratory test. A blood sample is taken from a patient and the test is used to determine whether the individual carries a *specific mutation* in a given gene with a yes/no answer. Barring technical problems resulting in false-positive results, a positive test indicates that a symptomatic individual has the disease. A positive result is independent of the accuracy of the clinical assessment.

As with all laboratory tests, clinical judgement is the final arbiter. It is possible that a patient has *symptomatic* multiple sclerosis and an inherited ataxia as indicated by a DNA test for which she is presymptomatic. There are two recent examples that illustrate this point. Cuda *et al.* (2002) describe a 5-year-old boy with dysarthria and a very severe hypertrophic cardiomyopathy. On molecular analysis, he was found to have a double mutation in two distinct genes, the frataxin gene and the cardiac troponin T gene. More importantly due to the therapeutic implications, Nicholl *et al.* (2001) describe a family with autosomal dominant tremor. One child developed a dystonic tremor and was found to have Wilson disease.

Examples of direct mutation testing include diseases caused by DNA repeats. The repeat is amplified by PCR and the allele with the expansion can be detected as a larger than normal fragment by gel electrophoresis. The same is true for other mutations that result in deletion or additions of DNA codons.

A negative test in a symptomatic individual needs to be interpreted realizing its limitations. Phenotype and gene test need to be correctly matched. If a patient with a dominant spinocerebellar ataxia (SCA) has a negative test for SCA1 or SCA3, this does not mean that this individual does not have a dominant SCA, because the mutation could be in one of the other SCA loci. Alternatively, the mutation could be in the respective gene, but it was not assayed by the gene test that was performed. For example, a progressive ataxia can be caused by expansion of a polyglutamine repeat in the $\alpha_{1A}$ voltage-dependent calcium channel subunit encoded by the CACNL1A gene on chromosome 19p. Specific missense mutations in this gene, which would not be detected by a PCR-based test for CAG repeat length, can also be associated with a progressive ataxia (see Chapter 22).

Direct sequence analysis is increasingly used to analyze genes for point mutations. The availability of chip technology will undoubtedly broaden the applicability of sequence analysis to larger genes with multiple different mutations (Hacia *et al.*, 1996). Direct sequence analysis is particularly useful when several different mutations are clustered in one exon and can be analyzed by sequencing of a single PCR product. Sequence analysis of entire genes is possible as well, initially only available in research laboratories, but now also performed commercially.

Most mutations are present in the heterozygous state. The detection of single base deletions or insertions is easy to detect, because instead of a single peak (or band) double peaks (or double bands) are seen downstream of the mutation. On the other hand, missense mutations are more difficult to detect, because a double signal is only present in a single position (Fig. 1.9). Therefore, the DNA sequence has to be of sufficient quality to recognize the presence of two different DNA bases. Since it is unknown where the mutation might occur, the quality of the sequence has to be high throughout the entire region that is being sequenced. Less than 100% sensitivity of mutation detection by sequencing has its causes in these technical limitations, but it can also be due to the presence of mutations outside the sequenced regions (such as intronic or promotor mutations or large scale deletions), or in nonallelic heterogeneity.

The differentiation of non disease-causing polymorphisms from pathogenic mutations may not be straightforward. This is illustrated by the fact that changes in the DRD2 receptor were found in patients with myoclonus dystonia, but it is not yet resolved whether these DNA changes are disease-causing (see Chapter 40). Similarly, some alleles may represent mutations associated with reduced penetrance. Examples include HD and SCA2 alleles with intermediate CAG expansion or Friedreich alleles associated with very late onset. Widespread application of sequence analysis to disease genes will undoubtedly result in the detection of sequence variants that at the time of detection may be of unknown biologic significance.

## B. Indirect Analysis Using Linked Genetic Markers

For indirect testing, samples from at least two affected first degree relatives are required. Using polymorphic DNA markers near or in the gene of interest, phase of the marker alleles and the disease are established (see also Fig. 1.2). Once phase is established, this information can be used to track the disease chromosome in other family members. Tracking of the disease chromosome (or more precisely the chromosomal *region* that contains the gene) can be performed independent of the precise mutation causing the disease. Thus, if a disease is caused by a great number of different mutations in a large gene, indirect testing is at the moment a feasible strategy.

In contrast to direct mutational testing, the accuracy of indirect molecular testing is dependent on the correctness of the diagnosis in all individuals used for testing. In addition, nonallelic heterogeneity must be absent for the phenotype being examined. In other words the phenotype must be caused by mutations in one locus only. Therefore, it would not be possible to use indirect testing in families with Parkinson's disease unless the pedigree size is sufficiently large to establish linkage to one of the PD loci prior to testing of specific individuals.

## V. ANIMAL MODELS

Genetic technologies have provided the tools to create animal models resembling human diseases even if the respective diseases do not naturally occur in these animals. In principle, these technologies can be applied to any animal, but *mice* (*Mus musculus*), zebra fish (*Danio rerio*), fruitfly (*Drosophila melanogoaster*), and the worm (*Cenorhabditis elegans*) are most commonly used. Two main approaches to alter the genetic constitution of animals are utilized. The first involves insertion of a novel, often mutated gene into the germline, the second an alteration of endogenous genes by gene targeting or random mutagenesis.

### A. Expression of Foreign Transgenes

Transgenes can be injected into fertilized oocytes or introduced into embryonic stem (ES) cells. Multiple copies of the transgene usually integrate into the host genome and after integration the transgene is transmitted to offspring. The expression of transgenes in mice does not necessarily mirror the expression of a mutated allele in human disease conditions. First, the transgene is expressed in addition to the two endogenous mouse alleles. Often multiple copies of the transgene insert into the mouse genome. Second, unless the endogenous promoter is used, expression may be higher than under physiologic conditions, expression may occur in different sets of cell types, and expression may not have the correct temporal profile. Even if the physiologic promoter is used the transgene construct may lack all the appropriate regulatory elements. Finally, by chance the transgene can integrate into an endogenous mouse gene and disrupt proper functioning of this gene (so called insertional mutagenesis).

Despite these shortcomings expression of transgenes with mutated human cDNAs can mirror important morphologic and functional aspects of human movement disorders as shown by animal models of polyglutamine diseases (for example, Burright *et al.*, 1995; Huynh *et al.*, 2000). It is important to compare phenotypes obtained by expression of the wild-type gene with those obtained by expression of a mutant gene to discern effects of the mutation versus those of overexpression of a normal protein.

## B. Gene Targeting (Knock-Out, Knock-In)

Using homologous recombination, a cloned gene or gene segment can be exchanged for the endogenous gene (Thomas *et al.*, 1986; Doetschman *et al.*, 1987). By inserting mutations into the cloned gene this mutation can now be transferred into the animal cell. When the cloned gene contains a nonsense mutation, the gene will be inactivated after recombination (knock-out).

Gene targeting can be achieved in somatic cells, but has its greatest value when it is performed with ES cells. By backcrossing chimeric animals, mice can be generated that are heterozygous for the mutation in all tissues. In these animals, the gene is now transcribed only from the one remaining normal allele. By mating heterozygous animals, homozygous offspring are obtained in which no functional gene product is made. For a gene with the symbol Gs, the animals thus produced are designated $Gs^{+/+}$ for the wildtype, $Gs^{+/-}$ for the heterozygote, and $Gs^{-/-}$ for the homozygously deleted animal.

Instead of a nonsense mutation, the gene fragment used for homologous recombination can also contain a missense mutation (knock-in). Such animals are potentially closer models of human disease than mice expressing a transgene, because the proper endogenous promoter is used, and gene dosage is not disturbed (see for example, Miranda *et al.*, 2002; Wheeler *et al.*, 2000).

A problem with homozygous knock-outs for many genes is that embryonic lethality results, because the respective genes have a function not only in adult tissues but also during embryogenesis. This can be circumvented by generating chimeric animals that carry a mixture of wild-type and homozygously deficient cells. By using the Cre-loxP recombination system, mice containing alleles with inserted loxP sites can be mated with mice expressing a Cre-recombinase transgene under the control of a tissue-specific promoter so that function of a gene is only abolished in a specific tissue (see for example, Puccio *et al.*, 2001).

## VI. WEB-BASED INFORMATION FOR GENETIC DIAGNOSIS AND TESTING

A large amount of information is contained in publicly available databases. Table 1.1 lists samples of different kinds of the databases that in turn contain links to other relevant sites. These may contain extensive information on the genetic and physical location of DNA markers. More recently, these databases also contain the location of genes, and the expression pattern of genes.

A classic resource is *Mendelian Inheritance in Man*. It is now available as a web-based version called *Online Mendelian Inheritance in Man* (OMIM). The site is easily searchable by disease name, gene symbol, or name of a protein. Its advantages are that it is comprehensive and very well updated. Several editors regularly perform literature searches and add information to disease categories. This is also one of its disadvantages in that information for some entries is lengthy and not revised in its entirety. Entries identified by searches of the Pubmed database are often linked to corresponding entries in OMIM.

GeneClinics (now combined into GeneTests-GeneClinics) is a site maintained by the University of Washington through funding from the NIH, HRSA, and DOE. It provides information on genetic diseases and associated tests including academic and commercial laboratories. At

TABLE 1.1 Sites on the Worldwide Web with Molecular-Genetic Databases

| | |
|---|---|
| *http://www.Geneclinics.org* | Online textbook of selected diseases and information on commercial and research laboratories |
| *http://www.ncbi.nlm.nihgov/Omim* | Online Mendelian Inheritance in Man |
| *http://www.gene.ucl.ac.uk/hugo* | Human Genome Organization |
| *http://www.ncbi.nlm.nih.gov/entrez/query.fcgi* | Online search of the medical literature |
| *http://www.neuro.wustl.edu/neuromuscular* | Clinical and research data on neuromuscular syndromes including ataxias |
| http://www.uwcm.ac.uk/uwcm/mg/hgmd0.html | Human Gene Mutation Database |
| *http://hgbase.interactiva.de* | Human gene polymorphisms |
| *http://websvr.mips.biochem.mpg.de/proj/medgen/mitop* | Mitochondrial information |
| *http://fruitfly.berkeley.edu* | Berkeley Drosophila Genome Project |
| *http://www.ataxia.org* | National Ataxia Foundation |
| *http://www.pcnet.cpm/~orphan/* | National Organization for Rare Disorders |
| *http://www.who.ch/ncd/hgn/hgn-home.htm* | WHO Human Genetics Programme |
| *http://www.wiley.com/genetherapy/* | Gene Therapy |
| *http://esg-www.mit.edu:8001/esgbio/7001main.html* | Biology Project |
| *http://www.biology.arizona.edu* | The Biology Project (site for basics in biology and genetics) |
| *http://www.mblab.gla.ac.uk~julian/Dict.html* | A dictionary of cell biological terms |
| *http://morgan.rutgers.edu/* | A well-illustrated genetics tutorial |
| *http://www.mc.vanderbilt.edu/gcrc/gene/index.html* | Introduction to Gene Therapy Home Page |

the time of publication the number of movement disorders covered in GeneClinics is still small. Disease reviews in the site are well organized, but not continuously updated.

A site primarily dedicated to neuromuscular diseases is maintained by the Neurology Department at Washington University. This site has some tabular information on inherited ataxias.

## Acknowledgments

The author thanks Dr. Hema Vakharia for critically reading the manuscript, Mr. Michael Goldsmith and Karla (Pattie) Figueroa, M.S., for help with illustrations, and Ms. Cloteal Moran for assistance with references.

## References

Adams, C., Starkman, S., and Pulst, S. M. (1997). Phenotype of SCA2 in a large kindred from Southern Italy. *Neurol.* **49**, 1163–1166.

Alarcon, M., Cantor, R. M., Liu, J., Gilliam, T. C., Geschwind, D. H. (2002). Evidence for a language quantitative trait locus on chromosome 7q in multiplex autism families. *Am. J. Hum. Genet.* **70**(1), 60–71.

Ausubel, F. M., Brent R., Kingston, R. E., and Moore, D. D. (1999). "Short Protocols in Molecular Biology," 4th Edition. John Wiley and Sons, New York.

Bennett, R. L., Steinhaus, K. A., Uhrich, S. B., O'Sullivan, C.K., Resta, R. G., Lochner-Doyle, D., Markel, D. S., Vincent, V., and Hamanishi, J. (1995). Recommendations for standardized human pedigree nomenclature. Pedigree Standardization Task Force of the National Society of Genetic Counselors. *Am. J. Hum. Genet.* **56**, 745–752.

Blumberg, R. B., Mendel Web Archive (2002). http://www.netspace.org/MendelWeb/MWtoc.html.

Botstein, D., White, R. L., Skolnick, M., and Davis, R. W. (1980). Construction of a genetic linkage map in man using restriction fragment length polymorphisms. *Am. J. Hum. Genet.* **32**, 314–331.

Burright, E. N., Clark, H. B., Servadio, A., Matilla, T., Federsen, R. M., Yunis, W. S., Duvick, L. A., Zoghbi, H. Y., and Orr, H. T. (1995). SCA1 transgenic mice; a model for neurodegeneration caused by an expanded CAG trinucleotide repeat. *Cell* **82**, 937–948.

Doetschman, T., Gregg, R. G., Maeda, N., Hooper, M. L., Melton, D. W., Thompson, S., and Smithies, O. (1987). Targeted correction of a mutant HPRT gene in mouse embryonic stem cells. *Nature* **330**, 576–578.

Gasser, T., Windgassen, K., Bereznai, B., Kabus, C., and Ludolph, A.C. (1998). Phenotypic expression of the DYT1 mutation: a family with writer"s cramp of juvenile onset. *Ann. Neurol.* **44**, 126–128.

Genetics Science learning center web site (2002.) http://gslc.genetics.utah.edu/.

George V., and Laud, P. W. (2002). A Bayesian approach to the transmission/disequilibrium test for binary traits. *Genet. Epidemiol.* **22**, 41–51.

Gusella, J. F., Wexler, N. S., Conneally, P. M., Naylor, S. L., Anderson, M. A., Tanzi, R. E., Watkins, P. C., Ottina, K., Wallace, M. R., Sakaguchi, A. Y., Younn, A. B., Shoulson, I., Bonilla, E., and Martin, J. B. (1983). A polymorphic DNA marker genetically linked to Huntington's disease. *Nature* **306**, 234–238.

Hacia, J. G., Brody, L. D., Chee, M. S., Fodor, S. P., and Collins, F. S. (1996). Detection of heterozygous mutations in BRCA1 using high density oligonucleotide arrays and two-color fluorescence analysis. *Nat. Genet.* **14**, 441–447.

Human Genome program web site (2002). http://www.ornl.gov/TechResources/Human_Genome

Huynh, D. P., Figueroa, K., Hoang, N., and Pulst S. M. (2000). Nuclear localization or inclusion body formation of ataxin-2 are not necessary for SCA2 pathogenesis in mouse or human. *Nat. Genet.* **26**, 44–50.

Jackson, J. F., Currier, R. D., Teresaki, P. I., and Morton, N. E. (1977). Spinocerebellar ataxia and HLA linkage: risk prediction by HLA typing. *N. Engl. J. Med.* **296**, 1138–1141.

Jorde, L. B., Carey, J. C., Bamshad, M. J., and White, R. L. (2000). "Medical Genetics." 2nd revised edition, Mosby-Year Book, St. Louis, MO.

Miranda, C. J., Santos, M. M., Ohshima, K., Smith, J., Li, L., Bunting, M., Cossee, M., Koenig, M., Sequeiros, J., Kaplan, J., and Pandolfo, M. (2002). Frataxin knockin mouse. *FEBS Lett.* **512**, 291–297.

Martin, E. R., Scott, W. K., Nance, M. A., Watts, R. L., Hubble, J. P., Koller, W. C., Lyons, K., Pahwa, R., Stern, M. B., Colcher, A., Hiner, B. C., Jankovic, J., Ondo, W. G., Allen, F. H., Goetz, C. G., Small, G. W., Masterman, D., Mastaglia, F., Laing, N. G., Stajich, J. M., Ribble, R. C., Booze, M. W., Rogala, A., Hauser M. A., Zhang, F., Gibson, R. A., Middleton, L. T., Roses, A. D., Haines J. L., Scott, B. L., Pericak-Vance, M.A., and Vance, J. M. (2001). Association of single-nucleotide polymorphisms of the tau gene with late-onset Parkinson disease. *JAMA* **286**, 2245–2250.

Mullis, K. B. (1990). The unusual origin of the polymerase chain reaction. *Sci. Am.* **262**, 56–65.

Nicholl, D. J., Ferenci, P., Polli, C., Burdon, M. B., and Pall, H. S. (2001). Wilson's disease presenting in a family with an apparent dominant history of tremor. *J. Neurol. Neurosurg. Psychiatry* **70**, 514–516.

Nicholls, R. D., Saitoh, S., and Horsthemke, B. (1998). Imprinting in Prader-Willi and Angelman syndromes. *Trends Genet.* **14**, 194–200.

Ott, J. (1999). "Analysis of Human Genetic Linkage." Baltimore, MD. Johns Hopkins Univ. Press, 3rd edition.

Papavassiliou, A. G. (1995). Transcription factors. *New Engl. J. Med.* **332**, 45–47.

Pauls, D. L., and Korczyn, A. D. (1990). Complex segregation analysis of dystonia pedigrees suggests autosomal dominant inheritance. *Neurology* **40**, 1107–1110.

Puccio, H., Simon, D., Crossee, M., Criqui-Filipe, P., Tiziano, F., Melki, J., Hindelang, C., Matyas, R., Rustin, P., and Koenig, M. (2001). Mouse models for Friedreich ataxia exhibit cardiomyopathy, sensory nerve defect and fe-S enzyme eficiency followed by intramitochondrial iron deposits. *Nat. Genet.* **27**, 181–186.

Pulst S. M. (2000). "Neurogenetics." Oxford University Press, New York.

Pulst, S. M. (1999). Genetic linkage analysis. *Arch. Neurol.* **56**, 667–672.

Saiki R. K., Scharf, S. J., Faloona, F., Mullis, K. B., Horn, G. T., Erlich, H. A., and Arnheim, N. (1985). Enzymatic amplification of beta-globin sequences and restriction site analysis for diagnosis of sickle cell anemia. *Science* **230**, 1350–1354.

Saito, T. (1988). An expected decrease in the incidence of autosomal recessive disease due to decreasing consanguineous marriages. *Genet. Epidemiol.* **5**, 421–432.

Sanger, F., Nicklen, S., and Coulson, A. R. (1977). DNA sequencing with chain-terminating inhibitors. *Proc. Natl. Acad. Sci. USA* **74**, 5463–5467.

Shimura, H., Schlossmacher, M. G., Hattori, N., Frosch, M. P., Trockenbacher, A., Schneider, R., Mizuno, Y., Kosik, K. S., and Selkoe, D. J. (2001). Ubiquitination of a new form of alpha-synuclein by parkin from human brain: implications for Parkinson's disease. *Science* **293**, 263–269.

Simon, D. K., Pulst, S. M., Sutton, J. P., Browne, S. E., Beal, M. F., and Johns, D.R. (1999). Familial multisystem degeneration with parkinsonism associated with the 11778 mitochondrial DNA mutation. *Neurology* **53**, 1787–1793.

Southern, E. M. (1975). Detection of specific sequences among DNA fragments separated by gel electrophoresis. *J. Mol. Biol.* **98**, 503–517.

Strachan, T., and Read, A. P. (1999). "Human Molecular Genetics." Bios Scientific Publishers Ltd., Oxford.

Sutton, J. P., and Pulst, S. M. (1997). Atypical parkinsonism in a family of Portuguese ancestry: absence of CAG repeat expansion in the MJD1 gene. *Neurology* **48**, 1285–1290.

Thomas, K. R., and Capecchi, M. R. (1986). Introduction of homologous DNA sequences into mammalian cells induces mutation in the cognate gene. *Nature* **324**, 34–38.

von Hippel, P. H. (1998). An integrated model of the transcription complex in elongation, termination, and editing. *Science* **281**, 660–665.

Wang, D. G., Fan, J. B., Siao, C. J. Berno, A., Young, P., Sapolsky, R., Ghandour, G., Perkins, N., Winchester, E., Spencer, J., Kruglyak, L., Stein, L., Hsie, L., Topaloglou, T., Hubbell, E., Robinson, E., Mittmann, M., Morris, M. S., Shen, N., Kilburn, D., Rioux, J., Nusbaum, C., Rozen, S., Hudson, T. J., Lipshultz, R., Chee, M., and Lander, E. S. (1998). Large-scale identification, mapping and genotyping of single-nucleotide polymorphisms in the human genome. *Science* **280**, 1077–1082.

Watson, J. D., Gilman, M., Witkowski, J., and Zoller, M. (1992). *Recombinant DNA*, 2nd ed., W. H. Freeman, New York.

Weber, J. L., and May, P. E. (1989). Abundant class of human DNA polymorphisms which can be typed using the polymerase chain reaction. *Am. J. Hum. Genet.* **44**, 388–396.

Wheeler, V. C., White, J. K., Gutekunst, C. A., Vrbanac, V., Weaver, M., Li, X. J., Li, S. H., Yi, H., Vonsattel, J. P., Gusella, J. F., Hersch, S., Auerbach, W., Joyner, A. L., and MacDonald, M. E. (2000). Long glutamine tracts cause nuclear localization of a novel form of huntingtin in medium spiny striatal neurons in HdhQ92 and HdhQ111 knock-in mice. *Hum. Mol. Genet* **9**, 503–513.

# CHAPTER 2

# Inherited Ataxias: An Introduction

STEFAN-M. PULST

*Division of Neurology, Cedars-Sinai Medical Center*
*Departments of Medicine and Neurobiology, UCLA School of Medicine*
*Los Angeles, California 90048*

I. Classification
   A. Autosomal Dominant Axis
II. Rencently Identified Ataxias
   A. New SCA Loci
   B. Spastic Ataxia
   C. Ataxia with Oculomotor Apraxia
   D. Other Recessive Ataxias
III. Worldwide Prevalence
IV. Mechanisms of Disease
   A. Anticipation
   B. DNA Testing
   C. Sporadic Ataxia
   D. Neuroimaging and Neuropathology
   E. Models
V. Genotype/Phenotype Correlations and Modifying Loci
VI. Progression and Treatment
   Acknowledgments
   References

As briefly reviewed in the previous chapter, the inherited ataxias have in many ways been at the forefront of the genetic revolution in neurology. What is now known as spinocerebellar ataxia type 1 (SCA1), was the first neurologic disorder (excluding enzyme defects) mapped to a human chromosome and the SCAs led the way in understanding novel mechanisms of disease pathogenesis. Notwithstanding the importance of the genotype in classifying and understanding movement disorders (Rosenberg, 1995), the ataxias already provide a glimpse into the future where different genotypes are again grouped together based on the effect of the mutation at the protein level. The term polyglutamine (polyQ) diseases was only possible through gene discovery, but points to a new mechanistic understanding to group diseases based on the effects on the transcriptome or proteome. Other ataxias, although caused by mutations in entirely unrelated genes (see Table 2.1), might in the future be grouped under mitochondrial iron homeostasis disorders (FRDA and ASAT) or ataxias with deficiencies in DNA repair (ATM and AOA1).

## I. CLASSIFICATION

Ataxias are characterized by variable degrees of dysfunction and degeneration of systems subserving motor coordination. Although many of these disorders lead to dysfunction of Purkinje cells, others involve the deep cerebellar nuclei, brainstem nuclei, and spinal sensory as well as spinocerebellar tracts. Prominent neuronopathy or neuropathy may further contribute to imbalance and incoordination.

The phenotypic classification of the ataxias has been confusing for many decades (reviewed in Pulst and Perlman, 2000). Few of the designations given to specific ataxias have withstood the test of time as disease entities. Phenotypic overlap among the different SCAs is significant. On the other hand some of the SCAs may not initially present with prominent ataxia. For example in SCA12 or SCA17, other symptoms such as tremor or dementia may predominate or may be first symptoms of the disease. One of the few entities that has withstood the test of time is that of Friedreich ataxia. But even with this disease entity the

TABLE 2.1  Recessive and Dominant Ataxias Excluding Most Metabolic and Mitochondrial Ataxias

| Disease name | Alternate names | Inheritance | OMIM number | Chromosomal location | Protein name |
|---|---|---|---|---|---|
| SCA1 | Spinocerebellar ataxia type 1 | AD | 164400 | 6p | Ataxin-1 |
| SCA2 | | AD | 183090 | 12q | Ataxin-2 |
| SCA3 | Machado-Joseph disease | AD | 109150 | 14q | Ataxin-3 |
| SCA4 | | AD | 600223 | 16q | |
| SCA5 | Lincoln ataxia | AD | 600224 | 11q | |
| SCA6 | | AD | 183086 | 19p | $\alpha_{1A}$-calcium channel subunit protein (CACNA1A) |
| SCA7 | | AD | 164500 | 3p | Ataxin-7 |
| SCA8 | | AD | 603680 | 13q | |
| SCA9 | | AD | Unassigned | | |
| SCA10 | | AD | 603516 | 22q | |
| SCA11 | | AD | 604432 | 15q | |
| SCA12 | | AD | 604326 | 5q | |
| SCA13 | | AD | 605259 | 19q | |
| SCA14 | | AD | 605361 | 19q | |
| SCA15 | | AD | 606658 | Unassigned | |
| SCA16 | | AD | 606364 | 8q | |
| SCA17 | | AD | 600075 | 6q27 | TATA Box-binding protein (TBP) |
| SCA18 | | AD | | | |
| SCA19 | | AD | | | |
| DRPLA | Myoclonic epilepsy with choreoathetosis | AD | 125370 | 12p | Atrophin-1 |
| SAX1 | Spastic ataxia type 1 | AD | | 12p | |
| SPAR | Spastic ataxia with mental retardation | AD | | | |
| EA1 | Episodic ataxia with myokymia | AD | 160120 | 12p | Potassium channel (KCNA1) |
| EA2 | Acetazolamide-responsive paroxysmal ataxia | AD | 108500 | 19p | |
| EA3 | Periodic vestibulocerebellar ataxia (PATX) | AD | 606552 | Unassigned | |
| FRDA | Friedreich ataxia | AR | 229300 | 9q | Frataxin |
| ARSACS | Autosomal recessive spastic ataxia of Charlevoix-Saguenay | AR | 270550 | 13q | Sacsin |
| AT | Ataxia telangiectasia | AR | 208900 | 11q | ATM |
| AOA1 | Ataxia with ocular apraxia and hypoalbuminemia | AR | 208920 | 9p | Aprataxin |
| AOA2 | Spinocerebellar ataxia recessive, non Friedreich type 1 (SCAR1) | AR | 606002 | 9q34 | |
| AVED | Ataxia, Friedreich-like, with selective vitamin E deficiency | AR | 277460 | 8q | $\alpha$-tocopherol transfer protein |
| ATCAY | Cerebellar ataxia, Cayman type | AR | 601238 | 19p | |
| IOSCA | Spinocerebellar ataxia, infantile, with sensory neuropathy | AR | 271245 | 10q | |
| ASAT | X-linked anemia, sideroblastic, and spinocerebellar ataxia (XLSA/A) | AR | 301310 | Xq | ATP-binding cassette transporter 7 (ABC7) |
| | Hypoceruloplasminemia | AD/AR | 604290 | 3q | Ceruloplasmin (CP) |
| | Ataxia with hypogonadotropic hypogonadism | AR | 212840 | | |
| | Ataxia with hearing impairment and optic atrophy | AR | | 6p | |

phenotypic spectrum has undergone revision and expansion, now including patients that without genotypic testing would not have entered the differential diagnosis of FRDA 10 years ago.

The genetic classification of the ataxias is in some ways easier to follow than the classification of the dystonias or of the inherited Parkinsonian syndromes. A clear distinction is made between autosomal dominant ataxias, which are referred to as SCAs, and the recessive ataxias (Table 2.1). The nomenclature committee of the Human Genome Organization (HUGO) assigns locus or gene names, such as SCA3 or SCA17. Out of necessity, some SCAs have received a number (such as SCA15) to distinguish it from other mapped SCAs before the actual locus is identified.

The infantile onset spinocerebellar ataxia (IOSCA) that is recessive and seen in the Finnish population was previously classified as SCA8, but has now been given the designation IOSCA. Unfortunately, SCA9 has remained unassigned at the time of publication.

The subsequent chapters first discuss the dominant SCAs, followed by the most common recessive ataxias. A separate chapter discusses disorders in which ataxia is usually part of a wider disease phenotype with typical associated features, suggestive of metabolic or mitochondrial disorders (Chapter 24). Early-onset ataxia with retained reflexes may fall into any of the aforementioned categories. Although multiple system atrophy (MSA) usually does not have an identifiable genetic basis, a separate chapter is devoted to it, because of the difficulties in differential diagnosis (Chapter 23). Since patients do not present to the physician with their genotype displayed on their forehead or on their MRI, Chapter 25 addresses the work-up of patients with ataxia that may in the end result in the correct genotypic assignment of the patient.

## A. Autosomal Dominant Ataxias

The late Anita Harding (1982) proposed a phenotypic classification of the autosomal dominant cerebellar ataxias (ADCAs), distinguishing three types (Fig. 2.1). Ashizawa and Pulst have suggested adding a fourth type, which would include ataxia and epilepsy (Fig. 2.1). This phenotype may also be seen in some patients with dentatorubro-pallidoluysian atrophy (DRPLA) and rarely in SCA17, although in DRPLA and SCA17 it is usually accompanied by other non-cerebellar neurologic signs.

| PHENOTYPE | | GENOTYPE |
|---|---|---|
| ADCA I | Ataxia plus degeneration of other neuronal systems +/- neuropathy | **SCA1, 2, 3**  SCA12, 17  SCA4, 8, 10, 13, DRPLA  SCA6, 7 |
| ADCA II | ADCA I phenotype as above and **retinal degeneration** | **SCA7**  SCA2 |
| ADCA III | Pure cerebellar ataxia often late onset | **SCA6**  SCA11, 14, 15, 16  SCA4, 5, 8, 10  SCA2, 3 |
| ADCA IV | Cerebellar ataxia with seizures | **SCA10**  SCA17, DRPLA |

**FIGURE 2.1** Relationship between modified Harding classification of autosomal dominant ataxias and genotype-bases classification. Boxed and bolded mutations are commonly and typically associated with the phenotype, bolded mutations show a typical association, but are overall rare. Mutations in smaller letters are either very rarely associated with the phenotype or are overall very rare.

The ADCAs share onset in adulthood (although rare infantile and juvenile cases are reported), a progressive course of the disease, and pathologic features that involve specific neuronal groups in the cerebellum and the brainstem. As Fig. 2.1 shows, each phenotype can be associated with mutations in multiple genes, and a mutation in a specific gene can give rise to different ADCA phenotypes, depending on the mutation, disease duration, and other factors, such as genetic background and environmental effects.

The correlation of modified Harding phenotypes with genotypes as shown in Fig. 2.1 has several problems. Some of the SCAs have only been described in few (SCA10, SCA12) or single pedigrees (SCA11, SCA13, SCA14, SCA15, SCA16). SCA4 shows an ADCA I phenotype in a large Utah kindred, but an ADCA III phenotype in a Japanese family. On formal grounds it is not yet conclusively clear that the mutation in this family is allelic to SCA4 described in the large Utah kindred (Flanigan et al., 1996). Outlier phenotypes can be seen in all the SCAs and in part reflect at what stage in the disease process a patient is examined.

As discussed in Chapter 1 (Table 1.1) the OMIM database represents an invaluable tool. Some entries in the OMIM database, however, reflect the problems of historic classifications. For example, entry 117210 (cerebellar ataxia, autosomal dominant pure) likely represents different SCAs, possibly including SCA6 and SCA11. The entry *117350 (Wadia-Swami syndrome) describes a cerebellar degeneration with slow eye movements. By genotyping it has now been established that some of these pedigrees have indeed SCA2, but others have mutations in the SCA1 gene (Wadia et al., 1998; Buttner et al., 1998; Burk et al., 1999). Unfortunately, OMIM editors are not always able to follow the development and links to newer references that would tie a phenotypic-based entry to genotypic-based entries.

Other ataxia phenotypes that appear genetically distinct have an OMIM number, but have not been assigned a HUGO designation, which is genetic locus based. For example, Melberg et al. (1995) described a pedigree (OMIM *604121) with a rather distinctive phenotype that included ataxia and variable association of deafness and narcolepsy. Despite having a number in OMIM, this pedigree has not been assigned a number in the "SCA" nomenclature by the HUGO nomenclature committee. A family with ataxia, extrapyramidal features, and marked anticipation was described by Devos et al. (2001), but neither an OMIM number nor SCA designations have been assigned.

In summary, the reader is faced with several classifications in the dominant ataxias: the ADCA types proposed by Harding, an OMIM number that is phenotype based, but may include genotypic information, and finally a locus/gene based designation approved by the HUGO nomenclature committee. In this book, we have followed a chapter structure based on the genotype.

## II. RECENTLY IDENTIFIED ATAXIAS

### A. New SCA Loci

Even within the short time of planning this book and its publication, several new ataxias have been described that did not receive their own chapters. SCA13, 14, and 16 are discussed in Chapter 14. SCA15 was assigned a locus designation in an Australian kindred with a dominantly inherited pure cerebellar ataxia (Storey et al., 2001). Progression appeared very slow. It has not been mapped except by exclusion of known ataxia loci. SCA17 had been described as a mutation in the TATA-binding protein in a single patient and subsequently in several German patients, but its assignment as SCA17 occurred only recently (see Chapter 15). SCA18 has only been reported in abstract form (Brkanac et al., 2001), and SCA19 has been assigned to chromosome 1 (Verbeek, in press). SCA9 has remained unassigned for several years for unknown reasons.

### B. Spastic Ataxia

Meijer et al. (2002a) examined the phenotypic spectrum of mutations in the spastin gene causing spastic paraplegia type 4 (SPG4). Although many patients had a pure hereditary spastic paraplegia (HSP), some had a complicated phenotype including ataxia. Hedera et al. (2002) further emphasized the co-existence of spastic paraplegia and ataxia in a family that showed three phenotypic patterns: "uncomplicated" spastic paraplegia, ataxia-spasticity, and a third phenotype combining spastic paraplegia, ataxia, and mental retardation. This phenotype was designated SPAR and was not linked to known ataxia or HSP loci. There was suggestive anticipation and a progression of the phenotype from spasticity in earlier generations to ataxia and mental retardation in the third generation.

When lower limb spasticity is accompanied by a general ataxia, Meijer et al. (2002b) suggested to group patients as hereditary spastic ataxia (HSA). They mapped a new HSA locus to chromosome 12p13. This locus was designated spastic ataxia type 1 (SPAX1).

### C. Ataxia with Oculomotor Apraxia

Two loci for this disorder have been identified one on chromosome 9p, the other one on 9q34. The gene for the 9p locus (AOA1) has now been identified (Date et al., 2001; Moreira et al., 2001). The 9q locus has been designated AOA2, although ocular apraxia appears to be a less consistent feature.

#### 1. AOA1

Aicardi et al. (1988) described an autosomal recessive syndrome that closely resembled ataxia telangiectasia (AT), but differed in important respects. They reported 14 patients in 10 families with a neurologic syndrome of oculomotor apraxia, ataxia, and choreoathetosis, who did not have extraneurologic features of AT and lacked alterations in immunoglobulins, alpha-fetoprotein, and T- and B-lymphocyte markers. Moreira et al. (2001) studied 13 Portuguese families with AOA and found linkage to 9p in five families. They showed homozygosity and haplotype sharing over a 2-cM region on 9p13, demonstrating a founder effect. They also analyzed two unrelated Japanese families with early-onset cerebellar ataxia with hypoalbuminemia (EOCA-HA). This disorder, previously only described in Japan, is characterized by marked cerebellar atrophy, peripheral neuropathy, mental retardation, and occasionally oculomotor apraxia. Both families appeared to show linkage to the AOA1 locus. Subsequently, the authors found hypoalbuminemia in all five Portuguese families with AOA1 with a long disease duration, suggesting that AOA1 and EOCA-HA correspond to the same entity that accounts for a significant proportion of all recessive ataxias.

Date et al. (2001) and Moreira et al. (2001) identified mutations in the *aprataxin (APRX)* gene as the cause of ataxia-oculomotor apraxia. Aprataxin has a highly conserved histidine-triad (HIT) motif (His-X-His-X-His-X-X, where X is a hydrophobic amino acid). The HIT family of proteins has been associated with nucleotide-binding and diadenosine polyphosphate hydrolase activities. At the moment, the designation of identical mutations is different between the two groups as a result of different base pair numbering. Insertion or deletion mutation resulted in a severe phenotype with childhood onset, whereas missense mutations resulted in a mild phenotype with relatively late age of onset. In one pedigree with compound heterozygous missense mutations the age of onset was 25 years.

#### 2. AOA2

Ataxia of later onset with inconsistent association of oculomotor apraxia may be designated ataxia-oculomotor apraxia-2. In the family reported by Nemeth et al. (2000) five brothers showed onset of ataxia in their childhood or teens. There was severe ocular apraxia, initially suggesting ataxia telangiectasia. Mild choreoathetosis with dystonic posturing was also seen. Cerebellar atrophy was not marked on MRI. A Japanese family with ataxia and elevated levels of α-fetoprotein, γ-globulin, and creatine kinase has been mapped to the same region (Bomont et al., 2000).

### D. Other Recessive Ataxias

There are several ataxias that are usually recessive, have an early onset, and are usually nonprogressive. Nystuen et al. (1996) mapped a recessive nonprogressive ataxia with psychomotor retardation to chromosome 19p13.3.

This ataxia is also referred to as Cayman ataxia due to the location of the pedigree. Allikmets et al. (1999) identified a mutation in the ABC7 gene leading to a nonprogressive ataxia with sideroblastic anemia. The protein localizes to the inner mitochondrial membrane and is involved in iron homeostasis. Bomont et al. (2000) mapped a recessive ataxia with hearing impairment and optic atrophy to human chromosome 6p21-23 by homozygosity mapping.

Marinesco-Sjogren syndrome (MSS, OMIM *248800) shows cerebellar ataxia, congenital cataracts, and retarded somatic and mental growth. Recently, Merlini et al. (2002) suggested that MSS might be allelic with the congenital cataract-facial dysmorphism-neuropathy (CCFDN) syndrome due to phenotypic overlap. Indeed, MSS mapped to the CCFDN locus on chromosome 18q.

## III. WORLDWIDE PREVALENCE

The prevalence of ataxias with known recessive or dominant inheritance or that of sporadic ataxia has received relatively little study. Epidemiological studies are conducted in defined geographic areas that may be subject to founder mutations. An example is the often quoted prevalence of 1 per 100,000 ADCAs in Iceland, an island with a population of under 300,000 inhabitants (Gudmundsson, 1969). Other studies are sparked by the interest of investigators to ascertain pedigrees and again may reflect the presence of unusual founder mutations in that particular region. Using a community-based prevalence study among 613,349 inhabitants in Tottori prefecture, Japan, Mori et al. (2001b) determined the genotype frequencies of patients with SCAs. The prevalence of SCA was 17.8 per 100,000 individuals with most cases representing SCA6. SCA6 patients shared a common haplotype explaining the high prevalence of ADCA (Mori et al., 2001a). Prevalences per 100,000 individuals were as follows: SCA6, 2.40; SCA1, 0.48; DRPLA, 0.32, and SCA3, 0.16.

In addition, most of the older studies suffer from the lack of genotypic definition of the ataxias (for review see van de Warrenburg et al., 2002). More data are available on the relative frequency of ataxia mutations as they present to academic laboratories (Table 2.2).

Filla et al. (1992) conducted an epidemiological survey of hereditary ataxias and paraplegias in Molise, Italy. Total prevalence was 7.5 per 100,000 inhabitants (95%

TABLE 2.2  Distribution of Ataxia Genotypes in Large Ataxia Clinics Worldwide

| Clinic location | SCA1 | SCA2 | SCA3 | SCA6 | SCA7 | FRDA | Other |
|---|---|---|---|---|---|---|---|
| **North America** | | | | | | | |
| Canada (Silveira et al., 1996) | | | | | | | |
|   All patients | 3% | — | 41% | — | — | — | DRPLA 1% |
|   Non-Portuguese | 10% | — | 17% | — | — | — | — |
| United States | | | | | | | |
|   Baylor | — | 18% (Lorenzetti et al., 2002) | — | — | — | 44% (Gunaratne and Richards, 1997) | — |
|   Minnesota (Moseley et al., 1998) | 5.6% | 15.2% | 20.8% | 15.2% | 4.5% | 16.6%[a] | Sporadic cases—SCA 4.4%, FRDA 5.2% |
|   UCLA (Geschwind et al., 1997a,b,c) | 6% | 13% | 23% | 12% | 5% | 62% | No SCA8, 12, or DRPLA found. (Cholfin, et al., 2001; Sobrido et al., 2001) |
| | 0% | 0% | 92%[b] | 0% | 2% | — | Sporadic cases—7.1% SCA8 |
| **South America** | | | | | | | |
| Brazil, (Jardim et al., 2001) | | | | | | | |
| **Northern Europe** | | | | | | | |
| UK-London (Giunti et al., 1998) | 37%[c] | 47%[d] | 15%[e] | — | 94%[f] | — | — |
| Cambridge (Leggo et al., 1997) | 9% | 27.3% | 4.5% | 31.8% | — | — | DRPLA 9% Sporadic cases (SCA 12%, FRDA 3%) |
| Germany, (Schols, et al., 1997a) (Riess et al., 1997) | 9% | 10–14% | 42% | 22% | — | — | Sporadic cases—0.8–1.8% (Schols et al., 1997b) 6.6–7.3% (Schols et al., 1998) 2.4% (Schols et al., 2000) |
| Russia (Illarioshkin et al., 1996) | 33% | — | 0% | — | — | — | |
| The Netherlands (van de Warrenburg et al., 2002) | 6% | 7% | 28% | 15% | 8% | — | Total adds up to 100%, because authors "assumed" that 36% of ataxias were not detected |

(continues)

TABLE 2.2 (continued)

| Clinic location | SCA1 | SCA2 | SCA3 | SCA6 | SCA7 | FRDA | Other |
|---|---|---|---|---|---|---|---|
| **Mediterranean** | | | | | | | |
| Italy | — | — | — | — | — | 11.8% | EOCARR 5.9% |
| Valle d'Aosta (Leone et al., 1995) | 41% | 29% | 0% | — | — | — | — |
| Milan (Pareyson et al., 1999) | 24% | 47% | 0% | 2% | 2% | — | DRPLA 2% |
| Naples (Filla et al., 2000) | | | | | | | |
| Sicily (Giuffrida et al., 1999) | — | — | — | — | — | — | Sporadic cases—28.6% SCA2 |
| Spain (Pujana et al., 1999) | 6% | 15% | 15% | 1% | 3% | — | DRPLA 1% Sporadic cases—0% SCA, 1.7% LOFA |
| Portugal (Silveira et al., 1998) | 0% | 4% | 74% | 0% | — | — | DRPLA 0% |
| **Africa** | | | | | | | |
| South Africa (Ramesar et al., 1997) | 42.9% | — | 0% | — | — | — | Sporadic cases—4.5% SCA1 |
| **Far East** | | | | | | | |
| China (Tang et al., 2000) | 4.7% | 5.9% | 48.2% | 0% | 0% | — | DRPLA 0% Sporadic cases—0% SCA |
| India (Basu et al., 2000) | 10.5% | 17.5% | 7% | 1.8% | 0% | — | DRPLA 0% |
| Hokkaido (Sasaki et al., 2000) | 9.75 | 7.7% | 23.9% | 29% | 0% | 0% | DRPLA 2.6% SCA8 0% |
| Kinki (Matsumura et al., 2000) | 0% | — | 46.8% | — | — | — | Sporadic cases—(Futamura et al., 1998) 22% SCA, mostly SCA6 DRPLA 19.8% |
| Nagoya (Watanabe et al 1998) | 0% | 5.9% | 33.7% | 5.9% | — | — | DRPLA 20.% |
| Niigata (Takano et al., 2000) | 24.8% | — | — | — | — | — | — |
| | | | | | | | DRPLA 5% |
| Tottori (Mori et al., 2001) | 15% | 0% | 5% | 25% | 0% | 0% | Sporadic cases—11% SCA6 |
| Korea (Jin et al., 1999) | | | | | | | DRPLA 3.4% |
| Singapore (Tan et al., 2000) | — | 7.1% (Malay) | 35.7% Chinese | — | — | — | |
| Taiwan (Hsien et al., 2000), Soong et al., 2001). | 1.4–5.4% | 9.6–10.8% | 27.3–47.3% | 10.8% | 1.4–2.7% | | DRPLA 1.4% SCA8 0% Sporadic cases—4.1% SCA6 |
| **Australia** | | | | | | | |
| (Storey et al., 2000) | 16% most of Anglo-Celtic descent | 6% (1/2 of Italian descent) | 12% (1/3 of Chinese descent) | 174 | 25 | | DRPLA 0% Sporadic cases—For every 3 families found with SCA 1,2,3, or 6, one more would be found without a family history |

Note: Percentages represent proportion of total ADCA or of total Friedrich ataxia phenotypes test.
[a]Patients with a phenotype typical of FRDA were excluded from this sample.
[b]This is felt to reflect an Azorean founder effect.
[c]When ethnicity was taken into account, 50% of Italian patients had SCA1.
[d]When ethnicity was taken into account, 37% of Italian patients and 44% of East Indian patients had SCA2.
[e]When ethnicity was taken into account, 44% of East Indian patients had SCA3.
[f]This the percentage relative to only families with the ADCA II phenotype.

confidence limits 4.8–11.1). There were seven patients with FRDA, five with EOCA with retained tendon reflexes, four with AT, two with dominant HSA and seven with recessive HSA. Surprisingly, there was no patient with ADCA, potentially owing to the lack of DNA based testing of patients.

Polo et al. (1991) conducted a survey of hereditary ataxias and paraplegias in Cantabria in Northern Spain from 1974–1986. The series comprised 48 index cases and 65 affected relatives. On prevalence day, 54 patients with cerebellar ataxia were alive, giving a prevalence of 10 cases per 100,000. FRDA accounted for half of the cases. Six patients (12%) from three families had autosomal dominant ataxia and eleven patients had a late-onset ataxia without known family history. The authors attributed the relatively high prevalence data to inclusion of many secondary cases. A frequency of consanguinity in FRDA cases was described leading to a pseudodominant pattern of inheritance in one family. In Portugal a survey of spastic paraplegia and ataxia was initiated in 1993. By 2001, 107 patients with autosomal recessive ataxia had been identified in a population of 9.8 million. FRDA accounted for a third of the cases, ataxia with ocular apraxia for 21% (Barbot et al., 2001).

Van de Warrenburg estimated the prevalence of ADCAs in The Netherlands using ascertainment of ADCAs by three molecular diagnostics laboratories. They identified 145 ADCA families with 391 affecteds. Assuming that mutation detection identifies approximately 64% of ADCA worldwide, they calculated a minimal prevalence of 3 per 100,000. Although this approach of calculating prevalence has some flaws due to the inherent assumptions regarding frequency of identifiable mutations and number of affecteds per family, it provides a higher minimal prevalence for ADCAs than previous estimates.

Ataxias defined by genotype (as opposed to phenotype) occur on every continent, although the precise distribution shows highly variable prevalence rates. Friedreich ataxia mutations are largely restricted to Caucasians (see Chapter 18). Premutation alleles are found in European populations, but also in subsaharan Africans. Based on haplotype analysis, most cases worldwide can be traced back to these founder alleles (see Chapter 18). In Finland, however, carrier frequency is only 1 in 500, not 1 in 90 as in most Caucasian populations (Juvonen et al., 2000).

On the other hand, it is quite apparent that repeat expansions in the SCA loci have arisen independently in many geographic regions. Actual prevalence rates are correlated with the allele distribution in the general population. Thus, the greater frequency of long normal alleles in the DRPLA gene predict a greater prevalence of DRPLA in the Japanese as compared with the prevalence in Europeans where long normal DRPLA alleles are less common (Takano et al., 1998).

Despite worldwide occurrence of the SCAs, haplotype analysis has also provided evidence for the existence of founder mutations. Patients with SCA2 on the island of Cuba all share the identical haplotype on disease chromosomes (Pulst, unpublished). On the other hand, SCA3/MJD mutations in the Azorian islands can be traced back to two founder haplotypes, and many patients worldwide share the same intragenic haplotype (Gaspar et al., 2001).

## IV. MECHANISMS OF DISEASE

Identification of ataxia genes has invigorated the studies aimed at understanding the basis of neuronal dysfunction and subsequent neuronal death. Much of this work at the basic science level reflects the genotypic splitting that has been so prominent for the dominant ataxias. Research groups usually "stick to their gene" and comparisons between the results obtained in different groups can be difficult due to subtle variations of the *in vitro* and animal models (see below).

It will be interesting to observe in the years to come what types of interaction will develop between the frataxin field and mitochondrial research. ASAT is caused by a mutation in a putative iron transporter providing yet another link. AT and AOA1 share phenotypic features, but the fields may also converge due to putative functions of aprataxin and ATM in DNA repair.

The emergence of the term polyglutamine (polyQ) disease has signaled a departure from genotype-based understanding of disease (Table 2.3). Although the ataxias contribute the greatest number of diseases to this category, a large amount of work is carried out in the Huntington disease (HD) field (Chapter 33). It is too early to discern whether we will be speaking of one or several polyQ diseases in the future. Large-scale analysis of changes in the transcriptome or proteome may unveil hidden common characteristics (for review see Kiehl et al., 2000).

Some diseases, however, stand alone at this time. The pathogenic function of the SCA8 mutation has remained enigmatic. If the mutation results in "toxic" RNA, then shared features with the mutation in myotonic dystrophy may emerge. The SCA10 mutation with a large intronic repeat expansion may function like the GAA intronic expansion resulting in reduced transcription, but reduced mRNA levels have not yet been demonstrated (see Chapter 11).

### A. Anticipation

The concept of anticipation, although now so readily and perhaps too easily accepted, was not widely embraced by the genetic community until the end of the last century. In 1947, Julia Bell applied quantitative genetic tools to the analysis of myotonic dystrophy. She determined that the correlation of age of onset (AO) was considerably stronger between sibs than between parents and children. Lionel Penrose thought to have disproved that theory by proposing an "index of anticipation" (D/2S, where D is the mean difference in age of onset between parent and child (years) and S is the General standard deviation of AO). Whereas the index was low for HD, it actually was relatively high for myotonic dystrophy. Penrose, however, discounted these results on the basis of an ascertainment bias and closed the door on serious discussion of this topic for many years (Fig. 2.2).

TABLE 2.3  Class of DNA Repeat, Disease, and Disease Mechanism

| Repeat | Location | Disease | Likely disease mechanism |
|--------|----------|---------|--------------------------|
| CAG | Coding | SCA1, 2, 3, 6, 7, 17 | PolyQ protein aggregation |
| CAG | Noncoding | SCA12 | Change in expression |
| CTG | Noncoding | SCA8 | ? Antisense function, ? toxic function |
| ATTCT | Noncoding | SCA10 | Unknown |
| GAA | Noncoding | FRDA | Reduced transcription |

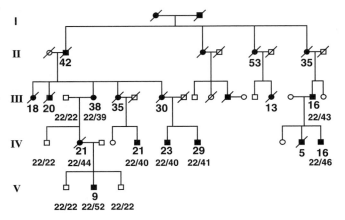

FIGURE 2.2 Anticipation of the age of onset in pedigree FS with SCA2. (Modified from Pulst et al., 1993). Mean age of onset is 24.9 with a standard deviation of 12.6. The Penrose index is 0.68 (compared with 0.36 for HD, 0.56 for diabetes mellitus, and 0.90 for DM1, from Harper (1998). Numbers under the age of onset indicate SCA2 genotype.

To quote Peter Harper (1998) in his excellent review of anticipation in myotonic dystrophy:

> "If any blame can be attributed for the 40-year neglect of anticipation following Penrose's 1948 paper, it is probably the tendency of workers to accept the verdict of scientific authority rather than to base conclusions on raw data, whether these be clinical or genetic in nature. The complete omission of the topic in successive editions of the most respected genetics textbooks (now hastily rectified in the latest editions!) should stand as a warning."

One can only hope that future generations will judge this book kindly.

A Pubmed search with the textwords "anticipation" and "ataxia" revealed 130 citations on 3–16–2002. The first citation was in 1991 in Chinese (abstract translated into English) describing in the abstract that anticipation was observed in 26 dominant families (Dai, 1991). Interestingly, the abstract describing the isolation of the SCA1 CAG repeat does not mention anticipation, although it provided for the first time the molecular basis of anticipation in the SCAs (Orr et al., 1993). The article itself contained an analysis of SCA1 repeat instability and the authors presciently suggested that "expansion of $(CAG)_n$ or other trinucleotide repeats is likely to be a predominant mechanism in these other forms of autosomal dominant ataxia." The next citation describes a single family with SCA2 that showed statistically significant anticipation (Pulst et al., 1993). A subsequent detailed analysis of 122 affected individuals from 36 unrelated families with ADCA type I indicated a mean anticipation of 9.4 years (Dürr et al., 1993).

The recognition of anticipation as an important aspect of any phenotype is now commonplace. But even today formal statistical methods to analyze its presence have not developed beyond the comparisons of parent/child and sib/sib correlations used by Julia Bell (1947), the "index of anticipation" proposed by Penrose (1948), a simple chi-square-based analysis used by Pulst et al. (1993), and comparison of AO means in parents and children (Raskind et al., 1997). Most authors use simple inspection of pedigrees to suggest the presence of anticipation.

Keeping Peter Harper's above comments in mind, one cannot overlook that anticipation has been described in disorders such as HSP type 4 and facioscapulohumeral dystrophy (FSHD) for which the causative genes have now been identified (Raskind et al., 1997; Flanigan et al., 2001). Neither mutation is an unstable DNA repeat, and one could ascribe the observation simply to ascertainment bias. But other explanations remain possible. For example, unstable DNA repeats in modifying alleles, which by themselves are not pathogenic, could influence the phenotype. Environmental changes and lifestyle choices affecting in aggregate the more recent generations could also lead to earlier disease onset.

Repeat instability is not identical for all repeats and shows marked differences for paternal and maternal transmission in some disorders. Stevanin et al. (2000) tabulated CAG repeat instability for SCA1, 2, 3, 7, and DRPLA. Average repeat gain for paternal transmissions was 5.1, for maternal transmission 1.5. SCA7 was the most unstable repeat, followed by DRPLA, and SCA2.

In the SCAs, the molecular basis for the phenotypic phenomenon of anticipation is the expansion of a DNA repeat in the causative gene. As a corollary, one would predict a correlation between age of onset and repeat length. This is indeed the case (Fig. 2.3), but the actual regression curves for each disorder may vary. Due to the recessive nature of the GAA expansion allele in FRDA, anticipation is not observed but a correlation between AO and repeat length is observed (see Fig 18.1).

Closer inspection of Fig. 2.3 can highlight several important points. For each protein, the precise lengths of the expansions of the polyQ tract that are disease causing are different. Furthermore, the length of the polyQ-tract that is pathogenic in ataxin-2 overlaps with the longest *normal* alleles in ataxin-3. This points to the importance of the protein context in which the polyQ tract is embedded, and to the differences in expression levels or relevant interactions with other proteins.

Regression curves are not linear, but give the best fit after logarithmic transformation of the age of onset. Slopes of regression lines after logarithmic transformation are different for each disorder. Figure 2.3 may suggest a clear threshold between normal repeats and pathologic repeats. For example, the shortest pathologic repeats for the SCA2 repeat shown in Fig. 2.3 contain 36 and 37 repeats. Extension of regression lines to 30 repeats suggest that if the underlying pathogenic mechanism remains similar, these alleles might still be pathogenic but associated with very late onset (see also Fig. 4.1). Indeed that was observed

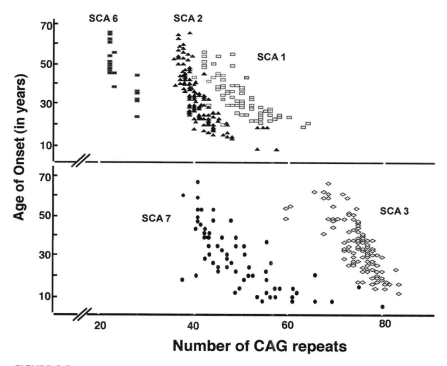

FIGURE 2.3 Scatterplot of CAG repeat number and age of ataxia onset for patients with SCA1, 2, 3, 6, and 7. Data were combined from Stevanin et al. (2000), Geschwind et al. (1997a,b), and Pulst et al. (1996).

subsequently in several SCA2 patients (for detailed discussion see Chapter 4). CAG repeats smaller than 21 in the CACNA1A gene are usually not considered pathogenic for causing SCA6. Recently, a patient with SCA6 and 20 CAG repeats was described (Komeichi et al., 2001). In addition, Mariotti et al. (2001) report a patient with SCA6 that was homozygous for a $CAG_{19}$ allele pointing to the importance of gene dosage. Thus, the concept of a pathologic threshold with clear separation between "normalcy" and "mutantcy" cannot be upheld any longer, and should be replaced with a probabilistic model specifying the likelihood of developing the disease given a certain repeat length.

Finally, although correlation between CAG length and AO is highly significant for all SCAs, there is significant variation in AO for a given repeat length. For example, in SCA6 patients the 23 repeat allele can result in AO as early as 35, but as late as 55 (Fig. 2.3). This has important implications for genetic counseling (see below).

## B. DNA Testing

For ataxia caused by DNA repeat expansions testing is readily available from commercial and research laboratories. Most repeats can be easily amplified by the polymerase chain reaction (PCR). Testing for the known SCAs and FRDA identifies between 50 and 80% of ataxias worldwide (see Table 3.2). As the subsequent chapters describe in detail, normal and pathologic repeat lengths are different for all the SCAs (see also Fig. 2.3).

Two reports from DNA testing laboratories have been published. Gunaratne and Richards (1997) summarized the experience of the Baylor DNA Diagnostics laboratory. Approximately 28% of these patients had an abnormal triplet repeat expansion in ataxia genes. The relative distribution was as follows: SCA1 3%, SCA2 8%, SCA3 11%, SCA6 2%, SCA7 3%, and 1.5% for FRDA. Of the 157 patients directly submitted for FRDA testing 69 (44%) had at least one GAA expansion in the frataxin gene.

Jones and Seltzer (2000) reported the experience at Athena Neurodiagnostics in abstract form summarizing testing on approximately 7000 samples. Not all samples, however, were tested for all mutations. The percentage of individuals with bi-allelic GAA frataxin expansions was 15.3%; an additional 3.7% had a heterozygous expansion. The relative distribution for the SCAs was as follows: SCA1 1.2%, SCA2 3.4%, SCA3 3.7%, SCA6 2%, and SCA7 1.3%. The differences in the relative distribution of positive results for SCA testing and FRDA testing between the two laboratories are noteworthy and may reflect differences in referral patterns.

Although PCR amplification for use in genetic testing may seem straightforward, specific challenges can be encountered in rare patients. For SCA1 long normal alleles have been reported that may show pathologic fragment sizes, but are interrupted by CAT repeats (see Chapter 3). CAT repeats do not code for glutamine, and the interrupted proteins do not appear to be pathogenic. Very large repeats may hinder amplification by PCR. Patients with large expansions appear as homozygotes. Therefore, patients with very early onset may require analysis by Southern blotting or use of special PCR conditions. In SCA10, Southern blot analysis is necessary due to the large number of ATTCT repeats on disease chromosomes (see Chapter 11).

The potential for large expansions is particularly relevant in the setting of prenatal diagnosis, where the potential age of disease onset in the fetus is unknown. The SCA7 and the SCA2 repeat appear to be particularly predisposed to large expansions, when the chromosome is paternal (see Chapters 4 and 9). Due to the prevalence of a common SCA2 allele with 22 repeats, the potential for homozygosity in a fetus is high (in the order of 80%). Unless heterozygosity can be clearly established in the fetus, DNA needs to be examined by Southern blot analysis for SCA2 prenatal diagnosis, when the father is affected.

FRDA is caused by GAA repeat expansion in 98% of cases, the remainder being point mutations. No patients with point mutations in both alleles have been identified. Thus, in patients with a typical phenotype (see Chapter 18) the presence of one allele with an expanded GAA repeat can be strongly supportive of the diagnosis.

The variation in AO for a given repeat length has important implications for counseling. This variation is seen not only in polyQ disorders (Fig. 2.3), but also in FRDA (Fig. 18.1). When presymptomatic individuals are counseled, precise prediction for a likely AO cannot be given. Future identification of modifying genes may make these estimates more precise. For other ataxias, mutation detection may be more complex and may involve sequence analysis of the entire gene, when multiple pathologic alleles exist. This is the case for ATT, AVED, AOA1, and ASAT. Measurement of metabolic parameters (such as vitamin E levels in AVED) may establish the diagnosis reliably even in the absence of identified mutations at the DNA level.

The ataxia chapters that follow also highlight some of the complexities of DNA testing. Not in all of the SCAs are normal and pathologic alleles separated by comfortable gaps (see also Fig. 2.3). Reduced penetrance alleles and short pathologic alleles associated with unusual phenotypes have been reported in the SCAs and in FRDA. Examples are the reports of an MSA-like phenotype in a Japanese patient with an "intermediate" SCA3 repeat of 56 CAG (Takiyama et al., 1997) and late-onset FRDA with short GAA expansions (see Chapter 18). In SCA2 and SCA6 normal and abnormal repeat sizes are close or even overlap leading to alleles with reduced penetrance (see Chapters 4 and 8).

Testing for repeat disorders can also identify premutation alleles. Although these alleles may not be pathogenic in the individuals carrying the respective allele, they are prone to further expansion. Thus, parents and siblings of probands need to be appropriately counseled regarding an increased recurrence risk.

## C. Sporadic Ataxia

DNA analysis should not be restricted to ataxia patients with a positive family history (see also Table 2.2). Obviously, recessive ataxias typically occur with normal parents and may only affect one sibling given smaller family sizes in the 21st century. Indeed, several series have shown that frataxin mutations can be found in a significant number of ataxia patients without family history including late-onset ataxias. Moseley et al. (1998) identified 7 of 124 (5.2%) sporadic patients with a frataxin mutation. It is noteworthy that they excluded patients with a typical FRDA phenotype from this analysis. Ten of 124 patients (8%) with sporadic ataxia had frataxin mutations with an AO between 13 and 36 years in the series by Schöls et al. (2000).

Mutations in ataxia patients without a family history are not restricted to recessive mutations. Due to anticipation, AO variation, or premature death of an at-risk parent, mutations in dominant SCA alleles are seen in sporadic ataxia patients as well as in families that show a suggestive recessive inheritance pattern (defined as normal parents, two or more affected children). The most commonly found mutations are SCA2 and SCA6 (Moseley et al., 1998; Schöls et al. 2000). Thirteen of 124 patients (8%) had SCA mutation in the series by Schöls et al. (2000), and 6 of 134 patients (4.4%) in the series by Moseley et al. (1998). Futamura et al. (1998) screened 85 Japanese ataxia patients without family history for mutations in the SCA1, SCA2, SCA3, SCA6, and DRPLA genes. Of the patients, 22% had expanded repeats, most frequently in the SCA6 gene. Other smaller or geographically more restricted series of sporadic ataxia patients are listed in Table 2.2.

Patients with an MSA phenotype, on the other hand, do not have a high frequency of SCA mutations (see also Chapter 23). Brandmann et al. (1997) examined the SCA1 and 3 genes in 80 MSA patients, but did not identify any repeat expansions. Schöls et al. (2000) examined 20 MSA patients for expansions in the SCA1, 2, 3, 6, 7, 8, and 12 genes, and in the GAA repeat of the frataxin gene. They did not detect any pathologic alleles. One patient with ataxia and significant orthostatic hypotension carrying an intermediate SCA3 repeat of 56 CAG has been reported (Takiyama et al., 1997).

## D. Neuroimaging and Neuropathology

Just as the clinical phenotype can be highly variable, so can imaging studies and pathologic examination (reviewed in Pulst and Perlman, 2000). Many of the dominant SCAs show similar imaging findings, although when seen as a group, differences between the different SCAs emerge. For example, SCA1 and 2 usually have prominent cerebellar and pontine atrophy, whereas atrophy in SCA6 is usually restricted to the cerebellum. Klockgether et al. (1998) performed a detailed comparison between patients with SCA1, 2, and 3. They were able to develop criteria that differentiated MRI presentations of SCA2 and SCA3, but found that SCA1 had an intermediate phenotype. FRDA is associated with spinal cord atrophy and normally does not involve the cerebellum. Late-onset cases, however, do show cerebellar atrophy. Many of the subsequent chapters provide examples of imaging studies.

### E. Models

Prior to positional or candidate identification of human ataxia genes, few models existed that closely resembled the human diseases. Naturally occurring mutations in mice had been described very often displaying a phenotype encompassing ataxia and other features such as epilepsy. Recent progress has defined some of the naturally occurring mouse mutants at the DNA level. Examples include the tottering mouse (*tg*) with homozygous point mutations in the $Ca_v2.1$ gene, whereas other mutations in this channel cause the phenotype leaner ($tg^{la}$) with ataxia and epilepsy (Fletcher *et al.*, 1996). *Stargazer* mice (*stg*) have epilepsy, ataxia, and unusual head postures. Stg mice have a mutation in stargazin which regulates trafficking of AMPA receptors and possible coupling with calcium channels (Chen *et al.*, 2000). Mutations in the *cacna2d2* gene encoding an α2-δ2 subunit that is strongly expressed in Purkinje cells were recently identified to cause epilepsy and ataxia in the recessive mouse mutants ducky *du* and $du^{2J}$ (Barclay *et al.*, 2001; Brodbeck *et al.*, 2002). For some mouse mutants, such as the Purkinje-cell-degeneration (*Pcd*) mouse, the mutation has just been identified, but for others (for example, the cerebellar deficient folia, (*cdf*, mouse), not yet (Fernandez-Gonzalez *et al.*, 2002; Park *et al.*, 2002).

The positional cloning of human ataxia genes has now led to a large number of mouse lines that have been genetically engineered to develop neurologic defects. Except for recently identified ataxia genes, mouse lines modeling the effects of dominant and recessive mutations are virtually for each disorder (see Chapters 3–24). For dominant disorders models usually utilize the expression of transgenes comparing the expression of normal alleles with that of mutant human alleles. The generation of lines in which the human mutation is introduced into the mouse gene (knock-in lines) has usually not led to deficits similar to the human disease (Lorenzetti *et al.*, 2000; Miranda *et al.*, 2002). For the modeling of recessive ataxias, a knock-out strategy has been the preferred approach (Yokota *et al.* 2001; Puccio *et al.*, 2001) including the study of metabolic ataxias such as Niemann Pick C disease (Voikar *et al.*, 2002).

The generation of animal models for FRDA and AVED was complex. Animals deficient for frataxin were embryonic lethals (Cossee *et al.*, 2000). This was presumably due to the fact that animals were completely frataxin-deficient, whereas human mutant alleles retain a small amount of function. Indeed, knock-in of the typical GAA mutation with 25% residual frataxin activity has generated a viable phenotype (Miranda *et al.*, 2002). A most useful FRDA mouse model was generated by conditional knock-out of frataxin using the cre/lox system leading to mouse lines with either a cardiac or neuronal phenotype (Puccio *et al.*, 2001).

Mouse models have helped to define necessary steps in pathogenesis. For example, expression of ataxin-1 requires nuclear localization for pathogenesis in Purkinje cells (Klement *et al.*, 1998), whereas for ataxin-2 cytoplasmic localization is sufficient to cause functional and morphologic changes (Huynh *et al.*, 2000). Pathogenesis is not due to loss of function because mice deficient in ataxin-1 or ataxin-2 do not show neurodegeneration (Matilla *et al.*, 1998; Kiehl *et al.*, 2001). These models are now also being used to assess the effect of expression of modifying alleles on pathogenesis (Shahbazian *et al.*, 2001; Vig *et al.*, 2001). Mouse models have also been useful in defining genotype-environment interactions. Mice deficient in the a-tocopherol transfer protein, a model for AVED, only developed a phenotype when fed a vitamin-E-deficient diet (Yokota *et al.*, 2001).

Animal models for the ataxias are not restricted to the use of mice. Expression of wild-type and mutant human transgenes has been targeted to the fly eye (Zhang *et al.*, 2002; Bonini, 2001; Link, 2001; Marsh *et al.*, 2000). As with mouse models, fly models may identify important steps in pathogenesis and permit the analysis of modifying alleles (Higashiyama *et al.*, 2002; Steffan *et al.*, 2001; Kazemi-Esfarjani and Benzer, 2000; Fernandez-Funez *et al.*, 2000). Parker *et al.* (2001) and Miller *et al.* (2001) have recently extended the analysis of polyQ diseases to the worm *Cenorhabditis elegans* and to zebrafish.

## V. GENOTYPE/PHENOTYPE CORRELATIONS AND MODIFYING LOCI

Even in Mendelian disorders where a particular allele has a significant effect on a phenotype, significant phenotypic variation can be observed. Phenotypic variation can be explained by several mechanisms that likely act in concert and some in a disease-specific fashion to modify the phenotype.

Clearly, a major part of phenotypic variance is explained by variation of the CAG length for the SCAs (reviewed in van de Warrenburg *et al.*, 2002) and by variation of the GAA repeat in FRDA (see Figure 18.1 and see Chapter 18). In other recessive ataxias, missense mutations that retain partial function are associated with milder phenotypes or later onset (see Chapters 19 and 24).

What explains the remainder of variance? Some of it may be due to variation of linked DNA polymorphsims such as variation in promoter sequences, some due to variation in the homologous allele, some of it due to variation at unlinked loci. Stochastic factors such as variation of DNA repeat length between different tissues or different cells in the CNS (somatic mosaicism) may play a role as well (for example see, Matsuura *et al.*, 1999).

Nongenetic contributions to phenotypic variation need to be considered as well. These may include factors as diverse as intrauterine development, perinatal events, diet and environmental toxins, and exposure to agents that influence the immune system. Identification of genetic modifiers and stratification of groups using these modifiers may facilitate the detection on nongenetic factors modifying the phenotype.

For the scientist interested in phenotypic variation, polyQ diseases provide an unusual experiment of nature. Repeat instability has generated a large mutational spectrum with a quantifiable increase in disease severity. Genetic modifiers of polyQ disease severity have been identified in animal models of polyQ ataxias (see especially Chapters 3 and 5) and in human patients (see Chapter 4).

Recently, Figueroa *et al.* (2001) examined variation in the polyglutamine tract in the human-calcium-activated potassium channel (*hSKCa3*) gene. Longer polyglutamine repeats were not causative for ataxias, but long normal alleles were overrepresented in patients with sporadic ataxia. As with all allelic association studies, independent confirmation of this finding is necessary.

## VI. PROGRESSION AND TREATMENT

Surprisingly little is known about the natural history of the ataxias with regard to progression of the disease. Klockgether *et al.* (1998) determined disease progression in genotypically defined ataxias. Progression was overall very similar for SCA1, 2, and 3 and was fastest for patients with MSA. Progression was correlated with CAG repeat size in SCA2 and SCA3. Lack of a correlation in SCA1 was probably a reflection of the smaller number of SCA1 patients and resulting loss of statistical power. Patients with FRDA and longer GAA repeats had earlier onset and faster disease progression. Onodera *et al.* (1998) studied progression of MRI findings in patients with SCA3. They found that progressive atrophy of cerebellum and brainstem was a function of age and the size of the expanded CAG repeat in the MJD1 gene in Machado-Joseph disease.

No long-term *prospectively* collected data for progression are available for any of the inherited ataxias. Establishment of progression rates and validation of clinical rating scales are necessary prerequisites for any controlled trials. It is likely that clinical rating scales will have to be complemented by quantitative measures of cerebellar function such as those proposed by Velazquez Perez *et al.* (2001).

The exponential increase in the understanding of inherited ataxias has not been paralleled by advances in the treatment of these disorders. Treatment of FRDA with the coenzyme Q10 analog idebenone has shown in promise for treatment of the cardiomyopathy (Rustin *et al.*, 1999), but has been disappointing for the treatment of ataxia in a small short-term trial (Schöls *et al.*, 2001). A short-term trial of branched amino acids showed some promise in SCA patients (Mori *et al.*, 2002), but treatment with trimethoprim-sulfamethoxazole did not (Schulte *et al.*, 2001). This contrasts with the sometimes dramatic response to acetazolamide in controlling episodes of ataxia with EA-2 and occasionally EA-1 (Chapter 22). Unfortunately, the treatment of cerebellar ataxia has remained primarily a neurorehabilitation challenge with occasional modest gains through the use of medications that can improve ataxia or dysarthria (reviewed in Perlman, 2000).

## Acknowledgments

The author thanks Dr. S. Perlman for providing Table 2.2 and Dr. Hema Vakharia for critically reading the manuscript. Mr. Michael Goldsmith and Ms. Cloteal Moran helped with illustrations and references.

## References

Aicardi, J., Barbosa, C., Andermann, E., Andermann, F., Morcos, R., Ghanem, Q., Fukuyama, Y., Awaya, Y., and Moe, P. (1988). Ataxia-ocular motor apraxia: a syndrome mimicking ataxia-telangiectasia. *Ann. Neurol.* **24**, 497–502.

Allikmets, R., Raskind, W. H., Hutchinson, A., Schueck, N. D., Dean, M., and Koeller, D. M. (1999). Mutation of a putative mitochondrial iron transporter gene (ABC7) in X-linked sideroblastic anemia and ataxia (XLSA/A). *Hum. Mol. Genet.* **8**, 743–749.

Bandmann, O., Sweeney, M. G., Daniel, S. E., Wenning, G. K., Quinn, N., Marsden, C. D., and Wood, N. W. (1997). Multiple-system atrophy is genetically distinct from identified inherited causes of spinocerebellar degeneration. *Neurology* **49**, 1598–1604.

Barbot, C., Coutinho, P., Chorao, R., Ferreira, C., Barros, J., Fineza, I., Dias, K., Monteiro, J., Guimaraes, A., Mendonca, P., do Ceu Moreira, M., and Sequeiros, J. (2001). Recessive ataxia with ocular apraxia: review of 22 Portuguese patients. *Arch. Neurol.* **58**, 201–205.

Barclay, J., Balaguero, N., Mione, M., Ackerman, S. L., Letts, V. A., Brodbeck, J., Canti, C., Meir, A., Page, K. M., Kusumi, K., Perez-Reyes, E., Lander, E. S., Frankel, W. N., Gardiner, R. M., Dolphin, A. C., and Rees, M. (2001). Ducky mouse phenotype of epilepsy and ataxia is associated with mutations in the Cacna2d2 gene and decreased calcium channel current in cerebellar Purkinje cells. *J. Neuroscii.* **21**, 6095–6104.

Basu, P., Chattopadhyay, B., Gangopadhaya, P. K., Mukherjee, S. C., Sinha, K. K., Das, S. K., Roychoudhury ,S., Majumder, P. P., and Bhattacharyya N. P. (2000). Analysis of CAG repeats in SCA1, SCA2, SCA3, SCA6, SCA7 and DRPLA loci in spinocerebellar ataxia patients and distribution of CAG repeats at the SCA1, SCA2 and SCA6 loci in nine ethnic populations of eastern India. *Hum. Genet.* **106**, 597–604.

Bell, J. (1947). Dystrophia myotonica and allied diseases. In "Treasury of Human Inheritance" (L.S. Penrose, Ed.). Vol. 4, part 5, pp. 343–410. Cambridge University Press, Cambridge.

Bomont, P., Watanabe, M., Gershoni-Barush, R., Shizuka, M., Tanaka, M., Sugano, J., Guiraud-Chaumeil, C., and Koenig, M. (2000). Homozygosity mapping of spinocerebellar ataxia with cerebellar atrophy and peripheral neuropathy to 9q33-34, and with hearing impairment and optic atrophy to 6p21-23. *Euro. J. Hum. Genet.* **8**, 986–990.

Bonini, N. M. (2001). Drosophila as a genetic approach to human neurodegenerative disease. *Parkinsonism Relat. Disord.* **7**, 171–175.

Brodbeck, J., Davies, A., Courtney, J. M., Meir, A., Balaguero, N., Canti, C., Moss, F. J., Page, K. M., Pratt, W. S., Hunt, S. P., Barclay, J., Rees, M., and Dolphin, A. C. (2002). The ducky mutation in Cacna2d2 results in altered Purkinje cell morphology and is associated with the expression of a truncated alpha 2 delta-2 protein with abnormal function. *J. Bio. Chem.* **277**, 7684–7693.

Burk, K., Fetter, M., Abele, M., Laccone, F., Brice A., Dichgans, J., and Klockgether, T. (1999). Autosomal dominant cerebellar ataxia type I: oculomotor abnormalities in families with SCA1, SCA2, and SCA3. *J. Neurol.* **246**, 789–797.

Buttner, N., Geschwind, D., Jen, J. C., Perlman, S., Pulst, S. M., and Baloh, R. W. (1998). Oculomotor phenotypes in autosomal dominant ataxias. *Arch. Neurol.* **55**, 1353–1357.

Chen, L., Chetkovich, D. M., Petralia, R. S., Sweeney, N. T., Kawasaki, Y., Wenthold, R. J., Bredt, D. S., and Nicoll, R. A. (2000). Stargazing regulates synaptic targeting of AMPA receptors by two distinct mechanisms. *Nature* **408**, 936–943.

Cholfin, J. A., Sobrido, M. J., Perlman, S., Pulst, S. M., and Geschwind, D. H. (2001). The SCA12 mutation is a rare cause of spinocerebellar ataxia. *Arch. Neuro.* in press.

Cossee, M., Puccio, H., Gansmuller, A., Koutnikova, H., Dierich, A., LeMeu, R. M., Fischbeck, K., Dolle, P., and Koenig, M. (2000). Inactivation of the Friedreich ataxia mouse gene leads to early embryonic lethality without iron accumulation. *Hum. Mol. Genet.* **9**, 1219–1226.

Dai, Z.H. (1991). Olivopontocerebellar atrophy:clinical analysis of 100 cases. (Article in Chinese). *Zhonghua. Shen. Jing. Jing. Shen. Ke. Za. Zhi.* **24**, 111–126.

Date, H., Onodera, O., Tanaka, H., Iwabuchi, K., Uekawa, K., Igarashi, S., Koike, R., Hiroi, T., Yuasa, T., Awaya, Y., Sakai, T.,Takahashi, T., Nagatomo, H., Sekijima, Y., Kawachi, I., Takiyama, Y., Nishizawa, M., Fukuhara, N., Saito, K., Sugano, S., and Tsuji, S. (2001). Early-onset ataxia with ocular motor apraxia and hypoalbuminemia is caused by mutations in a new HIT superfamily gene. *Nat. Genet.* **29**, 184–188.

Devos, D., Schraen-Maschke, S., Vuillaume, I., Dujardin, K., Naze, P., Willoteaux, C., Destee, A., Sablonniere, B. (2001). Clinical features and genetic analysis of a new form of spinocerebellar ataxia. *Neurology* **56**, 234–238.

Dürr, A., Chneiweiss, H., Khati, C., Stevanin, G., Cancel, G., Feingold, J., Agid, Y., Brice, A. (1993). Phenotypic variability in autosomal dominant cerebellar ataxia type I is unrelated to genetic heterogeneity. *Brain* **116**; 1497–1508.

Fernandez-Funez, P., Nino-Rosales, M. L., de Gouyon, B., She, W. C., Luchak, J. M., Martinez, P., Turiegano, E., Benito, J., Capovilla, M., Skinner, P. J., McCall, A., Canal, I., Orr, H. T., Zoghbi, H. Y., and Botas, J. (2000). Identification of genes that modify ataxin-1-induced neurodegeneration. *Nature* **408**, 101–106.

Fernandez-Gonzalez A., Spada, A. R., Treadaway, J., Higdon, J. C., Harris, B. S., Sidman, R. L., Morgan, J. I., and Zuo, J. (2002). Purkinje cell degeneration (pcd) Phenotypes Caused by Mutations in the Axotomy-Induced Gene, Nna1. *Science* **295**, 1904–1906.

Figueroa, K. P., Chan, P., Schöls, L., Tanner, C., Riess, O., and Perlman S., Geschwind, D. H., and Pulst S. M. (2001). Association of moderate polyglutamine tract expansions in the slow calcium-activated potassium channel type 3 with ataxia. *Arch. Neurol.* **58**, 1649–1653.

Filla, A., De Michele, G., Marconi, R., Bucci, L., Carillo, C., Castellano, A. E., Iorio, L., Kniahynicki, C., Rossi, F., and Campanella, G. (1992). Prevalence of hereditary ataxias and spastic paraplegias in Molise, a region of Italy. *J. Neurol.* **239**, 351–353.

Filla, A., Mariotti,C., Caruso, G., Coppola, G., Cocozza, S., Castaldo, I., Calabrese, O., Salvatore, E., De Michele, G., Riggio, M. C., Pareyson, D., Gellera, C., and Di Donato S. (2000). Relative frequencies of CAG expansions in spinocerebellar ataxia and dentatorubropallidoluysian atrophy in 116 Italian families. *Eur. Neurol.* **44**, 31–36.

Flanigan, K., Gardner, K., Alderson, K., Galster, B., Otterud, B., Leppert, M. F., Kaplan, C., and Ptacek, L. J. (1996). Autosomal dominant spinocerebellar ataxia with sensory axonal neuropathy (SCA4): clinical description and genetic localization to chromosome 16q22.1. *Am. J. Hum. Genet.* **59**, 392–399.

Flanigan, K. M., Coffeen, C.M., Sexton, L., Stauffer, D., Brunner, S., and Leppert, M. F. (2001). Genetic characterization of a large, historically significant Utah kindred with facioscapulohumeral dystrophy. *Neuromuscul. Disorder* **11**, 525–529.

Fletcher, C. F., Lutz, C. M., O'Sullivan, T. N., Shaughnessy, J. D. Jr., Hawkes, R., Frankel, W. N., Copeland, N. G., and Jenkins, N. A. (1996). Absence epilepsy in tottering mutant mice is associated with calcium channel defects. *Cell* **87**, 607–617.

Futamura, N., Matsumura, R., Fujimoto, Y., Horikawa, H., Suzumura A., and Takayanagi, T. (1998). CAG repeat expansions in patients with sporadic cerebellar ataxia. *Acta. Neurol. Scand.* **98**, 55–59.

Gaspar, C., Lopes-Cendes, I., Hayes, S., Goto, J., Arvidsson, K., Dias, A., Silveira, I., Maciel, P., Coutinho, P., Lima, M., Zhou, Y. X., Soong, B. W., Watanabe, M., Giunti, P., Stevanin, G., Riess, O., Sasaki, H., Hsieh, M., Nicholson, G. A, Brunt, E, Higgins, J. J., Lauritzen, M., Tranebjaerg, L., Volpini, V., Wood, N., Ranum, L., Tsuji S., Brice, A., Sequeiros, J., and Rouleau, G. A. (2001). Ancestral origins of the Machado-Joseph disease mutation: a worldwide haplotype study. *Am. J. Hum. Genet.* **68**, 523–528.

Geschwind, D. H., Perlman S., Figueroa C. P., Treiman L. J., and Pulst, S. M. (1997a). The prevalence and wide clinical spectrum of the spinocerebellar ataxia type 2 trinucleotide repeat in patients with autosomal dominant cerebellar ataxia. *Am. J. Hum. Genet.* **60**, 842–850.

Geschwind, D. H., Perlman, S., Figueroa, K. P., Karrim, J., Baloh, R. W., and Pulst, S. M. (1997b). Spinocerebellar ataxia type 6. Frequency of the mutation and genotype-phenotype correlations. *Neurology* **49**, 1247–1251.

Geschwind, D. H., Perlman, S., Grody, W. W., Telatar, M., Montermini, L., Pandolfo, M., and Gatti, R. A. (1997c). Friedreich's ataxia GAA repeat expansion in patients with recessive or sporadic ataxia. *Neurology* **49**, 1004–1009.

Giuffrida, S., Saponara, R., Trovato Salinaro, A., Restivo, D. A., Domina, E., Papotto, M., Le Pira F., Nicoletti A., Trovato, A., and Condorelli D. F. (1999). Identification of SCA2 mutation in cases of spinocerebellar ataxia with no family history in mid-eastern Sicily. *Ital. J. Neurol. Sci.* **20**, 217–221.

Giunti, P., Sabbadini, G., Sweeney, M. G., Davis, M. B., Veneziano, L., Mantuano, E., Federico, A., Plasmati, R., Frontali, M., and Wood, N. W. (1998). The role of the SCA2 trinucleotide repeat expansion in 89 autosomal dominant cerebellar ataxia families. Frequency, clinical and genetic correlates. *Brain* **121**, 459–467.

Gomez, C. M. (2001). Inherited cerebellar ataxia. In "Current Therapy in neurologic diseases" (J. T. Johnson, J. W. Griffin, and J. C. McArthur, eds.) 6th edition, 292–298. Mosby, St. Louis.

Gudmundsson, K. R. (1969). Prevalence and occurrence of some rare neurological diseases in Iceland. *Acta. Neurol. Scand.* **45**, 114–118.

Gunaratne, P. H., and Richards C. S. (1997). Estimated contribution of known ataxia genes in ataxia patients undergoing DNA testing. *Genet.*

*Test.* **1**, 275–278.

Harding, A. E. (1982). The clinical features and classification of the late onset autosomal dominant cerebellar ataxias. A study of 11 families, including descendants of the "the Drew family of Walworth." *Brain* **105**, 1–28.

Harper, P. S., (1998). Myotonic dystrophy as a trinucleotide repeat disorder-A clinical perspective. In "Genetic Instabilities and Hereditary Neurological Diseases" (R. Wells, and S. Warren, eds.). pp. 115–130, Academic Press, San Diego.

Hedera, P. (2002). Spastic paraplegia, ataxia, mental retardation (SPAR). A novel genetic disorder. *Neurology* **58**, 411–416.

Hsieh, M., Lin, S. J., Chen, J. F., Lin, H. M., Hsiao, K. M., Li, S. Y., Li ,C., and Tsai, C. J. (2000). Identification of the spinocerebellar ataxia type 7 mutation in Taiwan: application of PCR-based Southern blot. *J. Neurol.* **247**, 623–629.

Higashiyama, H., Hirose, F., Yamaguchi, M., Inoue, Y. H., Fujikake, N., Matsukage, A., and Kakizuka, A. (2002). Identification of ter94, Drosophila VCP, as a modulator of polyglutamine-induced neurodegeneration. *Cell Death Differ.* **9**, 264–273.

Huynh, D. P., Figueroa, K., Hoang, N., and Pulst, S. M. (2000). Nuclear localization or inclusion body formation of ataxin-2 are not necessary for SCA2 pathogenesis in mouse or human. *Nat. Genet.* **26**, 44–50

Illarioshkin, S. N., Slominsky, P. A., Ovchinnikov, I. V., Markova, E. D., Miklina, N. I., Klyushnikov, S. A., Shadrina, M., Vereshchagin, N. V., Limborskaya, S. A., and Ivanova-Smolenskaya I. A. (1996). Spinocerebellar ataxia type 1 in Russia. *J. Neurol.* **243**, 506–510

Jardim, L. B., Silveira, I., Pereira, M. L., Ferro, A., Alonso, I., do Ceu Moreira, M., Mendonca, P., Ferreirinha, F., Sequeiros, J., and Giugliani, R. (2001). A survey of spinocerebellar ataxia in South Brazil – 66 new cases with Machado-Joseph disease, SCA7, SCA8, or unidentified disease-causing mutations. *J. Neurol.* **248**, 870–876.

Jin, D. K., Oh, M. R., Song, S. M., Koh, S. W., Lee, M., Kim, G. M., Lee, W. Y., Chung, C. S., Lee, K. H., Im, J. H., Lee, M. J., Kim, J. W., and Lee, M. S. (1999). Frequency of spinocerebellar ataxia types 1,2,3,6,7 and dentatorubral pallidoluysian atrophy mutations in Korean patients with spinocerebellar ataxia. *J. Neurol.* **246**, 207–210.

Jones, J. G., and Seltzer, W. K. (2000). Mutation detection frequencies and allele size distributions for spinocerebellar and Friedreich's ataxia: A cumulative history of ataxia testing in a clinical reference laboratory. *Neurology* **54**, Suppl. 3, A357.

Juvonen, V., Hietala, M., Paivarinta, M., Rantamaki, M., Hakamies, L., Kaakkola, S., Vierimaa, O., Penttinen, M., and Savontaus, M. L. (2000). Clinical and genetic findings in Finnish ataxia patients with the spinocerebellar ataxia 8 repeat expansion. *Ann. Neurol.* **48**, 354–361.

Kazemi-Esfarjani, P., and Benzer, S. (2000). Genetic suppression of polyglutamine toxicity in Drosophila. *Science* **287**, 1837–1840.

Kiehl, R.T., Nechiporuk, A., Shibata, H., and Pulst, S. M. (2000). Knockout Models for Ataxin-2 and Ataxin-2-Binding-Protein: Gene Function, Developmental Genetics, and Clinical Relevance. *Neurology* **54**, A464.

Kiehl, T. R., Olson, J. M., and Pulst, S. M. ( 2001). The Hereditary Disease Array Group (HDAG)-Microarrays, Models and Mechanisms: A Collaboration Update. *Curr. Genomics* **2**, 221–229.

Klement, I. A., Skinner, P. J., Kaytor, M. D., Yi, H., Hersch, S. M., Clark, H. B., Zoghbi, H. Y., and Orr, H. T. (1998). Ataxin-1 nuclear localization and aggregation: role in polyglutamine-induced disease in SCA1 transgenic mice. *Cell* **95**, 41–53

Klockgether, T., Skalej, M., Wedekind, D., Luft, A. R., Welte, D., Schulz, J. B., Abele, M., Burk, K., Laccone, F., Brice, A., and Dichgans, J. (1998). Autosomal dominant cerebellar ataxia type I. MRI-based volumetry of posterior fossa structures and basal ganglia in spinocerebellar ataxia types 1, 2 and 3. *Brain* **121**, 1687–1693.

Komeichi, K., Sasaki, H., Yabe, I., Yamashita, I., Kikuchi, S., Tashiro, K. (2001). Twenty CAG repeats are sufficient to cause the SCA6 phenotype. *J. Med. Genet.* **38**, E38.

Leggo, J., Dalton, A., Morrison, P. J., Dodge, A., Connarty, M., Kotze, M. J., and Rubinsztein D. C. (1997). Analysis of spinocerebellar ataxia types 1, 2, 3, and 6, dentatorubral-pallidoluysian atrophy, and Friedreich's ataxia genes in spinocerebellar ataxia patients in the UK. *J. Med. Genet.* **34**, 982–985.

Leone, M., Bottacchi, E., D'Alessandro, G., and Kustermann, S. (1995). Hereditary ataxias and paraplegias in Valle d'Aosta, Italy: a study of prevalence and disability. *Acta. Neurol. Scand.* **91**, 183–187.

Link, C. D. (2001). Transgenic invertebrate models of age-associated neurodegenerative diseases. *Mech. Ageing Dev.* **122**, 1639–1649.

Lorenzetti, D., Watas, K., Xu, B., Matzuk, M. M., Orr, H. T., and Zoghbi, H. Y. Repeat instability and motor incoordination in mice with a targeted expanded CAG repeat in the Sca1 locus. (2002). *Hum. Mol. Genet.* **9**, 779–785.

Mariotti, C., Gellera, C., Grisoli, M., Mineri, R., Castucci, A., and Di Donato, S. (2001). Pathogenic effect of an intermediate-size SCA-6 allele (CAG)(19) in a homozygous patient. *Neurology* **57**, 1502–1504.

Marsh, J. J., Walker, H., Theisen, H., Zhu, Y. Z., Fielder, T., Purcell, J., and Thompson, L. M. (2000). Expanded polyglutamine peptides alone are intrinsically cytotoxic and cause neurogegeneration in Drosophila. *Hum. Mol. Genet.* **9**, 13–25.

Matilla, T., Volpini, V., Genis, D., Rosell, J., Corral, J., Davalos, A., Molins, A., and Estivill, X. (1993). Presymptomatic analysis of spinocerebellar ataxia type 1 (SCA1) via the expansion of the SCA1 CAG-repeat in a large pedigree displaying anticipation and parental male bias. *Hum. Mol. Genet.* **12**, 2123–2128.

Matilla, A., Roberson, E. D., Banfi, S., Morales, J., Armstrong, D. L., Burright, E. N., Orr, H. T., Sweatt, J. D., Zoghbi, H. Y., and Matzuk, M. M. (1998). Mice lacking ataxin-1 display learning deficits and decreased hippocampal paired-pulse facilitation. *J. Neurosci.* **18**, 5508–5516.

Matsuura, T, Sasaki, H, Yabe, I, Hamada, K, Hamada, T, Shitara, M, Tashiro, K. (1999). Mosaicism of unstable CAG repeats in the brain of spinocerebellar ataxia type 2. *J. Neurol.* **246**, 835–839.

Matsumura, R., Takayanagi, T., Murata, K., Futamura, N., and Fujimoto, Y. (1996). Autosomal dominant cerebellar ataxias in the Kinki area of Japan. *Jpn. J. Hum. Genet.* **41**, 399–406.

Meijer, I. A., Hand, C. K., Cossette, P., Figlewica, D. A., and Rouleau, G. A. (2002a). Spectrum of SPG4 mutations in a large collection of North American families with hereditary spastic paraplegia. *Arch. Neurol.* **59**, 281–286.

Meijer I., Hand, C. K., Grewal, K. K., Stefanelli, M. G., Ives, E. J., and Rouleau, G. A. (2002b). *Am. J. Hum. Genet.* **70**, 763–769.

Melberg, A., Hetta, J., Dahl, N., Nennesmo, I., Bengtsson, M., Wibom, R., Grant, C., Gustavson, K. H., and Lundberg, P. O. (1995). Autosomal dominant cerebellar ataxia deafness and narcolepsy. *J. Neurol. Sci.* **134**, 119–129.

Merlini, L., Gooding, R., Lochmuller, H., Muller-Felber, W., Walter, M. C., Angelicheva, D., Talim, B., Hallmayer, J., and Kalaydjieva, L. (2002). Genetic identity of Marinesco-Sjogren/myoglobinuria and CCFDN syndromes. *Neurology* **58**, 231–236.

Miranda, C. J., Santos, M. M., Ohshima, K., Smith, J., Li, L., Bunting, M., Cossee, M., Koenig, M., Sequeiros, J., Kaplan, J., and Pandolfo, M. (2002). Frataxin knockin mouse. *FEBS Lett.* **512**, 291–297.

Miller, V. M., Regagliati, M. R., Paulson, H. L. (2001). A zebafish model of polyglutamine disease. *Soc. Neurosci. Abstr.* **27**, 572.14.

Moreira, M. C., Barbot, C., Tachi, N., Kozuka, H., Uchida, E., Givson, T., Mendonca, P., Costa, M., Barros, J., Yanagisawa, T., Watanabe, M., Ideda, Y., Aoki, M., Nagata, T., Clutinho, P., Sequeiros, J., and Koenig, M. (2001). The gene mutated in ataxia-ocular apraxia 1 encodes the new HIT/Zn-finger protein aprataxin. *Nat. Genet.* **29**, 189–193.

Mori, M., Adachi, Y., Kusumi, M., and Nakashima, K. (2001a). Spinocerebellar ataxia type 6: founder effect in Western Japan. *J. Neurol. Sci.* **185**, 43–47.

Mori, M., Adachi, Y., Kusumi, M., and Nakashima, K. (2001b). A genetic epidemiological study of spinocerebellar ataxias in Tottori prefecture, Japan. *Neuroepidemiology* **20**, 144–149.

Mori, M., Adachi, Y., Mori, N., Kurihara, S., Kashiwaya, Y., Kusumi, M., Takeshima, T., Nakashima, K. (2002). Double-blind crossover study of branched-chain amino acid therapy in patients with spinocerebellar degeneration. *J. Neurol. Sci.* **195**, 149–152.

Moseley, M. L., Benzow, K. A., Schut, L. J., Bird, T. D., Gomez, C. M., Barkhaus, P. E., Blindauer, K. A., Labuda, M., Pandolfo, M., Koob, M. D., and Ranum L. P. (1998). Incidence of dominant spinocerebellar and Friedreich triplet repeats among 361 ataxia families. *Neurology* **51**, 1666–1671.

Nemeth, A. H., Bochukova, E., Dunne, E., Huson, S. M., Elston, J., Hanna, M. A., Jackson, M., Chapman, C. J., and Taylor, A. M. (2000). Autosomal recessive cerebellar ataxia with oculomotor apraxia (ataxia-telangiectasia-like syndrome) is linked to chromosome 9q34. *Am. J. Hum. Genet.* **67**, 1320–1326.

Nystuen, A., Benke, P. J., Merren, J., Stone, E. M., and Sheffield, V. C. (1996). A cerebellar ataxia locus identified by DNA pooling to search for linkage disequilibrium in an isolated population from the Cayman Islands. *Hum. Mol. Genet.* **5**, 525–531.

Onodera, O., Idezuka, J., Igarashi, S., Takiyama, Y., Endo, K., Takano, H., Oyake, M., Tanaka, H., Inuzuka, T., Hayashi, T., Yuasa, T., Ito, J., Miyatake, T., and Tsuji, S. (1998). Progressive atrophy of cerebellum and brainstem as a function of age and the size of the expanded CAG repeats in the MJD1 gene in Machado-Joseph disease. *Ann. Neurol.* **43**, 288–296.

Onodera, Y., Aoki, M., Tsuda, T., Kato, H., Nagata, T., Kameya, T., Abe, K., and Itoyama, Y. (2000). High prevalence of spinocerebellar ataxia type 1 (SCA1) in an isolated region of Japan. *J. Neurol. Sci.* **178**, 153–158.

Orr, H. T., Chung, M. Y., Banfi, S., Kwiatkowski, T. J. Jr., Servadio, A., Beaudet, A. L., McCall, A. E., Duvick, L. A., Ranum, L. P., and Zoghbi, H. Y. (1993). Expansion of an unstable trinucleotide CAG repeat in spinocerebellar ataxia type 1. *Nat. Genetics* **4**, 221–226.

Pareyson, D., Gellera, C., Castellotti, B., Antonelli, A., Riggio. M. C., Mazzucchelli, F., Girotti, F., Pietrini, V., Mariotti, C., and Di Donato, S. (1999). Clinical and molecular studies of 73 Italian families with autosomal dominant cerebellar ataxia type I: SCA1 and SCA2 are the most common genotypes. *J. Neurol.* **246**, 389–393.

Park, C., Finger, J. H., Cooper, J. A., and Ackerman, S. L. (2002). The cerebellar deficient folia (cdf) gene acts intrinsically in Purkinje cell migrations. *Genesis* **32**, 32–41.

Parker, J. A., Connolly, J. B., Wellington, C., Hayden, M., Dausset, J., and Neri, C. (2001). Expanded polyglutamines in Caenorhabditis elegans cause axonal abnormalities and severe dysfunction of PLM mechanosensory neurons without cell death. *Proc. Natl. Acad. Sci. U.S.A.* **98**, 13318–13323.

Penrose, L. S. (1948). The problem of anticipation in pedigrees of dystrophia myotonica. *Ann. Eugenics* **14**, 125–132.

Perlman, S. L. (2000). Cerebellar Ataxia. *Curr. Treat. Options Neurol.* **2**, 215–224.

Polo, J. M., Calleja, J., Combarros, O., Berciano, J. (1991). Hereditary ataxias and paraplegias in Cantabria, Spain. An epidemiological and clinical study. *Brain* **114**, 855–866.

Puccio, H., Simon, D., Cossee, M., Criqui-Filipe, P., Tiziano, F., Melki, J., Hindelang, C., Matyas, R., Rustin, P., and Koening, M. (2001). Mouse models for Friedreich ataxia exhibit cardiomyopathy, sensory nerve defect and Fe-S enzyme deficiency followed by intramitochondrial iron deposits. *Nat. Genet.* **27**, 181–186.

Pujana, M. A., Corral, J., Gratacos, M., Combarros, O., Berciano, J., Genis, D., Banchs, I., Estivill, X., and Volpini, V. (1999). Spinocerebellar ataxias in Spanish patients: genetic analysis of familial and sporadic cases. The Ataxia Study Group. *Hum. Genet.* **104**, 516–522.

Pulst, S. M., Nechiporuk, A., and Starkman, S. (1993). Anticipation in spinocerebellar ataxia type 2. *Nat. Genetics* **5**, 8–10.

Pulst, S. M., Nechiporuk, A., Nechiporuk, T., Gispert, S., Chen, X. N., Lopes-Cendes, I., Pearlman, S., Starkman, S., Orozco-Diaz, G., Lunkes, A., DeJong, P., and Rouleau, G. A., Auburger, G., Korenberg, J. R., Figueroa, C., and Sahba, S. (1996). Moderate expansion of a normally biallelic trinucleotide repeat in spinocerebellar ataxia type 2. *Nat. Genetics* **14**, 237–238.

Pulst, S. M., and Perlman, S. L. (2000). Hereditary ataxias. In "Neurogenetics" (S. M. Pulst, ed.), pp. 231–264. Oxford University Press, New York.

Ramesar, R. S., Bardien, S., Beighton, P., and Bryer, A. (1997). Expanded CAG repeats in spinocerebellar ataxia (SCA1) segregate with distinct haplotypes in South african families. *Hum. Genet.* **100**, 131–137.

Raskind, W. H., Pericak-Vance, M. A., Lennon, F., Wolff, J., Lipe, H. P., and Bird, T. D. (1997). Familial spastic paraparesis: evaluation of locus heterogeneity, anticipation, and haplotype mapping of the SPG4 locus on the short arm of chromosome 2. *Am. J. Med. Genet.* **74**, 26–36.

Riess, O., Laccone, F. A., Gispert, S., Schöls, L., Zuhlke, C., Vieira-Saecker, A. M., Herlt, S., Wessel, K., Epplen, J. T., Weber, B. H., Kreuz, F., Chahrokh-Zadeh, S., Meindl, A., Lunkes, A., Aguiar, J., Macek, M., Jr., Krebsova, A., Macek, M., Sr., Burk, K., Tinschert, S., Schreyer, I., Pulst, S. M., and Auburger, G. (1997). SCA2 trinucleotide expansion in German SCA patients. *Neurogenetics* **1**, 59–64.

Rosenberg, R. N. (1995). Autosomal dominant cerebellar phenotypes: the genotype has settled the issue. *Neurology* **45**, 1–5.

Rustin, P., von Kleist-Retzow, J. C., Chantrel-Groussard, K., Sidi, D., Munnich, A., and Rotig, A. (1999). Effect of idebenone on cardiomyopathy in Friedreich's ataxia: a preliminary study. *Lancet* **354**, 477–479.

Sasaki, H., Yabe, I., Yamashita, I., and Tashiro, K. (2000). Prevalence of triplet repeat expansion in ataxia patients from Hokkaido, the northernmost island of Japan. *J. Neurol. Sci.* **175**, 45–51.

Schöls, L., Amoiridis, G., Buttner, T., Przuntek, H., Epplen, J. T., and Riess, O. (1997a). Autosomal dominant cerebellar ataxia: phenotypic differences in genetically defined subtypes? *Ann. Neurol.* **42**, 924–932.

Schöls, L., Gispert, S., Vorgerd, M., Menezes Vieira-Saecker, A. M., Blanke, P., Auburger, G., Amoiridis, G., Meves, S., Epplen, J. T., Przuntek, H., Pulst, S. M., and Riess, O. (1997b). Spinocerebellar ataxia type 2. Genotype and phenotype in German kindreds. *Arch. Neurol.* **54**, 1073–1080.

Schöls, L., Kruger, R., Amoiridis, G., Przuntek, H., Epplen J. T., and Riess, O. (1998). Spinocerebellar ataxia type 6: genotype and phenotype in German kindreds. *J. Neurol. Neurosurg. Psychiatry* **64**, 67–73.

Schöls, L., Szymanski, S., Peters, H., Przuntek, H., Epplen, J. T., Hardt, C., and Riess, O. (2000). Genetic background of apparently idiopathic sporadic cerebellar ataxia. *Hum. Genet.* **107**, 132–137.

Schöls, L., Vorgerd, M., Schillings, M., Skipa, G., and Zange, J. (2001). Idebenone in patients with Friedreich ataxia. *Neurosci. Lett.* **29**, 169–172.

Schulte, T., Mattern, R., Berger, K., Szymansk, i. S., Klotz, P., Kraus, P. H., Przuntek, H., Schöls, L. (2001). Double-blind crossover trial of trimethoprim-sulfamethoxazole in spinocerebellar ataxia type 3/Machado-Joseph disease. *Arch. Neurol.* **58**, 1451–1457.

Shahbazian, M. D., Orr, H. T., and Zoghbi, H. Y. (2001). Reduction of Purkinje cell pathology in SCA1 transgenic mice by p53 deletion. *Neurobiol. Dis.* **8**, 974–981.

Silveira, I., Lopes-Cendes, I., Kish, S., Maciel, P., Gaspar, C., Coutinho, P., Botez, M. I., Teive H., Arruda, W., Steiner, C. E., Pinto-Junior W., Maciel, J. A., Jerin, S., Sack, G., Andermann, E., Sudarsky, L., Rosenberg, R., MacLeod, P., Chitayat, D., Babul, R., Sequeiros, J., and Rouleau, G. A. (1996). Frequency of spinocerebellar ataxia type 1, dentatorubro-pallidoluysian atrophy, and Machado-Joseph disease mutations in a large group of spinocerebellar ataxia patients. *Neurology* **46**, 214–218.

Silveira, I., Coutinho, P., Maciel, P., Gaspar, C., Hayes, S., Dias, A., Guimaraes, J., Loureiro, L., Sequeiros, J., and Rouleau, G. A. (1998). Analysis of SCA1, DRPLA, MJD, SCA2, and SCA6 CAG repeats in 48 Portuguese ataxia families. *Am. J. Med. Genet.* **81**, 134–138.

Sobrido, M. J., Cholfin, J. A., Perlman, S., Pulst, S. M., and Geschwind, D. H. (2001). SCA8 repeat length in patients with inherited and sporadic ataxia. *Neurology,* in press.

Soong, B. W., Lu, Y. C., Choo, K. B., and Lee H. Y. (2001). Frequency analysis of autosomal dominant cerebellar ataxias in Taiwanese patients and clinical and molecular characterization of spinocerebellar ataxia type 6. *Arch. Neurol.* **58**, 1105–1109.

Soto-Ares, G., Vinchon, M., Delmaire, C., Abecidan, E., Dhellemes, P., and Pruvo, J. P. (2001). Cerebellar atrophy after severe traumatic head injury in children. *Childs. Nerv. Syst.* **17**, 263–269.

Steffan, J. S., Bodai, L., Pallos, J., Poelmann, M., McCampbell, A., Apostol, B. L., Kazantsev, A., Schmidt, E., Zhu, Y. Z., Greenwald, M., Kurokawa, R., Housman, D. E., Jackson, G. R., Marsh, J. L., and Thompson, L. M. (2001). Histone deacetylase inhibitors arrest polyglutamine-dependent neurodegeneration in Drosophila. *Nature* **413**, 739–743.

Stevanin, S., Durr, A., and Brice, A. (2000). Clinical and molecular advances in autosomal dominant cerebellar ataxias: from genotype to phenotype and physiopathology. *Eur. J. Hum. Genet.* **8**, 4–18. Review.

Storey, E., du Sart, D., Shaw, J. H., Lorentzos, P., Kelly, L., McKinley, Gardner, R. J., Forrest, S. M., Biros, I., and Nicholson, G. A. (2000). Frequency of spinocerebellar ataxia types 1, 2, 3, 6, and 7 in Australian patients with spinocerebellar ataxia. *Am. J. Med. Genet.* **95**, 351–357.

Storey, E., Gardner, R. J. M., Knight, M. A., Kennerson, M. L., Tuck, R. R., Forrest, S. M., and Nichelson, G. A. (2001). A new autosomal dominant ataxia. *Neurology* **57**, 1913–1915.

Takano, H., Cancel, G., Ikeuchi, T., Lorenzetti, D., Mawad, R., Stevanin, G., Didierjean, O., Durr, A., Oyake, M., Shimohata, T., Sasaki, R., Koide, R., Igarashi, S., Hayashi, S., Takiyama, Y., Nishizawa, M., Tanaka, H., Zoghbi, H., Brice, A., and Tsuji, S. (1998). Close associations between prevalences of dominantly inherited spinocerebellar ataxias with CAG-repeat expansions and frequencies of large normal CAG alleles in Japanese and Caucasian populations. *Am. J. Hum. Genet.* **63**, 1060–1066.

Takiyama, Y., Sakoe, K., Nakano, I., and Nishizawa, M. (1997). Machado-Joseph disease: cerebellar ataxia and autonomic dysfunction in a patient with the shortest known expanded allele (56 CAG repeat units) of the MJD1 gene. *Neurology* **49**, 604–606.

Tan, E. K., Law, H. Y., Zhao, Y., Lim, E., Chan, L. L., Chang, H. M., Ng, I., and Wong, M. C. (2000). Spinocerebellar ataxia in Singapore: predictive features of a positive DNA test? *Eur. Neurol.* **44**, 168–171.

Tang, B., Liu, C., Shen, L., Dai, H., Pan ,Q., Jing, L., Ouyang, S., and Xia, J. (2000). Frequency of SCA1, SCA2, SCA3/MJD, SCA6, SCA7, and DRPLA CAG trinucleotide repeat expansion in patients with hereditary spinocerebellar ataxia from Chinese kindreds. *Arch. Neurol.* **57**, 540–544.

van de Warrenburg, B. P., Sinke, R. J., Verschuuren–Bemelmans, C. C., Scheffer, H., Brunt, E. R., Ippel, P. F., Maat–Kievit, J. A., Dooijes, D., Notermans, N. C., Lindhout, D., Knoers, N. V., and Kremer, H. P. (2002). Spinocerebellar ataxias in The Netherlands: Prevalence and age at onset variance analysis. *Neurology* **58**, 702–708.

Velazquez Perez, L., De La Hoz Oliveras, J., Perez Gonzalez, R., Hechevarria Pupo, R. R., and Herrera Dominguez, H. (2001). Quantitative evaluation of disorders of coordination in patients with cuban type 2 spinocerebellar ataxia. *Rev. Neurol.* **32**, 601–606.

Verbeek, D. S., Schelhaas, J. H., Ippel, E. F., Beemer, F. A., Pearson P. L., and Sinke, R. J. Identification of a novel SCA locus (SCA19) in Dutch Autosomal Dominant Cerebellar Ataxia (ADCA) family on chromosome region 1p21–q21. *Ann. Neurol.* in press.

Vig, J. P., Subramony, S. H., and McDaniel, D. O. (2001). Calcium homeostasis and spinocerebellar ataxia-1 (SCA-1). *Brain Res. Bull.* **56**, 221–225.

Voikar, V., Rauvala, H., and Ikonen, E. (2002). Cognitive deficit and development of motor impairment in a mouse model of Niemann-Pick type C disease. *Behav. Brain Res.* **132**, 1–10

Wadia, N., Pang, J., Desai, J., Mankodi, A., Desai, M., and Chamberlain, S. A. (1998). Clinicogenetic analysis of six Indian spinocerebellar ataxia (SCA2) pedigrees. The significance of slow saccades in diagnosis. *Brain,* **121**, 2341–2355.

Watanabe, H., Tanaka, F., Matsumoto, M., Doyu, M., Ando, T., Mitsuma, T., and Sobue, G. (1998). Frequency analysis of autosomal dominant cerebellar ataxias in Japanese patients and clinical characterization of spinocerebellar ataxia type 6. *Clin. Genet.* **53**, 13–19.

Yokota, T., Igarashi, K., Uchihara,T., Jishage, K., Tomita, H., Inaba, A., Li, Y., Arita, M., Suzuki, H., Mizusawa, H., and Arai, H. (2001). Delayed-onset ataxia in mice lacking alpha -tocopherol transfer protein: model for neuronal degeneration caused by chronic oxidative stress. *Proc. Natl. Acad. Sci. U.S.A.* **98**, 15185–15190.

Zhang, S., Xu, L., Lee, J., and Xu, T. (2002). Drosophila atrophin homolog functions as a transcriptional corepressor in multiple developmental processes. *Cell* **108**, 45–56.

Zoghbi, H. Y. (1997). The expansion of the CAG repeat in ataxin-2 is a frequent cause of autosomal dominant spinocerebellar ataxia. *Neurology* **49**, 1009–1013.

# CHAPTER 3

# Spinocerebellar Ataxia 1 (SCA1)

HARRY T. ORR

*Departments of Laboratory Medicine and Pathology, and Genetics, Cell Biology and Development*
*Institute of Human Genetics*
*University of Minnesota*
*Minneapolis, Minnesota 55455*

I. Summary
II. SCA1—The Phenotype
III. The *SCA1* Gene
  A. Diagnosis
IV. Models of Disease
  A. Cell Culture Studies
  B. Transgenic Mice
  C. SCA1 in Flies
V. Treatment
References

## I. SUMMARY

Spinocerebellar ataxia type 1 (SCA1) is an autosomal dominant neurodegenerative disease typically with mid-life onset characterized by motor symptoms in the absence of cognitive deficits (Zoghbi and Orr, 2001). SCA1 is a member of an intriguing group of neurodegenerative disorders known as the polyglutamine diseases. At the present, nine diseases been shown to result from expansion of CAG repeats coding for polyglutamine tracts in the respective proteins (Zoghbi and Orr, 2000; Nakamura et al., 2001). These disorders include spinobulbar muscular atrophy (SBMA), Huntington's disease (HD), and the spinocerebellar ataxias (SCA1, SCA2, SCA3/MJD, SCA6, SCA7, and SCA17), and dentatorubropallidoluysian atrophy (DRPLA). Expansion of the polyglutamine tract in the *SCA1* encoded protein, ataxin-1, results in an alteration in its folding. A key aspect of SCA1 pathogenesis is localization of mutant ataxin-1 to the nucleus.

## II. SCA1—THE PHENOTYPE

The clinical features seen in SCA1 patients usually include ataxia, dysarthria, and bulbar dysfunction. Additional features of the disease often vary depending on disease stage. Early in the course of SCA1, patients usually have a mild loss of limb and gait coordination, slurred speech, and poor handwriting skills. Some have hyperreflexia, hypermetric saccades, and nystagmus. As the disease progresses, the ataxia worsens, dysmetria, dysdiadochokinesis, and hypotonia develop and SCA1 patients often experience vibration and prorioceptive loss. At the advanced stage, the ataxia becomes very severe and brainstem dysfunction results in facial weakness, swallowing, and breathing problems. Patients typically will die from the loss of the ability to cough effectively, food aspiration, and respiratory failure. In SCA1, death usually occurs between 10 and 15 years after the onset of symptoms.

Schut and Haymaker (1951) published the first extensive neuropathological study on a kindred now known to have SCA1. They noted clinical variation within members of that kindred and attempted to subclassify these variants on the basis of their clinical and pathological features. A range of neuropathological features was identified in five patients who were examined at autopsy. Some patients had substantial olivopontocerebellar atrophy, while others had more cerebellar and/or sparing of the pons. The latter cases often had a clinical presentation that was more hyperreflexic with milder features of ataxia. There also was variability in the pathology of the spinal cord with most cases having significant axonal loss in the posterior columns and spinocerebellar tracts. Bulbar motor neurons, particularly the hypoglossal nuclei, were involved to some extent in all cases.

## III. THE *SCA1* GENE

The genetics of SCA1 has its roots in the late 1800s with the description of a recessive form of ataxia by Friedreich. In 1893, Pierre Marie described a different form of ataxia with an autosomal dominant pattern of inheritance. The designation SCA1 for one form of dominant ataxia reflects the fact that the genetic locus for this disease was the first to be localized to a specific chromosomal region, the shortarm of chromosome 6, by virtue of its linkage to the HLA complex (Yakura et al., 1974).

Using a positional cloning strategy, the *SCA1* gene was identified and isolated in 1993 (Orr et al., 1993). The *SCA1* gene consists of nine exons spanning over 450 kb of genomic DNA with the largest intron extending for at least 150 kb. The 2448 bp coding region is located within exons eight and nine. Thus, the 935 bp 5′UTR is contained within the first seven exons and a portion of the eighth exon. The 3″UTR is contained within the ninth exon and remarkably includes over 7000 bp yielding an *SCA1* transcript of 10,660 bp.

The disease-causing mutation is the expansion of an unstable CAG trinulceotide repeat within the coding region of the gene (Orr et al., 1993). Disease is caused by the expansion of this repeat producing an expanded tract of glutamine amino acids within the *SCA1* product, ataxin-1. There is an inverse relationship between the length of the CAG repeat tract on affected chromosomes and the age of onset and severity of disease. The longer the mutant CAG repeat, the earlier the age of onset and the more severe the disease (Ranum et al., 1994). The number of CAG repeats in *SCA1* is highly variable in the general population, ranging from 6–44. Interestingly, normal alleles of *SCA1* having more than 20 repeats typically have from one to three CAT triplet interruptions within the CAG tract (Chung et al., 1993). In contrast, affected alleles of *SCA1* have expansions between 39 and 82 of pure CAG repeat tract without a CAT interruption. The presence of the CAT interruptions is likely to be an important factor in the intergenerational stability of the longer normal alleles. Perhaps an early event in the conversion of a normal, stable allele of *SCA1* to an unstable allele is the loss of the CAT interruptions.

An immediate outcome of cloning *SCA1* was the development of a PCR-based DNA diagnostic test for the SCA1 mutation (Orr et al., 1993, Genis et al., 1995). This is a fairly straightforward test. However, care should be taken with repeat tracts in the range between 38 and 44 repeats. If these are determined to contain CAT interruptions then the allele should be diagnosed as a wild-type allele (Chung et al., 1993; Quan et al., 1995). The absence of a CAT interruption indicates that the allele is not a wild-type and is an affected allele. Either direct sequencing (Quan et al., 1995) or the restriction digestion of the PCR product with *Sfa*NI (Chung et al., 1993) can be used to assess the presence of CAT interruptions.

Ataxin-1 is a novel protein of about 800 amino acids, depending on the length of this polyglutamine tract. Initial searches of the sequence databases failed to reveal any homologies between ataxin-1 and other known proteins. Thus, the first efforts to reveal potential functions of ataxin-1 utilized antibodies to examine the regional and subcelluar patterns of ataxin-1 expression.

Expression of ataxin-1 in the normal human brain and peripheral tissues is widespread. The subcelluar pattern of ataxin-1 varies with cell type. In some cell types, such as lymphoblast cells, ataxin-1 is located primarily to cytoplasm. Interestingly, in neurons ataxin-1 is localized mainly within neuronal nuclei, although Purkinje cells also have a minor cytoplasmic component (Servadio et al., 1995). In brains from SCA1 patients and SCA1 transgenic mice, see below, single large ubiquitinated intranuclear inclusions have been described (Skinner et al., 1997; Duyckaerts et al., 1999). Specific antibodies against ataxin-1 (Skinner et al., 1997) and antibodies directed against polyglutamine expanses (Duyckaerts et al., 1999) also labeled these inclusions. Similar structures, containing the disease-specific protein, were described in another CAG-repeat disease, Huntington disease (DiFiglia et al., 1997), and also have been seen in other neurological diseases caused by expansions of CAG repeats (Paulson et al., 1997; Becher et al., 1998; Holmberg et al., 1998; Li et al., 1998). These data provide the first suggestion that the nucleus of neurons might have an important role in disease.

Another experimental strategy used to gain insight into the function of ataxin-1 has been the use of the yeast two-hybrid approach to seek out clones of proteins that might interact with ataxin-1. This analysis initially revealed that ataxin-1 has the ability to self-associate and that this ability requires amino acids 495–605 of the wild-type protein (Burright et al., 1997). This same amino acid stretch was also important for ataxin-1's ability to interact with many other cellular proteins using the yeast two-hybrid system.

The first protein identified that interacts with ataxin-1 was the cerebellar leucine-rich acidic nuclear (LANP) protein (Matilla et al., 1997). Several aspects of LANP and its interaction with ataxin-1 are intriguing and suggest that this interaction might be a critical aspect of SCA1 pathogenesis. First, the cellular pattern of LANP expression in the brain is more restricted than that of ataxin-1 and closely overlaps with the cellular pattern of SCA1 pathology. For example, LANP is prominently expressed in nuclei of Purkinje cells. Secondly, the interaction of LANP with ataxin-1 is directly related to the length of the polyglutamine tract in ataxin-1. LANP interacts significantly more with ataxin-1 having 82 glutamines than with a wild-type version of ataxin-1. Finally, co-expression of ataxin-1 and LANP in COS cells results in the normally soluble nuclear LANP being recruited to the insoluble nuclear matrix fraction along with ataxin-1.

Another ataxin-1-interacting protein that has been characterized is A1Up (Davidson *et al.*, 2000). Sequence analysis of A1Up suggests that it may link the chaperone and ubiquitin-proteasome pathways in a cell. The N-terminal region of A1Up contains a ubiquitin-like domain. In the C-terminal region A1Up is homologous to the human Chap1/Dsk2 protein that binds to the ATPase domain of HSP70. In the nucleus of transfected cells A1Up co-localized with mutant ataxin-1 further supporting that it is able to interact with ataxin-1.

The predominant localization of ataxin-1 to the nucleus of neurons prompted a search for peptide motifs that function in the transport of ataxin-1 from the cytoplasm, the site of protein synthesis, to the nucleus. Many nuclear proteins are directed to this subcellular site by the actions of short peptide sequence, a nuclear localization signal (NLS) within the protein. Many NLSs conform to a consensus cluster of four or five arginines and lysines. Inspection of the ataxin-1 sequence revealed two possible NLSs, one at the N terminus at lysine 15 and a second one at the C terminus at lysine 772 (Klement *et al.*, 1998). Using a single base mutational strategy that replaced each lysine with a threonine, the C-terminal NLS at lysine 772 was shown to be important for the localization of ataxin-1 to the nucleus of transfected cells. Thus, ataxin-1 appears to contain a single functional NLS at its C terminus. The type of NLS found in ataxin-1, a short stretch of basic amino acids, indicates that ataxin-1 is transported into the nucleus by the importin $\alpha/\beta$ pathway (Jans *et al.*, 2000).

To further examine the role of ataxin-1 in the nucleus, its ability to bind nucleic acids was examined *in vitro*. Ataxin-1 was shown to have the ability to bind specifically to homopolymers of poly(rG) (Yue *et al.*, 2001). Moreover, the RNA-binding capability of ataxin-1 decreased as the length of its polyglutamine tract was increased. These results raise the intriguing possibility that ataxin-1 is an RNA-binding protein and that this function is altered by the expansion of the polyglutamine tract. RNA-binding proteins are known to function in a wide variety of nuclear and cytoplasmic processes, including regulation of RNA splicing, mRNA stability and transport from the nucleus, and mRNA translation. It will be important to demonstrate that ataxin-1 is able to bind to and regulate the expression of a neuronal RNA.

## A. Diagnosis

### 1. DNA

The number of CAG repeats in *SCA1* is highly variable in the general population, ranging from 6–44. Assessing the length of the repeat tract in *SCA1* is readily accomplished using PCR (Chung *et al.*, 1993). Interestingly, normal alleles of *SCA1* having more than 20 repeats typically have from one to three CAT triplet interruptions within the CAG tract. In contrast, affected alleles of *SCA1* have expansions between 39 and 82 of pure CAG repeat tract without a CAT interruption (Chung *et al.*, 1993; Matilla *et al.* 1993; Jodice *et al.*, 1994; Quan *et al.*, 1995; Goldfarb *et al.*, 1996). Distinguishing normal from disease alleles requires additional evaluation when the allele size is in the 36 to 44 repeat range. To ascertain whether these alleles represent normal interrupted alleles or expanded uninterrupted alleles *Sfa*NI restriction digestion should be carried out (Chung *et al.*, 1993).

There is an inverse relationship between the length of the CAG repeat tract on affected chromosomes and the age of onset and severity of disease. The longer the mutant CAG repeat, the earlier the age of onset and the more severe the disease (Ranum *et al.*, 1994).

### 2. Neuroimaging

Imaging studies on genetically diagnosed SCA1 patients are limited. Magnetic resonance imaging (MRI) approaches have been used to examine structural changes in the brain associated with inherited and sporadic forms of ataxia. However, the changes seen occur late in disease progression and are relatively nonspecific (Wullner *et al.*, 1993). One study using single photon emission tomography to examine blood flow suggested that hypoperfusion appears to develop prior to anatomical alterations detectable by imaging (Michele *et al.*, 1998). The most extensive use of imaging to study the course of SCA1 was by Mascalchi *et al.* (1998) in a large Italian family. In this study, 10 individuals having the SCA1 mutation as determined by DNA analysis were examined by single-voxel proton magnetic resonance spectroscopy. These investigators used this approach to determine the *N*-acetylaspartate/creatine (NAA/Cr) ratio and thereby assess neuronal viability in brain regions. Compared with unaffected family members, asymptomatic SCA1 mutation carriers had decreased NAA/Cr ratios in the pons and to a lesser extent in the deep cerebellar hemisphere. Whether this change will be useful for monitoring treatments remains to be determined.

### 3. Neuropathology

A more recent pathological report (Robitaille *et al.*, 1995) on eleven additional patients from the same kindred examined by Schut and Haymaker (1951) emphasized the olivopontocerebellar degeneration in the majority of the cases. The most frequently seen and most severe alterations were loss of Purkinje cells in the cerebellar cortex, and loss of neurons in the inferior olivary nuclei, the cerebellar dentate nuclei, and the red nuclei. The cerebellar granular layer was spared or had slight loss of neurons. Involvement of the basal pontine nuclei ranged from mild to severe. The posterior columns and spinocerebellar tracts of the spinal cord typically were moderately to severely affected. Motor neurons of the anterior horns were decreased in number in all cases where the cord was examined. Nuclei of the third, tenth, and twelfth cranial nerves also had variable involvement, with the hypoglossal nuclei most frequently and severely affected. The striatonigral system was minimally affected or unaffected when examined by classical anatomical methods, but biochemical studies on the same population of patients revealed evidence of decreased dopaminergic terminals in the striatum without changes in the cellularity of the nigra (Kish et al., 1997). The cerebral cortex and hippocampus had slight loss of cellularity in some of the cases.

Neuropathological studies from other kindreds have shown findings similar to those listed above (Nino et al., 1980; Spadaro et al., 1992), although some studies, often with a limited number of cases, have shown variations such as more involvement of the striatum (Genis et al., 1995) or cerebral cortex (Bebdin et al., 1990). Morphometric studies showed a mean of 85% cell loss in the inferior olives and 60% loss of Purkinje cells in the cerebellar cortex in four cases of SCA1 (Bebdin et al., 1990). Some studies have made the clinical observation of optic atrophy (Landis et al., 1974), but neuropathological confirmation is lacking in nearly all instances. A review of the Japanese literature (Yagashita and Inoue, 1997) emphasizes the severe degeneration in the spinocerebellar tracts, posterior columns, and anterior horn neurons. In those cases the features of olivopontocerebellar degeneration were mild to moderate, but most had significant involvement of the dentatorubral system and some had changes in the external segment of the pallidum.

In general, SCA1 appears to have a defined neuropathological pattern that separates it from other dominantly inherited ataxias and from sporadic entities such as multiple system atrophy. Because of the range of neuropathological findings in SCA1 and the other forms of hereditary ataxia, there still is a possibility that individual cases of SCA1 will have neuropathological features that overlap with other diseases.

Investigations into the cellular changes associated with SCA1 have been limited primarily to the Purkinje cell, a principal target of the disease process. Golgi studies (Koeppen, 1991; Ferrer et al., 1994) have revealed reduced dendritic arborization in Purkinje cells with decreased numbers of small spiny dendritic processes. Axonal spheroids, or torpedo bodies, are frequent in Purkinje cells. Immunohistochemical assessments for nonphosphorylated neurofilaments and calbindin also delineate similar dendritic and axonal pathology of Purkinje neurons. Calbindin immunohistochemistry for Purkinje cells appears to remain intact, even when those cells show dendritic dwindling (Koeppen, 1991; Ferrer et al., 1994; Vig et al., 1996), but the preservation of Purkinje cell immunoreactivity for parvalbumin, another calcium-binding protein, is less consistent.

An electron microscopic study of cerebellar cortical biopsies was performed on two patients with SCA1 (Landis et al., 1974). A variety of mild ultrastructural abnormalities was identified although the pathogenic significance of these changes was uncertain. Some of the more severe alterations could be attributed to artifact, although there was evidence of axonal degenerative changes within afferent fibers in the granular layer as well as somatic and dendritic degeneration in some of the Purkinje cells.

## IV. MODELS OF DISEASE

### A. Cell Culture Studies

Two important aspects of SCA1 pathogenesis have been elucidated by cell culture studies. The first is that overexpression of mutant ataxin-1 alters the distribution of nuclear matrix-associated domains (Skinner et al., 1997). Expression of mutant ataxin-1 in transfected COS cells causes a specific redistribution of the nuclear matrix-associated nuclear domain containing the promyelocytic leukemia protein (PML). PML domains, or nuclear body, have associated with several nuclear functions (Zhong et al., 2000). Perhaps one of the more intriguing functions is their putative role as sites of protein degradation by proteasomes (Fabunmi et al., 2000).

The second important finding came from studies demonstrating that protein misfolding and proteolysis may have a role in SCA1 pathogenesis. Cummings et al. (1998) showed that ataxin-1 inclusions induce HSP 70 expression and result in its redistribution to the inclusions as well as another chaperone, HDJ-2/HSDJ and components of the proteasome. Overexpression of HDJ-2/HSDJ reduced the size and number of ataxin-1 nuclear inclusions in transfected cells. These studies indicate that ataxin-1 with an expanded polyglutamine tract is recognized by the cell as having an altered folding pattern. Furthermore, enhancing the cell's capability to either refold or clear the misfolded mutant ataxin-1 might mitigate its pathogenic potential.

### B. Transgenic Mice

The first transgenic mouse model of a polyglutamine disease utilized a strong Purkinje cell-specific promoter from the *Pcp2/L7* gene to direct expression of the human *SCA1* cDNA encoding full length ataxin-1 to model the Purkinje cell neuropathology of SCA1 (Burright et al., 1995). These lines expressed high levels of either a wild-type *SCA1* allele encoding ataxin-1 with a polyglutamine tract of 30 residues, ataxin-1[30Q], or an expanded allele, ataxin1[82Q]. Only the ataxin-1[82Q]-expressing transgenic mice developed severe ataxia and progressive Purkinje cell pathology. In contrast, transgenic mice expressing ataxin-1[30Q] failed to develop any signs of neurological or pathological abnormalities and were indistinguishable from nontransgenic littermates (Burright et al., 1995; Clark et al., 1997). These studies demonstrated that Purkinje cell pathological changes are induced specific to the expression of ataxin-1 with an expanded polyglutamine tract.

In transgenic mice from an ataxin-1[30Q] line, ataxin-1 localized to several ~0.5 μm nuclear inclusions. In contrast, in ataxin-1[82Q] mice ataxin-1 localized to a single ~2 μm ubiquitinated nuclear inclusion, similar to what was seen in tissue from patients with SCA1 (Skinner et al., 1997). The appearance of these inclusions, which stained immunohistochemically for the 20S proteasome and the HDJ-2/HSDJ (Hsp40) chaperone protein, preceded the onset of ataxia by approximately 6 weeks (Cummings et al., 1998). One notable difference between the Purkinje cell pathology observed in *SCA1* transgenic mice and that of SCA1 patients is that the mice lack axonal dilatations (torpedoes). *SCA1* transgenic animals from the ataxin-1[82Q] line had mild cerebellar impairment at 5 weeks of age; there was no evidence of gait abnormalities or balance problems at that age (Clark et al., 1997). By 12 weeks, the motor skill impairment progressed to overt ataxia, which worsened over time.

The first histological change detected was the development of cytoplasmic vacuoles within Purkinje cell bodies at postnatal day 25 (Clark et al., 1997; Skinner et al., 2001). Electron microscopic analysis indicated that these vacuoles are distended membranous structures with an aqueous interior. These vacuoles appear to originate as invaginations of the cell membrane and are sites of membrane protein degradation (Skinner et al., 2001). At 6 weeks of age a loss of proximal dendritic branches and a decrease in the number of dendritic spines became apparent, indicating that ataxin-1[82Q] impairs the maintenance of dendritic arborization (Clark et al., 1997). By 12–15 weeks, the complexity of the dendritic arborization of Purkinje cells was markedly reduced and the molecular layer was atrophic. Thus, transgenic mouse Purkinje cells that over-express a mutant allele of the SCA1 gene developed two morphological features seen in brain material from SCA1 patients; atrophy of the Purkinje cell dendritic tree and the presence of nuclear aggregates of ataxin-1. These observations indicate that the SCA1 mice are undergoing a disease process that is very similar to that seen in the humans. However, there are features of the disease process in the SCA1 transgenic mice that differ from those in SCA1 affected individuals. Most prominent of these features are the several heterotopic Purkinje cells seen within the molecular layer of the cerebellar cortex in the SCA1 mice by 12–15 weeks of age. The Purkinje cell heterotopia was not detected in young animals and, thus, is not a developmental abnormality. Most likely it reflects an attempt to preserve synaptic function in the presence of severely reduced proximal dendritic arborization. Importantly, Purkinje cell loss was minimal at the time of progressive gait abnormality. Thus, the *SCA1* transgenic mice indicated that the neurological impairment is due to neuronal dysfunction and not directly the result of neuronal loss.

Further, insight into the molecular basis of SCA1 pathogenesis was obtained by the observation that an expanded allele of SCA1 has the ability to alter Purkinje cell gene expression early on in the *SCA1* transgenic mice (Lin et al., 2000). Using the SCA1 transgenic mice and a PCR-based cDNA subtractive hybridization strategy, several genes, all expressed by Purkinje cells, were found to be down-regulated at an early stage of disease, prior to any detectable pathological or neurological alteration. Interestingly, a number of the genes found to be down-regulated encoded proteins involved in neuronal calcium signaling. These included inositol tripphosphate receptor type 1, sarcoplasmic endoplasmic reticulum calcium ATPase type 2, transient receptor potential type 3, and inositol polyphosphate 5-phosphatase type 1. Intriguingly, all of the genes whose expression was found to be altered early on in the *SCA1* transgenic mice were down-regulated. Furthermore, a decrease in the level of the calcium-binding protein parvalbumin has been seen by immunohistochemistry in Purkinje cells from the SCA1 transgenic mice (Vig et al., 1996). While the mechanism of this down-regulation could involve either transcription and/or post-transcription events, given that ataxin-1 is an RNA-binding protein *in vitro* and may bind RNA *in vivo*, it is possible to speculate that the down-regulation in gene expression seen in SCA1 involves a mechanism at the RNA level.

The *SCA1* transgenic mice support the concept that mutant ataxin-1 appears to gain a toxic function that underlies the development of disease. That disease is not caused by the loss of ataxin-1 function was confirmed by the generation and characterization of mice lacking expression of the murine homolog *Sca 1* (Matilla et al., 1998). Mice lacking expression of murine ataxin-1 are viable, fertile, and show no signs of the ataxia or Purkinje cell pathology seen in the *SCA1* over-expressing transgenic mice. However, the *Sca1* null mice provided additional insight into the normal function of ataxin-1. These mice demonstrated both motor and spatial learning deficits suggesting that ataxin-1 has a role in learning mediated by the cerebellum and the hippocampus, respectively.

While much has been learned from the *SCA1* transgenic mice, a more genetically accurate model of SCA1 is desirable. This can be accomplished by introducing an expanded CAG repeat tract into the murine *Sca1* locus via a homologous recombination strategy. In this case, the expanded allele would be under the control of its own promoter allowing an analysis of pathogenicity of mutant ataxin-1 when expressed at endogenous levels in a normal spatial and temporal pattern. To date, an expanded repeat tract with 78 CAG repeats has been introduced into the *Sca1* locus (Lorenzetti *et al.*, 2000). Mice homozygous for the mutant allele were not ataxic by cage behavior but did show a decreased performance level on the accelerating rotarod. These mice also failed to show any signs of Purkinje cell pathology as seen in the *SCA1* transgenic mice (Clark *et al.*, 1997). Thus, even with 78 repeats mutant ataxin-1 expressed at endogenous levels is unable to induce a disease within the lifespan of a mouse reminiscent of human SCA1. The combined data from all of the *SCA1* mouse studies indicates that pathogenesis is a function of polyglutamine tract length, protein levels, and duration of neuronal exposure to the mutant ataxin-1.

The *Sca1* 78 repeat knock-in mice did show one interesting feature not seen in the *SCA1* cDNA transgenic mice. Mice heterozygous for the expanded allele showed an intergenerational repeat instability ranging from an expansion of 2 repeats to contractions of 6 repeats (Lorenzetti *et al.*, 2000). As in humans (Chung *et al.*, 1993), the majority of contractions were the result of maternal transmissions of the mutant allele. These results indicate that chromosomal context of the repeat has a role in determining the intergenerational instability of a mutant allele.

A powerful aspect of transgenic mouse models of human disease is the ability to experimentally manipulate either the structure of the disease-causing protein or the genetic background in which the protein is expressed. These strategies enable important insights into the molecular and cellular basis of pathogenesis to be gained.

To ascertain whether ataxin-1 must be in the nucleus to cause disease, Klement *et al.* (1998) generated and characterized transgenic mice that express expanded ataxin-1[82Q] with a mutated nuclear localization sequence, ataxin-1$^{K772T}$. Although these mice expressed high levels of ataxin-1 in Purkinje cells, similar to those observed in the original SCA1, ataxin-1[82Q] transgenic mice, they never developed Purkinje cell pathology or motor dysfunction. Ataxin-1 was diffusely distributed throughout the cytoplasm and formed no aggregates, even when the mice were a year old. Nuclear localization is clearly critical for pathogenesis and ataxin-1 aggregation. Furthermore, the ataxin-1$^{K772T}$ mice expressed levels of *SCA1* mRNA as high as the ataxin-1[82] mice did and, yet, they failed to develop disease. Thus, these studies provided direct evidence that it is the ataxin-1 protein with an expanded number of glutamine residues that is pathogenic and not *SCA1* RNA with an expanded number of CAG triplet repeats.

To directly assess the role of nuclear inclusions in causing disease, transgenic mice using ataxin-1[77Q] with amino acids deleted from the self-association region found to be essential for ataxin-1 dimerization were generated (Klement *et al.*, 1998). These mice developed ataxia and Purkinje cell pathology similar to the original ataxin-1[82Q] transgenic mice, but without detectable nuclear ataxin-1 aggregates at either the light or electron microscopic levels. Thus, although nuclear localization of ataxin-1 is necessary, nuclear aggregation of ataxin-1 appears not to be required for initiation of Purkinje cell pathogenesis in transgenic mice. It is important to consider that the deletion of 122 amino acids might have compromised ataxin-1 in various ways, e.g., its folding, turn-over rate, or ability to interact with other cellular factors. However, this seems unlikely since this truncated ataxin-1 retained its ability to produce all of the neurobehavioral and unique pathologic features observed in the ataxin-1[82Q] mice.

Given that the nuclear inclusions of ataxin-1 seen in SCA1 patient brain material and *SCA1* transgenic mouse Purkinje cells also contained constituents of the ubiquitin proteasomal pathway (UPP), i.e., ubiquitin and components of the proteasome, Cummings *et al.* (2001) crossed ataxin-1[82Q] transgenic mice with mice lacking expression of the *Ube3a* gene. *Ube3a* encodes the E6-associated protein, E6-AP, which is a member of the E3 class of ubiquitin ligases (Huibregtse *et al.*, 1995). Two important observations were made in the mice expressing the full length ataxin-1[82Q] protein in the absence of *Ube3a* expression. The presence of nuclear inclusions of ataxin-1[82Q] was reduced significantly both in terms of their frequency and their size. Yet, the Purkinje cell pathology was considerably worse compared to that seen in the ataxin-1[82Q] mice. These studies demonstrated *in vivo* the importance of the ubiquitin pathway for the formation of the nuclear inclusions and indicated that inclusion formation is an active cellular process. Furthermore, they showed that pathology is not dependent on the formation of nuclear inclusions. Thus, the *SCA1* transgenic mice studies provided two distinct examples in which nuclear inclusions could be shown to be unnecessary for the development of a polyglutamine-induced neuronal disease.

The demonstration that over-expressing chaperones reduced ataxin-1 inclusion formation in transfected cells and neurodegeneration in invertebrate models of polyglutamine disease inspired an examination of whether enhancing chaperone activity in a mammalian model of SCA1 would have a similar effect. To accomplish this, the SCA1 transgenic mice were crossed with mice over-expressing the molecular chaperone inducible HSP70 (iHSP70). High levels of iHSP70 in Purkinje cells mitigated the pathogenic effects of ataxin-1[82Q] (Cummings *et al.*, 2001). Purkinje cell morphology in SCA1 mice was improved in the presence of high levels of iHSP70. In addition, the rotarod performance of the SCA1 transgenic mice, an assessment of motor function, was also significantly better with iHSP70 over-expression. Interestingly, suppression of pathology and improvement in motor function occurred in the absence of a detectable decrease in the formation of nuclear inclusions. These studies provide further hope that upregulating the ability of a neuron to handle a mutant polyglutamine protein has therapeutic potential.

### C. SCA1 in Flies

A powerful model system for the elucidation of genetic pathways that can modify polyglutamine-induced neurodegeneration is the fly *Drosophila*. Fernandez-Funez and colleagues (2000) directed the expression of human full-length *SCA1* gene to neurons in the retina of *Drosophila*. Flies expressing both wild type and mutant alleles were generated. While over-expression of wild-type ataxin-1 caused some signs of neurodegeneration, flies expressing a mutant allele of ataxin-1 showed a very dramatic and severe form of neurodegeneration. Genetic screens were conducted with these mutant flies to identify genes that modify the SCA1-induced pathogenesis. The modifiers identified highlight the importance of protein folding and clearance, nuclear transport, and RNA-binding in the development of disease. Thus, *Drosophila* as well as cell culture models of SCA1 provide high-throughput systems for the identification of potential therapeutic targets that are very likely to be applicable to the disease as it presents in the mammalian systems. One can anticipate that in the near future drug screens using these high-throughput SCA1 model systems will become a routine strategy.

#### 1. Genotype/Phenotype Correlations

A large percentage of phenotypic variation is attributable to variation in SCA1 CAG repeat length. Ranum *et al.* (1994) estimated that 66% of age of onset variance is explained by variation in CAG repeat length. Delayed or premature onset of disease appeared to cluster in families providing evidence for the existence of genetic modifiers.

As discussed above, *in vitro*, vertebrate, and invertebrate models have identified other modifying proteins that provide good candidates for modifying age of onset or disease severity in humans.

## V. TREATMENT

The data from the various models of SCA1 strongly support the notion that pathogenesis is induced by a toxic function gained by the mutant ataxin-1 protein as a result of the expanded glutamine tract. Polyglutamine expansion seems to cause mutant ataxin-1 to adopt an altered folding state, leading to its ubiquitination, inclusion formation, and resistance to proteasomal degradation. Yet the protein context in which the disease-causing polyglutamine tract is located plays an important role in defining the disease and its process. Most dramatic of these is the effect altering the subcellular localization has on pathogenesis in the SCA1 transgenic mice. A single amino acid substitution that alters the ability of ataxin-1 to be transported into the nucleus prevented the development of disease. Thus the disruption of nuclear structures and/or essential nuclear functions in a neuron appears to be key to SCA1 pathogenesis. Another role of protein context may be in determining the selective neuronal vulnerability seen in SCA1. Mutant ataxin-1 likely interacts abnormally with other proteins (e.g., LANP), which could contribute to selective neuronal dysfunction and degeneration. An important question for the future is whether there is therapeutic potential in altering the misfolding or nuclear transport of mutant ataxin-1?

## References

Becher, M. W., Kotzuk, J. A., Sharp, A. H., Davies, S. W., Bates, G. P., Price, D. L., and Ross, C. A. (1998). Intranuclear neuronal inclusions in Huntington's disease and dentatorubral and pallidoluysian atrophy: correlation between the density of inclusions and IT15 CAG triplet repeat length. *Neurobiol. Dis.* **4**, 387–397.

Bebin, E. M., Bebin, J., Currier, R. D., Smith, E. E., and Perry, T. L. (1990). Morphometric studies in dominant olivopontocerebellar atrophy. Comparison of cell losses with amino acid decreases. *Arch. Neurol.* **47**, 188–192.

Burright, E. N., Clark, H. B., Servadio, A., Matilla, T., Feddersen, R. M., Yunis, W. S., Duvick, L. A., Zoghbi, H. Y., and Orr, H. T. (1995). SCA1 transgenic mice: a model for neurodegeneration caused by an expanded CAG trinucleotide repeat. *Cell* **82**, 937–948.

Chung, M.-Y., Ranum, L. P. W., Duvick, L., Servadio, A., Zoghbi, H. Y., and Orr, H. T. (1993). Analysis of the CAG repeat expansion in spinocerebellar ataxia type I: evidence for a possible mechanism predisposing to instability. *Nat. Genet.* **5**, 254–258.

Clark, H. B., Burright, E. N., Yunis, W. S., Larson, S., Wilcox, C., Hartman, B., Matilla, A., Zoghbi, H. Y., and Orr, H. T. (1997). Purkinje cell expression of a mutant allele of SCA1 in transgenic mice leads to disparate effects on motor behaviors, followed by a progressive cerebellar dysfunction and histological alterations. *J. Neurosci.* **17**, 7385–7395.

Cummings, C. J., Mancini, M. A., Antalffy, B., DeFranco, D. B., Orr, H. T., and Zoghbi, H. Y. (1998). Chaperone suppression of aggregation and altered subcellular proteasome localization imply protein misfolding in SCA1. *Nat. Genet.* **19**, 148–154.

Cummings, C. J., Reinstein, E., Sun, Y., Antalffy, B., Jiang, Y.-H., Ciechanover, A., Orr, H. T., Beaudet, A. L., and Zoghbi, H. Y. (1999). Mutation of the E6-AP ubiquitin ligase reduces nuclear inclusion frequency while accelerating polyglutamine-induced pathology in SCA1 transgenic mice. *Neuron* **24**, 879–892.

Cummings, C. J., Sun, Y., Opal, P., Antalffy, B., Mestril, R., Orr, H. T., Dillmann, W. H., and Zoghbi, H. Y. (2001). Over-expression of inducible HSP70 chaperone suppresses neuropathology and improves motor function in SCA1 mice. *Hum. Mol. Genet.* **10**, 1511–1518.

Davidson, J. D., Riley, B., Burright, E. N., Duvick, L. A., Zoghbi, H. Y., and Orr, H. T. (2000). Identification and characterization of an ataxin-1 interacting protein: A1Up a ubiquitin-like nuclear protein. *Hum. Mol. Genet.* **9**, 2305–2312.

De Michele, G., Mainenti, P. P., Soricelli, A., Di Salle, F., Salvatore, E., Longobardi, M. R., Postiglione, A., Salvatore, M., and Filla, A. (1998).

Cerebral blood flow in spinocerebellar degenerations: a single photon emission tomography study in 28 patients. *J. Neurol.* **245**, 603–608.

DiFiglia, M., Sapp, E., Chase, K. O., Davies, S. W., Bates, G. P., Vonsattel, J.P., and Aronin, N. (1997). Aggregation of Huntingtin in neuronal intranuclear inclusions and dystrophic neurites in brain. *Science* **277**, 1990–1993.

Duyckaerts, C., Durr, A., Cancel, G., and Brice, A. (1999). Nuclear inclusions in spinocerebellar ataxia type 1. *Acta Neuropathol.* **97**, 201–207.

Fabunmi, R. P., Wigley, W. C., Thomas, P. J., and DeMartino, G. N. (2000). Interferon γ regulates accumulation of the proteasome activator PA28 and immunoproteasomes at nuclear PML bodies. *J. Cell Sci.* **114**, 29–36.

Ferrer, I., Genis, D., Davalos, A., Bernado, L., Sant, F., and Serrano, T. (1994). The Purkinje cell in olivopontocerebellar atrophy. A Golgi and immunocytochemical study. *Neuropathol. Appl. Neurobiol.* **20**, 38–46.

Genis, D., Matilla, T., Volpini, V., Rosell, J., Davalos, A., Ferrer, I., Molins, A., and Estivill, X. (1995). Clinical, neuropathologic, and genetic studies of a large spinocerebellar ataxia type 1 (SCA1). kindred: $(CAG)_n$ expansion and early premonitory signs and symptoms. *Neurology* **45**, 24–30.

Goldfarb, L.G., Vasconcelos, O., Platonov, F.A., Lunkes, A., Kipinis, V., Kononova, S., Chabrashvili, T., Vladimirtsev, V. A., Alexeev, V. P., and Gajdusek, D. C. (1996). Unstable triplet repeat and phenotypic variability of spinocerebellar ataxia type 1. *Ann. Neurol.* **39**, 500–506.

Holmberg, M., Duyckaerts, C., Dürr, A., Cance, G., Gourfinkel-An, I., Damier, P., Faucheux, B., Trottier, Y., Hirsch, E. C., Agid, Y., and Brice, A. (1998). Spinocerebellar ataxia type 7 (SCA7): a neurodegenerative disorder with neuronal intranuclear inclusions. *Hum. Mol. Genet.* **7**, 913–918.

Huibregtse, J. M., Scheffner, M., Beaudenon S., and Howley, P. M. (1995). A family of proteins structurally and functionally related to the E6-AP ubiquitin-protein ligase. *Proc. Natl. Acad. Sci. U.S.A.* **92**, 2563–2567.

Jans, D. A., Xiao, C.-Y., and Lam, M. H. C. (2000). Nuclear targeting signal recognition: a key control point in nuclear transport? *BioEssays* **22**, 532–544.

Jodice, C., Malaspina, P., Persichetti, F., Noveletto, A., Spadaro, M., Giunti, P., Morocutti, C., Terrenato, L., Harding, A. E., and Frontali, M. (1994). Effect of trinucleotide repeat length and parental sex on phenotypic variations in spinocerebellar ataxia 1. *Am. J. Hum. Genet.* **54**, 959–965.

Kish, S. J., Guttman, M., Robitaille, Y., el-Awar, M., Chang, L. J., and Levey, A. I. (1997). Striatal dopamine nerve terminal markers but not nigral cellularity are reduced in spinocerebellar ataxia type 1. *Neurology* **48**, 1109–1111.

Klement, I. A., Skinner, P. J., Kaytor, M. D., Yi, H., Hersch, S. M., Clark, H. B., Zoghbi, H. Y., and Orr, H. T. (1998). Ataxin-1 nuclear localization and aggregation: role in polyglutamine-induced disease in SCA1 transgenic mice. *Cell* **95**, 41–53.

Koeppen, A. H. (1991). The Purkinje cell and its afferents in human hereditary ataxia. *J. Neuropathol. Exp. Neurol.* **50**, 505–514.

Landis, D. M. D., Rosenberg, R. N., Landis, S. C., Schut, L. J., and Nyhan, W. L. (1974). Olivopontocerebellar degeneration. Clinical and ultrastructural abnormalities. *Arch. Neurol.* **31**, 295–307.

Li, M., Miwa, S., Kobayashi, Y., Merry, D. E., Yamamoto, M., Tanaka, F., Doyu, M., Hashizume, Y., Fischbeck, K. H., and Sobue, G. (1998). Nuclear inclusions of the androgen receptor in spinal and bulbar muscular atrophy. *Ann. Neurol.* **44**, 249–254.

Lin, X., Antalffy, B., Kang, D., Orr, H. T., Zoghbi, H. Y., (2000). Polyglutamine expansion downregulates specific neuronal genes before pathological changes in SCA1. *Nat Neurosci.* **3**, 157–163.

Lorenzetti, D., Watase, K., Xu, B., Matzuk, M. M., Orr, H. T., and Zoghbi, H. Y. (2000). Repeat instability and motor incoordination in mice with a targeted expanded CAG repeat in the Sca1 locus. *Hum. Mol. Genet.* **9**, 779–785.

Mascalchi, M., Tosetti, M., Plasmati, R., Bianchi, M. C., Tessa, C., Salvi, F., Frontali, M., Valzania, F., Bartolozzi, C., and Tassinari, C. A. (1998). Proton magnetic resonance spectroscopy in an Italian family with spinocerebellar ataxia type 1. *Ann. Neurol.* **43**, 244–252.

Matilla, T., Volpini, V., Genis, D., Rosell, J., Corral, J., Davalos, A., Molins, A., and Estivill, X. (1993). Presymptomatic analysis of spinocerebellar ataxia type 1 (SCA1). via the expansion of the SCA1 CAG-repeat in a large pedigree displaying anticipation and parental male bias. *Hum. Mol. Genet.* **2**, 2123–2128.

Matilla, A., Koshy, B., Cummings, C. J, Isobe, T., Orr, H. T., and Zoghby, H. Y. (1997). The cerebellar leucine rich acidic nuclear protein (LANP) interacts with ataxin-1. *Nature* **389**, 974–978.

Matilla, A., Robertson, E. D., Bonfi, S., Morales, J., Armstrong, D. L., Burright, E. N. Orr, H. T., Sweat, J. D., Zoghby, H. Y., and Matzuk, M. (1998). Mice lacking ataxin-1 display learning deficits and decreased hippocampal paired-pulse facilitation. *J. Neurosci.* **18**, 5508–5516.

Nakamura, K., Jeong, S.-Y., Uchihara, T., Anno, M., Nagashima, K., Nagashima, T., Ikeda, S.-I., Tsuji, S., and Kanazawa, I. (2001). SCA17, a novel sutosomal dominant cerebellar ataxia caused by an expanded polyglutamine in TATA-binding protein. *Hum. Mol. Genet.* **10**, 1441–1448.

Nino, H, E., Noreen, H. J., Dubey, D. P., Resch, J. A., Namboodiri, K., Elston, R. C., Yunis, E. J. (1980). A family with hereditary ataxia: HLA typing. *Neurology* **30**, 12–20.

Orr, H. T., Chung, M.-Y., Banfi, S. Kwiatkowski, T. J., Jr., Servadio, A, Beaudet, A. L., McCall, A. E., Duvick, L. A., Ranum, L. P. W., and Zoghbi, H. Y. (1993). Expansion of an unstable trinucleotide CAG repeat in spinocerebellar ataxia type 1. *Nat. Genet.* **4**, 221–226.

Paulson, H. L., Perez, M. K., Trottier, Y., Trojanowsk, J. Q., Subramony, S. H., Das, S. S., Vig, P., Mandel, J.-L., Fischbeck, K. H., and Pittman, R. N. (1997). Intranuclear inclusions of expanded polyglutamine protein in spinocerebellar ataxia Type 3. *Neuron* **19**, 333–334.

Quan, F., Janas, J., and Popovich, B. W. (1995). A novel CAG repeat configuration in the *SCA1* gene: implications for the molecular diagnosis of spinocerebellar ataxia type 1. *Hum. Mol. Gent.* **4**, 2411–2413.

Ranum, L. P. W., Chung, M.-Y., Banfi, S., Bryer, A., Schut, L. J., Ramesar, R., Duvick, L. A., McCall, A., Subramony, S. H., Goldfarb, L., Gomez, C., Sandkuijl, L. A., Orr, H. T., and Zoghbi, H. Y. (1994). Molecular and clinical correlations in spinocerebellar ataxia type 1: evidence for familial effects on the age at onset. *Am. J. Hum. Genet.* **55**, 244–252.

Robitaille, Y., Schut, L., and Kish, S. J. (1995). Structural and immunocytochemical features of olivopontocerebellar atrophy caused by the spinocerebellar ataxia type 1 (SCA-1) mutation define a unique phenotype. *Acta Neuropathol.* **90**, 572–581.

Schut, J. W., and Haymaker, W. (1951). Hereditary ataxia: A pathologic study of five cases of common ancestry. *J. Neuropathol. Clin. Neurol.* **1**, 183–213.

Servadio, A., Koshy, B., Armstrong, D., Antalffy, B., Orr, H. T., and Zoghby, H. Y. (1995). Expression analysis of the ataxin-1 protein in tissues from normal and spinocerebellar ataxia type 1 individuals. *Nat. Genet.* **10**, 94–98.

Skinner, P. J., Koshy, B. T., Cummings, C. J., Klement, I. A., Helin, K., Servadio, A, Zoghbi, H. Y., and Orr, H. T. (1997). Ataxin-1 with an expanded glutamine tract alters nuclear matrix-associated structures. *Nature* **389**, 971–74.

Skinner, P. J., Vierra-Green, C. A., Clark, H. B., Zoghbi, H. Y., and Orr, H. T. (2001). Altered trafficking of membrane proteins in Purkinje cells of SCA1 transgenic mice. *Am. J. Pathol.* **159**, 905–913.

Spadaro, M., Giunti, P., Lulli, P., Frontali, M., Jodice, C., Cappellacci, S., Morellini, M., Persichetti, F., Trabace, S., Anastasi, R., and Morocutti, C. (1992). HLA-linked spinocerebellar ataxia: a clinical and genetic study of large Italian kindreds. *Acta Neurol. Scand.* **85**, 257–265.

Wullner, U., Klockgether, T., Petersen, D., Naegele, T., and Dichgans, J. (1993). Magnetic resonance imaging in hereditary and idiopathic ataxia. *Neurology* **43**, 318–326.

Vig, P. J. S., Fratkin, J. D., Desaiah, D., Currier, R. D., and Subramony, S. H. (1996). Decreased parvalbumin immunoreactivity in surviving Purkinje cells of patients with spinocerebellar ataxia-1. *Neurology* **47**, 249–53.

Yagashita, S., and Inoue, M. (1997). Clinicopathology of spinocerebellar degeneration: Its correlation to the unstable CAG repeat of the affected gene. *Pathol. Int.* **47**, 1–15.

Yakura, H., Wakisaka, A., Fujimoto, S., and Itakura, K. (1974). Hereditary ataxia and HLA genotypes. *N. Engl. J. Med.* **291**, 154–155.

Yue, S., Serra, H., Zoghbi, H. Y., and Orr, H. T. (2001). The SCA1 protein, ataxin-1, has RNA-binding activity that is inversely affected by the length of its polyglutamine tract. *Hum. Mol. Genet.* **10**, 25–30.

Zoghbi, H. Y., and Orr, H. T. (2000). Glutamine repeats and neurodegeneratio. *Ann. Rev. Neurosci.* **23**, 217–247.

Zoghbi, H. Y., and Orr, H. T. (2001). Spinocerebellar ataxias. In: "The Metabolic and Molecular Bases of Inherited Disease" (C. S. Scriver, *et al.*, eds.), pp. 5741–575. 8th edition. McGraw-Hill, New York.

Zhong, S., Salomoni, P., and Pandolfi, P. P. (2000). The transcriptional role of PML and the nuclear body. *Nat. Cell Bio.* **2**, E85–E90.

# CHAPTER 4

# Spinocerebellar Ataxia 2 (SCA2)

STEFAN-M. PULST

*Division of Neurology, Cedars-Sinai Medical Center*
*Departments of Medicine and Neurobiology, UCLA School of Medicine*
*Los Angeles, California 90048*

I. Summary
II. Phenotype
   A. Ataxia
   B. Eye Movements and Retinal Changes
   C. Movement Disorders
   D. Neuropathy
   E. Dementia
   F. Progression Rate
   G. Epidemiology
III. Normal and Abnormal Gene Function
   A. Repeat Range
   B. Pathologic Alleles
   C. Permutation Alleles and Alleles with Reduced Penetrance
   D. Anticipation and Meiotic Instability of the SCA2 Repeat
   E. Normal Function
   F. Abnormal Function
IV. Diagnosis
   A. DNA
   B. Testing of Sporadic Patients
   C. Biochemistry
V. Neurophysiology
VI. Neuroimaging
VII. Neuropathology
VIII. Animal Models
   A. Mouse
   B. Drosophila
IX. Genotype/Phenotype/Modifying Alleles
X. Treatment
   Acknowledgments
   References

## I. SUMMARY

SCA2 is a neurodegenerative disease caused by expansion of an unstable CAG repeat in the SCA2 or ataxin-2 gene on human chromosome 12. SCA2 typically shows a phenotype of progressive ataxia accompanied by slow saccadic eye movements, but a wide phenotypic spectrum is observed including L-dopa-responsive Parkinsonism.

Ataxin-2 is a member of a novel protein family with putative RNA-binding domains that are evolutionarily conserved. The wide separation between normal and pathological repeat ranges seen in other CAG repeat disorders is not present in SCA2 and alleles with reduced penetrance exist. Animal models elucidating normal and mutant ataxin-2 function have been generated.

## II. PHENOTYPE

SCA2 has a worldwide distribution, but is very prevalent in the Cuban province of Holguin (Orozco et al., 1989, 1990; Santos et al., 1999). Most patients are of Caucasian Spanish ancestry in this founder population. In addition to ataxic gait and other cerebellar findings, many patients have slow saccadic eye movements. Tendon reflexes may be brisk during the first years of life, but absent several years later. In 1971, Wadia and Swami had already pointed to the importance of slowed saccadic eye movements in a subset of patients with inherited ataxias in India. Subsequent genotyping indicated that six of these families with slow eye movements carried mutations in the SCA2 gene (Wadia et al., 1998). In East Indian pedigrees slow eye movements do not appear to be a distinguishing feature (Chakravarty and Mukherjee, 2002).

Analysis of a large number of SCA2 pedigrees has indicated a wide range of phenotypic manifestations that make SCA2 indistinguishable from other SCAs in the individual patient. Some findings, however, are particularly common in SCA2. These are slow saccades, peripheral neuropathy, and dementia. Schols *et al.* (1997a) found that clinical features were highly variable within and between families. Although no specific single feature was sufficient to distinguish SCA2 from other SCAs, slowed saccades, postural and action tremors, myoclonus, and hyporeflexia were more common than in SCA1 and SCA3. Deficits of frontal executive function are also frequent in SCA2 even in early stages of the disease (Gambardella *et al.*, 1998; Burk *et al.*, 1999b; Storey *et al.*, 1999).

A neonatal phenotype reminiscent of neonatal SCA7 was reported in an infant born to a father with 43 repeats and an age of onset at 22 years (Babovic-Vuksanovic *et al.*, 1998). The infant had more than 200 repeats and presented with neonatal hypotonia, developmental delay, and dysphagia. Retinitis pigmentosa was noticed at 10 months of age.

### A. Ataxia

Ataxia is universally present and is usually a presenting sign, although some Cuban patients may present with muscle cramps. A subclinical neuropathy may be identified before any other clinical signs (Velazquez and Medina, 1998). Ataxia involves gait and stance, but is also prominent in appendicular functions. In the Cuban population a prominent truncal oscillation was seen, when patients were standing with their eyes open. Quantitative assessments have recently been used in the evaluation of these patients and may provide an important addition to ataxia rating scales (Velazquez Perez *et al*, 2001a).

### B. Eye Movements and Retinal Changes

Abnormal eye movements have been identified in all clinical studies of SCA2, some employing formal eye movement recordings (reviewed in Pulst and Perlman, 2000; Burk *et al.*, 1996, 1999; Rivaud-Pechoux *et al.*, 1998; Buttner *et al.*, 1998). Burk *et al.* (1996) examined several SCA2 patients defined by linkage analysis and compared them with SCA1 and SCA3 patients. SCA2 patients had significantly slower saccadic speed (138°/sec) than patients with SCA1 (244°/sec ) or SCA3 (347°/sec). All 8 SCA2 patients had saccadic velocities two standard deviations below the mean of a control group. In a follow-up study SCA2 patients were characterized by reduced saccadic velocity and the absence of square-wave jerks and gaze-evoked nystagmus (Burk *et al.*, 1999a).

Buttner *et al.* (1998) compared patients with SCA1, 2, 3, and 6 identified by direct mutation analysis. Patients with SCA2 had the slowest peak saccadic velocity ranging from 80 to 295°/sec (normal >400°/sec). Saccades were also slowed in SCA1 patients, but patients with SCA3 or SCA6 had normal saccades. Rivaud-Pecheaux (1998) suggested in addition that an increased saccade amplitude may help distinguish SCA1 from SCA2.

Retinal degeneration is common in SCA7 and was thought to point exclusively to a mutation in the *ataxin-7* gene (see Chapter 9). In SCA2 it has been described in the setting of infantile SCA2 associated with >200 repeats (Babovic-Vuksanovic *et al.*, 1998). Retinal pigmentary degeneration was also noticed in a 48-year-old woman who had developed night blindness at age 28, four years before the onset of ataxia (Rufa *et al.*, 2002). She had 41 CAG repeat in the SCA2 gene. Her two children with similar repeats sizes showed ataxia without retinal changes.

### C. Movement Disorders

Movement disorders are common in SCA2. Geschwind *et al.* (1997b) found a relatively high incidence of dystonia or chorea (38%). In a series of 111 patients from 32 families of diverse origins Cancel *et al.* (1997) described dystonia in 9%. Patients with dystonia had longer repeats than those without. Sasaki *et al.* (1998) point to the presence of choreiform movements in their patients in Japan.

Myoclonus is prominent in Cuban SCA2 patients, especially in those with early onset (Pulst, personal observation). Cancel *et al.* (1997) found that patients with myoclonus had longer repeats than those without. Schols *et al.* (2000) found a postural tremor as the most common extrapyramidal sign in SCA2.

Parkinsonism and even L-dopa responsive Parkinsonism without significant ataxia have recently been recognized as a feature of SCA2 mutations. Sasaki *et al.* (1998) described parkinsonism in a man homozygous for the SCA2 mutation. Recently, Gwinn-Hardy *et al.* (2000) described a Taiwanese family with several members that displayed prominent Parkinsonian signs that were responsive to L-dopa. Shan *et al.* (2001) confirmed this observation in another two Taiwanese patients including the reduction of 18F-dopa distribution in both the putamen and caudate nuclei.

### D. Neuropathy

Neuropathy is a common finding in all studies of SCA2 worldwide. In most studies hyperreflexia due to upper motor neuron dysfunction is followed by hyporeflexia. In Cuban SCA2 patients, a decrease in the amplitude of sensory nerve potentials was seen early and frequently, even in the absence of other clinical signs (Velazquez and Medina, 1998). Prolonged latencies of somatosensory and brainstem auditory potentials were also observed, but visual-evoked potentials remained normal. Subclinical involvement by a sensory neuropathy was also confirmed in two Japanese SCA2 pedigrees (Ueyama *et al.*, 1998).

Eighty percent of French SCA2 patients had a neuropathy (Kubis *et al.*, 1999). Cancel *et al.* (1997) observed fasciculations in 25% of SCA2 patients. Both CAG length and duration influenced the frequency of decreased reflexes and vibration sense in the lower extremities, amyotrophy, and fasciculations (Cancel *et al.*, 1997).

### E. Dementia

Formal mental status testing has indicated significant frontal executive dysfunction in SCA2. Even in nondemented subjects, verbal and executive dysfunction could be detected (Burk *et al.*, 1999b). Prevalence ranges from 25–37% in most studies (Dürr *et al.*, 1995; Geschwind *et al.*, 1997b; Burk *et al.*, 1999b). In a Northern Italian SCA2 pedigree, five of six individuals displayed frontal executive dysfunction despite a mini-mental status score in the nondemented range (Storey *et al.*, 1999). Gambardella *et al.* (1998) observed an early and selective impairment of conceptual reasoning ability as shown by the Wisconsin Card Sorting Test.

### F. Progression Rate

The rate of progression in SCA2 has not yet been studied in a prospective fashion or as a function of repeat length. In a retrospective study the rate of progression was similar for SCA1, 2, and 3 in one study (Klockgether *et al.*, 1998a). Presence of longer alleles correlated with faster progression in SCA2. Female gender was associated with shortened survival.

### G. Epidemiology

Direct prevalence rates for SCA2 are not known (see Chapter 2). In the group of dominant ataxias, SCA2 is one of the most frequent ataxias (see Table 2.3). In most studies SCA2 mutations represent about 15% of all SCA mutations and are slightly less common than SCA3 (Geschwind *et al.*, 1997b; Riess *et al.*, 1997; Cancel *et al.*, 1997; Lorenzetti *et al.*, 1997; Moseley *et al.*, 1998; Pujana *et al.*, 1999). SCA2 is frequent in southern Italy and Sicily (Giunti *et al.*, 1998; Filla *et al.*, 1999; Pareyson *et al.*, 1999), but rare in Taiwan (Hsieh *et al.*, 1999). In Cuba, the prevalence of SCA2 is 43 cases per 100,000 inhabitants in the province of Holguin; the highest rate was 503 cases per 100,000 inhabitants in part of the municipality of Baguanos (Velazquez Perez *et al.*, 2001b).

Most studies indicate that *SCA2* mutations have arisen on different founder chromosomes (Geschwind *et al.*, 1997b; Didierjean *et al.*, 1999). In Gunma prefecture, at least two founder haplotypes were identified with different CCG or CCGCCG interruption of the CAG repeat (Mizushima *et al.*, 1999). In contrast, some German, Serbian, and French families shared the same haplotype suggesting a common founder or a recurrent mutation on an at-risk chromosome (Didierjean *et al.*, 1999). An identical core haplotype established by alleles in the loci D12S1672 and D12S1333 in pedigrees of diverse ethnic origin from India, Japan, and England probably represent a haplotype common in these populations rather than indicating a common founder (Pang *et al.*, 1999).

## III. NORMAL AND ABNORMAL GENE FUNCTION

The *SCA2* gene was identified through a combination of positional cloning and candidate gene approaches (Pulst *et al.*, 1996; Sanpei *et al.*, 1996; Imbert *et al.*, 1996). The *SCA2* gene contains 25 exons encompassing 130 kb of genomic DNA (Nechiporuk *et al.*, 1997; Sahba *et al.*, 1998). The CAG repeat is located in exon 1 and codes for a polyglutamine tract in the N-terminal part of ataxin-2.

### A. Repeat Range

The *SCA2* CAG trinucleotide repeat is unusual in several aspects. The repeat is not highly polymorphic in normal individuals. Two alleles of 22 and 23 repeats account for >95% of alleles in most studies, although rare normal alleles ranging from 15–32 repeats have also been identified (Pulst *et al.*, 1996; Sanpei *et al.*, 1996; Riess *et al.*, 1997; Cancel *et al.*, 1997; Santos *et al.*, 1999). Normal alleles typically show one or two CAA interruptions. In contrast to the *SCA1* gene, which contains CAT interruptions coding for histidine, the CAA interruptions do not interrupt the glutamine tract at the protein level. Therefore, it was not surprising that disease alleles with CAA interruptions were reported (Costanzi-Porrini *et al.*, 2000). It is likely that interruptions in the repeat stabilize the repeat (Choudhry *et al.*, 2001).

Although initial studies appeared to suggest a gap between normal and abnormal repeat sizes, it is now apparent that similar to Huntington's disease alleles with reduced penetrance may exist.

### B. Pathologic Alleles

Expansions on disease chromosomes are relatively small compared with SCA1 and SCA3. The most common disease alleles contain 37–39 repeats and are thus smaller than the longest normal alleles seen in the SCA3/MJD gene (Fig. 4.1, see also Fig. 2.2). Initial analyses indicated that all symptomatic SCA2 patients had disease alleles of at least 36 repeats. In subsequent studies, several symptomatic individuals have been described carrying 34 or 35 repeat alleles (Malandrini *et al.*, 1998). Recently, Fernandez *et al.* (2000) described a family segregating 33 CAG repeat alleles in the SCA2 gene. These alleles were associated

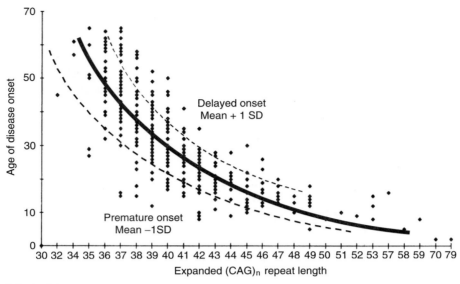

**FIGURE 4.1** Scattergram of CAG repeats in the *ataxin-2* gene and age of onset. Solid line indicates mean; interrupted lines indicate mean +/− one standard deviation for each repeat length after logarithmic transformation of age of onset years (Figueroa *et al.* 2002). The patients above the dotted line define the "delayed onset" group, those below the dotted line the "premature onset" group (for further details, see Section VB).

with very late onset, as late as 86 years in one patient. In the Holguin SCA2 population, a woman with a typical SCA2 phenotype and an age of onset of 48 years carried an SCA2 allele with 32 repeats (Santos *et al.*, 1999).

In contrast to normal alleles that are interrupted by one or several CAA repeats virtually all expanded alleles have a perfect repeat structure. Two patients carrying 34 repeat alleles containing one CAA interruption have been described (Costanzi-Porrini *et al.*, 2000).

### C. Premutation Alleles and Alleles with Reduced Penetrance

Premutation alleles likely come from a small reservoir of normal alleles that are uninterrupted. An allele of 34 repeats was seen in the asymptomatic mother of a woman with SCA2 (Riess *et al.*, 1997). An allele with 32 repeats has also been seen in an asymptomatic 19 year old whose symptomatic father carried an allele of 40 repeats (Cancel *et al.*, 1997). The contracted allele had no CAA interruptions. A Japanese unaffected male with 32 repeats had an affected son with 39 repeats (Futamura *et al.*, 1998).

### D. Anticipation and Meiotic Instability of the SCA2 Repeat

As in other diseases caused by unstable CAG DNA repeats, there is a clear inverse correlation between age of onset and repeat length. However, this correlation is not linear and is best approximated by a negative exponential fit (Pulst *et al.*, 1996). The widest range of age of onset is observed in patients with fewer than 40 repeats (Fig. 4.1, see also Fig. 2.2). For example, in one study the presence of 37 repeats was associated with ages of onset ranging from 20–60 years of age (Pulst *et al.*, 1996). Even in the Cuban population, which is genetically more homogeneous, a repeat of 37 repeats was associated with an age of onset from 15–65 years of age. For larger repeat sizes, the variability is less and repeat sizes of >45 are almost always associated with disease onset under 20 years of age (Pulst *et al.*, 1996; Sanpei *et al.*, 1996; Imbert *et al.*, 1996; Riess *et al.*, 1997; Cancel *et al.*, 1997; Geschwind *et al.*, 1997). Homozygosity for an expanded SCA2 allele does not appear to influence age of onset (Sanpei *et al.*, 1996), although the number of observations is small given the ranges observed with a specific repeat size. Animal studies, however, appear to suggest a clear dosage effect for mutant SCA2 alleles (Huynh *et al.*, 2000).

Initial observations in the FS pedigree from Southern Italy (Pulst *et al.*, 1993) did not point to consistent differences in the degree of anticipation depending on paternal or maternal inheritance. The lack of a paternal bias in expansion was also confirmed by Cancel *et al.* (1997). However, other studies have indicated that large expansions are almost exclusively observed, when the repeat is passed through the paternal germline (Riess *et al.*, 1997; Geschwind *et al.*, 1997b). Stevanin *et al.* (2000) summarized repeat instability from several studies and estimated that the SCA2 repeat increases by 3.5 (range −8 to +17) in paternal transmissions and by 1.7 (range −4 to +8) in maternal transmissions.

## E. Normal Function

The *SCA2* gene is transcribed from telomeric to centromeric and has 25 exons. The CAG repeat is contained in exon 1, which is also the largest exon with 15 kb. The identification of exon-intron boundaries allowed the identification of alternatively spliced transcripts. Transcripts missing either exons 9, 10, or 22 have been described (Affaitati *et al.*, 2001; Nechiporuk *et al.*, 1998; Sahba *et al.*, 1998) A transcript deleted in frame for exon 10 appears to be enriched in the cerebellum, although the functional consequences of this splicing event are unknown at this point (Sahba *et al.*, 1998).

The ataxin-2 cDNA sequence predicts a protein with 1312 amino acids with the CAG repeat coding for polyglutamine (Pulst *et al.*, 1996; Sanpei *et al.*, 1996). The 5′-sequence of the SCA2 cDNA is extremely GC rich and two potential ATG initiation codons can be identified. The most 5′ ATG is located 78 bp downstream of an in-frame stop codon. Usage of this translation initiation site predicts a protein of 140.1 kDa. The second ATG which has a better Kozak consensus sequence is located just 5′ to the CAG repeat and would result in a protein with relative MW of 125 kDa. Proteins observed by Western blot analysis and conservation of the 5′ ATG in the mouse (Nechiporuk *et al.*, 1998) suggest that the 5′ ATG is the predominant site of translation initiation.

Homology searches using the cDNA and amino acid sequences have identified homologies with several proteins (Fig. 4.2). Significant sequence homology was detected with a protein designated ataxin-2-related protein (A2RP) and the mouse SCA2 protein (Pulst *et al.*, 1996; Figueroa *et al.*, submitted). A2RP is also known under the name A2D (for ataxin-2 domain protein; Meunier *et al.*, 2002). A motif predicted to bind to poly(A)-binding protein is present in ataxin-2 and A2RP (Kozlov *et al.*, 2001).

Despite the significant homologies, the polyglutamine tract in human ataxin-2 is not present in A2RP or in mouse ataxin-2 suggesting that it may not be important for ataxin-2 function. However, all acidic amino acids comprising a highly acidic domain adjacent to the human polyglutamine domain are conserved in A2RP and mouse ataxin-2. Homologous proteins have also been identified in the worm and fruitfly (see Fig. 4.2). Knock-down of ataxin-2 expression in *Cenorhabditis elegans* using RNA interference results in embryonic lethality (Kiehl *et al.*, 2000).

Ataxin-2 interacts with a protein (A2BP) that is the human homolog of the *C. elegans* protein fox-1 (Shibata *et al.*, 2000). A2BP1 also shows homologies with poly(A)-binding protein. This is interesting in that both ataxin-2 and A2BP1 interact as do their respective homologous proteins

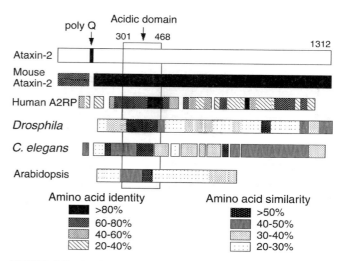

**FIGURE 4.2** Ataxin-2 homologs and orthologs. Amino acid sequence identity is indicated by degree of shading. Highly conserved domains are boxed. The corrected sequence of Drosophila ataxin-2 was provided by Drs. T. Satterfield and L. Pallanck. For invertebrates and plants, amino acid similarity is shown.

poly(A)-binding protein and poly(A)-binding protein interacting protein 1. A2BP1 and ataxin-2 co-localize in cultured cells and in human dentate and Purkinje cells (Shibata *et al.*, 2000).

### 1. Expression

The *SCA2* gene is widely expressed in human and mouse tissues. By Northern blot analysis, a 4.5-kb transcript is recognized in RNAs isolated from brain, heart, placenta, liver, skeletal muscle, and pancreas (Pulst *et al.*, 1996; Sanpei *et al.*, 1996; Imbert *et al.*, 1996). Little or no expression is seen in lung or kidney. The transcript is expressed throughout the brain. In RNAs isolated from SCA2 lymphoblastoid cell lines, expression of both, the normal and expanded, alleles is seen using reverse-transcribed polymerase chain reaction (PCR) (Pulst *et al.*, 1996). Antibodies to ataxin-2 recognize a 145-kDa protein in mouse and human brains (Nechiporuk *et al.*, 1998; Huynh *et al.*, 1999). The protein expression is not restricted to those neuronal populations that show the most severe degeneration. Ataxin-2 labeling is seen in a number of neuronal groups including brainstem neurons, cranial nerve nuclei, and hippocampal neurons, but also faintly in glial cells (Huynh *et al.*, 1999).

The *SCA2* transcript in the mouse is of identical size (Nechiporuk *et al.*, 1998). Expression during mouse embryonic development with strong expression at days E11 and E12 suggests that ataxin-2 may have a role in normal embryogenesis. Interestingly, little protein expression is seen, when mouse E11 embryos are stained with ataxin-2 antibodies and at E12 the predominant organ staining is detected in the liver. CNS staining is detected at later stages. The function of ataxin-2 in development may not be essential, since mouse lines homozygously deleted for ataxin-2 have apparently normal development (Kiehl and Pulst, submitted).

By Western blot analysis, a protein of ~145 kDa and several smaller proteins are recognized by anti-ataxin-2 antibodies (Huynh et al., 1999). A truncated protein of 41 kDa is seen that contains the polyQ domain (as shown by the 1C2 antibody that recognizes long polyQ tracts).

## F. Abnormal Function

In normal brains, ataxin-2 has a cytoplasmic localization. In SCA2 brain, antibodies to ataxin-2 or to expanded polyglutamine repeats show intense cytoplasmic staining that appears significantly stronger than in simultaneously stained control brains. Instead of a finely granular staining pattern, the entire cytoplasm is strongly immunoreactive in SCA2 brains. Huynh et al. (1999) did not detect intranuclear inclusions in Purkinje and dentate neuron cells. Although Huynh et al. (1999) did not identify intranuclear inclusion bodies in three cerebella from SCA2 patients, Koyano et al. (1999) reported ubiquitinated intranuclear inclusions in about 1–2% of neurons in affected areas. Purkinje neurons, however, did not show inclusions in Japanese patients suggesting that the formation of intranuclear inclusions is a late event and not necessary for pathogenesis.

Pang et al. (2002) detected intranuclear inclusions in a larger number of neurons in the brainstem and the cortex in two brains from SCA2 patients including neurons that are normally not involved in the neurodegenerative process. Unfortunately, the authors failed to use an antibody to ataxin-2 for the detection of the intranuclear inclusions relying instead on an antibody to polyglutamine repeats. These authors also reported on the presence of intranuclear inclusions in glial cells confirming the observations by Huynh et al. (1999) that ataxin-2 staining in glial cells was increased in SCA2 brains.

Thus, it appears that similar to other polyQ diseases, expansion of the polyQ tract results in aggregation of the protein. Although most of the aggregates appear to be cytoplasmic, a small number of neurons show intranuclear aggregates. The extent and distribution of neurons with intranuclear inclusions appear to vary between different groups. Purkinje cells, however, appear not to develop intranuclear inclusions in SCA2 patients. For further analysis of pathogenetic mechanisms the reader is referred Section V.

## IV. DIAGNOSIS

### A. DNA

The gold standard for diagnosis of SCA2 is a DNA test for presence of the SCA2 CAG repeat expansion. No other mutations in the ataxin-2 gene have been detected so far. PCR-based amplification of the SCA2 repeat followed by polyacrylamide gel electrophoresis has high specificity and sensitivity. Many laboratories perform the SCA2 test as a multiplex PCR with tests for other SCAs.

Although PCR-based diagnosis is highly sensitive and specific, particular problems arise in childhood-onset and prenatal cases. Repeats above 100 may not be efficiently amplified and escape detection. Due to the great frequency of the $CAG_{22}$ allele in the general population it is difficult to distinguish between homozygosity of this allele versus nonamplification of a greatly expanded allele. In these cases, Southern blot analysis should be performed.

Alleles with 30 or fewer repeats are not associated with the development of ataxia. Alleles with 32–34 repeats are associated with a normal phenotype or late to very-late onset ataxia. Alleles with 31 repeats are too rare for assessment of penetrance. One healthy octogenarian with 31 repeats has been identified (Riess et al., 1997), but also one middle-age individual with ataxia and 31 repeats (M. Pandolfo, personal communication). CAA interruption in alleles does not reduce the penetrance due the fact that the repeat is uninterrupted at the protein level.

Although a correlation exists between repeat length and age of onset, no clear counseling regarding likely age of onset can be given to presymptomatic probands that have elected to be tested. Individuals with uninterrupted repeats in the long normal or reduced penetrance range should be counseled regarding an increased risk of expansion to a pathologic range in their offspring.

### B. Testing of Sporadic Patients

Testing of DNA samples from patients with sporadic ataxia or without obvious family history of ataxia occasionally reveals SCA2 mutation. Only 2 of 842 sporadic ataxia patients in the series of Riess et al. (1997) had expansions of 41 and 49 repeats. In the series reported by Cancel et al. (1997) 2 out of 90 patients with sporadic olivo-ponto-cerebellar atrophy had alleles with 37 and 39 repeats. In a series by Moseley et al., a total of 5% of patients without obvious family history had a mutation in one of the known dominant ataxia genes. Of these, half were due to SCA2 mutations and half to SCA6 mutations (Moseley, 1998). SCA2 mutations were also found in one pedigree with two affected siblings, but parents without known disease. In Italy, two of seven patients with late-onset ataxia had an SCA2 mutation (Giuffrida et al., 1999a). In Japan, SCA2 was a rare cause of sporadic ataxia compared with SCA6 (Futamura et al., 1998)

Two parent-child pairs in which the asymptomatic parent carried a premutation allele have been described (Riess et al., 1997; Futamura et al., 1998).

### C. Biochemistry

Despite the widespread expression of ataxin-2 no biochemical abnormalities in serum or urine have been reported in SCA2 patients.

## V. NEUROPHYSIOLOGY

Nerve conduction studies show an axonal sensory neuropathy and appear to be more commonly abnormal in SCA2 than SCA1 or SCA3 (Kubis *et al.*, 1999). Reduced amplitude of sensory potentials, and prolonged latency of the central components of somatosensory and of brainstem auditory-evoked potentials are common in SCA2 (Abele *et al.*, 1997; Velazquez and Medina, 1998). Neurophysiological alterations can be observed even in the absence of clinical signs (Velazquez and Medina, 1998). Visual-evoked potentials and transcranial magnetic stimulation remained normal, but appear to be frequently altered in SCA1 (Abele *et al.*, 1997; Velazquez and Medina, 1998).

## VI. NEUROIMAGING

There are no CT or MR imaging features that are specific to SCA2. Typically, SCA2 leads to significant pontine atrophy. Compared with SCA1 and SCA3, cerebellar and brainstem atrophy appear to be more severe in SCA2 (Klockgether *et al.*, 1998b). SCA1 morphometry overlapped with that of SCA2 and SCA3. Volumes of the putaminal and caudate nuclei were reduced in SCA3, but not in SCA2. In a study of 20 Italian SCA2 patients, pontocerebellar atrophy did not appear to correlate with CAG repeat length or disease duration, but supratentorial atrophy did (Giuffrida *et al.*, 1999b). In cases with very-late onset, no atrophy or a purely cerebellar atrophy can be seen (Fernandez *et al.*, 2000).

## VII. NEUROPATHOLOGY

Post-mortem examinations have been reported in the Holguin population of Cuba, in Martinican pedigrees, and in Caucasian patients (Orozco *et al.*, 1989; Dürr *et al.*, 1995; Adams *et al.*, 1997, Huynh *et al.*, 1999; Estrada *et al.*, 1999; Pang *et al.*, 2002). Overall, autopsy findings are similar, although in keeping with marked phenotypic variation, some quantitative and qualitative variation is detected from brain to brain even within the same study.

In all autopsies cerebellar Purkinje cells were reduced in number. When performed, silver preparations show poor arborization of Purkinje cell dendrites and torpedo-like formation of their axons. Parallel fibers were scanty. Granule cells were decreased in number, whereas Golgi and basket cells were well preserved as well as neurons in the dentate and other cerebellar nuclei. In the brainstem, there was marked neuronal loss in the inferior olive and pontocerebellar nuclei. In contrast to SCA3, the dentate nucleus is not or only minimally affected in SCA2.

Degeneration in the nigro-luyso-pallidal system mainly involved the substantia nigra. Six of seven brains in the Cuban study had marked loss in the substantia nigra. In five spinal cords that were available for analysis marked demyelination was present in the posterior columns and to a lesser degree in the spinocerebellar tracts. Motor neurons and neurons in Clarke's column were reduced in size and number. Especially in lumbar and sacral segments, anterior and posterior roots were partially demyelinated. In some brains, severe gyral atrophy, most prominent in the fronto-temporal lobes, has been noted. The cerebral cortex was thinned, but without neuronal rarefaction. The cerebral white matter was atrophic and gliotic. One brain showed patchy loss in parts of the third nerve nuclei. Adams *et al.* (1997) reported similar findings in one member of the FS pedigree. Nerve biopsy has shown moderate loss of large myelinated fibers (Filla *et al.*, 1995).

Biochemical analysis for activity in glyceraldehyde-3-phosphate dehydrogenase (GAPDH) did not show region specific differences from normal controls in contrast to a slight reduction in Huntington's disease caudate nucleus and Alzheimer's disease temporal lobe (Kish *et al.*, 1998). Matsuura and colleagues (1999) described mosaicism in specific brain regions in a father-daughter pair. In the cerebellum, the repeat was three to eight repeats smaller than in other CNS regions.

In summary, SCA2 shares with SCA1 and SCA7 the significant involvement of the inferior olive, and the Purkinje cell loss. In contrast to SCA3, the dentate nucleus is spared, but SCA2 shows involvement of the substantia nigra that is similar to SCA3.

## VIII. ANIMAL MODELS

### A. Mouse

Animal models have begun to unravel normal and pathologic function of ataxin-2. Knock-down of expression of the ataxin-2 ortholog in *C. elegans* leads to early embryonic lethality. Knock-down of the A2BP1 ortholog, known as fox-1, leads to even more significant reduction in egg cell mass. Ataxin-2-deficient animals have been generated. These lines do not develop neurodegeneration and have no gross tissue or developmental changes supporting the hypothesis that polyQ expansion does not lead to a loss of function (Kiehl and Pulst, submitted).

Huynh *et al.* (2000) have expressed full-length human ataxin-2 under the control of the Purkinje cell specific Pcp2 promoter in C57BL/6JxDBA/2J mice. Only mice expressing ataxin-2 with 58 glutamine repeats (ataxin-2[Q58]), but not those with ataxin-2[Q22] showed a progressive motor deficit by stride analysis and by rotarod testing. For rotarod testing, mice are placed on a rod that rotates at increasing speeds from 4–40 rotations per minute for a total of 10 minutes (Fig. 4.3A). The time that mice remain on the rod is measured and compared between groups of mice (Fig. 4.3B).

Similar to human SCA2 brains, ataxin-2 staining was increased in mouse Purkinje cells from Q[58] lines, but intranuclear inclusions were not seen. Calbindin is an abundant protein in Purkinje cells and labels cytoplasm and the dendritic arbor strongly. In homozygous animals staining became abnormal at 4–6 weeks and showed a progressive disruption of the dendritic arborization, whereas mice expressing an ataxin-2[Q22] transgene or wild-type mice did not show any changes (Fig. 4.3C).

There was a clear dosage effect. Homozygous animals showed functional impairment earlier than heterozygous animals. This was paralleled by more severe morphologic changes in homozygous animals.

## B. Drosophila

Pallanck et al. (2002) have examined the function of the ataxin-2 Drosophila homolog (Datx-2). Mutations that reduce Datx2 activity, or transgenic overexpression of Datx2 result in locomotor defects, tissue degeneration, and lethality. Examination of tissues affected by altered Datx2 gene dosage revealed defects in actin filament organization and the appearance of polymerized actin aggregates. The Datx2 polypeptide is cytoplasmic, but fails to bind directly to actin filaments. Genetic studies demonstrated that Datx2 mutants exhibit interactions with defined components of actin filament formation pathways and with known and putative RNA-binding proteins.

## IX. GENOTYPE/PHENOTYPE/ MODIFYING ALLELES

As in other diseases caused by unstable DNA repeats, a significant proportion of phenotypic variation is explained

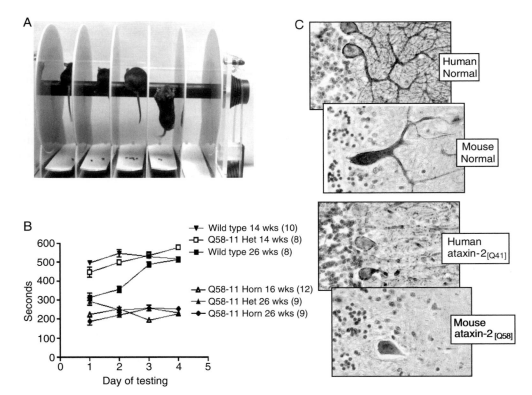

**FIGURE 4.3** Functional and morphologic analysis of ataxin-2[Q55] transgenic mouse lines. (A) Rotarod apparatus. (B) Performance of various mouse lines during 4 days of rotarod testing. Note that doubling transgene dosage already results in functional impairment at 16 weeks (open triangles) whereas heterozygous ataxin-2[Q58] animals only show functional abnormalities at 26 weeks (closed triangles). (C) Morphology of Purkinje cells in normal human brain and in a brain from a patient with 41 CAG repeats compared with the morphology seen in wild-type mice and a mouse brain expressing ataxin-2[Q58] at 27 weeks.

by length of the SCA2 CAG repeat. This is particularly apparent for age of disease onset. Approximately 60% of age of onset variability is explained by CAG repeat length in the Cuban population (Pulst and Velazquez Perez, unpublished). As in other polyQ diseases the phenotype is determined by an interaction between CAG repeat length and disease duration. In a very carefully conducted study, Cancel et al. (1997) found that the frequency of several clinical signs such as myoclonus, dystonia, and myokymia increased with the number of CAG repeats whereas the frequency of others was more related to disease duration.

Outlier phenotypes are observed in patients with pathological alleles at the lower end of the pathologic spectrum. Late-onset ataxia was seen with 32 repeats (Santos et al., 1999) and 33 repeats (Fernandez et al., 2000; Matsumura et al., 2001). Several pedigrees have now been reported in which SCA2 presents as a Parkinsonian syndrome that is L-dopa responsive (Gwinn-Hardy et al., 2000; Shan et al., 2001). One African-American pedigree had prominent dementia with onset before age 20 (Geschwind et al., 1997).

The effect of modifying alleles on age of disease onset has not been studied in detail for SCA2. Hayes et al. (2000) examined the effect of the normal polyQ tract length in several polyQ-containing proteins on age of onset in a diverse set of SCA2 patients. None of the polyQ tracts in disease-associated proteins showed any effect. However, the CAG repeat in RAI1 (retinoic acid induced gene 1) explained a small amount of the variance in age of onset.

We have attempted to identify modifier alleles in the Cuban SCA2 population using allelic association. We identified cases that had an age of onset outside one standard deviation of the mean after correction for the effect of the CAG repeat (Fig. 4.1). The two groups were designated as "premature onset" and "delayed onset" groups. We tested allelic association using $\chi^2$ analysis. There was no difference in the distribution of ApoE or RAI1 alleles between the two groups, although reduced allelic diversity reduced the power to detect differences. We also examined the contribution of polyQ tract length in all ataxia genes, the HD and DRPLA genes. Of these, only repeats in the CACNA1A gene (mutated in SCA6) modified age of onset with statistical significance (Figueroa et al., 2002). There was a trend for longer repeats in the ataxin-1 gene to be associated with earlier onset. This observation was biologically interesting, because ataxin-1 and the CACNA1A gene have strong expression in Purkinje cells and the calcium channel has a cytoplasmic expression.

Variation in SCA2 CAG repeat length has also been described as a modifier of other disease traits. Using transmission disequilibrium testing, the 22 repeat allele was preferentially transmitted to children with multiple sclerosis (Chataway et al., 1999). These findings, however, could not be confirmed in another study (Miterski et al., 2000).

## X. TREATMENT

No treatments are known that will change the underlying course of the disease. Rare patients with SCA2 and a Parkinsonian syndrome have shown response to treatment with L-dopa. Although no controlled trials have been performed, symptomatic treatment of ataxia, pseudobulbar symptoms, and tremor may show effects in selected patients. The role of rehabilitation in improving functional performance is undergoing active study in Cuban SCA2 patients (Velazquez Perez and Pulst, unpublished).

### Acknowledgments

This work was supported by the Carmen and Louis Warschaw Endowment for Neurology, F.R.I.E.N.D.s of Neurology, the National Ataxia Foundation, and grant RO1-NS33123 from the National Institutes of Health. The author thanks Dr. Hema Vakharia for critical comments and Karla (Pattie) Figueroa for performing amino acid comparisons shown in Fig. 4.2.

### References

Adams, C. R., Starkman, S., and Pulst S. M. (1997). Clinical and molecular analysis of a pedigree of southern Italian ancestry with spinocerebellar ataxia type 2. *Neurology* **49**, 1163–1166.

Abele, M., Burk, K., Andres, F., Topka, H., Laccone, F., Bosch, S., Brice, A., Cancel, G., Dichgans, J., and Klockgether, T. (1997). Autosomal dominant cerebellar ataxia type I, Nerve conduction and evoked potential studies in families with SCA1, SCA2 and SCA3. *Brain* **120**, 2141–2148.

Affaitati, A., de Cristofaro, T., Feliciello, A., and Varrone, S. (2001). Identification of alternative splicing of spinocerebellar ataxia type 2 gene. *Gene* **267**, 89–93.

Babovic-Vuksanovic, D., Snow, K., Patterson, M. C., and Michels, V. V. (1998). Spinocerebellar ataxia type 2 (SCA2) in an infant with extreme CAG repeat expansion. *Am. J. Med. Genet.* **79**, 383–387.

Burk, K., Abele, M., Fetter, M., Dichgans, J., Skalej, M., Laccone, F., Didierjean, O., Brice, A., and Klockgether, T. (1996). Autosomal dominant cerebellar ataxia type 1: clinical features and MRI in families with SCA1, SCA2 and SCA3. *Brain* **119**, 1497–1505.

Burk, K., Fetter, M., Abel, M., Laccone, F., Brice, A., Dichgans, J., and Klockgether, T.(1999a). Autosomal dominant cerebellar ataxia type I: oculomotor abnormalities in families with SCA1, SCA2, and SCA3. *J. Neurol. Sci.* **246**, 789–797.

Burk, K., Globas, C., Bosch, S., Graber, S., Abele, M., Brice, A., Dichgans, J., Daum, I., and Klockgether, T. (1999b). Cognitive deficits in spinocerebellar ataxia 2. *Brain* **122**, 769–777.

Buttner, N., Geschwind, D., Jen, J.C., Perlman, S., Pulst, S.M., and Baloh, R.W. (1998). Oculomotor phenotypes in autosomal dominant ataxias. *Arch. Neurol.* **55**, 1353–1357.

Cancel, G., Dürr, A., Didierjean, O., Imbert, G., Burk, K., Lezin, A., Belal, S., Benomar, A., Abada-Bendib, M., Vial, C., Guimaraes, J., Chneiweiss, H., Stevanin, G., Yvert, G., Abbas, N., Saudou, F., Lebre, A.S., Yahyaoui, M., Hentati, F., Vernant, J. C., Klockgether, T., Mandel, J. L, Agid, Y., and Brice A. (1997). Molecular and clinical correlations in spinocerebellar ataxia 2: a study of 32 families. *Hum. Mol. Genet.* **6**, 709–715.

Chakravarty, A., and Mukherjee, S. C. (2002). Autosomal dominant cerebellar ataxias in ethnic bengalees in West Bengal—an Eastern Indian state. *Acta Neurol. Scand.* **105**, 202–208.

Chataway, J., Sawcer, S., Coraddu, F., Feakes, R., Broadley, S., Jones, H. B., Clayton, D., Gray, J., Goodfellow, P. N., and Compston, A. (1999). Evidence that allelic variants of the spinocerebellar ataxia type 2 gene influence susceptibility to multiple sclerosis. *Neurogenetics* **2**, 91–96.

Choudhry, S., Mukerji, M., Srivastava, A. K., Jain, S., Brahmachari, S. K. (2001). CAG repeat instability at SCA2 locus: anchoring CAA interruptions and linked single nucleotide polymorphisms. *Hum. Mol. Genet.* **10**, 2437–46.

Costanzi-Porrini, S., Tssarolo, D., Abbruzzese, C., Liguori, M., Ashizawa, T., and Giacanelli, M. (2000). An interrupted 34-CAG repeat SCA-2 allele in patients with sporadic spinocerebellar ataxia. *Neurology* **54**, 491–493.

Didierjean, O., Cancel, G., Stevanin, G., Dürr, A., Burk, K., Benomar, A., Lezin, A., Belal, S., Abada-Bendid, M., Klockgether, T., and Brice A. (1999). Linkage disequilibrium at the SCA2 locus. *J. Med. Genet.* **36**, 415–417.

Dürr, A., Smadja, D., Cancel, G., Lezin, A., Stevanin, G., Mikol, J., Bellance, R., Buisson, G. G., Chneiweiss, H., and Dellanave, J. (1995). Autosomal dominant cerebellar ataxia type 1 in Martinique (French West Indies) Clinical and neuropathological analysis of 53 patients from three unrelated SCA2 families. *Brain* **118**, 1573–1581.

Estrada, R., Galarraga, J., Orozco, G., Nodarse, A., and Auburger G. (1999). Spinocerebellar ataxia 2 (SCA2): morphometric analyses in 11 autopsies. *Acta Neuropathol.* (Berlin) **97**, 306–310.

Fernandez, M., McClain, M. E., Martinez, R. A., Snow, K., Lipe, H., Ravits, J., Bird, T. D., and La Spada, A. R. (2000). Late-onset SCA2:33 CAG repeats are sufficient to cause disease. *Neurology* **55**, 569–572.

Figueroa, K., Santos, N., Velazque, L., and Pulst, S.M. (2002). Normal variation in SCA 1 and SCA6 CAG repeats modifies age of onset in spinocerebellar ataxia type 2. (SCA2). *Neurology* **58** (suppl. 3), A16.

Filla, A., DeMichele, G., Banfi, S., Santoro, L., Perretti, A., Cavalcanti, F., Pianese, L., Castaldo, I., Barbieri, F., Campanella, G. *et al.* (1995). Has spinocerebellar ataxia type 2 an distinct phenotype? Genetic and clinical study of an Italian family. *Neurology* **45**, 793–796.

Filla, A., De Michele, G., Santoro, L., Calabrese, O., Castaldo, I., Giuffrida, S., Restivo, D., Serlenga, L., Condorelli, D. F., Bonuccelli, U., Scala, R., Coppola, G., Caruso, G., and Cocozza, S. (1999). Spinocrebellar ataxia type 2 in southern Italy: a clinical and molecular study of 30 families. *J. Neurol.* **246**, 467–471.

Futamura, N., Matsumura, R., Fujimoto, Y., Horikawa, H., Suzumura, A., and Takayanagi, T. (1998). CAG repeat expansions in patients with sporadic cerebellar ataxia. *Acta Neurol. Scand.* **98**, 55–59.

Gambardella, A., Annesi, G., Bono, F., Spadafora, P., Valentino, P., Pasqua, A. A., Mazzei, R., Montesanti, R., Conforti, F. L., Oliveri, R. L., Zappia, M., Aguglia, U., and Quattrone, A. (1998). CAG repeat length and clinical features in three Italian families with spinocerebellar ataxia type 2 (SCA2): early impairment of Wisconsin Card Sorting Test and saccade velocity. *J. Neurol.* **245**, 647–652.

Geschwind, D. H., Perlman, S. B., Figueroa, K. P., Karrim, J., Baloh, R. W., and Pulst, S. M. (1997a). Spinocerebellar ataxia type 6. Frequency of the mutation and genotype-phenotype correlations. *Neurology* **49**, 1247–1251.

Geschwind, D. H., Perlman, S. B., Figueroa, C. P., Treiman, L. J., and Pulst S. M. ( 1997b). The prevalence and wide clinical spectrum of the spinocerebellar ataxia type 2 trinucleotide repeat in patients with autosomal dominant cerebellar ataxia. *Am. J. Hum. Genet.* **60**, 842–850.

Giuffrida, S., Lanza, S., Restivo, D. A., Saponara, R., Valvo, S. C., Le Pira, F., Trovato Salinaro, A., Spinella, F., Nicoletti, A., and Condorelli, D. F. (1999a). Clinical and molecular analysis of 11 Sicilian SCA2 families: influence of gender on age at onset. *Eur. J. Neurol.* **6**, 301–307.

Giuffrida, S., Saponara, R., Restivo, D. A., Trovato Salinaro, A., Tomarchio, L., Pugliares, P., Fabbri, G., and Maccagnano, C. (1999b). Supratentorial atrophy in spinocerebellar ataxia type 2: MRI study of 20 patients. *J. Neurol.* **246**, 383–388.

Giuffrida, S., Saponara, R., Trovato Salinaro, A., Restivo, D. A., Domina, E., Papotto, M., Le Pira, F., Nicoletti, A., Trovato, A., Condorelli, D. F. (1999). Identification of SCA2 mutation in cases of spinocerebellar ataxia with no family history in mid-eastern Sicily. *Ital. J. Neurol. Sci.* **20**, 217–221.

Giunti, P., Sabbadini, G., Sweeney, M. G., Davis, M. B., Veneziano, L., Mantuano, E., Federico, A., Plasmati, R., Frontali, M., and Wood, N. W. (1998). The role of the SCA2 trinucleotide repeat expansion in 89 autosomal dominant cerebellar ataxia families. Frequency, clinical and genetic correlates. *Brain* **121**, 459–467.

Gwinn-Hardy, K., Chen, J. Y., Liu, H. C., Liu, T. Y., Boss, M., Seltzer, W., Adam, A., Singleton, A., Koroshetz, W., Waters, C., Hardy, J., and Farrer, M. (2000). Spinocerebellar ataxia type 2 with parkinsonism in ethnic Chinese. *Neurology* **55**, 800–805.

Hayes, S., Turecki, G., Brisebois, K., Lopes-Cendes, I., Gaspar, C., Riess, O., Ranum, L. P., Pulst, S. M., and Rouleau, G. A. (2000). CAG repeat length in RAI1 is associated with age at onset variability in spinocerebellar ataxia type 2 (SCA2). *Hum. Mol. Genet.* **9**, 1753–1758.

Hsieh, M., Li, S. Y., Tsai, C. J., Chen, Y. Y., Liu, C. S., Chang, C. Y., Ro, L. S., Chen, D. F., Chen, S. S., and Li, C. (1999). Identification of five spinocerebellar ataxia type 2 pedigrees in patients with autosomal dominant cerebellar ataxia in Taiwan. *Acta Neurol. Scand.* **100**, 189–194.

Huynh, D. P., Del Bigio, M. R., Ho, D. H., and Pulst, S. M. (1999). Expression of ataxin-2 in brains from normal individuals and patients with Alzheimer's disease and spinocerebellar ataxia 2. *Ann. Neurol.* **45**, 232–241.

Huynh, D. P., Figueroa, K. P., Hoang, N., and Pulst, S. M. (2000). Nuclear localization or inclusion body formation are not necessary for SCA2 pathogenesis in man or mouse. *Nat. Genet.* **2**, 44–50.

Huynh, D., Tao, H., and Pulst, S. M. (2002). Expansion of the poly q traction ataxin 2 disrupts golgin localization. 54th American Academy of Neurology Annual Meeting, Denver, Colorado.

Imbert, G., Saudou, F., Yvert, G., Devys, D., Trottier, Y., Garnier, J. M., Weber, C., Mandel, J. L., Cancel, G., Abbas, N., Durr, A., Didierjean, O., Stevanin, G., Agid, Y., and Brice, A. (1996). Cloning of the gene forspinocerebellar ataxia 2 reveals a locus with high sensitivity to expanded CAG glutamine repeats. *Nat. Genet.* **14**, 285–291.

Kiehl, T. R., Shibata, H., and Pulst, S. M. (2000). The ortholog of human ataxin-2 is essential for early embryonic patterning in *C. elegans*. *J. Mol. Neurosci.* **15**, 231–241.

Kish, S. J., Lopes-Cendes, I., Guttman, M., Furukawa, Y., Pandolfo, M., Rouleau, G. A., Ross, B. M., Nance, M., Schut, L., Ang, L., and DiStefano, L. (1998). Brain glyceraldehyde-3-phosphate dehydrogenase activity in human trinucleotid repeat disorders. *Arch. Neurol.* **55**, 1299–1304.

Klockgether, T., Ludtke, R., Kramer, B., Abele, M., Burk, K., Schols, L., Riess, O., Laccone, F., Boesch, S., Lopes-Cendes, I., Brice, A, Inzelberg, R., Zilber, N., and Dichgans, J. (1998a). The natural history of degenerative ataxia: a retrospective study in 466 patients. *Brain* **121**, 589–600.

Klockgether, T., Skalej, M., Wedekind, D., Luft, A. R., Welte, D., Schulz, J. B., Abele, M., Burk, K., Laccone, F., Brice, A., Dichgans, J. (1998b). Autosomal dominant cerebellar ataxia type 1. MRI-based volumetry of posterior fossa structures and basal ganglia in spinocerebellar ataxia types 1,2, and 3. *Brain* **121**, 1687–1693.

Kozlov, G., Trempe, J. F., and Khaleghpour, K. (2001). Structure and function of the C-terminal |pABC domain of human Poly (A)-binding protein. *Proc. Natl. Acad. Sci. U.S.A.* **98**, 4409–4413.

Koyano, S., Uchihara, T., Fujigasaki, H., Nakamura, A., Yagishita, S., and Iwabuchi, K. (1999). Neuronal intranuclear inclusions in spinocerebellar ataxia type 2: triple-labeling immunofluorescent study. *Neurosci. Lett.* **273**, 117–20.

Kubis, N., Dürr, A., Gugenheim, M., Chneiweiss, H., Mazzetti, P., Brice, A., and Bouche, P. (1999). Polyneuropathy in autosomal dominant

Lorenzetti, D., Bohlega, S., and Zoghbi, H. Y. (1997). The expansion of the CAG repeat in ataxin-2 is a frequent cause of autosomal dominant spinocerebellar ataxia. *Neurology* **49**, 1009–1013.

cerebellar ataxias: Phenotype-genotype correlation. *Muscle Nerve* **22**, 712–717.

Malandrini, A., Galli, L., Villanova, M., Palmeri, S., Parrotta, E., DeFalco, D., Cappelli, M., Grieco, G.S., Renieri, A., and Guazzi, G. (1998). CAG repeat expansion in an Italian family with spinocerebellar ataxia type 2 (SCA2) a clinical and genetic study. *Eur. Neurol.* **40**, 164–168.

Matsumura, R., Futamura, N. (2001). Late-onset SCA2: 33 CAG repeats are sufficient to cause disease. *Neurology* **57**, 566.

Matsuura, T., Sasaki, H., Yabe, I., Hamada, K., Hamada, T., Shitara, M., and Tashiro, K. (1999). Mosaicism of unstable CAG repeats in the brain of spinocerebellar ataxia type 2. *J. Neurol.* **246**, 835–839.

Meunier, C., Bordereaux, D., Porteu, F., Gisselbrecht, S., Chretien, S., and Courtois, G. (2002). Cloning and characterization of a family of proteins associated with mpl. *J. Biol. Chem.* **277**, 9139–9147.

Miterski, B., Eppelen, J. T., Poehlau, D., Sindern, E., Haupts, M. Y. (2000). SCA2 alleles are not general predisposition factors for multiple sclerosis. *Neurogenetics* **2**, 235–236.

Mizushima, K., Watanabe, M., Kondo, I., Okamoto, K., Shizuka, M., Abe, K., Aoki, M., and Shoji, M. (1999). Analysis of spinocerebellar ataxia type 2 gene and haplotype analysis (CCG)1-2 polymorphism and contribution to founder effect. *J. Med. Genet.* **6**, 112–114.

Moseley, M. L., Benzow, K. A., Schut, L. J., Bird, T. D., Gomez, C. M., Barkhaus, P. E., Blindauer, K. A., Labuda, M., Pandolfo, M., Koob, M. D., and Ranum, L. P. (1998). Incidence of dominant spinocerebellar and Friedreich triplet repeats among 361 families. *Neurology* **51**, 1666–1671.

Nechiporuk, A., Lopes-Cendes, I., Nechiporuk, T., Starkman, S., Andermann, E., Rouleau, G. A., Weissenbach, J. S., Kort, E., Pulst, S. M. (1996). Genetic mapping of spinocerebellar ataxia type 2 gene on human chromosome 12. *Neurology* **46**, 1731–1735.

Nechiporuk, T., Nechiporuk, A., Sahba, S., Figueroa, K., Shibata, H., Chen, X. N., Korenberg, J. R., de Jong, P., and Pulst, S. M. (1997). A high-resolution PAC and BAC Map of the SCA2 Region. *Genomics* **44**, 321–329.

Nechiporuk, T., Huynh, D. P., Figueroa, K., Sahba, S., Nechiporuk, A., and Pulst, S. M. (1998). The mouse SCA2 gene: cDNA sequence, alternative splicing and protein expression. *Hum. Mol. Genet.* **8**, 1301–1309.

Orozco, G., Estrada, R., Perry, T. L., Arana, J., Fernandez, R., Gonzalez-Quevedo A, Galarraga, J., and Hansen, S. (1989). Dominantly inherited olivopontocerebellar atrophy from eastern Cuba: clinical, neuropathological, and biochemical findings. *J. Neurol. Sci.* **93**, 37–50.

Orozco, G., Fleites, A., Cordoves Sagaz, R., and Auburger, G. (1990). Autosomal dominant cerebellar ataxia: clinical analysis of 263 patients from a homogeneous population in Holguin, Cuba. *Neurology* **40**, 1369–1375.

Pallanck, L., et al. (2002). Ataxin-2 in the fly. 43rd Annual Drosophila Research Conference, Abstract 218B, April 10–14, San Diego, California. (Fly meeting 2002).

Pang, J., Allotey, R., Wadia, N., Sasaki, H., Bindoff, L., and Chamberlain, S. (1999). A common disease haplotype segregating in spinocerebellar ataxia 2 (SCA2) pedigrees of diverse ethnic origin. *Eur. J. Hum. Genet.* **7**, 841–845.

Pang, J. T., Giunti, P., Chamberlain, S., An, S.F., Vitaliani, R., Scaravilli, T., Martinian, L., Wood, N.W., Scaravilli, F., and Ansorge, O. (2002). Neuronal intranuclear inclusions in SCA2: a genetic, morphological and immunohistochemical study of two cases. *Brain* **125**, 656–663.

Pareyson, D., Gellera, C., Castellotti, B., Antonelli, A., Riggio, M.C., Mazzucchelli, F., Girotti, F., Pietrini,V., Mariotti, C., and Di Donato, S. (1999). Clinical and molecular studies of 73 Italian families with autosomal dominant cerebellar ataxia type I:SCA1 and SCA2 are the most common genotypes. *J. Neurol.* **246**, 389–393.

Pujana, M. A., Corral, J., Gratacos, M., Combarros, O., Berciano, J., Genis, D., Banchs, I., Estivill, X., and Volpini,V. (1999). Spinocerebellar ataxias in Spanish patients: genetic analysis of familial and sporadic cases. The Ataxia Study Group. *Hum. Genet.* **104**, 516–22.

Pulst, S.-M., Nechiporuk, A., and Starkman, S. (1993). Anticipation in spinocerebellar ataxia type 2. *Nat. Genet.* **5**, 8–10.

Pulst, S.-M., Nechiporuk, A., Nechiporuk, T., Gispert, S., Chen, X. N., Lopes-Cendes, I., Pearlman, S., Starkman, S., Orozco-Diaz, G., Lunkes, A., DeJong, P., Rouleau, G. A., Auburger, G., Korenberg, J. R., Figueroa, C., and Sahba, S. (1996). Moderate expansion of a normally biallelic trinucleotide repeat in spinocerebellar ataxia type 2. *Nat. Genet.* **14**, 269–276.

Pulst, S. M., and Perlman, S. (2000). Hereditary ataxias. In "Neurogenetics" (S. M. Pulst, (ed,). pp. 231–263. Oxford University Press, New York.

Riess, O., Laccone, F., Gispert, S., Schols, L., Zuhlke, C., Vieira-Saecker, A. M., Herlt, S., Wessel, K., Epplen, J. T., Weber, B. H., Kreuz, F., Chahrokh-Zadeh, S., Meindl, A., Lunkes, A., Aguiar, J., Macek, M. Jr., Krebsova, A., Macek, M. Sr., Burk, K., Tinschert, S., Schreyer, I., Pulst, S.M., and Auburger, G. (1997). SCA2 trinucleotide expansion in German SCA patients. *Neurogenetics* **1**, 59–64.

Rivaud-Pechoux, S., Dürr, A., Gaymard, B., Cancel, G., Ploner C.J., Agid, Y., Brice, A., and Pierrot-Deseilligny, C. (1998). Eye movement abnormalities correlate with genotype in autosomal dominant cerebellar ataxia type 1. *Ann. Neurol.* **43**, 297–302.

Rufa, A., Dotti, M. T., Galli, L., Orrico, A., Sicurelli, F., and Federico, A. (2002). Spinocerebellar ataxia type 2 (sca2) associated with retinal pigmentary degeneration. *Eur. Neurol.* **47**, 128–129.

Sahba, S., Nechiporuk, A., Figueroa, K. P., Nechiporuk, T., Pulst S. M. (1998). Genomic structure of the human gene for spinocerebellar ataxia type 2 (SCA2) on chromosome 12q24.1. *Genomics* **47**, 359–364.

Sanpei, K., Takano, H., Igarashi S., Sato, T., Oyake, M., Sasaki, H., Wakisaka, A., Tashiro, K., Ishida, Y., Ikeuchi, T., Koide, R., Saito, M., Sato, A., Tanaka, T., Hanyu, S., Takiyama, Y., Nishizawa, M., Shimizu, N., Nomura, Y., Segawa, M., Iwabuchi, K., Eguchi, I., Tanaka, H., Takahashi, H., and Tsuji, S. (1996). Identification of the spinocerebellar ataxia type 2 gene using a direct identification of repeat expansion and cloning technique, DIRECT. *Nat Genet.* **14**, 277–284.

Santos, N., Aguiar, J., Fernandez, J. et al. (1999). Molecular diagnosis of a sample of the Cuban population with Spinocerebellar ataxia type 2. *Biotecnol. Aplic.* **16**, 219–221.

Sasaki, H., Wakisaka, A., Sanpei, K., Takano, H., Igarashi, S., Ikeuchi, T., Iwabuchi, K., Fukazawa, T., Hamada, T., Yuasa, T., Tsuji, S., and Tashiro, K. (1998). Phenotype variation correlates with CAG repeat length in SCA2-a study of 28 Japanese patients. *J. Neurol. Sci.* **159**, 202–208.

Schols, L., Amoiridis, G., Buttner T., Przuntek, H., Epplen, J. T., and Riess O. (1997a). Autosomal dominant cerebellar ataxia: Phenotypic differences in genetically defined subtypes? *Annal. Neurol.* **42**, 924–932

Schols, L., Gispert, S., Vorgerd, M. et al. (1997b). Spinocerebellar ataxia type 2: Genotype and phenotype in German kindreds. *Arch. Neurol.* **54**, 1073–1080.

Schols, L., Peters, S., Szymanski, S., Kruger, R., Lange, S., Hardt, C., Riess, O., and Przuntek, H. (2000). Extrapyramidal motor signs in degenerative ataxias. *Arch. Neurol.* **57**, 1495–14500.

Shan, D. E., Soon, B. W., Sun, C. M., Lee, S. J., Liao, K. K., and Liu, R. S. (2001). Spinocerebellar ataxia type 2 presenting as familial levodopa-responsive parkinsonism. *Ann. Neurol.* **50**, 812–815.

Shibata, H., Huynh, D. P., and Pulst, S. M. (2000). A novel protein with RNA binding motifs interacts with ataxin-2. *Hum. Mol. Genet.* **9**, 1303–1313.

Stevanin, G., Durr, A., Brice, A. (2000). Clinical and molecular advances in autosomal dominant cerebellar ataxias: from genotype to phenotype and physiopathology. *Eur. J. Hum. Genet.* **8**, 4–18.

Storey, E., Forrest, S. M., Shaw, J. H., Mitchell, P., and Gardner, R. J. (1999). Spinocerebellar ataxia type 2: clinical features of a pedigree displaying prominent frontal-executive dysfunction. *Arch. Neurol.* **56**, 43–50.

Ueyama, H., Kumamoto, T., Nagao, S., *et al.* (1998). Clinical and genetic studies of spinocerebellar ataxia type 2 in Japanese kindreds. *Acta Neurol. Scand.* **98**, 427–432.

Velazquez Perez, L., De La Hoz Oliveras, J., Perez Gonzalez, R., Hechevarria Pupo, R. R., and Herrera Dominguez, H. (2001a). Quantitative evaluation of disorders of coordination in patients with Cuban type 2 spinocerebellar ataxia. *Rev. Neurol.* **32**, 601–606.

Velazquez Perez, L., Santos Falcon, N., Garcia Zaldivar, R., Paneque Herrera, M., and Hechevarria Pupo, R. R.(2001b). Epidemiology of Cuban hereditary ataxia. *Rev. Neurol.* **32**, 606–611.

Velazquez, L., and Medina, E. E. (1998). Electrophysiological characteristics of asymptomatic relatives of patients with type 2 spinocerebellar ataxia. *Rev. Neurol.* **160**, 955–963.

Wadia, N. H., and Swami, R. K. (1971). New form of heredo-familial spinocerebellar degeneration with slow eye movements (nine families). *Brain* **94**, 359–374.

Wadia, N., Pang, J., Desai, J., Mankodi, A., Desai, M., and Chamberlain, S. (1998). A clinicogenetic analysis of six Indian spinocerebellar ataxia (SCA2) pedigrees. The significance of slow saccades in diagnosis. *Brain* **121**, 2341–2355.

CHAPTER 5

# Spinocerebellar Ataxia 3—Machado-Joseph Disease (SCA3)

**HENRY PAULSON**
*Department of Neurology*
*University of Iowa Roy J. and Lucille A. Carver College of Medicine*
*Iowa City, Iowa 52242*

**S. H. SUBRAMONY**
*Department of Neurology*
*University of Mississippi Medical Center*
*Jackson, Mississippi 39216*

I. Historical Introduction
II. Prevalence of MJD
III. Phenotype
   A. Age of Onset
   B. Presenting Symptoms and Course
   C. Cerebellar Signs
   D. Oculomotor Signs
   E. Bulbar Signs
   F. Extrapyramidal and Upper Motor Neuron Signs
   G. Peripheral Nerve and Lower Motor Neuron Signs
   H. Cognitive Deficits
   I. Sleep Disturbances
   J. Autonomic Deficits
   K. Phenotypic Variability
IV. Diagnosis
   A. DNA
   B. Imaging Studies
   C. Evoked Potentials and Nerve Conduction Studies
V. Neuropathology
VI. Molecular Genetics
   A. CAG Repeat Expansion in the MJD1 Gene
   B. Somatic and Gametic Instability of the Expanded Repeat
VII. Phenotype-Genotype Correlation
   A. Other Genetic Issues
VIII. Pathogenic Mechanisms and Models
   A. Ataxin-3 the SCA3/MJD Disease Protein
   B. Protein Misfolding and Aggregation as a Central Feature of Disease
   C. Animal Models
   D. Possible Mechanisms of Neurotoxicity
IX. Treatments
References

## I. HISTORICAL INTRODUCTION

Nakano *et al.* (1972) first described a dominant ataxia among Azorean immigrants living in Massachussetts. The disorder in this family did not differ greatly from that seen in other families with dominant ataxia. Rosenberg and colleagues (1976) then reported another family from the Azores, descended from Anton Joseph, with a dominantly inherited neurodegenerative disease characterized by spasticity and rigidity. Neuropathologic examination in this family showed loss of nigral and spinal cord neurons. Other Azorean families were also reported to have a parkinsonian phenotype (Romanul *et al.*, 1977). Portuguese and American researchers noted a high incidence of a dominantly inherited neurodegenerative disorder among Azoreans, in which the different phenotypes noted in the U.S. families co-existed in the same family. The disease was variously named Machado-Joseph disease (MJD) and Azorean disease of the nervous system (Romanul *et al.*, 1977; Dawson, 1977). Though the disorder shared many features with dominantly inherited ataxia in non-Portuguese populations, the striking phenotypic variability was thought to reflect a disease prevalent only among Portuguese Azoreans. However, a similar clinical disorder was also reported among patients from other ethnic backgrounds such as African-Americans and the Japanese and it was speculated that these families had acquired the mutation from seafaring Portuguese (Healton *et al.*, 1980; Sakai *et al.*, 1983; Subramony *et al.*, 1993). In the early 1990s, the genetic locus for MJD was narrowed to chromosome 14q (Takiyama *et al.*, 1993; St.George-Hyslop *et al.*,1994; Sequiros *et al.*, 1994). The gene mutation was soon shown to be an unstable expansion of a CAG repeat sequence in the *MJD1* gene, coding for an expanded glutamine repeat in a novel protein (Kawaguchi *et al.*, 1994). Many families with dominantly inherited ataxia worldwide were subsequently documented to have the same CAG repeat expansion in the *MJD1* gene. In 1994, Stevanin and colleagues mapped what was thought to be an unrelated dominant ataxia, spinocerebellar ataxia 3 (SCA3), to the same region of chromosome 14 as MJD. SCA3 was soon shown to be caused by the same mutation as in MJD. Thus, MJD and SCA3 are genetically the same disorder. The Human Genome Organization (HUGO) now refers to this disease as MJD, though it has also been referred to as SCA3 or SCA3/MJD in the literature. It is now possible to diagnosis MJD retrospectively in families reported earlier in the literature. For example, the Drew family of Walworth, described in 1929 by Ferguson and Critchley, is now known to have MJD (Giunti *et al.*, 1995).

## II. PREVALENCE OF MJD

In many series, MJD is the most common molecularly defined dominant ataxia (Table 5.1), varying from about 20% in most U.S. series to close to 50% in German, Japanese, and Chinese series (Schols *et al.*, 1995; Durr *et al.*, 1996; Inoue *et al.*, 1996; Watanabe *et al.*, 1998; Moseley *et al.*, 1998; Soong 2001). In a comparative study of Japanese and Caucasian families, Takano *et al.* (1998) found a higher prevalence of MJD in Japan and speculated that this may be due to a higher prevalence of large normal alleles in the Japanese population, providing a "reservoir" for new expansions. MJD is not the most prevalent SCA in every ethnic group. For example, no MJD was found among dominant families from Italy and South Africa (Filla *et al.* 1996; Ramesar *et al.* 1997). In contrast to SCA2 and SCA6, the MJD mutation is rarely detected in patients with the diagnosis of "sporadic ataxia" (Moseley *et al.*, 1998; Schols *et al.*, 1995).

## III. PHENOTYPE

Many researchers have studied the clinical features of MJD in diverse populations (see Sequiros and Coutinho 1993; Takiyama *et al.*, 1994; Sasaki *et al.*, 1995; Higgins *et al.*, 1996; Matsumura *et al.*, 1996a; Durr *et al.*, 1996; Cancel *et al.*, 1995; Schols *et al.*, 1996; Watanabe *et al.*, 1996; Zhou *et al.*, 1997).

### A. Age at Onset

The age at onset of MJD varies from 5–70 years (Cancel *et al.*, 1995; Schols *et al.*, 1996; Watanabe *et al.*, 1996; Zhou *et al.*, 1997). The mean age at onset in a large Portuguese cohort of patients was 37 years (Sequiros and Coutinho, 1993). In many other series the mean age of onset is in the mid-30s.

### B. Presenting Symptoms and Course

It is important to recognize that MJD is clinically a highly variable disease, with ataxia simply being the most common feature (Table 5.2). The disease is characterized not only by ataxia related to cerebellar dysfunction, but also by numerous other signs related to dysfunction within the brainstem, oculomotor system, pyramidal system, extrapyramidal system, peripheral nervous system, and lower motor neurons. Patients usually present with gait imbalance and slurring of speech. Visual disturbances reflecting impaired ocular motility such as diplopia and difficulty

Table 5.1  Prevalence of MJD/SCA3 among Dominant Ataxias

| Series | Country | %MJD/SCA3 |
|---|---|---|
| Schols *et al.* (1995) | German | 49 |
| Durr *et al.* (1996) | European, North African | 28 |
| Inoue *et al.* (1996) | Japanese | 56 |
| Watanabe *et al.* (1996) | Japanese | 34 |
| Moseley *et al.* (1998) | African American, German | 21 |
| Takano *et al.* (1998) | Caucasian | 30 |
|  | Japanese | 43 |
| Pujana *et al.* (1999) | Spanish | 15 |
| Nakaoka *et al.* (1999) | Japanese | 20 |
| Saleem *et al.* (2000) | Indian | 5 |
| Soong *et al.* (2001) | Chinese | 47 |
| Zhou *et al.* (2001) | Chinese | 35 |

Table 5.2  Prevalence of Neurological Signs in MJD/SCA3*

| | |
|---|---|
| Age at onset (range) | 5–70 years |
| Age at onset (mean) | 30–38 years |
| Ataxia | 95–100% |
| Dysarthria | 62–100% |
| Dysphagia | 29–77% |
| Ophthalmoparesis | 40–100% |
| Nystagmus | 67–100% |
| Upper motor neuron signs | 25–100% |
| Lower motor neuron signs | 11–67% |
| Sensory loss | 19–80% |
| Basal ganglia signs: dystonia, akinesia | 7–80% |

*Note*: Other signs: bulging eyes and ocular stare, blepharospasm, perioral fasciculations, temporal atrophy, limb atrophy.

References: Takiyama *et al.* (1994); Sasaki *et al.* (1995); Cancel *et al.* (1995); Durr *et al.* (1996); Higgins *et al.* (1996); Schols *et al.* (1996); Matsumura *et al.* (1996); Watanabe *et al.* (1996); Zhou *et al.* (1997).

focusing are other complaints. Early in the course of the disease ataxia, dysarthria, spasticity, hyperreflexia, and nystagmus are frequent signs on examination. Later there is increasing motor dysfunction, associated with progressive bulbar symptoms and signs and oculomotor deficits. In advanced stages of disease, patients are chair-bound, have severe dysarthria and dysphagia, facial and temporal atrophy, ineffective cough, ophthalmoparesis, dystonic posturing, and amyotrophy. Death results from immobility, nutritional impairment, and respiratory compromise. Survival after disease onset ranges from 21–25 years and a wheelchair-bound state is usually seen 15–20 years after onset (Klockgether et al., 1998).

### C. Cerebellar Signs

Cerebellar signs are almost universally present, though occasional patients with MJD may have little ataxia for periods of time. Appendicular ataxia is usually less severe than gait ataxia, but patients typically exhibit the classic signs of cerebellar deficit such as finger-to-nose and heel-to-shin ataxia, kinetic and intention tremor, impaired tandem walking, and a broad-based ataxic gait.

### D. Oculomotor Signs

Early in the disease, frequently noted abnormalities are nystagmus and saccadic intrusions into pursuit eye movements. Later, there is slowing of saccades, disconjugate eye movements, and ophthalmoplegia with early limitation of upgaze. Ophthalmoplegia may have both supranuclear and infranuclear characteristics (Coutinho and Andrade, 1978). Ptosis, lid retraction, and apparent bulging of the eyes can be seen but are not early signs (Watanabe et al., 1996; Coutinho and Andrade, 1978; Zhou et al., 1997). Blepharospasm occurs in some patients, also late in the disease. Electro-oculographic examination of eye movements in MJD was characterized by gaze-evoked and rebound nystagmus, significant reduction in vestibulo-ocular reflex gain, and moderate pursuit gain abnormalities (Buttner et al., 1998). Peak saccade velocity was relatively preserved especially in comparison to that in patients with SCA2 (Burk et al., 1996; Buttner et al., 1998). MJD patients also have a high incidence of gaze nystagmus, hypometria of visually guided saccades, and abnormal smooth pursuit (Revaud-Pechoux et al., 1998). More recently, the presence of diplopia in MJD patients has been related to the presence of defective divergence mechanisms (Ohyagi et al., 2000).

### E. Bulbar Signs

Facial and temporal muscles may show atrophy. Perioral fasciculatory movements, often induced by lip movements, are frequently seen. The tongue may appear mildly atrophic and have fasciculations. Cough mechanisms may be compromised even early in the disease. A spastic-ataxic dysarthria occurs early and increasing dysphagia is a feature of more advanced disease.

### F. Extrapyramidal and Upper Motor Neuron Signs

Features suggesting extrapyramidal disease are frequent in MJD (Romanul et al., 1997; Subramony et al., 1996; Gwinn-Hardy et al., 2001). The occurrence of dystonia and chorea late in disease is common in several dominant ataxias, but early and prominent occurrence of such signs as akinesia, rigidity, and dystonia often responsive to dopaminergic therapy is particularly characteristic of MJD. Such signs often go hand in hand with evidence of upper motor neuron dysfunction such as brisk reflexes and extensor plantar responses and may occur as the predominant or sole clinical sign of the disease in some members of a family. Schols et al. (1997) noted that severe spasticity and pronounced peripheral neuropathy were more frequently associated with MJD than with the other dominant ataxias. Clinical signs referable to the motor system can change over the years so that a dopa-responsive akinetic rigid syndrome may give place to an ataxic syndrome that no longer responds to therapy.

### G. Peripheral Nerve and Lower Motor Neuron Signs

An associated peripheral neuropathy is characterized by loss of distal sensation and ankle areflexia. Amyotrophy and fasciculations of the limb muscles may occur later in disease but profound atrophy and weakness are rare. Previously brisk deep tendon reflexes may become hypoactive. In late-adult-onset MJD, peripheral neuropathy may be a more dominant feature with generalized areflexia. Klockgether et al. (1999) found the sural sensory response and tibial motor response to be of low amplitude in MJD patients compared to controls. Patient age at the time of the study best correlated with the degree of abnormality, suggesting that the normal age related attenuation of these responses was accentuated in MJD. There was no correlation with the CAG repeat number. Durr et al. (1996) also found an axonal neuropathy by nerve conduction studies in 60% of MJD patients. Soong et al. (1998) noted loss of large myelinated fibers in sural nerve biopsies; large repeat sizes correlated with loss of fiber density.

### H. Cognitive Deficits

Severe dementia is not a feature of MJD, even in the late stages. In six patients with MJD, Maruff et al. (1996) found abnormalities of visual attention characterized by slow processing of complex visual information and an inability to shift attention. They proposed that this reflected frontal subcortical dysfunction.

### I. Sleep Disturbances

Sleep problems are relatively common in MJD. Schols et al. (1998) found that MJD patients had greater trouble falling asleep and more nocturnal awakening, with sleep impairment being more common in older patients and patients with greater brainstem involvement. Restless leg syndrome (RLS) occurred in over half the patients with MJD. Though clinically evident polyneuropathy was more common among those with RLS, electrophysiological evidence for a neuropathy was found equally among those with and without RLS. Central sleep apnea has also been documented in some patients (Kitamura et al., 1989). In occasional patients with intermediate size expansions of 53–54 repeats, RLS and axonal polyneuropathy may be the only clinical deficits (Van Alfen et al., 2001).

### J. Autonomic Deficits

Sphincter disturbances occur in close to a third of the patients (Durr et al., 1996; Matsumura et al., 1996a). Impotence and orthostatic hypotension has been occasionally seen including in a patient with a short CAG expansion of 56 (Takiyama et al., 1997). Other evidence of autonomic dysfunction has included the absence of fungiform papillae in the tongue (Uchiyama et al., 2001).

### K. Phenotypic Variability

There may be startling phenotypic variability among affected persons in the same family that cannot be accounted for by duration of disease alone (Subramony et al., 1996). Portuguese workers classified MJD into several types based on this phenotypic variability (Table 5.3; Sequiros and Coutinho, 1993). Type I disease has a younger age of onset (mean about 25 years), is characterized by prominent spasticity and rigidity, bradykinesia and minimal ataxia (Rosenberg et al., 1976), and accounts for about 10% of cases. Type II disease has an intermediate age of onset (mean about 38 years), and is characterized by progressive ataxia and upper motor neuron signs. It is the most common phenotype (Sequiros and Coutinho 1993; Matsumura et al., 1996a). Type III disease has a mean age of onset closer to 50 years and is characterized by ataxia together with significant peripheral nerve involvement resulting in amyotrophy and generalized areflexia. A Type IV phenotype with Parkinsonian features has also been described. Parkinsonian features are especially prominent in some families (Romanul et al., 1977; Cancel et al., 1995; Gwinn-Hardy, 2001).

## IV. DIAGNOSIS

### A. DNA

The mutation in all MJD patients is the same: expansion of a CAG repeat in the *MJD1* gene on chromosome 14 (Kawaguchi et al., 1994). Normal alleles range from 12 to approximately 42 repeats, whereas expanded alleles are nearly always at least 60 repeats in length, ranging up to ~84 in length (e.g., Cancel et al., 1995; Durr et al., 1996; Maciel et al., 1995, 2001). Very rarely, intermediate alleles have been reported (Van Alfen et al., 2001). Thus, except for these rare intermediate alleles there is a significant gap between repeats lengths of normal and expanded alleles. This means that MJD differs from several other polyglutamine diseases in which the ranges for normal and expanded alleles are contiguous or narrowly separated.

Genetic testing for MJD is highly sensitive and specific. Except for the very rare intermediate alleles, all repeat lengths fall unequivocally into either the normal or disease distribution. Likewise, there is not a significant "zone of reduced penetrance" in MJD. All expanded alleles of ~60 or greater will result in signs of disease in a normal life span, whereas repeats of less than 42 do not cause disease. (The rare intermediate alleles, as described above, may manifest only with RLS and axonal neuropathy.) The large jump in repeat size that a normal allele would need to make to expand into the disease range reduces the likelihood of *de novo* expansions. Indeed, sporadic MJD due to *de novo* expansions is thought to be rare, much less common than it is in Huntington disease or SCA2, for example.

Table 5.3  Clinical Subtypes of MJD/SCA3[a]

| Type | Onset (year) | Repeat length | Features | Other findings |
|------|--------------|---------------|----------|----------------|
| I    | <25          | >75           | Dystonia, rigidity and ataxia | Oculomotor disturbance |
| II   | 20–50        | >73           | Ataxia, pyramidal and bulbar signs | Peripheral signs (late) |
| III  | >40          | <72           | Ataxia, peripheral signs | Distal wasting, hyporeflexia |

[a]The indicated CAG repeat lengths represent approximate transition points at which the clinical subtype is increasingly likely to occur.

As in all CAG/polyglutamine diseases, the size of the *MJD1* repeat correlates well with disease severity. Larger repeats cause earlier disease onset and are associated with faster disease progression than are smaller disease alleles. As described above, MJD had been classified into three clinical subtypes long before the genetic defect was known. It is now clear that CAG repeat length is the major determinant of the different clinical subtypes.

### B. Imaging Studies

Brain MRI abnormalities in MJD patients are consistent with the highly variable clinico-pathological findings. There is no single radiological feature that is specifically or consistently observed in MJD (Schols *et al.*, 1997; Murata *et al.*, 1998; Onodera *et al.*, 1998; Abe *et al.*, 1998). The most common abnormality may be marked dilatation of the fourth ventricle. This probably reflects atrophy of the superior and middle cerebellar peduncles, the pons and the cerebellar vermis. Other frequently noted MRI abnormalities include atrophy of the globus pallidus and midbrain. Although the cerebral cortex is relatively spared in MJD, frontal and temporal atrophy have been described. The severity of MRI abnormalities correlates with CAG repeat length: patients with longer repeats (and thus earlier disease) are more likely to show progressive MRI abnormalities at an early age. However, this correlation is not perfect because the severity of MRI findings also correlates with the patient's age *independent* of the age of onset and repeat length. In other words, elderly patients with late-onset disease often show surprisingly pronounced atrophy on MRI (Onodera *et al.*, 1998).

In planimetric and volumetric studies comparing brain MRIs in SCA1, 2, and MJD, Burk *et al.* (1996) and Klockgether *et al.* (1998) have shown atrophy of the brainstem and the cerebellum in all three types, but most prominent in SCA2. Caudate and putaminal volumes were decreased only in MJD, perhaps reflecting the more prominent basal ganglia signs in MJD. Recent positron emission tomographic (PET) studies have shown hypometabolism not only in the cerebellum and brainstem of these patients but also, unexpectedly, in the occipital cortex (Soong *et al.*, 1997). SPECT imaging studies also have documented diffuse abnormalities including in the cerebral cortex (Etchebehere *et al.*, 2001). Studies with 99m Tc TRODAT-1 SPECT imaging have shown reduction in the dopamine transporter density in the striatum in keeping with the presence of extrapyramidal signs (Yen *et al.*, 2000, 2002).

### C. Evoked Potentials and Nerve Conduction Studies

Increased latencies of visual- and brainstem-evoked responses have been noted in MJD patients by Durr *et al.* (1996). Nerve conduction studies are usually consistent with an axonal polyneuropathy.

## V. NEUROPATHOLOGY

Though there is considerable variability in the neuropathology of MJD, some findings are consistent enough that neuropathological criteria for diagnosing MJD can be considered (Sequiros and Coutinho, 1993). The following structures are consistently involved: substantia nigra and subthalamic nuclei; red nuclei; medial longitudinal fasciculus; pontine nuclei and the middle cerebellar peduncles; dentate nuclei; Clarke's column and spinocerebellar tracts; vestibular nuclei; anterior horn cells and motor cranial nerve nuclei; and posterior root ganglia and posterior columns. The cerebral cortex, striatum, cerebellar cortex, olivary nuclei, and corticospinal tract are usually not affected. Koeppen *et al.* (1999) also noted the relative sparing of the cerebellar Purkinje cells but found evidence of abnormal synaptophysin immunocytochemistry in the dentate nucleus. The cerebellar signs of MJD appear to be related to pontine and dentate pathology, while the less common akinetic-rigid syndrome may reflect substantia nigra pathology.

Until recently there were no known cytopathological hallmarks of MJD. Now, however, it is clear that the disease protein in MJD and most other polyglutamine diseases forms neuronal intranuclear inclusions or NI (Paulson *et al.*, 1997a; Schmidt *et al.*, 1998). NI are spherical, ubiquitinated aggregates that are found in neurons of select brain regions (Fig. 5.1). In MJD, they are abundant in pontine neurons but have also been observed in substantia nigra and certain brainstem neuronal populations. Extranuclear ubiquitin staining has also been noted in MJD brain. In fact, skein-like ubiquitin-positive deposits in the cytoplasm of motor neurons were described before the discovery of NI (Suenaga *et al.*, 1993).

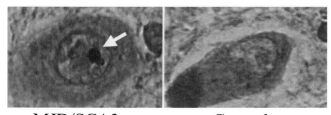

FIGURE 5.1 Nuclear inclusions in MJD/SCA3. Shown are pontine neurons from disease and control brain that have been immunohistochemically stained for ataxin-3. The arrow shows a typical nuclear inclusion in MJD/SCA3, a pathological hallmark of this and most other polyglutamine diseases.

## VI. MOLECULAR GENETICS

### A. CAG Repeat Expansion in the *MJD1* Gene

After Takiyama *et al.* (1993) localized the gene for MJD to chromosome 14q 24.3–q32, Kawaguchi *et al.* (1994) described the underlying mutation: an unstable CAG repeat expansion in the coding region of the *MJD1* gene. The gene contains eleven exons comprising a single open reading frame that encodes a novel protein, ataxin-3 (Ichikawa *et al.*, 2001). The CAG repeat is in the 10th exon, resulting in a glutamine repeat near the carboxy terminus of this ~42 kDa protein. The CAG repeat is highly polymorphic in normal individuals, varying from 12–43 in number (Mattila *et al.*, 1995; Ranum *et al.*, 1995; Sasaki *et al.*, 1995; Cancel *et al.*, 1995; Maciel *et al.*, 1995; Matsumura *et al.*, 1996a). Over 90% of normal alleles have fewer than 31 repeats, with most normal alleles distributed in peaks at 14, 21 to 24, and 27 repeats (Takiyama *et al.*, 1995; Limprasert *et al.*, 1996; Rubinsztein *et al.*, 1995). The nucleotide immediately following the CAG repeat is polymorphic, usually guanine in normal alleles. However, in normal alleles with repeat lengths of 20 or 21 and in over half the alleles with repeat lengths larger than 27, the guanine is replaced by cytosine (Limprasert *et al.*, 1996; Matsumura *et al.*, 1996b). This cytosine is also present in all expanded alleles and may play a role in the instability of the MJD1 repeat.

In individuals with MJD, the CAG repeat is expanded: disease-causing alleles range from 56–86 repeats and thus are easily distinguished from normal alleles which are no larger than 43 repeats. PCR-based genetic testing for MJD is highly sensitive and specific, firmly establishing the diagnosis in affected or at-risk individuals. The MJD mutation has now been documented in many families with dominant ataxia from diverse ethnic groups (Kawaguchi *et al.*, 1994; Mattila *et al.*, 1995; Ranum *et al.*, 1995; Maciel *et al.*, 1995; Schols *et al.*, 1996; Silviera *et al.*, 1996). In many regions of the world, it is the most common dominant ataxia.

### B. Somatic and Gametic Instability of the Expanded Repeat

Normal CAG repeat alleles in the *MJD1* gene are stably propagated in somatic tissues and upon transmission to the next generation (Limprasert *et al.*, 1996). In contrast, expanded repeats display somatic mosaicism and instability upon transmission (Maciel *et al.*, 1997). The somatic mosaicism is most notable in the cerebellar cortex, which typically has smaller expanded repeats compared to other areas of the CNS. This finding has also been observed in other CAG repeat diseases. Hashida *et al.* (1997) speculate that this may reflect smaller repeat sizes in the large number of cerebellar granule cells.

Intergenerational instability of the expanded *MJD1* repeat has been well documented (Takiyama *et al.*, 1995; Maciel *et al.*, 1995). Maciel *et al.* (1995) noted that 55% of 58 parent-child transmissions were unstable with over 75% of the changes causing further expansion. Though the frequency of contractions and expansions was similar in paternal and maternal transmissions, paternal transmission caused, on average, greater changes in repeat length. Sasaki *et al.* (1995) also noted slightly greater repeat instability with paternal transmission. In numerous studies, changes in repeat size from one generation to the next have varied from –8 to +9 repeats. In addition to this relatively modest paternal effect on instability, an intragenic CGG/GGG polymorphism just 3′ of the CAG repeat may contribute to intergenerational instability (Igarashi *et al.*, 1996).

The phenomenon of anticipation is well documented in MJD. With respect to age of onset, Durr *et al.* (1996) and Takiyama *et al.* (1994) noted a mean anticipation of 12 and 9 years, respectively. Anticipation is typically more prominent with paternal inheritance. Clearly the major factor underlying anticipation is a tendency for repeat sizes to become larger from one generation to the next. However, Takiyama *et al.* (1998) have reported significant anticipation upon maternal transmission with almost no change in repeat size, suggesting that other genetic or environmental factors contribute to anticipation. In addition, some observed anticipation may reflect ascertainment bias.

## VII. PHENOTYPE-GENOTYPE CORRELATION

There is an inverse correlation between age of onset and CAG repeat length in MJD, with the correlation coefficient varying from –0.67 to –0.92. For example, the largest MJD repeats, 83 and 86 reported by Zhou *et al.* (1997), occurred in children with ages of onset at 11 and 5 years, respectively. There is also a correlation between repeat length and the clinical phenotype (Table 5.3). Patients with type I disease typically have larger repeats than those with type II and III phenotypes. Consistent with this, a recent study showed that larger CAG repeats are associated with the presence of pyramidal signs and dystonia (Jardim, 2001). Sasaki *et al.* (1995) found that patients with type I, II, and III phenotypes had mean repeat sizes of 80, 76, and 73, respectively. Maciel's study (1995) noted a mean repeat size of 76 +/– 5 in type I patients. Schols *et al.* (1996) reported repeat sizes of 79 and 81 in 2 type I patients; but 4 patients with repeat size of over 80 were classified as type II, illustrating the imperfect correlation between phenotype and repeat size. Peripheral neuropathy is usually associated with a repeat size smaller that 73 (Schols *et al.*, 1996; Durr *et al.*, 1996). In their study, Durr *et al.* (1996) also found that many clinical signs, including amyotrophy, ophthalmoplegia, dysphagia, and sphincter difficulties, correlated more with disease duration than with CAG repeat size.

In a Yemenese family reported by Lerer et al. (1996), several homozygotes were reported to have an earlier onset and a more rapid progression than heterozygotes, similar to observations in Portugal (Rosenberg, 1984). The expansion size in this family was relatively small (66–72) and some heterozygotes remained asymptomatic well into their seventh decade. Some patients with small expansions and late adult onset may have a relatively "pure cerebellar" presentation.

### A. Other Genetic Issues

Stevanin et al. (1995) found several different haplotypes in patients from different geographic areas, suggesting multiple mutational origins for MJD worldwide. Gaspar et al. (1996) studied 64 unrelated patients, mostly from the Azores and Portugal, and nearly all shared a specific haplotype suggesting a founder effect in this population. However, Lima et al. (1998) identified two distinct haplotypes in Azorean families, suggesting more than one ancestral mutation in the Azores. Takiyama et al. (1995) found shared haplotypes between Japanese and Azorean patients with MJD, suggesting either a common founder or a haplotype that predisposes toward CAG repeat expansion. A recent worldwide study of many families with MJD showed that they shared two distinct intragenic polymorphisms suggesting a shared founder mutation (Gaspar, 2001).

Non-Mendelian contributions to the inheritance pattern in MJD have been suggested. Ikeuchi et al. (1996) noted that 73% of children from affected fathers inherited the disease allele, significantly in excess of the expected 50%. This pattern was not found among offspring of affected mothers. Among normal sized alleles, Rubinsztein et al. (1997) found that males transmitted larger and smaller alleles equally, but that females preferentially transmitted smaller alleles again suggesting meiotic drive. Further studies are needed to determine the significance of these findings.

## VIII. PATHOGENIC MECHANISMS AND MODELS

### A. Ataxin-3, the SCA3/MJD Disease Protein

The *MJD1* gene codes for a novel protein, ataxin-3, with a predicted molecular weight of 42 kDa (Fig. 5.2). The CAG repeat is translated into a polyglutamine tract of variable length near the carboxyl terminus. The *MJD1* gene is widely expressed in the CNS and elsewhere in the body. *In situ* hybridization shows relatively high expression in the cerebellum, hippocampus, substantia nigra, striatum, and cerebral cortex. Western blot and immunohistochemical studies indicate that the ataxin-3 protein is widely distributed in neuronal and non-neuronal tissues. In fact, ataxin-3 has

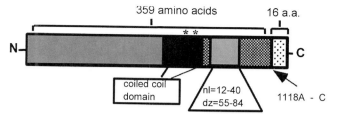

**FIGURE 5.2** The *MJD1* gene product, ataxin-3. Ataxin-3 is a small hydrophilic protein with glutamine repeat (Q) near the carboxyl terminus. The approximate ranges for normal and disease repeats are shown. The arrow indicates an intragenic polymorphism (1118 A-C) that alters the stop codon, extending the protein by 16 amino acids. Both an N-terminal conserved "josephin" domain and a predicted coiled-coil domain are shown. Two predicted ubiquitin interaction motifs are indicated by the asterisks. Splice variants for ataxin-3 are not shown.

been found in every mammalian tissue and cell line studied so far (Paulson et al., 1997a; Wang et al., 1997; Tait et al., 1998; Schmidt et al., 1998).

The rat ortholog of ataxin-3, which is otherwise nearly identical to human ataxin-3, contains a polyglutamine domain of only five glutamine residues interrupted by a histidine residue. This suggests that the glutamine repeat is not essential for normal ataxin-3 function. What the biological function of ataxin-3 is, however, remains unknown. The N terminus of the protein, termed the "josephin" domain, is conserved in orthologs from species as distant as *Arabidopsis*, *Drosophila*, and *Cenorhabditis*. Ataxin-3 is predicted to have a high degree of helical secondary structure, including a coiled-coil domain just upstream of the polyglutamine domain. Coiled-coil domains often mediate protein-protein interactions, but whether it does so in ataxin-3 is unknown. Protein motif analysis also suggests that ataxin-3 contains several putative ubiquitin interaction motifs flanking the glutamine repeat, indicating that ataxin-3 may be a ubiquitin-binding protein (Hofmann and Falquet, 2001).

The subcellular localization of ataxin-3 is complex. Some reports indicate predominantly cytoplasmic staining while others indicate nuclear staining (Paulson et al., 1997a; Wang et al., 1997; Tait et al., 1998; Schmidt et al., 1998; Trottier et al., 1998). Ataxin-3 is likely to be both a cytoplasmic and nuclear protein whose subcellular localization is regulated by one or more factors, including cell type. The most detailed study to date suggests multiple isoforms of ataxin-3 with heterogeneous patterns of subcellular localization (Trottier et al., 1998). In many cells a fraction of the cellular pool of ataxin-3 is intranuclear, bound to the nuclear matrix (Tait et al., 1998). This nuclear pool of ataxin-3 may be important to pathogenesis.

What controls the intracellular localization of ataxin-3? Although a potential nuclear localization signal (NLS) exists upstream of the polyglutamine domain (Tait et al.,

1998), this putative NLS is not sufficient to drive a fusion protein into the nucleus (H. Paulson, unpublished observations). Ataxin-3 may be small enough to enter the nucleus without assisted transport, or may be assisted in its transport by "piggybacking" on another protein that does undergo facilitated transport. Once in the nucleus, ataxin-3 binds the nuclear matrix and may assume a different conformation that makes the polyglutamine tract more accessible (Perez et al., 1999).

## B. Protein Misfolding and Aggregation as a Central Feature of Disease

The precise mechanism by which the CAG repeat expansion in *MJD1* causes selective neuronal degeneration is still uncertain. Growing evidence suggests that the pathogenic process occurs at the protein level. Dominantly inherited disease can, in principle, be caused by one of three mechanisms: partial deficiency of the protein (i.e., haploinsufficiency due to an inability of the mutant protein to perform its normal function), a dominant negative effect (interference by the mutant protein on the normal protein's function), or a "toxic gain of a function." Evidence is most consistent with the last of these. In particular, conformational abnormalities of the disease protein seem to be a central feature in MJD and other polyglutamine disorders. Recent studies of recombinant ataxin-3 support this view: mutant ataxin-3 adopts increased beta sheet secondary structure, directly demonstrating altered shape of the protein (Bevivino and Loll, 2001).

The subcellular distribution of ataxin-3 is altered in disease brain tissue. In brain regions known to be targets of MJD, surviving neurons develop intranuclear inclusions or NI that contain ataxin-3, ubiquitin, and presumably many other proteins (Paulson et al., 1997b). In MJD, these inclusions have been observed exclusively in neurons, most commonly in pontine neurons but also in the substantia nigra, globus pallidus, and inferior olive neurons. The presence of NI in disease brain argues that mutant ataxin-3 is prone to aggregate inside the nuclei of neurons.

## C. Animal Models

Aggregation by mutant ataxin-3 and its potential role in pathogenesis have been studied in transgenic animals and cellular models. Ikeda et al. (1996) expressed normal or mutant forms of ataxin-3 in transgenic mice using a Purkinje cell-specific promoter. Transgenic mice expressing the full-length mutant protein did not develop an ataxic phenotype. In contrast, mice expressing a truncated fragment of mutant ataxin-3 developed progressive ataxia with Purkinje cell degeneration. Truncated ataxin-3 has also been used to model polyglutamine neurodegeneration in the fruitfly, *Drosophila* (Warrick et al., 1998). This fly model recapitulates many important features of polyglutamine diseases, including inclusion formation and neurodegeneration. Aggregation and cytotoxicity of ataxin-3 have also been successfully modeled in transfected cells expressing truncated or full length forms of ataxin-3 (Ikeda et al., 1996; Paulson et al., 1997b; Yoshizawa et al. 2000); Chai et al., JBC 2001; Evert et al., 1999). Taken together, these studies have suggested that truncation of ataxin-3 accelerates aggregation and cytotoxocity, but do not prove that such a proteolytic event is critical to pathogenesis. Indeed, several labs have shown that full length ataxin-3 can form inclusions, though less efficiently than truncated ataxin-3 (Chai et al., JBC 2001; Evert et al., 1999). At least one study (Yoshizawa et al., 2000) showed that mutant ataxin-3 is more toxic to mitotically arrested cells, which may help explain the selective neurotoxicity of mutant polyglutamine proteins.

Whether nuclear inclusions are essential for pathogenesis in MJD remains unsettled. Interestingly, we have not observed nuclear inclusions in the cerebellar Purkinje cells of many MJD brains despite the fact that quantitative studies documented Purkinje cell loss in these same brains (Bebin et al., 1990; S. Subramony and P. Vig, unpublished obervations). Studies of other polyglutamine diseases further suggest that inclusions are not required for neuronal cell death. NI may instead represent an adaptive cellular response to the presence of abnormally folded, undegradable polyglutamine protein. Results in the *Drosophila* model of MJD shed light on the issue. In the fly, overexpressing specific heat shock protein chaperones markedly suppresses polyglutamine toxicity without noticeably changing the frequency or size of inclusions (Warrick et al., 1999). However, suppression of toxicity *is* accompanied by a marked increase in the solubility of mutant protein, arguing that misfolding and aggregation of disease protein is linked to pathogenesis in this model (Chan et al., 2000).

## D. Possible Mechanisms of Neurotoxicity

Whatever the pathogenic role of NI may be, their protein composition in MJD and other polyQ diseases continues to be a focus of study. Why? Because knowing this information may provide clues to interacting proteins and thereby to early steps in pathogenesis. These efforts have shown that the polyglutamine domain is important in mediating the recruitment of several other proteins, including *normal* ataxin-3 (Perez et al. 1998). Aggregates of the mutant protein can secondarily recruit full-length ataxin-3 as long as the glutamine tract of the latter has not been deleted. In studies employing green fluorescent protein (GFP) fusion proteins, simply adding nonpathogenic stretches of 19 or 35 glutamine repeats was sufficient to cause GFP to be recruited into NI (Perez et al., 1998). Consistent with this finding, endogenous proteins containing glutamine stretches often co-localize to NIs; examples include the eye absent protein (EYA) in the *Drosophila* model, and TATA-binding protein in human tissue (Perez *et al.*, 1998). Taken together, these data suggest that polyglutamine-polyglutamine interactions play an important role in the recruitment of various cellular proteins to inclusions. This is consistent with the polar-zipper hypothesis proposed by Perutz *et al.* (1994).

The ability of Hsp chaperones to suppress polyglutamine toxicity is just one piece of mounting evidence that the cell's "protein surveillance" machinery constitutes a first-line defense against mutant polyglutamine protein. A second major component of this cellular machinery, the ubiquitin-proteasome degradation pathway, has also been implicated in disease. The proteasome is a large, multi-subunit protease complex that degrades most cytosolic proteins in a ubiquitin-dependent manner. Ubiquitin and proteasomal components localize to NIs in SCA3/MJD disease brain, transgenic mice, and cellular models (Chai *et al.*, 1999). Moreover, inhibition of the proteasome increases polyglutamine aggregation in cellular models. Polyglutamine aggregates, in turn, inhibit proteasome activity (Bence *et al.*, 2001). This suggests a vicious cycle in which mutant polyglutamine proteins compromise the very cellular defenses that normally are marshaled against them. Some studies have shown decreased proteasome function in the aging brain, arguing that the age-dependent nature of polyglutamine disease may in part reflect age-related impairment of neuronal protein surveillance. It seems likely that as therapeutic compounds are discovered to suppress polyglutamine toxicity, some will act by modulating neuronal protein surveillance (e.g., facilitating the refolding, disaggregation, or degradation of the disease protein).

Transcriptional dysregulation as a pathogenic player is currently a hot topic in the polyglutamine disease field (see Steffan *et al.*, 2001). However, evidence for its involvement in MJD is still slim. The presence of misfolded, mutant ataxin-3 in the nucleus may alter gene transcription, perhaps by binding transcriptional factors like CREB-binding protein as has been shown in cellular models of Huntington disease (Nucifora *et al.*, 2001). Normal ataxin-3 has been shown to decrease CREB-dependent gene expression in transfected cells and mutant ataxin-3 can recruit essentially the entire cellular pool of CBP into inclusions (Chai *et al.*, 2001). Whether this occurs in humans with MJD, however, is currently unknown. Evert and colleagues have shown upregulation of cytokine genes in a cell line inducibly expressing mutant ataxin-3 (Evert *et al.*, 2001), but the molecular basis of these changes is also unknown. Research over the next few years will likely determine whether transcriptional dysregulation is important in MJD and whether it will represent a promising molecular target for the development of therapeutics.

## IX. TREATMENT

Presently, there is no preventive medicine for MJD or any other polyglutamine disease. Research is advancing at a remarkable rate, however, and several candidate compounds or genes have already been identified that suppress polyglutamine-mediated toxicity in cellular or animal models (e.g., Warrick et el., 1999). Ongoing, high-throughput drug screens in cell-based and invertebrate animal models should identify compounds that serve as the basis for preventive therapies in the future. The hope of every researcher and clinician is that an effective therapy for *any* polyglutamine disease will prove to be effective for *all* polyglutamine diseases.

However, despite this recent progress, the pipeline from lead compound to established preventive therapy for humans is slow. Fortunately, of all the SCAs MJD may be the most amenable to symptomatic therapy. Many patients with MJD have extrapyramidal signs, including parkinsonism, that respond to dopaminergic agents such as levodopa or dopamine agonists (Tuite *et al.*, 1995). Moreover, the RLS experienced by many MJD patients may respond to dopaminergic agents. Thus, any MJD patient with bradykinesia or dystonia deserves a trial of dopaminergic agents. Though earlier trials suggested that trimethoprim-sulfamethoxazole may benefit patients (Sakai *et al.*, 1996), a recent trial failed to confirm a benefit in MJD (Schulte *et al.*, 2001). Buspirone, a serotonin agonist, has also been described to provide modest benefit for the ataxic gait in hereditary ataxia, including in MJD (Friedman, 1997).

When it comes to the symptomatic treatment of MJD, it is difficult to practice evidence-based medicine because of the absence of well-controlled trials (in part due to the relative infrequency of the disease). But trials of medications like buspirone and amantadine in individual patients are certainly appropriate provided that the medication is stopped if no benefit is observed in a few weeks. Likewise, antioxidants such as vitamin E may be appropriate, provided the patient recognizes both that there is no definitive data showing a benefit in disease and that vitamins can have untoward effects when taken at higher than recommended doses. Most important is the recognition that supportive, nerurorehabilitative therapy is an essential part of the care of persons with degenerative ataxia like MJD (as reviewed by Perlman, 2000).

## References

Abe, Y., Tanaka, F., Matsumoto, M., Doyu, M., Hirayama, M., Kachi, T., and Sobue, G. (1998). CAG repeat number correlates with the rate of brainstem and cerebellar atrophy in Machado-Joseph disease. *Neurology* **51**, 882–884.

Bebin, E. M., Bebin, J., Currier, R. D., Smith, E. E., and Perry, T. L. (1990). Morphometric studies in dominant olivopontocerebellar atrophy. Comparison of cell losses with amini acid decreases. *Arch. Neurol.* **47**, 188–192.

Bence, N. F., Sampat, R. M., and Kopito, R. R. (2001). Impairment of the ubiquitin-proteasome system by protein aggregation. *Science* **292**, 1552–1555.

Bevivino, A. E., and Loll, P. J. (2001). An expanded glutamine repeat destabilizes native ataxin-3 structure and mediates formation of parallel beta-fibrils. *Proc. Natl. Acad. Sci. U.S.A.* **98**, 11955–11960.

Burk, K., Abele, M., Fetter, M., *et al.* (1996). Autosomal dominant cerebellar ataxia type I: clinical features and MRI in families with SCA 1, SCA 2 and SCA 3. *Brain* **119**, 1497–1505.

Buttner, N., Geschwind, D., Jen, J. C., Perlman, S., Pulst, S. M., and Baloh, R. W. (1998). Oculomotor phenotypes in autosomal dominant ataxias. *Arch. Neurol.* **55**, 1353–1357.

Cancel, G., Abbas, N., Stevanin, G., Durr, A., Chneiweiss, H., Neri, C., Duyckaerts, C., Penet, C., Cann, H. M., Agid, Y., and Brice, A. (1995). Marked phenotypic heterogeneity associated with expansion of a CAG repeat sequence at the spinocerebellar ataxia 3/Machado-Joseph disease locus. *Am. J. Hum. Genet.* **57**, 809–816.

Chai, Y., Koppenhafer, S. L., Shoesmith, S. J., Perez, M. K., and Paulson, H. L. (1999a). Evidence of proteasome involvement in polyglutamine disease: localization to nuclear inclusions in SCA3/MJD and suppression of polyglutamine aggregation in vitro. *Hum. Mol. Genet.* **8**, 673–682.

Chai, Y., Koppenhafer, S., Bonini, N., *et al.* (1999b). Analysis of the role of heat shock protein chaperones in polyglutamine disease. *J. Neurosci.* **19**, 10338–10347.

Chai, Y., Wu, L., Griffin, J., and Paulson, H. (2001). The role of protein context in specifying nuclear inclusion formation in polyglutamine disease. *J. Biol. Chem.* **276**, 44889–44897.

Chan, H. Y. E., Warrick, J. M., Gray-Board, G. L., Koppenhafer, S. L., Paulson, H. L., and Bonini, N. M. (2000). Mechanisms of chaperone suppression of polyglutamine disease: selectivity, synergy, and modulation of protein solubility in Drosophila. *Hum. Mol. Genet.* **9**, 2811–2820.

Coutinho, P., and Andrade, C. (1978). Autosomal dominant system degeneration in Portuguese families of the Azores Islands. *Neurology* **28**, 703–709.

Dawson, D. M. (1977). Ataxia in families from the Azores. *N. Eng. J. Med.* **296**, 1529–1530.

Durr, A., Stevanin, G., Cancel, G., Duyckaerts, C., Abbas, N., Didierjean, O., Chneiweiss, H., Benomar, A., Lyon-Caen, O., Julien, J., Serdaru, M., Penet, C., Agid, Y., and Brice, A. (1996). Spinocerebellar ataxia 3 and Machado-Joseph disease: clinical, molecular and neuropathological features. *Ann. Neurol.* **39**, 490–499.

Etchebehere, E., Fernando, C., Lopes-Cendes, I., *et al.* (2001). Brain single-photon emission computed tomography and magnetic resonance imaging in Machado-Joseph disease. *Arch. Neurology.* **58**, 1257–1263.

Evert, B. O., Wullner, U., Schulz, J. B., *et al.* (1999). High level expression of expanded full-length ataxin-3 in vitro causes cell death and formation of intranuclear inclusions in neuronal cells. *Hum. Mol. Genet.* **8**, 1169–1176.

Evert, B. O., Vogt, I. R., Kindermann, C., Ozimek, L., de Vos, R. A., Brunt, E. R., Schmitt, I., Klockgether, T., and Wullner, U. (2001). Inflammatory genes are upregulated in expanded ataxin-3-expressing cell lines and spinocerebellar ataxia type 3 brains. *J. Neurosci.* **21**, 5389–5396.

Ferguson, F. R., and Critchley, M. (1929). A clinical study of an heredofamilial disease resembling disseminated sclerosis. *Brain* **52**, 203–225.

Filla, A., De Michele, G., Campanella, G., Perretti, A., Santoro, L., Serlenga, L., Ragno, M., Calabrese, O., Castaldo, I., De Joanna, G., and Cocozza, S. (1996). Autosomal dominant cerebellar ataxia type I. Clinical and molecular study in 36 Italian families including a comparison between SCA1 and SCA2 phenotypes. *J. Neurol. Sci.* **142**, 140–147.

Friedman, J. H. (1997). Machado-Joseph disease/spinocerebellar ataxia 3 responsive to buspirone. *Movement Disorder* **12**, 613–614.

Gaspar, C., Lopes-Cendes, I., DeStefano, A. L., Maciel, P., Silveira, I., Coutinho, P., MacLeod, P., Sequeiros, J., Farrer, L. A., and Rouleau, G. A. (1996). Linkage disequilibrium analysis in Machado-Joseph disease patients of different ethnic origins. *Hum. Genet.* **98**, 620–624.

Gaspar, C., Jannatipour, M., Dion, P., Laganiere, J., Sequeiros, J., Brais, B., and Rouleau, G. A. (2000). CAG tract of MJD-1 may be prone to frameshifts causing polyalanine accumulation. *Hum. Mol. Genet.* **9**, 1957–1966.

Gaspar, C., Lopes-Cendes, I., Hayes, S., *et al.* (2001). Ancestral origins of the Machado-Joseph disease mutation: A worldwide haplotype study. *Am. J. Hum. Genet.* **68**, 523–528.

Giunti, P., Sweeney, M. G., and Harding, A. E. (1995). Detection of the Machado-Joseph disease/spinocerebellar ataxia three trinucleotide repeat expansion in families with autosomal dominant motor disorders, including the Drew family of Walworth. *Brain* **118**, 1077–1085

Gwinn-Hardy, K., Singleton, A., O'Suilleabhain, P., *et al.* (2001). Spinocerebellar ataxia type 3 phenotypically resembling parkinson disease in a black family. *Arch. Neurol.* **58**, 296–299.

Harding, A. E. (1982). The clinical features and classification of the late onset dominant cerebellar ataxias. *Brain* **105**, 1–28.

Hashida, H., Goto, J., Kurisaki, H., Mizusawa, H., and Kanazawa, I. (1997). Brain regional differences in the expansion of a CAG repeat in the Spinocerebellar ataxias: dentatorubral-pallidoluysian atrophy, Machado-Joseph disease, and spinocerebellar ataxia type I. *Ann. Neurol.* **41**, 505–511.

Healton, E. B., Brust, J. C., Kerr, D. L., Resor, S., and Penn, A. (1980). Presumable Azorean disease in a presumably non-Portuguese family. *Neurology* **30**, 1084–1089.

Higgins, J. J., Nee, L. E., Vasconcelos, O., Ide S. E., Lavedan, C., Goldfarb, L. G., Polyneropoulos, M. H. (1996). Mutations in American families with spinocerebellar ataxia (SCA) type 3: SCA 3 is allelic to Machado-Joseph disease. *Neurology* **46**, 4–8.

Hofmann, K., and Falquet, L. (2001). A ubiquitin-interacting motif conserved in components of the proteasomal and lysosomal protein degradation systems. *Trends. Biochem. Sci.* **6**, 347–350.

Ichikawa, Y., Goto, J., Hattori, M., Toyoda, A., Ishii, K., Jeong, S. Y., Hashida, H., Masuda, N., Ogata, K., Kasai, F., Hirai, M., Maciel, P., Rouleau, G. A., Sakaki, Y., and Kanazawa, I. (2001). The genomic structure and expression of MJD, the Machado-Joseph disease gene. *J. Hum. Genet.* **46**, 413–422.

Igarashi, S., Takiyama, Y., Cancel, G., Rogaeva, E. A., Sasaki, H., Wakisaka, A., Zhou, X-Y., Takano, H., *et al.* (1996). Intergenerational instability of the CAG repeat of the gene for Machado-Joseph disease (MJD1) is affected by the genotype of the normal chromosome: implications for the molecular mechanisms of the instability of the CAG repeat. *Hum. Mol. Genet.* **5**, 923–932.

Ikeda, H., Yamaguchi, M., Sugai, S., Aze, Y., Narumiya, S., and Kakizuka, A. (1996). Expanded polyglutamine in the Machado-Joseph disease protein induces cell death in vitro and in vivo. *Nat. Genet.* **13**, 196–202.

Ikeuchi, T., Igarashi, S., Takiyama, Y., Onodera, O., Oyake, M., Takano, H., Koide, R., Tanaka, H., and Tsuji, S. (1996). Non-mendelian transmission in dentatorubral-pallidoluysian atrophy and Machado-Joseph disease: the mutant allele is preferentially transmitted in male meiosis. *Am. J. Hum. Genet.* **57**, 730–733.

Inoue, K., Hanihara, T., Yamada, Y., Kosaka, K., Katsuragi, K., and Iwabuchi, K. (1996). Clinical and genetic evaluation of Japanese autosomal dominant cerebellar ataxias: is Machado-Joseph disease common in Japan? *J. Neurol. Neurosurg. Psychiatry* **60**, 697–698.

Jardim, L., Pereira, M., Silveira, I., *et al.* (2001). Neurologic findings in Machado-Joseph disease. *Arch. Neurol.* **58**, 899–904.

Kawaguchi, Y., Okamoto, T., Taniwaki, M., *et al.* (1994). CAG expansions in a novel gene for Machado-Joseph disease at chromosome 14q32.1. *Nat. Genet.* **8**, 221–228.

Kitamura, J., Kuvuki, Y., Tsuruta, K., Kurihara, T., and Matsukura, S. (1989). A new family with Joseph disease in Japan. Homovanillic acid, magnetic resonance and sleep apnea studies. *Arch. Neurol.* **46**, 425–428.

Klockgether, T., Schols, L., Abele, M., Burk, K., Topka, H., Andres, F., Amoiridis, G., Ludtke, R., Riess, L., Laccone, F., and Dichgans, J. (1996). Age related axonal neuropathy in spinocerebellar ataxia type 3/Machado-Joseph disease (SCA3/MJD). *J. Neurol. Neurosurg. Psychiatry* **66**, 222–224.

Klockgether, T., Skalej, M., Wedekind, D., Luft, A. R., Welte, D., Schulz, J. B., Abele, M., Burk, K., Laccone, F., Brice, A., and Dichgans, J. (1998). Autosomal dominant cerebellar ataxia type I. MRI based volumetry of posterir fossa structures and basal ganglia in spinocerebellar ataxia type 1, 2 and 3. *Brain* **121**, 1687–1693.

Klockgether, T., Ludtke, R., Kramer, B., *et al*. (1998). The natural history of degenerative ataxia: a retrospective study in 466 patients. *Brain* **121**, 589–600.

Koeppen, A. H., Dickson, A. C., Lamarche, J. B., *et al*. (1999). Synapses on the hereditary ataxias. *J. Neuropathol. Exp. Neurol.* **58**, 748–764.

Lerer, I., Merims, D., Abeliovich, D., Zlotogora, J., and Gadoth, N. (1996). Machado-Joseph disease: Correlation between clinical features, the CAG repeat length and homozygosity for the mutation. *Eur. J. Hum. Genet.* **4**, 3–7.

Lima, M., Mayer, F. M., Coutinho, P., and Abade, A. (1998). Origins of a mutation: population genetics of Machado-Joseph disease in the Azores (Portugal). *Hum. Biol.* **70**, 1011–1023.

Limprasert, P., Nouri, N., Heyman, R. A., Nopparatana, C., Kamonsilp, M., Deininger, P. L., and Keats, B. J. B. (1996). Analysis of CAG repeat of the Machado-Joseph gene in human, chimpanzee and monkey populations: a variant nucleotide is associated with the number of CAG repeats. *Hum. Mol. Genet.* **5**, 207–213.

Maciel, P., Gaspar, C., DeStefano, A. L., Silveira, I., Coutinho, P., Radvany, J., Dawson, D. M., Sudarsky, L., Guimaraes, J., Loureiro, J. E. L., Nezarati, M. M., Corwin, L. I., Lopes-Cendes, I., Rooke, K., Rosenberg, R., MacLeod, P., Farrer, L. A., Sequeiros, J., and Rouleau, G. A. (1995). Correlation between CAG repeat length and clinical features in Machado-Joseph disease. *Am. J. Hum. Genet.* **57**, 54–61.

Maciel, P., Lopes-Cendes, I., Kish, S., Sequeiros, J., and Rouleau, G. A. (1997). Mosaicism of the CAG repeat in CNS tissue in relation to age at death in spinocerebellar ataxia type 1 and Machado-Joseph disease patients. *Am. J. Hum. Genet.* **60**, 993–996.

Maciel, P., Costa, M. C., Ferro, A., Rousseau, M., Santos, C. S., Gaspar, C., Barros, J., Rouleau, G. A., Coutinho, P., and Sequeiros, J. (2001). Improvement in the molecular diagnosis of Machado-Joseph disease. *Arch. Neurol.* **58**, 1821–1827.

Maruff, P., Tyler, P., Burt, T., Currie, B., Burns, C., and Currie, J. (1996). Cognitive deficits in Machado-Jospeh disease. *Ann. Neurol.* **40**, 421–427.

Matilla, T., McCall, A., Subramony, S. H. *et al*. (1995). Molecular clinical correlations in spinocerebellar ataxia type 3 and Machado-Joseph disease. *Ann. Neurol.* **38**, 68–72.

Matsumura, R., Takayanagi, T., Fujimoto, Y., Murata, K., Mano, Y., Horikawa, H., and Chuma, T. (1996a). The relationship between trinucleotide repeat length and phenotypic variation in Machado-Joseph disease. *J. Neurol. Sci.* **139**, 52–57.

Matsumura, R., Takayanagi, T., Murata, K., Futamura, N., Hirano, M., and Ueno S. (1996b). Relationship between (CAG)n C configuration to repeat instability of the Machado-Joseph disease gene. *Hum. Genet.* **98**, 643–645.

Moseley, M. L., Benzow, K. A., Schut, L. J., Bird, T. D., Gomez, C. M., Barkhaus, P. E., Blindauer, K. A., Labuda, M., Pandolfo, M., Koob, M. D., and Ranum, L. P. (1998). Incidence of dominant spinocerebellar and Friedreich triplet repeats among 361 ataxia families. *Neurology* **51**, 1666–1671.

Murata, Y., Yamaguchi, S., Kawakami, H. *et al*. (1998). Characteristic magnetic resonance imaging findings in Machado-Joseph disease. *Arch. Neurol.* **55**, 33–37.

Nakano, K. K., Dawson, D. M., and Spence, A. (1972). Machado disease: a hereditary ataxia in Portuguese emigrants to Massachusetts. *Neurology* **22**, 49–55.

Nakaoka, U., Suzuki, Y., Kawanami, T. *et al*. (1999). Regional differences in genetic subgroup frequency in hereditary cerebellar ataxia, and a morphometrical study of brain MR images in SCA1, MJD and SCA6. *J. Neurol. Sci.* **164**, 187–194.

Nihiyama, K., Murayama, S., Goto, J., Watanabe, M., Hashida, H., Katayama, S., Numura, Y., Nakamura, S., and Kanazawa, I. (1996). Regional and cellular expression of the Machado-Joseph disease gene in brains of normal and affected individuals. *Ann. Neurol.* **40**, 776–781.

Nucifora, F. C. Jr., Sasaki, M., Peters, M. F., Huang, H., Cooper, J. K., Yamada, M., Takahashi, H., Tsuji, S., Troncoso, J., Dawson, V. L., Dawson, T. M., Ross, C. A. (2001). Interference by huntingtin and atrophin-1 with cbp-mediated transcription leading to cellular toxicity. *Science* **291**, 2423–2428.

Ohyagi, Y., Yamada, T., Okayama, A. *et al*. (2000). Vergence disorders in patients with spinocerebellar ataxia 3/Machado-Joseph disease: a synoptophore study. *J. Neurol. Sci.* **173**, 120–123.

Onodera, O., Idezuka, J., Igarashi, S. *et al*. (1998). Progressive atrophy of cerebellum and brainstem as a function of age and size of the expanded CAG repeats in the MJD1 gene in Machado-Joseph disease. *Ann. Neurol.* **43**, 1–9.

Paulson, H. L., Das, S. S., Crino, P. B., Perez, M. K., Patel, S. C., Gotsdiner, D., Fischbeck, K. H., and Pittman R. N. (1997a). Machado-Joseph disease gene product is a cytoplasmic protein widely expressed in brain. *Ann. Neurol.* **41**, 453–462.

Paulson, H. L., Perez, M. K., Trottier, Y., Trojanowski, J. Q., Subramony, S. H., Das, S. S., Vig, P., Mandel, J. L., Fischbeck, K. H., and Pittman, R. N. (1997b). Intranuclear inclusions of expanded polyglutamine protein in spinocerebellar ataxia type 3. *Neuron* **19**, 333–344.

Perez, K. P., Paulson, H. L., and Pittman, R. N. (1999). Ataxin-3 with an altered conformation that exposes the polyglutamine domain is associated with the nuclear matrix. *Hum. Mol. Genet.* **8**, 2377–2385.

Perez, M. K., Paulson, H. L., Pendse, S. J., Saionz, S. J., Bonini, N. M., and Pittman, R. N. (1998). Recruitment and the role of nuclear localization in polyglutamine-mediated aggregation. *J. Cell Biol.* **143**, 1457–1470.

Perlman, S. L. (2000). Cerebellar ataxia. *Curr. Treat. Options Neurol.* **2**, 215–224.

Perutz, M. F., Johnson, T., Suzuki, M., and Finch, J. T. (1994). Glutamine repeats as polar zippers: their possible role in inherited neurological diseases. *Proc. Natl. Acad. Sci. U.S.A.* **91**, 5355–5358.

Pujana, M. A., Corral, J., Gratacos, M. *et al*. (1999). Spinocerebellar ataxia in Spanish patients: genetic analysis of familial and sporadic cases. *Hum. Genet.* **104**, 516–522.

Ramesar, R. S., Bardlen, S., Beighton, P., and Bryer, A. (1997). Expanded CAG repeat in spinocerebellar ataxia (SCA 1) seggregates with distinct haplotypes in South African families. *Hum. Genet.* **100**, 131–137.

Ranum, L. P. W., Lundgren, J. K., Schut, L. J. *et al*. (1995). Spinocerebellar ataxia type 1 and Machado-Joseph disease: incidence of CAG expansions among adult-onset ataxia patients from 311 families with dominant, recessive or sporadic ataxia. *Am. J. Hum. Genet.* **57**, 603–608.

Rivaud-Pechoux, S., Durr, A., Gaymard, B., Cancel, G., Ploner, C. J., Agid, Y., Brice, A., and Pieroot-Deseilligny, C. (1998). Eye movement abnormalities correlate with genotype in autosomal dominant cerebellar ataxia type I. *Ann. Neurol.* **43**, 297–302.

Romanul, F. C. A., Fowler, H. L., Radvany, J. *et al*. (1977). Azorean disease of the nervous system. *N. Engl. J. Med.* **296**, 1505–1508.

Rosenberg, R. N., Nyhan, W. L., Bau, C., Shore, P. (1976). Autosomal dominant striatonigral degeneration. *Neurology* **26**, 703–714.

Rosenberg, R. N. (1984). Joseph Disease: an autosomal dominant motor system degeneration. *Adv. Neurol.* **41**, 179–194.

Rubinsztein, C., Leggo, J., Coetzee, G. A., Irvine, R. A., Buckley, M., and Ferguson-Smith, A. (1995). Sequence variation and size ranges of CAG repeats in the Machado-Joseph disease, spinocerebellar ataxia type 1 and androgen receptor genes. *Hum. Mol. Genet.* **4**, 1585–1590.

Rubinsztein, D. C., and Leggo, J. (1997). Non-mendelian transmission at the Machado-Joseph disease locus in normal females: preferential transmission of alleles with smaller CAG repeats. *J. Med. Genet.* **34**, 234–236.

Sakai, T., Ohta, M., and Ishino, H. (1983). Joseph disease in a non-Portuguese family. *Neurology* **33**, 74–80.

Sakai, T., Antoku, Y., Matsuishi, T., Iwashita, H. (1996). Tetrahydrobiopterin double-blind, crossover trial in Machado-Joseph disease. *J. Neurol. Sci.* **136**, 71–72.

Saleem, Q., Choudhry, S., Mukerji, M. *et al.* (2000). Molecular analysis if autosomal dominant hereditary ataxias in the Indian population: high frequency of SCA2 and evidence for a common founder mutation. *Hum. Genet.* **106**, 179–187.

Sasaki, H., Wakisaka, A., Fukazawa, T., Iwabuchi, K., Hamada, T., Takada, A., Mukai, E., Matsuura, T., Yoshiki, T., and Tashiro, K. (1995). CAG repeat expansion of Machado-Joseph disease in the Japanese: analysis of the repeat instability for parental transmission, and correlation with disease phenotype. *J. Neurol. Sci.* **133**, 128–133.

Schmidt, T., Landwehrmeyer, B., Schmitt, I. *et al.* (1998). An isoform of ataxin-3 accumulates in the nucleus of neuronal cells in affected brain regions of SCA 3 patients. *Brain Pathol.* **8**, 669–679.

Schols, L., Amoiridis, G., Epplen, J. T. *et al.* (1996). Relations between genotype and pheotype in German patients with the Machado-Joseph disease mutation. *J. Neurol. Neurosurg. Psychiatry* **61**, 466–470.

Schols, L. (1995). Machado-Joseph disease mutation as the genetic basis of most spinocerebellar ataxias in Germany. *J. Neurol. Neurosurg. Psychiatry* **59**, 49–50.

Schols, L., Amoirides, G., Buttner, T., Przuntek, H., Epplen, J., and Riess, O. (1997). Autosomal dominant cerebellar ataxia: phenotypic differences in genetically determined subtypes? *Ann. Neurol.* **42**, 924–932.

Schols, L., Haan, J., Riess, O., Amoiridis, G., and Przuntek, H. (1998). Sleep disturbance in spinocerebellar araxias. Is the SCA3 mutation a cause of restless legs syndrome? *Neurology* **51**, 1603–1607.

Schulte, T., Mattern, R., Berger, K., Szymanski, S., Klotz, P., Kraus, P. H., Przuntek, H., and Schols, L. (2001). Double-blind crossover trial of trimethoprim-sulfamethoxazole in spinocerebral ataxia type 3/Machado-Joseph disease. *Arch. Neurol.* **58**, 1451–1457.

Sequeiros, J., and Coutinho, P. (1993). Epidemiology and clinical aspects of Machado-Joseph disease. *Adv. Neurol.* **61**, 139–153.

Sequiros, J., Silviera, I., Maciel, P., Coutinho, P., Manaia, A., Gaspar, C., Burlet, P., Loureiro, L., Guimaraes, J., Tanaka, H., Takiyama, Y., Sakamoto, H., Nishizawa, M., Nomura, Y., Segawa, M., Tsuji, S., Melki, J., and Munnich, A. (1994). Genetic linkage studies of Machado-Joseph disease with chromosome 14q STRP's in 16 Portuguese-Azorean kindreds. *Genomics* **21**, 645–648.

Silveira, I., Lopes-Cendes, I., Kish S. *et al.* (1996). Frequency of spinocerebellar ataxia type I, dentatorubrual-pallidoluysian atrophy and Machado-Joseph disease mutations in a large group of spinocerebellar ataxia patients. *Neurology* **46**, 214–218.

Soong, B.-W., and Lin, K.-P. (1998). An electrophysiologic and pathologic study of peripheral nerves in individuals with Machado-Joseph disease. *Chin. Med.* **61**, 181–187.

Soong, B.-W., Cheng, C.-H., Liu, R.-S., and Shan, D.-E. (1997). Machado-Joseph disease: Clinical, molecular, and metabolic characterization in Chinese kindreds. *Ann. Neurol.* **41**, 446–452.

Soong, B., Lu ,Y., Choo, K. *et al.* (2001). Frequency analysis of autosomal dominant cerebellar ataxias in Taiwanese patients and clinical and molecular characterization of spinocerebellar ataxia type 6. *Arch. Neurol.* **58**, 1105–1109.

Steffan, J. S., Bodai, L., Pallos, J., Poelman, M., McCampbell, A., Apostol, B. L., Kazantsev, A., Schmidt, E., Zhu, Y. Z., Greenwald, M., Kurokawa, R., Housman, D. E., Jackson, G. R., Marsh, J. L., and Thompson, L. M. (2001). Histone deacetylase inhibitors arrest polyglutamine-dependent neurodegeneration in Drosophila. *Nature* **413**, 739–743.

Stevanin, G., Cancel, G., Durr, A. *et al.* (1995). The gene for spinal cerebellar ataxia 3 (SCA3) is located in a region of 3 cM on chromosome 14q24.3-q32.2. *Am. J. Hum. Genet.* **56**, 193–201.

St. George-Hyslop, P., Rogaeva, E., Huterer, J. *et al.* (1994). Machado-Joseph disease in pedigrees of Azorean descent is linked to chromosome 14. *Am. J. Hum. Genet.* **55**, 120–125.

Storey, E., du Sart, D., Shaw, J. *et al.* (2000). Frequency of Spinocerebellar Ataxia Types 1,2,3,6, and 7 in Australian Patients with Spinocerebellar Ataxia. *Am. J. Med. Genet.* **95**, 351–357.

Subramony, S. H., Bock, H. G., Smith, S. C., Currier, R. D. *et al.* (1993). Presumed Machado-Joseph disease: Four kindreds from Mississippi. In "Handbook of Cerebellar Disease" (R. Lechentenberg, ed.), pp. 353–362. Marcel Dekker, New York.

Subramony, S. H., and Currier, R. D. (1996). Intrafamilial variability in Machado-Joseph disease. *Movement Disorders* **11**, 741–743.

Suenaga, T., Matsushima, H. *et al.* (1993). Ubiquitin-immunoreactive inclusions in anterior horn cells and hypoglossal neurons in a case with Joseph's disease. *Acta Neuropathol.* **85**, 341–344.

Tait, D., Riccio, M., Sittler, A. *et al.* (1998). Ataxin-3 is transported into the nucleus and associates with the nuclear matrix. *Hum. Mol. Genet.* **7**, 991–997.

Takano, H., Cancel, G., Ikeuchi, T., Lorenzetti, D., Mawad, R., Stevanin, G., Didierjean, O., Durr, A., Oyake, M., Shimohata, T., Sasaki, R., Koide, R., Igarashi, S., Hayashi, S., Takiyama, Y., Nishizawa, M., Tanaka, H., Zoghbi, H. Y., Brice, A., and Tsuji, S. (1998). Close association between prevalence of dominantly inherited spinocerebellar ataxias with CAG-repeat expansions and frequencies of large normal CAG alleles in Japanaese and Caucasian populations. *Am. J. Hum. Genet.* **63**, 1060–1066.

Takiyama, Y., Nishizawa, S., Tanaka, H. *et al.* (1993). The gene for Machado-Joseph disease maps to human chromosome 14q. *Nat. Genet.* **4**, 300–305.

Takiyama, Y., Shimazaki, H., Morita, M., Soutome, M., Sakoe, K., Esumi, E., Muramatsu, S.-I., Yoshida, M. *et al.* (1998). Maternal anticipation in Machado-Joseph disease (MJD); some maternal factors independent of the number of CAG repeat units may play a role in genetic anticipation in a Japanese MJD family. *J. Neurol. Sci.* **155**, 141–145.

Takiyama, Y., Okynagi, S., Kawashima, S., Sakamoto, H., Saito, K., Yoshida, M., Tsuji, S., Mizuno, Y., and Nishizawa, M. (1994). A clinical and pathologic study of a large Japanese family with Machado-Joseph disease tightly linked to the DNA markers on chromosome 14q. *Neurology* **44**, 1302–1308.

Takiyama, Y., Sakoe, K., Nakano, I., and Nishizawa, M. (1997). Machado-Joseph disease: Cerebellar ataxia and autonomic dysfunction in a patient with the shortest known expanded allele (56 CAG repeat units) of the MJD1 gene. *Neurology* **49**, 604–606.

Takiyama, Y., Igarashi, S., Rogaeva, E. A., Endo, K., Rogaev, E. I., Tanaka, H., Sherrington, R., Sanpei, K., Liang, Y., Saito, M., Tsuda, T., Takano, H., Ikeda, M., Lin, C., Chi, H., Kennedy, J. L., Lang, A. E., Wherrett, J. R., Segawa, M., Nomura, Y., Yuasa, T., Weissenbach, J., Yoshida, M., Nishizawa, M., Kidd, K. K. *et al.* (1995). Evidence for intergenerational instability in the CAG repeat in the MJD1 gene and for conserved haplotypes at flanking markers amongst Japanese and Caucasian subjects with Machado-Joseph disease. *Hum. Mol. Genet.* **4**, 1137–1146.

Trottier, Y., Cancel, G., An-Gourfinkel, I., Lutz Y., Weber, C., Brice, A., Hirsch, E., and Mandel, J. L. (1998). Heterogeneous intracellular localization and expression of ataxin-3. *Neurobiol. Dis.* **5**, 335–347.

Tuite, P. J., Rogaeva, E. A., St. George-Hyslop, P. H., and Lang, A. E.

(1995). Dopa-responsive parkinsonism phenotype of Machado-Joseph disease: confirmation of 14q CAG expansion. *Ann. Neurol.* **38**, 684–687.

Uchiyama, T., Fukutake, T., Arai, K., Nakagawa, K., Hattori, T. (2001). Machado-Joseph disease associated with an absence of fungiform papillae on the tongue. *Neurology* **56**, 558–560.

Van Alfen, N., Sinke, R., Zwarts, M. *et al.* (2001). Intermediate CAG repeat lengths (53,54) for MJD/SCA3 are associated with an abnormal phenotype. *Ann. Neurol.* **49**, 805–808.

Wang, G., Keiko, I., Nukina, N. *et al.* (1997). Machado-Joseph Disease gene product identified in lymphocytes and brain. *Biochem. Biophys. Res. Commun.* **233**, 476–479.

Warrick, J. M., Paulson, H. L., Gray-Board, G. L., Bui, Q. T., Fischbeck, K. H., Pittman, R. N., and Bonini, N. M. (1998). Expanded polyglutamine protein forms nuclear inclusions and causes neural degeneration in drosophila. *Cell* **93**, 1–20.

Warrick J. M., Chan, E., Gray-Board, G. L. *et al.* (1999). Suppression of polyglutamine-mediated neurodegeneration in Drosophila by the molecular chaperone Hsp70. *Nat Genet* **23**, 425–442.

Watanabe, H., Tanaka, F., Matsumoto, M., Doyu, M., Ando, T., Mitsuma, T., and Sobue, G. (1998). Frequency analysis of autosomal dominant cerebellar ataxias in Japanese patietns and clinical characterization of spinocerebellar ataxia 6. *Clin. Genet.* **53**, 13–19.

Watanabe, M., Abe, K., Aoki, M., Kameya, T., Kaneko, J., Shoji, M., Ikeda, M., Shizuka, M., Ikeda, Y., Iizuke, T., Hirai, S., and Itoyama, Y. (1996). Analysis of CAG trinucleotide expansion associated with Machado-Joseph disease. *J. Neurol. Sci.* **136**, 101–107.

Yen, T. C., Tzen, K. Y., Chen, M. C., Chou, Y. H., Chen, R. S., Chen, C. J., Wey, S. P., Ting, G., Lu, C. S. (2002). Dopamine transporter concentration is reduced in asymptomatic Machado-Joseph disease gene carriers. *J. Nucl. Med.* **43**, 153–159.

Yen, T. C., Lu, C. S., Tzen, K. Y., Wey, S. P., Chou, Y. H., Weng, Y. H., Kao, P. F., Ting, G. (2000). Decreased dopamine transporter binding in Machado-Joseph disease. *J. Nucl. Med.* **41**, 994–998.

Yoshizawa, T., Yamagishi, Y., Koseki, N., Goto, J., Yoshida, H., Futoshi, S., Shoji, S., and Kanazawa, I. (2000). Cell cycle arrest enhances the in vitro cellular toxicity of the truncated Machado-Joseph disease gene product with an expanded polyglutamine stretch. *Hum. Mol. Genet.* **9**, 69–78.

Zhou, X. Y., Takiyama, Y., Igarashi, S., Li, Y. F., Zhou, B. Y., Gui, D. C., Endo, K., Tanaka, H., Chen, Z. H., Zhou, Y., Qiao, W., Gu, W. *et al.* (2001). Spinocerebellar Ataxia Type 1 in China. *Arch. Neurol.* **58**, 789–794.

Zhou, L. S., Fan, M. Z., Yang, B. X., Weissenbach, J., Wang, G. X., and Tsuji, S. (1997). Machado-Joseph disease in four Chinese pedigrees: molecular analysis of 15 patients including two juvenile cases and clinical correlations. *Neurology* **48**, 482–485.

CHAPTER 6

# Spinocerebellar Ataxia Type 4 (SCA4)

HIDEHIRO MIZUSAWA

*Department of Neurology and Neurological Science*
*Graduate School, Tokyo Medical and Dental University*
*Tokyo, 113-8519 Japan*

I. Summary
II. Phenotype
III. Gene Locus
IV. Diagnostic and Ancillary Tests
V. Neuropathology
VI. ADCCA or Pure Cerebellar Ataxia Linked to SCA4 Locus
References

## I. SUMMARY

SCA4 is a rare spinocerebellar ataxia characterized not only by cerebellar ataxia but also peripheral neuropathy and pyramidal tract involvement (Gardner *et al.*, 1994; Flanigan *et al.*, 1996). It was reported only in a large five-generation family of Scandinavian origin residing in Utah and Wyoming. The clinical manifestations include cerebellar ataxia such as ataxic gait, limb ataxia and dysarthria, loss of deep sensation and areflexia due to axonal sensory neuropathy, and occasional extensor plantar responses. The gene locus was mapped to chromosome 16q22.1. The same locus was demonstrated to show linkage in several Japanese families presenting with autosomal dominant pure cerebellar ataxia or cortical cerebellar atrophy (ADCCA). The 16q-linked ADCCA may be allelic to SCA4.

## II. PHENOTYPE

SCA4 patients start to suffer from gait ataxia typically in the fourth or fifth decade with a median age at onset of 39.9 (19–59) years (Flanigan *et al.*, 1996). Gait disturbance is followed by difficulty in fine motor tasks and dysarthria. On examination, in addition to ataxic gait (95%) and limb dysmetria (95%), loss of vibration and joint-position sense (100%) as well as minimal loss of pinprick sense (95%) is frequently demonstrated. Ankle and knee-jerk are lost in almost all patients and complete areflexia is found in 25%. Extensor plantar response is seen in 20%. Distal limb weakness is reported in 4 and proximal as well as distal limb weakness is in 2 out of 20 patients. The patients showing weakness have positive Babinski's sign suggesting pyramidal tract involvement. Dysarthria is present in 50% and there are rare eye signs including saccadic pursuit and spontaneous lateral movements. Genetic anticipation is not very clear since the median age at onset is 41.9 years for the fourth generation and 36.7 years for the fifth generation. The phenomenon is observed in some families and not in other families. Differential diagnosis may include Friedreich ataxia because both diseases share possible dorsal column involvement and areflexia. Friedreich ataxia, however, is characterized by autosomal recessive inheritance, juvenile onset, and the presence of various extra-neural signs such as pes carvus, scoliosis, and cardiomyopathy. Other autosomal dominant ataxias including even SCA1, SCA2, Machado-Joseph disease, and SCA7 do not have such a severe sensory disturbance. Ataxias such as SCA5 and SCA6 show pure cerebellar ataxia which is very distinct from SCA 4 phenotype.

## III. GENE LOCUS

The gene was reported to be tightly linked to the microsatellite marker D16S397 (lod score = 5.93 at theta = 0.00) in the 6-cM interval between D16S514 and D16S398.

## IV. DIAGNOSTIC AND ANCILLARY TESTS

Sural nerve sensory response is absent in 12 out of 13 patients examined and radial sensory response is lost in 3. No reports on neuroimaging, cellular and animal models and treatment were available.

## V. NEUROPATHOLOGY

The authors appeared to consider the phenotype of SCA4 similar to that of a French-German family with a late-onset ataxia and areflexia reported in 1954 (Biemond). In 1997, two sib cases with an SCA4 phenotype (Nachmanoff et al., 1997) were also reported. These reports included autopsy findings in which the dorsal column and dorsal roots of the spinal cord as well as the cerebellar cortex were affected. However, there has been no definite evidence that the cases of Biemond or Nachmanoff were SCA4.

## VI. ADCCA OR PURE CEREBELLAR ATAXIA LINKED TO SCA4 LOCUS

Recently a linkage study demonstrated many families of autosomal dominant pure cerebellar ataxia in Japan linked to a gene on SCA4 locus (Nagaoka, 2000). The gene locus was reported on 10.9 cM around D16S3107 (Fig. 6.1) which was later narrowed into 3.0 cM at the same region (Takashima et al., 2001). The mean age at onset is 55.9 (45–72) years with mild genetic anticipation (the age at onset is 4.9 years younger in offspring). The initial symptom is gait ataxia followed by cerebellar dysarthria (92.6%), limb ataxia (92.6%), hypotonus (92.6%), and horizontal gaze nystagmus (63.0%). Except for a mild decrease in vibration sense in a very old patient with long duration of the illness, there are no sensory deficits. Tendon reflexes are almost normal except that only 16.6% of the patients have slightly decreased ankle-jerks. No pyramidal tract signs including Babinski's sign are observed.

Sensory and motor nerve conduction studies and EEG are normal in all the patients examined. Brain MRI reveals cerebellar atrophy without apparent involvement of the brainstem and other structures (Fig. 6.2). Therefore the phenotype is that of pure cerebellar ataxia or autosomal dominant cerebellar ataxia type III (ADCA-III) of Harding's classification (Harding, 1982) and completely different from that of SCA4 which is characterized by profound sensory disturbance and classified into Harding's ADCA-I. The 16q-linked ADCCA shares almost the same phenotype as SCA6 and is distinguished from SCA6 only by gene analysis. Identification of the causative gene would address the question whether the 16q-linked ADCCA and SCA4 are allelic or not.

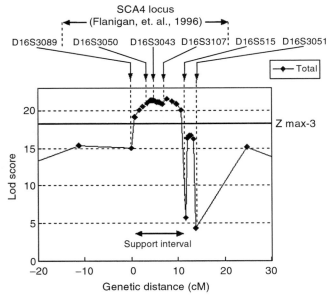

FIGURE 6.1 Multipoint lod scores of 16q-linked autosomal dominant pure cerebellar ataxia. The locus is supposed to be in the 10.9 cM region between 1.7 cM telomeric to D16S3089 and 1.3 cM centromeric to D16S515. The SCA4 locus lies in the region. (From Fig. 3, Nagaoka, U. et al., (2000). Neurology 54, 1971–1975. With permission.)

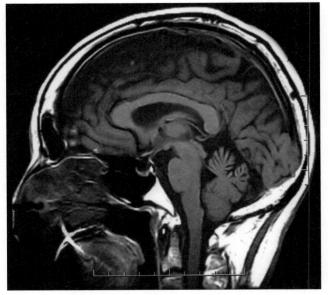

FIGURE 6.2 Brain MRI of a patient with 16q-linked ADCCA. The cerebellum, particularly in the vermis, is atrophic but the pons is well preserved.

## References

Biemond, A. (1954). La forme radiculo-cordonnale posteriure des degenerescences spinocerebelleuse. *Rev. Neurolog.* **91**, 2–21.

Flanigan, K., Gardner, K., Alderson, K., Galster, B., Otterud, B., Leppert, M. F., Kaplan, C., and Ptacek, L. J. (1996). Autosomal dominant spinocerebellar ataxia with sensory axonal neuropathy (SCA4): clinical description and genetic localization to chromosome 16q22.1. *Am. J. Hum. Genet.* **59(2)**, 392–399.

Gardner, K., Alderson, K., Galster, B., Kaplan, C., Leppert, M., Ptacek, L. (1994). Autosomal dominant spinocerebellar ataxia:Clinical description of a distinct hereditary ataxia and genetic localization to chromosome 16 (SCA4) in Utah kindred. *Neurology* **44**(Suppl.), 921S.

Harding, A. E. (1982). The clinical features and classification of the late onset autosomal dominant cerebellar ataxias. A study of 11 families, including descendants of "the Drew family of Walworth". *Brain* **105(Pt. 1)**, 1–28.

Nachmanoff, D. B., Segal, R. A., Dawson, D. M., Brown, R. B., and De Girolami, U. (1997). Hereditary ataxia with sensory neuronopathy: Biemond's ataxia. *Neurology* **48(1)**, 273–275.

Nagaoka, U., Takashima, M., Ishikawa, K., Yoshizawa, K., Yoshizawa, T., Ishikawa, M., Yamawaki, T., Shoji, S., and Mizusawa, H. (2000). A gene on SCA4 locus causes dominantly inherited pure cerebellar ataxia. *Neurology* **54(10)**, 1971–1975.

Takashima, M., Ishikawa, K., Nagaoka, U., Shoji, S., Mizusawa, H. (2001). A linkage disequilibrium at the candidate gene locus for 16q-linked autosomal dominant cerebellar ataxia type III in Japan. *J. Hum. Genet.* **46(4)**, 167–171.

Harding, A. E. (1982). The clinical features and classification of the late onset autosomal dominant cerebellar ataxias. A study of 11 families, including descendants of "the Drew family of Walworth". *Brain* **105(Pt. 1)**, 1–28.

# CHAPTER 7

# Spinocerebellar Ataxia 5 (SCA5)

**LAURA P.W. RANUM**
Department of Genetics, Cell Biology, and Development
Institute of Human Genetics
University of Minnesota
Minneapolis, Minnesota 55455

**KATHERINE A. DICK**
Department of Genetics, Cell Biology, and Development
Institute of Human Genetics
University of Minnesota
Minneapolis, Minnesota 55455

**J.W. DAY**
Department of Neurology
Institute of Human Genetics
University of Minnesota
Minneapolis, Minnesota 55455

I. Introduction
II. Anticipation
III. Genetic and Physical Mapping
IV. Repeat Expansion Detection and Rapid Cloning
V. Clinical Features
VI. Neuroimaging and Neuropathology
VII. Conclusions
    References

Spinocerebellar ataxia type 5 (SCA5) is a slowly progressive autosomal dominant ataxia that predominantly affects the cerebellum. The most common features of the disease are gait ataxia, incoordination of the upper extremities, and slurred speech. DNA samples from a 10-generation kindred descended from the paternal grandparents of President Lincoln were used to genetically map the SCA5 locus to chromosome 11q. Although the mutation has not yet been identified, dramatic intergenerational changes in severity and age of onset suggest the possibility that a microsatellite expansion may be involved.

## I. INTRODUCTION

In 1994, we described a 10-generation American kindred of European ancestry with a generally mild autosomal dominant form of spinocerebellar ataxia (SCA). (Ranum et al., 1994). The family has two major branches, which can be traced back to the paternal grandparents of President Abraham Lincoln (Fig. 7.1). Abbreviated pedigrees of the major branches are shown in Fig. 7.2. DNA was collected from 248 family members, including 77 affected individuals. After excluding linkage to the known ataxia loci, a genome-wide screen was performed to map the mutation to the long arm of chromosome 11 (Ranum et al., 1994).

Disease onset most often occurs in the third or fourth decade of life, but ranges from 10–68 years of age (Ranum et al., 1994; Stevanin et al., 1999). Symptoms, which progress over time, include incoordination of the upper extremities, slurred speech, and gait ataxia. SCA5 predominantly affects the cerebellum while largely sparing the brainstem and spinocerebellar tracts (Ranum et al., 1994; Schut et al., 2000; Liquori et al., 2001). The predominant cerebellar involvement and mild disease course of SCA5 are similar to SCA6 (Zhuchenko et al., 1997),

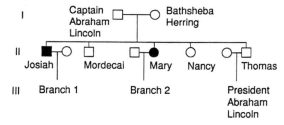

FIGURE 7.1 The common ancestry of the two branches of the family. The solid square and circle indicate President Lincoln's Uncle Josiah and Aunt Mary who passed the ataxia gene to their descendants. (From Ranum, L.P.W. et al. (1994). Nat. Genet. **8**, 280–294. With permission.)

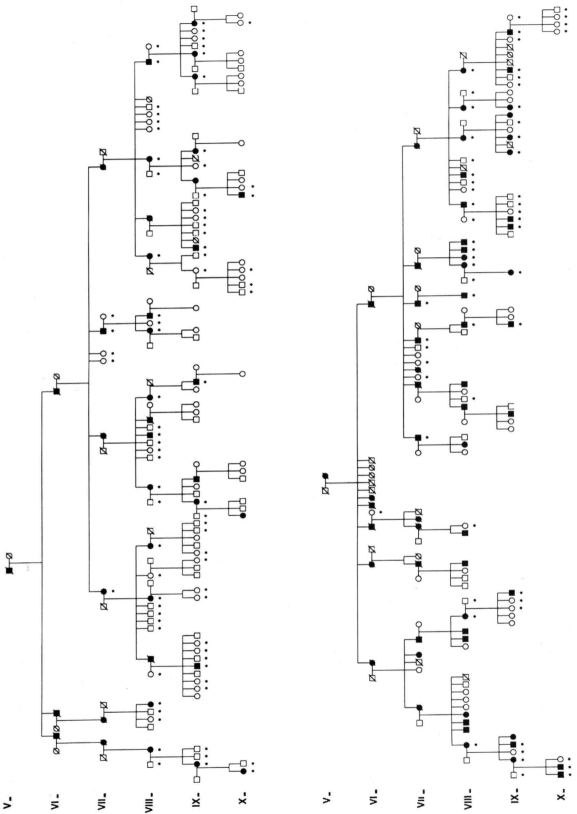

FIGURE 7.2 Abbreviated pedigrees of two branches of the family. To preserve confidentiality the identities of the branches are not given, generations three and four have been omitted, and changes have been made in the pedigree structure. While over 800 family members have been identified to date, only 275 individuals are represented. Family members who are not at first-degree risk and some younger family members who are at first-degree risk have been omitted. Furthermore, the number of siblings and the sex of some individuals have been changed. The portions of the two branches shown represent individuals from which most of the blood samples were obtained (black dot beneath symbol). Solid circles (females) and squares (males) represent affected individuals. Generations for both branches are numbered with reference to their common ancestors Captain Abraham Lincoln and Bathsheba Herring, of generation I. (From Ranum, L. P. W. et al. (1994). *Nat. Genet.* **8**, 280–284. With permission.)

SCA11 (Worth et al., 1999), and SCA15 (Storey et al., 2001). Unlike many of the other SCAs (Harding, 1983; Klockgether and Evert, 1998; Brice, 1998; Herman-Bert et al., 2000; O'Hearn et al., 2001; Rasmussen et al., 2001; Nakamura et al., 2001), SCA5 typically does not shorten life span (Ranum et al., 1994) probably because of the lack of bulbar paralysis in adult-onset patients. However, several patients with juvenile-onset SCA5 have been found to have some bulbar atrophy and to demonstrate a weakened ability to cough and swallow, suggesting that they may develop a predisposition for pneumonia that may shorten life span (Ranum et al., 1994). Stevanin et al. (1999) recently described SCA5 in a family of French decent with clinical features comparable to the American family.

## II. ANTICIPATION

To determine if anticipation was present in SCA5, we compared ages of onset for parent offspring pairs in the Lincoln family (Table 7.1). The mean ages of onset for the older (43.3 years) versus younger (29.9 years) generations are significantly different ($p < 0.001$). The average decreases in age of onset for maternally ($-15.7$ years) versus paternally ($-9.3$ years) inherited ataxia do not significantly differ. Several three-generation examples of anticipation were observed in which grandparents had onsets 10–40 years later in life than their children and grandchildren. In contrast to many forms of SCA in which juvenile-onset cases are preferentially associated with paternal transmission, five of the six juvenile-onset cases were maternally inherited (Ranum et al., 1994; Ranum, unpublished results).

Although an unstable trinucleotide or other microsatellite repeat expansion is an attractive explanation for the anticipation we see in the Lincoln family, the decrease in age of onset among the parent-offspring pairs could also be explained by an ascertainment bias. For example, the age-of-onset difference for the older versus the younger generations will be lessened when the currently unaffected offspring develop signs of the disease later in life and are included in the calculation (Ranum et al., 1994). Although the mean age of onset did not significantly decrease in the youngest generation of the French SCA5 family, anticipation may have not been apparent given the small size of the family (Stevanin et al., 1999).

## III. GENETIC AND PHYSICAL MAPPING

After mapping the SCA5 locus to chromosome 11 (Ranum et al., 1994) we developed a high-resolution genetic map of the region. Table 7.3 shows pairwise LOD scores for selected markers. Figure 7.4 summarizes the analyses of recombinant chromosomes, which place the SCA5 locus on 11q13, between PYGM and INT2 (Koob et al., 1995). All ten of the affected recombinants used to narrow the SCA5 critical region were maternally transmitted. The absence of paternally transmitted recombinant chromosome, consistent with the reduced male recombination rate for the region (Fain et al., 1996), limited the number of recombinant chromosomes available for analysis and slowed our efforts to genetically refine the region.

Raw sequence data, recently available from the Human Genome Project, has allowed us to develop hundreds of microsatellite markers and to create a more detailed map of the region. Currently, we are using a positional cloning approach to identify the gene.

TABLE 7.1 Age of Onset for Maternal versus Paternal Transmission

| Mother | Child | Age change | Father | Child | Age change |
|---|---|---|---|---|---|
| 68 | 50 | −18 | 50 | 32 | −18 |
| 68 | 28 | −40 | 50 | 37 | −13 |
| 68 | 40 | −28 | 50 | 20 | −30 |
| 40 | 30 | −10 | 40 | 42 | +2 |
| 22 | 25 | +3 | 40 | 29 | −11 |
| 59 | 20 | −39 | 50 | 30 | −20 |
| 59 | 36 | −23 | 27 | 38 | +11 |
| 36 | 18 | −18 | 38 | 28 | −10 |
| 55 | 33 | −22 | 45 | 50 | +5 |
| 26 | 24 | −2 | Average change | | −9.3 years |
| 39 | 25 | −14 | | | |
| 39 | 30 | −9 | | | |
| 29 | 28 | −1 | | | |
| 29 | 32 | +3 | | | |
| 28 | 13 | −15 | | | |
| 28 | 10 | −18 | | | |
| Average change | | −15.7 years | | | |

From Ranum, L.P.W. et al. (1994). Nat. Genet. **8**, 280–284. With permission.

TABLE 7.2 Pairwise LOD scores

| | LOD scores at θ = | | | | | |
| | 0.00 | 0.01 | 0.05 | 0.10 | 0.20 | 0.30 | 0.40 |
|---|---|---|---|---|---|---|---|
| GATA2A01 | −∞ | 8.76 | 10.25 | 10.09 | 8.30 | 5.68 | 2.69 |
| D11S913 | 14.66 | 14.39 | 13.28 | 11.85 | 8.83 | 5.65 | 2.48 |
| INT2 | −∞ | 12.60 | 14.40 | 14.07 | 11.66 | 8.18 | 4.05 |

From Liquori et al. (in press). Spinocerebellar ataxia type 5. In *The Cerebellum and its Disorders.* (M. Manto and M. Pandolfo, eds.), pp. 445–450. Cambridge University Press, Cambridge, U.K. Reprinted with the permission of Cambridge University Press.

**FIGURE 7.3** Genetic map of the SCA5 critical region summarizing recombinant information from affected individuals. The region of chromosome 11 common to all of the affected individuals in the Lincoln family is indicated in black. Regions in white indicate portions of chromsome 11 that do not contain the SCA5 mutation and regions in gray indicate the portions of the chromosome within which key recombination events defining the critical region have occurred.

## IV. REPEAT EXPANSION DETECTION AND RAPID CLONING

To investigate the possibility that the variable ages of onset and disease severity in the Lincoln family are caused by the expansion of an unstable CAG/CTG trinucleotide repeat expansion, we performed the Repeat Expansion Detection (RED) assay. RED is an elegant technique for detecting potentially pathological trinucleotide repeat expansions without requiring knowledge of chromosomal location (Schalling *et al.*, 1993). Human genomic DNA is used as a template for a two-step ligation cycling process, which results in the formation of ligated primers that correspond in length to the longest repeat in the genome (Schalling *et al.*, 1993). We detected a candidate CAG/CTG expansion by RED analysis in a juvenile-onset SCA5 patient. To determine if this CAG/CTG repeat expansion was the pathogenic cause of SCA5, we developed a method to clone and identify unique sequences surrounding expanded repeat tracts. Our method, called Repeat Analysis Pooled Isolation and Detection (RAPID) cloning uses an optimized RED protocol to follow an expanded repeat tract through a series of enrichment steps until a clone containing the expanded repeat is isolated, (Koob *et al.*, 1998). Nucleotide sequence flanking the repeat is then obtained and used to design a polymerase chain reaction (PCR) assay that can be used to determine if an expanded repeat co-segregates with the disease. We used RAPID cloning to identify both the SCA7 (Koob *et al.*, 1998) and SCA8 (Koob *et al.*, 1999) repeat expansions, and others have used this method to identify the mutations that cause SCA12 (Holmes *et al.*, 1999) and Huntington's disease-like 2 (Holmes *et al.*, 2001). Although RED analysis suggested the possibility that a CAG/CTG expansion causes SCA5, RAPID cloning and subsequent PCR analysis demonstrated that the expansion we detected was a nonpathogenic background expansion that did not co-segregate with the disease. Additional RED and RAPID analyses performed on genomic DNA from other family members with juvenile-onset SCA5, who would be expected to have the largest repeat expansions, also indicates that SCA5 is not caused by a CAG/CTG expansion greater than 40 repeats (Schut *et al.*, 2000; Liquori *et al.*, 2001). Our results suggest that if SCA5 is caused by a microsatellite expansion the mutation is most likely a trinucleotide motif other than CAG/CTG, or a variant repeat motif such as those that cause SCA10 (ATTCT; Matsuura *et al.*, 2000) or myotonic dystrophy type 2 (CCTG; Liquori *et al.*, 2001).

## V. CLINICAL FEATURES

Clinical features of affected members of the Lincoln family are summarized in Table 7.3. The mean age of onset for SCA5 is 33.0 ± 13 years (range = 10–68 years). Progressive ataxia of gait, in conjunction with truncal instability, has been the primary clinical feature within the family. Only 6 of 77 individuals examined had normal gait. Typically, onset is marked by excessive stumbling, difficulty negotiating stairs, or faltering while balancing on one foot in the shower. While symptoms progress over time, SCA5 only rarely leads to wheelchair dependence (Schut *et al.*, 2000; Liquori *et al.*, 2002).

While not as disabling as the gait ataxia, upper limb ataxia is present in over 90% of affected individuals. Only four individuals, all of whom were in their seventh to tenth decade of life, experienced severe upper extremity ataxia. Many subjects reported having no upper extremity symptoms, although ataxia was detectable during clinical testing. Ultimately, most affected patients encounter difficulties with activities that require fine finger dexterity, such as handwriting (Schut *et al.*, 2000; Liquori *et al.*, 2002).

**TABLE 7.3** Clinical Features of SCA5

| | |
|---|---|
| Limb ataxia | ++++ |
| Gait ataxia | ++++ |
| Truncal ataxia | +++ |
| Muscle weakness | – |
| Muscle atrophy | – |
| Deep tendon reflexes | |
|    Hypoactive | + |
|    Hyperactive | ++ |
| Bulbar abnormalities | +/– |
| Abnormal eye movements | +++ |
| Sensory | ++ |
| Dysarthria | +++ |
| Abnormal 3-cough sequence | +/– |
| Babinski sign | – |
| Abnormal romberg | +++ |

Legend: "++++" ≥ 90%; "+++" 50 to 89%; "++" 25 to 49%; "+" 10 to 24%; "+/–" 2 to 9%; "–" < 2%.

From Liquori *et al.* (2001). Spinocerebellar ataxia type 5. In *The Cerebellum and its Disorders*. (M. Manto and M. Pandolfo, eds.), pp. 445–450. Cambridge University Press, Cambridge, U.K. With permission.

The cerebellar dysarthria, detected in 75% of affected family members, lacks significant bulbar or spastic elements. While affected, articulation is generally not severely compromised, interfering with verbal communication in only one case. Mild gaze-evoked nystagmus, square wave-jerks, or saccadic intrusions during smooth pursuit were observed in half of the affected individuals. One third experienced mild sensory deficits. Tongue atrophy was absent, but fasciculations may have been detected in two individuals, one of whom developed symptoms at age 10. Brisk reflexes, present in one-third of affected family members, are suggestive of mild pyramidal tract involvement, though extensor plantar reflexes were observed in only one person (Schut *et al.*, 2000; Liquori *et al.*, 2002).

While onset is typically in the third or fourth decade of life, six individuals developed symptoms before 20 years of age and had signs of mild bulbar and pyramidal tract deterioration, in addition to the usual cerebellar findings. It remains unclear if atrophy in early-onset cases will progress and result in a more severe form of SCA5 (Schut *et al.*, 2000; Liquori *et al.*, 2002).

## VI. NEUROIMAGING AND NEUROPATHOLOGY

Brain magnetic resonance imaging (MRI) has been performed on twelve members of the Lincoln family and revealed dramatic cerebellar cortical atrophy, with widespread deterioration in the superior hemispheres and anterior vermis. The inferior vermis, tonsils, and brainstem appeared to be spared, as did the basal ganglia and cerebral hemispheres. These findings are illustrated in Fig. 7.4 by a sagittal MRI section of a 64-year-old affected woman who experienced disease onset at the age of 59 (Liquori *et al.*, 2002).

The brain of an elderly woman affected with SCA5 was examined at autopsy. Five years earlier she had been clinically examined and was found to have slight upper extremity ataxia, moderate truncal and gait ataxia, and minimal dysarthria. Sensory examination was normal, and motor and coordination testing showed no abnormality other than ataxia. For the last 10 years of her life, severe arthritis in addition to the ataxia made her unable to walk. Atypical of SCA5, this woman was also demented due to superimposed, pathologically confirmed Alzheimer's disease (Schut *et al.*, 2000; Liquori *et al.*, 2002).

The brain, examined after her death at age 89, weighed 940 g and displayed frontal and temporal atrophy, consistent with Alzheimer's disease. The cerebellum was extremely small (88 g), with marked folial atrophy in the anterior vermis and the anterior-superior portions of the hemispheres. In most areas of the cerebellar cortex there was significant Purkinje cell loss, molecular layer atrophy, loss of granular neurons, and frequent empty basket fibers.

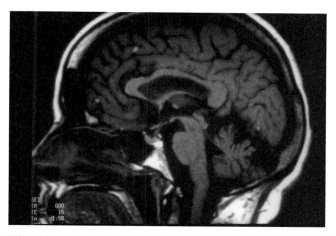

**FIGURE 7.4** Sagittal MRI scan from an affected individual at age 64. There is marked cerebellar atrophy, minimal brainstem atrophy, and no evidence of cerebral involvement. The relative preservation of the posterior vermis, posterior hemisphere, and tonsillar cortex is evident. (From Liquori *et al.* (2002). (Spinocerebellar ataxia type 5. In *The cerebellum and its Disorders*. (M. Manto and M. Pandolfo, eds.), pp. 445–450. Cambridge University Press, Cambridge, U.K.)

The tonsillar cortex appeared relatively spared. Deep cerebellar nuclei were gliotic without apparent neuronal loss, suggesting afferent loss from Purkinje cells. Although the basis pontis, red nuclei, cranial nerve nuclei, dorsal column nuclei, Clarke's nuclei, and spinocerebellar tracts were intact there was mild-to-moderate neuronal loss in the inferior olivary nuclei. Additionally, changes in the cerebral cortex were characteristic of the superimposed Alzheimer's disease. Pathologically, SCA5 appears to be primarily a cerebellar cortical degeneration, having predominant effects on Purkinje cells, and differing from dominantly inherited SCAs that affect afferent and efferent cerebellar connections as well as other brainstem nuclei (Schut *et al.*, 2000, Liquori *et al.*, 2002).

## VII. CONCLUSIONS

In contrast to many of the ataxias, SCA5 is a slowly progressive disease that usually does not shorten life span. Clinical, neuroimaging, and pathological data indicate that SCA5 primarily affects the cerebellum with little or no brainstem involvement, at least for adult-onset cases. Although onset typically occurs in mid-life, a broad range in the age of onset has been observed (10–68 years). Decreases in ages of onset with consecutive generations suggest that an unstable microsatellite expansion is responsible for the disease, however, RED and RAPID analysis indicate that the presence of a CAG/CTG expansion is unlikely. The SCA5 locus has been mapped to 11q13 and a positional cloning strategy is being used to identify the mutation. The eventual characterization of the SCA5 mutation will lead to a more comprehensive understanding of the genetic causes of ataxia and of the interdependence of the neuronal systems affected during SCA pathogenesis.

## Acknowledgments

Financial support from the National Ataxia Foundation and the National Institutes of Health (NS40389) is gratefully appreciated.

## References

Brice, A. (1998). Unstable mutations and neurodegenerative disorders. *J. Neurol.* **245**, 505–510.

Fain, P. R., Kort, E. N., Yousry, C., James, M. R., and Litt, M. (1996). A high resolution CEPH crossover mapping panel and integrated map of chromosome 11. *Hum. Mol. Genet.* **5**, 1631–1636.

Harding, A. (1983). Classification of the hereditary ataxias and paraplegias. *Lancet* **1**, 1151–1155.

Herman-Bert, A., Stevanin, G., Netter, J. C., Rascol, D. B., Calvas, P., Camuzat, A., Yuan, Q., Schalling, M., Durr, A., and Brice, A. (2000). Mapping of spinocerebellar ataxia 13 to chromosome 19q13.3–q13.4 in a family with autosomal dominant cerebellar ataxia and mental retardation. *Am. J. Hum. Genet.* **67**, 229–235.

Holmes, S. E., O'Hearn, E., McInnis, M. G., Gorelick-Feldman, D. A., Kleiderlein, J. J., Callahan, C., Kwak, N. G., Ingersoll-Ashworth, R. G., Sherr, M., Sumner, A. J., Sharp, A. H., Ananth, U., Seltzer, W. K., Boss, M. A., Vieria-Saecker, A. M., Epplen, J. T., Riess, O., Ross, C. A., and Margolis, R. L. (1999). Expansion of a novel CAG trinucleotide repeat in the 5′ region of PPP2R2B is associated with SCA12. *Nat. Genet.* **23**, 229–235.

Holmes, S. E., O'Hearn, E., Rosenblatt, A., Callahan, C., Hwang, H. S., Ingersoll-Ashworth, R. G., Fleisher, A., Stevanin, G., Brice, A., Potter, N. T., Ross, C. A., and Margolis, R. L. (2001). A repeat expansion in the gene encoding junctophilin-3 is associated with Huntington disease-like 2. *Nat. Genet.* **29**, 377–378.

Klockgether, T., and Evert, B. (1998). Genes involved in hereditary ataxias. *Trends Neurosci.* **21**, 413–418.

Koob, M. D., Lundgren, J. K., Nowak, N. J., Shows, T. B., Perlin, M. W., Schut, L. J., and Ranum, L. P. W. (1995). High-resolution genetic and physical mapping of spinocerebellar ataxia type 5 (SCA5) on 11q13. *Am. J. Hum. Genet.* **57**, A196.

Koob, M. D., Benzow, K. A., Bird, T. D., Day, J. W., Moseley, M. L., and Ranum, L. P. W. (1998). Rapid cloning of expanded trinucleotide repeat sequences from genomic DNA. *Nat. Genet.* **18**, 72–75.

Koob, M. D., Moseley, M. L., Schut, L. J., Benzow, K. A., Bird, T. D., Day, J. W., and Ranum, L. P. W. (1999). An untranslated CTG expansion causes a novel form of spinocerebellar ataxia (SCA8). *Nat. Genet.* **21**, 379–384.

Liquori, C. L., Ricker, K., Moseley, M. L., Jacobsen, J. F., Kress, W., Naylor, S. C., Day, J. W., and Ranum, L. P. W. (2001). Myotonic dystrophy type 2 caused by a CCTG expansion in intron 1 of ZNF9. *Science* **293**, 864–867.

Liquori, C. L., Schut, L. J., Clark, H. B., Day, J. W., and Ranum, L. P. W. (2002). Spinocerebellar ataxia type 5. In "The Cerebellum and it's Disorders" (M. Manto and M. Pandolfo, eds.), pp. 445–450. Cambridge University Press, Cambridge.

Matsuura, T., Yamagata, T., Burgess, D. L., Rasmussen, A., Grewal, R. P., Watase, K., Khajavi, M., McCall, A. E., Caleb, D. F., Zu, L, Achari, M., Pulst, S. M., Alonso, E., Noebels, J. L., Nelson, D. L., Zoghbi, H. Y., and Ashizawa, T. (2000). Large expansion of the ATTCT pentanucleotide repeat in spinocerebellar ataxia type 10. *Nat. Genet.* **26**, 191–194.

Nakamura, K., Jeong, S. Y., Uchihara, T., Anno, M., Nagashima, K., Nagashima, T., Ikeda, S., Tsuji, S., and Kanazawa, I. (2001). SCA17, a novel autosomal dominant cerebellar ataxia caused by an expanded polyglutamine in TATA-binding protein. *Hum. Mol. Genet.* **10**, 1441–1448.

O'Hearn, E., Holmes, S. E., Calvert, P. C., Ross, C. A., and Margolis, R. L. (2001). SCA–12: Tremor with cerebellar and cortical atrophy is associated with a CAG repeat expansion. *Neurology* **56**, 299–303.

Ranum, L. P. W., Schut, L. J., Lundgren, J. K., Orr, H. T., and Livingston, D. M. (1994). Spinocerebellar ataxia type 5 in a family descended from the grandparents of President Lincoln maps to chromosome 11. *Nat. Genet.* **8**, 280–284.

Rasmussen, A., Matsuura, T., Ruano, L., Yescas, P., Ochoa, A., Ashizawa, T., and Alonso, E. (2001). Clinical and genetic analysis of four Mexican families with spinocerebellar ataxia type 10. *Ann. Neurol.* **50**, 234–239.

Schalling, M., Hudson, T. J., Buetow, K. H., and Housman, D. E. (1993). Direct detection of novel expanded trinucleotide repeats in the human genome. *Nat. Genet.* **4**, 135–139.

Schut, L. J., Day, J. W., Clark, H. B., Koob, M. D., and Ranum, L. P. W. (2000). Spinocerebellar ataxia type 5. In "Handbook of Ataxia Disorders" (T. Klockgether, ed.), pp. 435–445. Marcel Dekker, New York.

Stevanin, G., Herman, A., Brice, A., and Durr, A. (1999). Clinical and MRI findings in spinocerebellar ataxia type 5. *Neurology* **53**, 1355–1357.

Storey, E., Gardner, R. J., Knight, M. A., Kennerson, M. L., Tuck, R. R., Forrest, S. M., and Nicholson, G. A. (2001). A new autosomal dominant pure cerebellar ataxia. *Neurology* **57**, 1913–1915.

Worth, P. F., Giunti, P., Gardner-Thorpe, C., Dixon, P. H., Davis, M. B., and Wood, N. W. (1999). Autosomal dominant cerebellar ataxia type III: linkage in a large British family to a 7.6-cM region on chromosome 15q14–21.3. *Nat. Genet.* **24**, 214–215.

Zhuchenko, O., Bailey, J., Bonnen, P., Ashizawa, T., Stockton, D. W., Amos, C., Dobyns, W. B., Subramony, S. H., Zoghbi, H. Y., and Lee, C. C. (1997). Autosomal dominant cerebellar ataxia (SCA6) associated with small polyglutamine expansions in the alpha-1A-voltage-dependent calcium channel. *Nat. Genet.* **15**, 62–69.

# CHAPTER 8

# Spinocerebellar Ataxia 6 (SCA6)

JOANNA C. JEN

*Department of Neurology*
*UCLA School of Medicine*
*University of California*
*Los Angeles, California 90095*

I. Introduction
II. Clinical Features
III. Genetics
IV. Diagnosis
V. Molecular Pathogenesis
VI. Neuropathology
VII. Animal Models
VIII. Treatment
References

## I. INTRODUCTION

The genetic characterization of glutamine-encoding CAG repeat expansions as the molecular basis of spinocerebellar ataxia types 1–3 prompted Zhuchenko and colleagues (1997) to search for CAG repeat expansions in other neural genes that might be associated with cerebellar ataxia syndromes. They found polymorphic CAG repeats in CACNA1A located on chromosome 19p, which was also involved in episodic ataxia type 2 (EA-2) and familial hemiplegic migraine (FHM); (Ophoff *et al.*, 1996). They then screened their ataxia patients for CAG repeat expansions in this gene. Indeed, in 233 ataxia patients and 475 normal controls who were screened, 8 ataxia patients each had an allele with 21 or more CAG repeats, while none of the normal controls had more than 16 repeats in either allele. Four of the patients with expanded CAG repeats were from families with additional ataxic individuals. In every family, expanded CAG repeats segregated with a slowly progressive ataxia of late onset, which was named spinocerebellar ataxia type 6 (SCA6). The association between modest CAG repeat expansions in CACNA1A and ataxia has since been reported in numerous families from around the world (Zoghbi, 1997).

## II. CLINICAL FEATURES

SCA6 typically presents with a slowly progressive truncal ataxia beginning in mid to late adulthood, with a mean age of onset in the late 40s to early 50s. Dysarthria, dysphagia, vertigo, and hypophonia are common symptoms (Geschwind *et al.*, 1997). Some patients experience recurrent position-dependent vertigo that preceded ataxic symptoms by decades (Jen *et al.*, 1998). Although SCA6 patients are bothered by a slowly progressive ataxia, some patients may complain of intermittent exacerbation triggered by stress, fatigue, or illness, reminiscent of similar triggers in EA-2 (Jodice *et al.*, 1997; Jen *et al.*, 1998).

Findings on neurologic examination are usually remarkable for dysarthria, gaze-evoked and downbeat nystagmus, and gait ataxia. Impaired smooth pursuit and dysmetric saccades are common oculomotor findings (Buttner *et al.*, 1998). Corticospinal tract findings such as hyperreflexia and extensor plantar responses occur in some patients, while other patients complain of mild sensory disturbances (Geschwind *et al.*, 1997).

## III. GENETICS

The CACNA1A gene encodes the pore-forming and voltage-sensing subunit of the Cav2.1 subunit of the P/Q-type voltage-gated calcium channel complex, so named because of its role in mediating the P-type calcium current first observed in the cerebellar Purkinje cells and Q-type current in the granule cells (Mori *et al.*, 1991). Mutations in the CACNA1A gene cause allelic disorders including EA-2, FHM, and SCA6.

In the initial report by Zhuchenko and colleagues (1997), individuals with 21–27 CAG repeats in the CACNA1A gene manifested progressive ataxia. The lowest repeat number reported to date was found in an individual homozygous for 19 repeats (Mariotti *et al.*, 2001), whose relatives heterozygous for the 19-repeat allele were asymptomatic. There is clearly limited penetrance, with numerous individuals with 22 repeats who remained asymptomatic into their 70s.

Contrary to the other SCAs, there appears to be remarkable stability in the trinucleotide expansion; intergenerational expansion has been noted in only a handful of cases. Furthermore, the modest expansion in SCA6 is well within the normal range in the other SCAs. Similar to the other SCAs, the number of repeats appears to be inversely correlated with the age of onset.

## IV. DIAGNOSIS

Because of the late onset and slow progression, some cases of SCA6 may appear to be sporadic. Familial cases have an autosomal dominant inheritance pattern.

Magnetic resonance imaging in SCA6 patients may reveal cerebellar atrophy, most prominent in the midline, as has been noted in some patients with EA-2.

Eye movement recordings have demonstrated an abnormal oculomotor profile unique to SCA6. (Buttner *et al.*, 1998) Moderately impaired in SCA1–3, saccade velocity remains normal in SCA6, although saccade dysmetria is common. Smooth pursuit, optokinetic responses, and suppression of vestibular-induced nystagmus are severely impaired.

Genetic testing for SCA6 is available in clinical laboratories. Specific markers have been developed to amplify the genomic sequence containing the polymorphic CAG repeats in CACNA1A. (Zhuchenko *et al.*, 1997)

## V. MOLECULAR PATHOGENESIS

When first characterized, the polymorphic CAG repeat in the CACNA1A gene was noted to reside in the 3′ untranslated region and did not correlate with either FHM or EA-2 (Ophoff *et al.*, 1996). In fact, CACNA1A was extensively alternatively spliced, with multiple isoforms. The longer transcripts contained a 5-bp insertion in the last exon, extending and shifting the reading frame to include the glutamine-encoding CAG repeats, along with additional amino acid residues in the carboxyl terminal (Zhuchenko *et al.*, 1997).

CACNA1A encodes Cav2.1, a neuronal calcium channel subunit that is the main pore-forming and voltage-sensing subunit of P/Q-type voltage-gated calcium channel complexes most abundantly expressed in the cerebellum but also found presynaptically at the neuromuscular junction (Mori *et al.*, 1991; Protti *et al.*, 1996). Cav2.1 contains four homologous domains (I–IV), each with six transmembrane segments (S1–S6), a motif shared by other members of the superfamily of voltage-gated ion channels (Fig. 8.1). S6 and the p loop, which interconnects S5 and S6 are thought to line the pore, while S4 is the voltage sensor. Proper channel

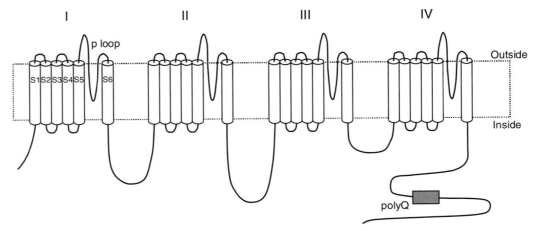

FIGURE 8.1 Putative topology of Cav2.1 channel subunit, with four internal homologous domains I–IV, each with 6 transmembrane segments S1–S6 and p loop interconnecting S5 and S6. The channel is made up of this subunit, along with an intracytoplasmic β subunit, a transmembrane γ subunit, and a largely extracellular α2δ subunit. Shaded box—polyglutamine expansion causing SCA6.

complex assembly and targeting to the plasma membrane require accessory subunits, each with several subtypes encoded by separate genes that are alternatively spliced.

## VI. NEUROPATHOLOGY

Several reports described detailed neuropathologic findings in SCA6, documenting marked Purkinje cell loss and less severe granule cell loss (Gomez et al., 1997; Ishikawa et al., 1999; Yang et al., 2000). Using antibodies directed against different epitopes on the Cav2.1 subunit, Ishikawa and colleagues (1999, 2001) demonstrated cytoplasmic and nuclear polyglutamine aggregates in SCA6 Purkinje cells, similar to findings in other polyglutamine diseases.

Curious whether SCA6 might be a channelopathy, others examined possible changes in the biophysical properties of Cav2.1 by expression analysis of clones harboring expanded CAG repeats. However, no consistent functional abnormalities have been demonstrated in the mutant channels. For example, Matsuyama et al. (1999) noted no difference in current density when mutant rabbit calcium channel constructs with CAG repeats of variable lengths were expressed in baby hamster kidney cells. In mutants with 30 or 40 glutamines they observed a subtle hyperpolarizing shift in the voltage dependence of inactivation, which predicted decreased overall calcium channel function. In contrast, Piedras-Renteria and colleagues (2001) found normal biophysical properties but markedly increased cell surface targeting and current density of mutant human calcium channel with expanded CAG repeats expressed in human embryonic kidney cells. Differences in the origin of the calcium channel clones, subtypes of accessory subunits co-transfected, and the heterologous expression system must contribute to the discrepancies in the experimental observations.

Since Cav2.1 is important in neurotransmission centrally and peripherally, we hypothesized that mutant Cav2.1 may impair neurotransmission to cause myasthenic symptoms. The long isoforms including glutamine-encoding CAG repeats were found to be highly expressed in the cerebellum and the spinal cord (Restituito et al., 2000). We examined in vivo neuromuscular transmission in three patients with SCA6 with episodic features and complaint of weakness. Contrary to our findings of jitter and blocking in EA-2 patients with point mutations in CACNA1A, single-fiber electromyography demonstrated no abnormality in neuromuscular transmission in patients with SCA6 (Jen et al., 2001).

## VII. ANIMAL MODELS

There are no known natural mutants with glutamine-encoding CAG repeat expansions in the CACNA1A orthologue in mice. There are two CACNA1A null mice that manifest dystonia and progressive ataxia with prominent cerebellar degeneration (Jun et al., 1999; Fletcher et al., 2001). The absence of P/Q-type currents in these null mice indicates that P/Q-type calcium channels are indeed splice isoforms of the CACNA1A gene (Bourinet et al., 1999; Jun et al., 1999; Fletcher et al., 2001). Heterozygotes are neurologically asymptomatic, despite decreased current density in cerebellar Purkinje and granule cells.

## VIII. TREATMENT

Patients are encouraged to remain physically and mentally active. Although in randomized clinical trials no medication has been shown to halt the progression of cerebellar degeneration in SCA6, some patients with episodic exacerbation have responded to acetazolamide. Acetazolamide, a carbonic anhydrase inhibitor that alters pH, can be dramatically effective in controlling episodes of ataxia in EA-2 (Griggs et al., 1978). Similarly, in several SCA6 families with slowly progressive ataxia with episodic features, acetazolamide (500–1000 mg daily in divided doses) decreased the frequency and severity of episodic ataxia (Jen et al., 1998). In another report, six SCA6 patients given 250–500 mg acetazolamide showed, during the first year of treatment, clinical improvement based on body sway analysis (Yabe et al., 2001). Typically, one may begin with acetazolamide 125 mg daily, then increase to 250 mg twice daily as tolerated. Some patients may require 1000 mg daily. Most patients experience paresthesias in the extremities, while some complain of anorexia because of metallic/abnormal taste with certain foods. These symptoms usually subside with time. The main concern with long-term use is increased risk of developing kidney stones, which can be prevented by drinking citrus juice.

## References

Bourinet, E., Soong, T. W., Sutton, K., Slaymaker, S., Mathews, E., Monteil, A., Zamponi, G. W., Nargeot, J., and Snutch, T. P. (1999). Splicing of alpha 1A subunit gene generates phenotypic variants of P- and Q- type calcium channels. *Nat. Neurosci.* **2**, 407–415.

Buttner, N., Geschwind, D., Jen, J., Perlman, S., Pulst, S. M., and Baloh, R. W. (1998). Oculomotor phenotypes in autosomal dominant ataxias. *Arch. Neurol.* **55**, 1353–1357.

Fletcher, C. F., Tottene, A., Lennon, V. A., Wilson, S. M., Dubel, S. J., Paylor, R., Hosford, D. A., Tessarollo, L., McEnery, M. W., Pietrobon, D., Copeland, N. G., and Jenkins, N. A. (2001). Dystonia and cerebellar atrophy in Cacna1a null mice lacking P/Q calcium channel activity. *FASEB J.* **5**, 5.

Geschwind, D., Perlman, S., Figueroa, K. P., Karrim, J., Baloh, R. W., and Pulst, S. M. (1997). Spinocerebellar ataxia type 6. Frequency of the mutation and genotype-phenotype correlations. *Neurology* **49**, 1247–1251.

Gomez, C. M., Thompson, R. M., Gammack, J. T., Perlman, S. L., Dobyns, W. B., Truwit, C. L., Zee, D. S., Clark, H. B., and Anderson,

J. H. (1997). Spinocerebellar ataxia type 6: gaze-evoked and vertical nystagmus, Purkinje cell degeneration, and variable age of onset. *Ann. Neurol.* **42**, 933–950.

Griggs, R. C., Moxley, R. T., Lafrance, R. A., and McQuillen, J. (1978). Hereditary paroxysmal ataxia: response to acetazolamide. *Neurology* **28**, 1259–1264.

Ishikawa, K., Fujigasaki, H., Saegusa, H., Ohwada, K., Fujita, T., Iwamoto, H., Komatsuzaki, Y., Toru, S., Toriyama, H., Watanabe, M., Ohkoshi, N., Shoji, S., Kanazawa, I., Tanabe, T., and Mizusawa, H. (1999). Abundant expression and cytoplasmic aggregations of [alpha]1A voltage- dependent calcium channel protein associated with neurodegeneration in spinocerebellar ataxia type 6. *Hum. Mol. Genet.* **8**, 1185–1193.

Ishikawa, K., Owada, K., Ishida, K., Fujigasaki, H., Shun Li, M., Tsunemi, T., Ohkoshi, N., Toru, S., Mizutani, T., Hayashi, M., Arai, N., Hasegawa, K., Kawakami, H., Kato, T., Makifuchi, T., Shoji, S., Tanabe, T., and Mizusawa, H. (2001). Cytoplasmic and nuclear polyglutamine aggregates in SCA6 Purkinje cells. *Neurology* **56**, 1753–1756.

Jen, J., Wan, J., Graves, M., Yu, H., Mock, A. F., Coulin, C. J., Kim, G., Yue, Q., Papazian, D. M., and Baloh, R. W. (2001). Loss-of-function EA2 mutations are associated with impaired neuromuscular transmission. *Neurology* **57**, 1843–1848.

Jen, J. C., Yue, Q., Karrim, J., Nelson, S. F., and Baloh, R. W. (1998). Spinocerebellar ataxia type 6 with positional vertigo and acetazolamide responsive episodic ataxia. *J. Neurol. Neurosurg. Psychiatry* **65**, 565–568.

Jodice, C., Mantuano, E., Veneziano, L., Trettel, F., Sabbadini, G., Calandriello, L., Francia, A., Spadaro, M., Pierelli, F., Salvi, F., Ophoff, R. A., Frants, R. R., and Frontali, M. (1997). Episodic ataxia type 2 (EA2) and spinocerebellar ataxia type 6 (SCA6) due to CAG repeat expansion in the CACNA1A gene on chromosome 19p. *Hum. Mol. Genet.* **6**, 1973–1978.

Jun, K., Piedras-Renteria, E. S., Smith, S. M., Wheeler, D. B., Lee, S. B., Lee, T. G., Chin, H., Adams, M. E., Scheller, R. H., Tsien, R. W., and Shin, H. S. (1999). Ablation of P/Q-type Ca(2+) channel currents, altered synaptic transmission, and progressive ataxia in mice lacking the alpha(1A)- subunit. *Proc. Natl. Acad. Sci. U.S.A.* **96**, 15245–15250.

Mariotti, C., Gellera, C., Grisoli, M., Mineri, R., Castucci, A., and Di Donato, S. (2001). Pathogenic effect of an intermediate-size SCA6 allele (CAG)19 in a homozygous patient. *Neurology* **57**, 1502–1504.

Matsuyama, Z., Wakamori, M., Mori, Y., Kawakami, H., Nakamura, S., Imoto, K. (1999). Direct alteration of the P/Q-type Ca2+ channel property by polyglutamine expansion in spinocerebellar ataxia 6. *J. Neurosci.* **19(21)**, RC14.

Mori, Y., Friedrich, I., Kim, M.-S., Mikami, A., Nakai, J., Ruth, P., Bosse, E., Hofmann, F., Flockerzi, V., Furuichi, T., Mikoshiba, K., Imoto, K., Tanabe, T., and Numa, S. (1991). Primary structure and functional expression from complementary DNA of a brain calcium channel. *Nature* **350**, 398–402.

Ophoff, R. A., Terwindt, G. M., Vergouwe, M. N., van Eijk, R., Oefner, P. J., Hoffman, S. M., Lamerdin, J. E., Mohrenweiser, H. W., Bulman, D. E., Ferrari, M., Haan, J., Lindhout, D., van Ommen, G. J., Hofker, M. H., Ferrari, M. D., and Frants, R. R. (1996). Familial hemiplegic migraine and episodic ataxia type-2 are caused by mutations in the Ca2+ channel gene CACNL1A4. *Cell* **87**, 543–552.

Piedras-Renteria, E. S., Watase, K., Harata, N., Zhuchenko, O., Zoghbi, H. Y., Lee, C. C., and Tsien, R. W. (2001). Increased expression of alpha1A Ca channel currents arising from expanded trinucleotide repeats in spinocerebellar ataxia type 6. *J. Neurosci.* **21**, 9185–9193.

Protti, D. A., Reisin, R., Mackinley, T. A., and Uchitel, O. D. (1996). Calcium channel blockers and transmitter release at the normal human neuromuscular junction. *Neurology* **46**, 1391–1396.

Restituito, S., Thompson, R. M., Eliet, J., Raike, R. S., Riedl, M., Charnet, P., and Gomez, C. M. (2000). The polyglutamine expansion in spinocerebellar ataxia type 6 causes a beta subunit-specific enhanced activation of P/Q-type calcium channels in Xenopus oocytes. *J. Neurosci.* **20**, 6394–6403.

Yabe, I., Sasaki, H., Yamashita, I., Takei, A., and Tashiro, K. (2001). Clinical trial of acetazolamide on spinocerebellar ataxia type 6 (SCA6). *Acta Neurol. Scand.* **104**, 44–47.

Yang, Q., Hashizume, Y., Yoshida, M., Wang, Y., Goto, Y., Mitsuma, N., Ishikawa, K., and Mizusawa, H. (2000). Morphological Purkinje cell changes in spinocerebellar ataxia type 6. *Acta Neuropathol* **100**, 371–376.

Zhuchenko, O., Bailey, J., Bonnen, P., Ashizawa, T., Stockton, D. W., Amos, C., Dobyns, W. B., Subramony, S. H., Zoghbi, H. Y., and Lee, C. C. (1997). Autosomal dominant cerebellar ataxia (SCA6) associated with small polyglutamine expansions in the alpha 1A-voltage-dependent calcium channel. *Nat. Genet.* **15**, 62–69.

Zoghbi, H. (1997). CAG repeats in SCA6. Anticipating new clues. *Neurology* **49**, 1196–1199.

# CHAPTER 9

# Spinocerebellar Ataxia 7 (SCA7)

**ANNE-SOPHIE LEBRE**
*Neurologie et Thérapeutique Expérimentale*
*INSERM U289*
*Département Génétique, Cytogénétique*
*et Embryologie*
*Group Hôspitalier Pitie-Salpêtrière*
*75651 Paris Cedex*
*France*

**GIOVANNI STEVANIN**
*Neurologie et Thérapeutique Expérimentale*
*INSERM U289*
*Group Hôspitalier Pitie-Salpêtrière*
*75651 Paris Cedex*
*France*

**ALEXIS BRICE**
*Neurologie et Thérapeutique Expérimentale*
*INSERM U289*
*Département Génétique, Cytogénétique*
*et Embryologie*
*Group Hôspitalier Pitie-Salpêtrière*
*75651 Paris Cedex*
*France*

I. Summary
II. Phenotype
   A. Visual Impairment
   B. Neurological Signs
III. Gene
   A. Normal Gene Function
   B. Abnormal Gene Function
IV. Diagnostic and Ancillary Tests
   A. DNA Tests
   B. Other Laboratory Tests
V. Neuroimaging
VI. Neuropathology
VII. Cellular and Animal Models of Disease
   A. *In Vitro*
   B. Invertebrates
   C. Mouse
VIII. Genotype/Phenotype Correlations/Modifying Alleles
   A. Genotype/Phenotype Correlations
   B. Modifying Alleles: Factors Influencing Clinical variability
IX. Treatment
X. Conclusion
   Acknowledgments
   References

## I. SUMMARY

Spinocerebellar ataxia 7 (SCA7) is a progressive autosomal dominant neurodegenerative disorder characterized clinically by cerebellar ataxia associated with progressive macular dystrophy. The disease affects primarily the cerebellum and the retina, but also many other CNS structures as the disease progresses. SCA7 is caused by expansion of an unstable trinucleotide CAG repeat encoding a polyglutamine tract in the corresponding protein, ataxin-7. Normal SCA7 alleles contain 4–35 CAG repeats, whereas pathological alleles contain from 36–306 CAG repeats. Ataxin-7 is a protein of unknown function expressed in many tissues including the CNS, but its mutation leads to selective neuronal death only in the brain. The expanded polyglutamine tract renders the protein toxic and leads to its aggregation in a subset of vulnerable neurons in SCA7 brain. These aggregates, named neuronal intranuclear inclusions (NIIs), also stain positively for ubiquitin, chaperones, and proteasome, as in other polyglutamine diseases, providing evidence that the pathogenesic mechanisms underlying these disorders have features in common. Cellular models and SCA7 transgenic mice have been generated which constitute valuable resources for studying the disease mechanism. They will also be useful for screening and evaluating possible therapeutic strategies.

## II. PHENOTYPE

SCA7 is an autosomal dominant cerebellar ataxia (ADCA), which has also been designated olivo-pontocerebellar-atrophy type III (Konigsmark and Weiner, 1970) or ADCA type II (Harding, 1982; 1993). This type of ADCA was initially described by Froment *et al.* (1937) and is characterized by very heterogeneous clinical presentations, disease severity, and age at onset (Benomar *et al.*,

1994; Enevoldson et al., 1994). Because of the retinopathy observed in most patients, this form of ADCA was considered to be a separate entity (OMIM database entry:164500).

## A. Visual Impairment

SCA7, unlike other ADCAs, is characterized by a progressive macular degeneration that can be visualized in most patients as a pigmented central core in the macula that can extend, in latter stages, into the periphery. Visual failure is progressive, bilateral, and symmetrical, and leads irreversibly to blindness (Enevoldson et al., 1994; Gouw et al., 1994). Central vision is affected first. Peripheral vision is preserved at early stages, explaining why patients do not complain of symptoms until failure is well advanced, and why night vision is not impaired. Interestingly, dyschromatopia in the blue-yellow axis is found years before visual failure becomes symptomatic (Gouw et al., 1994).

## B. Neurological Signs

The neurological signs of the ADCAs clearly overlap (Stevanin et al., 2000a). Studies of large groups of patients have, however, revealed a constellation of signs that is frequently found in SCA7 patients (Table 9.1). Cerebellar ataxia is always associated with dysarthria, but patients present variably with pyramidal signs (increased reflexes and/or extensor plantar reflexes and/or lower limb spasticity), decreased vibration sense, dysphagia, sphincter disturbances, and oculomotor abnormalities (supranuclear ophthalmoplegia and/or viscosity of eye movements). Extrapyramidal features (dystonia), myokymia, peripheral neuropathy and mental impairment are rare. The frequency of swallowing and sphincter disturbances significantly increases with disease duration (David et al., 1998). The association of cerebellar ataxia and dysarthria with pyramidal signs, supranuclear ophthalmoplegia, slow saccades, and decreased visual acuity is highly suggestive of SCA7.

## III. GENE

### A. Normal Gene Function

Three independent groups mapped the responsible gene, designated SCA7, to chromosome 3p12–p21.1 (Benomar et al., 1995; Gouw et al., 1995; Holmberg et al., 1995). As anticipation is particularly marked in SCA7, they postulated that a trinucleotide CAG repeat expansion might be involved in this disease. A pure $(CAG)_{10}$ repeat, located between markers D3S1600 and D3S1287, was found in a subclone of YAC 882_d_9 (David et al., 1997; Krols et al., 1997). PCR analysis with primers flanking this CAG repeat showed the presence of an expansion in SCA7 patients (David et al., 1997; Del-Favero et al., 1998), permitting cloning of the SCA7 cDNA. Lindblad et al. (1996), using the repeat expansion detection (RED) technique (Schalling et al., 1993), also demonstrated that long stretches of CAG repeats co-segregated with the disease in several ADCA II pedigrees. These results were confirmed by Koob et al. (1998) by a strategy of cloning derived from the RED technique.

A 7.5-kb transcript, detected by Northern-blot analysis, is expressed ubiquitously in adult and fetal tissues (David et al., 1997; Del-Favero et al., 1998). The 2727 bp open reading frame of the SCA7 cDNA encodes a protein of 892 amino acids designated ataxin-7 (David et al., 1997). Ataxin-7 contains several identifiable regions, such as polyalanine and polyglutamine repeats, a bipartite nuclear localization signal (NLS), and two serine-rich regions. Expression studies in several cellular models confirmed that the NLS directs normal and pathological ataxin-7 to the nucleus (Kaytor et al., 1999; Zander et al., 2001). Downstream of the polyglutamine tract, ataxin-7 contains four polyproline sequences predicted to be SH3-binding domains. Mushegian et al. identified a short motif in ataxin-7, homologous to a motif found in arrestins, that binds selectively to the phosphorylated activated forms of G-protein-coupled receptors and quenches their signaling (Gurevich et al., 1995; Mushegian et al., 2000). The arrestin-like sequence in ataxin-7 binds phosphate *in vitro* (Mushegian et al., 2000).

TABLE 9.1  Major Phenotypic Characteristics of 69 SCA7 Patients with a Mean Age at Onset of 29 ± 16 Years (1–70) and Carrying 51 ± 13 CAG repeats (38–130)

| Clinical signs | % |
| --- | --- |
| Cerebellar ataxia | 100 |
| Dysarthria | 98.5 |
| Decreased visual acuity | 81 |
| Brisk reflexes | 80 |
| Diminished or abolished reflexes | 3 |
| Babinski sign | 55 |
| Ophthalmoplegia | 54 |
| Slow saccades | 63 |
| Deep sensory loss | 60 |
| Sphincter disturbances | 55 |
| Amyotrophy | 25 |
| Auditory impairment | 24 |
| Axonal neuropathy | 18 |
| Facial myokimia | 13 |
| Dementia | 12 |
| Extrapyramidal rigidity | 14 |
| Dystonia | 9 |
| Bulging eyes | 6 |
| Nystagmus | 2 |

From Stevanin, et al. (2000b). Spinocerebellar ataxia 7 (SCA7). In *Handbook of Ataxia Disorders*. (T. Klockgether, ed.) Marcel Dekker, Inc., New York. With permission.

The role of ataxin-7 remains unknown, but expression studies can provide information on its normal and pathological functions. Lindenberg et al. studied the regional and cellular expression patterns of ataxin-7 at the mRNA level by *in situ* hybridization in normal human brain (Lindenberg et al., 2000). At the mRNA level ataxin-7 was preferentially expressed in neurons; the regional distribution reflected neuronal packing density.

Several groups have used monoclonal and polyclonal antibodies raised against the normal ataxin-7 to establish the distribution of this protein in brain, retina, and peripheral tissues (Lindenberg et al., 2000; Cancel et al., 2000; Einum et al., 2001). Immunoblotting demonstrated that ataxin-7 is widely expressed but that expression levels vary among tissues (Einum et al., 2001). Immunohistochemical studies have confirmed that ataxin-7 is widely expressed, and have shown that in most neurons, ataxin-7 immunoreactivity (IR) is preferentially found in cell bodies and processes. Ataxin-7 is found in the cytoplasmatic compartment although some nuclear ataxin-7 IR has also been detected in most neurons. Double immunolabeling coupled with confocal microscopy showed that some ataxin-7 co-localizes with BiP, a marker of the endoplasmic reticulum, but not with markers of mitochondria or the trans-Golgi network (Cancel et al., 2000). More intense and more prominently nuclear ataxin-7 IR was observed in neurons of the pons and the inferior olive, brain regions severly affected by the disease, suggesting that the subcellular localization and abundance of ataxin-7 is regulated in a region-specific manner. Lindenberg et al. suggested that since the neurons with more intense and more prominent nuclear ataxin-7 IR are those that are susceptible to SCA7, enrichment of normal ataxin-7 in the nuclear compartment may contribute to neurodegeneration (Lindenberg et al., 2000). However, strong cytoplasmatic and nuclear ataxin-7 IR is not limited to sites of SCA7 pathology and is not always related to the severity of neuronal loss. For example, IR is low in some vulnerable populations of neurons, such as Purkinje cells (Cancel et al., 2000). Ataxin-7 IR has also been detected throughout the retina and is particularly intense within the cell bodies and photosensitive outer segments of cone photoreceptors (Einum et al., 2001).

## B. Abnormal Gene Function

The mutated protein was first reported to be enriched in the nucleus of lymphoblasts of SCA7 patients (Trottier et al., 1995). The 1C2 monoclonal antibody, that specifically recognizes long polyglutamine stretches on Western blots, detected a specific 130-kDa protein in lymphoblasts (Trottier et al., 1995) and cerebral cortex (Stevanin et al., 1996) from SCA7 patients with predominantly early onset, which was not found in controls and patients with other ADCAs. The disease could result from a gain of function that occurs at the protein level and increases with repeat size after a threshold of approximately 36–40 glutamines. This is in agreement with the expression of both the mutated and the normal proteins as well as the dominant nature of the mutation. The expansion probably alters the conformation of the polyglutamine tract, as initially suggested by the specific detection of long repeats using the 1C2 antibody (Trottier et al., 1995; Stevanin et al., 1996). However, in Huntington's disease, recent results indicate that loss of the normal function of the protein might also contribute to the pathology (Cattaneo et al., 2001). This may also apply to SCA7, for which no knock-out models have yet to be reported.

Ataxin-7-containing aggregates, called NIIs, first detected in neurons of several brain regions in a juvenile SCA7 patient (Holmberg et al., 1998), are pathological structures characteristic of polyglutamine disorders (Zoghbi and Orr, 2000). NIIs have been detected in the brain and retina of SCA7 patients with the 1C2 antibody and with anti-ataxin-7 antibodies (Holmberg et al., 1998; Mauger et al., 1999 ; Cancel et al., 2000; Einum et al., 2001). NIIs were most frequent in the inferior olivary complex, a site of severe neuronal loss in SCA7 (Holmberg et al., 1998). They were also observed in other brain regions, including the cerebral cortex, not considered to be affected in the disease. In addition, cytoplasmic staining by the 1C2 antibody was also observed, particularly in the supramarginal gyrus, the hippocampus, the thalamus, the lateral geniculate body, and the pontine nuclei. These data confirm that, in SCA7, the NIIs are not restricted to the sites of severe neuronal loss. Conversely, no NIIs were detected in neurons of a SCA7 patient with 41 repeats after 40 years of disease duration (Einum et al., 2001). When observed by confocal microscopy, some inclusions were found to be ubiquitinated (Holmberg et al., 1998; Mauger et al., 1999), but to varying degrees, ranging from < 1% in the cerebral cortex to 60% in the inferior olive (Holmberg et al., 1998). In electron microscopy studies using immunogold labeling, ataxin-7 immunoreactive NIIs appear as dense aggregates containing a mixture of granular and filamentary structures (Mauger et al., 1999).

The inclusions may be only a hallmark of the diseases or reflect a cellular defense mechanism. Their implication in the degenerative process remains a matter of debate. Their presence in unaffected tissues in SCA7 patients does not seem sufficient to initiate the degenerative process (Holmberg et al., 1998).

## IV. DIAGNOSTIC AND ANCILLARY TESTS

### A. DNA Tests

Identification of SCA7 gene and its mutation enables direct detection of the mutation with routine laboratory tests in individuals who already present with symptoms of the disease. This molecular analysis is also useful to distinguish disorders which are clinically similar. Large series of controls and patients have now been analyzed and have shown that the SCA7 CAG repeat is polymorphic, with sizes ranging from 4–35 units in control, and from 36–306 in SCA7 and at-risk carrier chromosomes (Benton et al., 1998; David et al., 1998; Del-Favero et al., 1998; Gouw et al., 1998; Johansson et al., 1998; Koob et al., 1998; Lyoo et al., 1998; Stevanin et al., 1998; Nardacchione et al., 1999; Giunti et al., 1999). The largest expansions (between 114 and 306 CAG repeats) were found in juvenile or infantile cases (David et al., 1998; Benton et al., 1998; Johansson et al., 1998; Giunti et al., 1999).

The strong anticipation in SCA7 and the rarity of contractions should have led to its extinction within a few generations (see below). However, this tendency is counterbalanced by the occurrence of *de novo* SCA7 expansions from intermediate alleles (IAs) with 28–35 repeats in individuals that were still unaffected at ages 50–84 (Stevanin et al., 1998). Should these cases become affected with time, the intermediate size alleles may cause a disease with incomplete penetrance, as is the case for small expansions carrying 36–41 repeats at the HD locus (Andrew et al., 1997; Rubinsztein et al., 1996). Indeed, individuals carrying 34 (Gouw et al., 1998) and 35 (Koob et al., 1998) CAG repeats have been reported, but their clinical status and ages at onset were not described.

Care should therefore be taken to detect both intermediate size and large alleles. The latter are difficult to amplify but should be suspected in juvenile or infantile forms of the disease (Benton et al., 1998; David et al., 1998; Johansson et al., 1998; Giunti et al., 1999). Molecular diagnosis of isolated cases with a phenotype compatible with SCA7 is also crucial. Positive isolated cases rarely carry *de novo* mutation (Stevanin et al., 1998). More frequently, they reflect missing family histories if the transmitting parent died before the onset of symptoms or is still asymptomatic because of marked anticipation (Cancel et al., 1997; Schols et al., 1997)

Presymptomatic testing in SCA7 families is possible for adult at-risk individuals, but raises the same difficult ethical issues as other severe adult-onset disorders for which no treatment can be proposed and for which age at onset cannot be precisely predicted from the number of CAG repeats. The international guidelines established for Huntington's disease should also be followed for SCA7 (World Federation of Neurology Research Group on Huntington's Chorea, 1994).

### B. Other Laboratory Tests

Abnormal fundoscopy, initially consisting of a loss of the foveal reflex and progressive molting of pigment at the macula, is not constant and is often difficult to visualize at early stages of the disease, whereas dyschromatoptia in the blue-yellow axis is an early finding. Electroretinograms show abnormal photopic responses, but scotopic responses are preserved until late in the disease.

Somatosensory-evoked potentials are often abnormal. An infraclinical sensory and/or motor neuropathy on EMG and nerve conduction studies is found in a minority of patients. Visual-evoked potentials are not discriminative for diagnostic purposes (Enevoldson et al., 1994).

## V. NEUROIMAGING

Brain imaging shows marked atrophy in the cerebellum, particularly in the superior part of the vermis and the brainstem, which may be associated with moderate atrophy of the cerebral cortex (Fig. 9.1).

## VI. NEUROPATHOLOGY

Neuropathological examination reveals moderate to severe neuronal loss and gliosis in the cerebellum and associated structures (inferior olive, cerebellar cortex, dentate

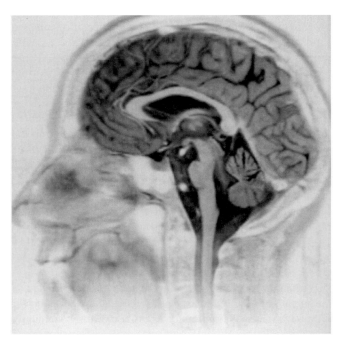

**FIGURE 9.1** Sagittal MRI of an SCA7 patient. T1 weighted sequence (TR = 450 ms, TE = 11 ms) showing obvious atrophy of the cerebellar vermis associated with mild atrophy of the pons (From David et al. (1998). In *Genetic Instabilities and Hereditary Neurological Diseases*. (R.D. Wells and S.T. Warren, eds.), pp. 273–282. Academic Press, San Diego. With permission.)

nucleus, pontine nuclei), the basal ganglia (*globus pallidus*, *substantia nigra*, subthalamic nucleus, red nucleus) and the spinal cord (Enevoldson *et al.*, 1994; Gouw *et al.*, 1994; Martin *et al.*, 1994; Holmberg *et al.*, 1998; Cancel *et al.*, 2000). Spinocerebellar, olivocerebellar, and efferent cerebellar tracts are severely affected. In the cerebellum, the vermis is more affected than the hemispheres where Purkinje cells, and to a lesser extent granule cells, degenerate. Mild cell loss also occurs in the dentate nucleus which, as the result of Purkinje cell degeneration, has a reduced mantel. Extensive neuronal loss is observed in the inferior olive, with marked astrocytic gliosis. Mild cell loss is also observed in the *substantia nigra* and the *basis pontis*, whereas the thalamus and the striatum are spared.

The distinctive neuropathological features of SCA7 are degeneration of optic pathways and the retina. The pregeniculate visual pathways and the optic nerve are affected, probably as a consequence of retinal degeneration. In juvenile cases presenting with blindness, those systems may not be altered, probably due to the rapid course of the disease. Pathological examination of the retina shows early degeneration of photoreceptors and of bipolar and granular cells, particularly in the foveal and parafoveal regions. Later, patchy loss of epithelial pigment cells and their ectopic migration into the retinal layers are observed (Martin *et al.*, 1994).

## VII. CELLULAR AND ANIMAL MODELS OF DISEASE

### A. *In Vitro*

Immunocytochemical studies have been carried out on neuronal and non-neuronal cells expressing both wild-type and mutated ataxin-7 (Kaytor *et al.*, 1999; Zander *et al.*, 2001). In both cases, intense labeling was observed almost exclusively in the nucleus. Nuclear transport was blocked by disruption of the nuclear localization signal (Kaytor *et al.*, 1999). In addition to diffuse distribution throughout the nucleus, ataxin-7 was associated with the nuclear matrix, PML promyelocytic leukemia protein (PML) bodies and nucleoli (Kaytor *et al.*, 1999). Thus, expanded ataxin-7 may exert its pathogenic effects in the nucleus by altering a matrix-associated nuclear structure or by disrupting nucleolar function.

Zander *et al.* (2001) have compared the inclusions in cells expressing mutant ataxin-7 and in human SCA7 brain tissue. The cells readily formed anti-ataxin-7 positive, fibrillar inclusions and small, nuclear electron dense structures. There were consistent signs of ongoing abnormal protein folding, including the recruitment of heat shock proteins and proteasome subunits (Fig. 9.2). Occasionally, transcription factors were sequestered in the inclusions. Activated caspase-

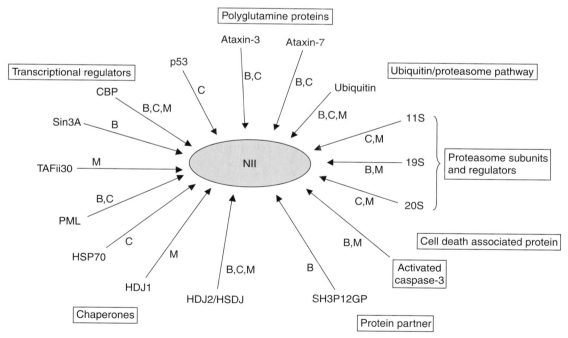

FIGURE 9.2 Composition of NIIs in human SCA7 brain (B), SCA7 cellular models (C), and SCA7 mouse models (M). CBP: CREB-binding protein; PML: promyelocytic leukemia protein; HSP: heat shock protein; HDJ1/2: proteins of the HSP40 family; and SH3P12GP: SH3P12 gene products (From Lebre *et al.* (2000) *Hum. Mol. Genet.* **10**, 1201–1213; Yvert *et al.* (2000). *Hum. Mol. Genet.* **9**, 2491–2506; Yvert *et al.* (2001). *Hum. Mol. Genet.* **10**, 1679–1692; Zander *et al.* (2001). *Hum. Mol. Genet.* **10**, 2569–2579.)

3 was recruited into the inclusions in both the cell models and human SCA7 brain, and its expression was upregulated in cortical neurons, suggesting that it may play a role in the disease process. Finally, on the ultrastructural level, there were signs of autophagy and nuclear indentations, indicative of a major stress response in cells expressing mutant ataxin-7.

Caspase cleavage of ataxin-7 might be implicated, since there are several potential cleavage sites in the ataxin-7 sequence (unpublished data). The presence of these sites in ataxin-7 is reminiscent of huntingtin, atrophin, ataxin-3, and the androgen receptor, which are suspected to be truncated by such proteases (Wellington et al., 1998) and, as a consequence, acquire a toxic property (Ellerby et al., 1999; Ikeda et al., 1996) or aggregate (Cooper et al., 1998; Martindale et al., 1998). Inhibition of caspases could, as in a Huntington's disease model (Ona et al., 1999), prevent the progression of SCA7 in animal and cellular models.

As in other polyglutamine diseases, the selective pattern of degeneration contrasts with the ubiquitous expression of ataxin-7. Several explanations have been proposed to explain this, as such a difference in the level of ataxin-7 expression or the presence of specific ataxin-7 binding proteins, but none is as yet satisfactory. Lebre et al. used a two-hybrid approach to screen a human retina cDNA library for ataxin-7 binding proteins (Lebre et al., 2001). A protein R85 was isolated and found to encode a splice variant of CAP (Cb1 associated protein; Ribon et al., 1998). R85 and CAP are generated by alternative splicing of the gene SH3P12 that was localized on chromosome 10q23–q24. The R85 cDNA contains a bipartite NLS, three SH3 domains, and interacts with normal and mutated ataxin-7. SH3P12 gene products (SH3P12GP) are expressed in Purkinje cells in the cerebellum. Ataxin-7 co-localizes with an SH3P12GP in NIIs in brain from an SCA7 patient. This interaction may be part of a physiological pathway related to the function or turnover of ataxin-7 and may be also involved in the pathophysiological process of SCA7 disease. Matilla et al. (2001) showed interaction of ataxin-7 with ATPase subunit S4 of the proteasomal 19S regulatory complex. Immunoblot analyses demonstrate reduced levels of S4 immunoreactivity in cerebellar protein extracts from two SCA7 patients. Authors suggest a role of S4 and proteasomal proteolysis in the molecular pathogenesis of SCA7.

### B. Invertebrates

No invertebrate models of SCA7 are available.

### C. Mouse

No SCA7 knock-out mice are available. Yvert et al. (2000, 2001) described transgenic SCA7 mice that overexpress full-length mutant ataxin-7 Q90 in Purkinje cells (P7 mice), in rod photoreceptors (R7 mice), or with a more widespread expression of mutant ataxin-7 (B7 mice), including neuronal cell types that are unaffected in SCA7 patients. P7 and R7 mice have deficiencies in motor coordination and vision, respectively, and display neurodegeneration with progressive ataxin-7 accumulation into ubiquitinated NIIs (Yvert et al., 2000). In R7 mice, severe degeneration of photoreceptors, but also secondary alterations of post-synaptic neurons was caused by overexpression of ataxin-7 Q90 in rods. Interestingly, ERG are abnormal in R7 mice before photoreceptor degeneration begins, thus providing a single means for assessing the protection of photoreceptors in future therapeutic trials (Yvert et al., 2000). Furthermore, these authors also suggest that proteolytic cleavage of mutant ataxin-7 and trans-neuronal responses are implicated in the pathogenesis of SCA7 (Yvert et al., 2000). In B7 mice, a similar handling of mutant ataxin-7, including a cytoplasm to nucleus translocation and accumulation of N-terminal fragments, was observed in all neuronal populations studied. An extensive screen for chaperones, proteasomal subunits, and transcription factors sequestered in NIIs disclosed no pattern unique to neurons undergoing degeneration in SCA7 (Fig. 9.2). In particular, the mouse TAF(II)30 subunit of the TFIID initiation complex accumulated markedly in NIIs, even though this protein does not contain a polyglutamine stretch. A striking discrepancy between mRNA and ataxin-7 levels in B7 mice expressing wild-type but not mutant ataxin-7 indicates selective stabilization of mutant ataxin-7, both in this model and in P7 mice. B7 mice, therefore, provide *in vivo* evidence that the polyglutamine expansion mutation can increase protein half-life (Yvert et al., 2001).

La Spada and collaborators (2001) also produced a transgenic SCA7 mice with a cone-rod dystrophy. NIIs were present, suggesting that the disease pathway involves the nucleus. Yeast two-hybrid assays indicated that a cone-rod homeobox protein (CRX) interacts with ataxin-7. Ataxin-7 and CRX co-localized and co-immunoprecipitated. In addition, polyglutamine-expanded ataxin-7 dramatically suppressed CRX transactivation. In this transgenic SCA7 model, electrophoretic mobility shift assays indicated reduced CRX binding activity, and RT-PCR analysis showed reduced expression of CRX-regulated genes. Authors suggest that interference with CRX-regulated transcription is responsible for retinal degeneration in this SCA7 model and may provide an explanation for how cell-type specificity is achieved in the polyglutamine disease.

## VIII. GENOTYPE/PHENOTYPE CORRELATIONS/ MODIFYING ALLELES

### A. Genotype/Phenotype Correlations

#### 1. First Sign at Onset

Cerebellar ataxia is usually the presenting symptom in adults with onset over thirty. In patients with earlier onset, however, decreased visual acuity, alone or associated with cerebellar ataxia, is the initial symptom (Benomar et al., 1994; Enevoldson et al., 1994). Some infantile cases may, however, result in early death without detectable retinal alterations (Martin et al., 1999). More than 45 years can elapse between the appearance of cerebellar symptoms and visual failure, while, in the reverse situation, the latency never exceeds 9 years (David et al., 1998; Giunti et al., 1999). In some patients with late onset, visual acuity may never decrease.

## 2. Age at Onset, Anticipation, and Disease Duration

The clinical manifestations typically begin in the third or fourth decade, with mean age at onset close to 30, but a range of 3 months or less to over 70 years (Martin et al., 1994; Holmberg et al., 1995; Jöbsis et al., 1997; Benton et al., 1998; David et al., 1998; Gouw et al., 1998; Giunti et al., 1999). Analysis of parent-child couples have revealed striking anticipation (~20 years/generation). Previous studies reported significantly greater anticipation in paternal than in maternal transmissions (Benomar et al., 1994, 1995; David et al., 1996). This was not confirmed in recent reports (Benton et al., 1998; David et al., 1998; Johansson et al., 1998; Del-Favero et al., 1998; Giunti et al., 1999), although all juvenile cases are paternally transmitted. This correlates well with the marked tendency to increased expansion in the sperm of patients compared to their blood (David et al., 1998). In a recent study using small pool PCR in sperm, very large expanded repeats were detected although they are under-represented in patients suggesting that a significant proportion of such alleles might be associated with embryonic lethality or dysfunctional sperm (Monckton et al., 1999).

There is a strong negative correlation between the size of the CAG expansion and age at onset. The former accounts for ~75 % of the variability of the latter, suggesting that other genetic and/or environmental factors play only a minor role in determining the SCA7 phenotype, unlike other disorders involving polyglutamine expansions (David et al., 1998; Gouw et al., 1998; Giunti et al., 1999). The CAG length/age at onset correlation, together with the increase in expansion size in successive generations, is consistent with the marked anticipation observed in ADCA II families (David et al., 1998).

In SCA7, anticipation is characterized by earlier onset, but also by more rapid disease progression and increased severity in successive generations. Disease duration until death is negatively correlated with the number of CAG repeats on the expanded allele and is limited to a few month or years in early onset patients (David et al., 1998). The mean duration in subjects carrying less than 49 repeats is 15 years and differs significantly from the 11-year duration when the number of repeats is 49 or more (Giunti et al., 1999). Longer disease durations, up to 30 years or more, are observed only in late onset patients (Neetens et al., 1990). Anticipation is also associated with increasing severity of symptoms in successive generations. The frequency of decreased visual acuity, ophthalmoplegia, scoliosis, and extensor plantar reflexes significantly increases with the size of the expansion (David et al., 1998). The phenotype of a given SCA7 patient partly depends on both the size of the mutation and disease duration at the time of examination. In some infantile cases with very large repeat expansions, progression is extremely rapid and the heart can be affected (Benton et al., 1998; Neetens et al., 1990). It is surprising that the retina may also be affected in juvenile SCA2 patients (Babovic-Vuksanovic et al., 1998). The retina and the cardiac muscle may be sensitive only to large or very large expansions, respectively. The pathological threshold of the polyglutamine expansion may therefore be tissue-dependent.

## B. Modifying Alleles: Factors Influencing Clinical Variability

No genetic factors influencing clinical variability have been identified so far, but several intragenic polymorphisms have been described. Polymorphisms in the number of GCN and CCG repeats, upstream and downstream, respectively, of the CAG repeat have been observed (Stevanin et al., 1998), as well as an A/G$^{3145}$ polymorphism resulting in a Met → Val substitution that does not affect function (Stevanin et al., 1999).

## IX. TREATMENT

There is no specific drug therapy for this neurodegenerative disorder. Currently, therapy remains purely symptomatic. Appropriate measures can reduce diplopia, swallowing, or sphincter disturbances. Dementia can be present (12%) and needs specific care. Therapy in gain of function diseases with adult onset is fraught with technical difficulties. Several major therapeutic avenues are being explored in polyglutamine diseases and could be tested in SCA7.

## X. CONCLUSION

SCA7 is a neurodegenerative disorder caused by expansion of an unstable trinucleotide CAG repeat encoding a polyglutamine tract, in which the degenerative process affects the retina in addition to other brain structures. SCA7 has a number of features in common with other diseases with polyglutamine expansions: (1) the appearance of clinical symptoms above a threshold number of CAG repeats (>35); (2) a strong negative correlation between the CAG repeat size and both age at onset and disease progression; (3) instability of the repeat sequence (~12 CAG/transmission) that accounts for the marked anticipation of ~20 years/generation; (4) ubiquitous expression of the gene; and (5) accumulation of the pathological protein in ubiquitinated NIIs. The expanded polyglutamine tract confers a toxic property on the proteins affected and leads to their aggregation in a subset of vulnerable neurons. These aggregates stain positively for ubiquitin, chaperones, and proteasome components.

SCA7 is characterized by marked instability of the responsible gene, a strong effect of CAG repeat size on the phenotype, and the wide variety of tissues affected, including the retina. The CAG repeat sequence is particularly unstable and *de novo* mutations can occur during paternal transmissions of intermediate size alleles (28–35 CAG repeats). This can explain the persistence of the disease in spite of the anticipation that should have resulted in its extinction. Molecular analysis can be used to confirm the clinical diagnosis and to identify gene carriers among at-risk individuals, in accordance with the now classical guidelines in use for Huntington's disease (World Federation of Neurology Research Group on Huntington's Chorea, 1994).

Cellular models and transgenic SCA7 mice are now available. They constitute valuable resources for studying the physiopathology of the disease, and will be useful for screening and evaluating the potential therapeutic strategies that are currently being explored in other polyglutamine diseases.

## Acknowledgments

The authors are grateful to Dr. Merle Ruberg for critical reading of the manuscript. The authors works are financially supported by the VERUM foundation, the Institut National de la Santé et de la Recherche Médicale (INSERM), the "Association Française de lutte contre les Myopathies" (AFM), the French "Ministère de la Recherche," and the "Association pour le Développement de la Recherche sur les Maladies Génétiques Neurologiques et Psychiatriques" (ADRMGNP).

## References

Andrew, S. E., Goldberg, Y. P., and Hayden, M. R. (1997). Rethinking genotype and phenotype correlations in polyglutamine expansion disorders. *Hum. Mol. Genet.* **6**, 2005–2010.

Babovic-Vuksanovic, D., Snow, K., Patterson, M. C., and Michels, V. V. (1998). Spinocerebellar ataxia type 2 (SCA2) in an infant with extreme CAG repeat expansion. *Am. J. Med. Genet.* **79**, 383–387.

Benomar, A., Le Guern, E., Dürr, A., Ouhabi, H., Stevanin, G., Yahyaoui, M., Chkili, T., Agid, Y., and Brice, A. (1994). Autosomal-dominant cerebellar ataxia with retinal degeneration (ADCA type II) is genetically different from ADCA type I. *Ann. Neurol.* **35**, 439–444.

Benomar, A., Krols, L., Stevanin, G., Cancel, G., Le Guern, E., David, G., Ouhabi, H., Martin, J. J., Dürr, A., Zaim, A., Ravise, N., Busque, C., Penet, C., Van Regemorter, N., Weissenbach, J., Yahyaoui, M., Chkili, T., Agid, Y., Van Broeckhoven, C., and Brice, A. (1995). The gene for autosomal dominant cerebellar ataxia with pigmentary macular dystrophy maps to chromosome 3p12–p21.1. *Nat. Genet.* **10**, 84–88.

Benton, C. S., de Silva, R., Rutledge, S. L., Bohlega, S., Ashizawa, T., and Zoghbi, H.Y. (1998). Molecular and clinical studies in SCA-7 define a broad clinical spectrum and the infantile phenotype. *Neurology* **51**, 1081–1086.

Cancel, G., Dürr, A., Didierjean, O., Imbert, G., Bürk, K., Lezin, A., Belal, S., Benomar, A., Abada-Bendib, M., Vial, C., Guimaraes, J., Chneiweiss, H., Stevanin, G., Yvert, G., Abbas, N., Saudou, F., Lebre, A.-S., Yahyaoui, M., Hentati, F., Vernant, J.-C., Klockgether, T., Mandel, J.-L., Agid, Y., and Brice, A. (1997). Molecular and clinical correlations in spinocerebellar ataxia 2: a study of 32 families. *Hum. Mol. Genet.* **6**, 709–715.

Cancel, G., Duyckaerts, C., Holmberg, M., Zander, C., Yvert, G., Lebre, A.S., Ruberg, M., Faucheux, B., Agid, Y., Hirsch, E., and Brice, A. (2000). Distribution of ataxin-7 in normal human brain and retina. *Brain* **123**, 2519–2530.

Cattaneo, E., Rigamonti, D., Goffredo, D., Zuccato, C., Squitieri, F., and Sipione, S. (2001). Loss of normal huntingtin function: new developments in Huntington's disease research. *Trends Neurosci.* **24**, 182–188.

Cooper, J. K., Schilling, G., Peters, M. F., Herring, W. J., Sharp, A. H., Kaminsky, Z., Masone, J., Khan, F. A., Delanoy, M., Borchelt, D. R., Dawson, V. L., Dawson, T. M., and Ross, C. A. (1998). Truncated N-terminal fragments of huntingtin with expanded glutamine repeats form nuclear and cytoplasmic aggregates in cell culture. *Hum. Mol. Genet.* **7**, 783–790.

David, G., Giunti, P., Abbas, N., Coullin, P., Stevanin, G., Horta, W., Gemmill, R., Weissenbach, J., Wood, N., Cunha, S., Drabkin, H., Harding, A. E., Agid, Y., and Brice, A. (1996). The gene for autosomal dominant cerebellar ataxia type II is located in a 5-cM region in 3p12–p13: genetic and physical mapping of the SCA7 locus. *Am. J. Hum. Genet.* **59**, 1328–1336.

David, G., Abbas, N., Stevanin, G., Dürr, A., Yvert, G., Cancel, G., Weber, C., Imbert, G., Saudou, F., Antoniou, E., Drabkin, H., Gemmill, R., Giunti, P., Benomar, A., Wood, N., Ruberg, M., Agid, Y., Mandel, J.-L., and Brice, A. (1997). Cloning of the SCA7 gene reveals a highly unstable CAG repeat expansion. *Nat. Genet.* **17**, 65–70.

David, G., Dürr, A., Stevanin, G., Cancel, G., Abbas, N., Benomar, A., Belal, S., Lebre, A.-S., Abada-Bendib, M., Grid, D., Holmberg, M., Yahyaoui, M., Hentati, F., Chkili, T., Agid, Y., and Brice, A. (1998). Molecular and clinical correlations in autosomal dominant cerebellar ataxia with progressive macular dystrophy (SCA7). *Hum. Mol. Genet.* **7**, 165–170.

Del-Favero, J., Krols, L., Michalik, A., Theuns, J., Löfgren, A., Goossens, D., Wehnert, A., van den Bossche, D., van Zand, K., Backhovens, H., Van Regenmorter, N., Martin, J. J., and Van Broeckhoven, C. (1998). Molecular genetic analysis of autosomal dominant cerebellar ataxia with retinal degeneration (ADCA type II) caused by CAG triplet repeat expansion. *Hum. Mol. Genet.* **7**, 177–186.

Einum, D. D., Townsend, J. J., Ptacek, L. J., and Fu, Y. H. (2001). Ataxin-7 expression analysis in controls and spinocerebellar ataxia type 7 patients. *Neurogenetics* **3**, 83–90.

Ellerby, L. M., Andrusiak, R. L., Wellington, C. L., Hackam, A. S., Propp, S. S., Wood, J. D., Sharp, A. H., Margolis, R. L., Ross, C. A., Salvesen, G. S., Hayden, M. R., and Bredesen, D. E. (1999). Cleavage of atrophin-1 at caspase site aspartic acid 109 modulates cytotoxicity. *J. Biol. Chem.* **274**, 8730–8736.

Enevoldson, T. P., Sanders, M. D., and Harding, A. E. (1994). Autosomal dominant cerebellar ataxia with pigmentary macular dystrophy. A clinical and genetic study of eight families. *Brain* **117**, 445–460.

Froment, J., Bonnet, P., and Colrat, A. (1937). Heredo-dégénérations rétinienne et spino-cérébelleuses: variantes ophtalmoscopiques et neurologiques présentées par trois générations successives. *J. Med. Lyon* **22**, 153–163.

Giunti, P., Stevanin, G., Worth, P., David, G., Brice, A., and Wood, N. W. (1999). Molecular and clinical study of 18 families with ADCA type II: evidence for genetic heterogeneity and de novo mutation. *Am. J. Hum. Genet.* **64**, 1594–1603.

Gouw, L. G., Digre, K. B., Harris, C. P., Haines, J. H., and Ptacek, L. J. (1994). Autosomal dominant cerebellar ataxia with retinal degeneration: clinical, neuropathologic, and genetic analysis of a large kindred. *Neurology* **44**, 1441–1447.

Gouw, L. G., Kaplan, C. D., Haines, J. H., Digre, K. B., Rutledge, S. L., Matilla, A., Leppert, M., Zoghbi, H. Y., and Ptacek, L. J. (1995). Retinal degeneration characterizes a spinocerebellar ataxia mapping to chromosome 3p. *Nat. Genet.* **10**, 89–93.

Gouw, L. G., Castaneda, M. A., McKenna, C. K., Digre, K. B., Pulst, S. M., Perlman, S., Lee, M. S., Gomez, C., Fischbeck, K., Gagnon, D., Storey, E., Bird, T., Jeri, F. R., and Ptacek, L. J. (1998). Analysis of the dynamic mutation in the SCA7 gene shows marked parental effects on CAG repeat transmission. *Hum. Mol. Genet.* **7**, 525–532.

Gurevich, V. V., Dion, S. B., Onorato, J. J., Ptasienski, J., Kim, C. M., Sterne-Marr, R., Hosey, M. M., and Benovic, J. L. (1995). Arrestin interactions with G protein-coupled receptors. Direct binding studies of wild type and mutant arrestins with rhodopsin, beta 2-adrenergic, and m2 muscarinic cholinergic receptors. *J. Biol. Chem.* **270**, 720–731.

Harding, A. E. (1982). The clinical features and classification of the late onset autosomal dominant cerebellar ataxias. A study of 11 families, including descendants of "the Drew family of Walworth" *Brain* **105**, 1–28.

Harding, A. E. (1993). Clinical features and classification of inherited ataxias. *Adv. Neurol.* **61**, 1–14.

Holmberg, M., Johansson, J., Forsgren, L., Heijbel, J., Sandgren, O., and Holmgren, G. (1995). Localization of autosomal dominant cerebellar ataxia associated with retinal degeneration and anticipation to chromosome 3p12–p21.1. *Hum. Mol. Genet.* **4**, 1441–1445.

Holmberg, M., Duyckaerts, C., Durr, A., Cancel, G., Gourfinkel-An, I., Damier, P., Faucheux, B., Trottier, Y., Hirsch, E. C., Agid, Y., and Brice, A. (1998). Spinocerebellar ataxia type 7 (SCA7): a neurodegenerative disorder with neuronal intranuclear inclusions. *Hum. Mol. Genet.* **7**, 913–918.

Ikeda, H., Yamaguchi, M., Sugai, S., Aze, Y., Narumiya, S., and Kakizuka, A. (1996). Expanded polyglutamine in the Machado-Joseph disease protein induces cell death in vitro and in vivo. *Nat. Genet.* **13**, 196–202.

Jöbsis, G. J., Weber, J. W., Barth, P. G., Keizers, H., Baas, F., van Schooneveld, M. J., van Hilten, J. J., Troost, D., Geesink, H. H., and Bolhuis, P. A. (1997). Autosomal dominant cerebellar ataxia with retinal degeneration (ADCA II): clinical and neuropathological findings in two pedigrees and genetic linkage to 3p12–p21.1. *J. Neurol. Neurosurg. Psychiatry* **62**, 367–371.

Johansson, J., Forsgren, L., Sandgren, O., Brice, A., Holmgren, G., and Holmberg, M. (1998). Expanded CAG repeat in Swedish Spinocerebellar ataxia type 7 (SCA7) patients: effect of CAG repeat length on the clinical manifestation. *Hum. Mol. Genet.* **7**, 171–176.

Kaytor, M. D., Duvick, L. A., Skinner, P. J., Koob, M. D., Ranum, L. P., and Orr, H. T. (1999). Nuclear localization of the spinocerebellar ataxia type 7 protein, ataxin-7. *Hum. Mol. Genet.* **8**, 1657–1664.

Konigsmark, B. W., and Weiner, L. P. (1970). The olivopontocerebellar atrophies: a review. *Medicine (Baltimore)* **49**, 227–241.

Koob, M. D., Benzow, K. A., Bird, T. D., Day, J. W., Moseley, M. L., and Ranum, L. P. W. (1998). Rapid cloning of expanded trinucleotide repeat sequences from genomic DNA. *Nat. Genet.* **18**, 72–75.

Krols, L., Martin, J. J., David, G., Van Regemorter, N., Benomar, A., Lofgren, A., Stevanin, G., Durr, A., Brice, A., and Van Broeckhoven, C. (1997). Refinement of the locus for autosomal dominant cerebellar ataxia type II to chromosome 3p21.1–14.1. *Hum. Genet.* **99**, 225–32.

La Spada, A. R., Fu, Y., Sopher, B. L., Libby, R. T., Wang, X., Li, L. Y., Einum, D. D., Huang, J., Possin, D. E., Smith, A. C., Martinez, R. A., Koszdin, K. L., Treuting, P. M., Ware, C. B., Hurley, J. B., Ptacek, L. J., and Chen, S. (2001). Polyglutamine-expanded ataxin-7 antagonizes crx function and induces cone-rod dystrophy in a mouse model of sca7. *Neuron* **31**, 913–927.

Lebre, A. S., Jamot, L., Takahashi, J., Spassky, N., Leprince, C., Ravise, N., Zander, C., Fujigasaki, H., Kussel-Andermann, P., Duyckaerts, C., Camonis, J. H., and Brice, A. (2001). Ataxin-7 interacts with a Cbl-associated protein that it recruits into neuronal intranuclear inclusions. *Hum. Mol. Genet.* **10**, 1201–1213.

Lindblad, K., Savontaus, M. L., Stevanin, G., Holmberg, M., Digre, K., Zander, C., Ehrsson, H., David, G., Benomar, A., Nikoskelainen, E., Trottier, Y., Holmgren, G., Ptacek, L. J., Anttinen, A., Brice, A., and Schalling, M. (1996). An expanded CAG repeat sequence in spinocerebellar ataxia type 7. *Genome Res.* **6**, 965–971.

Lindenberg, K. S., Yvert, G., Muller, K., and Landwehrmeyer, G. B. (2000). Expression analysis of ataxin-7 mRNA and protein in human brain: evidence for a widespread distribution and focal protein accumulation. *Brain Pathol.* **10**, 385–394.

Lyoo, C. H., Hun, K., Choi, Y. C., Lee, S. C., Stevanin, G., David, G., Brice, A., and Lee, M. S. (1998). CAG repeat expansion in the SCA7 gene in Korean families presenting with ADCA type II. *J. Kor. Neurol. Assoc.* **16**, 341–352.

Martin, J. J., Van Regemorter, N., Krols, L., Brucher, J. M., de Barsy, T., Szliwowski, H., Evrard, P., Ceuterick, C., Tassignon, M. J., Smet–Dieleman, H., Hayez-Delatte, F., Willems, P. J., and Van Broeckhoven, C. (1994). On an autosomal dominant form of retinal-cerebellar degeneration: an autopsy study of five patients in one family. *Acta Neuropathol. (Berlin)* **88**, 277–286.

Martin, J., Van Regemorter, N., Del-Favero, J., Lofgren, A., and Van Broeckhoven, C. (1999). Spinocerebellar ataxia type 7 (SCA7)—correlations between phenotype and genotype in one large Belgian family. *J. Neurol. Sci* **168**, 37–46.

Martindale, D., Hackam, A., Wieczorek, A., Ellerby, L., Wellington, C., McCutcheon, K., Singaraja, R., Kazemi-Esfarjani, P., Devon, R., Kim, S. U., Bredesen, D. E., Tufaro, F., and Hayden, M. R. (1998). Length of huntingtin and its polyglutamine tract influences localization and frequency of intracellular aggregates. *Nat. Genet.* **18**, 150–154.

Matilla, A., Gorbea, C., Einum, D. D., Townsend, T., Michalik, A., van Broeckhoven, C., Jensen, C. C., Murphy, K. J., Ptacek, L. J., and Fu, Y.-H., Association of ataxin-7 with the proteasome subunit S4 of the 19S regulatory complex. *Hum. Mol. Genet.* **10**, 2821–2831.

Mauger, C., Del-Favero, J., Ceuterick, C., Lubke, U., Van Broeckhoven, C., and Martin, J. (1999). Identification and localization of ataxin-7 in brain and retina of a patient with cerebellar ataxia type II using anti-peptide antibody. *Brain Res. Mol. Brain Res.* **74**, 35–43.

Monckton, D. G., Cayuela, M. L., Gould, F. K., Brock, G. J., Silva, R., and Ashizawa, T. (1999). Very large (CAG)(n) DNA repeat expansions in the sperm of two spinocerebellar ataxia type 7 males. *Hum. Mol. Genet.* **8**, 2473–2478.

Mushegian, A. R., Vishnivetskiy, S. A., and Gurevich, V. V. (2000). Conserved phosphoprotein interaction motif is functionally interchangeable between ataxin-7 and arrestins. *Biochemistry* **39**, 6809–6813.

Nardacchione, A., Orsi, L., Brusco, A., Franco, A., Grosso, E., Dragone, E., Mortara, P., Schiffer, D., and De Marchi, M. (1999). Definition of the smallest pathological CAG expansion in SCA7. *Clin. Genet.* **56**, 232–234.

Neetens, A., Martin, J. J., Libert, J., and Van Den Ende, P. (1990). Autosomal dominant cone-dystrophy-cerebellar atrophy (ADCoCA) (modified ADCA Harding II). *Neuroophthalmology* **10**, 261–275.

Ona, V. O., Li, M., Vonsattel, J. P., Andrews, L. J., Khan, S. Q., Chung, W. M., Frey, A. S., Menon, A. S., Li, X. J., Stieg, P. E., Yuan, J., Penney, J. B., Young, A. B., Cha, J. H., and Friedlander, R.M. (1999). Inhibition of caspase-1 slows disease progression in a mouse model of Huntington's disease. *Nature* **399**, 263–267.

Ribon, V., Printen, J. A., Hoffman, N. G., Kay, B. K., and Saltiel, A. R. (1998). A novel, multifuntional c-Cbl binding protein in insulin receptor signaling in 3T3-L1 adipocytes. *Mol. Cell Biol.* **18**, 872–879.

Rubinsztein, D. C., Leggo, J., Coles, R., Almqvist, E., Biancalana, V., Cassiman, J. J., Chotai, K., Connarty, M., Craufurd, D., Curtis, A., Curtis, D., Davidson, M. J., Differ, A. M., Dode, C., Dodge, A., Frontali, M., Ranen, N. G., Stine, O. C., Sherr, M., Abbott, M. H., Franz, M. L., Graham, C. A., Harper, P. S., Hedreen, J. C., Jackson, A., Kaplan, J. C., Losekoot, M., MacMillan, J. C., Trottier, Y., Novelleto, A., Simpson, S. A., Theilman, J., Whittaker, J. L., Folstein, S. E., Ross, C. A., and Hayden, M. R. (1996). Phenotypic characterization of individuals with 30–40 CAG repeats in the Huntington disease (HD) gene reveals HD cases with 36 repeats and apparently normal elderly individuals with 36–39 repeats. *Am. J. Hum. Genet.* **59**, 16–22.

Schalling, M., Hudson, T. J., Buetow, K. H., and Housman, D. E. (1993). Direct detection of novel expanded trinucleotide repeats in the human genome. *Nat. Genet.* **4**, 135–139.

Schols, L., Amoiridis, G., Buttner, T., Przuntek, H., Epplen, J. T., and Riess, O. (1997). Autosomal dominant cerebellar ataxia: phenotypic differences in genetically defined subtypes? *Ann. Neurol.* **42**, 924–932.

Stevanin, G., Trottier, Y., Cancel, G., Dürr, A., David, G., Didierjean, O., Bürk, K., Imbert, G., Saudou, F., Abada-Bendib, M., Gourfinkel-An, I., Benomar, A., Abbas, N., Klockgether, T., Grid, D., Agid, Y., Mandel, J.-L., and Brice, A. (1996). Screening for proteins with polyglutamine expansions in autosomal dominant cerebellar ataxias. *Hum. Mol. Genet.* **5**, 1887–1892.

Stevanin, G., Giunti, P., Belal, G. D. S., Durr, A., Ruberg, M., Wood, N., and Brice, A. (1998). De novo expansion of intermediate alleles in spinocerebellar ataxia 7. *Hum. Mol. Genet.* **7**, 1809–1813.

Stevanin, G., David, G., Durr, A., Giunti, P., Benomar, A., Abada-Bendib, M., Lee, M. S., Agid, Y., Brice, A. (1999). Multiple origins of the spinocerebellar ataxia 7 (SCA7) mutation revealed by linkage disequilibrium studies with closely flanking markers, including an intragenic polymorphism (G3145TG/A3145TG). *Eur. J. Hum. Genet.* **7**, 889–96.

Stevanin, G., Durr, A., and Brice, A. (2000a). Clinical and molecular advances in autosomal dominant cerebellar ataxias: from genotype to phenotype and physiopathology. *Eur. J. Hum. Genet.* **8**, 4–18.

Stevanin, G., Durr, A., and Brice, A. (2000b). Spinocerebellar ataxia 7 (SCA7). In "Handbook of Ataxia Disorders" (T. Klockgether, ed.). Marcel Dekker, Inc., New York.

Trottier, Y., Lutz, Y., Stevanin, G., Imbert, G., Devys, D., Cancel, G., Saudou, F., Weber, C., David, G., Laszlo, T., Agid, Y., Brice, A., and Mandel, J.-L. (1995). Polyglutamine expansion as a pathological epitope in Huntington's disease and four dominant cerebellar ataxias. *Nature* **378**, 403–406.

Wellington, C. L., Ellerby, L. M., Hackam, A. S., Margolis, R. L., Trifiro, M. A., Singaraja, R., McCutcheon, K., Salvesen, G. S., Propp, S. S., Bromm, M., Rowland, K. J., Zhang, T., Rasper, D., Roy, S., Thornberry, N., Pinsky, L., Kakizuka, A., Ross, C. A., Nicholson, D. W., Bredesen, D. E., and Hayden, M. R. (1998). Caspase cleavage of gene products associated with triplet expansion disorders generates truncated fragments containing the polyglutamine tract. *J. Biol. Chem.* **273**, 9158–9167.

World Federation of Neurology Research Group on Huntington's Chorea (1994). International Huntington Association and the World Federation of Neurology Research Group on Huntington's Chorea. Guidelines for the molecular genetics predictive test in Huntington's disease. *J. Med. Genet.* **31**, 555–559.

Yvert, G., Lindenberg, K. S., Picaud, S., Landwehrmeyer, G. B., Sahel, J. A., and Mandel, J. L. (2000). Expanded polyglutamines induce neurodegeneration and trans-neuronal alterations in cerebellum and retina of SCA7 transgenic mice. *Hum. Mol. Genet.* **9**, 2491–2506.

Yvert, G., Lindenberg, K. S., Devys, D., Helmlinger, D., Landwehrmeyer, G. B., and Mandel, J. L. (2001). SCA7 mouse models show selective stabilization of mutant ataxin-7 and similar cellular responses in different neuronal cell types. *Hum. Mol. Genet.* **10**, 1679–1692.

Zander, C., Takahashi, J., El Hachimi, K. H., Fujigasaki, H., Albanese, V., Lebre, A. S., Stevanin, G., Duyckaerts, C., and Brice, A. (2001). Similarities between spinocerebellar ataxia type 7 (SCA7) cell models and human brain: proteins recruited in inclusions and activation of caspase-3. *Hum. Mol. Genet.* **10**, 2569–2579.

Zoghbi, H. Y., and Orr, H. T. (2000). Glutamine repeats and neurodegeneration. *Annu. Rev. Neurosci.* **23**: 217–247.

# CHAPTER 10

# Spinocerebellar Ataxia 8 (SCA8)

MICHAEL D. KOOB

*Institute of Human Genetics*
*University of Minnesota*
*Minneapolis, Minnesota 55455*

I. Summary
II. Phenotype
III. Gene
IV. Diagnostic and Ancillary Tests
   A. DNA
   B. Neuroimaging
   C. Neuropathology
V. Cellular and Animal Models of Disease
VI. Genotype/Phenotype Correlation and Modifying Alleles
VII. Treatment
   References

## I. SUMMARY

Spinocerebellar ataxia type 8 (SCA8) patients typically have a slowly progressive, predominantly cerebellar disease involving dysarthria, limb and gait ataxia, impaired smooth pursuit, and nystagmus. A broad range of other clinical symptoms has also been reported, including tremor, spasticity, and various kinds of cognitive impairment. Although this clinical picture is not clearly distinguishable from that of other forms of inherited ataxia, SCA8 is unique among this group of diseases in many ways. The inheritance pattern for SCA8 in particular is perhaps the most complicated of all of the inherited spinocerebellar ataxias. SCA8 is caused by a CTG expansion mutation that has been shown to vary widely in size between generations. Unlike many of the other spinocerebellar ataxias, there is little direct relationship between expansion size and disease severity, and some individuals with large SCA8 CTG alleles never develop ataxia. For these reasons, it is not currently possible to define a clear "pathogenic size range" for SCA8 CTG expansions that is definitively predictive of either the onset or severity of the disease. The SCA8 CTG repeat is transcribed as part of a natural, untranslated antisense RNA, but the precise molecular pathogenic mechanism through which this repeat expansion causes disease is currently unknown.

## II. PHENOTYPE

Clinical evaluations of SCA8 ataxia patients have been published in detail for ataxia patients with large SCA8 CTG repeats identified from populations in Japan (Ikeda *et al.*, 2000a), Portugal (Silveira *et al.*, 2000), Finland (Juvonen *et al.*, 2000), and Italy (Cellini *et al.*, 2001) and for patients in a single large family from the United States (Day *et al.*, 2000; Koob *et al.*, 1999). A group of ataxia patients from Scotland with large SCA8 CTG expansions has also been identified, but details of this population have to date only been published in a brief abstract (Warner *et al.*, 1999).

The original report of SCA8 included a brief clinical description of patients from a single large family in which ataxia was genetically linked to the SCA8 locus on chromosome 13 (Koob *et al.*, 1999), and a more detailed clinical assessment of this same family was later reported (Day *et al.*, 2000). Clinical findings for affected family members most often included mild-to-moderate dysarthria, mild-to-severe truncal and limb ataxia, nystagmus and impaired smooth pursuit, with either gait ataxia or dysarthria being the initial clinical symptom. Many patients also had increased deep tendon reflexes, limb spasticity, and reduced vibratory sense. The severity of these clinical features was found to progress slowly.

The SCA8 patients that have been reported from the Japanese population (Ikeda et al., 2000a) had clinical features essentially identical to those of this family, as did most of the patients from the other populations that have been described (Cellini et al., 2001; Juvonen et al., 2000; Silveira et al., 2000). The clinical features of some of the ataxia patients with large SCA8 CTG expansions, however, did vary from those of the original SCA8 family. Tremor was present in most of the patients in the Finnish population (8/14 patients) and this was often the initial symptom (6/14 patients; Juvonen et al., 2000). Fluctuation of symptoms was also commonly observed in this population. One of the Finnish and one of the Italian ataxia patients with tremor were reported to have celiac disease (Cellini et al., 2001; Juvonen et al., 2000), one Italian ataxia patient had a vitamin E deficiency (Cellini et al., 2001), and one of the Portuguese patients had a severe congenital ataxia with myoclonic epilepsy and was mentally retarded (Silveira et al., 2000).

Cognitive impairments have been observed in many of the reported SCA8 patients, and this may represent a significant but variable clinical feature of this disease. Obvious cognitive difficulties were noted for 6 of the 15 Finnish patients, including two patients who had been treated for depression, one patient diagnosed with borderline personality disorder, and another with bipolar disorder (Juvonen et al., 2000). In the original SCA8 family 3 of the 13 affected family members and 2 of the 22 carriers of SCA8 expansion without ataxia had been treated for depression and one affected family member was being treated for a psychotic disorder (Day et al., 2000). Two of the 7 patients in the Portuguese study had mild-to-moderate memory impairment, and the patient with congenital ataxia was mentally retarded (Silveira et al., 2000). A detailed case study describing "a mother and son with the SCA8 expansion with executive, visuospatial and affective problems in addition to an ataxic syndrome" has been published (Stone et al., 2001). The authors of this study conclude by suggesting that clinical evaluations of even "apparently asymptomatic patients with putative SCA8 mutations should incorporate neuropsychological assessment." Interestingly, two studies have suggested that expanded SCA8 CTG repeats even in the absence of ataxia may be a susceptibility factor for developing bipolar affective disorder and schizophrenia, but these findings require larger sample sizes to reach statistical significance (Vincent et al., 2000a,b).

## III. GENE

The SCA8 CTG repeat tract is part of the natural antisense RNA of the Kelch-like 1 *(KLHL1)* gene (Koob et al., 1999; Nemes et al., 2000). *KLHL1* antisense transcripts *(KLHL1AS)* are transcribed from a promoter in the first intron of *KLHL1* across the splice donor site, the translation and the transcription start site of *KLHL1*, and are alternatively spliced and polyadenylated (Fig. 10.1). The SCA8 CTG repeat is transcribed in one of the two alternative 3′ terminal exons of *KLHL1AS*, and is located approximately 31 kb 5′ of the *KLHL1* transcription start site. No significant open reading frames are present in any of the *KLHL1AS* splice variants. We currently presume that the primary function of these antisense transcripts is to regulate the expression of the *KLHL1* gene.

*KLHL1* mRNA is primarily expressed in various brain tissues and encodes a 748 amino acid protein (Nemes et al., 2000). The proteins homologous to KLHL1 are actin-binding proteins that are, in general, thought to serve as actin-organizing proteins (Hernandez et al., 1997; Kim et al., 1998; Robinson and Cooley 1997; Soltysik-Espanola et al., 1999). We have shown that KLHL1 is also capable of both binding actin and of forming multimers, and so we

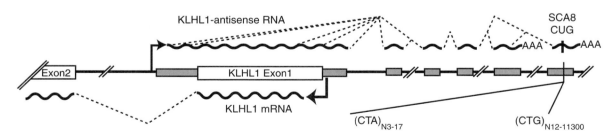

FIGURE 10.1 Organization of the SCA8 locus. The SCA8 CTG repeat is preceded by a polymorphic CTA repeat tract, and is transcribed as part of the KLHL1-antisense (KLHL1AS) RNA. The KLHL1AS RNA is variably spliced (as shown by dashed lines), has two alternative polyadenylation sites (AAA), and is transcribed across the first exon of the KLHL1 gene from a promoter in the first intron of KLHL1. Genomic DNA is represented by a solid line, KLHL1AS exons are indicated by gray boxes, KLHL1 exons are indicated by white boxes, transcription promoters are represented by bent arrows, and the waved line represents RNA.

speculate that KLHL1 may be involved in organizing the actin cytoskeleton of the brain cells in which it is expressed. We have developed monoclonal antibodies to the KLHL1 protein and have used them in immunohistochemical studies of the mouse brain to determine the cellular distribution of this protein (in preparation). We found that KLHL1 is present in the cytoplasm of many of major neurons of the brain, and in the cerebellum it is particularly prominent in the soma and dendritic arbor of Purkinje cells. Since loss of Purkinje cell dendritic function and structure has been shown to play an integral role in the pathogenesis of SCA1 (Clark *et al.*, 1997), we are actively pursuing the possibility that KLHL1 plays an important role in establishing or maintaining dendritic function in these cells.

## IV. DIAGNOSTIC AND ANCILLARY TESTS

### A. DNA

The size of the *SCA8* CTG tracts in a patient's genomic DNA can be determined by generating and electrophoretically sizing PCR products from both of the patient's *SCA8* alleles using an oligonucleotide primer pair that flanks this repeat tract. The primers SCA8-R4 (5′-GGTCCTTCATGTTAGAAAACCTGGCT-3′) and SCA8-F4 (5′-GTAAGAGATAAGCAGTATGAGGAAGTATG-3′); (Koob *et al.*, 1999) are most often used for this purpose, but other primer pairs designed from the unique genomic sequence flanking the repeat tract (GenBank accession AF126748) can also be used. The number of trinucleotide repeats present at this locus is extremely heterogeneous in the general population and, as a result, the genomic DNA from the vast majority of individuals will generate PCR products with two distinct sizes. In those instances where only a single size is detected, Southern analysis of the genomic DNA sample should be performed to differentiate between individuals with two SCA8 alleles of the same size and those who have one expanded allele that is too large to amplify by PCR (typically > ~200 CTG repeats). Digestion of the genomic DNA with the restriction endonuclease *Eco*RI will generate genomic fragments of the appropriate size for this Southern assay, and a PCR primer pair (e.g., 5′-GAATTCATTCCTTGCTTACAA-3′, and 5′-CTGCTGCATTTTTTAAAAATA3-′) can be used to generate a suitable probe from the flanking genomic sequence.

Determination of the SCA8 CTG repeat size by PCR analysis is complicated by the fact that the SCA8 CTG repeat tract is immediately adjacent to a polymorphic CTA repeat tract (Fig. 10.1). PCR products generated at this locus therefore contain both the CTA and the CTG repeat tracts. For this reason, the results of the SCA8 CTG PCR analysis are typically reported as a total combined size for both the CTA and CTG repeats. The CTA tract has been reported in one study to vary in size from 3 to 17 repeats (Koob *et al.*, 1999), although CTA allele sizes outside this range undoubtedly exist. No unstable expansions have been reported in the CTA portion of this combined repeat tract. We currently do not know if variations in the number of CTA repeats contribute to the pathogenic aspect of this locus, but individuals with a larger number of CTA repeats will appear by PCR analysis to have larger CTG expansions than individuals with a small number of CTA repeats. Precise determination of the actual SCA8 CTG repeat number requires either a second round of PCR using a modified assay (Warner *et al.*, 1999) or sequencing of isolated PCR products.

Interpretation of the SCA8 CTG DNA test is not straightforward. Individuals who do not have CTG expansions at the SCA8 locus do not have and will not develop this form of ataxia, and this test is useful for providing this information. On the other hand, detection of a large SCA8 CTG repeat and determination of its size cannot be used to directly predict either the severity or the onset of ataxia. At this point in time, we agree with the authors of the Finnish SCA8 population study that "the finding of an expanded allele should be regarded only as a direction to follow and not as a definitive diagnosis. Inclusion of more family members to obtain segregation data would be most helpful as would more information on the frequency of moderately sized expansions in the general populations to which the patient belongs" (Juvonen *et al.*, 2000).

We feel that the underlying problem with using the size of the SCA8 CTG repeat in a predictive manner is that the CTG expansion in the genomic DNA that is analyzed in a DNA test is not in itself the direct molecular cause of the SCA8 neuropathology. Rather, we currently believe that the cerebellar neurotoxicity of this expansion mutation is mediated through the transcription of this repeat into an RNA that in turn is cytotoxic either in a manner analogous to the mytonic dystrophy CUG repeat (Philips *et al.*, 1998) or by altering normal KLHL1 protein expression through abnormal antisense interactions. Because direct analysis of patient cerebellar RNA is not a practical option, definitive predictive testing for SCA8 will probably not be possible until the precise molecular mechanism that leads to neuropathology and the modifiers of this pathology are more fully understood.

A more complete discussion of the reduced penetrance of the *SCA8* CTG expansion mutation and of the sizes of *SCA8* repeats that have been found in both ataxia and control populations is included in Section VI.

### B. Neuroimaging

In each of the SCA8 clinical studies, ataxia patients analyzed with either magnetic resonance imaging (MRI) or computed tomography (CT) were typically found to have marked atrophy of the cerebellar vermis and hemispheres and relative preservation of the brainstem and other parts of the brain (Cellini et al., 2001; Day et al., 2000; Ikeda et al., 2000a; Juvonen et al., 2000; Stone et al., 2001). These findings are consistent with the classification of SCA8 as a "pure cerebellar" (ADCA III) type of ataxia (Harding, 1983). The MRI findings from SCA8 patients were compared to those from SCA6 patients, who are also considered to have a "pure cerebellar" ataxia, and no statistically significant differences were found between these two patient groups on MRI (Ikeda et al., 2000a). The slowly progressive nature of SCA8 was documented by MRI analysis in one study. MRI scans of an SCA8 patient taken 9 years and 18 years after disease onset were compared and showed little if any progression of the prominent cerebellar atrophy during those years (Day et al., 2000). The authors of the Finnish SCA8 study noted that significant atrophy was found in MRI scans taken very soon after disease onset, and suggested that atrophy was probably present during the preclinical stage of the disease (Juvonen et al., 2000). This suggestion is supported by a detailed report of the MRI analysis of one of the Japanese SCA8 families in which mild but obvious cerebellar atrophy was found in the asymptomatic SCA8 carrier father (Ikeda et al., 2000b). A neuroradiological finding of cerebellar atrophy may therefore be the first or only sign of SCA8 neuropathology in some patients.

### C. Neuropathology

Postmortem brain samples of individuals diagnosed with SCA8 have not been available for examination and so neuropathological studies of SCA8 patients have not yet been reported.

## V. CELLULAR AND ANIMAL MODELS OF DISEASE

SCA8 is the only known form of spinocerebellar ataxia caused by a CTG expansion (Klockgether et al., 2000), and we do not yet know how this expansion causes the cerebellar degeneration and ataxia found in SCA8 patients. The *KLHL1* and *KLHL1AS* transcripts were the only transcripts found in an exhaustive analysis of nearly 200 kb of genomic sequence flanking the SCA8 repeat (Nemes et al., 2000). Since both of these genes are expressed in the cerebellum, the pathogenic effect of the expansion may be mediated either directly or indirectly through one or both of these transcripts. The promoters for these transcripts are both located over 31 kb from the expanded repeat, so we do not think that expansions would directly alter the activity of these promoters. Rather, we believe that pathogenic *SCA8* CTG expansions that are transcribed in the *KLHL1AS* RNA are toxic to neuronal function either directly through a mechanism similar to that of the myotonic dystrophy CUG (Philips et al., 1998) or through altered antisense regulation of *KLHL1* expression. If the CUG expansions altered KLHL1AS RNA stability or processing, this could in turn affect the expression of the *KLHL1* gene in a dominant manner. For instance, if the expansion leads to an accumulation of *KLHL1AS* RNA, transcripts from both of the *KLHL1* alleles could potentially be negatively regulated through antisense interactions. Alternatively, if the CUG expansion prevents the *KLHL1AS* transcript from negatively regulating *KLHL1* expression, the resulting over-expression of the *KLHL1* protein could be toxic to cerebellar function and result in ataxia.

Mouse and cellular models to determine the roles of the *KLHL1* sense gene product, of the *KLHL1*-antisense RNA (SCA8 RNA), and of the CUG expansion in SCA8 neuropathology are being generated but results are not yet available from this work.

## VI. GENOTYPE/PHENOTYPE CORRELATION AND MODIFYING ALLELES

Before discussing the genotype/phenotype correlation for SCA8, it is critical to remember that the SCA8 DNA test is not a direct measure of a pathogenic agent. Since the pathological mechanism underlying SCA8 at the molecular level has not yet been determined, we do not know precisely how an expanded SCA8 CTG in the genome ultimately leads to neurodegeneration. We currently believe, however, that the neuropathology and resulting ataxic phenotype in SCA8 is mediated through the presence of the expanded SCA8 repeat in RNA in the cerebellum. A large number of factors could potentially influence both how much of the RNA containing the expansion is present in a particular individual's cerebellum and how efficiently this RNA initiates a neurodegenerative pathway. Although none of these hypothetical factors have as yet been identified, the data generated to this point strongly suggests that modifying alleles or environmental factors are in fact of critical importance in influencing the disease course of individuals who carry expanded SCA8 alleles.

For these reasons, defining a precise correlation between the size of the SCA8 trinucleotide repeat expansion mutation in a patient's genome and the clinical phenotype of the disease it causes may be currently impossible for SCA8. In general, only expanded alleles above a certain size appear to result in disease, but this pathogenic threshold seems to differ between SCA8 families and so no definitive "affected range" exists across all populations. In addition, there is also some evidence that expanded alleles above a certain size are also not often pathogenic (Moseley et al., 2000), but once again this upper pathogenic threshold is variable and not defined. Finally, SCA8 expansions in all size ranges are not fully penetrant, and so an unaffected individual who carriers an expanded SCA8 allele can never be certain that they will ever develop ataxia and should never be told that they will. In other words, although an expanded SCA8 allele appears to be a necessary component of the underlying genetic cause of the SCA8 form of ataxia, it is not necessarily in itself sufficient to always cause ataxia during an individual's lifetime.

To more fully appreciate the complexity of the SCA8 genotype/phenotype correlation, it is useful to review the findings from SCA8 genotyping studies of the one large family that has been genetically linked to the SCA8 locus, of collections of other smaller ataxia families, and of various control populations.

We originally cloned a genomic DNA fragment containing an expanded SCA8 CTG allele from the genome of an ataxia patient (Koob et al., 1999) using the RAPID cloning approach (Koob et al., 1998). The sequence of this clone revealed that this patient had the repeat configuration $(CTA)_{11}CTGCTA(CTG)_{80}$ (i.e., a combined CTG/CTA size of 93 with a tract of 80 uninterrupted CTG repeats). We used the genomic sequence flanking the repeats to develop a PCR assay for this expanded allele. PCR analysis of members from this original patient's family indicated that her affected mother also had an 80 CTG repeat expansion and two at-risk individuals, ages 38 and 35, had 87 and 113 CTG repeat expansions, respectively.

Because this family was too small to conclusively link the expansion to disease, SCA8 PCR analysis was performed on DNA samples from a collection of ataxia patients with apparently inherited forms of ataxia (Moseley et al., 1998) and a much larger SCA8 family (Fig. 10.2) was identified (Koob et al., 1999). PCR analyses indicated that all 11 affected members of this seven-generation kindred had expanded SCA8 alleles and identified 20 individuals with expanded alleles that were asymptomatic when clinically examined. Three CTA repeats preceded all of the expanded SCA8 CTG alleles in this family. The expansions often changed dramatically in size between generations, with changes from −11 to +33 seen in maternal transmissions and from −29 to +8 observed in paternal transmissions. Analysis of the linkage between ataxia and the expanded allele in this family gave a highly significant maximum lod score of 6.8 with no recombination between the expansion and the ataxia phenotype.

The asymptomatic carriers in this family were between 14 and 74 years old when examined. One of these carriers, age 42, had 140 CTG repeats and the other 19 carriers had CTG expansions ranging in size from 74 to 101 repeats. The affected family members had from 107–127 CTG repeats in their expanded allele, and were between 18 and 65 years old at the initial onset of disease symptoms. Although there was no significant correlation between repeat size and either the age of onset or disease severity among these affected individuals, there is in this family a perfect correlation between repeat size and affected status:

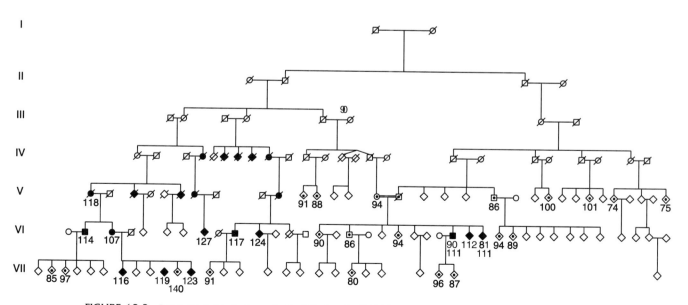

FIGURE 10.2 Large ataxia kindred genetically linked to the SCA8 locus. Filled symbols represent family members with ataxia and symbols with dots represent individuals that inherited the SCA8 CTG expansion but were not clinically affected by ataxia when examined. The lengths of expanded CTG alleles are indicated below the symbols. (Adapted from Koob et al. (1999). *Nat. Genet.* **21**, 379–384. With permission.)

all individuals with CTG repeats from 107–127 are affected and those with repeats outside this range are not affected. In this family the incomplete penetrance of the SCA8 CTG expansion can therefore apparently be explained by the size instability of this expansion when it is passed from one generation to the next. The sample size used to generate this observed "affected size range" is of course too small for it to be used for definitive predictive counseling for even this family, and it definitely should not be applied to other families for this purpose. This correlation would not exist, however, if the SCA8 CTG expansion was simply in linkage disequilibrium with "true causative mutations" of SCA8 as has at one point been suggested (Stevanin et al., 2000; Worth et al., 2000).

In the original SCA8 report (Koob et al., 1999), 102 families that had tested negative for other forms of ataxia were analyzed with the SCA8 PCR assay. Ataxia patients from 8 of these families (8%) were found to have expanded SCA8 alleles with combined CTA/CTG sizes of from 93 to about 250 repeats. Five of these families had been classified as having a dominant family history of ataxia, two with a probable recessive ataxia, and one individual did not have a family history of ataxia. Haplotype analyses identified at least seven different haplotypes associated with the expanded alleles, indicating that these expansions had arisen independently and were not in linkage disequilibrium with ataxia. In each family, the expanded allele was found in all affected family members.

Similar SCA8 genotyping studies have been reported for ataxia populations from Portugal (Silveira et al., 2000), Japan (Ikeda et al., 2000a), Finland (Juvonen et al., 2000), Italy (Cellini et al., 2001), Scotland (Warner et al., 1999), England (Worth et al., 2000), and Germany (Schols et al., 2000) and for a population described as European (Stevanin et al., 2000).

In the Portuguese SCA8 study (Silveira et al., 2000), 73 unrelated ataxia patients that had tested negative in the other SCA and DRPLA gene tests were analyzed by PCR for SCA8 expansions, and 5 of these patients (7%) had expansions with combined CTA/CTG repeats of from 100–152 repeats. All five of these families had distinct haplotypes associated with the expanded allele. Two of the families had one additional patient available for analysis, and in both cases these affected family members also had expanded alleles. A total of six unaffected member of these families (12–77 years old) were found to have expansions of from 128–170 combined repeats.

In the Japanese population study (Ikeda et al., 2000a), 9% (2/22) of the unrelated ataxia patients from families that had a family history of ataxia and 5% (2/41) of the sporadic cases of ataxia were found to have expanded SCA8 alleles. These patients had from 89–155 combined CTA/CTG repeats. The two SCA8 families each included two affected patients, and the expansion was present in both of the affected members of each family. Sequencing of the expanded alleles in one of these families showed that the affected father had 80 CTG repeats (with 9 CTA repeats) that had expanded to 96 CTG repeats (also with 9 CTA) in his affected daughter. The age of onset was 31 years earlier (age 42 vs. 73) in the daughter, who also had a faster disease progression and more severe cerebellar atrophy on MRI. In one of the apparently sporadic families (Ikeda et al., 2000b), the clinically unaffected father had 128 CTG repeats that contracted to 87 CTG repeats in his daughter with ataxia. MRI of these individuals showed mild but obvious cerebellar atrophy in the asymptomatic father (age 68) and significant cerebellar atrophy in the affected daughter (age 37 with onset at age 20).

In the Finnish study (Juvonen et al., 2000), 154 unrelated ataxia patients were tested and 9 families (6%) were found to have SCA8 expansions with combined CTA/CTG repeats of from 101–345 repeats. All of these families had different haplotypes associated with the expanded allele. Within this Finnish population, SCA8 expansions were more frequently found among the familial cases of ataxia, with 13% (7/54) of the families that had a family history of ataxia and 16% (5/32) of those families with a dominant inheritance pattern testing positive.

In the Italian study (Cellini et al., 2001), 5 of 167 genetically undefined ataxia patients (3%) had SCA8 expansions with from 90–320 combined CTA/CTG repeats. Although the Scottish study has not yet been reported in detail, analysis of over 100 individuals with cerebellar ataxia of unknown origin found that SCA8 "is currently the most common cause of spinocerebellar ataxia" in that population (Warner et al., 1999). In the English study (Worth et al., 2000), 1% (1/98) families with genetically undefined dominant ataxia had expanded repeats (152 and 141 combined repeats), 4.8% (1/21) families with recessive ataxias had an expanded allele (208 combined repeats), and 1 of the 145 isolated ataxia cases tested had a moderately large expansion (95 combined repeat). In a study of 104 apparently sporadic ataxia patients from the German population (Schols et al., 2000), 3 patients were found to have expanded SCA8 alleles. Finally, in a study of ataxia patients described as European (Stevanin et al., 2000), 6.6% (8/122) of the families with genetically undefined dominant ataxias had expanded SCA8 alleles of more than 93 combined repeats, 1 of 76 patients with apparent sporadic ataxia had 111 CTG repeats at the SCA8 allele, and none of the 26 patients with apparent recessive ataxia had expansions.

The sizes of the SCA8 CTG/CTA alleles found in individuals without ataxia were also reported in many of these SCA8 genotyping studies. In the original SCA8 study (Koob et al., 1999), DNA from 600 individuals with no reported neurological abnormalities was analyzed with the SCA8 PCR assay. Over 99% of the alleles found had combined CTG/CTA repeats of between 16 and 37 repeats. Three individuals had more than 75 combined repeats (0.3%), with the largest allele found by PCR being 92 combined repeats. Subsequent Southern blot analysis of the samples that were homozygous in the PCR assay, however, revealed that one of the Centre d'Etude Polymorphisme Humaine (CEPH) reference samples included in this study (pedigree 1334) has an extremely large SCA8 allele (about 800 repeats) that could not be detected by PCR. The SCA8 alleles found in this control group, therefore, were either smaller or much larger than the repeats found in the ataxia patients analyzed in this study.

Several other genotyping studies also reported that there was no overlap between the expanded SCA8 alleles found in the ataxia patients tested and those present in the control group from the same population (Warner et al., 1999; Ikeda et al., 2000a; Silveira et al., 2000; Cellini et al., 2001). Fifty-two individuals over the age of 79 with no known neurological abnormalities were used as a normal control group for the Japanese population study (Ikeda et al., 2000a), and the SCA8 alleles found in this group varied from 15–34 combined repeats. The control population for the Portuguese population study (Silveira et al., 2000) consisted of 909 samples taken from the Portuguese national screening for PKU. The vast majority of the alleles found in these individuals (99.3%) had between 15 and 37 combined repeats, and the remaining 0.7% had between 40 and 91 combined repeats. Similar results were obtained from the SCA8 genotyping of a large number of individuals without known neurological abnormalities from the Scottish population (Warner et al., 1999). For the Italian study (Cellini et al., 2001), DNA samples from 161 healthy individuals and from 125 patients with various psychotic diseases were genotyped, and SCA8 alleles with from 15–75 combined CTA/CTG repeats were found in these 286 individuals.

Some genotyping studies, however, have found large SCA8 alleles in individuals who neither have ataxia nor have a family history of the disease. In the English study (Worth et al., 2000), 5 out of 653 (0.7%) of the DNA samples analyzed from individuals without ataxia had 100–174 combined repeats. Three of these individuals had other neurological abnormalities other than ataxia. In the French population (Stevanin et al., 2000), expanded SCA8 alleles of 107, 111, and 123 combined alleles were found in 183 DNA samples from individuals with no family history of neurological disorders (1.6%), and in 2 of 46 individuals (4.3%) from families with neurological disorders other than ataxia. In the Finnish study (Juvonen et al., 2000), DNA from 448 individuals with no known ataxic disorders were analyzed, and 13 (2.9%) of these individuals had SCA8 expansions that overlapped in size with those found in the Finnish ataxia patients. Of these 13 people, 7 (1.6%) had large SCA8 expansions (100–165 combined repeats) detectable by the PCR assay, and 6 (1.3%) had extremely large repeats (about 320-675 repeats) detected by Southern analysis. Finally, in a large genotyping study of psychiatric patients (Vincent et al., 2000a), large SCA8 alleles (100–180 repeats) were found in 0.7% (8/1120) of the patient samples and in 0.3% (2/710, 103 and 117 repeats) of the unrelated controls without psychiatric disorders. Very large SCA8 alleles (about 250–1300 repeats) detected by Southern analysis were also found in 0.5% (6/1120) and 0.4% (3/710) of these two groups, respectively.

The authors of the SCA8 genotyping studies have suggested a variety of possible explanations for the broad range of clinical features associated with this expansion and for the presence of expanded SCA8 alleles in individuals without ataxia and without a family history of the disease. A pair of early papers suggested that the unstable CTG expansion is closely linked to another mutation that actually causes the SCA8 form of ataxia and may not be a direct cause of neuropathology (Stevanin et al., 2000; Worth et al., 2000). Other authors have suggested that the SCA8 CTG expansion may actually result in a broad range of neurological disease phenotypes (Cellini et al., 2001), including psychiatric disorders (Stone et al., 2001; Vincent et al., 2000b), and so "it is also possible that relevant neuropsychiatric features may have been overlooked in some 'controls' with expanded alleles" (Stone et al., 2001). Finally, some have hypothesized that the CTG expansion mutation is necessary for ataxia to develop, but that the precise configuration of the CTA/CTG allele (Moseley et al., 2000) or other components of the genetic background (i.e., modifiers) either enhance or suppress this disease phenotype (Juvonen et al., 2000; Sobrido et al., 2001; Vincent et al., 2000a).

## VII. TREATMENT

There is currently no known method for effectively treating SCA8 patients.

### References

Cellini, E., Nacmias, B., Forleo, P., Piacentini, S., Guarnieri, B. M., Serio, A., Calabro, A., Renzi, D., and Sorbi, S. (2001). Genetic and clinical analysis of spinocerebellar ataxia type 8 repeat expansion in Italy. *Arch. Neurol.* **58**, 1856–1859.

Clark, H. B., Burright, E. N., Yunis, W. S., Larson, S., Wilcox, C., Hartman, B., Matilla, A., Zoghbi, H. Y., and Orr, H. T. (1997). Purkinje cell expression of a mutant allele of SCA1 in transgenic mice leads to disparate effects on motor behaviors, followed by a progressive cerebellar dysfunction and histological alterations. *J. Neurosci.* **17**, 7385–7395.

Day, J. W., Schut, L. J., Moseley, M. L., Durand, A. C., and Ranum, L. P. (2000). Spinocerebellar ataxia type 8: clinical features in a large family. *Neurology* **55**, 649–657.

Harding, A. (1983). Classification of the hereditary ataxias and paraplegias. *Lancet* **1**, 1151–1155.

Hernandez, M. C., Andres-Barquin, P. J., Martinez, S., Bulfone, A., Rubenstein, J. L., and Israel, M. A. (1997). ENC-1: a novel mammalian kelch-related gene specifically expressed in the nervous system encodes an actin-binding protein. *J. Neurosci.* **17**, 3038–3051.

Ikeda, Y., Shizuka, M., Watanabe, M., Okamoto, K., and Shoji, M. (2000a). Molecular and clinical analyses of spinocerebellar ataxia type 8 in Japan. *Neurology* **54**, 950–955.

Ikeda, Y., Shizuka-Ikeda, M., Watanabe, M., Schmitt, M., Okamoto, K., and Shoji, M. (2000b). Asymptomatic CTG expansion at the SCA8 locus is associated with cerebellar atrophy on MRI. *J. Neurol. Sci.* **182**, 76–79.

Juvonen, V., Hietala, M., Paivarinta, M., Rantamaki, M., Hakamies, L., Kaakkola, S., Vierimaa, O., Penttinen, M., and Savontaus, M. L. (2000). Clinical and genetic findings in Finnish ataxia patients with the spinocerebellar ataxia 8 repeat expansion. *Ann. Neurol.* **48**, 354–361.

Kim, T. A., Lim, J., Ota, S., Raja, S., Rogers, R., Rivnay, B., Avraham, H., and Avraham, S. (1998). NRP/B, a novel nuclear matrix protein, associates with p110(RB) and is involved in neuronal differentiation. *J. Cell Biol.* **141**, 553–566.

Klockgether, T., Wullner, U., Spauschus, A., and Evert, B. (2000). The molecular biology of the autosomal-dominant cerebellar ataxias. *Movement Disorders* **15**, 604–612.

Koob, M. D., Benzow, K. A., Bird, T. D., Day, J. W., Moseley, M. L., and Ranum, L. P. W. (1998). Rapid cloning of expanded trinucleotide repeat sequences from genomic DNA. *Nat. Genet.* **18**, 72–75.

Koob, M. D., Moseley, M. L., Schut, L. J., Benzow, K. A., Bird, T. D., Day, J. W., and Ranum, L. P. (1999). An untranslated CTG expansion causes a novel form of spinocerebellar ataxia (SCA8). *Nat. Genet.* **21**, 379–384.

Moseley, M. L., Benzow, K. A., Schut, L. J., Bird, T. D., Gomez, C. M., Barkhaus, P. E., Blindauer, K. A., Labuda, M., Pandolfo, M., Koob, M. D., and Ranum, L. P. (1998). Incidence of dominant spinocerebellar and Friedreich triplet repeats among 361 ataxia families. *Neurology* **51**, 1666–1671.

Moseley, M. L., Schut, L. J., Bird, T. D., Koob, M. D., Day, J. W., and Ranum, L. P. (2000). SCA8 CTG repeat: en masse contractions in sperm and intergenerational sequence changes may play a role in reduced penetrance. *Hum. Mol. Genet.* **9**, 2125–2130.

Nemes, J. P., Benzow, K. A., and Koob, M. D. (2000). The SCA8 transcript is an antisense RNA to a brain-specific transcript encoding a novel actin-binding protein (KLHL1). *Hum. Mol. Genet.* **9**, 1543–1551.

Philips, A. V., Timchenko, L. T., and Cooper, T. A. (1998). Disruption of splicing regulated by a CUG-binding protein in myotonic dystrophy. *Science* **280**, 737–741.

Robinson, D. N., and Cooley, L. (1997). Drosophila kelch is an oligomeric ring canal actin organizer. *J. Cell Biol.* **138**, 799–810.

Schols, L., Szymanski, S., Peters, S., Przuntek, H., Epplen, J. T., Hardt, C., and Riess, O. (2000). Genetic background of apparently idiopathic sporadic cerebellar ataxia. *Hum. Genet.* **107**, 132–137.

Silveira, I., Alonso, I., Guimaraes, L., Mendonca, P., Santos, C., Maciel, P., Fidalgo De Matos, J. M., Costa, M., Barbot, C., Tuna, A., Barros, J., Jardim, L., Coutinho, P., and Sequeiros, J. (2000). High germinal instability of the (CTG)n at the SCA8 locus of both expanded and normal alleles. *Am. J. Hum. Genet.* **66**, 830–840.

Sobrido, M. J., Cholfin, J. A., Perlman, S., Pulst, S. M., and Geschwind, D. H. (2001). SCA8 repeat expansions in ataxia: a controversial association. *Neurology* **57**, 1310–1312.

Soltysik-Espanola, M., Rogers, R. A., Jiang, S., Kim, T. A., Gaedigk, R., White, R. A., Avraham, H., and Avraham, S. (1999). Characterization of Mayven, a novel actin-binding protein predominantly expressed in brain. *Mol. Biol. Cell* **10**, 2361–2375.

Stevanin, G., Herman, A., Durr, A., Jodice, C., Frontali, M., Agid, Y., and Brice, A. (2000). Are (CTG)n expansions at the SCA8 locus rare polymorphisms? *Nat. Genet.* **24**, 213.

Stone, J., Smith, L., Watt, K., Barron, L., and Zeman, A. (2001). Incoordinated thought and emotion in spinocerebellar ataxia type 8. *J. Neurol.* **248**, 229–232.

Vincent, J. B., Neves-Pereira, M. L., Paterson, A. D., Yamamoto, E., Parikh, S. V., Macciardi, F., Gurling, H. M., Potkin, S. G., Pato, C. N., Macedo, A., Kovacs, M., Davies, M., Lieberman, J. A., Meltzer, H. Y., Petronis, A., and Kennedy, J. L. (2000a). An unstable trinucleotide-repeat region on chromosome 13 implicated in spinocerebellar ataxia: a common expansion locus. *Am. J. Hum. Genet.* **66**, 819–829.

Vincent, J. B., Yuan, Q. P., Schalling, M., Adolfsson, R., Azevedo, M. H., Macedo, A., Bauer, A., DallaTorre, C., Medeiros, H. M., Pato, M. T., Pato, C. N., Bowen, T., Guy, C. A., Owen, M. J., O'Donovan, M. C., Paterson, A. D., Petronis, A., and Kennedy, J. L. (2000b). Long repeat tracts at SCA8 in major psychosis. *Am. J. Med. Genet.* **96**, 873–876.

Warner, J. P., Barron, L. H., and Porteous, M. E. (1999). SCA 8 in the Scottish population. *Am. J. Hum. Genet.* **65** (Suppl), A411.

Worth, P. F., Houlden, H., Giunti, P., Davis, M. B., and Wood, N. W. (2000). Large, expanded repeats in SCA8 are not confined to patients with cerebellar ataxia. *Nat. Genet.* **24**, 214–215.

# CHAPTER 11

# Spinocerebellar Ataxia 10 (SCA10)

**TOHRU MATSUURA**
*Department of Neurology*
*Baylor College of Medicine*
*Houston, Texas 77030*

**TETSUO ASHIZAWA**
*Department of Neurology*
*The University of Texas Medical Branch*
*Galveston, Texas 77555*

I. Summary
II. Phenotype
III. The SCA10 gene
IV. Instability of the Expanded ATTCT Repeat
V. Diagnosis
  A. DNA Testing
  B. Biochemistry
  C. Neuroimaging
  D. Neurophysiology
  E. Neuropathology
VI. Genotype-Phenotype Correlation
VII. Population Genetics
VIII. Models and Disease Mechanism of the ATTCT Expansion
IX. Treatment
  Acknowledgments
  References

## I. SUMMARY

Spinocerebellar ataxia type 10 (SCA10) is an autosomal dominant neurodegenerative disease characterized by ataxia and epilepsy (Grewal *et al.*, 1998; Matsuura *et al.*, 1999; Zu *et al.*, 1999) Within a given family, the disease typically shows progressively earlier onset of the disease in successive generations with increasing severity, a phenomenon known as anticipation. The combination of relatively pure cerebellar ataxia and epilepsy is unique among inherited ataxias. The clinical diagnosis of SCA10 can be suspected based on this unique clinical feature, especially in families of Mexican descent in which SCA10 appears to be prevalent.

In SCA10, the mutation is a very large expansion of an unstable ATTCT pentanucleotide repeat in intron 9 of the *SCA10* gene. In contrast, mutations of SCA1, SCA2, SCA3/MJD, SCA6, SCA7, SCA12, SCA17, and dentatorubropallidolysian atrophy (DRPLA) are expansions of polyglutamine-coding CAG trinucleotide repeats (reviewed in Rosa and Ashizawa, 2001). In SCA8, an expansion of an CTG repeat in the 3′ untranslated region (UTR) of the *SCA8* gene has been thought to be the disease-causing mutation. Furthermore, Friedreich's ataxia (FRDA1), the most common recessive ataxia, is caused by an expansion of an intronic GAA trinucleotide repeat in the *FRDA* gene (Campuzano *et al.*, 1996) . Thus, trinucleotide repeats had been the sole class of microsatellite repeat whose expansions cause inherited ataxias, until the pentanucleotide repeat expansion was discovered in SCA10. SCA10 is the only human disease caused by expansion of a pentanucleotide repeat.

We will describe clinical features and molecular genetics of this disease, and discuss potential pathogenic mechanisms of SCA10. Wherever appropriate, we will compare SCA10 with other diseases caused by expansions of short tandem repeats, and discuss their relevance to SCA10.

## II. PHENOTYPE

Clinical characteristics of SCA10, which is now genetically defined by the ATTCT repeat expansion, are currently based on data obtained from six Mexican families (Grewal *et al.*, 1998, Matsuura *et al.*, 1999; Zu *et al.*, 1999, Rasmussen *et al.*, 2001). The clinical phenotype of SCA10 is relatively homogeneous. The age of onset in reported cases ranges from 14–45 years. The central feature of the clinical phenotype is cerebellar ataxia that usually starts as poor balance on gait. The gait ataxia gradually worsens with an increasing number of falls, necessitating use of a cane, walker, and eventually wheelchair. In an advanced stage, the patient becomes unable to stand or sit without support. Scanning dysarthria appears within a few years after the onset of gait ataxia. Scanning speech is due to cerebellar ataxia involving the vocal cord, tongue, palate, cheek, and lip movements. Coordination of the diaphragm and other respiratory muscles are also impaired, contributing to the speech impairment. Poor coordination of tongue, throat, and mouth muscles also causes dysphagia in later stages of the disease. Dysphagia is not only a nuisance but often leads to life-threatening aspiration pneumonia. Severe dysphagia may require a percutaneous placement of a gastric tube for both prevention of aspiration and maintenance of nutritional intake. Hand coordination also starts deteriorating within a few years after the onset of gait ataxia. Fine motor tasks, such as handwriting and buttoning cuffs, are first to be impaired, followed by increasing difficulties with daily activities such as feeding, dressing, and personal hygiene. Tracking eye movements become abnormal, with fragmented pursuit, ocular dysmetria, and occasionally ocular flutter, which are all attributable to cerebellar dysfunction. Some patients with relatively severe ataxia show coarse gaze-induced nystagmus.

In addition to cerebellar ataxia, 20%–60% of affected members of SCA10 families have recurrent seizures (Grewal et al., 1998; Matsuura et al., 1999; Zu et al., 1999; Rasmussen et al., 2001). Most of these patients experience generalized motor seizures, but complex partial seizures have also been noted. An attack of complex partial seizure may occasionally be followed by a generalized motor seizure, suggesting secondary generalization of a focal seizure activity. In most cases, seizures are noted after the onset of gait ataxia. In untreated patients, generalized motor seizures could occur as frequently as daily, and complex partial seizures may be even more frequently–up to several times a day. Seizure characteristics do not appear to change with aging. However, seizure-related deaths have been noted in some affected members of SCA10 families (Grewal et al., 2002).

While there is no overt progressive dementia, some SCA10 patients exhibit mild cognitive dysfunctions (Rasmussen et al., 2001). Pyramidal and extrapyramidal dysfunction, visual impairment, hearing loss, and other nervous system abnormalities are usually absent, and if present, they are subtle. Although nerve conduction studies show evidence of mild sensory neuropathy in affected members of some SCA10 family, the patients seldom complain of neuropathic symptoms. In some SCA10 families, hepatic dysfunction and anemia appeared to co-segregate with the SCA10 mutation with some deaths from hepatic failure (Rasmussen et al., 2001). However, whether they are a part of the SCA10 phenotype remains to be determined. In spite of the potential existence of these associated features, the combination of relatively pure cerebellar ataxia and seizure remains the core phenotype of SCA10, which is unique to SCA10 and has not been seen in other ADCAs. Although some patients with DRPLA may show this combination, other neurological abnormalities such as cognitive or psychiatric impairments, involuntary movements with myoclonus and choreoathetosis, and pyramidal signs are far more conspicuous than SCA10. Although studies of additional SCA10 families are necessary to further define the clinical phenotype of SCA10, the unique combination of cerebellar ataxia and epilepsy justifies screening for the SCA10 mutation. A substantial number of patients with the SCA10 mutation have only cerebellar ataxia without seizures. Thus, other SCAs that cause pure cerebellar ataxia, such as SCA5, SCA6, SCA8, SCA11, and SCA16, should be considered among the differential diagnoses. It should also be noted that some patients with other SCAs may be presented with pure cerebellar ataxia, especially in the early stage of the disease. Furthermore, slight cognitive impairments, mild pyramidal sings, or subtle extrapyramidal signs in some patients with SCA10 may make the distinction from these SCAs complicated.

Anticipation was first noted by Grewal et al. (1998) in their large SCA10 family. While anticipation is striking in this family, it was less prominent in another larger family described by Matsuura et al. (1999). In small families, anticipation may be variable and difficult to evaluate (Rasmussen et al., 2001). It is also noteworthy that a severe early-onset phenotype has not been reported in SCA10, although juvenile cases with phenotype similar to those of adult onset disease have been seen in some SCA10 famlies.

## III. THE SCA10 GENE

SCA10 is an ADCA, which frequently shows anticipation (Grewal et al., 1995; Matsuura et al., 1999; Zu et al., 1999). Several other SCAs that show anticipation are caused by trinucleotide repeat expansions. In these SCAs, progressive increases in the expansion size are the molecular basis of anticipation. Therefore, we initially postulated that the mutation responsible for SCA10 is an expansion of an unstable trinucleotide repeat, and started to search for the mutation using a positional cloning strategy. Matsuura et al. (1999) and Zu et al. (1999) independently mapped the *SCA10* locus to the chromosome 22q13–qter region by linkage analyses in two large families. Two recombination events defined the critical *SCA10* region in a 3.8-cM interval between *D22S1140* and *D22S1160*. Studies using additional polymorphic markers narrowed the SCA10 region to a 2.7-cM region between *D22S1140* and *D22S1153* (Matsuura et al., 2000).

Chromosome 22 was the first human chromosome for which the Human Genome Project completed the nucleotide sequencing of the euchromatic parts (Dunham et al., 1999). However, there were still 11 heterochromatic gaps that remain to be sequenced at the time we were trying to clone the *SCA10* gene. Distal flanking markers, *D22S1160* and *D22S1153*, resided in one of these gaps. Consequently, we could not determine the exact physical size of the SCA10 candidate region. Nevertheless, two contigs composed of bacterial artificial chromosomes (BACs), phage P1-derived artificial chromosomes (PACs), and cosmids covered most of this region. The sequence data of these contigs enabled us to perform computer database searches for trinucleotide repeat sequences in this region.

Subsequently, we found four additional families with an autosomal dominant inheritance characterized by ataxia and seizures, which co-segregated with the SCA10 markers on 22q. Because all six families (two original large families and these four families) are of Mexican descent, we assumed a common ancestry and compared their haplotypes of the SCA10 region. We identified a common haplotype shared by these families within the region, although the telomeric end of this region could not be defined due to a gap of the available contigs. The chromosome 22 genome database at the Sanger Center showed 14 trinucleotide repeats, each consisting of >3 repeat units in length, in the contig sequences (Dunham et al., 1999). However, none of them showed larger repeat size in SCA10 patients than in normal subjects. Repeat expansion detection (RED) analysis failed to show evidence of a CAG or CAA expansion (Matsuura et al., 2000; Schalling et al., 1993). In Western blot analysis, a monoclonal antibody against expanded polyglutamine tracts failed to detect aberrant proteins in the extract of lymphoblastoid cells from SCA10 patients (Matsuura et al., 2000; Trottier et al., 1995). Nevertheless, there were still possibilities that a pathogenic trinucleotide repeat expansion exists in the telomeric gap. Small pathogenic expansions of less than 40 trinucleotide repeat units have been found in SCA6 and oculopharyngeal muscular dystrophy (Brais et al., 1998). Although small triplet repeat expansions are transmitted from generation to generation usually without repeat size instability, anticipation has been reported in multiple SCA6 families (Matsuyama et al., 1997; Watanabe et al., 1998; Soong et al., 2001; Sinke et al., 2001). Nevertheless, until the gap is cloned, it would be difficult to search for small triplet repeat expansions in the distal candidate region. Thus, we changed our strategy to search for an expansion of a nontriplet microsatellite repeat.

During our screening of such repeats, we encountered a pentanucleotide (ATTCT) repeat in intron 9 of a gene called *E46L*, which is now recognized as *SCA10* by the

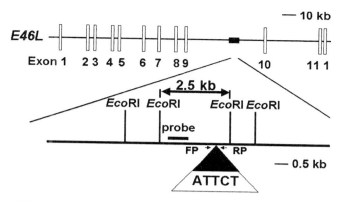

FIGURE 11.1 The physical map of the ATTCT pentanucleotide repeat region. (Adopted from Matsuura et al. (2000) *Nat. Genet.* **26**, 191–194. With modification). A schematic presentation of the structure of the *E46L* gene is shown at the top. Note that the ATTCT repeat is located in intron 9. The bottom part of the figure shows enlargement of the ATTCT repeat region with *Eco*RI restriction sites. "FP," "RP," and "Probe" indicate the positions of the forward and reverse PCR primers, and the probe used in the Southern analysis of the ATTCT repeat.

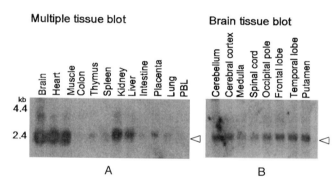

FIGURE 11.2 Northern blot analyses of multiple tissue blots. (A) Peripheral tissues. (B) Central nervous system tissues. (From Matsuura et al., (2000). *Nat. Genet.* **26**, 191–194. With permission.)

Human Genome Organization (HUGO. Fig. 11.1; (Matsuura et al., 2000). Multiple tissue Northern blots showed that this gene is expressed widely throughout the brain, as well as in the skeletal muscle, heart, liver, and kidney (Fig. 11.2). The widespread expression throughout the brain was confirmed by an *in situ* hybridization study using mouse brain sections (Fig. 11.3). PCR analysis showed polymorphisms in repeat number in normal individuals; the repeat size ranged from 10 to 22 ATTCTs with 82.1% heterozygosity in 604 chromosomes of three ethnic groups representing the Caucasian, Japanese and Mexican populations (Fig. 11.4). Sequence analysis of the alleles obtained from 20 normal individuals showed tandem repeats of ATTCT without interruption. The allele distributions in the three ethnic populations were similar to each other and consistent with Hardy-Weinberg equilibrium. In SCA10 families, PCR analysis demonstrated a lack of heterozygo-

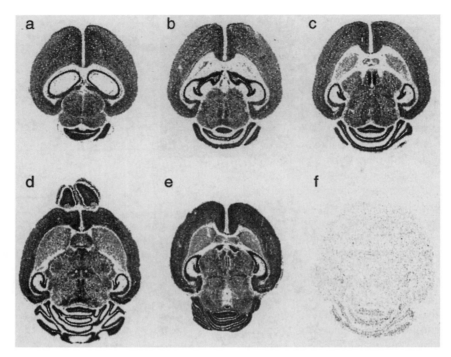

FIGURE 11.3  E46L is widely expressed in brain regions. E46L mRNA was detected by *in situ* hybridization of $^{32}$P-labeled probes to horizontal sections of 4-month-old adult (a–d) and 10-day-old juvenile (e) mouse brain. Expression was similar to the pattern of cell density determined by cresyl violet staining of the same sections (not shown). a–d, Dorsal to ventral progression; f, negative control for nonspecific hybridization to an adult brain section. (From Matsuura *et al.* (2000). *Nat. Genet.* **26**, 191–194. With permission.)

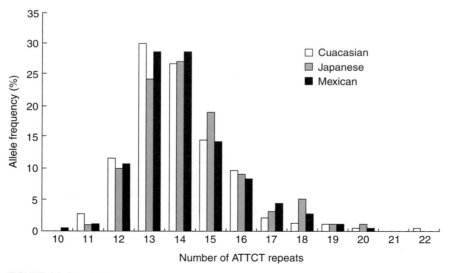

FIGURE 11.4  Distribution of the ATTCT repeat alleles in normal populations. Shown is a histogram of the normal ATTCT repeat alleles in Caucasian (n = 250), Japanese (n = 100), and Mexican (n = 254) chromosomes. (From Matsuura *et al.* (2000). *Nat. Genet.* **26**, 191–194. With permission.)

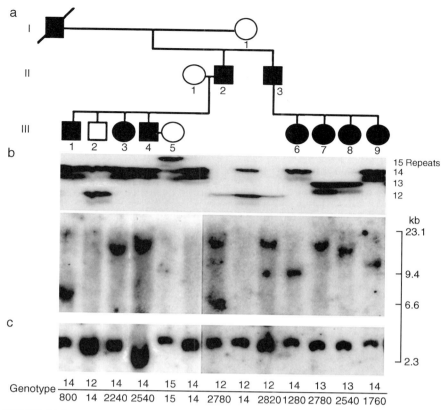

**FIGURE 11.5** Expansion mutations in SCA10 families. (Adopted from Matsuura et al., 2000). (a) The pedigree of an SCA10 family. Square and round symbols indicate male and female members, respectively. Open symbols are asymptomatic individuals, and filled symbols indicate affected members. A diagonal line across a symbol denotes a deceased individual. (b) PCR analysis of the ATTCT pentanucleotide repeat. All affected individuals showed a single allele (note that each band accompanies a shadow band underneath due to PCR artifact). Affected individuals in the second generation failed to transmit their 12-repeat allele to their affected offspring, while an unaffected offspring received this allele from the affected father. These data suggest that the affected individuals are apparently hemizygous for the ATTCT repeat. (c) Southern analysis of expansion mutations of the ATTCT repeat region. Southern blots of the genomic DNA samples digested with *Eco*RI using a 0.8% agarose gel show variably expanded alleles in affected members of the families shown above. All individuals examined have a normal allele (2.5 kb). The apparent variability of the normal allele size is attributable to gel-loading artifacts since additional analyses using the same (*Eco*RI) and different (*Eco*RV, *Hind*III and *Bgl*I) restriction enzymes did not show consistent variability of the normal allele size. The genotype of each individual is shown at the bottom, with an estimated number of pentanucleotide repeats. (From Matsuura et al. (2000). *Nat. Genet.* **26**, 191–194. With modifications).

sity of the amplifiable ATTCT repeat alleles in all affected individuals and carriers of the disease haplotype, with the single allele of the ATTCT repeat shared by their unaffected parent. The single allele amplified from the affected parent was never transmitted to any of the affected offspring (Fig. 11.5). Using multiple sets of PCR primer sets, we excluded the possibility that a mutation within the sequence where the PCR primers annealed was the cause of the lack of amplification. These data led us to postulate that the only allele on the normal (non-SCA10) chromosome is amplified and the SCA10 chromosome has an expansion or other rearragements at the repeat locus.

To investigate this hypothesis, Southern blots of *Eco*RI fragments of the genomic DNA obtained from normal and SCA10 patients were analyzed with a probe, which corresponds to the region immediately upstream of the ATTCT repeat (Matsuura et al., 2000). We detected only the expected 2.5-kb fragment in normal individuals, whereas a very large allele of variable size was detected in addition to the normal allele in each affected family member (Fig. 11.5). The expanded alleles were absent in over 600 normal chromosomes, suggesting that the expanded ATTCT repeats are specific for SCA10. The variable expansion allele size indicates that expanded ATTCT repeats are unstable.

## IV. INSTABILITY OF THE EXPANDED ATTCT REPEAT

Expansions of short tandem repeats have also been found in inherited neurodegenerative diseases without ataxia. Huntington's disease (The Huntington's Disease Collaborative Research Group, 1993), Kennedy's disease (La Spada *et al.*, 1991), myotonic dystrophy type 1 (DM1; Brook *et al.*, 1992; Fu *et al.*, 1992; Mahadevan *et al.*, 1992), fragile X syndrome (Fu *et al.*, 1991), FRAXE mental retardation (Knight *et al.*, 1994), and oculopharyngeal muscular dystrophy (Brais *et al.*, 1998) are caused by expansions of trinucleotide repeats. Recently, an expanded CCTG tetranucleotide repeat was identified as the mutation in myotonic dystrophy type 2 (DM2; Liquori *et al.*, 2001). This CCTG repeat expansion shows some similarities to the ATTCT repeat expansion in SCA10; both are intronic nontrinucleotide repeats that expand to disease-causing alleles by gaining thousands of repeat units. Although there are no other diseases known to be caused by expansions of microsatellite repeats, an expansion mutation of a dodecanucleotide repeat (CCCCGCCCCGCG repeat), which is classified as a minisatellite repeat, has been described in progressive myoclonic epilepsy (EPM1; Lafreniere *et al.*, 1997; Virtaneva *et al.*, 1997).

Instability of trinucleotide repeats have been extensively studied (Wells *et al.*, 1998), including exonic CAG repeats in SCA1 (Chung *et al.*, 1993; Chong *et al.*, 1995), SCA2 (Cancel *et al.*, 1997), SCA3/MJD (Cancel *et al.*, 1995, 1998), SCA7 Z(David *et al.*, 1998; Gouw *et al.*, 1998; Monckton *et al.*, 1999), DRPLA (Ueno *et al.*, 1995; Takano *et al.*, 1996; Takiyama *et al.*, 1999), Kennedy disease (Tanaka *et al.*, 1999), and Huntington disease (Telenius *et al.*, 1994; Leeflang *et al.*, 1995, 1999), 3′ UTR CTG repeats in DM1 (Ashizawa *et al.*, 1994, 1996; Monckton *et al.*, 1995; Wong *et al.*, 1995) and SCA8 (Moseley *et al.*, 2000), 5′ UTR CAG repeat in SCA12 (Holmes *et al.*, 1999), and nonpathogenic CAG repeats in ERDA1 (Ikeuchi *et al.*, 1994) and SEF2.1 (Breschel *et al.*, 1997). Repeat instability has also been investigated in 5′ UTR CGG/CCG repeats of fragile X syndrome (Eichler *et al.*, 1994) and FRAXE mental retardation (Knight *et al.*, 1994), and other untranslated CGG/CGG repeats at chromosomal fragile sites FRAXF (Parrish *et al.*, 1994), FRA11B (Jones *et al.*, 1995), and FRA16A (Nancarrow *et al.*, 1994), as well as the intronic GAA repeat in Friedreich's ataxia (Montermini *et al.*, 1997; De Michele *et al.*, 1998; Bidichandani *et al.*, 1999). All these repeats are variably unstable in somatic and germline cells, although small pathogenic expansions of trinucleotide repeats in SCA6 (Zhuchenko *et al.*, 1997) and oculopharyngeal muscular dystrophy (Brais *et al.*, 1998) show no or little repeat size instability. Diseases that involve large repeat size expansions such as DM1, fragile X syndrome, FRAXE mental retardation, and Friedreich's ataxia generally show greater degrees of instability.

The size of expanded repeat, however, is not the only determinant of the instability. The motif of each repeat unit (e.g., CAG vs. CGG vs. GAA) is clearly important; different repeats may form alternative DNA structures, which may be predisposed to different mechanisms of instability. Among CAG/CTG expansion diseases, the CG content in the region surrounding the repeat tract has been correlated with the degree of repeat size instability (Brock *et al.*, 1999). DNA mismatch repair gene *msh2* plays an important role in CAG repeat instability in Huntington's disease transgenic mice (Manley *et al.*, 1999). Yet, there appear to be other genetic, epigenetic and environmental factors that may influence the repeat instabilities. Additional studies on the CCTG repeat of DM2 (Liquori *et al.*, 2001) and minisatellites, including the dodecamer repeat involved in progressive EPM1 (Lafreniere *et al.*, 1997; Lalioti *et al.*, 1997; Virtaneva *et al.*, 1997) are also available in the literature.

Although these studies may provide insights about the instability of the ATTCT repeat, little is known about this novel class of disease-causing microsatellite repeat. In the genome database, there are over 80 loci where 10 or more ATTCTs show tandem repeats. Our preliminary ATTCT RED analysis suggests that ATTCT repeats are longer than 40 units in at least one locus in normal individuals (unpublished data). We are currently investigating the stability of various ATTCT repeats and exploring the possibility of instability and pathogenic involvement of these repeats in human diseases. The ATTCT repeat expansion in SCA10 is one of the largest microsatellite repeats involved in human diseases, perhaps, second to the CCTG repeat expansion in DM2. Our preliminary studies suggested the existence of age- and tissue-dependent instability in somatic and germline tissues (unpublished data). The expansion size often differs from one tissue to another, especially between the somatic tissues and sperm in patients with SCA10, and sperm appear to show a greater degree of instability. The expanded ATTCT repeats also show instability of the repeat size in lymphoblast cell lines derived from SCA10 patients. Instability in various experimental systems, such as transgenic animals and other cell culture models, may provide important data for understanding the mechanism of the ATTCT repeat instability.

## V. DIAGNOSIS

As discussed in the Section II, the clinical hallmark of SCA10 is a combination of relatively pure cerebellar ataxia and epilepsy. This autosomal dominant disease shows complete penetrance, and so far the SCA10 mutation has not been identified in patients without family history. However, it is conceivable that SCA10 could occur without family history especially if there was early parental death(s) or if the patient was adopted. Whether *de novo* mutations could occur in SCA10 remains unknown. SCA10 has been reported only in Mexican or Mexican-American families. In patients with a suspected SCA10 phenotype, the most accurate and cost-effective diagnosis is provided by DNA testing. Hispanic patients who have a familial disorder with a combination of relatively pure cerebellar ataxia and epilepsy should be considered for DNA testing for SCA10. The SCA10 phenotype in some of these patients has been attributed to alcoholism, which can cause both cerebellar ataxia and seizures, while in some other patients it has been mistaken for a seizure disorder complicated by cerebellar ataxia secondary to anti-epileptic drug therapy. Seizures in SCA10 patients have also been attributed to cysticercosis, which has a high prevalence in the Mexican population. It should also be noted that a substantial number of SCA10 patients show only cerebellar ataxia without epilepsy, particularly in the early phase of the disease. The indication of DNA testing for SCA10 should be determined taking these caveats into considerations.

## A. DNA Testing

The SCA10 DNA testing is now commercially available (see GeneTest web site at http://www.genetest.org). The testing is based on both PCR and Southern blot analyses (see Section III). PCR analysis of the ATTCT repeat should be performed first. If the subject's genomic DNA from somatic cells (e.g., blood leukocytes) shows two normal alleles, the molecular diagnosis will be "normal" and further DNA analysis will not be necessary. The normal range of the ATTCT repeat size has been reported as 10–21 repeats (Matsuura *et al.*, 2000). However, we have recently encountered a normal allele with 29 ATTCT repeats (unpublished data), defining the new normal range of 10–29 repeats. If the PCR analysis shows amplification of a single allele, the test subject may be either homozygous having two identical normal alleles or heterozygous having one normal allele and one nonamplifiable expanded allele. To distinguish these two possibilities, further analysis must be done using the Southern blot technique, which can detect the expanded allele. The expanded alleles have been found to range from 800 ATTCTs to 4500 ATTCT repeats. The size of SCA10 expansions have shown no overlapping with the normal allele range, providing 100% sensitivity and 100% specificity of the test. As an alternative to this two-step DNA diagnostic procedure, we recently developed a PCR assay that can amplify expanded repeats as a pentanucleotide ladder, using one primer placed upstream of the ATTCT repeat and the other primer within the repeat itself with a hanging unique sequence. This assay allows for quick screening of the ATTCT repeat expansion allele by a one-step PCR procedure (Matsuura and Ashizawa, 2002).

Although there have been no official guidelines for genetic testing of spinocerebellar ataxias, readers may refer to guidelines established for DNA testing of other repeat expansion diseases, such as Huntington's disease (International Huntington Association; IHA and the World Federation of Neurology; WFN Research Group on Huntington Chorea, 1994) and myotonic dystrophy (The International Myotonic Dystrophy Consortium, 2000). Genetic counseling is an indispensable part of presymptomatic and prenatal testing because of the complex ethical, legal, and socioeconomical consequences of the testing. Although the DNA diagnosis is relatively straightforward for confirmatory testing in symptomatic patients, genetic counseling is still recommended when appropriate. Asymptomatic testing is not recommended in juvenile at-risk subjects.

## B. Biochemistry

Biochemical changes in SCA10 are unknown. Some patients with SCA10 had abnormal hepatic functions. However, it is undetermined whether hepatic insufficiency is a part of the SCA10 phenotype or it is a coincidental finding.

## C. Neuroimaging

Magnetic resonance imaging (MRI; Fig. 11.6) and computerized tomography (CT) of the brain show pancerebellar atrophy without abnormalities in other regions (Rasmussen *et al.*, 2001).

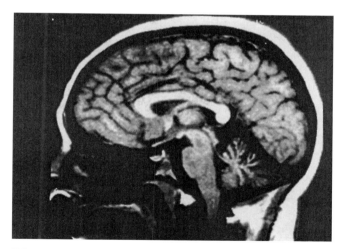

FIGURE 11.6 MRI (Sagittal T-1 weighted views) of a representative SCA10 patient showing atrophy of the cerebellar hemispheres and vermis. (From Rasmussen *et al.* (2001). *Ann. Neurol.* **50**, 234–239. With permission.)

## D. Neurophysiology

Interictal electoencephalography (EEG) often shows evidence of cortical dysfunctions with or without focal epileptiform discharges in some patients (Fig. 11.7; Rasmussen et al., 2001). Nerve conduction studies have shown decreased sensory conduction velocity and decreased amplitude or loss of sensory action potentials in some SCA10 patients.

## E. Neuropathology

Nothing is known about histopathology of the brain in SCA10 owing to a lack of autopsy or biopsy materials.

## VI. GENOTYPE-PHENOTYPE CORRELATION

We found that the size of expanded ATTCT repeat was inversely correlated with the age of onset ($p = 0.018$, $r^2 = 0.34$; Fig. 11.8), with the repeat number ranging from 800 to 4500 and the age of onset from 11 to 48 years (Matsuura et al., 2000). This relatively weak correlation implies that there are determinants other than the ATTCT repeat expansion size. The degree of correlation varies among different SCA10 families; in one large family the correlation was robust (Grewal et al., 1998; Zu et al., 1999), while small families may not clearly demonstrate anticipation (Rasmussen et al., 2001). Another large family clinically showed anticipation, but paternal transmissions exhibited

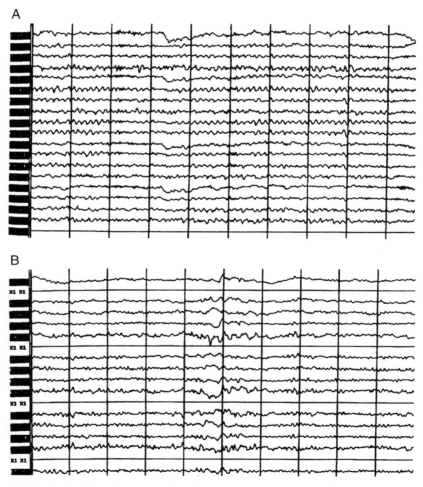

FIGURE 11.7 EEG of representative SCA10 patients showing (A) Disorganized basal activity with a 6–7Hz/15–25 μV irregular and unstable diffusely distributed theta rhythm; (B) generalized bouts of slow, sharp polymorphic waves, with predominance in the right frontocentral region. (From Rasmussen et al. (2001). Ann. Neurol. 50, 234–239. With permission.)

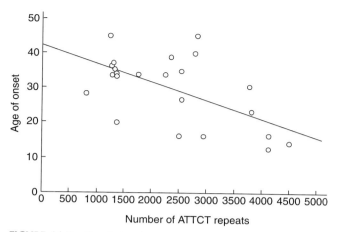

FIGURE 11.8 Correlation between the size of expanded SCA10 ATTCT repeat and the age of onset. (Adopted from Matsuura et al., 2000). A scatter plot shows an inverse correlation between the size of expansion and the age of onset in 26 SCA10 patients ($r^2 = 0.34$, $p = 0.018$). Each symbol represents an SCA10 patient, and the linear regression line is shown. (From Matsuura et al. (2000). *Nat. Genet.* **26**, 191–194. With permission.)

intergenerational contraction of the expanded repeat allele in spite of the clinically observed anticipation (Matsuura et al., 2000). Thus, family-dependent factors may have major effects on the age of onset. Paradoxical contractions of expanded repeats in the presence of anticipation have also been reported in paternal transmissions of DM1 (Ashizawa et al., 1992, 1994; Monckton et al., 1995), and was attributed to unusually strong somatic instability toward expansion in the father's leukocytes (Monckton et al., 1995). Studies on the somatic and germline instability of ATTCT repeats would allow us to investigate a similar mechanism that could explain the observed paradox in SCA10.

The frequency of patients with epilepsy in SCA10 patients varies from family to family. There has been no evidence of correlation between the expanded allele size and the seizure phenotype. Although we have not quantitatively assessed the severity of the disease in SCA10 patients, the expansion size appears to vary from family to family for patients with similar severity of the disease, suggesting the existence of family-dependent factors that influences the disease severity. Longitudinal clinical data should provide insights in understanding the correlation between the repeat size and the severity/progression of the disease in SCA10. Effects of the gender of the transmitting parent on the offspring's phenotype and the ATTCT repeat size also need further investigation.

## VII. POPULATION GENETICS

The exact prevalence of SCA10 is unknown. However, our study suggested that SCA10 is the second most common inherited ataxia in Mexico after SCA2 (Rasmussen et al., 2001). All SCA10 families that we have identified are Mexican nationals or Mexican-Americans. Whether SCA10 is unique to Mexicans or exists in other ethnic groups remains to be determined. However, screening of 169 patients with unassigned ataxia in U.S., Canada, Japan, Spain, and France showed no SCA10 mutations (Matsuura et al., 2002). The origin of the SCA10 mutation is also of interest. The mutation may have arisen in the local Mexican population, giving rise to a founder effect. Alternatively, the mutation may have arisen in ethnic South American Indians or may have been introduced into the current Mexican population by a Spanish conqueror. Further haplotype analysis may provide an answer to this question. Defining at-risk ethnic populations is an important issue for genetic counseling.

## VIII. MODELS AND DISEASE MECHANISM OF THE ATTCT EXPANSION

The molecular mechanism, through which the expanded ATTCT repeat leads to the SCA10 phenotype, is unknown. The challenge stems from the fact that *E46L* is a novel gene of unknown function and that the expanded ATTCT repeat is in an intron. *E46L* consists of 12 exons spanning 172.8 kb (Matsuura et al., 2000). The open reading frame of 1428 bp encodes 475 amino acids. Human *E46L* is highly conserved with its presumed mouse ortholog, *E46* (82% identity, 91% similarity over 475 amino acids). However, information is limited for *E46L* homologs of other species; the next most similar sequence found in the GenBank database is a putative plant protein of unknown function identified by the *Arabidopsis* genome project (24% identical, 41% similar over 409 amino acids). Analysis of the amino acid sequence of the human *E46L* protein suggests that it is a globular protein containing no transmembranous domains, nuclear localization signal, or other type of signal peptide (Golgi, peroxisomal, vacuolar, or endoplasmic reticulum-retention). Near the C terminus, the *E46L* protein contains armadillo repeats, which have been found in membrane structural proteins such as beta-catenin. However, it does not appear to contain any other known functional motifs, clusters, or unusual patterns of charged amino acids or internal repeats of specific amino acid runs (data not shown).

Intron 9 is large (66,420 bp). The location of the ATTCT repeat in this large intron raises the possibility that it might contain additional expressed sequences. This is of particular importance since some pentanucleotide repeats, such as (G/C)3NN repeats in *ETS-2* and dihydrofolate reductase genes (Wang and Griffith, 1996), may function as a *cis*-acting transcription regulator of the downstream gene. An antisense transcript to *E46L* could be disrupted by ATTCT repeat expansions. To investigate these possibilities, we carried out extensive sequence analysis of intron 9. Thirty-nine percent of intron 9 consists of interspersed repetitive elements; 11.39% short interspersed nuclear elements (SINEs), 21.64% long interspersed nuclear elements (LINEs), 3.51% long terminal repeat (LTR) elements, 2.69% medium reiteration frequency sequences (MER) elements, and 1.10% low complexity and simple-repeat sequences. We aligned the remaining intronic sequences to the nonredundant (NR) and expressed sequence tag (EST) subdivisions of the GenBank database using the BLASTN and BLASTX search programs. A processed pseudogene, apparently derived from the *CGI-47* gene on chromosome 3, was identified approximately 1.7 kb upstream of the ATTCT repeat. However, this pseudogene contains no introns and exhibits numerous frameshift and nonsense mutations relative to CGI-47. We also identified a total of 16 sequences with perfect identity to distinct ESTs in GenBank. Several lines of evidence strongly suggest that these ESTs represent hnRNA or DNA contamination artifacts: (1) none were represented more than once, (2) none exhibited evidence of splicing relative to genomic DNA, (3) 13 of the 16 were oriented on the sense strand relative to *E46L*, (4) several were terminated at a polyA nucleotide sequence, which could cause cryptic oligo-dT priming during cDNA synthesis, and (5) only 2 are derived from brain mRNA (both fetal). Finally, none of the EST sequences was positionally correlated with potential exons suggested by gene and exon identification algorithms (the NIX algorithms; http://www.hgmp.mrc.ac.uk/NIX), which include GRAIL, GENESCAN, FGENE, and HEXON.

At present, both loss-of-function and gain-of-function should be considered candidates for the pathogenic mechanism of SCA10. The large ATTCT expansion could affect transcription itself or post-transcriptional processing of *E46L*. Suppression of transcription by a large intronic repeat expansion has been documented in Friedreich ataxia, where an expanded GAA repeat interferes with transcription of the *FRDA* gene (Bidichandani et al., 1998; Ohshima et al., 1998; Sakamoto et al., 1999), presumably by forming triplex/sticky DNA structures (Sakamoto et al., 1998). Analysis of ATTCT repeat sequences suggests they could form a weak triplex structure. Currently, the only available tissue from SCA10 patients are transformed lymphoblastoid cells, and we have found no alterations in the level of SCA10 mRNA in these cells by Northern blot analysis. Allele-specific reverse-transcription PCR (RT-PCR) would be useful for comparing the levels of the E46L mRNA from normal and the SCA10 chromosomes. However, we have not been able to apply this method because of a lack of affected individuals who possess informative alleles at these polymorphic loci. We should also note that the lymphoblastoid cells were derived from B lymphocytes, which are not affected in SCA10, and the level of *E46L* expression is substantially lower in the blood compared with the tissues such as brain, muscle, heart, liver, and kidney. Thus, the mRNA level in affected tissues of SCA10 patients remains to be determined. Another possibility is that the expanded ATTCT repeat alters the splicing of the *E46L* transcript. Aberrant splicing may give rise to a product with a gain of toxic function. However, Northern blot and RT-PCR analyses of mRNA in the lymphoblastoid cells from patients with SCA10 have not shown convincing abnormalities of the isoforms. Investigating the functional role of *E46L* is critical for understanding the pathogenic mechanism of SCA10, since *E46L* is the prime candidate for the gene responsible for the disease. We are in the process of making *E46*-deficient and *E46L*-overexpressing mouse lines and cell lines, which may provide useful means to study the physiological functions of *E46L*. We are also exploring proteins that interact with the SCA10 protein by yeast two-hybrid and immuno-co-precipitation technologies.

Other possible mechanisms include *cis* and *trans* effects of the ATTCT expansion on genes other than *E46L*. In DM1, an unstable CTG repeat expands up to several thousand copies in the 3′ UTR of the *DMPK* gene. The DM1 CTG repeat expansions affect transcription of upstream and downstream genes, *DMWD* (Alwazzan et al., 1999) and *SIX5* (Klesert et al., 1997; Thornton et al., 1997). We searched for genes in the vicinity of the ATTCT repeats in available sequences in the human genome database. The ATTCT repeat is located on RP1–37M3 (the distal-end PAC clone of contig NT_001106). Contig NT_001106 extends 1452.3 kb upstream (toward the centromere) and 75.7 kb downstream of the ATTCT repeat. The downstream gap is estimated to be ~150 kb by FISH analyses and predicted to contain the 5′ end of a gene recognized with accession number AK023424 (from bases 1–167). AK023424 is a 2266 bp cDNA isolated from human placenta, and its remaining 3′ end is in genomic clone AL049853, which represents the centromeric end of the next contig NT_001113. This gene of unknown function is the closest known gene downstream of the *E46L*. The closest upstream gene to *E46L* is the fibulin 1 gene (*FBLN1*), of which 3′ end is over 200 kb from the ATTCT repeat. A CpG island is identified in the 5′ region of *E46L*. Although a putative CpG island is also found 70.8 kb downstream of the ATTCT repeat, no transcribed sequence has been identified from the corresponding region. There is a gene called *bK941F9.C22.6*, which starts about 168.3 kb upstream from the ATTCT repeat and extends toward the centromere (away from the ATTCT repeat), matching the EST AW134719. So far, two exons have been found in bK941F9. It will need further work to fully characterize this gene. Thus, the closest genes upstream and downstream are more than 200 kb away from the ATTCT repeat, which is far greater than the distances between the *DMPK* CTG repeat and the adjacent genes, *DMWD* (~13 kb) and SIX5 (~2 kb). However, we cannot exclude the transcriptional alteration of the neighboring genes in SCA10, since the *cis* effect of the ATTCT repeat expansion is difficult to predict.

Analogous to DM1, a gain of function in *trans* at the mRNA level may be a viable possibility for the pathogenic mechanism for SCA10. In DM1, the CUG repeat is located in the 3'UTR of the mature DMPK mRNA, and the mutant DMPK mRNA with expanded CUG repeats are retained in the nuclear foci (Taneja *et al.*, 1995). The gain of function by the mutant DMPK mRNA may be mediated by abnormal interactions of the expanded CUG repeat tract with CUG-binding proteins, such as CUGBP1, ETR3, MBLL, MBNL, and MBXL (Timchenko *et al.*, 1996; Philips *et al.*, 1998; Miller *et al.*, 2000). In SCA10, the ATTCT repeat is located in an intron, which is expected to be transcribed into AUUCU, but spliced out after transcription, and excluded from the mature mRNA. However, it is possible that the nuclear transcript with expanded AUUCU repeats may be processed differently in the nucleus. In this regard, SCA10 resembles DM2, in which a very large expansion of a CCTG repeat is located in intron 2 of the ZNF9 gene. Interestingly, in muscles obtained from DM2 patients, expanded CCUG repeats are accumulated within nuclear foci, similar to expanded CUG repeats in DM1 (Liquori *et al.*, 2001).

## IX. TREATMENT

There are no effective treatments for ataxia in SCA10. However, epilepsy can be effectively treated. Conventional anticonvulsants such as phenytoin, carbamazepine, and valproic acid usually bring the seizures under reasonable control, although occasional breakthrough seizures may be noted. With therapeutic serum levels, these anticonvulsants do not appear to aggravate ataxia.

## Acknowledgments

We thank the patients for their cooperation. Work in the authors' laboratories was supported by grants from the Oxnard Foundation/National Ataxia Foundation (T.A.), National Ataxia Foundation™, and NIH NS41547 (T.A.). We thank H.Y. Zoghbi, D.L. Nelson, D.L. Burgess, R.P. Grewal, J.F. Noebles, E. Alonso, and S.M. Pulst for their collaboration and for useful suggestions and comments.

## References

Alwazzan, M., Newman, E., Hamshere, M. G., and Brook, J. D. (1999). Myotonic dystrophy is associated with a reduced level of RNA from the DMWD allele adjacent to the expanded repeat. *Hum. Mol. Genet.* **8**, 1491–1497.

Ashizawa, T., Anvret, M., Baiget, M., Barceló, J. M., Brunner, H., Cobo, A. M., Dallapiccola, B., Fenwick, R. G. Jr., Grandell, U., Harley, H., Junien, C., Koch, M. C., Korneluk, R. G., Lavedan, C., Miki, T., Mulley, J. C., López de Munain, A., Novelli, G., Roses, A. D., Seltzer, W. K., Shaw, D. J., Smeets, H., Sutherland, G. R., Yamagata, H., and Harper, P. S. (1994). Characteristics of intergenerational contractions of the CTG repeat in myotonic dystrophy. *Am. J. Hum. Genet.* **54**, 414–423.

Ashizawa, T., Dubel, J. R., Dunne, C. J., Fu, Y-H., Pizzuti, A., Caskey, C. T., Boerwinkle, E., Perryman, M. B., Epstein, H. F., and Hejtmancik, J. F. (1992). Anticipation in myotonic dystrophy. II. Complex relationships between clinical findings and structure of the GCT repeat. *Neurology* **42**, 1877–1883.

Ashizawa, T., Monckton. D. G., Vaishnav, S., Patel, B. J., Voskova, A., and Caskey, C. T. (1996). Instability of the expanded (CTG)n repeats in the myotonin protein kinase gene in cultured lymphoblastoid cell lines from patients with myotonic dystrophy. *Genomics* **36**, 47–53.

Bidichandani, S. I., Ashizawa, T., and Patel, P. I. (1998). The GAA triplet-repeat expansion in Friedreich ataxia interferes with transcription and may be associated with an unusual DNA structure. *Am. J. Hum. Genet.* **62**, 111–112.

Bidichandani, S. I., Purandare, S. M., Taylor, E. E., Gumin, G., Machkhas, H., Harati, Y., Gibbs, R. A., Ashizawa, T., and Patel, P. I. (1999). Somatic sequence variation at the Friedreich ataxia locus includes complete contraction of the expanded GAA triplet repeat, significant length variation in serially passaged lymphoblasts and enhanced mutagenesis in the flanking sequence. *Hum. Mol. Genet.* **8**, 2425–2436.

Brais, B., Bouchard, J. P., Xie, Y. G., Rochefort, D. L., Chretien, N., Tome, F. M., Lafreniere, R. G., Rommens, J. M., Uyama, E., Nohira, O., Blumen, S., Korczyn, A. D., Heutink, P., Mathieu, J., Duranceau, A., Codere, F., Fardeau, M., Rouleau, G. A., and Korcyn, A. D. (1998). Short GCG expansions in the PABP2 gene cause oculopharyngeal muscular dystrophy. *Nat. Genet.* **18**, 164–167.

Breschel, T. S., McInnis, M. G., Margolis, R. L., Sirugo, G., Corneliussen, B., Simpson, S. G., McMahon, F. J., MacKinnon, D. F., Xu, J. F., Pleasant, N., Huo, Y., Ashworth, R. G., Grundstrom, C., Grundstrom, T., Kidd, K. K., DePaulo, J. R., and Ross, C. A. (1997). A novel, heritable, expanding CTG repeat in an intron of the SEF2-1 gene on chromosome 18q21.1. *Hum. Mol. Genet.* **6**, 1855–1863.

Brock, G. J., Anderson, N. H., and Monckton, D. G. (1999). Cis-acting modifiers of expanded CAG/CTG triplet repeat expandability: associations with flanking GC content and proximity to CpG islands. *Hum. Mol. Genet.* **8**, 1061–1067.

Brook, J. D., McCurrach, M. E., Harley, H. G., Buckler, A. J., Church, D., Aburatani, H., Hunter, K., Stanton, V. P., Thirion, J. P., and Hudson, T. (1992). Molecular basis of myotonic dystrophy: expansion of a trinucleotide (CTG) repeat at the 3' end of a transcript encoding a protein kinase family member. *Cell* **68**, 799–808.

Campuzano, V., Montermini, L., Molto, M. D., Pianese, L., Cossee, M., Cavalcanti, F., Monros, E., Rodius, F., Duclos, F., Monticelli, A., Zara, F., Cañizares, J., Koutnikova, H., Bidichandani, S. I., Gellera, C., Brice, A., Trouillas, P., De Michele, G., Filla, A., De Frutos, R., Palau, F., Patel, P. I., Di Donato, S., Mandel, J.-L., Cocozza, S., Koenig, M., and Pandolfo, M. (1996). Friedreich's ataxia: autosomal recessive disease caused by an intronic GAA triplet repeat expansion. *Science* **271**, 1423–1427.

Cancel, G., Abbas, N., Stevanin, G., Durr, A., Chneiweiss, H., Neri, C., Duyckaerts, C., Penet, C., Cann, H. M., Agid, Y., and Brice, A. (1995). Marked phenotypic heterogeneity associated with expansion of a CAG repeat sequence at the spinocerebellar ataxia 3/Machado-Joseph disease locus. *Am. J. Hum. Genet.* **57**, 809–816.

Cancel, G., Durr, A., Didierjean, O., Imbert, G., Burk, K., Lezin, A., Belal, S., Benomar, A., Abada-Bendib, M., Vial, C., Guimaraes, J., Chneiweiss, H., Stevanin, G., Yvert, G., Abbas, N., Saudou, F., Lebre, A. S., Yahyaoui, M., Hentati, F., Vernant, J. C., Klockgether, T., Mandel, J. L., Agid, Y., and Brice, A. (1997). Molecular and clinical correlations in spinocerebellar ataxia 2: a study of 32 families. *Hum. Mol. Genet.* **6**, 709–715.

Cancel, G., Gourfinkel-An, I., Stevanin, G., Didierjean, O., Abbas, N., Hirsch, E., Agid, Y., and Brice, A. (1998). Somatic mosaicism of the CAG repeat expansion in spinocerebellar ataxia type 3/Machado–Joseph disease. *Hum. Mutat.* **11**, 23–27.

Chung, M. Y., Ranum, L. P., Duvick, L. A., Servadio, A., Zoghbi, H. Y., and Orr, H.T. (1993). Evidence for a mechanism predisposing to intergenerational CAG repeat instability in spinocerebellar ataxia type I. *Nat. Genet.* **5**, 254–258.

Chong, S. S., McCall, A. E., Cota, J., Subramony, S. H., Orr, H. T., Hughes, M.R., and Zoghbi, H.Y. (1995). Gametic and somatic tissue-specific heterogeneity of the expanded SCA1 CAG repeat in spinocerebellar ataxia type 1. *Nat. Genet.* **10**, 344–350.

David, G., Durr, A., Stevanin, G., Cancel, G., Abbas, N., Benomar, A., Belal, S., Lebre, A. S., Abada-Bendib, M., Grid, D., Holmberg, M., Yahyaoui, M., Hentati, F., Chkili, T., Agid, Y., and Brice, A. (1998). Molecular and clinical correlations in autosomal dominant cerebellar ataxia with progressive macular dystrophy (SCA7). *Hum. Mol. Genet.* **7**, 165–170.

De Michele, G., Cavalcanti, F., Criscuolo, C., Pianese, L., Monticelli, A., Filla, A., and Cocozza, S., (1998). Parental gender, age at birth and expansion length influence GAA repeat intergenerational instability in the X25 gene: pedigree studies and analysis of sperm from patients with Friedreich's ataxia. *Hum. Mol. Genet.* **7**, 1901–1906.

Dunham, I., Shimizu, N., Roe, B. A., Chissoe, S., Hunt, A. R., Collins, J. E., Bruskiewich, R., Beare, D. M., Clamp, M., Smink, L. J., Ainscough, R., Almeida, J. P., Babbage, A., Bagguley, C., Bailey, J., Barlow, K., Bates, K. N., Beasley, O., Bird, C. P., Blakey, S., Bridgeman, A M., Buck, D., Burgess, J., Burrill, W. D., Burton, J., Carder, C., Carter, N. P., Chen, Y., Clark, G., Clegg, S. M., Cobley, V., Cole, C. G., Collier, R. E., Connor, R. E., Conroy, D., Corby, N., Coville1, G. J., Cox, A. V., Davis, R., Dawson, E., Dhami, P. D., Dockree, C., Dodsworth, S. J., Durbin, R. M., Ellington, A., Evans, K. L., Fey, J. M., Fleming, L., French, L., Garner, A. A., Gilbert, J. G. R., Goward, M. E., Grafham, D., Griffiths, M. N., Hall1, G., Hall, R., Hall-Tamlyn, G., Heathcott, R. W., Ho, S., Holmes, S., Hunt, S. E., Jones, M. C., Kershaw, K., Kimberley, A., King, A., Laird, G. K., Langford, C. F., Leversha, M. A., Lloyd, C., Lloyd, D. M., Martyn, I. D., Mashreghi-Mohammadi, M., Matthews, L., McCann, O. T., McClay, J., McLaren, S., McMurray, A. A., Milne, S. A., Mortimore, B. J., Odell, C. N., Pavitt, R., Pearce, A. V., Pearson, D., Phillimore, B. J., Phillips, S. H., Plumb, R. W., Ramsay, H., Ramsey, Y., Rogers, L., Ross, M. T., Scott, C. E., Sehra, H. K., Skuce, C. D., Smalley, S., Smith, M. L., Soderlund, C., Spragon, L., Steward, C. A., Sulston, J. E., Swann, R. M., Vaudin1, M., Wall, M., Wallis, J. M., , Whiteley, M. N., Willey, D., Williams, C., Williams, S., Williamson, H., Wilmer, T. E., Wilming, L., Wright, C. L., Hubbard, T., Bentley, D. R., Beck, S., Rogers, J., Shimizu, N., Minoshima, S., Kawasaki, K., Sasaki, T., Asakawa, S., Kudoh, J., Shintani, A., Shibuya, K., Yoshizaki, Y., Aoki, N., Mitsuyama, S., Roe, B. A., Chen, F., Chu, L., Crabtree, J., Deschamps, S., Do, A., Do, T., Dorman, A., Fang, F., Fu, Y., Hu, A., Hua, A., Kenton, S., Lai, H., Lao, H. I., Lewis, J., Lewis, S., Lin, S-P., Loh, P., Malaj, E., Nguyen, T., Pan, H., Phan, S., Qi, S., Qian, Y., Ray, L., Ren, Q., Shaull, S., Sloan, D., Song, L., Wang, Q., Wang, Y., Wang, Z., White, J., Willingham, D., Wu, H., Yao, Z., Zhan, M., Zhang, G., Chissoe, S., Murray, J., Miller, N., Minx, P., Fulton, R., Johnson, D., Bemis, G., Bentley, D., Bradshaw, H., Bourne, S., Cordes, M., Du, Z., Fulton, L., Goela, D., Graves, T., Hawkins, J., Hinds, K., Kemp, K., Latreille, P., Layman, D., Ozersky, P., Rohlfing, T., Scheet, P., Walker, C., Wamsley, A., Wohldmann, P., Pepin, K., Nelson, J., Korf, I., Bedell, J. A., Hillier, L., Mardis, E., Waterston, R., Wilson, R., Emanuel, B. S., Shaikh, T., Kurahashi, H., Saitta, S., Budarf, M. L., McDermid, H. E., Johnson, A., Wong, A. C. C., Morrow, B. E., Edelmann, L., Kim, U. J., Shizuya, H., Simon, M. I., Dumanski, J. P., Peyrard, M., Kedra, D., Seroussi, E., Fransson, I., Tapia, I., Bruder, C. E., and O'Brien, K. P. (1999). The DNA sequence of human chromosome 22. *Nature* **402**, 489–495.

Eichler, E. E., Holden, J. J., Popovich, B. W., Reiss, A. L., Snow, K., Thibodeau, S. N., Richards, C. S., Ward, P. A., and Nelson, D. L. (1994). Length of uninterrupted CGG repeats determines instability in the FMR1 gene. *Nat. Genet.* **8**, 88–94.

Fu, Y. H., Kuhl, D. P., Pizzuti, A., Pieretti, M., Sutcliffe, J. S., Richards, S., Verkerk, A. J., Holden, J. J., Fenwick, R. G., Jr., Warren, S. T., Oostra, B. A., Nelson, D. L., and Caskey, C. T. (1991). Variation of the CGG repeat at the fragile X site results in genetic instability: resolution of the Sherman paradox. *Cell* **67**, 1047–1058.

Fu, Y. H., Pizzuti, A., Fenwick, R. G., Jr., King, J., Rajnarayan, S., Dunne, P. W., Dubel, J., Nasser, G. A., Ashizawa, T., de Jong, P., Wieringa, B., Korneluk, R., Perryman, M. B., Epstein, H. F., and Caskey, C. (1992). An unstable triplet repeat in a gene related to myotonic muscular dystrophy. *Science* **255**, 1256–1258.

Gouw, L. G., Castaneda, M. A., McKenna, C. K., Digre, K. B., Pulst, S. M., Perlman, S., Lee, M. S., Gomez, C., Fischbeck, K., Gagnon, D., Storey, E., Bird, T., Jeri, F. R., and Ptacek, L. J. (1998). Analysis of the dynamic mutation in the SCA7 gene shows marked parental effects on CAG repeat transmission. *Hum. Mol. Genet.* **7**, 525–532.

Grewal, R. P., Tayag, E., Figueroa, K. P., Zu, L., Durazo, A., Nunez, C., and Pulst, S. M. (1998). Clinical and genetic analysis of a distinct autosomal dominant spinocerebellar ataxia. *Neurology*. **51**, 1423–1426.

Grewal, R. P. Achari, M., Matsuura, T., Durazo, A., Tayag, E., Zu, L., Pulst, S.-M., and Ashizawa, T. (2002). Clinical and genetic analysis of patients with spinocerebellar ataxia type 10. *Arch. Neurol.* (in press).

Holmes, S. E., O'Hearn, E. E., McInnis, M. G., Gorelick-Feldman, D. A., Kleiderlein, J. J., Callahan, C., Kwak, N. G., Ingersoll-Ashworth, R. G., Sherr, M., Sumner, A. J., Sharp, A. H., Ananth, U., Seltzer, W. K., Boss, M. A., Vieria-Saecker, A. M., Epplen, J. T., Riess, O., Ross, C. A., and Margolis, R. L. (1999). Expansion of a novel CAG trinucleotide repeat in the 5′ region of PPP2R2B is associated with SCA12. *Nat. Genet.* **23**, 391–392.

Ikeuchi, T., Sanpei, K., Takano, H., Sasaki, H., Tashiro, K., Cancel, G., Brice, A., Bird, T. D., Schellenberg, G. D., Pericak-Vance, M. A., Welsh-Bohmer, K. A., Clark, L. N., Wilhelmsen, K., and Tsuji, S. (1994). A novel long and unstable CAG/CTG trinucleotide repeat on chromosome 17q. *Genomics* 1998; 49, 321–326, Eichler, E. E., Holden, J. J. A., Popovich, B. W. et al. Length of uninterrupted CGG repeats determines instability in the FMR1 gene. *Nat. Genet.* **8**, 88–94.

International Huntington Association (IHA) and the World Federation of Neurology (WFN) Research Group on Huntington's Chorea. (1994). Guidelines for the molecular genetics predictive test in Huntington's disease. *Neurology* **44**, 1533–1536.

Jones, C., Penny, L., Mattina, T., Yu, S., Baker, E., Voullaire, L., Langdon, W. Y., Sutherland G. R., Richards, R. I., and Tunnacliffe, A. (1995). Association of a chromosome deletion syndrome with a fragile site within the proto-oncogene CBL2. *Nature* **376**, 145–149.

Klesert, T. R., Otten, A. D., Bird, T. D., and Tapscott, S. J. (1997). Trinucleotide repeat expansion at the myotonic dystrophy locus reduces expression of DMAHP. *Nat. Genet.* **16**, 402–406.

Knight, S. J., Voelckel, M. A., Hirst, M. C., Flannery, A. V., Moncla, A., and Davies, K. E. (1994). Triplet expansion at the FRAXE locus and X–linked mild mental handicap. *Am. J. Hum. Genet.* **55**, 81–86.

La Spada, A. R., Wilson, E. M., Lubahn, D. B., Harding, A. E., and Fischbeck, K. H. (1991). Androgen receptor gene mutations in X-linked spinal and bulbar muscular atrophy. *Nature* **352**, 77–79.

Lafreniere, R. G., Rochefort, D. L., Chretien, N., Rommens, J. M., Cochius, J. I., Kalviainen, R., Nousiainen, U., Patry, G., Farrell, K., Soderfeldt, B., Federico, A., Hale, B. R., Cossio, O. H., Sorensen, T., Pouliot, M. A., Kmiec, T., Uldall, P., Janszky, J., Pranzatelli, M. R., Andermann, F., Andermann, E., and Rouleau, G. A. (1997). Unstable insertion in the 5′ flanking region of the cystatin B gene is the most common mutation in progressive myoclonus epilepsy type 1. EPM1. *Nat. Genet.* **15**, 298–302.

Lalioti, M. D., Scott, H. S., Buresi, C., Rossier, C., Bottani, A., Morris, M. A., Malafosse, A., and Antonarakis, S. E. (1997). Dodecamer repeat expansion in cystatin B gene in progressive myoclonus epilepsy. *Nature* **386**, 847–851.

Leeflang, E. P., Tavare, S., Marjoram, P., Neal, C. O., Srinidhi, J., MacFarlane, H., MacDonald, M. E., Gusella, J. F., de, Young, M., Wexler, N. S., and Arnheim, N. (1999). Analysis of germline mutation spectra at the Huntington's disease locus supports a mitotic mutation mechanism. *Hum. Mol. Genet.* **8**, 173–183.

Leeflang, E. P., Zhang, L., Tavare, S., Hubert, R., Srinidhi, J., MacDonald, M. E., Myers, R. H., de Young, M., Wexler, N. S., Gusella, J. F., and Arnheim, N. (1995). Single sperm analysis of the trinucleotide repeats in the Huntington's disease gene: quantification of the mutation frequency spectrum. *Hum. Mol. Genet.* **4**, 1519–1526.

Liquori, C. L., Ricker, K., Moseley, M. L., Jacobsen, J. F., Kress, W., Naylor, S. L., Day, J. W., and Ranum, L. P. (2001). Myotonic dystrophy type 2 caused by a CCTG expansion in intron 1 of ZNF9. *Science* **293**, 864–867.

Mahadevan, M., Tsilfidis, C., Sabourin, L., Shutler, G., Amemiya, C., Jansen, G., Neville, C., Narang, M., Barcelo, J., O'Hoy, K., Leblond, S., Earle-Macdonald, J., de, Jong, P. J., Wieringa, B., and Korneluk, R. G. (1992). Myotonic dystrophy mutation: an unstable CTG repeat in the 3' untranslated region of the gene. *Science* **255**, 1253–1255.

Manley, K., Shirley, T. L., Flaherty, L., and Messer, A. (1999). Msh2 deficiency prevents in vivo somatic instability of the CAG repeat in Huntington disease transgenic mice. *Nat. Genet.* **23**, 471–473.

Matsuura, T., Achari, M., Khajavi, M., Bachinski, L. L., Zoghbi, H. Y., and Ashizawa, T. (1999). Mapping of the gene for a novel spinocerebellar ataxia with pure cerebellar signs and epilepsy. *Ann. Neurol.* **45**, 407–411.

Matsuura, T., Yamagata, T., Burgess, D. L., Rasmussen, A., Grewal, R. P., Watase, K., Khajavi, M., McCall, A. E., Davis, C. F., Zu, L., Achari, M., Pulst, S. M., Alonso, E., Noebels, J. L., Nelson, D. L., Zoghbi, H. Y., and Ashizawa T. (2000). Large Expansion of the ATTCT pentanucleotide repeat in spinocerebellar ataxia type 10. *Nat. Genet.* **26**, 191–194.

Matsuura, T., and Ashizawa, T. (2002), PCR amplification of expanded ATTCT repeat in SCA 10. *Ann. Neurol.* **51**, 271–272.

Matsuura, T., Ranum., L. P. W., Volpini, V., Pandolfo, M., Sasaki, H., Tashiro, K., Watase, K., Zoghbi, H. Y., and Ashizawa, T. (2002). Spinocerebellar ataxia type 10 is rare in populations other than Mexicans. *Neurology* **58**, 983–984.

Matsuyama, Z., Kawakami, H., Maruyama, H., Izumi, Y., Komure, O., Udaka, F., Kameyama, M., Nishio, T., Kuroda, Y., Nishimura, M., and Nakamura S. (1997). Molecular features of the CAG repeats of spinocerebellar ataxia 6 (SCA6). *Hum. Mol. Genet.* **6**, 1283–1287.

Miller, J. W., Urbinati, C. R., Teng-Umnuay, P., Stenberg, M. G., Byrne, B. J., Thornton, C. A., and Swanson M. S. (2000). Recruitment of human muscleblind proteins to (CUG)(n) expansions associated with myotonic dystrophy. *EMBO J.* **19**, 4439–4448.

Monckton, D. G., Cayuela, M. L., Gould, F. K., Brock, G. J., Silva, R., and Ashizawa T. (1999). Very large (CAG)$_n$ DNA repeat expansions in the sperm of two spinocerebellar ataxia type 7 males. *Hum. Mol. Genet.* **8**, 2473–2478.

Monckton, D. G., Wong, L. J., Ashizawa, T., and Caskey C. T. (1995). Somatic mosaicism. germline expansions. germline reversions and intergenerational reductions in myotonic dystrophy males: small pool PCR analyses. *Hum. Mol. Genet.* **4**, 1–8.

Montermini, L., Kish, S. J., Jiralerspong, S., Lamarche, J. B., and Pandolfos M. (1997). Somatic mosaicism for Friedreich's ataxia GAA triplet repeat expansions in the central nervous system. *Neurology* **49**, 606–610.

Moseley, M. L., Schut, L. J., Bird, T. D., Koob, M. D., Day, J. W., and Ranum L. P. (2000) SCA8 CTG repeat: en masse contractions in sperm and intergenerational sequence changes may play a role in reduced penetrance. *Hum. Mol. Genet.* **9**, 2125–2130.

Nancarrow, J. K., Kremer, E., Holman, K., Eyre, H., Doggett, N. A., Le Paslier, D., Callen, D. F., Sutherland, G. R., and Richards R. I. (1994). Implications of FRA16A structure for the mechanism of chromosomal fragile site genesis. *Science* **264**, 1938–1941.

Ohshima, K., Montermini, L., Wells, R. D., and Pandolfo M. (1998). Inhibitory effects of expanded GAA.TTC triplet repeats from intron I of the Friedreich ataxia gene on transcription and replication *in vivo. J. Biol. Chem.* **273**, 14588–14595.

Parrish, J. E., Oostra, B. A., Verkerk, A. J., Richards, C. S., Reynolds, J., Spikes, A. S., Shaffer, L. G., and Nelson D. L. (1994). Isolation of a GCC repeat showing expansion in FRAXF. a fragile site distal to FRAXA and FRAXE. *Nat. Genet.* **8**, 229–235.

Philips, A. V., Timchenko, L. T., and Cooper, T. A. (1998). Disruption of splicing regulated by a CUG-binding protein in myotonic dystrophy. *Science* **280**, 737–741.

Rasmussen, A., Matsuura, T., Ruano, L., Yescas, P., Ochoa, A., Ashizawa, T., and Alonso E. (2001). Clinical and genetic analysis of four Mexican families with spinocerebellar ataxia type 10. *Ann. Neurol.* **50**, 234–239.

Rosa, A. L., and Ashizawa, T. (2001). Genetic ataxia. In "Neurogenetics" (D. Lynch, ed.) Neurologic Clinic of North America. Academic Press, San Diego (in press).

Sakamoto, N., Chastain, P. D., Parniewski, P., Ohshima, K., Pandolfo, M., Griffith, J. D., and Wells R. D. (1999). Sticky DNA: self-association properties of long GAA. TTC repeats in R.R.Y triplex structures from Friedreich's ataxia. *Mol. Cell.* **3**, 465–475.

Schalling, M., Hudson, T. J., Buetow, K. H., and Housman, D. E. (1993). Direct detection of novel expanded trinucleotide repeats in the human genome. *Nat. Genet.* **4**, 135–139.

Sinke, R. J., Ippel, E. F., Diepstraten, C. M., Beemer, F. A., Wokke, J. H., van Hilten, B. J., Knoers, N. V., van Amstel, H. K., and Kremer H. P. (2001). Clinical and molecular correlations in spinocerebellar ataxia type 6: a study of 24 Dutch families. *Arch. Neurol.* **58**, 1839–1844.

Soong, B. W., Lu, Y. C., Choo, K. B., and Lee H. Y. (2001). Frequency analysis of autosomal dominant cerebellar ataxias in Taiwanese patients and clinical and molecular characterization of spinocerebellar ataxia type 6. *Arch. Neurol.* **58**, 1105–1109.

Takano, H., Onodera, O., Takahashi, H., Igarashi, S., Yamada, M., Oyake, M., Ikeuchi, T., Koide, R., Tanaka, H., Iwabuchi, K., and Tsuji S. (1996). Somatic mosaicism of expanded CAG repeats in brains of patients with dentatorubral-pallidoluysian atrophy: cellular population-dependent dynamics of mitotic instability. *Am. J. Hum. Genet.* **58**, 1212–1222.

Takiyama, Y., Sakoe, K., Amaike, M., Soutome, M., Ogawa, T., Nakano, I., and Nishizawa M. (1999). Single sperm analysis of the CAG repeats in the gene for dentatorubral-pallidoluysian atrophy (DRPLA): the instability of the CAG repeats in the DRPLA gene is prominent among the CAG repeat diseases. *Hum. Mol. Genet.* **8**, 453–457.

Tanaka, F., Reeves, M. F., Ito, Y., Matsumoto, M., Li, M., Miwa, S., Inukai, A., Yamamoto, M., Doyu, M., Yoshida, M., Hashizume, Y., Terao, Si., Mitsuma, T., and Sobue G. (1999). Tissue-specific somatic mosaicism in spinal and bulbar muscular atrophy is dependent on CAG-repeat length and androgen receptor-gene expression level. *Am. J. Hum. Genet.* **65**, 966–973.

Taneja, K. L., McCurrach, M., Schalling, M., Housman, D., and Singer R.H. (1995). Foci of trinucleotide repeat transcripts in nuclei of myotonic dystrophy cells and tissues. *J. Cell Biol.* **128**, 995–1002.

Telenius, H., Almqvist, E., Kremer, B., Spence, N., Squitieri, F., Nichol, K., Grandell, U., Starr, E., Benjamin, C., Castaldo, I., Calabrese, O., Anvret, M., Goldberg, Y. P., and Hayden M. R. (1995). Somatic mosaicism in sperm is associated with intergenerational (CAG)n changes in Huntington disease. *Hum. Mol. Genet.* **4**, 189–195.

Telenius, H., Kremer, B., Goldberg, Y. P., Theilmann, J., Andrew, S. E., Zeisler, J., Adam, S., Greenberg, C., Ives, E. J., Clarke, L. A., and Hayden M. R. (1994). Somatic and gonadal mosaicism of the Huntington disease gene CAG repeat in brain and sperm. *Nat. Genet.* **6**, 409–414.

The Huntington's Disease Collaborative Research Group. (1993). A novel gene containing a trinucleotide repeat that is expanded and unstable on Huntington's disease chromosomes. *Cell* **72**, 971–983.

The International Myotonic Dystrophy Consortium (IDMC) (2000). New nomenclature and DNA testing guidelines for myotonic dystrophy type 1 (DM1). *Neurology* **54**, 1218–1221.

Thornton, C. A., Wymer, J. P., Simmons, Z., McClain, C., and Moxley R. T. (1997). Expansion of the myotonic dystrophy CTG repeat reduces expression of the flanking DMAHP gene. *Nat. Genet.* **16**, 407–409.

Timchenko, L. T., Miller, J. W., Timchenko, N. A., DeVore, D. R., Datar, K. V., Lin, L., Roberts, R., Caskey, C. T., and Swanson M. S. (1996). Identification of a (CUG)n triplet repeat RNA-binding protein and its expression in myotonic dystrophy. *Nucleic Acids Res.* **24**, 4407–4414.

Trottier, Y., Lutz, Y., Stevanin, G., Imbert, G., Devys, D., Cancel, G., Saudou, F., Weber, C., David, G., Tora, L., Agid, Y., Brice, A., and Mandel J.-L. (1995). Polyglutamine expansion as a pathological epitope in Huntington's disease and four dominant cerebellar ataxias. *Nature* **378**, 403–406.

Ueno, S., Kondoh, K., Kotani, Y., Komure, O., Kuno, S., Kawai, J., Hazama, F., and Sano A. (1995). Somatic mosaicism of CAG repeat in dentatorubral-pallidoluysian atrophy (DRPLA). *Hum. Mol. Genet.* **4**, 663–666.

Virtaneva, K., D'Amato, E., Miao, J., Koskiniemi, M., Norio, R., Avanzini, G., Franceschetti, S., Michelucci, R., Tassinari, C.A., Omer, S., Pennacchio, L. A., Myers, R. M., Dieguez-Lucena, J. L., Krahe, R., de la Chapelle, A., and Lehesjoki, A. E. (1997). Unstable minisatellite expansion causing recessively inherited myoclonus epilepsy, EPM1. *Nat. Genet.* **15**, 393–396.

Wang, Y. H., and Griffith J. D. (1996). The [(G/C)3NN]n motif: a common DNA repeat that excludes nucleosomes. *Proc. Natl. Acad. Sci. U.S.A.* **93**, 8863–8867.

Watanabe, H., Tanaka, F., Matsumoto, M., Doyu, M., Ando, T., Mitsuma, T., and Sobue G. (1998). Frequency analysis of autosomal dominant cerebellar ataxias in Japanese patients and clinical characterization of spinocerebellar ataxia type 6. *Clin. Genet.* **53**, 13–19.

Wells, R. D., Warren, S.T., and Sarmiento M. (1998). "Genetic Instabilities and Hereditary Neurological Diseases." Academic Press, San Diego.

Wong, L. J., Ashizawa, T., Monckton, D. G., and Caskey C. T. (1995). Somatic heterogeneity of the CTG repeat in myotonic dystrophy is age and size dependent. *Am. J. Hum. Genet.* **56**, 114–122.

Zhuchenko, O., Bailey, J., Bonnen, P., Ashizawa, T., Stocton, D. W., Amos, C., Dobyns, W. B., Subramony, S. H., Zoghbi, H. Y., and Lee, C. C. (1997). Autosomal dominant cerebellar ataxia (SCA6) associated with small polyglutamine expansions in the alpha 1A-voltage-dependent calcium channel. *Nat. Genet.* **15**, 62–69.

Zu, L., Figueroa, K. P., Grewal, R., and Pulst S. M. (1999). Mapping of a new autosomal dominant spinocerebellar ataxia to chromosome 22. *Am. J. Hum. Genet.* **64**, 594–599.

CHAPTER 12

# Spinocerebellar Ataxia 11 (SCA11)

**HEMA VAKHARIA**
*Division of Neurology*
*Cedars-Sinai Medical Center*
*Los Angeles, California 90048*

**MIN-KYU OH**
*Division of Neurology*
*Cedars-Sinai Medical Center*
*Los Angeles, California 90048*

**STEFAN-M. PULST**
*Division of Neurology, Cedars-Sinai Medical Center*
*Departments of Medicine and Neurobiology,*
*UCLA School of Medicine*
*Los Angeles, California 90048*

I. Summary
II. Phenotype
III. Gene
IV. Neuroimaging and Ancillary Tests
References

## I. SUMMARY

Autosomal dominant cerebellar ataxia type III is characterized by a slowly progressing, often mild, and late-onset cerebellar syndrome without major involvement of other neurological systems. A disease allele mapping to chromosome 15q resulted in ADCA III. Although the nature of the mutation has not been identified, the clinical features included gait ataxia, dysarthria, horizontal nystagmus, and hyperreflexia. Worldwide, only one family with SCA11 has been reported so far.

## II. PHENOTYPE

Adult-onset ataxia with predominantly cerebellar features (also designated ADCA type III) has been associated with a number of gene mutations (see also Fig. 2.1). When familial, this phenotype is most commonly due to mutations in the CACNA1A gene (SCA6). But mutations in the SCA5 and 8 genes have also been associated with pure cerebellar ataxia, variably associated with pyramidal findings. Japanese families with linkage to the SCA4 locus (Nagaoka *et al.*, 2000) show an ADCA III phenotype instead of ataxia with sensory neuropathy SCA4 (Flanigan *et al.*, 1996). SCA10 most commonly presents as a cerebellar ataxia, but a significant number of patients also have epilepsy (Grewal *et al.*, 1998; Matsuura *et al.*, 1999; Zu *et al.*, 1999). SCA15, which has not been mapped, showed pure cerebellar degeneration in one Australian family (Storey *et al.*, 2001). SCA16 has an age of onset between 20 and 60 years of age and presents with ataxia and in some patients with head tremor (Miyoshi *et al.*, 2001).

When repeat expansions are large or disease duration is long (SCA6, see Chapter 8) or onset is early (SCA5, see Chapter 7), many of the above-named ataxias can have a phenotype that is more consistent with ADCA type I. On the other hand, patients with short pathologic expansions in SCA genes normally associated with an ADCA type I phenotype (such as SCA2) can present with late-onset pure cerebellar degeneration (Fernandez *et al.*, 2000; Ogawa *et al.*, 2001).

Worth *et al.* (1999) identified two British families with ADCA III. The first family identified in this study was a seven-generation pedigree (Fig. 12.1). Inheritance was consistent with autosomal dominant as male-to-male transmissions were seen. The average age of onset in this family was 24.7 ± 8.3 with normal life expectancy. There was no statistically significant evidence of anticipation; one child developed the disease before the mother.

In this family, all affected individuals had gait ataxia, jerky pursuit of eye movement, dysarthria, horizontal nystagmus, and hyperreflexia. In addition, 93% of the patients had limb ataxia. Of note 8 patients (57%) also had vertical nystagmus, and 1 patient had diplopia. There was neither supranuclear nor nuclear ophthalmoplegia, and saccadic speed was judged to be normal by clinical exam. Extrapyramidal signs, long tract signs, or fasciculation were absent.

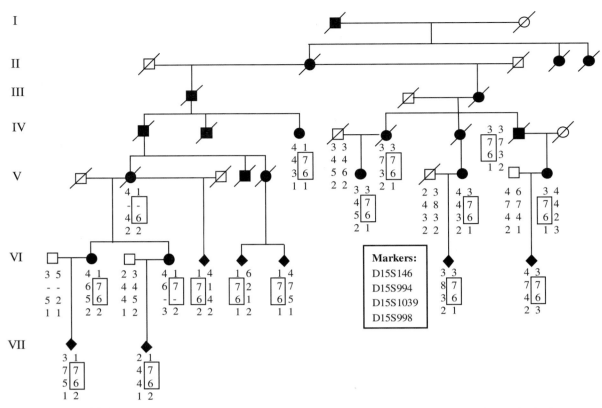

**FIGURE 12.1** Pedigree structure (unaffected members ommitted) and chromosome 15 marker haplotypes for family 1. Affected status is denoted by blackened symbols. Diamonds indicate concealed gender. The haplotypes cosegregating with the disease trait is boxed. Dashes are used if the marker could not be typed. (Modifed from Worth *et al*. 1999). *Am. J. Hum. Genet.* **65**, 420–426.

The second family (that ultimately did not show linkage to 15q) was a four-generation pedigree. Transmission in this family was also autosomal dominant with male-to-male transmission and the age of onset was 37.6 ± 18.2 years. Presence of anticipation could not be determined due to the small sample size. Similar to family 1, all affected individuals showed gait ataxia and jerky pursuit of eye movement. However, unlike family 1, only 40% of the patients from family 2 had horizontal nystagmus, 80% had dysarthria, and none had hyperreflexia or vertical nystagmus. In addition, 40% of the patients showed varying degrees of loss of deep sensation.

Anticipation, the characteristic of trinucleotide-repeat expansion, was not observed significantly in both families. However, there was not enough evidence to exclude the expansion from chromosome 15q due to the small number of subjects.

## III. GENE

Linkage and mutational analysis showed that the disease allele identified in this study was unlinked to any other known SCAs. After a genome-wide scan SCA11 was mapped to chromosome 15q14–21.3 in the first family. The SCA11 region is flanked by markers D15S146 and D15S998 and has a genetic size of approximately 8 cM. It is estimated to be of >8Mbp physical size. The lod score for marker D15S1039 was 4.67 (Table 12.1). The second family was excluded from the SCA11 locus.

The region of chromosome 15q14–21.3, where SCA11 was mapped, was not well defined. A number of expressed sequence tags (ESTs) and known genes were mapped to this region, but SCA11 has not yet been isolated.

**TABLE 12.1** Two-Point LOD Scores between Chromosome 15 Marker Alleles and the SCA 11 Phenotype

| Marker | LOD score at $\theta =$ | | | | | | | |
|---|---|---|---|---|---|---|---|---|
| | 0.00 | 0.001 | 0.01 | 0.05 | 0.100 | 0.200 | 0.300 | 0.400 |
| D15S146 | −0.94 | 0.09 | 0.99 | 1.40 | 1.34 | 0.95 | 0.53 | 0.21 |
| D15S994 | 4.41 | 4.40 | 4.31 | 3.93 | 3.45 | 2.46 | 1.49 | 0.65 |
| D15S1039 | 4.67 | 4.66 | 4.57 | 4.15 | 3.62 | 2.53 | 1.49 | 0.62 |
| D15S998 | −5.76 | −4.69 | −2.82 | −0.99 | −0.37 | 0.00 | 0.06 | 0.04 |

Adapted from Worth *et al.* (1999). *Am. J. Hum. Genet.* **65**, 420–426.

## IV. NEUROIMAGING AND ANCILLARY TESTS

Nerve-conduction studies and electromyography on three subjects in family 1 were normal. One patient at age 61 from the same family showed slightly reduced sensory nerve action potentials. Magnetic resonance imaging in the patients from family 1 showed isolated cerebellar atrophy without atrophy of the brainstem consistent with an ADCA III phenotype.

In summary, the SCA11 phenotype has been observed in only one pedigree. Three years after chromosomal assignment no other pedigrees with linkage to 15q have been described, although other SCA loci have been identified subsequently. Isolation of the causative gene will likely lead to the identification of additional patients and further definition of the phenotype.

## References

Fernandez, M., McClain, M. E., Martinez, R. A., Snow, K., Lipe, H., Ravits, J., Bird, T. D., and La Spada, A. R. (2000). Late-onset SCA2:33 CAG repeats are sufficient to cause disease. *Neurology* **55**, 569–572.

Flanigan, K., Gardner, K., Alderson, K., Galster, B., Otterud, B., Leppert, M. F., Kaplan, C., and Ptacek, L. J. (1996). Autosomal dominant spinocerebellar ataxia with sensory axonal neuropathy (SCA4): clinical description and genetic localization to chromosome 16q22.1. *Am. J. Hum. Genet.* **59**, 392–399.

Grewal, R. P., Tayag, E., Figueroa, K. P., Zu, L., Durazo, A., Nunez, C., and Pulst, S. M. (1998). Clinical and genetic analysis of a distinct autosomal dominant spinocerebellar ataxia. *Neurology* **51**, 1423–1426.

Matsuura, T., Achari, M., Khajavi, M., Bachinski, L. L., Zoghbi, H. Y., and Ashizawa, T. (1999). Mapping of the gene for a novel spinocerebellar ataxia with pure cerebellar signs and epilepsy. *Ann. Neurol.* **45**, 407–411.

Miyoshi, Y., Yamada, T., Tanimura, M., Taniwaki, T., Arakawa, K., Ohyagi, Y., Furuya, H., Yamamoto, K., Sakai, K., Sasazuki, T., and Kira, J. (2001). A novel autosomal dominant spinocerebellar ataxia (SCA16) linked to chromosome 8q22.1–24.1. *Neurology* **57**, 96–100.

Nagaoka, U., Takashima, M., Ishikawa, K., Yoshizawa, K., Yoshizawa, T., Ishikawa, M., Yamawaki, T., Shoji S., and Mizusawa, H. (2000). A gene on SCA4 locus causes dominantly inherited pure cerebellar ataxia. *Neurology* **54**, 1971–1975.

Ogawa, K., Suzuki, Y., Oishi, M., Mizutani, T., and Nakayama, T. (2000). A case of Machado-Joseph disease presenting pure cerebellar ataxia. *Rinsho Shinkeigaku* **41**, 512–514 (Article in Japanese).

Storey, E., Gardner, R. J. M., Knight, M. A., Kennerson, M. L., Tuck, R. R., Forrest, S. M., and Nicholson, G. A. (2001). A new autosomal dominant pure cerebellar ataxia. *Neurology* **57**, 1913–1915.

Worth, P. F., Giunti, P., Gardner-Thorpe, C., Dixon, P. H., Davis, M. B., and Wood, N. W. (1999). Autosomal dominant cerebellar ataxia type III: linkage in a large british family to a 7.6-cM region on chromosome 15q14–21.3. *Am. J. Hum. Genet.* **65**, 420–426.

Zu, L., Figueroa, K. P., Grewal, R., and Pulst, S.-M. (1999). Mapping of a new autosomal dominant spinocerebellar ataxia to chromosome 22. *Am. J. Hum. Genet.* **64**, 594–599.

CHAPTER 13

# Spinocerebellar Ataxia 12 (SCA12)

SUSAN E. HOLMES
*Department of Psychiatry
Division of Neurobiology
Johns Hopkins University School of Medicine
Baltimore, Maryland 21287*

ELIZABETH O'HEARN
*Departments of Neurology and Neuroscience
Johns Hopkins University School of Medicine
Baltimore, Maryland 21287*

SAMIR K. BRAHMACHARI
*Functional Genomics Unit
Center for Biochemical Technology
CSIR Delhi, India*

SHWETA CHOUDHRY
*Functional Genomics Unit
Center for Biochemical Technology
CSIR Delhi, India*

ACHAL K. SRIVASTAVA
*Department of Neurology
Neurosciences Center
All India Institute of Medical Sciences
New Delhi, India*

SATISH JAIN
*Department of Neurology
Neurosciences Center
All India Institute of Medical Sciences
New Delhi, India*

CHRISTOPHER A. ROSS
*Departments of Psychiatry and Neuroscience
Johns Hopkins University School of Medicine
Baltimore, Maryland 21205*

RUSSELL L. MARGOLIS
*Department of Psychiatry
Division of Neurobiology
Johns Hopkins University School of Medicine
Baltimore, Maryland 21287*

I. Introduction
II. Phenotype of SCA12
   A. Overview
   B. Case Reports
III. Normal and Abnormal Gene Function
   A. CAG Expansion Mutation
   B. CAG Repeat is within the Promoter Region of *PPP2R2B*
   C. *PPP2R2B* Promoter Region and CAG Repeat are Conserved in Nonhuman Primates
   D. *PPP2R2B* is a Subunit of PP2A, a Highly Conserved Constitutive Enzyme
IV. Diagnosis
   A. Clinical
   B. Laboratory Diagnosis
   C. Neuroimaging and Neuropathology
   D. Models of Pathogenesis
V. Treatment
Acknowledgments
References

Spinocerebellar ataxia type 12 (SCA12) is an autosomal dominant neurodegenerative disorder that has been described in pedigrees of German-American and Indian descent. The phenotype typically begins with tremor in the fourth decade, progressing to include ataxia and other cerebellar and cortical signs. SCA12 is associated with an expansion of a CAG repeat in the 5' region of the gene *PPP2R2B*, which encodes a brain-specific regulatory subunit of the protein phosphatase PP2A. The repeat size ranges from 55–78 triplets in the mutant allele of affected individuals, and from 7–31 triplets in normal alleles. It is possible that an expansion mutation in *PPP2R2B* may influence *PPP2R2B* expression, perhaps altering the activity of PP2A, an enzyme implicated in multiple cellular functions, including cell cycle regulation, tau phosphorylation, and apoptosis.

## I. INTRODUCTION

The spinocerebellar ataxias (SCAs) are a group of autosomal dominant progressive neurodegenerative disorders that have overlapping and variable phenotypes. They cannot usually be distinguished based on clinical features alone, but 16 distinct SCAs to date (SCAs 1–8 and 10–17) have been identified with the elucidation of the genetic basis (gene or linkage) for each (Evidente *et al.*, 2000; Herman-Bert *et al.*, 2000; Koide *et al.*, 1999; Matsuura *et al.*, 2000; Miyoshi *et al.*, 2001; Nakamura *et al.*, 2001; Subramony *et al.*, 1999; Worth *et al.*, 1999; Yamashita *et al.*, 2000; Zoghbi *et al.*, 2000). Still, 20–40% of adult dominant cerebellar ataxias (ADCAs) remain genetically unidentified (Grewal *et al.*, 1998; Moseley *et al.*, 1998). Cerebellar dysfunction is a hallmark of all the SCAs, but many also include abnormalities in other regions of the central and/or peripheral nervous system. At the genetic level, most of the SCAs in which the mutations have been identified (SCAs 1–3, 6, 7, and 17) are caused by translation of an expanded CAG repeat into an abnormally long polyglutamine tract within the corresponding protein. Three SCAs have now been associated with expansions of noncoding repeats. SCA10 was the first SCA shown to be caused by expansion of a nontriplet repeat (the pentamer ATTCT; Matsuura *et al.*, 2000). SCA8 is caused by an expanded CTG repeat within the 3′ region of an untranslated transcript (Koob *et al.*, 1999). Here we review SCA12, an unusual form of SCA that is associated with a CAG expansion at the 5′ end of the gene encoding PPP2R2B. It may provide a good model for study of noncoding repeat expansion SCAs, since the expansion is in a gene of known function.

## II. PHENOTYPE OF SCA12

### A. Overview

SCA12 is an autosomal dominant cerebellar ataxia described first in a single large pedigree ("R") of German descent (Holmes *et al.*, 1999) and subsequently identified in a second pedigree of Indian descent (Fujigasaki *et al.*, 2001). Though SCA12 appeared initially to be quite rare (Holmes *et al.*, 1999; Fujigasaki *et al.*, 2001; Worth and Wood, 2001) it has recently been reported to be more common in the Indian population (Srivastava *et al.*, 2001). The original pedigree, described in detail elsewhere (O'Hearn *et al.*, 2001) included 10 known living affected individuals (Fig. 13.1) and the second family included 9 living affected members. Seven smaller pedigrees were recently identified in India (Srivastava *et al.*, 2001; S. Jain, unpublished). Subjects typically present with action tremor of the upper extremities and progress to develop a wide range of signs and symptoms, including mild cerebellar dysfunction, hyperreflexia, subtle parkinsonian features, psychiatric symptoms, and in some of the oldest subjects, dementia. Table 13.1 lists the distribution of symptoms in pedigree R; the Indian subjects have a similar phenotype except for the paucity of parkinsonian features. Eight of the ten affected individuals of pedigree R have the combination of action tremor of head or arms, hyperreflexia, mild cerebellar dysfunction, subtle bradykinesia, and paucity of spontaneous movement. Age of onset ranges from 8–55 years,

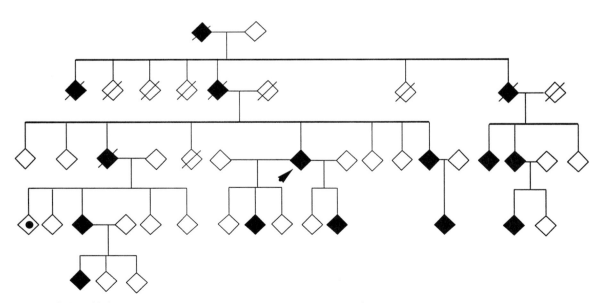

**FIGURE 13.1** Pedigree of the first described SCA12 family. The family is of American–German descent. The proband is indicated by an arrow. One subject carries the mutation but is clinically unaffected (dot within diamond). The pedigree structure has been slightly altered to preserve confidentiality. (From Holmes *et al.* (2001). *Brain Res. Bull.* **3–4**, 397–403. With permission from Elsevier Science.)

TABLE 13.1 Clinical Features of SCA12

| Clinical sign | Number affected, of 10 |
| --- | --- |
| Action tremor | 10 |
| Bradykinesia | 9 |
| Cerebellar signs, including ataxia | 8 |
| Hyperreflexia | 8 |
| Paucity of movement | 8 |
| Babinski present | 5 |
| Decreased tone | 4 |
| Psychiatric symptoms | 4 |
| Focal dystonia | 2 |
| Dementia | 2 |
| Incontinence | 2 |

with most individuals in both the American and Indian pedigrees presenting in the fourth decade. MRIs of affected individuals (Fig. 13.2) reveal generalized atrophy of the CNS, predominantly affecting the cerebral cortex and cerebellum.

## B. Case Reports

*Case 1.* The 64-year-old proband of pedigree R (Fig. 13.1, arrow) developed postural and kinetic tremor of the arms at 38 years of age. Head tremor developed and increased in magnitude over the next decade. Her posture became less upright and her gait developed a staggering quality during her 50s. On examination at age 60, she had moderate upper extremity dysmetria, mild dysarthria, and abnormally brisk reflexes. A head CT performed at age 62 (Fig. 13.2, A, B) revealed moderate, generalized atrophy of the cerebral hemispheres and cerebellum. Over the past two years, she has been unable to ambulate independently. She has been incontinent and cognitively impaired, and now requires assistance for most activities. She also has episodes of depression and irritability. On neurologic exam at age 64, the patient displayed disorientation, memory loss, perseveration, and mild left-sided neglect. She scored 8/28 on the Mini-Mental State Exam (MMSE). Eye movements were normal. The most notable sign was a large amplitude, low frequency tremor of the head. She also had a similar tremor of outstretched arms, persisting during moderately dysmetric finger-to-nose testing. She was diffusely hyper-reflexic with primitive reflexes, extensor plantar responses, and increased tone of arms and legs. Spontaneous movement was moderately reduced and she was mildly bradykinetic. Her speech was mildly dysarthric and hypophonic. When upright she was anteroflexed and had a tendency to retropulse. Her gait was ataxic and broad-based, and appeared to reflect a combination of cerebellar dysfunction, apraxia, and parkinsonian features.

*Case 2.* The 62-year-old proband of pedigree #6 of the set of Indian pedigrees (S. Jain, unpublished) provides an example of the phenotypic variability of the disease. He had presented with gradual onset and progressively worsening tremors of both hands for the last 8 years. His mother (dead) and sister (not examined) reportedly suffered from similar illness. On examination (at age 62) he had coarse action tremors in both hands with flinging of arms that resulted in severe functional disability. The tremors were violent on occasion and resembled ballistic movements or action myoclonus. (This patient's limb movements were similar to those seen in a 64-year-old member of pedigree R). He had mild dysarthria that was more prominent when he gestured with his hands. He had a normal MMSE, normal extraocular movements, normal reflexes with downgoing plantar responses, and had no gait ataxia or parkinsonian features. Brain MRI revealed mild cerebral and as well as cerebellar atrophy. Motor and sensory nerve conduction studies were normal.

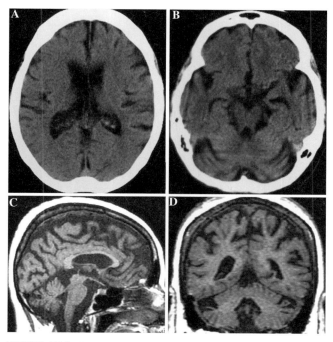

FIGURE 13.2 Neuroradiologic images from two patients with SCA12. (A, B) Head CTs of the proband at age 62 years reveals cerebellar and diffuse cerebral cortical atrophy. (C) sagittal, (D) coronal, T–1 weighted MRI scans of a 59-year-old affected subject also shows cerebellar and cortical atrophy. (From Holmes *et al.* (2001). *Brain Res. Bull.* **3–4**, 397–403. With permission from Elsevier Science.

## III. NORMAL AND ABNORMAL GENE FUNCTION

### A. CAG Expansion Mutation

DNA from the proband of pedigree R was used to clone a novel CAG repeat expansion using a variation of the

RAPID method (Koob *et al.*, 1998). This was the second SCA mutation cloned using this alternate strategy to traditional positional cloning. In brief, CAG RED analysis was performed according to Schalling *et al.* (1993) to identify a novel expanded repeat that segregated with disease status in the pedigree. To obtain flanking sequence of the expanded repeat, an *Eco*R1 digest of genomic DNA from the proband was subjected to electrophoresis on an LMP agarose gel. The gel was sectioned into 2-mm slices, and DNA was extracted from each slice. DNA was subjected to RED and the RED-positive sample was cloned into a lambda phage library and screened at high stringency (Margolis *et al.*, 1997) to detect clones containing long CAG repeats. Clones corresponding to the 24 strongest signals were partially or completely purified and isolated by *in vivo* excision. The single RED-positive clone, containing a 2.5-kb insert, was sequenced on both strands (ABI) and a repeat of 93 CAGs was identified. Primers were designed to amplify across the repeat.

This PCR assay was then used to show segregation of the repeat expansion with disease in pedigree R (Fig. 13.3).

Repeat length in affected individuals ranged from 66–78 triplets. No expansions were found in healthy individuals or patients with other movement disorders (Fig. 13.4). SCA12 was subsequently identified in a large pedigree of Indian descent (Fujigasaki *et al.*, 2001), with repeat lengths of affected individuals ranging from 55–61 triplets. No cases of SCA12 were found in several other sets of individuals with dominant familial ataxias (Fujigasaki *et al*, 2001; Worth and Wood, 2001; Silveira *et al.*, 2000; Holmes *et al.*, 1999; Cholfin *et al.*, 2001; Schols *et al.*, 2000), suggesting that it is a rare form of SCA.

### B. CAG Repeat is within the Promoter Region of *PPP2R2B*

The CAG repeat associated with SCA12 lies 133 nt upstream of the 5′ end of a cDNA of the gene *PPP2R2B* (GenBank accession M64930; Mayer *et al.*, 1991) and maps to 5q31–33 (Fig. 13.5). However, the precise relationship between the repeat and the PPP2R2B transcript

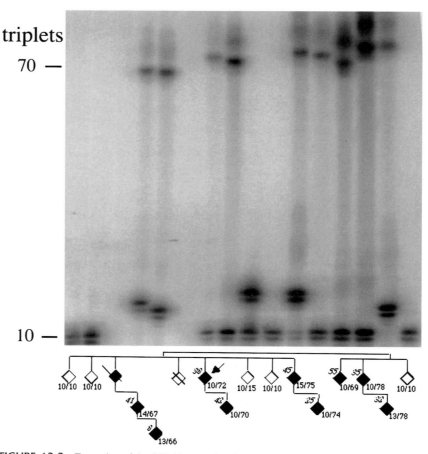

**FIGURE 13.3** Expansion of the SCA12 repeat in affected members of pedigree R. Allele sizes are noted to the bottom right of each individual and age of onset of affected members is at the top left of each filled diamond. (From Holmes *et al.* (1999). *Nat. Genet.* **23**, 391-392. With permission.)

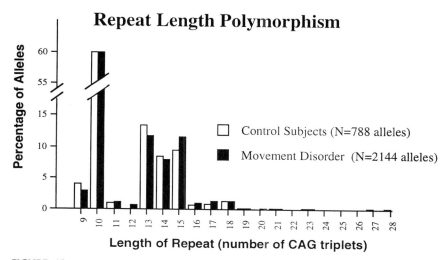

**FIGURE 13.4** Distribution of normal allele lengths in control subjects and movement disorder patients (including subjects with sporadic ataxia, familial ataxia, and Parkinson' disease) of European descent. (From Holmes et al. (2001). Brain Res. Bull. **3–4**, 397–403. With permission from Elsevier Science.

remains ambiguous. Several lines of evidence suggest that the actual start site of this gene is 5′ to that initially published, and that the expansion may in some instances lie within the 5′ UTR. First, 10 clones derived by the 5′ rapid amplification of cDNA ends (RACE) technique from several different brain cDNA samples were sequenced. The products extended variable distances 5′ to the published cDNA, ending as close as 9 nt 3′ to the CAG repeat (Fig. 13.5). Second, a mouse EST (AU051098) contains an interrupted CAG repeat $(CAG)_2GAG(CAG)_2CAC(CAG)_3$ that corresponds in position to the human repeat. The EST begins 22 nt 5′ to the human CAG repeat, extends 3′ into

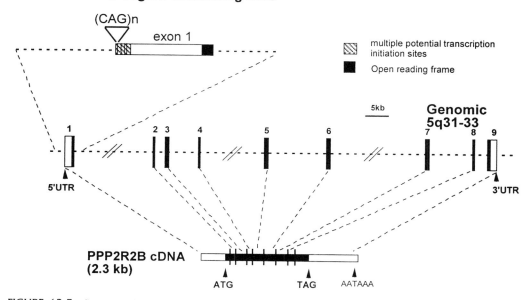

**FIGURE 13.5** Structure of PPP2R2B. Genomic sequence containing the nine known exons of *PPP2R2B* is shown aligned with the cloned cDNA. The gene stretches over approximately 290 kb (exons not to scale). Hatched box at the 5′ end of exon 1 indicates region containing multiple potential transcription start sites. The originally published PPP2R2B transcript is 3.4 kb long, but the majority of ESTs come from transcripts utilizing an upstream polyadenylation signal at approximately 2.3 kb. Preliminary evidence suggests the existence of additional upstream exons. (Modified from Holmes et al. (1999). Nat. Genet. **23**, 391–392. With permission.)

the protein-coding region, and has 91% homology to the human sequence. This suggests that the transcript, at least in some cases, includes the repeat. Third, two additional mouse ESTs (prepared using a Capfinder protocol) come to within 10 and 25 nts of the repeat. These results suggest the possibility of multiple alternative transcription start sites, indicated by hatching (Fig. 13.5), and are consistent with the presence of a (multiple start-site element downstream-1) (MED-1) sequence (GCTCCC; Ince and Scotto, 1995) 65 nt 3′ to the repeat. This promoter motif is typically found downstream of transcription initiation sites in genes that contain multiple initiation sites. Finally, analysis of human genome sequence and additional ESTs indicates several upstream exons and additional splice variants, suggesting a complex picture at the 5′ end of the gene. It now appears likely that the repeat is variably located within 5′ UTR, an intron, and 5′ flanking region, depending on alternate patterns of splicing and alternate sites of transcription initiation.

The possibility that the CAG repeat might encode polyglutamine, either within an unidentified ORF of *PPP2R2B*, or within another unidentified adjacent or overlapping gene, has been examined in several ways, and no evidence has emerged to support the existence of such a transcript. Polyglutamine expansions were not detected on Western blots of protein derived from lymphoblastoid cell lines of affected family members and probed with an antibody (1C2; Trottier *et al.*, 1995) that specifically detects long stretches of polyglutamine (Fig. 13.6). This does not eliminate the possibility of a brain-specific unidentified ORF of *PPP2R2B* in which the repeat encodes polyglutamine, but no patient brain tissue is available for analysis. Northern blots containing brain mRNA and mRNA from neuroblastoma cell line LA-N-1, which expresses high levels of *PPP2R2B*, were probed using oligonucleotides immediately 5′ and 3′ to the CAG repeat expansion, and no signal was detected (unpublished results). Finally, a genomic fragment of 49 kb that includes 27 kb upstream of the CAG repeat was analyzed using the program GENSCAN (Burge and Karlin, 1997). GENSCAN does predict an exon in which the (CAG)10 repeat encodes polyglutamine, but the prediction is of low probability. If the CAG repeat within the genomic sequence is reduced to a single triplet, no exon is predicted, suggesting that the prediction may be an artifact of a long triplet repeat.

Since the repeat apparently resides outside of an open reading frame, it does not appear that the pathogenesis of SCA12 involves a toxic effect of an expanded polyglutamine. A possible alternative mechanism for disease pathogenesis is an effect of the expansion on the level of gene expression, somewhat analogous to the 5′ UTR CGG expansion that causes the fragile X syndrome (Fu *et al.*, 1991; Verkerk *et al.*, 1991; Yu *et al.*, 1991). As the repeat lies in the putative promoter region of *PPP2R2B*, it is plausible that expansion of the repeat alters expression of this gene, which could in turn affect activity of the trimeric enzyme PP2A. Preliminary results indicate that the repeat-containing region does function as a promoter for *PPP2R2B*, and that expansion of the repeat causes a severalfold increase in gene expression (Holmes *et al.*, 2000). Further experiments are in process to determine if this effect is cell-specific and if it can be attributed to altered spacing of promoter elements, as with the expansion of the dodecamer in the cystatin B promoter leading to EPM1 (Lalioti *et al.*, 1999), or if it depends on the specific CAG sequence.

## C. PPP2R2B Promoter Region and CAG Repeat are Conserved in Nonhuman Primates

To better understand the role of the CAG repeats at the SCA12 locus, we analyzed the CAG repeat region in various nonhuman primates: chimpanzee, gorilla, langur, baboon, rhesus monkey, and bonnet monkey (unpublished results, S. Choudhry and S. K. Brahmachari). Sequence analysis of genomic DNA revealed that all the species examined possessed an uninterrupted CAG repeat configuration with variation of repeat length. The CAG repeat size variation seen in these species fell within the normal range observed in humans (Table 13.2). We also sequenced a region of 240 nts surrounding the CAG repeat. Except for a few base substitutions, this sequence is highly conserved among all primates, including humans (Fig. 13.7).

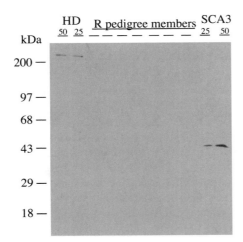

FIGURE 13.6 Absence of polyglutamine expansion in protein from lymphoblastoid cells of SCA12 subjects. Protein from lymphoblastoid cell lines of the R pedigree was probed with the 1C2 antibody, which preferentially detects stretches of greater than 35–40 glutamines. Lanes 1 and 2 contain 50 and 25 μg, respectively, of protein extract from a lymphoblastoid cell line of an HD patient with a (CAG)46 repeat, and lanes 10 and 11 contain 25 and 50 μg of protein from a cell line of an SCA3 subject with a (CAG)76 repeat. (From Holmes *et al.* (2001). *Brain Res. Bull.* **3–4**, 397–403. With permission from Elsevier Science.

TABLE 13.2 CAG Repeat Size Variation at the SCA12 Locus in Various Species of Nonhuman Primates

| Species | Scientific name | No. of chromosomes studied | CAG repeat length | GenBank Accession number |
|---|---|---|---|---|
| Chimpanzee | *Pan troglodytes* | 2 | 11, 14 | AF348396 |
| Gorilla | *Gorilla gorilla* | 2 | 8, 8 | AF348397 |
| Langur | *Presbytis entellus* | 4 | 10, 14 | AF348398 |
| Baboon | *Papio hamadryas* | 4 | 9–11 | AF348400 |
| Rhesus monkey | *Macaca mulatta* | 4 | 10, 11 | AF348399 |
| Bonnet monkey | *Macaca radiata* | 40 | 7–21 | AF348401 |

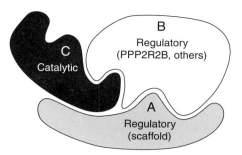

FIGURE 13.8 Schematic diagram of the trimeric holoenzyme PP2A (protein phosphatase 2A). A and C form a core dimer with which one of multiple B regulatory subunits may combine. B subunits are thought to confer substrate and cell-type specificity on the ubiquitous enzyme. *PPP2R2B* encodes one of the brain-specific B subunits. (From Holmes *et al.* (2001). *Brain Res. Bull.* **3–4**, 397–403. With permission from Elsevier Science.

## D. PPP2R2B is a Subunit of PP2A, a Highly Conserved Constitutive Enzyme

Protein phosphatase 2A (PP2A, also termed PP2) is a protein serine/threonine phosphatase that has been implicated in the regulation of many cellular processes involving protein phosphorylation, including cell growth and differentiation, DNA replication, cellular morphogenesis and cytokinesis, regulation of kinase cascades, ion channel function, neurotransmitter release, microtubule assembly, and apoptosis (Price and Mumby, 1999; Santoro *et al.*, 1998; Virshup, 2000). It is an essential enzyme expressed in all eukaryotic cells, and its constitutive components are among the most highly conserved proteins in evolution (Shenolikar, 1994). The designation PP2A actually refers to a subfamily of enzymes, all of which are composed of three subunits: a catalytic subunit (C), a structural subunit (A), and a regulatory (B) subunit (Fig. 13.8). A and C subunits form a complex to which one of multiple B regulatory subunits is recruited by subunit A. In mammals, two isoforms each of A and C exist, while there are at least 13 B regulatory subunits encoded by distinct genes. Theoretically, up to 40 different combinations of the three subunits could exist in mammalian cells. The A and C subunits of PP2A are expressed constitutively in all mammalian cells, whereas the B subunits are distributed differentially among different tissues and cell types, and are thought to modulate PP2A function through regulation of substrate specificity and intracellular targeting (e.g., nuclear, cytoplasmic, or mitochondrial; McCright *et al.*, 1996; Millward *et al.*, 1999; Tehrani *et al.*, 1996; Virshup, 2000).

```
        1             15 16            30 31            45 46            60 61            75 76            90
Human   GGAGGTTGGAAGAGC GCGTCTATGCGTTCC ACCTGAACCTGCAAG CTAGTCGCTTGCTGC CCAGGTGCCGATTCG CGGCTGCATGCGCCC
Chimp   --------------- --------------- --------------- --------------- --------------- ---------------
Gorilla --------------- A-------------- --------------- --------------- --------------- --------------T
Langur  --------------- ----------A---- --------------- --------------- --------------- ----------A----
Baboon  --------------- ----------A---- --------------- --------------- --------------- ---------------
Rhesus  --------------- A---------A---- --------------- --------------- --------------- ---------------
Bonnet  --------------- A---------A---- --------------- --------------- --------------- ---------------

        91           105 106           120 121           135 136           150 151           165 166           180
Human   CACTGCAGCAAAGAG CAGCCGCAGCCTCTG CCTCGGCCACCACTG CTGCTGGGAAAGAGT CGTGGGGCTGCTGAC GCGGTTGGGAGGAGC
Chimp   --------------- --------------- --------------- --------------- -C------------- ---------------
Gorilla --------------- --------------- --------------- --------------- -C------------- --T------------
Langur  -------T------- --A------------ --------------- ------------C-- -C------------- ---------------
Baboon  -------T------- --A------------ --------------- ------------C-- -C------------- ---------------
Rhesus  -------T------- --A------------ --------------- ------------CA- -C------------- ---------------
Bonnet  -------T------- --A------------ --------------- ------------CA- -C------------- ---------------

        181          195 196           210 211           225 226          240
Human   CTCGCCTTTAATGCA CCAGCCGCCTCCAGC CTCCTG(CAG)nCTGCGA GTGCGCGCGTGTGGG
Chimp   --------------- --------------- ------(CAG)n------ ---------------
Gorilla --------------- --------------- ------(CAG)n------ ---------------
Langur  --------------- --------------- ------(CAG)n------ ---------------
Baboon  --------------- --------------- ------(CAG)n------ ---------------
Rhesus  --------------- --------------- ------(CAG)n------ ---------------
Bonnet  --------------- --------------- ------(CAG)n------ ---------------
```

FIGURE 13.7 Conservation of the PPP2R2B promoter region in nonhuman primates. A region of 240 nts within the putative PPP2R2B promoter was analyzed in nonhuman primates by sequence of PCR products. Genbank accession numbers are indicated in Table 13.2.

One class of B subunits is referred to as PPP2R2, and includes PPP2R2A and PPP2R2B (also known as PR55β, B55β, and Bβ). PPP2R2B is brain-specific (Mayer et al., 1991) and widely expressed in neurons throughout the brain, including constitutive expression in Purkinje cells of the cerebellar cortex (Strack et al., 1998), a pattern consistent with the clinical and neuroradiologic involvement of multiple CNS regions in SCA12. As expected, the PPP2R2 class of regulatory units, including PPP2R2B, regulates PP2A dephosphorylation activity for specific substrates, including vimentin (Turowski et al., 1999), histone-1 (Ferrigno et al., 1993), and tau (Sontag et al., 1996). PPP2R2 subunits presumably regulate PP2A activity against many other still unidentified substrates. It is clear from a variety of studies on multiple organisms that mutant forms of PPP2R2 subunits produce a wide variety of deleterious effects. Mutations in the yeast PPP2R2 homolog *CDC55* result in defective cell separation and septation. The *Drosophila* mutant *aar¹* (*abnormal anaphase resolution*), in which a P element is integrated 4 base pairs downstream of the exon IV/intron 4 boundary of the *Drosophila* PPP2R2 homolog, has mitotic defects in anaphase. The nearly identical *Drosophila* mutant *twins^P* displays abnormal specification of neuronal identity during development; *twins^P* has reduced dephosphorylation activity against some substrates (caldesmon and histone H1), but not others (phosphorylase *a*). Recent evidence suggests that downregulation of PP2A containing either PPP2R2A or PPP2R2B in cultured mammalian brain slices results in Alzheimer's-like accumulation of hyperphosphorylated tau. The relative contribution of PPP2R2A vs. PPP2R2B *in vivo* remains to be determined.

In addition to substrate specificity, subcellular localization of PP2A is regulated by B subunits. PPP2R2-containing PP2A enzyme complexes are usually cytoplasmic and localized in neuronal cell bodies and dendrites. However, association of the PPP2R2 subunit with HIV-encoded protein Vpr results in targeting of PP2A to the nucleus, where it increases dephosphorylation of nuclear cdc25 (Hrimech et al., 2000).

Taken together, these findings demonstrate that PPP2R2B may be an important regulator of phosphatase activity in the brain. It therefore follows that quantitative or qualitative changes in PPP2R2B expression would be reflected in abnormal phosphatase activity.

## IV. DIAGNOSIS

### A. Clinical

SCA12 is the only SCA in which action tremor is the presenting and most common sign. The tremor resembles essential tremor, and in fact several SCA12 patients were initially diagnosed with essential tremor. In other patients, the tremor more closely resembles an intention tremor. As the disorder progresses, patients develop mild cerebellar dysfunction including gait ataxia and limb and eye movement dysmetria. They commonly display subtle parkinsonian signs, including increased muscle tone, a paucity of spontaneous movement, and bradykinesia. In common with other SCAs, patients often display hyperreflexia and in older patients, extensor plantar responses. Signs seen in a minority of patients that may be manifestations of SCA12 include early-onset nystagmus and dystonia, and late-onset dementia. Several patients have psychiatric disorders, including anxiety and depression. A diagnosis of SCA12 might therefore be considered for a patient with action tremor of the upper extremities as the presenting sign, with later development of subtle cerebellar dysfunction, parkinsonian features, hyperreflexia, or cognitive dysfunction. Only a single SCA12 pedigree has been identified in over 1600 pedigrees with neurological disorders, predominantly ataxias, from Europe or the United States. On the other hand, SCA12 may represent one of the more common forms of SCA in India (at least 6.5% of cases). Therefore, the level of suspicion that a patient with an SCA-12-like syndrome actually has SCA12 should be higher in Indian patients. The prevalence of SCA12 for different ethnic groups within India remains to be determined.

### B. Laboratory Diagnosis

PCR using primers flanking the CAG repeat is used to confirm the diagnosis of SCA12. Repeat length in the normal alleles of 2515 individuals has been published, and indicates that normal alleles contain from 7–32 triplets (Fujigasaki et al., 2001; Worth and Wood, 2001; Holmes et al., 1999; Cholfin et al., 2001; Schols et al., 2000, Srivastava et al., 2001). Expanded alleles from 10 members of the American family and 18 individuals from the 6 published Indian families contain 55–78 triplets (Holmes et al., 1999; Fujigasaki et al., 2001; Srivastava et al., 2001). In European, American, and Indian population samples, alleles with 10 triplets are the most common, and heterozygosity ranges from 60–74%. Two separate studies indicate that longer normal alleles (greater than 12 triplets) are more common in population samples from India than from Europe (Fujigasaki et al., 2001; Srivastava et al., 2001), consistent with the finding in other repeat expansion diseases that ethnic groups with a skew toward longer normal repeats have a higher frequency of expanded repeats. Our detection of an expansion of 93 repeats in cloned DNA from lymphocytes of the first SCA12 proband, taken together with the predominant signal of 78 triplets after PCR across the repeat in this patient, suggests that somatic mosaicism may be present in some situations (Holmes et al., 1999).

The extent of nonpenetrance of long alleles, or the presence of intermediate alleles that do not cause disease but are unstable in vertical transmission, has not been determined. However, a repeat expansion of 53 triplets has been identified in an Iranian woman with unipolar depression and her two monozygotic twin sons with schizophrenia (D. Wildenhauer, personal communication). While it is possible that a causal relationship exists between the psychiatric disorders in this family and the SCA12 mutation, it is also possible that this family has unrecognized but typical SCA12, or that the mutation is below the threshold for penetrance for SCA12 and the psychiatric disorders are coincidental. SCA12 expansions have not been detected in other subjects with bipolar disorder, schizophrenia, or Alzheimer's disease (Holmes, unpublished data). A 28-year-old unaffected individual from India with no known family history of degenerative neurological disorder has a repeat of 45 triplets, a finding of uncertain significance (Fujigasaki et al., 2001).

There does not appear to be a correlation between repeat expansion size and age of onset in SCA12, though the number of affected individuals identified to date is still small, and age of onset can be difficult to define precisely in this disorder. Repeat length is unstable. In Pedigree R, four of five vertical transmissions showed slight contractions (the four contractions were seen in maternal transmissions, and allele size was stable in one paternal transmission) and variability was seen in each of the two sibships (Holmes et al., 1999). In the largest Indian pedigree, differences of 3 and 5 repeats were found in the 2 sibships with paternal transmission, and no changes were detected in maternally transmitted alleles (Fujigasaki et al., 2001). In the five recently reported Indian pedigrees, two slight contractions and one slight expansion were detected in three paternal transmissions (Srivastava et al., 2001).

## C. Neuroimaging and Neuropathology

MRIs of affected individuals (Fig. 13.2) reveal generalized atrophy of the CNS, with the cerebral cortex most markedly affected, a more modest involvement of the cerebellum (somewhat more pronounced in the vermis than the hemispheres), and variable mild atrophy of other regions. This pattern of findings is not specific for SCA12, but in the presence of tremor and cerebellar signs and autosomal dominant inheritance, it would not be unreasonable to include SCA12 in the differential diagnosis. The relatively greater involvement of the cerebral cortex than the cerebellum may help distinguish SCA12 from other SCAs, and arguably suggests that SCA12 more closely resembles multisystem atrophy than classically defined SCA. Preliminary gross findings of the single SCA12 case that has come to autopsy are consistent with the MRI findings, with atrophy most marked in the cerebral cortex (O'Hearn and Troncoso, unpublished data).

## D. Models of Pathogenesis

As noted previously, the CAG repeat expansion associated with SCA12 appears to be located within the promoter region of the gene encoding PPP2R2B, which is a brain-specific regulatory subunit of the protein phosphatase PP2A, an enzyme implicated in a wide array of cellular processes, including apoptosis. Preliminary evidence suggests that repeat expansion alters the level of expression of PPP2R2B. However, various configurations of the 5' end of the transcript have been identified, and additional pathogenic mechanisms are possible. These include:

1. Alteration of the splicing pattern of nearby exons, as the repeat could be part of a splice enhancer or inhibitor.
2. A toxic effect at the RNA level, such as proposed in the case of myotonic dystrophy (Mankodi et al. 2000; Liquori et al., 2001), as the SCA12 repeat may in some instances be contained within the *PPP2R2B* transcript (5' UTR). The DM repeat expansion is substantially larger than that of SCA12, and the repeat is always found within the transcript, so this mechanism seems less likely in the case of SCA12.
3. Direct inhibition of transcription, as in the case of Friedreich's ataxia (Patel et al., 2001). Preliminary evidence suggests that alternately spliced *PPP2R2B* transcripts may contain exons upstream of the CAG repeat, with the repeat contained within an intron. However, the GAA expansion in Friedreich's ataxia is generally much longer than that of SCA12.
4. Association of PPP2R2B with proteins in addition to the PP2A complex, such that altered expression of PPP2R2B could lead to toxicity through effects on pathways unrelated to PP2A. While this has not yet been demonstrated for PPP2R2B, the PP2A A subunit, and at least one type of B subunit, associate with PP5 (Lubert et al., 2001), suggesting that B subunits may have other roles beyond regulating PP2A.

However, the most straightforward model of pathogenesis remains an alteration of PP2A function as follows (Fig. 13.9):

1. The CAG repeat expansion in *PPP2R2B* leads to abnormal levels of expression of PPP2R2B.
2. Abnormal levels of PPP2R2B, a key regulatory subunit of the enzyme, result in abnormal PP2A function.
3. Changes in PP2A function results in altered protein phosphorylation.
4. Altered phosphorylation of essential neuronal proteins leads to neurotoxicity.

Abnormal phosphatase activity would represent a novel pathogenic mechanism for the dominant spinocerebellar ataxias. Should this hypothesis of SCA12 pathogenesis prove correct, elucidation of the pathways that link PP2A to

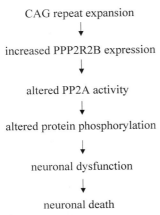

FIGURE 13.9 Hypothetical model of SCA12 pathogenesis.

neuronal death should prove of considerable importance in understanding SCA12 and other neurodegenerative disorders.

## V. TREATMENT

Therapy for essential tremor (beta blockers or phenobarbital derivatives) has been helpful in some cases. To our knowledge, dopaminergic agonists have not been used for parkinsonian features, but would be worth considering in appropriate patients. Psychiatric symptoms, which may cause significant morbidity in SCA12 and other SCAs, have been successfully managed in SCA12 with anxiolytic and antidepressant medicines. Other palliative treatments, including social support, physical therapy, and occupational therapy may help maximize patient functional capacity.

## Acknowledgments

We wish to thank the patients and their family members who have participated in our investigations of SCA12. We also thank C. Callahan, H. Hwang, J. Kleiderlein, D. Gorelick-Feldman, and A. Sharp for their technical assistance; R. Seeger for LA-N-1 cells; and P. R. McHugh for support and encouragement. This paper was reprinted in part from Brain Research Bulletin 56, Holmes, S. E., O'Hearn, E., Ross, C. A., and Margolis, R. L., SCA12: An unusual mutation leads to an unusual spinocerebellar ataxia, pages 397–403, Copyright 2001, with permission from Elsevier Science. This work was supported by grants from the Huntington's Disease Society of American, the National Ataxia Foundation, and NIH grants NS38054, NS16375, and MH01275.

## References

Burge, C., and Karlin, S. (1997). Prediction of complete gene structures in human genomic DNA. *J. Mol. Biol.* **268**, 78–94.

Cholfin, J. A., Sobrido, M. J., Perlman. S., Pulst, S, M., and Geschwind, D. H. (2001). The SCA12 Mutation as a Rare Cause of Spinocerebellar Ataxia. *Arch. Neurol.* **58**, 1833–1835.

Evidente, V. G., Gwinn-Hardy, K. A., Caviness, J. N., and Gilman, S. (2000). Hereditary ataxias. *Mayo Clin. Proc.* **75**, 475–490.

Ferrigno, P., Langan, T. A., and Cohen, P. (1993). Protein phosphatase 2A1 is the major enzyme in vertebrate cell extracts that dephosphorylates several physiological substrates for cyclin-dependent protein kinases. *Mol. Biol. Cell* **4**, 669–677.

Fu, Y.-H., Kuhl, D. P. A., Pizzuti, A., Pieretti, M., Sutcliffe, J. S., Richards, S., Verkerk, A. J. M. H., Holden, J. J. A., Fenmwick, R. G., Warren, S. T., Nelson, D. L., and Caskey, C. T. (1991). Variation of the CGG repeat at the Fragile X site results in genetic instability: resolution of the Sherman paradox. *Cell* **67**, 1047–1058.

Fujigasaki, H., Verma, I. C., Camuzat, A., Margolis, R. L., Zander, C., Lebre, A. S., Jamot, L., Saxena, R., Anand, I., Holmes, S. E., Ross, C. A., Durr, A., and Brice, A. (2001). SCA12 is a rare locus for autosomal dominant cerebellar ataxia: a study of an Indian family. *Ann. Neurol.* **49**, 117–121.

Gong, C. X., Lidsky, T., Wegiel, J., Zuck, L., Grundke-Iqbal, I., and Iqbal, K. (2000). Phosphorylation of microtubule-associated protein tau is regulated by protein phosphatase 2A in mammalian brain. Implications for neurofibrillary degeneration in Alzheimer's disease. *J. Biol. Chem.* **275**, 5535–5544.

Grewal, R. P., Tayag, E., Figueroa, K. P., Zu, L., Durazo, A., Nunez, C., and Pulst, S. M. (1998). Clinical and genetic analysis of a distinct autosomal dominant spinocerebellar ataxia. *Neurology* **51**, 1423–1426.

Healy, A. M., Zolnierowicz, S., Stapleton, A. E., Goebl, M., DePaoli-Roach, A. A., and Pringle, J. R. (1991). CDC55, a Saccharomyces cerevisiae gene involved in cellular morphogenesis: identification, characterization, and homology to the B subunit of mammalian type 2A protein phosphatase. *Mol. Cell. Biol.* **11**, 5767–5780.

Herman-Bert, A., Stevanin, G., Netter, J. C., Rascol, O., Brassat, D., Calvas, P., Camuzat, A., Yuan, Q., Schalling, M., Durr, A., and Brice, A. (2000). Mapping of spinocerebellar ataxia 13 to chromosome 19q13.3–q13.4 in a family with autosomal dominant cerebellar ataxia and mental retardation. *Am. J. Hum. Genet.* **67**, 229–235.

Holmes, S. E., O'Hearn, E. E., McInnis, M. G., Gorelick-Feldman, D. A., Kleiderlein, J. J., Callahan, C., Kwak, N. G., Ingersoll-Ashworth, R. G., Sherr, M., Sumner, A. J., Sharp, A. H., Ananth, U., Seltzer, W. K., Boss, M. A., Vieria-Saecker, A. M., Epplen, J. T., Riess, O., Ross, C. A., and Margolis, R. L. (1999). Expansion of a novel CAG trinucleotide repeat in the 5′ region of PPP2R2B is associated with SCA12. *Nat. Genet.* **23**, 391–392.

Holmes, S. E., Fujigasaki, H., O'Hearn, E., Antonarakis, S., Cooper, J. K., Callahan, C., Hwang, J., Gorelick-Feldman, D., Verma, I. C., Saxena, R., Durr, A., Brice, A., Ross, C. A., and Margolis, R. L. (2000). Spinocerebellar ataxia type 12 (SCA12): additional evidence for a causative role of the CAG repeat expansion in PPP2R2B. *Am. J. Hum. Genet.* **67**, Abstract 1074.

Hrimech, M., Yao, X. J., Branton, P. E., and Cohen, E. A. (2000). Human immunodeficiency virus type 1 Vpr-mediated G(2) cell cycle arrest: Vpr interferes with cell cycle signaling cascades by interacting with the B subunit of serine/threonine protein phosphatase 2A. *EMBO J.* **19**, 3956–3967.

Ince, T. A., and Scotto, K. W. (1995). A conserved downstream element defines a new class of RNA polymerase II promoters. *J. Biol. Chem.* **270**, 30249–30252.

Koide, R., Kobayashi, S., Shimohata, T., Ikeuchi, T., Maruyama, M., Saito, M., Yamada, M., Takahashi, H., and Tsuji, S. (1999). A neurological disease caused by an expanded CAG trinucleotide repeat in the TATA-binding protein gene: a new polyglutamine disease? *Hum. Mol. Genet.* **8**, 2047–2053.

Koob, M. D., Benzow, K. A., Bird, T. D., Day, J. W., Moseley, M. L., and Ranum, L. P. W. (1998). Rapid cloning of expanded trinucleotide repeat sequences from genomic DNA. *Nat. Genet.* **18**, 72–75.

Koob, M. D., Moseley, M. L., Schut, L. J., Benzow, K. A., Bird, T. D., Day, J. W., and Ranum, L. P. W. (1999). An untranslated CTG expansion causes a novel form of spinocerebellar ataxia (SCA8). *Nat. Genet.* **21**, 379–384.

Lalioti, M.D., Scott, H.S., and Antonarakis, S.E. (1999). Altered spacing of promoter elements due to the dodecamer repeat expansion contributes to reduced expression of the cystatin B gene in EPM1. *Hum. Mol. Genet.* **8**, 1791–1798.

Liquori, C. L., Ricker, K., Moseley, M. L., Jacobsen, J. F., Kress, W., Naylor, S. L., Day, J. W., and Ranum, L. P. W. (2001). Myotonic dystrophy type 2 caused by a CCTG expansion in intron 1 of ZNF9. *Science* **293**, 864–867.

Lubert, E. J., Hong, Y. I., and Sarge, K. D. (2001). Interaction between protein phosphatase 5 and the A subunit of protein phosphatase 2A. Evidence for a heterotrimeric form of protein phosphatase 5. *J. Biol. Chem.* **276**, 38582–38587.

Mankodi, A., Logigian, E., Callahan, McClain, C., White, R., Henderson, D., Krym, M., and Thornton, C. A. (2000). Myotonic dystrophy in transgenic mice expressing an expanded CUG repeat. *Science* **289**, 1701–1702.

Margolis, R. L., Abraham, M. A., Gatchell, S. B., Li, S.-H., Kidwai, A. S., Breschel, T. S., Stine, O. C., Callahan, C., McInnis, M. G., and Ross, C. A. (1997). cDNAs with long CAG trinucleotide repeats from human brain. *Hum. Genet.* **100**, 114–122.

Matsuura, T., Yamagata, T., Burgess, D. L., Rasmussen, A., Grewal, R. P., Watase, K., Khajavi, M., McCall, A. E., Davis, C. F., Zu, L., Achari, M., Pulst, S. M., Alonso, E., Noebels, J. L., Nelson, D. L., Zoghbi, H. Y., and Ashizawa, T. (2000). Large expansion of the ATTCT pentanucleotide repeat in spinocerebellar ataxia type 10. *Nat. Genet.* **26**, 191–194.

Mayer, R. E., Hendrix, P., Cron, P., Mattheis, R., Stone, S. R., Goris, J., Merlevede, W., Hofsteenge, J., Hemmings, B. A. (1991). Structure of the 55-kDa regulatory subunit of protein phosphatase 2A: evidence for a neuronal-specific isoform. *Biochemistry* **30**, 3589–3597.

Mayer-Jaekel, R. E., Ohkura, H., Gomes, R., Sunkel, C. E., Baumgartner, S., Hemmings, B. A., and Glover, D. M. (1993). The 55 kd regulatory subunit of Drosophila protein phosphatase 2A is required for anaphase. *Cell* **72**, 621–633.

Mayer-Jaekel R. E., Ohkura, H., Ferrigno, P., Andjelkovic, N., Shiomi, K., Uemura, T., Glover, D. M., and Hemmings, B. A. (1994). Drosophila mutants in the 55 kDa regulatory subunit of protein phosphatase 2A show strongly reduced ability to dephosphorylate substrates of p34cdc2. *J. Cell Sci.* **107**, 2609–2616.

McCright, B., Rivers, A. M., Audlin, S., and Virshup, D. M. (1996). The B56 family of protein phosphatase 2A (PP2A) regulatory subunits encodes differentiation-induced phosphoproteins that target PP2A to both nucleus and cytoplasm. *J. Biol. Chem.* **271**, 22081–22089.

Millward, T. A., Zolnierowicz, S., and Hemmings, B. A. (1999). Regulation of protein kinase cascades by protein phosphatase 2A. *Trends Biochem. Sci.* **24**, 186–191.

Miyoshi, Y., Yamada, T., Tanimura, M., Taniwaki, T., Arakawa, K., Ohyagi, Y., Furuya, H., Yamamoto, K., Sakai, K., Sasazuki, T., and Kira, J. (2001). A novel autosomal dominant spinocerebellar ataxia (SCA16) linked to chromosome 8q22.1–24.1. *Neurology* **57**, 96–100.

Moseley, M. L., Benzow, K. A., Schut, L. J., Bird, T. D., Gomez, C. M., Barkhaus, P. E., Blindauer, K. A., Labuda, M., Pandolfo, M., Koob, M. D., and Ranum, L. P. (1998). Incidence of dominant spinocerebellar and Friedreich triplet repeats among 361 ataxia families. *Neurology* **51**, 1666–1671.

Nakamura, K., Jeong, S., Uchihara, T., Anno, M., Nagashima, K., Nagashima, T., Ikeda, S., Tsuji, S., and Kanazawa, I. (2001). SCA17, a novel autosomal dominant cerebellar ataxia caused by an expanded polyglutamine in TATA-binding protein. *Hum. Mol. Genet.* **10**, 14.

O'Hearn, E., Holmes, S. E., Calvert, P. C., Ross, C. A., and Margolis, R. L. (2001). SCA12: Tremor with cerebellar and cortical atrophy is associated with a CAG repeat expansion. *Neurology* **56**, 299–303.

Patel, P. I., and Isaya, G. (2001). Friedreich ataxia: from GAA triplet-repeat expansion to frataxin deficiency. *Am. J. Hum. Genet.* **69**, 15–24.

Price, N. E., and Mumby, M. C. (1999). Brain protein serine/threonine phosphatases. *Curr. Opin. Neurobiol.* **9**, 336–342.

Santoro, M. F., Annand, R. R., Robertson, M. M., Peng, Y.-W., Brady, M. J., Mankovich, J. A., Hackett, M. C., Ghayur, T., Walter, G., Wong, W. W., and Giegel, D. A. (1998). Regulation of protein phosphatase 2A activity by caspase-3 during apoptosis. *J. Biol. Chem.* **273**, 13119–13128.

Schalling, M., Hudson, T. J., Buetow, K. W., and Housman, D. E. (1993). Direct detection of novel expanded trinucleotide repeats in the human genome. *Nat. Genet.* **4**, 135–139.

Schols, L., Szymanski, S., Peters, S., Przuntek, H., Epplen, J. T., Hardt, C., and Riess, O. (2000). Genetic background of apparently idiopathic sporadic cerebellar ataxia. *Hum. Genet.* **107**, 132–137.

Shenolikar, S. (1994). Protein serine/threonine phosphatases—new avenues for cell regulation. *Ann. Rev. Cell Biol.* **10**, 55–86.

Shiomi, K., Takeichi, M., Nishida, Y., Nishi, Y., and Uemura, T. (1994). Alternative cell fate choice induced by low-level expression of a regulator of protein phosphatase 2A in the Drosophila peripheral nervous system. *Development* **120**, 1591–1599.

Silveira, I., Miranda, C., Guimarães, L., Moreira, M.-C., Alonso, I., Mendonça, P., Ferro, A., Pinto-Basto, J., Coelho, J., Ferreirinha, F., Poirier, J., Parreira, E., Vale, J., Januário, C., Barbot, C., Tuna, A., Barros, J., Koide, R., Tsuji, S., Holmes, S. E., Margolis, R. L., Jardim, L., Pandolfo, M., Coutinho, P., and Sequeiros, J. (2002). Trinucleotide repeats in 202 families with ataxia: a small expanded (CAG)n allele at the SCA17 locus. *Arch. Neurol.* **59**, 623–629.

Sontag, E., Nunbhakdi-Craig, V., Lee, G., Bloom, G. S., and Mumby, M. C. (1996). Regulation of the phosphorylation state and microtubule-binding activity of tau by protein phosphatase 2A. *Neuron* **17**, 1201–1207.

Srivastava, A. K., Choudhry, S., Gopinath, M. S., Roy, S., Tripathi, M., Brahmachari, S. K., and Jain, S. (2001). Molecular and clinical correlation in five Indian families with spinocerebellar ataxia 12. *Ann. Neur.* **50**, 796–800.

Strack, S., Zaucha, J. A., Ebner, F. F., Colbran, R. J., and Wadzinski, B. E. (1998). Brain protein phosphatase 2A: developmental regulation and distinct cellular and subcellular localization by B subunits. *J. Comp. Neurol.* **392**, 515–527.

Subramony, S. H., Vig, P. J., and McDaniel, D. O. (1999). Dominantly inherited ataxias. *Sem. Neurol.* **19**, 419–425.

Tehrani, M. A., Mumby, M. C., and Kamibayashi, C. (1996). Identification of a novel protein phosphatase 2A regulatory subunit highly expressed in muscle. *J. Biol. Chem.* **271**, 5164–5170.

Trottier, Y., Lutz, Y., Stevanin, G., Imbert, G., Devys, D., Cancel, G., Saudou, F., Weber, C., David, G., Tora, L., Agid, Y., Brice, A., and Mandel, J.-L. (1995). Polyglutamine expansion as a pathological epitope in Huntington's disease and four dominant cerebellar ataxias. *Nature* **378**, 403–406.

Turowski, P., Myles, T., Hemmings, B. A., Fernandez, A., and Lamb, N. J. (1999). Vimentin dephosphorylation by protein phosphatase 2A is modulated by the targeting subunit B55. *Mol. Biol. Cell* **10**, 1997–2015.

Verkerk, A. J. M. H., Pieretti, M., Sutcliffe, J. S., Fu, Y.-H., Kuhl, D. P. A., Pizzuti, A., Reiner, O., Richards, S., Victoria, M. F., Zhang, F., Eussen, B. E., van Ommen, G.-J. B., Blonden, L. A. J., Riggins, G. J., Chastain, J. L., Kunst, C. B., Galjaard, H., Caskey, C. T., Nelson, D. L., Oostra, B. A., and Warren, S. T. (1991). Identification of a gene (FMR-1) containing CGG repeat coincident with a breakpoint cluster region exhibiting length variation in Fragile X syndrome. *Cell* **65**, 905–914.

Virshup, D. M. (2000). Protein phosphatase 2A: a panoply of enzymes. *Curr. Opin. Cell Biol.* **12**, 180–185.

Wera, S., and Hemmings, B. A. (1995). Serine/threonine protein phosphatases. *Biochem. J.* **311**, 17–29.

Worth, P. F., Giunti, P., Gardner-Thorpe, C., Dixon, P. H., Davis, M. B., and Wood, N. W. (1999). Autosomal dominant cerebellar ataxia type III: linkage in a large British family to a 7.6-cM region on chromosome 15q14-21.3. *Am. J. Hum. Genet.* **65**, 420–426.

Worth, P. F., and Wood, N. W. (2001). Spinocerebellar ataxia type 12 is rare in the United Kingdom. *Neurology* **56**, 419–420.

Yamashita, I., Sasaki, H., Yabe, I., Fukazawa, T., Nogoshi, S., Komeichi, K., Takada, A., Shiraishi, K., Takiyama, Y., Nishizawa, M., Kaneko, J., Tanaka, H., Tsuji, S., and Tashiro, K. (2000). A novel locus for dominant cerebellar ataxia (SCA14) maps to a 10.2-cM interval flanked by D19S206 and D19S605 on chromosome 19q13.4-qter. *Ann. Neurol.* **48**, 156–163.

Yu, S., Pritchard, M., Kremer, E., Lynch, M., Nancarrow, J., Baker, E., Holman, K., Mulley, J. C., Warren, S. T., Schlessinger, D., Sutherland, G. R., and Richards, R. I. (1991). Fragile X genotype characterized by an unstable region of DNA. *Science* **252**, 1179–1181.

Zoghbi, H. Y., and Orr, H. T. (2000). Glutamine repeats and neurodegeneration. *Ann. Rev. Neurosci.* **23**, 217–247.

CHAPTER 14

# Spinocerebellar Ataxia 13, 14, and 16

**HIROTO FUJIGASAKI**
*INSERM U289*
*Groupe Hôspitalier Pitié-Salpêtrière*
*75651 Paris Cedex*
*France*
*and*
*Department of Neurology and Neurological Science*
*Graduate School, Tokyo Medical and Dental University*
*Tokyo, Japan*

**GIOVANNI STEVANIN**
*Neurologie et Thérapeutique Expérimentale*
*INSERM U289*
*Groupe Hôspitalier Pitié-Salpêtrière*
*75651 Paris Cedex*
*France*

**ALEXANDRA DÜRR**
*INSERM U289*
*Département Génétique*
*Cytogénétique et Embryologie*
*Groupe Hôspitalier Pitié-Salpêtrière*
*75651 Paris Cedex*
*France*

**ALEXIS BRICE**
*Neurologie et Thérapeutique Expérimentale*
*INSERM U289*
*Département Génétique,*
*Cytogénétique, et Embryologie*
*Groupe Hôspitalier Pitié-Salpêtrière*
*75651 Paris Cedex*
*France*

I. Summary
II. Phenotype
   A. SCA13
   B. SCA14
   C. SCA16
III. Gene
   A. SCA13
   B. SCA14
   C. SCA16
IV. Diagnostic and Ancillary Tests
V. Neuroimaging
   A. SCA13
   B. SCA14
   C. SCA16
VI. Neuropathology
VII. Cellular and Animal Models of Disease
VIII. Genotype/Phenotype Correlation/Modifying Alleles
IX. Treatment
References

## I. SUMMARY

Autosomal ataxias (ADCA) are a highly heterogenous group of hereditary neurological diseases characterized by progressive cerebellar ataxia variably associated with other neurological deficits (Harding, 1993). The genetic etiologies for ADCA have been investigated in recent years, and at least 15 responsible loci, spinocerebellar ataxia (SCA) 1–8, 10, 11–4, 16, and 17, were identified (Orr *et al.*, 1993; Kawaguchi *et al.*, 1994; Ranum *et al.*, 1994; Flanigan *et al.*, 1996; Imbert *et al.*, 1996; Pulst *et al.*, 1996; Sanpei *et al.*, 1996; David *et al.*, 1997; Zuchenko *et al.*, 1997; Holmes *et al.*, 1999; Koob *et al.*, 1999; Worth *et. al.*, 1999; Herman-Bert *et al.*, 2000; Matsuura *et al.*, 2000; Yamashita *et al.*, 2000; Fujigasaki *et al.*, 2001; Miyoshi *et al.*, 2001; Nakamura *et al.*, 2001). SCA 13, 14, and 16 are the most recent. SCA13 maps to chromosome 19q. The clinical phenotype is slowly progressive ataxia with increased reflexes, mental retardation, and mild developmental delay in motor acquisition (Herman-Bert *et al.*, 2000). SCA14, which also maps to chromosome 19q close to the SCA13 locus, was identified in a Japanese family presenting cerebellar ataxia and axial myoclonus (Yamashita *et al.*, 2000). SCA16, linked to chromosome 8q, is a pure and mild form of cerebellar ataxia found in Japan (Miyoshi *et al.*, 2001). The genes responsible for these diseases have not yet been identified.

## II. PHENOTYPE

### A. SCA13

A single French family with nine affected individuals has been reported (Fig. 14.1). The clinical phenotype is characterized by progressive cerebellar ataxia and mental retardation. Surprisingly, onset occurs during childhood, but disease progression is very slow resulting in only mild disability despite the long disease duration. A single patient was wheelchair bound after a disease duration of 14 years. Seven of eight patients examined had dysarthria. Six patients had nystagmus. The most striking feature was the presence of mental retardation that was observed in all but one patient. Intellectual impairment is relatively mild but global, with IQs ranging from 62–76. Mild developmental delay in motor acquisition was also notable. Hyperreflexia, with or without Babinski's sign, was observed in all patients. Two older patients had urinary urgency. Petit mal epilepsy was associated in the only affected boy (IV–1). Electroencephalography on this patient showed a typical spike-wave complex at 3 cycles per second (Herman-Bert *et al.*, 2000). The clinical features in this family are summarized in Table 14.1.

### B. SCA14

Yamashita *et al.* (2000) mapped SCA14 to chromosome 19q in a single Japanese family. Twelve individuals were affected in this family. The main clinical feature was slowly progressive ataxia with a mean age at onset of 27.7±12.7 ranging from 12–42 years. The disease is characterized by very slow progression, a patient with a 30-year evolution was still able to walk independently. Five patients with early-onset (onset at age 27 or before) showed tremulous movements of the neck, induced or aggravated by tension or action, that were detected before ataxia became obvious. These movements were considered to be axial myoclonus with an irregular frequency between 5 and 7 Hz. Hyporeflexia or reduction of the Achilles tendon reflexes were observed in four of nine patients examined. Abnormalities of the pyramidal system, autonomic function, sensory system, or mental function were not a part of the phenotype (Table 14.1). SCA14 could be associated with reduced penetrance because the family included an asymptomatic obligate gene carrier who was examined at the age of 76. In addition, age at onset decreased in successive generations and disease severity increased in early onset patients, suggesting the existence of anticipation.

### C. SCA16

Recently, Miyoshi *et al.* (2001) reported a Japanese ADCA family with 9 affected patients linked to chromosome 8q. The clinical phenotype was relatively pure cerebellar ataxia with head tremor in three patients. Gaze-evoked nystagmus was constant. Age at onset ranged from 20–66 (mean 39.6±15.5). Anticipation was not apparent in this family. The initial symptom was cerebellar ataxia in eight patients and head tremor in one. Disease progression was slow, and disability was mild (Table 14.1).

## III. GENE

The responsible genes at these loci have not yet been identified. Identification might be difficult since the candidate intervals are large, due to the fact that only a single family is linked to each locus.

### A. SCA13

Linkage analyses in the French family localized the responsible gene on chromosome 19q. A maximal lod score

TABLE 14.1 Clinical Features of SCA13, 14, and 16

|  | SCA13 | SCA14 | SCA16 |
|---|---|---|---|
| Age at onset (years) | 0–45 | 12–42 | 20–66 |
| Cerebellar ataxia | Mild to severe | Slight to moderate | Slight to severe |
| Nystagmus | Absent to mild | Slight to moderate | Mild (constant) |
| Dysarthria | Absent to severe | Slight to moderate | Slight to severe |
| Reflex | Increase reflexes (constant) | Normal to increased reflexes | Normal |
| Mental retardation | Absent to mild | Absent | Absent |
| Additional signs | Petit mal seizure<br>Urinary urgency<br>Developmental delay | Axial myoclonus | Head tremor |

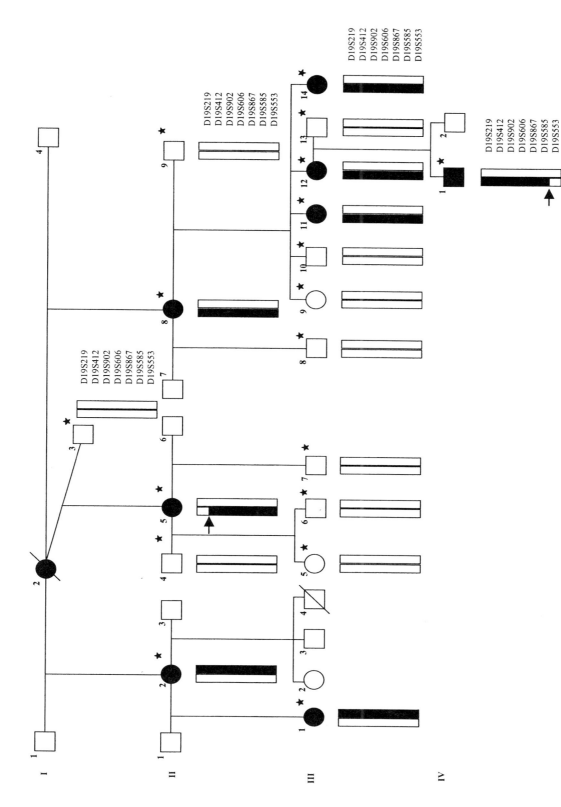

FIGURE 14.1 Partial pedigree of the SCA13 family. Haplotypes for seven microsatellite markers spanning ~8cM on chromosome 19q are shown. Black bars indicate the disease haplotype. Two recombination events (arrows) reduced the candidate interval. Stars indicate individuals sampled.

of 3.50 at θ = 0.00 was obtained for marker D19S867 by two-point linkage analysis. Multipoint linkage analyses with seven markers spanning the D19S219–D19S553 interval mapped the gene to the region between these markers and generated a maximal lod score of 3.85 at marker D19S867. Two recombination events observed in patients II–5 and IV–1 excluded markers D19S219 and D19S553, respectively, delimiting an –8 cM region on chromosome 19q 13.3–13.4 (Fig. 14.1). No evidence for linkage to this region has been obtained in other families with ADCA. To determine whether the mutation is a CAG repeat expansion, as in several SCAs, repeat expansion detection and Western blot analysis with the monoclonal antibody 1C2 were performed. No large CAG repeat expansions or proteins with expanded polyglutamine tracts were detected. Several genes in the 8-cM candidate interval, Bcl-2-associated X protein (BAX), phospholipase A2 (PLA2G4C), and calmodulin (CALM3), are potential candidates (Herman-Bert et al., 2000).

## B. SCA14

SCA14 was mapped to a 10.2-cM interval flanked by D19S206–D19S605 on chromosome 19q13.4–qter in a single Japanese family. A maximal lod score of 4.08 at θ = 0.00 was obtained for marker D19S924 by multipoint linkage analysis. No evidence of linkage to SCA14 locus has been found in 11 additional Japanese families (Yamashita et al., 2000). Although the SCA14 locus is very close to SCA13, these two loci might be independent since they are separated by 3.5 cM on the genetic map (Fig. 14.2, genetic map provided by Marshfield clinic, www.marshfieldclinic.org), that corresponds to approximately 1 Mb, according to the human genome sequence (Greengenes and Sanger Center).

## C. SCA16

The SCA16 locus was mapped to chromosome 8q22.1–24.1 in a large Japanese family. A maximal lod score of 3.06 at θ = 0.00 was obtained for marker D8S1804 by two-point analysis. Haplotype reconstruction restricted the candidate interval to a 37.6-cM region between markers D8S270 and D8S1720. The neuronal potassium channel α subunit gene (Kv8.1) and the KQT-like potassium channel gene (KCNQ3), which mapped to this region, are potential candidate genes (Miyoshi et al., 2001).

## IV. DIAGNOSTIC AND ANCILLARY TESTS

Since the responsible genes at these loci are unknown, molecular testing for these diseases can only be performed by linkage analysis in large informative families. The clinical features characteristic of each disease may help to make the diagnosis. They may not be reliable enough, however, since only a single family has been reported for each locus and phenotypic variants may be found.

## V. NEUROIMAGING

### A. SCA13

Brain MRI examination of a SCA13 patient at the age 5 showed cerebellar vermian atrophy, enlargement of the fourth ventricle, atrophy of the posterior pons, and bulbar thinning. Cerebral cortex was normal (Fig. 14.3a and b, from Herman-Bert et al., 2000).

### B. SCA14

Slight-to-moderate atrophy of the cerebellum was detected by MRI or computed tomography in the 8 patients examined (Fig. 14.3c, from Yamashita et al., 2000). A $^{123}$I-IMP SPECT study in 1 patient showed selectively reduced blood flow in the cerebellum.

### C. SCA16

Brain MRI showed marked atrophy of the cerebellum without brainstem involvement. $^{18}$F-fluoro-2-deoxyglucose PET was normal (Fig. 14.3d, from Miyoshi et al., 2001).

## VI. NEUROPATHOLOGY

No pathological examinations have been performed in patients with SCA13, 14, and 16.

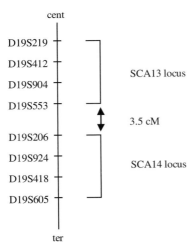

FIGURE 14.2 Partial genetic map of chromosome 19q. The map shows that the SCA13 and SCA14 loci are separated by a 3.5-cM interval.

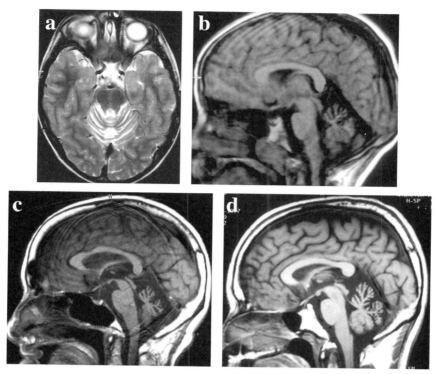

FIGURE 14.3 Brain MRI showing cerebellar vermian atrophy, enlargement of the fourth ventricle, slight atrophy of the pons and the medulla in SCA13 (a and b) (from Herman-Bert, et al. (2000). *Am. J. Hum. Genet.* **67**, 229–235. With permission); moderate cerebellar atrophy in SCA14 (c) (from Yamashita, et al., (2000). *Ann. Neurol.* **48**, 156–163. With permission); and cerebellar atrophy without brain stem involvement in SCA16 (d). (a, c, d [from Miyoshi, et al. (2001). *Neurology* **57**, 96–100. With permission]). T1-weighted sagittal image, (b) T2-weighted axial image.

## VII. CELLULAR AND ANIMAL MODELS OF DISEASE

No cellular or animal models are available.

## VIII. GENOTYPE/PHENOTYPE CORRELATION/ MODIFYING ALLELES

From the limited number of families described, some general features emerge:

1. In SCA13, onset is early and ataxia progresses slowly with a developmental delay in motor acquisition and mild mental retardation, cerebellar and brainstem atrophy is observed on MRI.
2. In SCA14, a relatively pure, slowly progressive cerebellar ataxia is preceded by axial myoclonus in patients with early onset. It is associated with moderate cerebellar atrophy, predominantly in the vermis. Anticipation and incomplete penetrance might be features of this disease.
3. In SCA16, pure cerebellar ataxia is always associated with gaze-evoked nystagmus. Sometimes head tremor is also observed. Marked cerebellar atrophy is present on MRI.

## IX. TREATMENT

No specific treatment for these diseases has been established. Physical and speech therapy is recommended. An anticonvulsant was effective for the petit mal seizures seen in one of the SCA13 patients. Clonazepam (0.5–2.0 mg/day) or valproic acid (400 mg/day) reduce axial myoclonus in SCA14 patients (Yamashita *et al.*, 2000).

### References

David, G., Abbas, N., Stevanin, G., Dürr, A., Yvert, G., Cancel, G., Weber, C., Imbert, G., Saudou, F., Antoniou, E., Drabkin, H., Gemmill, R., Giunti, P., Benomar, A., Wood, N., Ruberg, M., Agid, Y., Mandel, J. L., and Brice, A. (1997). Cloning of the SCA7 gene reveals a highly unstable CAG repeat expansion. *Nat. Genet.* **17**, 65–70.

Flanigan, K., Gardner, K., Alderson, K., Galster, B., Otterud, B., Leppert, M. F., Kaplan, C., and Ptacek, L. J. (1996). Autosomal dominant spinocerebellar ataxia with sensory axonal neuropathy (SCA4): clinical description and genetic localization to chromosome 16q22.1. *Am. J. Hum. Genet.* **59**, 392–399.

Fujigasaki, H., Martin, J. J., De Deyn, P. P., Camuzat, A., Deffond, D., Stevanin, G., Dermaut, B., Van Broeckhoven, C., Dürr, A., and Brice, A. (2001). CAG repeat expansion in the TATA box-binding protein gene causes autosomal dominant cerebellar ataxia. *Brain* **124**, 1939–1947.

Harding, A. E. (1993). Clinical features and classification of inherited ataxias. *Adv. Neurol.* **61**, 1–14

Herman-Bert, A., Stevanin, G., Netter, J. C., Rascol, O., Brassat, D., Calvas, P., Camuzat, A., Yuan, Q., Schalling, M., Dürr, A., and Brice, A. (2000) Mapping of spinocerebellar ataxia 13 to chromosome 19q13.3-q13.4 in a family with autosomal dominant cerebellar ataxia and mental retardation. *Am. J. Hum. Genet.* **67**, 229–235.

Holmes, S. E., O'Hearn, E. E., McInnis, M. G., Gorelick-Feldman, D. A., Kleiderlein, J. J., Callahan, C., Kwak, N. G., Ingersoll-Ashworth, R. G., Sherr, M., Sumner, A. J., Sharp, A. H., Ananth, U., Seltzer, W. K., Boss, M. A., Vieria-Saecker, A. M., Epplen, J. T., Riess, O., Ross, C. A., and Margolis, R. L. (1999). Expansion of a novel CAG trinucleotide repeat in the 5' region of PPP2R2B is associated with SCA12. *Nat. Genet.* **4**, 391–392.

Imbert, G., Saudou, F., Yvert, G., Devys, D., Trottier, Y., Garnier, J. M., Weber, C., Mandel, J. L., Cancel, G., Abbas, N., Dürr, A., Didierjean, O., Stevanin, G., Agid, Y., and Brice, A. (1996). Cloning of the gene for spinocerebellar ataxia 2 reveals a locus with high sensitivity to expanded CAG/glutamine repeats. *Nat. Genet.* **14**, 285–291.

Kawaguchi, Y., Okamoto, T., Taniwaki, M., Aizawa, M., Inoue, M., Katayama, S., Kawakami, H., Nakamura, S., Nishimura, M., Akiguchi, I., Kimura, J., Narumiya, S., and Kakizuka, A. (1994). CAG expansions in a novel gene for Machado-Joseph disease at chromosome 14q32.1. *Nat. Genet.* **8**, 221–228.

Koob, M. D., Moseley, M. L., Schut, L. J., Benzow, K. A., Bird, T. D., Day, J. W., and Ranum, L. P. W. (1999). An untranslated CTG expansion causes a novel form of spinocerebellar ataxia (SCA8). *Nat. Genet.* **21**, 379–384.

Matsuura, T., Yamagata, T., Burgess, D. L., Rasmussen, A., Grewal, R. P., Watase, K., Khajavi, M., McCall, A. E., Davis, C. F., Zu, L., Achari, M., Pulst, S. M., Alonso, E., Noebels, J. L., Nelson, D. L., Zoghbi, H. Y., and Ashizawa, T. (2000). Large expansion of the ATTCT pentanucleotide repeat in spinocerebellar ataxia type 10. *Nat. Genet.* **26**, 191–4

Miyoshi, Y., Yamada, T., Tanimura, M., Taniwaki, T., Arakawa, K., Ohyagi, Y., Furuya, H., Yamamoto, K., Sakai, K., Sasazuki, T., and Kira, J. (2001). A novel autosomal dominant spinocerebellar ataxia (SCA16) linked to chromosome 8q22.1–24.1. *Neurology* **57**, 96–100.

Nakamura, K., Jeong, S. Y., Uchihara, T., Anno, M., Nagashima, K., Nagashima, T., Ikeda, S., Tsuji, S., and Kanazawa, I. (2001). SCA17, a novel autosomal dominant cerebellar ataxia caused by an expanded polyglutamine in TATA-binding protein. *Hum. Mol. Genet.* **10**, 1441–1448.

Orr, H. T., Chung, M. Y., Banfi, S., Kwiatkowski, T. J. Jr, Servadio, A., Beaudet, A. L., McCall, A. E., Duvick, L. A., Ranum, L. P., and Zoghbi, H. Y. (1993). Expansion of an unstable trinucleotide CAG repeat in spinocerebellar ataxia type 1. *Nat. Genet.* **4**, 221–226.

Pulst, S. M., Nechiporuk, A., Nechiporuk, T., Gispert, S., Chen, X. N., Lopes-Cendes, I., Pearlman, S., Starkman, S., Orozco-Diaz, G., Lunkes, A., DeJong, P., Rouleau, G. A., Auburger, G., Korenberg, J. R., Figueroa, C., and Sahba, S. (1996). Moderate expansion of a normally biallelic trinucleotide repeat in spinocerebellar ataxia type 2. *Nat. Genet.* **14**, 269–276.

Ranum, L. P., Schut, L. J., Lundgren, J. K., Orr, H. T., and Livingston, D. M. (1994). Spinocerebellar ataxia type 5 in a family descended from the grandparents of President Lincoln maps to chromosome 11. *Nat. Genet.* **8**, 280–284.

Sanpei, K., Takano, H., Igarashi, S., Sato, T., Oyake, M., Sasaki, H., Wakisaka, A., Tashiro, K., Ishida, Y., Ikeuchi, T., Koide, R., Saito, M., Sato, A., Tanaka, T., Hanyu, S., Takiyama, Y., Nishizawa, M., Shimizu, N., Nomura, Y., Segawa, M., Iwabuchi, K., Eguchi, I., Tanaka, H., Takahashi, H., and Tsuji, S. (1996). Identification of the spinocerebellar ataxia type 2 gene using a direct identification of repeat expansion and cloning technique, DIRECT. *Nat. Genet.* **14**, 277–284.

Yamashita, I., Sasaki, H., Yabe, I., Fukazawa, T., Nogoshi, S., Komeichi, K., Takada, A., Shiraishi, K., Takiyama, Y., Nishizawa, M., Kaneko, J., Tanaka, H., Tsuji, S., and Tashiro, K. (2000). A novel locus for dominant cerebellar ataxia (SCA14) maps to a 10.2-cM interval flanked by D19S206 and D19S605 on chromosome 19q13.4-qter. *Ann. Neurol.* **48**, 156–163.

Worth, P. F., Giunti, P., Gardner-Thorpe, C., Dixon, P. H., Davis, M. B., and Wood, N. W. (1999). Autosomal dominant cerebellar ataxia type III: linkage in a large British family to a 7.6-cM region on chromosome 15q14-21.3. *Am. J. Hum. Genet.* **65**, 420–426.

Zhuchenko, O., Bailey, J., Bonnen, P., Ashizawa, T., Stockton, D. W., Amos, C., Dobyns, W. B., Subramony, S. H., Zoghbi, H. Y., and Lee, C. C. (1997). Autosomal dominant cerebellar ataxia (SCA6) associated with small polyglutamine expansions in the alpha 1A-voltage-dependent calcium channel. *Nat. Genet.* **15**, 62–69.

# CHAPTER 15

# Spinocerebellar Ataxia 17 (SCA17)

SHOJI TSUJI

*Department of Neurology*
*Brain Research Institute*
*Niigata University*
*Niigata 951, Japan*

and

*Department of Neurology*
*University of Tokyo*
*Tokyo 113-8655, Japan*

I. Summary
II. Phenotype
III. Gene
IV. Diagnosis
V. Neuropathology
VI. Neuroimaging
VII. Cellular and Animal Models of Disease
VIII. Treatment
References

## I. SUMMARY

Spinocerebellar ataxia type 17 (SCA17) is a rare autosomal dominant neurodegenerative disease caused by expansion of CAG repeats coding for polyglutamine stretches. Gait ataxia and dementia, progressing over several decades to include bradykinesia, dysmetria, dysdiadochokinesis, hyperreflexia, and paucity of movement characterize the clinical presentations of SCA17 diseases. In this chapter, clinical presentations and molecular genetic aspects of SCA17 are described.

## II. PHENOTYPE

Expansion of the CAG repeat of TATA-binding protein (TBP) gene was first described by Koide *et al.* (1999). The patient was a 14-year-old Japanese female. At the age of 6, she was noted to show gait disturbance and intellectual deterioration. At age 9, she showed truncal ataxia, spasticity, and muscle weakness. She also had a few episodes of atypical absence at age 9. The symptoms were slowly progressive, and she became confined to a wheelchair at age 13. On examination at age 14, she was found to be thin and short statured. She showed marked cerebellar ataxia of the limbs and the trunk, hyperreflexia, extensor plantar responses, cerebellar dysarthria, dysphagia, and severely impaired intellectual performance.

The patient was found to have an expanded CAG repeat of the TBP gene coding for 63 glutamines, exceeding the range of CAG repeats in normal individuals (25–42 repeats). The expanded CAG repeat consists of impure CAG repeat and was partially duplicated. Detailed nucleotide sequence analysis of the TBP genes of the parents as well as flanking markers revealed that *de novo* expansion involved the allele inherited from the father.

Subsequently three groups reported expansions in the TBP gene in familial cases (Fujigasaki *et al.*, 2001; Nakamura *et al.*, 2001; Zühlke *et al.*, 2001). In four Japanese pedigrees, CAG repeats in the *TBP* gene were expanded to 47–55 repeat units (Nakamura *et al.*, 2001). The mode of inheritance was autosomal dominant one with incomplete penetrance. The age at onset ranged from 19 to 48 years with the mean age of onset of 33.2 years. Including the above-described case with *de novo* expansion of the CAG repeat, a strong inverse correlation between the age at onset and the size of expanded CAG repeats was observed. The clinical presentations were characterized by gait ataxia and dementia, progressing over several decades to include bradykinesia, dysmetria, dysdiadochokinesis, hyperreflexia, and paucity of movement. The first symptom (ataxia, dementia, or parkinsonism) varied with the patients.

Zühlke et al. (2001) screened 469 sporadic and 135 familial cases with ataxia and gait disturbances without known SCA mutations and found repeat expansion in the TBP gene in four individuals in two families of Northern German origin with autosomal dominant inheritance of ataxia, dystonia, and intellectual decline. A marked intra- and interfamilial phenotypic variability was observed. One patient presented with a focal dystonia (writer's cramp) at age 20, but developed cerebellar dysfunction 3 years later. Elongated polyglutamine stretches between 50 and 55 residues were demonstrated in the four patients. Normal alleles in the German population ranged from 27–44 CAG repeats, the most common alleles with 37 and 38 repeats.

Fujigasaki et al. (2001) identified one index case after screening 162 ADCA (autosomal dominant cerebellar ataxia) families for expansion in the TBP gene. This individual belonged to a family with six affected individuals segregating a phenotype varying from ataxia and dementia to some patients presenting with psychosis.

## III. GENE

TATA-binding protein is a general transcription initiator factor. The mechanisms of neurodegeneration caused by expanded polyglutamine stretches of TBP, however, remain unclear. Expanded polyglutamine stretches of TBP may affect the function of TBP as a general transcriptional initiator factor, or, alternatively, the expanded polyglutamine stretches may exert toxic functions to neuronal cells similar to other polyglutamine diseases.

## IV. DIAGNOSIS

The CAG repeats of TBP gene in normal individuals code for 25–44 consecutive glutamines, while the expanded CAG repeats code for consecutive 47–63 glutamines. Although the number of SCA17 pedigrees described in literature is limited, mildly expanded CAG repeats of TBP gene are likely associated with reduced penetrance.

## V. NEUROPATHOLOGY

The conventional neuropathological examination revealed shrinkage and loss of small neurons in the caudate nucleus and putamen. Similar but moderate changes were detected in the thalamus, frontal cortex, and temporal cortex. Moderate Purkinje cell loss and an increase of Bergmann glia were seen in the cerebellum. Immunohistochemical examination of a postmortem brain of a patient with 48 CAG repeats detected neuronal intranuclear inclusions that are stained with anti-ubiquitin antibody, anti-TBP antibody, and 1C2 monoclonal antibody that preferentially recognizes expanded polyglutamine tracts. Furthermore, most neuronal nuclei are diffusely stained with 1C2 (Nakamura et al., 2001). Similar results were reported by Fujigasaki et al. (2001).

## VI. NEUROIMAGING

Magnetic resonance imaging (MRI) showed prominent cerebellar atrophy, and mild cerebral atrophy (Koide et al., 1999; Fujigasaki et al., 2001; Nakamura et al., 2001).

## VII. CELLULAR AND ANIMAL MODELS OF DISEASE

TATA-binding protein is an important general transcription initiation factor (Gostout et al., 1993; Kao et al., 1990) and is the DNA-binding subunit of RNA polymerase II transcription factor D (TFIID), the multisubunit complex crucial for the expression of most genes (Tan and Richmond, 1998; Green, 2000). Since expansion of the CAG repeat of the TBP gene has recently been described, pathophysiologic mechanisms of neurodegeneration caused by expanded CAG repeats remain to be elucidated. It is not clear at present whether the mutant TBP with expanded polyglutamine stretches alters the functions of TBP as a general transcription initiation factor.

## VIII. TREATMENT

There is no curative therapy for SCA17. Treatment for SCA17 is essentially palliative.

### References

Fujigasaki, H., Martin, J. J., De Deyn, P. P., Camuzat, A., Deffond, D., Stevanin, G., Dermaut, B., Van Broeckhoven, C., Dürr, A., and Brice, A. (2001). CAG repeat expansion in the TATA box-binding protein gene causes autosomal dominant cerebellar ataxia. Brain 124, 1939–1947.

Gostout, B., Liu, Q., and Sommer, S. S. (1993). "Cryptic" repeating triplets of purines and pyrimidines (cRRY(i)) are frequent and polymorphic: analysis of coding cRRY(i) in the proopiomelanocortin (POMC) and TATA-binding protein (TBP) genes. Am. J. Hum. Genet. 52, 1182–1190.

Green, M. R. (2000). TBP-associated factors (TAFIIs): multiple, selective transcriptional mediators in common complexes. Trends Biochem. Sci. 25, 59–63.

Kao, C. C., Lieberman, P. M., Schmidt, M. C., Zhou, Q., Pei, R., and Berk, A. J. (1990). Cloning of a transcriptionally active human TATA binding factor. *Science* **248**, 1646–1650.

Koide, R., Kobayashi, S., Shimohata, T., Ikeuchi, T., Maruyama, M., Saito, M., Yamada, M., Takahashi H., and Tsuji S. (1999). A neurological disease caused by an expanded CAG trinucleotide repeat in the TATA-binding protein gene: a new polyglutamine disease? *Hum. Mol. Genet.* **8**, 2047–2053.

Nakamura, K. *et al.* (2001). SCA17, a novel autosomal dominant cerebellar ataxia caused by an expanded polyglutamine in TATA-binding protein. *Hum. Mol. Genet.* **10**, 1441–1448.

Tan, S., and Richmond, T. J. (1998). Eukaryotic transcription factors. *Cur. Opin. Struct. Biol.* **8**, 41–48.

Zühlke, C., Hellenbroich, Y., Dalski, A., Kononowa, N., Hagenah, J., Vieregge, P., Riess, O., Klein, C., and Schwinger, E. (2001). Different types of repeat expansion in the TATA-binding protein gene are associated with a new form of inherited ataxia. *Eur. J. Hum. Genet.* **9**, 160–164.

CHAPTER 16

# Dentatorubral-Pallidoluysian Atrophy (DRPLA)

SHOJI TSUJI

*Department of Neurology*
*Brain Research Institute*
*Niigata University*
*Niigata 951, Japan*
*and*
*Department of Neurology*
*University of Tokyo*
*Tokyo 113-8655, Japan*

I. Phenotype
  A. Clinical Features of DRPLA
  B. Genotype/Phenotype Correlations
II. Molecular Genetics of DRPLA
III. Diagnostic and Ancillary Tests
  A. DNA
  B. Other Laboratory Tests
  C. Neuroimaging
  D. Neuropathology
IV. Cellular and Animal Models of Disease
  A. Mechanisms of Neurodegeneration caused by CAG Repeat Expansion
  B. Animal Models
V. Treatment
  References

Dentatorubral-pallidoluysian atrophy (DRPLA) and spinocerebellar ataxia type 17 (SCA17) are rare autosomal dominant neurodegenerative diseases caused by expansion of CAG repeats coding for polyglutamine stretches. Both diseases are characterized by broad spectra of clinical presentations as a function of the size of expanded polyglutamine stretches similarly as have been described in other polyglutamine diseases. In this chapter, clinical presentations and molecular genetic aspects of DRPLA are described.

## I. PHENOTYPE

Dentatorubral-pallidoluysian atrophy (DRPLA) is a rare autosomal dominant neurodegenerative disorder clinically characterized by various combinations of cerebellar ataxia, choreoathetosis, myoclonus, epilepsy, dementia, and psychiatric symptoms (MIM# 125370; Naito and Oyanagi, 1982). The term DRPLA was originally used by Smith *et al.* (1958) to describe a neuropathological condition associated with severe neuronal loss, particularly in the dentatorubral and pallidoluysian systems of the central nervous system, in a sporadic case without a family history (Smith, 1975). The hereditary form of DRPLA was first described in 1972 by Naito and his colleagues (1972).

### A. Clinical Features of DRPLA

The striking clinical characteristic features of DRPLA are the considerable heterogeneity in clinical presentation depending on the age of onset and the prominent genetic anticipation. Naito and Oyanagi reported that juvenile-onset patients (onset before the age of 20) frequently exhibit a phenotype of progressive myoclonus epilepsy (PME), characterized by ataxia, seizures, myoclonus, and progressive intellectual deterioration (Naito and Oyanagi, 1982). Epileptic seizures are a feature in all patients with onset before the age of 20, and the frequency of seizures decreases with age after 20. Occurrence of seizures in patients with onset after the age of 40 is rare. Various forms of generalized seizures including tonic, clonic, or tonic-clonic seizures are observed in DRPLA. Myoclonic epilepsy, and absence or atonic seizures are occasionally observed in patients with onset before the age of 20.

In contrast, patients with onset after the age of 20 tend to develop cerebellar ataxia, choreoathetosis, and dementia, thereby making this disease occasionally difficult to differentiate from Huntington disease (HD) and other spinocerebellar ataxias (Naito and Oyanagi, 1982). Some patients were occasionally diagnosed as having HD, since the main clinical presentations were involuntary movements and dementia, which masked the presence of ataxia. The evaluation of preceding ataxia and atrophy of the cerebellum and brainstem, in particular the pontine tegmentum, as detected by sagittal MRI scan is crucial for the differential diagnosis.

The mode of inheritance of DRPLA is an autosomal dominant one with a high penetrance. The prevalence rate of DRPLA in the Japanese population has been estimated to be 0.2–0.7 per 100,000, which is comparable with that of HD in the Japanese population (Inazuki et al., 1990).

## B. Genotype/Phenotype Correlations

The discovery of the gene for DRPLA has made it possible to analyze the diverse clinical presentations based on the size of expanded CAG repeats. There is an inverse correlation between the age at onset and the size of expanded CAG repeats. To clarify the clinical presentations of DRPLA, we analyzed the relationship between the common clinical features of DRPLA (ataxia, dementia or mental retardation, myoclonus, epilepsy, choreoathetosis, and psychiatric changes including character changes, delusions, or hallucinations) and the age at onset (Ikeuchi et al., 1995a–c). We found that ataxia and dementia are cardinal features irrespective of the age at onset. Patients with onset before the age of 20 frequently exhibit myoclonus and epilepsy in addition to ataxia and dementia. The combination of these clinical features corresponds to the PME phenotype. On the other hand, patients with onset after the age of 20 frequently exhibit choreoathetosis, and psychiatric disturbances in addition to ataxia and dementia. Since the age at onset is inversely correlated with the size of expanded CAG repeats, the above observations imply that the clinical presentation is strongly correlated with the expanded CAG repeat size. A similar correlation between the clinical features and the expanded CAG repeat size has been demonstrated in other diseases caused by CAG repeat expansions.

Although DRPLA has been reported to occur predominantly in Japanese individuals, several cases with quite similar clinical features have been described in other ethnic groups (Titica and van Bogaert, 1946; De Barsy et al., 1968; Farmer et al., 1989). Since the discovery of the gene for DRPLA (Koide et al., 1994; Nagafuchi et al., 1994b), expansion of the CAG repeat of the DRPLA gene has been demonstrated in patients with Haw River syndrome, confirming that DRPLA and Haw River syndrome are an identical disease (Burke et al., 1994). DRPLA has also been demonstrated in European families (Warner et al., 1994a–b; Norremolle et al., 1995; Potter, 1996; Connarty et al., 1996). Thus, DRPLA seems not to be as geographically exclusive as previously thought.

Analysis on the distribution of CAG repeats of DRPLA gene demonstrated that normal alleles with larger than 17 repeats were overrepresented in the Japanese population in comparison to that observed in Caucasian (Burke et al., 1994). This finding was further confirmed by analysis on a large data set (Takano et al., 1998). These findings support the hypothesis that frequencies of large normal alleles determine the frequencies of the corresponding diseases.

## II. MOLECULAR GENETICS OF DRPLA

DRPLA is characterized by prominent anticipation (Koide et al., 1994; Nagafuchi et al., 1994b; Ikeuchi et al., 1995a–c). Paternal transmission results in more prominent anticipation (26–29 years per generation) than does maternal transmission (14–15 years per generation). Given the strong parental bias on the degree of anticipation observed in HD (The Huntington's Disease Collaborative Research Group, 1993) and spinocerebellar ataxia type 1 (SCA1; Orr et al., 1993), we speculated that DRPLA must be a disease caused by unstable CAG repeat expansion of an as yet unidentified gene. Using cDNA clones known to carry CAG repeats as candidate genes, we and another study group independently discovered that the CAG repeat of a gene on chromosome 12, which had been reported as CTG-B37, was expanded in patients with DRPLA (Koide et al., 1994; Nagafuchi et al., 1994b). The CAG repeats in patients with DRPLA were expanded to 54–79 repeat units, as compared to 6–35 repeat units in normal individuals (Koide et al., 1994; Ikeuchi et al;l 1995a–c).

The DRPLA cDNA (Nagafuchi et al., 1994a; Onodera et al., 1995) is predicted to code for 1185 amino acids. The CAG repeat expansion in the DRPLA gene is located 1462 bp downstream from the putative methionine initiation codon and is predicted to code for a polyglutamine stretch. Putative nuclear localizing signals are present near the amino-terminus of DRPLA protein (Miyashita et al., 1998), which is compatible with recent observations that DRPLA protein is translocated into the nucleus preferentially in neuronal cells (Sato et al, 1999). The physiological functions of DRPLA protein, however, remain to be elucidated.

The gene for DRPLA was mapped to 12p13.31 by in situ hybridization (Takano et al., 1996). The human DRPLA gene spans approximately 20 kb and consists of 10 exons, with the CAG repeats located in exon 5 (Nagafuchi et al., 1994a).

Northern blot analysis revealed that a 4.7-kb transcript is widely expressed in various tissues including the heart, lung, kidney, placenta, skeletal muscle, and brain, without predilection for regions exhibiting neurodegeneration (Nagafuchi et al., 1994a; Onodera et al., 1995). The mutant DRPLA gene with expanded CAG repeats is expressed at levels comparable to the wild-type DRPLA gene, suggesting that CAG repeat expansion does not alter the transcriptional efficiency of the DRPLA gene (Onodera et al., 1995). The mutant DRPLA proteins are also expressed at levels similar to those of wild-type DRPLA proteins (Yazawa et al., 1995). Taken together, these studies strongly indicate that CAG repeat expansion does not alter the transcription or translation efficiency of the mutant DRPLA gene. Therefore, it seems likely that mutant DRPLA proteins with expanded polyglutamine stretches are toxic to neuronal cells, suggesting "gain of toxic functions."

As described above, DRPLA is characterized by a prominent genetic anticipation with a mean acceleration of age at onset of $25.6 \pm 2.4$ years in paternal transmission and $14.0 \pm 4.0$ years in maternal transmission (Koide et al., 1994; Nagafuchi et al., 1994b; Ikeuchi et al., 1995a–c). In accordance with the strong parental bias for genetic anticipation, a much larger intergenerational increase was observed for paternal transmission ($5.8 \pm 0.9$ repeat units per generation, n = 16) compared to maternal transmission ($1.3 \pm 1.6$ repeat units per generation, n = 4); (Koide et al., 1994; Ikeuchi et al., 1995a–c). This phenomenon has also been described for HD (Andrew et al., 1993; Duyao et al., 1993; Snell et al., 1993; The Huntington's Disease Collaborative Research Group, 1993). SCA1 (Chung et al., 1993; Orr et al., 1993), and SCA7 (David et al., 1997).

## III. DIAGNOSTIC AND ANCILLARY TESTS

### A. DNA

The size of the CAG repeats of DRPLA gene of healthy individuals ranges from 6 to 36, while that of the expanded CAG repeats of DRPLA patients range from 49–79. As have been observed for other CAG repeat diseases, strong inverse correlation between the size of expanded CAG repeats and the age at onset has been demonstrated in DRPLA as well (Koide et al., 1994; Nagafuchi et al., 1994b; Ikeuchi et al., 1995a–c). Interestingly, it has been shown that the haplotype associated with large normal alleles is identical to that associated with the mutant DRPLA gene (Yanagisawa et al., 1996), raising the possibility that the large normal alleles are the source for *de novo* mutation of the DRPLA gene. Furthermore, we found that the relative prevalence rates of dominant SCAs are in close association with the frequencies of large normal alleles in the corresponding ethnic population (Takano et al., 1998). This finding also supports the hypothesis that the large normal alleles are the source for *de novo* mutation of the DRPLA gene.

### B. Other Laboratory Tests

In patients with the age at onset before the age of 20, slowing of the background activity is observed in all cases, and epileptic discharges are frequently observed. The EEG abnormalities include burst of $\theta$ waves in frontal and central areas, and diffuse synchronized spike and wave complexes. Epileptic discharges are often provoked by photic stimulation. In patients with the age at onset in early adulthood (20–40), the epileptic discharges become less prominent compared with those of childhood-onset patients. In patients with onset in late adulthood (later than 40), slowing of the background activity is observed in substantial proportion of the patients. Epileptic discharges, however, have rarely been described.

### C. Neuroimaging

The MRI findings of DRPLA are characterized by cerebellar atrophy accompanied with dilatation of the fourth ventricle. Atrophy of brainstem structures, particularly midbrain tegmentum, and dilatation of cerebral aqueduct have been described as the characteristic findings. Cerebral atrophy accompanied with dilatation of the lateral ventricle has also been described, particularly in patients with long disease duration. In patients with long disease duration, high-intensity areas in pons, thalamus, and cerebral white matter, have also been described (Fig. 16.1).

A clear genotype-phenotype correlation was also observed in MRI findings of DRPLA patients (Koide et al., 1997). To clarify the relationship between the size of expanded CAG repeat of DRPLA gene and the atrophic changes of the brainstem and cerebellum, we quantitatively analyzed the MRI findings of 26 patients with DRPLA with the diagnosis confirmed by molecular analysis of the DRPLA gene. When the DRPLA patients were classified into two groups based on the size of the expanded CAG repeat of the DRPLA gene (group 1, number of CAG repeat units $\geq 66$; group 2, number of CAG repeat units $\leq 65$), we found strong inverse correlations between the age at MRI and the areas of midsagittal structures of the cerebellum and brainstem in group 1 but not in group 2, suggesting that a clear genotype-phenotype correlation was observed in patients with largely expanded CAG repeats. Furthermore, multiple regression analysis of the overall groups revealed that both the patient's age at MRI and the size of the expanded CAG repeat correlated with the areas of the midsagittal structures. Taken together, these results suggest that both the age and the size of expanded CAG repeats independently affect the atrophic changes in the midsagittal structures of the cerebellum and brainstem. Involvement of the cerebral white matter detected as areas with high-intensity signals on T2-weighted images was occasionally observed in DRPLA patients. We found that the involvement of cerebral white matter is more frequently observed in patients belonging to group 2 than in group 1 patients, suggesting that the disease duration is a major determinant for the white matter involvement.

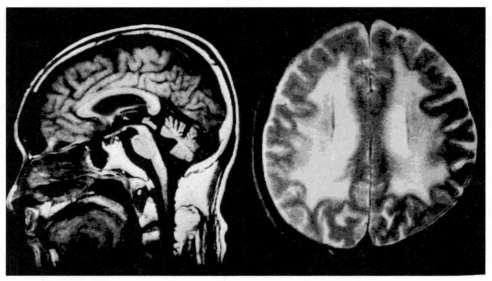

TR = 600 msec, TE = 15 msec      TR = 2500 msec, TE = 90 msec

**FIGURE 16.1** Representative MRI of a DRPLA patient with late-adulthood onset. MRI findings include atrophy of cerebellum and brainstem structures. High-intensity areas in cerebral white matter on T2-weighted images are often present in patients with long disease duration.

### D. Neuropathology

Neuropathological findings of DRPLA are characterized by combined degeneration of the dentatorubral and pallidoluysian systems. The globus pallidus, particularly the lateral segment, and the dentate nucleus are consistently involved, showing neuronal loss and astrocytosis (Naito and Oyanagi, 1982b; Takahashi et al., 1988). Detailed immunohistochemical analysis of autopsied brains of DRPLA patients using 1C2 monoclonal antibody that recognizes preferentially expanded polyglutamine stretches has demonstrated that diffuse accumulation of mutant DRPLA protein/atrophin-1 in the neuronal nuclei, rather than the formation of neuronal intranuclear inclusions (NIIs), was the predominant pathologic condition and involved a wide range of central nervous system regions far beyond the systems previously reported to be affected (Yamada et al., 2001).

## IV. CELLULAR AND ANIMAL MODELS OF DISEASE

### A. Mechanisms of Neurodegeneration caused by CAG Repeat Expansion

There is increasing evidence suggesting 'gain of toxic functions' of mutant proteins with expanded polyglutamine stretches, in particular, truncated mutant proteins containing expanded polyglutamine stretches. Such toxicities have been demonstrated not only in transient expression systems (Ikeda et al., 1996; Paulson et al., 1997; Cooper et al., 1998; Martindale et al., 1998; Igarashi et al., 1998), but also in transgenic mice (Ikeda et al., 1996; Mangiarini et al., 1996; Hodgson et al., 1999; Adachi et al., 2001).

Initial studies in transient expression systems demonstrated that expression of expanded polyglutamine stretches led to aggregate formation and concomitant apoptotic cell death (Ikeda et al., 1996; Paulson et al., 1997; Cooper et al., 1998; Martindale et al., 1998; Igarashi et al., 1998). Presence of aggregate bodies was subsequently confirmed as NIIs in HD and SCA1 transgenic mice (Davies et al., 1997; Skinner et al., 1997; Hodgson et al., 1999). Subsequent studies revealed NIIs in postmortem human brains, including cases of HD (Davies et al., 1997; Difiglia et al., 1997), SCA1 (Skinner et al., 1997), MJD/SCA3 (Paulson et al., 1997), DRPLA (Hayashi et al., 1998; Igarashi et al., 1998), SCA7 (Holmberg et al., 1998), spinal and bulbar muscular atrophy (SBMA) (Li et al., 1998) and SCA2 (Koyano et al., 1999). These results have led to the paradigm that expression of expanded polyglutamine stretches result in aggregate formation and apoptotic cell death.

Recent studies, however, suggest that this is not the case. Interestingly, neuronal death was not observed in transgenic mice for HD (Mangiarini et al., 1996; Turmaine et al., 2000) and SBMA (Adachi et al., 2001). Recent studies on a transgenic mouse model for HD using inducible promoters clearly demonstrated reversibility of neuropathology and phenotypic expressions (Yamamoto et al., 2000). These data strongly support that neuronal dysfunction but not apoptotic cell death is the primary event in polyglutamine diseases.

Although the physiological functions remain unclear, full-length wild-type DRPLA protein has recently been demonstrated to be localized predominantly in the nuclei of cultured cells (Miyashita *et al.*, 1998; Sato *et al.*, 1999). Such nuclear localization is presumably mediated by putative nuclear localization signals (NLS) in the DRPLA protein. In fact, we found that mutant DRPLA protein is expressed predominantly in the nucleus of neuronally differentiated PC12 cells using an adenovirus expression system. We furthermore demonstrated that intranuclear aggregate bodies are preferentially formed in neuronally differentiated PC12 cells and that these cells are more vulnerable than fibroblasts to the toxic effects of expanded polyglutamine stretches of the DRPLA protein (Sato *et al.*, 1998). These observations emphasize the importance of nuclear translocation of full-length or truncated DRPLA proteins with expanded polyglutamine stretches. The observations that transgenic mice expressing mutant ataxin-1 with a mutated NLS did not develop ataxia (Klement *et al.*, 1998), and that addition of a nuclear export signal (NES) to mutant huntingtin suppressed the formation of NIIs and apoptosis (Saudou *et al.*, 1998), further emphasize the role of nuclear translocation of mutant proteins with expanded polyglutamine stretches. These findings suggest that interaction of polyglutamine stretches and some nuclear proteins may be involved in the cytotoxicity caused by expanded polyglutamine stretches.

To date, a number of proteins have been found to associate with the gene products for polyglutamine diseases (Li *et al.*, 1995; Burke *et al.*, 1996; Kalchman *et al.*, 1996; Koshy *et al.*, 1996; Bao *et al.*, 1996; Matilla *et al.*, 1997; Onodera *et al.*, 1997; Skinner *et al.*, 1997; Wanker *et al.*, 1997; Sittler *et al.*, 1998; Faber *et al.*, 1998; Boutell *et al.*, 1998; Wood *et al.*, 1998; Steffan *et al.*, 2000; McCampbell *et al.*, 2000; Waragai *et al.*, 2000; Shimohata *et al.*, 2000; Nucifora, Jr. *et al.*, 2001). As described above, recent studies have emphasized the role of nuclear transport and nuclear accumulation of mutant proteins with expanded polyglutamine stretches prior to intranuclear inclusion formation in the pathogenesis of polyglutamine disease (Yamada *et al.*, 2001). These findings raise a new paradigm that interaction of mutant proteins containing expanded polyglutamine stretches with nuclear proteins may lead to neuronal dysfunction, in particular, transcriptional dysfunction. Recent discovery of sequestration of nuclear proteins by expanded polyglutamine stretches (Skinner *et al.*, 1997; Wood *et al.*, 1998; McCampbell *et al.*, 2000; Steffan *et al.*, 2000; Waragai *et al.*, 2000; Shimohata *et al.*, 2000; Nucifora, Jr. *et al.*, 2001) strongly supports this hypothesis. To further support this hypothesis, downregulation of selective genes has been demonstrated in the transgenic mice for HD (Luthi-Carter *et al.*, 2000) and SCA1 (Lin *et al.*, 2000). These results suggest that intranuclear accumulation of mutant proteins and transcriptional dysregulation are the primary pathogenic mechanisms of polyglutamine diseases. Given the reversibility of disease process in polyglutamine disease (Yamamoto *et al.*, 2000), there seems to be a wide time window for therapeutic approaches, and therapeutic strategies aimed against transcriptional dysregulation will be a challenging approach.

## B. Animal Models

Animal models have been established (Sato *et al.*, 1999; Schilling *et al.*, 1999). Schilling *et al.* (1999) created a transgenic mouse model inserting a full-length mutant DRPLA cDNA coding for 65 consecutive glutamines under the control of the prion promoter. The mice exhibit ataxia, tremors, abnormal movements, seizures, and premature death. They found intranuclear accumulation of atrophin-1 (DRPLA protein) and intranuclear inclusion bodies of multiple populations of neurons. They identified 120 kDa nuclear fragments of mutant atrophin-1, whose abundance increased with age and phenotypic severity. With these findings, they propose that the evolution of neuropathology in DRPLA involves proteolytic processing of mutant atrophin-1 and nuclear accumulation of truncated fragments.

Zhang *et al.* (2002) recently demonstrated that *Drosophila* atrophin genetically interacts with the transcription repressor even-skipped, and that *Drosophila* atrophin is required in diverse developmental processes including early embryonic patterning. They further demonstrated that both human atrophin-1 and *Drosophila* atrophin repress transcription *in vivo* when tethered to DNA, and poly-Q expansion in atrophin-1 reduces this repressive activity. These data suggest that atrophin proteins function as transcriptional co-repressors.

To investigate the molecular mechanisms of instability of CAG repeats, we generated transgenic mice harboring a single copy of a mutant DRPLA gene (Sato *et al.*, 1999). The transgenic mice, in fact, exhibited an age-dependent increase (+0.31 per year) in male transmission and an age-dependent contraction in female transmission (−1.21 per year). Such age-dependent increase in the intergenerational changes in the sizes of expanded CAG repeats in paternal transmission and age-dependent contraction in maternal transmission were also observed in 83 parent-offspring pairs of DRPLA patients (56 paternal and 27 maternal transmissions).

Based on a linear regression model and the continuous cell divisions required for spermatogenesis throughout adult life, the mean increase in the size of CAG repeats in male transmission in mice was calculated to be +0.31 per year and +0.0073 per spermatogenesis cycle. These values were comparable to those observed in DRPLA patients, which were calculated to be +0.27 and +0.012, respectively. These results strongly indicate that the difference in the actual intergenerational changes between humans and mice is due to the reproductive life span variations and that a common mechanism underlies the age-dependent increase in the sizes of CAG repeats both in humans and in mice.

In contrast to spermatogenesis, oogenesis occurs only during fetal life, and ceases at the diplotene stage of the first meiotic prophase by 5 days after birth, suggesting that age-dependent contraction of CAG repeats occurs after the cessation of meiotic DNA replication. Similar observations have been made in transgenic mice for SCA1 and SBMA (Kaytor et al., 1997; La Spada et al., 1998). These results strongly suggest that contraction of the CAG repeats occurs during the prolonged resting stage, and mechanisms such as repair of damaged DNA or selective degeneration of the primary oocyte with larger CAG repeats might be involved in the contraction process.

Manley et al. (1999) recently examined instability of the HD CAG repeat by crossing transgenic mice carrying exon 1 of human HD with Msh2−/− mice, and demonstrated that the mismatch repair enzyme MSH2 is required for somatic instability of the CAG repeat.

## V. TREATMENT

There is no curative therapy for DRPLA. Treatment for DRPLA is essentially palliative. Epilepsies should be treated with anticonvulsants. With availability of animal models, however, it is strongly hoped that the development of new therapeutic approaches will be tested using these animal models (Sato et al., 1999; Schilling et al., 1999).

## References

The Huntington's Disease Collaborative Research Group. (1993). A novel gene containing a trinucleotide repeat that is expanded and unstable on Huntington's disease chromosomes. Cell **72**, 971–983.

Adachi, H., Kume, A., Li, M., Nakagomi, Y., Niwa, H., Do, J., Sang, C., Kobayashi, Y., Doyu, M., and Sobue, G. (2001). Transgenic mice with an expanded CAG repeat controlled by the human AR promoter show polyglutamine nuclear inclusions and neuronal dysfunction without neuronal cell death. Hum. Mol. Genet. **10**, 1039–1048.

Andrew, S. E., Goldberg, Y. P., Kremer, B., Telenius, H., Theilmann, J., Adam, S., Starr, E., Squitieri, F., Lin, B., and Kalchman, M. A. (1993). The relationship between trinucleotide (CAG) repeat length and clinical features of huntington's disease. Nat. Genet. **4**, 398–403.

Bao, J., Sharp, A. H., Wagster, M. V., Becher, M., Schilling, G., Ross, C. A., Dawson, V. L., and Dawson, T. M. (1996). Expansion of polyglutamine repeat in huntingtin leads to abnormal protein interactions involving calmodulin. Proc. Natl. Acad. Sci. U.S.A. **93**, 5037–5042.

Boutell, J. M., Wood, J. D., Harper, P. S., and Jones, A. L. (1998). Huntingtin interacts with cystathionine beta-synthase. Hum. Mol. Genet. **7**, 371–378.

Burke, J. R., Enghild, J. J., Martin, M. E., Jou, Y. S., Myers, R. M., Roses, A. D., Vance, J. M., and Strittmatter, W. J. (1996). Huntingtin and DRPLA proteins selectively interact with the enzyme GAPDH. Nat. Med. **2**, 347–350.

Burke, J. R., Ikeuchi, T., Koide, R., Tsuji, S., Yamada, M., Pericak-Vance, M. A., and Vanve, J. M. (1994). Dentatorubral-pallidoluysian atrophy and Haw River syndrome. Lancet **344**, 1711–1712.

Chung, M. Y., Ranum, L. P., Duvick, L. A., Servadio, A., Zoghbi, H. Y., and Orr, H. T. (1993). Evidence for a mechanism predisposing to intergenerational CAG repeat instability in spinocerebellar ataxia type I. Nat. Genet. **5**, 254–258.

Connarty, M., Dennis, N. R., Patch, C., Macpherson, J. N., and Harvey, J. F. (1996). Molecular re-investigation of patients with Huntington's disease in Wessex reveals a family with dentatorubral and pallidoluysian atrophy. Hum. Genet. **97**, 76–78.

Cooper, J. K., Schilling, G., Peters, M. F., Herring, W. J., Sharp, A. H., Kaminsky, Z., Masone, J., Khan, F. A., Delanoy, M., Borchelt, D. R., Dawson, V. L., Dawson, T. S., and Ross, C. A. (1998). Truncated N-terminal fragments of huntingtin with expanded glutamine repeats form nuclear and cytoplasmic aggregates in cell culture. Hum. Mol. Genet. **7**, 783–790.

David, G., Abbas, N., Stevanin, G., Durr, A., Yvert, G., Cancel, G., Weber, C., Imbert, G., Saudou, F., Antoniou, E., Drabkin, H., Hemmill, R., Giunti, P., Benomar, A., Wood, N., Ruberg, M., Agid, Y., Mandel, J. L., and Brice, A. (1997). Cloning of the SCA7 gene reveals a highly unstable CAG repeat expansion. Nat. Genet. **17**, 65–70.

Davies, S. W., Turmaine, M., Cozen, B. A., DiFiglia, M., Sharp, A. H., Ross, C. A., Scherzinger, E., Wanker, E. E., Mangiarini, L., and Bates, G. P. (1997). Formation of neuronal intranuclear inclusions underlies the neurological dysfunction in mice transgenic for the HD mutation. Cell **90**, 537–548.

De Barsy, T. H., Myle, G., Troch, C., Matthys, R., and Martin, J. (1968). La Dyssynergie cerebelleuse myoclonique (R. Hunt): Affection autonome ou rariante du type degeneratif de l'epilepsie-myoclonie progressive (Unvericht-Lundborg) Appoche anatomo-alinique. J. Neurol. Sci. **8**, 111–127.

Difiglia, M., Sapp, E., Chase, K. O., Davies, S. W., Bates, G. P., Vonsattel, J. P., and Aronin, N. (1997). Aggregation of huntingtin in neuronal intranuclear inclusions and dystrophic neurites in brain. Science **277**, 1990–1993.

Duyao, M., Ambrose, C., Myers, R., Novelletto, A., Persichetti, F., Frontali, M., Folstein, S., Ross, C., Franz, M., and Abbott, M. (1993). Trinucleotide repeat length instability and age of onset in Huntington's disease. Nat. Genet. **4**, 387–392.

Faber, P. W., Barnes, G. T., Srinidhi, J., Chen, J., Gusella, J. F., and MacDonald, M. E. (1998). Huntingtin interacts with a family of WW domains proteins. Hum. Mol. Genet. **7**, 1463–1474.

Farmer, T. W., Wingfield, M. S., Lynch, S. A., Vogel, F. S., Hulette, C., and Katchinoff, B. (1989). Ataxia, chorea, seizures, and dementia. Pathologic features of a newly defined familial disorder. Arch. Neurol. **46**, 774–779.

Hayashi, Y., Kakita, A., Yamada, M., Egawa, S., Oyanagi, S., Naito, H., Tsuji, S., and Takahashi, H. (1998). Hereditary dentatorubral-pallidoluysian atrophy—ubiquitinated filamentous inclusions in the cerebellar dentate nucleus neurons. Acta Neuropathol. (Berlin) **95**, 479–482.

Hodgson, J. G., Agopyan, N., Gutekunst, C. A., Leavitt, B. R., LePiane, F., Singaraja, R., Smith, D. J., Bissada, N., McCutcheon, K., Nasir, J., Jamot, L., Li, X. J., Stevens, M. E., Rosemond, E., Roder, J. C., Phillips, A. G., Rubin, E. M., Hersch, S. M., and Hayden, M. R. (1999). A YAC mouse model for Huntington's disease with full-length mutant huntingtin, cytoplasmic toxicity, and selective striatal neurodegeneration. Neuron **23**, 181–192.

Holmberg, M., Duycjaerts, C., Durr, A., Cancel, G., Gourfinkel-An, I., Damier, P., Faucheux, B., Trottier, Y., Hirsch, E. C., Agid, Y., and Brice, A. (1998). Spinocerebellar ataxia type 7 (SCA7): a neurodegenerative disorder with neuronal intranuclear inclusions. Hum. Mol. Genet. **7**, 913–918.

Igarashi, S., Koide, R., Shimohata, T., Yamada, M., Hayashi, Y., Takano, H., Date, H., Oyake, M., Sato, T., Sato, A., Egawa, S., Ikeuchi, T., Tanaka, H., Nakano, R., Tanaka, K., Hozumi, I., Inuzuka, T., Takahashi, H., and Tsuji, S. (1998). Suppression of aggregate formation and apoptosis by transglutaminase inhibitors in cells expressing truncated DRPLA protein with an expanded polyglutamine stretch. *Nat. Genet.* **18**, 111–117.

Ikeda, H., Yamaguchi, M., Sugai, S., Aze, Y., Narumiya, S., and Kakizuka, A. (1996). Expanded polyglutamine in the Machado-Joseph disease protein induces cell death in vitro and in vivo. *Nat. Genet.* **13**, 196–202.

Ikeuchi, T., Koide, R., Onodera, O., Tanaka, H., Oyake, M., Takano, H., and Tsuji, S. (1995a). Dentatorubral-pallidoluysian atrophy (DRPLA). Molecular basis for wide clinical features of DRPLA. *Clin. Neurosci.* **3**, 23–27.

Ikeuchi, T., Koide, R., Tanaka, H. *et al.* (1995b). Dentatorubral-pallidoluysian atrophy (DRPLA): Clinical features are closely related to unstable expansions of trinucleotide (CAG) repeat. *Ann. Neurol.* **37**, 769–775.

Ikeuchi, T., Onodera, O., Oyake, M., Koide, R., Tanaka, H., and Tsuji, S. (1995c). Dentatorubral-pallidoluysian atrophy (DRPLA): close correlation of CAG repeat expansions with the wide spectrum of clinical presentations and prominent anticipation. *Sem. Cell. Biol.* **6**, 37–44.

Inazuki, G., Kumagai, K., and Naito, H. (1990). Dentatorubral-pallidoluysian atrophy (DRPLA): Its distribution in Japan and prevalence rate in Niigata. *Seishin Igaku* **32**, 1135–1138.

Kalchman, M. A., Graham, R. K., Xia, G., Koide, H. B., Hodgson, J. G., Graham, K. C., Goldberg, Y. P., Gietz, R. D., Pickart, C. M., and Hayden, M. R. (1996). Huntingtin is ubiquitinated and interacts with a specific ubiquitin-conjugating enzyme. *J. Biol. Chem.* **271**, 19385–19394.

Kaytor, M. D., Burright, E. N., Duvick, L. A., Zoghbi, H. Y., and Orr, H. T. (1997). Increased trinucleotide repeat instability with advanced maternal age. *Hum. Mol. Genet.* **6**, 2135–2139.

Klement, I. A., Skinner, P. J., Kaytor, M. D., Yi, H., Hersch, S. M., Clark, H. B., Zoghbi, H. Y., and Orr, H. T. (1998). Ataxin-1 nuclear localization and aggregation: role in polyglutamine-induced disease in SCA1 transgenic mice. *Cell* **95**, 41–53.

Koide, R., Ikeuchi, T., Onodera, O., Tanaka, H., Igarashi, S., Endo, K., Takahashi, H., Kondo, R., Ishikawa, A., Hayashi, T., Saito, M., Tomoda, A., Miike, T., Naito, H., Ikuta, F., and Tsuji, S. (1994). Unstable expansion of CAG repeat in hereditary dentatorubral-pallidoluysian atrophy (DRPLA). *Nat. Genet.* **6**, 9–13.

Koide, R., Onodera, O., Ikeuchi, T., Kondo, R., Tanaka, H., Tokiguchi, S., Tomoda, A., Miike, T., Isa, F., Beppu, H., Shimizu, N., Watanabe, Y., Horikawa, Y., Shimohata, T., Hirota, K., Ishikawa, A., and Tsuji, S. (1997). Atrophy of the cerebellum and brainstem in dentatorubral pallidoluysian atrophy. Influence of CAG repeat size on MRI findings. *Neurology* **49**, 1605–1612.

Koshy, B., Matilla, T., Burright, E. N., Merry, D. E., Fischbeck, K. H., Orr, H. T., and Zoghbi, H. Y. (1996). Spinocerebellar ataxia type-1 and spinobulbar muscular atrophy gene products interact with glyceraldehyde-3-phosphate dehydrogenase. *Hum. Mol. Genet.* **5**, 1311–1318.

Koyano, S., Uchihara, T., Fujigasaki, H., Nakamura, A., Yahishita, S., and Iwabuchi, K. (1999). Neuronal intranuclear inclusions in spinocerebellar ataxia type-2: triple-labelling immunofluorescent study. *Neurosci. Lett.* **274**, 117–120.

La Spada, A. R., Peterson, K. R., Meadows, S. A., McClain, M. E., Jeng, G., Chmelar, R. S., Haugen, H. A., Chen, K., Singer, M. J., Moore, D., Trask, B. J., Fischbeck, K. H., Clegg, C. H., and McKnight, G. S. (1998). Androgen receptor YAC transgenic mice carrying CAG 45 alleles show trinucleotide repeat instability. *Hum. Mol. Genet.* **7**, 959–967.

Li, M., Miwa, S., Kobayashi, Y., Merry, D. E., Yamamoto, M., Tanaka, F., Doyu, M., Hashizume, Y., and Fischbeck, K. H. (1998). Nuclear Inclusions of the androgen receptor protein in spinal and bulbar muscular atrophy. *Ann. Neurol.* **44**, 249–254.

Li, X. J., Li, S. H., Sharp, A. H., Nucifora, Jr., F. C., Schilling, G., Lanahan, A., Worley, P., Snyder, S. H., and Ross, C. A. (1995). A huntingtin-associated protein enriched in brain with implications for pathology. *Nature* **378**, 398–402.

Lin, X., Antalffy, B., Kang, D., Orr, H. T., and Zoghbi, H. Y. (2000). Polyglutamine expansion down-regulates specific neuronal genes before pathologic changes in SCA1. *Nat. Neurosci.* **3**, 157–163.

Luthi-Carter, R., Strand, A., Peters, N. L., Solano, S. M., Hollingsworth, Z. R., Menon, A. S., Frey, A. S., Spektor, B. S., Penney, E. B., Schilling, G., Ross, C. A., Borchelt, D. R., Tapscott, S. J., Young, A. B., Cha, J.-H. J., and Olson, J. M. (2000). Decreased expression of striatal signalling ggenes in a mouse model of Huntingtin's disease. *Hum. Mol. Genet.* **9**, 1259–1271.

Mangiarini, L., Sathasivam, K., Seller, M., Cozens, B., Harper, A., Hetherington, C., Lawton, M., Trottier, Y., Lehrach., H., Davies, S. W., and Bates, G. P. (1996). Exon 1 of the HD gene with an expanded CAG repeat is sufficient to cause a progressive neurological phenotype in transgenic mice. *Cell* **87**, 493–506.

Manley, K., Shirley, T. L., Flaherty, L., and Messer, A. (1999). Msh2 deficiency prevents in vivo somatic instability of the CAG repeat in Huntington disease transgenic mice. *Nat. Genet.* **23**, 471–473.

Martindale, D., Hackam, A., Wieczorek, A., Ellerby, L., Welllington, C., McCutcheon, K., Singaraja, R., Kazemi-Esfarjani, P., Devon, R., Kim, S. U., Bredesen, D. E., Tufaro, F., and Hayden, M. R. (1998). Length of huntingtin and its polyglutamine tract influences localization and frequency of intracellular aggregates. *Nat. Genet.* **18**, 150–154.

Matilla, A., Koshy, B. T., Cummings, C. J., Isobe, T., Orr, H. T., and Zoghbi, H. Y. (1997). The cerebellar leucine-rich acidic nuclear proteins interacts with ataxin-1. *Nature* **389**, 974–978.

McCampbell, A., Taylor, J. P., Taye, A. A., Robitschek, J., Li, M., Walcott, J., Merry, D., Chai, Y., Paulson, H., Sobue, G., and Fischbeck, K. H. (2000). CREB-binding protein sequestration by expanded polyglutamine. *Hum. Mol. Genet.* **9**, 2197–2202.

Miyashita, T., Nagao, K., Ohmi, K., Yanagisawa, H., Okamura-Oho, Y., and Yamada, M. (1998). Intracellular aggregate formation of dentatorubral-pallidoluysian atrophy (DRPLA) protein with the extended polyglutamine. *Biochem. Biophys. Res. Commun.* **249**, 96–102.

Nagafuchi, S., Yanagisawa, H., Ohsaki, E., Shirayama, T., Tadokoro, K., Inoue, T., and Yamada, M. (1994a). Structure and expression of the gene responsible for the triplet repeat disorder, dentatorubral and pallidoluysian atrophy (DRPLA). *Nat. Genet.* **8**, 177–182.

Nagafuchi, S., Yanagisawa, H., Sato, K., Sato, K., Shirayama, T., Ohsaki, E., Bundo, M., Takeda, T., Tadokoro, K., Kondo, I., Murayama, N., Tnaka, Y., Kikushima, H., Umino, K., Kurosawa, H., Furukawa, T., Nihei, K., Inoue, T., Sano, A., Komure, O., and Takahashi, M. (1994b). Expansion of an unstable CAG trinucleotide on chromosome 12p in dentatorubral and pallidoluysian atrophy. *Nat. Genet.* **6**, 14–18.

Naito, H., Izawa, K., Kurosaki, T., Kaji, S., and Sawa, M. (1972). Two families of progressive myoclonus epilepsy with Mendelian dominant heredity. *Psychiatr. Neurol. Jpn.* **74**, 871–897 (in Japanese).

Naito, H., and Oyanagi, S. (1982). Familial myoclonus epilepsy and choreoathetosis: hereditary dentatorubral-pallidoluysian atrophy. *Neurology* **32**, 798–807.

Norremolle, A., Nielsen, J. E., Sorensen, S. A., and Hasholt, L. (1995). Elongated CAG repeats of the B37 gene in a Danish family with dentato-rubro-pallido-luysian atrophy. *Hum. Genet.* **95**, 313–318.

Nucifora, Jr., F. C., Sasaki, M., Peters, M. F., Huang, H., Cooper, J. K., Yamada, M., Takahashi, H., Tsuji, S., Troncoso, J., Dawson, V. L., Dawson, T. M., and Ross, C. A. (20001). Interference by huntingtin and atrophin-1 with CBP-mediated transcription leading to cellular toxicity. *Science* **291**, 2423–2428.

Onodera, O., Burke, J. R., Miller, S. E., Hester, S., Tsuji, S., Roses, A. D., and Strittmatter, W. J. (1997). Oligomerization of expanded-polyglutamine domain fluorescent fusion proteins in cultured mammalian cells. *Biochem. Biophys. Res. Commun.* **238**, 599–605.

Onodera, O., Oyake, M., Takano, H., Ikeuchi, T., Igarashi, S., and Tsuji, S. (1995). Molecular cloning of a full-length cDNA for dentatorubral-pallidoluysian atrophy and regional expressions of the expanded alleles in the CNS. *Am. J. Hum. Genet.* **57**, 1050–1060.

Orr, H. T., Chung, M. Y., Banfi, S., Kwiatkowski, Jr., T. J., Servadio, A., Beaudet, A. L., McCall, A. E., Duvick, L. A., Ranum, L. P., and Zogbi, H. Y. (1993). Expansion of an unstable trinucleotide CAG repeat in spinocerebellar ataxia type 1. *Nat. Genet.* **4**, 221–226.

Paulson, H. L., Perez, M. K., Trottier, Y., Trojanowski, J. Q., Subramony, S. H., Das, S. S., Vig, P., Mandel, J. L., Fischbeck, K. H., and Pittman, R. N. (1997). Intranuclear inclusions of expanded polyglutamine protein in spinocerebellar ataxia type 3. *Neuron* **19**, 333–344.

Potter, N. T. (1996). The relationship between (CAG)n repeat number and age of onset in a family with dentatorubral-pallidoluysian atrophy (DRPLA): diagnostic implications of confirmatory and predictive testing. *J. Med. Genet.* **33**, 168–170.

Sato, T., Oyake, M., Nakamura, K., Nakao, K., Fukusima, Y., Onodera, O., Igarashi, S., Takano, H., Kikugawa, K., Ishida, Y., Shimohata, T., Koide, R., Ikeuchi, T., Tanaka, H., Futamura, N., Matsumura, R., Takayanagi, T., Tanaka, F., Sobue, G., Komure, O., Takahashi, M., Sano, A., Ichikawa, Y., Goto, J., and Kanazawa, I. (1999). Transgenic mice harboring a full-length human mutant DRPLA gene exhibit age-dependent intergenerational and somatic instabilities of CAG repeats comparable with those in DRPLA patients. *Hum. Mol. Genet.* **8**, 99–106.

Saudou, F., Finkbeiner, S., Devys, D., and Greenberg, M. E. (1998). Huntingtin acts in the nucleus to induce apoptosis but death does not correlate with the formation of intranuclear inclusions. *Cell* **95**, 55–66.

Schilling, G., Wood, J. D., Duan, K., Slunt, H. H., Gonzales, V., Yamada, M., Cooper, J. K., Margolis, R. L., Jenkins, N. A., Copeland, N. G., Takahashi, H., Tsuji, S., Price, D. L., Borchelt, D. R., and Ross, C. A. (1999). Nuclear accumulation of truncated atrophin-1 fragments in a transgenic mouse model of DRPLA. *Neuron* **24**, 275–286.

Shimohata, T., Nakajima, T., Yamada, M, Uchida, C., Onodera, O., Naruse, S., Kimura, T., Koide, R., Nozaki, K., Sano, Y., Ishiguro, H., Sakoe, K., Ooshima, T., Sato, A., Ikeuchi, T., Oyake, M., Sato, T., Aoyagi, Y., Hozumi, I., Nagatsu, T., Takiyama, Y., Nishizawa, M., Goto, J., Kanazawa, I., Davidson, I., Tanese, N., Takahashi, H., and Tsuji, S. (2000). Expanded polyglutamine stretches associated with CAG repeat diseases interact with TAFII130, interfering with CREB-dependent transcription. *Nat. Genet.* **26**, 29–36.

Sittler, A., Walter, S., Wedemeyer, N., Hasenbank, R., Scherzinger, E., Eickhoff, H., Bates, G. P., Lehrach, H., and Wanker, E. E. (1998). SH3GL3 associates with the Huntington exon 1 protein and promotes the formation of polygln-containing protein aggregates. *Mol. Cell* **2**, 427–436.

Skinner, P. J., Koshy, B. T., Cummings, C. J., Klement, I. A., Helin, K., Servadio A., Zoghbi, H. Y., and Orr, H. T. (1997). Ataxin-1 with an expanded glutamine tract alters nuclear matrix-associated structures. *Nature* **389**, 971–974.

Smith, J. K. (1975). Dentatorubropallidoluysian atrophy. In "Handbook of Clinical Neurology" (P. J. Vinken and G. W. Bruyn, eds.) pp. 519–534. North-Holland, Amsterdam.

Smith, J. K., Gonda, V. E., and Malamud, N. (1958). Unusual form of cerebellar ataxia: Combined dentato-rubral and pallido-Luysian degeneration. *Neurology* **8**, 205–209.

Snell, R. G., MacMillan, J. C., Cheadle, J. P., Fenton, I., Lazarou, L. P., Davies, P., MacDonald, M. E., Gusella, J. F., Harper, P. S., and Shaw, D. J. (1993). Relationship between trinucleotide repeat expansion and phenotypic variation in Huntington's disease. *Nat. Genet.* **4**, 393–397.

Steffan, J. S., Kazantsev, A., Spasic-Boskovic, O., Greenwald, M., Zhu, Y. Z., Gohler, H., Wanker, E. E., Bates, G. P., Housman, D. E., and Thompson, L. M. (2000). The Huntington's disease protein interacts with p53 and creb-binding protein and represses transcription. *Proc. Natl. Acad. Sci. U.S.A.* **97**, 6763–6768.

Takahashi, H., Ohama, E., Naito, H., Takeda, S., Nakashima, S., Makifuchi, T., and Ikuta, F. (1988). Hereditary dentatorubral-pallidoluysian atrophy: clinical and pathologic variants in a family. *Neurology* **38**, 1065–1070.

Takano, H., Cancel, G., Ikeuchi, T., Lorenzetti, D., Mawad, R., Stevanin, G., Didierjean, O., Durr, A., Oyake, M., Shimohata, T., Sasaki, R., Koide, R., Igarashi, S., Hayashi, S., Takiyama, Y., Nishizawa, M., Tanaka, H., Zoghbi, H., Brice, A., and Tsuji, S. (1998). Close associations between prevalences of dominantly inherited spinocerebellar ataxias with CAG-repeat expansions and frequencies of large normal CAG alleles in Japanese and Caucasian populations. *Am. J. Hum. Genet.* **63**, 1060–1066.

Takano, T., Yamanouchi, Y., Nagafuchi, S., and Yamada, M. (1996). Assignment of the dentatorubral and pallidoluysian atrophy (DRPLA) gene to 12p 13.31 by fluorescence in situ hybridization. *Genomics* **32**, 171–172.

Titica, J., and van Bogaert, L. (1946). Heredo-degenerative hemiballismus: A contribution to the question of primary atrophy of the corpus Luysii. *Brain* **69**, 251–263.

Turmaine, M., Raza, A., Mahal, A., Mangiarini, L., Bates, G. P., and Davies, S. W. (2000). Nonapoptotic neurodegeneration in transgenic mouse model of Huntington's disease. *Proc. Natl. Acad. Sci. U.S.A.* **97**, 8093–8097.

Wanker, E. E., Rovira, C., Scherzinger, E., Hasenbank, R., Walter, S., Tait, D., Colicelli, J., and Lehrach, H. (1997). HIP-I: a huntingtin interacting protein isolated by the yeast two-hybrid system. *Hum. Mol. Genet.* **6**, 487–495.

Waragai, M., Junn, E., Kajikawa, M., Takeuchi, S., Kanazawa, I., Shibata, M., Mouradian, M. M., and Okazawa, H. (2000). PQBP-1/Npw38, a nuclear protein binding to the polyglutamine tract, interacts with U5–15kD/dim1p via the carboxyl-terminal domain. *Biochem. Biophys. Res. Commun.* **273**, 592–595.

Warner, T. T., Lennox, G. G., Janota, I., and Harding, A. E. (1994a). Autosomal-dominant dentatorubropallidoluysian atrophy in the United Kingdom. *Movement Disorders* **9**, 289–296.

Warner, T. T., Williams, L., and Harding, A. E. (1994b). DRPLA in Europe. *Nat. Genet.* **6**, 225–225.

Wood, J. D., Yuan, J., Margolis, R. L., Colomer, V., Duan, K., Kushi, J, Kaminsky, Z., Kleiderlein, J. J., Sharp, A. H., and Ross, C. A. (1998). Atrophin-1, the DRPLA gene product, interacts with two families of WW domain-containing proteins. *Mol. Cell. Neurosci.* **11**, 149–160.

Yamada, M., Hayashi, S., Tsuji, S., and Takahashi, H. (2001). Involvement of the cerebral cortex and autonomic ganglia in Machado-Joseph disease. *Acta Neuropathol. (Berlin)* **101**, 140–144.

Yamamoto, A., Lucas, J. J., and Hen, R. (2000). Reversal of neuropathology and motor dysfunction and a conditional model of Huntington's disease. *Cell* **101**, 57–66.

Yanagisawa, H., Fujii, K., Nagafuchi, S., Nakahori, Y., Nakagome, Y., Akane, A., Nakamura, M., Sano, A., Komure, O., Kondo, I., Jin, D. K., Sorenson, S. A., Potter, N. T., Young, S. R., Nakamura, K., Nukina, N., Nagao, Y., Tadokoro, K., Okuyama, T., Miyashita, T., Inoue, T., Kanazawa, I., and Yamada, M. (1996). A unique origin and multistep process for the generation of expanded DRPLA triplet repeats. *Hum. Mol. Genet.* **5**, 373–379.

Yazawa, I., Nukina, N., Hashida, H., Goto, J., Yamada, M., and Kanazawa, I. (1995). Abnormal gene product identified in hereditary dentato-rubral-pallidoluysian atrophy (DRPLA) brain. *Nat. Genet.* **10**, 99–103.

Zhang, S., Xu, L., Lee, J., and Xu, T. (2002). Drosophila atrophin homolog functions as a transcriptional corepressor in multiple developmental processes. *Cell* **108**, 45–56.

# CHAPTER 17

# Ataxia in Prion Diseases

LEV G. GOLDFARB

*National Institute of Neurological Disorders and Stroke*
*National Institutes of Health*
*Bethesda, Maryland 20892*

I. The *PRNP* Gene and Protein Products
II. Phenotypes
   A. Sporadic TSE
   B. Horizontally Transmitted (Infectious) TSE
   C. Inherited TSE
III. Diagnostic and Ancillary Tests
   A. DNA Analysis
   B. CSF
   C. EEG
   D. Neuroimaging
   E. Neuropathology
IV. Cellular and Animal Models of Disease
V. Treatment and Management
   References

Human transmissible spongiform encephalopathies (TSEs) or prion diseases represent a unique group of neurodegenerative disorders that may be sporadic, infectious, or hereditary. Infectious forms of TSE may cause large outbreaks such as epidemics of kuru in New Guinea and variant Creutzfeldt-Jakob disease (vCJD) in the U.K. Inherited TSEs, familial Creutzfeldt-Jakob disease (fCJD), Gerstmann-Sträussler-Scheinker disease (GSS), and fatal familial insomnia (FFI), are associated with multiple missense, insertion, and deletion mutations in the *PRNP* gene. Sporadic form is defined as a TSE case with no apparent environmental source of infection or family history; sporadic TSE is evenly distributed around the world with a population frequency of around 1 per million per year. Kuru was the first human chronic noninflammatory neurodegenerative disorder proven to be transmissible to chimpanzees after incubation periods of 18–24 months. Sporadic and hereditary forms of TSE were also subsequently found to be experimentally transmissible to non-human primates and rodents with similarly long incubation times. Recent advances in cellular and molecular biology have yielded increasingly strong evidence that TSE results from the accumulation in the brain of an abnormal isoform of prion protein (PrP-sc), a post-translational modification of a normal host protein (PrP-c) encoded by the *PRNP* gene. This abnormal isoform and its toxic fragments gradually accumulate in neurons resulting in neuronal death and other pathogenic effects that are responsible for the TSE phenotypes.

## I. THE *PRNP* GENE AND PROTEIN PRODUCTS

*PRNP* gene is a single-copy gene in the human genome located on chromosome 20p12–pter (Sparkes *et al.*, 1986). The *PRNP* gene contains two exons separated by an intron of about 13 kb. The second exon includes the entire coding sequence (Fig. 17.1). The open reading frame of the wild-type *PRNP* gene is comprised of 759 nucleotides coding for a 253 amino acid product named prion protein (PrP; Kretzschmar *et al.*, 1986). The gene has five repeating sequences between codons 51 and 91, of which the first 27-nucleotide sequence is followed by four 24-bp variant repeats coding for a P(H/Q)GGGWGQ octapeptide. These repeats are remarkably conserved among species, which implies their important functional role. The 5′ promoter region includes two regulatory elements that strongly

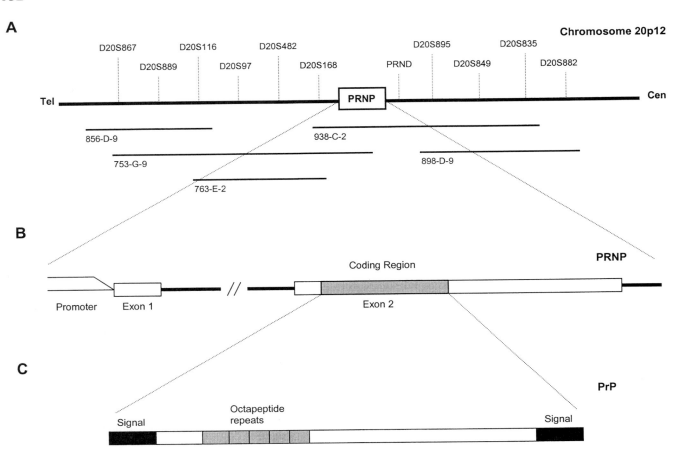

FIGURE 17.1 Genetic and physical maps of the PRNP region on chromosome 20p12-pter and diagrams of the human prion protein gene and prion protein. (A) genetic markers flanking the *PRNP* gene and human YAC clones covering the region; (B) *PRNP* gene consisting of a promoter and exons 1 and 2, with the coding sequence located entirely within exon 2; (C) prion protein precursor with signal peptides and nona/octapeptide repeat sequence.

enhance promoter activity (−131/−284 and −1303/−1543), and two repressing elements (−254/−567 and −567/−909; Funke-Kaizer et al., 2001). The protein product of the *PRNP* gene, the physiological cellular protein PrP-c is a 35-kDa sialoglycoprotein of unknown function. PrP-c is post-translationally processed to proteolytically remove a 22 amino acid amino-terminal signal peptide; a 23 amino acid signal peptide is removed from the C-terminal end as a glycophosphatidyl-inositol (GPI) anchor is added to serine at position 231. PrP-c contains two conserved disulfide-bonded cysteines at positions 179 and 217 and two asparagine side chains linked to large oligosaccharides at positions 181 and 197 (Goldmann, 1993). PrP-c is predominantly expressed in the upper cortical neurons of the neocortex and the Purkinje cells of the cerebellum, and located on the outer cell membrane, in Golgi and endosomal intracytoplasmic organelles. PrP-c is evenly distributed on the entire cell surface membrane, suggesting that it has a generalized cellular function (Laine et al., 2001).

The predominant view of TSE pathogenesis is currently based on the "protein only" hypothesis. According to this hypothesis, the infectious prion protein (PrP-sc), a predominantly beta-sheet molecule, is derived from a normal functional prion protein (PrP-c), a largely alpha-helical isoform (Prusiner, 1998). PrP-sc alone or in association with another as yet unidentified molecule (Telling et al., 1995) has the capacity to initiate the conformational conversion of PrP-c into PrP-sc if inoculated into a susceptible host. Proteases completely digest PrP-c, while the partially protease-resistant PrP-sc is reduced to a core protein with molecular weight of 27–30 kDa (McKinley et al., 1983) that has 43% beta-sheets and no alpha-helix. PrP27-30 exhibits remarkable resistance to physical and chemical agents, its turnover is approximately 5 times slower than that of the wild type PrP-c, and with disease progression it tends to form insoluble fibrillar aggregates in cell compartments. After cell death PrP-sc accumulates in the extracellular space forming amyloid plaques. Evidence indicating that PrP is the pathogenic protein responsible for TSE and transmission of this disease to other hosts comes from several thoroughly established experimental facts: accumulation of PrP-sc in the brain is observed in all forms of TSE; procedures destroying proteins inactivate infectivity; PrP-sc co-purifies with infectivity, and the concentration of PrP-sc is directly proportional to the infectivity titer; mice devoid of PrP gene are resistant to TSE; and no evidence was found for the presence of a virus-like particle or nucleic acid associated with TSE pathogenesis or transmissibility (Prusiner, 1998). PrP-c and PrP-sc are encoded by the same *PRNP* gene and have the same amino acid sequence, therefore they are antigenically indistinguishable by the infected host, thus accounting for the absence of immunologic and inflammatory changes in the TSE.

Multiple mutations in the *PRNP* gene have been linked to various phenotypes of human TSE; the two most frequent mutations are located at codons 178 and 200 (Gambetti *et al.*, 1999; Lee *et al.*, 1999). Currently, 26 pathogenic point mutations, 9 variants of insertional mutations, a single deletion, and 7 DNA polymorphisms have been identified in the *PRNP* gene. Many of the pathogenic mutations are located within or near one of the PrP alpha-helix domains, which is in agreement with the view that mutations alter the protein secondary structure predisposing the PrP molecule to acquire a beta-sheet configuration (Prusiner, 1998). Transgenic mice expressing mutant PrP analogous to the human P102L mutation spontaneously developed a transmissible neurologic illness characterized by ataxia, lethargy, and rigidity with characteristic vacuolar degeneration in the neocortex and PrP deposits in the cerebrum and cerebellum (Hsiao *et al.*, 1994). A specific association of mutant PrP with ataxia and neuronal death in the granular layer of the cerebellum has been observed in mice expressing PrP with deleted 32–121 and 32–134 residues (Shmerling *et al.*, 1998). The PrP methionine/valine polymorphism at position 129 has been the subject of intensive research. Accumulated evidence has shown that although this substitution does not by itself cause disease, it modifies TSE phenotype by influencing effects of other mutations in inherited forms (Goldfarb *et al.*, 1992) and predisposing to disease development in sporadic and iatrogenic CJD (Palmer *et al.*, 1991; Brown *et al.*, 2000), kuru (Lee *et al.*, 2001), and vCJD (Will, 2001).

## II. PHENOTYPES

The clinical and neuropathologic spectrum of the human TSEs is extremely diverse. The phenotype of an inherited form depends on the causative mutation, whereas the route of infection determines to a significant extent the features of an infectious form. Cerebellar ataxia is part of the clinical spectrum in each TSE form. In kuru, iatrogenic CJD with peripheral route of infection and several variants of GSS, ataxia is the presenting symptom that tends to dominate the clinical picture, while cognitive decline, psychiatric symptoms, or insomnia are more prominent in sporadic TSE, vCJD, and FFI. Spongiform change and astrocytic gliosis are found in most but not all forms of TSE. The presence and morphology of amyloid plaques and PrP-immunoreactive deposits is variable. Lesion topography has become an important basis for classification: cerebral cortex and subcortical ganglia are predominantly affected in all forms of CJD, the cerebellum is primarily affected in kuru and GSS, and the thalamus is the major site of degeneration in FFI.

### A. Sporadic TSE

Sporadic TSE is defined as a case of spongiform encephalopathy having no detectable cause. These cases lack family history or apparent environmental source of infection. Sporadic forms constitute an absolute majority of TSE cases varying between 83 and 93% in different countries and populations; they are evenly distributed throughout the world with a frequency of approximately 1 per million per year (Masters *et al.*, 1979). Sporadic TSE includes sporadic CJD and recently discovered sporadic forms of FFI (Parchi *et al.*, 1999). In sporadic CJD, the average age at disease onset is 60 years with a range from 16–82. The average duration of illness is 7 months. The onset of neurological disease is gradual, occurring over a period of several weeks. Initial symptoms are usually cognitive impairment manifested by memory loss, confusion, and behavioral changes. Cerebellar symptoms, eye movement abnormalities, or visual impairment may appear before, during, or after the onset of mental signs. As the illness evolves, the majority of patients develop deficits of higher cortical functions, overt dementia, gait ataxia, dysarthria, nystagmus, seizures, tremors, myoclonus, choreiform or athetoid movements, and pyramidal tract signs (Brown, 1994). Periodic sharp-wave discharge (PSD) on EEG appearing as 0.5–2 Hz generalized bi/triphasic periodic complexes is highly characteristic and considered to be diagnostic of sporadic CJD (Fig. 17.2). Magnetic resonance imaging (MRI) may show cerebral atrophy, but is most useful in excluding other conditions.

Characteristic appearances on MRI have been identified in several forms of CJD; sporadic CJD may be associated with high signal changes in the putamen and caudate head and vCJD is usually associated with hyperintensity of the pulvinar (posterior nuclei) of the thalamus. Using appropriate clinical and radiological criteria and tailored imaging protocols, MRI plays an important part in the *in vivo* diagnosis of this disease.

The clinical course is relentlessly progressive. Within 4–6 months, patients become globally demented, often mute,

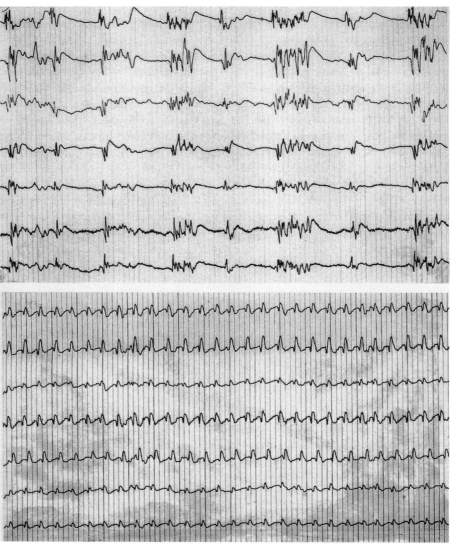

FIGURE 17.2 Electroencephalogram of two patients with sporadic Creutzfeldt-Jakob disease. Upper tracing: a characteristic of CJD but nonspecific "burst-suppression" pattern. Lower tracing: pathognomonic 1–2 cycles per second triphasic periodic sharp-wave discharge (PSD) pattern. (From Goldfarb et al. (2000). *Handbook of Ataxia Disorders* (T. Klockgether, ed.), pp. 523–543. Marcel Dekker, Inc., New York; Fig. 1, p. 532. With permission.)

and physically incapacitated by motor dysfunction, including severe ataxia, myoclonus, and rigidity (Brown 1994). The predominant histopathologic findings are spongiform degeneration and loss of neurons in the cerebral cortex, striatum, and the molecular layer of the cerebellum (Fig. 17.3). Reactive astrocytosis is a very prominent accompanying feature (Budka *et al.*, 1995b). An overt inflammatory response is not found, and the reaction of microglia and macrophages is minimal. Sporadic TSE cases do not show the presence of pathogenic mutations in the *PRNP* gene. A modifying polymorphism at *PRNP* codon 129 was found to be mildly influencing predisposition to sporadic CJD (Palmer *et al.*, 1991; The EUROCJD Group, 2001).

### B. Horizontally Transmitted (Infectious) TSE

#### 1. Kuru

Kuru is the prototype TSE that spread among the Fore people of New Guinea by serial transmission through ritual cannibalism. The origin of kuru in this community remains unknown, but the disease frequency grew dramatically from the beginning of the 20th century to reach epidemic proportions in the 1950s (Gajdusek, 1977). Kuru has since been disappearing gradually as a result of cessation of cannibalism. Since the incubation time may exceed three decades, a significant number of cases were recorded through the 1970s and 1980s. Kuru is characterized as a

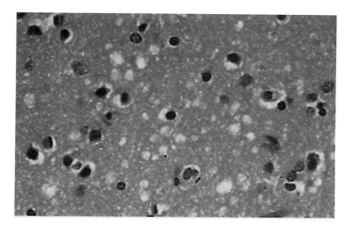

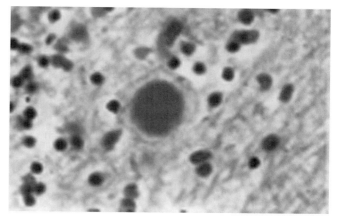

FIGURE 17.4 Neuropathology of kuru. Unicentric amyloid "kuru type" plaque in the granular layer of the cerebellum (PAS ×100). (From Brown et al. (1994). *Neurodegenerative Diseases* (D. B. Calne, ed.), pp. 839–876. W. B. Saunders, Philadelphia; Fig. 48, p. 851. With permission.)

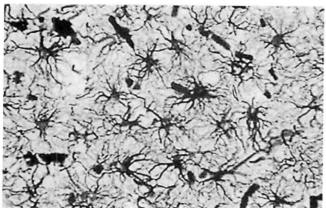

FIGURE 17.3 Neuropathology of sporadic TSE. Upper section: vacuolar degeneration in the cerebral cortex (H&E ×80). Lower section: proliferation and hypertrophy of astrocytes. Silver method ×80. (From Brown et al. (1994). *Neurodegenerative Diseases* (D. B. Calne, ed.), pp. 839–876. W. B. Saunders, Philadelphia; Fig. 48, p. 851. With permission.)

progressive fatal ataxia. The disease starts with unsteadiness and changes in personality and mood. Limb and truncal ataxia, dysarthria, and what has been described as "shivering tremor" progress to the point of inability to walk and sit or perform any activity. Myoclonus and choreiform and athetoid movements develop with disease evolution. Dementia may not be evident until a late stage of illness (Gajdusek, 1977). Death ensues 6–18 months after the onset of clinical illness. Consistent features of kuru neuropathology are diffuse astrocytic proliferation and gliosis throughout the gray matter with significant neuronal loss predominantly in the basal ganglia and pontine nuclei. The most characteristic feature is the presence of amyloid plaques having a delicate fibrillary structure surrounding a dark central core ("kuru-type" plaques; Fig. 17.4). The plaques are prominent in the dentate nucleus of the cerebellum (Klatzo et al., 1959). Spongiform degeneration is present in some cases. Brain tissue suspension of kuru patients transmitted the disease to chimpanzees after an initial incubation period of 22 months (Gajdusek et al., 1966).

## 2. Variant Creutzfeldt-Jakob Disease (vCJD)

TSE surveillance was introduced in the U.K. in 1990 when an epidemic of bovine spongiform encephalopathy (BSE) went out of control and the possibility of disease transmission to humans was strongly suspected. The epidemic of BSE in the U.K. started in 1986 and spread throughout the country to involve over 180,000 animals (Brown et al., 2001). It resulted from changes in rendering ovine and bovine carcasses for the production of animal food. Although the input of infectious carcasses ceased after a 1988 ban of the use of ruminant-derived protein to feed cattle, the BSE epidemic reached a peak only in late 1992 and subsided by 2001. In 1995 and early 1996, a number of human TSE cases with distinctive clinical presentations and neuropathologic features were reported in British adolescents and young adults (Will et al., 1996). Further observations suggested that this was a new variant of CJD perhaps related to the BSE epidemic. More than 100 cases of vCJD have so far occurred in Great Britain, three cases in France, and a single case in the Republic of Ireland and Hong Kong (R.G. Will, 2001, personal communication).

Distinctive features of vCJD include early age of onset, average 29 years (a single vCJD case in an older individual aged 74 was recently reported; Lorains et al., 2001), and a unique and consistent phenotype of early behavioral and personality change, depression, and memory loss. Ataxia is an early clinical feature in almost all patients. Dementia and myoclonus develop as the disease progress (Will et al., 1996). There are no periodic complexes on EEG. The average duration of illness is 14 months. The defining neuropathologic feature is a kuru-type amyloid plaque composed of a central amyloid core with a fibrillary periphery, surrounded by a rim of spongiform ("daisy" or "florid" plaques; Fig. 17.5). Clusters of PrP-immunoreactive deposits not associated with spongiform change are widespread in the cerebral and cerebellar cortices, the basal ganglia, thalamus, and brainstem (Ironside, 2000). Astrocytosis and neuronal loss characteristic of other CJD forms are also present.

### 3. Iatrogenic Creutzfeldt-Jakob Disease

CJD resulting from human-to-human transmission was first reported in the 1970s in a case of a contaminated corneal graft taken from a donor with neuropathologically confirmed CJD (Brown et al., 2000). Neurosurgical procedures involving contaminated instruments, EEG needles, and contaminated dura mater grafts were later found to be significant causes of CJD transmission. In 1985, the first case of CJD among the recipients of human growth hormone replacement therapy was reported, and to date, there have been 139 growth hormone-associated deaths. The length of incubation time and clinical features in iatrogenic CJD are determined to a significant extent by the route of infection. When the infectious agent is directly introduced into the brain, as with neurosurgical instruments or stereotactic EEG electrodes, the incubation time is 16–28 months and the predominant clinical feature is a rapidly progressing dementia. However, when the contaminated tissue is applied to the surface of the brain as with dura mater grafts, the incubation is on average 6 years and the clinical presentation may be either mental deterioration or cerebellar ataxia. With peripheral routes of infection as in hormone replacement therapy, the average incubation time increases to 12 years and the clinical presentation is invariably a cerebellar dysfunction with limb and truncal ataxia, dysarthria, and nystagmus. Neuropathological examination reveals widespread spongiform change, marked astrocytic gliosis, and neuronal loss closely resembling the changes seen in sporadic CJD.

## C. Inherited TSE

### 1. Familial Creutzfeldt-Jakob Disease

#### a. E200K Mutation

The *PRNP* E200K mutation is the most frequent cause of familial CJD accounting for approximately 70% of the fCJD-affected families worldwide. This mutation was identified in clusters of this disease in isolated populations of Slovakia, Libyan Jews, Chile, Italy, and Japan (Lee *et al.*, 1999). The mean age at onset is 55 years, and mean duration of illness 8 months (Brown, 1994, and Table 17.1). Patients present with cognitive impairment, behavioral change, and cerebellar signs. Limb and truncal ataxia and dysarthria are observed at the disease onset in approximately 50% of patients and may or may not be accompanied by mental deterioration. Incoordination gradually becomes disabling, and cortical blindness, startle-induced or spontaneous myoclonic jerks, and muscle rigidity develop during the course of illness. In the advanced stage of illness, all patients have dementia, 79% show cerebellar signs and 73% myoclonus. PSD on EEG is very characteristic of this form. CT scans show brain atrophy. Brain SPECT in one patient with a normal brain CT scan showed bilateral perfusion defects (Gambetti and Lugaresi, 1998). Histopathology is characterized by spongiosis, astrogliosis, and neuronal loss. These lesions are most severe and widely distributed in the cerebral cortex. PrP-immunostaining is consistently positive throughout the brain with the punctate or "synaptic" pattern, but deposits either in the form of amyloid or nonamyloid plaques are observed. Intracerebral

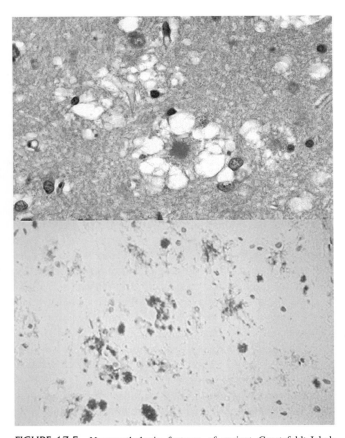

FIGURE 17.5 Neuropathologic features of variant Creutzfeldt-Jakob disease. Upper section: "daisy" amyloid plaques in the occipital cortex (PAS ×60). Lower section: PrP immunopositive aggregates in the cerebral cortex. Hydrolytic autoclaving method ×40. (Courtesy of Dr. James Ironside, National Creutzfeldt-Jakob Disease Surveillance Unit, Western General Hospital, Edinburgh EH4 2XU, U.K.).

TABLE 17.1  Genotype/Phenotype Relationship in Human Transmissible Spongiform Encephalopathies

| Allele[a] | Age at onset (years) | Duration of illness (months) | Signs and symptoms at onset | Signs and symptoms of advanced illness | Neuropathology |
|---|---|---|---|---|---|
| **Creutzfeldt-Jakob disease and fatal familial insomnia phenotypes** | | | | | |
| 24–nt Repeat expansion, 1–4 extra repeats | 67 | 4 | Abnormal behavior/ memory loss | Dementia/ataxia/spasticity/ PSD[b] cortex | Spongiosis/gliosis/neuronal loss in the celebral |
| 24–nt Repeat expansion, 5–9 extra repeats | 37 | 96 | Cognitive decline/ataxia/ spasticity | Dementia/ataxia/spasticity | Cortical atrophy/spongiosis/gliosis/ neuronal loss/widespread multicentric PrP plaques |
| D178N/129V | 45 | 22 | Memory loss | Dementia/ataxia/ myoclonus/spasticity | Spongiosis/gliosis/neuronal loss in the cerebral cortex |
| D178N/129M | 49 | 15 | Insomnia/ataxia | Insomnia/ataxia/ myoclonus/dementia | Preferential degeneration of thalamic nuclei/spongiosis |
| V180I | 79 | 15 | Rigidity/memory loss | Dementia/rigidity/tremor/ myoclonus | Spongiosis/gliosis/neuronal loss in the cerebral cortex |
| E200K | 55 | 8 | Cognitive decline | Dementia/ataxia/ myoclonus/spasticity/ PSD | Spongiosis/astrocytosis/neuronal loss in the cerebral cortex |
| V210I | 63 | 4 | Memory loss/ataxia | Dementia/ataxia/rigidity/ tremor/myoclonus/PSD | Spongiosis/astrocytosis/neuronal loss in the cerebral cortex |
| M232R | 65 | 4 | Memory loss | Dementia/myoclonus/PSD | Spongiosis/astrocytosis/neuronal loss in the cerebral cortex |
| **Gerstmann-Sträussler-Scheinker disease phenotypes** | | | | | |
| P102L | 48 | 72 | Ataxia/dysarthria | Ataxia/dysarthria/ spasticity/dementia | Uni- and multicentric amyloid plaques in the cerebellum/ spongiform change in the cerebral cortex |
| P105L | 44 | 108 | Cognitive decline/ spasticity | Spasticity/dementia | Cortical atrophy/gliosis/neuronal loss/multicentric PrP plaques in the cerebral cortex |
| A117V | 38 | 56 | Cognitive decline/ rigidity | Dementia/pseudobulbar/ rigidity/spasticity | Uni- and multicentric PrP plaques in cerebral cortex and subcortical ganglia |
| F198S | 52 | 72 | Cognitive decline/ ataxia | Dementia/ataxia/ dysarthria/rigidity | Uni- and multicentric neuritic amyloid plaques/neurofibrillary tangles in the cerebellum and cerebral cortex |
| Q217R | 52 | 72 | Abnormal behavior/ cognitive decline/ataxia | Dementia/ataxia/ dysarthria/rigidity | Uni- and multicentric neuritic amyloid plaques/neurofibrillary tangles in the cerebellum and cerebral cortex |

[a]Alleles causing TSE in at least three known kindreds are included.
[b]PSD—periodic synchronous discharge on EEG.

inoculation of brain homogenate from patients carrying the E200K mutation regularly transmitted the disease to primates (Brown, 1994).

### b. D178N Mutation

The PRNP D178N mutation has been linked to two separate familial disorders, fCJD and FFI. Considerable phenotypic distinction between them depends on coupling of the causative D178N mutation with the polymorphic sequence at PRNP codon 129 (Goldfarb et al., 1992). A PRNP allele with the D178N mutation and valine at polymorphic codon 129 (D178N/129V) is responsible for a dementing form of fCJD with an average age of onset of 45 years and duration of illness 22 months (Table 17.1). The most characteristic presentation is memory impairment often associated with depression, irritability, and abnormal behavior. Ataxia, dysarthria and aphasia, tremor, and myoclonus appear early during the course of illness. The 178/129V type of fCJD only rarely shows PSD on EEG (Brown, 1994). Most typical neuropathologic changes are severe and widely spread spongiosis, prominent gliosis, and neuronal loss. The frontal and temporal cortex is generally most severely affected, the thalamus is spared and minimal or no pathology is seen in the brainstem. The cerebellum shows minimal but definite PrP immunostaining despite the lack of structural changes (Gambetti et al., 1999). Disease has been transmitted to squirrel monkeys with brain tissue of seven patients from five kindreds (Brown, 1994).

*c. 24-bp Repeat Expansion*

Owen et al. (1989) reported insertion of an additional six 24-bp repeats in the *PRNP* repeat region in patients of a British family affected with an unusual dementia. Several further families with one to nine 24-bp extra repeats were subsequently reported in the U.S., several European countries, and Japan (Goldfarb et al., 1996). Each family carried a unique allele, which was apparently generated through unequal crossover and differed by the number of repeats as well as the order and composition of the repeat elements. The average age at disease onset in patients with 5 and more extra repeats was 37, significantly earlier than in patients with other types of familial TSE, and the mean duration of illness in these patients was 8 years (Table 17.1). In patients with 1–4 extra repeats, the age at onset was 67 years and the duration of illness 4 months. Thus, the age of disease onset correlates inversely with the repeat number, but an anticipation phenomenon is not recognized in the repeat-expansion families with TSE. The number of inserted sequences was exactly the same in the descendants of different lines in the six-generation British family (Poulter et al., 1992), as well as in four smaller families (Goldfarb et al., 1996). Clinical features were determined to a significant extent by the number of repeats. Patients with 1–4 repeats had rapidly progressive dementia often associated with ataxia and visual disturbances, myoclonus, and PSD on the EEG recordings. In patients with five or more octapeptide repeats the illness is characterized as a very slowly progressive mental deterioration with cerebellar and extrapyramidal signs often lacking the PSD complexes at the EEG examination. Patients with a low number of repeats showed histopathologic changes consistent with those of sporadic TSE, including spongiform degeneration or status spongiosus, astrogliosis, and neuronal loss. In contrast, autopsied patients having 7 or more octapeptide repeats show the presence of uni- or multicentric PrP amyloid plaques located in the molecular layer of the cerebellum and the cerebral gray matter, a distribution not seen in CJD and rather compatible with GSS (Gambetti et al., 1999). Brain suspension from three studied patients with 5, 7 and 8 extra repeats transmitted the disease to primates after intracerebral inoculation (Brown, 1994).

## 2. Fatal Familial Insomnia

Lugaresi et al. (1986) described a unique hereditary disease characterized by insomnia, dysautonomia, and motor deficits. This unique syndrome is associated with the D178N/129M haplotype. The disease starts between 20 and 71 years of age (mean, 49) and may have either a relatively short (6–13 months) or longer (24–48 months) duration (Gambetti et al., 1995). These variations in the duration of illness are, at least in part, genetically determined. The cardinal clinical features are insomnia (better defined as incapacity to generate sleep), dysautonomia, and motor signs. Rapid eye movement (REM) sleep and spindle activity during non-REM sleep are reduced and eventually disappear. Ataxia, dysarthria, and dysphagia are among the early signs, while cognitive functions remain relatively spared until late in the course of illness. Motor manifestations of advanced disease include dysarthria and limb ataxia. Cognitive functions show lack of vigilance and attention as part of selective impairment of memory. It has been noted that patients homozygous for the 129M allele follow a more rapid course of illness with prominent sleep and autonomic disturbances while signs of motor and cognitive dysfunction are mild. In contrast, the heterozygous 129M/V patients have a prolonged course, present with ataxia and dysarthria, and tend to have more prominent cognitive impairment and seizures while sleep disturbances and autonomic signs are less severe (Montagna et al., 1998). The histopathologic hallmark of FFI is the loss of neurons and astrogliosis in the medio-dorsal and anterior thalamic nuclei. Involvement of other thalamic nuclei varies. The inferior olives show neuronal loss and gliosis in most cases. The neocortex is virtually spared in patients with disease duration of less than one year, but focally affected by spongiosis and gliosis in those with a course between 12 and 20 months, and diffusely involved only in subjects with disease duration of more than 20 months (Lugaresi et al., 1998). Brain suspension of FFI patients transmitted the disease to wild-type (Tateishi et al., 1995) and transgenic mice (Telling et al., 1996b). Transgenic mice expressing chimeric human-mouse PrP showed degeneration and PrP accumulation in the thalamus.

## 3. Gerstmann-Sträussler-Scheinker Disease (GSS)

GSS was described as a familial disease with progressive limb and truncal ataxia, dysarthria, personality change, and cognitive decline. This is a disorder inherited with an autosomal dominant pattern and caused by mutations in the *PRNP* gene. The set of *PRNP* mutations responsible for GSS is different from those seen in familial CJD, suggesting that fCJD and GSS are allelic disorders (Goldfarb et al., 2000). GSS typically differs from CJD by more prominent cerebellar ataxia, longer duration of illness, and the presence of morphologically distinct multicentric amyloid plaques in the cerebellum.

*a. P102L Mutation*

The P102L mutation in the *PRNP* gene was first described in one family each from the U.S. and U.K. (Hsiao et al., 1989) and subsequently in many other families around the world. This is the most frequent GSS mutation causing the disease in about 80% of all known GSS families (Ghetti et al., 1995). In P102L mutation-associated GSS, the age of onset is on average 48 years and the duration of illness 6 years (Table 17.2). The clinical course is characterized by early development and slow progression of ataxia and dysarthria associated with pyramidal and extrapyramidal signs and pseudobulbar symptoms. Mental impairment is infrequent at the disease onset, but develops later in the course of illness manifesting as personality disorder and dementia. Cerebellar atrophy is documented by MRI (Fig. 17.6). Myoclonic movements and PSDs on EEG are rarely observed. The distinctive neuropathologic feature is the presence of multicentric PrP amyloid plaques in the cerebellum (Budka et al., 1995b). The plaques present as amorphous aggregates of spheroid bodies often consisting of a centrally located larger core encircled by several satellite plaques (Fig. 17.7). Spongiform degeneration varies from severe to absent (Piccardo et al., 1998).

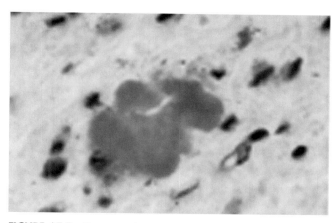

FIGURE 17.7 Representative section of the cerebellar cortex in a patient with Gerstmann-Sträussler-Scheinker disease. Large multicore amyloid plaques (PAS ×100). (From Brown et al. (1994). *Neurodegenerative Diseases* (D. B. Calne, ed.), pp. 839–876. W. B. Saunders, Philadelphia; Fig. 48, p. 851. With permission.)

*b. F198S and Q217R Mutations*

The large and well-studied Indiana kindred and another family segregating GSS with the F198S mutation (Farlow et al., 1989; Hsiao et al., 1992; Ghetti et al., 1995) and a Swedish family with GSS associated with the Q217R mutation (Hsiao et al., 1992) show a phenotype substantially different from the P102L patients. The age of onset is between 40 and 71 and disease duration between 2 and 12 years. Memory impairment and progressive ataxia, and dysarthria in association with bradykinesia and rigidity are observed in most patients. Distinct uni- and multicentric PrP amyloid plaques are widely spread in the cerebral, cerebellar, and midbrain parenchyma. In the neocortex, amyloid cores are surrounded by abnormal tau-positive neurites similar to neuritic plaques in Alzheimer's disease (Ghetti et al., 1995). To increase the similarity, neurofibrillary tangles are present in the same areas of the neocortex. Plaques and tangles are immunoreactive to PrP and not to A-beta protein.

## III. DIAGNOSTIC AND ANCILLARY TESTS

The diagnosis of TSE was traditionally based on clinical and neuropathologic data. Findings of rapidly progressing dementia, ataxia and/or insomnia and the presence of myoclonus and pyramidal tract signs may be the basis for a preliminary diagnosis in a living patient. The presence of spongiform change accompanied by neuronal loss and gliosis, and (in some TSE forms) PrP-immunoreactive deposits make the diagnosis definite (Budka et al., 1995a). With the discovery of TSE transmissibility to laboratory animals a successful transmission was considered a confirmation of TSE diagnosis since experimental transmission

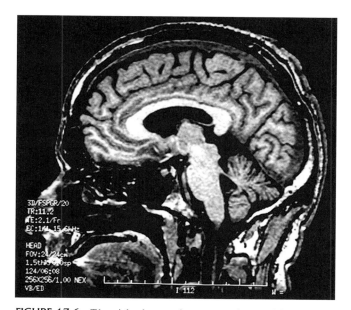

FIGURE 17.6 T1-weighted magnetic resonance image of the brain in a 38-year-old patient with Gerstmann-Sträussler-Scheinker disease, a sagittal projection. Marked diffuse cerebellar atrophy and enlargement of the fourth ventricle. (From Goldfarb et al., (2000). *Handbook of Ataxia Disorders* (T. Klockgether, ed.), pp. 523–543. Marcel Dekker, Inc., New York; Fig. 3, p. 538. With permission.)

rates in kuru, and sporadic and iatrogenic CJD is successful in 90–100% (Brown, 1994).

### A. DNA Analysis

In patients with inherited TSE (and sometimes in apparently sporadic TSE patients), pathogenic *PRNP* mutations can be identified. Genetic testing is available in research laboratories. Since multiple mutations may be responsible for similar disease phenotypes, sequencing of the entire *PRNP* coding sequence is the most reliable mutation detection strategy.

### B. CSF

The routine examination of the CSF is normal. Recent studies have indicated that the detection in the CSF of proteins 130 and 131 of a family of 14-3-3 brain proteinase inhibitors released from damaged neurons is extremely useful in the diagnosis of difficult cases (Zerr et al., 2000). This is a highly sensitive and specific test for the diagnosis of sporadic TSE, iatrogenic CJD, and hereditary CJD in patients with the E200K mutation. In sporadic CJD the estimated sensitivity is 94% and specificity 84% (Zerr et al., 2000). Other brain-specific proteins such as neuron-specific enolase (NSE), S-100b, and tau protein are also increased in the CSF of patients with sporadic CJD, but their diagnostic usefulness has not been fully investigated.

### C. EEG

Electroencephalography is routinely used in patients with CJD. The typical EEG abnormality in advanced disease consists of 1–2 cycles per second triphasic sharp waves superimposed on a depressed background (Fig. 17.2). They are usually asymmetric, may occur in synchrony with myoclonic jerks, and tend to become slower with the disease progression. Serial tracings reveal the characteristic triphasic sharp-wave pattern in up to 80% of patients with sporadic CJD at some time during its course, although PSD is rare in all inherited forms, with the exception of CJD associated with the E200K mutation (Brown, 1994).

### D. Neuroimaging

MRI reveals enlargement of the ventricles and widening of the cortical sulci, and in some cases, a symmetrical hyperintense signal in the basal ganglia. In GSS, significant cerebellar atrophy may be apparent (Fig. 17.6).

### E. Neuropathology

A new range of neuropathologic techniques such as PrP immunocytochemistry, PrP detection by the immunoblot or histoblot, *in situ* polymerase chain reaction (PCR), and determination of PrP "glycotypes" came into use as additions to the traditional neuropathologic examination and assessment of transmissibility.

## IV. CELLULAR AND ANIMAL MODELS OF DISEASE

Regular transmission of TSE to susceptible laboratory animals, including apes and monkeys, Syrian hamsters, and several strains of mice, was demonstrated in early TSE studies (Gajdusek, 1977). Neuropathology in the experimental animals was similar to human disease. Transmission of kuru by cannibalistic rituals was experimentally confirmed by studies of alimentary transmission of TSE in Syrian hamsters (who are natural cannibals) by neuropathologic confirmation and establishing infectivity of their brain tissue (Prusiner et al., 1985).

Bueler et al. (1993) produced knock-out mice by replacing 80% of the *PRNP* gene (codons 4 through 187) by a neocassette. Mice homozygous for the disrupted gene developed normally. They showed remarkable resistance to intracerebral inoculation of infectious prions surviving inoculation of $10^7$ $LD_{50}$ units of a mouse-adapted scrapie strain with no symptoms of disease, whereas the wild-type littermates all died after 130–158 days. Resistance of the *prnp*-null mice to TSE is explained by lack of prion propagation in the absence of the host PrP-c. Experiments with prion-infected grafts inserted into the brain tissue of knock-out mice show that the graft develops specific TSE changes while the surrounding PrP-deficient neurons remain intact (Brandner et al., 1996).

Transgenic mice expressing murine PrP with an introduced mutation equivalent to a human mutation at codon 102 spontaneously developed neurodegeneration, spongiform change, and astrogliosis (Hsiao et al., 1994). Brain extracts from these spontaneously ill mice transmitted disease to another strain of transgenic mice expressing PrP with the same mutation at a lower level (which did not cause the spontaneous disease by itself), but not to wild-type mice (Telling et al., 1996a). In another experiment, transgenic mice that expressed a mutant PrP containing 14 octapeptide repeats developed ataxia. Starting from birth, mutant PrP was converted into a protease-resistant and detergent-insoluble form that resembled the pathogenic isoform PrP-sc, and this isoform accumulates dramatically in many brain regions throughout the lifetime of the mouse. As PrP accumulates, there is massive apoptosis of cerebellar granule cells, as well as astrocytosis and deposition of PrP with a punctate pattern (Chiesa et al., 1998; Harris et al., 2000).

Transgenic technology also helped to demonstrate that patients carrying the same D178N mutation, but expressing different phenotypes actually generate distinct strains of prions. Limited digestion with proteinase K and deglycosylation of PrP-sc extracted from the brain tissue of FFI patients regularly exhibited a 19-kDa band on Western blot, while patients with fCJD associated with the same mutation showed a 21-kDa band (Monari et al., 1994). When mice expressing a chimeric mouse-human transgene were inoculated with protein extracts from FFI or fCJD patients, the PrP produced in these animals kept the same properties as the inoculated protein; the FFI "strain" was 19-kDa and the CJD "strain" was 21-kDa (Telling et al., 1996b), confirming that these strains had different conformations.

The pathogenesis of inherited TSEs was further successfully studied using cell models. In stably transfected lines of CHO cells that express mutant PrP, PrP-c molecules spontaneously convert to PrP-sc and cause profound and early changes in the metabolism and biophysical characteristics resulting from an altered conformation that the mutant PrP acquires co-translationally. The altered conformation confers characteristics of increased hydrophobicity, aggregability, and resistance to proteases reminiscent of those of the PrP-sc (Harris, 1999).

## V. TREATMENT AND MANAGEMENT

Early studies indicated that amphotericin B inhibits accumulation of infectivity in cell cultures and prevents or delays development of TSE in animal experiments (Pocchiari et al., 1989), but was ineffective in treatment of advanced human disease (Masullo et al., 1992). No effective treatment is currently available for TSE, but the progress of fundamental studies offers some hope. Prospective therapies most likely will be directed toward interruption of the conversion from normal prion protein to abnormal PrP-sc. Although PrP-c and PrP-sc are antigenically indistinguishable by the infected host, transgenic mice produce sustained anti-PrP titers that may be sufficient for protection against PrP-sc infection (Heppner et al., 2001). An antibody was created that recognizes a PrP-c epitope critical for PrP-c/PrP-sc interaction and binds PrP-c at the cell surface abolishing PrP-sc formation in a dose-dependent manner (Peretz et al., 2001). Promising results were obtained with some newly studied compounds such as porphyrins and phthalocyanines (Caughey et al., 1998) and tricyclic derivatives of acridine and phenothiazine, quinacrine, and chlorpromazine (Korth et al., 2001), that have in the past been used as antimalarial and antipsychotic drugs, respectively. Each shows anti-prion effects in experiments in cultured cells chronically infected with prions. Acridine and phenothiazine are currently being tested in clinical trials (Prusiner, 2001, personal communication).

PrP peptide encompassing residues 106 through 126 are highly toxic for cultured neurons and have a tendency to form stable beta-sheet structures and polymerize into amyloid-like fibrils (Forloni et al., 1993). The use of Chelex-100 treatment significantly reduced fibrillogenesis caused by this fragment due to limiting of copper and zinc access that is critical for PrP106–126 aggregation and neurotoxicity (Jobling et al., 2001). Conformational transition from PrP-c to PrP-sc may also be inhibited or prevented or even reversed by synthetic peptides homologous to PrP fragments implicated as transitional sites. These peptides named beta-sheet breakers reduced infectivity by 90–95% in mice with experimental TSE (Soto et al., 2000).

Precautions are recommended in the general care and management of hospitalized TSE patients. The infectious agent is not present in any external secretion, but it may be present in brain and spinal cord tissue, cerebrospinal fluid, eyeballs, pituitary gland, spleen, liver, kidneys, lymph nodes, and blood (Budka et al., 1995a). Penetrating injuries from potentially contaminated instruments carry risk of infection. Accidental contamination of intact skin should be treated with the application of fresh undiluted bleach or 1 N NaOH to the area for about 1 minute, followed by thorough washing with soap and water. Incidental transmission of vCJD with instruments and devices in general surgery and pathology has a greater risk as compared to other TSE forms because lymphoreticular tissues of vCJD patients appear to be more consistently infected (Bruce et al., 2001). Disposable instruments and other materials should be used whenever possible; if retained, instruments should be disinfected in two cycles of steam autoclaving at 134°C for 1 hour with subsequent soaking in 1N NaOH for 2 hours. Special precautions should be taken while handling pathological samples (Budka et al., 1995a). There is an ongoing discussion about safety of blood and blood products. Contamination of blood with the vCJD agent is causing most concern (Brown et al., 1999). Exclusion of at-risk donors may reduce or eliminate blood contamination. The identification of mutations in the *PRNP* gene opens a possibility of genetic counseling, as prenatal genetic testing can be done at the family's request (Brown et al., 1994). It is also expected that in the future the number of iatrogenic CJD cases will decrease as a result of the use of recombinant pituitary hormones replacing cadaveric hormones, and new highly efficient procedures are being introduced to sterilize dura mater grafts.

## References

Brandner, S., Isenmann, S., Raeber, A., Fischer, M., Sailer, A., Kobayashi, Y., Marino, S., Weissmann, C., and Aguzzi, A. (1996). Normal host protein necessary for scrapie-induced neurotoxicity. *Nature* **379**, 339–343.

Brown, P. (1994). Transmissible human spongiform encephalopathy (infectious cerebral amyloidosis): Creutzfeldt-Jakob disease, Gerstmann-Sträussler-Scheinker syndrome, and kuru. In: "Neurodegenerative Diseases" (D.B. Calne, ed.), pp. 839–876. W.B. Saunders, Philadelphia.

Brown, P., Cervenakova, L., Goldfarb, L. G., Gajdusek, D. C., Horwitz, J., Creacy, S. D., Bever, R. A., Wexler, P., Sujansky, E., and Bjork, R. J. (1994). Molecular genetic testing of a fetus at risk of Gerstmann-Sträussler-Scheinker's syndrome. *Lancet* **343**, 181–182.

Brown, P., Cervenakova, L., McShane, L. M., Barber, P., Rubenstein, R., and Drohan, W. N. (1999). Further studies of blood infectivity in an experimental model of transmissible spongiform encephalopathy, with an explanation of why blood components do not transmit Creutzfeldt-Jakob disease in humans. *Transfusion 1999* **39**, 1169–1178.

Brown, P., Preece, M., Brandel, J. P., Sato, T., McShane, L., Zerr, I., Fletcher, A., Will, R. G., Pocchiari, M., Cashman, N. R., d'Aignaux, J. H., Cervenakova, L., Fradkin, J., Schonberger, L. B., and Collins, S. J. (2000). Iatrogenic Creutzfeldt-Jakob disease at the millennium. *Neurology* **55**, 1075–1081.

Brown, P., Will, R. G., Bradley, R., Asher, D. M., and Detwiler, L. (2001). Bovine spongiform encephalopathy and variant Creutzfeldt-Jakob disease: background, evolution, and current concerns. *Emerg. Infect. Dis.* **7**, 6–16.

Bruce, M. E., McConnell, I., Will, R. G., and Ironside, J. W. (2001). Detection of variant Creutzfeldt-Jakob disease infectivity in extra-neural tissues. *Lancet* **358**, 208–209.

Budka, H., Aguzzi, A., Brown, P., Brucher, J. M., Bugiani, O., Collinge, J., Diringer, H., Gullotta, F., Haltia, M., Hauw, J.-J., Ironside, J. W., Kretzschmar, H. A., Lantos, P. L., Masullo, C., Pocchiari, M., Schlote, W., Tateishi, J., and Will, R. G. (1995a). Tissue handling in suspected Creutzfeldt-Jakob disease (CJD) and other human spongiform encephalopathies (prion diseases). *Brain Pathol.* **5**, 319–322.

Budka, H., Aguzzi, A., Brown, P., Brucher, J. M., Bugiani, O., Gullotta, F., Haltia, M., Hauw, J.-J., Ironside, J. W., Jellinger, K., Kretzschmar, H. A., Lantos, P. L., Masullo, C., Schlote, W., Tateishi, J., and Weller, R. O. (1995b). Neuropathological diagnostic criteria for Creutzfeldt-Jakob disease (CJD) and other human spongiform encephalopathies (prion diseases). *Brain Pathol.* **5**, 459–466.

Bueler, H., Aguzzi, A., Sailer, A., Greiner, R. A., Autenried, P., Aguet, M., and Weissmann, C. (1993). Mice devoid of PrP are resistant to scrapie. *Cell* **73**, 13391347.

Caughey, W. S., Raymond, L. D., Horiuchi, M., and Caughey, B. (1998). Inhibition of protease-resistant prion protein formation by porphyrins and phthalocyanines. *Proc. Natl. Acad. Sci. U.S.A.* **95**, 12117–12122.

Chiesa, R., Piccardo, P., Ghetti, B., and Harris, D. A. (1998). Neurological illness in transgenic mice expressing a prion protein with an insertional mutation. *Neuron* **21**, 1339–1351.

Collie, D. A., Sellar, R. J., Zeidler, M., Colchester, A. C., and Will, R. G. (2001). MRI of Creutzfeldt-Jakob disease: imaging features and recommended MRI protocol. *Clin. Radiol.* **56**, 726–739.

Farlow, M. R., Yee, R. D., Dlouhy, S. R., Conneally, P. M., Azzarelli, B., and Ghetti, B. (1989). Gerstmann-Sträussler-Scheinker disease. 1. Extending the clinical spectrum. *Neurology* **39**, 1446–1452.

Forloni, G., Angeretti, N., Chiesa, R., Monzani, E., Salmona, M., Bugiani, O., and Tagliavini, F. (1993). Neurotoxicity of a prion protein fragment. *Nature* **362**, 543–546.

Funke-Kaizer, H., Theis, S., Behrouzi, T., Thomas, A., Scheuch, K., Zollmann, F. S., Paterka, M., Paul, M., and Orzechowski, H.-D. (2001). Functional characterization of the human prion protein promoter in neuronal and endothelial cells. *J. Mol. Med.* **79**, 529–535.

Gajdusek, D. C. (1977). Unconventional viruses and the origin and disappearance of kuru. *Science* **197**, 943–960.

Gajdusek, D. C., Gibbs, C. J. Jr., and Alpers, M. (1966). Experimental transmission of kuru-like syndrome to chimpanzees. *Nature* **209**, 794–796.

Gambetti, P., and Lugaresi, E. (1998). Conclusions of the symposium. *Brain Pathol.* **8**, 571–575.

Gambetti, P., Parchi, P., Petersen, R. B., Chen, S. G., and Lugaresi, E. (1995). Fatal familial insomnia and familial Creutzfeldt-Jakob disease: clinical, pathological and molecular features. *Brain Pathol.* **5**, 43–51.

Gambetti, P., Petersen, R. B., Parchi, P., Chen, S. G., Capellari, S., Goldfarb, L., Gabizon, R., Montagna, P., Lugaresi, E., Piccardo, P., and Ghetti, B. (1999). Inherited prion diseases. In: "Prion Biology and Diseases" (S.B. Prusiner, ed.), pp. 509–583. Cold Spring Harbour Laboratory Press, New York.

Ghetti, B., Dlouhy, S. R., Giaccone, G., Bugiani, O., Frangione, B., Farlow, M. R., and Tagliavini, F. (1995). Gerstmann-Sträussler-Scheinker's disease and the Indiana kindred. *Brain Pathol.* **5**, 61–75.

Goldfarb, L. G., Petersen, R. B., Tabaton, M., Brown, P., LeBlanc, A. C., Montagna, P., Cortelli, P., Julien, J., Vital, C., Pendelbury, W. W., Haltia, M., Wills, P. R., Hauw, J.-J., McKeever, P. E., Monari, L., Schrank, B., Swergold, G. D., Autilio-Gambetti, L., Gajdusek, D. C., Lugaresi, E., and Gambetti, P. (1992). Fatal familial insomnia and familial Creutzfeldt-Jakob disease: disease phenotype determined by a DNA polymorphism. *Science* **258**, 806–808.

Goldfarb, L. G., Cervenakova, L., Brown, P., and Gajdusek, D. C. (1996). Genotype-phenotype correlations in familial spongiform encephalopathies associated with insert mutations. In: "Transmissible Subacute Spongiform Encephalopathies: Prion Disease" (L. Court and B. Dodet, eds.), pp. 425–431. Elsevier, Paris.

Goldfarb, L. G., Butefisch, C. M., and Brown, P. (2000). Ataxia in the transmissible spongiform encephalopathies. In: "Handbook of Ataxia Disorders" (T. Klockgether, ed.), pp. 523–543. Marcel Dekker, Inc., New York.

Goldmann, W. (1993). PrP gene and its association with spongiform encephalopathies. *Br. Med. Bull.* **49**, 839–859.

Haik S., Brandel J. P., Oppenheim C., Sazdovitch V., Dormont D., Hauw J. J., and Marsault, C. (2002). Sporadic CJD clinically mimicking variant CJD with bilateral increased signal in pulvinar. *Neurology* **58**, 148–149.

Harris, D. A. (1999). Cell biological studies of the prion protein. *Curr. Issues Mol. Biol.* **1**, 65–75.

Harris, D. A., Chiesa, R., Drisaldi, B., Quaglio, E., Migheli, A., Piccardo, P., and Ghetti, B. (2000). A transgenic model of a familial prion disease. *Arch. Virol. Suppl.* **16**, 103–112.

Heppner, F. L., Musahl, C., Arrighi, I., Klein, M. A., Rulicke, T., Oesch, B., Zinkernagel, R. M., Kalinke, U., and Aguzzi, A. (2001). Prevention of scrapie pathogenesis by transgenic expression of anti-prion protein antibodies. *Science* **294**, 178–182.

Hsiao, K. K., Baker, H. F., Crow, T. J., Poulter, M., Owen, F., Terwilliger, J., Westaway, D., Ott, J., and Prusiner, S. B. (1989). Linkage of a prion protein missense variant to Gerstmann-Sträussler syndrome. *Nature* **338**, 342–345.

Hsiao, K. K., Dlouhy, S. R., Farlow, M. R., Cass, C., DaCosta, M., Conneally, P. M., Hodes, M. E., Ghetti, B., and Prusiner, S. B. (1992). Mutant prion proteins in Gerstmann-Sträussler-Scheinker disease with neurofibrillary tangles. *Nat. Genet.* **1**, 68–71.

Hsiao, K. K., Groth, D., Scott, M., Yang, S.-L., Serban, H., Rapp, D., Foster, D., Torchia, M., DeArmond, S. J., and Prusiner, S. B. (1994). Serial transmission in rodents of neurodegeneration from transgenic mice expressing mutant prion protein. *Proc. Natl. Acad. Sci. U.S.A.* **91**, 9126–9130.

Ironside, J. W. (2000). Pathology of variant Creutzfeldt-Jakob disease. *Arch. Virol. Suppl.* **16**, 143–151.

Jobling, M. F., Huang, X., Stewart, L. R., Barnham, K. J., Curtain, C., Volitakis, I., Perugini, M., White, A. R., Cherny, R. A., Masters, C. L., Barrow, C. J., Collins, S. J., Bush, A. I., and Cappai, R. (2001). Copper and zinc binding modulates the aggregation and neurotoxic properties of the prion peptide PrP106-126. *Biochemistry* **40**, 8073–8084.

Klatzo, I., Gajdusek, D. C., and Zigas, V. (1959). Pathology of kuru. *Lab. Invest.* **8**, 799–847.

Korth, C., May, B. C., Cohen, F. E., and Prusiner, S. B. (2001). Acridine and phenothiazine derivatives as pharmacotherapeutics for prion disease. *Proc. Natl. Acad. Sci. U.S.A.* **98**, 9836–9841.

Kretzschmar, H. A., Stowring, L. E., Westaway, D., Stubblebine, W. H.,

Prusiner, S. B., and DeArmond, S. J. (1986). Molecular cloning of a human prion protein cDNA. *DNA* **5**, 315–324.

Laine, J., Marc, M. E., Sy, M. S., and Axelrad, H. (2001). Cellular and subcellular morphological localization of normal prion protein in rodent cerebellum. *Eur. J. Neurosci.* **14**, 47–56.

Lee, H. S., Sambuughin, N., Cervenáková, L., Chapman, J., Pocchiari, M., Litvak, S., Qi, H.-Y., Budka, H., del Ser, T., Furukawa, H., Brown, P., Gajdusek, D. C., Korczyn, A., and Goldfarb, L. G. (1999). Ancestral origins and worldwide distribution of the *PRNP* 200K mutation causing familial Creutzfeldt-Jakob disease. *Am. J. Hum. Genet.* **64**, 1063–1070.

Lee, H. S., Brown, P., Cervenakova, L., Garruto, R. M., Alpers, M. P., Gajdusek, D. C., and Goldfarb, L. G. (2001). Increased susceptibility to kuru of carriers of the PRNP 129 methionine/methionine genotype. *J. Infect. Dis.* **183**, 192–196.

Lorains, J. W., Henry, C., Agbamu, D. A., Rossi, M., Bishop, M., Will, R. G., and Ironside, J.W. (2001). Variant Creutzfeldt-Jakob disease in an elderly patient. *Lancet* **357**, 1339–1340.

Lugaresi, E., Medori, R., Montagna, P., Baruzzi, A., Cortelli, P., Lugaresi, A., Tinuper, P., Zucconi, M., and Gambetti, P. (1986). Fatal familial insomnia and dysautonomia with selective degeneration of thalamic nuclei. *N. Engl. J. Med.* **274**, 2079–2082.

Lugaresi, E., Tobler, I., Gambetti, P., and Montagna, P. (1998). The pathophysiology of fatal familial insomnia. *Brain Pathol.* **8**, 521–526.

Masters, C. L., Harris, J. O., Gajdusek, D. C., Gibbs, C. J. Jr., Bernoulli, C., and Asher, D. M. (1979). Creutzfeldt-Jakob disease: patterns of worldwide occurrence and the significance of familial and sporadic clustering. *Ann. Neurol.* **5**, 177–188.

Masullo, C., Macchi, G., Geng, X. Y., and Pocchiari, M. (1992). Failure to ameliorate Creutzfeldt-Jakob disease with amphotericin B therapy. *J. Infect. Dis.* **165**, 784–785.

McKinley, M. P., Bolton, D. C., and Prusiner, S. B. (1983). A protease-resistant protein is a structural component of the scrapie prion. *Cell* **35**, 57–62.

Monari, L., Chen, S. G., Brown, P., Parchi, P., Petersen, R. B., Mikol, J., Gray, F., Cortelli, P., Montagna, P., Ghetti, B., Goldfarb, L. G., Gajdusek, D. C., Lugaresi, E., Gambetti, P., and Autilio-Gambetti, L. (1994). Fatal familial insomnia and familial Creutzfeldt-Jakob disease: different prion proteins determined by a DNA polymorphism. *Proc. Natl. Acad. Sci. U.S.A.* **91**, 2839–2842.

Montagna, P., Cortelli, P., Avoni, P., Tinuper, P., Plazzi, G., Galassi, R., Portaluppi, F., Julien, J., Vital, C., Delise, M. B., Gambetti, P., and Lugaresi, E. (1998). Clinical features of fatal familial insomnia: phenotypic variability in relation to a polymorphism at codon 129 of the prion protein gene. *Brain Pathol.* **8**, 515–520.

Owen, F., Poulter, M., Lofthouse, R., Collinge, J., Crow, T. J., Risby, D., Baker, H. F., Ridley, R. M., Hsiao, K., and Prusiner, S. B. (1989). Insertion in prion protein gene in familial Creutzfeldt-Jakob disease. *Lancet* **1** (8628), 51–52.

Palmer, M. S., Dryden, A. J., Hughes, T., and Collinge, J. (1991). Homozygous prion protein genotype predisposes to sporadic Creutzfeldt-Jakob disease. *Nature* **352**, 340–342.

Parchi, P., Capellari, S., Chin, S., Schwarz, H. B., Schecter, N. P., Butts, J. D., Hudkins, P., Burns, D. K., Powers, J. M., and Gambetti, P. (1999). A subtype of sporadic prion disease mimicking fatal familial insomnia. *Neurology* **52**, 1757–1763.

Peretz, D., Williamson, R. A., Kaneko, K., Vergara, J., Leclerc, E., Schmitt-Ulms, G., Mehlhorn, I. R., Legname, G., Wormald, M. R., Rudd, P. M., Dwek, R. A., Burton, D. R., and Prusiner, S. B. (2001). Antibodies inhibit prion propagation and clear cell cultures of prion infectivity. *Nature* **412**, 739–743.

Piccardo, P., Dlouhy, S. R., Lievens, P. M. J., Young, K., Bird, T. D., Nochlin, D., Dickson, D.W., Vinters, H. V., Zimmerman, T. R., Mackenzie, I. R. A., Kish, S. J., Ang, L.-C., De Carli, C., Pocchiari, M., Brown, P., Gibbs, C. J., Jr., Gajdusek, D. C., Bugiani, O., Ironside, J., Tagliavini, F., and Ghetti, B. (1998). Phenotypic variability of Gerstmann-Sträussler-Scheinker disease is associated with prion protein heterogeneity. *J. Neuropathol. Exp. Neurol.* **57**, 979–988.

Pocchiari, M., Casaccia, P., and Ladogana, A. (1989). Amphotericin B: a novel class of antiscrapie drugs. *J. Infect. Dis.* **160**, 795–802.

Poulter, M., Baker, H. F., Frith, C. D., Leach, M., Lofthouse, R., Ridley, R. M., Shah, T., Owen, F., Collinge, J., Brown, J., Hardy, J., Mullan, M. J., Harding, A. E., Bennett, C., Doshi, R., and Crow, T. J. (1992). Inherited prion disease with 144 base pair gene insertion. 1. Genealogical and molecular studies. *Brain* **115**, 675–685.

Prusiner, S. B. (1998). The prion diseases. *Brain Pathol.* **8**, 499–513.

Prusiner, S. B., Cochran, S. P., and Alpers, M. P. (1985). Transmission of scrapie in hamsters. *J. Inf. Dis.* **152**, 971–978.

Shmerling, D., Hegyi, I., Fischer, M., Blattler, T., Brandner, S., Gotz, J., Rulicke, T., Flechsig, E., Cozzio, A., von Mering, C., Hangartner, C., Aguzzi, A., and Weissmann C. (1998). Expression of amino-terminally truncated PrP in the mouse leading to ataxia and specific cerebellar lesions. *Cell* **93**, 203–214.

Soto, C., Kascsak, R. J., Saborio, G. P., Aucouturier, P., Wisniewski, T., Prelli, F., Kascsak, R., Mendez, E., Harris, D. A., Ironside, J., Tagliavini, F., Carp, R. I., and Frangione, B. (2000). Reversion of prion protein conformational changes by synthetic beta-sheet breaker peptides. *Lancet* **355**, 192–197.

Sparkes, R. S., Simon, M., Cohn, V., Fournier, R. E. K., Lem, J., Klisak, I., Heinzmann, C., Blatt, C., Lucero, M., Mohandas, T., DeArmond, S. J., Westaway, D., Prusiner, S. B., and Weiner, L. P. (1986). Assignment of the human and mouse prion protein genes to homologous chromosomes. *Proc. Natl. Acad. Sci. U.S.A.* **83**, 7358–7362.

Tateishi, J., Brown, P., Kitamoto, T., Hoque, Z. M., Roos, R., Wollman, R., Cervenakova, L., and Gajdusek, D.C. (1995). First experimental transmission of fatal familial insomnia. *Nature* **376**, 434–435.

Telling, G. C., Scott, M., Mastriani, J., Gabizon, R., Torchia, M., Cohen, F. E., DeArmond, S. J., and Prusiner, S. B. (1995). Prion propagation in mice expressing human and chimeric PrP transgenes implicates the interaction of cellular PrP with another protein. *Cell* **53**, 79–90.

Telling, G. C., Haga, T., Torchia, M., Tremblay, P., DeArmond, S. J., and Prusiner, S. B. (1996a). Interactions between wild-type and mutant prion proteins modulate neurodegeneration in transgenic mice. *Genes Dev.* **10**, 1736–1750.

Telling, G. C., Parchi, P., DeArmond, S. J., Cortelli, P., Montagna, P., Gabizon, R., Mastrianni, J., Lugaresi, E., Gambetti, P., and Prusiner, S. B. (1996b). Evidence for the conformation of the pathologic isoform of the prion protein enciphering and propagating prion diversity. *Science* **274**, 2079–2082.

The EUROCJD Group. (2001). Genetic epidemiology of Creutzfeldt-Jakob disease in Europe. *Rev. Neurol.* (Paris) **157**, 633–637.

Will, R. G. (2001). The facts about vCJD. *Practitioner* **245** (1623), 469.

Will, R. G., Ironside, J. W., Zeidler, M., Cousens, S. N., Estibeiro, K., Alperovitch, A., Poser, S., Pocchiari, M., Hofman, A., and Smith, P. G. (1996). A new variant of Creutzfeldt-Jakob disease in the UK. *Lancet* **347**, 921–925.

Zerr, I., Pocchiari, M., Collins, S., Brandel, J. P., de Pedro Cuesta, J., Knight, R. S., Bernheimer, H., Cardone, F., Delasnerie-Laupretre, N., Cuadrado Corrales, N., Ladogana, A., Bodemer, M., Fletcher, A., Awan, T., Ruiz Bremon, A., Budka, H., Laplanche, J. L., Will, R. G., and Poser, S. (2000). Analysis of EEG and CSF 14-3-3 proteins as aids to the diagnosis of Creutzfeldt-Jakob disease. *Neurology* **55**, 811–815.

# CHAPTER 18

# Friedreich Ataxia

### MASSIMO PANDOLFO
Université Libre de Bruxelles-Hôpital Erasme
B-1070 Brussels, Belgium

I. Summary
II. Phenotype
   A. Epidermiology
   B. Onset
   C. Neurological Signs and Symptoms
   D. Neurological Examination
   E. Heart Disease
   F. Diabetes Mellitus
   G. Other Signs and Symptoms
   H. Neurophysiological Investigations
   I. Variant Phenotypes
   J. Prognosis
   K. Differential Diagnosis
III. Gene
   A. Gene Structure
   B. Gene Expression
   C. The Intronic GAA Triplet Repeat Expansion
   D. Frataxin Point Mutations
   E. Frataxin Function
   F. Pathogenesis of Friedreich Ataxia
IV. Diagnostic and Ancillary Tests
   A. DNA Tests
   B. Biochemical Investigations
   C. Neuroimaging
V. Pathology
   A. Central Nervous System
   B. Heart
   C. Other Organs
VI. Cellular and Animal Models of Disease
VII. Genetoype/Phenotype Correlations/Modifying Alleles
   A. Correlations between GAA Expansions and Phenotype
   B. Point Mutations in the frataxin Gene
VIII. Treatment
References

## I. SUMMARY

Friedreich ataxia is an inherited recessive disorder characterized by progressive neurological disability and heart abnormalities that may be fatal. The disease affects roughly 1 in 50,000 people, making it one of the most important genetic disorders of its kind. The first symptoms usually appear in childhood, but age of onset may vary from infancy to adulthood. Atrophy of sensory and cerebellar pathways causes ataxia, dysarthria, fixation instability, deep sensory loss, and loss of tendon reflexes. Corticospinal degeneration leads to muscular weakness and extensor plantar responses. A hypertrophic cardiomyopathy may contribute to disability and cause premature death. Other common problems include kyphoscoliosis, pes cavus, and, in 10% of patients, diabetes mellitus.

The Friedreich ataxia (FRDA) gene encodes a small mitochondrial protein, frataxin, which is produced in insufficient amounts in the disease, as a consequence of a GAA triplet repeat expansion in the first intron of the gene. Frataxin deficiency leads to excessive free radical production in mitochondria, progressive iron accumulation in these organelles, and dysfunction of Fe-S center containing enzymes (in particular respiratory complexes I, II, and III, and aconitase).

## II. PHENOTYPE

The disease was described by Friedreich in the 1860s, but rigorous diagnostic criteria were established only in the late 1970s and early 1980s (Geoffroy *et al.*, 1976; Harding *et al.*, 1981; Harding *et al.*, 1981). The phenotype of

Friedreich ataxia has been redefined and extended after the gene (FRDA) was discovered in 1996 (Campuzano *et al.*, 1996).

## A. Epidemiology

In Caucasians, Friedreich ataxia accounts for half of the overall hereditary ataxia cases and for three-quarters of those with onset before age 25 years (Harding, 1983). It has a prevalence of around $2 \times 10^{-5}$ (Harding, 1983; Skre, 1975; Romeo *et al.*, 1983; Leone *et al.*, 1990; Lopez-Arlandis *et al.*, 1995). There are local clusters of Friedreich ataxia due to a founder effect, as those observed in Rimouski, Québec (Bouchard *et al.*, 1979) and in Kathikas-Arodhes, Cyprus (Dean *et al.*, 1988). Friedreich ataxia may not exist in non-Caucasian populations (see below).

## B. Onset

The typical onset of Friedreich ataxia is around puberty (Geoffroy *et al.*, 1976; Harding, 1981; Filla *et al.*, 1990; Muller-Felber *et al.*, 1993; Campanella *et al.*, 1980; D'Angelo *et al.*, 1980), but it may be earlier (Ulku *et al.*, 1988; De Michele *et al.*, 1996a). After the gene was mapped and identified, it became clear that late-onset cases exist, even after 60 (late-onset Friedreich ataxia, LOFA; Klockgether *et al.*, 1993; De Michele *et al.*, 1994), contravening the clinical diagnostic criteria that fixed a limit at age 20 (Geoffroy *et al.*, 1976) or 25 (Harding, 1981). Age of onset may dramatically vary within a sibship, a phenomenon now in part explained by the dynamic nature of the underlying mutation (Campanuzo *et al.*, 1996).

The typical presentation at onset is with gait instability or generalized clumsiness. Most often the patient is a child who recently started to sway when walking and falls easily. As is the case in other degenerative diseases, an acute illness or injury may precede the appearance of the symptoms. Sometimes the child is active in sports and there were no clues as of the impending neurological illness, more commonly the child had been considered "clumsy" for some time before the overt appearance of symptoms. Non-neurological manifestations, in particular scoliosis, often considered to be idiopathic, may precede the onset of ataxia. Rare patients (5%) are diagnosed with idiopathic hypertrophic cardiomyopathy and treated as such for up to 2–3 years before neurological symptoms appear (Geoffroy *et al.*, 1976; Harding, 1981; Filla *et al.*, 1990; Muller-Felber *et al.*, 1993).

## C. Neurological Signs and Symptoms

Mixed cerebellar and sensory ataxia characterizes the disease. It begins as truncal ataxia causing swaying, imbalance, and falls. At the very beginning the gait abnormality is subtle and may reveal itself only as a difficulty in tandem gait. Subsequently, the ataxic nature of gait becomes evident, with irregular steps, veering, and difficulty in turning. Loss of upright stability initially consists of inability to stand on one foot, but soon becomes evident when standing with feet close together and is worsened by eye closure (positive Romberg sign). With further progression, gait becomes broad-based, with frequent loss of balance, requiring intermittent support (furniture, walls, an accompanying person's arm). Patients become completely unable to stand with feet close together, then need support even when standing with their feet apart. A cane, then a walker become necessary. Finally, on average 10–15 years after onset, patients lose the ability to walk, stand, and sit without support. Evolution is variable, however, with mild cases that are still ambulatory decades after onset.

Limb ataxia appears after truncal ataxia. Fine motor skills become impaired, with increasing difficulty in activities such as writing, dressing, and handling utensils. Dysmetria and intention tremor are evident. Ataxia is progressive and unremitting, though periods of stability are frequent at the beginning of the illness.

Dysarthria appears within 5 years from clinical onset (Geoffroy *et al.*, 1976; Harding, 1981; Filla *et al.*, 1990; Muller-Felber *et al.*, 1993), consisting of slow, jerky speech with sudden utterances (Gentil *et al.*, 1990; Cisneros *et al.*, 1995). It progresses until speech becomes almost unintelligible. Dysphagia, particularly for liquids, appears with advancing disease. Patients with very advanced disease frequently choke, requiring modified foods and eventually a nasogastric tube or gastrostomy feedings.

Cognitive functions are generally well preserved. Mental retardation and learning disabilities do not seem to be more prevalent in patients with Friedreich ataxia than in the general population. On the other hand, Friedreich ataxia, as any disease causing substantial physical disability, if not properly managed may have a substantial impact on academic, professional, and personal development. According to some authors, in-depth neuropsychological testing reveals a slowed information processing speed not accompanied by major prefrontal cortex and mood disorders (White *et al.*, 2000).

## D. Neurological Examination

The neurological examination of a typical case of Friedreich ataxia is very characteristic. We can group findings according to the underlying pathology.

Involvement of the central and peripheral sensory system results in deep sensory loss and abolished reflexes. Loss of position and vibration sense is invariable, but may not be evident at onset. Perception of light touch, pain, and temperature, initially normal, tends to decrease with advancing disease. Loss of tendon reflexes in the lower limbs was considered essential for the diagnosis (Geoffroy *et al.*, 1976; Harding, 1981). However, a minority of patients with a positive molecular test for Friedreich ataxia, referred to as Friedreich ataxia with retained reflexes (FARR), have elicitable, sometimes even exaggerated reflexes with spasticity.

Pyramidal involvement causes extensor plantar responses and progressive muscular weakness. The latter becomes severe only late in the progression of the disease. Ataxia and not weakness is the primary cause for loss of ambulation: even when patients become wheelchair-bound, they still maintain on average 70% of their normal strength in lower limbs (Beauchamp *et al.*, 1995). Some patients have flexor plantar responses, again contravening the older clinical diagnostic criteria (Geoffroy *et al.*, 1976; Harding, 1981). Probably, the relative impact of sensory neuropathy and of pyramidal tract degeneration varies from patient to patient, resulting most often in the typical picture of areflexia associated with extensor plantar responses, but sometimes one component prevails. Such partial pictures are usually observed in milder cases of the disease.

Muscle tone is often normal at onset. However, with advancing pyramidal involvement and particularly when ambulation becomes severely impaired, many patients complain of spasms in the lower limbs, mostly nocturnal. Despite limited involvement of lower motor neurons, distal amyotrophy in the lower limbs and in the hands is frequent even early in the course of the disease (Harding, 1981). When patients are wheelchair-bound, disuse atrophy occurs.

The typical oculomotor abnormality of Friedreich ataxia is fixation instability with square-wave jerks (Spieker *et al.*, 1995). Various combinations of cerebellar, vestibular, and brain stem oculomotor signs may be observed, but gaze-evoked nystagmus is uncommon and ophthalmoparesis does not occur. About 30% of the patients develop optic atrophy, with or without visual impairment (Geoffroy *et al.*, 1976; Harding, 1981; Muller-Febner *et al.*, 1993; Kirkham *et al.*, 1981; Livingstone *et al.*, 1981; Rabiah *et al.*, 1997). Sensorineural hearing loss affects about 20% of the patients (Ell *et al.*, 1984; Cassandro *et al.*, 1986). Optic atrophy and sensorineural hearing loss tend to be associated with each other and with diabetes (Harding, 1981; Montermini *et al.*, 1997).

### E. Heart Disease

Most commonly asymptomatic, heart disease may contribute to disability and cause premature death in a significant minority of patients, particularly those with earlier age of onset (Geoffroy *et al.*, 1976; Harding, 1981; Filla *et al.*, 1990; Muller-Felber *et al.*, 1993; Leone *et al.*, 1988; Hartman *et al.*, 1960; Boyer *et al.*, 1962; Harding *et al.*, 1983; Pentland *et al.*, 1983; Child *et al.*, 1986; Alboliras *et al.*, 1986; Maione *et al.*, 1997). Shortness of breath (40%) and palpitations (11%) are the most common symptoms (Harding, 1981). The electrocardiogram shows inverted T waves in essentially all patients, ventricular hypertrophy in most, conduction disturbances in about 10%, occasionally supraventricular ectopic beats and atrial fibrillation (Filla *et al.*, 1990; Muller-Felber *et al.*, 1993; Child *et al.*, 1986; Alboliras *et al.*, 1986). The latter is a negative prognostic sign (Harding, 1984). Electrocardiographic changes may vary in time with occasional normal recordings: this has led to underestimation of their frequency. Repeated recordings are the most sensitive test for the Friedreich ataxia cardiomyopathy. Echocardiography and Doppler echocardiography demonstrate concentric hypertrophy of the ventricles (62%) or asymmetric septal hypertrophy (29%), with diastolic function abnormalities (Harding, 1981; Morvan *et al.*, 1992).

### F. Diabetes Mellitus

About 10% of Friedreich ataxia patients develop diabetes mellitus, 20% have carbohydrate intolerance. The mechanisms are complex, with beta-cell dysfunction (Finocchiaro *et al.*, 1988) and atrophy (Schoenle *et al.*, 1989) as well as peripheral insulin resistance (Fantus *et al.*, 1993). Some cases may be initially controlled by oral hypoglycemic drugs, but insulin is eventually necessary. It is important to regularly check Friedreich ataxia patients for the development of diabetes, as it may increase the burden of disease, complicate the neurological picture, and promote potentially fatal complications.

### G. Other Signs and Symptoms

Kyphoscoliosis may cause pain and cardiorespiratory problems. Pes cavus and pes equinovarus may further affect ambulation. Autonomic disturbances, most commonly cold and cyanotic legs and feet, become increasingly frequent as the disease advances (Margalith *et al.*, 1984). Parasympathetic abnormalities such as decreased heart rate variability parameters have been reported (Pousset *et al.*, 1996). Urgency of micturition is rare.

### H. Neurophysiological Investigations

Sensory nerve action potentials (SNAPs) are severely reduced or, most often, lost (McLeod, 1971; Peyronnard *et al.*, 1976; Ackroyd *et al.*, 1984). Motor and sensory (when measurable) nerve conduction velocities (NCVs) are instead within or just below the normal range. These findings reflect the sensory axonal neuropathy of Friedreich ataxia and clearly distinguish an early case of this disease from a case of hereditary demyelinating sensorimotor neuropathy. The neurophysiological abnormalities related to the axonal sensory neuropathy do not appear to progress with time (Santoro *et al.*, 1999). Degeneration of both peripheral and central sensory fibers also results in dispersion and delay of somatosensory-evoked potentials (SEPs; Muller-Felber *et al.*, 1993). Brainstem auditory-evoked potentials (BAEPs), beginning from the most rostral component, wave V, progressively deteriorate in all patients (Muller-Felber *et al.*, 1993; Vanasse *et al.*, 1988). Visual-evoked potentials (VEPs) are commonly (50–90%) reduced in amplitude with a normal P 100 latency (Muller-Felber *et al.*, 1993); (Spieker *et al.*, 1995; Kirkham *et al.*, 1981; Livingstone *et al.*, 1981; Vanasse *et al.*, 1988). Central motor conduction velocity, determined by cortical magnetic stimulation, is slower than normal (Mondelli *et al.*, 1995), and, contrary to the sensory involvement, this slowing becomes more severe with increasing disease duration (Santoro *et al.*, 1999).

## I. Variant Phenotypes

A positive molecular test for Friedreich ataxia is found in up to 10% of patients with recessive or sporadic ataxia who do not fulfill the classical clinical diagnostic criteria for the disease (Moseley *et al.*, 1998). A re-analysis of these cases revealed that no clinical finding or combination of findings characterizes exclusively or is necessarily present in positive cases. Even neurophysiological evidence of axonal sensory neuropathy is very rarely absent in patients with a proven molecular diagnosis. Overall, however, the "classical" features of Friedreich ataxia, including cardiomyopathy, are highly predictive of a positive test. Absence of cardiomyopathy and moderate-to-severe cerebellar cortical atrophy, demonstrated by MRI, appear to be the best predictors of a negative test (Pandolfo, 1998). In the author's opinion, a molecular test for Friedreich ataxia, along with MRI and neurophysiological investigations, is indicated in the initial workup of all cases of sporadic or recessive degenerative ataxia, regardless of whether or not they fulfill the diagnostic criteria (Geoffroy *et al.*, 1976; Harding, 1981). More complex investigations, including a search for metabolic disorders such as those listed in sections II.K and IV.B, should be postponed until a negative Friedreich ataxia molecular test has been obtained.

Specific variant phenotypes due to the same genetic mutation causing typical Friedreich ataxia include LOFA, FARR, and Acadian ataxia. LOFA patients are defined as having a clinical onset after age 25 years. They exhibit an overall milder, slowly evolving disease with fewer skeletal abnormalities than typical cases. The frequency of cardiomyopathy in LOFA was found to be similar to that of typical Friedreich ataxia in some studies (De Michele *et al.*, 1994; Cassandro *et al.*, 1986), lower in others (Maione *et al.*, 1997). FARR patients are defined as having retained deep tendon reflexes in the lower limbs after the onset of neurological symptoms. They also tend to be mild cases with later onset and slower course than typical Friedreich ataxia patients (Cassandro *et al.*, 1986; Palau *et al.*, 1997). Acadian ataxia is found among the descendants of French colonizers who settled in Eastern Canada during the seventeenth century. After being expelled by the British in the eighteenth century, many moved to Louisiana, where they became known as "Cajuns" (a corruption of "Acadians"). Some later returned to the Canadian Atlantic provinces. The disease was introduced into this population by very few individuals, possibly a single couple, a so-called "founder effect." Acadian ataxia evolves more slowly than typical Friedreich ataxia, with less severe cardiomyopathy (Cassandro *et al.*, 1986; Barbeau *et al.*, 1984).

## J. Prognosis

Friedreich ataxia patients become wheelchair-bound on average 15 years after onset, but variability is very large (Harding, 1981; De Michele *et al.*, 1996b). Early onset and left ventricular hypertrophy predict a faster rate of progression (Cassandro *et al.*, 1986; De Michele *et al.*, 1996b). The burden of neurological impairment, cardiomyopathy, occasionally diabetes, shortens life expectancy (De Michele *et al.*, 1996b). Older studies found that most patients died in their thirties, but survival may be significantly prolonged with treatment of cardiac symptoms, particularly arrhythmias, by anti-diabetic treatment, and by preventing and controlling complications resulting from prolonged neurological disability. Carefully assisted patients may live several more decades.

## K. Differential Diagnosis

Nonhereditary diseases may cause early-onset, progressive ataxia with a chronic course. MRI will reveal posterior fossa tumors and malformations, including platybasia and basilar impression, as well as the lesions of multiple sclerosis. Detection of antigliadin and antiendomysial antibodies points to a diagnosis of gluten sensitivity, which may cause a progressive and treatable cerebellar syndrome even in the absence of gastrointestinal symptoms and signs of malabsorption (Hadjivassiliou *et al.*, 199; Pellecchia *et al.*, 1999).

Rare metabolic disorders may resemble Friedreich ataxia. In particular, a Friedreich ataxia-like clinical picture occurs in some cases of isolated vitamin E deficiency, an autosomal recessive disease due to defective conjugation of this vitamin to lipoproteins. Differentiating features of isolated vitamin E deficiency include a less severe peripheral neuropathy, the common occurrence of a nodding head tremor and of pigmentary retinopathy, and the lack of cardiomyopathy. A similar neurological picture may occur when vitamin E deficiency is secondary to a genetic disorder of lipoproteins (e.g., abetalipoproteinemia) or to malabsorption.

The so-called early-onset cerebellar ataxia with retained reflexes (EOCA) is a heterogeneous category of disorders that includes atypical Friedreich ataxia cases, that can be diagnosed by molecular testing, as well as different entities. Again, early prominent cerebellar atrophy is found only in non-Friedreich ataxia cases. The recent mapping of two ataxia-oculomotor apraxia loci and the cloning of one of them has started to provide a molecular basis for more EOCA cases.

## III. GENE

### A. Gene Structure

FRDA is localized in the proximal long arm of chromosome 9 (Chamberlain et al., 1988) and is composed of seven exons spread over 95 kb of genomic DNA (Campuzano et al., 1996). The major, and probably only functionally relevant mRNA, has a size of 1.3 kb, and corresponds to the first five exons, numbered 1 to 5a. The encoded protein, predicted to contain 210 amino acids, was called frataxin (Campuzano et al., 1996).

### B. Gene Expression

The gene is expressed in all cells, but at variable levels in different tissues and during development (Koutnikova et al., 1997; Jiralerspong et al., 1997). In adult humans, frataxin mRNA is most abundant in the heart and spinal cord, followed by liver, skeletal muscle, and pancreas. In mouse embryos, expression starts in the neuroepithelium at embryonic day 10.5 (E10.5), then reaches its highest level at E14.5 and into the postnatal period. In developing mice, the highest levels of frataxin mRNA are found in the spinal cord, particularly at the thoracolumbar level, and in the dorsal root ganglia. The developing brain is also very rich in frataxin mRNA, which is abundant in the proliferating neural cells in the periventricular zone, in the cortical plates, and in the ganglionic eminence (precursor of the basal ganglia).

### C. The Intronic GAA Triplet Repeat Expansion

Friedreich ataxia is the consequence of frataxin deficiency. Most patient are homozygous for the expansion of a GAA triplet repeat sequence (TRS) in the first intron, 5% are heterozygous for a GAA expansion and a point mutation in the frataxin gene. Repeats in normal chromosomes contain up to ~40 triplets, 90 to >1000 triplets in Friedreich ataxia chromosomes (Fig. 18.1). Expanded alleles show meiotic and mitotic instability. Lengths of the GAA TRS corresponding to pathological FRDA alleles can adopt a triple helical structure in physiological conditions and inhibit transcription in vivo (Ohshima et al., 1998). Two triplexes may associate to form a novel DNA structure called "sticky DNA" (Sakamoto et al., 1999), that was recently shown to inhibit transcription in vitro by sequestering RNA polymerase (Sakamoto et al., 2001).

### D. Frataxin Point Mutations

About 2% of Friedreich ataxia chromosomes carry GAA repeat of normal length, but have missense, nonsense, or

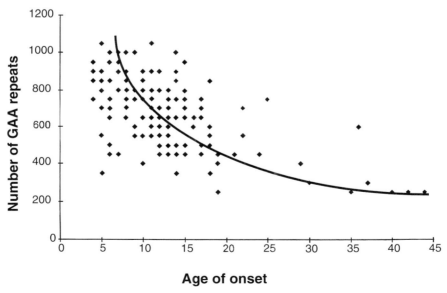

FIGURE 18.1 Scattergram showing the correlation between age of onset and size of the smaller GAA triplet repeat expansion in Friedreich ataxia.

splice site mutations ultimately affecting the frataxin coding sequence (Cossée et al., 1999). Friedreich ataxia patients carrying a frataxin point mutation are in all cases compound heterozygotes for a GAA repeat expansion. Therefore, as GAA expansion homozygotes, they express a small amount of normal frataxin, which is thought to be essential for survival (see below, mouse models).

### E. Frataxin Function

Frataxin does not resemble any protein of known function. Its amino acid sequence does not predict any transmembrane domain. It is highly conserved during evolution, with homologs in mammals, invertebrates, yeast, and plants. The protein is targeted to the mitochondrial matrix (Campuzano et al., 1997). Most of our knowledge about the functional role of frataxin comes from the investigation of yeast cells in which the frataxin homolog gene (YFH1) has been deleted (Babcock et al., 1997). $\Delta YFH1$ strains progressively lose respiratory competence and mitochondrial DNA (rho°). Iron accumulates in mitochondria, more then 10-fold in excess of wild-type, at the expense of cytosolic iron. These cells become hypersensitive to $H_2O_2$, indicating the occurrence of Fenton chemistry. Loss of respiratory competence requires the presence of iron in the culture medium, and occurs more rapidly as iron concentration is increased, suggesting that permanent mitochondrial damage is the consequence of iron toxicity. Normal human frataxin is able to complement the defect in $\Delta YFH1$ cells, while mutated human frataxin is unable to do so, indicating that the function of yfh1p is conserved in human frataxin. It is interesting that the expression of yeast and human frataxin is not regulated by iron levels and its mRNA does not contain an iron-responsive element. A series of publications described the ternary structure of frataxin (Dhe-Paganon et al., 2000; Musco et al., 2000) and of its bacterial homolog, CyaY (Cho et al., 2000). All studies agree that mature frataxin is a compact, globular protein containing an N-terminal α-helix, a middle β-sheet region composed of seven β strands, a second α-helix, and a C-terminal coil. The α-helices are folded upon the β-sheet, with the C-terminal coil filling a groove between the two α-helices. On the outside a ridge of negatively charged residues and a patch of hydrophobic residues are highly conserved suggesting that they interact with a large ligand, probably a protein. However, experiments aimed to identify a protein partner of frataxin, mostly by using the yeast two-hybrid method, have failed so far. Frataxin does not have any feature resembling known iron-binding sites and the NMR study failed to identify any structural change after iron addition. This is in contrast with biochemical data that suggested that frataxin binds iron in a high molecular weight complex, possibly protecting it from reactive oxygen species (ROS) (Adamec et al., 2000). This issue needs to be solved if the function of frataxin is to be understood (Fig. 18.2).

### F. Pathogenesis of Friedreich Ataxia

Altered iron metabolism, ROS damage, and mitochondrial dysfunction also occur in Friedreich ataxia. In Friedreich ataxia patients, iron accumulation occurs in myocardial cells and in the dentate nucleus. There is a moderate, but significant increase of mitochondrial iron in Friedreich ataxia fibroblasts (Delatycki et al., 1999). A high level of circulating transferrin receptor in Friedreich ataxia patients suggests a relative cytosolic iron deficit as observed in the yeast model (Wilson et al., 2000). Friedreich ataxia patients have increased plasma levels of lipid peroxidation (Emond et al., 2000) and oxidative DNA damage products and show hydroxy radical production (Schulz et al., 2000). Friedreich ataxia fibroblasts are hypersensitive to low doses of $H_2O_2$ (Wong et al., 1999). As for mitochondrial dysfunction, the same Fe-S enzyme defects found in $\Delta YFH1$ yeast (deficit of respiratory complexes I, II, and III, and of aconitase) are found in affected tissues from Friedreich ataxia patients (Rötig et al., 1997). In vivo, magnetic resonance spectroscopy analysis of skeletal muscle shows a reduced rate of ATP synthesis after exercise (Lodi et al., 1999), which is inversely correlated to GAA expansion sizes, and may be improved by coenzyme Q10 treatment (Lodi et al., 2001). Increased ROS production was directly shown in P19 cells engineered to produce reduced levels of frataxin (Santos et al., 2001). Furthermore, these cells as well as fibroblasts from Friedreich ataxia patients show abnormal antioxidant responses. In particular, an increase in mitochondrial superoxide dismutase (SOD) triggered by iron in normal cells is blunted in Friedreich ataxia cells (Jiralerspong et al., 2001). Finally, abnormal ROS production during differentiation of multipotent frataxin-deficient cells adversely affects neuronal differentiation and increases apoptosis, an observation that may have implication for the human disease (Santos et al., 2001).

## IV. DIAGNOSTIC AND ANCILLARY TESTS

### A. DNA Tests

DNA testing may be requested to confirm a clinical diagnosis, for carrier detection, and for prenatal diagnosis. A DNA test for Friedreich ataxia, along with MRI and neurophysiological investigations, is indicated in the initial workup of all cases of sporadic or recessive degenerative ataxia, regardless of whether or not they fulfill the diagnostic criteria (Geoffroy et al., 1976; Harding, 1981).

PCR or Southern blot analysis can be used to directly detect the intronic GAA expansion present on 98% of

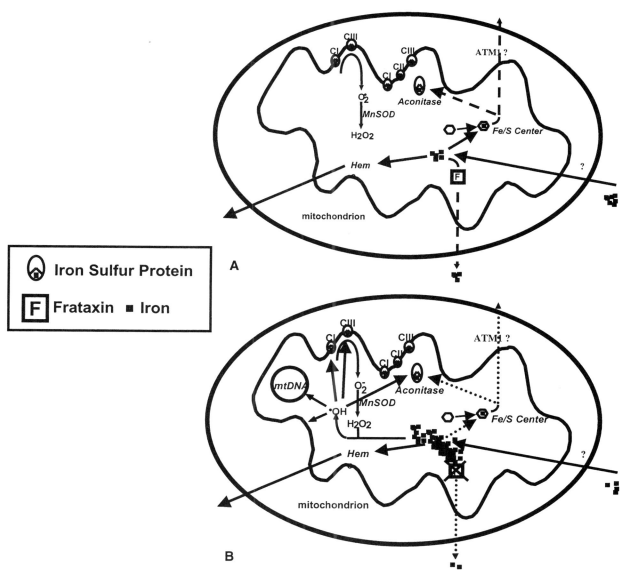

FIGURE 18.2 Simplified scheme showing the hypothetical function of frataxin and some consequences of its deficiency. (A) Normal mitochondrion in which frataxin prevents iron from accumulating and generating ROS and promotes its export to the cytosol. (B) Frataxin deficent mitochondrion in which iron reacts with repiratory chain-generated ROS to produce toxic hydroxyl radical and accumulates in the matrix.

Friedreich ataxia chromosomes. PCR requires only a small amount of DNA, but high quality DNA is extremely important. Ambiguous results, particularly in heterozygote detection, can be resolved by subsequent hybridization of PCR products with an oligonucleotide probe containing a GAA or CTT repeat. Southern blot analysis requires more DNA and is less accurate for determining the number of repeats, but it will not miss a heterozygote, therefore it is indicated when heterozygosity detection is critical.

The question of the lower limit for an expansion to be pathological, so far reported to be 66 triplets, is not so critical in Friedreich ataxia as in dominant unstable repeat diseases. *In vitro* data indicate that repeats exceeding about 50 triplets start to adopt a triplex structure and may form sticky DNA, suggesting that they may be pathological. The disease being recessive, when such a repeat is found on one chromosome, the diagnosis is strongly supported when the other chromosome carries a fully expanded GAA repeat or a frataxin point mutation. In general, length and sequence of the repeat must both be taken into account in uncertain cases. Interruptions of the GAA triplet repeat sequence prevent the formation of a triplex structure and sticky DNA *in vitro* (Sakamoto et al., 2001). *In vivo*, interrupted or variant repeats are occasionally found and may be non-pathological up to the equivalent of 130 triplets (Ohshima *et al.*, 1999). Therefore, sequence analysis of shorter expansions should be performed, particularly when testing for carrier status. In this case, the possibility of the catastrophic expansions of a parental repeat containing 34–60 pure GAA triplets should be kept into account, even though alleles in this size range are stably transmitted most of the time.

Testing gets more difficult when an affected individual is found to be heterozygous for a GAA expansion and a point mutation is suspected. This is done in research labs. No mutation in the coding sequence could be identified in up to 40% of compound heterozygous patients in some series (Cossée *et al.*, 1999). Some of them may actually have a different disease and carry a GAA expansion by chance, given its high frequency (1:90) in Caucasians. One such case has been documented (Smeyers *et al.*, 1996).

Individuals with a Friedreich ataxia-like phenotype but no GAA expansion most likely have mutations at a different locus. We obtained evidence against linkage to the FRDA locus in all families with such cases that we could test. Homozygosity for frataxin point mutations has never been reported. Based on the frequency of point mutations versus GAA expansion, in an equilibrium situation only 4 in 10,000 FRDA patients are expected to carry two point mutations. Therefore, if such cases exist, they should have consanguineous parents. However, biological reasons argue against the existence of these cases: mouse models demonstrated that complete absence of frataxin or the presence of only abnormal frataxin is incompatible with life.

Carrier detection can be requested by relatives of an affected individual and their spouses. A test for an expanded GAA repeat can be readily performed in such cases. In relatives of patients, a search for a previously identified point mutation segregating in the family is also relatively easy to perform. For spouses, exclusion of the expansion leaves an estimated 1/4000 risk of being a carrier. This risk may vary in different populations, for example, it is close to zero in French Canadians, as all Friedreich ataxia patients with this ancestry have two expanded GAA repeats, but it is not irrelevant in individuals from the Naples region, where the I154F mutation is relatively frequent (Cossée *et al.*, 1999). Prenatal diagnosis can be performed for a couple with affected children with a confirmed molecular diagnosis, proving either homozygosity for GAA repeat expansions or compound heterozygoty for a GAA expansion and an identified point mutation. It can also be performed when the parents have no affected children, but are known carriers.

Testing for a locally prevalent point mutation should accordingly be added to expansion analysis when determining carrier status in a spouse.

Presymptomatic diagnosis in a sibling of an affected individual is not currently performed, and is clearly excluded for minors. However, the situation may soon change with the identification of effective treatments that may delay or prevent certain disease manifestations.

## B. Biochemical Investigations

A number of biochemical tests are usually performed to exclude the diagnoses of abeta- or hypobetalipoproteinemia, isolated vitamin E deficiency, mitochondrial disorders, adrenoleukodystrophy or adrenomyeloneuropathy, Refsum disease, and a number of lysosomal storage diseases, all of which can be differential diagnoses, particularly at the early stage of the disease. These tests include lipoproteins, vitamin E, lactate, pyruvate, urinary organic acids, serum very long chain fatty acids (VLCFA), serum phytanic acid, and leukocyte and/or fibroblast lysosomal enzymes, which are all normal in Friedreich ataxia.

Attention is currently focused on iron metabolism and oxidative stress (Babcock *et al.*, 1997). Blood iron, iron-binding capacity, and ferritin are normal (Wilson *et al.*, 1998), but there is an increase in circulating transferrin receptors (Wilson *et al.*, 2000). In addition, several markers of oxidative stress are increased, including urinary 8-hydroxy-2′-deoxyguanosine (8OH2′dG), a marker of oxidative DNA damage (Schulz *et al.*, 2000), and plasma malondialdheyde, a marker of lipid peroxidation (Emond *et al.*, 2000).

## C. Neuroimaging

The characteristic neuroimaging finding in Friedreich ataxia is the thinning of the cervical spinal cord. It can be detected on sagittal and axial MRI images, along with signal abnormalities in the posterior and lateral columns (Riva *et al.*, 1995; Wessel *et al.*, 1989; Wullner *et al.*, 1993; Junck *et al.*, 1994; Mascalchi *et al.*, 1994). Brainstem, cerebellum, and cerebrum are apparently normal, but quantitative studies showed that all these structures are smaller in Friedreich ataxia patients compared to healthy controls. The extent of this diffuse atrophy correlates with clinical severity (Junck *et al.*, 1994). Vermian and lobar cerebellar atrophy only occurs in more severe and more advanced cases and is usually mild (Wullner *et al.*, 1993; Giroud *et al.*, 1994). In fact, marked cerebellar atrophy early in the course of the disease is a strong predictor of a negative molecular test for Friedreich ataxia. However, TC-HMPAO single positron emission tomography (SPECT) reveals a decrease in cerebellar blood flow, more than expected for the limited degree of atrophy (Giroud *et al.*, 1994).

Positron emission tomography (PET) scans with fluoro-deoxyglucose (FDG) as tracer reveal an increased glucose uptake in the brain of Friedreich ataxia patients who are still ambulatory (Gilman *et al.*, 1990). Glucose consumption decreases as disease progresses and the patients lose their ability to walk, eventually becoming subnormal (Junck *et al.*, 1994; Gilman *et al.*, 1990). Although not yet fully explained, this abnormality may relate to mitochondrial dysfunction (Babcock *et al.*, 1997).

Proton magnetic resonance spectroscopy ($^1$H-MRS) is starting to be used in Friedreich ataxia. It is a noninvasive technique not requiring the use of radioactive tracers, that is able quantify *in vivo* metabolites present in millimolar concentrations with a spatial resolution down to of a few cubic centimeters. The neuronal marker *N*-acetyl-aspartate (NAA), lactate, choline (Cho), and creatine-phosphocreatine (Cr) are metabolites that can be measured by this technique. In Friedreich ataxia, MRS analysis of skeletal muscle and heart showed a reduced rate of ATP synthesis after exercise (Lodi *et al.*, 1999), which is inversely correlated to GAA expansion sizes, and is improved by coenzyme Q10 treatment (Lodi *et al.*, 2001). There are not yet published data for the brain.

## V. PATHOLOGY

### A. Central Nervous System

The degeneration of the posterior columns of the spinal cord is the hallmark of the disease (Friedreich, 1863a–c; Friedreich, 1876; Friedreich, 1877). The posterior columns are shrunken, grayish, and translucent. Atrophy is more severe in the Goll than in the Burdach tract, therefore the longest fibers originating more caudally are most affected. The spinocerebellar tracts, the dorsal more than the ventral, and their neurons of origin in Clarke's column also become atrophic. In the brainstem, atrophy involves the gracilis and cuneate nuclei and the medial lemnisci, particularly in their ventral portion deriving from the gracile nuclei, as well as the cranial nerve sensory nuclei and entering roots of the V, IX, and X nerves, the descending trigeminal tracts, the solitary tracts, and the accessory cuneate nuclei, corresponding to Clarke's column in the spinal cord. Overall, Friedreich ataxia severely affects the sensory systems providing information to the brain and cerebellum about the position and speed of body segments. The motor system is directly affected as well—the long crossed and uncrossed corticospinal motor tracts are atrophic, more so distally, suggesting a "dying back" process (Said *et al.*, 1986). There is a variable loss of pyramidal neurons in the motor cortex, while motor neurons in the brainstem and in the ventral horns of the spinal cord are less affected.

In the cerebellum, cortical atrophy only occurs late and is mild, while atrophy of the dentate nuclei and of superior cerebellar pedunculi is prominent. Quantitative analysis of synaptic terminals indicates a loss of contacts over Purkinje cell bodies and proximal dendrites (Koeppen, 1991).

The auditory system and the optic nerves and tracts are variably affected (Carroll *et al.*, 1980). The external pallidus and subthalamic nuclei may show a moderate cell loss. Finally, since many patients with Friedreich ataxia die as a consequence of heart disease, widespread hypoxic changes and focal infarcts are often found in the CNS.

In the peripheral nervous system, the major abnormalities occur in the dorsal root ganglia (DRG), where a loss of large primary sensory neurons is observed, accompanied by proliferation of capsule cells that form clumps called "Residualknötchen" of Nageotte. Loss of large myelinated sensory fibers is prominent in peripheral nerves (Geoffroy *et al.*, 1976; Harding, 1981; Said *et al.*, 1986; Harding, 1984; Hughes *et al.*, 1968; Ouvrier *et al.*, 1982; Jitpimolmard *et al.*, 1993), while the fine, unmyelinated fibers are well preserved. Interstitial connective tissue is increased. This sensory axonal neuropathy is an early event in the course of the disease and, according to some authors is scarcely progressive (Santoro *et al.*, 1999).

In summary, patients with Friedreich ataxia have a deficit of deep sensory input to the brain and to the cerebellum as well as a deficient cerebellar output, determining a mixed cerebellar and sensory ataxia. The simultaneous presence of axonal sensory neuropathy and pyramidal tract degeneration causes the typical combination of absent deep tendon reflexes and extensor plantar responses.

### B. Heart

A hypertrophic cardiomyopathy is typically found in Friedreich ataxia (Geoffroy *et al.*, 1976; Harding, 1981; Filla *et al.*, 1990; Muller-Felber *et al.*, 1993; Hewer, 1968; Gottdiener *et al.*, 1982). Ventricular walls and interventricular septum are thickened (Gottdiener *et al.*, 1982; Pasternac *et al.*, 1980). Hypertrophic cardiomyocytes are found early in the disease and are intermingled with normal appearing ones. Then, atrophic, degenerating, even necrotic fibers progressively appear, there is diffuse and focal inflammatory cell infiltration and connective tissue increases. In the late stage of the disease, with extensive fibrosis, the cardiomyopathy becomes dilatative (Casazza *et al.*, 1996). A variable number of cardiomyocytes, from less than 1% to more than 10%, show intracellular iron deposits (Lamarche *et al.*, 1980, 1993). This is a specific finding in Friedreich ataxia and a direct consequence of the basic biochemical defect (Babcock *et al.*, 1997).

### C. Other Organs

More than three quarters of the patients have kyphoscoliosis, mostly as a double thoracolumbar curve resembling the idiopathic form (Schoenle *et al.*, 1989; Allard *et al.*, 1982; Labelle *et al.*, 1986). Pes cavus, pes equinovarus, and clawing of the toes are found in about half the cases

(Geoffroy et al., 1976; Harding, 1981). Patients with diabetes (about 10%) often show a loss if islet cells in the pancreas, which is not accompanied by the autoimmune inflammatory reaction found in type I diabetes (Aronsson et al., 1994).

## VI. CELLULAR AND ANIMAL MODELS OF DISEASE

The yeast model as well as some cellular models were described in a previous section.

The early embryonic lethality of frataxin KO mice (Cossée et al., 2000) has complicated the effort to generate a vertebrate animal model of the disease. A viable mouse model was eventually obtained through a conditional gene-targeting approach. A heart and striated muscle frataxin-deficient line and a line with more generalized, including neural, frataxin-deficiency were recently generated (Puccio et al., 2001). These mice reproduce important progressive pathophysiological and biochemical features of the human disease: cardiac hypertrophy without skeletal muscle involvement in the heart and striated muscle frataxin-deficient line, large sensory neuron dysfunction without alteration of the small sensory and motor neurons in the more generalized frataxin-deficient line, and deficient activities of complexes I–III of the respiratory chain and of the aconitases in both lines. Time-dependent intramitochondrial iron accumulation occurs in the heart of the line that is frataxin-deficient in the heart and striated muscle. These animals provide an important resource for pathophysiological studies and for testing of new treatments. However, they still do not mimic the situation occurring in the human disease because conditional gene targeting leads to complete loss of frataxin in some cells at a specific time in development, while Friedreich ataxia is characterized by partial frataxin deficiency in all cells and throughout life. Therefore, there is still a need to develop new animal models of the disease. An attempt was made to create a model by generating GAA expansion knock-in mice, but animals carrying a (GAA) expansion in heterozygosity with a KO allele still make enough frataxin (20% of normal) to present a normal phenotype (Miranda et al., 2002).

## VII. GENOTYPE/PHENOTYPE CORRELATIONS/ MODIFYING ALLELES

### A. Correlations between GAA Expansions and Phenotype

As expected by the fact that smaller expansions allow a higher residual expression of the frataxin gene (Cossée et al., 1997; Campuzano et al., 1997; Ohshima et al., 1998), the severity and age of onset of the disease are in part determined by the size of the expanded GAA repeat, in particular of the smaller allele. A number of studies (Montermini et al., 1997; Monros et al., 1997; Filla et al., 1996; Dürr et al., 1996; Lamont et al., 1997; Schols et al., 1997) have shown a direct correlation between the size of GAA repeats and earlier age of onset, earlier age of confinement to a wheelchair, more rapid rate of disease progression, and presence of nonobligatory disease manifestations indicative of more widespread degeneration. However, differences in GAA expansions account for only about 50% of the variability in age of onset. Other factors that influence the phenotype may include somatic mosaicism for expansion sizes, variations in the frataxin gene itself, modifier genes, and environmental factors.

### B. Point Mutations in the frataxin Gene

About 2% of the Friedreich's ataxia chromosomes have a normal GAA repeat but carry a missense, nonsense, or splice site mutation ultimately affecting the frataxin coding sequence (Campuzano et al., 1996; Cossé et al., 1997); (Bidichandani et al., 1997). All affected individuals with a point mutation so far identified are heterozygous for an expanded GAA repeat on the other homolog of chromosome 9. Mutations have been found frequently in the initiation ATG codon in exon 1, where changes can involve each of the nucleotides. A stretch of four Cs near the end of exon 1 is another hot spot for mutations, with insertions or deletions detected in several unrelated families (Cossée et al., 1999). Missense mutations have so far been identified only in the C-terminal portion of the protein corresponding to the mature intramitochondrial form of frataxin. Nonsense and most missense mutations result in a typical Friedreich's ataxia phenotype, while a few missense mutations are associated with milder atypical phenotypes with slow progression (Bidichandani et al., 1997; Cossée et al., 1999), suggesting that the mutated proteins preserve some residual function. The G130V mutation in particular is associated with early onset but slow progression, no dysarthria, mild limb ataxia, and retained reflexes (Bidichandani et al., 1997; Cossée et al., 1999). For unclear reasons, optic atrophy is more frequent in patients with point mutations of any kind (50%; Cossée et al., 1999).

## VIII. TREATMENT

There are no established treatments to stop or slow the degenerative process, but new knowledge derived from the characterization of the Friedreich ataxia gene product is starting to change this picture. Based on the hypothesis that iron-mediated oxidative damage plays a major role in the pathogenesis of Friedreich ataxia, removal of excess mitochondrial iron and/or antioxidant treatment may in principle be attempted. However, removal of excess mitochondrial iron is problematic with the currently available drugs. Desferioxamine (DFO) is effective in chelating iron in the extracellular fluid and cytosol, not directly in mitochondria. Furthermore, DFO toxicity may be higher when there is no overall iron overload. Thus, chelation therapy has a number of unknowns—it is probably better tested in pilot trials involving a small number of closely monitored patients. Iron depletion by phlebotomy, though less risky, presents the same uncertainties concerning possible efficacy. As far as antioxidants are concerned, these include a long list of molecules with specific mechanisms of action and pharmacokinetic properties. To have the potential to be effective in FRDA, an antioxidant must protect against the damage caused by the free radicals involved in this disease, in particular OH·, act in the mitochondrial compartment, and be able to cross the blood-brain barrier. At this time, co-enzyme Q (CoQ) derivatives, like its short chain analog idebenone, appear to be interesting molecules and are object of pilot studies (Rustin et al., 1999). These molecules seem to be particularly effective in treating the FRDA cardiomyopathy because they reduce heart size and activate ATP production, as shown by MRS (Lodi et al., 2001).

Attempts for symptomatic treatment have utilized drugs affecting neurotransmitters acting in the cerebellar circuitry. Tested compounds include cholinergic agonists, like physostigmine (Kark et al., 1981) and choline, neuropeptides as TRH (Le Witt et al., 1982; Filla et al., 1989), serotoninergic agonists as 5-OH tryptophan (Wessel et al., 1995; Trouillas et al., 1995) and buspirone, and the dopaminergic drug amantadine (Peterson et al., 1988; Botez et al., 1996). Overall, results have not been encouraging.

Rehabilitation programs should include exercises aimed at maximizing the residual capacity of motor control. Orthopedic interventions are sometimes necessary, including surgical correction of severe scoliosis, and, in patients who can still walk, of foot deformity.

New knowledge of frataxin function and pathogenesis is needed to progress toward an effective treatment of the disease. Pharmacological agents may be identified that counteract specific effects of frataxin deficiency. In the long term, gene replacement, protein replacement, or reactivation of the expression of endogenous frataxin could be cures and are all worth exploring.

## References

Ackroyd, R. S., Finnegan, J. A., and Green, S. H. (1984). Friedreich ataxia. A clinical review with neurophysiological and echocardiographic findings. *Arch. Dis. Child.* **59**, 217–221.

Adamec, J., Rusnak, F., Owen, W. G., Naylor, S., Benson, L. M., Gacy, A. M., and Isaya, G. (2000). Iron-dependent self-assembly of recombinant yeast frataxin: implications for Friedreich ataxia. *Am. J. Hum. Genet.* **67**, 549–562.

Alboliras, E. T., Shub, C., Gomez, M. R., Edwards, W. D., Hagler, D. J., Reeder, G. S., Seward, J. B., and Tajik, A. J. (1986). Spectrum of cardiac involvement in Friedreich ataxia: clinical, electrocardiographic and echocardiographic observations. *Am. J. Cardiol.* **58**, 518–524.

Allard, P., Dansereau, J., Thiry, P. S., Geoffroy, G., Raso, J. V., and Duhaime, M. (1982). Scoliosis in Friedreich ataxia. *Can. J. Neurol. Sci.* **9**, 105–111.

Aronsson, D. D., Stokes, I. A., Ronchetti, P. J., and Labelle, H. B. (1994). Comparison of curve shape between children with cerebral palsy, Friedreich ataxia, and adolescent idiopathic scoliosis. *Dev. Med. Child Neurol.* **36**, 412–418.

Babcock, M., de Silva, D., Oaks, R., Davis-Kaplan, S., Jiralerspong, S., Montermini, L., Pandolfo, M., and Kaplan, J. (1997). Regulation of mitochondrial iron accumulation by Yfh1, a putative homolog of frataxin. *Science* **276**, 1709–1712.

Barbeau, A., Roy, M., Sadibelouiz, M., and Wilensky, M. A. (1984). Recessive ataxia in Acadians and "Cajuns." *Can. J. Neurol. Sci.* **11**, 526–533.

Beauchamp, M., Labelle, H., Duhaime, M., and Joncas, J. (1995). Natural history of muscle weakness in Friedrich's Ataxia and its relation to loss of ambulation. *Clin. Orthop.* **311**, 270–275.

Bidichandani, S. I., Ashizawa, T., and Patel, P. I. (1997). Atypical Friedreich ataxia caused by compound heterozygosity for a novel missense mutation and the GAA triplet-repeat expansion. *Am. J. Hum. Genet.* **60**, 251–1256.

Botez, M. I., Botez-Marquard, T., Elie, R., Pedraza, O. L., Goyette, K., and Lalonde, R. (1996). Amantadine hydrochloride treatment in heredodegenerative ataxias: a double blind study. *J. Neurol. Neurosurg. Psychiatry* **61**, 259–264.

Bouchard, J. P., Barbeau, A., Bouchard, R., Paquet, M., and Bouchard, R. W. (1979). A cluster of Friedreich ataxia in Rimouski, Quebec. *Can. J. Neurol. Sci.* **6**, 205–208.

Boyer, S. H., Chisholm, A. W., and McKusick, V. A. (1962). Cardiac aspects of Friedreich ataxia. *Circulation* **25**, 493–505.

Campanella, G., Filla, A., De Falco, F., Mansi, D., Durivage, A., and Barbeau, A. (1980). Friedreich ataxia in the south of Italy: A clinical and biochemical survey of 23 patients. *Can. J. Neurol. Sci.* **7**, 351–357.

Campuzano, V., Montermini, L., Moltó, M. D., Pianese, L., Cossée, M., Cavalcanti, F., Monros, E., Rodius, F., Duclos, F., Monticelli, A., Zara, F., Cañizares, J., Koutnikova, H., Bidichandani S, Gellera, C., Brice, A., Trouillas, P., De Michele, G., Filla, A., de Frutos, R., Palau, F., Patel, P. I., Di Donato, S., Mandel, J. L., Cocozza, S., Koenig, M., and Pandolfo, M. (1996). Friedreich ataxia: autosomal recessive disease caused by an intronic GAA triplet repeat expansion. *Science* **271**, 1423–1427.

Campuzano, V., Montermini, L., Lutz, Y., Cova, L., Hindelang, C., Jiralerspong, S., Trottier, Y., Kish, S. J., Faucheux, B., Trouillas, P., Authier, F. J., Dürr, A., Mandel, J. L., Vescovi, A. L., Pandolfo, M., and Koenig, M. (1997). Frataxin is reduced in Friedreich ataxia patients and is associated with mitochondrial membranes. *Hum. Mol. Genet.* **6**, 1771–1780.

Carroll, W. M., Kriss, A., Baraitser, M., Barrett, G., and Halliday, A. M. (1980). The incidence and nature of visual pathway involvement in Friedreich ataxia. A clinical and visual evoked potential study of 22 patients. *Brain* **103**, 413–434.

Casazza, F., and Morpurgo, M. (1996). The varying evolution of Friedreich ataxia cardiomyopathy. *Am. J. Cardiol.* **77**, 895–898.

Cassandro, E., Mosca, F., Sequino, L., De Falco, F.A., and Campanella, G. (1986). Otoneurological findings in Friedreich ataxia and other inherited neuropathies. *Audiology* **25**, 84–91.

Chamberlain, S., Shaw, J., Rowland, A., Wallis, J., South, S., Nakamura, Y., von Gabain, A., Farrall, M., and Williamson, R. (1988). Mapping of mutation causing Friedreich ataxia to human chromosome 9. *Nature* **334**, 248–250.

Child, J. S., Perloff, J. K., Bach, P. M., Wolfe, A. D., Perlman, S., and Kark, R. A. (1986). Cardiac involvement in Friedreich ataxia: a clinical study of 75 patients. *J. Am. Coll. Cardiol.* **7**, 1370–1378.

Cho, S. J., Lee, M. G., Yang, J. K., Lee, J. Y., Song, H. K., and Suh, S. W. (2000). Crystal structure of Escherichia coli CyaY protein reveals a previously unidentified fold for the evolutionarily conserved frataxin family. *Proc. Natl. Acad. Sci. U.S.A.* **97**, 8932–8937.

Cisneros, E., and Braun, C. M. (1995). Vocal and respiratory diadochokinesia in Friedreich ataxia. Neuropathological correlations. *Rev. Neurol. (Paris)* **151**, 113–123.

Cossée, M., Campuzano, V., Koutnikova, H., Fischbeck, K. H., Mandel, J.-L., Koenig, M., Bidichandani, S., Patel, P. I., Moltó, M. D., Cañizares, J., de Frutos, R., Pianese, L., Cavalcanti, F., Monticelli, A., Cocozza, S., Montermini, L., and Pandolfo, M. (1997). Frataxin fracas. *Nat. Genet.* **15**, 337–338.

Cossée, M., Dürr, A., Schmitt, M., Dahl, N., Trouillas, P., Allinson, P., Kostrzewa, M., Nivelon-Chevallier, A., Gustavson, K.-H., Kohlschütter, A., Müller, U., Mandel, J.-L., Brice, A., Koenig, M., Cavalcanti, F., Tammaro, A., De Michele, G., Filla, A., Cocozza, S., Labuda, M., Montermini, L., Poirier, J., and Pandolfo, M. (1999). Frataxin point mutations and clinical presentation of compound heterozygous Friedreich ataxia patients. *Ann. Neurol.* **45**, 200–206.

Cossée, M., Puccio, H., Gansmuller, A., Koutnikova, H., Dierich, A., LeMeur, M., Fischbeck, K., Dolle, P., and Koenig, M. (2000). Inactivation of the Friedreich ataxia mouse gene leads to early embryonic lethality without iron accumulation. *Hum. Mol. Genet.* **9**, 1219–1226.

D'Angelo, A., Di Donato, S., Negri, G., Beulche, F., Uziel, G., and Boeri, R. (1980). Friedreich ataxia in northern Italy: I. Clinical, neurophysiological and in vivo biochemical studies. *Can. J. Neurol. Sci.* **7**, 359–365.

De Michele, G., Filla, A., Cavalcanti, F., Di Maio, L., Pianese, L., Castaldo, I., Calabrese, O., Monticelli, A., Varrone, S., Campanella, G., and Cocozza, S. (1994). Late onset Friedreich disease: clinical features and mapping of mutation to the FRDA locus. *J. Neurol. Neurosurg. Psychiatry* **57**, 977–979.

De Michele, G., Di Maio, L., Filla, A., Majello, M., Cocozza, S., Cavalcanti, F., Mirante, E., and Campanella, G. (1996a). Childhood onset of Friedreich ataxia: a clinical and genetic study of 36 cases. *Neuropediatrics* **27**, 3–7.

De Michele, G., Perrone, F., Filla, A., Mirante, E., Giordano, M., De Placido, S., and Campanella, G. (1996b). Age of onset, sex, and cardiomyopathy as predictors of disability and survival in Friedreich disease: a retrospective study on 119 patients. *Neurology* **47**, 1260–1264.

Dean, G., Chamberlain, S., and Middleton, L. (1988). Friedreich ataxia in Kathikas-Arodhes, Cyprus. *Lancet* **1**, 587.

Delatycki, M., Camakaris, J., Brooks, H., Evans-Whipp, T., Thorburn, D. R., Williamson, R., and Forrest, S. M. (1999). Direct evidence that mitochondrial iron accumulation occurs in Friedreich ataxia. *Ann. Neurol.* **45**, 673–675.

Dhe-Paganon, S., Shigeta, R., Chi, Y. I., Ristow, M., and Shoelson, S. E. (2000). Crystal structure of human frataxin. *J. Biol. Chem.* **275**, 30753–30756.

Dürr, A., Cossée, M., Agid, Y., Campuzano, V., Mignard, C., Penet, C., Mandel, J.-L., Brice, A., and Koenig, M. (1996). Clinical and genetic abnormalities in patients with Friedreich ataxia. *N. Engl. J. Med.* **335**, 1169–1175.

Ell, J., Prasher, D., and Rudge, P. (1984). Neuro-otological abnormalities in Friedreich ataxia. *J. Neurol. Neurosurg. Psychiatry* **47**, 26–32.

Emond, M., Lepage, G., Vanasse, M., and Pandolfo, M. (2000). Increased levels of plasma malondialdehyde in Friedreich ataxia. *Neurology* **55**, 1752–1753.

Fantus, I. G., Seni, M. H., and Andermann, E. (1993). Evidence for abnormal regulation of insulin receptors in Friedreich ataxia. *J. Clin. Endocrinol. Metab.* **76**, 60–63.

Filla, A., De Michele, G., Di Martino, L., Mengano, A., Iorio, L., Maggio, M.A., and Campanella, G. (1989). Chronic experimentation with TRH administered intramuscularly in spinocerebellar degeneration. Double-blind cross-over study in 30 subjects. *Riv. Neurol.* **59**, 83–88.

Filla, A., De Michele, G., Caruso, G., Marconi, R., and Campanella, G. (1990). Genetic data and natural history of Friedreich disease: a study of 80 Italian patients. *J. Neurol.* **237**, 345–351.

Filla, A., De Michele, G., Cavalcanti, F., Perretti, A., Santoro, L., Barbieri, F., D'Arienzo, G., and Campanella, G. (1990). Clinical and genetic heterogeneity in early onset cerebellar ataxia with retained tendon reflexes. *J. Neurol. Neurosurg. Psychiatry* **53**, 667–670.

Filla, A., De Michele, G., Cavalcanti, F., Pianese, L., Monticelli, A., Campanella, G., and Cocozza, S. (1996). The relationship between trinucleotide (GAA) repeat length and clinical features in Friedreich ataxia. *Am. J. Hum. Genet* **59**, 554–560.

Finocchiaro, G., Baio, G., Micossi, P., Pozza, G., and Di Donato, S. (1988). Glucose metabolism alterations in Friedreich ataxia. *Neurology* **38**, 1292–1296.

Friedreich, N. (1863a). Über degenerative Atrophie der spinalen Hinterstränge. *Virchows Arch. Pathol. Anat.* **27**, 1–26.

Friedreich, N. (1863b). Über degenerative Atrophie der spinalen Hinterstränge. *Virchows Arch. Pathol. Anat.* **26,** 433–459.

Friedreich, N. (1863c). Über degenerative Atrophie der spinalen Hinterstränge. *Virchows Arch. Pathol. Anat.* **26**, 391–419.

Friedreich, N. (1876). Über ataxie mit besonderer berücksichtigung der hereditären formen. *Virchows Arch. Pathol. Anat.* **68**, 145–245.

Friedreich, N. (1877). Über ataxie mit besonderer berücksichtigung der hereditären formen. *Virchows Arch. Pathol. Anat.* **70**, 140–142.

Gentil, M. (1990). Dysarthria in Friedreich disease. *Brain Lang.* **38**, 438–448.

Geoffroy, G., Barbeau, A., Breton, G., Lemieux, B., Aube, M., Leger, C., and Bouchard, J. P. (1976). Clinical description and roentgenologic evaluation of patients with Friedreich ataxia. *Can. J. Neurol. Sci.* **3**, 279–286.

Gilman, S., Junck, L., Markel, D. S., Koeppe, R. A., and Kluin, K. J. (1990). Cerebral glucose hypermetabolism in Friedreich ataxia detected with positron emission tomography. *Ann. Neurol.* **28**, 750–757.

Giroud, M., Septien, L., Pelletier, J. L., Dueret, N., and Dumas, R. (1994). Decrease in cerebellar blood flow in patients with Friedreich ataxia: A TC-HMPAO SPECT study of three cases. *Neurol. Res.* **16**, 342–344.

Gottdiener, J. S., Hawley, R. J., Maron, B.J., Bertorini, T. F., and Engle, W. K. (1982). Characteristics of the cardiac hypertrophy in Friedreich ataxia. *Am. Heart J.* **103**, 525–531.

Hadjivassiliou, M., Grünewald, R. A., Chattopadhyay, A. K., Davies-Jones, G. A., Gibson, A., Jarratt, J. A., Kandler, R. H., Lobo, A., Powell, T., and Smith, C.M.L. (1998). Clinical, radiological, neurophysiological, and neuropathological characteristics of gluten ataxia. *Lancet* **352**, 1582–1585.

Harding, A. E., and Zilkha, K. J. (1981). "Pseudo-dominant" inheritance in Friedreich ataxia. *J. Med. Genet.* **18**, 285–287.

Harding, A. E. (1981). Friedreich ataxia: a clinical and genetic study of 90 families with an analysis of early diagnosis criteria and intrafamilial clustering of clinical features. *Brain* **104**, 589–620.

Harding, A. E. (1983). Classification of the hereditary ataxias and paraplegias. *Lancet* **1**, 1151–1155.

Harding, A. E., and Hewer, R. L. (1983). The heart disease of Friedreich ataxia. A clinical and electrocardiographic changes in 30 cases. *Q. J. Med.* **52**, 489–502.

Harding, A. E. (1984). "The Hereditary Ataxias and Related Disorders." Churchill Livingstone, London.

Hartman, J. M., and Booth, R. W. (1960). Friedreich ataxia: a neurocardiac disease. *Am. Heart J.* **60**, 716–720.

Hewer, R. L. (1968). Study of fatal cases of Friedreich ataxia. *Br. Med. J.* **3**, 649–652.

Hughes, J. T., Brownell, B., and Hewer, R. L. (1968). The peripheral sensory pathway in Friedreich ataxia. An examination by light and electron microscopy of the posterior nerve roots, posterior root ganglia, and peripheral sensory nerves in cases of Friedreich ataxia. *Brain* **91**, 803–818.

Jiralerspong, S., Liu, Y., Montermini, L., Stifani, S., and Pandolfo, M. (1997). Frataxin shows developmentally regulated tissue-specific expression in the mouse embryo. *Neurobiol. Dis.* **4**, 103–113.

Jiralerspong, S., Ge, B., Hudson, T. J., and Pandolfo, M. (2001). Manganese superoxide dismutase induction by iron is impaired in friedreich ataxia cells. *FEBS Lett.* **509**, 101–105.

Jitpimolmard, S., Small, J., King, R. H., Geddes, J., Misra, P., McLaughlin, J., Muddle, J. R., Cole, M., Harding, A. E., and Thomas, P. K. (1993). The sensory neuropathy of Friedreich ataxia: an autopsy study of a case with prolonged survival. *Acta Neuropathol.* (Berlin). **86**, 29–35.

Junck, L., Gilman, S., Gebarski, S. S., Koeppe, R. A., Kluin, K., and Markel, D. S. (1994). Structural and functional brain imaging in Friedreich ataxia. *Arch. Neurol.* **51**, 349–355.

Kark, R. A., Budelli, M. M., and Wachsner, R. (1981). Double-blind, triple-crossover trial of low doses of oral physostigmine in inherited ataxias. *Neurology* **31**, 288–292.

Kirkham, T. H., and Coupland, S. G. (1981). An electroretinal and visual evoked potential study in Friedreich ataxia. *Can. J. Neurol. Sci.* **8**, 289–294.

Klockgether, T., Chamberlain, S., Wullner, U., Fetter, M., Dittmann, H., Petersen, D., and Dichgans, J. (1993). Late-onset Friedreich ataxia. Molecular genetics, clinical neurophysiology, and magnetic resonance imaging. *Arch. Neurol.* **50**, 803–806.

Koeppen, A. (1991). The Purkinje cell and its afferents in human hereditary ataxia. *J. Neuropathol. Exp. Neurol.* **50**, 505–514.

Koutnikova, H., Campuzano, V., Foury, F., Dollé, P., Cazzalini, O., and Koenig, M. (1997). Studies of human, mouse and yeast homologues indicate a mitochondrial function for frataxin. *Nat. Genet.* **16**, 345–351.

Labelle, H., Tohme, S., Duhaime, M., and Allard, P. (1986). Natural history of scoliosis in Friedreich ataxia. *J. Bone Joint Surg.* (Am) **68**, 564–572.

Lamarche, J. B., Côté, M., and Lemieux, B. (1980). The cardiomyopathy of Friedreich ataxia morphological observations in 3 cases. *Can. J. Neurol. Sci.* **7**, 389–396.

Lamarche, J. B., Shapcott, D., Côté, M., and Lemieux, B. (1993). Cardiac iron deposits in Friedreich ataxia. In "Handbook of Cerebellar Diseases" (R. Lechtenberg, ed.), pp. 453–458. Marcel Dekker, New York.

Lamont, P. J., Davis, M. B., and Wood, N. W. (1997). Identification and sizing of the GAA trinucleotide repeat expansion of Friedreich ataxia in 56 patients—clinical and genetic correlates. *Brain* **120**, 673–680.

Le Witt, P. A., and Ehrenkranz, J. R. (1982). TRH and spinocerebellar degeneration [letter]. *Lancet* **2**, 981.

Leone, M., Rocca, W. A., Rosso, M. G., Mantel, N., Schoenberg, B. S., and Schiffer, D. (1988). Friedreich disease: survival analysis in an Italian population. *Neurology* **38**, 1433–1438.

Leone, M., Brignolio, F., Rosso, M. G., Curtoni, E. S., Moroni, A., Tribolo, A., and Schiffer, D. (1990). Friedreich ataxia: a descriptive epidemiological study in an Italian population. *Clin. Genet.* **38**, 161–169.

Livingstone, I. R., Mastaglia, F. L., Edis, R., and Howe, J. W. (1981). Visual involvement in Friedreich ataxia and hereditary spastic ataxia. A clinical and visual evoked response study. *Arch. Neurol.* **38**, 75–79.

Lodi, R., Cooper, J. M., Bradley, J. L., Manners, D., Styles, P., Taylor, D., and Schapira, A. H. (1999). Deficit of in vivo mitochondrial ATP production in patients with Friedreich ataxia. *Proc. Natl. Acad. Sci. U.S.A.* **96**, 11492–11495.

Lodi, R., Hart, P. E., Rajagopalan, B., Taylor, D. J., Crilley, J. G., Bradley, J. L., Blamire, A. M., Manners, D., Sytles, P., Schapira, A. H. V., and Cooper, J. M. (2001). Antioxidant treatment improves in vivo cardiac and skeletal muscle bioenergetics in patients with Friedreich's ataxia. *Ann. Neurol.* **49**, 590–596.

Lopez-Arlandis, J. M., Vilchez, J. J., Palau, F., and Sevilla, T. (1995). Friedreich ataxia: an epidemiological study in Valencia, Spain, based on consanguinity analysis. *Neuroepidemiology* **14**, 14–19.

Maione, S., Giunta, A., Filla, A., De Michele, G., Spinelli, L., Liucci, G. A., Campanella, G., and Condorelli, M. (1997). May age onset be relevant in the occurrence of left ventricular hypertrophy in Friedreich ataxia? *Clin. Cardiol.* **20**, 141–145.

Margalith, D., Dunn, H. G., Carter, J. E., and Wright, J. M. (1984). Friedreich ataxia with dysautonomia and labile hypertension. *Can. J. Neurol. Sci.* **11**, 73–77.

Mascalchi, M., Salvi, F., Piacentini, S., and Bartolozzi, C. (1994). Friedreich ataxia: MR findings involving the cervical portion of the spinal cord. *AJR Am. J. Roentgenol.* **163**, 187–191.

McLeod, J. G. (1971). An electrophysiological and pathological study of peripheral nerves in Friedreich ataxia. *J. Neurol. Sci.* **12**, 333–349.

Miranda, C., Santos, M. M., Ohshima, K., Smith, J., Koenig, M., Sequeiros, J., Kaplan, J., and Pandolfo, M. (2002). Frataxin knock-in mice. *FEBS Lett.* **512**, 291–297.

Mondelli, M., Rossi, A., Scarpini, C., and Guazzi, G. C. (1995). Motor evoked potentials by magnetic stimulation in hereditary and sporadic ataxia. *Electromyogr. Clin. Neurophysiol.* **35**, 415–424.

Monros, E., Moltó, M. D., Martinez, F., *et al.* (1997). Phenotype correlation and intergenerational dynamics of the Friedreich ataxia GAA trinucleotide repeat. *Am. J. Hum. Genet.* **61**, 101–110.

Montermini, L., Richter, A., Morgan, K., Justice, C. M., Julien, D., Castelloti, B., Mercier, J., Poirier, J., Capazzoli, F., Bouchard, J. P., Lemieux, B., Mathieu, J., Vanasse, M., Seni, M. H., Graham, G., Andermann, F., Andermann, E., Melançon, S., Keats, B. J. B., Di Donato, S., and Pandolfo, M. (1997). Phenotypic variability in Friedreich ataxia: role of the associated GAA triplet repeat expansion. *Ann. Neurol.* **41**, 675–682.

Morvan, D., Komajda, M., Doan, L. D., Brice, A., Isnard, R., Seck, A., Lechat, P., Agid, Y., and Grosgogeat, Y. (1992). Cardiomyopathy in Friedreich ataxia: a Doppler-echocardiographic study. *Eur. Heart J.* **13**, 1393–1398.

Moseley, M. L., Benzow, K. A., Schut, L. J., Bird, T., Gomez, C., Barkhaus, P. E., Blindauer, K., Labuda, M., Pandolfo, M., Koob, M. D., and Ranum, L.P.W. (1998). Incidence of dominant spinocerebellar and Friedreich triplet repeats among 361 ataxia families. *Neurology* **51**, 1666–1671.

Muller-Felber, W., Rossmanith, T., Spes, C., Chamberlain, S., Pongratz, D., and Deufel, T. (1993). The clinical spectrum of Friedreich ataxia in German families showing linkage to the FRDA locus on chromosome 9. *Clin. Investig.* **71**, 109–114.

Musco, G., Stier, G., Kolmerer, B., Adinolfi, S., Martin, S., Frenkiel, T., Gibson, T., and Pastore, A. (2000). Towards a structural understanding of Friedreich ataxia: the solution structure of frataxin. *Structure Fold. Des.* **8**, 695–707.

Ohshima, K., Montermini, L., Wells, R. D., and Pandolfo, M. (1998). Inhibitory effects of expanded GAA•TTC triplet repeats from intron I of the Friedreich ataxia gene on transcription and replication in vivo. *J. Biol. Chem.* **273**, 14588–15595.

Ohshima, K., Sakamoto, N., Labuda, M., Poirier, J., Moseley, M. L., Montermini, L., Ranum, L. P. W., Wells, R. D., and Pandolfo, M. (1999). A non-pathogenic GAAGGA repeat in the frataxin gene: implications for the Friedreich ataxia GAA triplet repeat. *Neurology* **53**, 1854–1857.

Ouvrier, R. A., McLeod, J. G., and Conchin, T. E. (1982). Friedreich ataxia. Early detection and progression of peripheral nerve abnormalities. *J. Neurol. Sci.* **55**, 137–145.

Palau, F., De Michele, G., Vilchez, J. J., Pandolfo, M., Monros, E., Cocozza, S., Smeyers, P., Lopez-Arlandis, J., Campanella, G., Di Donato, S., and Filla, A. (1997). Early-onset ataxia with cardiomyopathy and retained tendon reflexes maps to Friedreich ataxia locus on chromosome 9q. *Ann. Neurol.* **37**, 359–362.

Pandolfo, M. (1998). A reappraisal of the clinical features of Friedreich ataxia: which indications for a molecular test? *Neurology* **52** (Suppl. 2), A260.

Pasternac, A., Krol, R., Petitclerc, R., Harvey, C., Andermann, E., and Barbeau, A. (1980). Hypertrophic cardiomyopathy in Friedreich ataxia: symmetric or asymmetric? *Can. J. Neurol. Sci.* **7**, 379–382.

Pellecchia, M. T., Scala, R., Filla, A., De Michele, G., Ciacci, C., and Barone, P. (1999). Idiopathic cerebellar ataxia associated with celiac disease: lack of distinctive neurological features. *J. Neurol. Neurosurg. Psychiatry* **66**, 32–35.

Pentland, B., and Fox, K. A. (1983). The heart in Friedreich ataxia. *J. Neurol. Neurosurg. Psychiatry* **46**, 1138–1142.

Peterson, P. L., Saad, J., and Nigro, M. A. (1988). The treatment of Friedreich ataxia with amantadine hydrochloride [see comments]. *Neurology* **38**, 1478–1480.

Peyronnard, J. M., Bouchard, J. P., and Lapointe, M. (1976). Nerve conduction studies and electromyography in Friedreich ataxia. *Can. J. Neurol. Sci.* **3**, 313–317.

Pousset, F., Kalotka, H., Durr, A., Isnard, R., Lechat, P., Le Heuzey, J. Y., Thomas, D., and Komajda, M. (1996). Parasympathetic activity in Friedrich's ataxia. *Am. J. Cardiol.* **78**, 847–850.

Puccio, H., Simon, D., Cossee, M., Criqui-Filipe, P., Tiziano, F., Melki, J., Hindelang, C., Matyas, R., Rustin, P., and Koenig, M. (2001). Mouse models for Friedreich ataxia exhibit cardiomyopathy, sensory nerve defect and Fe-S enzyme deficiency followed by intramitochondrial iron deposits. *Nat. Genet.* **27**, 181–186.

Rabiah, P. K., Bateman, J. B., Demer, J. L., and Perlman, S. (1997). Ophthalmologic findings in patients with ataxia. *Am. J. Ophthalmol.* **123**, 108–117.

Riva, A., and Bradac, G. B. (1995). Primary cerebellar and spinocerebellar ataxia an MRI study on 63 cases. *J. Neuroradiol.* **22**, 71–76.

Romeo, G., Menozzi, P., Ferlini, A., Fadda, S., Di Donato, S., Uziel, G., and Lucci, B. (1983). Incidence of Friedreich ataxia in Italy estimated from consanguineous marriages. *Am. J. Hum. Genet.* **35**, 52–529.

Rötig, A., deLonlay, P., Chretien, D., Foury, F., Koenig, M., Sidi, D., Munnich, A., and Rustin, P. (1997). Frataxin gene expansion causes aconitase and mitochondrial iron-sulfur protein deficiency in Friedreich ataxia. *Nat. Genet.* **17**, 215–217.

Rustin, P., von Kleist-Retzow, J. C., Chantrel-Groussard, K., Sidi, D., Munnich, A., and Rotig, A. (1999). Effect of idebenone on cardiomyopathy in Friedreich ataxia: a preliminary study. *Lancet* **354**, 477–479.

Said, G., Marion, M. H., Selva, J., and Jamet, C. (1986). Hypotrophic and dying-back nerve fibers in Friedreich ataxia. *Neurology* **36**, 1292–1299.

Sakamoto, N., Chastain, P. D., Parniewski, P., Ohshima, K., Pandolfo, M., Griffith, J. D., and Wells, R. D. (1999). Sticky DNA: self-association properties of long GAA•TTC repeats in R•R•Y triplex structures from Friedreich ataxia. *Mol. Cell* **3**, 465–475.

Sakamoto, N. K., Ohshima, K. L., Montermini, L. M., Pandolfo, M., and Wells, R. D. (2001). Sticky DNA, a self-associated complex formed at long GAA•TTC repeats in intron 1 of the Frataxin gene, inhibits transcription. *J. Biol. Chem.* **276**, 27171–27177.

Santoro, L., De Michele, G., Perretti, A. et al. (1999). Relation between trinucleotide GAA repeat length and sensory neuropathy in Friedreich ataxia. *J. Neurol. Neurosurg. Psychiatry* **66**, 93–96.

Santos, M., Ohshima, K., and Pandolfo, M. (2001). Frataxin deficiency enhances apoptosis in cells differentiating into neuroectoderm. *Hum. Mol. Genet.* **10**, 1935–1944.

Schoenle, E. J., Boltshauser, E. J., Baekkeskov, S., Landin Olsson, M., Torresani, T., and von Felten, A. (1989). Preclinical and manifest diabetes mellitus in young patients with Friedreich ataxia: no evidence of immune process behind the islet cell destruction. *Diabetologia* **32**, 378–381.

Schols, L., Amoiridis, G., Przuntek, H., Frank, G., Epplen, J. T., and Epplen, C. (1997). Friedreich ataxia. Revision of the phenotype according to molecular genetics. *Brain* **120**, 2131–2140.

Schulz, J. B., Dehmer, T., Schols, L., Mende, H., Hardt, C., Vorgerd, M., Burk, K., Matson, W., Dichgans, J., Beal, M. F., and Bogdanov, M. B. (2000). Oxidative stress in patients with Friedreich ataxia. *Neurology* **55**, 1719–1721.

Skre, H. (1975). Friedreich ataxia in western Norway. *Clin. Genet.* **7**, 287–298.

Smeyers, P., Monros, E., Vilchez, J., Lopez-Arlandis, J., Prieto, F., and Palau, F. (1996). A family segregating a Friedreich ataxia phenotype that is not linked to the FRDA locus. *Hum. Genet.* **97**, 824–828.

Spieker, S., Schulz, J. B., Petersen, D., Fetter, M., Klockgether, T., and Dichgans, J. (1995). Fixation instability and oculomotor abnormalities in Friedreich ataxia. *J. Neurol.* **242**, 517–521.

Trouillas, P., Serratrice, G., Laplane, D., Rascol, A., Augustin, P., Barroche, G., Clanet, M., Degos, C. F., Desnuelle, C., and Dumas, R. (1995). Levorotatory form of 5-hydroxytryptophan in Friedreich ataxia. Results of a double-blind drug-placebo cooperative study. *Arch. Neurol.* **52**, 456–460.

Ulku, A., Arac, N., and Ozeren, A. (1988). Friedreich ataxia: a clinical review of 20 childhood cases. *Acta Neurol. Scand.* **77**, 493–497.

Vanasse, M., Garcia-Larrea, L., Neuschwander, P., Trouillas, P., and Mauguiere, F. (1988). Evoked potential studies in Friedreich ataxia and progressive early onset cerebellar ataxia. *Can. J. Neurol. Sci.* **15**, 292–298.

Wessel, K., Schroth, G., Diener, H. C., Muller-Forell, W., and Dichgans, J. (1989). Significance of MRI-confirmed atrophy of the cranial spinal cord in Friedreich ataxia. *Eur. Arch. Psychiatry Neurol. Sci.* **238**, 225–230.

Wessel, K., Hermsdorfer, J., Deger, K., Herzog, T., Huss, G. P., Kompf, D., Mai, N., Schimrigk, K., Wittkamper, A., and Ziegler, W. (1995). Double-blind crossover study with levorotatory form of hydroxytryptophan in patients with degenerative cerebellar diseases. *Arch. Neurol.* **52**, 451–455.

White, M., Lalonde, R., and Botez-Marquard, T. (2000). Neuropsychologic and neuropsychiatric characteristics of patients with Friedreich's ataxia. *Acta Neurol. Scand.* **102**, 222–226.

Wilson, R. B., Lynch, D. R., and Fishbeck, K. H. (1998). Normal serum iron and ferritin concentrations in patients with Friedreich's ataxia. *Ann. Neurol.* **44**, 132–134.

Wilson, R. B., Lynch, D. R., Farmer, J. M., Brooks, D. G., and Fischbeck, K. H. (2000). Increased serum transferrin receptor concentrations in Friedreich ataxia. *Ann. Neurol.* **47**, 659–661.

Wong, A., Yang, J., Cavadini, P., Gellera, C., Lonnerdal, B., Taroni, F., and Cortopassi, G. (1999). The Friedreich ataxia mutation confers cellular sensitivity to oxidant stress which is rescued by chelators of iron and calcium and inhibitors of apoptosis. *Hum. Mol. Genet.* **8**, 425–430.

Wullner, U., Klockgether, T., Petersen, D., Naegele, T., and Dichgans, J. (1993). Magnetic resonance imaging in hereditary and idiopathic ataxia [see comments]. *Neurology* **43**, 318–325.

# CHAPTER 19

# Familial Ataxia with Isolated Vitamin E Deficiency (AVED)

**FAYCAL HENTATI**
Department of Neurology
Institut National de Neurologie
La Rabta, Tunis, Tunisia

**RIM AMOURI**
Department of Neurology
Institut National de Neurologie
La Rabta, Tunis, Tunisia

**MONCEF FEKI**
Hopital La Rabta
Servie Biochimie
Tunis, Tunisia

**SANA GABSI-GHERAIRI**
Department of Neurology
Institut National de Neurologie
La Rabta, Tunis, Tunisia

**SAMIR BELAL**
Department of Neurology
Institut National de Neurologie
La Rabta, Tunis, Tunisia

I. Introduction
II. Phenotype
    A. Epidemiological Data
    B. Clinical Phenotype
III. Gene
    A. Normal Gene Function
    B. Abnormal Gene Function
IV. Diagnostic and Ancillary Tests
    A. DNA Tests
    B. Other Laboratory Tests
    C. Neuroimaging
    D. Neuropathology
V. Cellular and Animal Models of the Disease
VI. Genotype/Phenotype Correlations—Modifying Alleles
VII. Treatment
    Acknowledgments
    References

Familial isolated vitamin E deficiency, also called ataxia with vitamin E deficiency (AVED), is a rare autosomal recessive neurodegenerative disease characterized, in the typical form, by a progressive cerebellar ataxia, areflexia, deep sensory disturbances, and pyramidal signs resembling Friedreich's ataxia with very reduced serum vitamin E levels in the absence of fat malabsorption. However, wide clinical variability could be observed ranging from a severe Friedreich's ataxia-like presentation to mild neurological impairment or very late onset of the disease. Sensory-evoked potentials (SEPs) show a severe involvement of the posterior columns, whereas peripheral nerve conduction studies give evidence of a mild axonal neuropathy. Pathological lesions consist of generalized neuronal lipofuscin storage associated with a marked degeneration of spinal tracts, predominantly the posterior columns.

The defective gene located on chromosome 8q13.1–q13.3, encodes a hepatic α-tocopherol transfer protein (α-TTP). This protein is involved in vitamin E transfer into circulating lipoproteins. Low serum vitamin E levels in AVED patients are caused by the failure to incorporate α-tocopherol (vitamin E) into very low density protein (VLDL) in liver because of α-TTP gene defects. The largest group of AVED patients is found in North Africa and shares a common genetic mutation, the 744delA mutation. However, different mutations have been found among various ethnic groups in Europe, North America, and Asia. Vitamin E supplementation may stop disease progression, especially when administered in the early stages of the disease.

## I. INTRODUCTION

Burck et al. (1981) first described a patient with progressive spinocerebellar syndrome and isolated vitamin E deficiency without evidence of lipid malabsorption. Subsequently, Laplante et al., (1984), Harding et al. (1985), Stumpf et al., Krendel et al., and Yokota et al. (1987), and

Sokol et al. (1988) reported further patients with cerebellar ataxia associated with selective vitamin E deficiency. The gene for Friedreich's ataxia was mapped to chromosome 9q in 1988 (Chamberlain et al., 1988). Ben Hamida et al. (1993) excluded genetic linkage of two Tunisian families with Friedreich's ataxia phenotype and selective vitamin E deficiency from the Friedreich's ataxia locus. This new entity was called "Friedreich ataxia with vitamin E deficiency" (FAVED); (Ben Hamida et al., 1993). In the same year, the FAVED locus of was mapped to chromosome 8q (Ben Hamida et al., 1993).

Arita et al. (1995) identified the gene sequence of hepatic α-tocopherol transfer protein (α-TTP) by cDNA cloning and mapped this gene to chromosome 8q (Arita et al., 1995). Ouahchi et al. (1995) demonstrated that mutations in this gene were responsible for FAVED. FAVED patients had no or reduced levels of α-TTP. This caused diminished hepatic secretion of vitamin E incorporated into VLDL, resulting in selective vitamin E deficiency and subsequent neurological disorder (Traber et al., 1990).

This disease was reported in the literature under different names: FAVED (Ben Hamida et al., 1993), AVED (Ouahchi et al., 1995), and familial isolated vitamin E deficiency (FIVE); (Burk et al., 1981; Harding et al., 1985; Traber et al. 1987, 1990).

## II. PHENOTYPE

### A. Epidemiological Data

AVED is a rare autosomal recessive inherited neurodegenerative disorder. Since the first patient reported by Burk et al. (1981), 40 patients have been reported in the literature. The majority of patients are from Mediterranean countries where consanguineous marriages are prevalent and in particular from Tunisia. Some patients are from the United States, Europe, and Japan.

### B. Clinical Phenotype

AVED patients usually develop progressive cerebellar ataxia. Phenotypic variability is very great ranging from severe Friedreich's ataxia-like presentation (Stumpf et al., 1987; Ben Hamida et al., 1993) to a mild neurological impairment (Sokol et al., 1988) and very late onset (Yokota et al., 1987). Most of the patients exhibiting the Friedreich's ataxia phenotype fulfill Harding's criteria (Harding et al., 1981) with some minor clinical differences.

#### 1. Age of Onset

Age of onset is variable. Usually, the age of onset is under 20 years in patients from the Mediterranean, North American, and European regions. The mean age of Tunisian patients is 13.6 ± 3.7 years (Gabsi et al., 2001). Some patients from Japan show a later age of onset occurring during the third, fourth or fifth decade (Shimohata et al., 1997, Yokota et al., 1997).

### 2. Symptoms at Onset

The onset is usually insidious with symptoms such as tiredness, gait instability, and rarely head tremor (Sokol et al., 1988). These symptoms worsen progressively, gait becomes difficult, and incoordination affects the limbs leading to a progressive difficulty in walking.

### 3. Clinical Features

#### a. Cerebellar Syndrome

Cerebellar ataxia is a constant feature and appears early in the course of the disease. Incoordination progressively affects the four limbs. The cerebellar syndrome includes posture and gait disturbances, dysmetria, asynergia, and adiadochokinesia. Cerebellar dysarthria is observed in almost all patients. The cerebellar signs worsen progressively leading to severe gait disturbances. Patients become wheelchair-bound after a mean disease duration of 11.18 ± 3.73 years (Gabsi et al., 2001). Head tremor is noted in about 28% of patients (Cavalier et al., 1998).

#### b. Sensory Disturbances

Proprioceptive sensory disturbances are present in all patients even in early stages of the disease. Vibratory sensation is usually completely absent distally in lower limbs and significantly diminished in the distal upper extremities. Romberg's sign is found in the majority of patients and position sensation is usually absent in legs and hand. Touch, pinprick, and temperature sensation are normal.

#### c. Pyramidal Syndrome

An upper motor neuron syndrome is usually clinically expressed by a spastic gait and a Babinski sign (Ben Hamida et al., 1993; Hentati et al., 1996; Hoshino et al., 1999; Krendel et al., 1987; Labauge et al., 1998; Stumpf et al., 1987; Yokota et al., 1997; Gabsi et al., 2001). However, plantar responses may be flexor in some patients (Hentati et al., 1996; Yokota et al., 1987; Harding et al., 1985; Laplante et al., 1984; Gotoda et al., 1995; Gabsi et al., 2001). Motor strength is usually well preserved until late stages of the disease.

#### d. Tendon Reflexes Changes

Deep tendon reflexes are usually abolished either in the four limbs (Harding et al., 1985; Hoshino et al., 1999; Jackson et al., 1996; Krendel et al., 1987; Labauge et al., 1998; Laplante et al., 1984; Martinello et al., 1998; Rayner et al., 1993; Shimohata et al., 1997; Stumpf et al., 1987;

Tamaru et al., 1997; Zouari et al., 1998; Gabsi et al., 2001) or only in the lower limbs (Ben Hamida et al., 1993; Yokota et al., 1997; Gabsi et al., 2001). Reflexes can be preserved (Stumpf et al., 1987; Sokol et al., 1988; Hentati et al., 1996) and can even be brisk (Cavalier et al., 1998).

*e. Ocular Signs*

Ocular signs are rare. Gaze paralysis (Krendel et al., 1987) and saccadic eye movements may occur (Tamaru et al., 1997). Retinitis pigmentosa was reported in five Japanese patients (Cavalier et al., 1998; Hoshino et al., 1999; Shimohata et al., 1997) and in only one Tunisian patient (Gabsi et al., 2001). A bilateral and horizontal nystagmus was frequently present (66.6%) in patients reported from Tunisia (Gabsi et al., 2001).

*f. Other Neurological Signs*

A distal amyotrophy of lower limbs is rare and could be observed in some patients with long disease duration (Laplante et al., 1984; Montermini et al., 1997; Shimohata et al., 1997; Gabsi et al., 2001). Deafness has been reported in only one patient from Japan (Shimohata et al., 1997). Patients may complain of urinary urgency or incontinence (Stumpf et al., 1987). Dystonic movements are observed in about 13% of AVED patients (Cavalier et al., 1998, Hentati et al., 1996; Krendel et al., 1987). Myoclonus was noted in one Japanese patient (Yokota et al., 1997). Tongue fasciculations may be observed (Krendel et al., 1987; Martinello et al., 1998).

*g. Cardiac Involvement*

The heart is less frequently involved in AVED than in Friedriech's ataxia (Cossée et al., 1999; Cruz-Martinez et al., 1997; Dürr et al., 1996; Montermini et al., 1997; Schölls et al., 1997).

Cardiac involvement affects about 19% of AVED patients (Cavalier et al., 1998). The most frequent cardiac abnormality is hypertrophic cardiomyopathy by echocardiogram. Congestive heart failure can occur in late stages of the disease and seems to be a frequent cause of death.

*h. Skeletal Abnormalities*

Pes cavus is often present at time of diagnosis (Krendel et al., 1987; Stumpf et al., 1987; Sokol et al., 1988; Gotoda et al., 1995; Jackson et al., 1996; Labauge et al., 1998; Rayner et al., 1993; Martinello et al., 1998; Gabsi et al., 2001). Kyphoscoliosis is less frequent and may appear later in the course of the disease (Laplante et al., 1984; Krendel et al., 1987; Yokota et al., 1997; Martinello et al., 1998; Gabsi et al., 2001).

*i. Other Clinical Signs*

Diabetes mellitus which is often associated with Friedreich ataxia (Cossée et al., 1999; Cruz-Martinez et al., 1997; Dürr et al., 1996; Montermini et al., 1997; Schölls et al., 1997) has not been notified in AVED patients. There is no mental retardation in AVED patients.

## III. GENE

### A. Normal Gene Function

#### 1. α-Tocopherol Transfer Protein (α-TTP) Gene Identification

Catignani et al. (1975) identified from rat liver, a cytosolic protein with a molecular weight of 31,000 daltons that binds α-tocopherol. The hepatic α-TTP was purified from rat liver and cDNA was cloned by Sato (1991). Subsequently, human α-TTP cDNA located on chromosome 8q13.1–q13.3 was isolated by Arita (1995). α-TTP consists of 278 amino acids and exhibits 94% identity with rat α-TTP. It has been demonstrated that α-TTP selectively recognizes α-tocopherol, the most biologically active form of vitamin E among tocopherol analogues.

The possible implication of TTPA in AVED was first investigated and established by Ouahchi et al. (1995). The gene consisting of five exons, encodes a 278 amino acid cytosolic protein showing structural similarity with human retinaldehyde binding protein.

#### 2. α-Tocopherol Transfer Protein Gene Function

The knowledge of structure, metabolism, and functions of vitamin E is a prerequisite for the comprehension of AVED pathogenic mechanism.

*a. Vitamin E Transport and Role*

**Transport**—Vitamin E applies to a family of eight structurally related compounds, the tocopherols and the tocotrienols. The four major forms of vitamin E (α, β, δ and γ tocopherol) differ by the number and position of methyl group substitution on the chromanol ring. Of these, α-tocopherol has the greatest biological activity. Natural α-tocopherol is referred to as RRR-α tocopherol or d-α tocopherol. Synthetic vitamin E is produced by coupling trimethylhydroquinone with isophytol resulting in a mixture of eight isomers, α-tocopherol is one of them and the seven other isomers have a less important biological activity. Diet contains different tocopherols (especially α- and γ-tocopherols) that are passively absorbed, incorporated in chylomicrons, and secreted by the intestine into the systemic circulation then to the liver. The hepatic α-TTP, which possesses stereospecificity toward the most abundant and active isomer RRR-α-tocopherol, specifically promotes its net mass transfer into VLDL. Most circulating vitamin E is found in low density lipoproteins (LDL). Tissue uptake of vitamin E requires the presence of a cell LDL-receptor.

Despite the fact that α-tocopherol is taken up by the LDL receptor pathway or by transfer during chylomicrons hydrolysis (Traber *et al.*, 1985), its incorporation into extrahepatic tissues could be related to other unknown processes.

**Role**—The most widely recognized function of vitamin E is its role as a scavenger of free radicals, which are highly reactive and are able to oxidize a variety of biologically important molecules with subsequent membrane and cell damage. Its antioxidant properties allow it to protect unsaturated fatty acids of membrane phospholipids from peroxydation and therefore maintain membrane stability. Moreover, many experimental studies suggest a role of α-tocopherol in the modulation of cell proliferation and gene expression (Azzi *et al.*, 2000). Anti-proliferative effects may be related to the inhibition of protein kinase C (Ricciarelli *et al.*, 1999) and/or the activation of the apoptosis signal Fas system (Israel *et al.*, 2000). Vitamin E has been shown to modulate the expression of some genes, such as α-tropomyosine, collagenase and CD 36 (Azzi *et al.*, 2000). These effects are probably mediated by the modulation of NF-κB and AP1 transcription factors (Nakamura *et al.*, 1999; Azzi *et al.*, 1999). The precise role of vitamin E in the nervous system is unknown and the mechanism of neurological dysfunction of vitamin E deficiency remains unclear. The antioxidant role of vitamin E may be crucial for neurological function, and the lack of this effect may cause structural and functional alterations of nervous tissue. Vitamin E deficiency may affect nervous tissue in other ways including production of increased amount of cytolytic phospholipids, via phospholipase A2 activation (Grau *et al.*, 1998), enhancement of glutamate-induced neuronal toxicity (Schubert *et al.*, 1992; Tirosh *et al.*, 2000) or disturbance of brain monoamine metabolism (Adachi *et al.*, 1999; Romero-Ramos *et al.*, 2000).

*b. Function of α-Tocopherol Transfer Protein in Liver*

The α-TTP mediates the secretion of α-tocopherol taken up by hepatocytes into systemic circulation and is necessary to maintain minimal levels of plasma α-tocopherol.

The α-TTP is critical in the hepatic handling of α-tocopherol by specifically binding and retaining α-tocopherol as well as possibly transferring α-tocopherol from an endocytic to a secretory compartment. α-TTP is required to discriminate between dietary forms of vitamin E for secretion in VLDL and thus regulate plasma α-tocopherol levels. Several studies have shown that α-TTP is expressed in significant amounts only by hepatocytes (Yoshida *et al.*, 1992; Arita *et al.*, 1995), but this protein was detected at extremely low levels in rat and human brain (Hosomi *et al.*, 1998, Coop *et al.*, 1999), suggesting a role of this protein in transport of α-tocopherol across the blood-brain barrier. α-TTP is implicated in vitamin E cellular uptake and/or intracellular trafficking, but the mechanisms by which vitamin E is transported, regulated within cells, and how it is involved in cellular signaling remain unclear.

### B. Abnormal Gene Function

#### 1. Abnormal α-Tocopherol Transfer Protein Function

The underlying biochemical defect in patients with AVED is the failure of α-tocopherol incorporation into VLDL in liver cells. As AVED patients lack α-TTP, they are not able to discriminate between stereoisomers of α-tocopherol. Thus, abnormal α-TTP prevents transfer of α-tocopherol to VLDL and its vascular secretion by the liver, and therefore impairs its delivery to peripheral tissues (Hosomi *et al.*, 1998).

#### 2. Gene Defects

The implication of α-TTP in familial ataxia with isolated vitamin E deficiency was first established by the identification of frameshift mutations in the TTPA gene (Ouahchi *et al.*, 1995; Hentati *et al.*, 1996). Cavalier (1998) reported the identification of 13 mutations in the TTPA gene in 27 families with AVED. Four mutations were found in two or more independent families: *744delA*, which is the major mutation in North Africa, and *513insTT*, *486delT*, and *400C>T*, in families of European origin. This study represents the largest group of patients and mutations reported for this often misdiagnosed disease and pointed to the need for an early differential diagnosis from Friedreich's ataxia in order to initiate therapeutic and prophylactic vitamin E supplementation before irreversible damage develops.

To date, 16 different mutations of α-TTP have been identified in AVED patients living in North Africa, Europe, North America, and Japan (Table 19.1).

TABLE 19.1  Mutations in the α-TTP gen

| Mutations | Authors |
| --- | --- |
| 205G>C | Cavalier *et al.*, 1998 |
| 306A>G | Cavalier *et al.*, 1998 |
| 486delT | Hentati *et al.*, 1996 |
| 552G>A | Tamaru *et al.*, 1997 |
| 513insTT | Hentati *et al.*, 1996 |
| 744delA | Ouahchi *et al.*, 1995 |
| 175C>T [R59W] | Cavalier *et al.*, 1998 |
| Arg134Ter | Cavalier *et al.*, 1998 |
| 303T>G [H101Q] | Gotoda *et al.*, 1995 |
| 358G>A [A120T] | Cavalier *et al.*,1998 |
| 421G>A [E141K] | Cavalier *et al.*, 1998 |
| 574G>A [R192H] | Hentati *et al.*, 1998 |
| 661C>T [R221W] | Cavalier *et al.*, 1998 |
| Met thr 1 | Hoshino *et al.*, 1999 |

The major reported mutations are

***744delA***—Deletion of 1 bp (A) at position 744. The mutation, referred to as Mediterranean, appeared to have spread in North Africa and Italy (Ouahchi *et al.*, 1995). This mutation results in the replacement of the last 30 amino acids by an aberrant 14 amino acid peptide. Although only the C-terminal tenth of the protein was altered, this mutation was associated with a severe phenotype.

***303T>G [His101Gln]***—A missense mutation in the TTPA gene was found by Gotoda (1995) in a Japanese family. This homozygous mutation corresponds to a T-to-G transversion at nucleotide 303 and results in replacement of histidine (CAT) with glutamine (CAG) at residue 101. This mutation was associated with a mild phenotype, very late onset, and retinitis pigmentosa.

***485delT***—Hentati (1996) found a severely affected American patient with ataxia and peripheral neuropathy who had homozygous deletion of nucleotide 485 in the TTPA gene. The deletion resulted in a frameshift and generation of a premature stop codon at residue 176.

***513insTT***—Hentati (1996) found a North American patient severely affected with ataxia and peripheral neuropathy who was homozygous for insertion of 2 thymine residues at nucleotide position 513 of their TTPA sequence, causing a frameshift and a premature stop codon.

***574 G>A***—Hentati *et al.* (1996) found a mildly affected patient with vitamin E deficiency who was a compound heterozygote for a 574G-A point mutation resulting in an arg192-to-his amino acid substitution, and the 513insTT mutation.

***400C>T***—In 2 independent Canadian families with AVED, Cavalier (1998) found a truncating arg134-to-ter mutation in homozygous state in 1 patient with consanguineous parents and in compound heterozygous state with the 486delT mutation in the second nonconsanguineous family.

***552G>A***—In a patient with ataxia and vitamin E deficiency, Schuelke *et al.* (1999) identified a homozygous 552G–A mutation in the TTPA gene. The missplicing caused a shift in the reading frame with an aberrant amino acid sequence from codon 120 leading to a premature stop at codon 134. The truncated protein completely lacked the domains encoded by exons 3–5.

## IV. DIAGNOSTIC AND ANCILLARY TESTS

### A. DNA Tests

A screening mutation by sequencing DNA from AVED patients may be useful to confirm a diagnosis or to identify at risk AVED families.

The 744delA deletion abolishes a MboII restriction site. This is used for direct mutation detection. The direct identification of this mutation by PCR and restriction digestion may constitute a useful diagnostic test in Tunisia and other North African countries where this mutation appears to be prevalent if not the unique one. Such a screening test may be more robust than direct determination of serum vitamin E levels and would allow early diagnosis and presymptomatic identification of patients in at-risk families.

### B. Other Laboratory Tests

#### 1. Blood Testing

Routine blood testing including complete blood count, creatine kinase, ceruloplasmin, copper, hepatic enzymes, bilirubin, glucose is normal. The normal serum vitamin E levels are about 5–15 mg/L. AVED patients display values always below 2 mg/L regardless of the specific clinical phenotype. Patients with a severe phenotype exhibit very low serum vitamin E levels usually below 0.5 mg/L.

Serum concentrations of cholesterol and triglycerides and lipoprotein electrophoresis profile are normal. There is no deficit in other lipo- or hydrosoluble vitamins. Biochemical liver and pancreatic tests, D-xylose test, and steatorrhea are also normal. No acanthocytes are present in peripheral blood (Shorer *et al.*, 1996). These results permit the exclusion of secondary vitamin E deficiency with confidence. Interestingly, absolute and lipid-adjusted vitamin E levels were found to be slightly reduced in unaffected family members of AVED families (Gotoda *et al.*, 1995). Unaffected heterozygotes for AVED showed significantly reduced lipid-adjusted vitamin E (Feki *et al.*, submitted), suggesting a slight decrease of $\alpha$-TTP function in presence of one muted allele.

Genetic testing is the gold standard for AVED diagnosis. Genetic testing, however, is expensive and is rarely available in developing countries for routine practice. Serum vitamin E levels should be tested for AVED diagnosis in the absence of genetic testing.

#### 2. Cerebrospinal Fluid

Cerebrospinal fluid examination is normal (Harding *et al.*, 1985; Jackson *et al.*, 1996; Krendel *et al.*, 1987; Stumpf *et al.*, 1987; Yokota *et al.*, 1987).

#### 3. Neurophysiological Findings

*a. Peripheral Nerve Conduction Study*

In AVED patients motor action potentials and motor nerve conduction velocities in peripheral nerves are usually normal. Sensory action potential amplitudes are normal or moderately reduced as well as sensory nerve conduction velocities, suggesting the presence of slight involvement of peripheral nerves (Harding *et al.*, 1985; Rayner *et al.*, 1993;

Stumpf *et al.*, 1987; Sokol *et al.*, 1988; Martinello *et al.*, 1998).

*b. Somatosensory-Evoked Potentials (SEP)*

SEP studies show a slight decrease of the N13 amplitude with normal N13 latency. The N20 is usually undetectable or has a significantly delayed latency and a prolonged value of the central sensory conduction time (N13–N20). P40 response after tibial nerve stimulation is usually absent. These findings document a severe involvement of somatosensory pathways in lumbar, thoracic, and cervical spinal cord. This involvement is present from early stages of the disease and is not correlated with disease duration (Harding *et al.*, 1985; Krendel *et al.*, 1987; Yokota *et al.*, 1987; Zouari *et al.*, 1998).

### 4. Nerve Biopsy

Nerve biopsy shows a mild-to-moderate loss of large myelinated fibers (MF), whereas small MF density is normal or increased. Regeneration features are frequently observed and result in the increase of small MF density. There are no onion bulb formations. Pi granules and dense bodies are frequently noted in Schwann cell cytoplasm (Zouari *et al.*, 1998).

These findings demonstrate the presence of a slight-to-moderate regenerative axonal neuropathy in AVED in contrast to Friedreich ataxia in which the peripheral sensory neuropathy is severe since early stages of the disease and without regeneration features.

## C. Neuroimaging

Cerebral CT scan is generally normal (Harding *et al.*, 1985; Jackson *et al.*, 1996; Kayden *et al.*, 1993; Laplante *et al.*, 1984; Martinello *et al.*, 1998; Rayner *et al.*,1993; Yokota *et al.*, 1987). Cerebral MRI was reported as normal in six AVED patients (Martinello *et al.*,1998; Yokota *et al.*, 1997; Jackson *et al.*, 1996), and showed marked dilatation of the cisterna magna in one patient (Shimohata *et al.*, 1997). Single-photon emission CT of the brain was reported as normal in four AVED patients (Yokota *et al.*, 1997).

## D. Neuropathology

A pathological study of the central nervous system in a patient with AVED was reported in 1997 (Larnaout *et al.*, 1997). It showed almost generalized neuronal lipofuscin storage associated with marked posterior column degeneration, moderate corticospinal tract involvement, and the presence of axonal spheroids in the spinal sensory system. The neuronal lipofuscin storage appeared as intracytoplasmic neuronal yellow fluorescent lipopigments markedly stained with periodic acid Schiff and Sudan black. These inclusions involved especially neurons of the third cortical layer, Purkinje cells, the dentate nucleus, and anterior horn cells as well as the neurons of twelfth and ambiguous cranial nuclei and inferior olive. At electron microscopic examination these inclusions appeared as numerous uniformly granular lipopigment bodies without intrinsic lamellae, filaments, membranes, or any other features. The brainstem and spinal cord were markedly atrophic with a marked pallor of posterior column affecting the gracile and cuneate fasciculi and moderate pallor of lateral corticospinal tract. These lesions suggest that there is a dying-back axonopathy underlying the pathological mechanism in AVED.

## V. CELLULAR AND ANIMAL MODELS OF THE DISEASE

Effects of vitamin E deficiency in animals have been studied since the early 1930s (Blackmore *et al.*, 1969; Ringsted., 1935). After 28 weeks of vitamin E-deficient diet, rats develop ataxic gait, muscle weakness, and wasting and die prematurely (Bertoni *et al.*, 1984; Goss-Sampson *et al.*, 1988; Sokol. *et al.*, 1990; Southam *et al.*, 1991; Towfighi, 1981). Kyphoscoliosis, anemia (Bertoni *et al.*, 1984; Towfighi, 1981), and retinitis pigmentosa may appear. Electromyographic studies showed signs of muscle denervation with normal motor nerve conduction velocities (Goss-Sampson *et al.*, 1988; Southam *et al.*, 1991). SEPs were disturbed (Goss-Sampson *et al.*, 1988; Southam *et al.*, 1991). These results are similar to those reported in man.

On muscle biopsy, there is evidence of denervation and autofluorescent bodies within muscular fibers (Goss-Sampson *et al.*, 1988). On nerve biopsy, large myelinated fibers are reduced in number. Pathologic studies of the central nervous system demonstrated degeneration of posterior columns and gracilis and cuneatus nuclei with dystrophic axons containing normal and impaired mitochondria, neurofilaments, lamellar dense bodies, and peroxysomes (Bertoni *et al.*, 1984; Goss-Sampson *et al.*, 1988; Sokol, 1990; Southam *et al.*, 1991; Towfighi, 1981). Degenerative lesions affect predominantly the gracilis nucleus in rats, while they are more important in the cuneatus nucleus in monkeys. Accumulation of perinuclear lipofuscin in dorsal root ganglion cells (Towfighi, 1981) and gracilis and cuneatus nuclei (Southam *et al.*, 1991) can be observed.

To explore the *in vivo* functions of α-TTP two groups generated α-TTP knock-out mice (Teresawa *et al.*, 2000; Yokota *et al.*, 2001). At 18 months of age, the α-TTP-deficient mice described by Teresawa *et al.* (2000) did not develop obvious signs of neurological disease in contrast to humans with α-TTP gene defects. A second report, however, did demonstrate a neurologic phenotype in Ttp(−/−)

mice with head tremor, ataxia, and dystonia (Yokota et al., 2001). This phenotype was only seen after one year of age and most pronounced when these mice were maintained on a diet deficient in α tocopherol. Teresawa et al. (2000) used 90 mg/kg, whereas Yokota et al. maintained mice on 35 mg/kg for the normal diet and 0 mg/kg for a deficient diet. This was also reflected in the plasma α-tocopherol levels which was not detectable in α-TTP mice with normal and deficient diet, suggesting that a severe deficiency of from the plasma and brain (less than 2% of wild-type mean value) was necessary to produce neurological phenotype.

## VI. GENOTYPE/PHENOTYPE CORRELATIONS— MODIFYING ALLELES

Clinical phenotypic variability in AVED patients has been correlated with allelic genetic heterogeneity (Cavalier et al., 1998). AVED patients from Mediterranean, European, and North American regions harbor mutations resulting in translation interruption (stop codon) or loss of transcription of one exon. This leads to a complete loss of function of α-TTP that may explain early onset of the disease and severe clinical phenotype in these patients. In Japanese AVED patients, mutations result in partial loss of α-TTP that is still able to incorporate α-tocopherol into VLDL. In these patients, age of onset of the disease is later and the course of the disease seems to be milder. In addition patients from Japan often develop visual loss related to retinitis pigmentosa probably because they have longer duration of the disease. Besides, genetic studies showed that retinitis pigmentosa was more frequently associated with the 303T>G mutation than with other mutations (Yokota et al., 1996). Hoshino et al. (1999) reported one Japanese patient with a mutation leading to complete loss of function of α-TTP and presenting with a clinical phenotype similar to that of Mediterranean, European, and North American AVED patients, Cardiomyopathy has only been reported in patients with frameshift or 661C>T mutations (Cavalier et al., 1998).

Patients with AVED exhibit a pronounced plasma vitamin E level decrease. This decrease seems to vary with the mutation type of α-TTP gene. Some mutations, such as the most common form 744delA (frequently observed in the Mediterranean basin) are associated with very low serum vitamin E levels (< 0.5 mg/L; Zouari et al., 1998; Gabsi et al., 2001), whereas the decrease is less pronounced in mutation 303T>G, reported in Japanese families (around 1.73 mg/L; Yokota et al., 1997). This 303T>G mutation resulting in His101Gln substitution, might be a conservative missense mutation that would allow the production of the partially functional protein leading to a mild phenotype.

## VII. TREATMENT

Results of vitamin E supplementation have been reported (Belal et al., 1995; Jackson et al., 1996; Kayden et al., 1993; Krendel et al., 1987; Labauge et al., 1998; Martinello et al., 1998; Rayner et al., 1993; Stumpf et al., 1987; Gabsi et al., 2001). Vitamin E is administered orally at mean dose of 800 mg/day and leads to a normalization of serum vitamin E levels. Over a one-year-treatment period, mean cooperative international ataxia rating scale (ARS) score, which assesses cerebellar ataxia, improved steadily. ARS score improvement was statistically significant, especially in AVED patients with a mean duration of the disease less than 15 years. Improvement concerned particularly action and head tremor. Deep sensory disturbances and deep tendon reflexes remained unchanged. However, no new neurological signs were elicited by periodic neurological examination over the period of the trial (Gabsi et al., 2001).

### Acknowledgments

This work has been supported by the Tunisian Ministry of Health and Tunisian Research Ministry.

### References

Adachi, K., Izumi, M., and Mitsuma, M. (1999). Effect of vitamin E deficiency on rat brain monoamine metabolism. *Neurochem. Res.* **24**, 1307–1311.

Aoki, K., Washimi, Y., Fujimori, N., Maruyama, K., and Yanagisawa, N. (1990). Familial idiopathic vitamin E deficiency associated with cerebellar atrophy. *Rinsho Shinkeigaku* **30**, 966–971.

Arita, M., Sato, Y., Miyata, A., Tanabe, T., Takahashi, E., Kayden, H., Arai, H., and Inoue, K. (1995). Human α-tocopherol transfer protein: cDNA cloning, expression and chromosomal localization. *Biochemistry Journal* **306**, 437–443.

Azzi, A., Boscoboinik, D., Clement, S., Ozer, N., Ricciarelli, R., and Stoker, A. (1999). Vitamin E mediated response of smooth muscle cell to oxidant stress. *Diabetes Res. Clin. Pract.* **45**, 191–198.

Azzi, A., and Stocker, A. (2000). Vitamin E: Non-antioxidant roles *Prog. Lip. Res.* **39**, 231–255.

Belal, S., Hentati, F., Ben Hamida, C., and Ben Hamida, M. (1995). Friedreich's ataxia-vitamin E responsive type. *Clin. Neurosci.* **3**, 39–42.

Belal, S., Panayides, K., Ionannou, P., Ben Hamida, C., Sirugo, G., Koenig, M., Mandel, J. L., Hentati, F., Ben Hamida, M., and Middleton, L. (1992). Genetic heterogeneity of Friedreich's ataxia in Tunisian families. *Neurology* **42**, (Suppl. 3) 4.

Ben Hamida, C., Doerflinger, N., Belal, S., Linder, C., Reutenauer, L., Dib, G., Gyapay, G., Vignal, A., Le Paslier, D., Cohen, D., Pandolfo, M., Mokini, V., Novelli, G., Hentati, F., Ben Hamida, M., Mandel, J. L., and Koenig, M. (1993). Localization of Friedreich ataxia phenotype with selective vitamin E deficiency to chromosome 8q by homozygosity mapping. *Nat. Genet.* **5**, 195–200.

Ben Hamida, M., Belal, S., Sirugo, G., Ben Hamida, C., Panayides, K., Ionannou, P., Beckmann, J., Mandel, J. L., Hentati, F., Koenig, M., and Middleton, L.T. (1993). Friedreich's ataxia phenotype not linked to chromosome 9 and associated with selective autosomal recessive vitamin E deficiency in two inbred Tunisian families. *Neurology* **43**, 2179–2183.

Bertoni, J. M., Abraham, F. A., Falls, H. F., and Itabashi, H. H. (1984). Small bowel resection with vitamin E deficiency and progressive spinocerebellar syndrome. *Neurology* **34**, 1046–1052.

Binder, H. J., Solitare, G. B., and Spiro, H. M. (1967). Neuromuscular disease in patients with steatorrhoea. *Gut* **8**, 605–611.

Blackmore, W. F., and Cavanagh, J. B. (1969). "Neuroaxonal dystrophy" occurring in an experimental " dying back " axonopathy process in the rat. *Brain* **92**, 789–804.

Burck, U., Goebel, H. H., Kuhlendahl, H. D., Meier, C., and Goebel, K. M. (1981). Neuromyopathy and vitamin E deficiency in man. *Neuropediatrics* **12**, 267–278.

Catignani, G. L., and Bieri, J. G. (1975). Rat liver a tocopherol binding protein. *Biochem. Biophys. Acta* **497**, 349–357.

Cavalier, L., Ouahchi, K., Kayden, H. J., Di Donato, S., Reutenauer, L., Mandel, J. L., and Koenig, M. (1998). Ataxia with isolated vitamin E deficiency: heterogeneity of mutations and phenotypic variability in a large number of families. *Am. J. Hum. Genet.* **62**, 301–310.

Chamberlain, S., Shaw, J., Rowland, A., Wallis, J., South, S., Nakamura, Y., and Williamson, R. (1988). Mapping of mutation causing Friedreich's ataxia to human chromosome 9. *Nature* **334**, 248–250.

Copp, R. P., Wisniewski, T., Hentati, F., Larnaout, A., Ben Hamida, M., and Kayden, H. J. (1999). Localization of alpha-tocopherol transfer protein in the brains of patients with ataxia with vitamin E deficiency and other oxidative stress related neurodegenerative disorders. *Brain Res.* **20**, 822, 80–87.

Cossée, M., Dürr, A., Schmitt, M., Dahl, N., Trouillas, P., Allinson, P., Kostrzewa, M., Nivelon-Chevallier, A., Gustavson, K. H., Kohlschutter, A., Muller, U., Mandel, J. L., Brice, A., Koenig, M., Cavalcanti, F., Tammaro, A., De Michele, G., Filla, A., Cocozza, S., Labuda, M., Montermini, L., Poirier, J., and Pandolfo, M (1999). Friedreich's ataxia: point mutations and clinical presentation of compound heterozygotes. *Ann. Neurol.* **45**, 200–206.

Cruz-Martinez, A., Anciones, B., and Palau, F. (1997). GAA trinucleotide repeat expansion in variant Friedreich's ataxia families. *Muscle Nerve* **20**, 1121–1126.

Doerflinger, N., Linder, C., Ouahchi, K., Gyapay, G., Weissenbach, J., Le Paslier, D., Rigault, P., Belal, S., Ben Hamida, C., Hentati, F., Ben Hamida, M., Pandolfo, M., DiDonato, S., Sokol, R., Kayden, H., Landrieu, P., Durr, A., Brice, A., Goutieres, F., Kohlschutter, A., Sabouraud, P., Benomar, A., Yahyaoui, M., Mandel, J. L., and Koenig, M. (1995). Ataxia with vitamin E deficiency: refinement of genetic localization and analysis of linkage disequilibrium by using new markers in 14 families. *Am. J. Hum. Genet.* **56**, 1116–1124.

Dürr, A., Cossée, M., Agid, Y., Campuzano, V., Mignard, C., Penet, C., Mandel, J. L., Brice, A., and Koenig, M. (1996). Clinical and genetic abnormalities in patients with Friedreich's ataxia. *N. Engl. J. Med.* **335**,1169–1175.

Feki, M., Belal, S., Feki, H., Souissi, M., Frih-Ayed, M., Kaabachi, N., Hentati, F., Ben Hamida, M., and Mebazaa, A. (2002). Serum vitamin E and lipid-adjusted vitamin E assessment in Friedreich ataxia phenotype patients and unaffected family members. *Clin. Chem.* **48(3)**, 577–579.

Gabsi, S., Gouider-Khouja, N., Belal, S., Fki, M., Kéfi, M., Turki, I., Ben Hamida M., Kayden, H., Mebazaa R., and Hentati, F. (2001). Effect of vitamin E supplementation in patients with ataxia with vitamin E deficiency. *Eur. J. Neurol.* **8**, 477–481.

Goss-Sampson, M. A., Kriss, A., Muddle, J. R., Thomas, P. K., and Muller, D. P. R. (1988). Lumbar and cortical somatosensory evoked potentials in rats with vitamin E deficiency. *J. Neurol. Neurosurg. Psychiatry*, **51**, 432–435.

Gotoda, T., Arita, M., Arai, H., Inoue, K., Yokota, T., Fukuo, Y., Yazaki, Y., and Yamada, N. (1995). Adult onset spinocerebellar dysfunction caused by a mutation in the a tocopherol transfer protein gene. *N. Eng. J. Med.* **333**, 1313–1318.

Grau, A., and Ortiz, A. (1998). Dissimilar protection of tocopherol isomers against membrane hydrolysis by phospholipase A2. *Chem. Phys. Lipids* **91**, 109–118.

Harding, A. E. (1981). Friedreich's ataxia: a clinical and genetic study of 90 families with analysis of early diagnostic criteria and intrafamilial clustering of clinical features. *Brain* **104**, 589–620.

Harding, A. E., Matthews, S., Jones, S., Ellis, C. J. K., Booth, I. W., and Muller, D. P. R. (1985). Spinocerebellar degeneration associated with a selective defect in vitamin E absorption. *N. Engl. J. Med.* **313**, 32–35.

Hentati, A., Deng, H.-X., Hung, W.-Y., Nayer, M., Ahmed, M. S., He, X., Tim, R., Stumpf, D. A., and Siddique, T. (1996). Human alpha-tocopherol transfer protein: gene structure and mutations in familial vitamin E deficiency. *Ann. Neurol.* **39**, 295–300.

Hoshino, M., Masuda, N., Ito, Y., Murato, M., Goto, J., Sakurai, and M., Kanazawa, I. (1999). Ataxia with isolated vitamin E deficiency: a Japanese family carrying a novel mutation in the α-tocopherol transfer protein gene. *Ann. Neurol.* **45**, 809–812.

Hosomi, A., Goto, K., Kondo, H., Iwatsubo, T., Yokota, T., et al. (1998). Localization of alpha-tocopherol transfer protein in rat brain. *Neurosci. Lett.* **256**, 159–162.

Israel, K., Yu, W., Sanders, B. G., and Kline, K. (2000). Vitamin E succinate induces apoptosis in human prostate cancer cell: role for Fas in vitamin E succinate-triggered apoptosis. *Nutr. Cancer* **36**, 90–100.

Jackson, C. E., Amato, A. A., and Barhon, R. J. (1996). Isolated vitamin E deficiency. *Muscle Nerve* **19**, 1161–1165.

Kayden, H. J. (1993). The neurologic syndrome of vitamin E deficiency: A significant cause of ataxia. *Neurology* **43**, 2167–2169.

Kohlschutter, A., Hubner, C., Jansen, W., and Lindner, S. G. (1988). A treatable familial neuromyopathy with vitamin E deficiency, normal absorption, and evidence of increased consumption of vitamin E. *J. Inherit. Metab. Dis.* **11**, 149–152.

Krendel, D. A., Gilchrist, J. M., Johnson, A. O., and Bossen, E. H. (1987). Isolated deficiency of vitamin E with progressive neurologic deterioration. *Neurology* **37**, 538–540.

Labauge, P., Cavalier, L., Ichalalène, L., and Castelnovo, G. (1998). Ataxie de type Friedreich et déficit héréditaire en vitamine E. *Rev. Neurol.* **4**, 339–341.

Laplante, P., Vanasse, M., Michaud, J., Geoffroy, G., and Brochu, P. (1984). A progressive neurological syndrome associated with an isolated vitamin E deficiency. *Can. J. Neurol. Sci.* **11**, 561–564.

Larnaout, A., Belal, S., Zouari, M., Fki, M., Ben Hamida, C., Goebel, H. H., Ben Hamida, M., and Hentati, F. (1997). Friedreich's ataxia with isolated vitamin E deficiency: A neuropathological study of a Tunisian patient. *Acta Neuropathol. (Berl)* **93(6)**, 633–637.

Martinello, F., Fardin, P., Ottina, M., Ricchieri, G.L., Koenig, M., Cavalier, L., and Trevisan, C. P. (1998). Supplemental therapy in isolated vitamin E deficiency improves the peripheral neuropathy and prevents the progression of ataxia. *J. Neurol. Sci.* **156**, 177–179.

Montermini, L., Richter, A., Morgan, K., Justice, C. M., Julien, D., Castellotti, B., Mercier, J., Poirier, J., Capozzoli, F., Bouchard, J. P., Lemieux, B., Mathieu, J., Vanasse, M., Seni, M. H., Graham, G., Andermann, F., Andermann, E., Melancon, S. B., Keats, B. J., Di Donato, S., and Pandolfo, M. (1997). Phenotypic variability in Friedreich ataxia: role of the associated GAA triplet repeat expansion. *Ann. Neurol.* **41**, 675–682.

Nakamura, T., Goto, M., Matsumoto, A., and Tanaka, I. (1998). Inhibition of NF-kappa B transcriptional activity by alpha-tocopherol succinate. *Biofactors* **7**, 21–30.

Ouahchi, K., Arita, M., Kayden, H., Hentati, F., Ben Hamida, M., Sokol, R., Arai, H., Inoue, K., Mandel, J.-L., and Koenig, M. (1995). Ataxia with isolated vitamin E deficiency is caused by mutations in the alpha-tocopherol transfer protein. *Nat. Genet.* **9**, 141–145.

Rayner, R. J., Doran, R., and Roussounis, S. H. (1993). Isolated vitamin E deficiency and progressive ataxia. *Arch. Dis. Child.* **69**, 602–603.

Ringsted, A. (1935). A preliminary note on the appearance of paresis in adult rats suffering from chronic avitaminosis E. *Biochemi. J.* **29**, 788–795.

Romero-Ramos, M., Venero, J. I., Santigo, M., Rodriguez-Gomez, J. A., Vizuete, M. L., Cano, J., and Machado, A. (2000). Decreased messenger RNA expression of key markers of the nigrostriatal dopaminergic system following vitamin E deficiency in the rat. *Neuroscience* **101**, 1029–1036.

Schölls, L., Amoiridis, G., Przuntek, H., Frank, G., and Epplen, J. T., Epplen, C. (1997). Friedreich's ataxia: revision of the phenotype according to molecular genetics. *Brain* **120**, 2131–2140.

Schubert, D., Kimura, H., and Maher, P. (1992). Growth factors and vitamin E modify neuronal glutamate toxicity. *Proc. Natl. Acad. Sci.* **89**, 8264–8267.

Schuelke, M., Mayatepek, E., Inter, M., Becker, M., Pfeiffer, E., Speer, A., Hubner, C., and Finckh, B. (1999). Treatment of ataxia in isolated vitamin E deficiency caused by alpha-tocopherol transfer protein deficiency. *J. Pediatr.* **134**, 240–244.

Shimohata, T., Date, H., Ishiguro, H., Suzuki, T., Takano, H., Tanaka, H., Tsuji, S., and Hirota, K. (1997). Ataxia with isolated vitamin E deficiency and retinitis pigmentosa. *Ann. Neurol.* **43**, 273.

Shiojiri, T., Yokota, T., Fujimori, N., and Mizusawa, H. (1999). Familial ataxia with isolated vitamin E deficiency not due to mutation of alpha-TTP (Letter). *J. Neurol.* **246**, 982.

Shorer, Z., Parvari, R., Brill, G., Sela, B. A., and Moses, S. (1996). Ataxia with isolated vitamin E deficiency in four siblings. *Pediatr. Neurol.* **15**, 340–343.

Sokol, R. J. (1990). Vitamin E and neurologic deficits. *Adv. Pediatr.* **37**, 119–148.

Sokol, R. J., Kayden, H. J., Bettis, D. B., Traber, M. G., Neville, H., Ringel, S., Wilson, W. B., and Stumpf, D. A. (1988). Isolated vitamin E deficiency in the absence of fat malabsorption-familial and sporadic cases: characterization and investigation of causes. *J. Lab. Clin. Med.* **111**, 548–559.

Southam, E., Thomas, P. K., King, H. M., Goss-Sampson, M. A., and Muller, D. P. R. (1991). Experimental vitamin E deficiency in rats. Morphological and functional evidence of abnormal axonal transport secondary to free radical damage. *Brain* **114**, 915–936.

Stumpf, D. A., Sokol, R. J., Bettis, D., Neville, H., Ringel, S., Angelini, C., and Bell, R. (1987). Friedreich's disease: V. Variant form with vitamin E deficiency and fat absorption. *Neurology* **37**, 68–74.

Tamaru, Y., Hirano, M., Kusaka, H., Ito, H., Imai, T., and Ueno, S. (1997). α-tocopherol transfer protein gene: Exon skipping of all transcripts causes ataxia. *Neurology* **49**, 584–588.

Terasawa, Y., Ladha, Z., Leonard, S. W., Morrow, J. D., Newland, D., Sanan, D., Packer, L., Traber, M. G., and Farese, R. V. Jr (2000). Increased atherosclerosis in hyperlipidemic mice deficient in alpha-tocopherol transfer protein and vitamin E. *Proc. Natl. Acad. Sci. U.S.A.* **97**, 13830–13834.

Tirosh, O., Sen, C. K., Roy, S., and Packer, L. (2000). Cellular and mitochondrial changes in glutamate-induced HT4 neuronal cell death. *Neuroscience* **97**, 531–541.

Towfighi, J. (1981), Effects of chronic vitamin E deficiency on the nervous system of the rat. *Acta Neuropathol.* **54**, 261–267.

Traber, M. G., and Kayden, H. J. (1987). Tocopherol distribution and intracellular localization in human adipose tissue. *Am. J. Clin. Nutr.* **46**, 488–495.

Traber, M. G., Olivecrona, T., and Kayden, H. J. (1985). Bovine milk lipoprotein lipase transfers tocopherol to human fibroblasts during triglyceride hydrolysis *in vitro. J. Clin. Invest.* **75**, 1729–1734.

Traber, M. G., Sokol, R. J., Burton, G. W., Ingold, K. U., Papas, A. M., Huffaker, J. E., and Kayden, H. J. (1990). Impaired ability of patients with familial isolated vitamin E deficiency to incorporate a tocopherol into lipoproteins secreted by the liver. *J. Clin. Invest.* **85**, 397–407.

Trouillas, P., Takayanagi, T., Hallett, M., Currier, R. D., Subramony, S. H., Wessel, K., and Bryer, A. *et al.* (1997). International cooperative ataxia rating scale for pharmacological assessment of the cerebellar syndrome. *J. Neurol. Sci.* **145**, 205–211.

Vatassery, G. T., Brin, M. F., Fahn, S., Kayden, H. J., and Traber, M. G. (1988). Effect of high doses of dietary vitamin E on the concentrations of vitamin E in several brain regions, plasma, liver and adipose tissue of rats. *J. Neurochem.* 621–623.

Yokota, T., Igarashi, K., Uchihara, T., Jishage, K., Tomita, H., Inaba, A., Li, Y., Arita, M., Suzuki, H., Mizusawa, H., and Arai, H. (2001). Delayed-onset ataxia in mice lacking-tocopherol transfer protein: Model for neuronal degeneration caused by chronic oxidative stress. *Proc. Natl. Acad. Sci. U.S.A.* **98**, 15,185–15,190.

Yokota, T., Shiojiri, T., Gotoda, T., Arita, M., Arai, H., Ohga, T., Kanda, T., Suzuki, J., Imai, T., Matsumoto, H., Harino, S., Kiyosawa, M., Mizusawa, H., and Inoue, K. (1997). Friedreich-like ataxia with retinitis pigmentosa caused by His101Gln mutation of the a-tocopherol transfer protein gene. *Ann. Neurol.* **41**, 826–832.

Yokota, T., Wada, Y., Furukawa, T., Tsukagoshi, H., Uchihara, T., and Watabiki, S. (1987). Adult-onset spinocerebellar syndrome with idiopathic vitamin E deficiency. *Ann. Neurol.* **22**, 84–87.

Yoshida, H., Yusin, M., Ren, I., Kuhlenkamp, J., Hirano, T., Stolz, A., and Kaplowitz, N. (1992). Identification, purification and immunochemical characterization of alpha tocopherol-binding protein in rat liver cytosol. *J. Lipid Res.* **33**, 343–350.

Zouari, M., Feki, M., Ben Hamida, C., Larnaout, A., Turki, I., Belal, S., Mebazaa, A., Ben Hamida, M., and Hentati, F. (1998). Electrophysiology and nerve biopsy: comparative study in Friedreich's ataxia and Friedreich's ataxia phenotype with vitamin E deficiency. *Neuromuscular Disord.* **8**, 416–425.

# CHAPTER 20

# Autosomal Recessive Spastic Ataxia of Charlevoix-Saguenay (ARSACS/SACS)—No Longer a Local Disease

ANDREA RICHTER

Service de Génétique Médicale
Hôpital Sainte-Justine
Département de Pédiatrie
Université de Montréal
Montréal, Québec, Canada

I. Phenotype
II. Gene
III. Diagnostic and Ancillary Tests
   A. DNA
   B. Other Ancillary Tests
IV. Neuroimaging
V. Neuropathology
VI. Genotype/Phenotype Correlations
   Acknowledgments
   References

Autosomal recessive spastic ataxia of Charlevoix-Saguenay (ARSACS/SACS, OMIM 270550) originally described in 1978 is a clinically homogeneous form of early-onset familial disease with prominent myelinated retinal nerve fibers (Bouchard et al., 1991). Over 300 patients have been identified and most of their families originated in the Charlevoix-Saguenay region of Northeastern Quebec in Canada. The frequency of several recessive diseases is increased in this region due to a founder effect caused by the settlement patterns of the late 17th to mid-19th centuries (Jetté et al., 1991, Gauvreau et al., 1991, DeBraekeleer, 1991). The gene carrier prevalence was estimated to be 1/22. (DeBraekeleer et al., 1993). Patients present in early childhood with spastic gait ataxia. The disease progresses rapidly in young adults and patients are wheelchair-bound by their fifth decade. The ARSACS locus was mapped to chromosome region 13q11 by noting increased homozygosity for locus *D13S787* in a genome-wide scan (Bouchard et al., 1998). Following extensive genetic, physical, and transcript mapping combined with directed sequencing, two mutations were detected in the Sacsin *(SACS)* gene in ARSACS families. (Richter et al., 1999, Engert et al., 1999, 2000). Both the single nucleotide deletion (g.6594delT, $\Delta$T) and the g.5254C > T (C > T) nonsense mutation cause the premature termination of the predicted sacsin protein. We calculated disease allele frequencies using data from more than 125 Québec ARSACS patients. Close to 94% of the disease alleles carried the $\Delta$T mutation, over 3% the C > T mutation. Interestingly close to 3% of the disease alleles carry unknown mutation(s), always in heterozygous form with $\Delta$T (Mercier et al., 2001). The sequencing of *SACS* is underway to identify these mutations.

There are descriptions of recessive spastic ataxias clinically very similar to ARSACS in France (Chaigne et al., 1993), Tunisia (Mrissa et al., 2000), Spain (Pascual-Castroviejo et al., 2000), and Turkey (Gücüyener et al., 2001). The availability of family material in two of the studies permitted linkage analysis. Results show that the disease linked to the *SACS* region on chromosome 13q in a large consanguineous kindred from Tunisia (Mrissa et al., 2000) and in two consanguineous families from Turkey (Gücüyener et al., 2001). These publications likely represent only the first few cases of recessive spastic ataxia where a diagnosis of ARSACS should be considered.

## I. PHENOTYPE

ARSACS patients show early-onset spasticity in the lower limbs, usually observed at gait initiation (12–18 months). At birth foot deformities are absent; however, in most but not all patients the progressive development of pes cavus, equines, and clawing of the toes has been observed in the first two decades of life. The clinical picture from early childhood is always that of gait ataxia, with a tendency to fall (Bouchard et al., 1978, Bouchard, 1991).

Passive movements of the lower limbs reveal increased tonus. Plantar response is always abnormal, deep tendon reflexes are usually increased, often with clonus at the ankles and patellae in the early adulthood. Walking is stiff at the beginning, with little movement of the often partially flexed knees. Early in the disease spasticity is already present even when there is little sign of cerebellar dysfunction. Speech can be slightly slurred in childhood and becomes explosive in adulthood. Hand and tongue dysdiadochokinesis are detected early. Overt dysmetria in upper limbs occurs later in the course of the disease, usually after the loss of ambulation (Bouchard et al., 2000).

The ocular signs are most typical of the disease. They appear early and are non-progressive. They include saccadic alteration of smooth ocular pursuit and prominent myelinated fibers radiating from the optic disc and embedding the retinal blood vessels at fundoscopy (Fig. 20.1). This highly unusual feature points to an early abnormal myelination process. Other ophthalmological examinations, including electroretinography, visual acuity and fields, color vision, slit lamp, and tonometry are normal. There is moderate to severe slowing of motor conduction velocities in the peripheral nerves in children, but this does not progress with age. (Bouchard et al., 1979; 1993).

No intellectual deficit is observed in early childhood in ARSACS patients despite the slower motor development.

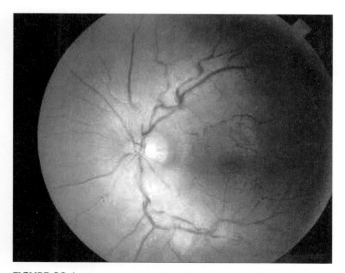

FIGURE 20.1 Presence of myelinated retinal nerve fibers that embed portions of blood vessels in ARSACS. Observed in all ascertained cases in Québec.

During elementary school, they are slow to learn and especially to learn to write. About half of the ARSACS patients drop out of school by the end of high school. However, despite their motor handicap, several patients have completed college and university degrees. Repeated neuropsychological testing in adult ARSACS patients shows that their verbal IQ is usually within normal limits, however, their handling of visual and spatial material is poor and tends to deteriorate. (Bouchard et al., 1978; 1979).

The progression of the disease is quite apparent in teenagers and young adults. There is a progressive increase in muscle tone and in deep tendon reflexes. The gait is often jerky, sometimes accompanied by scissoring. The presence of discrete-to-marked distal amyotrophy is usually a later feature, but in some families this appears early in the disease. Ankle jerks disappear around the age of 25, having been brisk and polyclonic in most patients. At this time, electromyography shows signs of denervation in the distal muscles, especially in the feet. These progressive signs are believed to result from the axonal degeneration in both the upper and lower motor neurons (Bouchard et al., 2000).

Some cases of Friedreich's ataxia (FRDA) with minimal GAA expansions present as spastic ataxia, but they are not likely to be confused with ARSACS. Although they may have disease onset similar to non-Québec ARSACS, the frequent presence of cardiomyopathy in FRDA should signal the need to investigate the frataxin gene. (Sorbi et al., 2000; Berciano et al., 2002).

The clinical findings in Québec ARSACS patients are as follows: initially a diffuse spastic syndrome that progresses throughout life. In contrast the signs of cerebellar disease are relatively sparse but increase slowly starting at adolescence. In the mid-20s, motor axonal polyneuropathy appears and exacerbates the deficits. Over a 20-year period, 320 ARSACS patients were identified and followed in the local neuromuscular clinics. ARSACS patients become wheelchair-bound at a mean age of 41, with a wide range of 17–57 years. The mean age at death for 34 patients is 51 (Bouchard, 1991); some survive into their 70s, but become bedridden by that time, and most of them die of recurrent infections. There is no cardiac involvement or associated disease. (Bouchard et al., 2000).

The clinical details in the Tunisian ARSACS patients are similar but not identical to those in Québec patients. In particular the age of onset is later and fundoscopy failed to reveal the presence of the myelinated retinal nerve fibers (Mrissa et al., 2000). In contrast, in the Turkish ARSACS families, age of onset is comparable to Québec patients, and myelinated retinal nerve fibers are present in at least one patient. However, one of the four patients in that study was described as having a mild intellectual handicap (Gücüyener et al., 2001). In both studies the neurophysiologic and nerve biopsy findings seem to be similar to the Québec findings. Presence of retinal fibers and psychomotor delay have been reported for the Spanish patient where the age of onset is in the pediatric period (Pascual-Castroviejo et al., 2000). The French cases also have early onset (Chaigne et al., 1993) and retinal myelinated fibers have been observed (J.-P. Bouchard, personal communication).

## II. GENE

We initially localized the ARSACS gene to chromosome 13q11 by genotyping 322 markers and noting a high degree of homozygosity at locus *D13S787*. Linkage analysis and the inspection of recombination events in multiplex families identified a candidate region of 5.5 cM between D13S175 and D13S787 (Bouchard et al., 1998). We then precisely mapped the ARSACS locus by location score analysis and constructed haplotypes to significantly narrow the candidate gene region to 1.58 cM between *D13S1275* and *D13S292*. Two groups of ARSACS-associated haplotypes were identified: one that represents an estimated 96% of ARSACS chromosomes and a small rare group. The haplotype groups did not appear to be closely related (Richter et al., 1999). A high resolution BAC and PAC map was constructed that includes the ARSACS critical region flanked by *D13S1275* and *D13S292* (Engert et al., 1999).

Six BAC/PAC clones forming a 450-kb contig between *D13S232* and *D13S787* were used for sample sequencing within the ARSACS critical region. Comparison of expressed sequence tag (EST) and genomic sequences suggested that many of the ESTs were derived from a single large genomic open reading frame (ORF). A 20-kb sequence contig revealed a genomic ORF of 11,487 bp that encodes 3829 amino acids and is preceded by an in-frame stop codon 75 bp upstream. We confirmed this by sequencing RT-PCR products spanning the coding sequence, which showed perfect correspondence with the DNA sequence. We called the gene sacsin (gene symbol: *SACS*). The sequencing DNA from ARSACS patients and controls detected a single-base deletion at position 6594 (ΔT) on all copies of the major ancestral haplotype examined. This mutation results in a frameshift and the introduction of a stop codon. A nonsense mutation at nucleotide 5254 (C > T) causing the substitution of a stop codon for an arginine was found on the minor ARSACS haplotype. The gene is expressed in many tissues and appears well conserved in the mouse (Engert et al., 2000).

*SACS* has a remarkable feature—it contains a large exon spanning at least 12,794 bp. The function of the protein is unknown but the presence of heat shock domains suggests a role in chaperone-mediated protein folding. *SACS* is expressed in a wide variety of tissues including the central nervous system.

## III. DIAGNOSTIC AND ANCILLARY TESTS

### A. DNA

We developed a simple allele-specific oligonucleotide (ASO) based test that permits efficient screening of the two *sacsin* mutations (Mercier et al., 2001). This is the current diagnosis confirmation mechanism in cases with known ancestry in the Charlevoix-Saguenay region and also for carrier testing in families of confirmed cases. The simplicity and low cost of the ASO test may also make population-based carrier screening practicable. For cases from other populations, failure to identify the known gene mutations should be followed by linkage analysis. If there is indication of linkage of the disease to the ARSACS region on chromosome 13q11, sequencing of the ORF should be undertaken.

### B. Other Ancillary Tests

After the initial clinical assessment, the basic investigation of new ARSACS patients includes ophthalmoscopic evaluation, electrophysiological studies, and head imaging. Initially, there were no abnormalities seen in endocrine functions, blood chemistry, and cerebrospinal fluid (Bouchard et al., 1978). In most patients, electromyography shows signs of severe denervation in distal muscles by the end of the 20s (Bouchard et al., 1979). Nerve conduction studies demonstrate signs of both early demyelination and progressive axonal neuropathy (Bouchard, 1991; Bouchard et al., 1979), the latter confirmed by nerve biopsy (Peyronnard et al., 1979). Motor nerve conduction velocities are moderately reduced but this is nonprogressive. In most cases, the distal motor latencies can no longer be recorded in the feet by the end of the third decade. Sensory action potentials are absent in the four limbs.

Somatosensory (SEP)-, brain stem auditory (BAEP)-, and pattern-reversal visual (VEP)-evoked potentials were recorded in 67 ARSACS patients (age: 5–56; mean 25). The results (DeLéan et al., 1989) showed a widespread axonal degeneration process in the primary sensory neurons as well as in the central nervous system.

Electroencephalographic (EEG) abnormalities occur in more than 60% of the cases (Bouchard et al., 1979). Bursts of generalized slow waves of subcortical origin are seen in most patients. Epileptic activity is rarely seen on the tracings, but 7% of the patients have generalized epilepsy, which starts in the late teens but can be easily controlled with an anti-epileptic drug monotherapy.

Electronystagmographic recordings of oculomotor and vestibular function were carried out in 11 patients (Dionne et al., 1979). Horizontal gaze nystagmus, marked impairment of smooth ocular pursuit, and optokinetic nystagmus, as well as defective fixation suppression of caloric nystagmus were found in all ARSACS patients. These findings suggest a diffuse cerebellar disease with particular involvement of the vermis and vestibulocerebellum.

## IV. NEUROIMAGING

Atrophy of the superior cerebellar vermis is always present on computed tomography scan or magnetic resonance imaging (Bouchard, 1991; Langelier et al., 1979), even in younger patients, and progresses slowly. The inferior vermis remains thicker throughout the disease, but there is a progressive cerebellar cortical atrophy. Cerebral atrophy becomes important in late life. Nevertheless, the cerebral white matter never shows abnormal signals, even late in the disease. The cervical spinal cord is flattened and markedly reduced. There is no scoliosis.

## V. NEUROPATHOLOGY

The pathological findings in a postmortem examination of a 21-year-old man were described previously (Bouchard, 1991). The superior cerebellar vermis was grossly atrophic, especially the anterior structures (central lobule and culmen). In this area Purkinje cells were practically absent, without significant Bergman's glia proliferation. In the spinal cord there was a severe bilateral symmetrical loss of myelin staining centered on the lateral corticospinal tracts, extending into the adjacent dorsal cerebellar tracts. There was no significant alteration of the posterior columns. It must be emphasized that the patient was young.

A second autopsy on a 59-year-old man revealed the presence of the same anomalies, but more pronounced with mild-to-moderate involvement of the gracilis and cuneiform fasciculi. Neuronal loss and gliosis were found in the anterior horn, thoracic nuclei, and globus palidus. Swollen perikarya were found in several of their neurons as well as in the thalamus, dentate nuclei, Purkinje cell layer, and, to a minor degree, in several other structures. In the thalamus, this swelling was occasionally associated with an increased amount of abnormal lipofuscin-like coarse granules. Tau-positive neurofibrillar degeneration was found in the hippocampus, the neocortex, the basal nucleus of Meynert, and rarely in other deep gray or brain stem structures. (Bouchard et al., 2000).

Currently we have no genotypic data on the samples that have undergone detailed neuropathologic studies.

Sural nerve biopsies were performed in six cases. An important loss of large myelinated fibers was seen at both the calf and ankle levels, and implied fibers of smaller size as well. On teasing, there was an increased variability of the internodal length along the same fiber, but this was quite variable. There was no sign of active myelin breakdown and only occasional changes suggestive of active axonal breakdown were observed (Bouchard, 1991; Peyronnard et al., 1979). Eight patients had proximal muscle biopsies. There was moderate-to-marked grouping of fiber types in five cases. Type I fiber hypotrophy was seen in five patients, and type II fiber hypotrophy in one. No sign of chronic or active denervation and reinnervation was observed (Peyronnard et al., 1979). In summary, the peripheral nervous system changes observed point to an early involvement and a possible developmental defect in this disease, as underlined by early abnormalities noted by motor and sensitive nerve conduction studies.

## VI. GENOTYPE/PHENOTYPE CORRELATIONS

Most Québec patients are homozygous for the $\Delta$T mutation. There are no phenotypic differences between them and the rare $\Delta$T/C > T compound heterozygotes. To date only one patient was found to be homozygous for the C > T mutation. The clinical picture was indistinguishable from the others. Even more intriguing, the clinical picture of the six Québec ARSACS patients heterozygous for the $\Delta$T mutation and an unknown mutation is also indistinguishable from the others. Finer genotype/phenotype subclassification may be possible when other samples with known genotype become available for neuropathology.

Mutation determination is underway for chromosome 13q linked cases of ARSACS in Turkish families (Topaloğlu, personal communication). The discovery of new sacsin mutations may permit us to start to understand the clinical differences between the commonly seen ARSACS cases described in Québec and the rare variants seen elsewhere.

ARSACS, a disease originally described in a population with demonstrated founder effect, is now taking its place among the diseases that should be considered in the differential diagnosis of recessive cerebellar ataxias with spasticity in other populations.

## Acknowledgments

I wish to thank Drs. J.-P. Bouchard, J. Mathieu, Y. Robitaille, and J. Michaud for their continued support in providing patient material, clinical information, results of neuropathology investigations, and especially for their interest in our efforts to try to understand ARSACS. The identification of sacsin would not have been possible without Drs. T. J. Hudson, K. Morgan, J. Rioux, J. C. Engert, and S. B. Melançon. I would like to thank Ms. J. Mercier, C. Prévost, and F. Gosselin for their prolonged contributions to this work. The investigation of ARSACS by our research group is currently supported by operating grant MOP14668 from the Canadian Institutes of Health Research. Previous funding has been obtained from the Muscular Dystrophy Association of Canada, the March of Dimes, the National Ataxia Foundation, and l'Association Canadienne de l'Ataxie de Friedreich.

## Refererences

Berciano, J., Mateo, I., De Pablos, C., Polo, J. M., and Combarros, O. (2002). Friedreich ataxia with minimal, GAA expansion presenting as adult-onset spastic ataxia. *J. Neurol. Sci.* **194**, 75–82.

Bouchard, J.-P. (1991). Recessive spastic ataxia of Charlevoix-Saguenay. In "Hereditary Neuropathies and Spinocerebellar Atrophies," *Handbook of Clinical Neurology*. (J.M.B.V. de Jong, ed.). Vol. 16 (60), pp. 451–459. Elsevier Science, Amsterdam.

Bouchard, J.-P., Barbeau, A., Bouchard, R., and Bouchard, R. W. (1978). Autosomal recessive spastic ataxia of Charlevoix-Saguanay. *Can. J. Neurol. Sci.* **5**, 61–69.

Bouchard, J.-P., Barbeau, A., Bouchard. R., and Bouchard, R. W. (1979). Electromyography and nerve conduction studies in Friedreich's ataxia and autosomal recessive spastic ataxia of Charlevoix-Saguenay (ARSACS). *Can. J. Neurol. Sci.* **6**, 185–189.

Bouchard, J.-P., Bouchard, R. W., Gagné, F., Richter, A., and Melançon. S. (1993). Recessive spastic ataxia of Charlevoix-Saguenay (RSACS): clinical, morphological and genetic studies. In: *Handbook of Cerebellar Disease*. (R. Lechtenberg, ed.), pp. 491–494. Marcel Dekker, New York.

Bouchard, J.-P., Richter, A., Mathieu, J., Brunet, D., Hudson, T. J., Morgan, K., and Melançon, S. B. (1998). Autosomal recessive spastic ataxia of Charlevoix-Saguenay. *Neuromusc. Disord.* **8**, 474–479.

Bouchard, J.-P., Richter, A., Melançon, S. B., Mathieu, J., and Michaud, J. (2000). Autosomal recessive spastic ataxia (Charlevoix-Saguenay). In: *Handbook of Ataxia Disorders*. (T. Klockgether, ed.), pp. 311–324. Marcel Dekker, New York.

Chaigne, D., Brauer, E., Ruh, D., Jouart, D., and Juif, J. G. (1993). L'ataxie spastique autosomique récessive: étude clinique, neurophysiologique, ophtalmologique et IRM de deux cas familiaux. *Rev. Neurol. (Paris)* **149**, 585. (Abstract).

De Braekeleer, M., Giasson, F., Mathieu, J., Roy, M., Bouchard, J.-P., and Morgan, K. (1993). Genetic epidemiology of autosomal recessive spastic ataxia of Charlevoix-Saguenay in northeastern Quebec. *Genet. Epidemiol.* **10**, 17–25.

De Braekeleer, M. (1991) Hereditary disorders in Saguenay-Lac-St-Jean (Quebec, Canada). *Hum. Heredity* **41**, 141–146.

Dionne, J., Wright, G., Barber, H., Bouchard, R., and Bouchard, J. P. (1979). Ocularmotor and vestibular findings in autosomal recessive spastic ataxia of Charlevoix-Saguenay. *Can. J. Neurol. Sci.* **6**, 177–184.

Engert, J. C., Dore, C,. Mercier, J., Ge, B., Betard, C., Rioux, J. D., Owen, C., Berube, P., Devon, K., Birren, B., Melancon, S. B., Morgan, K., Hudson. T. J., and Richter, A. (1999). Autosomal recessive spastic ataxia of Charlevoix-Saguenay (ARSACS): high-resolution physical and transcript map of the candidate region in chromosome region 13q11. *Genomics* **62**, 156–162.

Engert, J. C., Bérubé, P., Mercier, J., Doré, C., Lepage, P., Ge, B., Bouchard, J.-P., Mathieu, J., Melançon, S. B., Schalling, M., Lander, E. S., Morgan, K., Hudson, T.J., and Richter, A. (2000). ARSACS, a spastic ataxia common in northeastern Quebec, is caused by mutations in a new gene encoding an 11.5 kb ORF. *Nat. Genet.* **24**, 120–125.

Gauvreau, D., Guérien, M., and Hamel, M. (1991). De Charlevoix au Saguenay: mesure et caractéristiques du mouvement migratoire avant 1911. In *Histoire d'un Génome: Population et Génétique dans l'est du Québec*. (G. Bouchard and M. De Braekeleer, eds.), pp. 145–162. Presses de l'Université du Québec, Sillery.

Gücüyener, K., Özgül, K., Paternotte, C., Erdem, H., Prud'homme. J.-F., Özgüç, M., and Topaloğlu, H. (2001). Autosomal recessive spastic ataxia in two unrelated Turkish families. *Neuropediatrics* **32**, 142–146.

Jetté, R., Gauvreau, D., and Guérien, M. (1991). Aux origines d'une région: le peuplement fondateur de Charlevoix avant 1850. In *Histoire d'un Génome: Population et Génétique dans l'est du Québec*. (G. Bouchard and M. De Braekeleer, eds.), pp. 75–107. Presses de l'Université du Québec, Sillery.

Langelier, R., Bouchard, J.-P., and Bouchard, R. (1979). Computed tomography of posterior fossa in hereditary ataxias. *Can. J. Neurol. Sci.* **6**, 195–198.

Mercier, J., Prévost, C., Engert, J. C., Bouchard, J.-P., Mathieu, J., and Richter, A. (2001). Rapid detection of the sacsin mutations causing autosomal recessive spastic ataxia of Charlevoix-Saguenay. *Genet. Test.* **5**, 255–259.

Mrissa, N., Belal, S., Hamida, C. B., Amouri, R., Turki, I., Mrissa, R., Hamida, M. B., and Hentati, F. (2000). Linkage to chromosome 13q11-12 of an autosomal recessive cerebellar ataxia in a Tunisian family. *Neurology* **54**, 1408–1414.

Pascual-Castroviejo, I., Pascual-Pascual, S. I., Viano, J., and Martinez, V. (2000). Charlevoix-Saguenay type recessive spastic ataxia. A report of a Spanish case. *Rev. Neurol.* **31**, 36–38.

Peyronnard, J. M., Charron, L., and Barbeau, A. (1979). The neuropathy of Charlevoix-Saguenay ataxia: an electrophysiological and pathological study. *Can. J. Neurol. Sci.* **6**, 199–203.

Richter, A., Rioux, J. D., Bouchard, J.-P., Mercier, J., Mathieu, J., Ge, B., Poirier, J., Julien, D., Gyapay, G., Weissenbach, J., Hudson, T. J., Melançon, S. B., and Morgan, K. (1999). Location score and haplotype analysis of the locus for autosomal recessive spastic ataxia of Charlevoix-Saguenay in chromosome region 13q11. *Am. J. Hum. Genet.* **64**, 768–775.

Sorbi, S., Forleo, P., Cellini, E., Piacentini, S., Serio, A., Guarnieri. B., and Petruzzi, C. (2000). Atypical Friedreich ataxia with a very late onset and an unusual limited GAA repeat. *Arch. of Neurol.* **57**, 1380–1382.

# CHAPTER 21

# Ataxia–Telangiectasia

**KAI TREUNER**
The Laboratory of Genetics
The Salk Institute for Biological Studies
La Jolla, California 92037

**CARROLEE BARLOW**
The Laboratory of Genetics
The Salk Institute for Biological Studies
La Jolla, California 92037

I. Phenotype
  A. Neurological Phenotype
  B. Non-Neurological Phenotypes
II. ATM Gene and Function Based on Human Data
III. Biochemical Targets of ATM Kinase Activity
IV. Diagnosis
  A. Clinical
  B. Ancillary Tests
  C. Neuroimaging
  D. DNA and Cell-Based Testing
V. Cellular and Animal Models of the Disease
  A. The Role of ATM in DNA Repair
  B. The Role of ATM in Checkpoint Control
  C. Cytoplasmic Function of ATM
VI. Animal Models of A–T
VII. Genotype/Phenotype Correlations
VIII. Treatment
IX. Syndromes Related to A–T
  Acknowledgments
  References

Ataxia–Telangiectasia (A–T) is an autosomal recessive pleiotropic disease caused by mutations in the gene ataxia telangiectasia mutated (ATM). Affected children present as toddlers with truncal ataxia and oculomotor disturbances. The disease proceeds relentlessly with progressive neurodegeneration involving the cerebellum and multiple other brain regions leading to severe neurological compromise and death between 20–30 years of age. Patients also present with dilated blood vessels in the eyes and on the skin (telangiectasia), have immunodeficiency, are infertile, and more than one-third of the patients develop cancer. Patients are hypersensitive to ionizing radiation (IR) and reagents that cause DNA double strand breaks (DSBs).

It is interesting to note that while the clinical disorder affects only approximately 500 children in the U.S., the number of laboratories that intensively study the gene mutated in the disease is significant, and several mouse models of the disease exist. This likely stems from the pleiotropic nature of the disease and the insights that can be gained regarding the immune system, neural development, germ cell formation, and cancer. This work has led to major advances in our understanding of the disease and in our ability to detect patients with the disease. In this chapter we will review the basic clinical phenotypes and present an overview of the field paying particular attention to the neurological aspects of the disease, recent advances in the field, and areas that remain somewhat controversial.

## I. PHENOTYPE

The clinical syndrome was originally described by Boder and Sedgwick (1958), although the first report of progressive ataxia with telangiectasia dates back to 1926. There is extensive literature describing the major phenotypes of the disease (Gatti, 2001; Kastan and Lim, 2000; Spacey et al., 2000) and the reader is referred to these materials for detailed descriptions. Recently, several new insights have been gained regarding some phenotypes and we present a more detailed review of these areas.

## A. Neurological Phenotype

The most prominent clinical phenotype in A–T is progressive neurodegeneration. Several important points about the disease remain somewhat controversial, although as more studies are conducted the precise nature of the neurological phenotype is becoming clearer. Clinical studies as well as recent pathological analysis clearly demonstrate that in the majority of patients the phenotype is not manifest until 1–2 years of age. Based on these data, it is widely accepted by the field that the progressive neurodegeneration in A–T is not due to abnormal brain development. For example, the pathological analysis of autopsy material obtained from infants that were clinically normal but which had A–T were described. In these reports, the number of cells, the volume of the brain, and the morphology of the neurons were normal (Gatti, 2001 and references therein). However, the majority of pathological studies have been performed using autopsy materials obtained from older patients where it is difficult to determine the precise nature of the abnormalities as global neuronal cell loss is present, particularly in the cerebellum. In these cases, evidence of abnormal migration of Purkinje cells and/or abnormal Purkinje cell dendritic arborization exists (Vinters et al., 1985) as well as evidence of global degenerative changes affecting virtually all areas of the brain and spinal cord (Gatti, 2001). Clinically, the phenotype becomes apparent at approximately 1–2 years of age when there is clear evidence of truncal ataxia and oculomotor dysfunction consistent with cerebellar and/or basal ganglia dysfunction. Within several years of age the ataxia also involves peripheral coordination. Oculomotor apraxia and slurred speech are also noted early. Recently a quantitative neurologic assessment of patients was developed that allows the physician to more accurately follow the progression of the disease (Crawford et al., 2000). As the disease progresses, it is obvious that the cerebellum is involved and this is also evident by MRI at approximately 4–6 years of age. Importantly, A–T patients are not mentally retarded and motor impairment is the main manifestation of the disease. Children are able to learn to walk but due to progressive ataxia are generally wheelchair-bound by age 10. Choreoathetosis is found in almost all these patients followed by contractures in fingers and toes. Patients generally succumb to sinopulmonary infections as a result of neurological dysfunction in their 20s to 30s.

## B. Non-Neurological Phenotypes

### 1. Cancer

It is well established that A–T patients have an increased risk of developing cancer. Nearly one-third of A–T patients develop cancer during their lifetime (Khanna, 2000). The vast majority of these tumors are of lymphoid origin and mainly consist of aggressive and invasive lymphocytic leukemia and non-Hodgkin's lymphoma and a small number of patients develop nonlymphoid cancers (Boultwood, 2001; Haidar et al., 2000; Hecht and Hecht, 1990; Kastan, 1995; Morgan and Kastan, 1997; Starostik et al., 1998; Taylor et al., 1996).

### 2. Radiation Sensitivity

A–T patients cannot tolerate normal doses of chemotherapeutic agents that induce DSBs, nor radiation therapy at standard doses. Because patients are radiosensitive, conventional therapeutic regimens for the treatment of lymphoid malignancies are contraindicated. However, most patients respond well to a modified regimen (Abadir and Hakami, 1983; Kastan, 1995). The radiation sensitivity is also seen in cells isolated from A–T patients and several groups are working to exploit this finding for diagnostic purposes. We will discuss the details of the types of radiosensitivity found in A–T cell lines, as well as the potential to use the pattern of radiosensitivity as a diagnostic test for the disease, in Section VIII.

### 3. Infertility

Patients with A–T are likely infertile due to meiotic failure and germ cell degeneration. Because of the severity of the neurological aspect of the disease, reproduction is precluded and consequently, a systematic characterization of the phenotype has not been performed. There is some controversy as to whether or not male patients can produce viable sperm, but the prevailing view is that male A–T patients do not produce sperm although patients clearly have normal levels of testosterone (Gatti, 2001). In contrast, some patients have been documented that experience menses although there has never been a study to determine if ovulation occurs (Gatti, 2001). The mouse model of the disease has proven extremely useful in clarifying the nature of this aspect of the disease and we will discuss the role of ATM in germ cell development later in the chapter.

### 4. Telangiectasia

Approximately 95% of A–T patients develop telangiectasias within their lifetime. However, these are usually not detected until after 5 years of age. Telangiectasias are generally found on the conjunctiva and skin, mainly in sun-exposed areas, although there are anecdotal reports of their occurrence in non-sun-exposed areas. Radiation exposure greatly enhances their occurrence as does age. Generally it is accepted that the telangiectasias resemble the dilations of blood vessels characteristically found in normal aging or in patients with other disorders rather than an angiogenic response. The mechanism leading to the telangiectasias remains unclear but the most widely accepted view is that they occur as part of the normal cellular response to the increased DNA damage caused by aging, sun exposure, or irradiation.

## 5. Immune System Dysfunction

Many but not all patients have some degree of immune system impairment. Abnormalities range from mild IgA deficiencies to severe IgG deficiencies and T-cell dysfunction. Most patients at autopsy have no or minimal normal thymic tissue. There are significant cytogenetic abnormalities found in peripheral lymphoid cells in patients, included polyploidy (Naeim et al., 1994) and frequent translocations with aberrations of chromosome 14 including translocations to chromosome 7 (Taylor et al., 1996). In spite of these abnormalities, repeated studies have shown that both B- and T-cell function are virtually normal in most A–T patients. It is clear, however, that several different types of immune dysfunction can be found in some A–T patients. Importantly, no consistent immune system dysfunction has been found and even patients within the same family have widely varying immune function.

## 6. Diabetes

A recent finding is the increased incidence of diabetes in A–T patients, although the mechanism remains ambiguous. There have been several case reports demonstrating that A–T patients with hepatic dysfunction show evidence of insulin resistance, based on a blunted hypoglycemic response to exogenous insulin. The nature of the insulin resistance however, has differed in different clinical conditions. In some cases, clearly detectable insulin receptor antibodies (Bar et al., 1978) were present, whereas in other cases no antibodies were detected (Blevins and Gebhart, 1996). Importantly, however, Blevins and Gebhart conclusively showed a loss of the first phase of insulin response to intravenous glucose showing that A–T patients have clear evidence of pancreatic beta cell dysfunction (Blevins and Gebhart, 1996). It remains to be determined why A–T patients develop diabetes and several groups are beginning to study the role of ATM in insulin signaling pathways.

## II. ATM GENE AND FUNCTION BASED ON HUMAN DATA

A major breakthrough in A–T research came with the discovery of the gene mutated in the clinical syndrome. The original linkage analysis, performed by Richard Gatti and colleagues (1988) led to the localization of the gene to chromosome 11q22–23. The gene was then positionally cloned by an international consortium led by Yosef Shiloh and Francis Collins in 1995 (Savitsky, 1995). The gene, named ATM for ataxia telangiectasia mutated, spans over 150 kb of genomic DNA sequence and consists of 66 exons. The mRNA transcript is 13 kb in length and contains an open reading frame of 9.2 kb. It was shown that the ATM gene is ubiquitously expressed in various tissues and cell types, including brain, thymus, testis, and ovary. It shares a bi-directional promotor with the NPAT (E14/CAN3) gene, which interestingly encodes a substrate of the CyclinE/ CDK2 kinase involved in S-phase entry.

Over 400 mutations in the *ATM* gene have been identified thus far, and occur within almost every exon (for a list see http://www.vmresearch.org/atm.htm). Most mutations result in a truncated ATM protein. Only a few are due to missense mutations and these have been found to destabilize the mRNA or interfere with translation. One potential hot spot involves exon 54 in close proximity to the PI-3 Kinase domain, which might indicate an important structural function for this region. Interestingly most of A–T cell lines do not appear to contain any detectable ATM protein, indicating that the truncated ATM protein is unstable and is likely rapidly degraded in the cell.

The gene encodes a large protein of 3056 amino acids with a molecular weight of approximately 350 kDa. The ATM protein carries a carboxy terminal kinase domain with homology to the lipid kinase phosphatidylinositol-3 kinase (PI-3K), although ATM does not appear to be a lipid kinase but rather functions as a serine/threonine protein kinase. The protein belongs to the family of high molecular weight kinases that are conserved throughout eukaryotic evolution from yeast to mammals. These kinases regulate diverse processes such as proliferation, checkpoint control, meiotic progression, apoptosis, telomere length, and V(D)J-recombination. Members of this family besides ATM are the catalytic subunit of the mammalian DNA-dependent protein kinase (DNA-PKcs), ATR (ATM and Rad3-related), Tel1p, Mec1p, and Rad3p.

## III. BIOCHEMICAL TARGETS OF ATM KINASE ACTIVITY

It is well established that ATM kinase activity is induced after IR. ATM is known to phosphorylate target proteins at a sequence motif consisting of a serine or threonine residue followed by a glutamine residue. The optimal sequence for ATM phosphorylation was determined to be leucine-serine-glutamine-glutamic acid (LSQE). Several *in vivo* targets of ATM kinase activity have been identified. Targets include the tumor suppressor protein p53 and the regulator of its degradation MDM2, the breast cancer susceptibility gene BRCA1 and its interaction partner CtIP, the Nijmegen breakage syndrome protein Nbs1/p95, the telomeric protein TRF1/Pin2, the histone H2AX, and the checkpoint kinase chk2. ATM can also phosphorylate several other proteins *in vitro* or in an ATM-dependent manner. Among these are the nonreceptor tyrosine kinase c-Abl, replication protein A (RPA), the checkpoint kinase chk1, eIF-4E-binding protein 1 (4E-BP1), the human checkpoint protein Rad17, the NF-κB inhibitor IκBα and the IKK kinases (for a review see Kastan and Lim, 2000). The vast number of substrates in the cell reflects the importance of ATM in the regulation of diverse signaling pathways and repair mechanisms. Later in this review we will present in detail how ATM is involved in these processes based on *in vitro* results and studies using mouse models.

## IV. DIAGNOSIS

### A. Clinical

The diagnosis of A–T at present is based on clinical examination although with the advent of new sequencing methods and the cloning of the gene a few centers are working to establish reliable genetic tests for diagnosis. Clinical examination should reveal progressive cerebellar ataxia with onset between one and three years of age. A very sensitive measure of dysfunction was recently reported and consists of abnormalities in gaze (Lewis *et al.*, 1999) and ocular apraxia is almost universal by three years of age. The absence of telangiectasia is common and should not be considered in establishing a negative diagnosis but is often helpful in supporting the diagnosis. The telangiectasias generally occur between 4–6 years of age.

### B. Ancillary Tests

Supportive diagnostic tests include elevated serum alpha-fetoprotein (AFP) levels which are found in 95% of patients but which can vary markedly in the degree of elevation. Serum immunoglobulin levels can also be abnormal and the most consistent abnormality is IgA deficiency, although this can be highly variable.

### C. Neuroimaging

After four years of age MRI reveals cerebellar dystrophy. Functional MRI and PET are being evaluated as tools for both early diagnosis and for following the course of the disease. However, these methods are considered research tools because of several problems, including the risk of using radioactive tracers in A–T patients and the lack of cooperation with young patients.

### D. DNA and Cell-Based Testing

The majority of recent work has focused on establishing genetic testing and cell based diagnostic tests. For example, noncancerous lymphoid cells from A–T patients often show consistent karyotypic abnormalities such as T(7;14) translocations (Taylor *et al.*, 1996). Recent work from Dr. Gatti's laboratory has focused on optimizing a colony survival assay with isolated and transformed peripheral lymphocytes as a diagnostic method. This may prove particularly useful for diagnosing young A–T patients who present with cancer where there is often confusion as to the diagnosis and where treatment with conventional therapies is life threatening. The assay may also prove useful in detecting carriers of the disease mutation. Clearly the best diagnostic test would be to detect mutations at the DNA level. However, this approach is still not feasible, although several groups continue to work toward the goal.

## V. CELLULAR AND ANIMAL MODELS OF THE DISEASE

### A. The Role of ATM in DNA Repair

A variety of human diseases are caused by mutations in DNA repair proteins and cell cycle checkpoint proteins, several of which are known to be ATM targets (see Fig. 21.1A). IR causes DNA damage mainly in the form of DSBs, which can be repaired by either nonhomologous end joining (NHEJ) or homologous recombination (HR). NHEJ is thought to be the predominant pathway for DNA repair in humans and other mammals. Several proteins have been identified that are essential components of the NHEJ repair mechanism, among which are the catalytic subunit of DNA-PK, associated with the Ku70/Ku80 dimer and the DNA-ligase IV/XRCC4 complex. HR seems to be the dominant DSB repair pathway in yeast while its role in non-meiotic mammalian DNA repair remains unclear. HR requires the undamaged sister chromatid or sister chromosome to repair the broken DNA molecule and is mediated by a multimeric protein complex containing several members of the Rad protein family, including Rad51, Rad52, and Rad54.

One important question that remains unresolved is whether ATM regulates either one or possibly both DNA repair pathways and how it might exert this function. There have been reports of an increase in HR in mice deficient in ATM suggesting that NHEJ might be impaired and hence HR pathways are recruited in the absence of ATM (Bishop *et al.*, 2000). Alternatively, ATM might be critical for the normal regulation of HR. Other groups found evidence for a functional role of ATM in HR by analyzing the repair efficiency of either Ku70 or Rad54-deficient chicken DT40 cells in the absence of ATM. Ku70 and ATM deficient cells showed increased radiosensitivity and a dramatic inability to repair chromosome damage, while Rad54 and ATM-deficient cells show only slight abnormalities (Morrison *et*

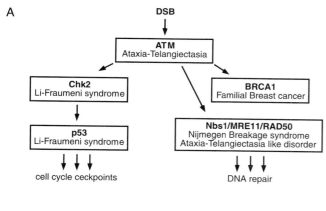

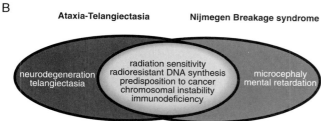

FIGURE 21.1 Defects in DNA repair trigger multiple diseases. (A) Shown are mutations in DNA repair proteins or cell cycle checkpoint proteins which have been documented in different human diseases. (B) Overlapping and nonoverlapping phenotypes of A–T patients and Nijmegen breakage syndrome patients. The various phenotypes shared by both patients are shown in yellow, the unique individual phenotypes are shown in red for A–T and green for NBS patients.

al., 2000). Further support for a role of ATM in HR comes from recent results showing defective assembly of the Rad51 complex in A–T cells in response to DSBs (Chen et al., 1999). Interestingly, the assembly of the Rad51 complex seems to be dependent on the phosphorylation of Rad51 by c-Abl, a nonreceptor tyrosine kinase which itself is activated by ATM in response to IR (Baskaran et al., 1997).

Importantly, multiple laboratories have shown that in the absence of ATM the kinetics of DSB repair is normal. This raises the possibility that the repair machinery is normal in the absence of ATM but that the ability to sense DNA damage may be abnormal. Consistent with this idea our laboratory has found that ATM deficiency causes no defect in the expression levels of various DNA repair proteins in neural progenitor cells of Atm–/– mice (Treuner et al., unpublished). It is currently unknown by which mechanism ATM and other kinases like ATR, are activated after DNA damage and how the cell senses the presence of DSBs. It has been suggested that several members of the Rad group of checkpoint proteins might function as DNA damage sensors. Three of these Rad proteins, Rad1, Rad9, and Hus1 form a complex which encircles the DNA and serves as a loading platform for the DNA repair machinery at sites of broken or damaged DNA. Further studies are required to understand the different signal transduction pathways, which are activated for the repair of DNA damage.

## B. The Role of ATM in Checkpoint Control

Cells derived from A–T patients, such as fibroblasts and lymphoblasts, display several characteristic abnormalities in culture including poor growth, high requirement for serum growth factors, chromosomal instability, reduced life span, hypersensitivity to IR, and failure to activate the $G_1$, S, and $G_2$ cell cycle checkpoints after DNA damage. The activation of these checkpoints leads to an arrest of the cell cycle providing time for the cell to repair the DNA and survive or induce apoptosis.

The IR-induced G1 checkpoint is the best characterized abnormal cellular checkpoint in ATM-deficient cells. The phosphorylation of p53 by ATM results in the stabilization and activation of p53. The stabilization of p53 is likely mediated by preventing its binding to the MDM2 protein, which targets p53 for ubiquitin-dependent degradation. Although p53 has multiple targets, the G1 checkpoint is mediated by the p53-dependent increase in cyclin-dependent kinase inhibitor p21. The transcriptional activation of p21 by p53 leads to the suppression of cyclinE/Cdk2 and prevents $G_1$- to S-phase progression of the cells.

The S-phase checkpoint is required to prevent the replication of damaged DNA. Failure to activate this checkpoint after IR leads to the radioresistant DNA synthesis (RDS) phenotype of A–T cells. It was shown that ATM activation of chk2 kinase is involved in this response by degrading Cdc25A, a protein tyrosine phosphatase that normally activates cyclinA/Cdk2 as cells progress from $G_1$ to S-phase. Additionally, Nbs1 also participates in this checkpoint. The phosphorylation of Nbs1 by ATM is well established, however, it is still unclear how the phosphorylation is linked to the function of Nbs1 and the RDS phenotype of NBS cells (Gatei et al., 2000; Lim et al., 2000; Wu et al., 2000; Zhao et al., 2000). Recent results support a model of two parallel, cooperating mechanisms which are both regulated by ATM (Falck et al., 2002).

The $G_2$ checkpoint is important to prevent the entry of DNA-damaged cells into mitosis. The current model postulates the activation of the chk1 and possibly chk2 kinase by either ATR and/or ATM leads to the phosphorylation and inhibition of the tyrosine phosphatase Cdc25C. The phosphorylation of Cdc25C creates a binding site for the 14–3–3 protein and results in a nuclear export and cytoplasmic sequestration of Cdc25C. As a result of this sequestration and possibly inactivation, Cdc25C can no longer dephosphorylate and activate the mitotic cyclinB/cdc2 kinase and the damaged cells can not enter mitosis. As outlined above, the role of ATM in the activation of this checkpoint is not completely understood, but is most likely mediated via the chk2-Cdc25C pathway.

## C. Cytoplasmic Function of ATM

Many functions of ATM such as its role in the DNA damage response and cell cycle control are consistent with a predominantly nuclear localization of the protein. However, a significant fraction of ATM is present in the cytoplasm and has been shown to be associated with peroxisomes and β-adaptin, a cytoplasmic protein involved in vesicle and protein transport (Lim et al., 1998). It was also recently shown that insulin can induce the kinase activity of ATM in a 3T3 L1 cell line after differentiation into adipocytes suggesting that ATM may also be activated by growth factors (Yang and Kastan, 2000). Finally, we and others have demonstrated that Atm−/− mice have a significant increase in the number of lysosomes in the brain and that ATM is found in the cytoplasm of Purkinje cells and other neurons (Barlow et al., 2000). It remains to be established whether the localization difference implicates an important role for ATM as a cytoplasmic protein in postmitotic neurons.

The participation of ATM in intracellular growth factor signaling opens a new and exiting field for future investigations regarding additional roles of ATM and may be of particular importance for the clinical phenotype of diabetes with pancreatic beta cell dysfunction and potentially neurological compromise.

## VI. ANIMAL MODELS OF A–T

The existence of a good animal model is of great importance to study a disease and to test potential treatments. Several laboratories including our own have generated a mouse model of A–T by targeted disruption of the murine Atm gene (Barlow et al., 1996). Atm−/− mice are viable and recapitulate most aspects of the disease. The mutant mice show growth retardation, infertility, reduced numbers of mature T- and B-cells, sensitivity to IR, and chromosomal instability. They develop aggressive thymic lymphomas between 2–4 months ultimately leading to death at age 4–5 months (the normal life span of a mouse is usually 2–3 years). Atm−/− mouse embryonic fibroblasts (MEFs) grow slowly in culture and reach early senescence. They also display high levels of chromosomal breaks similar to cells derived from A–T patients. The mouse model has proven particularly useful in understanding the cancer and germ cell phenotypes of the disease. Atm−/− mice are infertile and have very small testes and ovaries. They display complete absence of mature gametes. The arrest occurs as early as leptotene of prophase I, resulting in apoptotic degeneration of the germ cells. In addition, Rad51 localization on meiotic chromosomes is abnormal (Barlow et al., 1997). These results suggest that ATM may be involved in the regulation or surveillance of meiotic progression or prevents nicking or premature nicking in preparation for HR.

The mouse model of A–T also closely mimics the human condition with regard to the immune system and has been useful for defining the role of ATM in cancer (Barlow et al., 1996; Liyanage et al., 2000). Mutant mice display a reduction in mature CD4 or CD8 single-positive T lymphocytes. Virtually all Atm−/− mice succumb to aggressive T-cell lymphoblastic lymphoma (TCL) early in life, and these tumors closely resemble those found in A–T patients in several respect. (1) the vast majority of lymphoreticular cancers in A–T patients arise in childhood (Taylor et al., 1996), and the tumors also arise in very young Atm−/− mice (Barlow et al., 1996; Elson et al., 1996; Xu et al., 1996); (2) tumors in both humans (Boultwood, 2001; Hecht and Hecht, 1990; Starostik et al., 1998; Stoppa-Lyonnet et al., 1998) and mice (Barlow et al., 1996) are highly proliferative and invasive; (3) a large majority are of T-cell lineage and immature, CD4/CD8 double positive in the case of T-cell prolymphocytic leukemias arising in A–T patients (Taylor et al., 1996) and virtually all tumors arising in Atm−/− mice are also CD4/CD8 double positive; and (4) cytogenetically they both have translocations of chromosome 14 in humans and syntenic regions of mouse chromosomes 14 and 12 (Barlow et al., 1996; Liyanage et al., 2000; Petiniot et al., 2000). Chromosome 14 harbors the T-cell receptor alpha (TCRα) gene (located on chromosome 14q11.2 in humans and chromsome 14 at 19.5–7 cM in mice) which is often involved in the translocations and abnormalities of this loci. This has been detected using FISH in mice (Liyanage et al., 2000) and by karyotype analysis of lymphocytes in humans (Gatti, 2001). Because the loci involved normally undergo V(D)J recombination, it has been postulated that defects in V(D)J recombination are responsible for tumor formation in the absence of ATM. However, recently it was shown that V(D)J recombination is not required for lymphoma formation in the absence of Atm, as mice deficient in Rag1 or Rag2 (and therefore incapable of V(D)J recombination) and ATM still succumb to TCLs (Petiniot et al., 2000). An alternative possibility is that loss of ATM and the development of lymphoreticular malignancies occurs as a result of disruptions in signaling to cell cycle checkpoint machinery. Continued analysis of the mouse models may provide new insights and potentially new methods for treating the cancer found in the disease.

In contrast to the phenotypes described above, analysis of the mutant mice showed that the neurologic phenotype is markedly different in the animal as compared to the patients. The animal models do not show any evidence of neurodegeneration nor gross ataxia. The mice have a mild impairment in several motor function tests (Barlow et al., 1996). Extensive histological analysis has been performed by multiple laboratories and the overwhelming majority of work shows that the mutant mouse brain and spinal cord has normal architecture and no evidence of neuronal degeneration. A detailed examination of the cerebellum demonstrated normal Purkinje cell bodies, normal thickness of the granular cell and molecular layers, and normal migration and arborization as revealed by Golgi-Cox staining. However, the mice do show evidence of oxidative damage, particularly the Purkinje cells of the cerebellum, which are specifically affected in A–T (Barlow *et al.*, 1999). This suggests that increased oxidative stress may play a role in the death of neurons although the reason for the oxidative damage remains unknown. One potential source of oxidative stress is unrepaired DNA. In support of such a hypothesis, we recently showed that adult neurogenesis is abnormal in *Atm*–/– mice (Allen *et al.*, 2001; see Fig. 21.2). Neural progenitor cells of the dentate gyrus in *Atm*–/– mice have abnormally high rates of proliferation and genomic instability. In addition, they show a decrease in long-term survival and *in vitro* these cells are severely compromised in their ability to fully differentiate into oligodendrocytes and neurons. These findings suggest that neural progenitors are highly prone to incurring genome damage and that an inability to respond to such damage results in a substantial loss in viability. It is therefore possible that ATM deficiency allows damaged cells to be retained in the nervous system and that subsequent continuing genomic damage ultimately leads to neurodegeneration of these already compromised cells (Fig. 21.2). Further support for a role of ATM and DNA repair in the nervous system comes from a series of recent genetic studies in mice. Several different mutant mice have been crossed with the *Atm*–/– mice to generate mice deficient in ATM and other DNA repair proteins. Mice homozygous for mutations in the various components of the NHEJ repair pathway (DNA–PK-/- or SCID, Ku70-/-, Ku80-/-) and ATM die early during embryonic development. This suggests that ATM may be required for the proper function of the HR pathway if it serves as a backup pathway when NHEJ is impaired, or that ATM can compensate for the loss of function of these proteins in NHEJ. In contrast to the embryonic lethality found in the double mutants described above, loss of ATM rescues the embryonic lethality and neuronal apoptosis of both DNA Ligase 4 and XRCC4-deficient mice (Gao *et al.*, 1998; Lee *et al.*, 2000). Studies of these types will likely provide insight into the role of ATM in tissue, cell type, and process specific DNA repair. However, the mechanism and role of DNA repair in the adult nervous system and an explanation for the loss of neurons in the disease remains unclear.

One of the major drawbacks of the mouse model is the lack of progressive ataxia, which hinders the analysis of neurodegeneration in A–T. An important area of research will be to determine why the animal model shows preservation of the brain whereas all other organ systems are affected to the same degree and with the same phenotype as that of the patients. Studies in the animal model may also provide insight into the molecular events that occur early in the disease process, in spite of the lack of progressive abnormalities. However, the lack of an appropriate model with progressive neurodegeneration limits the usefulness of the model for testing therapies. Attempts are ongoing to "knockout" the ATM gene in primates, pigs, and cattle in an attempt to create new animal models of A–T. These experiments hold promise for studying the neurodegenerative aspect of A–T and might reveal new insights into the brain-specific function of ATM.

## VII. GENOTYPE/PHENOTYPE CORRELATIONS

As described earlier, several hundred different mutations have been found in the *ATM* gene (see web site http://www.vmresearch.org/atm.htm). The overwhelming majority of mutations results in complete loss of function and those that are missense mutations generally affect the kinase domain (for an extensive review of the mutations that have been found please refer to Gatti, 2001). An important

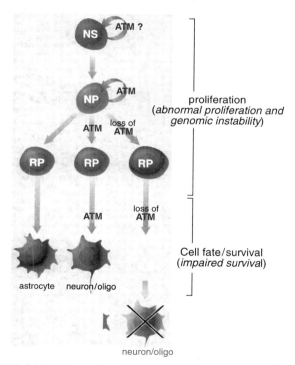

FIGURE 21.2 Presumptive role for ATM in the nervous system. During neural development ATM is required to maintain appropriate cell cycle response at baseline and in response to growth cues and also contributes to maintaining genomic stability. In the absence of ATM, neural progenitor cells (NP) have an increased rate of proliferation. Restricted progenitor cells (RP) with genomic aberrations are produced that are able to proliferate but have a decreased ability to differentiate along a nonastrocytic lineage and are less able to survive.

aspect of these studies that remains well established is that regardless of the mutation, the majority of patients have very similar phenotypes. Only a few rare patients have missense mutations and a slightly milder phenotype but at this stage the data are still controversial (McConville *et al.*, 1996; Taylor *et al.*, 1994). It remains to be established if there are any true genotype/phenotype correlations in A–T patients but as more efficient genotyping techniques become available these types of studies may prove extremely useful. In contrast to the lack of evidence of a genotype/phenotype correlation in A–T patients, mounting evidence suggests that there are phenotypic consequences that increase a persons risk for cancer in A–T carriers with particular mutations that give rise to missense mutations but a discussion of the heterozygous phenotype is outside the scope of this chapter.

## VIII. TREATMENT

The treatment of lymphoid leukemias with radiation therapy and radiomimetic drugs is problematic due to radiosensitivity, but treatment can be done using modified regimens. Several other symptoms of A–T such as frequent pulmonary or sinus infections can be treated conventionally with antibiotics. In contrast, there is currently no effective treatment for the neurodegeneration in A–T patients, and the underlying mechanisms causing neuronal cell death are poorly understood. In recent years patients have been treated with antioxidants like vitamin E, coenzyme Q10 and folic acid based on several studies demonstrating the presence of oxidative damage in A–T cells. The recent discovery of neural stem cells may be of importance because engraftment of these cells may eventually become a promising therapeutic approach. It remains to be seen whether retrovirus-based gene therapy might create new opportunities in the future. The challenge will be to achieve sufficient ATM expression in a high percentage of the cells in the brain, thymus, or other target organs. Most importantly, continued efforts to understand the role of ATM in the brain may help define new therapeutic options for the disease.

## IX. SYNDROMES RELATED TO A–T

Patients with Nijmegen breakage syndrome exhibit several similarities to A–T patients. They show extreme radiation sensitivity, radioresistant DNA synthesis (RDS), chromosomal instability, and predisposition to cancer. However, an important difference is the lack of neurodegeneration seen in A–T patients, showing that ATM and Nbs1 have nonoverlapping functions in the brain (see Fig. 21.1B). Interestingly, Nbs1, MRE11, and RAD50 form a trimeric protein complex, which participates in DNA repair and is important for the DNA damage checkpoint during S-phase. Hypomorphic mutations in the MRE11 gene give rise to a syndrome named ataxia-telangiectasia-like disorder (ATLD), because of the high similarity to A–T (see also Fig. 21.1A). ATLD patients do not have mutations in the ATM gene and have normal levels of the ATM protein. ATLD is characterized by radiation sensitivity, RDS, and progressive cerebellar degeneration, although the severity of the disease seems to markedly different. Unfortunately, the precise clinical characteristics of these patients remains to be established, in particular the nature of the progressive neurological dysfunction. These data provide evidence that ATM and the Nbs1/MRE11/RAD50 complex act in the same DNA damage response pathway. It will be important in the future to find out why A–T and ATLD patients show neurodegeneration and NBS patients do not and to characterize the molecular events leading to neuronal cell death.

### Acknowledgments

We would like to acknowledge R. Gatti for sharing unpublished work and for advice. We apologize to colleagues whose work we could not cite due to space restrictions. This work was supported by the Searle Family Trust Scholar Award and The V Foundation for Cancer Research to CB. KT is supported by the Deutsche Forschungsgemeinschaft.

### References

Abadir, R., and Hakami, N. (1983). Ataxia telangiectasia with cancer. An indication for reduced radiotherapy and chemotherapy doses. *Br. J. Radio.* **56**, 343–345.

Allen, D., van Praag, H., Ray, J., Weaver, Z., Winrow, C. J., Carter, T. A., Braquet, R., Harrington, E., Ried, T., Brown, K. D. *et al.* (2001). Ataxia telangiectasia mutated is essential during adult neurogenesis *Genes Dev.* **15**, 554–566.

Bar, R. S., Levis, W. R., Rechler, M. M., Harrison, L. C., Siebert, C., Podskalny, J., Roth, J., and Muggeo, M. (1978). Extreme insulin resistance in ataxia telangiectasia: defect in affinity of insulin receptors. *N. Engl. J. Med.* **298**, 1164–1171.

Barlow, C., Dennery, P. A., Shigenaga, M. K., Smith, M. A., Morrow, J. D., Roberts, L. J., 2nd, Wynshaw-Boris, A., and Levine, R. L. (1999). Loss of the ataxia-telangiectasia gene product causes oxidative damage in target organs. *Proc. Natl. Acad. Sci. U.S.A.* **96**, 9915–9919.

Barlow, C., Hirotsune, S., Paylor, R., Liyanage, M., Eckhaus, M., Collins, F., Shiloh, Y., Crawley, J., Ried, T., Tagle, D., and Wynshaw-Boris, A. (1996). Atm deficient mice: A paradigm of ataxia-telangiectasi. *Cell* **86**, 159–171.

Barlow, C., Liyanage, M., Moens, P., Deng, C. X., Ried, T., and Wynshaw-Boris, A. (1997). Partial rescue of the severe prophase I defects of Atm-deficient mice by p53 or p21 mutant alleles. *Nat. Genet.* **17**, 462–466.

Barlow, C., Ribaut-Barassin, C., Zwingman, T. A., Pope, A. J., Brown, K. D., Owens, J. W., Larson, D., Harrington, E. A., Haeberle, A. M., Mariani, J. *et al.* (2000). ATM is a cytoplasmic protein in mouse brain required to prevent lysosomal accumulation. *Proc. Natl. Acad. Sci. U.S.A.* **97**, 871–876.

Baskaran, R., Wood, L. D., Whitaker, L. L., Canman, C. E., Morgan, S. E., Xu, Y., Barlow, C., Baltimore, D., Wynshaw-Boris, A., Kastan, M. B., and Wang, J. Y. (1997). Ataxia telangiectasia mutant protein activates c-Abl tyrosine kinase in response to ionizing radiation. *Nature* **387**, 516–519.

Bishop, A. J., Barlow, C., Wynshaw-Boris, A. J., and Schiestl, R. H. (2000). Atm deficiency causes an increased frequency of intrachromosomal homologous recombination in mice. *Cancer Res.* **60**, 395–399.

Blevins, L. S., Jr., and Gebhart, S. S. (1996). Insulin-resistant diabetes mellitus in a black woman with ataxia-telangiectasia. *South Med. J.* **89**, 619–621.

Boder, E., and Sedgwick, R. P. (1958). Ataxia-telangiectasia: a familial syndrome of progressive cerebellar ataxia, oculocutaneous telangiectasia and frequent pulmonary infection. *Pediatrics* **21**, 526–554.

Boultwood, J. (2001). Ataxia telangiectasia gene mutations in leukaemia and lymphoma. *J. Clin. Pathol.* **54**, 512–516.

Chen, G., Yuan, S. S., Liu, W., Xu, Y., Trujillo, K., Song, B., Cong, F., Goff, S. P., Wu, Y., Arlinghaus, R. et al. (1999). Radiation-induced assembly of Rad51 and Rad52 recombination complex requires ATM and c-Abl. *J. Biol. Chem.* **274**, 12748–12752.

Crawford, T. O., Mandir, A. S., Lefton-Greif, M. A., Goodman, S. N., Goodman, B. K., Sengul, H., and Lederman, H. M. (2000). Quantitative neurologic assessment of ataxia-telangiectasia, *Neurology* **54**, 1505–1509.

Elson, A., Wang, Y., Daugherty, C. J., Morton, C. C., Zhou, F., Campos-Torres, J., and Leder, P. (1996). Pleiotropic defects in ataxia-telangiectasia protein-deficient mice. *Proc. Natl. Acad. Sci. U.S.A.* **93**, 13084–13089.

Falck, J., Petrini, J. H., Williams, B. R., Lukas, J., and Bartek, J. (2002). The DNA damage-dependent intra—S phase checkpoint is regulated by parallel pathways. *Nat. Genet.* **19**, 19.

Gao, Y., Sun, Y., Frank, K. M., Dikkes, P., Fujiwara, Y., Seidl, K. J., Sekiguchi, J. M., Rathbun, G. A., Swat, W., Wang, J. et al. (1998). A critical role for DNA end-joining proteins in both lymphogenesis and neurogenesis. *Cell* **95**, 891–902.

Gatei, M., Young, D., Cerosaletti, K. M., Desai-Mehta, A., Spring, K., Kozlov, S., Lavin, M. F., Gatti, R. A., Concannon, P., and Khanna, K. (2000). ATM-dependent phosphorylation of nibrin in response to radiation exposure [In Process Citation]. *Nat. Genet.* **25**, 115–119.

Gatti, R. A. (2001). "Ataxia-Telangiectasia." McGraw-Hill, New York.

Haidar, M. A., Kantarjian, H., Manshouri, T., Chang, C. Y., O'Brien, S., Freireich, E., Keating, M., and Albitar, M. (2000). ATM gene deletion in patients with adult acute lymphoblastic leukemia. *Cancer* **88**, 1057–1062.

Hecht, F., and Hecht, B. K. (1990). Cancer in ataxia-telangiectasia patients. *Cancer Genet. Cytogenet.* **46**, 9–19.

Kastan, M. (1995). Ataxia-telangiectasia—broad implications for a rare disorder. *N. Engl. J. Med.* **333**, 662–663.

Kastan, M. B., and Lim, D. S. (2000). The many substrates and functions of ATM. *Natl. Rev. Mol. Cell. Biol.* **1**, 179–86.

Khanna, K. K. (2000). Cancer risk and the ATM gene: a continuing debate, *J. Natl. Cancer Inst.* **92**, 795–802.

Lee, Y., Barnes, D. E., Lindahl, T., and McKinnon, P. J. (2000). Defective neurogenesis resulting from DNA ligase IV deficiency requires Atm. *Genes Dev.* **14**, 2576–2580.

Lewis, R. F., Lederman, H. M., and Crawford, T. O. (1999). Ocular motor abnormalities in ataxia telangiectasia. *Ann. Neurol.* **46**, 287–295.

Lim, D., Kirsch, D., Canman, C., Ahn, J., Ziv, Y., Newman, L., Darnell, R., Shiloh, Y., and Kastan, M. (1998). ATM binds to beta-adaptin in cytoplasmic vesicles. *Proc. Natl. Acad. Sci. U.S.A.* **18**, 10146–10151.

Lim, D. S., Kim, S. T., Xu, B., Maser, R. S., Lin, J., Petrini, J. H., and Kastan, M. B. (2000). ATM phosphorylates p95/nbs1 in an S-phase checkpoint pathway. *Nature* **404**, 613–617.

Liyanage, M., Weaver, Z., Barlow, C., Coleman, A., Pankratz, D. G., Anderson, S., Wynshaw-Boris, A., and Ried, T. (2000). Abnormal rearrangement within the alpha/delta T-cell receptor locus in lymphomas from Atm-deficient mice. *Blood* **96**, 1940–1946.

McConville, C. M., Stankovic, T., Byrd, P. J., McGuire, G. M., Yao, Q. Y., Lennox, G. G., and Taylor, M. R. (1996). Mutations associated with variant phenotypes in ataxia-telangiectasia. *Am. J. Hum. Genet.* **59**, 320–330.

Morgan, S. E., and Kastan, M. B. (1997). p53 and ATM. *Adv. Cancer Res.* **65**, 1–25.

Morrison, C., Sonoda, E., Takao, N., Shinohara, A., Yamamoto, K., and Takeda, S. (2000). The controlling role of ATM in homologous recombinational repair of DNA damage. *EMBO J.* **19**, 463–471.

Naeim, A., Repinski, C., Huo, Y., Hong, J. H., Chessa, L., Naeim, F., and Gatti, R. A. (1994). Ataxia-telangiectasia: flow cytometric cell-cycle analysis of lymphoblastoid cell lines in G2/M before and after gamma-irradiation. *Mod. Pathol.* **7**, 587–592.

Petiniot, L. K., Weaver, Z., Barlow, C., Shen, R., Eckhaus, M., Steinberg, S. M., Ried, T., Wynshaw-Boris, A., and Hodes, R. J. (2000). Recombinase-activating gene (RAG) 2-mediated V(D)J recombination is not essential for tumorigenesis in Atm-deficient mice. *Proc. Natl. Acad. Sci. U.S.A.* **97**, 6664–6669.

Savitsky, K., Bar-Shira, A., Gilad, S., Rotman, G., Ziv, Y., Vanagaite, L., Tagle, D.A., Smith, S., Uziel, T., Sfez, S., Ashkenazi, M., Pecker, I., Frydman, M., Harnik, R., Patanjali, S.R., Simmons, A., Clines, G. A., Sartiel, A., Gatti, R. A., Chessa, L., Sanal, O., Lavin, M.F., Jaspers, N. G. J., Taylor, A. M. R., Arlett, C.F., Miki, T., Weissman, S. M., Lovett, M., Collins, F. S., and Shiloh, Y. (1995). A single ataxia-telangiectasia gene with a product similar to PI-3 kinase. *Science* **268**, 1749–1753.

Spacey, S. D., Gatti, R. A., and Bebb, G. (2000). The molecular basis and clinical management of ataxia telangiectasia. *Can. J. Neurol. Sci.* **27**, 184–191.

Starostik, P., Manshouri, T, O'Brien, S., Freireich, E., Kantarjian, H., Haidar, M., Lerner, S., Keating, M., and M., Albitar. (1998). Deficiency of the ATM protein expression defines an aggressive subgroup of B-cell chronic lymphocytic leukemia. *Cancer Res.* **58**, 4552–4557.

Stoppa-Lyonnet, D., Soulier, J., Lauge, A., Dastot, H., Garand, R., Sigaux, F., and Stern, M. H. (1998). Inactivation of the ATM gene in T-cell prolymphocytic leukemias, *Blood* **91**, 3920–3926.

Taylor, A. M., McConville, C. M., Rotman, G., Shiloh, Y., and Byrd, P. J. (1994). A haplotype common to intermediate radiosensitivity variants of ataxia-telangiectasia in the UK. *Int. J. Radiat Biol.* **66**, S35–S41.

Taylor, A. M., Metcalfe, J. A., Thick, J., and Mak, Y. F. (1996). Leukemia and lymphoma in ataxia telangiectasia,. *Blood* **87**, 423–438.

Vinters, H. V., Gatti, R. A., and Rakic, P. (1985). Sequence of cellular events in cerebellar ontogeny relevant to expression of neuronal abnormalities in ataxia-telangiectasia. *Kroc. Found Ser.* **19**, 233–255.

Wu, X., Ranganathan, V., Weisman, D. S., Heine, W. F., Ciccone, D. N., O'Neill, T. B., Crick, K. E., Pierce, K. A., Lane, W. S., Rathbun, G. et al. (2000). ATM phosphorylation of Nijmegen breakage syndrome protein is required in a DNA damage response. *Nature* **405**, 477–482.

Xu, Y., Ashley, T., Brainerd, E., Bronson, R., Meyn, M., and Baltimore, D. (1996). Targeted disruption of ATM leads to growth retardation, chromosomal fragmentation during meiosis, immune defects, and thymic lymphoma. *Genes Dev.* **10**, 2411–2422.

Yang, D. Q., and Kastan, M. B. (2000). Participation of ATM in insulin signalling through phosphorylation of eIF-4E-binding protein 1. *Nat. Cell. Biol.* **2**, 893–898.

Zhao, S., Weng, Y. C., Yuan, S. S., Lin, Y. T., Hsu, H. C., Lin, S. C., Gerbino, E., Song, M. H., Zdzienicka, M. Z., Gatti, R. A. et al. (2000). Functional link between ataxia-telangiectasia and Nijmegen breakage syndrome gene products. *Nature* **405**, 473–477.

# CHAPTER 22

# Episodic and Intermittent Ataxias

**JOANNA C. JEN**
Department of Neurology
UCLA School of Medicine
University of California
Los Angeles, California 90095

**ROBERT W. BALOH**
Department of Neurology
UCLA School of Medicine
University of California
Los Angeles, California 90095

I. Clinical Features
  A. EA-1
  B. EA-2
  C. EA-3
  D. EA-4
II. Genetics
  A. *KCNA1*
  B. *CACNA1A*
III. Diagnosis
IV. *In Vivo* and *In Vitro* Models
  A. Molecular Pathogenesis
  B. Animal Models
V. Treatment
References

Several episodic ataxia syndromes have been described and recent genetic discoveries are providing insight into the molecular mechanisms of these dramatic clinical disorders. Episodic ataxia type 1 (EA-1) is characterized by brief episodes of ataxia and interictal myokymia while episodic ataxia type 2 (EA-2) is manifesed by longer episodes of ataxia with interictal nystagmus (Table 22.1). Episodic ataxia type 3 (EA-3, also known as periodic vestibulocerebellar ataxia) and episodic ataxia type 4 (EA-4) share features with EA-1 and EA-2, but so far the genes for these disorders have not been identified.

Litt and colleagues (1994) mapped EA-1 to chromosome 12q13 near a cluster of three potassium channel genes. Browne and colleagues (1994) discovered four different missense mutations in KCNA1 in four unrelated EA-1 pedigrees. This was the first report of a mutation in a human potassium channel gene and the first known ion channel mutation involving the brain. We and others localized the disease locus of EA-2 to a region on chromosome 19p previously shown to be the disease locus for familial hemiplegic migraine (FHM; Kramer *et al.*, 1995; Vahedi *et al.*, 1995). A calcium channel gene mapped to this locus on chromosome 19p, and Ophoff and colleagues (1996) analyzed the exons and flanking introns of CACNA1A, identifying point mutations that resulted in a premature stop codon or interfered with splicing in two families with EA-2 and missense mutations in four families with FHM. These findings confirm that EA-2 and FHM are allelic disorders. Damji *et al.* (1996) ruled out linkage to the 12q locus of EA-1 and the 19p locus of EA-2 in two large families with the EA-3 phenotype and Steckley *et al.* (2001) ruled out linkage to these two loci in a large family with EA-4. Because EA-3 and EA-4 share numerous features with other ion channel disorders, a mutation in a gene coding for an ion channel protein is expected with EA-3 and EA-4.

## I. CLINICAL FEATURES

### A. EA-1

EA-1 is characterized by sudden episodes of ataxia, lasting seconds to a few minutes, typically triggered by exercise, startle, or emotional upset (Browne *et al.*, 1994; Brunt and Van Weerden, 1990; Lubbers *et al.*, 1995; Van Dyke *et al.*, 1975). Aura-like symptoms typically include a feeling of weakness or falling, dizziness, and blurring of

TABLE 22.1  Clinical Features of Episodic Ataxia Syndromes

|  | EA–1 | EA–2 | EA–3 | EA–4 |
|---|---|---|---|---|
| Age of onset (years) | 3–20 | 3–30 | 15–50 | 1–42 |
| Duration of attacks | Seconds–minutes | Hours | Hours | Minutes |
| Common triggers | Stress Exercise | Stress Exercise | Stress Fatigue | Stress Fatigue |
| Associated symptoms |  | Vertigo, headaches | Vertigo, diplopia, tinnitus | Vertigo, tinnitus, visual blurring |
| Interictal examination | Myokymia | Nystagmus, ataxia | Nystagmus, ataxia | Myokymia, ataxia |
| Response to acetazolamide | Yes | Yes | No | No |
| Gene | KCNA1 | CACNA1A | Unknown | Unknown |

vision. During the episode, the ataxia involves the trunk and extremities and the speech is slurred. In between episodes, there is a continuous myokymia (muscle rippling) which may be either clinically evident or only detectable with EMG. The myokymia is most easily observed in the periorbital region and fingers. Several different types of epilepsy have been reported in families with EA–1 including complex partial, partial with generalization, and generalized motor seizures (Zuberi et al., 1999). The episodes of ataxia often diminish as the child becomes older and may completely disappear in the teens. Examination during an attack reveals a generalized ataxia and dysarthria but eye movements are normal. The interictal examination is typically normal except for myokymia.

## B. EA–2

EA–2 is characterized by episodes of ataxia lasting hours and interictal nystagmus and ataxia (Vahedi et al., 1995; Baloh et al., 1997; Denier et al., 1999; Jen et al., 1999). The episodes vary from pure ataxia to combinations of symptoms suggesting involvement of cerebellum and brain stem and even occasionally the cortex. Vertigo, nausea, and vomiting are the most common associated symptoms occurring in more than 50% of patients. The episodes are typically triggered by exercise and emotional stress and often relieved by treatment with acetazolamide. Other known triggers include alcohol, phenytoin, and caffeine. About half of the patients have headaches that meet the International Headache Society (IHS) criteria for migraine (Baloh et al., 1997).

On examination during an acute episode, patients typically show severe ataxia and dysarthria. They may exhibit spontaneous nystagmus not seen during the interictal examination. In between episodes, the most common finding is a gaze-evoked nystagmus with features typical of rebound nystagmus. Spontaneous vertical nystagmus, particularly downbeat nystagmus, is seen in about a third of cases. This may begin with a positional downbeat nystagmus in the head-hanging position but over time becomes a spontaneous downbeating nystagmus. Later in the course, a mild truncal ataxia may be seen along with impaired smooth pursuit and saccade dysmetria.

## C. EA–3

EA–3 is characterized by episodes of vertigo, ataxia, diplopia, and oscillopsia lasting hours (Farmer and Mustian, 1963; Damji et al., 1996). The episodes often increase in frequency over time and may eventually become constant. As with EA–1 and EA–2, the episodes are often triggered by stress and exercise, but unlike the latter two conditions, the episodes are not relieved by acetazolamide. In addition, some patients experience tinnitus and hearing loss suggesting additional involvement of the inner ear or central auditory pathways.

Interictal examination typically shows a gaze-evoked nystagmus with features of rebound nystagmus or downbeat nystagmus. Over time, most patients show a gradually progressing interictal truncal ataxia. Quantitative oculomotor testing shows features similar to those seen with EA–2, i.e., severely impaired smooth pursuit, optokinetic nystagmus, and fixation-suppression of vestibular nystagmus.

## D. EA–4

EA–4 is characterized by episodes of ataxia with vertigo and nausea, typically lasting minutes (Steckley et al., 2001). Also patients complain of a loud bilateral tinnitus during the attacks and they may also experience visual blurring or diplopia. The age of onset can vary from early infancy to as late as the fifth decade. As with EA–1 and EA–2, acetazolamide will decrease the frequency and severity of attacks. Interictal examination shows myokymia

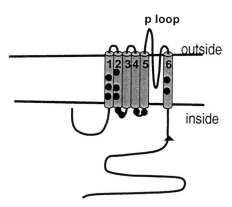

**FIGURE 22.1** Putative topology of Kv1.1 channel subunit, with six transmembrane segments S1–S6 and p loop connecting S5 and S6. The channel complex is made up of four such subunits. An overview of the location and nature of the mutations reported to date is presented: ●–, missense mutations; ▲–, nonsense mutation.

but no nystagmus. There may be a mild baseline ataxia of gait but no extremity ataxia.

## II. GENETICS

### A. KCNA1

The *KCNA1* gene codes for the six transmembrane segments (S1–S6) of the Kv1.1 potassium channel subunit (Fig. 22.1). Four of these subunits bond together along with other auxiliary units to form the Kv1.1 potassium channel. The four subunits are believed to be arranged in a ring like the staves of a barrel around a central pore. The four pore loops (between S5 and S6) reach into the barrel and confer the ion conduction properties. The *KCNA1* protein is localized in a variety of brain and peripheral nerve regions. It is heavily expressed in Purkinje cells and granular cells in the cerebellum and in the juxtaparanodal regions of the nodes of Ranvier in peripheral nerves (Rhodes *et al.*, 1997; Zhou *et al.*, 1998).

A wide range of missense mutations in *KCNA1* have been identified in families with EA–1 (Browne *et al.*, 1994; Bretschneider *et al.*, 1999). Most of the mutations involve either the transmembrane segments of the intracellular linkers affecting highly conserved amino acids. Otherwise there is no common pattern. (Fig. 22.1).

### B. CACNA1A

*CACNA1A* encodes for the $Ca_v2.1$ subunit of the P/Q calcium channel complex, which also contains accessory subunits β, γ, and α2δ. Each of the four homologous domains of the $Ca_v2.1$ calcium channel subunit has six putative alpha helical membrane-spanning segments (S1–S6) similar to the six transmembrane segments of the potassium channel Kv1.1 (Fig. 22.2). S5, S6, and the interconnecting p loop are thought to line the pore, while S4 is the putative voltage-sensing region. The selectivity filter is formed by the p loop from each of the four domains. *CACNA1A* is expressed throughout the brain but is particularly heavily expressed in the Purkinje and granular cells of the cerebellum (Mori *et al.*, 1991). It is also expressed in the neuromuscular junction where it is tightly coupled with neuromuscular transmission (Protti *et al.*, 1996).

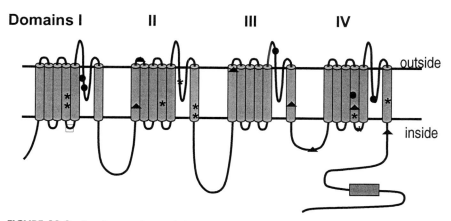

**FIGURE 22.2** Putative topology of $Ca_v2.1$ channel subunit, with four internal homologous domains I–IV, each with six transmembrane segments S1–S6 and p loop interconnecting S5 and S6. The channel complex is made up of this pore-forming and voltage-sensing subunit, along with an intracytoplasmic β subunit, a transmembrane γ subunit, and a largely extracellular α2δ subunit. ▲– EA–2-causing splice site, deletion, or nonsense mutations that disrupt the reading frame; ●– missense mutations causing EA–2. *– missense mutations causing FHM; box– polyglutamine expansion causing SCA6.

In the initial report of Ophoff et al. (1996), the two EA–2-causing mutations disrupted the reading frame of *CACNA1A*, thus predicting a truncated Ca$_v$2.1 subunit. Subsequently, numerous different mutations in *CACNA1A* have been identified in patients with EA–2, with the majority predicting truncated proteins due to premature stop codons (Table 22.2). However, a few missense mutations have been associated with the EA–2 phenotype (Denier et al., 2001; Guida et al., 2001; Jen et al., 2001). Although most cases have been familial, sporadic cases of EA–2 with *CACNA1A* mutations have been reported (Yue et al., 1998; Jen et al., 2001). By contrast, so far only missense mutations in *CACNA1A* have been associated with the FHM phenotype (Ducros et al., 2001). Some FHM patients with missense mutations in *CACNA1A* develop severe, even fatal, cerebral edema and coma following head trauma (Kors et al., 2001). Jouvenceau et al. (2001) reported an 11-year-old boy with primary generalized epilepsy, episodic and progressive ataxia, and mild learning difficulties due to a spontaneous nonsense mutation in *CACNA1A*.

There clearly is genetic heterogeneity with both FHM and EA–2. Some families with FHM have been linked to chromosome 1q while others are not linked to either 19p or the 1q locus (Ducros et al., 2001). We and others have identified families with EA–2 that are not linked to the 19p locus (Damji et al., 1996; Baloh and Jen, 2002). Escayg et al. (2000) identified a missense mutation in the calcium channel β4 subunit gene, CACNB4, in a family with generalized epilepsy and in one of our families with EA–2. This could represent a second genotype for EA–2 although, because of the different phenotypes in the two families, the mutation must be considered a candidate disease mutation until the *in vivo* consequences of the mutation can be determined.

How do mutations in the same gene lead to such varied clinical syndromes (Table 22.2)? The location and type of mutation are important but cannot be the only determining factors. Ducros and colleagues (2001) recently reported results of their studies on the clinical manifestations associated with mutations of *CACNA1A* in 28 families with FHM. Overall, they found nine mutations in *CACNA1A*, all of which were missense mutations. Eighty-nine percent of subjects with mutations had attacks of hemiplegic migraine. Six mutations were associated with hemiplegic migraine and cerebellar signs and 83% of the subjects with these 6 mutations had nystagmus, ataxia, or both. Only three mutations were associated with pure hemiplegic migraine. Although there appears to be a tendency for missense mutations to cause FHM and truncation mutations to cause EA–2, data from our laboratory and others suggest that this genotype/phenotype correlation is imperfect. We described a missense mutation in *CACNA1A* causing a severe progressive ataxia and episodic features in only a few family members (Yue et al., 1997). We also identified a missense mutation in a family with a typical EA–2 phenotype and associated episodes of hemiplegic migraine (Jen et al., 1999). Hoffman (2001) concluded that the clinical variability in patients with mutations in *CACNA1A* probably resulted from the combined result of different types of mutations, resulting biochemical defects in the channel, development abnormalities caused by channel dysfunction, and the effect of environment on all of the above.

## III. DIAGNOSIS

Magnetic resonance imaging (MRI) often will show atrophy of the cerebellum, particularly the midline, with EA–2. The degree of atrophy with EA–2 is most prominent in patients with longstanding symptoms and signs. Nonspecific paroxysmal slowing on electroencephalography (EEG) has been reported with both EA–1 and EA–2 (Zuberi et al., 1999; Feeney and Boyle, 1989; Zasorin et al., 1983). Continuous muscle rippling (myokymia) can be detected in patients with EA–1 and EA–4, particularly after provocation by limb ischemia. Repetitive duplets, triplets, or multiplets on EMG underlie the continuous muscle activity (Brunt and van Weerden, 1990; Zuberi et al., 1999). The spontaneous muscle activity typically occurs at a frequency of about 10 per second and is often made more prominent with hyperventilation.

Eye movement recordings demonstrate a unique, consistent oculomotor pattern in patients with EA–2 and EA–3 (Baloh et al., 1997; Damji et al., 1996; Table 22.3). Saccade velocity remains normal but saccade dysmetria

TABLE 22.2 Currently Identified Phenotypes Associated with Mutations in *CACNA1A*

| Phenotype | Type of mutation |
|---|---|
| Familial hemiplegic migraine (FHM) | Missense, usually S4–S6 regions |
| Episodic ataxia type 2 (EA–2) | Usually interrupts reading frame, can be missense |
| Episodic ataxia and epilepsy | Nonsense near carboxy terminus |
| Episodic and progressive ataxia | Missense in pore region |
| Cerebral edema and coma after trauma | Missense in I S–4–S5 link |
| Spinocerebellar ataxia type 6 (SCA6) | CAG repeat expansion in carboxy terminus |

TABLE 22.3 Summary of the Type and Degree of Oculographic Abnormalities Observed with EA-2 and EA-3

| | |
|---|---|
| Saccades | |
|   Peak velocity | Normal |
|   Reaction time | Normal |
|   Accuracy | Mildly impaired |
| Visual tracking | |
|   Smooth pursuit | Severely impaired |
|   Optokinetic nystagmus | Severely impaired |
|   Suppression of VOR | Severely impaired |
| Vestibulo-ocular reflex (VOR) | |
|   Gain | Normal |
|   Time constant | Normal |

occurs. Smooth pursuit and optokinetic responses are severely impaired and fixation of vestibular-induced nystagmus is severely impaired. On the other hand, the vestibulo-ocular reflex (VOR) gain is either normal or increased. This pattern of findings is localizing to the caudal midline cerebellum.

At the present time, genetic testing for EA-1 and EA-2 is not available in clinical laboratories. Since no single mutations have been prominent in either KCNA1 or CACNA1A, all of the coding regions of the genes would have to be screened for mutations. Since KCNA1 is only a single exon, this is relatively easy but the 47 exons of CACNA1A make screening it a much more difficult task. In the future, automated techniques will be available to rapidly screen these genes for the known point mutations.

## IV. IN VITRO AND IN VIVO MODELS

### A. Molecular Pathogenesis

#### 1. EA-1

KCNA1 encodes for a subunit of a delayed rectifier potassium channel Kv1.1. The delayed rectifier potassium channels are a family of potassium channels that allow a sustained K+ efflux with a delay after membrane depolarization. The outflow of potassium ions rapidly repolarizes the membrane. Kv1.1 channels play a key role in neurotransmitter release, action potential generation, and axonal impulse conduction (Zhou *et al.*, 1998). Studies of the biophysical properties of the mutant Kv1.1 channels in *Xenopus* oocytes or mammalian cell lines demonstrate physiologic consequences of the genetic mutations although no consistent pattern has emerged (Bretschneider *et al.*, 1999; Letts *et al.*, 1998; Adelman *et al.*, 1995). Boland *et al.* (1999) introduced seven different EA-1 mutations and studied the effect on the expressed channels in *Xenopus* oocytes. The voltage range of steady-state inactivation was altered in all cases and most changes were also associated with changes in activation gating. They concluded that the EA-1 mutations altered potassium channel function by two mechanisms: (1) reduced channel expression, and (2) altered channel gating. Bretschneider *et al.* (1999) studied the functional consequences of two EA-1 mutations by injecting cRNA coding for the mutations into mammalian cells and comparing currents with a patch-clamp technique in the mutant and wild-type channels. One of the mutant channels deactivated and inactivated faster compared to the wild-type while the others showed no change in maximum open probability, suggesting that its effect was due to a reduced number of functional channels on the cell surface. Interestingly, acetazolamide had no measurable effect on either the wild-type or mutant channels.

#### 2. EA-2

Kraus *et al.* (1998) introduced the four missense mutations reported in families with FHM by Ophoff *et al.* (1996) into *Xenopus laevis* oocytes and investigated possible changes in channel function after functional expression of the mutant subunits. Changes in channel gating were observed in three out of four of the mutants, but the time course of recovery from the channel inactivation was accelerated in two and slower in the third compared to the wild-type. Hans *et al.* (1999) introduced the four missense mutations causing FHM into human $Ca_v2.1$ and investigated their functional consequences after expression in human kidney cells. They also found a range of effects from shifts in the voltage range to activation toward more negative voltages, increases in both the open probability and the rate of recovery from inactivation, and decreases in the density of functional channels in the membrane. Interestingly, the reduction in single channel conductance induced by two of the mutations was not observed in some patches or periods of activity, suggesting that the abnormal channel may switch on and off due to some unknown factor.

We studied two nonsense mutations and a missense mutation causing EA-2 and found markedly decreased calcium currents when the mutant channels were expressed in mammalian cells (Jen *et al.*, 2001), similar to findings by Guida *et al.* (2001) on an EA-2-causing missense mutation. The dramatic reduction in calcium conductance in these EA-2-causing mutants was not observed in SCA6 or FHM mutations studied to date. Altered biophysical properties and decreased plasma membrane targeting of the mutant channels can both contribute to the decreased calcium currents.

Since $Ca_v2.1$ is important in neurotransmission centrally and peripherally, we hypothesized that mutant channels may cause impaired neurotransmission. Three EA-2 patients in whom we had identified a missense and two nonsense mutations had experienced fluctuating weakness in addition to episodic ataxia. We examined in vivo the neuromuscular junction in these three patients whose mutations had been studied in vitro. Indeed, we found increased jitter and blocking by single fiber EMG (Jen et al., 2001) due to presynaptic failure (unpublished observations). The abnormal neuromuscular transmission may reflect similarly impaired neurotransmission in the central nervous system in these patients.

## B. Animal Models

### 1. KCNA1

Sequencing of the Shaker locus in *Drosophila* identified the potassium channel that was the first of a large number of homologous genes in the Shaker family. KCNA1, the gene for EA–1, is a Shaker homolog in humans. There are no known natural mutants of the Shaker homolog in mice but Kv1.1 null mice generated by gene targeting in embryonic stem cells results in an epileptic phenotype but not ataxia (Smart et al., 1998). Kv1.1 null mice show a striking temperature-sensitive excitability change localized to the nerve terminals that can explain the myokymia seen with this disorder (Zhou et al., 1998).

### 2. CACNA1A

Mutations in genes coding for various calcium channel subunits have been identified in a number of recessive mouse mutants. Homozygous point mutations in the $Ca_v2.1$ gene (P1802L, domain II p loop) causes epilepsy and ataxia in the mutant mouse tottering (*tg*; Fletcher et al., 1996). Novel sequences in the intracellular carboxy terminus of the $Ca_v2.1$ subunit also result in a mutant mouse phenotype leaner (*tg^la*) with ataxia and epilepsy (Fletcher et al., 1996). Burgess and colleagues (1997) found a mutation in a calcium channel gene with a predicted deletion of the highly conserved α1 binding motif of the β4 subunit in the mutant mouse lethargic exhibiting both ataxia and seizures.

Characterization of the genetic defects in the stargazer and waggler mutants led to the identification of a new neuronal calcium channel subunit γ (Letts et al., 1998). *Stargazer* mice have epilepsy, ataxia, and an unusual head posturing suggesting inner ear vestibular damage. The mutant mouse *stg* showed no impairment in calcium channel function but a dramatic lack of functional AMPA (alpha-amino-3-hydroxyl-5-methyl-4-isoxazolepropionate) receptors in the cerebellar granule cells. With no modulatory effect on neuronal calcium channels, the protein stargazin was found to regulate synaptic targeting of AMPA receptors and possible coupling with calcium channels (Chen et al., 2000). Mutations in a gene *cacna2d2* encoding an α2δ subunit were recently identified to cause epilepsy and ataxia in recessive mouse mutants ducky *du* and *du^2J* (Barclay et al., 2001). Thus mutations in every subunit of the calcium channel complex can cause similar phenotypes in mice.

Lau and colleagues (1998) studied the expression of the $Ca_v2.1$ mRNA and $Ca_v2.1$ protein in the cerebellum from 20-day-old homozygous leaner mice and control mice using *in situ* hybridization, histochemistry, and immunocytochemistry. They found no difference in the messenger RNA or protein expression in the mutated $Ca_v2.1$ subunit in the leaner mice compared to controls. Thus, they showed that the $Ca_v2.1$ subunit splice donor consensus sequence carried by leaner mice does not result in any significant quantitative changes in either messenger RNA or protein expression. The data suggest that the leaner phenotype resulted from abnormal calcium channels that contain the altered $Ca_v2.1$ subunits.

There are *Drosophila* mutants harboring calcium channel mutations as well. Two temperature-sensitive *cacophony* mutants *cac* and *cac^T32* have been described (Peixoto and Hall, 1998; Dellinger et al., 2000), both involving a gene that encodes the calcium channel α1 subunit important in synaptic transmission, with shared homology with human calcium channel genes. The *cac* mutant showed locomotor defects and frequent convulsions when exposed to 37°C, with aberrant courtship song. The *cac^T32* mutant is paralyzed when exposed to higher temperatures.

## V. TREATMENT

Since emotional stress is often a trigger for attacks in patients with episodic ataxia syndromes, stress management techniques such as biofeedback and meditation can be helpful in controlling symptoms in some patients. Alcohol and caffeine should be avoided and regular modest exercise should be encouraged. Vigorous exercise will often trigger attacks of the episodic ataxia. Fluid and food intake should be regularly distributed throughout the day and binges should be avoided.

Acetazolamide can be dramatic in controlling episodes of ataxia with EA–2 (Griggs et al., 1978) and is occasionally beneficial with EA–1 (Lubbers et al., 1995) and EA–4 (Steckley et al., 2001), but not with EA–3. With EA–2 there can be a variable response to acetazolamide even within a single family with a known mutation. Acetazolamide presumably works by altering the pH within the cerebellum thus stabilizing the mutated ion channel (Bain et al., 1992). One typically begins with low doses (125 mg/day) and then works up to an average effective dose of between 500 and 750 mg/day. Most patients will experience paresthesias of the extremities after taking the drug, but these symptoms typically decrease over time. The main long-term side effect is development of kidney stones, which can be markedly decreased if the patient regularly drinks citrus juices. Patients with known allergies to sulfa-containing drugs may have an allergic reaction to acetazolamide. There is relatively little experience with other carbonic anhydrase inhibitors but these drugs are probably as effective as acetazolamide.

## References

Adelman, J. P., Bond, C. T., Pessia, M., and Maylie, J. (1995). Episodic ataxia results from voltage-dependent potassium channels with altered functions. *Neuron* **15**, 1449–1454.

Bain, P. G., O'Brien, M. D., Keevil, S. F., and Porter, D. A. (1992). Familial periodic cerebellar ataxia: a problem of molecular pH homeostasis. *Ann. Neurol.* **31**, 146–154.

Baloh, R. W, and Jen, J. C. (2002). Genetics of familial episodic vertigo and ataxia. *N.Y. Acad. Sci.* **956**, 338–345.

Baloh, R. W., Yue, Q., Furman, J. M., and Nelson, S. F. (1997). Familial episodic ataxia: clinical heterogeneity in four families linked to chromosome 19p. *Ann. Neurol.* **41**, 8–16.

Barclay, J., Balaguero, N., Mione, M., Ackerman, S. L., Letts, V. A., Brodbeck, J., Canti, C., Meir, A., Page, K. M., Kusumi, K., Perez-Reyes, E., Lander, E. S., Frankel, W. N., Gardiner, R. M., Dolphin, A. C., Rees, M. (2001) Ducky mouse phenotype of epilepsy and ataxia is associated with mutations in the Cacna2d2 gene and decreased calcium channel current in cerebellar Purkinje cells. *J. Neurosci.* **21**, 6095–6104.

Boland, L. M., Price, D. L., and Jackson, K. A. (1999). Episodic ataxia/myokymia mutations functionally expressed in the shaker potassium channel. *Neuroscience.* **91**, 1557–1564.

Bretschneider, F., Wrisch, A., Lehmann-Horn, F., and Grissmar, S. (1999). Expression in mammalian cells and electrophysiological characterization of two mutant Kv1.1 channels causing episodic ataxia type 1 (EA-1). *Eur. J. Neurosci.* **11**, 2403–2412.

Browne, D. L., Gancher, S. T., Nutt, J. G. *et al.* (1994) Episodic ataxia/myokymia syndrome is associated with point mutations in the human potassium channel, KCNA1. *Nat. Genet.* **8**, 136–140.

Brunt, E. R. and van Weerden, T. W. (1990). Familial paroxysmal kinesigenic ataxia and continuous myokymia. *Brain* **113**, 1361–1382.

Burgess, D. L., Jones, J. M., Meisler, M. H., and Noebels, J. L. (1997). Mutation in the Ca2+ channel beta subunit gene Cchb4 is associated with ataxia and seizures in the lethargic (lh) mouse. *Cell* **88**, 185–192.

Chen L., Chetkovich, D. M., Petralia, R. S., Sweeney, N. T., Kawasaki, Y., Wenthold, R. J., Bredt, D. S., Nicoll, R. A. (2000). Stargazing regulates synaptic targeting of AMPA receptors by two distinct mechanisms. *Nature* **408**, 936–943.

Damji, K. F., Allingham, R. R., Pollock, S. C., Small, K., Lewis, K. E. *et al.* (1996). Periodic vestibulocerebellar ataxia, an autosomal dominant ataxia with defective smooth pursuit, is genetically distinct from other autosomal dominant ataxias. *Arch. Neurol.* **53**, 338–344.

Dellinger, B., Felling, R., and Ordway, R. W. (2000) Genetic modifiers of the Drosophila NSF mutant, comatose, include a temperature-sensitive paralytic allele of the calcium channel alpha1-subunit gene, cacophony. *Genetics* **155**, 203–211.

Denier, C., Ducros, A., Durr, A. *et al.* (2001). Missense CACNA1A mutation causing episodic ataxia type 2. *Arch. Neurol.* **58**, 179–180.

Denier, C., Ducros, A., Vahedi, K. *et al.* (1999). High prevalence of CACNA1A truncations and broader clinical spectrum in episodic ataxia type 2. *Neurology* **52**, 1816–1821.

Ducros, A., Denier, C., Joutel, A., Cecillon, M., Lescoat, C. *et al.* (2001). The clinical spectrum of familial hemiplegic migraine with associated with mutations in a neuronal calcium channel. *N. Engl. J. Med.* **345**, 17–24.

Escayg, A., De Waard, M., Lee, D. D. *et al.* (2000). Coding and noncoding variation of the human calcium-channel β4-subunit gene CACNB4 in patients with idiopathic generalized epilepsy and episodic ataxia. *Am. J. Hum. Genet.* **66**, 1531–1539.

Farmer, T. W., and Mustian, V. M. (1963). Vestibulocerebellar ataxia. *Arch. Neurol.* **8**, 471–480.

Feeney, G. F. X., and Boyle, R. S. (1989). Paroxysmal cerebellar ataxia. *Aust. N.Z. J. Med.* **19**, 113–117.

Fletcher, C. F., Lutz, C. M., O'Sullivan, T. N. *et al.* (1996) Absence epilepsy in tottering mutant mice is associated with calcium channel defects. *Cell* **87**, 607–617.

Griggs, R. C., Moxley, R. T., Lafrance, R. A., and McQuillen, J. (1978). Hereditary paroxysmal ataxia: response to acetazolamide. *Neurology* **28**, 125–1264.

Guida, S., Trettel, F., Pagnutti, S. *et al.* (2001). Complete loss of P/Q calcium channel activity caused by CACNA1A missense mutation carried by patients with episodic ataxia type 2. *Am. J. Hum. Genet.* **68**, 759–764.

Hans, M., Luvisetto, S., Williams, M. E. *et al.* (1999). Functional consequences of mutations in the human $\alpha_{1A}$ calcium channel subunit linked to familial hemiplegic migraine. *J. Neurosci.* **19**, 1610–1619.

Hoffman, E. P. (2001). Hemiplegic migraine—downstream of a single-base change. *N. Engl. J. Med.* **345**, 57–59.

Jen, J. C., Wan, J., Graves, M., *et al.* (2001). Loss-of-function EA2 mutations are associated with impaired neuromuscular transmission. *Neurology* **57**, 1843–1848.

Jen, J. (1999). Calcium channelopathies in the central nervous system. *Curr. Opin. Neurobiol.* **9**, 274–280.

Jen, J., Yue, Q., Nelson, S. F., *et al.* (1999). A novel nonsense mutation in CACNA1A causes episodic ataxia and hemiplegia. *Neurology* **53**, 34–37.

Jouvenceau, A., Eunson, L. H., Spauchus, A. *et al.* (2001). Human epilepsy associated with dysfunction of the brain P/Q-type calcium channel. *Lancet* **358**, 801–807.

Kors, E. E., Terwindt, G. M, Vermeulen, F. L. M. G. *et al.* (2001). Delayed cerebral edema and fatal coma after minor head trauma: Role of the CACNA1A calcium channel subunit gene and relationship with familial hemiplegic migraine. *Ann. Neurol.* **49**, 753–760.

Kramer, P. L., Yue, Q., Gancher, S. T. *et al.* (1995) A locus for the nystagmus-associated form of episodic ataxia maps to an 11-cM region on chromosome 19p. *Am. J. Hum. Genet.* **57**, 182–185.

Kraus, R. L., Sinnegger, M. J., Glossman, H. *et al.* (1998). Familial hemiplegic migraine mutations change $\alpha_{1A}$ Ca2+ channel kinetics. *J. Biol. Chem.* **273**, 5586–5590.

Lau, F. C., Abbott, L. C., Rhyu, I. J. *et al.* (1998). Expression of calcium channel $\alpha_{1A}$ mRNA and protein in the leaner mouse ($tg^{la}/tg^{la}$) cerebellum. *Mol. Brain Res.* **59**, 93–99.

Letts, V. A., Felix, R., Biddlecome, G. H. *et al.* (1998). The mouse stargazer gene encodes a neuronal $Ca^{2+}$-channel g subunit. *Nat. Genet.* **19**, 340–347.

Litt, M., Kramer, P., Browne, D. *et al.* (1994). A gene for episodic ataxia/myokymia maps to chromosome 12p13. *Am. J. Hum. Genet.* **55**, 702–709.

Lubbers, W. J., Brunt, E. R., Scheffer, H. *et al.* (1995). Hereditary myokymia and paroxysmal ataxia linked to chromosome 12 is responsive to acetazolamide. *J. Neurol. Neurosurg. Psychiatry* **59**, 400–405.

Mori, Y., Friedrich, I., Kim, M. S. *et al.* (1991). Primary structure and functional expression from complementary DNA of a brain calcium channel. *Nature* **350**, 398–402.

Ophoff, R. A., Terwindt, G. M., Vergouwe, M. N. *et al.* (1996). Familial hemiplegic migraine and episodic ataxia type-2 are caused by mutations in the Ca2+ channel gene CACNA1A. *Cell* **87**, 543-552.

Peixoto, A. A., and Hall, J. C. (1998). Analysis of temperature-sensitive mutants reveals new genes involved in the courtship song of Drosophila. *Genetics* **148**, 827–838.

Protti, D. A., Reisin, R., Mackinley, T. A., and Uchitel, O. D. (1996). Calcium channel blockers and transmitter release at the normal human neuromuscular junction. *Neurology* **46**, 1391–1396.

Rhodes, K. J, Strassle, B. W., Monaghan, M. M. *et al.* (1997). Association and colocalization of the Kv$\beta_{1-}$ and Kv$\beta_{2-}$ subunits with Kv1 α-subunits in mammalian K$^+$ channel complexes. *J. Neurosci.* **17**, 8246–8258.

Smart, S. L., Lopantsev, V., Zhang, C. L. *et al.* (1998). Deletion of the Kv1.1 potassium channel causes epilepsy in mice. *Neuron* **20**, 809–819.

Steckley, J. L., Ebers, G. C., Cader, M. Z., and McLaclan, R. S. (2001). An autosomal dominant disorder with episodic ataxia, vertigo, tinnitus, and myokymia. *Neurology* **57**, 1499–1502.

Vahedi, K., Joutel, A., Van Bogaert, P. *et al.* (1995). A gene for hereditary paroxysmal cerebellar ataxia maps to chromosome 19p. *Ann. Neurol.* **37**, 289–293.

Van Dyke, D. H., Griggs, R. C., Murphy, M. J., and Goldstein, M. N. (1975). Hereditary myokymia and periodic ataxia. *J. Neurol. Sci.* **25**, 109–118.

Yue, Q., Jen, J. C., Nelson, S. F., and Baloh, R. W. (1997). Progressive ataxia due to a missense mutation in a calcium-channel gene. *Am. J. Hum. Genet.* **61**, 1078–1087.

Yue, Q., Jen, J. C., Thwe, M. M. *et al.* (1998). De novo mutation in CACNA1A caused acetazolamide-responsive episodic ataxia. *Am. J. Med. Genet.* **77**, 298–301.

Zasorin, N. L., Baloh, R. W., and Myers, L. B. (1983). Acetazolamide-responsive episodic ataxia syndrome. *Neurology* **33**, 1212–1214.

Zhou, L., Zhang, C.-L., Messing, A., and Chiu, S. Y. (1998). Temperature-sensitive neuromuscular transmission in Kv1.1 null mice: role of potassium channels under the myelin sheath in young nerves. *J. Neurosci.* **18**, 7200–7215.

Zuberi, S. M., Eunson, L. H., Spauschus, A. *et al.* (1999). A novel mutation in the human voltage-gated potassium channel gene (Kv1.1) associates with episodic ataxia type 1 and sometimes with partial epilepsy. *Brain* **122**, 817–825.

# CHAPTER 23

# Multiple System Atrophy

**CLIFF SHULTS**
Department of Neurosciences
University of California San Diego
and
Neurology Service
VA San Diego Healthcare System
La Jolla, California 92093

**SID GILMAN**
Department of Neurology
University of Michigan
Ann Arbor, Michigan 48109

I. Introduction
II. Clinical Features
   A. Multiple System Atrophy—Parkinson Type (MSA–P)
   B. Multiple System Atrophy—Cerebellar Type (MSA–C)
III. Diagnosis of MSA
   A. Diagnostic Criteria
   B. Supporting Diagnostic Studies
IV. Neuroimaging
   A. Structural Brain Imaging with MRI
   B. Functional Cardiac Imaging with SPECT
   C. Functional Brain Imaging with PET and SPECT
V. Pathology of MSA
VI. Epidemiology and Environmental Risk Factors
VII. Genetics
VIII. Management and Treatment
   A. Extrapyramidal Features
   B. Autonomic Disorders
   References

A progressive neurodegenerative disorder, multiple system atrophy (MSA) presents clinically with varying degrees of parkinsonism, autonomic dysfunction, and impaired cerebellar function. The current nomenclature utilizes the abbreviations MSA–P to indicate that parkinsonism dominates the clinical presentation, and MSA–C to indicate that cerebellar ataxia predominates. Current diagnostic criteria utilize clinical domains, features, and criteria to define possible, probable, and definite MSA. Diagnosis can be aided by anatomical imaging with magnetic resonance imaging (MRI), which reveals hypointensity in the putamen and a hyperintense rim lateral to the putamen in T2 weighted MRI scans in many cases of MSA. Functional imaging with positron emission tomography (PET) utilizing ligands to study cerebral glucose metabolic rates, striatal dopaminergic terminals, and striatal dopaminergic and opiate receptors can also assist diagnosis. Functional imaging with single photon emission computed tomography (SPECT) using striatal presynaptic and postsynaptic dopaminergic ligands also provides helpful diagnostic information. SPECT studies of cardiac innervation reveal postganglionic autonomic failure in Parkinson's disease but not in MSA, as autonomic failure occurs in the preganglionic neuron in MSA. The neuropathological features in MSA include neuronal degeneration affecting principally the nigrostriatal, olivopontocerebellar, and central autonomic structures and very large numbers of abnormal tubular structures termed glial cytoplasmic inclusions (GCIs). These abnormal structures also appear in oligodendrocyte cytoplasm, oligodendrocyte nuclei, neuronal cytoplasm, neuronal nuclei, and axons. Alpha-synuclein constitutes a major component of GCIs and neuronal cytoplasmic inclusions, and aggregates of α-synuclein also have been found in Parkinson's disease and related Lewy body disorders.

## I. INTRODUCTION

A progressive neurodegenerative disorder, MSA presents clinically with varying degrees of parkinsonism, autonomic dysfunction and impaired cerebellar function. The initial descriptions of the disorder focused on the most conspicuous clinical and neuropathological manifestations, hence it has been described as sporadic olivopontocerebellar atrophy (sOPCA; Dejerine and Thomas, 1900), Shy-Drager syndrome (Shy and Drager, 1960), and striatonigral degeneration (STN; Adams et al., 1961). Graham and Oppenheimer (1969) identified these clinical disorders as manifestations of a single entity that they termed multiple system atrophy. Identification of the cardinal neuropathological feature of the disorder, glial cytoplasmic inclusions in oligodendrocytes (Papp et al., 1989), verified the notion that these syndromes are manifestations of the same process. The current nomenclature utilizes the abbreviations MSA–P to indicate that parkinsonism domiates the clinical presentation, and MSA–C to indicate that cerebellar ataxia predominates. Published descriptions of the relative frequency of these clinical presentations vary with the interests of the investigators and the referral patterns to their clinics. As many of the currently published series of cases have been derived from movement disorders clinics, the focus has been on the parkinsonian presentations of the disorder. Nevertheless, MSA accounts for a substantial proportion of patients seen in ataxia clinics (Klockgether et al., 1998).

## II. CLINICAL FEATURES

### A. Multiple System Atrophy—Parkinsonian Type (MSA–P)

In one of the largest series of MSA cases from a Parkinson's disease (PD) clinic, Wenning et al. (1994) described the clinical features and natural history of 100 consecutive patients who fulfilled criteria that they had established for the diagnosis of clinically probable MSA. The cohort included 67 men and 33 women with a mean age at onset of the first reported symptom of 52.5 ± 8.1 years (mean ± SD). The initial, but not necessarily presenting, clinical feature consisted of parkinsonism in 46%, autonomic dysfunction in 41%, cerebellar dysfunction in 5%, and features of two or more systems in 7%. The parkinsonian features included tremor, which affected 66% of the cases, but a rest tremor affected only 29% of the cases and consisted of a classical parkinsonian tremor in only 9% of the cases. More commonly a postural (47%) or action (20%) tremor appeared. Akinesia (usually with rigidity) was considered necessary for the diagnosis of parkinsonism, hence all patients had this feature. Autonomic disorders appeared in 97% of the cases, including impotence in men (90%), postural faintness (53%), recurrent syncope (15%), urinary incontinence (71%), urinary retention (27%) and fecal incontinence (2%).

Cerebellar dysfunction affected 52% of the cases, and consisted of limb ataxia (47%), gait ataxia (37%), nystagmus (25%), and intention tremor (15%). Signs of corticospinal dysfunction appeared in 61% of subjects, consisting of hyperreflexia (54%) and the Babinski sign (49%). Other clinical features included inspiratory stridor (34%), which could be nocturnal, disproportionate antecollis (15%), and myoclonus (31%), which often affected the fingers. Certain clinical signs, when detected, should alert the clinician to the likelihood of MSA (Quinn, 1989, 1994; Klein et al., 1997; Schulz and Dichgans, 2000). These include emotional incontinence, rapid eye movement (REM) sleep behavior disorder, cold, dusky violaceous hands with poor circulatory return, and stimulus-sensitive myoclonus.

In a series of 38 autopsy-confirmed cases of MSA from the Parkinson's Disease Society Brain Bank in London, Wenning et al. (2000) found that sometime prior to death 95% of the patients had bradykinesia, 84% had signs of autonomic dysfunction, and 32% had signs of cerebellar dysfunction. They compared the findings in the MSA patients with those in 100 autopsy-confirmed cases of PD and found that the best discriminators of the two disorders consisted of a poor response to levodopa, autonomic features, speech or bulbar dysfunction, absence of dementia, and absence of levodopa-induced confusion and falls.

### B. Multiple System Atrophy—Cerebellar Type (MSA–C)

In a literature review of 203 pathologically proven cases in the literature up to 1995, Wenning et al. (1997) reported that 29% of patients developed cerebellar ataxia as the initial motor disturbance whereas 58% developed parkinsonism, and an additional 9% had parkinsonism combined with cerebellar ataxia. The individual clinical features in these cases included cerebellar ataxia in 54%, with gait ataxia in 49%, limb ataxia in 47%, intention tremor in 24% and nystagmus in 23%. For the entire group of 203 cases, the mean age at onset was 54.3 years (range 33–78) and the mean age at death was 60.3 (range 37–84). In the studies of Gilman et al. (2000), MSA–C presented as a sporadic adult-onset cerebellar degeneration of undetermined cause that was labeled sOPCA. Other authors have used various different terms for this disorder, including idiopathic late-onset cerebellar ataxia (ILOCA; Wenning et al., 1997) and idiopathic cerebellar ataxia (IDCA; Schulz and Dichgans, 2000). Appearing usually in the fifth or sixth decade or later, the initial symptoms consisted of difficulty in walking or loss of balance, followed soon by a dysarthria described as "slurring" of speech (Gilman et al., 2000). Subsequent incoordination of the hands and arms caused deterioration of handwriting, difficulty in sewing, and difficulty typing or playing musical instruments. The disease progressed steadily, but at highly variable rates between patients. In some, the symptoms increased slowly over many years, and in others, the disease progressed so rapidly from onset that the patient became confined to a wheelchair in 3–5 years. The cases with rapidly evolving cerebellar ataxia frequently progressed to develop autonomic failure, and later, features of parkinsonism (Klockgether et al., 1998; Gilman et al., 2000). The symptoms of autonomic failure began with sexual impotence in men, often occurring 5–10 years in advance of any other symptoms. Autonomic dysfunction affecting the urinary bladder presented with urgency and frequency initially, followed by incomplete bladder emptying and then urinary incontinence. Autonomic dysfunction affecting blood pressure control consisted of lightheadedness progressing to syncope with movement from the recumbent or seated position to the standing position. The symptoms of parkinsonism frequently appeared after the onset of autonomic failure, but at times before the onset, and consisted of difficulty turning over in bed, slowness in rising from a chair, and varying degrees of bradykinesia. Tremor as a component of parkinsonism in MSA–C occurred uncommonly. In a longitudinal study of the outcome of OPCA, approximately one-fourth of the patients evolved to MSA within five years, and this transition carried a poor prognosis for survival (Gilman et al., 2000) (Figs. 23.1–23.4). Older age at onset of ataxia and earlier presentation in a neurologic specialty clinic predicted transition to MSA (Figs. 23.3 and 23.4).

The clinical signs of MSA–C included a wide-based stance and an ataxic gait, with ataxia on testing with the finger-nose-finger and the heel-knee-shin maneuvers (Gilman et al., 2000; Schulz and Dichgans, 2000). Muscle strength usually remained preserved. Many abnormalities of extraocular movements appeared, principally square-wave jerks, gaze paretic nystagmus, saccadic intrusions into smooth pursuit movements, ocular dysmetria, and nystagmus in the primary position. The speech disturbances were principally ataxic, with excess and equalized stress patterns and abnormal modulations of pitch and volume (Kluin et al., 1996). Spasticity often accompanied the ataxia of speech, resulting in a strained-strangled quality, and hypokinetic components could appear early in the course, even before the diagnosis of MSA–C became apparent. The autonomic disorder responsible for orthostatic symptoms was detected by taking the blood pressure and pulse with the patient recumbent for 2–3 minutes, then repeating the measurements after the patient had stood up for 2–3 minutes. With standing, a decline of 30 mmHg systolic pressure and of 20 mmHg diastolic pressure occurred and in some patients faintness accompanied the decreased blood pressure. Signs of corticospinal disease appeared

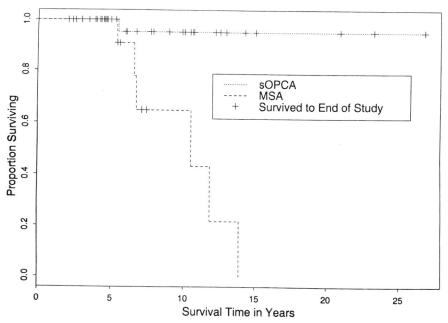

FIGURE 23.1 Estimated distribution of survival for patients with sporadic olivopontocerebellar atrophy (sOPCA) who evolved to develop multiple system atrophy (MSA; dashed line) as compared with patients who did not develop MSA (dotted lines). Crosses represent patients still surviving at the end of the study. Survival times are measured from the onset of symptoms of ataxia to the time of death or the last visit. (From Gilman, S. et al. (2000). Neurology, **55**, 527–532. With permission.)

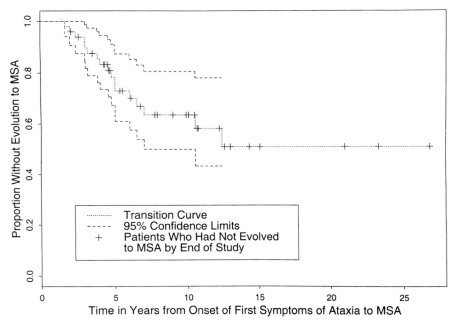

**FIGURE 23.2** Estimated distribution of time from the onset of ataxia to the transition to MSA (dotted line). Crosses represent patients who had not evolved to develop MSA by the end of the study. Dashed lines indicate the 95% confidence intervals, which are terminated at 12 years since the sample size of cases with more than 12 years from onset is small. (From Gilman, S. et al. (2000). *Neurology*, **55**, 527–532. With permission.)

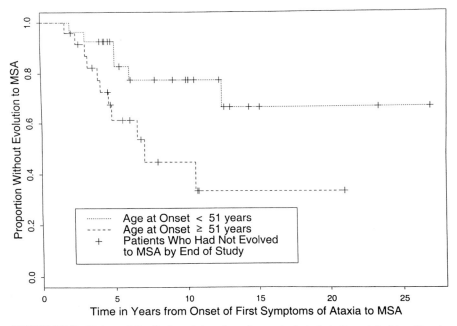

**FIGURE 23.3** Estimated distribution of time from the onset of ataxia to the point of transition to MSA, classified into two groups based upon the median age at onset of the initial symptoms of ataxia (51 years). Data for patients with onset younger than 51 years are shown with dotted lines and those with onset older than 51 years with dashed lines. Crosses represent patients who had not evolved to develop MSA by the end of the study. (From Gilman, S. et al. (2000). *Neurology*, **55**, 527–532. With permission.)

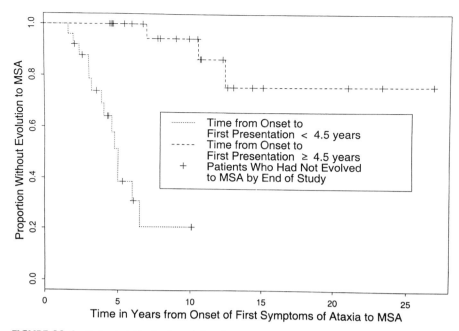

**FIGURE 23.4** Estimated distribution of time from onset of ataxia to point of transition to MSA, classified into two groups based upon the median time from the onset of ataxia to presentation in a neurological specialty clinic (4.5 years). Data for patients who first presented below 4.5 years are shown with dotted lines and for those above 4.5 years with dashed lines. Crosses represent patients who had not evolved to MSA by the end of the study. (From Gilman, S. *et al.* (2000). *Neurology*, **55**, 527–532. With permission.)

frequently, with limb spasticity, usually most markedly in the legs than the arms, hyperreflexia, and extensor plantar responses (Wenning *et al.*, 1996). Muscle strength usually remained preserved. In some cases parkinsonism preceded the symptoms of autonomic failure, but more commonly appeared after the onset of postural hypotension and urinary frequency, urgency, and incontinence. The clinical features included stooped posture, bradykinesia, masked face with infrequent blinking, slowness of movement, and short steps with a shuffling gait. In patients presenting with cerebellar ataxia, the severity of the ataxia diminished as the parkinsonian features became more prominent (Gilman *et al.*, 2000). Table 23.1 provides a comparison of the symptoms and signs typically seen in MSA as compared with Parkinson's disease and idiopathic late onset cerebellar ataxia.

## III. DIAGNOSIS OF MSA

### A. Diagnostic Criteria

Two sets of diagnostic criteria have been proposed, first by Quinn (1989, 1994) and then by a consensus panel (Gilman *et al.*, 1999b). Neither of these sets of diagnostic criteria has been applied prospectively. Quinn advanced the first criteria for the diagnosis of MSA in 1989 and later proposed a modification (Quinn, 1994). The criteria utilized the older terms SND type for the predominantly parkinsonian presentation and OPCA type for the predominantly cerebellar presentation. He used three levels of certainty for diagnosis—possible, probable, and definite (Table 23.2).

Quinn (1994) subsequently modified the criteria shown in Table 23.2. He created a category of possible MSA of the OPCA type with a sporadic adult-onset cerebellar syndrome in combination with parkinsonism. He changed the criteria for probable MSA of the SND type by adding a pathological sphincter electromyogram (EMG) as an alternative finding. He also permitted the diagnosis of probable MSA of the SND type without autonomic failure if the patient had signs of cerebellar or pyramidal disease or a pathological sphincter EMG. (Unfortunately, this resulted in an inconsistency, as the combination of a cerebellar syndrome with parkinsonism could then qualify as *possible* MSA of the OPCA type or *probable* MSA of the SND type). Probable MSA of the OPCA type required a sporadic adult-onset cerebellar syndrome (with or without parkinsonism or pyramidal signs) plus severe symptomatic autonomic failure or a pathological sphincter EMG. The term sporadic was redefined so that there could be no other case of MSA among first- or second-degree relatives. The down-gaze supranuclear palsy as an exclusion factor was

TABLE 23.1 Symptoms and Signs Characteristic of MSA, PD, and Late-Onset Ataxia

| Symptoms and signs | MSA | PD | Late-onset cerebellar ataxia |
|---|---|---|---|
| Symmetry of motor disorders | Symmetric | Asymmetric (tremor and bradykinesia typically worse on one side) | Symmetric |
| Tremor | Uncommon at rest; postural and kinetic tremor common | Distal tremor frequently found at rest | Uncommon at rest; postural and kinetic tremor common |
| Rigidity | Present | Present | Absent |
| Bradykinesia | Present | Present | Absent |
| Response to levodopa | Poorly sustained, poor or absent | Excellent | Not applicable |
| Hypokinetic speech | Present in MSA–P and late in the course of MSA–C | Present | Absent or mild |
| Ataxic extraocular movements | Frequent in MSA–C, absent in MSA–P | Absent | Frequently present |
| Nystagmus | Frequent in MSA–C, absent in MSA–P | Absent | Frequently present |
| Saccadic interruptions of smooth ocular pursuit movements | Frequently present in both MSA–C and MSA–P | Frequently present | Frequently present |
| Truncal ataxia | Frequent in MSA–C, absent in MSA–P | Absent | Frequently present |
| Limb ataxia | Frequent in MSA–C, absent in MSA–P | Absent | Present |
| Ataxic speech | Frequent in MSA–C, often absent in MSA–P | Absent | Present |
| Postural hypotension | Common | Uncommon early, mild in late stages | Absent |
| Urinary urgency, incomplete emptying, or incontinence | Common | Uncommon | Uncommon |
| Antecollis | Present | Absent | Absent |
| Inspiratory stridor | Present | Absent | Absent |
| Dusky hands | Present | Absent | Absent |
| Myoclonus | Present | Absent | Absent |
| Rapid eye movement sleep behavior disorder | Present | Present | Frequently present |
| Obstructive sleep apnea | Frequently present | Frequently absent | Frequently present |

TABLE 23.2 Quinn Criteria for the Diagnosis of MSA

**MSA of the striatonigral type**

Possible MSA—Sporadic, adult-onset, parkinsonism unresponsive or poorly responsive to treatment with carbidopa/levodopa. The features could not include dementia, predominant down-gaze progressive supranuclear palsy (PSP), or other identifiable causes of a parkinsonian syndrome. The cases were required to be sporadic, defined as no more than one other case of typical idiopathic PD among first- or second-degree relatives. The onset was required to be at age 30 years or above.

Probable MSA—Cases meeting the above criteria plus severe symptomatic autonomic failure and/or cerebellar signs and/or pyramidal signs. Severe symptomatic autonomic failure was defined as postural syncope or presyncope and/or marked urinary incontinence or retention not due to other causes.

Definite MSA—Required postmortem confirmation.

**MSA of the olivopontocerebellar type**

Possible MSA—This category was not included in the diagnostic criteria.

Probable MSA—A sporadic adult-onset cerebellar syndrome with or without a pyramidal syndrome, with severe symptomatic autonomic failure and/or parkinsonism. The features could not include dementia, predominant down-gaze PSP, or other identifiable causes of a cerebellar disorder.

Definite MSA—Required postmortem confirmation.

From Quinn, N. (1989). *J. Neurol. Neurosurg. Psychiatry* **52**, Special Suppl. 78–89. With permission.

termed "prominent" rather than "predominant." For the diagnosis of probable MSA of the SND type, a moderate or good but often waning response to levodopa could occur, but multiple "atypical features" would be required for the diagnosis.

In 1998, a consensus panel convened in Minneapolis to develop new criteria for the diagnosis of MSA (Gilman et al., 1999b). These criteria employed the terms clinical domains, features, and criteria in the diagnosis of MSA (Table 23.3). Clinical domains referred to the predominant clinical feature involved, autonomic failure, and parkinsonian or cerebellar ataxia. Features (A) referred to a characteristic of the disease and criteria (B) indicated a defining feature or composite of features required for diagnosis.

TABLE 23.3  Consensus Criteria

**Clinical domains, features, and criteria**
  I. Autonomic and urinary dysfunction
     A. Autonomic and urinary features
        1. Orthostatic hypotension (by 20 mmHg systolic or 10 mmHg diastolic*)
        2. Urinary incontinence or incomplete bladder emptying
     B. Criterion for autonomic failure or urinary dysfunction in MSA
        1. Orthostatic fall in blood pressure (by 30 mmHg systolic or 15 mmHg diastolic) or urinary incontinence (persistent, involuntary partial or total bladder emptying, accompanied by erectile dysfunction in men) or both.
     * (Note the different figures for orthostatic hypotension depending on whether it is used as a feature or a criterion).
 II. Parkinsonism
     A. Parkinsonian features
        1. Bradykinesia (slowness of voluntary movement with progressive reduction in speed and amplitude)
        2. Rigidity
        3. Postural instability (not caused by primary visual, vestibular, cerebellar, or proprioceptive dysfunction)
        4. Tremor (postural, resting, or both)
     B. Criterion for Parkinsonism in MSA
        1. Bradykinesia plus at least one of items IIA2–A4
III. Cerebellar dysfunction
     A. Cerebellar features
        1. Gait ataxia (wide-based stance with steps of irregular length and direction)
        2. Ataxic dysarthria
        3. Limb ataxia
        4. Sustained gaze-evoked nystagmus
     B. Criterion for cerebellar dysfunction in MSA
        1. Gait ataxia plus at least one of items IIIA2–A4
 IV. Corticospinal tract dysfunction
     A. Corticospinal tract features
        1. Extensor plantar responses with hyperreflexia

**Diagnostic categories of MSA**
  I. Possible MSA: One criterion plus two features from separate other domains. When the criterion is parkinsonism, a poor levodopa response qualifies as one feature (hence only one additional feature is required)
 II. Probable MSA: Criterion for autonomic failure/urinary dysfunction plus poorly levodopa responsive parkinsonism or cerebellar dysfunction
III. Definite MSA: Pathologically confirmed by the presence of a high density of glial cytoplasmic inclusions in association with a combination of degenerative changes in the nigrostriatal and olivopontocerebellar pathways

**Exclusion criteria for the diagnosis of MSA**
  I. History
     Symptomatic onset under 30 years of age
     Family history of a similar disorder.
     Systemic diseases or other identifiable causes for features listed in the inclusion criteria
     Hallucinations unrelated to medication
 II. Physical examination
     DSM criteria for dementia
     Prominent showing of vertical saccades or vertical supranuclear gaze palsy
     Evidence of focal cortical dysfunction such as aphasia, alien limb syndrome, and parietal dysfunction
III. Laboratory investigation
     Metabolic, molecular genetic, and imaging evidence of an alternative cause of features listed in criteria.

From Gilman, S. *et al.* (1999b) *J. Neurol. Sci.* **163**, 94–98. With permission.

## B. Supporting Diagnostic Studies

Investigators have evaluated a number of ancillary tests to assist in the diagnosis of MSA. These have focused primarily on studies of the autonomic system and neuroimaging. Please refer to Section IV for a review of neuroimaging in MSA.

Comparative studies of progressive autonomic failure and MSA have revealed differences in the sites of autonomic dysfunction (Cohen et al., 1987). The two groups showed similar responses in tests of postganglionic sudomotor function, vagal function, and standing plasma norepinephrine levels. The groups differed, however, in the supine plasma-free norepinephrine level. These findings indicated that autonomic failure in MSA results primarily from a preganglionic disorder. Abnormalities in anal and urethral sphincter EMGs have been reported to identify patients with MSA with good sensitivity and specificity (Palace et al., 1997; Tison et al., 2000). However, patients with advanced PD and progressive supranuclear palsy (PSP) may also have abnormal sphincter EMG (Valldeoriola et al., 1995; Giladi et al., 2000; Libelius and Johansson, 2000), hence the test has limited specificity.

## IV. NEUROIMAGING

### A. Structural Brain Imaging with MRI

The initial finding in MSA of hypointensity in the putamen and a hyperintense rim lateral to the putamen in T2-weighted MRI scans (Drayer et al., 1986; Pastakia et al., 1986) prompted investigators to assess the sensitivity of MRI in the diagnosis of MSA. In one study, two neuroradiologists blindly and independently rated axial T2-weighted and proton density MRI scans of 44 clinically probable MSA cases (28 MSA–P and 16 MSA–C), 47 clinically diagnosed idiopathic PD patients, and 45 age-matched control subjects (Schrag et al., 1998). The investigation demonstrated that putaminal atrophy, a hyperintense putaminal rim, and infratentorial signal change had high specificity but low sensitivity for MSA. The infratentorial changes included atrophy of the cerebellum and middle cerebellar peduncles, and midbrain and signal change in the pons ("hot cross bun") and cerebellar peduncles. On 1.5 Tesla scans the overall sensitivity was 88% and the specificity of the findings for MSA in comparison to idiopathic PD and controls was 93 and 91%, respectively. Putaminal isointensity or hypointensity relative to globus pallidus, absolute putaminal hypointensity, and altered size of the olives were not useful discriminators.

In another study, slit-hyperintensity in the outer margin of the putamen on T2-weighted MRI scans appeared in 17 of 28 clinically diagnosed MSA patients and in none of 25 clinically diagnosed PD (Konagaya et al., 1994).

A third study evaluated the frequency and specificity of hypointense MRI signal changes in the putamen in parkinsonism of various origins (Kraft et al., 1999). This finding was compared with the frequency and specificity of a pathological MRI pattern consisting of a hyperintense lateral rim and a hypointense signal in the dorsolateral putamen on T2-weighted MRIs. The study group included 15 MSA patients of the STN type (MSA–P), 65 with PD, and 10 with PSP. Of the 15 MSA patients 9 showed the putaminal pattern of hyperintense lateral rim and a dorsolateral hypointense signal on T2-weighted images. This pattern occurred in none of the 65 patients with PD and none of the 10 patients with PSP. Exclusively hypointense changes in the putamen were found in 5 MSA–P patients (36%), 6 PD patients (9%), and 4 PSP patients (40%). These data suggest that a hyperintense lateral rim and a hypointense signal in the dorsolateral putamen is a highly specific (100%) but only moderately sensitive (60%) finding in MSA. In contrast, putaminal hypointensity alone is a sensitive but nonspecific MRI sign of MSA.

A fourth investigation evaluated the sensitivity of MRI in the diagnosis of MSA, PSP and corticobasal degeneration (CBD; Schrag et al., 2000). Two neuroradiologists independently and blindly rated axial T2-weighted and proton density MR images of 54 MSA patients (30 MSA–P and 24 MSA–C), 35 clinically probable PSP patients (diagnosis according to NINDS criteria), 5 CBD patients (autopsy confirmed in 4), and 44 normal control subjects. The findings considered characteristic of MSA–P included a hyperintense putaminal rim, putaminal atrophy and hyperintensity, atrophy and decreased signal in the globus pallidus, thinning or smudging of the substantia nigra, and infratentorial signal increase and atrophy. All MSA–C cases had infratentorial atrophy and signal increase. No CBD case showed infratentorial signal increase, a hyperintense putaminal rim, putaminal atrophy and hyperintensity, or atrophy and decreased signal in the globus pallidus. Findings typical of PSP included a midbrain diameter less than 17 mm, signal increase in the midbrain, dilatation of the third ventricle, frontal or temporal atrophy, increased signal in the globus pallidus, and atrophy or increased signal in the red nucleus. These features helped distinguish PSP from MSA. More than 70% patients with PSP and more than 80% of patients with MSA–C could be classified correctly, and no patient in these groups became misclassified. In the remaining patients an unequivocal differentiation could not be made. Only approximately 50% of patients with MSA–P could be classified correctly, and 19% of them (all of whom had been evaluated with 0.5-T scans) were misclassified.

The studies reviewed above suggest that in T2-weighted MRI scans, a hyperintense rim lateral to the putamen and hypointensity or atrophy in the putamen are relatively

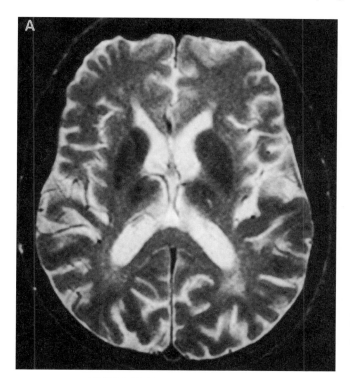 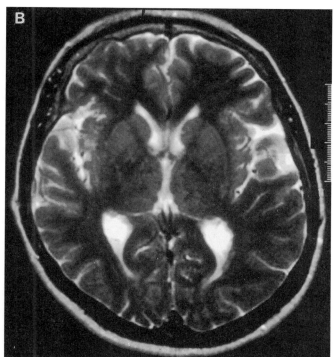

FIGURE 23.5 (A) Axial T2-weighted MRI TR/TE 3500/90 msec in a 59-year-old male with probable MSA of the parkinsonian type. The image shows decreased signal intensity of the putamen and caudate nucleus bilaterally, slit-like hyperintensity along the outer margin of the putamen, slightly increased size of the lateral ventricles, and volume loss in the cerebral cortex with increased size of the sulci. (B) Axial T2-weighted MRI TR/TE 4687/18 msec in a 71-year-old male with progressive supranuclear palsy. The image shows mild volume loss that is compatible with age but no other abnormalities.

specific for MSA. These findings can be particularly helpful in differentiating MSA from PD, PSP, and normal control subjects (Fig. 23.5). Also, infratentorial changes can help to distinguish MSA from PD and other forms of atypical parkinsonism such as PSP (Fig. 23.6) and CBD. These findings have not been used thus far in attempts to differentiate MSA–C from sOPCA patients. None of these studies was carried out prospectively, consequently the utility of MRI in the early diagnosis of MSA remains to be determined.

## B. Functional Cardiac Imaging with SPECT

SPECT studies with [$^{123}$I]metaiodobenzylguanidine (MIBG) have been used to compare the autonomic innervation of the heart in PD and MSA. Structurally similar to norepinephrine, MIBG enters and labels postganglionic adrenergic neurons. In 15 PD patients with autonomic failure, the heart-to-mediastinum ratio of [$^{123}$I]MIBG was pathologically impaired independent of the duration and severity of parkinsonian and autonomic symptoms (Braune et al., 1999). In contrast, the five MSA subjects had normal ratios, demonstrating that autonomic failure occurs in postganglionic neurons in PD but in preganglionic neurons in MSA. These findings were extended in a study attempting to differentiate between the early stages of PD and MSA by examining patients who did not require levodopa therapy (Druschky et al., 2000). The median cardiac [$^{123}$I]MIBG uptake was significantly decreased in 10 PD and 20 MSA patients compared to controls. The uptake was significantly lower in PD than in MSA, but some overlap occurred between the two groups. Even in PD patients without clinical signs of autonomic failure, [$^{123}$I]MIBG uptake was significantly lower than in MSA. These studies indicate that [$^{123}$I]MIBG-SPECT may be able to differentiate MSA from PD, even in early stages of the diseases. These studies in patients with early disease need to be replicated and extended to other disorders such as PSP and CBD.

## C. Functional Brain Imaging with PET and SPECT

Many functional imaging studies with PET and SPECT have demonstrated abnormalities in the nigrostriatal pathway and in the striatum of MSA patients in comparison with patients with PD and other parkinsonian syndromes (Booji et al., 1999, Schulz and Dichgans, 2000).

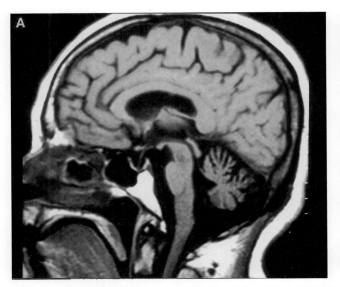 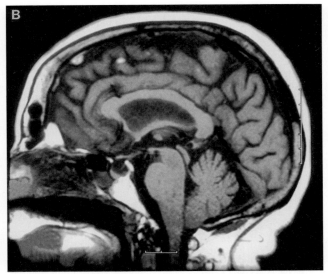

FIGURE 23.6 (A) Sagittal T1-weighted MRI TR/TE 500/20 msec 5 mm thick in a 47-year-old female with probable MSA of the cerebellar type. The image shows decreased volume of the pons, particularly its caudal, enlargement of the fourth ventricle, and decreased volume of the cerebellar vermis. (B) Sagittal T1-weighted MRI TR/TE 550/16 msec 5 mm thick in a 75-year-old woman with progressive supranuclear palsy. The image shows mild generalized volume loss compatible with age but no other abnormality.

### 1. Metabolic Markers

Studies utilizing PET with [$^{18}$F]fluorodeoxyglucose have consistently shown decreased local cerebral metabolic rates for glucose in the striatum of MSA–P patients and in the cerebellum and brainstem of MSA–C patients. The first study examined 7 patients with probable STN as compared with 16 normal, age-matched control subjects (De Volder et al., 1989). The STN subjects had marked hypometabolism in the caudate and putamen. A later study compared 10 patients with clinically likely STN to 10 age-matched normal control subjects, 10 disease severity-matched PD patients, and 10 disease duration-matched PD subjects (Eidelberg et al., 1993). The STN subjects had significantly reduced glucose metabolic rates in the caudate nucleus and putamen compared to the normal controls and both PD groups. The reduction was more striking in the putamen than in the caudate nucleus. In the first study of MSA–C, marked hypometabolism appeared not only in the brainstem and cerebellum, but also in the caudate nucleus, putamen, thalamus, and cerebral cortex (Gilman et al., 1994). A subsequent investigation revealed hypometabolism primarily in the putamen/pallidum complex in MSA–P and principally in the cerebellum in MSA–C (Perani et al., 1995). A later study demonstrated decreased glucose metabolic rates in the frontal and temporal cortex, caudate nucleus, putamen, cerebellum, and brainstem in MSA as compared with PD and normal control subjects, but with overlap between MSA cases and normal controls (Otsuka et al., 1997). Another investigation revealed that reduced striatal glucose metabolic rates could distinguish MSA–P cases from PD and normal control subjects (Antonini et al., 1997), however, striatal hypometabolism has also been reported in CBD (Nagahama et al., 1997).

### 2. Markers of Striatal Dopaminergic Terminals

Although studies utilizing PET with [$^{18}$F]fluorodopa have shown consistently decreased striatal dopaminergic terminals in MSA–P, a similar pattern has been found in other parkinsonian syndromes. In a comparison of 25 MSA subjects with 28 clinically probable PD, 10 PSP and 27 normal subjects, the brain ligand inflow constants ($K_i$) in both the caudate nucleus and putamen allowed differentiation of PD from the normal controls (Burn et al., 1994). The $K_i$ value in the caudate nucleus allowed differentiation of PSP from PD, but did not distinguish MSA from either PD or PSP. The $K_i$ value in the putamen did not distinguish PD, MSA, or PSP from each other. A subsequent investigation also showed that [$^{18}$F]fluorodopa could distinguish between normal controls and parkinsonian subjects (Antonini et al., 1997). A third investigation compared 9 MSA patients (5 with MSA–C, 2 with STN, and 2 with MSA–P) with15 PD subjects (Otsuka et al., 1997). In keeping with the previous studies, both MSA and PD subjects had reduced uptake in the caudate and putamen compared to control subjects.

Other investigations of the striatal monoaminergic terminals in MSA have utilized PET with (+)-

[$^{11}$C]dihydrotetrabenazine, a ligand for the type 2 vesicular monoamine transporter (Gilman et al., 1996a, 1999a). The studies compared inflow constants and binding in 7 normal controls with 8 MSA–P, 8 MSA–C, and 6 sOPCA patients. All patient groups had significantly reduced specific binding in the putamen in the order MSA–P < MSA–C < sOPCA, as compared with controls. The brain ligand transport ($K_1$) was significantly decreased in the putamen of all patient groups and in the cerebellar hemispheres of MSA–C and sOPCA but not MSA–P groups, as compared with controls. These findings suggested fundamental differences between MSA–P and MSA–C groups reflecting differential severity of degeneration of nigrostriatal and cerebellar systems in these two forms of MSA. The findings also showed that some sOPCA patients have subclinical nigrostriatal dysfunction and are at risk of developing MSA with disease progression. This finding is particularly intriguing in light of a later report from this group that 24% of sOPCA subjects evolve into an MSA phenotype within 5 years of onset of symptoms (Gilman et al., 2000).

SPECT has been used with [$^{123}$I] beta-CIT, a ligand for the dopamine transporter, to compare MSA with sOPCA and parkinsonian syndromes. One study demonstrated reduced dopaminergic terminals in sOPCA as well as in MSA (Kim et al., 2000), suggesting that these sOPCA patients may evolve into MSA–C. Another investigation of 9 MSA, 4 PSP, and 113 PD patients demonstrated a reduction in the dopamine transporter of all three groups compared to controls (Brucke et al., 1997).

### 3. Markers of Striatal Postsynaptic Dopaminergic Receptors

Several investigations utilizing PET with [$^{11}$C]raclopride or SPECT with [$^{123}$I]-iodobenzamide have demonstrated decreased density of striatal dopaminergic $D_2$ receptors in MSA, indicating striatal pathology. The first investigation used both [$^{18}$F] fluorodopa and [$^{11}$C]raclopride in nine levodopa naive patients with clinically diagnosed parkinsonism (Sawle et al., 1993). In eight of the subjects the putamen with the lower [$^{18}$F]fluorodopa uptake had higher [$^{11}$C]raclopride binding on that side, suggesting more upregulation of $D_2$ receptors on the side of greater dopaminergic denervation. In one patient, however, the putamen on the side with the lower [$^{18}$F]fluorodopa uptake had lower [$^{11}$C]raclopride binding, suggesting postsynaptic dysfunction as well as presynaptic dysfunction. This patient subsequently developed a course typical of MSA–P.

In a study utilizing [$^{11}$C]raclopride (in addition to [$^{18}$F]fluorodeoxyglucose and [$^{18}$F]fluorodopa) in 10 PD patients, 9 MSA patients, and 10 normal control subjects, the MSA cases had significantly lower binding than both the healthy control subjects and the PD patients (Antonini et al., 1997).

SPECT scanning with [$^{123}$I]-iodobenzamide has also demonstrated decreased binding to the $D_2$ receptor in MSA. In an investigation of 11 patients with MSA–P and 21 patients with MSA–C, binding was more than two standard deviations below the normal range in approximately two-thirds of the MSA cases (Schulz et al., 1994). The reduction appeared to be equivalent in both MSA–P and MSA–C cases. In a longitudinal investigation, 9 PD and 9 Parkinson plus syndrome patients were followed over 11–53 months, with scans obtained at the beginning and end (Hierholzer et al., 1998). The Parkinson plus syndrome patients had significantly lower binding than the PD patients at both the initial and the final scan. Also, binding declined over the course of the study in the Parkinson plus but not in the PD patients. SPECT studies with this ligand in PSP also show reduced striatal binding (van Royen et al., 1993; Arnold et al., 1994), hence studies of the $D_2$ receptor may not differentiate MSA from other parkinsonian syndromes.

Two PET studies of postsynaptic opioid receptors with the nonselective ligand [$^{11}$C]diprenorphin have shown decreased binding in MSA, further documenting the extensive striatal pathology in this disorder. In one (Burn et al., 1995), binding was no different in 8 PD as compared with 8 normal control subjects. In the same study, binding was significantly reduced in the putamen but not the caudate nucleus of 7 SND patients compared to the controls, and in both the caudate nucleus and putamen of 6 PSP cases compared to the controls. In the other study, (Rinne et al., 1995), 10 MSA–C patients were studied with both [$^{11}$C]diprenorphine and [$^{18}$F]fluorodopa. The binding of both ligands was significantly reduced in the putamen of the patients as compared to a group of normal controls.

The above studies indicate that PET studies of glucose metabolic rates can differentiate MSA from PD and normal control subjects, but perhaps not MSA from PSP or CBD patients, although a definitive study has not been undertaken. PET or SPECT with markers of presynaptic dopaminergic terminals cannot differentiate among PD, MSA, and PSP cases, but may be able to identify subjects with sOPCA who will evolve into MSA. Similarly, PET or SPECT ligands for the $D_2$ receptor can distinguish MSA from controls and untreated PD patients, but perhaps not MSA from PSP, CBD, or PD patients receiving dopaminergic medications (which downregulate $D_2$ receptors), although a rigorous comparative study has not been reported.

## V. PATHOLOGY OF MSA

Papp et al. (1989) first described the cardinal neuropathological feature of MSA, the GCI, by using a silver impregnation technique in 11 patients with various combinations of STN, OPCA, and the Shy-Drager syndrome. They found GCIs in oligodendrocytes. Subsequently, Papp and Lantos (1992, 1994) found these abnormal tubular structures in five locations: oligodendrocyte cytoplasm, oligodendrocyte nuclei, neuronal cytoplasm, neuronal nuclei and axons. GCIs appeared commonly in the cortical motor regions, putamen, caudate nucleus, pallidum, reticular formation of the brainstem, basis pontis, and intermediate zone of the spinal cord. GCIs were found particularly in the motor systems (pyramidal, extrapyramidal, and corticocerebellar) and in the supraspinal autonomic systems, but not in the visual, auditory, olfactory, somatosensory systems, and the limbic cortical and subcortical systems. They noted that oligodendrocyte degeneration (as demonstrated by density of GCIs) exceeded neuronal degeneration (as demonstrated by neuronal cytoplasmic and nuclear inclusions, degenerated neurons, and loss of nerve cells). This observation suggests (but does not prove) that GCIs may develop in advance of neuronal loss and that the formation of GCIs does not require neuronal degeneration.

Initial immunohistochemical studies revealed in GCIs the presence of ubiquitin, tau and α- and β-tubulin-like material but not actin, vimentin, desmin, cytokeratin, glial fibrillary acidic protein, and neurofilament-like material. Subsequently, several groups discovered that α-synuclein constitutes a major component of GCIs and neuronal cytoplasmic inclusions (Gai *et al.*, 1998; Spillantini *et al.*, 1998; Tu *et al.*, 1998; Wakabayashi *et al.*, 1998). Gai *et al.* (1998) found GCIs located mainly in white matter, but also demonstrated α-synuclein-IR inclusions in cytoplasmic and intranuclear inclusions of neurons in the putamen and pons. Degenerating neurites also were immunoreactive for α-synuclein. Spillantini *et al.* (1998) used antibodies that recognize amino and carboxyl epitopes, and reported that full-length α-synuclein appeared to be present.

In a biochemical study of the components of GCIs, Gai *et al.* (1999) used density gradient enrichment and anti-α-synuclein immunomagnetic techniques to isolate GCIs from the white matter of MSA patients. Immunoblotting revealed that the inclusions contained not only α-synuclein, but also αB crystallin, tubulins, ubiquitin, and a species of possibly truncated α-synuclein as high molecular weight aggregates. Gai *et al.* (1999) did not find evidence of neurofilaments, β-amyloid or β-synuclein.

Several recent excellent reviews have described the role of α-synuclein in degenerative neurological disorders (Dickson, 2001; Galvin *et al.*, 2001; Goedert, 2001; Goldberg and Lansbury, 2000). Alpha-synuclein is a highly soluble, 140 amino acid long protein with acidic regions in the C terminus and 6 degenerate KTKEGV motifs within residues 10–86 of the N terminus (Dickson, 1999; Lavedan, 1998; Souza *et al.*, 2000). Homologous proteins have been identified and cloned from several species. The synuclein family now includes α-synuclein, β-synuclein (also designated phosphoneuronoprotein 14 or PNP14), and γ-synuclein (also called breast cancer specific gene 1 or BCSG1 and persyn). The subsequent identification of a new γ-synuclein-like retinal protein (synoretin) suggests that not all members of the synuclein family have been identified. Separate genes on chromosome 4q21.3–q22, 5q35, and 10q23 encode α-synuclein, β-synuclein, and γ-synuclein, respectively. Although brain tissue expresses all synucleins, primarily in axon terminals and presynaptic sites, other tissues (breast, ovary, skin, retina) also express these proteins. Alpha-synuclein develops very little structure in aqueous solutions, but in tissues it associates with presynaptic and unilamellar phospholipid vesicles. Alpha-synuclein also undergoes fast axonal transport, modulates the neurofilament network, and inhibits phospholipase D. Despite fragmentary information on the normal biology of α-, β- and γ-synuclein, it appears that α-synuclein may participate in synaptic functions and plasticity.

Reports of an association of two mutations in the gene encoding α-synuclein with autosomal dominantly inherited parkinsonism has spurred research into α-synuclein (Kruger *et al.*, 1998; Polymeropoulos *et al.*, 1997). This research has shown that aggregates of α-synuclein appear not only in the GCIs characteristic of MSA, but also in glia as well as neurons in PD and related Lewy body disorders (Arai *et al.*, 1999; Kruger *et al.*, 1999; Dickson, 2001; Gwinn-Hardy *et al.*, 2000; Wakabayashi *et al.*, 2000). Recent studies have demonstrated selective nitration of α-synuclein in GCIs of MSA and other α-synucleinopathies, implicating nitrative/oxidative damage in the mechanisms of brain degeneration of these neurodegenerative disorders (Giasson *et al.*, 2000).

## VI. EPIDEMIOLOGY AND ENVIRONMENTAL RISK FACTORS

Epidemiological studies of MSA have been carried out only recently. The scarcity of studies stems from the relatively recent recognition of MSA as a distinct clinicopathological entity. The studies were performed principally to ascertain the frequency of other movement disorders, particularly PD. This method of ascertainment neglects cases with primary cerebellar or autonomic presentations and thus underreports the true incidence and prevalence of MSA. Trenkwalder *et al.* (1995) carried out a door-to-door survey of the total population older than 65 years in two rural Bavarian villages to determine the prevalence of various forms of parkinsonism in elderly Germans. They found among the 1190 persons investigated a prevalence of MSA of 0.31% (310/100,000). Schrag *et al.* (1999) studied the prevalence of PSP and MSA by screening the computerized records of 15 general practices in London. They found an age-adjusted prevalence of MSA of 4.4/100,000 with a prevalence of 28/100,000 in persons over the age of 65. In an earlier study carried out in the Faroe Islands, Wermuth et al. (1997) found an overall prevalence of Shy-Drager disease of 2.3/100,000. Chio et al. (1998) found a prevalence of 4.9/100/000 in a survey of persons taking anti-parkinsonian drugs in Northwestern Italy. Only one study, Bower et al. (1997), has examined the incidence of MSA. Incidence, which is a measure of the number of new cases of an illness during a time period, is a better index of the frequency of a disease than prevalence, as it is less affected by survival. Bower et al. (1997) reviewed medical records to study the incidence of MSA in Olmstead County, MN, from 1976 to 1990. They found an annual incidence of 3.0/100,000 in persons over the age of 50.

Most of the epidemiological studies of MSA have been carried out in Europe, hence we do not know the distribution of MSA throughout the world. Also, there was a broad range in the prevalence of MSA among the studies. While this variability could reflect methodological differences in the studies, the variability stems from variations in genetic and environmental exposures that underlie MSA.

Knowledge regarding the environmental risk factors associated with MSA is sadly lacking, and the few studies done to date have yielded only rudimentary information about potential causative agents. Two very small case-control studies of MSA have been reported. An initial study (Nee et al., 1991) showed a significantly higher proportion of MSA patients than controls with prior exposures to metal dusts and fumes (Odds ratio, OR – 14.75), inorganic dusts (OR – 7.79), pesticides (OR – 5.8), plastic monomers and additives (OR – 5.25), organic dusts (OR – 3.14) and organic solvents (OR – 2.41). Vanacore et al. (2000) reported that in comparison to a control population, cigarette smoking was significantly less common in MSA and PD, but not PSP. Frumkin (1998) reported a single case of MSA, principally with cerebellar dysfunction, following chronic exposure to carbon disulfide in a rayon factory. Shill and Fife (2000) reported a single case of valproic acid toxicity mimicking MSA.

In a review of the medical records of 100 patients with clinically diagnosed MSA Hanna et al. (1999) discovered a history of exposure to high levels of toxins such as pesticides and solvents in 11 of the cases. Although this study is intriguing, it lacks a control population, and the results need to be evaluated in a case control study.

## VII. GENETICS

MSA typically occurs sporadically. To date there are no reports of families with MSA in multiple generations suggesting an inherited disorder. In her study of parkinsonism in twin pairs, Tanner (2001) identified a pair of monozygotic twins with parkinsonism, and at autopsy one of the brothers had MSA–P.

Some of the genetically identified dominantly inherited spinocerebellar ataxias (SCAs) can be associated with parkinsonism. Gilman et al. (1996b) described a family with SCA 1 that exhibited facial akinesia, limb rigidity, dystonia, and peripheral neuropathy in addition to severe ataxia. At neuropathological examination, two of the family members exhibited neuronal degeneration and gliosis in the basal ganglia, cerebellum, brainstem, and spinal autonomic nuclei, with argyrophilic glial cytoplasmic inclusions resembling those seen in MSA (Gilman et al., 1996b). Schols et al. (2000) found that patients with SCA 3 and SCA 6, but not SCA 1, SCA 2, or Friedrich's ataxia, often exhibited bradykinesia and/or rigidity. The prevalence of autonomic dysfunction in subjects with the SCAs is not well defined. Soong et al. (1997) found autonomic dysfunction in 6 of 25 patients with SCA3. Filla et al. (1999) found that 55% of 72 Italian patients with SCA2 had symptoms of sphincteric dysfunction. In patients who have cerebellar dysfunction and evidence of parkinsonism it seems reasonable to screen for SCA 3 and 6, and in patients with cerebellar and autonomic dysfunction, screening for SCA 2, particularly if there is evidence of a peripheral neuropathy.

A small number of studies have investigated possible genetic contributions to the development of MSA. Ozawa et al. (1999) studied the entire coding region for α-synuclein in 11 cases of autopsy-verified MSA and found no mutations. They added the caveat that there might be mutations in regulatory regions or introns. This possibility has been underscored by the recognition that a dinucleotide repeat polymorphism, REP1, in the promoter region of α-synuclein has been linked in some studies to parkinsonism (Izumi et al., 2001; Kruger et al., 1999; Parsian et al., 1998; Touchman et al., 2001; Tan et al., 2000). A study of the tau gene (Bennett et al., 1998) confirmed the higher frequency of the A0 allele in PSP, but reported that the frequency of the allele in 35 cases of MSA was not increased. Cairns et al. (1997) found no difference in the frequency of the apolipoprotein e4 allele in 22 cases of autopsy-proven MSA as compared with 66 neurologically normal control cases.

The sporadic occurrence of MSA suggests that environmental factors participate in the development of MSA but does not negate the possible role of genetic factors in MSA. The situation may be similar to that of PSP, which also typically occurs as a sporadic disorder. Recent studies, however, have shown that homozygous tau A0 alleles can be found more frequently among PSP patients than normal control subjects, suggesting that genetic factors may be important in PSP (Conrad et al., 1997; Morris et al., 1999).

## VIII. MANAGEMENT AND TREATMENT

### A. Extrapyramidal Features

In MSA patients with bradykinesia, rigidity, and tremor, levodopa can be helpful, but usually the effect is small and poorly sustained (Hughes et al., 1992). Nevertheless, treatment should be initiated with carbidopa/levodopa 25/100 three times daily, with an escalation of dose every other day as tolerated to a total of eight tablets per day. The most important side effects are nausea, involuntary movements of the choreiform or dystonic type, postural hypotension, and psychological disturbances such as depression, paranoid ideation, and hallucinations. Some patients need and tolerate larger doses than most patients with PD. If levodopa/carbidopa proves ineffective, other dopaminergic agonist agents usually do not help. It is worth attempting a trial of anticholinergic agents such as trihexyphenidyl 1 mg once daily with an increase in dose by 2 mg every 3–5 days to a total daily dose of 10–12 mg in three divided doses daily as tolerated. The main side effects are dry mouth, blurred vision, and nausea. Amantadine 100 mg twice daily can also be utilized if both dopaminergic and anticholinergics agents fail.

### B. Autonomic Disorders

Symptomatic orthostatic hypotension must be monitored and treated. Patients should sit down or lie down as soon as symptoms appear, which usually occurs when they rise from the recumbent or seated position. Patients should avoid extreme heat because of reflex peripheral vasodilatation and also avoid overeating and straining at stool, which increase vagal activity. Pressure stockings should be worn constantly to increase central venous volume. The head of the patient's bed should be elevated 6 inches to enhance renin secretion. Persistent symptomatic orthostatic hypotension requires administration of fludrocortisone at 0.1 mg daily, increasing progressively to a maximum of 0.4 mg per day in two divided doses. If this is not helpful, administration of indomethacin 25 mg three times daily with meals increasing to 50 mg three times daily may be helpful by inhibition of prostaglandin synthesis, which causes vasodilatation. An alternative to this is to prescribe the alpha adrenergic agonist clonidine 0.1 mg twice daily increasing to 0.3 mg twice daily. Ephedrine given at a dose of 25 mg three times daily can be helpful by enhancing peripheral vasoconstriction. Midodrine can also be useful, beginning with a dose of 2.5 mg tid and a maximal dose of 10 mg tid. When using midodrine, the clinician should be careful to avoid supine hypertension.

MSA causes three abnormalities of lower urinary tract function: (1) involuntary detrusor contraction results from bladder filling, perhaps owing to loss of inhibitory influences from the corpus striatum and substantia nigra; (2) the capacity to initiate a voluntary micturition reflex is lost, probably owing to degeneration of neurons in pontine and medullary nuclei and in sacral intermediolateral columns of the sacral spinal cord and a reduction in the density of acetylcholinesterase-containing nerves innervating the bladder musculature; and (3) severe urethral dysfunction occurs, partly from dysfunction of the striated urethral sphincter musculature, leading to bladder neck incompetence. Urinary symptoms of frequency and incontinence often respond to a peripherally acting anticholinergic agent such as oxybutynin 5–10 mg at bedtime. An alternative therapy is propantheline 15–30 mg at bedtime, which is helpful in treating detrusor hyperreflexia. Anticholinergic medications can worsen the constipation that frequently occurs in MSA.

Constipation can be treated effectively with a high-fiber diet, including daily administration of one to three doses of psyllium, a high-fiber cereal at breakfast, an apple and at least one serving of broccoli or similar high-fiber vegetable. Stool softening is readily accomplished with one eight-ounce glass of prune juice and additional servings of other fruit juices as needed.

Treatment of ataxia in MSA has not been effective. A trial of clonazepam can be used at a dose of 0.05 mg (1/2 tablet) at bedtime, increasing to 0.1 mg.

## References

Adams, R. D., van Bogaert, L., and van der Eecken, H. (1961). Dégénérescences nigro-striées et cérébello-nigro-striées. *Psychiat. Neurol.* **142**, 219–259.

Antonini, A., Leenders, K. L., Vontobel, P., Maguire, R. P., Missimer, J., Psylla, M., Gunther, I. (1997). Complementary PET studies of striatal neuronal function in the differential diagnosis between multiple system atrophy and Parkinson's disease. *Brain* **120**, 2187–2195.

Arai, T., Ueda, K., Ikeda, K., Akiyama, H., Haga, C., Kondo, H., Kuroki, N., Niizato, K., Iritani, S., and Tsuchiya, K. (1999). Argyrophilic glial inclusions in the midbrain of patients with Parkinson's disease and diffuse Lewy body disease are immunopositive for NACP/alpha–synuclein. *Neurosci. Lett.* **259**, 83–86.

Arnold, G., Tatsch, K., Oertel, W. H., Vogl, T., Schwarz, J., Kraft, E., and Kirsch, C. M. (1994). Clinical progressive supranuclear palsy: differential diagnosis by IBZM-SPECT and MRI. *J. Neural Transm. Suppl.* **42**, 111–118.

Bennett, P., Bonifati, V., Bonuccelli, U., Colosimo, C., De Mari, M., Fabbrini, G., Marconi, R., Meco, G., Nicholl, D. J., Stocchi, F., Vanacore, N., Vieregge. P., and Williams, A. C. (1998). Direct genetic evidence for involvement of tau in progressive supranuclear palsy. European Study Group on Atypical Parkinsonism Consortium. *Neurology* **51**, 982–985.

Booji, J., Tissingh, G., Winogrodzka, A., and van Royen, E. A. (1999). Imaging of the dopaminergic neurotransmission system using single-photon emission tomography and positron emission tomography in patients with parkinsonism. *Eur. J. Nucl. Med.* **26**, 171–182.

Bower, J. H., Maraganore, D. M., McDonnell, S. K., and Rocca, W. A. (1997). Incidence of progressive supranuclear palsy and multiple system atrophy in Olmsted County, Minnesota, 1976 to 1990. *Neurology* **49**, 1284–1288.

Braune, S., Reinhardt, M., Schnitzer, R., Riedel, A., and Lucking, C. H. (1999). Cardiac uptake of [123I]MIBG separates Parkinson's disease from multiple system atrophy. *Neurology* **53**, 1020–1025.

Brucke, T., Asenbaum, S., Pirker, W., Djamshidian, S., Wenger, S., Wober, C., Muller, C., and Podreka, I. (1997). Measurement of the dopaminergic degeneration in Parkinson's disease with [$^{123}$I] beta-CIT and SPECT. Correlation with clinical findings and comparison with multiple system atrophy and progressive supranuclear palsy. *J. Neural Transm. Suppl.* **50**, 9–24.

Burn, D. J., Sawle, G. V., and Brooks, D. J. (1994). Differential diagnosis of Parkinson's disease, multiple system atrophy and Steele-Richardson-Olszewski syndrome: discriminant analysis of striatal18F-dopa PET data. *J. Neurol. Neurosurg. Psychiatry* **57**, 278–84.

Burn, D. J., Rinne, J. O., Quinn, N. ., Lees, A. J., Marsden, C. D., and Brooks D. J. (1995). Striatal opioid receptor binding in Parkinson's disease, STN and Steele-Richardson-Olszewski syndrome, a [$^{11}$C]diprenorphine PET study. *Brain* **118**, 951–958.

Cairns, N. J., Atkinson, P. F., Kovacs, T., Lees, A. J., Daniel, S. E., and Lantos, P. L. (1997). Apolipoprotein E e4 allele frequency in patients with multiple system atrophy. *Neurosci. Lett.* **221**, 161–164.

Chio, A., Magnani, C., and Schiffer, D. (1998). Prevalence of Parkinson's disease in Northwestern Italy: comparison of tracer methodology and clinical ascertainment of cases. *Mov. Disord.* **13**, 400–405.

Cohen, J., Low, P., Fealey, R., Sheps, S., and Jiang, N. S. (1987). Somatic and autonomic function in progressive autonomic failure and multiple system atrophy. *Ann. Neurol.* **22**, 692–699.

Conrad, C., Andreadis, A., Trojanowski, J. Q., Dickson, D. W., Kang, D., Chen, X, Wiederholt, W., Hansen, L., Masliah, E., Thal, L. J., Katzman, R., Xia, Y., and Saitoh, T. (1997). Genetic evidence for the involvement of tau in progressive supranuclear palsy. *Ann. Neurol.* **41**, 707–708.

Dejerine, J. M., and Thomas, A. A. (1900). L'atrophie olivo-ponto-cérébelleuse. *Nouv. Icono. Salpêtrière* **13**, 330–370.

De Volder, A. G., Francart, J., Laterre, C., Dooms, G., Bol, A., Michel, C., and Goffinet, A. M. (1989). Decreased glucose utilization in the striatum and frontal lobe in probable striatonigral degeneration. *Ann. Neurol.* **26**, 239–247.

Dickson, D. W. (1999). Multiple system atrophy: a sporadic synucleinopathy. *Brain Pathol.* **9**, 721–732.

Dickson, D. W. (2001). Alpha-synuclein and the Lewy body disorders. *Curr. Opin. Neurol.* **14**, 423–432.

Drayer, B. P., Olanow, W., Burger, P., Johnson, G. A., Herfkens, R., and Riederer, S. (1986). Parkinson plus syndrome: diagnosis using high field MR imaging of brain iron. *Radiology* **159**, 493–498.

Druschky, A., Hilz, M. J., Platsch, G., Radespiel-Troger, M., Druschky, K., Kuwert, T., and Neundorfer, B. (2000). Differentiation of Parkinson's disease and multiple system atrophy in early disease stages by means of I-123-MIBG-SPECT. *J. Neurol. Sci.* **175**, 3–12.

Eidelberg, D., Takikawa, S., Moeller, J. R., Dhawan, V., Redington, K., Chaly, T., Robeson, W., Dahl, J. R., Margouleff, D., and Fazzini, E. (1993). Striatal hypometabolism distinguishes striatonigral degeneration from Parkinson's disease. *Ann. Neurol.* **33**, 518–527.

Filla, A., De Michele, G., Santoro, L., Calabrese, O., Castaldo, I., Giuffrida, S., Restivo, D., Serlenga, L., Condorelli, D. F., Bonuccelli, U., Scala, R., Coppola, G., Caruso, G., and Cocozza, S. (1999). Spinocerebellar ataxia type 2 in southern Italy: a clinical and molecular study of 30 families. *J. Neurol.* **246**, 467–471.

Frumkin, H. (1998). Multiple system atrophy following chronic carbon disulfide exposure. *Environ. Health Perspect.* **106**, 611–613.

Gai, W. P., Power, J. H., Blumbergs, P. C., and Blessing, W. W. (1998). Multiple-system atrophy: a new alpha-synuclein disease? *Lancet* **352**, 47–48.

Gai, W. P., Power, J. H., Blumbergs, P. C., Culvenor, J. G., and Jensen, P. H. (1999). Alpha-synuclein immunoisolation of glial inclusions from multiple system atrophy brain tissue reveals multiprotein components. *J. Neurochem.* **73**, 2093–2100.

Galvin, J. E., Lee, V. M., and Trojanowski, J. Q. (2001). Synucleinopathies: clinical and pathological implications. *Arch. Neurol.* **58**, 186–190.

Giasson, B. I., Duda, J. E., Murray, I. V., Chen, Q., Souza, J. M., Hurtig, H. I., Ischiropoulos, H., Trojanowski, J. Q., and Lee, V. M. (2000). Oxidative damage linked to neurodegeneration by selective alpha-synuclein nitration in synucleinopathy lesions. *Science* **290**, 985–989.

Giladi, N., Simon, E. S., Korczyn, A. D., Groozman, G. B., Orlov, Y., Shabtai, H., and Drory, V. E. (2000). Anal sphincter EMG does not distinguish between multiple system atrophy and Parkinson's disease. *Muscle Nerve* **23**, 731–734.

Gilman, S., Koeppe, R. A., Junck, L., Kluin, K. J., Lohman, M., and St. Laurent, R. T. (1994) Patterns of cerebral glucose metabolism detected with positron emission tomography differ in multiple system atrophy and olivopontocerebellar atrophy. *Ann. Neurol.* **36**, 166–175.

Gilman, S., Frey, K. A., Koeppe, R. A., Junck, L., Little, R., Vander Borght, T. M., Lohman, M., Martorello, S., Lee, L. C., Jewett, D. M., and Kilbourn, M. R. (1996a). Decreased striatal monoaminergic terminals in olivopontocerebellar atrophy and multiple system atrophy demonstrated with positron emission tomography. *Ann. Neurol.* **40**, 885–892.

Gilman, S., Sima, A. A. F., Junck, L., Kluin, K. J., Koeppe, R. A., Lohman, M. E., and Little, R. (1996b). Spinocerebellar ataxia type 1 with multiple system degeneration and glial cytoplasmic inclusions. *Ann. Neurol.* **39**, 241–255.

Gilman, S., Koeppe, R. A., Junck, L., Little, R., Kluin, K. J., Heumann, M., Martorello, S., and Johanns, J. (1999a). Decreased striatal monoaminergic terminals in multiple system atrophy detected with PET. *Ann. Neurol.* **45**, 769–777.

Gilman, S., Low, P. A., Quinn, N., Albanese, A., Ben-Shlomo, Y., Fowler, C. J., Kaufmann, H., Klockgether, T., Lang, A. E., Lantos, P. L., Litvan, I., Mathias, C. J., Oliver, E., Robertson, D., Schatz, I., and Wenning, G. K. (1999b). Consensus statement on the diagnosis of multiple system atrophy. *J. Neurol. Sci.* **163**, 94–98.

Gilman, S., Little, R., Johanns, J., Heumann, M., Kluin, K. J., Junck, L., Koeppe, R. A., and An, H. (2000). Evolution of sporadic olivopontocerebellar atrophy into multiple system atrophy. *Neurology* **55**, 527–532.

Goedert, M. (2001). Alpha-synuclein and neurodegenerative diseases. *Natl. Rev. Neurosci.* **2**, 492–501.

Goldberg, M. S., and Lansbury, P. T., Jr. (2000). Is there a cause-and-effect relationship between alpha-synuclein fibrillization and Parkinson's disease? *Nat. Cell. Biol.* **2**, E115–119.

Graham, J. G., and Oppenheimer, D. R. (1969). Orthostatic hypotension and nicotine sensitivity in a case of multiple system atrophy. *J. Neurol. Neurosurg. Psychiatry* **32**, 28–34.

Gwinn-Hardy, K., Mehta, N. D., Farrer, M., Maraganore, D., Muenter, M., Yen, S. H., Hardy, J., and Dickson, D. W. (2000). Distinctive neuropathology revealed by alpha-synuclein antibodies in hereditary parkinsonism and dementia linked to chromosome 4p. *Acta Neuropathol.* **99**, 663–672.

Hanna, P. A., Jankovic, J., and Kirkpatrick, J. B. (1999). Multiple system atrophy, the putative causative role of environmental toxins. *Arch. Neurol.* **56**, 90–94.

Hierholzer, J., Cordes, M., Venz, S., Schelosky, L., Harisch, C., Richter, W., Keske, U., Hosten, N., Maurer, J., Poewe, W., and Felix, R. (1998). Loss of dopamine-D2 receptor binding sites in Parkinsonian plus syndromes. *J. Nucl. Med.* **39**, 954–960.

Hughes, A. J., Colosimo, C., Kleedorfer, B., Daniel, S. E., and Lees, A. J. (1992). The dopaminergic response in multiple system atrophy. *J. Neurol. Neurosurg. Psychiatry* **55**, 1009–1013.

Izumi, Y., Morino, H., Oda, M., Maruyama, H., Udaka, F., Kameyama, M., Nakamura, S., and Kawakami, H. (2001). Genetic studies in Parkinson's disease with an alpha-synuclein/NACP gene poly-

morphism in Japan. *Neurosci. Lett.* **300**, 125–127.
Kim, G. M., Kim, S. E., and Lee, W. Y. (2000). Preclinical impairment of the striatal dopamine transporter system in sporadic olivopontocerebellar atrophy: studied with [(123)I]beta-CIT and SPECT. *Eur. Neurol.* **43**, 23–29.
Klein, C., Brown, R., Wenning, G., and Quinn, N. (1997). The "cold hands sign" in multiple system atrophy. *Mov. Disord.* **12**, 514–518.
Klockgether, T., Lüdtke, R., Kramer, B., Abele, M., Bürk, K., Schöls, L., Riess, O., Laccone, R., Boesch, S., Lopes-Cendes, I., Brice, A., Inzelberg, R., Zilber, N., and Dichgans, J. (1998). The natural history of degenerative ataxia: a retrospective study in 466 patients. *Brain* **121**, 589–600.
Kluin, K. J., Gilman, S., Lohman, M., and Junck, L. (1996). Characteristics of the dysarthria of multiple system atrophy. *Arch. Neurol.* **53**, 545–548.
Konagaya, M., Konagaya, Y., and Iida, M. (1994). Clinical and magnetic resonance imaging study of extrapyramidal symptoms in multiple system atrophy. *J. Neurol. Neurosurg. Psychiatry* **57**, 1528–1531.
Kraft, E., Schwarz, J., Trenkwalder, C., Vogl, T., Pfluger, T., and Oertel, W. H. (1999). The combination of hypointense and hyperintense signal changes on T2-weighted magnetic resonance imaging sequences: a specific marker of multiple system atrophy? *Arch. Neurol.* **56**, 225–228.
Kruger, R., Kuhn, W., Muller, T., Woitalla, D., Graeber, M., Kosel, S., Przuntek, H., Epplen, J. T., Schols, L., and Riess, O. (1998). Ala30Pro mutation in the gene encoding alpha-synuclein in Parkinson's disease. *Nat. Genet.* **18**, 106–108.
Kruger, R., Vieira-Saecker, A. M., Kuhn, W., Berg, D., Muller, T., Kuhnl, N., Fuchs, G. A., Storch, A., Hungs, M., Woitalla, D., Przuntek, H., Epplen, J. T., Schols, L., and Riess, O. (1999). Increased susceptibility to sporadic Parkinson's disease by a certain combined alpha-synuclein/apolipoprotein E genotype. *Ann. Neurol.* **45**, 611–617.
Lavedan, C. (1998). The synuclein family. *Genome Res.* **8**, 871–80.
Libelius, R., and Johansson, F. (2000). Quantitative electromyography of the external anal sphincter in Parkinson's disease and multiple system atrophy. *Muscle Nerve* **23**, 1250–1256.
Morris, H. R., Lees, A. J., and Wood, N. W. (1999). Neurofibrillary tangle parkinsonian disorders—tau pathology and tau genetics. *Mov. Disord.* **14**, 731–736.
Nagahama, Y., Fukuyama, H., Turjanski, N., Kennedy, A., Yamauchi, H., Ouchi, Y., Kimura, J., Brooks, D. J., and Shibasaki, H. (1997). Cerebral glucose metabolism in corticobasal degeneration: comparison with progressive supranuclear palsy and normal controls. *Mov. Disord.* **12**, 691–696.
Nee, L. E., Gomez, M. R., Dambrosia, J., Bale, S., Eldridge, R., and Polinsky, R. J. (1991). Environmental-occupational risk factors and familial associations in multiple system atrophy: a preliminary investigation. *Clin. Autonom. Res.* **1**, 9–13.
Otsuka, M., Kuwabara, Y., Ichiya, Y., Hosokawa, S., Sasaki, M., Yoshida, T., Fukumura, T., Kato, M., and Masuda, K. (1997). Differentiating between multiple system atrophy and Parkinson's disease by positron emission tomography with 18F-dopa and 18F-FDG. *Ann. Nucl. Med.* **11**, 251–257.
Ozawa, T., Takano, H., Onodera, O., Kobayashi, H., Ikeuchi, T., Koide, R., Okuizumi, K., Shimohata, T., Wakabayashi, K., Takahashi, H., and Tsuji, S. (1999). No mutation in the entire coding region of the alpha–synuclein gene in pathologically confirmed cases of multiple system atrophy. *Neurosci. Lett.* **270**, 110–112.
Palace, J., Chandiramani, V. A., and Fowler, C. J. (1997). Value of sphincter electromyography in the diagnosis of multiple system atrophy. *Muscle Nerve* **11**, 1396–1403.
Papp, M. I., Kahn, J. E., and Lantos, P. L. (1989). Glial cytoplasmic inclusions in the CNS of patients with multiple system atrophy (striatonigral degeneration, olivopontocerebellar atrophy and Shy-Drager syndrome). *J. Neurol. Sci.* **94**, 79–100.
Papp, M. I., and Lantos, P. L. (1992). Accumulation of tubular structures in oligodendroglial and neuronal cells as the basic alteration in multiple system atrophy. *J. Neurol Sci.* **107**, 172–82.
Papp, M. I., and Lantos, P. L. (1994). The distribution of oligodendroglial inclusions in multiple system atrophy and its relevance to clinical symptomatology. *Brain* **117**, 235–243.
Parsian, A., Racette, B., Zhang, Z. H., Chakraverty, S., Rundle, M., Goate, A., and Perlmutter, J. S. (1998). Mutation, sequence analysis, and association studies of alpha-synuclein in Parkinson's disease. *Neurology.* **51**, 1757–1759.
Pastakia, B., Polinsky, R., Di Chiro, G., Simmons, J. T., Brown, R., and Wener, L. (1986). Multiple system atrophy (Shy-Drager syndrome): MR imaging. *Radiology* **159**, 499–502.
Perani, D., Bressi, S., Testa, D., Grassi, F., Cortelli, P., Gentrini, S., Savoiardo, M., Caraceni, T., and Fazio, F. (1995). Clinical/metabolic correlations in multiple system atrophy. A fludeoxyglucose F 18 positron emission tomographic study. *Arch. Neurol.* **52**, 179-185.
Polymeropoulos, M. H., Lavedan, C., Leroy, E., Ide, S. E., Boyer, R., Stenroos, E. S., Chandrasekharappa, S., Athanassiadou, A., Papapetropoulos, T., Johnson, W. G., Lazzarini, A. M., Duvoisin, R. C., Di Iorio, G., Golbe L. I., and Nussbaum, R. L. (1997). Mutation in the alpha-synuclein gene identified in families with Parkinson's disease. *Science* **27**, 2045–2047.
Quinn, N. (1989). Multiple system atrophy-the nature of the beast. *J. Neurol. Neurosurg. Psychiatr.* **52**, Special Suppl 78–89.
Quinn, N. (1994). Multiple system atrophy. In "Movement Disorders 3" (C. D. Marsden and S. Fahn, eds.), pp. 262–281 Butterworth-Heinemann, London.
Rinne, J. O., Burn, D. J., Mathias, C. J., Quinn, N. P., Marsden, C. D., and Brooks, D. J. (1995). Positron emission tomography studies on the dopaminergic system and striatal opioid binding in the olivopontocerebellar atrophy variant of multiple system atrophy. *Ann. Neurol.* **37**, 568–573.
Sawle, G. V., Playford, E. D., Brooks, D. J., Quinn, N., and Frackowiak, R. S. (1993). Asymmetrical pre-synaptic and post-synaptic changes in the striatal dopamine projection in dopa naive parkinsonism. Diagnostic implications of the D2 receptor status. *Brain* **116**, 853–867.
Schols, L., Peters, S., Szymanski, S., Kruger, R., Lange, S., Hardt, C., Riess, O., and Przuntek, H. (2000). Extrapyramidal motor signs in degenerative ataxias. *Arch. Neurol.* **57**, 1495–1500.
Schrag, A., Kingsley, D., Phatouros, C., Mathias, C. J., Lees, A. J., Daniel, S. E., and Quinn, N. P. (1998). Clinical usefulness of magnetic resonance imaging in multiple system atrophy. *J. Neurol. Neurosurg. Psychiatry* **65**, 65–71.
Schrag, A., Ben-Shlomo, Y., and Quinn, N. P. (1999). Prevalence of progressive supranuclear palsy and multiple system atrophy: a cross-sectional study. *Lancet* **354**, 1771–1775.
Schrag, A., Good, C. D., Miszkiel, K., Morris, H. R., Mathias, C. J., Lees, A. J., and Quinn, N. P. (2000). Differentiation of atypical parkinsonian syndromes with routine MRI. *Neurology* **54**, 697–702.
Schulz, J. B., Klockgether, T., Petersen, D., Jauch, M., Muller-Schauenburg, W., Spieker, S., Voigt, K., and Dichgans, J. (1994). Multiple system atrophy: natural history, MRI morphology, and dopamine receptor imaging with 123IBZM-SPECT. *J. Neurol. Neurosurg Psychiatry* **57**, 1047–1056.
Schulz, J. B., and Dichgans, J. (2000). Idiopathic cerebellar degeneration. In "Handbook of Ataxia Disorders" (T. Klockgether, ed.). Chapter 27, pp. 545–569. Marcel Dekker, New York.
Shill, H. A., and Fife, T. D. (2000). Valproic acid toxicity mimicking multiple system atrophy. *Neurology* **55**, 1936–1937.
Shy, G. M., and Drager, G. A. (1960). A neurological syndrome associated with orthostatic hypotension. *Arch. Neurol.* **2**, 511–527.
Soong, B., Cheng, C., Liu, R., and Shan, D. (1997) Machado-Joseph disease: clinical, molecular, and metabolic characterization in Chinese kindreds. *Ann. Neurol.* **41**, 446–452.

Souza, J. M., Giasson, B. I., Lee, V. M., and Ischiropoulos, H. (2000). Chaperone-like activity of synucleins. *FEBS Lett.* **474**, 116–119.

Spillantini, M. G., Crowther, R., Jakes, R., Cairns, N. J., Lantos, P. L., and Goedert, M. (1998). Filamentous alpha-synuclein inclusions link multiple system atrophy with Parkinson's disease and dementia with Lewy bodies. *Neurosci. Lett.* **251**, 205–208.

Tan, E. K., Matsuura, T., Nagamitsu, S., Khajavi, M., Jankovic, J., and Ashizawa, T. (2000). Polymorphism of NACP-Rep1 in Parkinson's disease: an etiologic link with essential tremor? *Neurology* **54**, 1195–1198.

Tanner, C. (2001). Personal communication.

Tison, F., Arne, P., Sourgen, C., Chrysostome, V., and Yeklef, F. (2000). The value of external anal sphincter electromyography for the diagnosis of multiple system atrophy. *Mov. Disord.* **15**, 1148–1157.

Touchman, J. W., Dehejia, A., Chiba-Falek, O., Cabin, D. E., Schwartz, J. R., Orrison, B. M., Polymeropoulos, M. H., and Nussbaum, R. L. (2001). Human and mouse alpha–synuclein genes: comparative genomic sequence analysis and identification of a novel gene regulatory element. *Genome Res.* **11**, 78–86.

Trenkwalder, C., Schwarz, J., Gebhard, J., Ruland, D., Trenkwalder, P., Hense, H. W., and Oertel, W. H. (1995). Starnberg trial on epidemiology of Parkinsonism and hypertension in the elderly. Prevalence of Parkinson's disease and related disorders assessed by a door-to-door survey of inhabitants older than 65 years. *Arch. Neurol.* **52**, 1017–1022.

Tu, P.-H., Galvin, J. E., Baba, M., Giasson, B., Tomita, T., Leight, S., Nakajo, S., Iwatsubo, T., Trojanowski, J. Q., and Lee, V. M.-Y. (1998). Glial cytoplasmic inclusions in white matter oligodendrocytes of multiple system atrophy brains contain insoluble α-synuclein. *Ann. Neurol.* **44**, 415–422.

Valldeoriola, F., Valls-Sole, J., Tolosa, E. S., and Marti, M. J. (1995). Striated anal sphincter deneration in patients with progressive supranuclear palsy. *Mov. Disord.* **10**, 550–555.

van Royen, E., Verhoeff, N. F., Speelman, J. D., Wolters, E. C., Kuiper, M. A., and Janssen, A. G. (1993). Multiple system atrophy and progressive supranuclear palsy. Diminished striatal D2 dopamine receptor activity demonstrated by 123I-IBZM single photon emission computed tomography. *Arch. Neurol.* **50**, 513–516.

Vanacore, N., Bonifati, V., Fabbrini, G., Colosimo, C., Marconi, R., Nicholl, D., Bonuccelli, U., Stocchi, F., Lamberti, P., Volpe, G., De Michele, G., Iavarone, I., Bennett, P., Vieregge, P., and Meco, G. (2000). Smoking habits in multiple system atrophy and progressive supranuclear palsy. European Study Group on Atypical Parkinsonisms. *Neurology* **54**, 114–119.

Wakabayashi, K., Yoshimoto, M., Tsuji, S., Takahashi, H. (1998). Alpha-synuclein immunoreactivity in glial cytoplasmic inclusions in multiple system atrophy. *Neurosci. Lett.* **249**, 180–182.

Wakabayashi, K., Hayashi, S., Yoshimoto, M., Kudo, H., and Takahashi, H. (2000). NACP/alpha-synuclein-positive filamentous inclusions in astrocytes and oligodendrocytes of Parkinson's disease brains. *Acta Neuropathol. (Berlin)* **99**, 14–20.

Wenning, G. K., Ben Shlomo, Y., Magalhães, M., Daniel, S. E., and Quinn, N. P. (1994). Clinical features and natural history of multiple system atrophy. An analysis of 100 cases. *Brain* **117**, 835–845.

Wenning, G. K., Tison, F., Elliott, L., Quinn, N. P., and Daniel, S. E. (1996). Olivopontocerebellar pathology in multiple system atrophy. *Mov. Disord.* **11**, 157–162.

Wenning, G. K., Tison, F., Ben Shlomo, Y., Daniel, S. E., and Quinn, N. P. (1997). Multiple system atrophy: a review of 203 pathologically proven cases. *Mov. Disord.* **12**, 133–147.

Wenning, G. K., Ben-Shlomo, Y., Hughes, A., Daniel, S. E., Lees, A., and Quinn, N. P. (2000). What clinical features are most useful to distinguish definite multiple system atrophy from Parkinson's disease? *J. Neurol. Neurosurg. Psychiatry* **68**, 434–440.

Wermuth, L., Joensen, P., Bunger, N., and Jeune, B. (1997). High prevalence of Parkinson's disease in the Faroe Islands. *Neurology* **49**, 426–432.

# CHAPTER 24

# Metabolic and Mitochondrial Ataxias

ENRICO BERTINI
*Department of Neurosciences*
*Laboratory of Molecular Medicine*
*Bambino Gesu' Children Research Hospital*
*Rome, Italy*

CARLO DIONISI-VICI
*Department of Neurosciences*
*Division of Metabolic Disorders*
*Bambino Gesu' Children Research Hospital*
*Rome, Italy*

MASSIMO ZEVIANI
*Division of Biochemistry and Genetics, and*
*Division of Child Neurology*
*Istituto Nazionale Neurologico "C. Besta"*
*Milano, Italy*

I. Ataxia in Mitchondrial Disorders
   A. Definition and Classification of Mitochondrial Disorders
   B. Genetics of mtDNA Mutations
   C. Ataxia Associated with mtDNA Mutations
   D. Ataxia Associated with Nuclear Disease Genes
   E. Defects of Structural Components or Assembly Factors of Respiratory Chain Complexes
   F. Defects in Factors Controlling the Stability of mtDNA
   G. Defects in Factors Involved in Metabolic Pathways Influencing the Biogenesis of Mitochondria, including OXPHOS
   H. Factors Involved in Metabolic Pathways Influencing the Biogenesis of Mitochondria, including OXPHOS
   I. Other Disorders of Mitochondrial Energy Metabolism
II. Ataxia in Lipid Disorders
   A. Abetalipoproteinemia
   B. Cerebrotendineous Xanthomatosis
   C. Niemann-Pick Disease Type C
III. Ataxia in Lysosomal Disorders
   A. Tay-Sachs Disease
   B. Neuraminidase Deficiency
   C. Sialurias
IV. Ataxias Associated with Other Metabolic Disorders
   A. Refsum Disease
   B. Adrenoleukodystrophy (ALD)
   C. Mevalonic Aciduria
   D. Pyroglutamicaciduria
   E. 1-2-hydroxyglutaricacidemia
   F. Congenital Disorders of Glycosylation
   G. 4-Hydroxybutyricaciduria
   H. Galactosemia
   I. Harnup Disease
   J. Cerebellar Hypoplasia and Neonatal Diabetes
   K. Guidance to the Work-up when Metabolic or a Mitochondrial Disorder is Considered
References

After the comprehensive review written by Anita E. Harding (1984), metabolic determinants of ataxia have conspicuously expanded in the last decade, particularly in the field of mitochondrial encephalopathies. Moreover, there have been numerous advances in understanding the pathogenetic mechanisms of other metabolic conditions that were already known before, especially for lysosomal disorders, for Niemann-Pick type C, for some organic acidurias, and for congenital disorders of glycosilation. We will consider mitochondrial encephalopathies in more detail because they are a frequent and rapidly expanding cause of metabolic derangement causing ataxia.

## I. ATAXIA IN MITOCHONDRIAL DISORDERS

### A. Definition and Classification of Mitochondrial Disorders

The term "mitochondrial disorders" is to a large extent applied to the clinical syndromes associated with abnormalities of the common final pathway of the mitochondrial energy metabolism, i.e., the oxidative phosphorylation (OXPHOS). OXPHOS is carried out in the inner mitochondrial membrane by the five enzymatic complexes of the respiratory chain (Zeviani and Antozzi, 1997).

From a genetic standpoint, the respiratory chain is unique, since it is formed through the complementation of two distinct genetic systems, the nuclear and the mitochondrial genomes. Nuclear genes provide most of the protein subunits of the respiratory complexes, the factors that control their intramitochondrial transport, assembly, and turnover, as well as the enzymes for the synthesis of prosthetic groups. In addition, most of the components of the mitochondrial DNA (mtDNA)—replication and mtDNA—expression systems are encoded by genes localized in the nucleus. In turn, human mtDNA, a maternally transmitted circular minichromosome, is present in 2–10 copies per organelle (polyploidy). It is composed of mRNA genes encoding 13 subunits of respiratory complexes I, III, IV, and V, as well as of tRNA and rRNA genes that are part of the RNA apparatus needed for intraorganellar mtDNA translation (Anderson et al., 1981).

In the past, the diagnosis of mitochondrial (encephalo) myopathies was largely based on the detection of ragged red fibers in the muscle biopsy (Rowland et al., 1991) and/or biochemical defects of the respiratory complexes. The discovery of mutations of mtDNA and, later, of several OXPHOS-related nuclear genes associated with mitochondrial disease has ushered in a new era of research. However, in several cases, mitochondrial disorders, defined on the basis of morphological or biochemical findings, still lack a molecular-genetic definition.

A widely accepted classification of mitochondrial disorders links the clinical and biochemical features to genetic abnormalities. Accordingly, mitochondrial diseases can be divided into three main groups: (1) genetically defined defects of the mitochondrial genome; (2) disorders due to abnormalities of nuclear genes; and (3) biochemically defined disorders.

Given the scope of the present review only the clinical presentations including ataxia as a prominent and consistent feature will be considered.

## B. Genetics of mtDNA Mutations

In normal conditions the mitochondrial genotype of an individual is composed of a single mtDNA species, a condition known as homoplasmy. Mutations of mtDNA can produce a transitory condition known as heteroplasmy, where the wild-type and the mutant genomes co-exist intracellularly. Because of mitochondrial polyploidy, during mitosis the two mtDNA species are stochastically distributed to subsequent cell generations. By a form of genetic drift called "mitotic segregation" (Marchington et al., 1998), the relative proportion between mutant and wild-type genomes can vary widely among cells, and, consequently, among tissues and individuals. This phenomenon can explain the extreme variability of the phenotypic expression of a given mtDNA mutation, as is often observed in mitochondrial disorders. Heteroplasmic mutations will only be expressed phenotypically as a cellular dysfunction leading to disease when mutated gene copies accumulate over a certain threshold. At that point the deleterious effects of the mutation will no longer be complemented by the co-existing wild-type mtDNA. In addition, phenotypic expression will depend upon the nature of the mutation, i.e., its intrinsic pathogenicity, its tissue distribution, and the relative reliance of each organ system on the mitochondrial energy supply.

## C. Ataxia Associated with mtDNA Mutations

Of the very many mtDNA-related syndromes associated with ataxia, four are relatively frequent and well characterized both clinically and genetically. These are (1) Kearns-Sayre syndrome (KSS), (2) mitochondrial encephalopathy, lactic acidosis, and stroke-like episodes (MELAS), (3) myoclonus-epilepsy with ragged-red fibers (MERRF), and (4) neurogenic weakness, ataxia, and retinitis pigmentosa (NARP). KSS is almost invariably associated with a heteroplasmic mtDNA deletion or duplication, while MELAS, MERRF, and NARP are due to different point mutations.

### 1. Kearns-Sayre Syndrome (KSS)

KSS is a sporadic, severe disorder characterized by the invariant triad of (1) progressive external ophthalmoplegia (PEO), (2) pigmentary retinopathy, and (3) onset before age 20. Frequent additional symptoms are poor growth, a progressive cerebellar syndrome, heart block, and increased protein content in the cerebrospinal fluid (CSF). Cerebellar ataxia can for long be the only sign of central nervous system involvement in KSS. A partial and milder variant of KSS, sporadic PEO, is an adult-onset disease characterized by bilateral ptosis and ophthalmoplegia, frequently associated with variable degrees of proximal muscle weakness and wasting, and exercise intolerance. PEO is variably associated with signs of central nervous system involvement, among which mild ataxia is frequent. Pearson's bone marrow-pancreas syndrome (PS) is a rare disorder of early infancy characterized by sideroblastic anemia with pancytopenia and exocrine pancreatic insufficiency. It is mentioned here because of the observation that infants surviving into childhood may develop the clinical features of KSS (Rotig et al., 1990; McShane et al., 1991).

KSS is characterized by distinctive neuroradiological abnormalities in the cerebellum and brainstem, but also in supratentorial structures including the basal ganglia, thalami, and subcortical white matter (Barkovich et al., 1993). In the brainstem, the mesencephalon can be affected diffusely, but occasionally the red nuclei are selectively involved. In the cerebellum, the most severely affected structures are the dentate nuclei and the dentato-rubral fibers in the superior cerebellar peduncle. Loss of Purkinje cells has been reported in KSS, along with severely reduced expression of mtDNA-encoded proteins in neurons of the dentate nucleus (Tanji *et al.*, 1999). These findings are rather specific to KSS, and contribute to explain why cerebellar ataxia is a prominent, and occasionally the only CNS symptom.

KSS, PEO, and PS are all associated with single large-scale heteroplasmic rearrangements of mtDNA. Usually, deletions are easily detected by Southern blot analysis of muscle DNA, while they can be absent in lymphocyte or fibroblast DNA, especially in the milder cases (e.g., PEO).

## 2. MELAS

MELAS is defined by the presence of (1) stroke-like episodes due to focal brain lesions often localized in the parieto-occipital lobes and (2) lactic acidosis and/or RRF. Other signs of central nervous system involvement include dementia, recurrent headache and vomiting, focal or generalized seizures, and deafness. Ataxia can be observed in some patients (Hirano *et al.*, 1992).

MELAS was first associated with a heteroplasmic point mutation in the $tRNA^{Leu(UUR)}$, an A→G transition at position 3243 (Goto *et al.*, 1990). Other MELAS-associated point mutations were later reported (see also Mitomap: http://infinity.gen.emory.edu/mitomap.html), although the A3243G remains by far the most frequent. The genotype-phenotype correlation of the A3243G mutation is rather loose since the observed clinical manifestations are not limited to the full-blown MELAS syndrome. For instance, the A3243G mutation has been detected in several patients (and families) with maternally inherited PEO, isolated myopathy alone, cardiomyopathy, or in pedigrees with maternally inherited diabetes mellitus and deafness.

## 3. MERRF

MERRF is a maternally inherited neuromuscular disorder characterized by myoclonus, epilepsy, muscle weakness and wasting, cerebellar ataxia, deafness, and dementia (Fukuhara *et al.*, 1980; Wallace *et al.*, 1988; Berkovic *et al.*, 1989 ).

Cerebellar lesions are prominent neuropathological features of MERRF syndrome, as originally described by Fukuhara (1991). Neurodegenerative changes involve the dentate nuclei, and the posterior columns and spinocerebellar tracts of the spinal cord. Neuronal loss and gliosis of the cerebellar dentate nuclei and inferior olives have been reported in patients carrying the 8344 mtDNA mutation in association with MERRF (Lombes *et al.*, 1989; Oldfors *et al.*, 1995) or Leigh syndrome (Lindboe *et al.*, 1995). These neuropathological findings have been confirmed by few neuroimaging studies on MERRF cases. Diffuse cerebral and cerebellar atrophy and calcifications in the basal ganglia can be associated with signal abnormalities in the dentate nuclei, superior cerebellar peduncles, and inferior olives (Barkovich *et al.*, 1993).

The most commonly observed mutation of mtDNA associated with MERRF is an A→G transition at nt 8344 in the $tRNA^{Lys}$ gene (Shoffner *et al.*, 1990). A second mutation has been reported in the same gene, at position 8356 (Silvestri *et al.*, 1992; Zeviani *et al.*, 1993).

MERRF must be considered in the differential diagnosis of progressive myoclonus epilepsies, including Ramsay-Hunt syndrome and Unverricht-Lundborg disease, in which cerebellar signs are prominent (Berkovic *et al.*, 1993).

The 8344 A→G mutation has been reported in association with other phenotypes, including MELAS, Leigh syndrome, and a variant neurologic syndrome characterized by ataxia, myopathy, hearing loss, and neuropathy (Austin *et al.*, 1998).

## 4. NARP

NARP is a maternally inherited syndrome in which ataxia is the cardinal manifestation of central nervous system involvement.

Magnetic resonance imaging (MRI) examination of NARP patients has revealed the presence of moderate, diffuse cerebral and cerebellar atrophy, and, in the most severely affected patients, symmetric lesions of the basal ganglia (Uziel *et al.*, 1997).

NARP is associated with a heteroplasmic T→G transversion at position 8993 in the ATPase 6 subunit gene (Holt *et al.*, 1990). A transition in the same position (T8993C) has later been described in other NARP patients (deVries *et al.*, 1993). RRF fibers are consistently absent in the muscle biopsy. The degree of heteroplasmy is correlated with the severity of the disease. For instance, when the percentage of mutant mtDNA is more than 95% patients show the clinical, neuroradiologic and neuropathologic findings of maternally inherited Leigh's syndrome (hence called MILS) (Tatuch *et al.*, 1992). NARP/MILS phenotypes have been described in association with other mutations of the ATPase 6 gene, e.g., mutation 9176T→C (Thagarajan *et al.*, 1995; Dionisi-Vici *et al.*, 1998), and 9176 T > G (Carrozzo *et al.*, 2001). NARP and MILS may co-exist in the same family.

## 5. Hearing Loss, Ataxia, and Myoclonus

Finally, a brief mention deserves a syndrome characterized by hearing loss, ataxia, and myoclonus, originally found in a large Italian pedigree (Tiranti *et al.*, 1995). The responsible mutation, 7472insC, affects the $tRNA^{Ser(UCN)}$

gene. This mutation has later been reported in several families, in which affected members showed a wide range of clinical manifestations, from isolated hearing loss (Verhoeven et al., 1999), to epilepsia partialis continua and ataxia (Schuelke et al., 1998), to overt MERRF (Jaksch et al., 1998). Given the increasing frequency at which the 7472insC has been found, the search for this mutation should become part of the routine screening of mitochondrial ataxias.

### D. Ataxia Associated with Nuclear Disease Genes

Three groups of nuclear gene defects are related to mitochondrial disorders.

1. Structural components or assembly factors of respiratory chain complexes
2. Factors controlling the stability of mtDNA
3. Factors involved in metabolic pathways influencing the biogenesis of mitochondria, including OXPHOS

### E. Defects of Structural Components or Assembly Factors of Respiratory Chain Complexes

The most relevant recent contribution in this group of mitochondrial disorders has been the (partial) elucidation of the molecular basis of Leigh syndrome.

*Leigh syndrome* (LS) is one of the most common mitochondrial disorders of the respiratory chain in infancy and childhood. Affected infants show severe psychomotor delay, cerebellar and pyramidal signs, dystonia, respiratory abnormalities, incoordination of ocular movements, and recurrent vomiting. RRF are absent. The MRI picture reflects the typical neuropathological findings which define this condition (Medina et al., 1990; Savoiardo et al., 1995; Valanne et al., 1998). Symmetric lesions usually involve the medulla, the pontine tegmentum, and the periaqueductal region, and, in the cerebellum, the dentate nuclei and the deep white matter surrounding these nuclei (Figure 24.1). Basal ganglia and posterior fossa structures may be involved simultaneously.

LS is clearly a genetically heterogeneous entity. In some cases it is attributable to mtDNA mutations, as in the case of NARP/MILS, in others the defect is X-linked or sporadic, as in the case of the defect of the E1α subunit of pyruvate dehydrogenase (PDH, Section I.I below). In still other cases it is attributable to an autosomal recessive defect of a nuclear gene. Defects of complex I, IV, or, more rarely, complex II, have been reported in autosomal recessive LS. Defects of nuclear genes encoding different subunits of complex I and complex II have been found in LS associated with biochemical deficiency of the corresponding enzyme (Bourgeron et al., 1995; Smeitink et

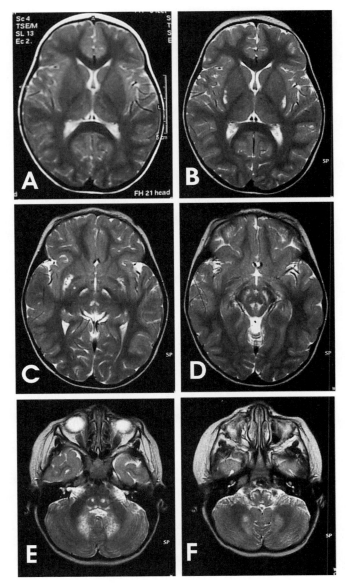

**FIGURE 24.1** MRI (T2-weighted images) of a 2-year-old girl with Leigh syndrome and citochrome-c-oxidase deficiency (Surf-1 mutation). The girl had an almost normal MRI at age 1 year (A) with typical abnormal hyperintense signals only in the subthalamic nuclei. Six months later clinical conditions worsened and MRI showed hyperintense areas in putamina (B), and subthalamic nuclei (C), dorsal and ventral mesencephalon (D), pontine corticospinal pathways (E), lower bulbar olives, and dentate cerebellar nuclei (F).

al., 1998; Triepels et al., 1999). In addition, most of the cases of LS due to a defect of complex IV (cytochrome c oxidase, COX) are caused by loss-of-function mutations of SURF-1. Interestingly, the product of SURF-1 is not a subunit of COX, but, like its yeast homolog, is an integral component of the mitochondrial inner membrane, probably involved in the assembly of the complex (Tiranti et al., 1998; Zhu et al., 1998).

## F. Defects in Factors Controlling the Stability of mtDNA

Two clinical syndromes of this group are occasionally associated with ataxia: autosomal dominant progressive external ophthalmoplegia (adPEO), and mitochondrial neurogastrointestinal encephalomyopathy (MNGIE). Both are rare human diseases that show a Mendelian inheritance pattern, but are characterized by large-scale mtDNA deletions.

AD PEO has been attributed to mutations in different genes, namely ANT1, TWINKLE, and POLG1, whose protein products are all related to mtDNA replication. ANT1 encodes the muscle-specific adenine-nucleotide translocator, which regulates the supply of intramitochondrial nucleotides, i.e., the precursors of mtDNA; TWINKLE encodes the mtDNA helicase; POLG1 encodes the main subunit of polymerase gamma, the mtDNA-specific polymerase.

MNGIE is an autosomal recessive disease characterized by the unusual combination of six features: (1) PEO; (2) severe gastrointestinal dysmotility; (3) cachexia; (4) peripheral neuropathy; (5) diffuse leukoencephalopathy on brain MRI; (6) evidence of mitochondrial dysfunction (histological, biochemical, or genetic abnormalities of the mitochondria); (Nishino et al., 2001). The gene responsible for MNGIE encodes thymidine phosphorylase (TP), which is involved in the catabolism of pyrimidines.

## G. Defects in Factors Involved in Metabolic Pathways Influencing the Biogenesis of Mitochondria, Including OXPHOS

Neurodegenerative disorders not obviously linked to overt OXPHOS defects have been attributed to mutations in mitochondrial proteins indirectly related to respiration and energy production. This observation further broadens the concept of mitochondrial disease and extends the possible involvement of mitochondrial energy metabolism in a previously unsuspected large number of important clinical phenotypes. This group includes paraplegin, a mitochondrial metalloprotease associated with autosomal recessive (or dominant) spastic paraplegia (Casari et al., 1998) and DDP1, a component of the import machinery for mitochondrial carrier proteins, which is responsible for X-linked deafness-dystonia syndrome (Mohr-Tranebjaerg syndrome); (Tranebjaerg et al., 2000) More recently, mutations in OPA1, a gene on chr. 3 encoding a dynamin-related protein, have been found in an autosomal dominant optic neuropathy causing decreased visual acuity, color vision deficits, a centrocecal scotoma, and optic nerve pallor (Alexander et al., 2000). Dynamin-related proteins identified in lower eukaryotes are essential for the maintenance and inheritance of mitochondria, possibly controlling their proliferation and cytoplasmic distribution. Interestingly, the pathophysiology and clinical symptoms observed in autosomal dominant optic atrophy show overlap with those occurring in Leber's optic atrophy. However, the two disease genes more consistently associated with ataxia are frataxin, a mitochondrial iron-storage protein, which is responsible for Friedreichs' ataxia (FA) (Ristow et al., 2000), and ABC7, an iron mitochondrial exporter, which controls the generation of cytosolic iron-sulfur proteins. The role of frataxin in FA is discussed in Chapter 18 of this book. Mutations in ABC7 are linked to X-linked sideroblastic anemia with ataxia (ASAT; 301310), a recessive disorder characterized by an infantile to early childhood onset of nonprogressive cerebellar ataxia and mild anemia with hypochromia and microcytosis (Allikmets et al., 1999).

## H. Factors Involved in Metabolic Pathways Influencing the Biogenesis of Mitochondria, Including OXPHOS

### 1. Ataxia Associated with Biochemically Defined Mitochondrial Disorders

Other phenotypes associated with ataxia still lack a molecular genetic definition, and are classified based on biochemical and/or morphological findings only. For instance, a peculiar autosomal recessive syndrome characterized by severe sensory neuropathy, PEO, ataxia, and myoclonus epilepsy, has been described in six adult patients from three separate families. The presence of RRF in muscle biopsies and elevation of CSF lactate suggest a mitochondrial etiology of the multisystem degeneration in these patients (van Domburg et al, 1996).

Another example of this broad category of defects is coenzyme Q10 muscle deficiency (Boitier et al., 1998). A syndrome associated with CoQ10 deficiency in muscle, characterized by the triad of recurrent myoglobinuria, brain involvement (seizures, ataxia, mental retardation), and RRF/lipid storage in muscle has been described in a few unrelated cases (Sobreira et al., 1997). More recently, several patients with unexplained cerebellar ataxia, pyramidal signs, and seizures, but with only unspecific myopathic change and no myoglobinuria, have been found to have very low levels of CoQ10 in muscle (26–35% of normal). All patients responded to CoQ10 supplementation with improvement of ataxia and seizures were reduced in frequency (Musumeci et al., 2001). MRI shows frequently a global cerebellar atrophy in this condition (Fig. 24.2).

A severe and unusual pontoneocerebellar hypoplasia has been reported in a child with altered kinetic properties of COX in fibroblasts; parental consanguinity suggested autosomal recessive inheritance in this condition (Nijtmans et al., 1995).

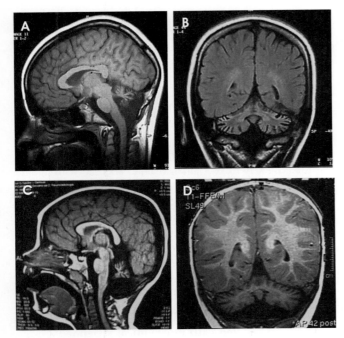

**FIGURE 24.2** In A and B, longitudinal and coronal T1-weighted images of the brain in a 9-year-old-girl with a CDG type 1a and a mutation in the PMM gene, showing a prominent and global cerebellar atrophy. Two years earlier, the girl had a cerebral infarction with a lesion in the left temporal lobe. In C and D, T1-weighted MRI of a 7-year-old-girl with a severe congenital ataxia, showing a global cerebellar atrophy. The girl has never been able to walk and was affected by a moderate mental deficiency and occasional epileptic seizures. Muscle biopsy showed reduced coenzyme Q10.

## I. Other Disorders of Mitochondrial Energy Metabolism

### 1. Ataxia in Pyruvate Dehydrogenase Deficiency

An important cause of progressive or intermittent ataxia is pyruvate dehydrogenase (PDH) deficiency, another mitochondrial disorder affecting the energy metabolic pathway. Intermittent ataxia and PDH deficiency has been reported by Blass et al. (1970) in an 8-year-old boy who had suffered 2–6 episodes of ataxia each year since the age of 16 months. Most attacks followed nonspecific febrile illness. Choreoathetosis as well as cerebellar ataxia were present during the episodes. Serum pyruvic acid and alanine levels were elevated. Thiamine in large doses seemed to benefit. Other similar cases with more detailed molecular informations have been reported later, due to E1a deficiency (Bindoff et al., 1989). Genetic defects in the PDH complex are the most common cause of primary lactic acidosis. PDH complex has a molecular weight of approximately 7 million and is composed of multiple copies of 4 principal components: pyruvate decarboxylase (E1), dihydrolipoyl transacetylase (E2), and dihydrolipoamidedehydrogenase (E3), and protein X. Sixty E2 subunits form the core of the complex and the other enzymes are attached to the surface. The E1 enzyme is itself a complex structure, a heterotetramer of 2 α- and 2 β-subunits, whereas E2 and E3 have a single type of polypeptide chain. The E1α subunit plays a key role in the function of the PDH complex since it contains the E1 active site. The functional gene locus for the E1α subunit of the PDH complex is located in the Xp22.2–p22.1 region, and the E1β is located on 4q22.

Although the PDH complex plays an important role in all metabolically active tissues, it plays a critical role in the brain under normal conditions. This is because the brain usually obtains most of its energy from the aerobic oxidation of glucose. Brown et al. (1994) reviewed all aspects of PDH deficiency, including prenatal diagnosis and treatment. They stated that treatment of PDH deficiency rarely influences the course of the disease, although a more favorable outcome can be expected in the very rare patients with a thiamine-responsive form. A short therapeutic trial of thiamine is indicated in all cases. Most patients with PDH deficiency have mutations in the E1α subunit gene and it is one of the few X-linked diseases in which a high proportion of heterozygous females manifest severe symptoms and the pathogenesis is understood. The clinical spectrum of PDH deficiency is broad, ranging from fatal lactic acidosis in the newborn to chronic or intermittent neurological dysfunction with intermittent acidosis (Robinson et al., 1987; Brown et al., 1988). Heterogeneity is also in the severity of the biochemical abnormality that can be detected in fibroblasts and muscle. Intermittent ataxia has been reported in detail in a 3-year-old boy affected by a 36 bp insertion in exon 10 of the E1a gene. Paradoxically his sister was more severely affected. The boy showed hypotonia at birth and delayed milestones during development. From the age of 3 months he had several episodes of ptosis lasting 1–2 days that were associated later with swallowing disturbances, paralysis of lateral gaze, tachypnea, episodes of acute ataxia with weakness between the ages 2 and 3 years. Metabolic acidosis was found. MRI of the brain demonstrated bilateral pallidal lesions and demyelinating pons lesions. The more severely affected younger sister presented at birth with severe hypotonia and dysmorphia and by the age of 10 months developed spastic quadriplegia with areflexia and severe mental retardation (De Merleir et al., 1998).

PDH deficiency and abnormalities in protein X have been related with onset of spastic ataxia by Marsac et al. (1993) in one out of two affected brothers, offspring of first cousin parents.

E3 deficiency has also been reported to occur with recurrent attacks of ataxia, vomiting, abdominal pain, increased transaminases, and encephalopathy (Shaag et al., 1999). This disorder is also classified as the E3-deficient form of maple syrup urine disease, or MSUD type III, because it presents a combined deficiency of branched-chain alpha-ketoacid dehydrogenase, PDH, and alpha-ketoglutarate dehydrogenase complexes. It can be detected by gas chromatography/mass spectroscopy (GC/MS) in blood and urine showing increased pyruvate, lactate, alpha-ketoglutarate, and branched-chain amino acids. This is the result of E3 being a common component of the three mitochondrial multienzymes. Hypoglycemia is intermittently observed and thiamine therapy is generally of no benefit.

## 2. Pyruvate Carboxylase Deficiency

Clinical features of pyruvate carboxylase (PC; locus: 11q13.4–q13.5) deficiency are similar to PDH deficiency, and patients generally present with lactic acidosis, increased serum alanine and pyruvate levels. Human pyruvate carboxylase is expressed in the liver, is a tetramer composed of identical subunits, and is a key regulatory enzyme in gluconeogenesis, lipogenesis, and neurotransmitter synthesis.

Two distinct clinical presentations of PC deficiency have been identified. An infantile form presents soon after birth with chronic lacticacidemia, delayed neurologic development in survivors, and normal lactate-to-pyruvate ratio despite acidemia. The second form, reported particularly from France (Saudubray et al., 1976), also presents early with lactic acidosis but shows elevated blood levels of ammonia, citrulline, proline, and lysine, and the lactate-to-pyruvate ratio is elevated as is the ratio of acetoacetate to 3-hydroxybutyrate. As in PDH deficiency ataxia has also been reported in PC deficiency with onset in childhood as well as LS (256000).

Other conditions that can provoke *intermittent ataxia* as a prominent symptom are

1. 4-hydroxyphenylpyruvate dioxygenase (locus 12q24–qter) deficiency, described in a 17-month-old girl with tirosinemia, no signs of liver failure, and drowsiness (Giardini et al., 1983).
2. Holocarboxylase synthetase (21q22.1; 253270) deficiency (Sander et al., 1980) that causes a multiple carboxylase deficiency because holocarboxylase synthetase is the enzyme that covalently links biotin to propionyl-CoA-carboxylase, PC, and beta-methylcrotonyl-CoA carboxylase. The neonatal form of multiple caroxylase deficiency (MCD) presents with severe manifestations of lactic acidosis, lethargy, vomiting, alopecia, keratoconjunctivitis, perioral erosions, seizures, hypotonia and ataxia is evident in babies with infantile-onset disease. These symptoms are generally corrected by biotin supplementation (20–100 mg/d).

Care must be taken to differentiate the inherited multiple carboxylase deficiencies from acquired biotin deficiencies, such as those that develop after excessive dietary intake of avidin, an egg-white glycoprotein that binds specifically and essentially irreversibly to biotin (Sweetman et al., 1981) or prolonged parenteral alimentation without supplemental biotin (Mock et al., 1981).

## 3. Biotinidase Deficiency

Biotin is an essential water-soluble vitamin and is the coenzyme for 4 carboxylases necessary for normal metabolism in humans. Biotinidase (253260; chr. 3p25) functions to recycle biotin in the body by cleaving biocytin (biotin-epsilon-lysine), a normal product of carboxylase degradation resulting in regeneration of free biotin. Therefore, biotinidase deficiency causes multiple carboxylase deficiency from impaired generation of free biotin from biotinyl residues of dietary protein. The most frequent initial symptom is seizures, either alone or with other neurologic or cutaneous findings. Hypotonia, ataxia, hearing loss, and optic atrophy are neurologic features; skin rash and alopecia are cutaneous features. Ketolactic acidosis and a typical organic aciduria are important clues for diagnosis. If untreated, symptoms usually become progressively worse, and coma and death may occur. Treatment with massive doses of biotin reverses the symptoms of alopecia, skin rash, ataxia, and developmental delay, which typically appear at about 3 months of age.

Muhl et al. (2001) characterized the spectrum of biotinidase mutations. The authors concluded that, based on mutation analysis, it is not predictable whether or not an untreated patient will develop symptoms; however, they found it essential to differentiate biochemically between patients with residual biotinidase activity lower or higher than 1%.

# II. ATAXIA IN LIPID DISORDERS

## A. Abetalipoproteinemia

Abetalipoproteinemia (200100) or Bassen-Kornzweig syndrome is a rare autosomal recessive disorder due to a deficiency of microsomal triglyceride transfer protein (MTP; locus 4q22–q24), which catalyzes the transport of triglyceride, cholesteryl ester, and phospholipid from phospholipid surfaces (Shoulders et al., 1993). MTP is required for lipoprotein assembly, and is a heterodimer composed of the multifunctional protein, protein disulfide isomerase, and a unique large subunit with an apparent molecular mass of 88 Kd. It isolated as a soluble protein from the lumen of the microsomal fraction of liver and intestine (Wetterau et al., 1992). The basic defect provokes inability to synthesize the apoB peptide of LDL and VLDL. (Apo B is the sole apoprotein of LDL; VLDL has a complex composition: apoC, about 50% of VLDL protein;

apoB, about 35%; apoA, about 5%; apoE, about 10%.). Many of the manifestations of this disorder are the consequence of vitamin E deficiency, and treatment with vitamin E is recommended. Features are celiac syndrome, retinitis pigmentosa, progressive ataxic neuropathy, and acanthocytosis. It somewhat resembles Friedreich's ataxia, and the peripheral nerve biopsy shows extensive central and peripheral demyelination. Intestinal absorption of lipids is defective, serum cholesterol very low, and serum beta lipoprotein absent. Many but not all of the patients reported are Jews. Hypobetalipoproteinemia and normotriglyceridemic abetalipoproteinemia are not allelic disorders, they can also rarely provoke cerebellar ataxia, and are caused by mutations in the apolipoprotein B gene (APOB; Araki et al., 1991).

A targeting mouse gene knock for MTP has been performed by Raabe et al. (1998). In the heterozygous knock-out mice, the activity levels of MTP were reduced by 50% in both liver and intestine with reduced plasma levels of low density lipoprotein cholesterol and a 28% reduction in plasma apoB-100 levels. All homozygous knock-out embryos died during embryonic development.

Spinocerebellar degeneration due to vitamin E may also occur with various forms of chronic intestinal malabsorption, including that of cholestatic liver disease and of Crohn disease (Harding et al., 1982).

## B. Cerebrotendineous Xanthomatosis

Cerebrotendinous xanthomatosis (CTX; 213700), first reported by Van Bogaert (1937), is a rare, lipid-storage disease characterized clinically by progressive neurological dysfunction of cerebellar ataxia beginning in the teens, spinal cord involvement progressing to a pseudobulbar syndrome, dementia, premature atherosclerosis, cataracts, and death. It is inherited as an autosomal recessive trait. Large deposits of cholesterol and cholestanol are found in virtually every tissue, particularly the Achilles tendons, brain, and lungs. Cholestanol, the 5-α-dihydro derivative of cholesterol, is enriched relative to cholesterol in all tissues. Bilateral juvenile cataract associated with chronic diarrhea may represent the earliest clinical manifestations of CTX together with tendon xanthomata and neurological impairment is generally a late development.

CTX is due to lack of a hepatic mitochondrial C27-steroid 26-hydroxylase (CYP27A1; locus: 2q33–qter), involved in the normal biosynthesis of cholic acid and chenodeoxycholic acid (CDCA) (Bjorkhem et al., 1983). CYP27A1 is a mitochondrial cytochrome P–450 which, together with two protein cofactors, adrenodoxin and adrenodoxin reductase, hydroxylates a variety of sterols at the C27 position. In the bile acid synthesis pathway, sterol 27-hydroxylase catalyzes the first step in the oxidation of the side chain of sterol intermediates.

### 1. Diagnosis

The diagnosis can be made by demonstrating cholestanol in abnormal amounts in the serum and tendon of persons suspected of being affected. Diagnosis can also be based on determination of urinary bile alcohols, in particular 5-β-cholestane-3-α,7-α,12-α,23,25-pentol, by means of capillary gas chromatography (Koopman et al., 1988). Plasma cholesterol concentrations are low normal in CTX patients, and there is an abnormality of high density lipoproteins (HDL).

### 2. Neuroimaging

Dotti et al., (1994) described the CT and MR findings in brain and spinal cord of 10 patients with cerebrotendinous xanthomatosis. All patients were aged 35 years or older and had cerebral and/or cerebellar atrophy. The majority had focal lesions distributed through the cerebrum, cerebellum, brainstem, or basal nuclei. Some of these lesions appeared to be xanthomata. Treatment with CDCA produces a substantial reduction in cholestanol synthesis and lowers the cholestanol levels (Salem et al., 1987).

### 3. Animal Models

By targeted disruption Rosen et al. (1998) has generated a mouse model deficient in sterol 27-hydroxylase. They found that mice with disrupted Cyp27A1 had normal plasma levels of cholesterol, retinol, tocopherol, and 1,25-dihydroxy vitamin D. Excretion of fecal bile acids was decreased (less than 20% of normal), and formation of bile acids from tritium-labeled 7-α-hydroxycholesterol was less than 15% of normal. Compensatory upregulation of hepatic cholesterol 7-α-hydroxylase and hydroxymethylglutaryl-CoA reductase (9- and 2- to 3-fold increases in mRNA levels, respectively) was found. No CTX-related pathologic abnormalities were observed.

## C. Niemann-Pick Disease Type C

Niemann-Pick disease type C (257220) is an inherited lipid storage disorder that affects the viscera and central nervous system. It must be distinguished from other types of Niemann-Pick disease due to a primary sphingomyelinase deficiency: type A, acute neuronopathic form; type B, chronic form without nervous system involvement; and type E, adult, non-neuronopathic form. Patients with type C usually appear normal for 1 or 2 years and sometimes even longer. They start with mental deficiency for some years and gradually develop neurological abnormalities, which are initially manifested by ataxia, grand mal seizures, and loss of previously learned speech. Hepatosplenomegaly is less striking than in types A and B. Early onset and transitory cholestatic jaundice occurs in some patients. Foamy Niemann-Pick cells and "sea blu hysticocytes" (Fig. 24.3)

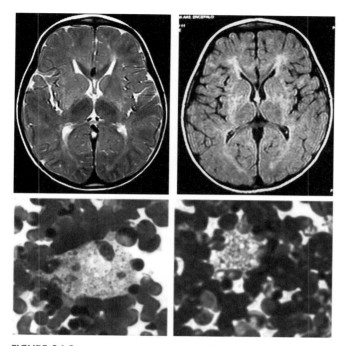

FIGURE 24.3 MRI of a patient with a Niemann-Pick type C (T2-weighted on the upper left and FLAIR on the upper right image), showing nonspecific hyperintense abnormal signals in the periventricular anterior and posterior areas. The girl had a normal development and started with a spastic-ataxic gait at age 9 years. Vertical ophthalmoplegia was observed at age 12 years and addressed the diagnosis. Niemann-Pick type C that was suggested by the presence of "sea blu histiocytes" (lower left) and foamy cells (lower right) in the bone marrow. The diagnosis was confirmed by the filipin test in fibroblasts and mutations in the NPC1 gene.

are found in the bone marrow. Death usually occurs at age 5–15. Most patients have vertical ophthalmoplegia. Three phenotypes have been proposed: (1) an early-onset, rapidly progressive form associated with severe hepatic dysfunction and psychomotor delay during infancy and later with supranuclear vertical gaze paresis, ataxia, marked spasticity, and dementia; (2) a delayed-onset, slowly progressive form heralded by the appearance, usually in early childhood, of mild intellectual impairment, supranuclear vertical gaze paresis, and ataxia, and later associated with dementia and, variably, seizures and extrapyramidal deficits; and (3) a late-onset, slowly progressive form distinguished from the second pattern by later age of onset (adolescence or adulthood) and a much slower rate of progression. The form of type C Niemann-Pick disease that starts in adolescence or adulthood shows a slower evolution than does the infantile form. Psychomotor retardation is a consistent feature and generally precedes other neurological symptoms by several years. Cerebellar ataxia and extrapyramidal manifestations are often found rather than pyramidal manifestations. Supranuclear ophthalmoplegia with paralysis of down-gaze is nearly constant. Cataplexy and other types of seizures may be found during the evolution of the disease. In some cases a psychosis may be the only manifestation for several years; the treatment by psychotropic drugs raises the question of a superimposition of a drug-induced lipidosis. Although hepatosplenomegaly is a consistent finding in children in the infantile form of the disease, hepatomegaly is often absent in the adult forms and splenomegaly, although generally present, is not pronounced. Phenotypes 1 and 2 have been observed in the same sibship.

### 1. Diagnosis

Assay for the deficiency, which shows impaired ability of cultured fibroblasts to esterify exogenously supplied cholesterol, is useful for confirming the diagnosis in patients (Fink *et al.*, 1989).

Nerve cells demonstrate not only storage of cholesterol but also neurofibrillary tangles containing paired helical filaments, similar to the neurofibrillary tangles present in Alzheimer's disease, Kufs disease, Down syndrome, tuberous sclerosis, progressive supranuclear palsy, and Hallervorden-Spatz disease, among others. The presence of neurofibrillary tangles in Niemann-Pick type C distinguishes it from other types of Niemann-Pick disease.

Incubation of fibroblasts from patients with type C Niemann-Pick disease with low density lipoprotein results in excessive intracellular accumulation of unesterified cholesterol not only in lysosomes but also at an early stage in the Golgi complex. Therefore the Golgi complex may play a role in the intracellular translocation of exogenously derived cholesterol and that disruptions of the cholesterol transport pathway at the Golgi may, in part, be responsible for the deficiency in cholesterol utilization in this disorder (Blanchette-Mackie *et al.*, 1988). This is consistent with the view that the primary defect is one affecting the cellular transport and/or processing of free cholesterol and that it is the intracellular storage of cholesterol that causes a marked but secondary attenuation of lysosomal sphingomyelinase activity.

Vanier *et al.* (1991) indicated that the diagnosis is best reached by the combined demonstration of a deficient induction of esterification of cholesterol and of an intravesicular cholesterol storage by cytochemistry after filipin staining. Genetic heterogeneity was later provided by Vanier *et al.* (1996) from complementation studies by somatic cell hybridization and linkage analysis, using the filipin stain. Crosses between various cell lines revealed a major complementation group in 27 unrelated patients and a second minor group of 5 patients. Linkage analysis in a family belonging to the minor complementation group showed that the mutated gene did not map to 18q11–q12 where the major gene is located (NPC1). No consistent clinical or biochemical phenotype was associated with a second complementation group. Three of the 5 patients in group 2, however, presented with a new rare phenotype associated with severe pulmonary involvement leading to death with the first year of life. No biochemical abnormality specific to either group could be demonstrated.

## 2. Gene

By positional cloning, Carstea et al. (1997) identified the NPC1 gene, that was found to have insertion, deletion, and missense mutations in NPC patients. NPC1 is a member of a family of genes encoding membrane-bound proteins containing putative sterol-sensing domains. The cellular and subcellular localization and regulation of NPC1was investigated by light and electron microscopic immunocytochemistry, using a specific antipeptide antibody to human NPC1. NPC1 was expressed predominantly in presynaptic astrocytic glial processes, and at a subcellular level, NPC1 localized to vesicles with the morphologic characteristics of lysosomes and to sites near the plasma membrane. Analysis of the temporal and spatial pattern of neurodegeneration in the NPC BALB/c mouse, a spontaneous mutant model of human NPC, by aminocupric-silver staining, showed that the terminal fields of axons and dendrites are the earliest sites of degeneration that occur well before the appearance of a neurological phenotype (Patel et al., 1999). These studies show that NPC1 in brain is predominantly a glial protein present in astrocytic processes closely associated with nerve terminals, the earliest site of degeneration in NPC. Given the vesicular localization of NPC1 and its proposed role in mediating retroendocytic trafficking of cholesterol and other lysosomal cargo. These results suggest that disruption of NPC1-mediated vesicular trafficking in astrocytes may be linked to neuronal degeneration in NPC. Davies et al. (2000) demonstrated that the NPC1 protein has homology with the resistance-nodulation-division (RND) family of prokaryotic permeases and may normally function as a transmembrane (TM) efflux pump. Studies of acriflavine loading in normal and NCP1 fibroblasts indicate that NPC1 uses a proton motive force to remove accumulated acriflavine from the endosomal/lysosomal system. Expression of NPC1 in E. coli facilitates the transport of acriflavine across the plasma membrane, causing cytosolic accumulation, and results in transport of oleic acid but not cholesterol or cholesterol-oleate across the plasma membrane (Ioannou et al., 2000). The NPC1 gene product has been described as a large polytopic glycoprotein with a cytoplasmic tail containing a dileucine endosome-targeting motif. The NPC1 protein sequence shares strong homology with NPC1-like-1, or NPC1L1, and the morphogen receptor Patched. In addition, a group of 5 NPC1 transmembrane domains share homology with the sterol-sensing domain of proteins involved in cellular cholesterol homeostasis. Subcellular localization studies have shown NPC1 to reside in late endosomes and to transiently associate with lysosomes and the trans-Golgi network. Analysis of its topologic arrangement in membranes suggests that NPC1 contains 13 transmembrane domains and 3 large, hydrophilic, lumenal loops. A number of observations suggested that NPC1 may be related to a family of prokaryotic efflux pumps and thus may also act as a molecular pump (Ioannou et al., 2000).

The biosynthesis and trafficking of NPC1 has been tracked using green fluorescent protein-fused NPC1. Newly synthesized NPC1 was exported from the endoplasmic reticulum and required transit through the Golgi before it was targeted to late endosomes. NPC1-containing late endosomes then moved by a dynamic process involving tubulation and fission, followed by rapid retrograde and anterograde migration along microtubules. Cell fusion studies with normal and mutant NPC1 cells showed that exchange of contents between late endosomes and lysosomes depended upon ongoing tubulovesicular late endocytic trafficking. In turn, rapid endosomal tubular movement required an intact NPC1 sterol-sensing domain and was retarded by elevated endosomal cholesterol content. The authors concluded that the neuropathology and cellular lysosomal lipid accumulation in NPC1 disease results, at least in part, from striking defects in late endosomal tubulovesicular trafficking (Zhang et al., 2001).

## 3. Animal Models

Mutant mouse models have played a pivotal role in delineating the biochemical basis of NPC. One is a BALB/c mouse with clinical and biochemical features of NPC (Pentchev et al., 1984); the other is the C57BL/Ks mouse characterized by attenuated sphingomyelinase activity and excess sphingomyelin accumulation (Miyawaki et al., 1986). Niemann-Pick disease type C has also been described in a domestic short-hair kitten (Lowenthal et al., 1990).

Naureckiene et al. (2000) found that NPC2 is caused by deficiency of HE1 (human epididymis-1), a ubiquitously expressed lysosomal protein identified previously as a cholesterol-binding protein. HE1 was undetectable in fibroblasts from NPC2 patients but present in fibroblasts from unaffected controls and NPC1 patients. Mutations in the HE1 gene were found in NPC2 patients but not in controls. Treatment of NPC2 fibroblasts with exogenous recombinant HE1 protein ameliorated lysosomal accumulation of LDL-derived cholesterol. By radiation hybrid analysis and by inclusion within mapped clones, the HE1 gene was mapped to chromosome 14q24 (Naureckiene et al., 2000).

## III. ATAXIA IN LYSOSOMAL DISORDERS

### A. Tay-Sachs Disease

GM2 gangliosidosis or Tay-Sachs disease (272800) is an autosomal recessive, progressive neurodegenerative disorder, which in the classic infantile form, is usually fatal by age 2 or 3 years, resulting from marked deficiency of the enzyme hexosaminidase A (HexA; locus: 15q23–q24). This condition is generally characterized by the onset in infancy with developmental retardation, followed by paralysis, seizures, dementia, and blindness, with death in the second or third year of life. A gray-white area around the retinal fovea centralis, due to lipid-laden ganglion cells, leaving a central "cherry-red" spot is a typical funduscopic finding. The finding of the typically ballooned neurons in the central nervous system provides pathologic verification. An early and persistent extension response to sound ("startle reaction") is useful for recognizing the disorder.

Other allelic variants are juvenile- and adult-onset forms with slowly progressive deterioration of gait and posture, muscle atrophy beginning distally, pes cavus, foot drop, spasticity, ataxia of limbs and trunk, dystonia, and dysarthria. Intelligence is little affected, vision and optic fundi are sometimes normal, and no seizures occur. Autopsy shows diffuse neuronal storage with zebra bodies and increased GM2-ganglioside. Conzelmann et al. (1983) have been able to demonstrate a correlation between the level of residual activity and clinical severity: Tay-Sachs disease, 0.1% of normal; late-infantile, 0.5%; adult GM2-gangliosidosis, 2–4%; healthy persons with "low hexosaminidase," 11 and 20%. Some juvenile- or adult-onset patients with HexA deficiency have a marked cerebellar atrophy, dementia, denervation motor neuron disease, and axonal (neuronal) motor-sensory peripheral neuropathy (Mitsumoto et al. 1985).

A spinocerebellar syndrome mimicking Friedreich ataxia has been reported as a prominent and isolated initial symptom in some adult patients (Johnson et al., 1977; Willner et al., 1981).

Atypical features are prominent muscle cramps, postural and action tremor, recurrent psychosis, incoordination, corticospinal and corticobulbar involvement, dysarthria, dystonia, dementia, amyotrophy, choreoathetosis (Oates et al., 1986; Table 24.1).

TABLE 24.1  GM2 Gangliosidoses

| Clinical and biochemical forms | Age at onset | Symptoms | Laboratory examinations |
| --- | --- | --- | --- |
| **Infantile acute encephalopathy** Tay-Sachs disease, Sandhoff disease, GM2 activator deficiency | 3–5 months | Increased startle response at onset. Regression of mental and motor skills (6–10 months) and progression to decerebrate posturing. *Cherry spot*, progressive visual inattention with blindness by the second year of life. Seizures. Macrocephaly in the second year. Death by age 2–4 years. | MRI: T2-weighted hyperintensity of caudate and white matter. ERG: normal PEV: abnormal Increased AST and ALT |
| Late infantile and juvenile (subacute encephalopathy) | 2–10 years | CNS manifestations Progressive ataxia, choreoathetosis, and dysarthria Progressive spasticity leading to decerebrate rigidity by 10-12 years of age Progressive dementia → vegetative state Seizures Ophthalmologic manifestations Cherry red spot not a consistent finding Optic atrophy Retinitis pigmentosa Progressive loss of vision → blindness Death by age 10–15 years. | ERG: abnormal PEV: abnormal |
| Adult (chronic encephalopathy) | 2–20 years | Extreme variability, also intrafamilial CNS manifestations A wide range of symptoms and abnormal findings Symptoms of spinocerebellar and lower motor neuron dysfunction are most prominent: Motor weakness (motor neuron disease) Supranuclear gaze paralysis and nystagmus Tremor Personality changes Psychoses Depression Normal intelligence | EMG: chronic denervation (motor neuron disease) ERG: normal PEV: abnormal Rectal biopsy: concentrically arraged, membranous cytoplasmic bodies MRI: severe cerebellar atrophy |

Hex-A has the structure (alpha-beta)$^3$, while Hex-B is beta$^6$ (a homopolymer of beta chains); Tay-Sachs disease is an alpha-minus mutation, which Sandhoff disease, that is phenotypically similar to Tay-Sachs disease, is a beta-minus mutation. In the absence of beta subunits there is increased polymerization of alpha units to form Hex-S, which is a normal constituent of plasma and probably has a structure of alpha-6 (Beutler et al., 1975). Neufeld (1989) provided a review of the disorders related to mutations in the HexA and HexB genes. Biochemical complementation studies have clarified that the B variant and the B1 variant correspond to different mutations in the alpha subunit, while the 0 variant to mutations in the beta subunit (Sandhoff disease) and the AB variant to abnormalities in the Gm2-activator protein.

The frequency of the condition is much higher in Ashkenazi Jews of Eastern European origin than in others. Parental consanguinity is frequent in non-Jewish cases, relatively infrequent in the Jewish cases, facts that also emphasize the difference in gene frequency in the two groups. According to estimates in the 1960s, the gene frequency in New York City Jews was between 0.013 and 0.016 and that in non-Jews was only about 1/100 of this value (Kaback et al., 1977).

### 1. Animal Models

Taniik et al. (1995) produced a mouse model of Tay-Sachs disease by targeted disruption of the HexA gene. The mice were devoid of β-hexosaminidase A activity, accumulated GM2 ganglioside in the central nervous system, and displayed neurons with membranous cytoplasmic bodies identical to those of Tay-Sachs disease in humans. Unlike human Tay-Sachs disease in which all neurons store GM2 ganglioside, no storage was evident in the olfactory bulb, cerebellar cortex, or spinal anterior horn cells of these mice. Sango et al. (1995) likewise found that disruption of the HexA gene in mouse embryonic stem cells resulted in mice that showed no neurological abnormalities, although they exhibited biochemical and pathologic features of the disease. In contrast, mice in whom the Hex gene was disrupted as a model of Sandhoff disease were severely affected. The authors suggested that the phenotypic differences between the two mouse models were the result of differences in the ganglioside degradation pathway between mice and humans. The authors postulated that alternative ganglioside degradative pathway revealed by the hexosaminidase-deficient mice may be significant in the analysis of other mouse models of the sphingolipidoses, as well as suggesting novel therapies for Tay-Sachs disease.

### 2. Treatment

Platt et al. (1997) evaluated N-butyldeoxynojirimycin, an inhibitor of glycosphingolipid (GSL) biosynthesis as a strategy for treatment of the disorder. When Tay-Sachs mice were treated with this agent, the accumulation of GM2 in the brain was prevented, with the number of storage neurons and the quantity of ganglioside stored per cell markedly reduced. Thus, the authors concluded that limiting the biosynthesis of the substrate for the defective Hexa enzyme prevented GSL accumulation and the neuropathology associated with its storage in lysosomes.

Guidotti et al. (1999) determined the in vivo strategy leading to the highest HexA activity in the maximum number of tissues in Hexa-deficient knock-out mice. They demonstrated that intravenous co-administration of adenoviral vectors coding for both alpha and beta subunits, resulting in preferential liver transduction, was essential to obtain the most successful results. Only the supply of both subunits allowed for HexA overexpression, leading to massive secretion of the enzyme in serum, and full or partial restoration of enzymatic activity in all peripheral tissues tested. These results emphasized the need to overexpress both subunits of heterodimeric proteins in order to obtain a high level of secretion in animals defective in only one subunit. Otherwise, the endogenous nondefective subunit is limiting.

## B. Neuraminidase Deficiency

Neuraminidase deficiency or sialidase deficiency (256550) is a lysosomal storage disease that is classified in two clinical forms: *sialidosis type 1* or normosomatic type (also known as the "cherry-red spot-myoclonus syndrome") and *sialidosis type 2* or dysmorphic type (Lowden and O'Brien, 1979). There are two sialidases: one for glycoprotein and one for glycolipid (ganglioside). The glycoprotein sialidase, a lysosomal enzyme, is deficient in the sialidoses. The dysmorphic type 2 form has juvenile (mucolipidosis I; lipomucopolysaccharidosis) and infantile (Goldberg syndrome) forms. Type 1 form presents with cherry-red macular spots in childhood, progressive debilitating myoclonus, bilateral perinuclear cataracts, ataxia, insidious visual loss, and normal intelligence. Deep tendon reflexes are increased. Onset may be in the second decade and is not associated with dementia. Somatic and bony abnormalities are absent. Histologic findings show neuronal lipidosis and vacuolated Kupffer cells. The diagnosis can be suggested by chromatographic screening of the urine for sialyloligosaccharides, which are normally cleaved by the neuraminidase. Diagnosis is confirmed by low activity of sialidase (locus: 6p21.3) in fibroblasts (Lowden and O'Brien, 1979). Sialidosis type II presents as an infantile sialidosis simulating GM1 gangliosidosis by clinical features, with Hurleroid features, progressive kyphosis, scoliosis, and pectus carinatum with a protracted course (juvenile) or short lasting disease (infantile).

Neuraminidase levels are much lower in sialidosis type 2 than in type 1 (O'Brien and Warner, 1980). Young et al. (1987) have suggested the following classifications: (1) primary neuraminidase deficiency without dysmorphism (type 1); (2) primary neuraminidase deficiency with dysmorphism, congenital form (type 2, congenital); (3) primary neuraminidase deficiency with dysmorphism, childhood-onset (type 2, juvenile); (3) combined neuraminidase/beta-galactosidase deficiency, infantile-onset; (4) combined neuraminidase/beta-galactosidase deficiency, juvenile-onset. The sialurias, which include Salla disease, are distinguished from the sialidoses discussed here by the accumulation and excretion of free (unbound) sialic acid and by normal or elevated activity of sialidase (N-acetylneuraminidase). The form of deficiency of glycoprotein neuraminidase activity unassociated with beta-galactosidase deficiency (sialidosis I and II) has a defect in the structural gene for neuraminidase, whereas the form which combines neuraminidase and beta-galactosidase deficiencies appears to have a defect in a 32,000-dalton glycoprotein, protective protein/cathepsin A (PPCA, chr. 20), necessary for activation or proteolytic protection of these two enzymes.

In summary, neuraminidase (sialidase) has an essential role in the removal of terminal sialic acid residues from sialoglycoconjugates, and this lysosomal enzyme occurs in a complex with beta-galactosidase and PPCA. It is deficient in two genetic disorders: sialidosis, caused by a structural defect in the neuraminidase gene, and galactosialidosis, in which the loss of neuraminidase activity is secondary to a deficiency of PPCA.

### 1. Diagnosis

Lysosomal neuraminidase occurs in a high molecular weight complex with beta-galactosidase and PPCA. Association of the enzyme with PPCA is crucial for its correct targeting and lysosomal activation. On the basis of the subcellular distribution and residual catalytic activity of the mutant neuraminidases, three mutation groups have been suggested (Bonten et al., 1996): (1) catalytically inactive and not in lysosome; (2) catalytically inactive, but localized in lysosome; and (3) catalytically active and lysosomal. In general, there was a close correlation between the residual activity of the mutant enzymes and the clinical severity of disease. Patients with the severe infantile type II disease had mutations from the first group, whereas patients with a mild form of type I disease had at least one mutation from the third group. Mutations from the second group were mainly found in juvenile type II patients with intermediate clinical severity.

Imaging studies (CT/MRI) generally show generalized cerebral atrophy, and skeletal X-rays show dysostosis multiplex (type II) or stippling of epiphyses and periosteal cloaking (neonatal form). Increased secretion of sialyloligosaccharides and sialylglycopeptides in the urine can be detected by thin-layer chromatography. The diagnosis is accomplished by finding a deficiency of α-N-acetylneuraminidase activity in leukocytes or cultured fibroblasts and prenatal diagnosis by analyzing α-N-acetylneuraminidase activity in cultured chorionic villi or amniocytes.

### 2. Animal Models

The naturally occurring inbred strain mouse SM/J shows a number of phenotypic abnormalities attributable to reduced neuraminidase activity. The mice were originally characterized by their altered sialylation of several lysosomal glycoproteins. The defect was linked to the Neu1 gene on mouse chromosome 17, which was mapped by linkage analysis to the H–2 locus. SM/J mice have a leu209-to-ile (L209I) amino acid substitution in the Neu1 protein that is responsible for the partial deficiency of lysosomal neuraminidase (Rottier et al., 1998).

## C. Sialurias

Sialic acid storage disease (269920) is an autosomal recessive neurodegenerative disorder that may present as a severe infantile form, i.e., infantile sialic acid storage disease (ISSD), or as a slowly progressive adult form that is prevalent in Finland (Salla disease). The main symptoms are hypotonia, cerebellar ataxia, mental retardation, visceromegaly, and coarse features. MRI has documented progressive cerebellar atrophy and dysmyelination in the Finnish subtype, Salla disease. ISSD presents with severe visceral involvement, dysostosis multiplex, psychomotor retardation, and early death. Nephrotic syndrome is frequent. Unlike Salla disease, ISSD has no particular ethnic prevalence. Seven patients have been studied by PET using 2-fluoro-2-deoxy-D-glucose (FDG) as a tracer. Local cerebral metabolic rates for glucose (LCMRGlc) in individual brain regions were measured and compared with controls. The FDG PET showed significantly increased LCMRGlc values in the frontal and sensory-motor cortex and especially in the basal ganglia of the patients. Cerebellar hypometabolism was present in all seven patients with marked ataxia, whereas the less severely affected patients without obvious ataxia had normal or even high glucose uptake in the cerebellum (Suhonen-Polvi et al., 1999). A neuropathological study was performed on two patients with Salla disease, one male and one female, from different families. Macroscopically the cerebral white matter was severely reduced. Histologically marked loss of axons and myelin sheaths was accompanied by pronounced astrocytic proliferation. The remaining axons frequently showed ovoid swellings surrounded by a myelin sheath. Many cortical nerve cells displayed in relation to age an abnormal amount of lipofuscin. Neurofibrillary tangles were observed in nerve cells of the neocortex, nucleus basalis of Meynert and locus ceruleus. Cerebellum showed moderate loss of Purkinje cells. In the spinal cord axonal degeneration was observed in both ascending and descending tracts (Autio-Harmainen et al., 1988).

### 1. Diagnosis

Enlarged lysosomes are seen on electron microscopic studies, and patients excrete large amounts of free sialic acid in the urine. Electron microscopy of skin, and conjunctiva show generalized lysosomal storage of a polysaccharide-like material, with vacuolated structures resembling honeycombs. Biochemical analyses of urine and cultured fibroblasts show that levels are particularly increased for the unbound (free) sialic acid with no increase in labeled, bound sialic acid (Tondeur et al., 1982).

Using a positional candidate gene approach, Verheijen et al. (1999) identified a gene, SLC17A5, encoding a protein, sialin, with a predicted transport function in the lysosomes, belonging to a family of anion/cation symporters (ACS). This transporter is common for a variety of acidic sugars, such as sialic acid or glucuronic acid. They found a mutation in five Finnish patients with Salla disease and different SLC17A5 mutations in six patients of different non-Finnish ethnic origins with ISSD.

## IV. ATAXIAS ASSOCIATED WITH OTHER METABOLIC DISORDERS

### A. Refsum Disease

The cardinal clinical features of Refsum disease (266500) are retinitis pigmentosa, chronic polyneuropathy, and cerebellar signs. Most cases have electrocardiographic changes, and some have nerve deafness and/or ichthyosis. Multiple epiphyseal dysplasia is a conspicuous feature in some cases (Steinberg, 1978).

Phytanic acid, an unusual branched-chain fatty acid (3,7,11,15-tetramethyl-hexadecanoic acid), accumulates in tissues and body fluids of patients with Refsum disease. The patients are unable to metabolize phytanic acid, which is exclusively derived from exogenous sources. Phytanic acid normally undergoes alpha-oxidation in which the chain is shortened by one carbon atom, yielding pristanic acid and carbon dioxide (Jansen et al., 1997). Pristanic acid can be degraded by beta-oxidation to yield three molecules of acetyl-coenzyme A (CoA), three of propionyl-CoA, and one of isobutyryl-CoA. Patients with Refsum disease have deficient alpha-oxidation of (14)C-phytanic acid to pristanic acid, whereas the subsequent beta-oxidation of pristanic acid is normal. The gene, that codes for phytanoyl-CoA hydroxylase (PAHX; chr. 10pter–p11.2 ), has been characterized and mutations have been found confirming that this disease can be classified as a true peroxisomal disorder (Mihalik et al.,1997; Jansen et al., 1997).

### B. Adrenoleukodystrophy (ALD)

Ataxia may be the first clinical presentation of metabolic leukodystrophies, particularly in the juvenile-onset phenotypes of metachromatic and Krabbe leukodystrophies, and of adrenoleukodystrophy (300100). We will deal only with ALD because spastic ataxia is a particularly frequent clinical presentation in this disorder. ALD is an X-linked disorder (Xq28) in which four diferent phenotypes have been reported. In addition to the classic childhood-onset, subacute ALD form, and the juvenile-adult adreno-myeloneuropathy (AMN), there is the cerebral adult form, and a fourth spastic-ataxic phenotype observed in heterozygous females. The neurologic picture of ANM is dominated by a chronic and slowly progressive spastic paraplegia and associated with an endocrinologic disorder of adrenal insufficiency (Moser et al., 1995).

An unusual ALD phenotype of spinocerebellar degeneration has been described in two adult male first cousins with progressive limb and truncal ataxia, slurred speech, and spasticity of the limbs. Brain CT scans showed atrophy of the pons and cerebellum. Very long-chain fatty acids (VLCFA) were elevated in the plasma and red cell membranes of the affected patients and were increased to intermediate levels in the female carriers (Kobayashi et al., 1994).

### 1. Gene

The ALD protein is a peroxisomal transporter protein involved in the import or anchoring of VLCFA-CoA synthetase (Aubourg et al., 1993).

### 2. Diagnosis

Very high levels of VLCFA in plasma (high levels of C26, abnormal ratio of C26:C22 and C24:C22 fatty acids), RBC, and cultured skin fibroblasts is diagnostic and positive in 100% of affected males, and in 85% of carrier females. Adrenal insufficiency is detected by the presence of hyponatremia, hyperkalemia, mild metabolic acidosis, low serum cortisol level with elevated ACTH levels, and impaired cortisol response to ACTH stimulation in 85% of patients.

### 3. Treatment

Baclofen can be used for acute episodes of painful muscle spasms, anticonvulsants for seizure activity, chloral hydrate for changes in sleep-wake cycle, and soft diet, pureed foods, and gastrostomy for loss of bulbar muscular control. Therapeutic trials with VLCFA-restricted diet with Lorenzo's Oil (4 parts of Glyceryl Trierucate Oil/1 part Glyceryl Trioleate Oil) may normalize C26:0 levels within 4 weeks, but has not been demonstrated to stop progression of the disease. Bone marrow transplant is experimental in selected cases and gives poor results in severe cases.

### C. Mevalonic Aciduria

Mevalonic aciduria is the first recognized defect in the biosynthesis of cholesterol and isoprenoids, and is a consequence of a deficiency of mevalonate kinase (ATP:mevalonate 5-phosphotransferase; locus: 12q24; 251170). Transmission is autosomal recessive.

Mevalonic aciduria is generally associated with severe phenotypes starting in the first years of life with failure to thrive, developmental delay, anemia, hepatosplenomegaly, central cataracts, and dysmorphic facies. In mild forms the principal symptom is cerebellar ataxia (Berger et al., 1985; Gibson et al., 1988). Hyperimmunoglobulinemia D and periodic fever syndrome is an allelic variant in mevalonate kinase deficiency.

#### 1. Diagnosis

Urinary organic acid analysis by GC/MS invariably demonstrates a high urinary excretion of mevalonic acid, and concentrations of CoQ10 in plasma are decreased in most patients. Mevalonic acid accumulates because of failure of conversion to 5-phosphomevalonic acid, which is catalyzed by mevalonate kinase. Mevalonic acid is synthesized from 3-hydroxy-3-methylglutaryl-CoA, a reaction catalyzed by HMG-CoA reductase. The activity of mevalonate kinase, the enzyme that catalyzes the first step in mevalonate metabolism, is severely deficient in the patient's fibroblasts, lymphocytes, and lymphoblasts, and is about half-normal in each parent (Gibson et al., 1988).

### D. Pyroglutamicaciduria

5-Oxoprolinuria (266130; pyroglutamicaciduria), resulting from glutathione synthetase deficiency (chr. 20q11.2), is an autosomal recessive disorder characterized, in its severe form, by massive urinary excretion of 5-oxoproline, metabolic acidosis, hemolytic anemia, and central nervous system damage. The metabolic defect results in low GSH levels, that can be measured in erythrocytes. 5-Oxoprolinuria is presumably a product of feedback overstimulation of gamma-glutamylcysteine synthesis and its subsequent conversion to 5-oxoproline. The overproduction of pyroglutamate is probably caused by increased *in vivo* activity of gamma-glutamyl-cysteine synthetase, which in turn is caused by absence of normal feedback inhibition by glutathione with resulting superabundance of substrates available for gamma-glutamyl cyclotransferase. Lack of glutathione in the erythrocytes is apparently tolerable, but in nonrenewable neurons leads to serious neurological problems of a progressive nature.

In a big series of 28 patients with GSS deficiency (Ristoff et al., 2001), a clinical classification has been proposed into 3 types based on severity of clinical signs: mild (hemolytic anemia only), moderate (neonatal acidosis), and severe (neurological involvement).

#### 1. Diagnosis

Patients have severe metabolic acidosis with a wide anion gap, elevated levels of 5-oxoproline, and reduced glutathione content of RBCs (0–25% of normal) in serum. In the urine there generally is massive 5-oxoprolinuria (up to 40 g/day), and reduced excretion of urea. In erythrocytes there are reduced levels of glutathione, increased amounts of 5-oxoproline, and low glutathione synthetase activity. The diagnosis is obtained by the demonstration of a deficiency of glutathione synthetase activity in erythrocytes (both forms) and cultured skin fibroblasts.

#### 2. Treatment

It was concluded that early supplementation with vitamins C and E may improve the long-term clinical outcome of these patients. Other have proposed that *N*-acetylcysteine may be of value in increasing the low intracellular glutathione concentrations and cysteine availability in patients with this disorder. Drugs and oxidants that may cause hemolysis, i.e., ascorbic acid, should be avoided.

Ataxia is frequent in the neurological presentation and the disease can also appear in young adults with mental and neurological deterioration, and epileptic seizures. Erythrocytes contained no detectable glutathione, and glutathione synthetase activity was less than 2% of normal in an adult patient (Marstein et al., 1976).

### E. L-2-hydroxyglutaricacidemia

L-2-hydroxyglutaricacidemia (236792) is a disorder of unknown pathogenesis characterized by multiple neurological symptoms and growth deficiency associated with increased levels of L-2-hydroxyglutaric (2-OH-glu) acid in serum, urines, and CSF. It was first reported by Barth et al. (1992) in a total of eight (four males and four females) mentally retarded children from five unrelated families, including three pairs of sibs. In addition to mental retardation, patients had seizures and a definite cerebellar dysfunction was identified in seven. Besides L-2-hydroxyglutaric acid, lysine was increased likewise in plasma and CSF. In all patients, MRI revealed an identical abnormal pattern with subcortical leukoencephalopathy, cerebellar atrophy, and signal changes in the putamina and dentate nuclei. No specific biochemical function or catabolic pathway involving L-2-hydroxyglutaric acid was known in mammals, including humans. Preliminary loading and dietary studies failed to reveal the origin of the compound. One of the two families from Turkey had consanguineous parents and two affected children. This disorder has also been reported in two adult sisters (age 57 and 47 years) with nonspecific mild mental deficiency that started to show cerebellar speech, horizontal nystagmus and saccadic eye movement, dystonia of the arms, mild limb and truncal ataxia, and ataxic gait. Deep tendon reflexes were hyperactive. Manifestations were slowly progressive over the next years (Fujitake *et al.*, 1999).

## F. Congenital Disorders of Glycosylation

Congenital disorders of glycosylation (CDG) outline a heterogeneous number of expanding genetic disorders that are related to enzymatic defects affecting pathways for *N*-glycosylation of secretory proteins. The proteins are covalently modified with carbohydrates, which are important for their stability and folding, and which mediate diverse recognition events in growth and development.

Defects in the attachment of carbohydrate to protein give rise to multisystemic disorders with mental and psychomotor retardation, dysmorphism, and blood coagulation defects. CDGs can be divided into two types, depending on whether they impair lipid-linked oligosaccharide (LLO) assembly and transfer (CDG I), or affect trimming of the protein-bound oligosaccharide or the addition of sugars to it (CDG II) (Orlean, 2000).

The CDG disorder with prominent early onset ataxic symptoms is caused by mutations in the gene for phosphomannomutase-2 (PMM2; chr. 16p13.3–p13.2). It is now classified as CDG Ia (212065). CDG Ib is caused by mutations in the gene for phosphomannose isomerase-1 but does not produce ataxia. It was first described by Jaeken and Stiebler (1989) as a neurologic syndrome with cerebellar atrophy (Fig. 24.2) and peripheral demyelination. Further observations have added numerous symptoms to this condition: severe hypertrophic cardiomyopathy, failure to thrive since the neonatal period, feeding difficulties, bilateral pleural and pericardial effusions, hepatic insufficiency, olivopontocerebellar atrophy, micronodular cirrhosis, renal tubular microcysts, retinitis pigmentosa, psychomotor retardation, hypotonia, esotropia, inverted nipples, and lipodystrophy. Because coagulation factors and inhibitors are glycoproteins, patients show decreased activity of factor XI and of the coagulation inhibitors antithrombin III and protein C. This combined coagulation inhibitor deficiency may explain the stroke-like episodes occurring in children with this disorder. The most severe clinical form of CDG is olivopontocerebellar atrophy of neonatal onset (OPCA), reported for the first time by Harding *et al.*, (1988) with characteristic neuropathology. Serial CT and MRI have been described on three Japanese patients with this syndrome at different ages. A small cerebellum, with peculiar enlargement of the cisterna magna, and a small brain stem were already present in infancy with atrophy of the anterior vermis and these abnormalities slowly progressed in the cerebellum and brain stem after birth (Akaboshi *et al.*, 1995).

### 1. Gene

Matthijs *et al.* (1997a) cloned the PMM1 gene and demonstrated that it is located on 22q13, thus excluding it as the cause of CDG1a, which maps to 16p. In a later report, Matthijs *et al.* (1997b) identified a second human PMM gene, called PMM2, which was located on 16p13. The PMM2 gene was found to encode a protein with 66% identity to PMM1. In 16 CDG I patients from different geographic origins and with a documented phosphomannomutase deficiency, Matthijs *et al.* (1997b) found 11 different missense mutations in PMM2. Additional mutations include point mutations, deletions, intronic mutations, and exon-skipping mutations.

### 2. Diagnosis

The activity of phosphomannomutase, the enzyme that converts mannose 6-phosphate to mannose 1-phosphate, is markedly deficient (10% or less of control activity) in fibroblasts, liver, and/or leukocytes of patients with CDG Ia.

A different CDG (CDG1c; 603142) with the same CDGS type I serum sialotransferrin pattern but normal levels of PMM activity was reported by Burda *et al.* (1998) and Korner *et al.* (1998). These patients had mostly a neurological phenotype: muscular hypotonia, ataxia, mental and motor developmental retardation, and seizures occurring during infections in some. Nerve conduction velocity was always normal and an MRI showed a slight general atrophy of the cerebrum and cerebellum. In one patient there was an atrophic retinal pigmentation, reduction of retinal vascularization, and hyperopia. Fibroblasts from patients showed a specific deficiency in the assembly of the dolichol-linked oligosaccharide (chr. 1p22.3), namely an impaired glycosylation of the dolichol-linked oligosaccharide leading to accumulation of dolichylpyrophosphate-linked Man(9)GlcNAc(2).

## G. 4-Hydroxybutyricaciduria

Deficiency of succinic semialdehyde dehydrogenase (SSADH; ALDH5A1, chr. 6p22; 271980), also known as 4-hydroxybutyricaciduria, is a rare inborn error in the metabolism of the neurotransmitter 4-aminobutyric acid (GABA) inherited as an autosomal recessive trait. The phenotype includes ataxia, psychomotor retardation, mild autism, hypotonia, hyperreflexia, lethargy, refractory seizures, and electroencephalographic abnormalities (Gibson et al., 1998). 4-Hydroxybutyric acid is of special interest because of the accumulation of a compound of known neurotoxicity.

### 1. Diagnosis

Increased concentrations of GABA are found in the urine and CSF. Deficiency of the enzyme can be detected in lymphocyte lysates of patients as well as reduced levels of enzyme activity consistent with heterozygosity in parents (Gibson et al., 1998).

### 2. Animal Models

Hogema et al. (2001) developed Aldh5a1-deficient mice that displayed, at postnatal days 16–22, ataxia and generalized seizures leading to rapid death. They showed increased amounts of gamma-hydroxybutyric acid and total GABA in urine, brain, and liver homogenates, and significant gliosis was detected in the hippocampus of Aldh5a1$^{-/-}$ mice. Intervention with phenobarbital or phenytoin was ineffective, whereas intervention with vigabatrin or with a GABA(B) receptor antagonist prevented tonic/clonic convulsions and significantly enhanced survival of mutant mice. Indeed, treatment of mutant mice with the amino acid taurine rescued Aldh5a1$^{-/-}$ mice. The findings provided insight into pathomechanisms and may have therapeutic relevance for human SSADH deficiency disease and 4-hydroxybutyricaciduria overdose and toxicity.

## H. Galactosemia

The cardinal features of galactosemia are failure to thrive, hepatomegaly, cataracts, and mental retardation and the defect concerns galactose-1-phosphate uridyltransferase (230400; chr. 9p13). Vomiting or diarrhea usually begins within a few days of milk ingestion. Jaundice of intrinsic liver disease may be accentuated by the severe hemolysis occurring in some patients. Cataracts have been observed within a few days of birth.

These may be found only on slit-lamp examination and missed with an ophthalmoscope, since they consist of punctate lesions in the fetal lens nucleus. There appears to be a high frequency of neonatal death due to *E. coli* sepsis, with a fulminant course. Long-term results of treatment with lactose restriction have been disappointing; IQ is low in many despite early and seemingly adequate therapy. Ataxia has been reported in the long-term follow-up of two siblings, a 27-year-old man and his 24-year-old sister, who were diagnosed with classic transferase deficiency galactosemia at birth and were treated with strict lactose restriction. Despite well-documented dietary management, both siblings were mentally retarded and manifested a progressive neurological condition characterized by hypotonia, hyperreflexia, dysarthria, ataxia, and a postural and kinetic tremor. Magnetic resonance imaging revealed moderate cortical atrophy, a complete lack of normal myelination, and multifocal areas of increased signal in the periventricular white matter on T2-weighting (Koch et al., 1992).

## I. Hartnup Disease

Hartnup disorder (234500) is characterized by a pellagra-like light-sensitive rash, cerebellar ataxia, emotional instability, and aminoaciduria (Scriver et al., 1965). The defect involves the intestinal and renal transport of certain neutral alpha-amino acids (particularly methionine and tryptophan but also lysine and glycine). Stool indoles and urinary indican are elevated after oral tryptophan loading. Inheritance is autosomal recessive and the gene has recently been localized to human chromosome 5p15 (Nozaki et al., 2001).

## J. Cerebellar Hypoplasia and Neonatal Diabetes

A severe, cerebellar hypoplasia/agenesis has been reported in three cases of neonatal diabetes mellitus and dysmorphism occurring within a highly consanguineous family (Hoveyda et al., 1999). Two of these cases are sisters and the third case is a female first cousin. The pattern of inheritance suggests an autosomal recessive disorder. Prenatal diagnosis of the condition in this family was possible by demonstration of the absence of the cerebellum and severe intrauterine growth retardation.

## K. Guidance to the Work-Up When Metabolic or a Mitochondrial Disorder is Considered

First of all, it is unquestionable that inherited metabolic disorders have to be well known in order to be diagnosed. However, in view of the large number of inherited metabolic disorders, it might appear that their diagnosis requires precise knowledge of a large number of biochemical pathways. As a matter of fact, an adequate diagnostic approach can be based on the proper use of only a few screening tests (Table 24.2).

For disorders with intermittent symptomatology, involving *energy metabolism* and the *intermediary metabolism*, there are several investigations that can be performed. In urine these tests are acetone (Acetest, Ames), reducing substances (Clinitest, Ames), keto acids (DNPH),

TABLE 24.2  Metabolic Causes of Adolescent or Adult-Onset Ataxia where Ataxia is Prominent as Opposed to Syndromes where Ataxia is Present, but Secondary to Other Prominent Findings

| Prominent ataxia syndrome | Metabolic syndromes associated with ataxia |
|---|---|
| **Pyruvate dehydrogenase (PDH) deficiency** (intermittent ataxia) | **Niemann-Pick C** (mental retardation, dystonia) |
| **Pyruvate carboxylase deficiency (PC)** (intermittent ataxia) | **NARP** (neuropathy, mental retardation, pigmental retinopathy) |
| **Cerebrotendinous xanthomatosis** | **MERRF** (lactic acidosis, myoclonic epilepsy, neuropathy, myopathy)<br>**Kearns-Sayre S.** (pigmental retinopathy, progressive external ophthalmoplegia, cardiomyopathy, myopathy)<br>**MELAS** (failure to thrive, myopathy, stroke like episodes, cardiomyopathy)<br>**Hearing loss, ataxia, myoclonus** (7472insC, tRNA$^{Ser(UCN)}$)<br>**GM2 gangliosidosis** (motor neuron disease, spasticity, dystonia)<br>**Adrenoleukodystrophy** (spasticity, adrenal failure, leukodystrophy)<br>**SPG7** (spasticity, optic atrophy)<br>**X-linked sideroblastic anemia with ataxia**<br>**Ataxia and CoQ10 deficiency** (epilepsy)<br>**Abetalipoproteinemia** (retinitis pigmentosa, progressive ataxic neuropathy, and acanthocytosis).<br>**Sialidosis type 1** (cherry red macular spots, progressive myoclonus, cataracts, visual loss)<br>**Refsum disease** (retinitis pigmentosa, chronic polyneuropathy, and cerebellar signs) |

pH (pHstix, Merck), sulfites (Merck), Brandt reaction, electrolytes (Na, K), urea, creatine, uric acid; in blood, they are blood cell count, electrolytes (anion gap), glucose, calcium, blood gases (pH, $pCO_2$, $HCO_3H^-$, $pO_2$), uric acid, prothrombin time, transaminases (and other liver tests), CK, ammoniemia, lactic and pyruvic acids, 3-hydroxybutyrate, acetoacetate, free fatty acids, organic acids GC/MS, amino acids; in the CSF, i.e., lactic and pyruvic acids, organic acids GC/MS; chest X-ray; cardiac echography; cerebral ultrasound (in infants), EEG, and MRI. Fresh samples of urine and CSF can be collected and stored in the refrigerator.

When suspecting a mitochondrial disorder it is mandatory to search for increased lactate and pyruvate in biological fluids, and perform a muscle biopsy (morphology, biochemistry). However, muscle morphology can be normal, for example in NARP syndrome or in complex I deficiency. MRI can be crucial to start a work-up on *energy metabolism disorders* when showing symmetric necrosis of the basal ganglia, of subthalamic nuclei, dorsal and ventral mesencephalon, or dentate nuclei.

For disorders involving disturbance of the synthesis of *complex molecules*, such as lysosomal, and peroxisomal disorders, disorders of intracellular trafficking and processing (congenital defects of glycosylation, cholesterol synthesis), where symptoms are permanent, progressive, independent of intercurrent events, a preliminary clinical and neuroimaging assessment is essential. Urinary oligosaccharides and enzymatic lysosomal assays in leukocytes or in fibroblasts should be performed for diagnosis. For peroxisomal disorders screening of plasma VLCFA, pristanic, phytanic, and plasmalogens in erythrocytes can be useful for a first diagnostic approach. For glycosylation disorders it is important to ask for isoelectric focusing of the sialotransferrin pattern.

## References

Akaboshi, S., Ohno, K., and Takeshita, K. (1995). Neuroradiological findings in the carbohydrate-deficient glycoprotein syndrome. *Neuroradiology* **37**, 491–495.

Alexander, C., Votruba, M., Pesch, U. E. A., Thiselton, D. L., Mayer, S., Moore, A., Rodriguez, M., Kellner, U., Leo-Kottler, B., Auburger, G., Bhattacharya, S. S., and Wissinger, B. (2000). OPA1, encoding a dynamin-related GTPase, is mutated in autosomal dominant optic atrophy linked to chromosome 3q28. *Nat. Genet.* **26**, 211–215.

Allikmets, R., Raskind, W. H., Hutchinson, A., Schueck, N. D., Dean, M., and Koeller, D. M. (1999). Mutation of a putative mitochondrial iron transporter gene (ABC7) in X-linked sideroblastic anemia and ataxia (XLSA/A). *Hum. Mol. Genet.* **8**, 743–749.

Anderson, S., Bankier A. T., Barrell, B. G., de Bruijn, M. H., Coulson, A. R., Drouin, J., Eperon, I. C., Nierlich, D. P., Roe, B. A., Sanger, F., Schreier, P. H., Smith, A. J., Staden, R., and Young, I. G. (1981). Sequence and organization of the human mitochondrial genome. *Nature* **290**, 457–465.

Araki, W., Hirose, S., Mimori, Y., Nakamura, S., Kimura, J., Ohno, K., and Shimada, T. (1991). Familial hypobetalipoproteinaemia complicated by cerebellar ataxia and steatocystoma multiplex. *J. Intern. Med.* **229**, 197–199.

Aubourg, P., Mosser, J., Douar, A. M., Sarde, C. O., Lopez, J., and Mandel, J. L. (1993). Adrenoleukodystrophy gene: unexpected homology to a protein involved in peroxisome biogenesis. *Biochimie* **75**, 293–302.

Austin, S. A., Vriesendorp, F. J., Thandroyen, F. T., Hecht, J. T., Jones, O. T., and Johns, D. R. (1998). Expanding the phenotype of the 8344 transfer RNAlysine mitochondrial DNA mutation. *Neurology* **51**, 1447–1450.

Autio-Harmainen, H., Oldfors, A., Sourander, P., Renlund, M., Dammert, K., and Simila, S. (1988). Neuropathology of Salla disease. *Acta Neuropathol (Berlin)* **75**, 481–490.

Barkovich, A. J., Good, W. V., Koch, T. K., and Berg, B. O. (1993). Mitochondrial disorders: analysis of their clinical and imaging characteristics. *Am. J. Neuroradiol.* **14**, 1119–1137.

Barth, P. G., Hoffmann, G. F., Jaeken, J., Lehnert, W., Hanefeld, F., van Gennip, A. H., Duran, M., Valk, J., Schutgens, R. B. H., Trefz, F. K., Reimann, G., and Hartung, H.-P. (1992). L-2-hydroxyglutaric acidemia: a novel inherited neurometabolic disease. *Ann. Neurol.* **32**, 66–71.

Berger, R., Smit, G. P. A., Schierbeek, H., Bijsterveld, K., and le Coultre, R. (1985). Mevalonic aciduria: an inborn error of cholesterol biosynthesis? *Clin. Chim. Acta* **152**, 219–222.

Berkovic, S. F., Carpenter, S., Evans, A., Karpati, G., Shoubridge, E. A., Andermann, F., Meyer, E., Tyler, J. L., Diksic, M., and Arnold, D. et al. (1989). Myoclonus epilepsy and ragged-red fibres (MERRF). 1. A clinical, pathological, biochemical, magnetic resonance spectrographic and positron emission tomographic study. *Brain* **112**, 1231–1260.

Berkovic, S. F., Cochius, J., Andermann, E., and Andermann, F. (1993). Progressive myoclonus epilepsies: clinical and genetic aspects. *Epilepsia* **34** Suppl. 3, S19–30. Review.

Beutler, E., Kuhl, W., and Comings, D. (1975). Hexosaminidase isozyme in type O Gm2 gangliosidosis (Sandhoff-Jatzkewitz disease). *Am. J. Hum. Genet.* **27**, 628–638.

Bindoff, L. A., Birch-Machin, M. A., Farnsworth, L., Gardner-Medwin, D., Lindsay, J. G., and Turnbull, D. M. (1989). Familial intermittent ataxia due to a defect of the E1 component of pyruvate dehydrogenase complex. *J. Neurol. Sci.* **93**, 311–318.

Bjorkhem, I., Fausa, O., Hopen, G., Oftebro, H., Pedersen, J. I., and Skrede, S. (1983). Role of the 26-hydroxylase in the biosynthesis of bile acids in the normal state and in cerebrotendinous xanthomatosis: an in vivo study. *J. Clin. Invest.* **71**, 142–148.

Blanchette-Mackie, E. J., Dwyer, N. K., Amende, L. M., Kruth, H. S., Butler, J. D., Sokol, J., Comly, M. E., Vanier, M. T., August, J. T., Brady, R. O., and Pentchev, P. G. (1988). Type-C Niemann-Pick disease: low density lipoprotein uptake is associated with premature cholesterol accumulation in the Golgi complex and excessive cholesterol storage in lysosomes. *Proc. Nat. Acad. Sci. U.S.A.* **85**, 8022–8026.

Blass, J. P., Avigan, J., and Uhlendorf, B. W. (1970). A defect in pyruvate decarboxylase in a child with an intermittent movement disorder. *J. Clin. Invest.* **49**, 423–432.

Boitier, E., Degoul, F., Desguerre, I., Charpentier, C., Francois, D., Ponsot, G., Diry, M., Rustin, P., and Marsac, C. (1998). A case of mitochondrial encephalomyopathy associated with a muscle coenzyme Q10 deficiency. *J. Neurol. Sci.* **156**, 41–46.

Bonten, E., van der Spoel, A., Fornerod, M., Grosveld, G., and d'Azzo, A. (1996). Characterization of human lysosomal neuraminidase defines the molecular basis of the metabolic storage disorder sialidosis. *Genes Dev.* **10**, 3156–3169.

Bourgeron, T., Rustin P., Chretien, D., Birch-Machin, M., Bourgeois, M., Viegas-Pequignot, E., Munnich, A., and Rotig, A. (1995). Mutation of a nuclear succinate dehydrogenase gene results in mitochondrial respiratory chain deficiency. *Nat. Genet.* **11**, 144–149.

Brown, G. K., Haan, E. A., Kirby, D. M., Scholem, R. D., Wraith, J. E., Rogers, J. G., and Danks, D. M. (1988). Cerebral' lactic acidosis: defects in pyruvate metabolism with profound brain damage and minimal systemic acidosis. *Eur. J. Pediatr.* **147**, 10–14.

Brown, G. K., Otero, L. J., LeGris, M., and Brown, R. M. (1994). Pyruvate dehydrogenase deficiency. *J. Med. Genet.* **31**, 875–879.

Burda, P., Borsig, L., de Rijk-Andel, J., Wevers, R., Jaeken, J., Carchon, H., Berger, E. G., and Aebi, M. (1998). A novel carbohydrate-deficient glycoprotein syndrome characterized by a deficiency in glucosylation of the dolichol-linked oligosaccharide. *J. Clin. Invest.* **102**, 647–652.

Carrozzo, R., Tessa, A., Vazquez-Memije., M. E., Piemonte, F., Patrono, C., Malandrini, A., Dionisi-Vici, C., Vilarinho, L., Villanova, M., Schagger, H., Federico, A., Bertini, E., and Santorelli, F. M. (2001). The T9176G mtDNA mutation severely affects ATP production and results in Leigh syndrome. *Neurology* **56**, 687–690.

Carstea, E. D., Morris, J. A., Coleman, K. G., Loftus, S. K., Zhang, D., Cummings, C., Gu, J., Rosenfeld, M. A., Pavan, W. J., Krizman, D. B., Nagle, J., Polymeropoulos, M. H. et al. (1997). Niemann-Pick C1 disease gene: homology to mediators of cholesterol homeostasis. *Science* **277**, 228–231.

Casari, G., De Fusco, M., Ciurmatori, S., Zeviani, M., Mora, M., Fernandez, P., De Michele, G., Filla, A., Cocuzza, S., Marconi, R., Durr, A., Fontane, B., and Ballabio, A. (1998). Spastic paraplegia and OXPHOS impairment caused by mutations in paraplegin, a nuclear-encoded mitochondrial metalloprotease. *Cell* **93**, 973–983.

Conzelmann, E., Kytzia, H.-J., Navon, R., and Sandhoff, K. (1983). Ganglioside GM2 N-acetyl-beta-D-galactosaminidase activity in cultured fibroblasts of late-infantile and adult GM2 gangliosidosis patients and of healthy probands with low hexosaminidase level. *Am. J. Hum. Genet.* **35**, 900–913.

Davies, J. P., Chen, F. W., and Ioannou, Y. A. (2000). Transmembrane molecular pump activity of Niemann-Pick C1 protein. *Science* **290**, 2295–2298.

De Meirleir, L., Specola, N., Seneca, S., and Lissens, W. (1998). Pyruvate dehydrogenase E1-alpha deficiency in a family: different clinical presentation in two siblings. *J. Inherit. Metab. Dis.* **21**, 224–226.

de Vries, D. D., van Engelen, B. G. M., Gabreels, F. J. M., Ruitenbeek, W., and van Oost, B. A. (1993). A second missense mutation in the mitochondrial ATPase 6 gene in Leigh's syndrome. *Ann. Neurol.* **34**, 410–412.

Dionisi-Vici, C., Seneca, S., Zeviani, M., Fariello, G., Rimoldi, M., Bertini, E., and De Meirleir, L. (1998). Fulminant Leigh syndrome and sudden unexpected death in a family with the T9176C mutation of the mitochondrial ATPase 6 gene. *J. Inherit. Metab. Dis.* **21**, 2–8.

Dotti, M. T., Federico, A., Signorini, E., Caputo, N., Venturi, C., Filosomi, G., and Guazzi, G. C. (1994). Cerebrotendinous xanthomatosis (van Bogaert-Scherer-Epstein disease): CT and MR findings. *Am. J. Neuroradiol.* **15**, 1721–1726.

Fink, J. K., Filling-Katz, M. R., Sokol, J., Cogan, D. G., Pikus, A., Sonies, B., Soong, B., Pentchev, P. G., Comly, M. E., Brady, R. O., and Barton, N. W. (1989). Clinical spectrum of Niemann-Pick disease type C. *Neurology* **39**, 1040–1049.

Fujitake, J., Ishikawa, Y., Fujii, H., Nishimura, K., Hayakawa, K., Inoue, F., Terada, N., Okochi, M., and Tatsuoka, Y. (1999). L-2-hydroxyglutaric aciduria: two Japanese adult cases in one family. *J. Neurol.* **246**, 378–382.

Fukuhara, N. (1991). MERRF: a clinicopathological study. Relationships between myoclonus epilepsies and mitochondrial myopathies. *Rev. Neurol. (Paris)* **147**, 476–479. Review.

Fukuhara, N., Tokiguchi, S., Shirakawa, K., and Tsubaki, T. (1980). Myoclonus epilepsy associated with ragged-red fibers (mitochondrial abnormalities): disease entity or a syndrome? Light- and electron-microscopic studies of two cases and a review of the literature. *J. Neurol. Sci.* **47**, 117–133.

Giardini, O., Cantani, A., Kennaway, N. G., and D'Eufemia, P. (1983). Chronic tyrosinemia associated with 4-hydroxyphenylpyruvate dioxygenase deficiency with acute intermittent ataxia and without visceral and bone involvement. *Pediatr. Res.* **17**, 25–29.

Gibson, K. M., Hoffmann, G. F., Hodson, A. K., Bottiglieri, T., Jakobs, C. (1998). 4-Hydroxybutyric acid and the clinical phenotype of succinic semialdehyde dehydrogenase deficiency, an inborn error of GABA metabolism. *Neuropediatrics* **29**, 14–22.

Gibson, K. M., Hoffmann, G., Nyhan, W. L., Sweetman, L., Berger, R., le Coultre, R., and Smit, G. P. A. (1988). Mevalonate kinase deficiency in a child with cerebellar ataxia hypotonia and mevalonic aciduria. *Eur. J. Pediatr.* **148**, 250–252.

Gibson, K. M., Sweetman, L., Nyhan, W. L., Lenoir, G., Divry, P. (1984). Defective succinic semialdehyde dehydrogenase activity in 4-hydroxybutyric aciduria. *Eur. J. Pediatr.* **142**, 257–259,

Goto, Y., Nonaka, I., and Horai, S. (1990). A mutation in the tRNA-leu(UUR) gene associated with the MELAS subgroup of mitochondrial encephalomyopathies. *Nature* **348**, 651–653

Guidotti, J. E., Mignon, A., Haase, G., Caillaud, C., McDonell, N., Kahn, A., and Poenaru, L. (1999). Adenoviral gene therapy of the Tay-Sachs disease in hexosaminidase A-deficient knock-out mice. *Hum. Mol. Genet.* **8**, 831–838.

Harding, A. E., Muller, D. P. R., Thomas, P. K., and Willison, H. J. (1982). Spinocerebellar degeneration secondary to chronic intestinal malabsorption: a vitamin E deficiency syndrome. *Ann. Neurol.* **12**, 419–424.

Harding, A. E. (1994). Ataxia associated with metabolic disorders. In "The Hereditary Ataxias and Related Disorders"(A.E. Harding, ed), pp. 23–56. Churchill Livingstone, London.

Harding, B. N., Dunger, D. B., Grant, D. B., and Erdohazi, M. (1988). Familial olivopontocerebellar atrophy with neonatal onset: a recessively inherited syndrome with systemic and biochemical abnormalities. *J. Neurol. Neurosurg. Psychiatry* **51**, 385–390,.

Hirano, M., Ricci, E., Koenigsberger, M. R., Defendini, R., Pavlakis, S. G., DeVivo, D. C., DiMauro, S., and Rowland, L. P. (1992). MELAS: an original case and clinical criteria for diagnosis. *Neuromuscular Disord.* **2**, 125–135

Hogema, B. M., Gupta, M., Senephansiri, H., Burlingame, T. G., Taylor, M., Jakobs, C., Schutgens, R. B. H., Froestl, W., Snead, O. C., Diaz-Arrastia, R., Bottiglieri, T., Grompe, M., and Gibson, K. M. (2001). Pharmacologic rescue of lethal seizures in mice deficient in succinate semialdehyde dehydrogenase. *Nat. Genet.* **29**, 212–216.

Holt, I. J., Harding, A. E., Petty, R. K., and Morgan-Hughes, J. A. (1990). A new mitochondrial disease associated with mitochondrial DNA heteroplasmy. *Am. J. Hum. Genet.* **46**, 428–33.

Hoveyda, N., Shield, J. P. H., Garrett, C., Chong, W. K. 'Kling', Beardsall, K., Bentsi-Enchill, E., Mallya, H., and Thompson, M. H. (1999). Neonatal diabetes mellitus and cerebellar hypoplasia/agenesis: report of a new recessive syndrome *J. Med. Genet.*, **36**, 700–704.

Ioannou, Y. A. (2000). The structure and function of the Niemann-Pick C1 protein. *Mol. Genet. Metab.* **71**, 175–181.

Jaeken, J. (1989). A not-previously described hereditary neurological disease with a deficiency of sialic acid, galactose and N-acetylglucosamine of plasma glycoproteins. (Dutch). *Verh. K. Acad. Geneeskd. Belg.* **51**, 377–406.

Jaeken, J., Stibler, H., and Hagberg, B. (eds.) (1991). The carbohydrate-deficient glycoprotein syndrome: a new inherited multisystemic disease with severe nervous system involvement. *Acta Paediatr. Scand. Suppl.* **375**, 1–71.

Jaksch, M., Hofmann, S., Kleinle, S., Liechti-Gallati, S., Pongratz, D. E., Muller-Hocker, J., Jedele, K. B., Meitinger, T., and Gerbitz, K.-D. (1998). A systematic mutation screen of 10 nuclear and 25 mitochondrial candidate genes in 21 patients with cytochrome c oxidase (COX) deficiency shows tRNA-ser(UCN) mutations in a subgroup with syndromal encephalopathy. *J. Med. Genet.* **35**, 895–900.

Jansen, G. A., Ofman, R., Ferdinandusse, S., Ijlst, L., Muijsers, A. O., Skjeldal, O. H., Stokke, O., Jakobs, C., Besley, G. T. N., Wraith, J. E., and Wanders, R. J. A. (1997). Refsum disease is caused by mutations in the phytanoyl-CoA hydroxylase gene. *Nat. Genet.* **17**, 190–193,.

Johnson, W. G., Chutorian, A., and Miranda, A. (1977). A new juvenile hexosaminidase deficiency disease presenting as cerebellar ataxia. Clinical and biochemical studies. *Neurology* **27**,1012–8.

Kaback, M. M., Rimoin, D. L., and O'Brien, J. S. (1977). *Tay-Sachs Disease: Screening and Prevention.* Alan R. Liss, New York.

Kobayashi, T., Yamada, T., Yasutake, T., Shinnoh, N., Goto, I., and Iwaki, T. (1994). Adrenoleukodystrophy gene encodes an 80 kDa membrane protein. *Biochem. Biophys. Res. Commun.* **201**, 1029–1034.

Koch, T. K., Schmidt, K. A., Wagstaff, J. E., Ng, W. G., and Packman, S. (1992). Neurologic complications in galactosemia. *Pediatr. Neurol.* **8**, 217–220.

Koopman, B. J., Wolthers, B. G., van der Molen, J. C., van der Slik, W., Waterreus, R. J., and van Spreeken, A. (1988). Cerebrotendinous xanthomatosis: a review of biochemical findings of the patient population in The Netherlands. *J. Inherit. Metab. Dis.* **11**, 56–75.

Korner, C., Knauer, R., Holzbach, U., Hanefeld, F., Lehle, L., and von Figura, K. (1998). Carbohydrate-deficient glycoprotein syndrome type V: deficiency of dolichyl-P-Glc:Man(9)GlcNAc(2)-PP-dolichyl glucosyltransferase. *Proc. Nat. Acad. Sci. U.S.A.* **95**, 13200–13205.

Lindboe, C. F., Lie, A. K., Aase, S. T., Schjetne, O. B., and Haave, I. (1995). Neuronal degeneration in subacute necrotizing encephalomyelopathy (Leigh's disease). Case report. *APMIS* **103**, 54–58.

Lombes, A., Mendell, J. R., Nakase, H., Barohn, R. J., Bonilla, E., Zeviani, M., Yates, A. J., Omerza, J., Gales, T. L., and Nakahara, K. *et al.* (1989). Myoclonic epilepsy and ragged-red fibers with cytochrome oxidase deficiency: neuropathology, biochemistry, and molecular genetics. *Ann. Neurol.* **26(1)**, 20–33.

Lowden, J. A., and O'Brien, J. S. (1979). Sialidosis: a review of human neuraminidase deficiency. *Am. J. Hum. Genet.* **31**, 1–18.

Lowenthal, A. C., Cummings, J. F., Wenger, D. A., Thrall, M. A., Wood, P. A., and de Lahunta, A. (1990). Feline sphingolipidosis resembling Niemann-Pick disease type C. *Acta Neuropathol.* **81**, 189–197

Marchington, D. R., Macaulay, V., Hartshorne, G. M., Barlow, D., and Poulton, J. (1998). Evidence from human oocytes for a genetic bottleneck in an mtDNA disease. *Am. J. Hum. Genet.* **63**, 769–777.

Marsac, C., Stansbie, D., Bonne, G., Cousin, J., Jehenson, P., Benelli, C., Leroux, J.-P., and Lindsay, G. (1993). Defect in the lipoyl-bearing protein X subunit of the pyruvate dehydrogenase complex in two patients with encephalomyelopathy. *J. Pediatr.* **123**, 915–920.

Marstein, S., Jellum, E., Halpern, B., Eldjarn, L., and Perry, T. L. (1976). Biochemical studies of erythrocytes in a patient with pyroglutamic acidemia (5-oxoprolinemia). *N Eng. J. Med.* **295**, 406–412.

Matthijs, G., Schollen, E., Pardon, E., Veiga-Da-Cunha, M., Jaeken, J., Cassiman, J.-J., Van Schaftingen, E. (1997a). Mutations in PMM2, a phosphomannomutase gene on chromosome 16p13, in carbohydrate-deficient glycoprotein type I syndrome (Jaeken syndrome). *Nat. Genet.* **16**, 88–92.

Matthijs, G., Schollen, E., Pirard, M., Budarf, M. L., Van Schaftingen, E., and Cassiman, J.-J. (1997b). PMM (PMM1), the human homologue of SEC53 or yeast phosphomannomutase, is localized on chromosome 22q13. *Genomics* **40**, 41–47.

McShane, M. A., Hammans, S. R., Sweeney, M., Holt, I. J., Beattie, T. J., Brett, E. M., and Harding, A. E. (1991). Pearson syndrome and mitochondrial encephalomyopathy in a patient with a deletion of mtDNA *Am. J. Hum. Genet.* **48**, 39–42

Medina, L., Chi, T. L., DeVivo, D. C., and Hilal, S. K. (1990). MR findings in patients with subacute necrotizing encephalomyelopathy (Leigh syndrome): correlation with biochemical defect. *AJR Am. J. Roentgenol.* **154**, 1269–1274.

Mihalik, S. J., Morrell, J. C., Kim, D., Sacksteder, K. A., Watkins, P. A., and Gould, S. J. (1997). Identification of PAHX, a Refsum disease gene. *Nat. Genet.* **17**, 185–189,.

Mitsumoto, H., Sliman, R. J., Schafer, I. A., Sternick, C. S., Kaufman, B., Wilbourn, A., and Horwitz, S. J. (1985). Motor neuron disease and adult hexosaminidase A deficiency in two families: evidence for multisystem degeneration. *Ann. Neurol.* **17**, 378–385.

Miyawaki, S., Yoshida, H., Mitsuoka, S., Enomoto, H., and Ikehara, S. (1986). A mouse model for Niemann-Pick disease: influence of genetic background on disease expression in spm/spm mice. *J. Hered.* **77**, 379–384.

Mock, D. M., deLorimer, A. A., Liebman, W. M., Sweetman, L., and Baker, H. (1981). Biotin deficiency: an unusual complication of parenteral alimentation. *N Eng. J. Med.* **304**, 820–822.

Moser, H. W., Smith, K. D., and Moser, A. B. (1995). X-linked adrenoleukodystrophy. In: Scriver, C. R., Beaudet, A. L., Sly, W. S., Valle, D. (eds.) *The Metabolic and Molecular Bases of Inherited Disease*, pp. 2325–2349. McGraw-Hill, New York.

Muhl, A., Moslinger, D., Item, C. B., and Stockler-Ipsiroglu, S. (2001). Molecular characterisation of 34 patients with biotinidase deficiency ascertained by newborn screening and family investigation. *Eur. J. Hum. Genet.* **9**, 237–243.

Musumeci, O., Naini, A., Slonim, A. E., Skavin, N., Hadjigeorgiou, G. L., Krawiecki, N., Weissman, B. M., Tsao, C. Y., Mendell, J. R., Shanske, S., De Vivo, D. C., Hirano, M., and DiMauro, S. (2001). Familial cerebellar ataxia with muscle coenzyme Q10 deficiency. *Neurology* **56**, 849–855.

Naureckiene, S., Sleat, D. E., Lackland, H., Fensom, A., Vanier, M. T., Wattiaux, R., Jadot, M., and Lobel, P. (2000). Identification of HE1 as the second gene of Niemann-Pick C disease. *Science* **290**, 2298–2301.

Neufeld, E. F. (1989). Natural history and inherited disorders of a lysosomal enzyme, beta-hexosaminidase. *J. Biol. Chem.* **264**, 10927–10930.

Nijtmans, L. G., Barth, P. G., Lincke, C. R., Van Galen, M. J., Zwart, R., Klement, P., Bolhuis, P. A., Ruitenbeek, W., Wanders, R. J., and Van den Bogert, C. (1995). Altered kinetics of cytochrome c oxidase in a patient with severe mitochondrial encephalomyopathy. *Biochim. Biophys. Acta* **1270**, 193–201

Nishino, I., Spinazzola, A., and Hirano, M. (2001). MNGIE, from nuclear DNA to mitochondrial DNA. *Neuromuscular Disord.* **11**, 7–10.

Nozaki, J., Dakeishi, M., Ohura, T., Inoue, K., Manabe, M., Wada, Y., and Koizumi, A. (2001). Homozygosity mapping to chromosome 5p15 of a gene responsible for Hartnup disorder. *Biochem. Biophys. Res. Commun.* **284**, 255–260.

Oates, C. E., Bosch, E. P., and Hart, M. N. (1986). Movement disorders associated with chronic GM(2) gangliosidosis: case report and review of the literature. *Eur. Neurol.* **25**, 154–159.

O'Brien, J. S., and Warner, T. G. (1980). Sialidosis: delineation of subtypes by neuraminidase assay. *Clin. Genet.* **17**, 35–38.

Oldfors, A., Holme, E., Tulinius, M., and Larsson, N. G. (1995). Tissue distribution and disease manifestations of the tRNA(Lys) A→G(8344) mitochondrial DNA mutation in a case of myoclonus epilepsy and ragged red fibres. *Acta Neuropathol (Berlin)* **90**, 328–333.

Orlean, P. (2000). Congenital disorders of glycosylation caused by defects in mannose addition during N-linked oligosaccharide assembly. *J. Clin. Invest.* **105**, 131–132.

Patel, S. C., Suresh, S., Kumar, U., Hu, C. Y., Cooney, A., Blanchette-Mackie, E. J., Neufeld, E. B., Patel, R. C., Brady, R. O., Patel, Y. C., Pentchev, P. G., and Ong, W.-Y. (1999). Localization of Niemann-Pick C1 protein in astrocytes: implications for neuronal degeneration in Niemann-Pick type C disease. *Proc. Nat. Acad. Sci. U.S.A.* **96**, 1657–1662.

Pentchev, P. G., Boothe, A. D., Kruth, H. S., Weintroub, H., Stivers, J., and Brady, R. O. (1984). A genetic storage disorder in BALB/C mice with a metabolic block in esterification of exogenous cholesterol. *J. Biol. Chem.* **259**, 5784–5791.

Platt, F. M., Neises, G. R., Reinkensmeier, G., Townsend, M. J., Perry, V. H., Proia, R. L., Winchester, B., Dwek, R. A., and Butters, T. D. (1997). Prevention of lysosomal storage in Tay-Sachs mice treated with N-butyldeoxynojirimycin. *Science* **276**, 428–431.

Raabe, M., Flynn, L. M., Zlot, C. H., Wong, J. S., Veniant, M. M., Hamilton, R. L., and Young, S. G. (1998). Knockout of the abetalipoproteinemia gene in mice: reduced lipoprotein secretion in heterozygotes and embryonic lethality in homozygotes. *Proc. Nat. Acad. Sci. U.S.A.* **95**, 8686–8691.

Ristoff, E., Mayatepek, E., and Larsson, A. (2001). Long-term clinical outcome in patients with glutathione synthetase deficiency. *J. Pediatr.* **139**, 79–84.

Ristow, M., Pfister, M. F., Yee, A. J., Schubert, M., Michael, L., Zhang, C. Y., Ueki, K., Michael, M. D. 2nd, Lowell, B. B., and Kahn, C.R. (2000). Frataxin activates mitochondrial energy conversion and oxidative phosphorylation. *Proc. Natl. Acad. Sci. U.S.A.* **97**, 12239–12243.

Robinson, B. H., MacMillan, H., Petrova-Benedict, R., and Sherwood, W. G. (1987). Variable clinical presentation in patients with defective E1 component of pyruvate dehydrogenase complex. *J. Pediatr.* **111**, 525–533.

Rosen, H., Reshef, A., Maeda, N., Lippoldt, A., Shpizen, S., Triger, L., Eggertsen, G., Bjorkhem, I., and Leitersdorf, E. (1998). Markedly reduced bile acid synthesis but maintained levels of cholesterol and vitamin D metabolites in mice with disrupted sterol 27-hydroxylase gene. *J. Biol. Chem.* **273**, 14805–14812.

Rotig, A., Cormier, V., Blanche, S., Bonnefont, J. P., Ledeist, F., Romero, N., Schmitz, J., Rustin, P., Fischer, A., Saudubray, J. M., and Munnich, A. (1990). Pearson's marrow-pancreas syndrome. A multisystem mitochondrial disorder in infancy. *J. Clin. Invest.* **86**, 1601–1608.

Rottier, R. J., Bonten, E., and d'Azzo, A. (1998). A point mutation in the neu-1 locus causes the neuraminidase defect in the SM/J mouse. *Hum. Mol. Genet.* **7**, 313–321.

Rowland, L. P., Blake, D. M., Hirano, M., Di Mauro, S., Schon, E. A., Hays, A. P., and Devivo, D. C. (1991). Clinical syndromes associated with ragged red fibers. *Rev Neurol (Paris)* **147**, 467–473.

Salem, G., Berginer, V., Shore, V., Horak, I., Horak, E., Tint, G. S., and Shefer, S. (1987). Increased concentrations of cholestanol and apolipoprotein B in the cerebrospinal fluid of patients with cerebrotendinous xanthomatosis: effect of chenodeoxycholic acid. *N Eng. J. Med.* **316**, 1233–1238.

Sander, J. E., Malamud, N., Cowan, M. J., Packman, S., Amman, A. J., and Wara, D. W. (1980). Intermittent ataxia and immunodeficiency with multiple carboxylase deficiencies: a biotin-responsive disorder. *Ann. Neurol.* **8**, 544–547.

Sango, K., Yamanaka, S., Hoffmann, A., Okuda, Y., Grinberg, A., Westphal, H., McDonald, M. P., Crawley, J. N., Sandhoff, K., Suzuki, K., and Proia, R. L. (1995). Mouse models of Tay-Sachs and Sandhoff diseases differ in neurologic phenotype and ganglioside metabolism. *Nat. Genet.* **11**, 170–176.

Saudubray, J. M., Marsac, C., Cathelineau, C. L., Besson Leaud, M., Leroux, J. P. (1976). Neonatal congenital lactic acidosis with pyruvate carboxylase deficiency in two siblings. *Acta Paediatr. Scand.* **65**, 717–724.

Savoiardo, M., Ciceri, E., D'Incerti, L., Uziel, G., and Scotti, G. (1995). Symmetric lesions of the subthalamic nuclei in mitochondrial encephalopathies: an almost distinctive Mark of Leigh disease with COX deficiency. *Am. J. Neuroradiol.* **16 (8)**, 1746–1747.

Schuelke, M., Bakker, M., Stoltenburg, G., Sperner, J., and von Moers, A. (1998). Epilepsia partialis continua associated with a homoplasmic mitochondrial tRNA(Ser(UCN)) mutation. *Ann. Neurol.* **44**, 700–704.

Scriver, C. R. (1965). Hartnup disease: a genetic modification of intestinal and renal transport of certain neutral alpha-amino acids. *N Eng. J. Med.* **273**, 530–532.

Shaag, A., Saada, A., Berger, I., Mandel, H., Joseph, A., Feigenbaum, A., and Elpeleg, O. N. (1999). Molecular basis of lipoamide dehydrogenase deficiency in Ashkenazi Jews. *Am. J. Med. Genet.* **82**, 177–182.

Shoffner, J. M., Lott, M. T., Lezza, A. M., Seibel, P., Ballinger, S. W., and Wallace, D. C. (1990). Myoclonic epilepsy and ragged-red fiber disease (MERRF) is associated with a mitochondrial DNA tRNA(Lys). mutation. *Cell* **61**, 931–937.

Shoulders, C. C., Brett, D. J., Bayliss, J. D., Narcisi, T. M. E., Jarmuz, A., Grantham, T. T., Leoni, P. R. D., Bhattacharya, S., Pease, R. J., Cullen, P. M., Levi, S., Byfield, P. G. H., Purkiss, P., and Scott, J. (1993) Abetalipoproteinemia is caused by defects of the gene encoding the 97 kDa subunit of a microsomal triglyceride transfer protein. *Hum. Mol. Genet.* **2**, 2109–2116.

Silvestri, G., Moraes, C. T., Shanske, S., Oh, S. J., and DiMauro, S. (1992). A new mtDNA mutation in the tRNA(Lys) gene associated with myoclonic epilepsy and ragged-red fibers (MERRF). *Am. J. Hum. Genet.* **51**, 1213–1217.

Smeitink, J., Loeffen, J., Smeets, R., Triepels, Ruitenbeek, W., Trijbels, F., and van den Heuvel, L. (1998). Molecular characterization and mutational analysis of the human B17 subunit of the mitochondrial respiratory chain complex I. *Hum. Genet.* **103**, 245–250.

Sobreira, C., Hirano, M., Shanske, S., Keller, R. K., Haller, R. G., Davidson, E., Santorelli, F. M., Miranda, A. F., Bonilla, E., Mojon, D. S., Barreira, A. A., King, M. P., and DiMauro, S. (1997). Mitochondrial encephalomyopathy with coenzyme Q10 deficiency. *Neurology* **48**, 1238–1243.

Steinberg, D. (1978). Phytanic acid storage disease: Refsum's syndrome. In: *Metabolic Basis of Inherited Disease.* (J. B. Stanbury, J. B. Wyngaarden, and D. S. Fredrickson, eds), pp. 688–706. McGraw-Hill, New York.

Suhonen-Polvi, H., Varho, T., *et al.* (1999). Increased brain glucose utilization in Salla disease (free sialic acid storage disorder). *J. Nucl. Med.* **40**, 12–18.

Sweetman, L., Surh, L., Baker, H., Peterson, R. M., and Nyhan, W. L. (1981) Clinical and metabolic abnormalities in a boy with dietary deficiency of biotin. *Pediatrics* **68**, 553–558.

Taniike, M., Yamanaka, S., Proia, R. L., Langaman, C., Bone-Turrentine, T., and Suzuki, K. (1995). Neuropathology of mice with targeted disruption of Hexa gene, a model of Tay-Sachs disease. *Acta Neuropathol.* **89**, 296–304.

Tanji, K. Vu, T. H., Schon, E. A., DiMauro, S., and Bonilla, E. (1999). Kearns-Sayre syndrome: unusual pattern of expression of subunits of the respiratory chain in the cerebellar system. *Ann. Neurol.* **45**, 377–383.

Tatuch, Y., Christodoulou, J., Feigenbaum, A., Clarke, J. T. R., Wherret, J., Smith, C., Rudd, N., Petrova-Benedict, R., and Robinson, B. H. (1992). Heteroplasmic mtDNA mutation (T-to-G) at 8993 can cause Leigh disease when the percentage of abnormal mtDNA is high. *Am. J. Hum. Genet.* **50**, 852–858.

Thyagarajan, D., Shanske, S., Vazquez-Memije, M., De Vivo, D., and DiMauro, S. (1995). A novel mitochondrial ATPase 6 point mutation in familial bilateral striatal necrosis. *Ann. Neurol.* **38**, 468–472.

Tiranti, V., Chariot, P., Carella, F., Toscano, A., Soliveri, P., Girlanda, P., Carrara, F., Fratta, G. M., Reid, F. M., Mariotti, C., and Zeviani, M. (1995). Maternally inherited hearing loss, ataxia and myoclonus associated with a novel point mutation in mitochondrial tRNA-ser(UCN) gene. *Hum. Mol. Genet.* **4**, 1421–1427.

Tiranti, V., Hoertnagel, K., Carrozzo, R., Galimberti, C., Munaro, M., Granatiero, M., Zelante L., Gasparini P., Marzella R., Rocchi, M., Bayona-Bafaluy, M. P., Enriquez, J. A., Uziel, G., Bertini, E., Dionisi-Vici C., Franco, B., Meitinger, T., and Zeviani, M. (1998). Mutations of SURF-1 in Leigh disease associated with cytochrome c oxidase deficiency. *Am. J. Hum. Genet.* **63**, 1609–1621.

Tondeur, M., Libert, J., Vamos, E., Van Hoof, F., Thomas, G. H., and Strecker, G. (1982). Infantile form of sialic acid storage disorder: clinical, ultrastructural, and biochemical studies in two siblings. *Eur. J. Pediatr.* **139**, 142–147.

Tranebjaerg, L., Hamel, B. C. J., Gabreels, F. J. M., Renier, W. O., Van and Ghelue, M. (2000). A de novo missense mutation in a critical domain of the X-linked DDP gene causes the typical deafness-dystonia-optic atrophy syndrome. *Eur. J. Hum. Genet.* **8**, 464–467.

Triepels, R. H., van den Heuvel, L. P., Loeffen, J. L., Buskens, C. A., Smeets, R. J., Rubio Gozalbo, M. E., Budde, S. M., Mariman, E. C., Wijburg, F. A., Barth, P. G., Trijbels, J. M., and Smeitink, J. A. (1999). Leigh syndrome associated with a mutation in the NDUFS7 (PSST) nuclear encoded subunit of complex I. *Ann. Neurol.* **45**, 787–790.

Uziel, G., Moroni, I., Lamantea, E., Fratta, G. M., Ciceri, E., Carrara, F., and Zeviani, M. (1997). Mitochondrial disease associated with the T8993G mutation of the mitochondrial ATPase 6 gene: a clinical, biochemical, and molecular study in six families." *J. Neurol. Neurosurg. Psychiatry* **63**, 16–22.

Valanne L., Ketonen L., Majander A., Suomalainen A., and Pihko H. (1998). Neuroradiologic findings in children with mitochondrial disorders. *Am. J. Neuroradiol.* **19**, 369–377.

Van Bogaert, L., Scherer, H. J., and Epstein, E. (1937). *Une Forme Cerebrale de la Cholesterinose generalisee.* Masson, Paris.

van Domburg, P. H., Gabreels-Festen, A. A., Gabreels, F. J., de Coo, R., Ruitenbeek, W., Wesseling, P., and ter Laak, H. (1996). Mitochondrial cytopathy presenting as hereditary sensory neuropathy with progressive external ophthalmoplegia, ataxia and fatal myoclonic epileptic status. *Brain* **119** (Pt. 3), 997–1010.

Vanier, M. T., Duthel, S., Rodriguez-Lafrasse, C., Pentchev, P., and Carstea, E. D. (1996). Genetic heterogeneity in Niemann-Pick C disease: a study using somatic cell hybridization and linkage analysis. *Am. J. Hum. Genet.* **58**, 118–125.

Vanier, M. T., Rodriguez-Lafrasse, C., Rousson, R., Duthel, S., Harzer, K., Pentchev, P. G., Revol, A., and Louisot, P. (1991). Type C Niemann-Pick disease: biochemical aspects and phenotypic heterogeneity. *Dev. Neurosci.* **13**, 307–314.

Verheijen, F. W., Verbeek, E., Aula, N., Beerens, C. E. M. T., Havelaar, A. C., Joosse, M., Peltonen, L., Aula, P., Galjaard, H., van der Spek, P. J., and Mancini, G. M. S. (1999). A new gene, encoding an anion transporter, is mutated in sialic acid storage diseases. *Nat. Genet.* **23**, 462–465.

Verhoeven, K., Ensink, R. J. H., Tiranti, V., Huygen, P. L. M., Johnson, D. F., Schatteman, I., Van Laer, L., Verstreken, M., Van de Heyning, P., Fischel-Ghodsian, N., Zeviani, M., Cremers, C. W. R. J., Willems, P. J., and Van Camp, G. (1999). Hearing impairment and neurological dysfunction associated with a mutation in the mitochondrial tRNA-ser(UCN) gene. *Eur. J. Hum. Genet.* **7**, 45–51.

Wallace, D. C., Zheng, X. X., Lott, M. T., Shoffner, J. M., Hodge, J. A., Kelley, R. I., Epstein, C. M., and Hopkins, L. C. (1988). Familial mitochondrial encephalomyopathy (MERRF): genetic, pathophysiological, and biochemical characterization of a mitochondrial DNA disease. *Cell* **55**, 601–610.

Wetterau, J. R., Aggerbeck, L. P., Bouma, M.-E., Eisenberg, C., Munck, A., Hermier, M., Schmitz, J., Gay, G., Rader, D. J., and Gregg, R. E. (1992). Absence of microsomal triglyceride transfer protein in individuals with abetalipoproteinemia. *Science* **258**, 999–1001.

Willner J. P., Grabowski G. A., Gordon R. E., Bender A. N., and Desnick R. J. (1981). Chronic GM2 gangliosidosis masquerading as atypical Friedreich ataxia: clinical, morphologic, and biochemical studies of nine cases. *Neurology* **31**, 787–98.

Young, I. D., Young, E. P., Mossman, J., Fielder, A. R., Moore, J. R. (1987). Neuraminidase deficiency: Case report and review of the phenotype. *J. Med. Genet.* **24**, 283–290.

Zeviani, M., and Antozzi, C. (1997). Mitochondrial disorders. *Mol. Hum. Reprod.* **3**, 133–148.

Zeviani, M., Muntoni, F., Bavarese, N., Serra, G., Tiranti, V., Carrara, F., Mariotti, C., and DiDonato, S. (1993). A MERRF/MELAS overlap syndrome associated with a new point mutation in the mitochondrial DNA tRNA(Lys) gene. *Eur. J. Hum. Genet.* **1**, 80–87.

Zhang, M., Dwyer, N. K., Love, D. C., Cooney, A., Comly, M., Neufeld, E., Pentchev, P. G., Blanchette-Mackie, E. J., and Hanover, J. A. (2001). Cessation of rapid late endosomal tubulovesicular trafficking in Niemann-Pick type C1 disease. *Proc. Nat. Acad. Sci. U.S.A.* **98**, 4466–4471.

Zhu Z., Yao J., Johns T., Fu K., De Bie I., Macmillan C., Cuthbert A. P., Newbold R. F., Wang, J., Chevrette, M., Brown, G. K., Brown, R. M., and Shoubridge, E. A. (1998). SURF1, encoding a factor involved in the biogenesis of cytochrome c oxidase, is mutated in Leigh syndrome. *Nat. Genet.* **20**, 337–343.

# CHAPTER 25

# Diagnostic Evaluation of Ataxic Patients

SUSAN L. PERLMAN

*Department of Neurology*
*UCLA School of Medicine*
*University of California*
*Los Angeles, California 90095*

I. Defining the Neurological Phenotype in Patients with Ataxia as the Primary Symptom
   A. The History and Physical Examination
   B. Confirming the Phenotype with Imaging and Electrophysiological Studies
   C. Constructing a Template for the Differential Diagnosis Based on Neural Phenotype
II. Determining Whether the Disease is Genetic
   A. Is the Neurologic Phenotype Consistent with a Genetic Cause?
   B. Constructing an Accurate Pedigree
   C. Choosing Appropriate Gene Tests or Obtaining Other Confirmation of an Inherited Disorder
   D. What to do when the Initial Round of Testing is Uninformative
   References

"Ataxia is not a foreign cab" (National Ataxia Foundation, 1999). It is a neurologic symptom, defined as a lack of accuracy or coordination of movement (DeJong, 1979). It is the predominant symptom of cerebellar disease, but "inaccuracy of movement" may also be caused by muscle weakness or alterations in tone, intrusion of involuntary movements, large-fiber/posterior column sensory loss, central or peripheral vestibular problems, visual changes, or any combination of them, and so may be an accompanying symptom of other noncerebellar, neurologic conditions. The spectrum of cerebellar disease extends from the pure or predominantly cerebellar syndromes through complex conditions that affect many parts of the neuraxis.

The steps in accurately diagnosing a patient presenting with the chief complaint of "ataxia" (with features of imbalance or falling, hand "clumsiness" or "shakiness," slurred speech, choking, double vision, "dizziness") include:

1. Defining the neural systems causing ataxia, by neurologic examination, imaging, and electrophysiological tests
2. Identifying any associated neural or medical features
3. Establishing the time course of the ataxic complaint
4. Constructing an accurate family history and history of other illnesses, medications, or nutritional supplement use that could explain at least part of the ataxic etiology
5. Performing appropriate additional laboratory studies to confirm a suspected acquired or genetic cause.

Molecular genetic research in the past decade has identified a growing number of gene mutations that result in inherited ataxia, which has greatly improved efforts to classify these disorders, explain their pathophysiology, and develop gene-specific therapies. It has also simplified and reduced the cost of the diagnostic work-up and improved the accuracy of prognostication and genetic counseling for patients and families (details of these specific genetic conditions can be found in earlier chapters of this book). However, even in a family with a genetically defined ataxia, specific individuals may also have acquired factors or a second genetic condition that contributes to the primary disease, or is responsible for all the symptoms, in someone who may not carry the defective gene at all.

## I. DEFINING THE NEUROLOGIC PHENOTYPE IN PATIENTS WITH ATAXIA AS THE PRIMARY SYMPTOM

### A. The History and Physical Examination

#### 1. The Neurological Examination—Is it Ataxia? Is it Cerebellar?

A careful neurological history and examination should enable the clinician to identify all the neural systems contributing to the ataxic complaint. Wide-based or unsteady casual stance and gait, trouble on turns or stairs, and difficulty with single-limb balance or tandem gait are usually the presenting complaints in the ataxic patient, but are not specific to the cerebellum (Table 25.1).

Cerebellar symptoms and signs, that occur due to involvement of the cerebellum and its outflow tracts, include decreased muscle tone with pendular tendon reflexes; limb dysmetria or hypermetria (inaccurate trajectory, overshoot, loss of check reflex); slowing and loss of rhythm of movement; postural and kinetic tremor and titubation; dysarthria (scanning speech, decreased rapid alternating movements of speech); and eye movement abnormalities (fixation instability, gaze-evoked nystagmus with blurring or doubling of vision or "dizziness," saccade overshoot, impaired smooth pursuit, and inability to suppress the vestibulo-ocular reflex). Deep tendon reflexes might be depressed. Fatigue is a common complaint.

The presence of muscle atrophy or weakness, spasticity or rigidity, involuntary movements (resting tremor, choreoathetosis, dystonia), postural instability or motor impersistance, sensory loss, vestibular features (vertigo, oscillopsia, rotatory nystagmus, past-pointing), or vision loss or distortion can cause instability in the absence of cerebellar disease or may be important secondary features in a predominantly cerebellar syndrome.

TABLE 25.1 Neurologic Phenotypes in Predominantly Ataxic Syndromes

Cerebellar
- Midline/vermian—primarily truncal and gait ataxia
- Hemispheric—primarily limb ataxia with decreased tone, dysmetria, and action tremor dysarthria
- "Flocculo-nodular"—notable eye movement changes (nystagmus, loss of vestibulo-ocular reflex)

Spinocerebellar
- Signs similar to those seen in midline or hemispheric cerebellar processes
- Suggested by symptom development (caudal to rostral) or associated spinal long tract signs
- Differentiated from intrinsic cerebellar origin by imaging or electrophysiological testing

Cerebral cortical or subcortical
- Alone or in combination with cerebellar cortical changes (combined cortical syndromes)
- Differentiated by imaging or electrophysiological testing

Vestibular
- Peripheral—rotatory nystagmus, stationary or paroxysmal positional
  - Suppressed by visual fixation
  - Past-pointing or stepping to side of "weaker" vestibular signal
  - If bilaterally weak vestibular activity, oscillopsia often present
- Central—rotatory and gaze-paretic nystagmus
  - Incompletely suppressed by fixation
- Dysfunction in either location may need to be confirmed with electrophysiologic testing

Somatosensory
- Peripheral large-fiber—stocking-glove defects in vibration and joint position sense
- Posterior column—deficits may progress from caudal to rostral
  - Spinal level for vibration may be present
- Both with positive Romberg test
- Dysfunction in either location may need to be confirmed with electrophysiologic testing

Secondary or associated neurologic features
- Intrinsic brainstem signs
- Basal ganglia signs
- Upper motor neuron signs
- Motor unit signs (anterior horn cell, motor axon, muscle)
- Small-fiber sensory signs
- Autonomic features
- Special sensory changes (retina or optic nerve, auditory pathways)
- Dementia
- Seizures, myoclonus

## 2. Associated System Involvement—Is it Pure Cerebellar or Cerebellar-Plus?

The neurological examination might be consistent with a pure or predominantly cerebellar syndrome or it might have revealed signs of both cerebellar and extracerebellar dysfunction, involving upper motor neurons or motor unit structures, basal ganglia, large-fiber/posterior columns or small-fiber/spinothalamic tracts, and vestibular or visual pathways. Additional neural systems involvement that can help refine the phenotypic diagnosis include dementia (severe or subtle), seizures, myoclonus, sensorineural hearing loss, intrinsic brainstem changes (nuclear or supranuclear ophthalmoplegia with slowed or absent saccades; facial, pharyngeal, laryngeal, neck, or tongue weakness with depressed jaw jerk and gag reflexes; parasomnias or central or obstructive sleep apnea), and autonomic dysfunction (orthostatic hypotension, lower motor neuron bladder or bowel signs, impotence).

Non-neurologic systems might be involved in some cerebellar conditions, and identifying them on physical exam could expedite making the specific diagnosis. These might include conjunctival/ocular or cutaneous features (telangiectasias, Kayser-Fleischer rings, cataracts, optic atrophy, retinitis pigmentosa, cherry-red spot; angiokeratomas, lipomas, tendon xanthomas), skeletal abnormalities (platybasia, scoliosis, pes cavus, arthritic joint changes or history of gout), liver or splenic enlargement, signs of cardiomyopathy (arrhythmias, cardiac enlargement—confirmed by EKG, Holter monitor, echocardiogram), or a history of anemia, thyroid disease, diabetes, diabetes insipidus, GI tract or kidney problems, recurrent infection, or cancer.

Table 25.2 profiles non-neurologic findings in the history and physical examination that might focus the evaluation

**TABLE 25.2** Non-Neurologic Features that Suggest a Specific Diagnosis

| Feature | Possible diagnosis | Follow-up laboratory studies |
|---|---|---|
| Anemia—hemolytic sideroblastic | Gamma-glutamyl cysteine synthetase deficiency<br>ATP-binding cassette7 transporter | Glutathione levels<br>Bone marrow aspirate |
| Cardiomyopathy, history of diabetes | Friedreich's ataxia<br>Mitochondrial disorders | FRDA gene test<br>See below |
| Cataracts, tendon xanthomas | Cerebrotendinous xanthomatosis | Urine cholestanol |
| Cherry-red spot<br>Lens or corneal opacities<br>Angiokeratomas | Sialidosis (mucolipidosis type 1) | Urine oligosaccharides<br>Fibroblast α-neuraminidase assay |
| Diabetes insipidus | Langerhans' cell histiocytosis<br>Wolfram's syndrome | Skull X-rays, brain MRI |
| GI malabsorption/diarrhea | Coeliac disease<br>Whipple's disease | Antigliadin antibodies, GI biopsy<br>CSF PCR |
| Gout, renal stones | Hypoxanthine-guanine phosphoribosyl transferase (HGPRT) deficiency, partial (Lesch-Nyhan syndrome) | Plasma urate, urine urate/cr |
| Kayser-Fleischer rings<br>Evidence of cirrhotic liver disease | Wilson's disease | Plasma copper, ceruloplasmin<br>Urine copper (post-penicillamine)<br>Liver function studies, biopsy |
| Lipomas | Mitochondrial disorders | See below |
| Optic atrophy (with or without hearing loss)<br>Retinitis pigmentosa | Abeta/hypobetalipoproteinemia<br>Alpha-tocopherol transfer protein deficiency (AVED) | Vitamin E, lipoproteins, lipids |
| | Mitochondrial disorders (Kearns-Sayre syndrome, Leber's hereditary optic neuropathy, neuropathy-ataxia-retinitis pigmentosa, and others); Wolfram's syndrome | 2-hour postprandial glucose, lactate, pyruvate, aminoacids, ammonia; serum coenzyme Q10; CSF lactate; muscle or skin biopsy; LHON and NARP mtDNA tests |
| | Neuronal ceroid lipofuscinosis | Conjunctival, skin, rectal biopsy |
| | Peroxisomal disorders (Refsum's disease, adrenomyeloneuropathy) | Phytanic acid, very long chain fatty acids |
| | SCA7 | SCA7 gene test |
| Photosensitive skin rash | Hartnup's disease | Urine aminoacid chromatography |
| Platybasia | Arnold-Chiari malformation | Skull X-ray, brain/C-S MRI |
| Splenomegaly | Niemann-Pick type C | Bone marrow aspirate |
| Telangiectases<br>History of sinopulmonary infection<br>History of leukemia, lymphoma, cancer | Ataxia-telangiectasia | Alpha-fetoprotein, quantitative immunoglobulins, T-cell studies, chromosomal breakage studies, radiation sensitivity studies |

toward a specific diagnosis and lead to confirmatory laboratory testing.

## B. Confirming the Phenotype with Imaging and Electrophysiological Studies

### 1. Imaging Studies can Provide an Anatomic Basis for the Observed Phenotype and Identify Some Etiologies

The presence of platybasia (Milhorat et al., 1999; Sonstein et al., 1996), a history of head or neck trauma or surgery, a history of diabetes insipidus (Jarquin-Valdivia and Buchhalter, 2001; Birnbaum et al., 1989), or findings on examination that localize to the cervical spine or cord may lead to performance of plain X-rays of the skull or C-spine, suggesting an acquired cause for the cerebellar or spinocerebellar signs (see Table 25.2), which should be confirmed with more sensitive imaging technology. Computed tomographic scanning (CT) is widely available and relatively inexpensive; preferred for assessment of skull base, calvaria, and the presence of bleeding; and sensitive to a variety of space-occupying lesions and gross atrophy of intracranial structures. However, magnetic resonance imaging (MRI), with its exceptional contrast resolution, multiplanar capacity, choice of sequences to emphasize different tissue characteristics, and lack of harmful effects, remains the preferred imaging technique for patients with neurologic disease. Patients with contraindications to MRI (metal implants, claustrophobia) usually undergo CT.

Structural imaging (MRI) can confirm the presence of atrophy in cerebellum (midline or hemispheres), brainstem, cerebrocortical or subcortical structures, and cervical cord. It can reveal hypo- or hyperintense changes, that may be specific for certain conditions (see Table 25.3), as well as subcortical white matter disease of the elderly (Baloh et al., 1995; Baloh and Vinters, 1995), frank cerebrovascular disease, hydrocephalus (normal pressure or partial obstructive), congenital malformations (cerebellar hypoplasia, Dandy-Walker and Arnold-Chiari malformations), mass lesions, and changes diagnostic of infection, inflammation, vasculitis, and multiple sclerosis. Hyperintensities are not usually seen in the genetic spinocerebellar ataxias (SCAs).

TABLE 25.3  Hyper- or Hypointensities on MRI that Suggest a Specific Diagnosis

| Feature | Possible diagnosis | Follow-up laboratory studies |
| --- | --- | --- |
| Cortical abnormalities with diffusion-weighted (DWI) and fluid attenuated inversion recovery (FLAIR) sequences[a] Hyperintensities in striatum, thalamus, hippocampus, cerebellum[b] | Creutzfeldt-Jakob disease | EEG CSF 14-3-3 protein[c,d] Blood prion protein gene analysis[e] |
| High signal in thalamus, peri-aqueductal midbrain, and superior cerebellar vermis/hemispheres[f,g] | Acute Wernicke's encephalopathy | Give intravenous thiamin and monitor neurologic exam and blood transketolase activity |
| High signal in pons and middle cerebellar peduncles ("the cross sign")[h,i] | Sporadic OPCA or MSA Also reported in SCA3[j] | Consider 18F-fluorodopa PETscan Consider autonomic testing SCA3 gene test |
| Hyperintensities in putamenal lateral rim with hypointensities in dorsolateral putamen[k,l] | MSA, differentiated from other L-dopa unresponsive akinetic-rigid syndromes | Consider 18F-fluorodopa PETscan Consider autonomic testing |
| Signs of demyelination | Multiple sclerosis or other immunologic processes. Infection (PML, SSPE). Leukodystrophy (metachromatic, globoid) Peroxisomal disorders | Evoked potential and CSF analysis CSF analysis, brain biopsy Lysosomal hydrolases assay Very long chain fatty acid assay |

[a]Jacobs et al. (2001)
[b]Collie et al. (2001)
[c]Poser et al. (1999)
[d]Chapman et al. (2000)
[e]This test is done on a research basis only. Research labs involved in this testing can be accessed at www.geneclinics.org. (Collinge, 2001).
[f]Antunez et al. (1998)
[g]Murata et al. (2001)
[h]Adachi et al. (2000)
[i]Burk et al. (2001)
[j]Murata et al. (1998)
[k]Kraft et al. (1999)
[l]Macia et al. (2001)

Atrophy seen on MRI may be helpful in defining the phenotype, but is not specific for etiology. Pure cerebellar atrophy (see Table 25.4) can be seen from (1) congenital causes; (2) in certain genetic syndromes (ataxia-telangiectasia, SCA4, 5, 6, 8, 10, 11, 14, 15, and 16); and as a result of (3) infectious or postinfectious processes (usually viral, e.g., varicella, measles, rubella, Epstein-Barr, the enteroviruses, HTLV1).

Isolated cerebellar atrophy can also be a feature of conditions associated with a number of identified autoantibodies. Because of the potential reversibility with treatment, immune-mediated causes should be sought in all sporadic cases of ataxia, in patients with subacute, waxing-waning, or stepwise progression of symptoms, and in individuals with signs of associated neuropathy, white matter or vasculitic change on MRI, or elevated CSF protein. Prior history of preceding flu-like illness, of other autoimmune disease (rheumatologic, endocrine) or of cancer in the patient or family increase suspicion of involvement of the immune system.

TABLE 25.4  Causes of Isolated Cerebellar Atrophy on MRI

Congenital, inherited
   Ataxia-telangiectasia
   SCA4, 5, 6, 8, 10, 11, 14, 15, 16
Infectious/postinfectious
   Ebstein-Barr enterovirus
   HTLV1
   Measles
   Rubella
   Varicella
Metabolic
   Acute thiamin deficiency
   Chronic vitamin B12 and E deficiencies
   Autoimmune thyroiditis and hypothyroidism
Paraneoplastic
   Anti-Yo, Hu, Ri, MaTa, CV2 (most common)
   Anti-calcium channel, CRMP-5, ANNA-1,2,3, mGluR1
Other autoantibodies
   Anti-gliadin (most common—reported also in the inherited syndromes as a possible secondary factor)
   Anti-GluR2, GAD, MPP1, GQ1b ganglioside
Post-anoxia, hyperthermia, trauma and in chronic epilepsy
Toxic
Drug reactions
   Amiodarone
   Cytosine arabinoside
   5-Fluorouracil
   Phenytoin
   Valproic acid
Environmental
   Acrylamide
   alcohol
   Organic solvents
   Organolead/mercury/tin
   Inorganic bismuth/mercury/thallium

Paraneoplastic anti-cerebellar antibodies seen in the context of occult carcinoma include anti-Yo, Hu, Ri (Bradwell, 2000), MaTa (Rosenfeld et al., 2001), CV2 (de la Sayette et al., 1998), calcium channel (Lennon et al., 1995), CRMP–5 (Yu et al., 2001), ANNA–1,2,3 (Chan et al., 2001), and mGluR1 (Smitt et al., 2001)). Non-cancer related antibodies are anti-GluR2 (Gahring et al., 1997), anti-GAD (Abele et al., 1999; Honnorat et al., 2001), anti-MPP1 (Fritzler et al., 2000), anti-gliadin (Burk et al., 2001; Bushara et al., 2001; Luostarinen et al., 2001), and anti-GQ1b ganglioside (Odaka et al., 2001). The related case reports suggest that the presence of these antibodies may contribute to or cause the associated cerebellar syndrome and that immune-modulating therapies (high-dose steroid taper, plasmapheresis, intravenous immunoglobulin infusion, immunosuppressant drugs) may reverse some of the symptoms (Burk, 2001b). The long-term benefits of these treatments are not yet known. The presence of anti-gliadin antibodies in individuals with one of the known genetic ataxias (Bushara, 2001) raises the possibility that the products of neural degeneration may trigger an immune response that accelerates further nerve cell loss and that treatment of the inherited ataxias might well include immunomodulatory strategies. Further research in this area is needed before these treatments can be recommended.

Antibody testing that is clinically available through routine diagnostic laboratories includes several of the paraneoplastic markers (anti-Yo, Hu/ANNA-1, Ri/ANNA-2, MaTa, and CV2), associated with occult breast, gynecologic, small cell lung, colon, and testicular carcinoma. Non-cancer related antibodies for which testing can be ordered include anti-GAD, anti-gliadin, and anti-GQ1b ganglioside.

Cerebellar atrophy is also characteristic of cerebellar syndromes associated with (1) postanoxia, hyperthermia (Fujino et al., 2000) and trauma (Soto-Ares et al., 2001); (2) chronic epilepsy (Crooks et al., 2000; Sandok et al., 2000); and (3) various acquired metabolic states (acute thiamin and chronic B12 and E deficiencies, autoimmune thyroiditis and hypothyroidism; Selim and Drachman, 2001) adverse drug reactions (amiodarone, cytosine arabinoside, 5-fluorouracil, phenytoin, valproic acid; Shill and Fife 2000), and toxic exposures (acrylamide, alcohol, organic solvents, organolead/mercury/tin compounds, inorganic bismuth/mercury/thallium).

Functional imaging (positron emission tomography, PET; single proton emission computed tomography, SPECT) is becoming increasingly available for clinical use and allows the imaging of physiologic features of brain tissue, including perfusion patterns, metabolic activity (glucose and oxygen utilization), and location and density of specific neurotransmitters and receptors. Decreases in perfusion or glucose metabolism in the cerebellum, brainstem, cerebral cortex, and other brain regions may precede atrophy, and SPECT or PET can be used to confirm the suspicion of abnormalities, in the face of a still normal MRI scan (Soong and Liu, 1998; Matsuda et al., 2001). Decreased nigrostriatal glucose metabolism, dopamine receptor density (Gilman, 2001), and dopamine transporter activity (Kim et al., 2000) have been observed in individuals with multiple system atrophy (MSA) and in others with apparent sporadic cerebellobrainstem syndromes without basal ganglia features (sOPCA), suggesting that the latter group may be at risk to evolve into MSA and that functional imaging may be helpful in differentiating early MSA from sOPCA. Proton magnetic resonance spectroscopy (MRS) has been reported to be more sensitive than MRI or SPECT for detection of early neuronal degeneration (Konaka et al., 2000).

## 2. Electrophysiological Studies can Provide an Anatomic Basis for the Observed Phenotype and Identify Some Etiologies

Electrodiagnostics should be obtained when the neurological examination either does not confirm or does not adequately define a certain neural system. If suspicion is high for a dementia or if a question of epileptiform activity or myoclonus is raised, an electroencephalogram (EEG) can add clarification (Table 25.5). The presence of periodic, 1–2 per second, generalized di- or triphasic sharp waves in a patient with a rapidly progressive dementia and ataxia is felt to be diagnostic of Creutzfeldt-Jakob disease (CJD). Early optic neuropathy or pigmentary retinopathy can be revealed with visual-evoked response testing (VER) or

TABLE 25.5 Electrophysiological Test Results that Suggest a Specific Diagnosis

| Feature | Possible diagnosis | Follow-up laboratory studies |
|---|---|---|
| EEG evidence of periodic triphasic sharp waves | Creutzfeldt-Jakob disease | MRI with DWI and FLAIR<br>CSF 14-3-3 protein (see Table 25.3)<br>Blood prion protein gene test (see Table 25.3) |
| VER or ERG abnormalities (even subclinical) | Abeta/hypobetalipoproteinemia | Vitamin E, lipoproteins, lipids |
| | Aminoacid disorders | 2 hour postprandial amino acids |
| | Friedreich's ataxia | FRDA gene test |
| | Mitochondrial disorders | 2 hour postprandial glucose, lactate, pyruvate; CSF lactate; muscle or skin biopsy; LHON and NARP mtDNA tests |
| | Neuronal ceroid lipofuscinosis | Conjunctival, skin, rectal biopsy |
| | Peroxisomal disorders | Phytanic acid, very-long chain fatty acids |
| | Sialidosis | Urine oligosaccharides; fibroblast α-neuraminidase assay |
| | SCA7 | SCA7 gene test |
| ENG abnormalities | | |
| Saccade intrusions[a] | Friedreich's ataxia | FRDA gene test |
| Downbeat nystagmus[b] | Craniocervical junction syndrome | MRI of posterior fossa/C-spine |
| | SCA6 | SCA6 gene test |
| Loss of VOR[c] | SCA3 | SCA3 gene test |
| Periodic alternating nystagmus/gaze deviation and slowed vertical saccades[d] | Creutzfeldt-Jakob disease | See above |
| Delayed saccade initiation with normal velocity (oculomotor apraxia)[e] | Ataxia-telangiestasia | See Table 25.2 |
| | Ataxia-oculomotor apraxia 1 (AOA1)[f] | EMG/NCV; serum albumin (decr.), cholesterol (incr) |
| | Ataxia-oculomotor apraxia 2 (AOA2)[g] | Serum IgG, AFP, CPK (incr.) |
| | AT-like disorder (ATLD)[h] | Linkage studies for hMre11 |
| EMG showing myopathic change | Mitochondrial disorders | See above; MERRF mtDNA test |
| sphincter dennervation | MSA | See Table 25.3 |
| Sympathetic skin response change | MSA | See Table 25.3 |
| | SCA3 | SCA3 gene test |

[a]Moschner et al. (1994)
[b]Gomez et al. (1997)
[c]Buttner et al. (1998)
[d]Grant et al. (1993)
[e]Baloh et al. (1978)
[f]Moreira et al. (2001)
[g]Nemeth et al. (2000)
[h]Stewart et al. (1999)

an electroretinogram (ERG). Brainstem auditory-evoked responses (BAER) can confirm the presence of disease in peripheral and central auditory pathways. Electronystagmography (ENG, with recordings for pathologic nystagmus, visual ocular control, and vestibulo-ocular reflex activity) can explain unusual eye movements, reveal disease in vestibular pathways, and confirm the presence of disease in specific cerebellar and brainstem regions (Baloh and Honrubia, 1990). The use of BAER and ENG are valuable when one is trying to decide if a particular "pure cerebellar" patient has subclinical involvement of the brainstem, that might ultimately progress to a more complex cerebello-brainstem or even MSA picture. ENG can also be of help in ruling in or out cerebellar and vestibular involvement in an individual with nonspecific gait instability and a normal MRI. Somatosensory-evoked response testing can confirm the presence of spinal cord and supraspinal abnormalities along posterior column pathways. Electromyography and nerve conduction studies (EMG/NCV) can confirm the presence of pathology in muscle, motor or sensory peripheral nerve, or anterior horn cell and can define the type of neuropathy present (axonal versus demyelinating).

When MSA is suspected, autonomic studies (measures of heart rate variability, tilt-table response, sympathetic skin response (De Marinis et al., 2000; Holmberg et al., 2001; Kitae et al., 2001), and cardiac I$^{123}$-MIBG-SPECT (Druschky et al., 2000) may be sought, to show changes in pre-ganglionic sympathetic or parasympathetic pathways, before the clinical features of orthostatic hypotension, bowel and bladder dysfunction, and impotence are apparent. However, Machado-Joseph disease (SCA 3), which shares the phenotype of cerebello-brainstem, basal ganglia, and corticospinal dysfunction with MSA, is also noted to have abnormal results on the sympathetic skin response and cardiac SPECT measures (Kazuta et al., 2000). Videourodynamic and sphincter EMG studies (Vodusek, 2001; Sakakibara et al., 2001) are felt to show changes of denervation that are fairly specific for MSA, but not foolproof (Colosimo et al., 2000).

Sleep-disordered breathing and parasomnias (central and obstructive sleep apneas, periodic leg movements of sleep, sleep talking, nightmares) and pharyngeal/laryngeal symptoms (aspiration, stridor, hypophonia) are often present in patients with cerebellar/lower brainstem dysfunction and may be life-threatening. Identifying them (with the use of polysomnography and video swallowing and laryngoscopic testing) not only confirms the presence of pathology, but also opens up the opportunity for therapeutic interventions that can greatly improve the patient's quality of life (Asahina et al., 2000; Maurer et al., 1999; Munschauer et al., 1990; Plazzi et al., 1997); (Silber and Levine 2000).

Imaging and electrophysiologic studies can be helpful in differentiating the various inherited and acquired ataxias, but will not of themselves confirm the diagnosis, in the absence of neuropathologic or molecular genetic studies (see Tables 25.6 and 25.7).

## C. Constructing a Template for the Differential Diagnosis Based on Neural Phenotype

### 1. Factoring in Disease Course

Occasionally the neurologic phenotype, obtained by history and physical examination with back-up imaging and electrodiagnostic testing, can be diagnostic (for instance, in ataxia-telangiectasia, Friedreich's ataxia, Creutzfeldt-Jakob disease, and MSA) or highly suggestive of a certain etiology (for instance, the mitochondrial disorders or a typical dominant ataxia in a family with a known genotype), usually then confirmed by appropriate additional studies. Rarely, in these cases, the confirmatory testing will not prove the apparent diagnosis, and better diagnostic studies must be sought or the search for other causes widened. Most often, the neurologic phenotype is not diagnostic and is without distinguishing features (pure cerebellar syndrome or cerebello-brainstem syndrome, without known familial cause), and other factors must be considered. The tempo of disease can be of help. Acute-onset cerebellar conditions are more likely to be traumatic, vascular, metabolic/toxic (including intoxications with prescription medications), infectious, inflammatory, or one of the episodic disorders. Static cerebellar dysfunction will be seen in congenital, post-traumatic, postvascular, or other postacute diseases that have been stabilized, with residual cerebellar damage. A subacute course (progression over weeks, rather than months) can also suggest a post-traumatic, metabolic/toxic, or infectious/inflammatory process, but may represent a neoplastic or paraneoplastic etiology. Slowly progressive ataxia, while it can be seen with metabolic/toxic, infectious/inflammatory, and neoplastic/paraneoplastic conditions, usually results from genetic or incompletely understood degenerative conditions.

### 2. Ruling Out Acquired Causes

Everyone deserves a screen for thyroid dysfunction, low vitamin B12 and E, commonly implicated infections (syphilis, Lyme disease, Epstein-Barr virus, HTLV1) and immune-mediated conditions (lupus, paraneoplastic and other autoantibodies), urine copper and heavy metal presence, and CSF signs of demyelination, infection, and neoplasm. Even an individual with a gene-proven SCA is not immune to other complicating acquired conditions, and there are reports of individuals with complex phenotypes and symptomatic genes for two separate inherited syndromes (McNeil et al., 2001; Petzinger et al., 2000; Rubio et al., 1996, and one individual seen by us with both paternally transmitted Huntington's disease and maternally-transmitted dominant acanthocytosis).

TABLE 25.6  Comparison of Imaging and Electrical Data in Some of the Inherited Ataxias

| Genotype | MRI pattern of atrophy and other imaging abnormalities | Evoked Potentials | ENG[a] | Other |
|---|---|---|---|---|
| SCA1 | MRI—cerebellum, brainstem<br>MRS—early pontine change[b] | Motor—very abnl[c]<br>BAER, SSER—abnl<br>VER—occ. abnl[d] | Saccade overshoot<br>Less often, decreased saccade velocity<br>?Abnl VOR | EMG/NCV—mixed neuropathy[e,f] usually with larger CAG burden<br>ERG—occ. abnl[g] |
| SCA2 | MRI—cerebellum, pons, supratentorial with disease duration[h]<br>MRS—diffuse changes[i] | BAER—abnl<br>SSER—later abnl<br>Motor—nl | Saccade velocity decreased | EMG/NCV—lg fiber axonal sensory neuropathy, occ. fasciculations[f]<br>ERG—occ. abnl with large CAG burden[j] |
| SCA3 | MRI—enlarged fourth ventricle, atrophy of cerebellum, brainstem (worse with disease duration, larger CAG burden),[k] frontoparietal,[l] caudate and putamen[m]<br>SPECT—basal ganglia, frontotemporal[l]; diffusely decreased BDZ/GABA binding[n] and dopamine transporter binding[o]<br>PET—striatonigral,[p] cortex[q] | BAER, SSER—abnl<br>VER—occ. abnl<br>Motor—nl | Gaze-evoked nystagmus<br>Saccade hypometria<br>Impaired VOR (worse with larger CAG burden) | EMG/NCV—axonal neuropathy,[f] usually with smaller CAG burden<br>ERG-occ. abnl[r] |
| SCA4 | MRI—cerebellum[s] | | | EMG/NCV—sensory axonal neuropathy[t] |
| SCA5 | MRI—cerebellum[u] | | | |
| SCA6 | MRI—cerebellum, middle peduncle, red nucleus[v]<br>PET—cortex, basal ganglia, brainstem[w] | BAER—abnl[z] | Vertical and horizontal nystagmus<br>Impaired smooth pursuit<br>?abnl VOR[x,y] | EMG/NCV—mild axonal neuropathy[z] |
| SCA7 | MRI—cerebellum, brainstem[qq,bb] | VER—abnl<br>BAER—abnl | Early slowed saccades[cc] | ERG—abnl<br>EMG/NCV—nl |
| SCA8 | MRI—cerebellum[dd] | | Gaze-paretic nystagmus<br>Impaired smooth pursuit | EMG/NCV—occ. large fiber axonal sensory neuropathy |
| SCA10 | MRI—cerebellum[ee] | | Gaze-paretic nystagmus | EEG—abnl if seizures present[ee] |
| SCA11 | MRI—cerebellum[ff] | | Gaze-paretic nystagmus | NCV—nl |
| SCA12 | MRI—cerebellum, cortex[gg] | | | |
| SCA13 | MRI—cerebellum, pons[hh] | | | |
| SCA14 | MRI—cerebellum[ii] | | | |
| SCA15 | MRI—cerebellum[jj] | | | |
| SCA16 | MRI—cerebellum[kk] | | | |
| DRPLA | MRI—atrophy in cerebellum, pons, midbrain, cortex; with later onset high signal in thalamus, subcortical white matter[ll,mm,nn] | BAER—nl | | EEG—abnl<br>EMG/NCV—nl |
| FRDA1,2<br>FRDA–like | MRI-spinal cord atrophy; late, mild cerebellar atrophy<br>PET-early cerebral hyper-metabolism[oo] | Motor-early abnl<br>SSER-early abnl<br>VER-late abnl<br>BAER-late abnl[pp] | Saccade intrusions<br>Impaired VOR[qq] | EMG/NCV-axonal neuropathy<br>ERG-abnl in the vitamin E associated phenocopies[rr] |
| ATM[ss]<br>ATMlike<br>(Table 25.4) | MRI-cerebellum | | Delayed saccade initiation with normal velocity (oculomotor apraxia)[tt] | EMG/NCV-late neuropathy; in AOA1-early axonal motor neuropathy, AOA2-sensory |
| Other EOCAs[uu] with retained reflexes | MRI-cerebellum > brainstem[vv] | BAER-abnl<br>SSER-abnl<br>VER-abnl[ww]<br>(nl in LOTS) | Impaired saccades | EMG/NCV-lg fiber axonal motor > sensory neuropathy[ww] |

[a]Buttner N. et al. (1998), Rivaud-Pechoux et al. (1998), and Burk et al. (1999)
[b]Mascalchi et al. (1998)
[c]Yokota et al. (1998)
[d]Illarioshkin et al. (1996)
[e]Schols et al. (1995)
[f]Kubis et al. (1999)
[g]Abe et al. (1997)
[h]Giuffrida et al. (1999)
[i]Boesch et al. (2001)
[j]Babovic-Vuksanovic et al. (1998)
[k]Onodera et al. (1998)
[l]Etchebehere et al. (2001)
[m]Klockgether et al. (1998)
[n]Ishibashi et al. (1998)
[o]Yen et al. (2000)
[p]Shinotoh et al. (1997)
[q]Taniwaki et al. (1997)
[r]Isashiki et al. (2001)
[s]Nagaoka et al. (2000)
[t]Flanigan et al. (1996)
[u]Stevanin et al. (1999)
[v]Murata et al. (1998)
[w]Soong et al. (2001)
[x]Buttner et al. (1998)
[y]Gomez et al. (1997)
[z]Kumagai et al. (2000)
[aa]Benomar et al. (1994)
[bb]Gouw et al. (1994)
[cc]Oh et al. (2001)
[dd]Ikeda et al. (2000)
[ee]Matsuura et al. (1999)
[ff]Worth et al. (1999)
[gg]O'Hearn et al. (2001)
[hh]Herman-Bert et al. (2000)
[ii]Yamashita et al. (2000)
[jj]Storey et al. (2001)
[kk]Miyoshi et al. (2001)
[ll]Farmer et al. (1989)
[mm]Uyama et al. (1995)
[nn]Tomiyasu et al. (1998)
[oo]Klockgether et al. (1993)
[pp]Nuwer et al. (1983)
[qq]Moschner et al. (1994)
[rr]Abeta/hypobetalipoproteinemia, alpha-tocopherol transfer protein deficiency (AVED; Cavalier et al. (1998)
[ss]Becker-Catania et al. (2000)
[tt]Baloh et al. (1978)
[uu]Early-onset cerebellar ataxia, including autosomal recessive spastic ataxia of Charlevoix-Saguenay (ARSACS; Engert et al. (2000) and late-onset Tay-Sachs disease (LOTS; Navon, et al. (1986)
[vv]De Michele et al. (1995)
[ww]Pal et al. (1995)

TABLE 25.7 Phenotypic Templates for use in Categorizing the Progressive Ataxias

| Typical phenotype | Common features | Less common features | Unlikely features |
|---|---|---|---|
| Friedreich ataxia-like With gene-positive FRDA, the smaller the GAA repeat, the more likely the less common features will be seen, subject to the presence of mosaicism or heterozygous point mutations | Onset between 5–25y/o<br>Recessive or sporadic inheritance[a]<br>Caudal > rostral ataxia<br>Absent lower limb DTRs and upgoing toes with later muscle atrophy and spasticity.<br>Decreased vibration<br>Later hearing or vision loss<br>Scoliosis, pes cavus<br>Cord atrophy on MRI<br>Early abnormalities on motor/sensory EPs<br>Axonal neuropathy on EMG/NCV<br>Saccade intrusions on electronystagmography<br>Cardiomyopathy<br>Diabetes mellitus | Onset >25y/o (LOFA[b])<br>Retained reflexes in lower limbs (FARR[c]) or spasticity without ataxia[d]<br>Chorea[e]<br>Normal EKG | Onset <2y/o<br>Dominant inheritance<br>Sub-Saharan African, Amerindian, Chinese, Japanese, or Southeast Asian ethnicity[f]<br>Early muscle atrophy, spasticity, dystonia<br>Normal sensation<br>Early hearing/vision loss, retinal degeneration<br>Ophthalmoplegia<br>Dementia or mental retardation<br>Seizures<br>Early cerebellar atrophy on MRI<br>Demyelinating neuropathy on NCV<br>Diabetes insipidus |
| Early-onset cerebellar ataxia with retained reflexes (EOCARR), not FARR. The presence of unusual, distinctive features in a patient with EOCARR may provide a "hook" for fishing in the growing body of rare EOCAs that are now being genetically defined[g] | Onset between 2–20y/o<br>Recessive or sporadic<br>Retained/brisk DTRs<br>Cerebellar atrophy on MRI | Sensory neuropathy<br>Absent ankle jerks | Optic atrophy<br>Scoliosis<br>Cardiomyopathy |
| Mitochondrial* | Maternal inheritance | Dominant inheritance | |

*Distinguishing neurological features include: dementia, dystonia, exercise intolerance, hearing loss, myelopathy, myoclonus, myopathy, neuropathy, ophthalmoplegia, optic neuropathy, pigmentary retinopathy, seizures, stroke-like episodes or other episodicity, vascular headache.
Distinguishing systemic features include: cardiomyopathy or conduction defects, cataracts, diabetes mellitus, exocrine pancreas dysfunction, hyperaldosteronism, hypoadrenalism, hypogonadism/ovarian failure, hypoparathyroidism, hypothyroidism, intestinal pseudo-obstruction, lactic acidosis, lipomatosis, liver disease, pancytopenia, renal tubulopathies, rhabdomyolysis, short stature, sideroblastic anemia.
The presence of any of these features in a patient with ataxia may warrant a screen for mitochondrial disease.

| Typical phenotype | Common features | Less common features | Unlikely features |
|---|---|---|---|
| Autosomal Dominant Cerebellar Ataxia (ADCA) Type 1, or Typical Dominant Ataxia or SCAn The observed phenotype will vary with genotype, triplet repeat length, and disease duration, most notably with SCA2,3,[h] 7 | Onset 4th–6th decade<br>Anticipation<br>Pyramidal features<br>Extrapyramidal, "mild"<br>Neuropathy<br>Ophthalmoplegia<br>Optic atrophy<br>Dementia | Onset <25y/o<br>Early slow saccades/SCA2<br>Downbeat nystagmus/SCA6<br>Akinetic rigidity/SCA3,12<br>Chorea/SCA1,[i]2<br>Prominent tremor/SCA1,12,16 (ADCA V)<br>Areflexia/SCA2,4, later onset SCA3 | Onset <10 years/ SCA1,2,3,7,8,10,12,13<br>Early chorea, seizures, dementia/SCA2,7,17 mental retardation/SCA13<br>Myoclonus/ SCA14, DRPLA (ADCA IV)<br>Maculopathy/SCA7 (ADCA II) |

ADCA III—"pure cerebellar"/SCA5,6,8,10,11,15
This phenotype is typically slowly progressive with a variable age of onset, occasional mild pyramidal features (SCA8), or rarely seizures (SCA10). MRI shows only cerebellar atrophy. To be distinguished from parenchymatous cerebellar cortical atrophy by familial features and molecular genetic studies.

| Typical phenotype | Common features | Less common features | Unlikely features |
|---|---|---|---|
| Sporadic Olivoponto-cerebellar Atrophy (sOPCA) or Multiple System Atrophy (MSA), with features of ataxia, parkinsonism (bradykinetic rigidity with poor levodopa response), and autonomic failure (othostatic hypotension, incontinence, impotence)[j] | Onset 5th–7th decade<br>More rapid disease progression than SCA<br>About 1/4 of sOPCA will evolve into MSA within 5years, especially if onset was >50y/o[k,l]<br>80% Parkinson-onset (MSA-P)<br>20% Ataxic-onset (MSA-C) | Dementia[o]<br>Motor > sensory neuropathy[p] (MSA-C)<br>Neck extensor myopathy[q] | Familial features Levodopa responsive Parkinsonian symptoms (SCA3 should be considered)<br>Ophthalmoplegia[r]<br>Chorea[s]<br>Motor-evoked potential abnl[t] |

(continues)

## TABLE 25.7 (continued)

| Typical phenotype | Common features | Less common features | Unlikely features |
|---|---|---|---|
| | Later autonomic sxx, with the exception of impotence, which may precede ataxia/rigidity by 5–10 years | | |
| | Pyramidal features | | |
| | Motor > sensory neuropathy,[m] especially sphincter dennervation (MSA-P) | | |
| | VER, SSER abnl BAER abnl (MSA-C)[n] | | |
| | Sympathetic skin response abnl | | |
| | REM sleep disturbance | | |
| | Obstructive sleep apnea | | |
| | Day time stridor | | |

[a] Pseudodominant inheritance may occur in populations with a high carrier frequency
[b] Late-onset Friedreich ataxia
[c] Friedreich ataxia with retained reflexes
[d] Castelnovo et al. (2000)
[e] Hanna et al. (1998)
[f] Labuda et al. (2000)
[g] Bomont et al. (2000)
[h] Jardim et al. (2001)
[i] Namekawa et al. (2001)
[j] Gilman et al. (1999)
[k] Gilman et al. (2000)
[l] DopaPET scan may show changes in basal ganglia before MSA-C shows clinical parkinsonism
[m] Abele et al. (2000a)
[n] Abele et al. (2000b)
[o] Wenning et al. (2000)
[p] Abele et al. (2000a)
[q] Askmark et al. (2001)
[r] Burk et al. (1997)
[s] Berciano (1982)
[t] Abele et al. (2000b)

### 3. Generic Chronic, Progressive Ataxias

Harding (1984) greatly simplified earlier phenotypic classifications of the chronic, progressive ataxias by eliminating the use of eponyms. The spinocerebellar degenerations were grouped by age of onset (under 5 years, between 5 and 25 years, over 25 years), then subclassified by known metabolic defect (inborn error of metabolism or DNA repair—weighted toward the earlier onset ataxias) or unknown cause. Those with unknown cause were further divided into the predominantly cerebellar forms and the "pure" and "complicated" spastic paraplegias. The cerebellar forms were categorized as (1) Friedreich's ataxia-like, (2) other "early-onset" ataxias (with specific features different from Friedreich's—retained reflexes, X-linked inheritance, unusual associated signs), (3) autosomal dominant cerebellar ataxias of late onset (ADCA I, II, III, IV, V, and those with episodic features), and (4) idiopathic late-onset cerebellar ataxia (type A—frequently with dementia, type B—with prominent tremor, and type C—with clear brainstem involvement; the latter more consistent with sporadic olivopontocerebellar atrophy, sOPCA or MSA, as observed clinically). The ADCAs were type I—"typical"—with variable inclusion of optic atrophy/ophthalmoplegia/dementia/extrapyramidal features/amyotrophy; type II—with pigmentary retinal degeneration; type III—"pure cerebellar"; type IV—with myoclonus and deafness type V—with essential tremor. Features of the dominant ataxias that further set them apart were the presence of episodic features (type VI) or severe dementia (which group would later come to include the familial prion disease, Gerstmann-Strauussler-Scheinker disease; (Ghetti et al., 1995). When known acquired causes have been ruled out, Harding's classification has continued to be the phenotypic "hook" with which one can dip into the multiple possible presentations of progressive ataxia and pull out an organized diagnostic strategy (see Tables 25.7 and 25.8).

TABLE 25.8  Phenotypic Templates in Designing the Work-Up of the Ataxic Patient

| Phenotype | Representative diseases/syndromes |
|---|---|
| Congenital, etiology unknown after imaging, EEG | These are usually associated with mental retardation and more generalized developmental and neurologic dysfunction (e.g., Marinesco-Sjogren syndrome). Some may be infantile onset forms of ADCA I or II (SCA2,[a] SCA7[b]). |
| Early (or later) onset, with no diagnosis on MRI | |
|   Friedreich ataxia-like (spinal recessive or sporadic) | Friedreich ataxia (FRDA) |
| | Late-onset Tay-Sachs (LOTS) |
| | Vitamin E associated syndromes (AVED, abeta/hypobetalipoproteinemia) |
|   EOCARR (central or spinal dominant, recessive, or X-linked) | Ataxia-telangiectasia and related syndromes |
| | Complicated hereditary spastic paraplegia (e.g. ARSACS) |
| | Various inborn and acquired metabolic defects (lysosomal, peroxisomal)[c] |
|   With mitochondrial features | Kearns-Sayre, LHON, NARP, MELAS, MERRF, and others not yet known to be mitochondrially based (infantile-onset spinocerebellar ataxia with sensory neuropathy—IOSCA[d]) |
|   With dementia or seizures | Mitochondrial disorders |
| | Neuronal ceroid lipofuscinosis |
| | Niemann-Pick type C, sialidosis |
| | Prion-related diseases |
| | SCA2, 10, 13, 17, DRPLA |
| | Whipple disease |
| | Wilson disease, hereditary hemochromatosis |
|   With episodic features | Episodic ataxia (EA)1,2,3,4 (dominant inheritance) |
| | Inborn errors of pyruvate/lactate, amino acid, or ammonia metabolism |
| | Labyrinthine abnormalities |
| | Vertebrobasilar insufficiency or migraine |
| Late (or earlier) onset, with no diagnosis on MRI | |
|   Dominant—ADCA I (typical) | SCA1, 2, 3, 4, 18; Gerstmann-Straussler-Scheinker |
|            ADCA II (retinal degeneration) | SCA7 (in about 94% with the ADCA II phenotype5) |
|            ADCA III (pure cerebellar) | SCA5, 6, 8, 10, 11, 15 |
|            ADCA IV (myoclonus, deafness) | SCA14, DRPLA, mitochondrial disorders |
|            ADCA V (essential tremor) | SCA12, 16 |
|            ADCA VI (episodic) | EA1, 2, 3, 4; early SCA6 |
|   Idiopathic—Type A (dementia) | Any of the ADCA Is or ADCA IIIs; parenchymatous cerebellar cortical atrophy |
|            Type B (tremor) | Any of the ADCA Vs |
|            Type C (sOPCA/MSA) | The true sOPCA/MSA group or SCA3 |

[a] Babovic-Vuksanovic et al. (1998)
[b] Benton, et al. (1998)
[c] Arylsulfatase C deficiency, cholestanolosis, partial HGPRT deficiency, leukodystrophies (metachromatic, globoid, adrenomyeloneuropathy, Refsum), Niemann-Pick type C, Wilson disease.
[d] Lonnqvist et al. (1998)
[e] Giunti et al. (1999)

With the advent of molecular genetic testing, we now recognize that not all Friedreich' ataxia or other recessively inherited metabolic disease (Gray et al., 2000) is early onset, that some "idiopathic" late-onset cerebellar ataxias are genetic (Moseley et al., 1998), and that some ataxia genotypes can have variable phenotypes. The balancing act between "lumping" and "splitting" will continue to be refined by the use of molecular diagnosis as the bottom line.

## II. DETERMINING WHETHER THE DISEASE IS GENETIC

### A. Is the Neurologic Phenotype Consistent with a Genetic Cause?

Hereditary ataxias are usually slowly progressive and associated with gait ataxia and cerebellar atrophy on imaging. There are few distinguishing features that allow their differentiation from the acquired ataxias or from each other. Even the family history (or lack thereof) can be misleading, for instance, in a family with more than one type of ataxia present (MSA in one generation and FRDA in the next) or in conditions of recessive or X-linked inheritance with only one affected individual identified. Even in families with a known dominant ataxia, an affected individual may actually have an acquired problem (we have one patient from a Machado-Joseph's family, carried with the dual diagnosis of Joseph's disease and multiple sclerosis, who was found after the advent of SCA3 testing to have only multiple sclerosis). It is appropriate for all ataxic patients to obtain a detailed medical and neurologic history, a complete physical and neurologic examination, MRI brain and cervical cord imaging, and supportive electrophysiological studies as needed. A detailed family history can provide important direction for further DNA-based testing, but an apparent "sporadic" presentation should not exclude the use of gene tests.

### B. Constructing an Accurate Pedigree

A family history should be obtained with at least three generations represented and detailed attention paid to family members with neurologic symptoms. Even commonplace complaints like migraine, motion sickness, Parkinson's disease, stroke, or neuropathy can be meaningful, when the gene being considered has variable penetrance. Confirmation (or exclusion) of neurologic concerns can be accomplished by meeting with and examining those individuals or by reviewing their medical records, imaging studies, gene test reports, or autopsy findings. Problems in constructing an accurate pedigree include unavailable information (loss of contact with other branches of the family, paternity issues, adoption, or willful hiding of the family history by the informant), incomplete penetrance of the disease, or anticipation of a triplet-repeat-based disease from the indeterminate range into the symptomatic range ("new dominant mutation"). Other explanations for an apparent sporadic presentation of a possibly gene-based ataxia are recessive or maternal inheritance with only one affected known (careful screening for consanguinity or ethnic factors can be helpful). Truly sporadic ataxias remain more common than all the inherited ataxias combined. See Table 25.9 for the distribution of specific ataxia genotypes worldwide.

### C. Choosing Appropriate Gene Tests or Obtaining Other Confirmation of an Inherited Disorder

Non-gene-based tests are available to help confirm the diagnosis in a number of the recessively inherited ataxias, primarily the inborn errors of metabolism with biochemical markers in blood, urine, or CSF (see Table 25.8). Even when the mutant gene is known, screening for a causative point mutation is difficult for most recessive alleles, outside of specific ethnic groups with founder effects, due to the possible presence of multiple disease-causing alleles (Concannon and Gatti, 1997). Among the recessive ataxias, Friedreich's ataxia is the exception, with the screenable GAA repeat expansion causing disease in 97% of patients. DNA sequencing as a way to confirm the presence of a specific point mutation has limited value, due to issues of cost and availability (many sequencing tests, for instance, ataxia-telangiectasia, episodic ataxia type 2(EA-2), and Friedreich's ataxia are done only in a few research laboratories).

DNA testing on blood is clinically available for ten of the ADCAs (SCA1, 2, 3, 6, 7, 8, 10, 17, DRPLA), for Friedreich's ataxia type 1, for Baltic myoclonic epilepsy (EPM1), and for four mitochondrial disorders (LHON, MELAS, MERRF, and NARP). Muscle tissue can be obtained and sent for clinical analysis for mitochondrial DNA rearrangements (Kearns-Sayre syndrome) or mitochondrial enzyme biochemistry. Individual DNA tests cost between $250–500, with some usual reduction in total cost if a "panel" is ordered. The first DNA test to become clinically available for hereditary spastic paraplegia is spastin/SPG4, at a cost of over $2000, as it is a DNA sequencing assay. Health insurance companies may not always pay for genetic testing, and patients frequently do not want their insurance companies to know they are being screened for a genetic disease, due to concerns about confidentiality and continued insurability. Cost of DNA testing may be a significant factor in lack of compliance with a recommended ataxia evaluation.

Clinical DNA screening for the ADCAs can reveal a specific diagnosis in about half the cases where a dominant inheritance pattern is known, but in less than a third where inheritance is not noted. Submissions (classic phenotype or late-onset sporadic ataxia) for Friedreich's ataxia X25 screening yielded 20–44% confirmation (Gunaratne and Richards, 1997; Jones and Seltzer, 2000). Not all clinical molecular genetic laboratories use the same ranges for normal, intermediate, and expanded nucleotide repeat alleles, so borderline results might require consultation with a specialist or review of the literature, to determine the significance of the result in an individual patient (Potter and Nance, 2000).

A useful (although not comprehensive) list of clinical and research laboratories performing various DNA tests for the inherited ataxias can be found at www.geneclinics.org, sponsored by the School of Medicine and Children's Hospital of the University of Washington at Seattle.

Because of the clinical overlap in the ADCAs, DNA testing for a patient with a family history consistent with autosomal dominant inheritance usually includes the typical SCAs (SCA1, 2, 3, 6, 8, 10, 17) and often SCA7 and DRPLA. A clinician (or the testing laboratory) may request

TABLE 25.9  Distribution of Ataxia Genotypes in Large Ataxia Clinics Worldwide

| Clinic location | SCA1 | SCA2 | SCA3 | SCA6 | SCA7 | FRDA | Other |
|---|---|---|---|---|---|---|---|
| North America |  |  |  |  |  |  |  |
| Canada[a] |  |  |  |  |  |  |  |
|   All patients | 3% | — | 41% | — | — | — | DRPLA 1% |
|   Non–Portuguese | 10% | — | 17% | — | — | — | — |
| United States |  |  |  |  |  |  |  |
|   Baylor | — | 18%[d] | — | — | — | 44%[e] | — |
|   Minnesota[b] | 5.6% | 15.2% | 20.8% | 15.2% | 4.5% | 16.6% | Sporadic cases 4.4% SCA+ 5.2% FRDA+ |
|   UCLA[c] | 6% | 13% | 23% | 12% | 5% | 62% | No SCA8, 12, or DRPLA found[g] |
| South America |  |  |  |  |  |  |  |
|   Brazil, south[h] | 0% | 0% | 92%[i] | 0% | 2% | — | Sporadic cases 7.1% SCA8 |
| Northern Europe |  |  |  |  |  |  |  |
|   UK–London[j] | 37%[p] | 47%[q] | 15%[r] | — | 94%[s] | — | — |
|   Cambridge[k] | 9% | 27.3% | 4.5% | 31.8% | — | — | DRPLA 9% Sporadic cases 12% SCA+ 3% FRDA+ |
|   Germany[l,m] | 9% | 10–14% | 42% | 22% | — | — | Sporadic cases 0.8%–1.8% SCA2[t] 6.6–7.3% SCA6[u] 2.4% SCA8[v] |
|   Netherlands[n] | 10.9% | 10.2% | 46.7% | 20.4% | 11.7% | — | — |
|   Russia[o] | 33% | — | 0% | — | — | — | — |
| Mediterranean |  |  |  |  |  |  |  |
|   Italy |  |  |  |  |  |  |  |
|     Valle d'Aosta[w] | — | — | — | — | — | 11.8% | EOCARR 5.9% |
|     Milan[x] | 41% | 29% | 0% | — | — | — | — |
|     Naples[y] | 24% | 47% | 0% | 2% | 2% | — | DRPLA 2% |
|     Sicily[z] | — | — | — | — | — | — | Sporadic cases 28.6% SCA2 |
|   Spain[aa] | 6% | 15% | 155 | 1% | 3% | — | DRPLA 1% |
|   Portugal[bb] | 0% | 4% | 74% | 0% | — | — | Sporadic cases 0% SCA 1.7% LOFA DRPLA 0% |
| Africa |  |  |  |  |  |  |  |
|   South Africa[cc] | 42.9% | — | 0% | — | — | — | Sporadic cases 4.5% SCA1 |
| Far East |  |  |  |  |  |  |  |
|   China[dd] | 4.7% | 5.9% | 48.2% | 0% | 0% | — | DRPLA 0% Sporadic cases 0% SCA |
|   India[ee] | 10.5% | 17.5% | 7% | 1.8% | 0% | — | DRPLA 0% |
|   Japan |  |  |  |  |  |  |  |
|     Hokkaido[ff] | 9.7% | 7.7% | 23.9% | 29% | 0% | 0% | DRPLA 2.6% SCA8 0% |
|     Kinki[gg] | 0% | — | 46.8% | — | — | — | Sporadic cases[oo] 22% SCA, most SCA6 |
|     Nagoya[hh] | 0% | 5.9% | 33.7% | 5.9% | - | — | DRPLA 19.8% |
|     Niigata[ii] | 3% | 5% | 43% | 11% | — | — | DRPLA 20% |
|     Tohoku[jj] | 24.8% | — | — | — | — | — | — |
|     Tottori[kk] | 15% | 0% | 5% | 25% | 0% | 0% | DRPLA 5% 11% SCA6 |
|   Korea[ll] | 0% | 12.6% | 4.6% | 6.9% | 0% | — | DRPLA 3.4% |
|   Singapore[mm] | — | 7.1% (Malay) | 35.7% Chinese | — | — | — | — |

(continues)

**TABLE 25.9** (continued)

| Clinic location | SCA1 | SCA2 | SCA3 | SCA6 | SCA7 | FRDA | Other |
|---|---|---|---|---|---|---|---|
| Taiwan[nn] | 1.4–5.4% | 9.6–10.8% | 27.3–47.3% | 10.8% | 1.4–2.7% | — | DRPLA 1.4%<br>SCA8 0%<br>Sporadic cases<br>4.1% SCA6 |
| Australia[pp] | 16%<br>most of<br>Anglo-Celtic<br>descent | 6%<br>half of<br>Italian<br>descent | 12%<br>1/3 of<br>Chinese<br>descent | 17% | 2% |  | DRPLA 0%<br>Sporadic cases<br>For every 3 families found with SCA1, 2, 3,<br>or 6, one more would be found without a<br>family history |

Percentages represent proportion of total ADCA or of total Friedreich ataxia phenotypes tested.
[a]Silveira et al. (1996).
[b]Moseley et al. (1998)
[c]Geschwind et al. (1997a,b,c)
[d]Lorenzetti et al. (1997)
[e]Gunaratne and Richards (1997)
[f]Sample these tests were drawn from may have met fewer criteria for phenotypic Friedreich ataxia.
[g]Cholfin et al. (2001), Sobrido et al. (2001)
[h]Jardim et al. (2001)
[i]This is felt to reflect an Azorean founder effect.
[j]Giunti et al. (1998)
[k]Leggo et al. (1997). Percentages are calculated relative to the total of SCAs identified (#ADCAs in total sample was not given, due to incomplete family history data).
[l]Schols et al. (1997a)
[m]Riess et al. (1997)
[n]van de Warrenburg (2001)
[o]Illarioshkin et al. (1996)
[p]When ethnicity was taken into account, 50% of Italian patients had SCA1.
[q]When ethnicity was taken into account, 37% of Italian patients and 44% of East Indian patients had SCA2.
[r]When ethnicity was taken into account, 44% of East Indian patients had SCA3.
[s]This is the percentage relative to only families with the ADCA II phenotype.
[t]Schols et al. (1997b)
[u]Schols et al. (1998)
[v]Schols et al. (2000)
[w]Leone et al. (1995)
[x]Pareyson et al. (1999). Northern Italian descent showed more SCA1, southern Italian more SCA2.
[y]Filla et al. (2000). Northern Italian descent showed more SCA1, southern Italian more SCA2.
[z]Giuffrida et al. (1999)
[aa]Pujana et al. (1999)
[bb]Silveira et al. (1998)
[cc]Ramesar et al. (1997)
[dd]Tang et al. (2000)
[ee]Basu et al. (2000)
[ff]Sasaki et al. (2000)
[gg]Matsumura et al. (1996)
[hh]Watanabe et al. (1998)
[ii]Takano et al. (1998)
[jj]Onodera et al. (2000)
[kk]Mori et al. (2001)
[ll]Jin et al. (1999)
[mm]Tan et al. (2000)
[nn]Hsieh et al. (2000), Soong et al. (2001)
[oo]Futamura et al. (1998)
[pp]Storey et al. (2000)

that the more common SCAs (SCA1, 2, 3, and 6) be screened first, and the others only after negative results for the first group have been found. If there is a strong clinical indicator present (retinopathy, suggesting SCA7) or a known genotype in the family, single gene testing can be requested. Most clinical laboratories will hold the DNA sample for 6 months (and many research labs will bank DNA indefinitely), leaving open the option for additional gene testing in the future (Tan and Ashizawa, 2001).

When there are cases present in only one generation or in just one individual, suggesting recessive or maternal inheritance or sporadic occurrence, the phenotype might suggest a set of diagnostic studies (see Table 25.8). Where there is no family history or phenotypic clue, after acquired causes have been ruled out (see Section I.C.2.), DNA testing for SCAs and Friedreich's ataxia should be considered, due to the 5–10% probability that a patient with a sporadic ataxia may have one of these (Moseley et al., 1998).

When one of the episodic ataxias (1–4) is suspected, a therapeutic trial of either phenytoin (EA-1) or acetazolamide (EA-2 and 4) can further define the condition, in the absence of readily available molecular genetic testing (Steckley et al., 2001).

## D. What to do when the Initial Round of Testing is Uninformative

If extensive history and examination, imaging and electrical studies; and laboratory, biopsy, and DNA testing have not given an answer as to the cause of the patient's ataxia, it is worthwhile to reconsider the phenotype, look for incompatible findings, and search the literature for additional clues. Good sites for this are the various on-line medical libraries—Medline or PubMed; Online Mendelian Inheritance in Man (OMIM) at the National Center for Biotechnology Information, www.ncbi.nlm.nih.gov; the GeneReviews/Hereditary Ataxia Overview service of the University of Washington at www.geneclinics.org; and the National Ataxia Foundation web site at www.ataxia.org, with its links to various medical and scientific sites dealing with ataxia. This could lead to additional testing or biopsy, or to the decision to pursue a relationship with a research laboratory involved with the patient's phenotype. It is often useful to watch, wait, and manage symptoms (Perlman, 2000), as the phenotype may evolve with the passage of time and the diagnosis may become more obvious. This also gives the clinician time to establish a supportive relationship with the patient and family, which encourages communication and the sharing of information and enhances the management of diseases that are still too often relentlessly progressive and disabling.

## References

Abe, T., Abe, K., Aoki, M., Itoyama, Y., and Tamai, M. (1997). Ocular changes in patients with spinocerebellar degeneration and repeated trinucleotide expansion of spinocerebellar ataxia type 1 gene. *Arch. Ophthalmol.* **115**, 231–236.

Abele, M., Weller, M., Mescheriakov, S., Burk, K., Dichgans, J., and Klockgether T. (1999). Cerebellar ataxia with glutamic acid decarboxylase autoantibodies. *Neurology* **52**, 857–859.

Abele, M., Schulz, J. B., Burk, K., Topka, H., Dichgans, J., and Klockgether, T. (2000a). Nerve conduction studies in multiple system atrophy. *Eur. Neurol.* **43**, 221–223.

Abele, M., Schulz, J. B., Burk, K., Topka, H., Dichgans, J., and Klockgether, T. (2000b). Evoked potentials in multiple system atrophy (MSA). *Acta Neurol. Scand.* **101**, 111–115.

Adachi, M., Hosoya, T., Yamaguchi, K., Kawanami, T., and Kato, T. (2000). Diffusion- and T2-weighted MRI of the transverse pontine fibres in spinocerebellar degeneration. *Neuroradiology* **42**, 803–809.

Antunez, E., Estruch, R., Cardenal, C., Nicolas, J. M., Fernandez-Sola, J., and Urbano-Marquez, A. (1998). Usefulness of CT and MR imaging in the diagnosis of acute Wernicke's encephalopathy. *Am. J. Roentgenol.* **171**, 1131–1137.

Asahina, M., Yamaguchi, M., Fukutake, T., and Hattori, T. (2000). Sleep apnea in multiple system atrophy. *Nippon Rinsho* **58**, 1722–1727.

Askmark, H., Eeg-Olofsson, K., Johansson, A., Nilsson, P., Olsson, Y., and Aquilonius S. (2001). Parkinsonism and neck extensor myopathy: a new syndrome or coincidental findings? *Arch. Neurol.* **58**, 232–237.

Babovic-Vuksanovic, D., Snow, K., Patterson, M. C., and Michels, V. V. (1998). Spinocerebellar ataxia type 2 (SCA 2) in an infant with extreme CAG repeat expansion. *Am. J. Med. Genet.* **79**, 383–387.

Baloh, R. W., and Honrubia, V. (1990). "*Clinical Neurophysiology of the Vestibular System.*" F.A. Davis Company, Philadelphia.

Baloh, R. W., and Vinters, H. V. (1995). White matter lesions and disequilibrium in older people. II. Clinicopathologic correlation. *Arch. Neurol.* **52**, 975–981.

Baloh, R. W., Yee, R. D., and Boder, E. (1978). Eye movements in ataxia-telangiectasia. *Neurology* **28**, 1099–1104.

Baloh, R. W., Yue, Q., Socotch, T. M., and Jacobson, K. M. (1995). White matter lesions and disequilibrium in older people. I. Case-control comparison. *Arch. Neurol.* **52**, 970–974.

Basu, P., Chattopadhyay, B., Gangopadhaya, P. K., Mukherjee, S. C., Sinha, K. K., Das, S. K., Roychoudhury, S., Majumder, P. P., and Bhattacharyya N. P. (2000). Analysis of CAG repeats in SCA1, SCA2, SCA3, SCA6, SCA7 and DRPLA loci in spinocerebellar ataxia patients and distribution of CAG repeats at the SCA1, SCA2 and SCA6 loci in nine ethnic populations of eastern India. *Hum. Genet.* **106**, 597–604.

Becker-Catania, S. G., Chen, G., Hwang, M. J., Wang, Z., Sun, X., Sanal, O., Bernatowska-Matuszkiewicz, E., Chessa, L., Lee, E. Y., and Gatti, R. A. (2000). Ataxia-telangiectasia: phenotype/genotype studies of ATM protein expression, mutations, and radiosensitivity. *Mol. Genet. Metab.* **70**, 122–133.

Benomar, A., Le Guern, E., Durr, A., Ouhabi, H., Stevanin, G., Yahyaoui, M., Chkili, T., Agid, Y., and Brice, A. (1994). Autosomal-dominant cerebellar ataxia with retinal degeneration (ADCA type II) is genetically different from ADCA type I. *Ann. Neurol.* **35**, 439–444.

Benton, C. S., de Silva, R., Rutledge, S. L., Bohlega, S., Ashizawa, T., and Zoghbi, H. Y. (1998). Molecular and clinical studies in SCA-7 define a broad clinical spectrum and the infantile phenotype. *Neurology* **51**, 1081–1086.

Berciano, J. (1982). Olivopontocerebellar atrophy. A review of 117 cases. *J. Neurol. Sci.* **53**, 253–272.

Birnbaum, D. C., Shields, D., Lippe, B., Perlman, S., and Phillipart, M. (1989). Idiopathic central diabetes insipidus followed by progressive spastic cerebral ataxia. Report of four cases. *Arch. Neurol.* **46**, 1001–1003.

Boesch, S. M., Schocke, M., Burk, K., Hollosi, P., Fornai, F., Aichner, F. T., Poewe, W., and Felber, S. (2001). Proton magnetic resonance spectroscopic imaging reveals differences in spinocerebellar ataxia types 2 and 6. *J. Magn. Reson. Imaging* **13**, 553–559.

Bomont, P., Watanabe, M., Gershoni-Barush, R., Shizuka, M., Tanaka, M., Sugano, J., Guiraud-Chaumeil, C., and Koenig, M. (2000). Homozygosity mapping of spinocerebellar ataxia with cerebellar atrophy and peripheral neuropathy to 9q33-34, and with hearing impairment and optic atrophy to 6p21-23. *Eur. J. Hum. Genet.* **8**, 986–990.

Bradwell, A. R. (2000). Paraneoplastic neurological syndromes associated with Yo, Hu, and Ri autoantibodies. *Clin. Rev. Allergy Immunol.* **19**, 19–29.

Burk, K., Fetter, M., Skalej, M., Laccone, F., Stevanin, G., Dichgans, J., and Klockgether, T. (1997). Saccade velocity in idiopathic and autosomal dominant cerebellar ataxia. *J. Neurol. Neurosurg. Psychiatry* **62**, 662–664.

Burk, K., Fetter, M., Abele, M., Laccone, F., Brice, A., Dichgans, J., and Klockgether, T. (1999). Autosomal dominant cerebellar ataxia type I: oculomotor abnormalities in families with SCA1, SCA2, and SCA3. *J. Neurol.* **246**, 789–797.

Burk, K., Skalej, M., and Dichgans, J. (2001a). Pontine MRI hyperintensities ("the cross sign") are not pathognomonic for multiple system atrophy (MSA). *Mov. Disord.* **16**, 535.

Burk, K., Bosch, S., Muller, C. A., Melms, A., Zuhlke, C., Stern, M., Besenthal, I., Skalej, M., Ruck, P., Ferber, S., Klockgether, T., and Dichgans, J. (2001b). Sporadic cerebellar ataxia associated with gluten sensitivity. *Brain* **124**, 1013–1019.

Bushara, K. O., Goebel, S. U., Shill, H., Goldfarb, L. G., and Hallett, M. (2001). Gluten sensitivity in sporadic and hereditary cerebellar ataxia. *Ann. Neurol.* **49**, 540–543.

Buttner, N., Geschwind, D., Jen, J. C., Perlman, S., Pulst, S. M., and Baloh, R. W. (1998). Oculomotor phenotypes in autosomal dominant ataxias. *Arch. Neurol.* **55**, 1353–1357.

Castelnovo, G., Biolsi, B., Barbaud, A., Labauge, P., and Schmitt, M. (2000). Isolated spastic paraparesis leading to diagnosis of Friedreich's ataxia. *J. Neurol. Neurosurg. Psychiatry* **69**, 693.

Cavalier, L., Ouahchi, K., Kayden, H. J., Di Donato, S., Reutenauer, L., Mandel, J. L., and Koenig, M. (1998). Ataxia with isolated vitamin E deficiency: heterogeneity of mutations and phenotypic variability in a large number of families. *Am. J. Hum. Genet.* **62**, 301–310.

Chan, K. H., Vernino, S., and Lennon, V. A. (2001). ANNA-3 anti-neuronal nuclear antibody: marker of lung cancer-related autoimmunity. *Ann. Neurol.* **50**, 301–311.

Chapman, T., McKeel, D. W., Jr., and Morris, J. C. (2000). Misleading results with the 14-3-3 assay for the diagnosis of Creutzfeldt-Jakob disease. *Neurology* **55**, 1396–1397.

Cholfin, J. A., Sobrido, M. J., Perlman, S., Pulst, S. M., and Geschwind, D. H. (2001). The SCA12 mutation is a rare cause of spinocerebellar ataxia. *Arch. Neurol.* in press.

Collie, D. A., Sellar, R. J., Zeidler, M., Colchester, A. C., Knight, R., and Will, R. G. (2001). MRI of Creutzfeldt-Jakob disease: imaging features and recommended mri protocol. *Clin. Radiol.* **56**, 726–739.

Collinge, J. (2001). Prion diseases of humans and animals: their causes and molecular basis. *Annu. Rev. Neurosci.* **24**, 519–550.

Colosimo, C., Inghilleri, M., and Chaudhuri, K. R. (2000). Parkinson's disease misdiagnosed as multiple system atrophy by sphincter electromyography. *J. Neurol.* **247**, 559–561.

Concannon, P., and Gatti, R. A. (1997). Diversity of ATM gene mutations detected in patients with ataxia-telangiectasia. *Hum. Mutat.* **10**, 100–107.

Crooks, R., Mitchell, T., and Thom, M. (2000). Patterns of cerebellar atrophy in patients with chronic epilepsy: a quantitative neuropathological study. *Epilepsy Res.* **41**, 63–73.

de la Sayette, V., Bertran, F., Honnorat, J., Schaeffer, S., Iglesias, S., and Defer, G. (1998). Paraneoplastic cerebellar syndrome and optic neuritis with anti-CV2 antibodies: clinical response to excision of the primary tumor. *Arch. Neurol.* **55**, 405–408.

De Marinis, M., Stocchi, F., Gregori, B., and Accornero, N. (2000). Sympathetic skin response and cardiovascular autonomic function tests in Parkinson's disease and multiple system atrophy with autonomic failure. *Mov. Disord.* **15**, 1215–1220.

De Michele, G., Di Salle, F., Filla, A., D'Alessio, G., Ambrosio, G., Viscardi, L., Scala R., and Campanella, G. (1995). Magnetic resonance imaging in "typical" and "late onset" Friedreich's disease and early onset cerebellar ataxia with retained tendon reflexes. *Ital. J. Neurol. Sci.* **16**, 303–308.

DeJong, R. N. (1979). *The Neurological Examination,* 4th Edition. Harper & Row, Hagerstown, MD.

Druschky, A., Hilz, M. J., Platsch, G., Radespiel-Troger, M., Druschky, K., Kuwert, T. and Neundorfer B. (2000). Differentiation of Parkinson's disease and multiple system atrophy in early disease stages by means of I-123-MIBG-SPECT. *J. Neurol. Sci.* **175**, 3–12.

Engert, J. C., Berube, P., Mercier, J., Dore, C., Lepage, P., Ge B., Bouchard, J. P., Mathieu, J., Melancon, S. B., Schalling, M., Lander, E. S., Morgan, K., Hudson, T. J., and Richter, A. (2000). ARSACS, a spastic ataxia common in northeastern Quebec, is caused by mutations in a new gene encoding an 11.5-kb ORF. *Nat. Genet.* **24**, 120–125.

Etchebehere, E. C., Cendes, F., Lopes-Cendes, I., Pereira, J. A., Lima, M. C., Sansana, C. R., Silva,C. A., Camargo, M. F., Santos, A. O., Ramos, C. D., and Camargo, E. E. (2001). Brain single-photon emission computed tomography and magnetic resonance imaging in Machado-Joseph disease. *Arch. Neurol.* **58**, 1257–1263.

Farmer, T. W., Wingfield, M. S., Lynch, S. A., Vogel, F. S., Hulette, C., Katchinoff, B., and Jacobson, P. L. (1989). Ataxia, chorea, seizures, and dementia. Pathologic features of a newly defined familial disorder. *Arch. Neurol.* **46**, 774–779.

Filla, A., Mariotti, C., Caruso, G., Coppola, G., Cocozza, S., Castaldo, I., Calabrese, O., Salvatore, E., De Michele, G., Riggio, M. C., Pareyson, D., Gellera, C., and Di Donato, S. (2000). Relative frequencies of CAG expansions in spinocerebellar ataxia and dentatorubropallidoluysian atrophy in 116 Italian families. *Eur. Neurol.* **44**, 31–36.

Flanigan, K., Gardner, K., Alderson, K., Galster, B., Otterud, B., Leppert, M. F., Kaplan, C., and Ptacek, L. J. (1996). Autosomal dominant spinocerebellar ataxia with sensory axonal neuropathy (SCA4): clinical description and genetic localization to chromosome 16q22.1. *Am. J. Hum. Genet.* **59**, 392–399.

Fritzler, M. J., Kerfoot, S. M., Feasby, T. E., Zochodne, D. W., Westendorf, J. M., Dalmau, J. O., and Chan, E. K. (2000). Autoantibodies from patients with idiopathic ataxia bind to M-phase phosphoprotein-1 (MPP1). *J. Investig. Med.* **48**, 28–39.

Fujino, Y., Tsuboi, Y., Shimoji, E., Takahashi, M., and Yamada, T. (2000). Progressive cerebellar atrophy following acute antidepressant intoxication. *Rinsho Shinkeigaku* **40**, 1033–1037.

Futamura, N., Matsumura, R., Fujimoto, Y., Horikawa, H., Suzumura, A., and Takayanagi, T. (1998). CAG repeat expansions in patients with sporadic cerebellar ataxia. *Acta. Neurol. Scand.* **98**, 55–59.

Gahring, L. C., Rogers, S. W., and Twyman, R. E. (1997). Autoantibodies to glutamate receptor subunit GluR2 in nonfamilial olivopontocerebellar degeneration. *Neurology* **48**, 494–500.

Geschwind, D. H., Perlman, S., Figueroa, C. P., Treiman, L. J., and Pulst, S. M. (1997a). The prevalence and wide clinical spectrum of the spinocerebellar ataxia type 2 trinucleotide repeat in patients with autosomal dominant cerebellar ataxia. *Am. J. Hum. Genet.* **60**, 842–850.

Geschwind, D. H., Perlman, S., Figueroa, K. P., Karrim, J., Baloh, R. W., and Pulst, S. M. (1997b). Spinocerebellar ataxia type 6. Frequency of the mutation and genotype-phenotype correlations. *Neurology* **49**, 1247–1251.

Geschwind, D. H., Perlman, S., Grody, W. W., Telatar, M., Montermini, L., Pandolfo, M., and Gatti, R. A. (1997c). Friedreich's ataxia GAA repeat expansion in patients with recessive or sporadic ataxia. *Neurology* **49**, 1004–1009.

Ghetti, B., Dlouhy, S. R., Giaccone, G., Bugiani, O., Frangione, B., Farlow, M. R., and Tagliavini, F. (1995). Gerstmann-Straussler-Scheinker disease and the Indiana kindred. *Brain. Pathol.* **5**, 61–75.

Gilman, S. (2001). Biochemical changes in multiple system atrophy detected with positron emission tomography. **7**, 253–256.

Gilman, S., Little, R., Johanns, J., Heumann, M., Kluin, K. J., Junck, L., Koeppe, R. A., and An, H. (2000). Evolution of sporadic olivopontocerebellar atrophy into multiple system atrophy. *Neurology* **55**, 527–532.

Gilman, S., Low, P. A., Quinn, N., Albanese, A., Ben-Shlomo, Y., Fowler, C. J., Kaufmann, H., Klockgether, T., Lang, A. E., Lantos, P. L., Litvan, I., Mathias, C. J., Oliver, E., Robertson, D., Schatz, I., and Wenning, G. K. (1999). Consensus statement on the diagnosis of multiple system atrophy. *J. Neurol. Sci.* **163**, 94–98.

Giuffrida, S., Saponara, R., Restivo, D. A., Trovato Salinaro, A., Tomarchio, L., Pugliares, P., Fabbri, G., and Maccagnano, C. (1999a). Supratentorial atrophy in spinocerebellar ataxia type 2: MRI study of 20 patients. *J. Neurol.* **246**, 383–388.

Giuffrida, S., Saponara, R., Trovato Salinaro, A., Restivo, D. A., Domina, E., Papotto, M., Le Pira, F., Nicoletti, A., Trovato, A., and Condorelli, D. F. (1999b). Identification of SCA2 mutation in cases of spinocerebellar ataxia with no family history in mid-eastern Sicily. *Ital. J. Neurol. Sci.* **20**, 217–221.

Giunti, P., Stevanin, G., Worth, P. F., David, G., Brice, A., and Wood, N. W. (1999). Molecular and clinical study of 18 families with ADCA type II: evidence for genetic heterogeneity and de novo mutation. *Am J Hum Genet* 64, 1594–1603.

Giunti, P., Sabbadini, G., Sweeney, M. G., Davis, M. B., Veneziano, L., Mantuano, E., Federico, A., Plasmati, R., Frontali, M., and Wood, N. W. (1998). The role of the SCA2 trinucleotide repeat expansion in 89 autosomal dominant cerebellar ataxia families. Frequency, clinical and genetic correlates. *Brain* **121** ( Pt. 3), 459–467.

Gomez, C. M., Thompson, R. M., Gammack, J. T., Perlman, S. L., Dobyns, W. B., Truwit, C. L., Zee, D. S., Clark, H. B., and Anderson, J. H. (1997). Spinocerebellar ataxia type 6: gaze-evoked and vertical nystagmus, Purkinje cell degeneration, and variable age of onset. *Ann. Neurol.* **42**, 933–950.

Gouw, L. G., Digre, K. B., Harris, C. P., Haines, J. H., and Ptacek, L. J. (1994). Autosomal dominant cerebellar ataxia with retinal degeneration: clinical, neuropathologic, and genetic analysis of a large kindred. *Neurology* **44**, 1441–1447.

Grant, M. P., Cohen, M., Petersen, R. B., Halmagyi, G. M., McDougall, A., Tusa, R. J., and Leigh, R. J. (1993). Abnormal eye movements in Creutzfeldt-Jakob disease. *Ann. Neurol.* **34**, 192–197.

Gray, R. G., Preece, M. A., Green, S. H., Whitehouse, W., Winer, J., and Green, A. (2000). Inborn errors of metabolism as a cause of neurological disease in adults: an approach to investigation. *J. Neurol. Neurosurg. Psychiatry* **69**, 5–12.

Gunaratne, P. H., and Richards, C. S. (1997). Estimated contribution of known ataxia genes in ataxia patients undergoing DNA testing. *Genet. Test.* **1**, 275–278.

Hanna, M. G., Davis, M. B., Sweeney, M. G., Noursadeghi, M., Ellis, C. J., Elliot, P., Wood, N. W., and Marsden, C. D. (1998). Generalized chorea in two patients harboring the Friedreich's ataxia gene trinucleotide repeat expansion. *Mov. Disord.* **13**, 339–340.

Harding, A. E. (1984). *The Hereditary Ataxias and Related Disorders.* Churchill Livingstone, Edinburgh.

Herman-Bert, A., Stevanin, G., Netter, J. C., Rascol, O., Brassat, D., Calvas, P., Camuzat, A., Yuan, Q., Schalling, M., Durr A., and Brice, A. (2000). Mapping of spinocerebellar ataxia 13 to chromosome 19q13.3-q13.4 in a family with autosomal dominant cerebellar ataxia and mental retardation. *Am. J. Hum. Genet.* **67**, 229–235.

Holmberg, B., Kallio, M., Johnels, B., and Elam, M. (2001). Cardiovascular reflex testing contributes to clinical evaluation and differential diagnosis of Parkinsonian syndromes. *Mov. Disord.* **16**, 217–225.

Honnorat, J., Saiz, A., Giometto, B., Vincent, A., Brieva, L., de Andres, C., Maestre, J., Fabien, N., Vighetto, A., Casamitjana, R., Thivolet, C., Tavolato, B., Antoine, J., Trouillas, P., and Graus, F. (2001). Cerebellar ataxia with anti-glutamic acid decarboxylase antibodies: study of 14 patients. *Arch. Neurol.* **58**, 225–230.

Hsieh, M., Lin, S. J., Chen, J. F., Lin, H. M., Hsiao, K. M., Li, S. Y., Li, C. and Tsai, C. J. (2000). Identification of the spinocerebellar ataxia type 7 mutation in Taiwan: application of PCR-based Southern blot. *J. Neurol.* **247**, 623–629.

Ikeda, Y., Shizuka, M., Watanabe, M., Okamoto, K., and Shoji, M. (2000). Molecular and clinical analyses of spinocerebellar ataxia type 8 in Japan. *Neurology* **54**, 950–955.

Illarioshkin, S. N., Slominsky, P. A., Ovchinnikov, I. V., Markova, E. D., Miklina, N. I., Klyushnikov, S. A., Shadrina, M., Vereshchagin, N. V., Limborskaya, S. A., and Ivanova-Smolenskaya, I. A. (1996). Spinocerebellar ataxia type 1 in Russia. *J. Neurol.* **243**, 506–510.

Isashiki, Y., Kii, Y., Ohba, N., and Nakagawa, M. (2001). Retinopathy associated with Machado—Joseph disease (spinocerebellar ataxia 3) with CAG trinucleotide repeat expansion. *Am. J. Ophthalmol.* **131**, 808–810.

Ishibashi, M., Sakai, T., Matsuishi, T., Yonekura, Y., Yamashita ,Y., Abe, T., Ohnishi, Y., and Hayabuchi, N. (1998). Decreased benzodiazepine receptor binding in Machado-Joseph disease. *J. Nucl. Med.* **39**, 1518–1520.

Jacobs, D. A., Lesser, R. L., Mourelatos, Z., Galetta, S. L., and Balcer, L. J. (2001). The Heidenhain variant of Creutzfeldt-Jakob disease: clinical, pathologic, and neuroimaging findings. *J. Neuroophthalmol.* **21**, 99–102.

Jardim, L. B., Pereira, M. L., Silveira, I., Ferro, A., Sequeiros, J., and Giugliani, R. (2001a). Neurologic findings in Machado-Joseph disease: relation with disease duration, subtypes, and (CAG)n. *Arch. Neurol.* **58**, 899–904.

Jardim, L. B., Silveira, I., Pereira, M. L., Ferro, A., Alonso, I., do Ceu Moreira, M., Mendonca, P., Ferreirinha, F., Sequeiros, J., and Giugliani, R. (2001b). A survey of spinocerebellar ataxia in South Brazil—66 new cases with Machado-Joseph disease, SCA7, SCA8, or unidentified disease-causing mutations. *J. Neurol.* **248**, 870–876.

Jarquin-Valdivia, A. A., and Buchhalter, J. (2001). Delayed diagnosis of pediatric Langerhans' cell histiocytosis: case report and retrospective review of pediatric cases seen at Mayo Clinic. *J. Child. Neurol.* **16**, 535–538.

Jin, D. K., Oh, M. R., Song, S. M., Koh, S. W., Lee, M., Kim, G. M., Lee, W. Y., Chung, C. S., Lee, K. H., Im, J. H., Lee, M. J., Kim, J. W., and Lee, M. S. (1999). Frequency of spinocerebellar ataxia types 1,2,3,6,7 and dentatorubral pallidoluysian atrophy mutations in Korean patients with spinocerebellar ataxia. *J. Neurol.* **246**, 207–210.

Jones, J. G., and Seltzer, W. K. (2000). Mutation detection frequencies and allele size distributions for spinocerebellar and Friedreich's ataxia: a cumulative history of ataxia testing in a clinical reference laboratory. American Academy of Neurology, AAN meetings, San Diego.

Kazuta, T., Hayashi, M., Shimizu, T., Iwasaki, A., Nakamura, S., and Hirai, S. (2000). Autonomic dysfunction in Machado-Joseph disease assessed by iodine123-labeled metaiodobenzylguanidine myocardial scintigraphy. *Clin. Auton. Res.* **10**, 111–115.

Kim, G. M., Kim, S. E., and Lee, W. Y. (2000). Preclinical impairment of the striatal dopamine transporter system in sporadic olivopontocerebellar atrophy: studied with [(123)I]beta-CIT and SPECT. *Eur. Neurol.* **43**, 23–29.

Kitae, S., Murata, Y., Tachiki, N., Okazaki, M., Harada, T., and Nakamura, S. (2001). Assessment of cardiovascular autonomic dysfunction in multiple system atrophy. *Clin. Autonom. Res.* **11**, 39–44.

Klockgether, T., Wullner, U., Dichgans, J., Grodd, W., Nagele, T., Petersen, D., Schroth, G., Schmidt, O., and Voigt, K. (1993). Clinical and imaging correlations in inherited ataxias. *Adv. Neurol.* **61**, 77–96.

Klockgether, T., Skalej, M., Wedekind, D., Luft, A. R., Welte, D., Schulz, J. B., Abele, M., Burk, K., Laccone, F., Brice, A., and Dichgans, J. (1998). Autosomal dominant cerebellar ataxia type I. MRI-based volumetry of posterior fossa structures and basal ganglia in spinocerebellar ataxia types 1, 2 and 3. *Brain* **121** (Pt. 9), 1687–1693.

Konaka, K., Kaido, M., Okuda, Y., Aoike, F., Abe, K., Kitamoto, T., and Yanagihara, T. (2000). Proton magnetic resonance spectroscopy of a patient with Gerstmann-Straussler-Scheinker disease. *Neuroradiology* **42**, 662–665.

Kraft, E., Schwarz, J., Trenkwalder, C., Vogl, T., Pfluger, T., and Oertel, W. H. (1999). The combination of hypointense and hyperintense signal changes on T2-weighted magnetic resonance imaging sequences: a specific marker of multiple system atrophy? *Arch. Neurol.* **56**, 225–228.

Kubis, N., Durr, A., Gugenheim, M., Chneiweiss, H., Mazzetti, P., Brice, A., and Bouche, P. (1999). Polyneuropathy in autosomal dominant cerebellar ataxias: phenotype-genotype correlation. *Muscle Nerve* **22**, 712–717.

Kumagai, R., Kaseda, Y., Kawakami, H., and Nakamura, S. (2000). Electrophysiological studies in spinocerebellar ataxia type 6: a statistical approach. *Neuroreport* **11**, 969–972.

Labuda, M., Labuda, D., Miranda, C., Poirier, J., Soong, B. W., Barucha, N. E., and Pandolfo, M. (2000). Unique origin and specific ethnic distribution of the Friedreich ataxia GAA expansion. *Neurology* **54**, 2322–2324.

Leggo, J., Dalton, A., Morrison, P. J., Dodge, A., Connarty, M., Kotze, M. J., and Rubinsztein, D. C. (1997). Analysis of spinocerebellar ataxia types 1, 2, 3, and 6, dentatorubral-pallidoluysian atrophy, and Friedreich's ataxia genes in spinocerebellar ataxia patients in the UK. *J. Med. Genet.* **34**, 982–985.

Lennon, V. A., Kryzer, T. J., Griesmann, G. E., O'Suilleabhain, P. E., Windebank, A. J., Woppmann, A., Miljanich, G. P., and Lambert, E. H. (1995). Calcium-channel antibodies in the Lambert-Eaton syndrome and other paraneoplastic syndromes. *N. Engl. J. Med.* **332**, 1467–1474.

Leone, M., Bottacchi, E., D'Alessandro, G., and Kustermann, S. (1995). Hereditary ataxias and paraplegias in Valle d'Aosta, Italy: a study of prevalence and disability. *Acta Neurol. Scand.* **91**, 183–187.

Lonnqvist, T., Paetau, A., Nikali, K., von Boguslawski, K., and Pihko, H. (1998). Infantile onset spinocerebellar ataxia with sensory neuropathy (IOSCA): neuropathological features. *J. Neurol. Sci.* **161**, 57–65.

Lorenzetti, D., Bohlega, S., and Zoghbi, H. Y. (1997). The expansion of the CAG repeat in ataxin-2 is a frequent cause of autosomal dominant spinocerebellar ataxia. *Neurology* **49**, 1009–1013.

Luostarinen, L. K., Collin, P. O., Peraaho, M. J., Maki, M. J., and Pirttila, T. A. (2001). Coeliac disease in patients with cerebellar ataxia of unknown origin. *Ann. Med.* **33**, 445–449.

Macia, F., Yekhlef, F., Ballan, G., Delmer, O., and Tison, F. (2001). T2-hyperintense lateral rim and hypointense putamen are typical but not exclusive of multiple system atrophy. *Arch. Neurol.* **58**, 1024–1026.

Mascalchi, M., Tosetti, M., Plasmati, R., Bianchi, M. C., Tessa, C., Salvi, F., Frontali, M., Valzania, F., Bartolozzi, C., and Tassinari, C. A. (1998). Proton magnetic resonance spectroscopy in an Italian family with spinocerebellar ataxia type 1. *Ann. Neurol.* **43**, 244–252.

Matsuda, M., Tabata, K., Hattori, T., Miki, J., and Ikeda, S. (2001). Brain SPECT with 123I-IMP for the early diagnosis of Creutzfeldt-Jakob disease. *J. Neurol. Sci.* **183**, 5–12.

Matsumura, R., Takayanagi, T., Murata, K., Futamura, N., and Fujimoto, Y. (1996). Autosomal dominant cerebellar ataxias in the Kinki area of Japan. *Jpn. J. Hum. Genet.* **41**, 399–406.

Matsuura, T., Achari, M., Khajavi, M., Bachinski, L. L., Zoghbi, H. Y., and Ashizawa, T. (1999). Mapping of the gene for a novel spinocerebellar ataxia with pure cerebellar signs and epilepsy. *Ann. Neurol.* **45**, 407–411.

Maurer, J. T., Juncker, C., Baker-Schreyer, A., and Hormann, K. (1999). Sleep apnea syndromes in multiple system atrophy. *Hno.* **47**, 117–121.

McNeil, D. E., Linehan, W. M., and Glenn, G. M. (2001). Comorbid genetic diseases, von Hippel-Lindau disease and spinocerebellar ataxia type 2, confounding the diagnosis of cerebellar dysfunction in an adolescent. *Clin. Neurol. Neurosurg.* **103**, 216–219.

Milhorat, T. H., Chou, M. W., Trinidad, E. M., Kula, R. W., Mandell, M., Wolpert, C., and Speer, M. C. (1999). Chiari I malformation redefined: clinical and radiographic findings for 364 symptomatic patients. *Neurosurgery* **44**, 1005–1017.

Miyoshi, Y., Yamada, T., Tanimura, M., Taniwaki, T., Arakawa, K., Ohyagi, Y., Furuya, H., Yamamoto, M., Sakai, K., Sasazuki, T., and Kira, J. (2001). A novel autosomal dominant spinocerebellar ataxia (SCA16) linked to chromosome 8q22.1-24.1. *Neurology* **57**, 96–100.

Moreira, M. C., Barbot, C., Tachi, N., Kozuka, N., Uchida, E., Gibson, T., Mendonca, P., Costa, M., Barros, J., Yanagisawa, T., Watanabe, M., Ikeda, Y., Aoki, M., Nagata, T., Coutinho, P., Sequeiros, J., and Koenig M. (2001). The gene mutated in ataxia-ocular apraxia 1 encodes the new HIT/Zn-finger protein aprataxin. *Nat. Genet.* **29**, 189–193.

Mori, M., Adachi, Y., Kusumi, M., and Nakashima, K. (2001). A genetic epidemiological study of spinocerebellar ataxias in Tottori Prefecture, Japan. *Neuroepidemiology* **20**, 144–149.

Moschner, C., Perlman, S., and Baloh, R. W. (1994). Comparison of oculomotor findings in the progressive ataxia syndromes. *Brain* **117** (Pt. 1), 15–25.

Moseley, M. L., Benzow, K. A., Schut, L. J., Bird, T. D., Gomez, C. M., Barkhaus, P. E., Blindauer, K. A., Labuda, M., Pandolfo, M., Koob, M. D., and Ranum, L. P. (1998). Incidence of dominant spinocerebellar and Friedreich triplet repeats among 361 ataxia families. *Neurology* **51**, 1666–1671.

Munschauer, F. E., Loh, L., Bannister, R., and Newsom-Davis, J. (1990). Abnormal respiration and sudden death during sleep in multiple system atrophy with autonomic failure. *Neurology* **40**, 677–679.

Murata, T., Fujito, T., Kimura, H., Omori, M., Itoh, H., and Wada, Y. (2001). Serial MRI and 1H-MRS of Wernicke's encephalopathy: report of a case with remarkable cerebellar lesions on MRI. *Psychiatry Res.* **108**, 49–55.

Murata, Y., Kawakami, H., Yamaguchi, S., Nishimura, M., Kohriyama, T., Ishizaki, F., Matsuyama, Z., Mimori, Y., and Nakamura, S. (1998a). Characteristic magnetic resonance imaging findings in spinocerebellar ataxia 6. *Arch. Neurol.* **55**, 1348–1352.

Murata, Y., Yamaguchi, S., Kawakami, H., Imon, Y., Maruyama, H., Sakai, T., Kazuta, T., Ohtake, T., Nishimura, M., Saida, T., Chiba, S., Oh-i, T., and Nakamura, S. (1998b). Characteristic magnetic resonance imaging findings in Machado-Joseph disease. *Arch. Neurol.* **55**, 33–37.

Nagaoka, U., Takashima, M., Ishikawa, K., Yoshizawa, K., Yoshizawa, T., Ishikawa, M., Yamawaki, T., Shoji, S., and Mizusawa, H. (2000). A gene on SCA4 locus causes dominantly inherited pure cerebellar ataxia. *Neurology* **54**, 1971–1975.

Namekawa, M., Takiyama, Y., Ando, Y., Sakoe, K., Muramatsu, S., Fujimoto, K., Nishizawa, M., and Nakano, I. (2001). Choreiform movements in spinocerebellar ataxia type 1. *J. Neurol. Sci.* **187**, 103–106.

National Ataxia Foundation (1999). T-shirt Logo, in *Generations*, p. 44.

Navon, R., Argov, Z., and Frisch, A. (1986). Hexosaminidase A deficiency in adults. *Am. J. Med. Genet.* **24**, 179–196.

Nemeth, A. H., Bochukova, E., Dunne, E., Huson, S. M., Elston, J., Hannan, M. A., Jackson, M., Chapman, C. J., and Taylor, A. M. (2000). Autosomal recessive cerebellar ataxia with oculomotor apraxia (ataxia-telangiectasia-like syndrome) is linked to chromosome 9q34. *Am. J. Hum. Genet.* **67**, 1320–1326.

Nuwer, M. R., Perlman, S. L., Packwood, J. W., and Kark, R. A. (1983). Evoked potential abnormalities in the various inherited ataxias. *Ann. Neurol.* **13**, 20–27.

Odaka, M., Yuki, N., and Hirata, K. (2001). Anti-GQ1b IgG antibody syndrome: clinical and immunological range. *J. Neurol. Neurosurg. Psychiatry* **70**, 50–55.

Oh, A. K., Jacobson, K. M., Jen, J. C., and Baloh, R. W. (2001). Slowing of voluntary and involuntary saccades: an early sign in spinocerebellar ataxia type 7. *Ann. Neurol.* **49**, 801–804.

O'Hearn, E., Holmes, S. E., Calvert, P. C., Ross, C. A., and Margolis, R. L. (2001). SCA-12: Tremor with cerebellar and cortical atrophy is associated with a CAG repeat expansion. *Neurology* **56**, 299–303.

Onodera, O., Idezuka, J., Igarashi, S., Takiyama, Y., Endo, K., Takano, H., Oyake, M., Tanaka, H., Inuzuka, T., Hayashi, T., Yuasa, T., Ito, J., Miyatake, T., and Tsuji S. (1998). Progressive atrophy of cerebellum and brainstem as a function of age and the size of the expanded CAG repeats in the MJD1 gene in Machado-Joseph disease. *Ann. Neurol.* **43**, 288–296.

Onodera, Y., Aoki, M., Tsuda, T., Kato, H., Nagata, T., Kameya, T., Abe, K., and Itoyama, Y. (2000). High prevalence of spinocerebellar ataxia type 1 (SCA1) in an isolated region of Japan. *J. Neurol. Sci.* **178**, 153–158.

Pal, P., Taly, A. B., Nagaraja, D., and Jayakumar, P. N. (1995). Early onset cerebellar ataxia with retained tendon reflexes: a clinical, electrophysiological and computed tomographic study. *J. Assoc. Physicians India* **43**, 608–613.

Pareyson, D., Gellera, C., Castellotti, B., Antonelli, A., Riggio, M. C., Mazzucchelli, F., Girotti, F., Pietrini, V., Mariotti, C., and Di Donato, S. (1999). Clinical and molecular studies of 73 Italian families with autosomal dominant cerebellar ataxia type I: SCA1 and SCA2 are the most common genotypes. *J. Neurol.* **246**, 389–393.

Perlman, S. L. (2000). Cerebellar Ataxia. *Curr. Treat. Options. Neurol.* **2**, 215–224.

Petzinger, G. M., Figueroa, K. P., Jakowec, M. W., Fahn, S., and Pulst, S. M. (2000). Double mutants for Huntington and spinocerebellar ataxia type 2 (SCA2) in a large pedigree segregating both mutations. American Academy of Neurology, AAN meetings, San Diego.

Plazzi, G., Corsini, R., Provini, F., Pierangeli, G., Martinelli, P., Montagna, P., Lugaresi, E., and Cortelli, P. (1997). REM sleep behavior disorders in multiple system atrophy. *Neurology* **48**, 1094–1097.

Poser, S., Mollenhauer, B., Kraubeta, A., Zerr, I., Steinhoff, B. J., Schroeter, A., Finkenstaedt, M., Schulz-Schaeffer, W. J., Kretzschmar, H. A., and Felgenhauer, K. (1999). How to improve the clinical diagnosis of Creutzfeldt-Jakob disease. *Brain* **122** (Pt. 12), 2345–2351.

Potter, N. T., and Nance, M. A. (2000). Genetic testing for ataxia in North America. *Mol. Diagn.* **5**, 91–99.

Pujana, M. A., Corral, J., Gratacos, M., Combarros, O., Berciano, J., Genis, D., Banchs, I., Estivill, X., and Volpini, V. (1999). Spinocerebellar ataxias in Spanish patients: genetic analysis of familial and sporadic cases. The Ataxia Study Group. *Hum. Genet.* **104**, 516–522.

Ramesar, R. S., Bardien, S., Beighton, P., and Bryer, A. (1997). Expanded CAG repeats in spinocerebellar ataxia (SCA1) segregate with distinct haplotypes in South african families. *Hum. Genet.* **100**, 131–137.

Riess, O., Laccone, F. A., Gispert, S., Schols, L., Zuhlke, C., Vieira-Saecker, A. M., Herlt, S., Wessel, K., Epplen, J. T., Weber, B. H., Kreuz, F., Chahrokh-Zadeh, S., Meindl, A., Lunkes, A., Aguiar, J., Macek, M., Jr., Krebsova, A., Macek, M., Sr., Burk, K., Tinschert, S., Schreyer, I., Pulst, S. M., and Auburger, G. (1997). SCA2 trinucleotide expansion in German SCA patients. *Neurogenetics* **1**, 59–64.

Rivaud-Pechoux, S., Durr, A., Gaymard, B., Cancel, G., Ploner, C. J., Agid, Y., Brice, A., and Pierrot-Deseilligny, C. (1998). Eye movement abnormalities correlate with genotype in autosomal dominant cerebellar ataxia type I. *Ann. Neurol.* **43**, 297–302.

Rosenfeld, M. R., Eichen, J. G., Wade, D. F., Posner, J. B., and Dalmau, J. (2001). Molecular and clinical diversity in paraneoplastic immunity to Ma proteins. *Ann. Neurol.* **50**, 339–348.

Rubio, A., Steinberg, K., Figlewicz, D. A., MacDonald, M. E., Greenamyre, T., Hamill, R., Shoulson, I., and Powers, J. M. (1996). Coexistence of Huntington's disease and familial amyotrophic lateral sclerosis: case presentation. *Acta Neuropathol. (Berlin)* **92**, 421–427.

Sakakibara, R., Hattori, T., Uchiyama, T., and Yamanishi, T. (2001). Videourodynamic and sphincter motor unit potential analyses in Parkinson's disease and multiple system atrophy. *J. Neurol. Neurosurg. Psychiatry* **71**, 600–606.

Sandok, E. K., O'Brien, T. J., Jack, C. R., and So, E. L. (2000). Significance of cerebellar atrophy in intractable temporal lobe epilepsy: a quantitative MRI study. *Epilepsia* **41**, 1315–1320.

Sasaki, H., Yabe, I., Yamashita, I., and Tashiro, K. (2000). Prevalence of triplet repeat expansion in ataxia patients from Hokkaido, the northernmost island of Japan. *J. Neurol. Sci.* **175**, 45–51.

Schols, L., Amoiridis, G., Buttner, T., Przuntek, H., Epplen, J. T., and Riess, O. (1997a). Autosomal dominant cerebellar ataxia: phenotypic differences in genetically defined subtypes? *Ann. Neurol.* **42**, 924–932.

Schols, L., Gispert, S., Vorgerd, M., Menezes Vieira-Saecker, A. M., Blanke, P., Auburger, G., Amoiridis, G., Meves, S., Epplen, J. T., Przuntek, H., Pulst, S. M., and Riess, O. (1997b). Spinocerebellar ataxia type 2. Genotype and phenotype in German kindreds. *Arch. Neurol.* **54**, 1073–1080.

Schols, L., Kruger, R., Amoiridis, G., Przuntek, H., Epplen, J. T., and Riess, O. (1998). Spinocerebellar ataxia type 6: genotype and phenotype in German kindreds. *J. Neurol. Neurosurg. Psychiatry* **64**, 67–73.

Schols, L., Szymanski, S., Peters, S., Przuntek, H., Epplen, J. T., Hardt, C., and Riess, O. (2000). Genetic background of apparently idiopathic sporadic cerebellar ataxia. *Hum. Genet.* **107**, 132–137.

Schols, L., Riess, O., Schols, S., Zeck, S., Amoiridis, G., Langkafel, M., Epplen, J. T., and Przuntek, H. (1995). Spinocerebellar ataxia type 1: Clinical and neurophysiological characteristics in German kindreds. *Acta Neurol. Scand.* **92**, 478–485.

Selim, M., and Drachman, D. A. (2001). Ataxia associated with Hashimoto's disease: progressive non-familial adult onset cerebellar degeneration with autoimmune thyroiditis. *J. Neurol. Neurosurg. Psychiatry* **71**, 81–87.

Shill, H. A., and Fife, T. D. (2000). Valproic acid toxicity mimicking multiple system atrophy. *Neurology* **55**, 1936–1937.

Shinotoh, H., Thiessen, B., Snow, B. J., Hashimoto, S., MacLeod, P., Silveira, I., Rouleau, G. A., Schulzer, M., and Calne, D. B. (1997). Fluorodopa and raclopride PET analysis of patients with Machado-Joseph disease. *Neurology* **49**, 1133–1136.

Silber, M. H., and Levine, S. (2000). Stridor and death in multiple system atrophy. *Mov. Disord.* **15**, 699–704.

Silveira, I., Coutinho, P., Maciel, P., Gaspar, C., Hayes, S., Dias, A., Guimaraes, J., Loureiro, L., Sequeiros, J., and Rouleau, G. A. (1998). Analysis of SCA1, DRPLA, MJD, SCA2, and SCA6 CAG repeats in 48 Portuguese ataxia families. *Am. J. Med. Genet.* **81**, 134–138.

Silveira, I., Lopes-Cendes, I., Kish, S., Maciel, P., Gaspar, C., Coutinho, P., Botez, M. I., Teive, H., Arruda, W., Steiner, C. E., Pinto-Junior, W., Maciel, J. A., Jerin, S., Sack, G., Andermann, E., Sudarsky, L., Rosenberg, R., MacLeod, P., Chitayat, D., Babul, R., Sequeiros, J., and Rouleau, G. A. (1996). Frequency of spinocerebellar ataxia type 1, dentatorubropallidoluysian atrophy, and Machado-Joseph disease mutations in a large group of spinocerebellar ataxia patients. *Neurology* **46**, 214–218.

Smitt, P. S., Kinoshita, A., De Leeuw, B., De Zeeuw, C., Jaarsma, D., Nakanishi, S., Coesmans, M., Frens, M., Linden, D., and Shigemoto, R. (2001). Paraneoplastic autoantibodies cause ataxia by functional blocking of the metabotropic glutamate receptor mGluR1. *N. Engl. J. Med.* in press.

Sobrido, M. J., Cholfin, J. A., Perlman, S., Pulst, S. M., and Geschwind, D. H. (2001). SCA8 repeat length in patients with inherited and sporadic ataxia. *Neurology* in press.

Sonstein, W. J., LaSala, P. A., Michelsen, W. J., and Onesti, S. T. (1996). False localizing signs in upper cervical spinal cord compression. *Neurosurgery* **38**, 445–448; discussion 448–449.

Soong, B., Liu, R., Wu, L., Lu, Y., and Lee, H. (2001a). Metabolic characterization of spinocerebellar ataxia type 6. *Arch. Neurol.* **58**, 300–304.

Soong, B. W., Lu, Y. C., Choo, K. B., and Lee, H. Y. (2001b). Frequency analysis of autosomal dominant cerebellar ataxias in Taiwanese patients and clinical and molecular characterization of spinocerebellar ataxia type 6. *Arch. Neurol.* **58**, 1105–1109.

Soong, B. W., and Liu, R. S. (1998). Positron emission tomography in asymptomatic gene carriers of Machado-Joseph disease. *J. Neurol. Neurosurg. Psychiatry* **64**, 499–504.

Soto-Ares, G., Vinchon, M., Delmaire, C., Abecidan, E., Dhellemes, P., and Pruvo, J. P. (2001). Cerebellar atrophy after severe traumatic head injury in children. *Childs Nerv. Syst.* **17**, 263–269.

Steckley, J. L., Ebers, G. C., Cader, M. Z., and McLachlan, R. S. (2001). An autosomal dominant disorder with episodic ataxia, vertigo, and tinnitus. *Neurology* **57**, 1499–1502.

Stevanin, G., Herman, A., Brice, A., and Durr, A. (1999). Clinical and MRI findings in spinocerebellar ataxia type 5. *Neurology* **53**, 1355–1357.

Stewart, G. S., Maser, R. S., Stankovic, T., Bressan, D. A., Kaplan, M. I., Jaspers, N. G., Raams, A., Byrd, P. J., Petrini, J. H., and Taylor, A. M.

(1999). The DNA double-strand break repair gene hMRE11 is mutated in individuals with an ataxia-telangiectasia-like disorder. *Cell* **99**, 577–587.

Storey, E., Gardner, R. J., Knight, M. A., Kennerson, M. L., Tuck, R. R., Forrest, S. M., and Nicholson, G. A. (2001). A new autosomal dominant pure cerebellar ataxia. *Neurology* **57**, 1913–1915.

Storey, E., du Sart, D., Shaw, J. H., Lorentzos, P., Kelly, L., McKinley Gardner, R. J., Forrest, S. M., Biros, I., and Nicholson, G. A. (2000). Frequency of spinocerebellar ataxia types 1, 2, 3, 6, and 7 in Australian patients with spinocerebellar ataxia. *Am. J. Med. Genet.* **95**, 351–357.

Takano, H., Cancel, G., Ikeuchi, T., Lorenzetti, D., Mawad, R., Stevanin, G., Didierjean, O., Durr, A., Oyake, M., Shimohata, T., Sasaki, R., Koide, R., Igarashi, S., Hayashi, S., Takiyama, Y., Nishizawa, M., Tanaka, H., Zoghbi, H., Brice, A., and Tsuji, S. (1998). Close associations between prevalences of dominantly inherited spinocerebellar ataxias with CAG-repeat expansions and frequencies of large normal CAG alleles in Japanese and Caucasian populations. *Am. J. Hum. Genet.* **63**, 1060–1066.

Tan, E. K., and Ashizawa, T. (2001). Genetic testing in spinocerebellar ataxias: defining a clinical role. *Arch. Neurol.* **58**, 191–195.

Tan, E. K., Law, H. Y., Zhao, Y., Lim, E., Chan, L. L., Chang, H. M., Ng, I., and Wong, M. C. (2000). Spinocerebellar ataxia in Singapore: predictive features of a positive DNA test? *Eur. Neurol.* **44**, 168–171.

Tang, B., Liu, C., Shen, L., Dai, H., Pan, Q., Jing, L., Ouyang, S., and Xia, J. (2000). Frequency of SCA1, SCA2, SCA3/MJD, SCA6, SCA7, and DRPLA CAG trinucleotide repeat expansion in patients with hereditary spinocerebellar ataxia from Chinese kindreds. *Arch. Neurol.* **57**, 540–544.

Taniwaki, T., Sakai, T., Kobayashi, T., Kuwabara, Y., Otsuka, M., Ichiya, Y., Masuda, K., and Goto, I. (1997). Positron emission tomography (PET) in Machado-Joseph disease. *J. Neurol. Sci.* **145**, 63–67.

Tomiyasu, H., Yoshii, F., Ohnuki, Y., Ikeda, J. E., and Shinohara, Y. (1998). The brainstem and thalamic lesions in dentatorubral-pallidoluysian atrophy: an MRI study. *Neurology* **50**, 1887–1890.

Uyama, E., Kondo, I., Uchino, M., Fukushima, T., Murayama, N., Kuwano, A., Inokuchi, N., Ohtani, Y., and Ando, M. (1995). Dentatorubral-pallidoluysian atrophy (DRPLA): clinical, genetic, and neuroradiologic studies in a family. *J. Neurol. Sci.* **130**, 146–153.

van de Warrenburg, B. P. (2001). Autosomal dominant cerebellar ataxias in the Netherlands: a national inventory. *Ned. Tijdschr. Geneeskd.* **145**, 962–967.

Vodusek, D. B. (2001). Sphincter EMG and differential diagnosis of multiple system atrophy. *Mov. Disord.* **16**, 600–607.

Watanabe, H., Tanaka, F., Matsumoto, M., Doyu, M., Ando, T., Mitsuma, T., and Sobue, G. (1998). Frequency analysis of autosomal dominant cerebellar ataxias in Japanese patients and clinical characterization of spinocerebellar ataxia type 6. *Clin. Genet.* **53**, 13–19.

Wenning, G. K., Ben-Shlomo, Y., Hughes, A., Daniel, S. E., Lees, A., and Quinn, N. P. (2000). What clinical features are most useful to distinguish definite multiple system atrophy from Parkinson's disease? *J. Neurol. Neurosurg. Psychiatry* **68**, 434–440.

Worth, P. F., Giunti, P., Gardner-Thorpe, C., Dixon, P. H., Davis, M. B., and Wood, N. W. (1999). Autosomal dominant cerebellar ataxia type III: linkage in a large British family to a 7.6-cM region on chromosome 15q14-21.3. *Am. J. Hum. Genet.* **65**, 420–426.

Yamashita, I., Sasaki, H., Yabe, I., Fukazawa, T., Nogoshi, S., Komeichi, K., Takada, A., Shiraishi, K., Takiyama, Y., Nishizawa, M., Kaneko, J., Tanaka, H., Tsuji, S., and Tashiro, K. (2000). A novel locus for dominant cerebellar ataxia (SCA14) maps to a 10.2-cM interval flanked by D19S206 and D19S605 on chromosome 19q13.4-qter. *Ann. Neurol.* **48**, 156–163.

Yen, T. C., Lu, C. S., Tzen, K. Y., Wey, S. P., Chou, Y. H., Weng, Y. H., Kao, P. F., and Ting, G. (2000). Decreased dopamine transporter binding in Machado-Joseph disease. *J. Nucl. Med.* **41**, 994–998.

Yokota, T., Sasaki, H., Iwabuchi, K., Shiojiri, T., Yoshino, A., Otagiri, A., Inaba, A., and Yuasa, T. (1998). Electrophysiological features of central motor conduction in spinocerebellar atrophy type 1, type 2, and Machado-Joseph disease. *J. Neurol. Neurosurg. Psychiatry* **65**, 530–534.

Yu, Z., Kryzer, T. J., Griesmann, G. E., Kim, K., Benarroch, E. E., and Lennon, V. A. (2001). CRMP-5 neuronal autoantibody: marker of lung cancer and thymoma-related autoimmunity. *Ann. Neurol.* **49**, 146–154.

# CHAPTER 26

# Parkinson's Disease: Genetic Epidemiology and Overview

**CONNIE MARRAS**
*The Parkinson's Institute*
*Sunnyvale, California 94089*

**CAROLINE TANNER**
*The Parkinson's Institute*
*Sunnyvale, California 94089*

I. Introduction
II. The Clinical and Pathological Features of Parkinson's Disease
III. Diagnosis of Parkinson's Disease
IV. The Treatment of Parkinson's Disease
V. Challenges Investigating the Etiology of Parkinson's Disease
VI. Studies of Familial Aggregation
  A. Studies in Populations not Selected by Familial Patterns
  B. Studies in Multiplex Families
VII. Twin Studies
VIII. Single Gene Associations
IX. Conclusion
  References

Parkinson's disease is a progressive neurodegenerative disorder characterized by tremor, rigidity, bradykinesia, and loss of postural reflexes. The etiology of Parkinson's disease is unknown. Theories of genetic, environmental, and complex gene/environment interactions have each predominated at various times in the past. While a family history of Parkinson's disease does appear to be more common in people with Parkinson's disease than in controls, most patients do not report affected family members. Genetic factors appear to be more important in young-onset cases, however. Causative genes have been found in a few families with parkinsonism inherited in a Mendelian fashion. The phenotype in these families tends to include some atypical features in at least a subset of affected individuals, so that the relationship of these cases to typical Parkinson's disease is uncertain. No single genes have been shown to play a significant role in the etiology of typical sporadic Parkinson's disease. Researchers have begun to investigate interactions between specific environmental factors and specific genes, guided by current theories of the pathophysiology of Parkinson's disease.

## I. INTRODUCTION

Parkinson's disease is a common neurodegenerative disorder of unknown etiology. Since first proposed by Gowers in 1888 and subsequently supported by many anecdotal reports, a hereditary contribution to the etiology of Parkinson's disease has been actively investigated. Theories of a genetic etiology to Parkinson's disease have not always dominated, however. During the past century, predominant etiologic theories for Parkinson's disease have shifted from heritability (Gowers, 1888), to an infectious etiology (Poskanzer and Schwab, 1963), back to heritability (Kondo et al., 1973; Martin et al., 1973; Barbeau and Pourcher, 1982), to an environmental etiology (Langston et al., 1983; Duvoisin, 1993), to a purely genetic etiology (Duvoisin and Golbe, 1995) to multifactorial theories of gene-environment interaction (Tanner and Langston, 1990; Semchuk et al., 1993; Sim et al., 1995; Tanner and Goldman, 1996). If the cause of Parkinson's disease were primarily genetic, then it would be expected that Parkinson's disease cases would cluster primarily within at-risk pedigrees, but few such pedigrees have been observed. A primary environmental cause, such as infection or toxicant exposure, classically causes clustering of cases in time or space, yet this is also rarely observed. If Parkinson's disease results from a combination of genetic and environmental factors then familial, temporal, and spatial disease clusters might not be observed. Although there is much support for genetic contributions to Parkinson's disease etiology, the precise determinants of most Parkinson's disease cases remain uncertain.

Recently, genetic defects responsible for the occurrence of some cases of parkinsonism have been identified (Polymeropoulos et al., 1996; Hattori et al., 1998b; Kitada et al., 1998). In many of these cases, the clinical features are indistinguishable from idiopathic Parkinson's disease. However, within affected families there are commonly clinical features in other individuals with clinical features which are unusual for Parkinson's disease. The relationship of these genetically defined entities to Parkinson's disease is uncertain, and to date they represent only a small fraction of recognized cases of parkinsonism. There is also support for genetic contributions to the etiology of typical Parkinson's disease through three major research strategies: familial studies, twin studies, and single gene associations. This chapter discusses the information that can be gained through each strategy, and summarizes the conclusions from familial and twin studies. The associations between Parkinson's disease and individual genes will be discussed in detail in Chapters 27–29. First, we will provide a brief introduction to the clinical features of Parkinson's disease and highlight the unique challenges presented by Parkinson's disease to the investigation of etiologic factors.

## II. THE CLINICAL AND PATHOLOGICAL FEATURES OF PARKINSON'S DISEASE

The cardinal symptoms of Parkinson's disease are tremor at rest, bradykinesia, rigidity, and postural instability. Parkinson's disease is largely a disorder of the elderly. The average age of symptom onset is in the late fifties in clinic series, later in community-based series. Both the incidence and prevalence increase dramatically beyond age 50 (Marras and Tanner, 2002). Pathologically, the most notable features of Parkinson's disease are the degeneration of dopaminergic neurons in the substantia nigra pars compacta and the formation of Lewy bodies within the degenerating neurons. Lewy bodies are eosinophilic inclusions which also stain positively with antibodies against α-synuclein, a protein which localizes to presynaptic nerve terminals. Abnormalities of α-synuclein have been associated with a rare genetic form of Parkinson's disease (Polymeropoulos et al., 1997).

The degeneration of the neurons of the substantia nigra results in a loss of dopaminergic input to the striatum. Loss of neurons projecting largely to the dorsal putamen are thought to account for the symptoms of akinesia and rigidity. There is also degeneration of selected brain stem nuclei (locus coeruleus, raphe nuclei, dorsal motor nucleus of the vagus), cortical neurons (particularly within the cingulate gyrus and the entorhinal cortex), the nucleus basalis of Meynert and preganglionic sympathetic and parasympathetic neurons. Degeneration in these regions likely accounts for the common non-motoric features of autonomic dysfunction, mood disturbance, and cognitive decline. Symptoms are generally slowly progressive, and the age at death of patients with Parkinson's disease is comparable or only slightly less than that of the general population (Roos et al., 1996). Nonetheless, symptoms usually become severe enough to interfere significantly with quality of life, and often independence (Guillard et al., 1986; Karlsen et al., 2000; Schrag et al., 2000).

There is much heterogeneity in the symptoms, age of onset, and rate of progression of Parkinson's disease (Marttila and Rinne, 1991; Payami et al., 1995). Such heterogeneity could be due to differences in genetic or environmental influences on the same underlying pathologic process. Alternatively, it may reflect different pathologic processes producing overlapping symptom complexes. Our understanding is currently limited by the lack of a diagnostic test for Parkinson's disease during life. This is compounded by the lack of pathological confirmation in most cases. As we begin to understand the genetic heterogeneity of Parkinson's disease the causes of clinical heterogeneity may become clearer. For example, in the late 1990s mutations in the *parkin* gene were found to cause an autosomal recessive disorder characterized by symptoms of parkinsonism which in many cases are indistinguishable from idiopathic Parkinson's disease. The clinical feature of this disorder that is most distinct from idiopathic Parkinson's disease is the young age at onset (Hattori et al., 1998a; Kitada et al., 1998). The discovery of this genetic defect has shed some light on the heterogeneity of the clinical disorder we call Parkinson's disease. Known genetic forms of parkinsonism constitute only a very small fraction (5% or less) of the cases, however (Lin et al., 1999; Chan et al., 2000; Oliveri et al., 2001). The determinants of most cases of Parkinson's disease, whether genetic or environmental, are yet to be determined. These determinants are a subject of intense investigation.

## III. DIAGNOSIS OF PARKINSON'S DISEASE

The diagnosis of Parkinson's disease is based on symptoms and physical signs. At this time, there is no diagnostic test for Parkinson's disease during life. Clinical diagnostic criteria have been developed, but these have been developed based on clinical experience of typical cases. Because there is no gold standard, the number of cases with the same pathological process excluded by these criteria because of mild or unusual manifestations is unknown. As structural imaging is usually normal in Parkinson's disease, conventional neuroimaging techniques such as CT or MRI are only helpful to rule out alternative causes of parkinsonism, such as cerebrovascular disease. Functional imaging, however, does reveal some characteristic abnormalities. Positron emission tomography (PET) and single photon emission computed tomography (SPECT) are imaging techniques that detect and display the distribution of radiolabelled tracers within the body. They can be used to provide quantitative or semiquantitative estimates of regional cerebral blood flow or receptor binding. Ligands that bind to presynaptic dopamine reuptake sites, such as fluorodopa or β-CIT, are taken up into nerve terminals projecting from the substantia nigra to the striatum. The amount of radiolabeled tracer accumulation in the striatum can thus be used as an indicator of the integrity of the nigrostriatal system. Patients with Parkinson's disease show reduced tracer accumulation in the striatum contralateral to the affected limbs using markers of the presynaptic dopaminergic system, with normal levels of accumulation of tracers with affinity for postsynaptic dopaminergic neurons. Reductions in activity with presynaptic ligands can be seen even before the onset of symptoms (Sawle et al., 1992; Piccini et al., 1999).

These functional neuroimaging techniques have the potential to provide a diagnostic test for Parkinson's disease, and even provide a presymptomatic test for individuals perceived to be at risk. However, the ability to distinguish normal from abnormal and to distinguish Parkinson's disease from other forms of parkinsonism have not yet been developed to the extent that these techniques can be used outside the research setting. Their limited availability is also a barrier. However, with further experience such techniques may be of value in diagnosis. These techniques have begun to be implemented in epidemiologic research in Parkinson's disease. This is relevant to the study of genetic and environmental risk factors, where incorrect classification of cases and controls can significantly alter results.

## IV. THE TREATMENT OF PARKINSON'S DISEASE

Treatments are currently available which considerably reduce the symptoms of Parkinson's disease. The most effective treatment strategies augment dopaminergic transmission, presumably compensating for the reduced dopaminergic innervation of the striatum by neurons from the substantia nigra. This can be achieved with levodopa, which is converted into dopamine within the central nervous system. The other mainstay of treatment is with dopamine receptor agonists, which have varying affinities for different subgroups of dopamine receptors. Most patients eventually require treatment with levodopa, which is the most potent pharmacologic treatment for Parkinson's disease. In the early stages of Parkinson's disease, these treatments are very effective and can significantly improve or completely abolish symptoms and signs. However, with time the duration of symptomatic improvement with each dose of levodopa is progressively reduced, causing parkinsonism to re-emerge before the next dose of medication. Such fluctuations in symptom control can also become sudden and unpredictable. Another type of motor complication, dyskinesia, is also experienced by many patients with Parkinson's disease during treatment with dopaminergic therapy. Dyskinesias are choreic or dystonic involuntary movements that most commonly occur when medications are working most effectively in reducing parkinsonism. Both dyskinesias and fluctuations can be disabling, and there is currently intensive research into the pathophysiology and determinants of both of these complications of therapy.

Individual genotype may prove important in choosing a therapeutic regimen (Maimone et al., 2001). A reduced risk of developing levodopa-induced dyskinesias has been suggested in patients carrying certain polymorphisms of the dopamine receptor D2 gene. (Oliveri et al., 1999) In these patients, initial therapy with levodopa may be acceptable, while a dopamine agonist may be preferred in patients who do not carry these protective polymorphisms. In the future, knowledge of the genotypes of dopamine receptor subtypes or variants in other genes may guide the choice of dopamine receptor agonists. At the present time, however, the choice of therapeutic agent remains largely an exercise in trial and error.

## V. CHALLENGES INVESTIGATING THE ETIOLOGY OF PARKINSON'S DISEASE

All clinical studies of Parkinson's disease are limited by the fact that there is no diagnostic test during life. This introduces uncertainty into any investigation relying entirely or primarily on clinical findings. In considering conclusions derived from genetic studies of Parkinson's disease, the following potential limitations should be remembered:

1. There is considerable evidence for the existence of "presymptomatic Parkinson's disease," yet there is no established biomarker for either a presymptomatic state nor for susceptibility. Some asymptomatic individuals who have been found to have reduced striatal fluorodopa uptake during PET scanning have later developed symptoms of Parkinson's disease (Sawle et al., 1992; Piccini et al., 1999). This suggests that degeneration of the nigro-striatal pathways supplying dopamine to the striatum begins well before symptoms appear. The finding of Lewy bodies in the brains of persons not known to have clinical evidence of Parkinson's disease

during life lends further support to the existence of a presymptomatic phase of disease (Forno, 1969; Gibb and Lees, 1988). If these incidental Lewy body cases and asymptomatic striatal dopamine depletion represent preclinical Parkinson's disease, then many persons with these pathologic or biochemical changes will be missed using current diagnostic methods. If these cases represent the earliest stages of Parkinson's disease, then misclassifying these presymptomatic individuals as unaffected can introduce bias into any study of disease etiology.

2. Parkinson's disease is a disorder of late life, and this further increases the chances of misclassification. First, individuals destined to be affected may never manifest symptoms before death. Second, individuals who develop parkinsonism shortly before death may never come to medical attention. Third, investigators often must rely on proxy reports by medical records, family, or close acquaintances to guide assessments of diagnosis in very unwell or deceased individuals.

3. There is significant inaccuracy in the diagnosis of Parkinson's disease, even by specialists in movement disorders. The symptoms of other disorders, such as progressive supranuclear palsy, multiple system atrophy, and corticobasal degeneration overlap with those of Parkinson's disease. Given that there is no way to confirm the diagnosis during life, this overlap of symptoms compromises the accuracy of diagnosis. Surveys of postmortem findings in cases clinically diagnosed as having Parkinson's disease found typical Parkinson's disease neuropathology in only 80–90% of cases (Rajput and Rozdilsky, 1991; Hughes et al., 1993; 2001). These results probably overestimate the actual misdiagnosis rate, however, since autopsy is most likely to be performed when there is a question regarding the clinical diagnosis (Maraganore et al., 1999a). Nonetheless, any series of subjects may have a significant number of patients with other diseases which may not share the same genetic determinants as Parkinson's disease. This makes studies susceptible to spurious results.

4. Variations in diagnostic criteria make it difficult to compare studies. For example, persons with "atypical" parkinsonism may have been classified as having Parkinson's disease in some earlier reports. Thus, some studies of "familial Parkinson's disease" may concern a different disorder with prominent parkinsonism. An example is frontotemporal dementia with parkinsonism, which appears to have genetic susceptibility factors distinct from Parkinson's disease. Inclusion of patients with this disorder in investigations of the determinants of Parkinson's disease could lead to false conclusions. Even when symptoms are not atypical, persons with isolated tremor or with both parkinsonism and dementia may be classified as either primary disorder, depending on the diagnostician.

With all of these factors combined, there can be significant uncertainty in the classification of many study subjects as affected or unaffected. This uncertainty increases the need for confirmation of results with additional studies in different populations or samples and, ultimately, by careful postmortem evaluation.

## VI. STUDIES OF FAMILIAL AGGREGATION

Familial patterns of disease can provide valuable clues to the importance of genetic factors to Parkinson's disease. Genetic influences can be studied in samples unselected by family history of disease, measuring the frequency of affected family members in cases versus controls or measuring the degree of "relatedness" of affected individuals versus the general population. Such investigations can give important information regarding the overall importance of genetic influences (when the population with Parkinson's disease is considered as a whole), however, they do not give insight into the specific genetic factors involved. Alternatively, familial patterns can be studied in families with multiple affected individuals (multiplex families) looking for patterns of chromosomal segregation which may point to the locations of causative genes. This strategy has successfully revealed several important loci and, ultimately, genes and proteins. However, results from this type of investigation (involving multiplex families) have uncertain generalizability to the whole population with Parkinson's disease. We will summarize the information provided to date by these two strategies.

### A. Studies in Populations not Selected by Familial Patterns

Multiple case-control studies have found that persons with Parkinson's disease more frequently report affected family members than do controls (Table 26.1). The odds of having Parkinson's disease if a family member has Parkinson's disease reported to be from 2.3–14.6 times greater than if no relatives are affected. Most of these studies involved patients in specialty clinics (Morano et al., 1994; Payami et al., 1994; Bonifati et al., 1995; Vieregge and Heberlein, 1995; De Michele et al., 1996; Taylor et al., 1999; Werneck and Alvarenga, 1999; Autere et al., 2000; Preux et al., 2000). Since attendance at such clinics may be more likely if there are unusual characteristics to the illness, including a familial pattern, these findings may not be generalizable to the whole population. Community-based studies have also shown an increased risk, but of lower magnitude (odds ratios 2.3–3.7; Semchuk et al., 1993; Marder et al., 1996; Elbaz et al., 1999). Consideration of relatives of different degree (first only, or first and second) may also account for some of the variability in odds ratios that has been observed.

TABLE 26.1  Risk of Parkinson's Disease Associated with a Family History of Disease: Case/Control studies

| Study | # Cases/controls | Odds ratio (95% confidence interval) | Study population characteristics |
|---|---|---|---|
| Semchuck et al., 1993 | 130/260 | 2.4 (1.03, 5.4)[b] | Population-based |
|  |  | 3.7 (1.7, 7.9)[c] |  |
| Morano et al., 1994 | 74/148 | 3.9 (1.3, 11.6) | Movement disorders clinic |
| Payami et al., 1994 | 114/114 | 3.5 (1.3, 9.4)[b] | Movement disorders clinic |
| Bonifati et al., 1995 | 100/100 | 4.9 (2.0, 11.9)[a] | Movement disorders clinic |
| Vieregge and Heberlein, 1995 | 66/72 | 7.1 (0.35, 25.2)[c] | Movement disorders clinic |
| Marder et al., 1996 | 233/1172 | 2.3 (1.3, 4.0)[b] | Population-based |
| De Michele et al., 1996 | 116/232 | 14.6 (7.2, 29.6)[a] | Movement disorders clinic |
| Elbaz et al., 1999 | 175/481 | 3.2 (1.6–6.6)[b] | Population-based |
| Taylor et al., 1999 | 140/147 | 3.9 (1.7, 8.9)[a] | Movement disorders clinic |
| Werneck and Alvarenga, 1999 | 92/110 | 14.5 (2.98, 91.38)[a] | Hospital neurology clinic |
| Preux et al., 2000 | 140/280 | 10.1 (2.9–35.0)[b] | Hospital clinic |
| Autere et al., 2000 | 268/210 | 2.9 (1.3–6.4)[b] | Hospital neurology clinic |

[a] Any relative.
[b] First-degree relatives.
[c] First- and second-degree relatives.

While these studies suggest a genetic contribution to the occurrence of Parkinson's disease, each has its limitations. First, it is unusual for family history data to be confirmed by examination. This leaves such studies susceptible to bias caused by affected individuals being more likely to recognize and report disease in their relatives than unaffected individuals. Relatives of an affected individual may also be more likely to come to medical attention because of heightened awareness of the significance of parkinsonian symptoms. Second, few have assessed the validity of family history data for Parkinson's disease. In one of the best studies, Marder et al (1996) found 100% specificity and sensitivity for family history when 57 siblings of Parkinson's disease patients were examined, but this was based on only three siblings with Parkinson's disease. The validity for more distant relatives is even more uncertain. Third, even if Parkinson's disease is more common among relatives of patients with Parkinson's disease, families share behaviors and environments as well as genes, therefore familial clustering can have nongenetic causes.

The role of genetic factors in the risk of Parkinson's disease has been investigated at the population level in Iceland and Finland. The population of Iceland provides a unique means of investigating the role of genetics in Parkinson's disease because of their long-standing database of genealogical information in a relatively genetically isolated population. After identifying 772 living and dead persons with parkinsonism over the preceding 50 years, genealogical information over 11 centuries was used to determine relatedness of those with parkinsonism and a control group (Sveinbjornsdottir et al., 2000). Parkinson's disease cases were more closely related than the controls, and Parkinson's disease risk was increased most in siblings (6.3 times); less so in offspring (3.0 times). The increased risk in siblings compared to children of cases supports a role for a common early environmental factor or, alternatively, recessive-inheritance. The overall influence of genetic factors found, however, was weak. This suggests that genetic influences play a modifying role in the risk of Parkinson's disease along with other, more important determining factors. The demographics of the population with Parkinson's disease identified in this Icelandic study differs from "typical" Parkinson's disease observed in other countries. Remarkably, most cases had a relatively early age of onset. The onset of symptoms was before the age of 50 in 212 of the 772 patients in their study. Perhaps a greater proportion of Icelandic parkinsonism is made up of inherited familial forms, as these typically have a younger age of onset.

Analysis of familial patterns of Parkinson's disease in Finland, a similarly isolated population, was performed through ascertainment of family history from 265 specialist-confirmed Parkinson's disease cases registered in a hospital database. A complex segregation analysis concluded that genetic factors were important in the pathogenesis of Parkinson's disease, more so in early-onset disease, and that contributing genetic factors were likely to be heterogeneous (Moilanen et al., 2001). The failure to find strong genetic determinants in older onset "typical" Parkinson's disease once again supports an important environmental determinant(s). Although the genetically isolated nature of the Finnish and Icelandic populations makes such extensive genealogical investigation possible, it also creates uncertainty about the degree to which these findings can be extended to other populations. Investigation in other populations, with careful clinical assessment of cases, will be an important next step.

Patterns of familial aggregation described in the studies to date support a model of gene-environment and/or gene-gene interactions contributing to the Parkinson's disease phenotype, as opposed to a simple Mendelian pattern of inheritance.

### B. Studies in Multiplex Families

Valuable information regarding genetic risk factors has been gained from studying families with Parkinson's disease occurring in several members (multiplex families). Parkinsonism occurring in at least three generations, seemingly inherited in an autosomal dominant fashion, has been described in a number of families (Golbe et al., 1990; Duvoisin, 1993; Wszolek et al., 1995; Kruger et al., 1998; Farrer et al., 2000; Gwinn-Hardy et al., 2000). However, potential limitations include: (1) the lack of diagnostic confirmation of many "affected" individuals by examination, leaving the diagnosis to rest on family-derived historical information; (2) the lack of autopsy confirmation in most cases; (3) the unusual clinical or pathological features of at least some of the cases in most families. Nonetheless, in many individual cases the disorder is indistinguishable from idiopathic Parkinson's disease, and studying these families may provide important clues to the cause of typical, apparently sporadic cases of Parkinson's disease through genetic linkage analysis. To date, genetic linkage analysis has revealed causative mutations in only a few of the families. The familial forms have been designated PARK 1 though 7. These are mentioned briefly here and are discussed in greater detail in Chapters 27–29.

### 1. PARK 1 (Mutation in α-Synuclein, 4q21–23)

An Italian-American family originating near Salerno, Italy, and referred to as the Contursi kindred has many members with parkinsonism. The inheritance pattern is consistent with autosomal dominant transmission with high penetrance (Golbe et al., 1990, 1996). Although the symptoms of many cases are indistinguishable from typical Parkinson's disease, at least 20% have clinically atypical features (Golbe et al., 1996). The pathology of this disorder has thus far shown all of the classical features of Parkinson's disease, although additional cortical pathology has been reported, and α-synuclein deposits appear to have a slightly different molecular composition (Langston et al., 1998; Spira et al., 2001). The specific causative mutation was identified in the gene encoding α-synuclein, which localizes to 4q21–23 (Polymeropoulos et al., 1997). A number of families of Greek origin are reported to have the same Ala53Thr mutation as the original family (Papadimitriou et al., 1999). A second mutation (Ala30Pro) in the α-synuclein gene has been identified in a German family, also with an autosomal dominant pattern of inheritance (Kruger et al., 1998). Similar to the Contursi kindred, the disease phenotype is somewhat atypical.

Searches for mutations in the α-synuclein gene in other families and in sporadic Parkinson's disease (Chan et al., 1998a, b, 2000; Farrer et al., 1998, 2000; Vaughan et al., 1998a, b; Lin et al., 1999; Nagar et al., 2001) have failed to identify additional cases, suggesting that α-synuclein mutations are a rare cause of Parkinson's disease. Nonetheless, the discovery of the association may give important clues to the pathophysiology of the more common sporadic form.

Recent support for this is suggested by the report of an association between a polymorphic variation in the promoter region of this gene and Parkinson's disease. Among 319 cases and 196 controls drawn from a Caucasian tertiary care clinic population, alleles of the α-synuclein promoter were distributed differently between cases and controls (Farrer et al., 2001). These polymorphisms appear to influence the expression of α-synuclein (Chiba-Falek and Nussbaum, 2001). Another case-control study of sporadic cases, however, has not found an association between α-synuclein genotype and Parkinson's disease (Khan et al., 2001). Investigation of the protein itself led to the discovery that α-synuclein is a component of Lewy bodies, another piece of evidence suggesting that this protein may play an important role in typical Parkinson's disease.

### 2. PARK 2 (Mutation in the *parkin* Gene, 6q25.2–27)

For years, Japanese neurologists have described a form of levodopa responsive "juvenile parkinsonism" (Yamamura et al., 1973; Yokochi et al., 1984). These patients typically have disease onset before the age of 30 and a positive family history compatible with autosomal recessive inheritance. They also have a high incidence of foot dystonia, levodopa-induced dyskinesias, and motor fluctuations, but otherwise their disease is quite similar to typical Parkinson's disease. Mutations of a gene on chromosome 6, called the *parkin* gene, have now been identified in several of these Japanese families (Hattori et al., 1998a; Kitada et al., 1998), as well as in European, Middle Eastern, and North African families (Lucking et al., 1998; Nisipeanu et al., 1999; van de Warrenburg et al., 2001). Few autopsies have been reported. Most do not find Lewy bodies (Yamamura et al., 1993; Takahashi et al., 1994), but atypical cases with both parkin mutations and Lewy bodies have been described (Farrer et al., 2001). The parkin gene product is a ubiquitin-protein ligase, functioning in protein degradation. (Mizuno et al., 2001) This finding may hold clues to the pathophysiology of typical, late-onset Parkinson's disease. Abnormalities in the cellular protein degradation system involving ubiquitin may be a unifying pathophysiologic feature in the cause(s) of Parkinson's disease.

*Parkin* mutations appear to account for a significant proportion of cases of very early onset parkinsonism. Of 73 families with at least one member having onset before age 46, 49% had *parkin* mutations. The age of onset can be as high as 58, although this is unusual (Lucking *et al.*, 2000). Even in early-onset patients without a family history, *parkin* mutations may be a relatively common finding, in keeping with its recessive mode of inheritance. Lucking *et al.* (2000) reported that of 100 cases without a family history beginning before age 46, 12 had *parkin* mutations. In contrast, no *parkin* mutations were found in a sample of 118 patients with typical later-onset (over 45 years of age) Parkinson's disease, even though 23 of these subjects reported a family history suggesting autosomal recessive inheritance (Oliveri *et al.*, 2001). Since at least 95% of parkinsonism in the community has onset after age 50 (Nelson *et al.*, 1997), *parkin* gene mutations remain a rare cause of parkinsonism.

## 3. PARK 3 (Chromosome 2p13)

In a study of six different families, all from a relatively small area in Northern Germany and Southern Denmark, parkinsonism showed linkage to 2p13, but with low penetrance (40%; Gasser *et al.*, 1998). Autopsies have been obtained in members from three different families and in each case Lewy bodies and nigral cell degeneration were observed. Dementia was present in cases from two out of the six families and neurofibrillary tangles and senile plaques were present on pathological examination. This gene remains to be identified. As with the α-synuclein gene, linkage studies in other families with multiple affected individuals have not found an association, suggesting that mutations at this locus are not a common cause of parkinsonism. (Farrer *et al.*, 2000; Gwinn-Hardy *et al.*, 2000).

## 4. PARK 4 (Chromosome 4p14–16.3)

In a family known as the Iowa kindred, the causative genetic abnormality for parkinsonism inherited in an autosomal dominant pattern has been mapped to 4p14–16.3 (Farrer *et al.*, 1999). Affected family members have a younger than average age of onset (mean 34.1 ± 8.8 years), a more rapid clinical course, and dementia early in the course of disease. Response to levodopa therapy is not always observed. Most remarkably, in this family, linkage is also observed for individuals with a single sign—a postural tremor resembling essential tremor. In the cases with parkinsonism, the few autopsied cases have demonstrated Lewy bodies in substantia nigra and hippocampus (Muenter *et al.*, 1998).

## 5. PARK 5 (Mutation in Ubiquitin Carboxy-Terminal Hydrolase-1, Chromosome 4p14–15)

A second gene has been tentatively associated with a form of young-onset parkinsonism, although clinical characterization is limited (Leroy *et al.*, 1998). This is a missense mutation in the ubiquitin carboxy-terminal hydrolase-L1 (UCH-L1) gene located on chromosome 4p, identified in two younger-onset patients with progressive parkinsonism. Studies in others with familial parkinsonism have not found this mutation (Harhangi *et al.*, 1999; Lincoln *et al.*, 1999). In a case-control investigation of 139 Parkinson's disease cases, no association with this mutation was found (Maraganore *et al.*, 1999b), suggesting that mutations in this gene are not a common cause of Parkinson's disease.

## 6. PARK 6 (Chromosome 1p35–36)

A Sicilian family has been identified with early-onset parkinsonism and autosomal recessive inheritance. Slow progression of disease and marked response to levodopa characterize the phenotype, much like parkinsonism associated with mutations of the *parkin* gene. No autopsy findings have yet been reported. The gene has yet to be identified. *Parkin* gene mutations were excluded, however, and linkage analysis identified a single region on the short arm of chromosome 1 as causative (Valente *et al.*, 2001).

## 7. PARK7 (Chromosome 1p36)

A family with early-onset autosomal recessive parkinsonism has been identified from The Netherlands (van Duijn *et al.*, 2001). As with PARK 6 and PARK 2, slow progression and levodopa responsiveness was noted. No autopsy information is yet available. Homozygosity mapping showed linkage to 1p36, near but distinct from the locus identified for PARK 6. Once again, the gene has not yet been identified.

The significance of PARK 5, 6, and 7 in relation to Parkinson's disease awaits verification of the underlying neuropathology of the disorders and the identification of additional affected families.

## 8. Genome-Wide Scans

To comprehensively examine chromosomal regions that may influence risk for Parkinson's disease, two genetic linkage studies have been reported which each used over 300 genetic markers spanning the genome (DeStefano *et al.*, 2001; Scott *et al.*, 2001). One study reported results based on 870 individuals from 174 multiplex families. These families were identified from 378 probands in clinics throughout the United States and Australia. Evidence for linkage to five chromosomal regions was found: chromosome 6 within the *parkin* gene in families with at least one individual with onset younger than 40 years, and chromosomes 17q, 8p, 5q, and 9q in families with later-onset disease (Scott *et al.*, 2001). The second study investigated linkage in 113 sibling pairs affected by Parkinson's disease, finding suggestive evidence for four different sites on chromosomes 1, 9, 10, and 16 (DeStefano *et al.*, 2001). Both sets of results suggest that multiple genetic factors may be involved in the etiology of Parkinson's disease. If the two study cohorts were comprised of somewhat different subsets of patients with parkinsonism, this could partially explain the differing results. For example, unintentional inclusion of patients with frontotemporal dementia and parkinsonism could explain linkage to a marker on 17q close to the tau gene, the strongest result obtained. Careful cognitive testing is now being done to explore this possibility. Neither study found strong evidence of linkage to any other site. The results are compatible with multiple contributing genetic factors to Parkinson's disease. More studies with large samples and careful examination of subjects will be required to demonstrate which results will be reproducible. Determining the generalizability of these findings to those without affected relatives will also be an important next step.

### 9. Mitochondrial Inheritance

Mitochondrial dysfunction has been proposed to underlie the pathogenesis of nerve cell death in Parkinson's disease (Schapira *et al.*, 1998, 1999; Kosel *et al.*, 1999). Complex I activity has been found to be decreased in various tissues from individuals with Parkinson's disease (Parker and Swerdlow, 1998). Morphological abnormalities in mitochondria within hybrid cell lines containing mitochondria from patients with Parkinson's disease have also been found (Trimmer *et al.* 2000). Because mitochondrial DNA is maternally inherited, a pattern of cases related through maternal lines is expected in disorders caused by defects in the mitochondrial genome. A few such families have been reported (Wooten *et al.*, 1997b; Swerdlow *et al.*, 1998) and one study found that in families with multiple affected siblings the affected parent was more often the mother (Wooten *et al.*, 1997a). Complex I activity in mitochondria from one multigenerational kindred with maternal inheritance was found to be lower than in mitochondria from persons with paternal inheritance. (Swerdlow *et al.*, 1997) Despite these suggestive findings, most families do not show evidence of matrilineal inheritance (Zweig *et al.*, 1992; Tanner *et al.*, 1999). Therefore inherited mitochondrial dysfunction may be a contributing factor in the etiology of Parkinson's disease but is unlikely to be an important cause in most cases. This does not, however, rule out acquired mitochondrial dysfunction as an important factor in the pathogenesis of Parkinson's disease.

All of the preceding strategies for investigating familial aggregation in Parkinson's disease have the potential to provide insight into the determinants of familial parkinsonism. However, at least 80% of those with Parkinson's disease do not report having other affected family members. Therefore, all investigations selecting cases on the basis of belonging to a family with two or more affected members are studying a select group of individuals whose disease determinants may be different from most cases of Parkinson's disease. To understand the relevance of these findings to typical Parkinson's disease, it will be important to test hypotheses of genetic linkage generated through studies in multiplex families on samples composed of typical, sporadic cases.

## VII. TWIN STUDIES

Monozygotic (MZ) twins provide a unique opportunity to assess the relative roles of genetic and environmental factors in disease etiology because they have identical nuclear DNA at conception. If genetic factors are solely responsible for determining disease occurrence, MZ twins should have identical susceptibility and should have 100% concordance for developing Parkinson's disease. Exceptions would be diseases due to mitochondrial DNA mutations, and diseases caused by acquired mutations in nuclear DNA. In diseases such as Parkinson's disease where genetic factors do not appear to be solely responsible, comparison of concordance rates in MZ and dizygotic (DZ) twins allows the relative influence of genes and environment to be determined. Intrapair concordance rates should be higher in MZ twins than in DZ twins if genetic factors are etiologically important in Parkinson's disease.

The first twin studies consistently showed similar low rates of clinical concordance in MZ and DZ twin pairs (Ward *et al.*, 1983; Marsden, 1987; Marttila *et al.*, 1988; Vieregge *et al.*, 1992). However, these studies had several limitations, including small, selected samples, lack of follow-up of discordant twin pairs and often, lack of confirmation of diagnosis. One study by Zimmerman *et al.* (1991) which performed a second follow-up of 20 monozygotic twin pairs highlighted these limitations. They reported an autopsy diagnosis of striatonigral degeneration in one twin initially classified as having Parkinson's disease, and found normal fluorodopa PET scan imaging in another subject who had been classified as affected by clinical examination. Clinical Parkinson's disease was also observed in one monozygotic co-twin after a discordance interval of 26 years. Studies have been performed more recently to overcome these limitations.

To investigate clinical concordance in a large, unselected sample of twins, a registry of approximately 32,000 Caucasian male twins (formed from World War II Veterans) was used to identify twin pairs with Parkinson's disease in one or both twins. Beginning in, 1993, the first effort to directly contact all living members of the cohort was begun. Those with suspected Parkinson's disease, as identified by telephone screening interviews, and their twin brothers were asked to undergo a standardized diagnostic evaluation performed by neurologists with expertise in Parkinson's disease (Tanner et al., 1999). 14,436 of 19,842 twins were interviewed; the remainder were either deceased or unable to be located with no proxy information obtainable. Of these, 193 cases of Parkinson's disease were identified in 163 pairs. MZ and DZ concordance did not differ overall. For twins with onset after age 50, concordance rates were essentially identical in MZ and DZ twins (11% each). In the young-onset cases (under age 50), however, MZ concordance was much increased (100% in 4 MZ twin pairs versus 16.7% in 12 DZ pairs). The overall similarity of concordance in monozygotic and dizygotic pairs is not consistent with a significant genetic cause of Parkinson's disease. In the less common young-onset parkinsonism, however, genetic factors appear to be primary. This is consistent with the family studies already described, in which younger age at onset is a more common clinical finding than in unselected series of patients with Parkinson's disease (Payami et al., 1995).

If some of the clinically unaffected twins actually had preclinical disease, the results may have been substantially altered if those cases had been investigated by imaging techniques or longer follow-up studies. Discrepancies of up to 28 years in age of onset between MZ twins and 31 years in DZ twins have been reported (Tanner et al., 1999). There have been a number of small twin studies investigating striatal dopamine function using neuroimaging. In several studies of twin pairs discordant for Parkinson's disease clinically, the asymptomatic twins had significantly decreased putamenal fluorodopa uptake measured by PET scan (Mark et al., 1991; Burn et al., 1992; Holthoff et al., 1994; Laihinen et al., 2000). These results suggest a higher concordance for dopamine system deficits than for clinical expression of parkinsonism. These studies did not find a higher concordance for fluorodopa PET scan abnormalities in MZ than DZ twins, however, arguing against classical Mendelian inheritance as a cause for the concordance. In one study of 34 twin pairs, decreased putamenal fluorodopa uptake was higher in MZ than DZ twins and the higher MZ concordance was most evident in the 19 pairs in whom longitudinal follow-up was possible (Piccini et al., 1999). A separate clinical and PET study of 23 twin pairs followed for 8 years found similar clinical concordance for Parkinson's disease between MZ and DZ twins (Vieregge et al., 1999). Twins with decreased fluorodopa uptake on prior PET scanning had not developed clinical Parkinson's disease, calling into question the clinical relevance of the finding of decreased putamenal fluorodopa uptake in an asymptomatic individual, at least over an eight-year period. Imaging was not repeated to know if the dopaminergic deficits previously found were progressive. These apparently contradictory findings highlight the importance of uniform long-term follow-up. A final interpretation of these different findings will require clarification in a larger, unselected population with longitudinal neuroimaging follow-up. Such an investigation is now underway in the large, unselected veteran twin cohort described above.

## VIII. SINGLE GENE ASSOCIATIONS

Single genes have been investigated for their association with Parkinson's disease based on their involvement in the metabolism of toxic xenobiotics or dopamine, their role in mitochondrial function, or their involvement in other neurodegenerative disorders. If disease susceptibility is found to be modified by certain genes this may give important clues to pathophysiology, and guide research into new treatment strategies.

Polymorphisms of many genes have been found to be associated with an increase or decrease in risk for Parkinson's disease in at least one or more studies (Table 26.2), although for most the results are conflicting. The conflicting results may be due to a number of factors:

1. Genetic susceptibility factors may vary according to patient characteristics (e.g., ethnicity, gender, early- versus late-onset disease)
2. Susceptibility factors which increase risk to a mild or moderate degree may require large samples to demonstrate positive results
3. Unknown gene-gene or gene-environment interactions may produce misleading results if cases and controls are not matched with respect to these factors
4. Subjects may be misclassified in the absence of postmortem evaluation
5. Studies may consider different polymorphisms of the same gene
6. Some studies may analyze allelic frequencies while others analyze genotype frequency
7. Studies may use subjects with familial or sporadic parkinsonism or mixed samples, and familial and sporadic disease may have different genetic risk factors
8. Low frequency alleles may be under or over-represented in small samples.

Genes and environmental factors may interact to determine the expression of Parkinson's disease in an individual. Several environmental factors have been consistently shown to modify the risk for Parkinson's disease in epidemiologic studies, including pesticide exposure (increasing risk), cigarette smoking, and caffeine

TABLE 26.2  Some Genes with Polymorphisms Associated with Risk of Parkinson's Disease in One or More Studies

| Function/rationale | Gene or gene product (Reference)[a] |
|---|---|
| Detoxification | Cytochrome P450 2D6, 1A1 (Checkoway et al., 1998a; Riedl et al., 1998)<br>Paraoxonase (Kondo and Yamamoto, 1998; Akhmedova et al., 1999, 2001; Taylor et al., 2000; Wang and Liu, 2000)<br>Glutathione transferase (Tan et al., 2000)<br>Superoxide dismutase (Checkoway et al., 1998a; Grasbon-Frodl et al., 1999b)<br>Alcohol dehydrogenase (Buervenich et al., 2000; Tan et al., 2001)<br>N-acetyl transferase (Tan et al., 2000) |
| Dopamine metabolism or function | Dopamine transporter (DAT1) (Tan et al., 2000)<br>Dopamine D2 receptor (Tan et al., 2000)<br>Dopamine D4 receptor (Tan et al., 2000)<br>Tyrosine hydroxylase (Kurth and Kurth, 1992; Plante-Bordeneuve et al., 1994)<br>Monoamine oxidase A and B (Tan et al., 2000)<br>Catechol-O-methyl transferase (Tan et al., 2000) |
| Causative in familial parkinsonism or other neurodegenerative disorders | Parkin (Satoh and Kuroda, 1999; Wang et al., 1999; Hu et al., 2000)<br>α-synuclein (Farrer et al., 2001)<br>Ubiquitin carboxy-terminal hydrolase-1 (Maraganore et al., 1999; Mellick and Silburn, 2000; Wintermeyer et al., 2000)<br>Apolipoprotein E (Tan et al., 2000)<br>Tau (Pastor et al., 2000; Martin et al., 2001)<br>α-1-antichymotrypsin (Yamamoto et al., 1997; Grasbon-Frodl et al., 1999a; Munoz et al., 1999; Wang et al., 2001) |
| Mitochondrial metabolism | tRNA genes (Checkoway et al., 1998; Tan et al., 2000)<br>Mitochondrial DNA-encoded complex 1 (ND2) (Checkoway et al. 1998a; Tan et al., 2000) |
| Other | Serotonin transporter (McCann et al., 2000)<br>TNF-alpha (Kruger et al., 2000) |

[a]References emphasize reviews where available.

intake (both decreasing risk; Marras and Tanner, 2002). Numerous other potential environmental factors have been identified, but consistency across studies has yet to be confirmed. If Parkinson's disease results from a combination of genetic and environmental factors, then an interaction of genetic factors with these exposures could result in a high level of disease risk. For example, increased risk from an environmental toxin could be influenced by the genetically determined level of activity of metabolizing enzymes. Few gene-environment interactions have been investigated. One case-control study suggests that smoking history modifies the effect of family history on the risk for Parkinson's disease, such that the odds ratio is highest in those with a history of smoking and a family history of Parkinson's disease (odds ratio 10.0; Elbaz et al., 2000). This is a surprising finding given the increasing body of evidence that smoking is negatively associated with the occurrence of Parkinson's disease. A possible interaction between monoamine oxidase B gene polymorphisms and smoking behavior has also been reported. A reduced risk of Parkinson's disease with increasing number of pack-years of smoking was found in the presence of the G allele, while Parkinson's disease risk *decreased* with increasing pack-years smoked in the presence of the A allele (Checkoway et al., 1998b). Interactions between xenobiotic metabolizing enzyme genotype and pesticide exposure in the risk of Parkinson's disease have also been studied (Menegon et al., 1998; Taylor et al., 2000). The association between pesticide exposure and Parkinson's disease may be modified by glutathione transferase P1 polymorphisms (Menegon et al., 1998). In their study of 96 patients and 95 controls, no overall difference in the distribution of GST P1 genotypes was found between cases and controls. In those with pesticide exposure, however, the GST P1 AA genotype was associated with the lowest risk for Parkinson's disease. Confirmation of each of these observations in additional populations is necessary.

Only a few gene-gene interaction investigations have been reported. Conflicting results have been found for an interaction between α-synuclein and apolipoprotein E polymorphisms. Among 158 patients with Parkinson's disease and 172 controls, the combined presence of a polymorphism within the α-synuclein promoter region and the ApoE4 allele was associated with an odds ratio for Parkinson's disease of 12.8 (Khan et al., 2001). No significant association between this combination of alleles and Parkinson's disease was found in a separate sample of 305 patients and 330 controls, however (Kruger et al., 1999). A study in Taiwanese patients studied genotypes for two enzymes involved in dopamine metabolism; catechol-O-methyl transferase (COMT) and monoamine oxidase type B (MAO-B). Their results suggested an association between MAO-B polymorphisms and the risk for Parkinson's disease that was modified by the COMT genotype (Wu et al., 2001). Patients carrying the combination of MAO-B G and COMT L alleles had significantly greater risk than other allelic combinations. One report has suggested a somewhat higher risk of Parkinson's disease (odds ratio = 3) in the presence of the combined MAO-B G allele and DRD2 A1 or B1 alleles (Costa-Mallen et al., 2000). If several different genes and environmental factors influence risk for Parkinson's disease, then systematic investigation of a large number of possible interactions will be required to fully understand the determinants of susceptibility.

## IX. CONCLUSION

The etiology of most cases of Parkinson's disease remains uncertain. Most studies do not support a major genetic contribution except for rare families and the small subgroup of patients with a younger age of onset. This field of research is complicated by the uncertain relationship between these unusual subgroups and most cases of Parkinson's disease (which are of older onset, in general, and do not have a strong family history). Nonetheless, multiple genes and various environmental factors may combine to determine the level of risk for Parkinson's disease in any one individual. Continued investigation of single gene associations and interactions between genes and environmental factors may provide important clues to disease etiology. Such investigations may ultimately lead to the discovery of new strategies for prevention or treatment. Investigations of genetic variations that determine treatment responsiveness could also provide guidance for the optimal treatment strategy for an individual, allowing us to maximize symptom control using the agents available to us today.

## References

Akhmedova, S., Anisimov, S. et al. (1999). Gln to Arg 191 polymorphism of paraoxonase and Parkinson's disease. *Hum. Hered.* 178–180.

Akhmedova, S. N., Yakimovsky, A. K. et al. (2001). Paraoxonase 1 Met—Leu 54 polymorphism is associated with Parkinson's disease. *J. Neurol. Sci.* **184(2)**, 179–182.

Autere, J. M., Moilanen, J. S. et al. (2000). Familial aggregation of Parkinson's disease in a Finnish population. *J. Neurol. Neurosurg. Psychiatry* **69**(1), 107–109.

Barbeau, A., and Pourcher, E. (1982). New data on the genetics of Parkinson's disease. *Can. J. Neurol. Sci.* **9**(1), 53–60.

Bonifati, V., Fabrizio, E., et al. (1995). Familial Parkinson's disease: a clinical genetic analysis. *Can. J. Neurol. Sci.* **22**, 272–279.

Buervenich, S., Sydow, O. et al. (2000). Alcohol dehydrogenase alleles in Parkinson's disease. *Mov. Disord.* **15**(5), 813–818.

Burn, D. J., Mark, M. H. et al. (1992). Parkinson's disease in twins studied with $^{18}$F-dopa and positron emission tomography. *Neurology* **42**, 1894–1900.

Chan, D. K., Mellick, G. et al. (2000). The alpha-synuclein gene and Parkinson disease in a Chinese population. *Arch. Neurol.* **57**(4), 501–503.

Chan, P., Jiang, X. et al. (1998a). Absence of mutations in the coding region of the alpha-synuclein gene in pathologically proven Parkinson's disease. *Neurology* **50**(4), 1136–1137.

Chan, P., Tanner, C. M. et al. (1998b). Failure to find the alpha-synuclein gene missense mutation (G209A) in 100 patients with younger onset Parkinson's disease. *Neurology* **50**(2), 513–514.

Checkoway, H., Farin, F. M. et al. (1998a). Genetic polymorphisms in Parkinson's disease. *Neurotoxicology* **19**(4–5), 635–643.

Checkoway, H., Franklin, G. M. et al. (1998b). A genetic polymorphism of MAO-B modifies the association of cigarette smoking and Parkinson's disease. *Neurology* **50**, 1458–1461.

Chiba-Falek, O., and Nussbaum, R. L. (2001). Effect of allelic variation at the NACP-Rep1 repeat upstream of the alpha-synuclein gene (SNCA) on transcription in a cell culture luciferase reporter system. *Hum. Mol. Genet.* **10**(26), 3101–3109.

Costa-Mallen, P., Costa, L. G. et al. (2000). Genetic polymorphism of dopamine D2 receptors in Parkinson's disease and interactions with cigarette smoking and MAO-B intron 13 polymorphism. *J. Neurol. Neurosurg. Psychiatry* **69**(4), 535–537.

De Michele, G., Filla, A. et al. (1996). Environmental and genetic risk factors in Parkinson's disease: a case-control study in southern Italy. *Mov. Disord.* **11**, 17–23.

DeStefano, A. L., Golbe, L. I. et al. (2001). Genome-wide scan for Parkinson's disease: the GenePD Study. *Neurology* **57**(6), 1124–1126.

Duvoisin, R. C. (1993). The genetics of Parkinson's disease—a review. *Adv. Neurology* **60**, 306–315.

Duvoisin, R. C., and Golbe, L. I. (1995). Kindreds of dominantly inherited Parkinson's disease: keys to the riddle. *Ann. Neurol.* **38**(3), 355–356.

Elbaz, A., Grigoletto, F. et al. (1999). Familial aggregation of Parkinson's disease: a population-based case-control study in Europe. EUROPARKINSON Study Group. *Neurology* **52**(9), 1876–1882.

Elbaz, A., Manubens-Bertran, J. M. et al. (2000). Parkinson's disease, smoking, and family history. EUROPARKINSON Study Group. *J. Neurol.* **247**(10), 793–798.

Farrer, M., Chan, P. et al. (2001). Lewy bodies and parkinsonism in families with parkin mutations. *Ann. Neurol.* **50**(3), 293–300.

Farrer, M., Destee, T. et al. (2000). Linkage exclusion in French families with probable Parkinson's disease. *Mov. Disord.* **15**(6), 1075–1083.

Farrer, M., Gwinn-Hardy, K. et al. (1999). A chromosome 4p haplotype segregating with Parkinson's disease and postural tremor. *Hum. Mol. Genet.* **8**(1), 81–85.

Farrer, M., Maraganore, D. M. et al. (2001). alpha-Synuclein gene haplotypes are associated with Parkinson's disease. *Hum. Mol. Genet.* **10**(17), 1847–1851.

Farrer, M., Wavrant-De Vrieze, F. et al. (1998). Low frequency of alpha-synuclein mutations in familial Parkinson's disease. *Ann. Neurol.* **43**(3), 394–397.

Forno, L. S. (1969). Concentric hyalin intraneuronal inclusions of Lewy type in brains of elderly persons (60 incidental cases): relationships to parkinsonism. *J. Am. Ger. Soc.* **17**, 557–575.

Gasser, T., Muller-Myhsok, B. et al. (1998). A susceptibility locus for Parkinson's disease maps to chromosome 2p13. *Nat. Genet.* **18**, 262–265.

Gibb, W. R. G., and Lees, A. J. (1988). The relevance of the Lewy body to the pathogenesis of idiopathic Parkinson's disease. *J. Neurol. Neurosurg. Psychiatry* **51**, 745–752.

Golbe, L. I., Di Iorio, G. et al. (1996). Clinical genetic analysis of Parkinson's disease in the Contursi kindred. *Ann. Neurol* **40**, 767–775.

Golbe, L. I., Di Torio, G. et al. (1990). A large kindred with autosomal dominant Parkinson's disease. *Ann. Neurol* **26**, 276–282.

Gowers, W. (1888). *Diseases of the Nervous System.* P. Blakiston, Son & Co, Philadelphia, PA.

Grasbon-Frodl, E. M., Egensperger, R. et al. (1999a). The alpha1-antichymotrypsin A-allele in German Parkinson disease patients. *J. Neural. Transm.* **106(7–8)**, 729–736.

Grasbon-Frodl, E. M., Kosel, S. et al. (1999b). Analysis of mitochondrial targeting sequence and coding region polymorphisms of the manganese superoxide dismutase gene in German Parkinson disease patients. *Biochem. Biophys. Res. Commun.* **255(3)**, 749–752.

Guillard, A., Chastang, C. et al. (1986). Long-term study of 416 cases of Parkinson disease. Prognostic factors and therapeutic implications. *Rev. Neurol.* **142(3)**, 207–214.

Gwinn-Hardy, K. A., Crook, R. et al. (2000). A kindred with Parkinson's disease not showing genetic linkage to established loci. *Neurology* **54(2)**, 504–507.

Harhangi, B. S., Farrer, M. J. et al. (1999). The Ile93Met mutation in the ubiquitin carboxy-terminal-hydrolase-L1 gene is not observed in European cases with familial Parkinson's disease. *Neurosci. Lett.* **270(1)**, 1–4.

Hattori, N., Kitada, T. et al. (1998a). Molecular genetic analysis of a novel Parkin gene in Japanese families with autosomal recessive juvenile parkinsonism: evidence for variable homozygous deletions in the Parkin gene in affected individuals. *Ann. Neurol.* **44(6)**, 935–941.

Hattori, N., Matsumine, H. et al. (1998b). Point mutations (Thr240Arg and Gln311Stop) [correction of Thr240Arg and Ala311Stop] in the Parkin gene. *Biochem. Biophys. Res. Commun.* **249(3)**, 754–758.

Holthoff, V., Vieregge, P. et al. (1994). Discordant twins with Parkinson's disease: positron emission tomography and early signs of impaired cognitive circuits. *Ann. Neurol.* **36**, 176–182.

Hu, C. J., Sung, S. M. et al. (2000). Polymorphisms of the parkin gene in sporadic Parkinson's disease among Chinese in Taiwan. *Eur. Neurol.* **44(2)**, 90–93.

Hughes, A. J., Daniel, S. E. et al. (1993). A clinicopathologic study of 100 cases of Parkinson's disease. *Arch. Neurol.* **50**, 140–148.

Hughes, A. J., Daniel, S. E. et al. (2001). Improved Accuracy of Clinical Diagnosis of Lewy Bocy Parkinson's Disease. *Neurology* **57**, 1497–1499.

Karlsen, K. H., Tandberg, E. et al. (2000). Health related quality of life in Parkinson's disease: a prospective longitudinal study. *J. Neurol. Neurosurg. Psychiatry* **69(5)**, 584–589.

Khan, N., Graham, E. et al. (2001). Parkinson's disease is not associated with the combined alpha-synuclein/apolipoprotein E susceptibility genotype. *Ann. Neurol.* **49(5)**, 665–658.

Kitada, T., Asakawa, S. et al. (1998). Mutations in the parkin gene cause autosomal recessive juvenile parkinsonism [see comments]. *Nature* **392(6676)**, 605–608.

Kondo, I., and Yamamoto, M. (1998). Genetic polymorphism of paraoxonase 1 (PON1) and susceptibility to Parkinson's disease. *Brain Res.* 271–273.

Kondo, K., Kurland, L. et al. (1973). Parkinson's disease: genetic analysis and evidence of a multifactorial etiology. *Mayo Clin. Proc.* **48**, 465–475.

Kosel, S., Hofhaus, G. et al. (1999). Role of mitochondria in Parkinson disease. *Biol Chem* **380(7–8)**, 865–870.

Kruger, R., Hardt, C. et al. (2000). Genetic analysis of immunomodulating factors in sporadic Parkinson's disease. *J. Neural. Transm.* **107(5)**, 553–562.

Kruger, R., Kuhn, W. et al. (1998). Ala30Pro mutation in the gene encoding alpha-synuclein in Parkinson's disease [letter]. *Nat. Genet.* **18(2)**, 106–108.

Kruger, R., Vieira-Saecker, A. M. et al. (1999). Increased susceptibility to sporadic Parkinson's disease by a certain combined alpha-synuclein/apolipoprotein E genotype. *Ann. Neurol.* **45(5)**, 611–617.

Kurth, J. H., and Kurth, M. C. (1992). Allelic frequencies of tyrosine hydroxylase in Parkinson's disease patients. *Mov. Disord.* **7(3)**, 290.

Laihinen, A., Ruottinen, H. et al. (2000). Risk for Parkinson's disease: twin studies for the detection of asymptomatic subjects using [18F]6-fluorodopa PET. *J. Neurol.* **247** (Suppl 2), II110–II113.

Langston, J. W., Ballard, P. A. et al. (1983). Chronic parkinsonism in humans due to a product of meperidine analog synthesis. *Science* **219**, 979–980.

Langston, J. W., Sastry, S. et al. (1998). Novel alpha-synuclein-immunoreactive proteins in brain samples from the Contursi kindred, Parkinson's, and Alzheimer's disease. *Exp. Neurol.* **154(2)**, 684–690.

Leroy, E., Boyer, R. et al. (1998). The ubiquitin pathway in Parkinson's disease. *Nature* **395(6701)**, 451–452.

Lin, J., Yueh, K. et al. (1999). Absence of G209A and G88C mutations in the alpha-synuclein gene of Parkinson's disease in a Chinese population. *Eur. Neurol.* **42(4)**, 217–220.

Lincoln, S., Crook, R. et al. (1999). No pathogenic mutations in the beta-synuclein gene in Parkinson's disease. *Neurosci. Lett.* **269(2)**, 107–109.

Lucking, C. B., Abbas, N. et al. (1998). Homozygous deletions in parkin gene in European and North African families with autosomal recessive juvenile parkinsonism. The European Consortium on Genetic Susceptibility in Parkinson's Disease and the French Parkinson's Disease Genetics Study Group. *Lancet* **352(9137)**, 1355–1356.

Lucking, C. B., Durr, A. et al. (2000). Association between early-onset Parkinson's disease and mutations in the parkin gene. French Parkinson's Disease Genetics Study Group. *N. Engl. J. Med.* **342(21)**, 1560–1567.

Maimone, D., Dominici, R. et al. (2001). Pharmacogenomics of neurodegenerative diseases. *Eur. J. Pharmacol.* **413(1)**, 11–29.

Maraganore, D. M., Anderson, D. W. et al. (1999a). Autopsy patterns for Parkinson's disease and related disorders in Olmsted County, Minnesota. *Neurology* **53(6)**, 1342–1344.

Maraganore, D. M., Farrer, M. J. et al. (1999b). Case-control study of the ubiquitin carboxy-terminal hydrolase L1 gene in Parkinson's disease. *Neurology* **53(8)**, 1858–1860.

Marder, K., Tang, M. et al. (1996). Risk of Parkinson's disease among first-degree relatives: a community-based study. *Neurology* **47**, 155–160.

Mark, M., Burn, D. J. et al. (1991). Parkinson's Disease and Twins: An 18F-Dopa PET study. *Neurology* **41** (Suppl. 1), 255.

Marras, C., and Tanner, C. M. (2002). The epidemiology of Parkinson's disease. In *Movement Disorders Neurologic Principles and Practice*, 2nd ed. (R. L. Watts and W. C. Koller, eds.) McGraw Hill, New York.

Marsden, C. D. (1987). Twins and Parkinson's disease. *J. Neurol. Neurosurg. Psychiatry* **50**, 105–106.

Martin, E. R., Scott, W. K. et al. (2001). Association of single-nucleotide polymorphisms of the tau gene with late-onset Parkinson disease. *JAMA* **286(18)**, 2245–2250.

Martin, W., Young, W. et al. (1973). Parkinson's disease: A genetic study. *Brain* **96**, 495–506.

Marttila, R. J., Kaprio, J. et al. (1988). Parkinson's disease in a nationwide twin cohort. *Neurology* **3817**, 1217–1219.

Marttila, R. J., and Rinne, U. K. (1991). Progression and survival in Parkinson's disease. *Acta Neurolog. Scand.* **84** (Suppl. 136), 24–28.

McCann, S. J., McManus, M. E. et al. (2000). The serotonin transporter gene and Parkinson's disease. *Eur. Neurol.* **44(2)**, 108–111.

Mellick, G. D., and Silburn, P. A. (2000). The ubiquitin carboxy-terminal hydrolase-L1 gene S18Y polymorphism does not confer protection against idiopathic Parkinson's disease. *Neurosci. Lett.* **293(2)**, 127–130.

Menegon, A., Board, P. G. et al. (1998). Parkinson's disease, pesticides, and glutathione transferase polymorphisms. *Lancet* **352(9137)**, 1344–1346.

Mizuno, Y., Hattori, N. *et al.* (2001). Parkin and Parkinson's disease. *Curr. Opin. Neurol.* **14(4)**, 477–482.

Moilanen, J. S., Autere, J. M. *et al.* (2001). Complex segregation analysis of Parkinson's disease in the Finnish population. *Hum. Genet.* **108(3)**, 184–189.

Morano, A., Jimenez-Jimenez, F. J. *et al.* (1994). Risk-factors for Parkinson's disease: case-control study in the province of Caceres, Spain. *Acta Neurol. Scand.* **89(3)**, 164–170.

Muenter, M. D., Forno, L. S. *et al.* (1998). Hereditary form of parkinsonism—dementia. *Ann. Neurol.* **43(6)**, 768–781.

Munoz, E., Obach, V. *et al.* (1999). Alpha1-antichymotrypsin gene polymorphism and susceptibility to Parkinson's disease. *Neurology* **52(2)**, 297–301.

Nagar, S., Juyal, R. C. *et al.* (2001). Mutations in the alpha-synuclein gene in Parkinson's disease among Indians. *Acta Neurol. Scand.* **103(2)**, 120–122.

Nelson, L., Van Den Eeden, S. *et al.* (1997). Incidence of idiopathic Parkinson's disease (PD) in a health maintenance organziation (HMO): variations by age, gender and race/ethnicity. *Neurology* A334.

Nisipeanu, P., Inzelberg, R. *et al.* (1999). Autosomal-recessive juvenile parkinsonism in a Jewish Yemenite kindred: Mutation of Parkin gene. *Neurology* **53**, 1602–1604.

Oliveri, R. L., Annesi, G. *et al.* (1999). Dopamine D2 receptor gene polymorphism and the risk of levodopa-induced dyskinesias in PD. *Neurology* **53(7)**, 1425–1430.

Oliveri, R. L., Zappia, M. *et al.* (2001). The parkin gene is not involved in late-onset Parkinson's disease. *Neurology* **57(2)**, 359–362.

Papadimitriou, A., Veletza, V. *et al.* (1999). Mutated alpha-synuclein gene in two Greek kindreds with familial PD: incomplete penetrance? *Neurology* **52(3)**, 651–654.

Parker, W. D., Jr., and Swerdlow, R. H. (1998). Mitochondrial dysfunction in idiopathic Parkinson disease. *Am. J. Hum. Genet* **62(4)**, 758–762.

Pastor, P., Ezquerra, M. *et al.* (2000). Significant association between the tau gene A0/A0 genotype and Parkinson's disease. *Ann Neurol* **47(2)**, 242–245.

Payami, H., Bernard, S. *et al.* (1995). Genetic anticipation in Parkinson's disease. *Neurology* **45**, 135–138.

Payami, H., Larsen, K. *et al.* (1994). Increased risk of Parkinson's disease in parents and siblings of patients. *Ann. Neurol* **36**, 659–661.

Piccini, P., Burn, D. J. *et al.* (1999). The role of inheritance in sporadic Parkinson's disease: evidence from a longitudinal study of dopaminergic function in twins. *Ann. Neurol* **45(5)**, 577–582.

Plante-Bordeneuve, V., Davis, M. B. *et al.* (1994). Tyrosine hydroxylase polymorphism in familial and sporadic Parkinson's disease. *Mov. Disord.* **9(3)**, 337–339.

Polymeropoulos, M., Lavedan, C. *et al.* (1997). Mutation in the α-synuclein gene identified in families with Parkinson's disease. *Science* 2045–2047.

Polymeropoulos, M. H., Higgins, J. J. *et al.* (1996). Mapping of a gene for Parkinson's disease to chromosome 4q21–q23. *Science* **274**, 1197–1199.

Poskanzer, D. C., and Schwab, R. S. (1963). Cohort analysis of Parkinson's syndrome. Evidence for a single etiology related to subclinical infection about 1920. *J. Chron. Dis.* **16**, 961–973.

Preux, P. M., Condet, A. *et al.* (2000). Parkinson's disease and environmental factors. Matched case-control study in the Limousin region, France. *Neuroepidemiology* **19(6)**, 333–337.

Rajput, A. H., and Rozdilsky, B. (1991). Accuracy of clinical diagnosis in parkinsonism—a prospective study. *Can. J. Neurol. Sci.* **18**, 275–278.

Riedl, A. G., Watts, P. M. *et al.* (1998). P450 enzymes and Parkinson's disease: the story so far. *Mov. Disord.* 212–220.

Roos, R. A. C., Jongen, J. C. F. *et al.* (1996). Clinical course of patients with Parkinson's disease. *Mov. Disord.* **11(3)**, 236–242.

Satoh, J., and Kuroda, Y. (1999). Association of codon 167 Ser/Asn heterozygosity in the parkin gene with sporadic Parkinson's disease. *Neuroreport* **10(13)**, 2735–2739.

Sawle, G. V., Wroe, S. J. *et al.* (1992). The identification of presymptomatic parkinsonism: clinical and [$^{18}$F]dopa positron emission tomography studies in an Irish kindred. *Ann. Neurol.* **32**, 609–617.

Schapira, A. H. (1999). Mitochondrial involvement in Parkinson's disease, Huntington's disease, hereditary spastic paraplegia and Friedreich's ataxia. *Biochim. Biophys. Acta* **1410(2)**, 159–170.

Schapira, A. H., Gu, M. *et al.* (1998). Mitochondria in the etiology and pathogenesis of Parkinson's disease. *Ann. Neurol.* **44(3)** (Suppl. 1), S89–S98.

Schrag, A., Jahanshahi, M. *et al.* (2000). How does Parkinson's disease affect quality of life? A comparison with quality of life in the general population. *Mov. Disord.* **15(6)**, 1112–1118.

Scott, W. K., Nance, M. A. *et al.* (2001). Complete genomic screen in Parkinson disease: evidence for multiple genes. *JAMA* **286(18)**, 2239–2244.

Semchuk, K., Love, E. *et al.* (1993). Parkinson's disease: A test of the multifactorial etiologic hypothesis. *Neurology* **43**, 1173–1180.

Sim, E., Stanley, L. A. *et al.* (1995). Xenogenetics in multifactorial disease susceptibility. *Trends Genet.* **11(12)**, 509–512.

Spira, P. J., Sharpe, D. M. *et al.* (2001). Clinical and pathological features of a Parkinsonian syndrome in a family with an Ala53Thr alpha-synuclein mutation. *Ann. Neurol.* **49(3)**, 313–319.

Sveinbjornsdottir, S., Hicks, A. A. *et al.* (2000). Familial aggregation of Parkinson's disease in Iceland. *N. Engl. J. Med.* **343(24)**, 1765–1770.

Swerdlow, R. H., Parks, J. K. *et al.* (1997). Mitochondrial electron transport function in cybrid cells from members of a Parkinson's disease kindred spanning three generations. *Soc. Neurosci.* **23**, 1907.

Swerdlow, R. H., Parks, J. K. *et al.* (1998). Matrilineal inheritance of complex I dysfunction in a multigenerational Parkinson's disease family. *Ann. Neurol.* **44(6)**, 873–881.

Takahashi, H., Ohama, E. *et al.* (1994). Familial juvenile parkinsonism: clinical and pathologic study in a family. *Neurology* **44**, 437–441.

Tan, E. K., Khajavi, M. *et al.* (2000). Variability and validity of polymorphism association studies in Parkinson's disease. *Neurology* **55(4)**, 533–538.

Tan, E. K., Nagamitsu, S. *et al.* (2001). Alcohol dehydrogenase polymorphism and Parkinson's disease. *Neurosci. Lett.* **305(1)**, 70–72.

Tanner, C., and Goldman, S. (1996). Epidemiology of Parkinson's disease. *Neurol. Clin.* **14**, 317–335.

Tanner, C. M., and Langston, J. W. (1990). Do environmental toxins cause Parkinson's disease? A critical review. *Neurology* **40(10)**, Suppl. 3, Suppl. 17–30; Discussion 30–31.

Tanner, C. M., Ottman, R. *et al.* (1999). Parkinson disease in twins: an etiologic study. *JAMA* **281(4)**, 341–346.

Taylor, C. A., Saint-Hilaire, M. H. *et al.* (1999). Environmental, medical, and family history risk factors for Parkinson's disease: a New England-based case control study. *Am. J. Med. Genet.* **88(6)**, 742–749.

Taylor, M. C., Le Couteur, D. G. *et al.* (2000). Paraoxonase polymorphisms, pesticide exposure and Parkinson's disease in a Caucasian population. *J. Neural. Transm.* **107(8–9)**, 979–983.

Trimmer, P. A., Swerdlow, R. H. *et al.* (2000). Abnormal mitochondrial morphology in sporadic Parkinson's and Alzheimer's disease cybrid cell lines. *Exp. Neurol.* **162(1)**, 37–50.

Valente, E. M., Bentivoglio, A. R. *et al.* (2001). Localization of a novel locus for autosomal recessive early-onset parkinsonism, PARK6, on human chromosome 1p35–p36. *Am. J. Hum. Genet.* **68(4)**, 895–900.

van de Warrenburg, B. P. C., Lammens, M. *et al.* (2001). Clinical and pathologic abnormalities in a family with parkinsonism and parkin gene mutations. *Neurology* **56**, 555–557.

van Duijn, C. M., Dekker, M. C. *et al.* (2001). Park7, a novel locus for autosomal recessive early-onset parkinsonism, on chromosome 1p36. *Am. J. Hum. Genet.* **69(3)**, 629–634.

Vaughan, J., Durr, A. *et al.* (1998a). The alpha-synuclein Ala53Thr mutation is not a common cause of familial Parkinson's disease: a study of 230 European cases. European Consortium on Genetic

Susceptibility in Parkinson's Disease. *Ann. Neurol.* **44(2)**, 270–273.

Vaughan, J. R., Farrer, M. J. *et al.* (1998b). Sequencing of the alpha-synuclein gene in a large series of cases of familial Parkinson's disease fails to reveal any further mutations. The European Consortium on Genetic Susceptibility in Parkinson's Disease (GSPD). *Hum. Mol. Genet.* **7(4)**, 751–753.

Vieregge, P., Hagenah, J. *et al.* (1999). Parkinson's disease in twins: a follow-up study. *Neurology* **53(3)**, 566–572.

Vieregge, P., and Heberlein, I. (1995). Increased risk of Parkinson's disease in relatives of patients. *Ann. Neurol.* **37**, 685.

Vieregge, P., Schiffke, K. A. *et al.* (1992). Parkinson's disease in twins. *Neurology* **42**, 1453–1461.

Wang, J., and Liu, Z. (2000). No association between paraoxonase 1 (PON1) gene polymorphisms and susceptibility to Parkinson's disease in a Chinese population. *Mov. Disord.* **15(6)**, 1265–1267.

Wang, M., Hattori, N. *et al.* (1999). Polymorphism in the parkin gene in sporadic Parkinson's disease. *Ann. Neurol.* **45(5)**, 655–658.

Wang, Y. C., Liu, H. C. *et al.* (2001). Genetic association anaiysis of alpha-1-antichymotrypsin polymorphism in Parkinson's disease. *Eur. Neurol.* **45(4)**, 254–256.

Ward, C., Duvoisin, R. *et al.* (1983). Parkinson's disease in 65 pairs of twins and in a set of quadruplets. *Neurology* **33**, 815–824.

Werneck, A. L., and Alvarenga, H. (1999). Genetics, drugs and environmental factors in Parkinson's disease. A case-control study. *Arq. Neuropsiquiatr.* **57(2B)**, 347–355.

Wintermeyer, P., Kruger, R. *et al.* (2000). Mutation analysis and association studies of the UCHL1 gene in German Parkinson's disease patients. *Neuroreport* **11(10)**, 2079–2082.

Wooten, G. F., Currie, L. J. *et al.* (1997a). Maternal inheritance in Parkinson's disease. *Ann. Neurol.* **41(2)**, 265–268.

Wooten, G. F., Currie, L. J. *et al.* (1997b). Maternal inheritance in two large kindreds with Parkinson's disease. *Neurology* **48** (Suppl 3), A333.

Wszolek, Z., Pfeiffer, R. *et al.* (1995). Western Nebraska family (family D) with autosomal dominant parkinsonism. *Neurology* **45**, 502–505.

Wu, R. M., Cheng, C. W. *et al.* (2001). The COMT L allele modifies the association between MAOB polymorphism and PD in Taiwanese. *Neurology* **56(3)**, 375–382.

Yamamoto, M., Kondo, I. *et al.* (1997). Genetic association between susceptibility to Parkinson's disease and alpha1-antichymotrypsin polymorphism. *Brain Res* **759(1)**, 153–155.

Yamamura, Y., Arihiro, K. *et al.* (1993). Early-onset Parkinsonism with diurnal fluctuation: clinical and pathological studies. *Clin. Neurol.* **33**, 491–196.

Yamamura, Y., Sobue, I. *et al.* (1973). Paralysis agitans of early onset with marked diurnal fluctuations of symptoms. *Neurology* **23**, 239–244.

Yokochi, M., Narabayashi, H. *et al.* (1984). Juvenile parkinsonism—some clinical, pharmacological, and neuropathological aspects. In *Advances in Neurology, Volume 40*. (R. G. Hassler and J. F. Christ, eds), pp. 407–413. Raven Press, New York.

Zimmerman, T. R., Jr., Bhatt, M. *et al.* (1991). Parkinson's disease in monozygotic twins: A follow-up study. *Neurology* **41** (Suppl. 1), 255.

Zweig, R. M., Singh, A. *et al.* (1992). The familial occurrence of Parkinson's disease. Lack of evidence for maternal inheritance. *Arch. Neurol.* **49**, 1205–1207.

CHAPTER 27

# PARK1 and α-Synuclein: A New Era in Parkinson's Research

J. WILLIAM LANGSTON
*Parkinson's Institute*
*Sunnyvale, California 94089*

LAWRENCE I. GOLBE
*UMDNJ-Robert Wood Johnson Medical School*
*New Brunswick, New Jersey 08901*

SEUNG-JAE LEE
*Parkinson's Institute*
*Sunnyvale, California 94089*

I. Introduction
II. The Contursi Kindred
   A. Historical Background
III. The Clinical Phenotype of the A53T Mutation
   A. Similarities to Idiopathic Parkinson's Disease
   B. Differences from Idiopathic Parkinson's Disease
   C. The Neuropathological Features of A53T Parkinsonism
IV. Parkinsonism Due to a A30P Mutation
   A. A Second Mutation is Discovered in the α-Synuclein Gene
   B. Clinical Features of A30P Parkinsonism
   C. Impact on Parkinson's Disease Research
V. Gene Function
   A. The Synuclein Protein Family
   B. Normal Function of α-Synuclein
   C. Proteins that Interact with α-Synuclein
   D. α-Synuclein as a Chaperone
VI. Aggregation of α-Synuclein
   A. Fibrillation and Structural Changes of α-Synuclein
   B. Do α-Synuclein Aggregates Cause Neurodegeneration?
   C. Potential Modulators of the α-Synuclein Aggregation Process
VII. Diagnosis
   A. DNA
VIII. Neuroimaging
   A. A53T
   B. A30P
IX. Animal Models of α-Synucleinopathies
X. Is the A53T Phenotype Parkinson's Disease?
XI. Treatment
XII. Conclusions
   References

In this chapter, we describe the first of the genetic parkinsonisms in which the actual causative mutation has been identified, an A53T mutation in the region encoding for the protein α-synuclein. While families harboring this mutation, and a subsequently identified A30P mutation in the same coding region, are quite rare, these original studies led to the discovery that α-synuclein is a major component of Lewy bodies, a cardinal pathological feature of typical, sporadic Parkinson's disease. This in turn has dramatically changed the research landscape in Parkinson's disease, and given birth to a focus on the disease as a protein aggregation or "handling" disorder. The clinical phenotypes of these mutations are reviewed, and similarities and differences between sporadic Parkinson's disease and this form of genetic parkinsonism are highlighted. While the normal function of α-synuclein is not known, there is a rapidly expanding body of basic research on this protein and ways that its abnormal aggregation or removal could cause disease. While the transgenic mouse models for this disorder have been somewhat disappointing, the fly model has been surprisingly reflective of that which is seen in human Parkinson's disease. Although this form of parkinsonism has proved to be an exceptionally rare cause of parkinsonism in humans, it represents a powerful example of how a rare genetic variant of a common disease can shed tremendous light on the sporadic form of the disorder that it tends to model.

## I. INTRODUCTION

The second half of the 20th century was a fertile time for advancements in Parkinson's disease research. The opening shot was sounded when, in 1959, Arvid Carlsson (Carlsson,

1959) suggested that a dopamine deficiency in the central nervous system might be responsible for the clinical manifestations of Parkinson's disease. He based this hypothesis on experiments showing reserpine-induced akinesia in rabbits responded dramatically to systemic administration of L-dopa. In a remarkable leap of deductive reasoning, he suggested that (1) the L-dopa was replenishing a central dopamine depletion that was causing the akinesia in these animals, and (2) that this might be similar to Parkinson's disease. A year later, 1960, Ehringer and Hornykiewicz showed for the first time that there was a profound deficiency of striatal dopamine in patients who had died with Parkinson's disease. This in turn eventually led to successful trials of the dopamine precursor L-dopa in patients with Parkinson's disease (Cotzias, 1968), which was no less dramatic in humans than it had been in reserpinized rabbits. This, of course, revolutionized the treatment of Parkinson's disease.

Since that time there have been numerous additions to our pharmacologic armamentarium for Parkinson's disease, including dopamine agonists, monoamine oxidase B and catechol-*O*-methyl transferase inhibitors (Factor, 2002). However, advances have not been limited just to new therapies. Indeed several have had important implications for research on the cause of the disease. The first of these occurred when it was reported that a very simple molecule (MPTP) was capable of inducing virtually all of the motor signs of Parkinson's disease in humans who had self-administered the compound under the impression it was "synthetic heroin" (Langston *et al.*, 1983). This observation led to a renaissance of research in Parkinson's disease, and reawakened interest in the hypothesis that Parkinson's disease might be due primarily to environmental causes. It also provided the first good animal model for the disease, a highly useful tool to study mechanisms of neurodegeneration, and ways to test new neuroprotective strategies (Langston, 2002).

The second discovery, and the one which forms the basis for this chapter, occurred literally as the 20th century drew to a close, and that discovery tilted the scientific compass back in the direction of inheritance as an important factor in the investigation of Parkinson's disease. The major focus of this chapter will be on that discovery, and its impact on our understanding of the inherited parkinsonism and Parkinson's disease itself. Indeed, this discovery has opened a huge new line of scientific inquiry into the mechanisms of neuronal degeneration that occur in all forms of parkinsonism, including the idiopathic disease, as well as all diseases which have intraneuronal inclusions known as Lewy bodies as part of their neuropathological substrate. In this chapter, we will detail the history of this discovery, and summarize our current understanding of this form of genetic parkinsonism, including its clinical features, as well as the tremendous amount of basic research that its recognition has engendered.

## II. THE CONTURSI KINDRED

### A. Historical Background

The events leading up to the discovery of this first mutation directly linked to a form of parkinsonism began in 1990, when the clinical details of a large family with an autosomal dominant pattern of heritance were published by Golbe and colleagues (1990). Exploring the genealogy of this family led these investigators back to the common ancestor in this family (a member of Generation I of the kindred) who was born in approximately 1700 in Contursi, a village in the Salerno province of South Central Italy. While three of the six extant affected families of this "Contursi" kindred, as it has come to be known, still remained in the area, many members had emigrated, mostly to the United States, but also to Northern Italy, Germany, and Argentina. The first quantitative clinical genetic analysis of this kindred was published in early 1996 (Golbe *et al.*, 1996), along with an expanded genealogical and clinical description. A linkage analysis was performed using the single kindred comprising 592 members of 6 families. Affected individuals were examined and/or their DNA collected from subjects in all of the locales mentioned.

While the disease status of the common ancestor, like that of all members of the first five generations of the Contursi kindred, was unknown, among the members of Generations VI through X, 60 affected members could be characterized in regard to disease status. The families have been genealogically distinct for three generations (i.e., since Generation VII) and when this investigation took place, none of them were aware of the others. DNA samples from 16 of these family members were used for an "affecteds-only" linkage analysis. Linkage was established in late 1996 (Polymeropoulos *et al.*, 1996) when genetic markers on chromosome 4q21–q23 were found to be linked to the parkinsonian phenotype. The actual sequence was reported a few months later (Polymeropoulos *et al.*, 1997), revealing an A53T mutation in the gene encoding the protein α-synuclein. Ironically, this protein had been investigated previously in relationship to the other major neurodegenerative disease of aging, Alzheimer's disease, but had never before been associated with Parkinson's disease.

This same group also searched for the A53T mutation using PCR in 314 chromosomes from patients diagnosed as having Parkinson's disease, both sporadic and familial, but not known to be related to the Contursi kindred (Polymeropoulos *et al.*, 1997). Three families, all of Greek origin, were found to carry the A53T mutation.

As might have been expected, identification of the A53T mutation set off an explosion of interest in the possibility that many patients with Parkinson's disease might harbor a similar mutation. In a relatively brief period of time well over one thousand patients, including both familial and sporadic cases of Parkinson's disease, had been examined for this mutation (Chan et al., 1998; El-Agnaf et al., 1998; Farrer et al., 1998; Ho and Kung, 1998; Muñoz et al., 1997; Pastor et al., 2001; Scott et al., 1999; Vaughan et al., 1998a, b; Wang et al., 1998; Zareparsi et al., 1998). Surprisingly, only nine more families, all of Greek ancestry, were found to have the A53T mutation (Athanassiadou et al., 1999; Bostantjopoulou et al., 2001; Markopoulou et al., 1995; Papadimitriou et al., 1999; Papapetropoulos et al., 2001). Although no genealogical relationship is known among these families or between any of them and the Contursi kindred, the genealogies of these families are known only for three or four generations. Just as the families in the Contursi kindred were related to one another only via more remote generations (and were unaware of this), a similar genealogical relationship between the 12 Greek families with A53T parkinsonism has not been excluded. Furthermore, Athanassiadou et al., (1999) found that the Contursi kindred and the three Greek A53T families reported by Polymeropoulos et al. (1997) share a haplotype at 4q21. This makes a founder effect quite likely.

From a historical perspective, it is interesting to note that southern Italy was colonized by Greece in the 4th Century BC and considered part of "Magna Graeca." This may explain the observation that all known carriers of the A53T mutation are of either Italian or Greek ancestry. Contursi has been a tourist destination since ancient times because of its warm, sulfurous springs mentioned in Virgil's writings. The town, in fact, continues to promote medical tourism by styling itself as Il Paese della Salute ("Healthtown"). This stimulates the speculation that the A53T gene was brought to Contursi by a visitor, but perhaps it is too much to speculate that he was seeking an aqueous cure for his parkinsonism.

## III. THE CLINICAL PHENOTYPE OF THE A53T MUTATION

### A. Similarities to Idiopathic Parkinson's Disease

#### 1. Motor Features

Most if not all cases harboring the A53T mutation who have been examined to date have exhibited bradykinesia and rigidity, two of the classical triad of parkinsonian signs. While tremor (the third sign of this classic triad) appears to be less common (see discussion below), there is little doubt that some patients appear to have the full clinical constellation of features that typify sporadic Parkinson's disease.

#### 2. Non-motor Features

Bostantjopoulou and colleagues (2001) have performed the largest and most systematic study of neuropsychological features of patients with the A53T mutation. These investigators studied 8 patients from 6 different Greek families with a mean onset age of 40 and mean disease duration of 5.4 years. The results from 12 neuropsychological test batteries were compared to a group of 11 patients with sporadic Parkinson's disease. The groups were matched for educational attainment, onset age, symptom duration, and motor disability. As a group, the A53T patients displayed slightly more impairment of memory and spatial abilities. However, tests of overall cognition, executive function, and naming were no different between the two groups. Only one of the A53T patients was demented, and the dementia was mild in degree. Others displayed low scores in many of the tests and the authors predicted that they would meet criteria for dementia a few years hence. But overall, the hypothesis that individuals with A53T parkinsonism are more likely to develop dementia than patients with sporadic Parkinson's disease was not convincingly supported by this study.

Bostantjopoulou and colleagues (2001) also measured depression, and found it to be no more severe among the A53T patients than among Parkinson's disease controls. The mean Beck Depression score was 5.3 in A53T patients, not significantly different from the mean score of 7.3 in Parkinson's disease patients. Only one patient met criteria for depression, and in that case it was relatively mild in degree. This result is consistent with informal observation of affected members of the Contursi Kindred (Golbe et al., 1996).

### B. Differences from Idiopathic Parkinson's Disease

Just as there are many similarities between parkinsonism associated with the A53T mutation and the more common idiopathic form of parkinsonism, there are also a number of differences between these two conditions. We next will discuss some of these differences, including age of onset, disease duration, tremor, psychological status, and cognition. In addition, several cases with atypical clinical features are briefly described.

#### 1. Age of Onset

The features that most strikingly differentiate patients with the A53T mutation from sporadic Parkinson's disease are its young age of onset and shortened period of subsequent survival. In the Contursi kindred, the mean onset age is only 45.6 years (standard deviation: 13.48 years). In the series reported by Bostantjopoulou and colleagues (2001), it was even less (39.7 years with a standard deviation of 7.6 years). Thus the age of onset is about 13–15 years younger than the mean onset age of Parkinson's disease (Granieri et al., 1991). On the other hand, there is substantial variability in the age of onset in the Contursi kindred as noted above, which is similar to that seen in most community- or clinic-based series of Parkinson's disease. This is an interesting phenomenon, since this wide variation in onset age occurs despite the apparently unitary etiology of the A53T for parkinsonism. This suggests that other factors influence the expressivity of the principal mutation, a possibility that is under investigation in many laboratories, including our own.

## 2. Survival

The mean survival for individuals with sporadic Parkinson's disease is approximately 13–15 years (Hughes *et al.*, 1993). In the Contursi kindred, it is shorter, with an average of only 9.2 years (standard deviation 4.87 years). In the family reported by Markopoulou *et al.* (1995), the average survival time was 9.0 years, and in the series of Bostantjopoulou *et al.* (2001) it was only 5.4 years (standard deviation: 2.1 years), a duration that differs dramatically from that which is typically seen in sporadic Parkinson's disease. An observation by Hughes *et al.* (1993) amplifies this difference even more. In their series of patients with apparently typical, sporadic disease but with an onset in the mid-40s, the mean survival was approximately 25 years. If one accepts these data, A53T parkinsonism appears to be approximately three times as malignant as typical Parkinson's disease.

## 3. Tremor

This cardinal feature of Parkinson's disease seems to occur less commonly in patients with A53T parkinsonism. For example, it was observed in only 2 of 15 (13%) patients in the series reported by Papapetropoulos *et al.* (2001). The presence of tremor was 50% (4 of 8) in the series of Bostanjopoulos *et al.* (2001), 58% in the Contursi kindred (18 of 31 in whom tremor data was available; Golbe *et al.*, 1996) and 69% (11 of 16) of the patients reported by Markoupoulou and colleagues (1995). While the physiology of the tremor was not described in detail in any of these reports, informal observation in the Contursi kindred fails to reveal more than a rare individual with the "typical" parkinsonian rest tremor, which is characterized by low frequency, multiple joint involvement, and multiple planes of movement.

It is possible that the differences reported on the prevalence of tremor reflect the use of different definitions of tremor and varying periods of observation. But there is also the possibility that closely related individuals may share a tremor-causing mutation that modifies the action of the A53T mutation. This possibility has been examined in the Contursi kindred (Golbe *et al.*, 1996). There were 25 affected individuals in 9 sibships with two or more affected members for whom data on the presence of tremor was available. Of these 25, 19 had a tremor status that was concordant with the majority of his/her sibs. While this suggests a trend toward intrasibship clustering, it was not statistically different from the figure of 12, a result expected under the null hypothesis. This analysis, or other cluster analysis of tremor, has not been applied to nuclear family units in other reports of kindreds with the A53T mutation. However, it is quite possible that this may prove to be a useful approach for the study of the genetic determinants of tremor.

## 4. Atypical Clinical Features

While many of the patients in the Contursi kindred, and those with the A30P mutation, appear to have a diseae phenotype that is similar to typical sporadic Parkinson's disease, at least some patients have features that are clearly atypical. For example, a patient followed during life by one of the authors developed a severe fluent aphasia midway through her illness (a sign that is not seen in typical Parkinson's disease), culminating in a "word salad" late in her illness, and eventually a terminal apathetic dementia. The patient described by Spira *et al.*, (2001) developed severe central hypoventilation and prominent myclonus, again features that are not characteristic of sporadic Parkinson's disease. Such cases hint at a more extensive or diverse pathology than is seen in typical Parkinson's disease.

## C. The Neuropathological Features of A53T Parkinsonism

Examining the neuropathologic features of the human phenotype of the A53T mutation is critical for assessing its relevance to Parkinson's disease. This is because Parkinson's disease is a clinicopathological entity, and there is no clinical feature that accurately and consistently predicts the typical pathology of the disorder. However, there are only a few cases of A53T parkinsonism that have been studied pathologically to date, and only one of these is published in any detail. In this section, we will summarize what is known about the pathological features of the disease at the current time, and compare them with the typical neuropathological features of Parkinson's disease.

The first mention of the neuropathological features of A53P parkinsonism is contained in the original report by Golbe and colleagues (1990) in which two cases are briefly described. In one patient, who died at age 66, only limited sections of the substantia nigra were available, but in the locus ceruleus cell loss was clearly evident, and two Lewy bodies were seen. The second patient (the nephew of the first patient) died of accidental causes after being symptomatic for 11 years. In this case, severe nerve cell loss was seen in the substantia nigra. There was also mild cell loss in several other areas typically affected in sporadic Parkinson's disease, including the locus ceruleus, and Lewy bodies were seen in many of these areas. An atypical feature was the loss of nerve cells in Sommer's sector and the subiculum of the hippocampus. A brief description of the pathological findings in a third member of this kindred has been reported separately (Langston et al., 1998); (this case will be reported in more detail elsewhere). Briefly, this third patient had findings very similar to the second described by Golbe above, but the patient did not have cell loss in the hippocampus. The most striking finding was the predominance of synuclein immunostaining in neuronal processes (Lewy neurites) rather than cell bodies (Fig. 27.1), and microvacuolation in the temporal cortex, a finding that is more common in dementia with Lewy bodies than in Parkinson's disease. Finally, Spira et al. (2001) reported the neuropathological findings in an affected member of a family of Greek origin. In addition to many of the typical neuropathological features of Parkinson's disease, they also noted an excess of α-synuclein-positive neurites (Lewy neurites) and temporal lobe vacuolation, similar to our patient. Subcortical basal ganglia gliosis and nerve cell loss was also observed, leading the authors to conclude that "the A53T α-synuclein mutation can produce a more widespread disorder than found in typical idiopathic Parkinson's disease."

Finally, a 35 kDa band on Western blotting that is synuclein immunoreactive has been identified in the brain of an affected member of the Contursi kindred, which was not present in either controls or in typical Parkinson's disease brain (Langston et al., 1998). While this could be a dimer of α-synuclein, its exact identity remains unknown, and it could be providing a clue regarding the differences in pathophysiology between Contursi parkinsonism and Parkinson's disease.

## IV. PARKINSONISM DUE TO A A30P MUTATION

### A. A Second Mutation is Discovered in the α-Synuclein Gene

After the identification of the A53T mutation, investigators searched not only for other patients in the population who might have this mutation, but also for any other mutations in the region encoding for α-synuclein. Surprisingly, only one other such mutation has been reported to date, and it was found quite quickly. This A30P mutation was identified by the combination of single strand conformation polymorphism analysis and DNA sequencing by Kruger and colleagues (1998). As of this writing, no other examples of this mutation have been reported. Because this mutation was also associated with a form of typical parkinsonism, these observations, together with the landmark observation that Lewy bodies, the pathologic hallmark of Parkinson's disease, are laden with α-synuclein (to be discussed later in Section VI), cemented the notion that the A53T mutation in the Contursi kindred and Greek families was pathogenic rather than an innocent allelic variant.

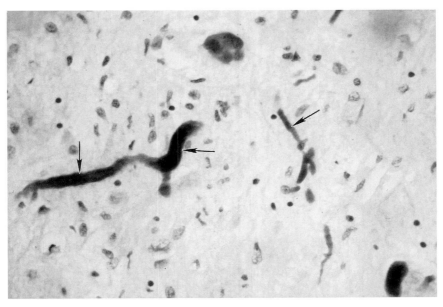

FIGURE 27.1 Two Lewy bodies (arrows) are present in a nerve cell in substantia nigra from a case of Parkinson's disease. Neuromelanin (NM) can clearly be seen between the Lewy bodies, indicating this is a catecholaminergic neuron. The Lewy bodies were immunostained with antibody to α-synuclein. (×1170; courtesy of Dr. Lysia Forno.)

## B. Clinical Features of A30P Parkinsonism

Because so few cases have been reported, it is difficult to make too many generalizations regarding the phenotype in this disorder. The five symptomatic individuals with the A30P mutation who have been reported to date presented between the ages 58–76 (average age: 60) with signs and symptoms that are reported to be similar to those of sporadic Parkinson's disease. But once again progression is said to be more rapid, and rest tremor less common, mirroring the feature of A53T parkinsonism, and reflecting differences with sporadic Parkinson's disease.

## C. Impact on Parkinson's Disease Research

While identification of mutations that were responsible for a parkinsonism phenotype generated tremendous excitement when first reported, a reality check came about when it became clear that this was not a widespread phenomenon. However, this disappointment was quickly dispelled with the first report that α-synuclein is a major component of Lewy bodies, a cardinal neuropathological feature of Parkinson's disease (Spillantini *et al.*, 1998; Fig. 27.2). This is because many investigators feel that unraveling the nature of the Lewy body is a key to understanding Parkinson's disease itself. Indeed this appeared to be the first big break in understanding the Lewy body, and perhaps the pathophysiologic process underlying neurodegeneration of Parkinson's disease. For this reason, much of the second half of this chapter is dedicated to what has been learned about α-synuclein, its relationship to parkinsonism, and other Lewy body disorders, and the animal models that have been developed to date based on these mutations.

## V. GENE FUNCTION

### A. The Synuclein Protein Family

Three human proteins have been identified in this family: α-synuclein, β-synuclein, and γ-synuclein. The genes for these proteins are located on chromosome 4q21(Campion *et al.*, 1995; Spillantini *et al.*, 1995; Chen *et al.*, 1995), 5q35 (Spillantini *et al.*, 1995), and 10q23 (Lavedan *et al.*, 1998), respectively. The synucleins are highly conserved in amino acid sequence and have similar domain organizations. The N termini of these proteins are composed of imperfect repeats that contain the consensus sequence KTKEGV (Fig. 27.3). This region is predicted to have helical conformation and mediate the interaction to phospholipid membranes (Davidson *et al.*, 1998; Perrin *et al.*, 2000). The mid-domain is highly hydrophobic, and the 12 amino acid motif in this region, which is absent in

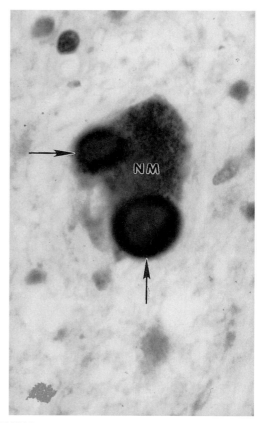

**FIGURE 27.2** Two Lewy neurites (arrows) are seen in the dorsal motor nucleus of the vagus. This section was taken from the brain of an affected member of the Contursi kindred who was documented to have an A53T mutation in the encoding region for α-synuclein. The neurites were immunostained with an antibody to α-synuclein. (×470; courtesy of Dr. Lysia Forno.)

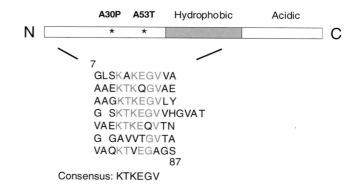

**FIGURE 27.3** Three domains are distinguished based on their sequence characteristics. N terminal half of the protein is composed of 7 imperfect 11 amino acid repeats with the core consensus sequence of KTKEGV. Both A53T and A30P mutations are located outside the core consensus sequence in this repeat region. The central region of this protein is rich in hydrophobic amino acids and thought to be essential for aggregation. This region was originally isolated as a non-amyloid component of senile plaques, but whether this fragment is a core component of senile plaques is controversial. Finally, the C-terminal region shows a high density of negatively charged amino acids.

β-synuclein, appears to be critical for aggregation (Giasson et al., 2001). The C terminus is the least homologous domain among these proteins, but shares the acidic characteristics due to high contents of negatively charged amino acids. The C-terminal domain may be particularly important, as it could be responsible for maintaining the solubility of the protein, thus preventing aggregation (Crowther et al., 1998).

## B. Normal Function of α-Synuclein

Although little is known about the function of α-synuclein, it has most frequently been implicated in synaptic function. Expression of synelfin, the zebra finch homolog of α-synuclein, is induced in the lateral magnocellular nucleus during a critical period when song learning occurs (George et al., 1995), suggesting a role in synaptic plasticity and learning. Other evidence for a role of α-synuclein in synaptic transmission comes from work in primary hippocampal cultures, where it has been shown that a reduction in α-synuclein expression by antisense oligonucleotide causes a decrease in the size of the presynaptic vesicular pool (Murphy et al., 2000). Also consistent with this proposed role are immunohistochemical studies indicating that α-synuclein is enriched in neuronal presynaptic terminals (Iwai et al., 1995), and observations in biochemical fractionations of brain homogenate showing the presence of α-synuclein in synaptosomal fractions (George et al., 1995; Irizarry et al., 1996; Kahle et al., 2000; Lee et al., 2002a). Surprisingly, given the potential importance of this protein, a targeted null mutation of the α-synuclein gene in mice resulted in only subtle abnormalities in nigrostriatal synaptic transmission (Abeliovich et al., 2000). Without doubt more work is needed on the relationship between this protein and synaptic function, in terms of both its normal role, and how it is affected in an abnormal setting, such as in the face of neuronal degeneration.

As noted above, α-synuclein has been associated with synaptic vesicles. It appears that this protein interacts with acidic phospholipid liposomes through the N-terminal repeat region (Clayton and George, 1998). Initial binding of α-synuclein to artificial vesicles is sensitive to ionic strength, but the pre-bound proteins cannot be washed off by high ionic strength solutions (Davidson et al., 1998; Jo et al., 2002; Lee et al., 2002a), suggesting the involvement of hydrophobic interactions to fatty acyl chains of phospholipids in the membrane in the stably associated state. Consistent with this, the interaction between α-synuclein and fatty acids has been demonstrated in solution (Sharon et al., 2001). In a more recent study, α-synuclein has been shown to accumulate on the surface of triglyceride-rich lipid droplets in cells treated with high concentration of fatty acids (Cole et al., 2002). Furthermore, the binding of α-synuclein to the lipid droplets appears to protect triglyceride from hydrolysis. Together, these findings suggest that α-synuclein might be involved in cytoplasmic lipid trafficking and/or metabolism, which could in turn influence the lipid-mediated signaling pathways and vesicle trafficking. Interestingly, the A30P mutant protein appears to be less capable of binding to both bilayer membranes and triglyceride-rich lipid droplets, and the A53T mutant, although its lipid binding ability does not seem to be affected, is less efficient in inhibiting triglyceride turnover (Cole et al., 2002; Jensen et al., 1998; Jo et al., 2002). Therefore, alteration of lipid binding and subsequent action on the lipids might be a part of the pathogenic processes of synucleinopathies.

## C. Proteins that Interact with α-Synuclein

Identification of interacting proteins often leads to critical clues for predicting the physiological role of the proteins with unknown function, and this has certainly been the case with α-synuclein. There have been several attempts to identify interacting proteins, which in turn have yielded proteins involved in a variety of cytophysiological functions. Synphilin-1 was the first to be identified as an α-synuclein-binding protein by yeast two-hybrid screening (Engelender et al., 1999). The function of synphilin is unknown, but the fact that it has multiple protein-protein interaction motifs, such as ankyrin-like repeats and a coiled-coil domain, suggests a role as an adaptor protein (Engelender et al., 1999). Synphilin-1 promotes α-synuclein aggregation when the two are co-expressed, and is ubiquitinated by parkin (see Chapter 28), raising the possibility that synphilin-1 might be involved in the formation of Lewy bodies (Chung et al., 2001). In fact an immunohistochemical study has shown the presence of synphilin-1 in Lewy bodies (Wakabayashi et al., 2000).

In addition to synphilin, several signaling proteins have been identified as α-synuclein interacting proteins. These include extracellular signal regulated kinases (ERK), protein kinase C (PKC) isozymes, dephosphorylated BAD, 14-3-3, and phospholipase D (PLD) isozymes (Ahn et al., 2002; Jenco et al., 1998; Ostrerova et al., 1999). The role of α-synuclein interaction with these proteins is not yet fully understood, but it may result in the inhibition of both PKC and PLD activities (Ahn et al., 2002; Jenco et al., 1998; Ostrerova et al., 1999). Furthermore, several kinases have been shown to phosphorylate α-synuclein. Casein kinase 1 and 2 efficiently phosphorylate α-synuclein at serine 129, and to a lesser extent at serine 87 (Okochi et al., 2000), while G protein-coupled receptor kinases 2 and 5 exclusively phosphorylate the protein at serine 129 (Pronin et al., 2000). Tyrosine kinases, c-src and fyn, also phosphorylate α-synuclein at tyrosine 125 (Ellis et al., 2001; Nakamura et al., 2001). Although the role of phosphorylation in α-synuclein function is not clear at present, both serine and tyrosine phosphorylations seem to affect the ability of α-synuclein to inhibit PLD activity (Ahn *et al.*, 2002; Pronin *et al.*, 2000). These results suggest that α-synuclein functions as a regulator of signal transduction by relaying the upstream kinase activities to the downstream effector molecules. Whether this function of α-synuclein is related to the pathogenesis of Parkinson's disease is unknown at present.

Another set of interacting proteins, including tubulin heterodimer, microtubule associated protein (MAP) 1B, and tau, suggested that α-synuclein has a role in microtubule organization (Alim *et al.*, 2002; Jensen *et al.*, 1999, 2000; Payton *et al.*, 2001). Association of α-synuclein with microtubules has been confirmed using purified microtubule preparations from rat brain (Alim *et al.*, 2002). Furthermore, α-synuclein can promote protein kinase A-mediated tau phosphorylation in a cell-free system (Jensen *et al.*, 1998), which would decrease the ability of tau to bind microtubules. Given that the microtubule is a key cytoskeletal element in the transport of vesicles and macromolecular complexes through neuritic processes, alteration of α-synuclein function in the microtubule stability might be detrimental to neurons. However, exact function of α-synuclein with respect to the microtubule dynamics remains to be elucidated.

Finally, Shimura *et al.* (2001) have recently shown that parkin specifically binds to and ubiquitinates an *O*-glycosylated form of α-synuclein in human brain. Since this interaction leads to the ubiquitination of α-synuclein, it is proposed to be a part of the degradation mechanism. However, whether parkin-mediated destruction represents a major metabolic pathway for α-synuclein is unknown, but this possibility is certainly worthy of further exploration.

### D. α-Synuclein as a Chaperone

This ability of α-synuclein to interact to the variety of unrelated proteins described above and its flexible structure (which could bind to the exposed hydrophobic patches of partially unfolded proteins) has prompted a hypothesis that this protein may function as a molecular chaperone. This hypothesis has been tested in cell-free settings by two different groups (Kim *et al.*, 2000; Souza *et al.*, 2000). Both studies showed that α-synuclein suppressed the aggregation of thermally or chemically denatured proteins, a hallmark property of molecular chaperones. Although this hypothesis still needs to be tested by extensive structural and cell biological studies, there are several other pieces of indirect evidence that support this hypothesis. First, expression of α-synuclein can be induced by neurotoxicants, such as MPTP and paraquat, *in vivo* (Manning-Bog *et al.*, 2002; Vila *et al.*, 2000). Secondly, induction of α-synuclein expression protects cells from apoptotic stimuli and oxidative insults (da Costa *et al.*, 2000; Hashimoto *et al.*, 2002; Lee *et al.*, 2001).

Usually, the action of molecular chaperones is an ATP-dependent process. Therefore, it is noteworthy that one of the α-synuclein binding partners, synphilin-1, contains a predicted ATP/GTP-binding domain (Engelender *et al.*, 1999). Together, these results are consistent with the hypothesis that α-synuclein protects neurons from various stresses by maintaining the protein folding homeostasis. Assuming that the inclusion bodies are the manifestation of the cellular accumulation of misfolded proteins, impaired function of α-synuclein as a molecular chaperone might have direct implication in Lewy body formation.

## VI. AGGREGATION OF α-SYNUCLEIN

A substantial amount has been learned about α-synuclein structure, fibrillation, and aggregation properties since this protein burst on the Parkinson's disease scene in the late 1990s. This wealth of new knowledge is presented here because of its potential importance to the neurodegenerative process that underlies Parkinson's disease.

### A. Fibrillation and Structural Changes of α-Synuclein

Purified recombinant α-synuclein forms fibrillar aggregates spontaneously in solution. α-Synuclein fibrils are typical amyloids with cross β-sheet conformation (Conway *et al.*, 2000a; Serpell *et al.*, 2000) and form in a nucleation-dependent manner (Wood *et al.*, 1999). These fibrils have similar morphology with the ones found in the Lewy bodies, suggesting that the mechanism of cell-free fibrillation could reflect what actually happens as part of a pathogenic process. Recent studies have also identified several prefibrillar oligomeric aggregates, called protofibrils, with different sizes and shapes (Fig. 27.4). Protofibrils are also rich in β-sheet conformation (Volles *et al.*, 2001), indicating that they might be structurally related to the fibrils. In cell-free fibrillation, protofibrils appear before the fibrils, and are consumed as fibrils form, implicating the precursor-product relationship. These results suggest that protofibrils are kinetically unstable intermediates of fibrillation.

Unlike compactly folded fibrillar and protofibrillar forms, monomeric wild-type α-synuclein is largely unstructured in solution, and so are both A53T and A30P mutant forms. Therefore, the transition from monomer to fibril is a process of acquiring structure through a multistep process. It seems very likely that structural characterization of intermediate states and transitions that link them will lead to a greater understanding of the molecular details of the α-synuclein fibrillation process. In a recent study, a

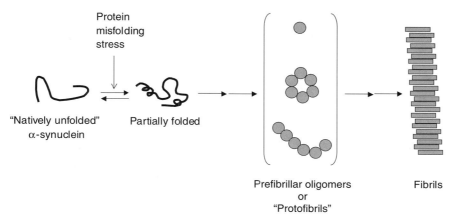

FIGURE 27.4  Working model of α-synuclein aggregation process. While the native state of α-synuclein is characterized by unfolded conformation, it could gain a partial conformation after exposure to protein misfolding stresses. This structural intermediate is highly unstable in a monomeric state, but self-association between the molecules to form dimers or oligomers stabilizes the conformation. The small oligomers with partial conformation undergo structural transitions to gain high-density β-sheet structure, while they evolve into a prefibrillar large oligomeric state, called a protofibril. Various shapes of protofibrils have been observed and they seem to constitute the intermediate state in the fibrillation process. However, the relationship among different protofibrils or the ability of each protofibrillar species to turn into fibrils is not yet understood.

partially folded conformation was described in increased temperatures and low pH solutions, and the increase of this intermediate structure tightly correlated with the enhanced formation of α-synuclein fibrils (Uversky et al., 2001b; Fig. 27.4). Although this structural transition is rapidly reversible in the monomer state, oligomerization of the protein stabilizes the partially folded conformation (Uversky et al., 2001a). Establishing methods to either populate or isolate individual oligomeric intermediates will be a crucial next step for the structural analyses of the fibrillation process.

## B. Do α-Synuclein Aggregates Cause Neurodegeneration?

While the role of impaired α-synuclein function in neurodegenerative processes remains elusive, several lines of evidence support the hypothesis that α-synuclein causes neurodegeneration through a toxic gain-of-function mechanism. First, the mutations in the α-synuclein gene that are linked to familial parkinsonism show an autosomal dominant rather than a recessive pattern of inheritance (Kruger et al., 1998; Polymeropoulos et al., 1997). Secondly, the pathological features of Parkinson's disease, dopaminergic neuronal loss and Lewy body formation can be replicated by simply overexpressing α-synuclein in the transgenic fly (Feany and Bender, 2000) (see Section IX for further discussion of this model). Finally, the lack of a neurodegenerative phenotype in the knock-out mouse suggests that neurodegeneration is not a loss-of-function phenotype of α-synuclein (Abeliovich et al., 2000). These observations are consistent with the idea that α-synuclein exerts its role in causing parkinsonism by acquiring a toxic function, probably through abnormal aggregation, rather than a loss of normal function.

On the other hand, while deposition of fibrillar α-synuclein is a characteristic pathological feature of many neurodegenerative diseases, the potential neurotoxic effect of α-synuclein fibrils is controversial. Some recent observations suggest that the fibril itself may not be pathogenic. The brain of the α-synuclein transgenic mouse that displays degeneration of nigral nerve terminals and motor deficit does not contain fibrillar α-synuclein, but rather contains nonfibrillar aggregates (Masliah et al., 2000). Another piece of evidence against the fibril as being pathogenic comes from the kinetic comparison of cell-free fibrillation among wild-type and two mutant forms of α-synuclein. In this study, both of the familial parkinsonism-linked α-synuclein mutations accelerated the disappearance of monomer but not the appearance of the end-product fibril (Conway et al., 2000b), suggesting that some form of aggregation is accelerated by the mutations but the rate of fibril formation per se is not. Because of these studies, the hypothesis has emerged that the nonfibrillar oligomeric form of α-synuclein may be a potential pathogenic species. Interestingly, one of the biochemical properties of protofibrils would be consistent with the protofibril-specific toxicity model. This property is that protofibrillar α-synuclein, but not the monomer or fibril, has the ability to disrupt the integrity of phospholipid membranes when mixed with artificial liposomes (Volles et al., 2001). Although cell-free experiments have introduced the protofibrillar α-synuclein as a pathogenic species, the testing of this hypothesis has been hampered by the transient nature of this form. Establishing methods to accumulate or deplete the protofibrils in cell or animal models will allow us to determine the cytotoxicity of this form.

Although no direct evidence is available for the toxicity of protofibrillar α-synuclein at the moment, cytotoxicity of protofibrillar aggregates has been observed in other amyloidogenic proteins, such as Aβ in Alzheimer's disease. For example, the cytotoxicity of protofibrillar Aβ has been reported in experiments utilizing primary neuronal cultures (Hartley et al., 1999). Moreover, in Huntington's disease and the spinocerebellar ataxias, both disorders characterized by the abnormal accumulation of polyglutamine proteins, the formation of detectable fibrillar inclusions and cell death have been dissociated using both cell and animal model systems (Klement et al., 1998; Saudou et al., 1998; Cummings et al., 1999; Faber et al., 1999; Kazemi-Esfarjani and Benzer, 2000), raising the possibility that protofibrillar aggregates might be the key to neuronal degeneration in these diseases. Thus, cytotoxicity of prefibrillar oligomeric aggregates is emerging as a general mechanism of neurodegeneration mediated by endogenous proteotoxins. Identification of regulatory factors that act at a specific stage of the α-synuclein aggregation process, thus accumulating or depleting specific aggregate species in cells, will be a critical step toward determining the cytotoxic form of α-synuclein.

## C. Potential Modulators of the α-Synuclein Aggregation Process

### 1. Pathogenic Mutations in α-Synuclein

Many researchers have independently reported that both A53T and A30P mutations promoted α-synuclein aggregation in cell-free experiments. Little difference in secondary structures has been observed among wild-type and mutant forms of α-synuclein using low-resolution optical methods (Conway et al., 2000a; Li et al., 2001). However, high-resolution analysis using NMR spectroscopy has shown that the wild-type protein has a region with residual helical structure and that the mutant proteins have different degrees of preference for helical conformation in the same region (Bussell and Eliezer, 2001). The diversity of in vitro aggregation kinetics displayed by the mutants might be due to the slight difference in the residual structure. Interestingly, A30P mutation promoted the formation of prefibrillar protofibrils, but not the fibrils, while A53T mutation accelerated the formation of both protofibrils and fibrils (Conway et al., 2000b). Assuming that the cytotoxic function of α-synuclein is enhanced by both mutations, this result suggests that protofibrils, rather than the fibrils, might be responsible for the toxicity of the protein. Although expression of mutant α-synuclein increased cytotoxicity in several studies (Kanda et al., 2000; Lee et al., 2001; Ostrerova et al., 1999; Ostrerova-Golts et al., 2000; Tabrizi et al., 2000; Zhou et al., 2000, 2002), the accelerated protofibril formation by the mutant forms has yet to be verified in cell systems.

### 2. β- and γ-Synucleins

In a recent study, Masliah and colleagues (Hashimoto et al., 2001) generated doubly transgenic mice that express both human α- and β-synuclein and showed that β-synuclein, a non-amyloidogenic homolog of α-synuclein, inhibited α-synuclein aggregation and neurodegeneration. The inhibitory activity of β-synuclein, as well as γ-synuclein, against α-synuclein aggregation has also been shown in cell-free experiments with purified proteins (Hashimoto et al., 2001; Uversky et al., 2002). These studies suggest that β- and γ-synucleins might be natural negative regulators of α-synuclein aggregation. The inhibitory action of β-synuclein might impose on the very early stage of α-synuclein aggregation, since the monomeric proteins seem to interact with each other in vitro (Hashimoto et al., 2001). Such anti-amyloidogenic properties of β- and γ-synucleins might provide therapeutic strategies to inhibit α-synuclein aggregation and neuronal cell death.

### 3. Parkin

Parkin is a 55-kDa protein (465 amino acid) with a ubiquitin-like domain in the N terminus and C terminal RING domains, a structural feature shared by a family of E3 ubiquitin ligases (Giasson and Lee, 2001). While parkin, which was first identified as the cause of an autosomal recessive, juvenile form of parkinsonism (ARJP; Kitada et al., 1998) is covered in detail in Chapter 28, in this volume, it will be briefly reviewed here because of its relevance to α-synuclein. The E3 ligase function of parkin, which involves an E2-dependent transfer of ubiquitin to target proteins, has been demonstrated by several groups (Imai et al., 2000; Shimura et al., 2000; Zhang et al., 2000). It is currently thought that parkin-mediated ubiquitination determines the degradation of target proteins by the proteasomal system. Therefore, defects in parkin function could result in the accumulation of misfolded proteins, which then could lead to protein aggregation. Parkin has been linked to Lewy bodies and synuclein in several ways. For example, it has now been shown that Lewy bodies contain parkin, even in sporadic Parkinson's disease cases (Shimura et al., 2001), suggesting that the role of parkin may not be limited to the rare ARJP cases. Interestingly, Lewy body pathology does not occur in ARJP cases with homozygous null mutations in the parkin gene (Hayashi et al., 2000). There is one reported case that showed typical Lewy pathology (Farrer et al., 2001), but this case contained an R275W mutation in one allele, which results in a parkin protein with a reduced but substantial amount of ubiquitin ligase activity (Chung et al., 2001). Therefore, loss-of-parkin function might be responsible not only for the neuronal cell loss but also for the lack of inclusions in many ARJP cases. Chung et al. (2001) showed that co-expression of parkin did not change the efficiency of synphilin-mediated α-synuclein aggregation, but was required to form ubiquitin-positive aggregates.

Given these findings, it can be proposed that parkin might play a role in building the unique architecture of Lewy bodies by putting small nonfibrillar aggregates together and assisting conformational transitions within the aggregates through its E3 ligase activity. However, whether parkin is involved in α-synuclein fibrillation and the formation of Lewy pathology is still speculative. Nevertheless, given the pathological characteristics of ARJP cases and the interaction with a rare glycosylated form of α-synuclein, it would be worthwhile to investigate the exact role of parkin in the formation of fibrillar inclusion, especially in the protofibril-to-fibril conversion.

## 4. Dopamine

In a screening of drug-like molecules that interfere with α-synuclein fibrillation, catecholamines, including dopamine and its precursor L-dopa, were found effective in inhibiting fibrillation (Conway et al., 2001). The inhibitory activity of dopamine appears to be mediated by the formation of a dopamine-α-synuclein adduct. Importantly, this inhibition occurs in the protofibril-to-fibril conversion step through the stabilization of protofibrillar intermediates. These observations raise the possibility that dopamine might play a role in cell type-selective protein aggregation and neuronal degeneration in Parkinson's disease. Although evidence that dopamine has the same inhibitory activity in biological systems is lacking at the moment, discovery of such inhibitory molecules could provide important tools to control the conversion of protofibrils to fibrils experimentally and should in theory allow the assessment of their relative toxicity. Since this study shows that it is possible to control the fibrillation process by a "small molecule," it seems possible that similar approaches might be used to develop neuroprotective strategies in Parkinson's disease.

## 5. Environmental Factors

Epidemiological studies have suggested the involvement of environmental risk factors, such as pesticides and heavy metals, in the etiology of Parkinson's disease (Altschuler, 1999; Good et al., 1992; Gorell et al., 1999; Hertzman et al., 1994; Hirsch et al., 1991; Le Couteur et al., 1999; Tanner, 1989; Yasui et al., 1992). After the discovery of fibrillar α-synuclein in the pathology of Parkinson's disease, the role of these risk factors in α-synuclein fibrillation has been investigated in various experimental systems. In cell-free experiments, heavy metals and pesticides promoted conformational changes and fibrillation of α-synuclein (Uversky et al., 2001c, d). In rodents, chronic administration of rotenone induced Lewy body-like inclusions (Betarbet et al., 2000), and paraquat administration caused thioflavin S-positive amorphous α-synuclein deposits (Manning-Bog et al., 2002). Rotenone-induced formation of α-synuclein-positive inclusions has also been confirmed in cultured mammalian cells (Lee et al., 2002b). Due to the broad and rapidly evolving array of new model systems, it seems quite possible that we could witness dramatic progress in the not too distant future with respect to the molecular mechanisms of α-synuclein aggregation under the influence of such environmental factors.

## 6. Other Factors

Some interacting proteins have been shown to promote α-synuclein aggregation. For example, tubulin acts like a seed for α-synuclein aggregation at a very low molar ratio, and self-assembly of tubulin itself appears to increase the seeding activity (Alim et al., 2002). However, the seeding activity of tubulin has yet to be reproduced in a biological system. On the other hand, an aggregation-promoting effect of synphilin-1 has been demonstrated in mammalian cells that co-express α-synuclein and synphilin-1 (Chung et al., 2001). Interestingly, immunohistochemical studies have shown that both tubulin and synphilin-1 co-localize with α-synuclein in Lewy bodies, at least raising the possibility of pathogenic interactions between these proteins (Alim et al., 2002; Wakabayashi et al., 2000).

In addition to the interactions with other proteins, interactions with the membrane also seem to modulate the aggregation property of α-synuclein. In a recent cell-free experiment using isolated rat brain vesicles, membrane-bound α-synuclein exhibited a higher aggregation tendency than the cytosolic form, and the aggregates formed in the membranes appeared in turn capable of seeding aggregation of the cytosolic form (Lee et al., 2002a). The aggregation in the membrane appears to be mediated by oxidative modification of lipids (Lee et al., 2002a). This study suggests that even though the membrane-bound form represents only 15–20% of total brain α-synuclein, a shift of α-synuclein distribution to the membranes and/or a slight increase of lipid oxidation could accelerate the seed formation. Certainly these observations appear to warrant further investigation.

Finally, using an antibody specific to phosphorylated α-synuclein at serine 129, Iwatsubo and colleagues (Fujiwara et al., 2002) have shown that Lewy bodies contain phospho-α-synuclein. They also estimated that approximately 90%

of sarcosyl-insoluble, urea-soluble α-synuclein from the brains of patients with dementia with Lewy bodies is phosphorylated at serine 129, whereas in normal rat brain, only about 4% of total α-synuclein has the same modification. In a cell-free system, the phospho-form of α-synuclein produced both fibrils and protofibrils much more rapidly than the non-phosphorylated form (Fujiwara et al., 2002).

## VII. DIAGNOSIS

### A. DNA

As described in detail in the earlier sections of this chapter, two mutations that are associated with a parkinsonian phenotype (A53T and A30P) have been identified in the region encoding for α-synuclein (Fig. 27.3). Both of these mutations were found by direct sequencing of all the exons of the α-synuclein gene, confirmed by restriction enzyme digestion of PCR products. The subsequent screening by other investigators has typically been done simply by restriction digestion of the PCR product of corresponding exons. The possibility that these were polymorphisms was excluded by examining an appropriate number of control samples.

Because so few examples of either of these mutations have been found in the general population (see earlier discussion) screening for such mutations is probably warranted only when there is a very strong family history (particularly if the pattern of inheritance appears to be autosomal dominant) or for research purposes. In regard to the latter, there is a tantalizing hint of a relationship between polymorphisms in the promotor region of the α-synuclein gene and sporadic Parkinson's disease. In 1999 Kruger and colleagues (1999) reported a significant difference in the allelic distributions of a polymorphism in the promoter region of the α-synuclein gene (NACP-Rep1) between Parkinson's disease patients and controls with a risk ratio of 2.2 (interestingly, when they combined the APOE4 allele with allele 1 of the NACP-Rep1, a 12.8-fold increase in the relative risk was found in patients harboring this genotype). To date, however, this potentially important observation has not been confirmed by others (Izumi et al., 2001; Khan et al., 2001; Parsian et al., 1998; Tan et al., 2000).

## VIII. NEUROIMAGING

### A. A53T

Although little information is available on imaging studies in patients who harbor the A53T mutation, at least one such study has been performed. Samii and colleagues (1999) examined four affected members of a Greek family using positron emission tomography (PET) to assess fluorodopa uptake. The PET scans in these patients were reported to be very similar to that seen in idiopathic Parkinson's disease. This study indicates that the distribution of neuronal terminal damage in this form of parkinsonism follows a pattern quite close to that seen in Parkinson's disease.

### B. A30P

Only one imaging study has been reported in regard to this mutation as well. Recently, Kruger and colleagues (2001) performed fluorodopa, raclopride and fluorodeoxyglucose (FDG) PET in a presymptomatic A30P carrier. They found no clear abnormality except for impaired FDG uptake in temporal and frontomedial cortices and in the caudate nucleus areas, all on one side, findings of little relevance to the issue of providing a prognosis to the patient. On the other hand, fluorodopa PET studies in two symptomatic individuals gave results similar to those expected in sporadic Parkinson's disease, except for greater cortical involvement on the FDG scan of one demented individual.

## IX. ANIMAL MODELS OF α-SYNUCLEINOPATHIES

After the mutation was unraveled in the Contursi kindred, it appeared that we were just steps away from having the ideal neuropathological model for Parkinson's disease. The MPTP model has served us extraordinarily well in terms of elucidating the function and circuitry of the basal ganglia in the presence of a nigrostriatal deficit, and has been widely used to test new treatments for the disease, but it does not typically manifest all of the pathologic features of the disease, particularly Lewy bodies (although at least one study in non-human primates has shown Lewy body-like inclusions in aged animals (Forno et al., 1986)).

Since humans with the A53T mutation clearly exhibit Lewy bodies, it seemed a natural step to develop transgenic mouse models that would have all of the pathologic features of Parkinson's disease. Such a model would be invaluable in exploring mechanisms of neurodegeneration and new neuroprotective strategies, and should even allow us to determine if Lewy bodies are indeed a key causative factor in neurodegeneration, or, as suggested by some, could actually be protective. And indeed, transgenic mouse lines that overexpress wild-type or mutant forms of α-synuclein in neurons were quickly developed (Kahle et al., 2000; Masliah et al., 2000; Matsuoka et al., 2001; van der Putten et al., 2000). However, only one group has reported what appears to be α-synuclein-positive intraneuronal inclusions, although even these were of a granular nature, not of the more typical fibrillar composition as seen in typical Lewy bodies (Masliah et al., 2000). Although this mouse line showed degeneration of dopaminergic presynaptic terminals, the neuronal loss was not observed at the cell body level (Masliah et al., 2000). Thus, attempts to create mouse models using a transgenic approach have not consistently lead to a biologic recapitulation of pathological and behavioral features of the human disease, much to the frustration of many. Hopefully, it will be only a matter of time before more consistently successful efforts prevail.

Surprisingly, in contrast to the experience with the transgenic mouse, transgenic flies that express human wild-type or mutant α-synuclein under control of either a pan-neuronal promoter (*elav*) or a dopaminergic neuron-specific promoter (DOPA-decarboxylase) have produced dramatic, if not stunning, results, with fibrillar intraneuronal inclusions and the loss of dopaminergic neurons (Auluck et al., 2002; Feany and Bender, 2000). These pathological phenotypes are even accompanied by a decline of motor function. According to Feany and Bender (2000), the timing of inclusion appearance depends on the level of α-synuclein expression, but they did not address the dose effects on dopaminergic degeneration and behavioral deficits. The fact that the simple expression of α-synuclein produces pathogenic phenotypes in flies but not in mice raised the possibility that defense mechanisms against α-synuclein aggregation/toxicity may exist, which is far more efficient in mice than in flies. In support of this idea, co-expression of human chaperone hsp70 suppresses the pathogenic phenotypes of α-synuclein transgenic flies (Auluck et al., 2002). However, this could also be a function of amount of overexpression.

Careful analysis of several lines with different levels of expression should be useful in addressing relative tolerance of different species to α-synuclein overexpression. Interestingly, in all the transgenic animal models generated so far, overexpression of wild-type α-synuclein was as efficient as that of mutant forms in producing pathogenic phenotypes (Feany and Bender, 2000; Masliah et al., 2000). This underscores the idea that pathogenic function of α-synuclein is not restricted to the rare familial Parkinson's disease cases, but is likely to contribute to the far more frequent idiopathic Parkinson's disease.

Finally, we should also note that in addition to the transgenic approaches, the chronic systemic administration of rotenone, a common pesticide and an inhibitor of mitochondrial complex I, caused the selective degeneration of nigral dopaminergic neurons in rats (Betarbet et al., 2000). These rats also produced Lewy body-like fibrillar cytoplasmic inclusions that contain α-synuclein and ubiquitin. These pathological abnormalities are associated with hypokinesia and rigidity. Thus, we should also keep in mind that transgenic models are not always the only game in town. Indeed, all avenues should be pursued to better understand the relationship between α-synuclein and the neurodegenerative process that underlies Parkinson's disease.

## X. IS THE A53T PHENOTYPE PARKINSON'S DISEASE?

The advent of genetic forms of parkinsonism has led to some interesting questions in the field of Parkinson's disease, which, while on the surface appear to be simple issues of nomenclature, actually go very deep. These issues are also causing more than a minor amount of confusion in the field. While they have been thrown into relief by each of the emerging forms of genetic parkinsonism, nowhere are they more evident than in regard to the parkinson-inducing mutations in α-synuclein. We will focus on the A53T mutation, because so much is known about its clinical and pathological phenotype. The issue to be addressed here can be simply put in the form of a question: Is Contursi A53T parkinsonism actually Parkinson's disease?

To really answer this question, one must first deal with the definition of "disease" first. *Stedman's Medical Dictionary* provides the following definition: "A disease entity is characterized usually by at least two of these criteria—a recognized etiologic agent, an identifiable group of signs and symptoms or a consistent anatomic alteration." Both forms of parkinsonism have similar signs and symptoms, but, using these criteria, a clear difference between sporadic Parkinson's disease and Contursi parkinsonism is the presence of what is generally accepted as a clear-cut etiology in the form of the A53T mutation in the Contursi kindred (and other identified families with this mutation). *Dorland's Medical Dictionary* further defines Parkinson's disease as "a form of parkinsonism of unknown etiology, usually occurring in later life," but goes on to note that a juvenile form has also been described. We now know that at least some of these juvenile cases harbor parkin mutations.

What about anatomic location? Certainly the key features are similar if not identical in both Contursi parkinsonism and sporadic Parkinson's disease, but there are differences, which are described in Section III. C. Whether or not these are adequate to clearly differentiate the two diseases, at least by a standard neuropathologic definition of Parkinson's disease, could probably be argued either way. However, the case described by Spira and colleagues (2002) seems to have a feature (subcortical gliosis) that extends beyond the typical pathologic boundaries of Parkinson's disease. As noted earlier in this chapter, there are significant differences in age of onset and disease duration, again suggesting that these two conditions are not identical disorders. Finally, while the phenotype of these two disorders can be virtually identical, at least some cases have features that take A53T parkinsonism clearly into the realm of atypical disease.

Given these observations, it seems reasonable to conclude that these are very similar but different diseases. There still remain a large number of cases of apparently sporadic disease (certainly constituting the vast majority of what we call Parkinson's disease at the moment) that have an older onset and a typical clinicopathologic picture. It seems wisest at the moment to continue to refer to these patients as having idiopathic Parkinson's disease, and refer to the clearly defined genetic parkinsonisms as just that. The evolving nomenclature for these genetic forms of parkinsonism may gradually gain widespread acceptance, or the research and clinical community may simply refer to them by mutation. With time, as we learn more about these disorders, should it prove that there are hundreds of etiologies for what we now call Parkinson's disease (genetic and nongenetic) there may be a point where we should reconsider even the use of the term that honors James Parkinson's name, but at the moment we still seem far from reaching that point.

## XI. TREATMENT

The short patient survival in Section III. B.2 cannot be explained by a lack of responsiveness to levodopa. Patients with A53T parkinsonism respond as well as patients with typical Parkinson's disease, both early on and late in the course of the course of the disease (Bostantjopoulou *et al.*, 2001; Golbe *et al.*, 1996; Markopoulou *et al.*, 1995; Spira *et al.*, 2001). However, this response is often complicated by hallucinations and/or choreiform dyskinesias. While this is not dissimilar from typical Parkinson's disease, these complications seem to occur disproportionately early in the course of the illness, hinting at yet another clinical difference between these cases and sporadic, older-onset Parkinson's disease.

## XII. CONCLUSIONS

As should be evident from reading this chapter, in many ways the discovery of the Contursi kindred, and the subsequent linkage analysis and identification of the causative mutation could be considered a seminal event in the evolution of Parkinson's disease research. While the A53T mutation has proven to be exceedingly rare as a cause of genetic parkinsonism, the scientific repercussions have been widespread and are likely to continue to be so for some time. From a practical standpoint, however, given the rarity of this mutation, screening for the mutation in the general population of parkinsonian patients is hard to justify, other than as part of a specific research program and/or when there is a strong family history of the disorder.

From a broader research prospective, the research impact of this rare form of parkinsonism has been impressive. As noted at the beginning of this chapter, this was the last of a series of discoveries and advances that occurred during the second half of the 20th century, and one that came near the end of its closing decade. As a result, this discovery (and those that quickly followed) have ushered Parkinson's disease research into the 21st century in a manner that few would have guessed, and that is as a protein aggregation (or "handling") disorder. This is an arena that has been familiar to researchers in the Alzheimer's field for some time now, but relatively new to Parkinson's disease. As a result, it has led to a burst of research energy as techniques in the Alzheimer's research area are applied to Parkinson's disease. It is probably naive to assume that protein aggregation will prove to be the whole story in sporadic Parkinson's disease (after all, most patients don't harbor mutations in the α-synuclein gene, and therefore there must be other reasons for protein aggregation in those cases). Nonetheless, before the discovery of the relevance of α-synuclein (and other proteins, including parkin), we were probably approaching the problem of neurodegeneration in Parkinson's disease in at least a partially blinded manner. Now it will be interesting to see how far this new research vision will take us in the future.

## References

Abeliovich, A., Schmitz, Y., Farinas, I., Choi-Lundberg, D., Ho, W. H., Castillo, P. E., Shinsky, N., Verdugo, J. M., Armanini, M., Ryan, A., Hynes, M., Phillips, H., Sulzer, D., and Rosenthal, A. (2000). Mice lacking alpha-synuclein display functional deficits in the nigrostriatal dopamine system. *Neuron* **25**, 239–252.

Ahn, B. H., Rhim, H., Kim, S. Y., Sung, Y. M., Lee, M. Y., Choi, J. Y., Wolozin, B., Chang, J. S., Lee, Y. H., Kwon, T. K., Chung, K. C., Yoon, S. H., Hahn, S. J., Kim, M. S., Jo, Y. H., and Min, D. S. (2002). Alpha-synuclein interacts with phospholipase D isozymes and inhibits pervanadate induced phospholipase D activation in human embryonic kidney 293 cells. *J. Biol. Chem.* **30**, 30.

Alim, M. A., Hossain, M. S., Arima, K., Takeda, K., Izumiyama, Y., Nakamura, M., Kaji, H., Shinoda, T., Hisanaga, S., and Ueda, K. (2002). Tubulin seeds alpha-synuclein fibril formation. *J. Biol. Chem.* **277**, 2112–2117.

Altschuler, E. (1999). Aluminum-containing antacids as a cause of idiopathic Parkinson's disease. *Med. Hypotheses* **53**, 22–23.

Athanassiadou, A., Voutsinas, G., Psiouri, L., Leroy, E., Polymeropoulos, M. H., Ilias, A., Maniatis, G. M., and Papapetropoulos, T. (1999). Genetic analysis of families with Parkinson disease that carry the Ala53Thr mutation in the gene encoding alpha-synuclein. *Am. J. Hum. Genet.* **65**, 555–558.

Auluck, P. K., Chan, H. Y., Trojanowski, J. Q., Lee, V. M., and Bonini, N. M. (2002). Chaperone suppression of alpha-synuclein toxicity in a Drosophila model for Parkinson's disease. *Science* **295**, 865–868.

Betarbet, R., Sherer, T. B., MacKenzie, G., Garcia-Osuna, M., Panov, A. V., and Greenamyre, J. T. (2000). Chronic systemic pesticide exposure reproduces features of Parkinson's disease. *Nat. Neurosci.* **3**, 1301–1306.

Bostantjopoulou, S., Katsarou, Z., Papadimitriou, A., Veletza, V., Hatzigeorgiou, G., and Lees, A. (2001). Clinical features of parkinsonian patients with the alpha-synuclein (G209A) mutation. *Mov. Disord.* **16**, 1007–1013.

Bussell, R., Jr., and Eliezer, D. (2001). Residual structure and dynamics in Parkinson's disease-associated mutants of alpha-synuclein. *J. Biol. Chem.* **276**, 45996–46003.

Campion, D., Martin, C., Heilig, R., Charbonnier, F., Moreau, V., Flaman, J. M., Petit, J. L., Hannequin, D., Brice, A., and Frebourg, T. (1995). The NACP/synuclein gene: chromosomal assignment and screening for alterations in Alzheimer disease. *Genomics* **26**, 254–257.

Carlsson, A. (1959). The occurrence, distribution and physiological role of catecholamines in the nervous system. *Pharmacol. Rev.* **11**, 490–493.

Chan, P., Tanner, C. M., Jiang, X., and Langston, J. W. (1998). Failure to find the alpha-synuclein gene missense mutation (G209A) in 100 patients with younger onset Parkinson's disease. *Neurology* **50**, 513–514.

Chen, X., de Silva, H. A., Pettenati, M. J., Rao, P. N., St. George-Hyslop, P., Roses, A. D., Xia, Y., Horsburgh, K., Ueda, K., and Saitoh, T. (1995). The human NACP/alpha-synuclein gene: chromosome assignment to 4q21.3–q22 and TaqI RFLP analysis. *Genomics* **26**, 425–427.

Chung, K. K., Zhang, Y., Lim, K. L., Tanaka, Y., Huang, H., Gao, J., Ross, C. A., Dawson, V. L., and Dawson, T. M. (2001). Parkin ubiquitinates the alpha-synuclein-interacting protein, synphilin-1: implications for Lewy-body formation in Parkinson disease. *Nat. Med.* **7**, 1144–1150.

Clayton, D. F., and George, J. M. (1998). The synucleins: a family of proteins involved in synaptic function, plasticity, neurodegeneration and disease. *Trends Neurosci.* **21**, 249–254.

Cole, N. B., Murphy, D. D., Grider, T., Rueter, S., Brasaemle, D., and Nussbaum, R. L. (2002). Lipid droplet binding and oligomerization properties of the Parkinson's disease protein alpha-synuclein. *J. Biol. Chem.* **277**, 6344–6352.

Conway, K. A., Harper, J. D., and Lansbury, P. T., Jr. (2000a). Fibrils formed in vitro from alpha-synuclein and two mutant forms linked to Parkinson's disease are typical amyloid. *Biochemistry* **39**, 2552–2563.

Conway, K. A., Lee, S. J., Rochet, J. C., Ding, T. T., Williamson, R. E., and Lansbury, P. T., Jr. (2000b). Acceleration of oligomerization, not fibrillization, is a shared property of both alpha-synuclein mutations linked to early-onset Parkinson's disease: implications for pathogenesis and therapy. *Proc. Natl. Acad. Sci. U.S.A.* **97**, 571–576.

Conway, K. A., Rochet, J. C., Bieganski, R. M., and Lansbury, P. T., Jr. (2001). Kinetic stabilization of the alpha-synuclein protofibril by a dopamine- alpha-synuclein adduct. *Science* **294**, 1346–1349.

Cotzias, G. C. (1968). L-Dopa for Parkinsonism. *N. Engl. J. Med.* **278**, 630.

Crowther, R. A., Jakes, R., Spillantini, M. G., and Goedert, M. (1998). Synthetic filaments assembled from C-terminally truncated alpha-synuclein. *FEBS Lett.* **436**, 309–312.

Cummings, C. J., Reinstein, E., Sun, Y., Antalffy, B., Jiang, Y., Ciechanover, A., Orr, H. T., Beaudet, A. L., and Zoghbi, H. Y. (1999). Mutation of the E6-AP ubiquitin ligase reduces nuclear inclusion frequency while accelerating polyglutamine-induced pathology in SCA1 mice. *Neuron* **24**, 879–892.

da Costa, C. A., Ancolio, K., and Checler, F. (2000). Wild-type but not Parkinson's disease-related alα-53 → Thr mutant alpha-synuclein protects neuronal cells from apoptotic stimuli. *J. Biol. Chem.* **275**, 24065–24069.

Davidson, W. S., Jonas, A., Clayton, D. F., and George, J. M. (1998). Stabilization of alpha-synuclein secondary structure upon binding to synthetic membranes. *J. Biol. Chem.* **273**, 9443–9449.

Ehringer, H., and Hornykiewicz, O. (1960). Distribution of noradrenaline and dopamine (3-hydroxytyramine) in the human brain and their behavior in diseases of the extrapyramidal system. *Klin. Wochenschr.* **38**, 1236–1239.

El-Agnaf, O. M., Curran, M. D., Wallace, A., Middleton, D., Murgatroyd, C., Curtis, A., Perry, R., and Jaros, E. (1998). Mutation screening in exons 3 and 4 of alpha-synuclein in sporadic Parkinson's and sporadic and familial dementia with Lewy bodies cases. *Neuroreport* **9**, 3925–3927.

Ellis, C. E., Schwartzberg, P. L., Grider, T. L., Fink, D. W., and Nussbaum, R. L. (2001). Alpha-synuclein is phosphorylated by members of the Src family of protein-tyrosine kinases. *J. Biol. Chem.* **276**, 3879–3884.

Engelender, S., Kaminsky, Z., Guo, X., Sharp, A. H., Amaravi, R. K., Kleiderlein, J. J., Margolis, R. L., Troncoso, J. C., Lanahan, A. A., Worley, P. F., Dawson, V. L., Dawson, T. M., and Ross, C. A. (1999). Synphilin-1 associates with alpha-synuclein and promotes the formation of cytosolic inclusions. *Nat. Genet.* **22**, 110–114.

Faber, P. W., Alter, J. R., MacDonald, M. E., and Hart, A. C. (1999). Polyglutamine-mediated dysfunction and apoptotic death of a *Caenorhabditis elegans* sensory neuron. *Proc. Natl. Acad. Sci. U.S.A.* **96**, 179–184.

Factor, S. A., Weiner, W J. (2002). "Parkinson's Disease—Diagnosis and Clinical Management," pp. 685. Demos Medical Publishing, Inc., New York.

Farrer, M., Chan, P., Chen, R., Tan, L., Lincoln, S., Hernandez, D., Forno, L., Gwinn-Hardy, K., Petrucelli, L., Hussey, J., Singleton, A., Tanner, C., Hardy, J., and Langston, J. W. (2001). Lewy bodies and parkinsonism in families with parkin mutations. *Ann. Neurol.* **50**, 293–300.

Farrer, M., Wavrant-De Vrieze, F., Crook, R., Boles, L., Perez-Tur, J., Hardy, J., Johnson, W. G., Steele, J., Maraganore, D., Gwinn, K., and Lynch, T. (1998). Low frequency of alpha-synuclein mutations in familial Parkinson's disease. *Ann. Neurol.* **43**, 394–397.

Feany, M. B., and Bender, W. W. (2000). A Drosophila model of Parkinson's disease. *Nature* **404**, 394–398.

Forno, L. S., Langston, J. W., DeLanney, L. E., Irwin, I., and Ricaurte, G. A. (1986). Locus ceruleus lesions and eosinophilic inclusions in MPTP-treated monkeys. *Ann. Neurol.* **20**, 449–455.

Fujiwara, H., Hasegawa, M., Dohmae, N., Kawashima, A., Masliah, E., Goldberg, M. S., Shen, J., Takio, K., and Iwatsubo, T. (2002). Alpha-Synuclein is phosphorylated in synucleinopathy lesions. *Nat. Cell Biol.* **4**, 160–164.

George, J. M., Jin, H., Woods, W. S., and Clayton, D. F. (1995). Characterization of a novel protein regulated during the critical period for song learning in the zebra finch. *Neuron* **15**, 361–372.

Giasson, B. I., and Lee, V. M. (2001). Parkin and the molecular pathways of Parkinson's disease. *Neuron* **31**, 885–888.

Giasson, B. I., Murray, I. V., Trojanowski, J. Q., and Lee, V. M. (2001). A hydrophobic stretch of 12 amino acid residues in the middle of alpha-synuclein is essential for filament assembly. *J. Biol. Chem.* **276**, 2380–2386.

Golbe, L. I., Di Iorio, G., Bonavita, V., Miller, D. C., and Duvoisin, R. C. (1990). A large kindred with autosomal dominant Parkinson's disease. *Ann. Neurol.* **27**, 276–282.

Golbe, L. I., Di Iorio, G., Sanges, G., Lazzarini, A. M., La Sala, S., Bonavita, V., and Duvoisin, R. C. (1996). Clinical genetic analysis of Parkinson's disease in the Contursi kindred. *Ann. Neurol.* **40**, 767–775.

Good, P. F., Olanow, C. W., and Perl, D. P. (1992). Neuromelanin-containing neurons of the substantia nigra accumulate iron and aluminum in Parkinson's disease: a LAMMA study. *Brain Res.* **593**, 343–346.

Gorell, J. M., Johnson, C. C., Rybicki, B. A., Peterson, E. L., Kortsha, G. X., Brown, G. G., and Richardson, R. J. (1999). Occupational exposure to manganese, copper, lead, iron, mercury and zinc and the risk of Parkinson's disease. *Neurotoxicology* **20**, 239–247.

Granieri, E., Carreras, M., Casetta, I., Govoni, V., Tola, M. R., Paolino, E., Monetti, V. C., and De Bastiani, P. (1991). Parkinson's disease in Ferrara, Italy, 1967 through 1987. *Arch. Neurol.* **48**, 854–857.

Hartley, D. M., Walsh, D. M., Ye, C. P., Diehl, T., Vasquez, S., Vassilev, P. M., Teplow, D. B., and Selkoe, D. J. (1999). Protofibrillar intermediates of amyloid beta-protein induce acute electrophysiological changes and progressive neurotoxicity in cortical neurons. *J. Neurosci.*

19, 8876–8884.

Hashimoto, M., Hsu, L. J., Rockenstein, E., Takenouchi, T., Mallory, M., and Masliah, E. (2002). α-Synuclein protects against oxidative stress via inactivation of the C-jun N-terminal kinase stress signaling pathway in neuronal cells. *J. Biol. Chem.* **14**, 14.

Hashimoto, M., Rockenstein, E., Mante, M., Mallory, M., and Masliah, E. (2001). Beta-Synuclein inhibits alpha-synuclein aggregation: a possible role as an anti-parkinsonian factor. *Neuron* **32**, 213–223.

Hayashi, S., Wakabayashi, K., Ishikawa, A., Nagai, H., Saito, M., Maruyama, M., Takahashi, T., Ozawa, T., Tsuji, S., and Takahashi, H. (2000). An autopsy case of autosomal-recessive juvenile parkinsonism with a homozygous exon 4 deletion in the parkin gene. *Mov. Disord.* **15**, 884–888.

Hertzman, C., Wiens, M., Snow, B., Kelly, S., and Calne, D. (1994). A case-control study of Parkinson's disease in a horticultural region of British Columbia. *Mov. Disord.* **9**, 69–75.

Hirsch, E. C., Brandel, J. P., Galle, P., Javoy-Agid, F., and Agid, Y. (1991). Iron and aluminum increase in the substantia nigra of patients with Parkinson's disease: an X-ray microanalysis. *J. Neurochem.* **56**, 446–451.

Ho, S. L., and Kung, M. H. (1998). G209A mutation in the alpha-synuclein gene is rare and not associated with sporadic Parkinson's disease. *Mov. Disord.* **13**, 970–971.

Hughes, A. J., Daniel, S. E., Blankson, S., and Lees, A. J. (1993). A clinicopathologic study of 100 cases of Parkinson's disease. *Arch. Neurol.* **50**, 140–148.

Imai, Y., Soda, M., and Takahashi, R. (2000). Parkin suppresses unfolded protein stress-induced cell death through its E3 ubiquitin-protein ligase activity. *J. Biol. Chem.* **275**, 35661–355664.

Irizarry, M. C., Kim, T. W., McNamara, M., Tanzi, R. E., George, J. M., Clayton, D. F., and Hyman, B. T. (1996). Characterization of the precursor protein of the non-A beta component of senile plaques (NACP) in the human central nervous system. *J. Neuropathol. Exp. Neurol.* **55**, 889–895.

Iwai, A., Masliah, E., Yoshimoto, M., Ge, N., Flanagan, L., de Silva, H. A., Kittel, A., and Saitoh, T. (1995). The precursor protein of non-A beta component of Alzheimer's disease amyloid is a presynaptic protein of the central nervous system. *Neuron* **14**, 467–475.

Izumi, Y., Morino, H., Oda, M., Maruyama, H., Udaka, F., Kameyama, M., Nakamura, S., and Kawakami, H. (2001). Genetic studies in Parkinson's disease with an alpha-synuclein/NACP gene polymorphism in Japan. *Neurosci. Lett.* **300**, 125–127.

Jenco, J. M., Rawlingson, A., Daniels, B., and Morris, A. J. (1998). Regulation of phospholipase D2: selective inhibition of mammalian phospholipase D isoenzymes by alpha- and beta-synucleins. *Biochemistry* **37**, 4901–4909.

Jensen, P. H., Hager, H., Nielsen, M. S., Hojrup, P., Gliemann, J., and Jakes, R. (1999). Alpha-synuclein binds to Tau and stimulates the protein kinase A-catalyzed tau phosphorylation of serine residues 262 and 356. *J. Biol. Chem.* **274**, 25,481–25,489.

Jensen, P. H., Islam, K., Kenney, J., Nielsen, M. S., Power, J., and Gai, W. P. (2000). Microtubule-associated protein 1B is a component of cortical Lewy bodies and binds alpha-synuclein filaments. *J. Biol. Chem.* **275**, 21,500–21,507.

Jensen, P. H., Nielsen, M. S., Jakes, R., Dotti, C. G., and Goedert, M. (1998). Binding of alpha-synuclein to brain vesicles is abolished by familial Parkinson's disease mutation. *J. Biol. Chem.* **273**, 26,292–26,294.

Jo, E., Fuller, N., Rand, R. P., St. George-Hyslop, P., and Fraser, P. E. (2002). Defective membrane interactions of familial Parkinson's disease mutant A30P alpha-synuclein. *J. Mol. Biol.* **315**, 799–807.

Kahle, P. J., Neumann, M., Ozmen, L., Muller, V., Jacobsen, H., Schindzielorz, A., Okochi, M., Leimer, U., van Der Putten, H., Probst, A., Kremmer, E., Kretzschmar, H. A., and Haass, C. (2000). Subcellular localization of wild-type and Parkinson's disease-associated mutant alpha-synuclein in human and transgenic mouse brain. *J. Neurosci.* **20**, 6365–6373.

Kanda, S., Bishop, J. F., Eglitis, M. A., Yang, Y., and Mouradian, M. M. (2000). Enhanced vulnerability to oxidative stress by alpha-synuclein mutations and C-terminal truncation. *Neuroscience* **97**, 279–284.

Kazemi-Esfarjani, P., and Benzer, S. (2000). Genetic suppression of polyglutamine toxicity in Drosophila. *Science* **287**, 1837–1840.

Khan, N., Graham, E., Dixon, P., Morris, C., Mander, A., Clayton, D., Vaughan, J., Quinn, N., Lees, A., Daniel, S., Wood, N., and de Silva, R. (2001). Parkinson's disease is not associated with the combined alpha-synuclein/apolipoprotein E susceptibility genotype. *Ann. Neurol.* **49**, 665–668.

Kim, T. D., Paik, S. R., Yang, C. H., and Kim, J. (2000). Structural changes in alpha-synuclein affect its chaperone-like activity in vitro. *Protein Sci.* **9**, 2489–2496.

Kitada, T., Asakawa, S., Hattori, N., Matsumine, H., Yamamura, Y., Minoshima, S., Yokochi, M., Mizuno, Y., and Shimizu, N. (1998). Mutations in the parkin gene cause autosomal recessive juvenile parkinsonism. *Nature* **392**, 605–608.

Klement, I. A., Skinner, P. J., Kaytor, M. D., Yi, H., Hersch, S. M., Clark, H. B., Zoghbi, H. Y., and Orr, H. T. (1998). Ataxin-1 nuclear localization and aggregation: role in polyglutamine- induced disease in SCA1 transgenic mice. *Cell* **95**, 41–53.

Kruger, R., Kuhn, W., Leenders, K. L., Sprengelmeyer, R., Muller, T., Woitalla, D., Portman, A. T., Maguire, R. P., Veenma, L., Schroder, U., Schols, L., Epplen, J. T., Riess, O., and Przuntek, H. (2001). Familial parkinsonism with synuclein pathology: clinical and PET studies of A30P mutation carriers. *Neurology* **56**, 1355–1362.

Kruger, R., Kuhn, W., Muller, T., Woitalla, D., Graeber, M., Kosel, S., Przuntek, H., Epplen, J. T., Schols, L., and Riess, O. (1998). Ala30Pro mutation in the gene encoding alpha-synuclein in Parkinson's disease. *Nat. Genet.* **18**, 106–108.

Kruger, R., Vieira-Saecker, A. M., Kuhn, W., Berg, D., Muller, T., Kuhnl, N., Fuchs, G. A., Storch, A., Hungs, M., Woitalla, D., Przuntek, H., Epplen, J. T., Schols, L., and Riess, O. (1999). Increased susceptibility to sporadic Parkinson's disease by a certain combined alpha-synuclein/apolipoprotein E genotype. *Ann. Neurol.* **45**, 611–617.

Langston, J. W. (2002). The Impact of MPTP on Parkinson's Disease: Past, Present, and Future. In Parkinson's Disease: Diagnosis and Clinical Management. (W. J. Weiner, ed.), pp. 299–327. Demos Medical Publishing, Inc., New York.

Langston, J. W., Ballard, P., Tetrud, J. W., and Irwin, I. (1983). Chronic Parkinsonism in humans due to a product of meperidine-analog synthesis. *Science* **219**, 979–980.

Langston, J. W., Sastry, S., Chan, P., Forno, L. S., Bolin, L. M., and Di Monte, D. A. (1998). Novel alpha-synuclein-immunoreactive proteins in brain samples from the Contursi kindred, Parkinson's, and Alzheimer's disease. *Exp. Neurol.* **154**, 684–690.

Lavedan, C., Leroy, E., Dehejia, A., Buchholtz, S., Dutra, A., Nussbaum, R. L., and Polymeropoulos, M. H. (1998). Identification, localization and characterization of the human gamma-synuclein gene. *Hum. Genet.* **103**, 106–112.

Le Couteur, D. G., McLean, A. J., Taylor, M. C., Woodham, B. L., and Board, P. G. (1999). Pesticides and Parkinson's disease. *Biomed. Pharmacother.* **53**, 122–130.

Lee, H. J., Choi, C., and Lee, S. J. (2002a). Membrane-bound alpha-synuclein has a high aggregation propensity and the ability to seed the aggregation of the cytosolic form. *J. Biol. Chem.* **277**, 671–678.

Lee, H. J., Shin, S. Y., Choi, C., Lee, Y. H., and Lee, S. J. (2002b). Formation and removal of alpha-synuclein aggregates in cells exposed to mitochondrial inhibitors. *J. Biol. Chem.* **277**, 5411–5417.

Lee, M., Hyun, D., Halliwell, B., and Jenner, P. (2001). Effect of the overexpression of wild-type or mutant alpha-synuclein on cell susceptibility to insult. *J. Neurochem.* **76**, 998–1009.

Li, J., Uversky, V. N., and Fink, A. L. (2001). Effect of familial

Parkinson's disease point mutations A30P and A53T on the structural properties, aggregation, and fibrillation of human alpha-synuclein. *Biochemistry* **40**, 11604–11613.

Manning-Bog, A. B., McCormack, A. L., Li, J., Uversky, V. N., Fink, A. L., and Di Monte, D. A. (2002). The herbicide paraquat causes up-regulation and aggregation of alpha-synuclein in mice: paraquat and alpha-synuclein. *J. Biol. Chem.* **277**, 1641–1644.

Markopoulou, K., Wszolek, Z. K., and Pfeiffer, R. F. (1995). A Greek-American kindred with autosomal dominant, levodopa-responsive parkinsonism and anticipation. *Ann. Neurol.* **38**, 373–378.

Masliah, E., Rockenstein, E., Veinbergs, I., Mallory, M., Hashimoto, M., Takeda, A., Sagara, Y., Sisk, A., and Mucke, L. (2000). Dopaminergic loss and inclusion body formation in alpha-synuclein mice: implications for neurodegenerative disorders. *Science* **287**, 1265–1269.

Matsuoka, Y., Vila, M., Lincoln, S., McCormack, A., Picciano, M., LaFrancois, J., Yu, X., Dickson, D., Langston, W. J., McGowan, E., Farrer, M., Hardy, J., Duff, K., Przedborski, S., and Di Monte, D. A. (2001). Lack of nigral pathology in transgenic mice expressing human alpha-synuclein driven by the tyrosine hydroxylase promoter. *Neurobiol. Dis.* **8**, 535–539.

Munoz, E., Oliva, R., Obach, V., Marti, M. J., Pastor, P., Ballesta, F., and Tolosa, E. (1997). Identification of Spanish familial Parkinson's disease and screening for the Ala53Thr mutation of the alpha-synuclein gene in early onset patients. *Neurosci. Lett.* **235**, 57–60.

Murphy, D. D., Rueter, S. M., Trojanowski, J. Q., and Lee, V. M. (2000). Synucleins are developmentally expressed, and alpha-synuclein regulates the size of the presynaptic vesicular pool in primary hippocampal neurons. *J. Neurosci.* **20**, 3214–3220.

Nakamura, T., Yamashita, H., Takahashi, T., and Nakamura, S. (2001). Activated Fyn phosphorylates alpha-synuclein at tyrosine residue 125. *Biochem. Biophys. Res. Commun.* **280**, 1085–1092.

Okochi, M., Walter, J., Koyama, A., Nakajo, S., Baba, M., Iwatsubo, T., Meijer, L., Kahle, P. J., and Haass, C. (2000). Constitutive phosphorylation of the Parkinson's disease associated alpha-synuclein. *J. Biol. Chem.* **275**, 390–397.

Ostrerova, N., Petrucelli, L., Farrer, M., Mehta, N., Choi, P., Hardy, J., and Wolozin, B. (1999). Alpha-Synuclein shares physical and functional homology with 14-3-3 proteins. *J. Neurosci.* **19**, 5782–5791.

Ostrerova-Golts, N., Petrucelli, L., Hardy, J., Lee, J. M., Farer, M., and Wolozin, B. (2000). The A53T alpha-synuclein mutation increases iron-dependent aggregation and toxicity. *J. Neurosci.* **20**, 6048–6054.

Papadimitriou, A., Veletza, V., Hadjigeorgiou, G. M., Patrikiou, A., Hirano, M., and Anastasopoulos, I. (1999). Mutated alpha-synuclein gene in two Greek kindreds with familial PD: incomplete penetrance? *Neurology* **52**, 651–654.

Papapetropoulos, S., Paschalis, C., Athanassiadou, A., Papadimitriou, A., Ellul, J., Polymeropoulos, M. H., and Papapetropoulos, T. (2001). Clinical phenotype in patients with alpha-synuclein Parkinson's disease living in Greece in comparison with patients with sporadic Parkinson's disease. *J. Neurol. Neurosurg. Psychiatry* **70**, 662–665.

Parsian, A., Racette, B., Zhang, Z. H., Chakraverty, S., Rundle, M., Goate, A., and Perlmutter, J. S. (1998). Mutation, sequence analysis, and association studies of alpha-synuclein in Parkinson's disease. *Neurology* **51**, 1757–1759.

Pastor, P., Munoz, E., Ezquerra, M., Obach, V., Marti, M. J., Valldeoriola, F., Tolosa, E., and Oliva, R. (2001). Analysis of the coding and the 5′ flanking regions of the alpha-synuclein gene in patients with Parkinson's disease. *Mov. Disord.* **16**, 1115–1119.

Payton, J. E., Perrin, R. J., Clayton, D. F., and George, J. M. (2001). Protein-protein interactions of alpha-synuclein in brain homogenates and transfected cells. *Brain Res. Mol. Brain. Res.* **95**, 138–145.

Perrin, R. J., Woods, W. S., Clayton, D. F., and George, J. M. (2000). Interaction of human alpha-synuclein and Parkinson's disease variants with phospholipids. Structural analysis using site-directed mutagenesis. *J. Biol. Chem.* **275**, 34,393–34,398.

Polymeropoulos, M. H., Higgins, J. J., Golbe, L. I., Johnson, W. G., Ide, S. E., Di Iorio, G., Sanges, G., Stenroos, E. S., Pho, L. T., Schaffer, A. A., Lazzarini, A. M., Nussbaum, R. L., and Duvoisin, R. C. (1996). Mapping of a gene for Parkinson's disease to chromosome 4q21–q23. *Science* **274**, 1197–1199.

Polymeropoulos, M. H., Lavedan, C., Leroy, E., Ide, S. E., Dehejia, A., Dutra, A., Pike, B., Root, H., Rubenstein, J., Boyer, R., Stenroos, E. S., Chandrasekharappa, S., Athanassiadou, A., Papapetropoulos, T., Johnson, W. G., Lazzarini, A. M., Duvoisin, R. C., Di Iorio, G., Golbe, L. I., and Nussbaum, R. L. (1997). Mutation in the alpha-synuclein gene identified in families with Parkinson's disease. *Science* **276**, 2045–2047.

Pronin, A. N., Morris, A. J., Surguchov, A., and Benovic, J. L. (2000). Synucleins are a novel class of substrates for G protein-coupled receptor kinases. *J. Biol. Chem.* **275**, 26515–226512.

Samii, A., Markopoulou, K., Wszolek, Z. K., Sossi, V., Dobko, T., Mak, E., Calne, D. B., and Stoessl, A. J. (1999). PET studies of parkinsonism associated with mutation in the alpha-synuclein gene. *Neurology* **53**, 2097–2102.

Saudou, F., Finkbeiner, S., Devys, D., and Greenberg, M. E. (1998). Huntingtin acts in the nucleus to induce apoptosis but death does not correlate with the formation of intranuclear inclusions. *Cell* **95**, 55–66.

Scott, W. K., Yamaoka, L. H., Stajich, J. M., Scott, B. L., Vance, J. M., Roses, A. D., Pericak-Vance, M. A., Watts, R. L., Nance, M., Hubble, J., Koller, W., Stern, M. B., Colcher, A., Allen, F. H., Jr., Hiner, B. C., Jankovic, J., Ondo, W., Laing, N. G., Mastaglia, F., Goetz, C., Pappert, E., Small, G. W., Masterman, D., Haines, J. L., and Davies, T. L. (1999). The alpha-synuclein gene is not a major risk factor in familial Parkinson disease. *Neurogenetics* **2**, 191–192.

Serpell, L. C., Berriman, J., Jakes, R., Goedert, M., and Crowther, R. A. (2000). Fiber diffraction of synthetic alpha-synuclein filaments shows amyloid-like cross-beta conformation. *Proc. Natl. Acad. Sci. U.S.A.* **97**, 4897–4902.

Sharon, R., Goldberg, M. S., Bar-Josef, I., Betensky, R. A., Shen, J., and Selkoe, D. J. (2001). Alpha-Synuclein occurs in lipid-rich high molecular weight complexes, binds fatty acids, and shows homology to the fatty acid-binding proteins. *Proc. Natl. Acad. Sci. U.S.A.* **98**, 9110–9115.

Shimura, H., Hattori, N., Kubo, S., Mizuno, Y., Asakawa, S., Minoshima, S., Shimizu, N., Iwai, K., Chiba, T., Tanaka, K., and Suzuki, T. (2000). Familial Parkinson disease gene product, parkin, is a ubiquitin-protein ligase. *Nat. Genet.* **25**, 302–305.

Shimura, H., Schlossmacher, M. G., Hattori, N., Frosch, M. P., Trockenbacher, A., Schneider, R., Mizuno, Y., Kosik, K. S., and Selkoe, D. J. (2001). Ubiquitination of a new form of alpha-synuclein by parkin from human brain: implications for Parkinson's disease. *Science* **293**, 263–269.

Souza, J. M., Giasson, B. I., Lee, V. M., and Ischiropoulos, H. (2000). Chaperone-like activity of synucleins. *FEBS Lett* **474**, 116–119.

Spillantini, M. G., Crowther, R. A., Jakes, R., Hasegawa, M., and Goedert, M. (1998). Alpha-Synuclein in filamentous inclusions of Lewy bodies from Parkinson's disease and dementia with Lewy bodies. *Proc. Natl. Acad. Sci. U.S.A.* **95**, 6469–6473.

Spillantini, M. G., Divane, A., and Goedert, M. (1995). Assignment of human alpha-synuclein (SNCA) and beta-synuclein (SNCB) genes to chromosomes 4q21 and 5q35. *Genomics* **27**, 379–381.

Spira, P. J., Sharpe, D. M., Halliday, G., Cavanagh, J., and Nicholson, G. A. (2001). Clinical and pathological features of a Parkinsonian syndrome in a family with an Ala53Thr alpha-synuclein mutation. *Ann. Neurol.* **49**, 313–319.

Tabrizi, S. J., Orth, M., Wilkinson, J. M., Taanman, J. W., Warner, T. T., Cooper, J. M., and Schapira, A. H. (2000). Expression of mutant alpha-synuclein causes increased susceptibility to dopamine toxicity. *Hum. Mol. Genet.* **9**, 2683–2689.

Tan, E. K., Matsuura, T., Nagamitsu, S., Khajavi, M., Jankovic, J., and

Ashizawa, T. (2000). Polymorphism of NACP-Rep1 in Parkinson's disease: an etiologic link with essential tremor? *Neurology* **54**, 1195–1198.

Tanner, C. M. (1989). The role of environmental toxins in the etiology of Parkinson's disease. *Trends Neurosci.* **12**, 49–54.

Uversky, V. N., Lee, H. J., Li, J., Fink, A. L., and Lee, S. J. (2001a). Stabilization of partially folded conformation during {alpha}-synuclein oligomerization in both purified and cytosolic preparations. *J. Biol. Chem.* **5**, 5.

Uversky, V. N., Li, J., and Fink, A. L. (2001b). Evidence for a partially folded intermediate in alpha-synuclein fibril formation. *J. Biol. Chem.* **276**, 10737–10744.

Uversky, V. N., Li, J., and Fink, A. L. (2001c). Metal-triggered structural transformations, aggregation, and fibrillation of human alpha-synuclein. A possible molecular link between Parkinson's disease and heavy metal exposure. *J. Biol. Chem.* **276**, 44284–44296.

Uversky, V. N., Li, J., and Fink, A. L. (2001d). Pesticides directly accelerate the rate of alpha-synuclein fibril formation: a possible factor in Parkinson's disease. *FEBS Lett.* **500**, 105–108.

Uversky, V. N., Li, J., Souillac, P., Jakes, R., Goedert, M., and Fink, A. L. (2002). Biophysical properties of the synucleins and their propensities to fibrillate: inhibition of alpha-synuclein assembly by beta- and gamma-synucleins. *J. Biol. Chem.* **25**, 25.

van der Putten, H., Wiederhold, K. H., Probst, A., Barbieri, S., Mistl, C., Danner, S., Kauffmann, S., Hofele, K., Spooren, W. P., Ruegg, M. A., Lin, S., Caroni, P., Sommer, B., Tolnay, M., and Bilbe, G. (2000). Neuropathology in mice expressing human alpha-synuclein. *J. Neurosci.* **20**, 6021–6029.

Vaughan, J., Durr, A., Tassin, J., Bereznai, B., Gasser, T., Bonifati, V., De Michele, G., Fabrizio, E., Volpe, G., Bandmann, O., Johnson, W. G., Golbe, L. I., Breteler, M., Meco, G., Agid, Y., Brice, A., Marsden, C. D., and Wood, N. W. (1998a). The alpha-synuclein Ala53Thr mutation is not a common cause of familial Parkinson's disease: a study of 230 European cases. European Consortium on Genetic Susceptibility in Parkinson's Disease. *Ann. Neurol.* **44**, 270–273.

Vaughan, J. R., Farrer, M. J., Wszolek, Z. K., Gasser, T., Durr, A., Agid, Y., Bonifati, V., DeMichele, G., Volpe, G., Lincoln, S., Breteler, M., Meco, G., Brice, A., Marsden, C. D., Hardy, J., and Wood, N. W. (1998b). Sequencing of the alpha-synuclein gene in a large series of cases of familial Parkinson's disease fails to reveal any further mutations. The European Consortium on Genetic Susceptibility in Parkinson's Disease (GSPD). *Hum. Mol. Genet.* **7**, 751–753.

Vila, M., Vukosavic, S., Jackson-Lewis, V., Neystat, M., Jakowec, M., and Przedborski, S. (2000). Alpha-synuclein up-regulation in substantia nigra dopaminergic neurons following administration of the parkinsonian toxin MPTP. *J. Neurochem.* **74**, 721–729.

Volles, M. J., Lee, S. J., Rochet, J. C., Shtilerman, M. D., Ding, T. T., Kessler, J. C., and Lansbury, P. T., Jr. (2001). Vesicle permeabilization by protofibrillar alpha-synuclein: implications for the pathogenesis and treatment of Parkinson's disease. *Biochemistry* **40**, 7812–7819.

Wakabayashi, K., Engelender, S., Yoshimoto, M., Tsuji, S., Ross, C. A., and Takahashi, H. (2000). Synphilin-1 is present in Lewy bodies in Parkinson's disease. *Ann. Neurol.* **47**, 521–523.

Wang, W. W., Khajavi, M., Patel, B. J., Beach, J., Jankovic, J., and Ashizawa, T. (1998). The G209A mutation in the alpha-synuclein gene is not detected in familial cases of Parkinson disease in non-Greek and/or Italian populations. *Arch. Neurol.* **55**, 1521–1523.

Wood, S. J., Wypych, J., Steavenson, S., Louis, J. C., Citron, M., and Biere, A. L. (1999). Alpha-synuclein fibrillogenesis is nucleation-dependent. Implications for the pathogenesis of Parkinson's disease. *J. Biol. Chem.* **274**, 19509–19512.

Yasui, M., Kihira, T., and Ota, K. (1992). Calcium, magnesium and aluminum concentrations in Parkinson's disease. *Neurotoxicology* **13**, 593–600.

Zareparsi, S., Kaye, J., Camicioli, R., Kramer, P., Nutt, J., Bird, T., Litt, M., Payami, H., and Kay, J. (1998). Analysis of the alpha-synuclein G209A mutation in familial Parkinson's disease. *Lancet* **351**, 37–38.

Zhang, Y., Gao, J., Chung, K. K., Huang, H., Dawson, V. L., and Dawson, T. M. (2000). Parkin functions as an E2-dependent ubiquitin-protein ligase and promotes the degradation of the synaptic vesicle-associated protein, CDCrel-1. *Proc. Natl. Acad. Sci. U.S.A.* **97**, 13354–13359.

Zhou, W., Hurlbert, M. S., Schaack, J., Prasad, K. N., and Freed, C. R. (2000). Overexpression of human alpha-synuclein causes dopamine neuron death in rat primary culture and immortalized mesencephalon-derived cells. *Brain Res.* **866**, 33–43.

Zhou, W., Schaack, J., Zawada, W. M., and Freed, C. R. (2002). Overexpression of human alpha-synuclein causes dopamine neuron death in primary human mesencephalic culture. *Brain Res.* **926**, 42–50.

# CHAPTER 28

# Parkin Mutations (Park2)

**YOSHIKUNI MIZUNO**
*Department of Neurology*
*Juntendo University School of Medicine*
*Tokyo 113-8421, Japan*

**NOBUTAKA HATTORI**
*Department of Neurology*
*Juntendo University School of Medicine*
*Tokyo 113-8421, Japan*

**HIROYO YOSHINO**
*Department of Neurology*
*Juntendo University School of Medicine*
*Tokyo 113-8421, Japan*

**SHIUCHI ASAKAWA**
*Department of Molecular Biology*
*Keio University School of Medicine*
*Tokyo 160-8582, Japan*

**SHINSEI MINOSHIMA**
*Department of Molecular Biology*
*Keio University School of Medicine*
*Tokyo 160-8582, Japan*

**NOBUYOSHI SHIMIZU**
*Department of Molecular Biology*
*Keio University School of Medicine*
*Tokyo 160-8582, Japan*

**TOSHIAKI SUZUKI**
*Department of Molecular Oncology*
*The Tokyo Metropolitan Institute of Medical Science*
*Tokyo, Japan*

**TOMOKI CHIBA**
*Department of Molecular Oncology*
*The Tokyo Metropolitan Institute of Medical Science*
*Tokyo, Japan*

**KEIJI TANAKA**
*Department of Molecular Oncology*
*The Tokyo Metropolitan Institute of Medical Science*
*Tokyo, Japan*

I. Introduction
II. Gene
   A. Normal Gene Function
   B. Abnormal Gene Function
III. Diagnostic and Ancillary Tests
   A. DNA Test
IV. Neuroimaging
   A. MRI
   B. PET, fMRI, and MR Spectroscopy
V. Neuropathology
VI. Cellular and Animal Models
VII. Genotype and Phenotype Correlations/Modifying Alleles
VIII. Treatment
   Acknowledgments
   References

Park2 (autosomal recessive-juvenile parkinsonism, AR-JP) presents young-onset parkinsonism, consisting of gait disturbance, rest tremor, cogwheel rigidity, and bradykinesia. Clinical features are essentially similar to those of late-onset sporadic Parkinson's disease. They respond to levodopa well. Progression is slow. Pathologic features include extensive nigral and locus coeruleus degeneration and gliosis without Lewy body formation. The disease gene has been identified and named *parkin*, which is located on the long arm of chromosome 6 at 6q25–27.2. Varieties of deletion mutations and point mutations of *parkin* have been found in patients with Park2. Also compound heterozygotes were found. Parkin protein functions as a ubiquitin ligase and a number of candidate substrates for Parkin have been reported including CDCrel 1, α-synuclein 22, Pael receptor, synphilin-1, and CDCrel 2A. Accumulation of one or more of the candidate substrates appears to be the cause of nigral degeneration. Transgenic and knock-out animals of *parkin* have not been reported in the literature. Park2 has been considered to represent the most common form of familial Parkinson's disease.

## I. INTRODUCTION

Park2 was first delineated as a distinct clinical entity by Yamamura *et al.* (1973). They reported 16 patients (13 familial cases in 5 unrelated families and 3 sporadic cases);

clinical features of 11 patients from the initial 4 families were essentially identical; only in 1 of those 11 patients, the disease started at age 42; in the remaining 10 patients the initial symptoms appeared between 17 and 28 years; female preponderance was noted (M:F = 1:10); all the patients showed tremor, rigidity, bradykinesia, and postural instability. Sleep benefit, i.e., temporary improvement in parkinsonism after a nap or sleep, was a characteristic feature of their patients. Dementia and autonomic failures were absent. Consanguineous marriages were seen in two families; and none of the affected parents had parkinsonism indicating autosomal recessive mode of inheritance. The progression was slow.

Subsequently, Ishikawa and Tsuji (1996) and Yamamura et al. (1998) summarized the clinical features (Table 28.1). Clinical features of this entity were thought to be uniform initially, i.e., onset usually before 40 years of age, more often gait disturbance than tremor as an initial symptom (tremor is the most common initial symptom in sporadic Parkinson's disease), manifesting four cardinal symptoms of Parkinson's disease, good response to levodopa, and high incidence of levodopa-induced dyskinesia and motor fluctuations.

With mapping of Park2 to the long arm of chromosome 6 at 6q25–27.2 (Matsumine et al., 1997), it became apparent that the age of onset could be as late as seventies and that the initial symptom was more often tremor in late onset patients (Klein et al., 2000a). Clinical features were essentially identical to those of idiopathic sporadic Parkinson's disease.

## II. GENE

### A. Normal Gene Function

#### 1. Character of *parkin*

*Parkin* is a megagene of approximately 1.4 Mb, the second known largest gene after the dystrophin gene. The cDNA consists of 2960 bp, of which 1395 bp constitute the coding region encoding a protein composed of 465 amino acids of 52 kDa (Kitada et al., 1998). The salient features of *parkin* are summarized in Table 28.2 and nucleotide and amino acid numbers in Table 28.3. The promoter region of parkin was analyzed recently by West et al. (2001). The *parkin* promoter lacks TATA or CAAT boxes and appears to share homology to the α-synuclein promoter. In addition, it is interesting to note that a neighboring novel gene is located in a head-to-head direction with *parkin* with only a 198-bp interval (Asakawa et al., 2001).

Parkin has a ubiquitin-like domain in the amino terminal region and two RING finger-like structures on the carboxy terminal side. Between the two RINGs, there is an in-between-RING (IBR) structure comprising a RING-box in this region (Fig. 28.1).

TABLE 28.1  Clinical Features of Park 2

|  | Yamamura et al., 1998 | Ishikawa and Tsuji 1996 |
|---|---|---|
| Number of families | 22 | 12 |
| Consanguinity | 10 | 11 |
| Number of patients | 43 | 17 |
| Male | 16 | 5 |
| Female | 27 | 12 |
| Age of onset | 26.1 ± 7.8 | 27.8 ± 9.0 |
| Sleep benefit | 95.3% | 100% |
| Initial symptom |  | nd |
| Dystonic gait | 18 |  |
| Parkinsonian gait | 8 |  |
| Rest tremor | 13 |  |
| Bradykinesia | 3 |  |
| Upper limb dystonia | 1 |  |
| Dystonia | 77.5% | 62.5% |
| Hyperreflexia | 92.5% | 64.7% |
| Levodopa response | 100% | 100% |

Note: nd; not described

TABLE 28.2  Profile of *parkin*

| Name | Parkin |
|---|---|
| Chromosome locus | 6q25.2–q27 |
| Total size | 1.4 Mb |
| Nunber of exons | 12 |
| cDNA | 2960 bp |
| Coding region | 1395 bp |
| N-terminal | 30% Homology to ubiquitin |
| C-terminal | 2 RING-finger motives and in-between RINGs |
| Gene product | 465 amino acids |
| Molecular weight | 51,652 |

TABLE 28.3  Nucleotide and Amino Acid Number of *parkin* cDNA and Parkin

|  | Nucleotide number | Amino acid number |
|---|---|---|
| Exon 1 | 1–7 | 1–2 |
| Exon 2 | 8–171 | 3–57 |
| Exon 3 | 172–412 | 58–137 |
| Exon 4 | 413–534 | 138–178 |
| Exon 5 | 535–618 | 179–206 |
| Exon 6 | 619–734 | 207–245 |
| Exon 7 | 735–871 | 246–290 |
| Exon 8 | 872–933 | 291–311 |
| Exon 9 | 934–1093 | 312–361 |
| Exon 10 | 1094–1167 | 362–389 |
| Exon 11 | 1168–1285 | 390–428 |
| Exon 12 | 1286–1398 | 428–465 |

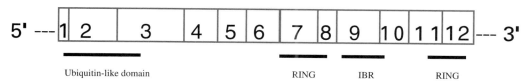

FIGURE 28.1  Structure of *parkin*. RING: Rare Interesting New Gene, IBR: In-between RINGs.

## 2. *parkin* in Animals

Kitada *et al.* (2001) isolated mouse cDNA clones that were homologous to human *parkin*. One of those cDNA clones had a 1392-bp open reading frame encoding a protein of 464 amino acids with a predicted molecular weight of 51,615. The amino acid sequence of mouse parkin exhibited 83.2% identity to human Parkin, including the ubiquitin-like domain at the N terminus and the RING finger-like domain near the C terminus. Two other clones had a 783-bp open reading frame encoding a truncated protein without RING-finger-like domain. This was a novel splicing variant. Northern blot analysis revealed that mouse *parkin* was expressed in various tissues including the brain, heart, liver, skeletal muscle, kidney, and the testis.

Stichel *et al.* (2000) cloned a mouse cDNA and generated polyclonal antisera against the N-terminal part of mouse Parkin. They studied Parkin expression profile using immunohistochemistry and Western blot analysis, and compared with that of the mRNA yielded by in situ hybridization and RT-PCR analysis. Parkin protein was widely distributed in all subdivisions of the mouse brain. The brain stem contained a large number of cells heavily expressing Parkin. Dopaminergic cells of the substantia nigra pars compacta exhibited high levels of Parkin mRNA but no Parkin protein, while striatum contained immuno-positive profiles but no mRNA signals. In human brains, nigral neurons do express Parkin heavily (Shimura *et al.*, 1999).

Gu *et al.* (2001) cloned a rat parkin cDNA by screening a rat hypothalamus cDNA library with a $^{32}$P-labeled probe containing the entire open reading frame of the human *parkin* cDNA. The rat parkin cDNA consisted of 1576 bp, which encoded a 465 amino acid protein (the same number of amino acids as human Parkin). There were 85 and 95% identity to the human and mouse parkin cDNA, respectively. Both the N-terminal ubiquitin and the RING-IBR-RING finger domains were highly conserved among rat, human, and mouse Parkin. Immunostaining with an antibody raised against a synthetic peptide corresponding to amino acids 295–311 of the parkin sequence showed cytoplasmic location and Parkin was widely distributed in the rat brain. Glia cells expressing Parkin were also detected. The authors concluded that the role of Parkin might be much more global than previously thought.

D'Agata *et al.* (2000a) isolated and sequenced a partial cDNA coding for the rat homolog of parkin by RT-PCR. They identified a 1.46 kb rat cDNA clone containing a 1376-bp coding sequence, which showed strong similarity with the human parkin cDNA. *In situ* hybridization revealed widespread expression of parkin in the rat brain. Thus parkin is widely expressed both in the messenger level and the protein level in the brain.

## 3. Expression of *parkin* Messenger

Parkin mRNA is expressed ubiquitously among general organs and brain regions (Kitada *et al.*, 1998; Solano *et al.*, 2000; Wang *et al.*, 2001). Solano *et al.* (2000) reported more robust expression of parkin mRNA in the human substantia nigra by the digoxygenin *in situ* hybridization method. Parkin messenger is also expressed in peripheral leukocytes, but a shorter messenger, in which exons 3–5 were spliced out, was a predominant form (Sunada *et al.*, 1999). We recently compared the distribution of *parkin* messenger and ubiquitin conjugating enzyme R7 (UbcR7) messenger in rat brains, as the latter is one of the interacting proteins of Parkin. We found essentially identical distribution of these two messengers (Wang *et al.*, 2001).

Solano *et al.* (2000) examined expression of mRNA for α-synuclein, parkin, and ubiquitin carboxy-terminal hydrolase-L1 (UCH-L1) by in situ hybridization in human brain tissues. Expression of α-synuclein and parkin messages were quite similar and robust in the melanin-containing neurons in the substantia nigra, while expression of UCH-L1 was more uniform. They suggested that those two proteins might be involved in common pathways contributing to the pathophysiology of Parkinson's disease.

## 4. Expression of Parkin Protein

Parkin protein is expressed in many structures of the brain (Shimura *et al.*, 1999; Huynh *et al.*, 2000; Zarate-Lagunes *et al.*, 2001), but it is more intensely expressed in the substantia nigra (Shimura *et al.*, 1999), hippocampal formation, the pallidal complex, the red nucleus, and the cerebellum (Zarate-Lagunes *et al.*, 2001) in human brains. Parkin immunoreactivity was reported to be absent in the glial-fibrillary acidic protein-positive astrocytes in rats (D'Agata *et al.*, 2001b).

At cellular level, Parkin is expressed mainly in the Golgi apparatus and synaptic vesicles in cultured cells

(Kubo et al., 2001). In human brains, Parkin was also found in the cytoplasmic fraction and the microsome fraction, but not in the nuclear and the mitochondrial fractions (Shimura et al., 1999).

Parkin is a highly conserved protein. It is expressed in brains of rat, mouse, bird, frog, and *Drosophila* brains (D'Agata et al., 2001b; Horowitz et al. 2001a, b).

## 5. Function of Parkin

As expected from its unique structure, Parkin was found to be a ubiquitin ligase, an E3 of the ubiquitin system (Shimura et al., 2000). Choi et al. (2000) reported di-ubiquitination of Parkin itself and suggested proteolysis of Parkin by 26S proteasome.

Imai et al. (2000) also showed that Parkin is a RING-type E3 ubiquitin-protein ligase which binds to E2 ubiquitin-conjugating enzymes (UbcH7 and UbcH8), through its RING-IBR-RING motif. They also found that unfolded protein stress-induced up-regulation of both the mRNA and the protein level of Parkin. Furthermore, over-expression of Parkin specifically suppressed unfolded protein stress-induced cell death. Their results would indicate that Parkin contributes to protection from neurotoxicity induced by unfolded protein stresses.

Recently, several candidate substrates for Parkin have been described including CDCrel-1 (Zhang et al., 2000), 22-kDa α-synuclein (Shimura et al., 2001), Pael receptor (Imai et al., 2001), and synphilin-1 (Chung et al., 2001). Accumulation of 22-kDa α-synuclein and Pael receptor in autopsied brains of Park2 patients was also reported. But the exact molecular mechanism of nigral degeneration of Park2 is still unknown. Interestingly, iron content in the substantia nigra was markedly increased in Park2 patients; it was more pronounced than that seen in sporadic Parkinson's disease patients (Takanashi et al., 2001).

Huynh et al. (2000) reported co-localization of parkin with actin filaments in COS 1 kidney cells and nerve growth factor-induced PC12 neurons using an antibody against Parkin and phalloidin, a strong actin filament binder.

### B. Abnormal Gene Function

We have studied Parkin immunoreactivity in four autopsied patients with exon 4 deletion of *parkin* (Shimura et al., 1999). No immunoreactivity was found in the brains. Thus loss of Parkin function appears to be responsible for the selective nigral neuronal degeneration in Park2. It is not known yet whether or not other mutated parkin proteins such as those with small deletions or point mutations are expressed in brains. Thus the functions of abnormal *parkin* or Parkin have yet to be studied.

## III. DIAGNOSTIC AND ANCILLARY TESTS

### A. DNA Test

Parkin mutations can be tested using peripheral blood or tissue specimens as shown in the next sections.

#### 1. PCR for Homozygous Deletion Mutations

The primer set that we are using to amplify each exon of parkin is shown in Table 28.4. If a homozygous deletion mutation of any exon or exons of *parkin* is found, there is no question about the genetic diagnosis of Park2. We have not found apparently normal persons carrying a homozygous deletion mutation of *parkin*. Thus the specificity is very high. The sensitivity of this method is still under investigation.

TABLE 28.4  Primer Sets for Amplifying and Sequencing of *parkin* Exons

|       | Forward (5'–3')            | Reverse (5'–3')            |
|-------|----------------------------|----------------------------|
| Ex 1  | GCGCGGCTGGCGCCGCTGCGCGCA   | GCGGCGCAGAGAGGCTGTAC       |
| Ex 2  | ATGTTGCTATCACCATTTAAGGG    | AGATTGGCAGCGCAGGCGGCATG    |
| Ex 3  | ACATGTCACTTTTGCTTCCCT      | AGGCCATGCTCCATGCAGACTGC    |
| Ex 4  | ACAAGCTTTTAAAGAGTTTCTTGT   | AGGCAATGTGTTAGTACACA       |
| Ex 5  | ACATGTCTTAAGGAGTACATTT     | TCTCTAATTTCCTGGCAAACAGTG   |
| Ex 6  | AGAGATTGTTTACTGTGGAAACA    | GAGTGATGCTATTTTTAGATCCT    |
| Ex 7  | TGCCTTTCCACACTGACAGGTACT   | TCTGTTCTTCATTAGCATTAGAGA   |
| Ex 8  | TGATAGTCATAACTGTGTGTAAG    | ACTGTCTCATTAGCGTCTATCTT    |
| Ex 9  | GGGTGAAATTTGCAGTCAGT       | AATATAATCCCAGCCCATGTGCA    |
| Ex10  | ATTGCCAAATGCAACCTAATGTC    | TTGGAGGAATGAGTAGGGCATT     |
| Ex11  | ACAGGGAACATAAACTCTGATCC    | CAACACACCAGGCACCTTCAGA     |
| Ex12  | GTTTGGGAATGCGTGTTTT        | AGAATTAGAAAATGAAGGTAGACA   |

## 2. Sequencing of *parkin*

If there is no apparent homozygous deletions in parkin, then the next step is sequencing of parkin to detect small deletions and point mutations. The set of primers that we use for the sequence is shown in Table 28.4. The sensitivity and the specificity of this method are still under investigation. If there is a homozygous point mutation (except for normal polymorphisms) or a small deletion, there is no question about genetic diagnosis of Park2.

But at times, heterozygous point mutations are found in patients whose clinical pictures are consistent with Park2. Such patients are likely to have a deletion mutation on the other chromosome. This hemizygous deletion mutation can be detected by the real time PCR. If there is no hemizygous deletion mutation in the coding region, then the analysis of the promotor region and introns are necessary.

## 3. Real Time PCR for the Detection of Compound Heterozygotes

There are variable combinations of deletion mutations and point mutations in Park2 patients. The detection of such patients is by no means easy. Hemizygous exonic deletion mutations can be detected by quantifying the amount of each exon by real time PCR. But still not all the combinations of parkin mutations can be detected with the current method.

## 4. Haplotype Analysis

In case a parkin mutation is not found and still the clinical features are consistent with early-onset autosomal recessive Parkinson's disease, the next step is the haplotype analysis to see the linkage to the Park2 locus. Microsatellite markers for haplotype analysis are shown in Table 28.5. Multiple affected and nonaffected subjects are necessary to discuss the linkage in any given family. If the linkage to the Park2 locus is excluded, the next study is the haplotype analysis using microsatellite markers of the Park6 and Park7 loci in the short arm of chromosome 1 (Duijin *et al.*, 2001; Valente *et al.*, 2001).

## 5. Mutations Found in *parkin*

Mutations found in our laboratories and those reported in the literature are summarized in Table 28.6 (Kitada *et al.*, 1998; Hattori *et al.*, 1998a, b; Leroy *et al.*, 1998; Abbas *et al.*, 1999; Lucking *et al.*, 2000; Klein *et al.*, 2000a; Maruyama *et al.*, 2000; Munoz *et al.*, 2000; Tassin *et al.*, 2000; Periquet *et al.*, 2001; Ujiki *et al.*, 2001).

## 6. Parkin Mutations in Apparently Sporadic Patients

Kobayashi *et al.* (2000) analyzed parkin gene mutations in a cohort of sporadic PD patients. They screened 200 patients (103 women and 97 men) for parkin gene mutations, and 4 out of the 200 patients had homozygous exonic deletions in *parkin*. The ages of onset of these patients were 33, 38, 47, and 48, respectively. These four patients comprised 6.3% of the patients with an age of onset below 50 years. Clinical features were essentially similar to those without parkin gene mutations. Lucking *et al.* (2000) also reported apparently sporadic patients with *parkin* mutation with later onset of ages.

## 7. Pseudoautosomal Dominant Inheritance

Klein *et al.* (2000a) reported *parkin* deletions in a family with adult-onset, tremor-dominant parkinsonism. They presented a large pedigree from South Tyrol with adult-onset and clinically typical tremor-dominant parkinsonism of apparently autosomal dominant inheritance. Compound heterozygous deletions in the parkin gene (one large and one truncating) were identified in four affected male siblings. Clinical features were essentially identical to those of idiopathic PD. The age of onset was 31–75, but 8 out of 11 patients had the disease onset after 50 years. None of them displayed any of the clinical hallmarks described in patients with previously reported parkin mutations. They suggested that the parkin gene might be important in the etiology of the more frequent late-onset typical PD. The reason for pseudo-autosomal dominant inheritance in this report appears to be extensive consanguineous marriages in their family. In addition, carrier states may have been common among those people living in a small restricted region without much movement in the population.

Maruyama *et al.* (2000) also reported a Japanese family with pseudo-autosomal dominant inheritance with *parkin* mutations. In this family, the age of onset was in the young range.

## 8. Polymorphisms of *parkin*

Three polymorphisms of *parkin*, i.e., S/N167, R/W366, and V/L380, have been reported to date (Wang *et al.*, 1999; Hu *et al.*, 2000; Satoh and Kuroda 2000).

Satoh and Kuroda (1999) studied a codon 167 serine/asparagine (S/N167) polymorphism located in exon 4 in 71 patients with sporadic Parkinson's disease and 109 age-matched non-Parkinson's disease controls. The

TABLE 28.5 Microsattelite Markers for the Study of Linkage to the Park2 Locus

| | | | |
|---|---|---|---|
| D6S225 | D6S253 | D6S263 | D6S294 |
| D6S305 | D6S364 | D6S411 | D6S434 |
| D6S437 | D6S444 | SOD2 | |

TABLE 28.6  Mutations Found in *parkin*

**Exonic deletions**
| | | | |
|---|---|---|---|
| Exon 2 | | | Lucking *et al.*, 2000, Tassin *et al.*, 2000 |
| Exon 3 | | | Hattori *et al.*, 1998a, Maruyama *et al.*, 2000 |
| Exon 4 | | | Hattori *et al.*, 1998a, Kobayashi *et al.*, 2000 |
| Exon 5 | | | Kitada *et al.*, 1998, Hattori *et al.*, 1998a |
| Exon 7 | | | |
| Exon 8 | | | Lucking *et al.*, 2000 |
| Exons 1–7/8 | | | Klein *et al.*, 2000a |
| Exons 2–3 | | | Lucking *et al.*, 2000 |
| Exons 2–4 | | | Maruyama *et al.*, 2000, Lucking *et al.*, 2000 |
| Exons 3–4 | | | Hattori *et al.*, 1998a, Kobayashi *et al.*, 2000 |
| Exons 3–5 | | | Kobayashi *et al.*, 2000 |
| Exons 3–6 | | | Lucking *et al.*, 2000 |
| Exons 3–7 | | | Kitada *et al.*, 1998 |
| Exons 3–9 | | | Lucking *et al.*, 2000 |
| Exons 4–6 | | | Portman *et al.*, 2001 |
| Exons 5–6 | | | Lucking *et al.*, 2000, Periquet *et al.*, 2001 |
| Exons 5–7 | | | Leroy *et al.*, 1998 |
| Exons 6–7 | | | Lucking *et al.*, 2000, Maruyama *et al.*, 2000 |
| Exons 7–9 | | | Lucking *et al.*, 2000 |
| Exons 8–9 | | | Lucking *et al.*, 2000, Tassin *et al.*, 2000 |

**Duplication**
| | | | |
|---|---|---|---|
| Exon 3 | | | Lucking *et al.*, 2000, Periquet *et al.*, 2001 |
| Exon 6 | | | Lucking *et al.*, 2000 |
| Exon 7 | | | Lucking *et al.*, 2000, Periquet *et al.*, 2001 |
| Exon 11 | | | Lucking *et al.*, 2000 |

**Triplication**
| | | | |
|---|---|---|---|
| Exon 2 | | | Lucking *et al.*, 2000 |

**Small deletions**
| | | | |
|---|---|---|---|
| 202-203delAG | 2 | Frame shift | Abbas *et al.*, 1999, Lucking *et al.*, 2000 |
| 255delA | 2 | Frame shift | Abbas *et al.*, 1999, Lucking *et al.*, 2000 |
| 535/6delG | intron 4 | Frame shift | Hattori *et al.*, 1998a |
| 871delG | 7 | Frame shift | Munoz *et al.*, 2000 |
| 970delC | 9 | Frame shift | Abbas *et al.* 1999 |
| 1072del? | 9 | Frame shift | Klein *et al.*, 2000a |
| 1142-3delGA | 9 | Frame shift | Lucking *et al.*, 2000 |

**Small insertions**
| | | | |
|---|---|---|---|
| 321-2insGT | 3 | Frame shift | Abbas *et al.*, 1999, Periquet *et al.*, 2001 |

**Point mutations**
| | | | |
|---|---|---|---|
| Arg33Stop | 2 | Nonsense | Maruyama *et al.*, 2000 |
| Arg33X | 2 | Missense | Maruyama *et al.*, 2000 |
| Arg42Pro | 2 | Missense | Terreni *et al.*, 2001 |
| Lys161Asn | 4 | Missense | Abbas *et al.*, 1999, Lucking *et al.*, 2000 |
| Lys211Asn | 6 | Missense | Lucking *et al.*, 2000 |
| Thr240Arg | 7 | Missense | Hattori *et al.*, 1998b |
| Arg256Cys | 7 | Missense | Abbas *et al.*, 1999, Lucking *et al.*, 2000 |
| Cys268Stop | 7 | Nonsense | Lucking *et al.*, 2000 |
| Arg275Trp | 7 | Missense | Abbas *et al.*, 1999, Lucking *et al.*, 2000 |
| Asp280Asn | 7 | Missense | Lucking *et al.*, 2000 |
| Cys289Gly | 7 | Missense | Lucking *et al.*, 2000 |
| Ala311Stop | 9 | Nonsense | Hattori *et al.*, 1998b |
| Gly328Glu | 9 | Missense | Lucking *et al.*, 2000 |
| Arg334Cys | 9 | Missense | Lucking *et al.*, 2000 |
| Thr415Asn | 11 | Missense | Abbas *et al.*, 1999, Lucking *et al.*, 2000 |
| Gly430Asp | 12 | Missense | Lucking *et al.*, 2000, Periquet *et al.*, 2001 |
| Cys431Phe | 12 | Missense | Maruyama *et al.*, 2000 |
| Trp453Stop | 12 | Nonsense | Abbas *et al.*, 1999, Lucking *et al.*, 2000 |

Mutations without reference are mutations found in our laboratories.

frequency of either 167S or 167N allele was not statistically different between PD patients and controls, while the frequency of 167S/N heterozygotes was significantly higher in PD patients (62.0% versus 45.9%), compared with that of both 167S/S and 167N/N homozygotes combined ($\chi^2$ = 4.467, $p$ = 0.0346; odds ratio = 1.92, 95% confidence interval = 1.05–3.54). They suggested that the heterozygosity at codon 167 in the *parkin* gene might represent a genetic risk factor for development of sporadic PD.

Wang *et al.* (2000) examined three polymorphisms of the *parkin* gene, i.e., G-to-A transition in exon 4 (S/N167), C-to-T transition in exon 10 (R/W366), and G-to-C transition in exon 10 (V/L380), in 160 Japanese sporadic Parkinson's disease patients and 160 age- and sex-matched control subjects. The C-to-T transition (R/W366) was significantly lower in PD patients (1.2 versus 4.4%), suggesting that it might be a protective factor against Parkinson's disease.

Hu *et al.* (2001) also studied the same *parkin* polymorphism (S/N167, R/W366, V/L380) in 92 sporadic PD patients and in 98 controls in the Chinese population in Taiwan. They found no association between these polymorphic mutations and sporadic Parkinson's disease.

Klein *et al.* (2000b) studied polymorphism of the *parkin*. They studied 100 Parkinson's disease patients of Central European origin and 100 ethnically matched controls. They analyzed the allele frequencies of S/N167, R/W366, and V/L380 of the *parkin* gene. They did not find significant difference in the allele frequencies of these polymorphisms between their Parkinson's patients and the controls. Then they studied 79 sporadic Parkinson's disease patients with onset before age 50 years. They detected a significant difference in allele frequencies for V/L380 between these early-onset Parkinson's disease patients and controls ($p$ = 0.02). Carriers of this polymorphism were at increased risk to develop early-onset parkinsonism.

### 9. Other Laboratory Tests

There are no other laboratory tests that would help the diagnosis of Park2. Gene analysis is the essential part.

## IV. NEUROIMAGING

### A. MRI

MRI is essentially normal in Park2 patients.

### B. PET, fMRI, and MR Spectroscopy

There are only three reports on PET. Broussolle *et al.* (2000) reported [$^{18}$F]-dopa PET study in patients with juvenile-onset PD and *parkin* gene mutations. They found a profound decrease of [$^{18}$F-]dopa uptake, representing 28% in putamen and 44% in caudate nucleus compared to their control subject values. PET findings were essentially similar to those of sporadic PD patients.

Hilker *et al.* (2001) reported a PET study on five adult-onset patients in Tyrol with apparently autosomal dominant transmission harboring compound heterozygous parkin mutations reported by Klein *et al.* (2000a). What they found was presynaptic striatal 18-fluorodopa (FDOPA) storage decrease in those patients with the most prominent reduction in the posterior part of the putamen. They also found a uniform reduction of the striatal 11C-raclopride binding index in them indicating the postsynaptic lesions as well. In asymptomatic carriers of a single parkin mutation with an apparently normal allele, they found a mild but statistically significant decrease of mean FDOPA uptake compared to control subjects.

Portman *et al.* (2001) studied two brothers with a parkin mutation (deletion of exons 4–6) by PET. Fluorodopa uptake was markedly diminished both in the caudate nucleus and the putamen. Raclopride binding and fluorodeoxyglucose metabolism was within normal range. Thus more extensive decrease in FDOPA uptake appears to be characteristic of Park2.

To date no report has been published on functional MRI or MR spectroscopy on Park2 patients.

## V. NEUROPATHOLOGY

Pathologic features of Park2 consist of neuronal degeneration in the substantia nigra pars compacta and the locus coeruleus (Takahashi *et al.*, 1994; Mori *et al.*, 1998; Yamamura *et al.*, 1998; Hayashi *et al.*, 2000). The substantia nigra is involved more extensively. Not only the lateral part of the substantia nigra but also the medial part shows neuronal loss. Gliosis is also extensive in these structures. In these reports, no Lewy bodies were seen and absence of Lewy bodies appears to be a pathologic characteristic of Park2. However, recently Farrer *et al.* (2001) reported a Park2 patient who showed Lewy bodies in the substantia nigra at autopsy. This patient had compound heterozygous mutations of Arg275Try at exon 7 and 40-bp deletion in exon 3. Thus the specificity of absence of Lewy bodies in Park2 should further be studied.

Additional interesting findings include accumulation of tau protein in the remaining nigral neurons in a patient with exon 4 deletion (Mori *et al.*, 1998) and neuronal loss and fibrillary gliosis in the substantia nigra pars reticulata in addition to the pars compacta in a patient with exon 4 deletion (Hayashi *et al.*, 2000).

There appears to be a reason for the absence of Lewy bodies in Park2. Alpha-synuclein is a major component of Lewy bodies (Spillantini *et al.*, 1997). Parkin is also expressed in Lewy bodies of sporadic Parkinson's disease patients (Shimura et al., 2001; Schlossmacher et al., 2002). Furthermore, synphilin-1 interacts with both α-synuclein (Engelender et al., 1999) and Parkin (Chung et al., 2001), and present in Lewy bodies (Wakabayashi et al., 2000). Interaction of Parkin, synphilin-1, and α-synuclein may be essential for the formation of Lewy bodies. In the absence of interaction between Parkin and α-synuclein and/or synphilin-1, Lewy bodies may not be formed. If mutated Parkin is still able to bind with α-synuclein and/or synphilin-1, Lewy bodies may be formed. The patient reported by Farrer et al. (2001) may be such an example. They observed expression of mutated Parkin and Lewy body formation in a Park2 patient at autopsy.

## VI. CELLULAR AND ANIMAL MODELS

At the time of the preparation of this manuscript, there is no published report on transgenic or knock-out models of *parkin* at cellular or animal (invertebrates and vertebrates) levels.

## VII. GENOTYPE AND PHENOTYPE CORRELATIONS/MODIFYING ALLELES

To date, no clear correlation exists between types of parkin mutations and clinical features or pathologic features. Abbas et al. (1999) examined a large cohort of *parkin* mutations and found a tendency for patients with point mutations to have higher ages of onset compared to those with deletion mutations. But there is an overlap. Further studies are needed to make a definite conclusion on the clinico-genetic correlation.

## VIII. TREATMENT

Treatment is essentially similar to that of sporadic Parkinson's disease. But as the age of Park2 patients is usually young and the progression is slow, most of those patients treated with levodopa develop dyskinesias and motor fluctuations. Therefore, treatment with a dopamine agonist is recommended as long as possible. When levodopa is instituted, it is important to use a small dose (50–100 mg of levodopa with a dopa-decarboxylase inhibitor) at frequent doses in a day (every 2–3 hours) to minimize pulsatile stimulation of dopamine receptors.

## Acknowledgments

This study was in part supported by Grant-in-Aid for Scientific Research on Priority Areas from Ministry of Education, Science, Sports, and Culture, Japan, Grant-in-Aid for Health Science Promotion, and Grant-in-Aid for Neurodegenerative Disorders from Ministry of Health and Welfare, Japan, and by "Center of Excellence" Grant from National Parkinson Foundation, Miami.

## References

Abbas, N., Lücking, C. B., Ricard, S., Dürr, A., Bonifati, V., De Michele, G., Bouley, S., Vaughan, J. R., Gasser, T., Marconi, R., Broussolle, E., Brefel-Courbon, C., Harhangi, B. S., Oostra, B. A., Fabrizio, E., Böhme, G. A., Pradier, L., Wood, N. W., Filla, A., Meco, G., Denefle, P., Agid, Y., Brice, A., and the French Parkinson's Disease Genetics Study Group and the European Consortium on Genetic susceptibility in Parkinson's Disease. (1999). A wide variety of mutations in the parkin gene are responsible for autosomal recessive parkinsonism in Europe. *Hum. Mol. Gen.* **8**, 567–574.

Asakawa, S., Tsunematsu, K., Takayanagi, A., Sasaki, T., Shimizu, A., Shintani, A., Kawasaki, K., Mungall, A.J., Beck, S., Minoshima, S., and Shimizu, N. (2002). The genomic structure and promoter region of the human parkin gene. *Biochem. Biophys. Res. Commun.* **286**, 863–868.

Broussolle, E., Lucking, C. B., Ginovart, N., Pollak, P., Remy, P., and Durr, A. (2000). [$^{18}$F]-dopa PET study in patients with juvenile-onset PD and parkin gene mutations. *Neurology* **55**, 877–879.

Choi, P., Ostrerova-Golts, N., Sparkman, D., Cochran, E., Lee, J. M., Wolozin, B. (2000). Parkin is metabolized by the ubiquitin/proteosome system. *Neuroreport* **11**, 2635–2638.

Chung, K. K. K., Zhang, Y., Lim, K. L., Tanaka, Y., Huang, H., Gao, J., Ross, C. A., Dawson, V. L., and Dawson, T. M. (2001). Parkin ubiquitinates the α-synuclein-interacting protein synphilin-1: implications for Lewy-body formation in Parkinson disease. *Nat. Med.* **7**, 1144–1150.

D'Agata, V., Zhao, W., and Cavallaro, S. (2000a). Cloning and distribution of the rat parkin mRNA. *Brain Res. Mol. Brain Res.* **75**, 345–349.

D'Agata, V., Grimaldi, M., Pascal, A., and Cavallaro, S. (2000b). Regional and cellular expression of the parkin gene in the rat cerebral cortex. *Eur. J. Neurosci.* **12**, 3583–3588.

Duijin, C. M. V., Dekker, M. C. J., Bonifati, V., Galijaad, R. J., Houwing-Duistemaat, J., Nijders, P. J. L. M., Testers, L., Breedveld, G. J., Horstink, M., Sandkuijl, L. A., Swieten, J. C. V., Oostra, B. A., and Heutink, P. (2001). PARK7, a novel locus for autosomal recessive early-onset parkinsonism, on chromosome 1p36. *Am. J. Hum. Genet.* **69**, 629–634.

Engelender, S., Kaminsky, Z., Guo, X., Sharp, A. H., Amaravi, R. K., Kleiderlein, J. J., Margolis, R. L., Troncoso, J. C., Lanahan, A. A., Worley, P. F., Dawson, V. L., Dawson, T. M., and Ross, C. A. (1999). Synphilin-1 associates with alpha-synuclein and promotes the formation of cytosolic inclusions. *Nat. Genet.* **22**, 110–114.

Farrer, M., Chan, P., Chen, R., Tan, L., Lincoln, S., Hernandez, D., Forno, L., Gwinn-Hardy, K., Petrucelli, L., Hussey, J., Singleton, A., Tanner, C., Hardy, J., and Langston, J. W. (2001). Lewy bodies and parkinsonism in families with parkin mutations. *Ann. Neurol.* **50**, 293–300.

Gu, W. G., Abbas, N., Lagunes, M. A., Parent, A., Pradier, L., Bohme, G. A., Agid, Y., Nirsch, E. C., Raisoman-Vozari, R., and Brice, A. (2000). Cloning of rat parkin cDNA and distribution of parkin in rat brain. *J. Neurochem.* **74**, 1773–1776.

Hattori, N., Matsumine, H., Kitada, T., Asakawa, S., Yamamura, Y., Kobayashi, T., Yokochi, M., Yoshino, H., Wang, M., Kondo, T., Kuzuhara, S., Nakamura, S., Shimizu, N., and Mizuno, Y. (1998a). Molecular analysis of a novel ubiquitin-like protein (PARKIN) gene in Japanese families with AR-JP: evidence of homozygous deletions in the PARKIN gene in affected individuals. *Ann. Neurol.* **44**, 935–941.

Hattori, N., Matsumine, H., Asakawa, S., Kitada, T., Yoshino, H., Elibol, B., Brooks, A. J., Yamamura, Y., Kobayashi, T., Wang, M., Yoritaka, A., Minoshima, S., Shimizu, N., and Mizuno, Y. (1998b). Point mutations (Thr240Arg and Gln311Stop) in the Parkin gene. *Biochem. Biohys. Res. Commun.* **249**, 754–758.

Hayashi, S., Wakabayashi, K., Ishikawa, A., Nagai, H., Saito, M., Maruyama, M., Takahashi, T., Ozawa, T., Tsuji, S., and Takahashi, H. (2000). An autopsy case of autosomal-recessive juvenile parkinsonism with a homozygous exon 4 deletion in the parkin gene. *Mov. Disord.* **15**, 884–884.

Hilker, R., Klein, C., Ghaemi, M., Kis, B., Strotmann, T., Ozelius, L. J., Lenz, O., Vieregge, P., Herholz, K., Heiss, W. D., and Pramstaller, P. P. (2001). Positron emission tomographic analysis of the nigrostriatal dopaminergic system in familial parkinsonism associated with mutations in the parkin gene. *Ann. Neurol.* **49**, 367–376.

Horowitz, J. M., Vernace, V. A., Myers, J., Hanlon, D. W., Fraley, G. S., Stachowiak, M. K., and Torres, G. (2001a). Immunodetection of Parkin protein in vertebrate and invertebrate brains: a comparative study using specific antibodies. *J. Chem. Neuroanat.* **21**, 75–93.

Horowitz, J. M., Nyers, J., Vernace, V. A., Stachosiak, M. K., and Torres, G. (2001b). Spatial distribution, cellular integration and stage development of Parkin protein in Xenopus brain. *Dev. Brain Res.* **126**, 31–41.

Hu, C. J., Sung, S. M., Liu, H. C., Lee, C.C., Tsai, C. H., and Chang, J. G. (2000). Polymorphisms of the parkin gene in sporadic Parkinson's disease among Chinese in Taiwan. *Eur. Neurol.* **44**, 90–93.

Huynh, D. P., Scoles, D. R., Ho, T. H., Bel Bigio, M. R., and Pulst, S. M. (2000). Parkin is associated with acting filaments in neuronal and nonneuronal cells. *Ann. Neurol.* **48**, 737–744.

Imai, Y., Soda, M., and Takahashi, R. (2000). Parkin suppresses unfolded protein stress-induced cell death through its E3 ubiquitin-protein ligase activity. *J. Biol. Chem.* **275**, 35661–35664.

Imai, Y., Soda, M., Inoue, H., Hattori, N., Mizuno, Y., and Takahashi, R. (2001). An unfolded putative transmembrane polypeptide, which can lead to endoplasmic reticulum stress, is a substrate of Parkin. *Cell* **105**, 891–902.

Ishikawa, A., and Tsuji, S. (1996). Clinical analysis of 17 patients in 12 Japanese families with autosomal-recessive type juvenile parkinsonism. *Neurology* **47**, 160–169.

Kitada, T., Asakawa, S., Hattori, N., Matsumine, H., Yamamura, Y., Minoshima, S., Yokochi, M., Mizuno, Y., and Shimizu, N. (1998). Deletion mutation in a novel protein "Parkin" gene causes autosomal recessive juvenile parkinsonism (AR-JP). *Nature* **392**, 605–608.

Kitada, T., Asakawa, S., Minoshima, S., Mizuno, Y., and Shimizu, N. (2000). Molecular cloning, gene expression, and identification of a splicing variant of the mouse parkin gene. *Mamm. Genome* **11**, 417–421.

Klein, C., Pramstaller, P. P., Kis, B., Page, C. C., Kann, M., Leung, J., Woodward, H., Castellan, C. C., Sherer, M., Vieregge, P., Breakfield, X. O., Kramer, P. L., and Ozelius, L. J. (2000a). Parkin deletions in a family with adult-onset, tremor-dominant parkinsonism: expanding the phenotype. *Ann. Neurol.* **48**, 65–71.

Klein, C., Schumacher, K., Jacobs, H., Hagenah, J., Kris, B., Garrels, J., Schwinger, E., Ozelius, L., Pramistaller, P., Vieregge, P., and Kramer, P. L. (2000b). Association studies of Parkinson's disease and *parkin* polymorphisms. *Ann. Neurol.* **48**, 126–127.

Kobayashi, T., Wang, M., Hattori, N., Matsumine, H., Kondo, T., and Mizuno, Y. (2000). Exonic deletion mutations of the parkin gene among sporadic patients with Parkinson's disease. *Parkinsonism Relat. Disord.* **6**, 129–131.

Kubo, S., Kitami, T., Noda, S., Shimura, H., Uchiyama, Y., Asakawa, S., Minoshima, S., Shimizu, N., Mizuno, Y., and Hattori, N. (2001). Parkin is associated with cellular vesicles. *J. Neurochem.* **78**, 42–54.

Leroy, E., Anastasopoulos, D., Konitsiotis, S., Lavedan, C., and Polymeropoulos, M. (1998). Deletions in the parkin gene and genetic heterogeneity in a Greek family with early onset Parkinson's disease. *Hum. Genet.* **103**, 424–427.

Lucking, C. B., Durr, A., Bonifati, V., Vaughan, J., De Michele, G., Gasser, T., Harhangi, B. S., Meco, G., Denefle, P., Wood, N. W., Agid, Y., and Brice, A. (2000). Association between early-onset Parkinson's disease and mutations in the parkin gene. French Parkinson's Disease Genetics Study Group. *N. Engl. J. Med.* **342**, 1560–1567.

Maruyama, M., Ikeuchi, T., Saito, M., Ishikawa, A., Yuasa, T., Tanaka, H., Hayashi, S., Wakabayashi, K., Takahashi, H., and Tsuji, S. (2000). Novel mutations, pseudo-dominant inheritance, and possible familial affects in patients with autosomal recessive juvenile parkinsonism. *Ann. Neurol.* **48**, 245–250.

Matsumine, H., Saito, M., Shimoda-Matsubayashi, S., Tanaka, H., Ishikawa, A., Nakagawa-Hattori, Y., Yokochi, M., Kobayashi, T., Igarashi, S., Takano, H., Sanpei, K., Koike, R., Mori, H., Kondo, T., Mizutani, Y., Schaffer, A. A., Yamamura, Y., Nakamura, S., Kuzuhara, S., Tsuji, S., and Mizuno, Y. (1997). Localization of a gene for autosomal recessive form of juvenile parkinsonism (AR-JP) to chromosome 6q25.2–27. *Am. J. Hum. Genet.* **60**, 588–596.

Mori, H., Kondo, T., Yokochi, M., Matsumine, H., Nakagawa-Hattori, Y., Miyake, T., Suda, K., and Mizuno, Y. (1998). Pathologic and biochemical studies of juvenile parkinsonism linked to chromosome 6q. *Neurology* **51**, 890–892.

Munoz, E., Pastor, P., Marti, M. J., Oliva, R., and Tolosa, E. (2000). A new mutation in the parkin gene in a patient with atypical autosomal recessive juvenile parkinsonism. *Neurosci. Lett.* **289**, 66–68.

Periquet, M., Lucking, C., Vaughan, J., Bonifati, V., Durr, A., De Michele, G., Horstink, M., Farrer, M., Illarioshkin, S. N., Pollak, P., Borg, M., Brefel-Courbon, C., Denefle, P., Meco, G., Gasser, T., Breteler, M. M., Wood, N., Agid, Y., and Brice, A. (2001). French Parkinson's Disease Genetics Study Group. The European Consortium on Genetic Susceptibility in Parkinson's Disease. Origin of the mutations in the parkin gene in Europe: exon rearrangements are independent recurrent events, whereas point mutations may result from founder effects. *Am. J. Hum. Genet.* **68**, 617–626.

Portman, A. T., Gilade, N., Leender, K. L., Maguire, P., Veenma-van der Duin, L., Swart, J., Pruim, J., Simon, E. S., Hassin-Baer, S., and Korczyn, A. D. (2001). The nigrostriatal dopaminergic system in familial early onset parkinsonism with *parkin* mutations. *Neurology* **56**, 1759–1762.

Satoh, J., and Kuroda, Y. (1999). Association of codon 167 Ser/Asn heterozygosity in the parkin gene with sporadic Parkinson's disease. *Neuroreport* **10**, 2735–2739.

Schlossmacher, M. G., Frosch, M. P., Gai, W. P., Medina, M., Sharma, N., Forno, L., Ochiishi, T., Shimura, H., Hattori, N., Langston, J. W., Mizuno, Y., Hyman, B. T., Selkoe, D. J., and Kosik, K. S. (2002). Parkin localizes to the Lewy bodies of Parkinson's disease and dementia with Lewy bodies. *Am. J. Pathol.* **160**, 1665–1667.

Shimura, H., Hattori, N., Kubo, S., Yoshikawa, M., Kitada, T., Matsumine, H., Asakawa, S., Minoshima, S., Yamamura, Y., Shimizu, N., and Mizuno, Y. (1999). Immunohistochemical and subcellular localization of Parkin: absence of protein in AR-JP. *Ann. Neurol.* **45**, 668–672.

Shimura, H., Hattori, N., Kubo, S., Mizuno, Y., Asakawa, S., Minoshima, S., Shimizu, N., Iwai, K., Chiba, T., Tanaka, K., and Suzuki, T. (2000). Familial Parkinson's disease gene product, Parkin, is a ubiquitin-protein ligase. *Nat. Genet.* **25**, 302–305.

Shimura, H., Schlossmacher, M. G., Hattori, N., Frosch, M. P., Trockenbacher, A., Schneider, R., Mizuno, Y., Kosik, K. S., and Selkoe, D. J. (2001). Ubiquitination of a new form of alpha-synuclein by parkin from human brain: implication for Parkinson's disease. *Science* **293**, 263–269.

Spillantini, M.G., Schmidt, M.L., Lee, A.M.Y., Trajanowski, J.Q., Jakes, R., Goedert, M. (1997). α-Synuclein in Lewy bodies. *Nature* **388**, 839–840.

Solano, S. M., Miller, D. W., Augood, S. J., Young, A. B., and Penney, J. B. Jr. (2000). Expression of alpha-synuclein, parkin, and ubiquitin carboxy-terminal hydrolase L1 mRNA in human brain: genes associated with familial Parkinson's disease. *Ann. Neurol.* **47**, 201–210.

Stichel, C. C., Augustin, M., Kuhn, K., Shu, X. R., Engels, P., Ullmer, C., and Lubbert, H., (2000). Parkin expression in the adult mouse brain. *Eur. J. Neurosci.* **12**, 4148–4198.

Sunada, Y., Saito, F., Matsumura, K., and Shimizu, T. (1999). Differential expression of the parkin gene in the human brain and peripheral leukocytes. *Neurosci. Lett.* **254**, 180–182.

Takahashi, H., Ohama, E., Suzuki, S., Horikawa, Y., Ishikawa, A., Morita, T., Tsuji, S., and Ikuta, F. (1994). Familial juvenile parkinsonism: clinical and pathologic study in a family. *Neurology* **44**, 437–441.

Takanashi, M., Mochizuki, H., Yokomizo, K., Hattori, N., Mori, H., Yamamura, Y., and Mizuno, Y. (2001). Iron accumulation in the substantia nigral of autosomal recessive juvenile parkinsonism (ARJP). *Parkinsonsm Rel. Disord.* **7**, 311–314.

Tassin, J., Durr, A., Bonnet, A. M., Gil, R., Vidailhet, M., Lucking, C. B., Goas, J. Y., Durif, F., Abada, M., Echenne, B., Motte, J., Lagueny, A., Lacomblez, L., Jedynak, P., Bartholome, B., Agid, Y., and Brice, A. (2000). *Brain* **123**, 1112–1121.

Terrini, L., Cababress, E., Calella, A. M., Forloni, G., and Mariani, C. (2001). New mutation (R42P) of the parkin gene in the ubiquitinlike domain associated with parkinsonism. *Neurology* **56**, 463–466.

Ujike, H., Yamamoto, M., Kanzaki, A., Okumura, K., Takaki, M., and Kuroda, S. (2001). Prevalence of homozygous deletions of the parkin gene in a cohort of patients with sporadic and familial Parkinson's disease. *Mov Disord* **16**, 111–113.

Valente, E. M., Bentivolglio, A. R., Dixon, P. H., Ferraris, A., Ialongo, T., Frontali, M., Albanese, A., and Wood, N. W. (2001). Localization of a novel locus for autoromal recessive early-onset parkinsonims, PARK6, on human chromosome 1p35–36. *Am. J. Hum. Genet.* **68**, 895–900.

Wakabayashi, K., Engelender, S., Yoshimoto, M., Tsuji, S., Ross, C. A., and Takahashi, H. (2000). Synphilin-1 is present in Lewy bodies in Parkinson's disease. *Ann. Neurol.* **47**, 521–523.

Wang, M., Hattori, N., Matsumine, H., Kobayashi, T., Yoshino, H., Morioka, A., Kitada, T., Asakawa, S., Minoshima, S., Shimizu, N., and Mizuno, Y. (1999). Polymorphism in the Parkin gene in sporadic Parkinson's disease. *Ann. Neurol.* **45**, 655–658.

Wang, M., Suzuki, T., Kitada, T., Asakawa, S., Minoshima, S., Shimizu, N., Tanaka, K., Mizuno, Y., and Hattori, N. (1999). Developmental changes in the expression of parkin and UbcR7, a parkin-interacting and ubiquitin-conjugating enzyme, in rato brain. *J. Neurochem.* **77**, 1561–1568.

West, A., Farrer, M., Petrucelli, L., Cookson, M., Lockhart, P., and Hardy, J. (2001). Identification and characterization of the human parkin gene promoter. *J. Neurochem.* **78**, 1146–1152.

Yamamura, Y., Sobue, I., Ando, K., Iida, M., Yanagi, T., and Kono, C. (1973). Paralysis agitans of early onset with marked diurnal fluctuation of symptoms. *Neurology* **23**, 239–244.

Yamamura, Y., Kuzuhara, S., Kondo, K., Matsumine, H., and Mizuno, Y. (1998). Clinical, pathologic, and genetic studies on autosomal recessive early-onset Parkinsonism with diurnal fluctuation. *Parkinsonism Related. Disord.* **4**, 65–72.

Zarate-Lagunes, M., Gu, W. J., Blanchard, V., Francois, C., Muriel, M. P., Mouatt-Prigent, A., Bonici, B., Parent, A., Hartmann, A., Yelnik, J., Boehme, G. A., Pradier, L., Moussaoui, S., Faucheux, B., Raisman-Vozari, R., Agid, Y., Brice, A., and Hirsch, E. C. (2001). Parkin immunoreactivity in the brain of human and non-human primates: an immunohistochemical analysis in normal conditions and in Parkinsonian syndromes. *J. Comp. Neurol.* **432**, 184–196.

Zhang, Y., Gao, J., Chung, K. K., Huang, H., Dawson, V. L., and Dawson, T. M. (2000). Parkin functions as an E2 dependent ubiquitin-protein ligase and promotes the degradation of the synaptic vesicle associated protein, CDCrel-1. *Proc. Natl. Acad. Sci. U.S.A.* **21**, 13354–13359.

# CHAPTER 29

# PARK3, Ubiquitin Hydrolase-L1 and Other PD Loci

REJKO KRÜGER
Department of Neurology
University of Tübingen
Tübingen, Germany

OLAF RIESS
Department of Medical Genetics
University of Tübingen
Tübingen, Germany

I. *PARK3* (MIM 602404)
II. *PARK4* (MIM 605543)
III. *PARK5* (MIM 191342)
IV. *PARK6* (MIM 605909)
V. *PARK7* (MIM 606324)
VI. *PARK8* and *PARK9*
VII. Other Linked PD Loci
References

Recent progress in the genetic characterization of Parkinson's disease (PD) provides increasing evidence for genetic heterogeneity. Linkage analysis in some large families with autosomal dominant inheritance of the disease allowed the identification of several gene loci (PARK1, PARK3, and PARK4) with subsequent characterization of the first disease gene, α-synuclein (PARK1). In other families with early-onset PD and autosomal recessive inheritance additional gene loci were identified (PARK2, PARK6, PARK7). Functional studies implicated that the PARK2 gene product, parkin, encodes a ubiquitin ligase. Based on the fact that parkin ubiquitinates α-synuclein a model of altered protein degradation as the cause for PD has been proposed. This model currently stimulates the genetic analysis of other candidate genes and led to the identification of a mutation in the ubiquitin C terminal hydrolase-L1 gene (UCH-L1). The identification of rare mutations in PD patients in potentially involved gene products of this pathway as in the neurofilament genes or in the α-synuclein interacting protein synphilin-1 is, in particular, intriguing. These studies are complemented by the first successful whole genome mapping approaches with indications for linkage on chromosome 5q, 8p, and 17q, respectively. Clinically it becomes obvious that the phenotype is broader than previously anticipated.

## I. *PARK3* (MIM 602404)

Using a whole genome mapping approach in six families with an autosomal dominantly inherited parkinson's disease (ADPD) the *PARK3* locus (Table 29.1) has been initially mapped to a 10.6-cM region on chromosome 2p13 (Gasser et al., 1998). Four of these families showed positive lod scores and in two families (family B and C), that were traced back to the border region of Southern Denmark and Northern Germany, a common haplotype has been found.

Clinically patients in these families display typical signs of PD indistinguishable from sporadic forms of the disease including an average age at disease onset of 61 years. Diagnosis of definite parkinsonism was based on the criteria of Calne (1994) and supported by positron emission tomography (PET) and by autopsy findings demonstrating the presence of Lewy bodies in the substantia nigra of affected individuals (Wszolek et al., 2001). In addition some affected individuals presented with signs of dementia, which neuropathologically corresponded to the presence of neurofibrillary tangles and amyloid plaques (Denson and Wszolek, 1995; Wszolek et al., 1999). These results further support a clinical and pathological overlap between Alzheimer's disease (AD) and PD and indicate the phenotypic variability of the underlying mutation.

TABLE 29.1 Gene Loci for PD

| Locus name | MIM | Chromosomal localization | Gene product | Mode of inheritance | Lewy body pathology | Special clinical features[a] |
|---|---|---|---|---|---|---|
| PARK1 | 601508 | 4q23 | α-Synuclein | Ad | Yes | Dementia |
| PARK2 | 600116 | 6 | Parkin | Ar | No | Early onset, levodopa-induced dyskinesia, sleep benefit, foot dystonia |
| PARK3 | 602404 | 2p13 | Unknown | Ad | Yes | Dementia |
| PARK4 | 605543 | 4p15 | Unknown | Ad | Yes | Dementia, postural tremor |
| PARK5 | 191342 | 4p14 | UCH–L1 | Ad | Unknown | Not described |
| PARK6 | 605909 | 1p35–36 | Unknown | Ar | Unknown | Early onset |
| PARK7 | 606324 | 1p36 | Unknown | Ar | Unknown | Early onset, dystonia, psychosis |
| LUBAG | 314250 | Xq13.1 | Unknown | X-linked | Not described | Dystonia |

[a] In addition to typical PD symptoms, i.e., tremor, rigidity, postural instability.

The fact that affected individuals in two families share the same haplotype suggested the possibility of a founder effect. Since the majority of individuals carrying the risk haplotype were not affected a reduced penetrance of about 40% has been postulated for the disease-causing mutation. Reduced penetrance might mask a positive family history and therefore the 2p13 locus was implicated in sporadic PD. However, in two studies based on sporadic German PD patients residing in the area of Germany, where one of the PARK3 families were traced, no evidence for a common haplotype was found (Gasser et al., 1999; Klein et al., 1999). These results argue against a major role of the PARK3 locus in the pathogenesis of sporadic PD.

Subsequent genotyping of polymorphisms in the candidate region on chromosome 2p13 allowed to reduce the common haplotype to a physical distance of 2.5 Mb. To date 15 genes in this candidate region have been screened for mutations in the coding region, including the transforming growth factor alpha, which is known to function as a neurotrophic factor for dopaminergic nerve cells. However, no disease-causing mutation has been found in the coding region of the respective genes (West et al., 2001; Zink et al., 2001). However, flanking intronic sequences have not been investigated for most of the characterized genes. Also, the critical PARK3 region might harbor other than the currently known genes for which disease-causing mutation(s) will have to be excluded.

## II. PARK4 (MIM 605543)

The PARK4 locus has been linked to human chromosome 4p15 in two families with an autosomal dominantly inherited form of PD known as the "Spellman-Muenter" and the, "Waters-Miller" family (Table 29.1). Subsequent genealogical studies allowed the identification of a common ancestor of these families which led to naming these families collectively as the Iowa kindred (Farrer et al., 1999).

The phenotype of this family shows a wide range of clinical symptoms and includes typical PD as well as essential tremor (ET) and dementia with Lewy bodies (DLB). The parkinsonian symptoms in the Iowa kindred include typical clinical signs as rigidity, resting tremor, and impaired postural stability, which were responsive to levodopa therapy. However, additional atypical signs as an early age at disease onset (mean 34.1 ± 8.8 years), early weight loss, rapid disease progression, and dementia have been reported (Spellman, 1962; Waters and Miller, 1994; Muenter et al., 1998). Interestingly all family members presenting with postural tremor shared the chromosome 4p risk haplotype. Since these affecteds present with probable or possible ET according to Louis (1998), it may be speculated that ET represents one end of a broad and overlapping clinical spectrum of the parkinsonian syndrome. To date two loci responsible for familial ET have been mapped to chromosome 2p22–25 and 3q13. However, in these families associated parkinsonism was not reported (Higgins et al., 1998; Gulcher et al., 1997).

Neuropathological examinations of deceased affecteds of the Iowa kindred confirmed the presence of Lewy bodies (LB) in affected brain regions. However, also atypical pathological features as pleomorphic cortical LB, glial α-synuclein positive inclusions, and widespread severe neuritic dystrophy were found (Gwinn-Hardy et al., 2000) reflecting a pathological correlate for the observed dementia in the respective individual. These features together with clinical aspects in some affecteds argue in favor of the diagnosis of DLB. In summary, clinical data of the Iowa kindred suggest that the disease entities ET, PD, and DLB may represent different stages based on the same pathomechanism.

The chromosome 4 haplotype is shared by all affected individuals in the Iowa pedigree. However, as for other families with ADPD (i.e., *PARK*1 and *PARK*3), there are individuals older than the expected age at disease onset which display no neurological symptoms. This might reflect reduced penetrance of the disease-causing mutation and it has to be established, which factors, environmental or genetic modifying factors, are responsible for the variability of clinical symptoms and age at disease onset.

The *PARK4* locus on chromosome 4p15 encompasses several candidate genes based on the hypothesis of disturbed dopaminergic transmission and protein degradation, i.e., the dopamine D5 receptor and UCH-L1 genes (Edwards *et al.*, 1991), respectively. The latter gene was recently identified as the *PARK5* locus (see Section III) in a German family with ADPD (Leroy *et al.*, 1998). Due to the short distance between the *PARK4* and the *PARK5* locus it has been suspected that UCH-L1 mutations might be responsible for the *PARK4* phenotype. However, detailed mutation analyses failed to demonstrate a disease-causing mutation in the *UCH-L1* gene and further mapping approaches definitively excluded the *UCH-L1* gene from the candidate region. Therefore additional efforts are necessary to unravel the disease-causing gene for ADPD in the Iowa kindred.

## III. *PARK5* (MIM 191342)

The identification of α-synuclein protein as the major component of LB in brains of PD patients and of two mutations in the *α-synuclein* gene in some rare forms of ADPD led to the hypothesis of disturbed protein degradation and pathological aggregation as the major pathogenetic mechanism causing PD. The ubiquitin proteasome system (UPS) represents an important cellular machinery for the elimination of misfolded proteins. The fact that proteins aggregated in LB are ubiquitinated suggested an involvement of UPS dysfunction in PD (Fig. 29.2). Based on a candidate gene approach, Leroy and co-workers screened the gene encoding UCH-L1 for mutations in 72 index patients with familial PD. UCH-L1 is abundant in the brain (Doran *et al.*, 1983) and represents 1–2% of total soluble brain protein (Wilkinson *et al.*, 1989). UCH-L1 has been identified as a component of LB in PD brains (Dickson *et al.*, 1994). Most strikingly UCH-L1 is involved in the UPS being responsible for the reutilization of ubiquitin-monomers from degraded protein fragments. Subsequently, an Ile93Met mutation was found in one German family with autosomal dominant inheritance (Leroy *et al.*, 1998). This mutation was absent in a total of 500 chromosomes of control individuals. Therefore the authors concluded that the Ile93Met mutation is responsible for ADPD in this family. Two living affected sibs displayed typical clinical signs of PD with an age at disease onset of 51 and 49 years, respectively, and a good response to levodopa therapy. The fact that the deceased father of the index patient showed no neurological symptoms might indicate reduced penetrance of the Ile93Met mutation.

Due to the limited size of the pedigree the possibility of a rare variant which co-segregates by chance with the disease in the two sibs has been discussed. Evidence in favor of a causative role of the Ile93Met mutation came from *in vitro* studies showing a 50% reduced catalytic activity of mutated UCH-L1. Moreover a homozygous deletion of exons 7 and 8 in the *UCH-L1* gene was found to be responsible for degeneration of sensory and motor neurons in so-called *gad* (gracile axonal dystrophy) mice (Saigoh *et al.*, 1999). Interestingly immunohistochemistry revealed intraneuronal protein aggregates in affected neurons, which could be stained with antibodies against ubiquitin and proteasome. In addition a role of UCH-L1 in PD has been supported by association studies in sporadic PD. Several independent genetic studies found a protective effect of a Ser18Tyr polymorphism in the *UCH-L1* gene for PD. Individuals carrying a tyrosine in position 18 of the amino acid sequence were significantly more frequent in healthy controls compared to sporadic PD patients (Maraganore *et al.*, 1999; Wintermeyer *et al.*, 2000; Zhang *et al.*, 2000; Satoh and Kuroda, 2001).

Some doubt concerning the pathogenic relevance of the Ile93Met mutation in the *UCH-L1* gene came from additional studies in families with ADPD. Farrer and colleagues (2000) identified a Met124Leu mutation in one index patient with ADPD. Although this mutation was not observed in controls it did not co-segregate with the disease in the respective family. This may indicate the presence of rare sequence variations in the *UCH-L1* gene without pathogenic relevance. Therefore, based on present data, the presence of a disease gene closely linked to *UCH-L1* cannot be completely ruled out. Further studies in familial PD are required to confirm the *UCH-L1* gene as the PARK5 locus responsible for one rare form of ADPD.

## IV. *PARK6* (MIM 605909)

Based on the identification of mutations in the *parkin* gene (Fig. 29.1), autosomal recessively inherited juvenile PD (ARJP) has been identified as a distinct clinical and genetical subgroup of parkinsonian syndromes. Affected individuals display characteristic early-onset of first clinical signs of PD and an early occurrence of levodopa-induced motor fluctuations. However some affected individuals present clinical features indistinguishable from idiopathic PD. Although the prevalence of mutations in the *parkin* gene was suprisingly high, in more than 50% of families with ARJP no linkage to the *PARK2* locus could be established and no pathogenic mutations were identified in the *parkin* gene. This strongly supported the presence of further genes causing an ARJP phenotype. Indeed, Valente

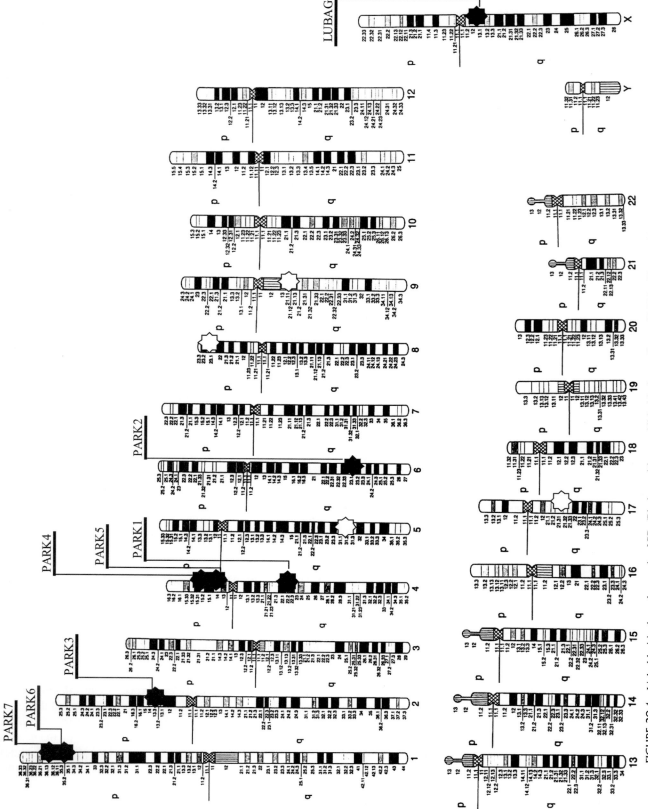

FIGURE 29.1 Linked loci in the pathogenesis of PD. (Black asterisks—established loci according to OMIM; white asterisks—candidate loci based on whole genome mapping approach.)

and colleagues (2001) identified a novel locus on chromosome 1p35–p36 in a large Italian family with early-onset autosomal recessive parkinsonism. Linkage was established to a 12.5-cM region with a maximum lod score of 4.01 based on calculations using a reduced penetrance of 90% as shown for the *parkin* gene.

Clinically affected individuals showed typical signs of PD except for an early age at disease onset (range 32–48 years) and the early occurrence of levodopa-associated dyskinesia. Other signs typical of ARJP in families with parkin mutations, like foot dystonia and sleep benefit, were absent. Pathological data on the presence of LB are not yet available.

In order to evaluate the relevance of the *PARK6* locus for early-onset parkinsonism, Valente and co-workers (2002) performed linkage analysis in additional 25 families with early-onset parkinsonism that are not linked to the *PARK2* locus. In eight families positive linkage to the *PARK6* locus was established reducing the candidate region to a 9-cM interval. Interestingly affected individuals from these families do not share a common haplotype indicating independent mutational events. Therefore present data suggest a major relevance for the *PARK6* locus in early-onset parkinsonism being responsible for about one third of non-parkin early-onset parkinsonism families.

## V. *PARK7* (MIM 606324)

Another locus that has been identified in the pathogenesis of autosomal recessive early-onset parkinsonism also maps to the short arm of chromosome 6 in one Dutch family (van Duijn et al., 2001). Initially it has been suspected that the Dutch family might share the *PARK6* defining region on chromosome 1p35–36. However, van Duijn and colleagues definitely excluded the *PARK6* locus in their family and mapped *PARK7* more than 25 cM distal to *PARK6* on chromosome 1p36. Based on a model with 100% penetrance by the age of 40 years a maximum lod score of 4.3 was calculated. Recently, additional pedigrees have been mapped to this locus, although the authors point to the difficulty in distinguishing between *PARK6* and *PARK7* due to the proximity of the two loci (Bonifati et al., 2002).

The age at disease onset in the Dutch family was ≤40 years. All four affected individuals displayed typical signs of PD. As for parkin mutation carriers affected individuals showed significant dystonic features, i.e., blepharospasm and laterocollis. In addition affected family members present with neurotic signs in two individuals and psychotic episodes in an untreated affected, which might indicate specific symptoms of parkinsonism linked to the *PARK7* locus. As for the Sicilian family (*PARK6*) no pathological data are available. Therefore it remains to be established, if early-onset autosomal recessively inherited parkinsonism is associated with Lewy bodies in affected brain regions like typical idiopathic PD.

These findings confirm that autosomal recessive early-onset parkinsonism represents a genetically and clinically heterogenous group in the large spectrum of parkinsonian phenotypes.

## VI. *PARK8* AND *PARK9*

Recently a new locus for autosomal dominantly inherited PD was mapped on chromosome 12p11.2–q13.1 in a Japanese family (*PARK8*; Funayama et al., 2002). The phenotype largely reflects common idiopathic PD. Low penetrance of the expected disease-causing mutation is reflected by the presence of unaffected family members sharing the risk haplotype.

Another approach based on a genome-wide scan on 117 Icelandic PD patients from 51 families with more than one affected individual identified a susceptibility gene for parkinson's disease on chromosome 1p32 (Hicks et al., 2001). Patients present with typical signs of idiopathic PD including a late disease onset. Thus this locus is a good candidate for being assigned as *PARK9*.

## VII. OTHER LINKED PD LOCI

Using complete genome screening in large cohorts of familial PD patients substantial effort has been undertaken in order to define susceptibility loci for PD. Based on this approach a recent study defined five distinct chromosomal regions (Table 29.2) containing loci linked to PD (Scott et al., 2001). For this study affected individuals of 174 families with PD including at least 1 affected sib pair were screened. The average age at onset of PD was 59.9 years (SD 12.6) and all affecteds were selected presenting with at least two of the cardinal signs of PD (tremor, rigidity, bradykinesia) and no atypical clinical features.

TABLE 29.2 Candidate Loci for PD

| Chromosomal region | Marker | max. lod score | Candidate gene |
|---|---|---|---|
| 5q | D5S816 | 2.39 | Synphilin–1 |
| 8p | D8S520 | 2.22 | Neurofilament–L, –M |
| 9q | D9S301 | 2.59[a] | Torsin A |
| 17q | D17S921 | 1.92[b] | Tau |

[a]Levodopa nonresponsive families.
[b]Late-onset PD.

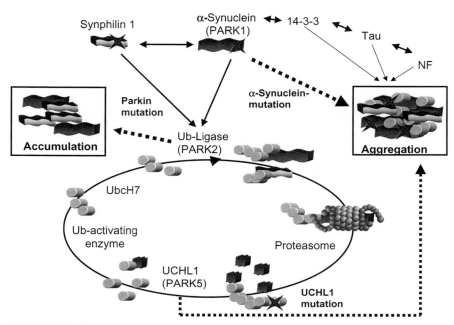

FIGURE 29.2 The functional implication of a disturbed ubiquitin proteasome system (UPS) in the pathogenesis of PD leading to abnormal protein accumulation and aggregation.

Markers on chromosome 5q, 8p, and 17q gave lod scores greater than 1.5 (Fig. 29.2). Among these loci, strongest evidence for linkage was found on chromosome 8p. This chromosomal region contains the genes for the neurofilament (NF) subunits L and M. Neurofilaments are cytoskleletal proteins responsible for neuronal integrity and maintainance of neuronal function, which are abundantly expressed in the nervous system (reviewed in Julien and Mushinski, 1998). Evidence for pathogenic relevance of NF came from neuropathological studies identifying NF as a major component of Lewy bodies in PD brains (Hill et al., 1991). Interestingly these NF aggregates are ubiquitinated and hyperphosphorylated (Fig. 29.1), which parallels pathological alterations seen in another cytoskeletal protein, the microtubuli-associated protein tau (Heutink, 2000). Moreover dysregulation of *NF* gene expression has been found in brains of PD patients (Hill et al., 1993). Recently, Polymeropoulos and colleagues reported a Gly336Ser mutation in the *NF-M* gene in one 22-year-old index patient of a family with EOPD (Lavedan et al. 2002). Our group independently screened 328 German PD patients for mutations in the *NF-M* gene. We identified a novel del830Val mutation, which was not present in more than 400 chromosomes of control individuals. However this mutation did not co-segregate with the disease in the respective family (Krüger et al., 2001, Abstract). Therefore we conclude that mutations in the *NF-M* gene are not a frequent cause for PD in Caucasians.

Another linked locus for PD could be observed on the long arm of chromosome 5. This chromosomal region harbors *synphilin–1*, a gene which has already been linked to the pathogenesis of PD for several reasons (Fig. 29.1): (1) synphilin–1 protein has been identified as an α-synuclein (*PARK1*) interacting protein using yeast two hybrid screening (Engelender et al., 1998), (2) it could be shown that co-expression of synphilin-1 and α-synuclein results in the formation of LB-like aggregations in cultured cells (Engelender et al., 1998; Kawamata et al., 2001), and (3) most recently synphilin–1 has been identified as substrate of the ubiquitin-ligase parkin (*PARK2*). Since synphilin–1 is a component of LB in brains of sporadic PD patients (Wakabayashi et al., 2000), we performed mutation analysis in the *synphilin–1* gene in a large sample of German sporadic PD patients. We identified a Arg621Cys exchange in two independent PD patients, which were not observed in healthy controls (Holtzmann et al., 2001 Abstract). If this sequence variation represents a functional relevant mutation or only a rare variant has still to be determined.

In the whole genome mapping effort by the group of Pericak-Vance, stratification of the patients group for late-onset PD (mean age at onset 62.7 years) revealed strongest linkage to chromosome 17q, a region containing the *tau* gene (Scott et al., 2001). Mutations in the *tau* gene have already been identified as responsible for frontotemporal dementia with parkinsonism (FTDP) and a certain haplotype of the *tau* gene has been implicated in progressive supranuclear palsy (PSP; Hutton et al., 1998; Baker et al., 1999). Subsequently the same authors identified a haplotype of the *tau* gene conferring an increased risk for idiopathic PD (Martin *et al.*, 2001, Fig. 29.2). This indicates the relevance of detailed mutation analyses in PD patients to demonstrate a possible causative role of tau in the pathogenesis of PD.

Another locus on chromosome 9q was identified in the subgroup of families with variable response to levodopa therapy. Interestingly this region contains the *torsin A* gene, in which deletion mutations cause torsion dystonia (DYT1).

Finally linkage studies in the subgroup of early-onset PD (EOPD) families (mean age at disease onset 39.7 years) revealed strong linkage with markers on chromosome 6q, the region containing the *parkin* gene, responsible for an ARJP PD (Kitada *et al.*, 1998). This result was confirmed by the identification of causative mutations in the *parkin* gene in 8 of 18 linked EOPD families. For the other ten families additional loci might be involved possibly referring to the PARK6 and PARK7 loci, respectively (Fig. 29.2).

Another whole genome mapping approach in PD identified a locus on chromosome X containing a possible disease-causing gene (Pankratz *et al.*, 2001, Abstract). In former studies, a chromosome Xq13.1 locus has already been linked to a parkinsonian phenotype (DYT3; MIM 314250) in the Filipino population (Kupke *et al.*, 1990). An abnormally high frequency of a movement disorder including dystonic features and parkinsonian features consistent with autosomal recessive X-linked inheritance has been identified by Lee and colleagues (1976) in Filipino kindreds. Affected individuals show primarily dystonic symptoms tending to generalize, which led Wilhelmsen and colleagues (1991) to refer this disorder as "lubag," a term used by the families to describe intermittent twisting movements. However, about one-third of patients display additional parkinsonian features as bradykinesia, resting tremor, or loss of postural reflexes which accompany or precede the dystonic symptoms. The mean age at disease onset was described as 38.6 years ranging from 12–56 years (Fahn and Moskowitz, 1988). Clinical symptoms are not responsive to levodopa therapy and yet no effective treatment has been found. Typical neuropathological features include neuronal loss and multifocal astrocytosis in the caudate nucleus and lateral putamen (Waters *et al.*, 1993). If the phenotypic spectrum of this locus might include typical parkinsonism as seen in patients used for linkage studies by Pankratz and colleagues (2001) has to be established.

## References

Baker, M., Litvan, I., Houlden, H., Adamson, J., Dickson, D., Perez-Tur, J., Hardy, J., Lynch, T., Bigio, E., and Hutton, M. (1999). Association of an extended haplotype in the tau gene with progressive supranuclear palsy. *Hum. Mol. Genet.* **8**, 711–715.

Bonifati, V., Breedveld, G. J., Squitieri, F., Vanacore, N., Brustenghi, P., Harhangi, B. S., Montagna, P., Cannella, M., Fabbrini, G., Rizzu, P., van Duijn, C. M., Oostra, B. A., Meco, G., and Heutink, P. (2002). Localization of autosomal recessive early-onset parkinsonism to chromosome 1p36 (PARK7) in an independent dataset. *Ann Neurol.* **51**, 253–256.

Caine, D. B. (1994). Initiating treatment for ideopathic parkinsonism. *Neurology* **44**, S19–22.

Denson, M. A., and Wszolek, Z. K. (1995). Familial parkinsonism: our experience and review. *Parkinsonism Related Disord.* **1**, 35–46.

Dickson, D. W., Schmidt, M. L., Lee, V. M., Zhao, M. L., Yen, S. H., and Trojanowski, J. Q. (1994). Immunoreactivity profile of hippocampal CA2/3 neurites in diffuse Lewy body disease. *Acta Neuropathol. (Berlin)* **87**, 269–276.

Doran, J. F., Jackson P., Kynoch, P., and Thompson, R. J. (1983). Isolation of PGP9.5, a new human neurone-specific protein detected by high resolution two-dimensional electrophoresis. *J. Neurochem.* **40**, 1542–1547.

Edwards, Y. H., Fox, M. F., Povey, S., Hinks, L. J., Day, I. N. M., and Thompson, R. J. (1991). The gene for human neuron specific ubiquitin C-terminal hydrolase maps to chromosome 4p14. *Cytogenet. Cell Genet.* **58**, 1886–1887.

Engelender, S., Wanner, T., Kleiderlein, J. J., Wakabayashi, K., Tsuji, S., Takahashi, H., Ashworth, R., Margolis, R. L., and Ross, C. A. (2000). Organization of the human synphilin-1 gene, a candidate for parkinson's disease. *Mamm. Genome* **11**, 763–766.

Engelender, S., Kaminsky, Z., Guo, X., Sharp, A. H., Amaravi, R. K., Kleiderlein, J. J., Margolis, R. L., Troncoso, J. C., Lanahan, A. A., Worley, P. F., Dawson, V. L., Dawson, T. M., and Ross, C. A. (1999). Synphilin-1 associates with alpha-synuclein and promotes the formation of cytosolic inclusions. *Nat. Genet.* **22**, 110–114.

Fahn, S., and Moskowitz, C. (1988). X-linked recessive dystonia and parkinsonism in Filipino males. *Ann. Neurol.* **24**, 179 (Abstract).

Farrer, M., Gwinn-Hardy, K., Muenter, M., DeVrieze, F. W., Crook, R., Perez-Tur, J., Lincoln, S., Maraganore, D., Adler, C., Newman, S., MacElwee, K., McCarthy, P., Miller, C., Waters, C., and Hardy, J. (1999). A chromosome 4p haplotype segregating with parkinson's disease and postural tremor. *Hum. Mol. Genet.* **8**, 81–85.

Farrer, M., Destee, T., Becquet, E., Wavrant-De, Vrieze, F., Mouroux, V., Richard, F., Defebvre, L., Lincoln, S., Hardy, J., Amouyel, P., and Chartier-Harlin, M. C. (2000). Linkage exclusion in French families with probable parkinson' s disease. *Mov. Disord.* **15**, 1075–1083.

Funayama, M., Hasegawa, K., Kowa, H., Saito, M., Tsuji, S., and Obata, F. (2002). A new locus for parkinson's disease (PARK8) maps to chromosome 12p11.2-q13.1. *Ann. Neurol.* **51**, 296–301.

Gasser T. H., Bereznai, B., Wieditz, G., Wszolek, Z. K., Müller-Myhsok, B. (1999). Evaluation of a Danish/German founder haplotype at the PARK3-locus on chromosome 2p13. *Mov. Disord.* **13**, 103 (Abstract).

Gasser, T., Muller-Myhsok, B., Wszolek, Z. K., Oehlmann, R., Calne, D. B., Bonifati, V., Bereznai, B., Fabrizio, E., Vieregge, P., and Horstmann, R. D. (1998). A susceptibility locus for parkinson's disease maps to chromosome 2p13. *Nat. Genet.* **18**, 262–265.

Gulcher, J. R., Jonsson, P., Kong, A., Kristjansson, K., Frigge, M. L., Karason, A., Einarsdottir, I. E., Stefansson, H., Einarsdottir, A.S., Sigurthoardottir, S., Baldursson, S., Bjornsdottir, S., Hrafnkelsdottir, S. M., Jakobsson, F., Benedickz, J., and Stefansson, K. (1997). Mapping of a familial essential tremor gene, FET1, to chromosome 3q13. *Nat. Genet.* **17**, 84–87.

Gwinn-Hardy, K., Mehta, N. D., Farrer, M., Maraganore, D., Muenter, M., Yen, S. H., Hardy, J., and Dickson, D. W. (2000). Distinctive neuropathology revealed by alpha-synuclein antibodies in hereditary parkinsonism and dementia linked to chromosome 4p. *Acta Neuropathol. (Berlin)* **99**, 663–672.

Heutink, P. (2000). Untangling tau-related dementia. *Hum. Mol. Genet.* **9**, 979–86.

Hicks, A., Petursson, H., Jonsson, T., Stefansson, H., Johannsdottir, H., Sainz, J., Frigge, M. L., Kong, A., Gulcher, J. R., Stefansson K., and Sveinbjörnsdottir, S. (2001). A susceptibility gene for late-onset idiopathic parkinson's disease succesfully mapped. *Am. J. Hum. Genet.* **69**, 200 (Abstract).

Higgins J. J., Loveless, J. M., Jankovic, J., and Patel, P. I. (1998). Evidence that a gene for essential tremor maps to chromosome 2p in four families. *Mov. Disord.* **13**, 972–927.

Hill, W. D., Lee, V. M., Hurtig, H. I., Murray, J. M., and Trojanowski, J. Q. (1991). Epitopes located in spatially separate domains of each neurofilament subunit are present in parkinson's disease Lewy bodies. *J. Comp. Neurol.* **309**, 150–160.

Hill, W. D., Arai, M., Cohen, J. A., and Trojanowski, J. Q. (1993). Neurofilament mRNA is reduced in parkinson's disease substantia nigra pars compacta neurons. *J. Comp. Neurol.* **329**, 328–336.

Holtzmann C., Li, L., Krüger, R., Berger, K., and Riess, O. (2001). Identification of an Arg621Cys exchange in the synphilin-1 gene in German patients with sporadic parkinson's disease. *Med. Genet.* **3**, 318 (Abstract).

Hughes A. J., Daniel, S. E., Blankson, S., and Lees, A. J. (1993) A clinicopathologic study of 100 cases of parkinson's disease. *Arch. Neurol.* **50**, 140–148.

Hutton, M., Lendon, C. L., Rizzu, P., Baker, M., Froelich, S., Houlden, H., Pickering-Brown, S., Chakraverty, S., Isaacs, A., Grover, A., Hackett, J., Adamson, J., Lincoln, S., Dickson, D., Davies, P., Petersen, R. C., Stevens, M., de Graaff, E., Wauters, E., van Baren, J., Hillebrand, M., Joosse, M., Kwon, J. M., Nowotny, P., Heutink, P. *et al.* (1998). Association of missense and 5'-splice-site mutations in tau with the inherited dementia FTDP-17. *Nature* **393**, 702–705.

Julien, J. P., and Mushynski, W. E. (1998). Neurofilaments in health and disease. *Prog. Nucleic Acid Res. Mol. Biol.* **61**, 1–23.

Kawamata, H., McLean, P. J., Sharma, N., and Hyman, B. T. (2001). Interaction of a-synuclein and synphilin-1: effect of parkinson's disease-associated mutations. *J. Neurochem.* **77**, 929–934.

Kitada, T., Asakawa, S., Hattori, N., Matsumine, H., Yamamura, Y., Minoshima, S., Yokochi, M., Mizuno, Y., and Shimizu, N. (1998). Mutations in the parkin gene cause autosomal recessive juvenile parkinsonism. *Nature* **392**, 605–608.

Klein, C., Vieregge, P., Hagenah, J., Sieberer, M., Doyle, E., Jacobs, H., Gasser, T., Breakefield, X. O., Risch, N. J., and Ozelius, L. J. (1999). Search for the PARK3 founder haplotype in a large cohort of patients with parkinson's disease from northern Germany. *Ann. Hum. Genet.* **63**, 285–91.

Krüger, R., Rahner, N., Fischer, C., Schulte, T., Holzmann, C., Epplen, J. T., Schöls, L., and Riess, O. (2001). Mutation analysis and association studies of the neurofilament L, M and H genes in German parkinson's disease patients. *Am. J. Hum. Genet.* **69**, 567 (Abstract).

Kupke, K. G., Lee, L. V., and Muller, U. (1990). Assignment of the X-linked torsion dystonia gene to Xp21 by linkage analysis. *Neurology* **40**, 1438–42.

Lavedan, C., Buchholtz, S. Nussbaum, R. L., Albin, R. L., and Polymeropoulos, M. H. (2002). A mutation in the human neurofilament M gene in Parkinson's disease that suggests a role for cytoskeleton in neuronal degeneration. *Neurosci. Lett.* **322**, 57–61.

Lee, L. V., Pascasio, F. M., Fuentes, F. D., and Viterbo, G. H. (1976). Torsion dystonia in Panay, Philippines. *Adv. Neurol.* **14**, 137–151.

Leroy, E., Boyer, R., Auburger, G., Leube, B., Ulm, G., Mezey, E., Harta, G., Brownstein, M. J., Jonnalagada, S., Chernova, T., Dehejia, A., Lavedan, C., Gasser, T., Steinbach, P. J., Wilkinson, K. D., and Polymeropoulos, M. H. (1998). The ubiquitin pathway in parkinson's disease. *Nature* **395**, 451–452.

Louis, E. D., Ford, B., Lee, H., Andrews, H., and Cameron, G. (1998). Diagnostic criteria for essential tremor: a population perspective. *Arch. Neurol.* **55**, 823–828.

Lücking, C. B., Durr, A., Bonifati, V., Vaughan, J., De Michele, G., Gasser, T., Harhangi, B. S., Meco, G., Denefle, P., Wood, N. W., Agid, Y., and Brice, A. (2000). Association between early-onset parkinson's disease and mutations in the parkin gene. *N. Engl. J. Med.* **342**, 1560–1567.

Maraganore, D. M., Farrer, M. J., Hardy, J.A., Lincoln, S. J., McDonnell, S. K., and Rocca, W. A. (1999). Case-control study of the ubiquitin carboxy-terminal hydrolase L1 gene in parkinson's disease. *Neurology* **53**, 1858–1860.

Martin, E. R., Scott, W. K., Nance, M. A., Watts, R. L., Hubble, J. P., Koller, W. C., Lyons, K., Pahwa, R., Stern, M. B., Colcher, A., Hiner, B. C., Jankovic, J., Ondo, W. G., Allen, F. H., Jr., Goetz, C. G., Small, G. W., Masterman, D., Mastaglia, F., Laing, N. G., Stajich, J. M., Ribble, R. C., Booze, M. W., Rogala, A., Hauser, M. A., Zhang, F., Gibson, R. A., Middleton, L. T., Roses, A. D., Haines, J. L., Scott, B. L., Pericak-Vance, M. A., and Vance, J. M. (2001). Association of single-nucleotide polymorphisms of the tau gene with late-onset parkinson disease. *JAMA* **286**, 2245–2250.

Muenter, M. D., Forno, L. S., Hornykiewicz, O., Kish, S. J., Maraganore, D. M., Caselli, R. J., Okazaki, H., Howard, F. M. Jr., Snow, B. J., and Calne, D. B. (1998). A familial parkinson-dementia syndrome. *Ann. Neurol.* **43**, 768–781.

Pankratz, N. D., Nichols, W. C., Uniacke, S. K., Rudolph, A., Halter, C., Siemers, E., Hubble, J. P., Conneally, P. M., Foroud, T., and parkinson Study Group. (2001). Genome screen to identify loci contributing to susceptibility for parkinson's disease. Abstract. *Am. J. Hum. Genet.* **69**, 535.

Saigoh, K., Wang, Y. L., Suh, J. G., Yamanishi, T., Sakai, Y., Kiyosawa, H., Harada, T., Ichihara, N., Wakana, S., Kikuchi, T., and Wada, K. (1999). Intragenic deletion in the gene encoding ubiquitin carboxy-terminal hydrolase in gad mice. *Nat. Genet.* **23**, 47–51.

Satoh, J., and Kuroda, Y. (2001) A polymorphic variation of serine to tyrosine at codon 18 in the ubiquitin C-terminal hydrolase-L1 gene is associated with a reduced risk of sporadic parkinson's disease in a Japanese population. *J. Neurol. Sci.* **189**, 113–117

Scott, W. K., Nance, M. A., Watts, R. L., Hubble, J. P., Koller, W. C., Lyons, K., Pahwa, R., Stern, M. B., Colcher, A., Hiner, B. C., Jankovic, J., Ondo, W. G., Allen, F. H., Jr., Goetz, C. G., Small, G. W., Masterman, D., Mastaglia, F., Laing, N. G., Stajich, J. M., Slotterbeck, B., Booze, M. W., Ribble, R.C., Rampersaud, E., West, S. G., Gibson, R. A., Middleton, L. T., Roses, A. D., Haines, J. L., Scott, B. L., Vance, J. M., and Pericak-Vance, M. A. (2001). Complete genomic screen in parkinson's disease. *JAMA* **286**, 2239–2244.

Spellman, G. G. (1962). Report of familial cases of parkinsonism. *JAMA* **179**, 160–162.

Valente, E. M., Bentivoglio, A. R., Dixon, P. H., Ferraris, A., Ialongo, T., Frontali, M., Albanese, A., and Wood, N. W. (2001). Localization of a novel locus for autosomal recessive early-onset parkinsonism, PARK6, on human chromosome 1p35–p36. *Am. J. Hum. Genet.* **68**, 895–900.

Valente, E. M., Brancati, F., Ferraris, A., Graham, E. A., Davis, M. B., Breteler, M. M., Gasser, T., Bonifati, V., Bentivoglio, A. R., De Michele, G., Durr, A., Cortelli, P., Wassilowsky, D., Harhangi, B. S., Rawal, N., Caputo, V., Filla, A., Meco, G., Oostra, B. A., Brice, A., Albanese, A., Dallapiccola, B., and Wood, N. W. (2002). PARK6-linked parkinsonism occurs in several European families. *Ann. Neurol.* **51**, 14–18.

van Duijn, C. M., Dekker, M. C., Bonifati, V., Galjaard, R. J., Houwing-Duistermaat, J. J., Snijders, P. J., Testers, L., Breedveld, G. J., Horstink, M., Sandkuijl, L. A., van Swieten, J. C., Oostra, B. A., and Heutink, P. (2001). Park 7, a novel locus for autosomal recessive early-onset parkinsonism, on chromosome 1p36. *Am. J. Hum. Genet.* **69**, 629–634.

Wakabayashi, K., Engelender, S., Yoshimoto, M., Tsuji, S., Ross, C. A., and Takahashi, H. (2000). Synphilin-1 is present in Lewy bodies in parkinson's disease. *Ann. Neurol.* **47**, 521–523.

Waters, C. H., and Miller, C. A. (1994). Autosomal dominant Lewy body parkinsonism in a four-generation family. *Ann. Neurol.* **35**, 59–64.

Waters, C. H., Faust, P. L., Powers, J., Vinters, H., Moskowitz, C., Nygaard, T., Hunt, A. L., and Fahn, S. (1993). Neuropathology of Lubag (X-linked dystonia-parkinsonism). *Mov. Disord.* **8**, 387–390.

West A. B., Zimprich, A., Lockhart, P. J., Farrer, M., Singleton, A., Holtom, B., Lincoln, S., Hofer, A., Hill, L., Muller-Myhsok, B., Wszolek, Z. K., Hardy, J., and Gasser, T. (2001). Refinement of the PARK3 locus on chromosome 2p13 and the analysis of 14 candidate genes. *Eur. J. Hum. Gen*et. **9**, 659–666.

Wilhelmsen, K. C., Weeks, D. E., Nygaard, T. G., Moskowitz, C. B., Rosales, R. L., dela Paz, D. C., Sobrevega, E. E., Fahn, S., and Gilliam, T. C. (1991). Genetic mapping of 'Lubag' (X-linked dystonia-parkinsonism) in a Filipino kindred to the pericentromeric region of the X chromosome. *Ann. Neurol.* **29**, 124–131.

Wilkinson, K. D, Lee, K. M., Deshpande, S., Duerksen-Hughes, P., Boss, J. M., and Pohl, J. (1989). The neuron-specific protein PGP 9.5 is a ubiquitin carboxyl-terminal hydrolase. *Science* **246**, 670–672.

Wintermeyer, P., Kruger, R., Kuhn, W., Muller, T., Woitalla, D., Berg, D., Becker, G., Leroy, E., Polymeropoulos, M., Berger, K., Przuntek, H., Schols L., Epplen, J. T., and Riess, O. (2000). Mutation analysis and association studies of the UCHL1 gene in German Parkinson's disease patients. *Neuroreport* **11**, 2079–2082.

Wszolek, Z. K., Gwinn-Hardy, K. A., and Muenter, M. D. (1999). Family C (German-American) with late onset parkinsonism: longitudinal observations including autopsy. *Neurology* **52**, A221.

Wszolek, Z. K., Uitti, R. J., and Markopoulou, K. (2001). Familial Parkinson's disease and related conditions. In *Parkinson's Disease*. (Calne, E., Calne, S., eds.), Vol. 86, pp. 33–43. Advances in Neurology, Philadelphia.

Wszolek, Z. K., Pfeiffer, B., Fulgham, J. R., Parisi, J. E., Thompson, B. M., Uitti, R. J., Calne, D. B., and Pfeiffer, R. F. (1995). Western Nebraska family (family D) with autosomal dominant parkinsonism. *Neurology* **45**, 502–505.

Zhang, J., Hattori, N., Leroy, E., Morris, H. R., Kubo, S., Kobayashi, T., Wood, N. W., Polymeropoulos, M. H., and Mizuno, Y. (2000). Association between a polymorphism of ubiquitin carboxy-terminal hydrolase L1 gene in sporadic Parkinson's disease. *Parkinsonian Related Disord.* **6**, 195–197.

Zink, M. Grim, L., Wszolek, Z. K., and Gasser, T. (2001). Autosomal-dominant parkinson's disease linked to 2p13 is not caused by mutations in the transforming growth factor alpha (TGF alpha). *J. Neurol. Transm.* **108**, 1029–1034.

# CHAPTER 30

# tau Genetics in Frontotemporal Lobe Dementia, Progressive Supranuclear Palsy, and Corticobasal Degeneration

JOSEPH J. HIGGINS

*Center for Human Genetics and Child Neurology*
*Mid-Hudson Family Health Institute*
*New Paltz, New York 12561*

I. Introduction
II. Anatomy of the *tau* Gene
   A. Regulation of *tau* Gene Transcription
   B. *tau* Gene Transcription and RNA Splicing
   C. *tau* Gene Translation
III. *tau* Genetics and Molecular Function
   A. *tau* Gene Exon 10 Splicing Defects
   B. *tau* Gene Missense Mutations
   C. Translation Defects
IV. Clinical Phenotypes Caused by *tau* Gene Mutations
   A. FTDP-17
   B. Familial PSP
   C. Familial CBD
   D. MAPT Pathology
   E. Neuroimaging
V. Clinical Genetics in FTD, PSP, and CBD
   A. Genetic Testing
   B. Clinical Implications
   C. Ethical Dilemmas
VI. *tau* Genetics and Transgenic Models of Disease
VII. Treatment
   Acknowledgments
   References

## I. INTRODUCTION

The presence of neuronal and glial, fibrillary, intracellular inclusions that contain aberrant precipitates of the microtubule-associated protein tau (MAPT) led to the hypothesis that mutations in the *tau* gene were responsible for the neuropathology of frontotemporal lobe dementia (FTD), progressive supranuclear palsy (PSP), and corticobasal degeneration (CBD). This hypothesis remained unproved until 1998 when mutations in the coding and splice-site sequences of the *tau* gene were identified in families with dominantly inherited FTD with parkinsonism linked to chromosome 17 (FTDP-17; Hutton *et al.*, 1998; Poorkaj *et al.*, 1998; Spillantini *et al.*, 1998a). The mutations can impair MAPT function, modify the alternative splicing of *tau* exons, and promote the intracellular aggregation of MAPT. This notable discovery firmly established the role of the *tau* gene as an etiopathogenetic factor in FTDP-17 and continues to facilitate genetic research in other disorders with MAPT pathology.

The molecular classification of FTD, PSP, and CBD is controversial. Although there is considerable neuropathological overlap between these disease entities (Feany *et al.*, 1996), a significant number of cases with the FTD phenotype do not have MAPT pathology (Higgins and Mendez, 2000; Mann *et al.*, 2000) or *tau* mutations (Kertesz *et al.*, 2000). Conversely, there are some families with *tau* gene mutations and an absence of MAPT pathology (Heutink *et al.*, 1997; Hutton *et al.*, 1998). Overall, mutations in the *tau* gene are an uncommon cause of sporadic FTD (Rizzu *et al.*, 1999; Poorkaj *et al.*, 2001), PSP (Bonifati *et al.*, 1999; Higgins *et al.*, 1999b), and CBD (Higgins *et al.*, 1999a; Houlden *et al.*, 2001). Despite these shortcomings, FTD, PSP, CBD, and other genetic disorders (Table 30.1) are sometimes classified as the "tauopathies" because of their distinctive MAPT migration patterns on Western blots and their immunostaining characteristics. This classification is limited because it may incorrectly exclude many cases

TABLE 30.1 Genetic Disorders with MAPT Pathology

| Disorder | Chromosome | Gene mutations | Ref. |
|---|---|---|---|
| Frontotemporal dementia with parkinson linked to chromosome 17 | 17q21.1 | tau | Hutton et al., 1988; Poorkaj et al., 1988; Spillantini et al., 1988 |
| Pallidopontonigral degeneration | 17q21.1 | tau | Clark et al., 1988; Wszolek et al., 1992 |
| Pick-like disease dementia | 17q21.1 | tau | Murrel et al., 1999 |
| Progressive supranuclear palsy-like | 17q21.1 | tau | Stanford et al., 2000; Wszolek et al., 2001 |
| Niemann-Pick type C | 18q11–q12 | NPC1 | Auer et al., 1995; Carstea et al., 1997 |
| Pantothenate kinase-associated neurodegeneration | 20p13 | Pantothenate kinase | Saito et al., 2000; Zhou et al., 2001 |
| Down syndrome | 21q21 | Beta-APP[a] overexpression | Hanger et al., 1991 |
| Familial Alzheimer's disease | 21q21 | Beta-APP | Goate et al., 1991 |
|  | 14q24.3 | Presenilin 1 | Sherrington et al., 1995 |
|  | 1q31 | Presenilin 2 | Rogaev et al., 1995; Levy-Lahad et al., 1995 |

[a]APP = amyloid protein precursor.

based solely on their neuropathological phenotype. However, the classification has clinical utility when it is integrated with genetic data. In this chapter, tau genetics will be examined in the context of three disorders with MAPT neuropathology namely, FTD, PSP, and CBD. First, the structural anatomy of the tau gene and its relationship to fundamental genetic processes (i.e., transcription, translation, and post-translational modification) will be reviewed to assist the reader in interpreting the molecular characteristics of FTD, PSP, and CBD. This knowledge will serve as the basis for understanding the genotype-phenotype correlations and variability that occur in these disorders.

## II. ANATOMY OF THE tau GENE

The human tau gene is a single copy gene on chromosome 17 (Neve et al., 1986) that is organized into 16 exons (Fig. 30.1) including an exon that is transcribed but not translated (designated as exon –1); (Andreadis et al., 1992). Tau gene expression follows a pattern called alternative splicing because it gives rise to more than one mRNA sequence. Six of the exons are alternatively excised (exons 2, 3, 4A, 6, 8, and 10) with complex splicing patterns that are anatomically specific, developmentally regulated, and demonstrate interspecies variation (Himmler et al., 1989; Nelson et al., 1996). The organizational structure of the tau gene can be defined by its regulation during the basic processes of transcription, translation, and post-translational modification.

### A. Regulation of tau Gene Transcription

The regulation of tau gene expression at the transcriptional level is not completely understood. The 5'-untranslated exon –1 (5'-UTR) is rich in guanine and cytosine, and contains a promoter region with binding sites for one specific (GCF) and two general (AP2, Sp1) transcription factors (Andreadis et al., 1996). One of the binding sites, AP2, is commonly found in cells of neural crest origin and is activated by retinoic acid (Williams and Tijan, 1991; Falconer et al., 1992). This 5'-UTR promoter's composition

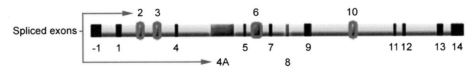

FIGURE 30.1 The genomic organization of the tau gene. The tau gene is organized into 16 coding regions that are labeled numerically as exons 1–14 plus two exons, –1 and 4A. The exons are black or gray and the introns are white. Constitutive exons are in black and the six alternatively spliced exons (2, 3, 4A, 6, 8, and 10) are shown in gray. Exon 6 is expressed at very low levels in the brain. However, exons 4A and 8 (unframed gray) are not transcribed in the human central nervous system. The intron between exons 13 and 14 is retained in human tau mRNA and results in a long 3'-UTR. Exon –1 is untranslated.

is atypical for a regulated gene (i.e., absent TATA and CAAT boxes, GC rich) and gives rise to a 2-kb mRNA species that lacks the neuronal specificity of the other *tau* gene mRNA transcripts that migrate at 6 and 8-kb mRNA on Northern blots (Andreadis *et al.*, 1996). The promoters for these other *tau* transcripts are not well defined but are presumably strictly regulated because of their restricted expression in neurons (6-kb mRNA), and the peripheral nervous system (8-kb mRNA). Expression of the neuronal *tau* transcript (6-kb RNA) is induced by thyroid hormone, nerve growth factor (NGF), and retinoic acid (Drubin *et al.*, 1988; Aniello *et al.*, 1991; Falconer *et al.*, 1992). The peripheral nervous system transcript (8-kb mRNA) is insensitive to NGF (Drubin *et al.*, 1988; Couchie *et al.*, 1992; Goedert *et al.*, 1992; Gache *et al.*, 1993). Although the precise regulatory mechanisms that operate at the transcriptional level are ill-defined, factors must exist to repress, or alter, *tau* transcription in non-neuronal cells or, in rare, central nervous system transcripts (Wei *et al.*, 2000). These elements may include other promoters or neuronal-specific silencers or enhancers.

## B. *tau* Gene Transcription and RNA Splicing

As noted above, the pre-mRNA *tau* gene transcript undergoes a complex alternative splicing pattern that produces several bands on Northern blot analysis at approximately 2, 6, and 8 kilobases (Drubin *et al.*, 1988). As depicted in Fig. 30.1, 6 of the exons are regulated (exons 2, 3, 4A, 6, 8, and 10). Two polyadenylation sites are used to generate the two major brain mRNA *tau* gene transcripts that migrate at 2 and 6-kb on Northern blots (Goedert *et al.*, 1989a; Himmler, 1989; Himmler *et al.*, 1989; Andreadis *et al.*, 1992). The intron between exons 13 and 14 is retained and results in a long 3′-UTR (Andreadis *et al.*, 1992; Fig. 30.1). The 2-kb transcript is localized to the nucleus (Wang *et al.*, 1993). The 8-kb transcript termed "big *tau*" is expressed in the retina, peripheral nervous system, and muscle and includes exon 4A (Nuñez and Fisher, 1997; Wei and Andreadis, 1998). Exons 4A and 8 are not transcribed in the human central nervous system and exon 6 is expressed at very low levels (Wei *et al.*, 1998). As shown in Figs. 30.2 and 30.3, this differential inclusion or exclusion of exons 2, 3, and 10 is the basis for the major expression patterns in the human brain. The translation of this alternatively transcript results in six MAPT isoforms as depicted in Fig. 30.4 (Andreadis *et al.*, 1992; Couchie *et al.*, 1992; Goedert *et al.*, 1992; Himmler *et al.*, 1989). Although present in adult human brain, exons 2 and 3 are absent from fetal human brain. The expression profile in adult human cerebellum, hippocampus, and whole brain are the same with a pattern that shows a predominance of exons $2^-3^-$ and $2^+3^-$, with a minor component of exons $2^+3^+$ (Wei *et al.*, 1998; Fig. 30.2).

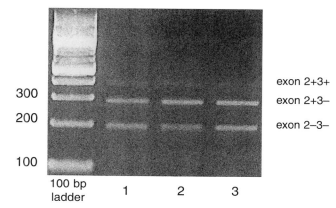

**FIGURE 30.2** The expression profile of *tau* gene exons 2 and 3. This figure shows the results of RT-PCR performed on samples from the midbrain (lane 1), and the cerebral cortex (lane 2) of an individual with PSP, and a normal adult cerebral cortex (lane 3). Exon 3 alone but never exon 2 alone is deleted (e.g., $2^-3^+$) in *tau* transcripts. The differential inclusion of exons 2 and 3 produces three of four combinations: $2^-3^-$ (196 bp), $2^+3^-$ (283 bp), $2^+3^+$ (370 bp). The product ratio is approximately $2^-3^- \geq 2^+3^- >> 2^+3^+$. (The primer used in reverse transcription of RNA to cDNA was 5′-AGGGACCCAATCTTCGACTG-3′. In the PCR reaction, the sense primer was 5′-TCCTCGCCTCTGTCGACTATC-3′, and the antisense primer was 5′-TCTCCAATGCCTGCTTCTTC-3′). (Unpublished data, Higgins and DeBiase, 2000).

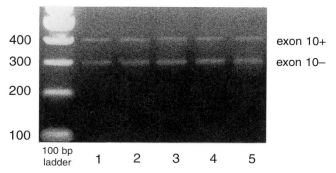

**FIGURE 30.3** The expression profile of *tau* gene exon 10. This figure shows the results of RT-PCR performed on samples from the midbrain (lanes 1 and 2), and the cerebral cortex (lane 3 and 4) of two individuals with PSP, and a normal adult cerebral cortex (lane 5). The differential inclusion of exons 10 ($10^+ = 388$ bp, $10^- = 296$ bp), is a factor in the formation of the six MAPT isoforms shown in Fig. 30.4. (The primer used in reverse transcription of RNA to cDNA was 5′-AGGGACCCAATCTTCGACTG -3′. In the PCR reaction, the sense primer was 5′-TGGTCCGTACTCCACCCAAG-3′, and the antisense primer was 5′-AGGGACCCAATCTTCGACTG-3′). (Unpublished data, Higgins and DeBiase, 2000).

## C. *tau* Gene Translation

In the human brain, the alternative splicing of *tau* mRNA translates into six distinct MAPT isoforms with molecular weights between 60 and 72 kDa, but isoforms of 110 kDa are found in some neural tissues (Drubin *et al.*, 1986, 1988). The alternative splicing of three exons (2, 3,

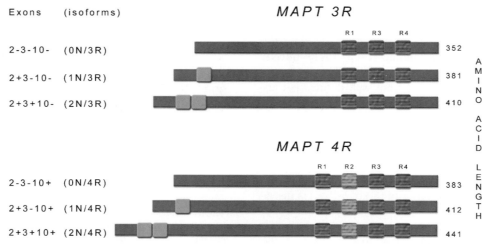

FIGURE 30.4 MAPT isoforms. The six MAPT isoforms are translation products from the alternative splicing of exons 2, 3, and 10. Depending on the splicing patterns of exons 2 and 3, the N terminus of MAPT has 0 (0N), 29 (1N), or 58 (2N) amino acid inserts. At the C terminus of MAPT either three (3R) or four (4R) tandem amino acid repeats of 31 or 32 amino acids are encoded by exons 9 (R1), 10 (R2), 11 (R3), and 12 (R4). The inclusion of exon 10 results in the 4R isoforms. The amino acid inserts produced by exons 2 and 3 are in light gray, and the 31 amino acid tandem repeat produced by exon 10 is stippled dark gray.

and 10) accounts for six mRNA species (2⁻3⁻10⁻, 2⁺3⁻10⁻, 2⁺3⁺10⁻, 2⁻3⁻10⁺, 2⁺3⁻10⁻, 2⁺3⁺10⁺) (Fig. 30.4). These isoforms differ from each other at the amino terminus (N terminus) and the carboxy terminus (C terminus). The N terminus has 0 (0N, exons 2⁻3⁻), 29 (1N, exons 2⁺3⁻), or 58 (2N, exons 2⁺3⁺) amino acid inserts encoded by exons 2 and 3 (Goedert et al., 1989a). At the C terminus, either three (3R) or four (4R) tandem amino acid repeats of 31 or 32 amino acids are encoded by exons 9, 10, 11, and 12 (Goedert et al., 1989b; Andreadis et al., 1992). The ratio of 3R to 4R in adult human brain is approximately one, but 0N, 1N, and 2N vary (Goedert and Jakes, 1990; Hong et al., 1998). The three N terminus and the two C-terminus variations give rise to three MAPT isoforms with 3R each and three isoforms have 4R each (Fig. 30.4).

### D. Post-Translational Modification of MAPT

Normal MAPT is modified by phosphorylation at approximately 30 of 79 potential serine and threonine phosphate receptor sites (Buee et al., 2000). These sites are clustered in the regions around the microtubule binding domains and may be phosphorylation sites for proline-directed kinases (e.g., glycogen synthase kinase-3, GSK-3; extracellular signal-related kinase-1 and -2; and mitogen-activated protein kinases, p38 kinase, and c-jun N-terminal kinase; Anderton et al., 2001). At least one tau gene mutation, R406W, affects phosphorylation of MAPT by GSK-3 (Dayanandan et al., 1999; Connell et al., 2001). This finding defines a mechanism where an alteration in the DNA sequence of the tau gene can have a long-range conformational effect on the post-translational modification of MAPT. Distinct patterns of hyperphosphorylated MAPT on Western blots help define the effects of tau gene mutations (Fig. 30.5).

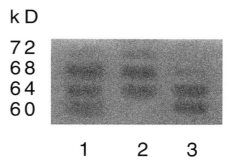

FIGURE 30.5 Representation of Western blot banding patterns of insoluble MAPT. This drawing shows the banding pattern of non-dephosphorylated insoluble MAPT from the brains of individuals with aberrant MAPT deposition. Lane 1 shows three dominant bands at 60, 64, and 68 kDa with a minor band at 72 kDa. This pattern is seen in the tau gene missense mutations, VAL337MET and ARG406TRP, that are found in exons 12 and 13, respectively. These mutations do not affect splicing. Lane 2 shows two major bands at 64 and 68 kDa with a minor band at 72 kDa. This pattern is seen in tau gene missense (ASN279LYS, SER305ASN), silent (LEU284LEU, ASN296ASN, SER305SER), and intronic splice site mutations. This pattern is also seen in the PRO301LEU tau gene mutation but further analysis of dephosphorylated insoluble MAPT and soluble MAPT, demonstrate that there is selective aggregation of the 4R isoform. Lane 3 shows two major bands at 60 and 64 kDa with a minor band at 72 kDa. This pattern is seen in the tau gene missense mutations, LYS257THR and GLY389ARG, that are found in exons 9 and 13, respectively.

## III. tau GENETICS AND MOLECULAR FUNCTION

Now that the organizational structure of the *tau* gene has been reviewed, this section will discuss the molecular dysfunction that is caused when a mutation in the *tau* gene is expressed in the human brain. Although there are exceptions, the postmortem examinations of individuals from families with *tau* gene mutations demonstrate the presence of abundant MAPT filaments composed of hyperphosphorylated protein. A general rule is that the morphology of pathological MAPT filaments is determined by whether the *tau* gene mutation affects mRNA splicing of exon 10 or not. Missense *tau* mutations (Table 30.2) appear to cause pathological consequences at the protein level by reducing the ability of MAPT to interact with microtubules and stimulating MAPT filament formation.

### A. *tau* Gene Exon 10 Splicing Defects

The intronic mutations, IVS10, A-G, +13; IVS10, C-U, +14 and IVS10, C-U, +16, (Table 30.2) increase the production of exon 10 *tau* mRNA (Hutton *et al.*, 1998; Spillantini *et al.*, 1998a). These mutations cause a two- to six-fold increase in the ratio of exon 10+ to 10– *tau* mRNA and consequently cause a preponderance of the 4R MAPT repeat isoform in the brain (Hutton *et al.*, 1998; Spillantini *et al.*, 1998a; D'Souza *et al.*, 1999; Grover *et al.*, 1999). The putative molecular mechanisms that disrupt splicing is novel in humans. As shown in Fig. 30.6, splicing occurs in two stages in an *in vitro* model (Lewin, 1994). In the first stage, the 5' end of the intron is cut separating the left exon and the right intron-exon molecule. The left exon takes the form of a linear molecule, but the right intron-exon is linked at a branch site to form a lariat or loop. The second stage involves cutting the 3'-splice site to free the intron in lariat form and ligating the right and left exons together. The lariat is subsequently debranched and degraded. The mechanism of splice site recognition can proceed by the involvement of independent small nuclear RNA molecules (snRNA; Fig. 30.6). These RNA species exist as small ribonucleoproteins (snRNP) and often contain splicing factors. The snRNP pair to the intronic 5'-splice site and mutations within the introns of the pre-mRNA can affect overall splicing efficiency (Lewin, 1994). By this mechanism, mutations in the intronic 5'-splice site of *tau* exon 10 can cause a decrease in the splicing efficiency of exon 10+ *tau* mRNA. Exonic mutations at the 5'-(ASN279LYS, and ΔLYS280) and the 3'-(SER305ASN, and SER305SER) ends of *tau* exon 10 are though to destabilize the lariat structure and either abolish (ΔLYS280) or inhibit *tau* exon 10 splicing (D'Souza *et al.*, 1999; Grover *et al.*, 1999; Varani *et al.*, 1999; D'Souza and Schellenberg; 2000, Gao,

TABLE 30.2 *Tau* Gene Mutations in Families with FTD, PSP, and CBD

| Type | Allelic variant | Location | Ref. |
|---|---|---|---|
| Missense | LYS257THR | Exon 9 | Rizzini, *et al.*, 2000 |
| | GLY272VAL | Exon 9 | Heutink *et al.*, 1997; Hutton *et al.*, 1998; Matsumura *et al.*, 1999 |
| | ASN279LYS[a] | Exon 10 | Clark *et al.*, 1998; Delisle *et al.*, 1999 |
| | PRO301LEU | Exon 10 | Heutink *et al.*, 1997; Hutton *et al.*, 1998; Hong *et al.*, 1998; Clark *et al.*, 1998 |
| | SER305ASN | Exon 10 | Iijima *et al.*, 1999 |
| | VAL337MET | Exon 12 | Poorkaj *et al.*, 1998 |
| | GLU342VAL | Exon 12 | Lippa *et al.*, 2000 |
| | LYS369ILE | Exon 12 | Neumann *et al.*, 2001 |
| | GLY389ARG | Exon 13 | Murrel *et al.*, 1999 |
| | ARG406TRP[a] | Exon 13 | Hutton *et al.*, 1998; Higgins *et al.*, 1998a; Perez *et al.*, 2000; Miyasaka *et al.*, 2001 |
| Deletion | ΔLYS280 | Exon 10 | Rizzu *et al.*, 1999; Hasegawa *et al.*, 1998 |
| | ΔASN296 homozygous[a] | | Pastor *et al.*, 2001 |
| Intronic | IVS9, G-A, +33 | Splicing, intron 9? | Rizzu *et al.*, 1999 |
| | IVS10, G-A, +3 | Splice-site, intron 10 | Spillantini *et al.*, 1998; Tolnay *et al.*, 2000 |
| | IVS10, T-C, +11 | Splice-site, intron 10 | Miyamoto *et al.*, 2001 |
| | IVS10, C-T, +12 | Splice-site, intron 10 | Yasuda *et al.*, 2000 |
| | IVS10, A-G, +13 | Splice-site, intron 10 | Hutton *et al.*, 1998 |
| | IVS10, C-U, +14 | Splice-site, intron 10 | Hutton *et al.*, 1998; Wilhelmsen *et al.*, 1994 |
| | IVS10, C-U, +16 | Splice-site, intron 10 | Hutton *et al.*, 1998 |
| Silent | SER305SER[a] | Exon 10 | Stanford *et al.*, 2000; Wszolek *et al.*, 2001 |
| | ASN296ASN[b] | Exon 10 | Spillantini *et al.*, 2000 |
| | LEU284LEU | Exon 10 | D'Souza *et al.*, 1999 |

[a]PSP-like phenotype.
[b]CBD-like phenotype.

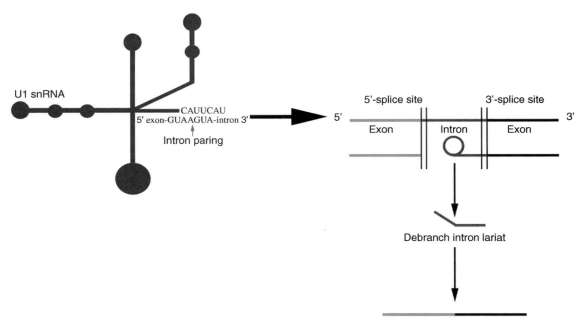

FIGURE 30.6  Splicing mechanisms and snRNA. Independent RNA molecules such as small nuclear RNAs (snRNA) recognize splicing sites in the form of active ribonucleoprotein particles. The secondary structure of the human U1 snRNA is depicted with emphasis on the 5′ terminal region labeled as the intron pairing site. This site has a string of single-stranded nucleotides that is complementary to the consensus sequence at the 5′ end of the intron. The exon splicing reaction proceeds through two basic stages within a complex called the spliceosome. In the first stage, the 5′ end of the intron is excised and forms a loop or lariat at a branch site that is located upstream of the 3′ end of the intron. In the second stage, the 3′ end of the intron is excised and the right and left exons are ligated or spliced together. The intron that is released as a lariat, debranched, and subsequently degraded.

et al., 2000; Jiang, et al., 2000). Although splicing is essentially viewed as a controlled deletion of intron sequences, *cis*-acting regulatory elements may also enhance or inhibit splicing efficiency. The silent mutations, LEU284LEU and ASN296ASN, are thought to affect either silencer or enhancer elements (D'Souza and Schellenberg, 2000) that modify the utilization of the 5′-splice site in *tau* exon 10. The recent discoveries that *tau* gene mutations disrupt the alternative splicing of exon 10 indicates that *tau* gene regulation is a complex process involving interactions between *cis*- and, perhaps even, *trans*-acting factors. These mechanisms may explain the temporal and regional differences in the ratios of 4R to 3R MAPT isoforms during human development and in pathological states.

## B. *tau* Gene Missense Mutations

The missense mutation, PRO301LEU, is the most common mutation in FTDP-17 (Heutink et al., 1997; Bird et al., 1999; Mirra et al., 1997). Missense mutations cluster in the region that contains the microtubule-binding domains at the C terminus of MAPT. These alterations interfere with the interaction of MAPT and microtubules by several mechanisms (Poorkaj et al., 1998; Hong et al., 1998; Hutton et al., 1998). Some missense (PRO301LEU, VAL337MET, and ARG406TRP) and the ΔLYS280 deletion mutations reduce the binding capacity and the affinity of MAPT for microtubules (Hong et al., 1998). These mutations, as well as the mutation, GLY272VAL, reduce the ability of 4R and 3R MAPT isoforms to polymerize tubulin (Rizzu et al., 1999; Hong et al., 1998; Hasegawa et al., 1998). The largest effect on microtubule binding and tubulin polymerization is caused by the deletion mutation, ΔLYS280 (Hong et al., 1998; Hasegawa et al., 1998). As noted above in Section III.A, the missense mutations in exon 10 (ASN279LYS, SER305ASN, and SER305SER) perturb the alternative splicing of exon 10 and do not directly affect MAPT-microtubule binding.

## C. Translation Defects

In family with a form of FTDP-17 known as hereditary dysphasic disinhibition dementia 2, there are no mutations in the *tau* gene and *tau* mRNA is expressed in a normal pattern. However, there is a selective loss of all six MAPT isoforms (Lendon et al., 1998; Zhukareva et al., 2001). These findings suggests that MAPT is controlled either at the level of *tau* mRNA translation or through mechanisms that regulate *tau* mRNA stability. The mechanisms that precipitate the dysregulation of *tau* mRNA are unknown but may involve intrinsic or extrinsic factors that affect the ribosomal initiation, elongation, or the termination of protein synthesis. This type of defect may be the cause of the more common, sporadic forms of dementia and represents a novel mechanism that can cause neurodegeneration.

## IV. CLINICAL PHENOTYPES CAUSED BY tau GENE MUTATIONS

Mutations in the *tau* gene have neuropathological and clinical consequences that overlap and vary considerably (Hulette *et al.*, 1999; Nasreddine *et al.*, 1999; van Swieten *et al.*, 1999; Kertesz *et al.*, 2000). There are many exceptions to the clinical descriptions that are summarized in Table 30.3, but certain tau mutations have distinct clinico-pathologic effects. Phenotypic features vary considerably and the age of disease onset and disease duration can differ within and between families with the same *tau* gene mutation. In general, individuals with an earlier onset have a more protracted, severe clinical course but this also varies. Most families with *tau* gene mutations have a predominantly FTD phenotype except those designated with a superscript a or b in Table 30.2. Parkinsonism is present in almost half of the families with *tau* gene mutations with ocular motility abnormalities or dystonia in some individuals. The phenotype in families with the PRO301LEU mutation may include CBD or epilepsy late in the disease. In some families with VAL337MET the early onset of psychotic features is uncommon but later become apparent. Sporadic forms of PSP and CBD have not been associated with *tau* gene mutations (Morris *et al.*, 2002) but clinically atypical familial forms of these disorders have been described (Spillantini *et al.*, 2000; Stanford *et al.*, 2000; Pastor *et al.*, 2001). However, a common feature of both familial and sporadic forms of FTD, PSP, and CBD is that the clinical onset is variable and the disease is often not recognized until its later stages (deYebenes *et al.*, 1995; Spillantini, *et al.*, 1998b; Bird *et al.*, 1999).

### A. FTDP-17

The gross pathological hallmarks of FTD are the selective lobar degeneration of the frontal and temporal lobes of the cerebral cortex and their clinical accompaniments (Table 30.3). The lobar degeneration is sometimes unilateral and additional neurodegenerative changes can occur in the subcortical areas such as the substantia nigra. The initial symptoms typically occur before age 65 and are characterized by a marked change in personality and social behavior. Memory is preserved until late in the course of the disease. In 1996, an international conference was

TABLE 30.3 Common Clinical Manifestations of FTDP-17, PSP, and CBD

|  | FTD-17 | PSP | CBD |
| --- | --- | --- | --- |
| Neurobehavioral | Apathy/disinhibition<br>Personal neglect<br>Impulsivity<br>Perseveration<br>Aggression<br>Psychosis/depression<br>Hyperorality/Klüver-Bury<br>Indifference | Apathy/disinhibition<br>Inappropriate jocularity<br>Dysphoria<br>Anxiety<br>Emotional incontinence | Depression<br>Inappropriate jocularity<br>Emotional lability |
| Cortical | Dementia<br>Poor executive functioning<br>Memory loss<br>Echolalia, mutism<br>Nonfluent aphasia<br>Visuospatial dysfunction | Dementia<br>Poor executive functioning | Dementia<br>Alien limb syndrome[b]<br>Ideomotor apraxia[b]<br>Mutism<br>Memory loss<br>Cortical sensory loss |
| Motor/sensory | Parkinsonism<br>Amyotrophy<br>Dysarthria, dysphagia<br>Dystonia<br>Pyramidal tract signs | Postural instability and falls[a]<br>Bradykinesia<br>Dysarthria, dysphagia<br>Tremor minimal or absent<br>Pyramidal tract signs<br>Axial > limb rigidity<br>Procerus sign (worried look)[c] | Late gait and balance disorder<br>Asymmetric akinetic-rigidity<br>Focal myoclonias<br>Limb dystonia |
| Ophthalmology | Eyelid apraxia | Vertical supranuclear palsy[a] | Impaired vertical gaze |

[a]Occurs within the first year and strongly indicates a diagnosis of PSP.
[b]Prominent in CBD.
[c]Romano and Colosimo, 2001.

convened at Ann Arbor, Michigan, to clarify the phenotype of dominantly inherited FTD and to agree on a nosological classification based on genetic linkage data (Foster *et al.*, 1997). The phenotype was inherited as an autosomal dominant condition with age-dependent penetrance. FTDP-17 was the term designated to define the clinical and neuropathological features of 13 family pedigrees that had their disease locus assigned to a minimal critical region of 2 cM between *D17S791* and *D17S800*. The prime positional candidate gene in this region, *tau*, was later found to possess distinct mutations that caused FTD in these families (Table 30.2). These families demonstrated phenotypic variability but common signs were found between the families. Most families presented with severe behavioral or psychiatric manifestations progressing to dementia, while others first manifested a parkinsonian-plus syndrome. In the original description of FTDP-17 (Lynch *et al.*, 1994), the psychiatric prodromal symptoms in affected individuals were personality and behavioral changes. Early in the course of the disease, affected individuals would become disinhibited and display deviant behaviors such as alcoholism, inappropriate sexual behavior, Klüver-Bucy syndrome, religiosity, and stealing. One peculiar behavior that appeared distinctive was the hoarding and craving of sweets. Rigidity, bradykinesia, postural instability, and primitive reflexes were common neurological signs that appeared later. Eventually, all affected family members developed frontal release signs and dementia. At first, the changes in behavior and personality suggested psychiatric diagnoses such as psychosis and depression, but the inheritance pattern and progressive clinical course eventually suggested FTDP-17.

### B. Familial PSP

Several *tau* gene mutations cause a phenotype similar to PSP (Table 30.2). The neuropathology is similar to sporadic cases of PSP with neurofibrillary tangles concentrating within the subcortical regions of the basal ganglia. The presenting neurological signs can vary within family members. Some individuals present with dementia and later develop supranuclear vertical gaze palsy and extrapyramidal signs. The initial symptoms in others can be asymmetric dystonia, dysarthria, and later supranuclear gaze palsy and frequent falls (Stanford *et al.*, 2000; Pastor *et al.*, 2001). In a Spanish family, two individuals were identified with a homozygous deletion ($\Delta$ASN296) with a PSP phenotype, and heterozygotes had a Parkinson disease phenotype (Pastor *et al.*, 2001).

### C. Familial CBD

A single family with a silent *tau* gene mutation (ASN296ASN) pathologically resembles CBD but has clinical features similar to FTDP-17 (Spillantini *et al.*, 2000). The common clinical features of sporadic CBD are found in Table 30.3.

### D. MAPT Pathology

The gross neuropathologic findings in the tauopathies include lobar and generalized brain atrophy. Typically, FTDP-17 demonstrates frontotemporal atrophy with neuronal loss and spongiform changes in the cerebral cortex (Fig. 30.7). Microscopically, neuronal loss and gliosis occurs in the substantia nigra and amygdala. In FTDP-17

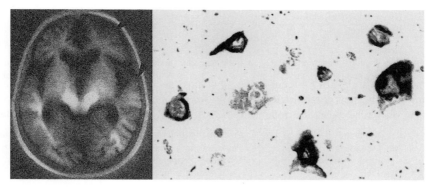

**FIGURE 30.7** Clinicopathological consequences of *tau* gene mutations. The T1-weighted, magnetic resonance image of the brain shows the typical neuroimaging findings in an individual with lobar atrophy. The arrows show the marked atrophy in the frontal and temporal lobes. The accompanying neuropathology can be diverse, but in FTDP-17 common features include neuronal loss, gliosis, and spongiosis and neuronal loss in the substantia nigra. (For more details see Foster *et al.* (1997). *Ann. Neurol.* **41**, 706–715. The MAPT immunolabeled composite image on the left shows MAPT+ neurons that are irregularly shaped, and swollen and some contain neurofibrillary tangle-like inclusions.

patients with amyotrophy, changes can be found in the spinal cord anterior horn cells. However, the spinal cord changes in humans are not as extensive as the pathology described in some transgenic murine models for the tauopathies (Section VI). In typical PSP, a high density of neurofibrillary tangles and neuropil threads are found in the globus pallidus, subthalamic nucleus, substantia nigra, or pons and to a lesser degree in the striatum, oculomotor complex, medulla, and dentate nucleus. Patients with CBD have lobar atrophy in the parietal or frontoparietal areas with swollen achromatic neurons (Litvan et al., 1996, 1997). Until recently, these neuropathological criteria when integrated with the clinical history were the sole means of categorizing FTD, PSP, and CBD. The nosology of these disorders has now expanded to include specific genetic and biochemical abnormalities. There are several neurodegenerative diseases that share a common biochemical aberration in MAPT (Table 30.1) in the form of neurofibrillary lesions within certain regions of the brain. The most common of these disorders is Alzheimer disease, in which MAPT deposits are found in neurofibrillary tangles, neuropil threads, and neurites. The MAPT findings in FTD-17, PSP, and CBD overlap with Alzheimer's disease. The absence of beta-amyloid deposits helps differentiate Alzheimer's disease from these disorders and the MAPT deposits are usually twisted filaments that differ in diameter and periodicity from the paired helical filaments of Alzheimer disease (Spillantini et al., 1998a). MAPT extracted from the filaments show varying patterns on Western blotting that depend on the specific tau gene mutation (Fig. 30.5). In the sporadic forms of FTD, PSP, and CBD, the MAPT isoform composition of neurons and oligodendroglia differ but the astrocytic MAPT isoforms are similar (Arai et al., 2001). In sporadic PSP and CBD the 4R MAPT isoform is predominant. However, in sporadic FTD both 3R and 4R MAPT are present with the 3R isoform predominating over the 4R. In genetic terms, this is mainly the consequence of the aggregation of specific sets of MAPT isoforms dictated by the lack or presence of exon 10. The MAPT intraneuronal inclusions found in PSP and CBD contains exon 10 where there is a general deficit in FTD (Sergeant et al., 1999). Therefore, it seems that the neuropathological and clinical phenotypes in FTD, PSP, and CBD are related to specific sets of MAPT isoforms expressed by certain vulnerable neuronal populations.

### E. Neuroimaging

The lobar atrophy seen on MRI in FTD and CBD may aid the clinician in diagnosing these disorders (Fig. 30.7). Studies using positron emission tomography (PET) using $^{18}$F-dopa and $^{18}$fluorodeoxyglucose can evaluate regional cerebral dopaminergic function and glucose metabolism. PET scans in a Japanese family with FTD and a PRO301LEU tau gene mutation demonstrate bilateral frontotemporal hypometabolism (Kodama et al., 2000). In familial PSP, a reduction in caudate and putamen $^{18}$F-dopa uptake along with a significant reduction in striatal, lateral, and medial premotor area and dorsal prefrontal cortex glucose metabolism is found (Piccini et al., 2001). Cerebral glucose utilization and fluorodopa metabolism in CBD demonstrate specific abnormalities with a marked asymmetry in the parietal cortex (the primary motor and sensory cortex and the lateral parietal cortex), the thalamus, the caudate nucleus, and the putamen of the dominantly affected hemisphere (Nagasawa et al., 1996).

## V. CLINICAL GENETICS IN FTD, PSP, AND CBD

In general, tau gene mutations are rare. There are no sporadic cases of PSP or CBD with tau gene mutations (Baker et al., 1999; Higgins et al., 1999a, b; Houlden et al., 2001) and only a single sporadic case of FTD is reported with a three base pair deletion in exon 10 (Rizzu et al., 1999). In cases with a family history of FTD, tau gene mutations are found infrequently in mixed populations (Houlden et al., 1999, Poorkaj et al., 2001) and in less than half of patients in more homogeneous populations (Rizzu et al., 1999). Although PSP is generally thought of as a sporadic disease, dominant inheritance is sometimes found in families that have affected individuals with a PSP-like phenotype (de Yebenes et al., 1995; Rojo et al., 1999). Recessive inheritance was first postulated in PSP based on a statistical model used in genetic association studies (Higgins et al., 1998) and corroborated in a single family with homozygous tau gene mutations (Pastor et al., 2001). The clinical genetics of CBD are not well characterized but at least one family with similar neuropathological features has a silent mutation in the tau gene (Spillantini et al., 2000).

### A. Genetic Testing

Table 30.2 categorizes most of the various mutations that have been identified in the tau gene in FTD, and, in rare instances, PSP and CBD-like disorders. With advances in our knowledge of the genetic mechanisms underlying the inherited tauopathies, genetic testing may eventually have an important role in clarifying diagnostic discrepancies. A high level of test sensitivity and specificity is required for the clinical utility of such testing. The feasibility of genetic testing for sporadic forms of PSP and CBD using an extended haplotype consisting of single nucleotide polymorphisms (SNPs) has significant limitations. An extended haplotype (HapA) in the 5' portion of tau gene (Fig. 30.8) can detect patients with PSP with a sensitivity of 98% and a specificity of 67% (Higgins, et al. 2000). Extended haplotypes using SNPs in the 5'-promoter region have a similar

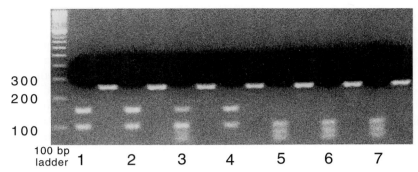

FIGURE 30.8 Extended haplotype analysis using polymorphisms at the 5′ end of the *tau* gene. An extended haplotype (HapA) using four polymorphisms in the *tau* gene exons 1, 4A, and 8 demonstrate co-segregation of a homozygous allele in patients with typical PSP (Higgins et al. (1999b,). *Neurology*, **53**, 1421–1424; Higgins et al. (2000). *Neurology*, **54**, 784–785. With permission.) This means that the analysis of one of these polymorphisms predicts the presence of the other three polymorphisms. In exon 4A, a T/T nucleotide substitution at position 554 (GenBank accession number AF047858) eliminates a *BsgI* restriction site. Digestion of the 254–bp PCR product (uncut products are to the right of each digested sample) yields four bands at 137, 117, 68, and 69 bp. The T/T homozygotes in lanes 1, 2, and 4 are affected with sporadic PSP and have one *BsgI* site that yields products of 137 and 117 bp. The other samples in lanes 3, 5, 6, and 7 are from age-matched controls. (Unpublished data, Higgins and DeBiase, 2000.) For a discussion of the relevance of extended haplotype analyses see Section V.-A.

sensitivity (de Silva et al., 2001; Ezquerra et al., 1999), but a haplotype (H1) spanning the entire *tau* gene was found in 87.5% of PSP patients compared to 62.8% of normal individuals (Baker et al., 1999). This haplotype has no defined effect on the pathological or biochemical phenotype of PSP (Liu et al., 2001). In sporadic CBD, the results are similar but less striking (Di Maria et al., 2000, Houlden et al., 2001). The lack of specificity of these analyses precludes their sole use as diagnostic tests in sporadic PSP and CBD.

## B. Clinical Implications

Almost all cases of *tau* gene mutation-associated diseases are in large family pedigrees. To identify these cases, the clinical evaluation of an affected individual must include a carefully constructed, four-generation pedigree of both maternal and paternal sides of the patient's family. The medical history of each person in the pedigree should be elicited from a knowledgeable family historian. Disease features that may represent atypical forms of disorder should be taken into account. For example, what may be considered "typical senility" in a family may otherwise go unnoticed if the appropriate questions are not asked. Because reduced penetrance could occur in mild cases, the disease might appear to skip one or several generations. The age at death should be noted in the family pedigree because an early death of an obligate mutation carrier may obscure the inheritance pattern of dominant late-onset disease. Once the pedigree is assembled and thoroughly checked, family members who are either at-risk or unaffected carriers can be identified. In these cases, genetic counseling is appropriate and can provide education about the natural history and genetics of the condition. In those cases where the family history suggests a dominantly segregating mutation, DNA studies of the patient or other affected family member may be warranted. Unless there is a strong family history of dementia and MAPT-related neuropathology, the yield of genetic testing is expected to be low. On the other hand, identifying a family member with a *tau* gene mutation has significant implications for other family members. Currently, *tau* gene DNA studies are provided on a research basis and are not available for clinical diagnostic testing.

## C. Ethical Dilemmas

The ultimate goal of research on *tau* genetics is to identify the metabolic perturbations that cause neurodegeneration and to develop therapeutic strategies for the prevention and cure of diseases that involve MAPT. The ability to identify the *tau* gene mutations that are associated with FTD, PSP, and CBD raises the future possibility of using DNA testing to identify individuals who have a mutation or susceptibility allele. These tests may confirm the diagnosis or predict whether an individual will develop the disease later in life. In the situation of testing for *tau* gene mutations in families, the problems are similar to other late-onset genetic disorders, such as Huntington disease, Alzheimer's disease, and the spinocerebellar ataxias. Testing for disease susceptibility based on a *tau* haplotype analysis raises different concerns because of its uncertainty and lack of specificity. At present, genetic testing for *tau* gene mutations is available on a research basis and is not commercially available. In many situations, state and federal regulations do not permit the results of research tests to be used diagnostically. Even so, *tau* genetic testing is not appropriate for most individuals with sporadic forms of dementia. However, predictive or diagnostic genetic testing for *tau* gene mutations may be warranted in the rare instances when a family is identified with an autosomal dominant pattern of inheritance. In these instances, genetic counseling permits individuals to make their own informed decisions.

## VI. *tau* GENETICS AND TRANSGENIC MODELS OF DISEASE

The overall goal of transgenic technology is to create an animal model to aid in the development of successful treatments for a human disease. Before the discovery that *tau* gene mutations caused human phenotypes, the results of transgenic models for MAPT pathology were confusing. Interestingly since transgenic mice lacking MAPT appear to be histologically normal. However, these null mutants had decreased microtubule stability in small caliber axons and behavioral changes (Harada et al., 1994). It is now thought that this phenotype bears resemblance to the family described in Section III.C without MAPT isoforms despite the presence of *tau* mRNA. Several murine transgenic models using models for overexpressing MAPT isoforms show age-dependent deposition of insoluble, hyperphosphorylated MAPT and intraneuronal inclusions. MAPT inclusions are usually present in cortical and brainstem neurons but unlike the human tauopathies, they are most abundant in spinal cord neurons causing amyotrophy and motor weakness (Ishihara et al., 1999; Spittaels et al., 1999; Duff et al., 2000; Probst et al., 2000). The discovery of *tau* gene mutations in FTDP-17 led to the idea that expression of these mutations could cause similar changes in transgenic animals. Expression of the most common *tau* gene mutation, PRO301LEU, causes motor and behavioral deficits in transgenic mice (Götz et al. 2001; Lewis et al. 2000). As in the overexpression models, the pathological changes occur in an age-dependent manner. There is central nervous system involvement but the spinal cord, peripheral nerves, and muscle were also affected. As in the human PRO301LEU mutation, a prominent insoluble 64-kDa MAPT isoform was found. Interestingly, the injection of beta-amyloid into the brains of these PRO301LEU mutant mice accelerates the production of neurofibrillary tangles (Götz et al., 2001). Cross-breeding experiments using PRO301LEU *tau* gene mutants and mice expressing mutant beta-amyloid precursor protein, produce double mutants with increased neurofibrillary tangle pathology (Lewis et al., 2001). These results suggest that an interaction between beta-amyloid and MAPT may be part of a fundamental pathway that leads to the degenerative changes in Alzheimer disease. However, the mechanisms underlying MAPT-mediated neurotoxicity remain unclear.

An intriguing genetic model of MAPT-related neurodegeneration involves expressing wild-type and mutant forms of the human *tau* gene in the fruit fly, *Drosophila melanogaster*. The mutant flies demonstrate an adult onset, and progressive neurodegeneration with a shortened life span. There is an accumulation of MAPT but neurodegeneration occurs without the neurofibrillary tangle formation that is seen in human and transgenic murine models (Wittmann et al., 2001). Although none of the transgenic models to date precisely mimic the myriad of neuropathology in FTDP-17, PSP, or CBD, they serve as experimental systems to investigate the cascade of events that leads to MAPT-induced neurodegeneration. These models are also valuable resources in expediously evaluating the effects of therapeutic interventions. There are certain limitations to this technology. For example, it is possible that murine neurons may be less susceptible to degeneration than human neurons. However, the major shortcoming so far in the transgenic mice models is that *tau* gene expression is not under the control of the endogenous *tau* promoter. As discussed in Sections II and III, *tau* gene mutations may disrupt *tau* gene regulation by a complex process involving the interaction between *cis*- and perhaps even *trans*-acting factors. The transgene expression pattern in the current mice models is dependent on promoter choice, copy number, and the transgene integration site. Therefore, it is not possible to determine the relationship between the temporal and regional differences in the ratios of 4R to 3R MAPT isoforms during human development or in pathological states.

## VII. TREATMENT

Unfortunately there are no curative therapies for the FTD, PSP, or CBD but some patients transiently respond to L-dopa therapy (Golbe, 2001). Because the behavioral manifestations of these disorders can be debilitating, psychiatric treatment is usually rendered in the early stages. Although acetylcholinesterase inhibitors such as donepezil has produced marginal benefit in Alzheimer's disease, its efficacy is lacking in PSP (Fabbrini et al., 2001; Litvan et al., 2001). Currently, the neurologist can assist patients and their family by providing prognostic information, genetic counseling, and by prescribing palliative measures to minimize aspiration and falling.

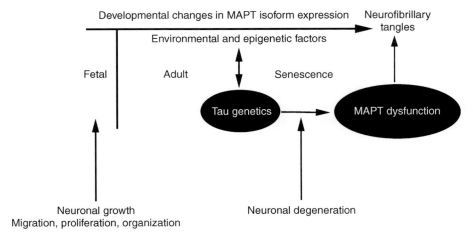

FIGURE 30.9 MAPT isoform expression scheme. MAPT isoform expression changes in the fetal period and remains constant during adulthood. It is not until adulthood that the known *tau* gene mutations cause clinically significant effects. *Tau* genetics and the influence of environmental and epigenetic factors play a role in creating aberrant MAPT isoform production, formation of neurofibrillary tangles, and subsequent neurodegeneration. This final pathway is common to several neurodegenerative disorders including FTDP-17, PSP, CBD, and Alzheimer's disease.

Defining the developmental events that lead to MAPT dysfunction will facilitate the design of rational therapeutic strategies that target the decisive stages that herald the inexorable progression of FTD, PSP, and CBD (Fig. 30.9). Although agents that reduce or prevent the pool of functionally impaired MAPT may be of therapeutic value, therapeutic compounds may be designed to act at several levels. Examples of potential targets include extrinsic factors such as the prolyl isomerase, Pin1, that restores the ability of phosphorylated MAPT to bind microtubules (Lu *et al.*, 1999) or intrinsic *cis-* and *trans-* acting factors that are involved in *tau* gene regulation.

## Acknowledgments

The author wishes to acknowledge Dr. Jonathan Clyman of the Center for Human Genetics & Child Neurology in New Paltz, New York, for his contribution on genetic counseling. The author is indebted to his wife, Teresa, and his children Joseph, Kalyn, Ryan, and John for their patience and support. The unpublished research by the author was supported in part by a grant from the Society for Progressive Supranuclear Palsy.

## References

Anderton, B. H., Betts, J., Blackstock, W. P., Brion, J. P., Chapman, S., Connell, J., Dayanandan, R., Gallo, J. M., Gibb, G., Hanger, D. P., Hutton, M., Kardalinou, E., Leroy, K., Lovestone, S., Mack, T., Reynolds, C. H., and Van Slegtenhorst, M. (2001). Sites of phosphorylation in tau and factors affecting their regulation. *Biochem. Soc. Symp.* **67**, 73–80.

Andreadis, A., Brown, W. M., and Kosik, K. S. (1992). Structure and novel exons of the human t gene. *Biochemistry* **31**, 10,626–10,633.

Andreadis, A., Wagner, B. K., Broderick, J. A., and Kosik, K. S. (1996). A τ promoter region without neuronal specificity. *J. Neurochem.* **66**, 2257–2263.

Aniello, F., Couchie, D., Bridoux A.-M., and Nunez, J. (1991). Splicing of juvenile and adult tau mRNA variants is regulated by thyroid hormone. *Proc. Natl. Acad. Sci. U.S.A.* **88**, 4035–4039.

Arai, T., Ikeda, K., Akiyama, H., Tsuchiya, K., Yagishita, S., and Takamatsu, J. (2001) Intracellular processing of aggregated tau differs between corticobasal degeneration and progressive supranuclear palsy. *Neuroreport.* **12**, 935–938.

Auer, I. A., Schmidt, M. L., Lee, V. M.-Y., Curry, B., Suzuki, K., Shin, R.-W., Pentchev, P. G., Carstea, E. D., and Trojanowski, J. Q. (1995). Paired helical filament tau (PHFtau) in Niemann-Pick type C disease is similar to PHFtau in Alzheimer's disease. *Acta Neuropathol.* **90**, 547–551.

Baker, M., Litvan, I., Houlden, H., Adamson, J., Dickson, D., Perez-Tur, J., Hardy, J., Lynch, T., Bigio, E., and Hutton, M. (1999). Association of an extended haplotype in the tau gene with progressive supranuclear palsy. *Hum. Mol. Genet.* **8**, 711–715.

Bird, T. D., Nochlin, D., Poorkaj, P., Cherrier, M., Kaye, J., Payami, H., Peskind, E., Lampe, T. H., Nemens, E., Boyer, P. J., and Schellenberg, G. D. (1999). A clinical pathological comparison of three families with frontotemporal dementia and identical mutations in the tau gene (P301L). *Brain* **122**, 741–756.

Bonifati, V., Joosse, M., Nicholl, D. J., Vanacore, N., Bennett, P., Rizzu, P., Fabbrini, G., Marconi, R., Colosimo, C., Locuratolo, N., Stocchi, F., Bonuccelli, U., De Mari, M., Wenning, G., Vieregge, P., Oostra, B., Meco, G., and Heutink, P. (1999). The tau gene in progressive supranuclear palsy: exclusion of mutations in coding exons and exon 10 splice sites, and identification of a new intronic variant of the disease-associated H1 haplotype in Italian cases. *Neurosci. Lett.* **274**, 61–65.

Buee, L., Bussiere, T., Buee-Scherrer, V., Delacourte, A., and Hof, P. R. (2000). Tau protein isoforms, phosphorylation and role in neurodegenerative disorders. *Brain Res. Rev.* **33**, 95–130.

Carstea, E. D., Morris, J. A., Coleman, K. G., Loftus, S. K., Zhang, D., Cummings, C., Gu, J., Rosenfeld, M. A., Pavan, W. J., Krizman, D. B., Nagle, J., Polymeropoulos, M. H., Sturley, S. L., Ioannou, Y. A., Higgins, M. E., Comly, M., Cooney, A., Brown, A., Kaneski, C. R., Blanchette-Mackie, E. J., Dwyer, N. K., Neufeld, E. B., Chang, T. Y., Liscum, L., and Tagle, D. A. (1997). Niemann-Pick C1 disease gene:

homology to mediators of cholesterol homeostasis. *Science* **277**, 228–231.
Clark, L. N., Poorkaj, P., Wszolek, Z.; Geschwind, D. H., Nasreddine, Z. S., Miller, B., Li, D., Payami, H., Awert, F., Markopoulou, K., Andreadis, A., D'Souza, I., Lee, V. M.-Y., Reed, L., Trojanowski, J. Q., Zhukareva, V., Bird, T., Schellenberg, G., and Wilhelmsen, K. C. (1998). Pathogenic implications of mutations in the tau gene in pallido-ponto-nigral degeneration and related neurodegenerative disorders linked to chromosome 17. *Proc. Natl. Acad. Sci. U.S.A.* **95**, 13103–13107.
Connell, J. W., Gibb, G. M., Betts, J. C., Blackstock, W. P., Gallo, J., Lovestone, S., Hutton, M., and Anderton, B. H. (2001). Effects of FTDP-17 mutations on the *in vitro* phosphorylation of tau by glycogen synthase kinase 3beta identified by mass spectrometry demonstrate certain mutations exert long-range conformational changes. *FEBS Lett.* **493**, 40–44.
Couchie, D., Mavilia, C., Georgieff, I.S., Liem, R. K. H., Shelanski, M. L., and Nuñez, J. (1992). Primary structure of high molecular weight tau present in the peripheral nervous system. *Proc. Natl. Acad. Sci. U.S.A.* **89**, 4378–4381.
Dayanandan, R., Van Slegtenhorst, M., Mack, T. G., Ko, L., Yen, S. H., Leroy, K., Brion, J. P., Anderton, B. H., Hutton, M., and Lovestone, S. (1999). Mutations in tau reduce its microtubule binding properties in intact cells and affect its phosphorylation. *FEBS Lett.* **446**, 228–232.
Delisle, M. B., Murrell, J. R., Richardson, R., Trofatter, J. A., Rascol, O., Soulages, X., Mohr, M., Calvas, P., and Ghetti, B. (1999). A mutation at codon 279 (N279K) in exon 10 of the Tau gene causes a tauopathy with dementia and supranuclear palsy. *Acta. Neuropathol.* **98**, 62–77.
de Silva, R., Weiler, M., Morris, H. R., Martin, E. R., Wood, N. W., and Lees, A. J. (2001). Strong association of a novel Tau promoter haplotype in progressive supranuclear palsy. *Neurosci. Lett.* **311**, 145–148.
de Yebenes J. G., Sarasa, J. L., Daniel, S. E., and Lees, A. J. (1995). Familial progressive supranuclear palsy. Description of a pedigree and review of the literature. *Brain* **118**, 1095–1103.
Di Maria, E., Tabaton, M., Vigo, T., Abbruzzese, G., Bellone, E., Donati, C., Frasson, E., Marchese, R., Montagna, P., Munoz, D. G., Pramstaller, P.P., Zanusso, G., Ajmar, F., and Mandich, P. (2000). Corticobasal degeneration shares a common genetic background with progressive supranuclear palsy. *Ann. Neurol.* **47**, 374–377.
D'Souza, I., Poorkaj, P., Hong, M., Nochlin, D., Lee, V. M., Bird, T. D., and Schellenberg, G. D. (1999). Missense and silent tau gene mutations cause frontotemporal dementia with parkinsonism-chromosome 17 type, by affecting multiple alternative RNA splicing regulatory elements. *Proc. Natl. Acad. Sci. U.S.A.* **96**, 5598–5603.
D'Souza, I., and Schellenberg, G. D. (2000). Determinants of 4-repeat tau expression. Coordination between enhancing and inhibitory splicing sequences for exon 10 inclusion. *J. Biol. Chem.* **275**, 17700–17709.
Drubin, D. G., Kobayashi, S., and Kirschner, M. (1986). Association of tau protein with microtubules in living cells. *Ann. N.Y. Acad. Sci.* **466**, 257–268.
Drubin, D. G., Kobayashi, S., Kellog, D., and Kirschner, M. (1988). Regulation of microtubule protein levels during cellular morphogenesis in nerve growth factior-treated PC12 cells. *J. Cell Biol.* **106**, 1583–1591.
Duff, K., Knight, H., Refolo, L. M., Sanders, S., Yu, X., Picciano, M., Malester, B., Hutton, M., Adamson, J., Goedert, M., Burki, K., and Davies, P. (2000). Characterization of pathology in transgenic mice over-expressing human genomic and cDNA tau transgenes. *Neurobiol. Dis.* **7**, 87–98.
Ezquerra, M., Pastor, P., Valldeoriola, F., Molinuevo, J. L., Blesa, R., Tolosa, E., and Oliva, R. (1999). Identification of a novel polymorphism in the promoter region of the tau gene highly associated to progressive supranuclear palsy in humans. *Neurosci. Lett.* **275**, 183–186.
Fabbrini, G., Barbanti, P., Bonifati, V., Colosimo, C., Gasparini, M., Vanacore, N., and Meco, G. (2001). Donepezil in the treatment of progressive supranuclear palsy. *Acta Neurol. Scand.* **103**, 123–125.

Falconer, M. M., Echeverri, C. J., and Brown, D. L. (1992). Differential sorting of β-tubulin isotypes into colchicine-stable microtubules during neuronal and muscle differentiation of embryonic carcinoma cells. *Cell Motil. Cytoskel.* **21**, 1583–1591.
Feany, M. B., Mattiace, L. A., and Dickson, D. W. (1996). Neuropathologic overlap of progressive supranuclear palsy, Pick's disease and corticobasal degeneration. *J. Neuropathol. Exp. Neurol.* **55**, 53–67.
Foster, N. L., Wilhelmsen, K., Sima, A. A., Jones, M. Z., D'Amato, C. J., Gilman, S., and conference participants (1997). Frontotemporal dementia and parkinsonism linked to chromosome 17: a consensus conference. *Ann. Neurol.* **41**, 706–715.
Gache, Y., Guilleminot, J., Bridoux, A.-M., and Nuñez, J. (1993). Heterogeneity of the high molecular weight τ proteins in N115 neuroblastoma cells. *J. Neurochem.* **61**, 873–880.
Gao, Q. S., Memmott, J., Lafyatis, R., Stamm, S., Screaton, G., and Andreadis, A. (2000). Complex regulation of tau exon 10, whose missplicing causes frontotemporal dementia. *J. Neurochem.* **74**, 490–500.
Goate, A., Chartier-Harlin, M.-C., Mullan, M., Brown, J., Crawford, F., Fidani, L., Giuffra, L., Haynes, A., Irving, N., James, L., Mant, R., Newton, P., Rooke, K., Roques, P., Talbot, C., Pericak-Vance, M., Roses, A., Williamson, R., Rossor, M., Owen, M., and Hardy, J. (1991). Segregation of a missense mutation in the amyloid precursor protein gene with familial Alzheimer's disease. *Nature* **349**, 704–706.
Goedert, M., Spillantini, M. G., Potier, M. C., Ulrich, J., and Crowther, R. A. (1989a). Cloning and sequencing the cDNA encoding an isoform of microtubule-associated protein tau containing four tandem repeats: differential expression of tau protein mRNAs in human brain. *EMBO J.* **8**, 393–399.
Goedert, M., Spillantinin, M. G., Jakes, R., Rutherford, D., and Crowther, R. A. (1989b). Multiple isoforms of human microtubule-associated protein tau: sequences and localization in neurofibrillary tangles of Alzheimer's disease. *Neuron* **3**, 519–526.
Goedert, M, and Jakes, R. (1990). Expression of separate isoforms of human tau protein: correlation with the tau pattern in brain and effects on tubulin polymerization. *EMBO J.* **9**, 4225–4230.
Goedert, M., Spillantini, M. G., and Crowther R.A. (1992). Cloning of a big tau microtubule-associated protein characteristic of the peripheral nervous system. *Proc. Natl. Acad. Sci. U.S.A.* **89**, 1983–1987.
Golbe, L. I. (2001). Progressive supranuclear palsy. *Curr. Treat. Options Neurol.* **3**, 473–477.
Götz, J., Chen, F., van Dorpe, J., and Nitsch, R. M. (2001). Formation of neurofibrillary tangles in P301L tau transgenic mice induced by A-beta42 fibrils. *Science* **293**, 1491–1495.
Grover, A., Houlden, H., Baker, M., Adamson, J., Lewis, J., Prihar, G., Pickering-Brown, S., Duff, K., and Hutton, M. (1999). 5′ splice site mutations in tau associated with the inherited dementia FTDP-17 affect a stem-loop structure that regulates alternative splicing of exon 10. *J. Biol. Chem.* **274**, 15,134–15,143.
Hanger, D. P., Brion, J. P., Gallo, J. M., Cairns, N. J., Luthert, P. J., and Anderton, B. H. (1991). Tau in Alzheimer's disease and Down's syndrome is insoluble and abnormally phosphorylated. *Biochem. J.* **275**, 99.
Harada, A., Oguchi, K., Okabe, J., Kuno, S., Terada, T., Ohshima, R., Sato-Yoshitake, Y., Takei, Y., Noda, T., and Hirokawa, N. (1994). Altered microtubule organization in small-calibre axons of mice lacking tau protein. *Nature* **369**, 488–491.
Hasegawa, M., Smith, M. J., and Goedert, M. (1998). Tau proteins with FTDP-17 mutations have a reduced ability to promote microtubule assembly. *FEBS Lett.* **437**, 207–210.
Heutink, P., Stevens, M., Rizzu, P., Bakker, E., Kros, J. M., Tibben, A., Niermeijer, M. F., van Duijn, C. M., Oostra, B. A., and van Swieten, J. C. (1997). Hereditary frontotemporal dementia is linked to chromosome 17q21–q22: a genetic and clinicopathological study of three Dutch families. *Ann. Neurol.* **41**, 150–159.

Higgins, J. J., Litvan, I., Pho, L. T., Li, W., and Nee, L. E. (1998). Progressive supranuclear gaze palsy is in linkage disequilibrium with the tau and not the alpha-synuclein gene. *Neurology* **50**, 270–273.

Higgins, J. J., Litvan, I., Nee, L. E., and Loveless, J. M. (1999a). A lack of the R406W tau mutation in progressive supranuclear palsy and corticobasal degeneration. *Neurology* **52**, 404–406.

Higgins, J. J., Adler, R. L., and Loveless, J. M. (1999b). Mutational analysis of the tau gene in progressive supranuclear palsy. *Neurology* **53**, 1421–1424.

Higgins, J. J., and Mendez, M. F. (2000). Roll over Pick and tell Alzheimer the news! *Neurology* **54**, 784–785.

Higgins, J. J., Golbe, L. I., DeBiase, A., Jankovic, J., Factor, S. A., and Adler, R. L. (2000). An extended 5′-tau susceptibility haplotype in progressive supranuclear palsy. *Neurology* **55**, 1364–1367.

Himmler, A. (1989). Structure of the bovine gene: alternatively spliced transcripts generate a gene family. *Mol. Cell. Biol.* **9**, 1389–1396.

Himmler, A., Drechsel, D., Kirschner, M. W., and Martin, D. W. (1989). Tau consists of a set of proteins with repeated C-terminal microtubule binding domains and variable N-terminal domains. *Mol. Cell. Biol.* **9**, 1381–1388.

Hong, M., Zhukareva, V., Vogelsberg-Ragaglia, V., Wszolek, Z., Reed, L., Miller, B. I., Geschwind, D. H., Bird, T. D., McKeel, D., Goate, A., Morris, J. C., Wilhelmsen, K. C., Schellenberg, G. D., Trojanowski, J. Q., and Lee, V. M. (1998). Mutation-specific functional impairments in distinct tau isoforms of hereditary FTDP-17. *Science* **282**, 1914–1917.

Houlden, H., Baker, M., Adamson, J., Grover, A., Waring, S., Dickson, D., Lynch, T., Boeve, B., Petersen, R.C., Pickering-Brown, S., Owen, F., Neary, D., Craufurd, D., Snowden, J., Mann, D., and Hutton, M. (1999). Frequency of tau mutations in three series of non-Alzheimer's degenerative dementia. *Ann. Neurol.* **46**, 243–248.

Houlden, H., Baker, M., Morris, H. R., MacDonald, N., Pickering-Brown, S., Adamson, J., Lees, A. J., Rossor, M. N., Quinn, N. P., Kertesz, A., Khan, M. N., Hardy, J., Lantos, P. L., St George-Hyslop, P., Munoz, D. G., Mann, D., Lang, A. E., Bergeron, C., Bigio, E. H., Litvan, I., Bhatia, K. P., Dickson, D., Wood, N. W., and Hutton, M. (2001). Corticobasal degeneration and progressive supranuclear palsy share a common tau haplotype. *Neurology* **56**, 1702–1706.

Hulette, C. M., Pericak-Vance, M. A., Roses, A. D., Schmechel, D. E., Yamaoka, L. H., Gaskell, P. C., Welsh-Bohmer, K. A., Crowther, R. A., and Spillantini, M. G. (1999). Neuropathological features of frontotemporal dementia and parkinsonism linked to chromosome 17q21-22 (FTDP-17): Duke Family 1684. *J. Neuropathol. Exp. Neurol.* **58**, 859–866.

Hutton, M., Lendon, C. L., Rizzu, P., Baker, M., Froelich, S., Houlden, H., Pickering-Brown, S., Chakraverty, S., Isaacs, A., Grover, A., Hackett, J., Adamson, J., Lincoln, S., Dickson, D., Davies, P., Petersen, R. C., Stevens, M., de Graaff, E., Wauters, E., van Baren, J., Hillebrand, M., Joosse, M., Kwon, J. M., Nowotny, P., Che, L. K., Norton, J., Morris, J. C., Reed, L. A., Trojanowski, J., Basun, H., Lannfelt, L., Neystat, M., Fahn, S., Dark, F., Tannenberg, T., Dodd, P. R., Hayward, N., Kwok, J. B. J., Schofield, P. R., Andreadis, A., Snowden, J., Craufurd, D., Neary, D., Owen, F., Oostra B.A., Hardy, J., Goate, A., van Swieten J., Mann, D., Lynch, T., and Heutink, P. (1998). Association of missense and 5′-splice-site mutations in tau with the inherited dementia FTDP-17. *Nature* **39**, 702–705.

Hutton, M. (2001). Missense and splice site mutations in *tau* associated with FTDP-17. Multiple pathogenic mechanisms. *Neurology* **56**, S21–S25.

Iijima, M., Tabira, T., Poorkaj, P., Schellenberg, G. D., Trojanowski, J. Q., Lee, V. M., Schmidt, M. L., Takahashi, K., Nabika, T., Matsumoto, T., Yamashita, Y.,Yoshioka, S., and Ishino, H. (1999). A distinct familial presenile dementia with a novel missense mutation in the tau gene. *Neuroreport* **10**, 497–501.

Ishihara, T., Hong, M., Zhang, B., Nakagawa, Y., Lee, M. K., Trojanowski, J. Q., and Lee, V. M. (1999). Age-dependent emergence and progression of a tauopathy in transgenic mice overexpressing the shortest human tau isoform. *Neuron* **24**, 751–762.

Jiang, Z., Cote, J., Kwon, J. M., Goate, A. M., and Wu, J. Y. (2000). Aberrant splicing of tau pre-mRNA caused by intronic mutations associated with the inherited dementia frontotemporal dementia with parkinsonism linked to chromosome 17. *Mol. Cell Biol.* **20**, 4036–4048.

Kertesz, A., Kawarai, T., Rogaeva, E., St. George-Hyslop, P., Poorkaj, P., Bird, T. D., and Munoz, D. G. (2000). Familial frontotemporal dementia with ubiquitin-positive, tau-negative inclusions. *Neurology* **54**, 818–827.

Kodama, K., Okada, S., Iseki, E., Kowalska, A., Tabira, T., Hosoi, N., Yamanouchi, N., Noda, S., Komatsu, N., Nakazato, M., Kumakiri, C., Yazaki, M., and Sato, T. (2000). Familial frontotemporal dementia with a P301L tau mutation in Japan. *J. Neurol. Sci.* **176**, 57–64.

Lendon, C. L., Lynch, T., Norton, J., McKeel D. W., Jr., Busfield, F., Craddock, N., Chakraverty, S., Gopalakrishnan, G., Shears, S. D., Grimmett, W., Wilhelmsen, K. C., Hansen, L., Morris, J. C., and Goate, A. M. (1998). Hereditary dysphasic disinhibition dementia: a frontotemporal dementia linked to 17q21–22. *Neurology* **50**, 1546–1555.

Levy-Lahad, E., Wasco, W., Poorkaj, P., Romano, D. M., Oshima, J., Pettingell, W. H., Yu, C., Jondro, P. D., Schmidt, S. D., Wang, K., Crowley, A. C., Fu, Y.-H., Guenette, S. Y., Galas, D., Nemens, E., Wijsman, E. M., Bird, T. D., Schellenberg, G. D., and Tanzi, R. E. (1995). Candidate gene for the chromosome 1 familial Alzheimer's disease locus. *Science* **269**, 973–975.

Lewin, B. (1994). *Genes V*. Oxford University Press, New York.

Lewis, J., McGowan, E., Rockwood, J., Melrose, H., Nacharaju, P., Van Slegtenhorst, M., Gwinn-Hardy, K., Paul Murphy, M., Baker, M., Yu, X., Duff, K., Hardy, J., Corral, A., Lin, W. L., Yen, S. H., Dickson, D. W., Davies, P., and Hutton, M. (2000). Neurofibrillary tangles, amyotrophy and progressive motor disturbance in mice expressing mutant (P301L) tau protein. *Nat. Genet.* **25**, 402–405.

Lewis, J., Dickson, D. W., Lin, W.-L., Chisholm, L., Corral, A., Jones, G., Yen, S.-H., Sahara, N., Skipper, L.; Yager, D., Eckman, C., Hardy, J., Hutton, M., and McGowan, E. (2001). Enhanced neurofibrillary degeneration in transgenic mice expressing mutant tau and APP. *Science* **293**, 1487–1491.

Lippa, C.F., Zhukareva, V., Kawarai, T., Uryu, K., Shafiq, M., Nee, L. E., Grafman, J., Liang, Y., St. George-Hyslop, P. H., Trojanowski, J. Q., and Lee, V. M. (2000). Frontotemporal dementia with novel tau pathology and a Glu342Val tau mutation. *Ann. Neurol.* **48**, 850–858.

Litvan, I., Hauw, J., Bartko, J.J., Lantos, P.L., Daniel, S.E., Horoupian, D.S., McKee, A., Dickson, D., Bancher, C., Tabaton, M., Jellinger, K., and Anderson, D.W. (1996). Validity and reliability of the preliminary NINDS neuropathologic criteria for progressive supranuclear palsy and related disorders. *J. Neuropathol. Exp. Neurol.* **55**, 97–105.

Litvan, I., Agid, Y., Goetz, C., Jankovic, J., Wenning, G. K., Brandel, J. P., Lai, E. C., Verny, M., Ray-Chaudhuri, K., McKee, A., Jellinger, K., Pearce, R. K., and Bartko, J. J. (1997). Accuracy of the clinical diagnosis of corticobasal degeneration: a clinicopathologic study. *Neurology* **48**, 119–125.

Litvan, I., Phipps, M., Pharr, V. L., Hallett, M., Grafman, J., and Salazar, A. (2001). Randomized placebo-controlled trial of donepezil in patients with progressive supranuclear palsy. *Neurology* **57**, 467–473.

Liu, W. K., Le, T. V., Adamson, J., Baker, M., Cookson, N., Hardy, J., Hutton, M., Yen, S. H., and Dickson, D. W. (2001). Relationship of the extended tau haplotype to tau biochemistry and neuropathology in progressive supranuclear palsy. *Ann. Neurol.* **50**, 494–502.

Lu, P.-J., Wulf, G., Zhou, X. Z., Davies, P., and Lu, K. P. (1999). The prolyl isomerase Pin1 restores the function of Alzheimer-associated phosphorylated tau protein. *Nature* **399**, 784–788.

Lynch, T., Sano, M., Marder, K. S., Bell, K. L., Foster, N. L., Defendini, R. F., Sima, A. A., Keohane, C., Nygaard, T. G., Fahn, S., Mayeux, R., Rowland, L. P., and Wilhelmsen, K. C. (1994). Clinical characteristics of a family with chromosome 17-linked disinhibition-dementia-parkinsonism-amyotrophy complex. *Neurology* **44**, 1878–1884.

Mann, D. M., McDonagh, A. M., Snowden, J., Neary, D., and Pickering-Brown, S. M. (2000) Molecular classification of the dementias. *Lancet* **355**, 626.

Matsumura, N., Yamazaki, T., and Ihara, Y. (1999). Stable expression in Chinese hamster ovary cells of mutated tau genes causing frontotemporal dementia and parkinsonism linked to chromosome 17 (FTDP-17). *Am. J. Pathol.* **154**, 1649–1656.

Mirra, S. S., Murrell, J. R., Gearing, M., Spillantini, M. G., Goedert, M., Crowther, R. A., Levey, A.I., Jones, R., Green, J., Shoffner, J. M., Wainer, B. H., Schmidt, M. L., Trojanowski, J. Q., and Ghetti, B. (1997). Tau pathology in a family with dementia and a P301L mutation in tau. J *Neuropathol. Exp. Neurol.* **58**, 335–345.

Miyamoto, K., Kowalska, A., Hasegawa, M., Tabira, T., Takahashi, K., Araki, W., Akiguchi, I., and Ikemoto, A. (2001). Familial frontotemporal dementia and parkinsonism with a novel mutation at an intron 10+11-splice site in the tau gene. *Ann. Neurol.* **50**, 117–120.

Miyasaka, T., Morishima-Kawashima, M., Ravid, R., Heutink, P., van Swieten, J.C., Nagashima, K., and Ihara, Y. (2001) Molecular analysis of mutant and wild-type tau deposited in the brain affected by the FTDP-17 R406W mutation. *Am. J. Pathol.* **158**, 373–379.

Morris, H. R., Katzenschlager, R., Janssen, J. C., Brown, J. M., Ozansoy, M., Quinn, N., Revesz, T., Rossor, M. N., Daniel, S. E., Wood, N.W. and Lees, A. J. (2002). Sequence analysis of tau in familial and sporadic progressive supranuclear palsy. *J. Neurol. Neurosurg. Psychiatry* **72**, 388–390.

Murrell, J. R., Spillantini, M. G., Zolo, P., Guazzelli, M., Smith, M. J.; Hasegawa, M., Redi, F., Crowther, R. A., Pietrini, P., Ghetti, B., and Goedert, M. (1999). Tau gene mutation G389R causes a tauopathy with abundant Pick body-like inclusions and axonal deposits. *J. Neuropath. Exp. Neurol.* **58**, 1207–1226.

Nagasawa, H., Tanji, H., Nomura, H., Saito, H., Itoyama, Y., Kimura, I., Tuji, S., Fujiwara, T., Iwata, R., Itoh, M., and Ido, T. (1996). PET study of cerebral glucose metabolism and fluorodopa uptake in patients with corticobasal degeneration. *J. Neurol. Sci.* **139**, 210–217.

Nasreddine, Z. S., Loginov, M., Clark, L. N., Lamarche, J., Miller, B. L, Lamontagne, A., Zhukareva, V., Lee, V. M., Wilhelmsen, K. C., and Geschwind, D. H. (1999). From genotype to phenotype: a clinical pathological, and biochemical investigation of frontotemporal dementia and parkinsonism (FTDP-17) caused by the P301L tau mutation. *Ann. Neurol.* **45**, 704–715.

Nelson, P. T., Stefansson, K., Gulcher, J., and Spear, C. B. (1996). Molecular evolution of t protein: implications for Alzheimer's disease. *J. Neurochem.* **67**, 1622–1632.

Neumann, M., Schulz-Schaeffe, W., Crowther, R. A., Smith, M. J., Spillantini, M. G,. Goedert, M., and Kretzschmar, H. A. (2001). Pick's disease associated with the novel tau gene mutation K369I. *Ann. Neurol.* **50**, 503–513.

Neve, R. L., Harris, R., Kosik, K. S., Kurnit, D. M., and Donlan, T. A. (1986). Identification of cDNA clones for the human MAP tau and chromosomal localization for the genes for tau and MAP2. *J. Mol. Brain Res.* **1**, 271–280.

Nuñez, J., and Fisher I. (1997). Microtubule-associated proteins (MAPs) in the peripheral nervous system during development and regeneration. *J. Mol. Neurosci.* **8**, 207–222.

Pastor, P., Pastor, E., Carnero, C., Vela, R., Garcia, T., Amer, G., Tolosa, E., and Oliva, R. (2001). Familial atypical progressive supranuclear palsy associated with homozygosity for the delN296 mutation in the tau gene. *Ann. Neurol.* **49**, 263–267.

Perez, M., Lim, F., Arrasate, M., and Avila, J. (2000). The FTDP-17-linked mutation R406W abolishes the interaction of phosphorylated tau with microtubules. J. *Neurochem.* **74**, 2583–2589.

Piccini, P., de Yebenez, J., Lees, A. J., Ceravolo, R., Turjanski, N., Pramstaller, P., and Brooks, D. J. (2001). Familial progressive supranuclear palsy: detection of subclinical cases using 18F–dopa and 18fluorodeoxyglucose positron emission tomography. *Arch. Neurol.* **58**, 1846–1851.

Poorkaj, P., Bird, T. D., Wijsman, E., Nemens, E., Garruto, R. M., Anderson, L., Andreadis, A., Wiederholt, W. C., Raskind, M., and Schellenberg, G. D. (1998). Tau is a candidate gene for chromosome 17 frontotemporal dementia. *Ann. Neurol.* **43**, 815–825.

Poorkaj, P., Grossman, M., Steinbart, E., Payami, H., Sadovnick A., Nochlin D., Tabira, T., Trojanowski, J.Q., Borson, S., Galasko, D, Reich, S., Quinn, B., Schellenberg, G., and Bird, T. (2001). Frequency of tau gene mutations in familial in sporadic cases of non-Alzheimer dementia. *Arch. Neurol.* **58**, 383–387.

Probst, A., Götz, J., Wiederhold, K. H., Tolnay, M., Mistl, C., Jaton, A. L., Hong, M., Ishihara, T., Lee, V. M., Trojanowski, J. Q., Jakes, R., Crowther, R. A., Spillantini, M. G., Burki, K., and Goedert, M. (2000). Axonopathy and amyotrophy in mice transgenic for human four-repeat tau protein. *Acta Neuropathol.* **99**, 469–481.

Rizzini, C., Goedert, M., Hodges, J. R., Smith, M. J., Jakes, R., Hills, R., Xuereb, J. H., Crowther, R. A., and Spillantini, M. G. (2000). Tau gene mutation K257T causes a tauopathy similar to Pick's disease. *J. Neuropathol. Exp. Neurol.* **59**, 990–1001.

Rizzu, P., Van Swieten, J. C., Joosse, M., Hasegawa, M., Stevens, M., Tibben, A., Niermeijer, M. F., Hillebrand, M., Ravid, R., Oostra, B. A., Goedert, M., van Duijn, C. M., and Heutink, P. (1999). High prevalence of mutations in the microtubule-associated protein tau in a population study of frontotemporal dementia in the Netherlands. *Am. J. Hum. Genet.* **64**, 414–421.

Rogaev, E. I., Sherrington, R., Rogaeva, E. A., Levesque, G., Ikeda, M., Liang, Y., Chi, H., Lin, C., Holman, K., Tsuda, T., Mar, L., Sorbi, S., Nacmias, B., Placentini, S., Amaducci, L., Chumakov, I., Cohen, D., Lannfelt, L., Fraser, P. E., Rommens, J. M., and St. George-Hyslop, P. H. (1995). Familial Alzheimer's disease in kindreds with missense mutations in a gene on chromosome 1 related to the Alzheimer's disease type 3 gene. *Nature* **376**, 775–778.

Rojo, A., Pernaute, R. S., Fontan, A., Ruiz, P. G., Honnorat, J., Lynch, T., Chin, S., Gonzalo, I., Rabano, A., Martinez, A., Daniel, S., Pramstaller, P., Morris, H., Wood, N., Lees, A., Tabernero, C., Nyggard, T., Jackson, A. C., Hanson, A., and de Yebenes, J. G. (1999). Clinical genetics of familial progressive supranuclear palsy. *Brain* **122**, 1233–1245.

Romano, S., and Colosimo, C. (2001). Procerus sign in progressive supranuclear palsy. *Neurology* **57**, 1928.

Saito, Y., Kawai, M., Inoue, K., Sasaki, R., Arai, H., Nanba, E., Kuzuhara, S., Ihara, Y., Kanazawa, I., and Murayama, S. (2000). Widespread expression of alpha-synuclein and tau immunoreactivity in Hallervorden-Spatz syndrome with protracted clinical course. *J. Neurol. Sci.* **177**, 48–59.

Sergeant, N., Wattez, A., and Delacourte, A. (1999). Neurofibrillary degeneration in progressive supranuclear palsy and corticobasal degeneration: tau pathologies with exclusively "exon 10" isoforms. *J. Neurochem.* **72**, 1243–1249.

Sherrington, R., Rogaev, E. I., Liang, Y., Rogaeva, E. A., Levesque, G., Ikeda, M., Chi, H., Lin, C., Li, G., Holman,,K., Tsuda, T., Mar, L., Foncin, J.-F., Bruni, A. C., Montesi, M. P., Sorbi, S., Rainero, I., Pinessi, L., Nee, L., Chumakov, I., Pollen, D., Brookes, A., Sanseau, P., Polinsky, R. J., Wasco, W., Da Silva, H. A. R., Haines, J. L., Pericak-Vance, M. A., Tanzi, R. E., Roses, A. D., Fraser, P. E., Rommens, J. M., and St. George-Hyslop, P. H. (1995). Cloning of a gene bearing mis-sense mutations in early-onset familial Alzheimer's disease. *Nature* **375**, 754–760.

Spillantini, M. G., Murrell, J. R., Goedert, M., Farlow, M. R., Klug, A., and Ghetti, B. (1998a) Mutation in the tau gene in familial multiple system tauopathy with presenile dementia. *Proc. Natl. Acad. Sci. U.S.A.* **95**, 7737–7741.

Spillantini, M. G., Bird, T. D., and Ghetti, B. (1998b). Frontotemporal dementia and Parkinsonism linked to chromosome 17: a new group of tauopathies. *Brain Pathol.* **8**, 387–402.

Spillantini, M. G., Yoshida, H., Rizzini, C., Lantos, P. L., Khan, N., Rossor, M. N., Goedert, M., and Brown, J. (2000). A novel tau mutation (N296N) in familial dementia with swollen achromatic neurons and corticobasal inclusion bodies. *Ann. Neurol.* **48**, 939–943.

Spittaels, K., Van den Haute, C., Van Dorpe, J., Bruynseels, K., Vandezande, K., Laenen, I., Geerts, H., Mercken, M., Sciot, R., Van Lommel, A., Loos, R., and Van Leuven, F. (1999) Prominent axonopathy in the brain and spinal cord of transgenic mice overexpressing four-repeat human tau protein. *Am. J. Pathol.* **155**, 2153–2165.

Stanford, P. M., Halliday, G. M., Brooks, W. S., Kwok, J. B., Storey, C. E., Creasey, H., Morris, J. G., Fulham, M. J., and Schofield, P. R. (2000). Progressive supranuclear palsy pathology caused by a novel silent mutation in exon 10 of the tau gene: expansion of the disease phenotype caused by tau gene mutations. *Brain* **123**, 880–893.

Tolnay, M., Spillantini M. G., Rizzini, C., Eccles, D., Lowe, J., and Ellison, D. (2000). A new case of frontotemporal dementia and parkinsonism resulting from an intron 10 +3-splice site mutation in the tau gene: clinical and pathological features. *Neuropathol. Appl. Neurobiol.* **26**, 368–378.

van Swieten, J. C., Stevens, M., Rosso, S. M., Rizzu, P., Joosse, M., de Koning, I., Kamphorst, W., Ravid, R., Spillantini, M. G., Niermeijer M. F., and Heutink, P. (1999). Phenotypic variation in hereditary frontotemporal dementia with tau mutations. *Ann. Neurol.* **46**, 617–626.

Varani, L., Hasegawa, M., Spillantini, M. G., Smith, M. J., Murrell, J. R., Ghetti, B., Klug, A., Goedert, M., and Varani, G. (1999). Structure of tau exon 10 splicing regulatory element RNA and destabilization by mutations of frontotemporal dementia and parkinsonism linked to chromosome 17. *Proc. Natl. Acad. Sci. U.S.A.* **96**, 8229–8234.

Wang, Y., Loomis, P. A., Zinkowski, R. P., and Binder, L. I. (1993). A novel tau transcript in cultured human neuroblastoma cells expressing nuclear tau. *J. Cell Biol.* **121**, 257–267.

Wei, M.-L., and Andreadis, A. (1998). Splicing of a regulated exon reveals additional complexity in the axonal microtubule-associated protein tau. *J. Neurochem.* **70**, 1346–1356.

Wei, M. L., Memmott, J., Screaton, G., and Andreadis, A. (2000). The splicing determinants of a regulated exon in the axonal MAP tau reside within the exon and in its upstream intron. *Brain Res. Mol. Brain Res.* **80**, 207–218.

Wilhelmsen, K. C., Lynch, T., Pavlou, E., Higgins, M., and Hygaard, T. G. (1994). Localization of disinhibition-dementia-parkinsonism-amyotrophy complex to 17q21–22. *Am. J. Hum. Genet.* **55**, 1159-1165.

Williams, T., and Tijan, R. (1991) Analysis of the DNA binding and activation properties of the human transcription factor AP2. *Genes Dev.* **5**, 670–682.

Wittmann, C. W., Wszolek, M. F., Shulman, J. M., Salvaterra, P. M., Lewis, J., Hutton, M., and Feany, M. B. (2001). Tauopathy in Drosophila: neurodegeneration without neurofibrillary tangles. *Science* **293**, 711–714.

Wszolek, Z. K., Pfeiffer, R. F., Bhatt, M. H., Schelper, R. L., Cordes,M., Snow, B. J., Rodnitzky, R. L., Wolters, E. C., Arwert, F., and Calne, D. B. (1992). Rapidly progressive autosomal dominant parkinsonism and dementia with pallido-ponto-nigral degeneration. *Ann. Neurol.* **32**, 312–320.

Wszolek, Z. K., Tsuboi, Y., Uitti, R. J., Reed, L., Hutton, M. L., and Dickson D. W. (2001). Progressive supranuclear palsy as a disease phenotype caused by the S305S tau gene mutation. *Brain* **124**, 1666–1670.

Yasuda, M., Takamatsu, J., D'Souza, I., Crowther, R.A., Kawamata, T., Hasegawa, M., Hasegawa, H., Spillantini, M. G., Tanimukai, S., Poorkaj, P., Varani, L., Varani, G., Iwatsubo, T., Goedert, M., Schellenberg, D. G., and Tanaka, C. (2000). A novel mutation at position +12 in the intron following exon 10 of the tau gene in familial frontotemporal dementia (FTD-Kumamoto). *Ann. Neurol.* **47**, 422–429.

Zhou, B., Westaway, S. K., Levinson, B., Johnson, M. A., Gitschier, J., and Hayflick, S. J. (2001). A novel pantothenate kinase gene (PANK2) is defective in Hallervorden-Spatz syndrome. *Nat. Genet.* **28**, 345–349.

Zhukareva, V., Vogelsberg-Ragaglia, V., Van Deerlin, V. M., Bruce, J., Shuck, T., Grossman, M., Clark, C. M., Arnold, S. E., Masliah, E., Galasko, D., Trojanowski, J. Q., and Lee V. M. (2001). Loss of brain tau defines novel sporadic and familial tauopathies with frontotemporal dementia. *Ann. Neurol.* **49**, 165–175.

# CHAPTER 31

# Wilson Disease

JOHN H. MENKES
*Division of Pediatric Neurology*
*Cedars-Sinai Medical Center*
*Los Angeles, California 90048*

I. Introduction
II. Phenotype
III. Normal and Abnormal Gene Function
IV. Diagnosis
V. Neuroimaging
VI. Pathologic Anatomy
VII. Animal Models
VIII. Treatment
    References

## I. INTRODUCTION

Wilson Disease (WD) is an autosomal recessive disorder of copper metabolism caused by mutations in the ATP7B gene. The WD gene product, a P-type ATPase (ATP7B), is necessary for both the incorporation of copper into ceruloplasmin and its excretion into bile. The disease is associated with increased hepatic copper storage leading to a clinically heterogeneous illness marked by cirrhosis of the liver and degenerative changes in the basal ganglia. Several animal models of the disease, including the Long-Evans cinnamon (LEC) rat, and canine copper toxicosis have been described.

During the second half of the 19th century, a condition termed "pseudosclerosis" was distinguished from multiple sclerosis by the lack of ocular signs. In 1902, Kayser observed green corneal pigmentation in one such patient (Kayser, 1902); in 1903, Fleischer who had also noted the green pigmentation of the cornea in 1903, commented on the association of the corneal rings with pseudosclerosis (Fleischer, 1912). In 1912, Wilson gave the classic description of the disease and its pathologic anatomy, but failed to mention the abnormal corneal pigmentation (Wilson, 1912).

In 1913, one year after Wilson's report, Rumpel found unusually large amounts of copper in the liver of a patient with hepatolenticular degeneration (Rumpel, 1913). Although this finding was confirmed and an elevated copper concentration was detected in the basal ganglia by Luthy (1931), the implication of these reports went unrecognized until 1945, when Glazebrook demonstrated abnormally high copper levels in serum, liver, and brain in a patient with WD (Glazebrook, 1945). In 1952, five years after the discovery of ceruloplasmin, several groups of workers simultaneously found it to be low or absent in patients with WD, and for many years it was believed that WD was due to ceruloplasmin deficiency. It has now become evident that absence of ceruloplasmin (aceruloplasminemia) results in a severe disorder of iron metabolism (Mukhopadhyay *et al.*, 1998; Gitlin, 1998).

## II. PHENOTYPE

WD is a progressive condition with a tendency toward temporary clinical improvement and arrest (Scheinberg and Sternlieb, 1984). The presenting symptoms can be neurological, hepatic, psychiatric, or less frequently hematological. Symptoms at onset are depicted in Table 31.1. The condition occurs in all races, with a particularly high incidence in Eastern European Jews, Italians from Southern Italy and

TABLE 31.1 Clinical Manifestations at Onset of Wilson Disease

| Symptoms | Percentage |
|---|---|
| Hepatic or hematologic abnormalities | 35 |
| Behavioral abnormalities | 25 |
| Neurological abnormalities | 40 |
| "Pseudosclerotic" form: one or more of the following | 40 |
|   Resting or intentional tremor | |
|   Dysarthria or scanning speech | |
|   Diminished dexterity or mild clumsiness | |
|   Unsteady gait | |
|     Tremor alone | 33 |
|     Dysarthria alone | 5 |
| "Dystonic form: one or more of the following: | 60 |
|   Hypophonic speech or mutism | |
|   Drooling | |
|   Rigid mouth, arms, or legs | |
| Seizures | 1 |
| Chorea or small-amplitude twitches | <1 |

Figures are approximate. Table prepared with the assistance of Drs. I.H. Scheinberg and I Sternlieb, Department of Medicine, Albert Einstein College of Medicine, Bronx, New York.

Sicily, and in people from some of the smaller islands of Japan—groups with a high rate of inbreeding.

In the experience of Gow and co-workers (2000) 22% of symptomatic subjects, aged 7–58 years, presented with fulminant hepatic failure, 54% presented with liver abnormalities, 10% with neurological features, and 4% with hemolysis. In 10% of subjects hepatic and neurological symptoms were conjoined.

In a fair number of cases, primarily in young children, initial symptoms can be hepatic, such as jaundice or portal hypertension, and the disease can assume a rapidly fatal course without any detectable neurological abnormalities (Ferlan-Marolt et al., 1999; Arima et al., 1977; Slovis et al. 1971). In many of these patients, an attack of what appears to be an acute viral hepatitis heralds the onset of the illness (Scott et al., 1978). The presentation of WD with hepatic symptoms is common among affected children in the United States. In the series of Werlin and associates (1978), who surveyed patients in the Boston area, the primary mode of presentation was hepatic in 61% of patients younger than age 21 years. In about 10% of affected children in the United States, WD presents as an acute or intermittent, Coombs'-test-negative, nonspherocytic anemia that is accompanied by leukopenia, and thrombocytopenia (Werlin et al., 1978).

When neurological symptoms predominate, the manifestations are so varied that it is impossible to describe a characteristic clinical picture. In the past, texts have distinguished between pseudosclerotic and dystonic forms of the disease: the former dominated by tremor, the latter by rigidity and contractures. In actuality, most patients, if untreated, ultimately develop both types of symptoms. As a rule, the appearance of neurological symptoms is delayed until 10–20 years of age, and the disease progresses at a slower rate than in the hepatic form. Symptoms of basal ganglia damage usually predominate but cerebellar symptoms are occasionally in the foreground. Tremors and rigidity are the most common early signs. The tremor may be of the intention type, or it may be the alternating resting tremor of parkinsonism. More commonly, especially when the disease is in its more advanced stages, the tremor is localized to the arms and is best described by the term "wing beating." The tremor is often absent when the arms are at rest; it develops after a short latent period when the arms are extended. The beating movements may be confined to the muscles of the wrist, but it is more common for the arm to be thrown up and down in a wide arc. The movements increase in severity; at times they reach a point at which the patient is thrown off balance. The tremor may affect both arms but is usually more severe in one.

Rigidity and spasms of the muscles are often present. In some instances a typical parkinsonian rigidity involves all muscles. Torticollis, tortipelvis, and other dystonic postures are not uncommon. Many patients have a fixed open mouth smile with drooping of the lower jaw and excess salivation. Spasticity of the laryngeal and pharyngeal muscles can lead to dysarthria and dysphagia. A nearly pure Parkinson-like syndrome and progressive choreoathetosis or hemiplegia has also been described (Lingan et al., 1987).

Tendon reflexes are increased, but extensor plantar responses are exceptional. In essence, WD is a disorder of motor function; despite often widespread cerebral atrophy, there are no sensory symptoms or reflex alterations.

When WD presents during childhood, the first signs are usually bulbar; these can include indistinct speech and difficulty in swallowing. A rapidly progressive dystonic syndrome is not unusual. Such patients can present with acute dystonia, rigidity, and fever, with an occasional elevation of serum CPK (Kontaxakis et al., 1988).

In the experience of Arima and his group (1977), who studied pediatric patients with WD, 33% of children presented with hepatic symptoms, mainly jaundice or ascites. They were 4–12 years old at the time of medical attention. Cerebral symptoms, notably dystonia, drooling, or gait disturbances, were the presenting symptoms in 30% of children. These patients were 9–13 years of age. The remainder had a mixed hepatocerebral picture and were 6–12 years old at the time of medical attention.

Behavioral or personality disorders were noted in Wilson's (1912) original description of the disease. These are almost invariable, and predominate in about one-third of subjects (Akil and Brewer, 1995; Portala et al., 2001). These include impaired school or work performance, depression, mood swings, and frank psychosis. In those patients who show primarily neurologic symptoms about two-thirds have had psychiatric problems before the diagnosis of WD was made. Minor intellectual impairment can also be observed, but seizures or mental deterioration are not prominent features of the disease.

The intracorneal, ring-shaped pigmentation, first noted by Kayser and Fleischer (Kayser, 1902; Fleischer, 1912), might be evident to the naked eye or might appear only with slit-lamp examination. The color of the Kayser-Fleischer ring varies from yellow to green to brown. It is the consequence of copper deposition close to the endothelial surface of Descemet's membrane. Copper deposition in this area occurs in two or more layers, with particle size and distance between layers influencing the ultimate appearance of the ring (Uzman and Jakus, 1957). The ring can be complete or incomplete. It is present in 79% of subjects who present with hepatic symptoms, and in all subjects who present with cerebral or a combination of cerebral and hepatic symptoms (Arima et al., 1977; Gow et al., 2000). The Kayser-Fleischer ring can antedate overt symptoms of the disease and has been detected in the presence of normal liver functions. In the large pediatric series of Arima it was never present prior to 7 years of age (Arima et al., 1977). "Sunflower" cataracts are less commonly encountered, as are azure lunulae of the fingernails (Cairns et al., 1969). Hypercalciuria and nephrocalcinosis are not uncommon, and can be presenting signs of WD (Azizi, et al., 1989).

Without treatment, death ensues within 1–3 years of the onset of neurologic symptoms and is usually a result of hepatic insufficiency.

## III. NORMAL AND ABNORMAL GENE FUNCTION

Because the derangement of copper metabolism is one of the important features of WD, it is pertinent to review briefly our present knowledge of the field (Cuthbert 1998; Culotta and Gitlin, 2000).

Copper homeostasis is an important biological process. By balancing intake and excretion the body avoids copper toxicity on the one hand, and on the other hand ensures the availability of adequate amounts of the metal for a variety of vital enzymes, notably superoxide dismutase, cytochrome oxidase, ceruloplasmin, lysyl oxidase, and dopamine-β-hydroxylase. The daily dietary intake of copper ranges between 1 and 5 mg. Healthy persons who consume a free diet absorb 150–900 µg/day, or about 40% of dietary copper (Vulpe and Packman, 1995). The concentration of copper in normal liver ranges from 20–50 µg/g dry weight.

Cellular copper transport consists of three processes: copper uptake, intracellular distribution and utilization, and copper excretion.

The site of copper absorption is in the proximal portion of the gastrointestinal tract, notably the stomach and duodenum. Several copper transporters (CTR1, CTR2, CTR3) have been described in yeast, and a human gene hCTR1, homologous to CTR1, has been identified (Andrews, 2001). The mechanism of copper transport appears to differ under conditions of high and low copper intake (Rolfs and Hediger, 2001). Once inside the intestinal epithelial cell, copper is exported bound to albumin or amino acids, notably histidine, and is transported into hepatocytes. There it is guided to various intracellular locations for incorporation into copper-containing proteins by one or more copper chaperones. Several copper chaperones have been described. Thus the Atx1/Atox1 Cu chaperone interacts with the cytosolic Cu-binding domains of the Cu-transporting ATPases to provide copper to the secretory compartment (Hoffman and O'Halloran, 2001; Lee et al., 2001). In hepatocytes, copper combines with apoceruloplasmin to form ceruloplasmin, which then enters the circulation (Fig. 31.1A). Excess copper is either excreted into bile or is stored in liver lysosomes in what is probably a polymeric form of metallothionein (MT), a family of low molecular weight metal-binding proteins containing large amounts of reduced cysteine. The WD gene product, ATP7B, is a P-type ATPase required both for the incorporation of copper into ceruloplasmin and for its excretion into bile (Fig. 31.1B; Cuthbert, 1998). The exact localization of ATP7B in normal hepatic cells is still under dispute. According to Harada et al. (2000), ATP7B is localized to the late endosomes in both the steady and the copper-loaded states with ATP7B translocating copper from the cytosol to the late endosomal lumen. In this manner ATP7B participates in the biliary excretion of copper (Harada et al., 2000). Other investigators have reported that ATP7B is redistributed from the trans-Golgi network to the cytoplasmic vesicular department under conditions of high cellular copper concentrations, and returns to the trans-Golgi network as the cytoplasmic copper concentration falls (Schaefer et al., 1999; Culotta and Gitlin, 2000). It is also becoming evident that the cellular localization of ATP7B and its responsiveness to intracellular copper concentrations differ between various WD mutants (Harada et al., 2001). Another copper-transporting, P-type ATPase, ATP7A, is deficient in Menkes disease, but fully active in WD. It is involved in the export of copper from the intestinal epithelial cell, and in the transport of copper across the placenta and blood-brain barrier (Menkes, 1999; Culotta and Gitlin, 2000).

Ceruloplasmin is an α-2-glycoprotein with a single continuous polypeptide chain and a molecular weight of 132 kDa; it has six or seven cupric ions per molecule. More than 95% of serum copper is in the form of ceruloplasmin. The protein has multiple functions, principally it is involved in peroxidation of ferrous to ferric transferrin and

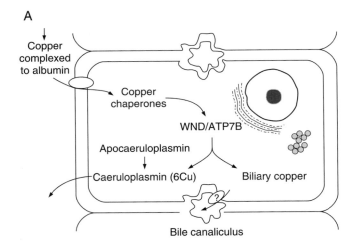

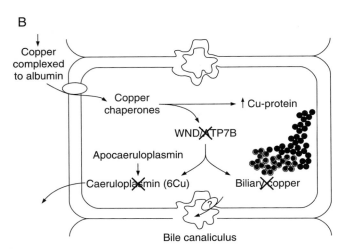

FIGURE 31.1 Normal hepatocyte copper metabolism and effects of WD mutation. (A) Copper complexed to albumin is transported into hepatocytes and then escorted to sites of incorporation into copper-containing enzymes and proteins. The WND/ATP7B copper transporter is necessary for incorporation of copper into ceruloplasmin and also for excretion of copper into bile. (B). In WD excess copper accumulates. Reactive species generated by copper ions damage hepatocytes. The copper released from hepatocytes then leads to damage of other tissues and organs. (Courtesy of Dr. Jennifer A. Cuthbert, Department of Internal Medicine, Southwestern Medical School, the University of Texas Southwestern Medical Center, Dallas, Texas.)

is believed to control the release of iron into plasma from cells, in which the metal is stored in the form of ferritin. A membrane-bound ceruloplasmin homolog, hephaestin, which functions as a ferroxidase, is necessary for iron egress from intestinal enterocytes into the circulation. (Frazer et al., 2001; Vulpe et al., 1999). Whether its activity is defective in WD is still unknown.

The concentration of ceruloplasmin in plasma is normally between 20 and 40 mg/dl. A familial component, independent of WD genotype, appears to be a major factor accounting for the variation in ceruloplasmin levels in unaffected subjects. The concentration of ceruloplasmin is less in normal newborns than in adults and remains less than normal up to approximately two months of age. In serum of the normal adult ceruloplasmin exists in two molecular isoforms, one of 125 kDa, also found in bile, the other of 132 kDa. In the neonate, by contrast, the biliary isoform is markedly reduced, an indication that the formation of biliary ceruloplasmin is transiently inoperative, perhaps due to a delay in the expression of the WD gene until after birth (Chowrimootoo et al., 1997; Cuthbert, 1998). As a consequence, fetal copper concentration is markedly increased, to the point where it is up to twenty times that of normal adult liver. In presymptomatic neonates homozygous for WD both the biliary and the plasma isoforms of ceruloplasmin are probably reduced or absent. (Chowrimootoo et al., 1998).

Ceruloplasmin is reduced in children in Menkes disease (Menkes, 1999), and in infants who suffer from a combined iron and copper deficiency anemia. In the nephrotic syndrome, low levels are caused by the vast renal losses of ceruloplasmin.

Ceruloplasmin is elevated in a variety of circumstances, including pregnancy or other conditions with high estrogen concentrations, infections, cirrhosis, malignancies, hyperthyroidism, and myocardial infarction. It is also elevated in Indian childhood cirrhosis, a condition that results from high environmental copper ingestion, and that like WD can present with a chronic liver disorder (Pandit and Bhave, 1996).

The concentrations of none of the other copper-containing proteins isolated from mammalian tissues, all of which are deficient in Menkes disease are altered in WD.

The gene for ATP7B has 21 expressed exons spanning over 80 kb of genomic sequence with tissue-specific alternate splicing of exons 6–8 and 12–13 (Fig. 31.2). More than 200 mutations of the WD gene have been characterized and most patients are compound heterozygotes (Thomas et al., 1995; Shah et al., 1997; Butler et al., 2001). Reported mutations include missense, nonsense, deletion, splice site, and promotor region mutations. Most mutations are found at low frequencies, and there are innumerable "private" mutations. Polymorphisms in the WD gene are common, and are probably encountered in almost every normal individual (Cox, 2002). The more common mutations and the populations in which they are commonly encountered are presented in Table 31.2. A complete list of mutations is being maintained by Cox at http://www.medgen.med.ualberta.ca/database.html. Although several workers have attempted to correlate genotype and phenotype, the same mutation can present vastly different clinical pictures (Palsson et al., 2001). For example, patients homozygous for the common his1069-to-gln mutation show a wide range of clinical presentations (Duc et al., 1998). It is therefore becoming apparent that several other genetic

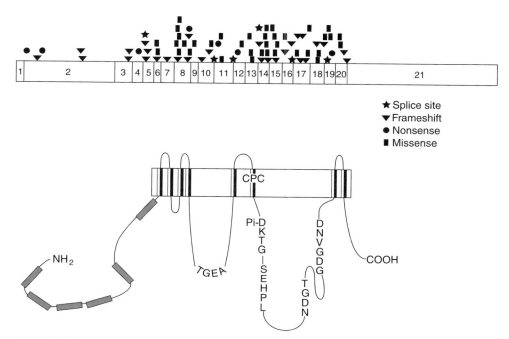

**FIGURE 31.2** Summary of some of the WD mutations in exonic and splice-type junction sequences. (Courtesy of Dr. T. Conrad Gillian, College of Physicians and Surgeons, Columbia University, New York, and the American Society of Human Genetics.)

TABLE 31.2 Common Mutations Detected in Wilson Disease Chromosomes

| Mutation | Exon | Domain | Population |
|---|---|---|---|
| Missense | | | |
| Arg778Leu | 8 | Tm 4 | Japan, China, Korea, Taiwan |
| Arg778Gln | 8 | Tm 4 | China |
| His1069Gln | 14 | SEHPL | Northern, Eastern Europe, U.S. |
| Frameshift | | | |
| 2007del7 | 7 | Tm 1 | Iceland |
| 2302insC | 8 | Tm 4 | Sardinia, Turkey, Albania |
| 2533delA | 10 | Td[1] | Italy, Turkey, Albania |
| 2464delC | 10 | Td[1] | Sardinia |
| 2871delC | 13 | Tm 6 | Eastern Japan |
| Promoter Region | | | |
| −441/−427 del | | | Sardinia |

and environmental factors determine the presentation and course of the disease.

The genetic mutation in WD induces extensive changes in copper homeostasis. Normally the amount of copper in the body is kept constant through excretion of copper from the liver into bile. The two fundamental defects in WD are a reduced biliary transport and excretion of copper, and an impaired formation of ceruloplasmin (Cuthbert, 1998). Biliary excretion of copper is between 20 and 40% of normal, and fecal output of copper is also reduced (Frommer, 1974).

One of the most important features of WD is the accumulation of copper within liver. In the earliest stages of the disease the metal is firmly bound to copper proteins such as ceruloplasmin and superoxide dismutase, or is in the cupric form complexed with MT. As the disease progresses the copper load overwhelms the binding capacity of MT and cytotoxic cupric copper is released, causing damage to hepatocyte mitochondria and peroxisomes (Iyengar *et al.*, 1988) (Fig. 31.1b). Ultimately, copper leaks from liver into blood, where it is taken up by other tissues, including brain, which in turn are damaged by copper (Scheinberg and Sternlieb, 1984).

Two other biochemical abnormalities are consistently found in patients with WD:

1. *Low to low-normal levels of plasma iron-binding globulin.* This abnormality can be demonstrated in the asymptomatic carrier, and suggests that WD may also encompass a disorder of iron metabolism. This may result from the deficiency of hephaestin, a ceruloplasmin homolog, which has been implicated in intestinal iron transport and may be involved in the transfer of iron from tissue cells to plasma transferrin (Vulpe *et al.*, 1999). Iron deposition in the liver is increased (Shiono *et al.*, 2001), and an increase in cerebral iron uptake can be demonstrated by PET scanning (Bruehlmeier, *et al.*, 2000).

2. *Persistent aminoaciduria.* This is most marked during the later stages but may be noted in some asymptomatic

patients. The presence of other tubular defects (e.g., impaired phosphate resorption in patients without aminoaciduria) suggests that a toxic action of the metal on renal tubules causes the aminoaciduria. Since the WD gene is expressed in kidney, these manifestations could either be primary or secondary to copper overflow from the liver (Cuthbert, 1998; Bull et al., 1993).

## IV. DIAGNOSIS

When WD presents with neurological manifestations some of the diagnostic features are progressive extrapyramidal symptoms commencing after the first decade of life, abnormal liver function, aminoaciduria, cupriuria, and absent or decreased serum ceruloplasmin. In the experience of Gow et al. (2000), and Steindl et al. (1997), the diagnosis is more difficult in patients who present with hepatic symptoms, particularly those who present in fulminant hepatic failure.

The presence of a Kayser-Fleischer ring is the single most important diagnostic feature; its absence in a subject with neurological symptoms virtually rules out the diagnosis of WD. The ring is not seen in the majority of presymptomatic patients, nor is it seen in 33% of patients who present with hepatic symptoms (Gow et al., 2000).

An absent or low serum ceruloplasmin level is of lesser diagnostic importance; some 5–20% of patients with WD have normal levels of the copper protein. In the experience of Gow et al. (2000) all patients with WD had low ceruloplasmin concentrations when measured by the oxidase method, which recognizes only the copper-containing form of ceruloplasmin, whereas 28% of patients had a normal ceruloplasmin when the immunoprecipitation method was used. The latter method also recognizes apoceruloplasmin. Ceruloplasmin is abnormally low in about 10% of heterozygotes for WD. In affected families, the differential diagnosis between heterozygotes and presymptomatic homozygotes is of utmost importance, inasmuch as it is generally accepted that presymptomatic homozygotes should be treated preventively (Walshe, 1988).

Since 90% of serum copper is attributable to ceruloplasmin, serum copper is generally low, but it can be normal or even high (Brewer, 1998; Ogihara et al., 1995).

Several workers have stressed the diagnostic value of measuring 24-hour urine copper concentrations (Brewer and Yusbasiyan-Gurkan, 1992). In the experience of Gow et al. (2000), however, 41% of patients who had a nonfulminant presentation of WD, including all those with a neurologic presentation, had normal urinary copper (less than 100 µg per 24 hours).

The gold standard for the diagnosis of WD is liver biopsy (Brewer and Yusbasiyan-Gurkan, 1992). When this procedure is undertaken it is advisable to perform histological studies with stains for copper and copper-associated proteins and chemical quantitation for copper. In all confirmed cases of WD, hepatic copper is greater than 3.9 µmol/g dry weight (237.6 µg/g) as compared to a normal range of 0.2–0.6 µmol/g (20–50 µg/g dry tissue).

The presymptomatic child with WD can have a Kayser-Fleischer ring (seen in 33% of presymptomatic patients in Walshe's series), an increased 24-hour urine copper, hepatosplenomegaly (seen in 38%), and abnormal neuroimaging (seen in about one-quarter). Should the diagnosis in a child at risk be in doubt, an assay of liver copper is indicated (Walshe, 1988). A low ceruloplasmin level in an asymptomatic family member of a subject with WD is not necessarily diagnostic of the presymptomatic stage of the illness; some 10%–20% of heterozygotes have ceruloplasmin levels below 15 mg/dl. Since gene carriers comprise about 1% of the population, gene carriers with low ceruloplasmin values are seen forty times more frequently than patients with WD and low ceruloplasmin values. Therefore when low ceruloplasmin levels are found on routine screening and are unaccompanied by any abnormality of hepatic function or copper excretion, the subject is much more likely to be a heterozygote for WD than a presymptomatic WD patient (Scheinberg and Sternlieb, 1984). However, this supposition should be confirmed by liver biopsy.

Linkage analysis has been used for the identification of the presymptomatic patient in families with one or more informative family members. A variety of microsatellite markers flanking the WD locus have been used (Curtis et al., 1999). Because of the large number of mutations causing the disease, PCR-based screening of mutations can be laborious. About one-third of North American and United Kingdom WD subjects have a point mutation (His 1069Glu), and screening for this mutation can be performed rapidly (Shah et al., 1997; Curtis et al., 1999). A combination of mutation and linkage analysis is required for prenatal diagnosis (Loudianos et al., 1994).

Yamaguchi et al. (1999) have used a monoclonal antibody specific to holoceruloplasmin for a WD mass screening trial. In view of the low serum ceruloplasmin levels in neonates they believe that the best time to institute such a program is not until 3 years of age.

## V. NEUROIMAGING

CT scans usually reveal ventricular dilatation and diffuse atrophy of the cortex, cerebellum, and brain stem. About half the patients have symmetric hypointensities in the head of caudate, pallidum, substantia nigra, and red nuclei. The histopathology of WD suggests that these hypointensities are secondary to the presence of protein-bound copper in the thalamus and basal ganglia. Cortical atrophy and focal lesions in cortical white matter are also noted. Increased density owing to copper deposition is not observed.

The MRI demonstrates symmetric areas of increased signal intensity on T2-weighted images in the putamen, particularly in its outer rim, the thalami, the head of the caudate nucleus, and the globus pallidus (Fig. 31.3A,B). In the series of King *et al.* (1996) the midbrain was abnormal in 77% of patients with WD (Fig. 31.3C,D). The tegmentum was primarily involved, but also the substantia nigra, and the mesencephalic tectum. Abnormalities are also seen in the pons and the cerebellum. These areas are hypointense on T1-weighted images. (van Wassenaer, *et al.*, 1996). Correlation with the clinical picture is not good in that patients with neurological symptoms can have a normal MRI, and other subjects with no neurological symptoms can have an abnormal MRI (King *et al.*, 1996; Prayer, *et al.*, 1990).

Positron emission tomography (PET) demonstrates a widespread depression of glucose metabolism, with the

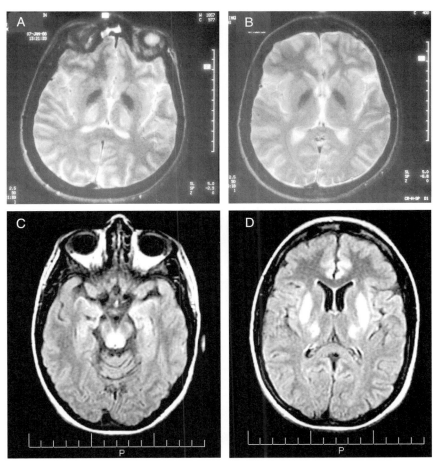

FIGURE 31.3 (A) Coronal T2-weighted MRI images of a 22-year-old woman with WD. Three months after the disease had been diagnosed and at start of penicillamine therapy, there are bilateral hyperintense thalamic lesions that were hypointense on T1-weighted images. (Courtesy of Dr. I. Prayer, Zentral Institut für Radiodiagnose und Ludwig Boltzmann Institut, University of Vienna, Austria). (B) The same patient as in Figure 4 after 13 months of penicillamine therapy shows a significant regression of the thalamic lesions. (Courtesy of Dr. I Prayer, Zentral Institut fur Radiodiagnose und Ludwig Boltmann Institut, University of Vienna, Austria). (C) Coronal FLAIR MRI in a 22-year-old woman with one-year history of dysphagia, foot dystonia, and mild personality changes. There is signal change in the midbrain surrounding the aqueduct and the upper pons dorsally. (D) Symmetrical signal abnormality involving the putamina. There is also signal abnormality involving the anterior portion of the thalami.

greatest focal hypometabolism being seen in the lenticular nucleus. This abnormality precedes any alteration seen on CT scan (Hawkins *et al.*, 1987), and improves with chelation therapy (Schlaug *et al.*, 1996). In addition, a reduction in dopamine D2 receptor binding, and loss of striatal dopamine transporters has been documented (Oder *et al.*, 1996; Jeon *et al.*, 1998).

## VI. PATHOLOGIC ANATOMY

The abnormalities in copper metabolism result in a deposition of the metal in several tissues. Anatomically, the liver shows a focal necrosis that leads to a coarsely nodular, postnecrotic cirrhosis. The nodules vary in size and are separated by bands of fibrous tissues of different widths. Some hepatic cells are enlarged and contain fat droplets, intranuclear glycogen, and clumped pigment granules; other cells are necrotic with regenerative changes in the surrounding parenchyma (Strohmeyer and Ishak, 1980).

Electron microscopic studies indicate that copper is initially spread diffusely within cytoplasm, probably as the monomeric MT complex. Later in the course of the disease, the metal is sequestered within lysosomes, which become increasingly sensitive to rupture (Scheinberg and Sternlieb, 1984). Copper probably initiates and catalyzes oxidation of the lysosomal membrane lipids, resulting in lipofuscin accumulation. Within the kidneys, the tubular epithelial cells can degenerate, and their cytoplasm can contain copper deposits.

In the brain, particularly in patients whose symptoms commenced prior to the onset of puberty, the basal ganglia show the most striking alterations. They have a brick-red pigmentation; spongy degeneration of the putamen frequently leads to the formation of small cavities (Wilson, 1912). Lesser degenerative changes are seen in the brain stem, the dentate nucleus, the substantia nigra, and the convolutional white matter.

Microscopic studies reveal a loss of neurons, axonal degeneration, and large numbers of protoplasmic astrocytes, including giant forms termed Alzheimer cells. These cells are not specific for WD; they can also be seen in the brains of patients dying in hepatic coma or as a result of argininosuccinic aciduria or other disorders of the urea cycle. Opalski cells, also seen in WD, are generally found in gray matter. They are large cells with a rounded contour and finely granular cytoplasm. They probably represent degenerating astrocytes. In about 10% of patients, cortical gray matter and white matter are more affected than the basal ganglia. Here, too, there is extensive spongy degeneration and proliferation of astrocytes (Richter, 1948). Copper is deposited in the pericapillary area and within astrocytes, but it is uniformly absent from neurons and ground substance.

## VII. ANIMAL MODELS

Several animal models have been described. These include the LEC rat, the toxic milk mouse mutant, and canine copper toxicosis seen in Bedlington terriers. In addition, a strain of mice have been generated that are homozygous null mutants for the WD gene.

In the LEC rat acute hepatitis develops some four months after birth. Survivors of the attack continue to suffer from chronic hepatitis. Hepatic copper is elevated, and hepatitis can be prevented by treatment with penicillamine (Klein *et al.*, 2000). This strain has a large deletion in the coding region for ATP7B at its 3' end (Wu *et al.*, 1994).

The toxic milk (tx) mouse mutant has a point mutation in murine homolog of the gene for ATP7B (Theophilos *et al.*, 1996). As a consequence copper translocation within the cell is defective, and the metal accumulates within liver (Voskoboinik *et al.*, 2001). The milk of mothers carrying the mutant gene is markedly deficient in copper and pups are born with a marked copper deficiency (Huang and Gitschier, 1997).

Canine copper toxicosis, which clinically expresses itself by the accumulation of copper in liver and the evolution of a fatal hepatic and neurologic disease, is not due to a defect in ATP7B. Rather it may involve one of the various copper chaperones (Nanji and Cox, 1999). The disease can be diagnosed by its linkage to a microsatellite marker and is effectively treated by the various regimens used for human WD (Brewer, 1998).

Buiakova *et al.* (1999) have developed a strain of mice that are homozygous null mutants for the gene for WD. These mice have a gradual accumulation of hepatic copper to a level that is 60 times greater than normal, and a histological picture of nodular cirrhosis. Brain and kidney copper are also increased. Milk from the mutant females is copper deficient, and offspring develop growth retardation and neurological symptoms characteristic for copper deficiency.

## VIII. TREATMENT

Even though it is clear that all patients with WD whether symptomatic or asymptomatic, require treatment there is no current consensus as to the optimal means of treating WD. The aims of treatment are initially to remove the toxic amounts of copper and secondarily to prevent tissue reaccumulation of the metal (Brewer, 1995; Sturniolo *et al.*, 1999).

Treatment can be divided into two phases: the initial phase, when toxic copper levels are brought under control, and maintenance therapy. There is no currently agreed upon regimen for the treatment of the new patient with neurologic or psychiatric symptoms (Brewer, 1999; Walshe, 1999). In the past, most centers recommended starting patients on D-penicillamine (600–3000 mg/day). Although this drug is effective in promoting urinary excretion of copper, adverse reactions during both the initial and maintenance phases of treatment are seen in about 25% of patients. In a significant proportion of patients there is worsening of neurologic symptoms during the initial phases of treatment. This is frequently irreversible (Brewer, 1999). Systemic lupus erythematosus and immune complex nephritis are seen in about 5% of patients on penicillamine (Walshe, 1999). Skin rashes, gastrointestinal discomfort, and hair loss are also encountered. During maintenance therapy, one may see polyneuropathy, polymyositis, and nephropathy. Some of these adverse effects can be prevented by treatment with pyridoxine (25 mg/day).

Because of these side effects most institutions now advocate initial therapy with ammonium tetrathiomolybdate (60–300 mg/day, administered in six divided doses, three with meals and three between meals; Brewer, 1998, 1999). Tetrathiomolybdate forms a tripartite complex with protein and copper and experimental work indicates that it acts by transferring MT copper to high molecular weight copper fractions, probably albumin (McQuaid and Mason, 1991). The time of administration of ammonium tetrathiomolybdate in relation to meals is important. When the drug is given with food it blocks the absorption of copper. When given between meals the drug complexes nonceruloplasmin copper to albumin and makes it unavailable for cellular uptake. The major drawbacks to the widespread use of this drug are bone marrow suppression and skeletal growth disturbances that can occur with long-term therapy (Cuthbert, 1998). The drug has not been approved for general use in the United States (Brewer et al., 1996).

Triethylene tetramine dihydrochloride (trientine); (250 mg four times a day, given at least one hour before or two hours after meals) is another chelator, which increases urinary excretion of copper. Its effectiveness is less than that of penicillamine, but the incidence of toxicity and hypersensitivity reactions is lower, and it therefore has been recommended for patients who are intolerant to penicillamine. Adverse reactions include a colitis that improved after withdrawal of the drug. Additionally deterioration of neurological symptoms has been encountered (Dahlman et al., 1995)

Zinc acetate (50 mg of elemental zinc acetate three times a day, given on an empty stomach) acts by inducing intestinal MT by as much as 15-fold (Sturniolo et al., 1999). MT has a high affinity for copper and prevents its entrance into blood. Brewer (1998) recommends zinc for the treatment of the presymptomatic patient, and because of its low teratogenicity for the treatment of the symptomatic pregnant patient. When zinc is given to children for the treatment of presymptomatic WD, subjects aged between one and five years are given 25 mg twice a day, subjects aged 6–15 years, if under 125 lbs body weight are given 25 mg three times a day (Brewer et al., 2001). Zinc is far less toxic than penicillamine but is much slower acting, and therefore because of its slow control of copper toxicity, treatment with zinc predisposes a patient to the risk of further progression of the disease and therefore is not the optimal drug for initial treatment of the patient with symptomatic WD.

Zinc is the optimum drug for maintenance therapy, and for the treatment of the presymptomatic patient. Trientine in combination with zinc acetate has been suggested for patients who present with hepatic disease (Brewer, 2000).

Liver transplantation can be helpful in the patient who presents with fulminant hepatic failure or who is in end-stage liver disease. The procedure appears to correct the metabolic defect and can reverse neurological symptoms and neuroimaging abnormalities (Bax et al., 1998; Wu et al., 2000). Schumache et al. (1997) have also recommended its use for patients with normal liver function but whose neurological symptoms have not responded to the various chelating agents.

Diet does not play an important role in the management of WD disease, although Brewer (1995) recommends restriction of liver and shellfish during the first year of treatment. Evidence has recently accumulated that oxidative damage due to free radical formation can play a significant role in producing cell damage in WD. Gu et al. (2000) have noted severe mitochondrial dysfunction in the liver of patients with WD, with significant reduction in the activities of all enzyme complexes involved in oxidative phosphorylation. Other findings pointing to oxidative stress are the reduction in the plasma levels of various antioxidants such as ascorbate and urate, and the increase of allantoin, a possible marker of free radical generation (Ogihara et al., 1995). These findings provide experimental support for the addition of antioxidants such as ascorbate or vitamin E to the therapeutic regimen of all patients with WD.

On the above regimens, there is gradual improvement in neurological symptoms. The Kayser-Fleischer ring begins to fade within 6–10 weeks of the onset of therapy and disappears completely in a couple of years (Mitchell and Heller, 1968). Improvement of neurological symptoms starts 5–6 months after therapy has begun and is generally complete in 24 months. As shown by serial neuroimaging studies there is a significant regression of lesions within thalamus and basal ganglia. Successive biopsies show a reduction in the amount of hepatic copper, but it rarely returns to normal levels (Mason et al., 1989) . Both total serum copper and ceruloplasmin levels fall, and the aminoaciduria and phosphaturia diminish. Evoked potentials have also been used to assess the response to treatment. Auditory-evoked potentials improve within one month of the onset of therapy, with the somatosensory evoked responses being somewhat slower to return to normal (Grimm et al., 1990).

As a rule, patients who are started on therapy prior to the evolution of symptoms remain normal. Children who have had hepatic disease exclusively do well, and in 80%, hepatic functions return to normal. About 40% of children who present with neurological symptoms become completely asymptomatic and remain so for 10 or more years. Children with the mixed hepatocerebral picture do poorly. Less than 25% recover completely, and about 25% continue to deteriorate, often with the appearance of seizures. In all forms of the disease, the earlier the start of therapy, the better the outlook (Arima et al., 1977).

When symptom-free patients with WD discontinue chelation therapy, their hepatic function deteriorates within nine months to three years, a rate that is far more rapid than deterioration following birth (Walshe and Dixon, 1986). Scheinberg et al. (1987) postulate that penicillamine not only removes copper from tissue but also detoxifies the metal by inducing MT synthesis.

Several variants of WD have been recognized. One type begins in adolescence and is marked by progressive tremor, dysarthria, disturbed eye movements, and dementia. It is characterized biochemically by low serum copper and ceruloplasmin. Kayser-Fleischer rings are absent, and liver copper concentrations are low. Metabolic studies using labeled copper suggest a failure in copper absorption from the lower gut (Godwin-Austin et al., 1978).

In another type, the patient developed extrapyramidal movements, but no liver disease. There were no Kayser-Fleischer rings. Blood copper levels were low, but hepatic copper was markedly elevated, with the metal stored in cytoplasm (Heann and Saffer, 1988).

In familial apoceruloplasmin deficiency, an autosomal recessive disorder, the clinical presentation is with adult-onset dementia, retinal degeneration, ataxia, and a variety of movement disorders (Gitlin, 1998; Miyajima et al., 2001). MR imaging and neuropathological examinations demonstrate iron deposition in the basal ganglia. The clinical and pathologic findings confirm the essential role of ceruloplasmin in iron metabolism and in brain iron homeostasis. The relationship between this condition and Hallervorden-Spatz syndrome has been clarified by the recent demonstration that the latter condition is the result of a defect in the metabolism of pantothenic acid (Zhou et al., 2001).

## References

Akil, M., and Brewer, G. J. (1995). Psychiatric and behavioral abnormalities in Wilson's disease. In *Behavioral Neurology of Movement Disorders*. (W.J. Weiner and A.E. Lang, eds.) Vol. 65, pp. 171–178. Raven press, New York.

Andrews, N. C. (2001). Mining copper transport genes. *Proc. Nat. Acad. Sci. U.S.A.* **98**, 6543–6545.

Arima, M., Takeshita, K., Yoshino, K. et al. (1977). Prognosis of Wilson's disease in childhood. *Eur. J. Pediatr.* **126**, 147–154.

Azizi, E., Eshel, G., and Aladjem, M. (1989). Hypercalciuria and nephrolithiasis as a presenting sign in Wilson disease. *Euro. J. Pediatr.* **148**, 548–549.

Bax, R. T., Hässler, A., Luck, W. et al. (1998). Cerebral manifestation of Wilson's disease successfully treated with liver transplantation. *Neurology* **51**, 863–865.

Brewer, G. J. (1995). Practical recommendations and new therapies for Wilson's disease. *Drugs* **50**, 240–249.

Brewer, G. J. (1998). Wilson disease and canine copper toxicosis. *Am. J. Clin. Nutr.* **67** (Suppl.), 1087S–1090S.

Brewer, G. J. (1999). Penicillamine should not be used as initial therapy in Wilson disease. *Mov. Disord.* **14**, 551–554.

Brewer, G. J. (2000). Wilson's disease. *Curr. Treat. Options Neurol.* **2**, 193–204.

Brewer, G. J., Dick, R. D., Johnson, V. D. et al. (2001). Treatment of Wilson's disease with zinc. XVI: treatment during the pediatric years. *J. Lab. Clin. Med.* **137**, 191–198.

Brewer, G. J., Johnson, V., Dick, R. D. et al. (1996). Treatment of Wilson disease with ammonium tetrathiomolybdate. II. Initial therapy in 33 neurologically affected patients and follow-up with zinc therapy. *Arch. Neurol.* **53**, 1017–1025.

Brewer, G. J., and Yusbasiyan-Gurkan, V. (1992). Wilson disease. *Medicine* **71**, 139–164.

Bruehlmeier, M., Leenders, K. L., Vontobel, P. et al. (2000). Increased cerebral iron uptake in Wilson's disease: a $^{52}$Fe-citrate PET study. *J. Nucl. Med.* **41**, 781–787.

Buiakova, O. I., Xu, J., Lutsenko, S. et al. (1999). Null mutation of the murine ATP7B (Wilson disease) gene results in intracellular copper accumulation and late-onset hepatic nodular transformation. *Hum. Mol. Genet.* **8**, 1665–1671.

Bull, P. C., Thomas, G. R., Rommens, J. M. et al. (1993). The Wilson disease gene is a putative copper transporting P-type ATPase similar to the Menkes gene. *Nature. Genet.* **5**, 327–337.

Butler, P., McIntyre, N., and Mistry, P. K. (2001). Molecular diagnosis of Wilson disease *Mol. Genet. Metab.* **72**, 223–230.

Cairns, J. E., Williams, H. P., and Walshe, J. M. (1969). "Sunflower cataract" in Wilson's disease. *Br. Med. J.* **3**, 95–96.

Chowrimootoo, G. F., Andoh, J., and Seymour, C. A. (1997). Western blot analysis in patients with hypocaeruloplasminaemia. *Q. J. Med.* **90**, 197–202.

Chowrimootoo, G. F., Scowcroft, H., and Seymour, C. A. (1998). Caeruloplasmin isoforms in Wilson's disease in neonates. *Arch. Dis. Child Fetal Neonatal Ed.* **79**, F198–201.

Cox, D. W., (2002). Personal Communication.

Culotta, V. C., and Gitlin, J. D. (2000). Disorders of copper transport. In *The Metabolic and Molecular Bases of Inherited Disease*. 8th edition. (C. R. Scriver, A. L. Beaudet, W. S. Sly, D. Valle, B. Vogelstein, B. Chids, eds.), pp. 3105–3126. McGraw-Hill, New York.

Curtis, D., Durkie, M., Balac (Morris), P. et al. (1999). A study of Wilson disease mutations in Britain. *Hum. Mutat.* **14**, 303–311.

Cuthbert, J. A. (1998). Wilson's disease. Update of a systemic disorder with protean manifestations. *Gastroenterol. Clin. North. Am.* **27**, 655–81.

Dahlman, T., Hartvig, P., Lofholm, M. et al. (1995). Long-term treatment of Wilson's disease with triethylene tetramine dihydrochlordie (trientine). *Q. J. Med.* **88**, 609–616.

Duc, H. H., Hefter, H., Stremmel, W. et al. (1998). His1069Gln and six novel Wilson disease mutations: analysis of relevance for early diagnosis and phenotype. *Eur. J. Hum. Genet.* **6**, 616–623.

Ferlan-Marolt, V., and Stepec, S. (1999). Fulminant Wilsonian hepatitis unmasked by disease progression: report of a case and review of the literature. *Dig. Dis. Sc.* **44**, 1054–1058

Fleischer, B. (1912); Über einer der "Pseudosklerose" nahestehende bisher unbekannte Krankheit (gekennzeichnet durch Tremor, psychische Störungen, bräunliche Pigmentierung bestimmter Gewebe, insbesondere auch der Hornhautperipherie, Lebercirrhose). *Deutsch. Z. Nervenheilk.* **44**, 179–201.

Frazer, D. M., Vulpe, C. D., McKie, A. T. et al. (2001). Cloning and gastrointestinal expression of rat hephaestin: relationship to other iron transport proteins. *Am. J. Physiol. Gastrointest. Liver Physiol.* **281**, G931–939.

Frommer, D. J. (1974). Defective biliary excretion of copper in Wilson's disease. *Gut* **15**, 125–129.

Gitlin, J. D. (1998). Aceruloplasminemia. *Pediatr. Res.* **44**, 271–276.

Glazebrook, A. J. (1945). Wilson's disease. *Edinburgh Med. J.* **52**, 83–87.

Godwin-Austin, R. B., Robinson, A., Evans, K., and Lascelles, P. T. (1978). An unusual neurological disorder of copper metabolism clinically resembling Wilson's disease but biochemically a distinct entity. *J. Neurol. Sci.* **39**, 85–98.

Gow, P. J., Smallwood, R. A., Angus, P. W., Smith, A. L. Wall, A. J., and Sewell, R. B. (2000). Diagnosis of Wilson's disease: an experience over three decades. *Gut* **46**, 415–419.

Grimm, G., Oder, W., Praye, L. et al. (1990). Evoked potentials in assessment and follow-up of patients with Wilson's disease. *Lancet* **336**, 963–964.

Gu, M., Cooper, J. M, Butler, P. et al. (2000). Oxidative-phosphorylation defects in liver of patients with Wilson's disease. *Lancet* **356**, 469–474.

Harada, M., Sakisaka, S., Kawaguchi, T. et al. Copper does not alter the intracellular distribution of ATP7 copper-transporting ATPase. *Biochem. Biophys. Res. Commun.* **275**, 871–876.

Harada, M., Sakisaka, S., Terada, K. et al. (2001). A mutation of the Wilson disease protein, ATP7B, is degraded in the proteasomes and forms protein aggregates. *Gastroenterology* **120**, 967–974.

Hawkins, R. A., Mazziotta, J. C., and Phelps, M. E. (1987). Wilson's disease studied with FDG and positron emission tomography. *Neurology* **37**, 1707–1711.

Heann, J., and Saffer, D. (1988). Abnormal copper metabolism: Another "non-Wilson's" case. *Neurology* **38**,1493–1495.

Hoffman, D. L., and O'Halloran, T. V. (2001). Function, structure, and mechanisms of intracellular copper trafficking proteins. *Ann. Rev. Biochem.* **70**, 677–701.)

Huang, I., and Gitschier, J. (1997). A novel gene involved in zinc transport is deficient in the lethal milk mouse. *Nat. Genet.* **17**, 292–297.

Iyengar, V., Brewer, G. J., Dick, R. D., and Chung O. Y. (1988). Studies of cholecystokinin-stimulated biliary secretions reveal a high molecular weight copper-binding substance in normal subjects that is absent in patients with Wilson's disease. *J. Lab. Clin. Med.* **111**, 267–274.

Jeon, B., Kim, J. M., Jeong, J. M. et al. (1998). Dopamine transporter imaging with [123I]-beta-CIT demonstrates presynaptic nigrostriatal dopaminergic damage in Wilson's disease. *J. Neurol. Neurosurg. Psychiatry* **65**, 60–64.

Kayser, B. (1902). Ueber einen Fall von angeborener grünlicher Verfärbung der Cornea. *Klin. Monatsbl. Augenheilkd.* **40**, 22–25.

King, A. D., Walshe, J. M., Kendall, B. E. et al. (1996). Cranial MR imaging in Wilson's disease. *Am. J. Roentgenol.* **167**, 1579–1584.

Klein, D., Lichtmannegger, J., Heinzmann, U., and Summer, K. H. (2000). Dissolution of copper-rich granules in hepatic lysosomes by penicillamine prevents the development of fulminant hepatitis in Long-Evans cinnamon rats. *J. Hepatol.* **32**, 193–201.

Kontaxakis, V., Stefanis, C., Markidis, M. et al. (1988). Neuroleptic malignant syndrome in a patient with Wilson's disease [Letter]. *J. Neurol. Neurosurg. Psychiatry* **51**, 1001–1002.

Lee, J., Prohaska, J. R., and Thiele, D. J. (2001). Essential role for mammalian copper transporter Ctr1 in copper homeostasis and embryonic development. *Proc. Natl. Acad. Sci. U.S.A.* **98**, 6842–6847.

Lingan, S., Wilson, J., Nazer, H., and Mowat, A. P. (1987). Neurological abnormalities in Wilson's disease are reversible. *Neuropediatrics* **18**, 11–12.

Loudianos, G., Figus, A. L., Loi, A. et al. (1994). Improvement of prenatal diagnosis of Wilson disease using microsatellite markers. *Prenatal Diagn.* **14**, 999–1002.

Lüthy, F. (1931). Über die hepato-lentikuläre Degeneration (Wilson-Westphal-Strümpell). *Deutsch. Z. Nervenheilk.* **123**, 101–181.

Mason, J., McQuaid, A., and Pheiffer, H. (1989) Can patients with Wilson's disease be decoppered? *Lancet* **1**, 1455.

McQuaid, A., and Mason, J. (1991). A. comparison of the effects of penicillamine, trientine, and trithiomolybdate on [35S]-labeled metallothionein *in vitro*; implications for Wilson's disease therapy. *J. Inorg. Biochem.* **41**, 87–92.

Menkes, J. H. (1999). Menkes disease and Wilson disease: Two sides of the same copper coin. Part I. Menkes disease. *Eur. J. Paed. Neurol.* **3**, 147–158.

Mitchell A. M., and Heller G. L. (1968). Changes in Kayser-Fleischer ring during treatment of hepatolenticular degeneration. *Arch. Ophthalmol.* **80**, 622–631.

Miyajima, H., Kono, S., Takahashi, Y. et al. (2001). Cerebellar ataxia associated with heteroallelic ceruloplasmin gene mutation. *Neurology* **57**, 2205–2210.

Mukhopadhyay, C. K., Attieh, Z. K., and Fox, P. L. (1998).Role of ceruloplasmin in cellular iron uptake. *Science* **279**, 714–716.

Nanji, M. S., and Cox, D. W. (1999). The copper chaperone Atox1 in canine copper toxicosis in Bedlington terriers. *Genomics* **62**, 108–112.

Oder, W., Brucke, T., Kollegger, H. et al. (1996) Dopamine D2 receptor binding is reduced in Wilson's disease: A correlation of neurological deficits with striatal 123I-iodobenzamide binding. *J. Neural. Transm.* **103**, 1093–1103.

Ogihara, H., Ogihara, T., Miki, M. et al. (1995). Plasma copper and antioxidant status in Wilson's disease. *Pediatr. Res.* **37**, 219–226.

Palsson, R., Jonasson, J. G., Kristjansoonk M. et al. (2001). Genotype-phenotype interactions in Wilson's disease: insights from an Icelandic mutation. *Eur. J. Gasroenterol. Hepatol.* **13**, 433–436.

Pandit, A., and Bhave, S. (1996). Present interpretation of the role of copper in Indian childhood cirrhosis. *Am. J. Clin. Nutr.* **63**, 830S–835S.

Portala, K., Westermark, K., Ekselius, L., and von Knorring, L. (2001). Personality traits in treated Wilson's disease determined by means of the Karolinska Scales of Personality (KSP). *Eur. Psychiatry* **16**, 362–371.

Prayer, L., Wimberger, D., Kramer, J. et al. (1990). Cranial MRI in Wilson's disease. *Neuroradiology* **32**, 211–214.

Richter, R. (1948). The pallial component in hepatolenticular degeneration. *J. Neuropathol. Exp. Neurol.* **7**, 1–18.

Rolfs, A., and Hediger, M. A. (2001). Intestinal metal ion absorption: an update. *Curr. Opin. Gastroenterol.* **17**, 177–183.

Rumpel, A. (1973). Über das Wesen und die Bedeutung der Leberveränderungen und der Pigmentierungen bei den damit verbundenen Fällen von Pseudosklerose, zugleich ein Beitrag zur Lehre der Pseudosklerose (Westphal-Strümpell). *Deutsch. Z. Nervenheilk.* **49**, 54–73.

Schaefer, M., Hopkins, R. G., Failla, M. L., and Gitlin, J. D. (1999). Hepatocyte-specific localization and copper-dependent trafficking of the Wilson's disease protein in the liver. *Am. J. Physiol.* **276**, G 639–646.

Scheinberg, I. H., and Sternlieb, I. (1984). *Wilson's Disease.* W.B. Saunders, Philadelphia, PA.

Scheinberg, I. H., Sternlieb, I., Schilsky, M., and Stockert, R. J. (1987). Penicillamine may detoxify copper in Wilson's disease. *Lancet* **2**, 95.

Schlaug, G., Hefter, H., Engelbrecht, V. et al. (1996). Neurological impairment and recovery in Wilson's disease: evidence from PET and MRI. *J. Neurol. Sci.* **136**, 129–139.

Schumacher, G., Platz, K. P., Mueller, A. R. et al. (1997). Liver transplantation: treatment of choice for hepatic and neurological manifestations of Wilson's disease. *Clin. Transplant.* **11**, 217–224.

Scott, J., Golan, J. L., Samourian, S., and Sherlock, S. (1978). Wilson's disease presenting as chronic active hepatitis. *Gastroenterology* **74**, 645–651.

Shah, A. B., Chernov, I., Zhang, H. T. et al. (1997). Identification and analysis of mutation in the Wilson disease gene (ATP7B): population frequencies, genotype-phenotype-correlation, and functional analyses. *Am. J. Hum. Genet.* **61**, 317–328.

Shiono, Y., Hayashi, H., Wakusawa, S., and Yano, M. (2001). Ultrastructural identification of iron and copper accumulation in the liver of a male patient with Wilson disease. *Med. Electron. Microsc.* **34**, 54–60.

Slovis, T. L., Dubois, R. S., Rodgerson, D. O., and Silverman A. (1971). The varied manifestations of Wilson's disease. *J. Pediatr.* **78**, 578–584.

Steindl, P., Ferenci, P., Dienes, H. P. *et al.* (1997). Wilson's disease in patients presenting with liver disease: a diagnostic challenge. *Gastroenterology* **113**, 212–218.

Strohmeyer, F. W., and Ishak, K. G. (1980). Histology of the liver in Wilson's disease: a study of 34 cases. *Am. J. Clin. Pathol.* **73**, 12–24.

Sturniolo, G. C., Mestriner, C., Irato, P. *et al.* (1999). Zinc therapy increases duodenal concentrations of metallothionein and iron in Wilson's disease patients. *Am. J. Gastroenterol.* **94**, 334–338.

Theophilos, M. B., Cox, D. W., and Mercer, J. F. B. (1996). The toxic milk mouse is a murine model of Wilson disease. *Hum. Molec. Genet.* **5**, 1619–1624.

Thomas, G. R., Forbes, J. R., Roberts, E. A. *et al.* (1995). The Wilson disease gene: spectrum of mutations and their consequence. *Nat. Genet.* **9**, 210–217.

Uzman, L. L., and Jakus, M. A. (1957). The Kayser-Fleischer ring: A histochemical and electron microscope study. *Neurology* **7**, 341–355.

Van Wassenaer-van Hall, H. N., van den Heuvel, A. G., Algra, A. *et al.* (1996). Wilson disease: findings at MR imaging and CT of the brain: clinical correlation. *Radiology* **198**, 531–536.

Voskoboinik, I., Greenough, M., La Fontaine, S. *et al.* (2001). Functional studies on the Wilson copper P-type ATPase and toxic milk mouse mutant. *Biochem. Biophys. Res. Commun.* **281**, 966–970.

Vulpe, C. D., and Packman, S. (1995). Cellular copper transport. *Annu. Rev. Nutr.* **15**, 293–322.

Vulpe, C. D., Kuo, Y. M., Murphy, T. L. *et al.* (1999). Hephaestin, a ceruloplasmin homologue implicated in intestinal transport, is defective in the sla mouse. *Nature. Genet.* **21**, 195–199.

Walshe, J. M. (1988). Diagnosis and treatment of presymptomatic Wilson's disease. *Lancet* **2**, 435–437.

Walshe, J. M. (1999). Penicillamine: the treatment of first choice for patients with Wilson's disease. *Mov. Disord.* **14**, 545–550.

Walshe, J. M., and Dixon, A. K. (1986). Dangers of non-compliance in Wilson's disease. *Lancet* **1**, 845–847.

Werlin, S. L., Grand, R. J., Perman, J. A., and Watkins, J. B. (1978). Diagnostic dilemmas of Wilson's disease: Diagnosis and treatment. *Pediatrics* **62**, 47–51.

Wilson, S. A. K. (1912). Progressive lenticular degeneration: A familial nervous disease associated with cirrhosis of the liver. *Brain* **34**, 295–509.

Wu, J., Forbes, J. R., Chen, H. S. *et al.* (1994). The LEC rat has a deletion in the copper-transporting ATPase gene homologous to the Wilson disease gene. *Nat. Genet.* **7**, 541–545.

Wu, J. C., Huang, C. C., Jeng, L. B., and Chu, N. S. (2000). Correlation of neurological manifestations and MR images in a patient with Wilson's disease after liver transplantation. *Acta. Neurol. Scand.* **102**, 135–139.

Yamaguchi, Y., Aoki, T., Arashima, S. *et al.* (1999). Mass screening for Wilson's disease: results and recommendations. *Pediatr. Int.* **41**, 405–408.

Zhou, B., Westaway, S. K., Levinson B. *et al.* (2001). A novel pantothenate kinase gene (PANK2) is defective in Hallervorden-Sptaz syndrome. *Nat. Genet.* **28**, 299–300.

# CHAPTER 32

# Essential Tremor

**ELAN D. LOUIS**
*G.H. Sergievsky Center and Department of Neurology*
*College of Physicians and Surgeons*
*Columbia University*
*New York, New York 10032*

**RUTH OTTMAN**
*G.H. Sergievsky Center*
*College of Physicians and Surgeons*
*Columbia University*
*New York, New York 10032*

I. Phenotype
II. Gene
III. Diagnostic and Ancillary Tests
IV. Neuroimaging
V. Neuropathology
VI. Cellular and Animal Models of the Disease
VII. Genotype/Phenotype Correlations/Modifying Alleles
VIII. Treatment
References

Essential tremor (ET) is a progressive neurological disorder that is characterized by a 4- to 12-Hz kinetic tremor in the hands and/or head (Louis, 2001a; Louis and Greene, 2000; Hubble *et al.*, 1989; Findley and Koller, 1987; Critchley, 1949). Patients also may have signs of more widespread cerebellar involvement (e.g., intention tremor, ataxia), (Stolze *et al.*, 2000; Deuschl *et al.*, 2000; Singer *et al.*, 1994) abnormalities referable to the basal ganglia (e.g., rest tremor, subclinical signs of bradykinesia); (Rajput *et al.*, 1993, Cohen *et al.*, 2002) and cognitive deficits (Gasparini *et al.*, 2001; Lombardi *et al.*, 2001). There are both familial and nonfamilial forms of the disease, (Louis and Ottman, 1996; Bain *et al.*, 1994) and the contribution of genetic factors to the etiology of this disease is not known (Louis and Ottman, 1996). In a small number of families with apparently autosomal dominant inheritance, genetic linkage has been established to regions on chromosomes 2p and 3q (Gulcher *et al.*, 1997; Higgins *et al.*, 1997, 1998), but no specific susceptibility gene has been identified yet. Animal models for action tremor, involving the administration of chemicals to laboratory animals, include the harmaline (Elble, 1998) and the penitrem A models (Cavanagh *et al.*, 1998).

## I. PHENOTYPE

ET is the most prevalent adult-onset movement disorder (Louis *et al.*, 1998), and the most common disease characterized by tremor. It is as much as twenty times more prevalent than Parkinson's disease (Rautakorpi *et al.*, 1982). Prevalence estimates in population-based studies have ranged from 0.4–6% (Louis *et al.*, 1998, 2001c; Rautakorpi *et al.*, 1982). Both the incidence and prevalence of the disease increase with advancing age (Rajput *et al.*, 1984; Louis *et al.*, 1995). Although cases may arise in childhood, most incident and prevalent cases of ET are in their 60s or older (Louis *et al.*, 1998; Rautakorpi *et al.*, 1982; Louis *et al.*, 2001c).

The most common finding in ET is a kinetic tremor (i.e., a tremor that occurs during voluntary movement), and this tremor has a frequency, depending on age, between 4 and 12 Hz (Brennan *et al.*, 2002). The tremor occurs when the hands are being used actively to perform voluntary activities such as writing, pouring, eating, and other daily activities. Patients with severe ET also have a postural tremor, which appears when the arms are held outstretched in front of the body (Brennan *et al.*, 2002). Tremor in ET most commonly affects the arms, but it may also affect additional regions, including the head, voice, and occasionally the trunk and lower extremities (Critchley, 1949).

In groups of cases ascertained from adult practices, the proportion of subjects with both arm and head tremor has ranged from 34–53% (Bain *et al.*, 1994; Lou *et al.*, 1991; Hubble *et al.*, 1997; Ashenhurst, 1973).

The tremor probably originates from an abnormality in cerebellar-thalamic loops, suggesting that the disease is a cerebellar outflow disease (Elble, 1998). There is some evidence as well that cerebellar input from the inferior olive may be abnormal in ET (Hallett and Dubinsky, 1993). Aside from the kinetic and postural tremors, which are referable to cerebellar pathways, patients with ET may have signs of involvement of other systems. Two recent studies have demonstrated cognitive abnormalities in ET, and more specifically, problems with verbal fluency, recent memory, working memory, and mental set-shifting, suggesting that there is cortical (especially frontal) involvement (Gasparini *et al.*, 2001; Lombardi *et al.*, 2001). Cerebellar-thalamic loops ultimately involve the motor cortex, perhaps accounting for some of this involvement of the frontal cortex in ET. Several studies have demonstrated postural instability and ataxia in ET, suggesting a more widespread cerebellar involvement (Stolze *et al.*, 2000; Deuschl *et al.*, 2000; Singer *et al.*, 1994), although not as widespread as that occurring in the spinocerebellar ataxias. Finally, as many as 20% of ET cases may have rest tremor, which could be a manifestation of subclinical involvement of the basal ganglia (Cohen *et al.*, 2002).

In the past, the tremor of ET was often labeled "benign," but we now know that the tremor usually is progressive (Critchley, 1949), producing disabilities with basic daily activities such as eating, writing, body care, and driving (Bain *et al.*, 1994; Rautakorpi, 1978). More than 90% of clinic patients report disability (Louis *et al.*, 2001a), and severely affected end-stage patients are physically unable to feed or dress themselves. In these, the tremor prevents any normal activity, resulting in a substantial loss of independence and even incapacitation (Koller, 1986). Between 15 and 25% of clinic patients are forced to retire prematurely, and 60% choose not to apply for a job or promotion because of uncontrollable shaking (Bain *et al.*, 1994; Rautakorpi, 1978).

The etiologies of ET are both genetic and nongenetic (Tanner *et al.*, 2001; Louis, 2001b). Differences between the genetic and nongenetic forms of the disease have not been identified, other than a possibly younger age of onset in genetic forms of the disease (Larsson and Sjogren, 1960). In this chapter, the primary focus is the genetic form of the disease, or hereditary ET.

One important question is the magnitude of a genetic contribution to the etiology of this disease on a population level. It is commonly stated in the literature that 50% of ET cases are attributed to genetic causes. This estimate appears to be based on the proportion of cases who report a family history. However, estimates of the proportion with a family history range from as low as 17% to as high as 100% (Table 32.1; Louis and Ottman, 1996).

With this degree of variability, it is difficult to get a sense of the true proportion, and the commonly cited value of 50% appears highly questionable. Furthermore, the proportion with a positive family history does not accurately reflect the proportion with a genetic etiology. Most of the studies examining this question have not enrolled control subjects. Many of the positive family histories could be explained by chance co-occurrence of a highly prevalent disorder rather than a genetic predisposition for tremor (Louis and Ottman, 1996; Louis *et al.*, 1997b). Studies suggest that as many as 18% of families may contain an affected individual, even if ascertained through an unaffected control (Louis *et al.*, 2001c).

Additional methodologic problems with previous studies raise further doubt about the proportion of genetic cases. Most studies have reported the percentage of probands with a family history rather than the proportion of at-risk relatives who are affected with ET, making it impossible to test consistency with genetic models or to control for characteristics of the relatives (e.g., age, sex, history of environmental exposures; Louis and Ottman, 1996). Most studies have selected probands from clinical care settings (hospitals, doctors' offices, clinics) rather than from the community. ET cases who are seen in clinics and doctors' offices probably represent a very small proportion of all ET cases in the population (as few as 0.5%; Larsson and Sjogren, 1960), and these cases are five times more likely to report a positive family history than are those who never make it to clinics (Louis *et al.*, 2001b). Clinic populations might be self-selected to over-represent familial and genetic forms of ET, and possibly autosomal dominant forms as well (Louis *et al.*, 2001b). Finally, most studies have

TABLE 32.1 Proportion of Cases with a Positive Family History of ET

| Study | % of Cases with a positive family history of ET |
|---|---|
| Kulcke | 17.4 |
| Hornabrook and Nagurney | 18.0 |
| Salemi *et al.* | 35.5 |
| Critchley | 38.1 |
| Rajput *et al.* | 38.7 |
| Martinelli *et al.* | 43.2 |
| Aiyesimoju *et al.* | 60.0 |
| Lou and Jankovic | 62.6 |
| Rautakorpi *et al.* | 70.0 |
| Mengano *et al.* | 79.0 |
| Latinen | 79.3 |
| Busenbark *et al.* | 96.0 |
| Larsson and Sjogren | 100.0 |

Modified from Louis and Ottman (1996). *Neurology* **46**, 1200–1205.

obtained family history information by interviewing the probands rather than by examining the relatives themselves, and the sensitivity of probands' reports may be as low as 16% (Louis et al., 1999).

In the Washington Heights-Inwood Genetic Study of Essential Tremor, a population-based family study of ET that enrolled relatives of ET cases and relatives of control subjects, we found that a first-degree relative of an ET case was 4.7 times more likely to have ET than was a first-degree relative of a control subject (Louis et al., 2001c). In addition, the magnitude of increased risk in relatives of ET patients versus. controls was greater in relatives of ET cases with onset ≤ 50 years than in relatives of those with older onset (relative risk, RR = 10.38 versus 4.82; Louis et al., 2001c).

In a recent twin study (Tanner et al., 2001), 3 of 5 (60%) monozygotic twins were concordant for ET, compared with only 3 of 8 (27%) dizygotic twins. Although concordance in monozygotic twins was approximately two times that in dizygotic twins, the monozygotic concordance was not 100%.

## II. GENE

Specific genes for ET have not yet been identified. Given the high prevalence of this disorder, one would expect that multiple genetic loci may contribute to this disease. Linkage has been reported on two different chromosomes (3q13 and 2p22; Gulcher et al., 1997; Higgins et al., 1997), suggesting that ET may be genetically heterogeneous.

Gulcher et al. (1997) reported the results of a genome scan for familial ET (FET) genes in 16 Icelandic families containing 75 affected individuals. The average age of family members was 49.9 years. The scan revealed significant evidence for linkage to chromosome 3q13 under either an autosomal dominant model assuming 1% disease prevalence and 90% penetrance (lod score = 3.71), or a "nonparametric" model (NPL score = 4.70, $p = 6.4 \times 10^{-6}$). In that study, the average age of onset was 26.7 years, and all patients had had bilateral postural tremor with or without kinetic tremor of the hands for at least five years. Neither the proportion with head tremor nor the response to medications was reported.

In that same year, Higgins et al. (1997) reported the results of a linkage analysis in a large American family of Czech descent (Fig. 32.1). Data were available on 67 family members, among whom 18 were affected. Evidence was obtained for linkage to chromosome 2p22–25 with a maximum lod score of 5.92. For this study, they assumed a disease prevalence of 1%, an autosomal dominant model of inheritance, and a penetrance of 100%. Higgins et al. (1998) subsequently reported that the same locus may be responsible for ET in other families, although the contribution of this or other loci to disease etiology on a population level is not known. In that study (Higgins et al., 1998), all patients had bilateral postural tremor with 2- to 4-cm excursions in at least one arm. Neither the proportion with head tremor nor the response to medications was indicated.

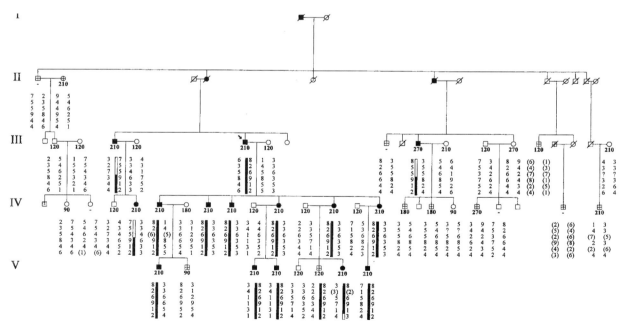

**FIGURE 32.1** Linkage to chromosome 2p22–25 in a large pedigree with familial tremor. From Higgins et al. (1997). *Mov. Disord.* **12**, 859–864. With permission.)

The mean age of affected individuals was 50 years, and the mean age of onset was 32 years.

One interesting question is whether the mode of inheritance of ET, on a population level, is autosomal dominant, as it seems to be in these pedigrees (Gulcher et al., 1997; Higgins et al., 1997). There is evidence that this pattern of inheritance may not typify all families with ET. First, these large families with multiple affected individuals over several generations have been difficult to identify for genetic linkage studies. Second, some of the data from the Washington Heights-Inwood Genetic Study of Essential Tremor may not be consistent with an autosomal dominant model either. In a simple autosomal dominant model, the RR in second-degree relatives is expected to be one-half of the RR in first-degree relatives. For example, if the RR among first-degree relatives is 4.7, then the RR among second-degree relatives would be expected to be one-half of that (i.e., 2.35). We found a relative risk of 4.7 in first-degree relatives, but only 0.9 in second-degree relatives (although confidence intervals did overlap). These findings could be consistent with a model involving multiple interacting genes (epistasis; Weiss et al., 1982). Third, as in other highly complex diseases, genetic heterogeneity is likely, with variable mode of inheritance across families (Weiss et al., 1982).

## III. DIAGNOSTIC AND ANCILLARY TESTS

The diagnostic approach in patients with ET includes a history, physical examination, and laboratory tests (Louis, 2001a; Louis and Greene, 2000). Because a gene for ET has not yet been identified, there is no diagnostic genetic test for ET. Moreover, the utility of such a test would be limited by a variety of factors. First, it is unknown what proportion of ET cases have an important genetic contribution, but this proportion may be much lower than 50% (Louis and Ottman, 1996; Louis, 2001b). Second, in those with a major genetic effect, multiple genes may be involved, either interacting within individuals, or as alternative etiologies in different patients. Third, even in those with a major genetic effect, an as-yet-identified environmental factor may be required to trigger onset of symptoms.

The following is an outline of the diagnostic approach to a patient who is suspected of having ET. During the history, information on the age of onset, type, and progression of the tremor should be assessed, along with family history information. Caffeine, cigarettes, and several medications, including lithium, prednisone, thyroxine, asthma inhalers, valproate, and selective serotonin re-uptake inhibitors commonly result in enhanced physiological tremor, which can resemble ET, and hence a complete inventory of all current medications, as well as caffeine and smoking habits, is important in order to exclude these as the cause of tremor. Patients with tremor due to other disorders usually have symptoms of these disorders at presentation. Thus, patients with hyperthyroidism may complain of diarrhea or weight loss, and patients with Parkinson's disease may note a loss of normal facial expression or a change in normal arm swing.

During the physical examination, the clinician should carefully assess the characteristics of the tremor. Although many patients with Parkinson's disease manifest a mild postural tremor (Koller et al., 1989), rest tremor is also present, and affects approximately 85% (Louis et al., 1997a) of patients with autopsy-proven Parkinson's disease. Approximately 20–30% of patients with dystonia have a postural tremor that resembles ET, but on closer inspection, the tremor differs from that of ET (e.g., the tremor is often irregular and jerky rather than regularly oscillatory, and there may be a null point, which is a hand or arm position that they can find which will result in temporary tremor resolution). If symptoms or signs of hyperthyroidism are present, then thyroid function tests should be performed.

In any patient with action tremor who is under the age of 40 years, the possibility of Wilson's disease should be explored with a serum ceruloplasmin. The ceruloplasmin is low (<20 mg/dl) in 95% of patients with Wilson's disease (Scheinberg and Sternlieb, 1984). If other clinical features suggestive of Wilson's disease are present (e.g., dysarthria, dystonia, parkinsonism), then a careful slit lamp examination of the eye by an experienced ophthalmologist should be performed. Kayser-Fleischer rings on Descemet's membrane are detectible in 99.3% of Wilson's patients with neurological abnormalities (Walshe and Yealland, 1992).

Although positron emission tomography (PET) has revealed a relative increase in cerebellar and red nuclear blood flow in some ET cases (Colebatch et al., 1990; Hallett and Dubinsky, 1993; Wills et al., 1994; Boecker et al., 1996; Jenkins et al., 1993), its diagnostic validity has not been established. Quantitative computerized tremor analysis, with accelerometers attached to the arms, is available at some tertiary care centers, and it may guide the clinician in distinguishing ET from other tremors, but its diagnostic validity has not been established. At present, the diagnosis of ET is clinical; there are no serological, radiological, or pathological markers. Because of the absence of a gold standard, one cannot assess the validity of clinical diagnoses. However, one can assess the agreement between clinicians using the same clinical criteria. High reliability has been established for one set of clinical criteria (Louis et al., 2001c).

## IV. NEUROIMAGING

Standard MRI has not demonstrated structural radiological abnormalities in ET cases, although MRI is not used routinely in the initial diagnosis or subsequent treatment of ET patients, so that most ET patients never have MRI studies. One study, using high-resolution proton density- and T2-weighted images of 12 ET cases and 15 control subjects, showed no structural abnormalities of the brain (Bucher et al., 1997). Volumetric analyses of the cerebellum and other pertinent structures, however, were not performed. Although the pathology of ET has been limited, several of the postmortem cases have shown extensive Purkinje cell loss (Mylle and Van Bogaert, 1940, 1948; Hassler, 1939), and at least one of these, a notable amount of cerebellar cortical atrophy. One functional MRI (fMRI) study demonstrated bilateral cerebellar activation (Bucher et al., 1997), PET imaging studies have documented increased regional blood flow in the cerebellar hemispheres and red nucleus of ET cases during certain activation procedures (arm extension), and also occasionally at rest (Colebatch et al., 1990; Hallett and Dubinsky, 1993; Wills et al., 1994; Boecker et al., 1996; Jenkins et al., 1993). These studies suggest that cerebellar circuitry is involved in the initiation of tremor, and hence, blood flow to these regions is enhanced. An unusual feature of this cerebellar activation is that it is both ipsilateral and contralateral to the tremulous arm. In normal individuals without ET, the cerebellar hemispheres modulate movement predominantly in the ipsilateral and, to a far smaller extent, the contralateral limbs (Grodd et al., 2001). This suggests that in ET, the tremor in each arm is due to an abnormal pattern of input from both cerebellar hemispheres. In contrast to fMRI or PET, which assess cerebral blood flow, [$^1$H] magnetic resonance spectroscopic imaging (MRSI) measures the concentrations of several intracellular metabolites. Among these is $N$-acetylaspartate (NAA), often expressed as a ratio to total creatine (tCR). NAA/tCR levels are decreased in the setting of neuronal damage or cell death (Tedeschi et al., 1996; Constantinidis, 1998; Ross and Bluml, 1999), and as such, serve as a marker for disturbed neuronal integrity. Using [$^1$H] MRSI, we demonstrated a reduction in NAA/tCR within the cerebellar cortex in ET cases, suggesting that there is cell damage or cell death in this region of the brain in ET (Louis et al., 2002). We demonstrated that the reduction in cerebellar cortical NAA/tCR seen in cases of successively older ages seemed to exceed the age-dependent decline in cerebellar cortical NAA/tCR seen in normal control subjects (Louis et al., 2002). This is consistent with the view that damage to or loss of cerebellar neurons progresses at an accelerated rate in ET cases compared to the normal age-dependent rate of loss observed in control subjects. This suggests that, like Parkinson's disease and Alzheimer's disease, ET may be neurodegenerative. In other words, the disease may be the result of a progressive accelerated loss of a specific neuronal population.

## V. NEUROPATHOLOGY

Autopsy studies are particularly important in ET since there are no natural animal models for this condition. Despite the widespread occurrence of ET, postmortem studies are limited to 15 cases, with many of these published 50–100 years ago (Mylle and Van Bogaert, 1940, 1948; Hassler, 1939; Herskovitz and Blackwood, 1969; Bergamasco, 1907; Lapresle et al., 1974; Rajput et al., 1991; Frankl-Hochwart, 1903; Boockvar et al., 2000). No consistent pathological abnormality was reported in these studies. There was, however, occasional loss of cerebellar Purkinje cells in one brain and more marked loss in three others, however, quantitative cell counts were not performed (Mylle and Van Bogaert, 1940, 1948; Hassler, 1939; Herskovitz and Blackwood, 1969). With normal aging, Purkinje cells die at rate of 2.5–5% per decade (Ellis, 1920; Hall et al., 1975), and these four cases ranged in age at the time of death from 61–80 years (mean = 70.3), making the loss of Purkinje cells difficult to interpret in the absence of age-matched control brains for comparison. Formal quantitative studies of cerebellar Purkinje or olivary neurons were not performed in any, nor were control brains studied for comparison. There was no immunohistochemistry. Moreover, in many of the cases, the diagnosis of ET is highly questionable because patients also had other involuntary movements (chorea, athetosis, parkinsonism) or were members of families in which individuals variably expressed a mixed phenotype of action tremor with or without parkinsonism. A concerted effort is needed to collect postmortem tissue on a larger number of ET cases, to study these cases using a standardized approach involving quantitative cell counts and immunohistochemistry, and to compare this tissue to that of control subjects of similar age.

## VI. CELLULAR AND ANIMAL MODELS OF DISEASE

There are no cellular models for ET. There are no transgenic or knock-out models of the disease. There are, however, two chemically induced animal models for ET, the harmaline model (Elble, 1998) and the penitrem A model (Cavanagh et al., 1998).

Acute exposure to the tremorogenic beta-carbolines, such as harmaline, results in an acute to subacute (hours to days) generalized and intense action tremor in a broad range of laboratory species including mice, cats, and monkeys (O'Hearn and Molliver, 1997). In human volunteers exposed to intravenous doses of 150–200 mg of harmine, neurological effects, including an acute coarse tremor, become apparent (Pennes and Hoch, 1957; Lewin, 1928). In animals, the tremor shares many features with ET including clinical features (Milner *et al.*, 1995; Trouvin *et al.*, 1987; Fuentes and Longo, 1971), pharmacological responsiveness to benzodiazepines, alcohol, and barbiturates, which facilitate gamma amino butyric acid (GABA)-mediated inhibitory neurotransmission (Cross *et al.*, 1993; Sinton *et al.*, 1989; Rappaport *et al.*, 1984), and underlying pathogenesis in cerebellar and GABAergic systems. The tremor-producing property of the beta-carboline alkaloids is related to their ability to enhance rhythmic firing in olivocerebellar neurons with secondary Purkinje cell degeneration (O'Hearn and Molliver, 1997; Sinton *et al.*, 1989).

Penitrem A, which is produced by the fungus *Penicillium crustosum*, is one of a group of tremor-producing agents produced by several fungi species (Cavanagh *et al.*, 1998). Within 10 minutes of an intraperitoneal injection, laboratory animals develop severe generalized 6–8 Hz tremors and ataxia that peaks within 6 hours but can persist to some extent for as long as 2 days. Despite widespread loss of Purkinje cells, particularly in the vermis and paravermis, the animals become normal within one week of the injection (Cavanagh *et al.*, 1998).

## VII. GENOTYPE/PHENOTYPE CORRELATIONS/ MODIFYING ALLELES

Intrafamilial variability in age of onset and anatomical distribution of tremor has been reported, but has not been investigated systematically. A study in rural Sweden (Larsson and Sjogren, 1960) is the most helpful in this regard. In that study, age at onset within families varied considerably (e.g., onset age ranging within some families between 20 and 60 years). There was variable involvement of the head, arm, tongue, legs, and trunk within many of the families as well. One group reported younger age of disease onset in successively younger generations (Higgins *et al.*, 1997, 1998), and suggested that there might be genetic anticipation in ET. However, true age at onset, rather than age at initial recognition, may be difficult to ascertain because the disease has an insidious onset. Once families have come to medical attention, there is greater potential for recognition of mild (i.e., young) cases.

Although the penetrance of ET is generally considered to be complete by age 65–70 years, there are few data. One study in Sweden (Larsson and Sjogren, 1960) suggested that the penetrance was complete at age 70 years because the highest registered age of onset in their cohort was 70 years, although it is apparent from their data that there were many unaffected relatives who were over the age of 70, and it is not known how many of these carried a genetic predisposition for ET. Perhaps the most thorough study of penetrance (Bain *et al.*, 1994) suggested that the penetrance was complete by age 65 years because 46% (i.e., nearly one-half) of relatives of familial ET cases had developed tremor by that age. Assuming an autosomal dominant model of inheritance and complete penetrance, 50% of first-degree relatives of familial cases would be expected to develop the disease. However, these calculations do not account for non-genetic causes of ET among relatives. We reported that 11.1% of control subjects' first-degree relatives had developed ET by age 60 years, and 22.2% by age 80 years (Louis *et al.*, 2001c). Nongenetic causes of ET should be similar in relatives and cases and relatives of controls. Therefore, in an autosomal dominant model with complete penetrance, more than 50% of first-degree relatives of familial cases would be expected to develop the disease by age 80 years.

Data from the Washington Heights-Inwood Genetic Study of Essential Tremor indicate that subclinical ET may be present and penetrance may not be complete even among older relatives (Louis *et al.*, 2001d). Data were analyzed on 201 case relatives and 212 control relatives who did not meet diagnostic criteria for ET. All subjects were examined. Clinically detectable tremor was present in 96.0% of these case relatives and 97.6% of control relatives, and tremor severity (as measured by a total tremor score) was higher in first-degree case relatives than in first-degree control relatives. Among first-degree relatives who were ≥ 60 years of age (mean age among case relatives = 72.5 years and mean age among control relatives = 71.8 years), a larger proportion of case relatives had higher total tremor scores. One possible explanation for the observed difference in distribution of tremor scores among case and control relatives is that relatives of ET cases have mild partially expressed forms of ET. The mean total tremor score was higher in relatives of cases than in relatives of controls among first-degree relatives but not among more distantly related (second-degree) relatives, suggesting that a genetic predisposition for tremor may have contributed to the observed difference.

## VIII. TREATMENT

The treatment of ET includes pharmacotherapy as well as surgery. Pharmacotherapy may be used to improve functional disability or to reduce the embarrassment that may be associated with ET. In mild cases without dysfunction or embarrassment, medications are not indicated. Surgery is indicated for severe cases that are refractory to medications. The following discussion focuses primarily on results from double-blind placebo-controlled trials.

Several issues deserve special highlighting. First, the ultimate evidence of efficacy is a reduction in functional impairment rather than a reduction in tremor amplitude. Second, a large proportion of patients, perhaps as many as 25–55%, are non-responders to multiple medications. Unfortunately, the factors that predict response to particular medications have not been identified. This can lead to frustration among patients and loss of confidence in their treating physician. Several factors may account for this high nonresponse rate. There is evidence that ET is not a homogeneous condition. There may be different clinical subtypes (young-onset ET, or ET with head tremor) that may differ with regard to their underlying etiology and pathogenesis (Louis et al., 2000). There is probably genetic heterogeneity (Gulcher et al., 1997; Higgins et al., 1997, 1998). It is possible that some subtypes may respond to a particular medication while others may not. In addition, little is known about the underlying pathophysiology of ET. Most current medications were discovered due to serendipity (Winkler and Young, 1971; Obrien et al., 1981) rather than an understanding of the disease mechanisms. The metabolic pathways that are involved in ET are not clear, although GABAergic and catecholamine pathways are a possibility (Louis, 1999; Rajput et al., 2001). If and where in these metabolic pathways the problem lies is not known. This information would be useful in terms of conceiving therapeutic interventions aimed at specific points in the cascade of biochemical events that may be occurring in ET. One final issue that deserves highlighting is that the patients who do respond to medications are often only partial responders, and the tremor is rarely reduced to asymptomatic levels. The treatments are outlined in Table 32.2.

Propranolol and primidone are the first line agents in the treatment of ET (Louis, 2001a). Peripheral β-adrenergic receptors (Jefferson et al., 1979) most probably mediate the effects of β-adrenergic blocking agents, although there has been some discussion as to whether central mechanisms may be involved as well. Several studies have demonstrated that propranolol, given in doses of 120 mg/day or more, results in a significant reduction in tremor compared with placebo (Tolosa and Loewenson, 1975), and subjectively, 45–75% of study patients report that propranolol is more effective than the placebo in reducing their tremor. Although propranolol is generally well-tolerated, relative contraindications include asthma, congestive heart failure, diabetes mellitus, and atrioventricular block. Propranolol is a nonselective β-adrenergic receptor antagonist, and it is more effective than antagonists with relatively selective β1 activity (Jefferson et al., 1979). Trials of β1-selective agents such as atenolol and metroprolol have shown mixed results (Jefferson et al., 1979; Dietrischson and Epsen, 1981). Propranolol-LA is as effective as conventional propranolol (Cleeves and Findley, 1988).

Primidone, which is an anticonvulsant medication, is metabolized to phenylethylmalonamide and phenobarbitone. Phenylethylmalonamide itself has no therapeutic effect. While the barbiturate metabolite may contribute to the sustained therapeutic effect, the parent compound itself (primidone) is thought to mediate most of the acute and sustained effect (Findley et al., 1985; Sasso et al., 1990; Gorman et al., 1986). Primidone is superior to phenobarbital in reducing tremor (Sasso et al., 1988). In doses of up to 750 mg/day, primidone resulted in a significant reduction in tremor compared with placebo effect (Findley et al., 1985; Sasso et al., 1990; Gorman et al., 1986), however, tolerability is a common problem. Even at starting doses as low as 62.5 mg/day, an acute toxic reaction, consisting of nausea, vomiting, or ataxia has been reported in 22.7–72.7% (Sasso et al., 1990) of patients, requiring discontinuation of primidone in approximately 20% of patients in some studies (Findley et al., 1985).

The two front line agents have been compared to one another in several studies. Primidone, at a dose of 750 mg/day, was compared with propranolol, at a dose of 120 mg/day, and each was significantly better than placebo (Gorman et al., 1986); however, neither has been conclusively shown to be superior to the other. The mean reduction in tremor amplitude was 75.6% while patients were taking primidone compared with 59.7% on propranolol, but given the modest sample size of 14 cases, this difference was not significant (Gorman et al., 1986). Interestingly, when the study patients were asked which of the two medications they preferred, a larger proportion chose primidone (Gorman et al., 1986). In another study, a 60% mean improvement occurred in patients taking primidone (250 mg/day in 9 patients) versus a 35% improvement in 22 patients on maximum effective dose of propranolol (p value not reported; Koller and Royse, 1986). While initial tolerability is a problem with primidone, one study provides tentative evidence that long-term tolerability of primidone is superior to that of propranolol. In a study of 25 ET patients, acute adverse reactions occurred in 8% with propranolol and 32% with primidone; however, what the authors referred to as "significant" side effects after one year occurred in 0% with primidone compared with 17% taking propranolol (Koller and Vetere-Overfield, 1989).

TABLE 32.2 Medications Used to Treat ET

| Agent | Usual therepeutic dose | Side effects |
|---|---|---|
| Propranolol | 160–320 mg per day | Fatigue, impotence, headache, breathlessness, bradycardia, depression |
| Primidone | 62.5–1000 mg per day | Sedation, nausea, vomiting |
| Gabapentin | 1200–3600 mg per day | Drowsiness, fatigue, slurred speech, imbalance, nausea, dizziness |
| Topiramate | 75–400 mg per day | Appetite suppression, weight loss, parethesias |
| Alprazolam | 0.75–2.75 mg per day | Sedation, fatigue |
| Nimodipine | 120 mg per day | Orthostatic hypotension |
| Theophylline | 150–300 mg per day | None reported |

Modified from Louis, E.D. (2001a). N. Engl. J. Med. 345, 887–891.

Several other agents have been used with variable efficacy in the treatment of ET. Among these is gabapentin, an anticonvulsant medication that is structurally similar to the inhibitory neurotransmitter GABA. There is some evidence that there is a disturbance of the GABAergic system in ET (Louis, 1999). There have been three gabapentin trials, enrolling in combination a total of 54 patients. In two of the three, gabapentin (doses ranging from 1200–3600 mg/day) resulted in a significant reduction in tremor compared with placebo, and in one of the two, its effect was similar to that of propranolol (Gironell et al., 1999; Pahwa et al., 1998; Ondo et al., 2000). Gabapentin is generally well-tolerated.

In a double-blind placebo-controlled trial, topiramate was administered to 24 patients at a single center (Connor, 2000). The imputed intent-to-treat analyses using baseline observations carried forward for discontinuations demonstrated a significant effect of topiramate in most but not all outcomes. While promising, the results were published in abstract form. The proportion of patients who demonstrated a reduction in tremor amplitude and the final tremor scores are not documented. These results will need to be reproduced at other centers.

Benzodiazepines potentiate the effect of GABA by binding to the $GABA_A$ receptor. Although a variety of benzodiazepines, including diazepam and clonazepam, are used to treat ET, alprazolam is the only benzodiazepine shown to be effective in ET in controlled trials. In one trial, alprazolam (Huber and Paulson, 1988); (dose ranging from 0.75–2.75 mg/day) resulted in a significant reduction in tremor compared with placebo, and 75% of patients demonstrated at least some improvement. However, one problem was that drowsiness or sedation occurred in 50% of the patients. Another agent, clonazepam did not result in a significant reduction in tremor severity compared with placebo (Thompson et al., 1984). One problem with the benzodiazepines in general as a treatment for ET is that their antitremor effect often comes at a dose that is associated with sedation and/or cognitive slowing.

Calcium channel blockers have had variable success in treating ET. Trials of flunarizine, given at a dose of 10 mg/day, have reported mixed results. In one trial, flunarizine resulted in a significant reduction in tremor compared with placebo; 13 of 15 patients improved; however, none of the patients improved in a second trial (Curran and Lang, 1993). Moreover, flunarazine is not available in the United States. In one trial (Biary et al., 1995), nimodipine, at a daily dose of 30 mg, resulted in a significant reduction in tremor compared with placebo; 8 of 15 patients improved. Some calcium channel blockers may actually worsen tremor. Tremor was acutely worsened by 71% on average in a study of eight patients (Topaktas et al., 1987) on a single oral dose of 10 mg of nifedipine.

Intramuscular botulinum toxin injections may reduce tremor by producing weakness or by blockade of gamma motor efferents and muscle spindle efferents. Patients with ET received one to two injections of botulinum toxin or placebo into the intrinsic muscles of the dominant hand in one trial (Jankovic et al., 1996). Although there was a significant reduction in tremor amplitude, no significant improvement in function was observed. Also, the complication of muscle weakness made a truly blinded study impossible (Jankovic et al., 1996).

Action tremor may be mediated by neuronal loops which pass from the cerebellum to the cortex by way of the ventral intermediate nucleus of the thalamus. Two surgical approaches have been used for tremor reduction. These involve continuous deep-brain stimulation through an electrode implanted in the ventral intermediate nucleus of the thalamus and surgical lesioning of the ventral intermediate nucleus of the thalamus (i.e., thalamotomy). The two methods were compared in a prospective randomized single-blinded study (Schuurman et al., 2000). Both procedures were equally effective in reducing tremor (Schuurman et al., 2000). All patients had moderate-to-severe tremor at baseline and a complete resolution of tremor six months after treatment (Schuurman et al., 2000). Thalamic stimulation was reported to result in greater improvement in self-reported measures of function. Adverse events were significantly more common in the thalamotomy group. These events included cognitive deterioration, dysarthria, and gait or balance disturbances (Schuurman et al., 2000). Thalamic stimulation is the surgery of choice because there are fewer adverse events and the clinician has the ability to adjust the stimulator settings during follow-up care (Hariz et al., 2002).

In summary, treatment of ET should be reserved for patients who experience functional disability or embarrassment. Both primidone and propranolol are superior to placebo, and there is some evidence that primidone may be more effective than propranolol. However, this needs to be studied further. The long-term tolerability of primidone may be superior, and therefore, primidone is a reasonable first choice, and if not tolerated, then propranolol is the second choice. Gabapentin and benzodiazepines are alternatives, although the latter may result in sedation at the doses required to treat tremor. If the patient does not respond to medications and is significantly disabled, then unilateral implantation of a deep brain thalamic stimulator is reasonable. The stimulator should be implanted in the thalamus contralateral to the more functionally disabled arm.

## References

Ashenhurst, E. M. (1973). The nature of essential tremor. *CMAJ* **109**, 876–878.

Bain, P. G., Findley, L. J., Thompson, P. D., Gresty, M. A., Rothwell, J. C., Harding, A. E., and Marsden, C. D. (1994). *Brain* **117**, 805–824.

Bergamasco, I. (1907). Intorno ad un caso di tremore essenziale simulant in parte il quadro della sclerosi multipla. *Riv. Pat. Nerv. Ment.* **115**, 80–90.

Biary, N., Bahou, Y., Sofi, M. A., Thomas, W., and Al Deeb, S. M. (1995). The effect of nimodipine on essential tremor. *Neurology* **45**, 1523–1525.

Boecker, H., Wills, A. J., Ceballos-Baumann, A., Samuel, M., Thompson, P. D., Findley, L. J., and Brooks, D. J. (1996). The effect of ethanol on alcohol-responsive essential tremor: A positron emission tomography study. *Ann. Neurol.* **36**, 650–658.

Boockvar, J., Telfeian, A., and Baltuch, G. H. (2000). Long-term deep brain stimulation in a patient with essential tremor: Clinical response and postmortem correlation with stimulator termination sites in ventral thalamus. *J. Neurosurg.* **93**, 140–144.

Brennan, K. C., Jurewicz, E., Ford, B., Pullman, S. L., and Louis, E. D. (2002). Is essential tremor predominantly a kinetic or a postural tremor? A clinical and electrophysiological study. *Mov. Disord.* **17**, 313–316.

Bucher, S. F., Seelos, K. C., Dodel, R. C., Reiser, M., and Oertel, W. H. (1997). Activation mapping in essential tremor with functional magnetic resonance imaging. *Ann. Neurol.* **41**, 32–40.

Busenbark, K. L., Nash, J., Nash, S., Hubble, J. P., and Koller, W. C. (1991). Is essential tremor benign? *Neurology* **41**, 1982–1983.

Cavanagh, J. B., Holton, J. L., Nolan, C. C., Ray, C. C., Naik, D. E., and Mantle, P. G. (1998). The effects of the tremorogenic mycotoxin penitrem A on the rat cerebellum. *Vet. Pathol.* **35**, 53–63.

Cleeves, L., and Findley, L. J. (1988). Propranolol and propranolol-LA in essential tremor: a double blind comparative study. *J. Neurol. Neurosurg. Psychiatry* **51**, 379–384.

Cohen, O., Pullman, S., Jurewicz, E., Watner, D., and Louis E. D. (2002). Rest tremor in essential tremor patients: Prevalence, clinical correlates, and electrophysiological characteristics. *Neurology* **58**(suppl. 3), A253.

Colebatch, J. G., Findley, L. J., Frackowiak, R. S. J., Marsden, C. D., and Brooks, D. J. (1990). Preliminary report: Activation of the cerebellum in essential tremor. *Lancet* **336**, 1028–1030.

Connor, G. S. (2000). Efficacy of topiramate in treatment of essential tremor: a randomized, double-blind, placebo-controlled, cross-over study. *Ann. Neurol.* **48**, 486.

Constantinidis, I. (1998). Magnetic resonance spectroscopy and the practicing neurologist. *Neurologist* **4**, 77–98.

Critchley M. (1949). Observations on essential tremor (heredofamilial tremor). *Brain* **72**, 113–139.

Cross, A. J., Misra, A., Sandilands, A., Taylor, M. J., and Green, A. R. (1993). Effect of chlormethiazole, dizocilpine and pentobarbital on harmaline-induced increase of cerebellar cyclic GMP and tremor. *Psychopharmacology* **111**, 96–98.

Curran, T., and Lang, A. E. (1993). Flunarizine in essential tremor. *Clin. Neuropharmacol.* **16**, 460–463.

Deuschl, G., Wenzelburger, R., Loffler, K., Raethjen, J., and Stolze, H. (2000). Essential tremor and cerebellar dysfunction. Clinical and kinematic analysis of intention tremor. *Brain* **123**, 1568–1580.

Dietrischson, P., and Espen, E. (1981). Effects of timolol and atenolol on benign essential tremor: Placebo-controlled studies based on quantitative recording. *J. Neurol. Neurosurg. Psychiatry* **44**, 677–683.

Elble, R. J. (1998). Animal models of action tremor. *Mov. Disord.* **13** (Suppl. 3), 35–39.

Ellis, R. S. (1920). Norms for some structural changes in the human cerebellum from birth to old age. *J. Comp. Neurol.* **32**, 1–33.

Findley, L. H., Cleeves, L., and Calzetti, S. (1985). Primidone in essential tremor of the hands and head: A double blind controlled clinical study. *J. Neurol. Neurosurg Psychiatry* **48**, 911–915.

Findley, L. J., and Koller, W. C. (1987). Essential tremor: a review. *Neurology* **37**, 1194–1197.

Frankl-Hochwart. (1903). "La degenerescence hepato-lenticulaire (maladie de Wilson, pseudo-sclerose)." Masson et Cie, Paris.

Fuentes, J. A., and Longo, V. G. (1971). An investigation on the central effects of harmine, harmaline and related B-carbolines. *Neuropharmacology* **10**, 15–23.

Gasparini, M., Bonifati, V., Fabrizio E. *et al.* (2001). Frontal lobe dysfunction in essential tremor. A preliminary study. *J. Neurol.* **248**, 399–402.

Gironell, A., Kulisevsky, J., Barbanoj, M., Lopez-Villegas, D., Hernandez, G., and Pascual-Sedano, B. (1999). A randomized placebo-controlled comparative trial of gabapentin and propranolol in essential tremor. *Arch. Neurol.* **56**, 475–480.

Gorman, W. P., Cooper, R., Pocock, P., and Campbell, M. J. (1986). A comparison of primidone, propranolol, and placebo in essential tremor, using quantitative analysis. *J. Neurol. Neurosurg. Psychiatry* **49**, 64–68.

Grodd, W., Halsmann, E., Lotze, M., Wildgruber, and Erb, M. (2001). Sensorimotor mapping of human cerebellum: fMRI evidence of somatotopic organization. *Hum. Brain Map* **13**, 55–73.

Gulcher, J. R., Jonsson, P., Kong, A. *et al.* (1997). Mapping of a familial essential tremor gene, FET1, to chromosome 3q13. *Nat. Genet.* **17**, 84–87.

Hall, T. C., Miller, A. K. H., and Corsellis, J. A. N. (1975). Variations in the human Purkinje cell population according to age and sex. *Neuropathol. Appl. Neurobiol.* **1**, 267–292.

Hallet, M., and Dubinsky, R. M. (1993). Glucose metabolism in the brain of patients with essential tremor. *J. Neurol. Sci.* **114**, 45–48.

Hariz, G. M., Lindberg, M., and Bergenheim, A. T. (2002). Impact of thalamic deep brain stimulation on disability and health-related quality of life in patients with essential tremor. *J. Neurol. Neurosurg. Psychiatry* **72**, 47–52.

Hassler, R. (1939). Zur pathologischen anatomie des senilen und des parkinsonistischen Tremor. *J. Psychol. Neurol.* **49**, 193–230.

Herskovitz, E., and Blackwood, W. (1969). Essential (familial, hereditary) tremor: a case report. *J. Neurol. Neurosurg. Psychiatry* **32**, 509–511.

Higgins, J. J., Loveless, J. M., Jankovic, J., and Patel, P. I. (1998). Evidence that a gene for essential tremor maps to chromosome 2p in four families. *Mov. Disord.* **13**, 972–977.

Higgins, J. J., Pho, L. T., and Nee, L. E. (1997). A gene (ETM) for essential tremor maps to chromosome 2p22-p25. *Mov. Disord.* **12**, 859–864.

Hubble, J. P., Busenbark, K. L., and Koller, W. C. (1989). Essential tremor. *Clin. Neuropharmacol.* **12**, 453–482.

Hubble, J. P., Busenbark, K. L., Pahwa, R., Lyons, K., and Koller, W. C. (1997). Clinical expression of essential tremor: Effects of gender and age. *Mov. Disord.* **12**, 969–972.

Huber, S. J., and Paulson, G. W. (1988). Efficacy of alprazolam for essential tremor. *Neurology* **38**, 241–243.

Jankovic, J., Schwartz, K., Clemence, W., Aswad, A., and Mordunt, J. (1996). A randomized, double-blind, placebo-controlled study to evaluate botulinum toxin type A in essential hand tremor. *Mov. Disord.* **11**, 250–256.

Jefferson, D., Jenner, P., and Marsden, C. D. (1979). B-Adrenoreceptor antagonists in essential tremor. *J. Neurol. Neurosurg. Psychiatry* **42**, 904–909.

Jenkins, I. H., Bain, P. G., Colebatch, J. G. *et al.* (1993). A positron emission tomography study of essential tremor: Evidence for overactivity of cerebellar connections. *Ann. Neurol.* **34**, 82–90.

Koller, W. C., Biary, N., and Cone, S. (1986). Disability in essential tremor: Effect of treatment. *Neurology* **36**, 1001–1004.

Koller, W. C., and Royse, V. L. (1986). Efficacy of primidone in essential tremor. *Neurology* **36**, 121–124.

Koller, W. C., and Vetere-Overfield, B. (1989). Acute and chronic effects of propranolol and primidone in essential tremor. *Neurology* **39**, 1587–1588.

Koller, W. C., Vetere-Overfield, B., and Barter, R. (1989). Tremors in early Parkinson's disease. *Clin. Neuropharmacol.* **12**, 293–297.

Lapresle, J., Rondot, P., and Said, G. (1974). Tremblement idopathique de repos, d'attitude et d'action. Etude anatomo-clinique d'une observation. *Rev. Neurol.* **130**, 343–348.

Larsson, T., and Sjogren, T. (1960). Essential tremor: A clinical and genetic population study. *Acta Psychiatrica Neurol. Scand.* **36** (S 144), 1–176.

Lewin, L. (1928). Untersuchungen Uber Banisteria caapi. *Sp. Arch. Exp. Pathol. Pharmacol.* **129**, 133–149.

Lombardi, W. J., Woolston D. J., Roberts, W. J., and Gross, R. E. (2001). Cognitive deficits in patients with essential tremor. *Neurology* **57**, 785–790.

Lou, J. S., and Jankovic, J. (1991). Essential tremor: Clinical correlates in 350 patients. *Neurology* **41**, 234–238.

Louis, E. D. (1999). A new twist for stopping the shakes? Revisiting GABAergic therapy for essential tremor. *Arch. Neurol.* **56**, 807–808.

Louis, E. D. (2001a). Essential tremor. *N. Engl. J. Med.* **345**, 887–891.

Louis, E. D. (2001b). Etiology of essential tremor: Should we be searching for environmental causes? *Mov. Disord.* **16**, 822–829.

Louis, E. D., Barnes, L. F., Albert, S. M., Cote, L., Schneier, F., Pullman, S. L., and Yu, Q. (2001a). Correlates of functional disability in essential tremor. *Mov. Disord.* **16**, 914–920.

Louis, E. D., Barnes, L. F., Ford, B., and Ottman, R. (2001b). Family history information on essential tremor: Potential biases related to the source of the cases. *Mov. Disord.* **16**, 320–324.

Louis, E. D., Ford, B., and Barnes, L. F. (2000). Clinical subtypes of essential tremor. *Arch. Neurol.* **57**, 1194–1198.

Louis, E. D., Ford, B., Frucht, B., Barnes, L. F., Tang, M.-X., and Ottman, R. (2001c). Risk of tremor and impairment from tremor in relatives of patients with essential tremor: A community based family study. *Ann. Neurol.* **49**, 761–769.

Louis, E. D., Ford, B., Frucht, S., and Ottman, R. (2001d). Mild tremor in relatives of patients with essential tremor: What does this tell us about the penetrance of the disease? *Arch. Neurol.* **58**, 1584–1589.

Louis, E. D., Ford, B., Ottman, R., and Wendt, K.J. (1999). Validity of family history data in essential tremor. *Mov. Disord.* **14**, 456–461.

Louis, E. D., and Greene, P. (2000). Essential tremor. In *Merritt's Textbook of Neurology, 10th Edition* (L.P. Rowland, ed.), pp. 678–679. Lea & Febiger, Philadelphia.

Louis, E. D., Klatka, L. A., Lui, Y., and Fahn, S. (1997a). Comparison of extrapyramidal features in 31 pathologically confirmed cases of diffuse Lewy body disease and 34 pathologically confirmed cases of Parkinson's disease. *Neurology* **48**, 376–380.

Louis, E. D., Marder, K., Cote, L., Pullman, S., Ford, B., Wilder, D., Tang, M.-X., Lantigua, R., Gurland, B., and Mayeux, R. (1995). Differences in the prevalence of essential tremor among elderly African-Americans, Whites and Hispanics in Northern Manhattan, NY. *Arch. Neurol.* **52**, 1201–1205.

Louis, E. D., and Ottman, R. (1996). How familial is familial tremor?: Genetic epidemiology of essential tremor. *Neurology* **46**, 1200–1205.

Louis, E. D., Ottman, R. A., Ford, B., Pullman, S., Martinez, M., Fahn, S., and Hauser, W. A. (1997b). The Washington Heights Essential Tremor Study: Methodologic issues in essential-tremor research. *Neuroepidemiology* **16**, 124–133.

Louis, E. D., Ottman, R., and Hauser, W. A. (1998). How common is the most common adult movement disorder?:Estimates of the prevalence of essential tremor throughout the world. *Mov. Disord.* **13**, 5–10.

Louis, E. D., Shungu, D., Chan, S., Mao, X., Jurewicz, E. C., and Watner, D. (2002). Metabolic abnormality in essential tremor: A $^1$H Magnetic Resonance Spectroscopic Imaging Study. *Neurology* **58** (suppl. 3), A253.

Milner, T. E., Cadoret, G., Lessard, L., and Smith, A. M. (1995). EMG analysis of harmaline-induced tremor in normal and three strains of mutant mice with Purkinje cell degeneration and the role of the inferior olive. *J. Neurophysiol.* **73**, 2568–2577.

Mylle, G., and Van Bogaert, L. (1940). Etudes anatomo-cliniques de syndromes hypercinetiques complexes. I. Sur le tremblement familial. *Mschr. Psychiatr. Neurol.* **103**, 28–43.

Mylle, G., and Van Bogaert, L. (1948). Du tremblement essentiel non familial. *Mschr. Psychiatr. Neurol.* **115**, 80–90.

Obrien, M. D., Upton, A. R., and Toseland, P. A. (1981). Benign familial tremor treated with primidone. *BMJ* **282**, 178–180.

O'Hearn, E., and Molliver, M. E. (1997). The olivocerebellar projection mediates ibogaine-induced degeneration of Purkinje cells: A model of indirect, trans-synaptic excitotoxicity. *J. Neurosci.* **17**, 8828–8841.

Ondo, W., Hunter, C., Dat Vuong, K., Schwartz, K., and Jankovic, J. (2000). Gabapentin for essential tremor: A multiple-dose, double-blind, placebo-controlled trial. *Mov. Disord.* **15**, 678–682.

Pahwa, R., Lyons, K., Hubble, J. P. et al. (1998). Double-blind controlled trial of gabapentin in essential tremor. *Mov. Disord.* **13**, 465–467.

Pennes, H. H., and Hoch, P H. (1957). Psychotomimetics, clinical and theoretical considerations: harmine, Win-2299 and Nalline. *Am. J. Psychiatry* **113**, 885–892.

Rajput, A. H., Hornykiewicz, O., Deng, Y., Birdi, S., Miyashita, H., and Macaulay, R. (2001). Increased noradrenaline levels in essential tremor. *Neurology* **56** (Suppl. 3), A302.

Rajput, A. H., Offord, K. P., Beard, C. M., and Kurland, L. T. (1984). Essential tremor in Rochester, Minnesota: A 45-year study. *J. Neurol. Neurosurg. Psychiatry* **47**, 466–470.

Rajput, A. H., Rozdilsky, B., Ang, L., and Rajput, A. (1991). Clinico-pathological observations in essential tremor: Report of six cases. *Neurology* **41**, 1422–1424.

Rajput, A. H., Rozdilsky, B., Ang, L., and Rajput, A. (1993). Significance of Parkinsonian manifestations in essential tremor. *Can. J. Neurol. Sci.* **20**, 114–117.

Rappaport, M. S., Gentry, R. T., Schneider, D. R., and Dole, V. P. (1984). Ethanol effects on harmaline-induced tremor and increase of cerebellar cyclic GMP. *Life Sci.* **34**, 49–56.

Rautakorpi, I. (1978). Essential Tremor. An epidemiological, clinical and genetic study. Research Reports from the Department of Neurology, No. 12, University of Turku, Finland.

Rautakorpi, I., Takala, J., Martilla, R. J., Sievers, K., and Rinne, U. K. (1982). Essential tremor in a Finnish population. *Acta. Neurol. Scand.* **66**, 58–67.

Ross, B. D., and Bluml, S. (1999). Neurospectroscopy. In *Neuroimaging*. (J.O. Greenberg, ed.), McGraw-Hill, New York.

Sasso, E., Perucca, E., and Calzetti, S. (1988). Double-blind comparison of primidone and phenobarbitol in essential tremor. *Neurology* **38**, 808–810.

Sasso, E., Perucca, E., Fava, R., and Calzetti, S. (1990). Primidone in the long-term treatment of essential tremor: A prospective study with computerized quantitative analysis. *Clin. Neuropharmacol.* **13**, 67–76.

Scheinberg, I. H., and Sternlieb, I. (1984). *Wilson's Disease*. W.B. Saunders, Philadelphia.

Schuurman, P. R., Bosch, D. A., Bossuyt, P. M. M. et al. (2000). A comparison of continuous thalamic stimulation and thalamotomy for suppression of severe tremor. *N. Engl. J. Med.* **432**, 461–468.

Singer, C., Sanchez-Ramos, J., and Weiner, W. J. (1994). Gait abnormality in essential tremor. *Mov. Disord.* **9**, 193–196.

Sinton, C. M., Krosser, B. I., Walton, K. D., and Llinas R. R. (1989). The effectiveness of different isomers of octanol as blockers of harmaline-induced tremor. *Pfluegers Arch.* **414**, 31–36.

Stolze, H., Petersen, G., Raethjen, J., Wenzelburger, R., and Deuschl, G. (2000). Gait analysis in essential tremor—further evidence for a cerebellar dysfunction. *Mov. Disord.* **15** (Suppl. 3), 87.

Tanner, C. M., Goldman, S. M., Lyons, K. E., Aston, D. A., Tetrud, J. W., Welsh, M. D., Langston, J. W., and Koller, W. C. (2001). Essential tremor in twins: An assessment of genetic vs. environmental determinants of etiology. *Neurology* **57**, 1389–1391.

Tedeschi, G., Bertolini, A., Campbell, G. et al. (1996). Reproducibility of proton MR spectroscopic imaging findings. *Am. J. Neuroradiol.* **17**, 1871–1879.

Thompson, C., Lang, A., Parkes, J. D., and Marsden, C. D. (1984). A double-blind trial of clonazepam in benign essential tremor. *Clin. Neuropharmacol.* **7**, 83–88.

Tolosa, E. S., and Loewenson, R. B. (1975). Essential tremor: Treatment with propranolol. *Neurology* **25**, 1041–1044.

Topaktas, S., Onur, R., and Dalkara, T. (1987). Calcium channel blockers and essential tremor. *Eur. Neurol.* **27**, 114–119.

Trouvin, J. H., Jacqmin, P., Rouch, C., Lesne, M., and Jacquot, C. (1987). Benzodiazepine receptors are involved in taberanthine-induced tremor: *in vitro* and *in vivo* evidence. *Eur. J. Pharmacol.* **140**, 303–309.

Walshe, J. M., and Yealland, M. (1992). Wilson's disease: The problem of delayed diagnosis. *J. Neurol. Neurosurg. Psychiatry* **55**,692–696.

Weiss, K. M., Chakraborty, R., and Majumder, P. P. (1982). Problems in the assessment of relative risk of chronic disease among biological relatives of affected individuals. *J. Chron. Dis.* **35**, 539–551.

Wills, A. J., Jenkins, I. H., Thompson, P. D., Findley, L. J., and Brooks, D. J. (1994). Red nuclear and cerebellar but no olivary activation associated with essential tremor: A positron emission tomographic study. *Ann. Neurol.* **36**, 636–642.

Winkler, G. F., and Young, R. R. (1971). The control of essential tremor by propranolol. *Trans. Am. Neurol. Assoc.* **96**, 66–68.

CHAPTER

# 33

# Molecular Biology of Huntington's Disease (HD) and HD-Like Disorders

DAVID C. RUBINSZTEIN
*Department of Medical Genetics*
*Cambridge Institute for Medical Research*
*Addenbrooke's Hospital*
*Cambridge, CB2 2XY, United Kingdom*

I. Summary
II. Symptomatology of HD
III. Neuropathology
IV. Neuroimaging
V. Genetics of HD
VI. Diagnostic and Predictive Testing
VII. Gene, Normal Gene, and Abnormal Gene Function
  A. Evidence for "Gain of Function" Caused by Mutation
  B. Function of Normal Huntingtin
VIII. Huntingtin Aggregates
  A. Evidence for Protein Aggregation and the Role of Neuronal Intracellular Inclusions in the Pathogenesis of HD
  B. The Role of Ubiquitin, Proteasomes, and Chaperones
  C. Proposed Mechanisms of Protein Aggregation
IX. Cell Death in HD
  A. Is there Apoptosis in HD?
  B. The Role of Caspases
  C. Cell Death versus Cell Dysfunction
  D. Selective Cell Death and Protein-Protein Interactions
X. Early Changes in Gene Expression
XI. Excitotoxicity and Impaired Energy Production
XII. Animal Models
XIII. Genotype/Phenotype/Modifying Alleles
XIV. Treatment
XV. HD-Like Disorders
  A. Benign Hereditary Chorea
  B. Dominant Adult-Onset Basal Ganglia Disease
  C. Neuroacanthocytosis
  D. McLeod Syndrome
  E. Familial Prion Disease (also Known as HD-Like-1)
  F. HD Like-2
  G. Autosomal Recessive Huntington-Like Disorder (HD-Like-3)
References

## I. SUMMARY

Huntington's disease (HD) is an autosomal dominant neurodegenerative disorder characterized by movement disorders, cognitive deterioration, and psychiatric symptoms. It is caused by a $(CAG)_n$ trinucleotide repeat expansion close to the 5' end of a gene encoding the huntingtin protein. The repeats are translated into an abnormally expanded polyglutamine tract and are thought to cause disease by conferring a novel deleterious function on huntingtin. The pathology of HD is characterized by a selectivity of neuronal loss with greatest severity in the striatum and the deeper layers of the cerebral cortex. The disease is associated with the formation of intraneuronal inclusions or aggregates. The role of these aggregates, which can be found in both the nucleus and cytoplasm, is controversial and the debates are reviewed in this chapter, drawing on experimental data from animal and cellular models of HD and related polyglutamine diseases. This chapter considers the possibility that early changes in cAMP-mediated transcription may play a role in the early changes seen in HD. The potential roles of excitotoxicity and mitochondrial dysfunction in disease pathogenesis are discussed and form a backdrop for a review of some of the approaches that are being used as experimental therapies for this currently incurable disease. Finally, the chapter briefly reviews the genetics and biology of some diseases that can mimic HD symptomatology.

## II. SYMPTOMATOLOGY OF HD

HD is associated with a triad of symptoms, including movement disorders, cognitive deterioration, and psychiatric disturbances. Symptoms of the disease begin insidiously, most commonly between the ages of 35 to 50, but the age of onset can vary from early childhood until old age. The disease is relentlessly progressive and fatal some 15–20 years after the onset of symptoms. Juvenile cases often progress more rapidly and death occurs within 7–10 years from the onset of symptoms.

Disturbances of motor function are classical features of the disease. They include choreiform involuntary movements of the proximal and distal muscles, and progressive impairment of coordination of voluntary movements (Brandt *et al.*, 1984; Folstein *et al.*, 1987; Penney *et al.*, 1990). In patients with juvenile HD, the signs and symptoms are somewhat different; they include bradykinesia, rigidity and dystonia, and chorea may be completely absent. Involuntary movements may take the form of tremor, and affected children often develop epileptic seizures (Van Dijk *et al.*, 1986). Where cerebellar atrophy is a feature of the disease, as seen in some juvenile cases, ataxia may be a prominent presenting feature (Rodda, 1981).

The cognitive difficulties usually begin with a slowing of intellectual processes. HD is classified as a subcortical dementia. In contrast to Alzheimer's disease, patients with HD have problems with retrieval of established memory rather than formation of new memories. Cognitive losses accumulate progressively and late-stage patients with HD have profound dementia. Emotional disturbances in the form of depression and manic-depressive behavior are common features (Folstein *et al.*, 1983). Personality changes, including irritability, apathy and sexual disturbances often accompany the psychiatric syndrome (Dewhurst *et al.*, 1970; Jensen *et al.*, 1993). The rate of suicide has been estimated at 5–10% (Lipe *et al.*, 1993; Di Maio *et al.*, 1993).

## III. NEUROPATHOLOGY

HD is characterized by a striking specificity of neuronal loss. The regions most sensitive to the mutation are the striatum, where there is typically about 57% loss of cross-sectional area from the caudate nucleus and about 65% loss of the putamen (de la Monte *et al.*, 1988). There is about 30% loss of white matter volume from these regions. There is loss of cortical volume, particularly in cases with more advanced disease, which affects predominantly the large pyramidal neurons in layers III, V, and VI. In advanced cases, there is also neuronal loss in the thalamus, substantia nigra pars reticulata, and in the subthalamic nucleus. There is a reduction in the volume of the globus pallidus, which is primarily due to loss of neuropil volume. Thus, globus pallidus volume loss is likely to be due primarily to loss of striatal fiber connections, rather than to loss of neurons. Cerebellar atrophy is most frequently reported in cases with juvenile-onset disease (reviewed in Vonsattel and DiFiglia, 1998).

The pathological hallmark of HD is gradual atrophy of the neostriatum, which occurs in an ordered and topographical distribution. The tail of the caudate nucleus shows more degeneration than the body, which is more affected than the head. Likewise, the caudal putamen is more affected than the rostral portion. Along the coronal axis, the dorsal neostriatum is more affected compared to the ventral regions. As the disease progresses, degeneration spreads caudo to rostral, dorsal to ventral and medially to laterally. Fibrillary astrogliosis mirrors the loss of neurons as the disease progresses (reviewed in Vonsattel and DiFiglia, 1998).

Within the striatum, the most sensitive cell population is the medium spiny neurons, which show dendritic changes including recurving of dentrites and altered density, shape, and size of spines. Among spiny neurons, enkephalin-containing neurons projecting to the external globus pallidus externa are more susceptible compared to substance-P-containing neurons projecting to the internal globus pallidus.

Vonsattel and colleagues (1985) described a grading system for standardization of HD pathology based on gross and microscopical examinations. It is interesting that the mildest category, grade 0, is grossly indistinguishable from normal brains but cell counting can show up to 40% loss of neurons in the head of the caudate nucleus in such brains.

## IV. NEUROIMAGING

Structural MRI studies suggest that overt loss of striatal volume precedes clinical signs of HD, as Aylward and colleagues (1996, 2000) have reported that individuals with the HD mutation show decreased basal ganglia volumes years before the onset of disease. Functional imaging studies suggest reduced D1 and D2 receptor binding in HD striatum precedes disease onset (Antonini *et al.*, 1996; Weeks *et al.*, 1996). Basal ganglia atrophy continues as the disease progresses (Aylward *et al.*, 2000). Interestingly, no changes of frontal lobe volume were seen in mildly affected cases, although frontal lobe volumes do decrease as the disease becomes more severe (Aylward *et al.*, 1998).

In the early stages of the disease there is loss of opioid and benzodiazdepine receptor binding in the striatum (Antonini *et al.*, 1996; Sedvall *et al.*, 1994; Weeks *et al.*, 1997). There is a continuing decrease in striatal and cortical glucose metabolism and dopamine binding as the disease progresses (Kuwert *et al.*, 1990; Turjanski *et al.*, 1995; Martin *et al.*, 1992; Sedvall *et al.*, 1994).

## V. GENETICS OF HD

The gene and the mutation responsible for the disease were identified in 1993 (Huntington's Disease Collaborative Research Group, 1993). The gene is located on the short arm of chromosome 4 (4p16.3) and encodes a large protein called huntingtin that contains more than 3000 residues. Exon 1 of the wild-type gene contains a stretch of uninterrupted CAG trinucleotide repeats, which is translated into a series of consecutive glutamine residues, a polyglutamine (polyQ) tract. Asymptomatic individuals have 35 or fewer CAG repeats and HD is caused by expansions of 36 or more repeats (Rubinsztein *et al.*, 1996). There is an inverse relationship between CAG repeat number and the age of onset of symptoms; the greater the number of CAG repeats, the earlier the age of onset (Fig. 33.1); (Ross and Hayden, 1998). Most adult-onset cases have CAG repeats ranging from 40–50, whereas expansions of >55 repeats frequently cause the juvenile form of the disease. Incomplete penetrance has been observed in individuals with 36–39 repeats —some individuals in their 9th and 10th decades with alleles in this size range have no signs, symptoms, or gross neuropathological features of HD (Rubinsztein *et al.*, 1996; McNeil *et al.*, 1997).

Previous studies suggested that almost all patients with HD had a positive family history. New mutations in symptomatic individuals without a previous family history have been observed but were believed to be very rare (Kremer *et al.*, 1995). However, recent data from Falush *et al.* (2001) suggest that the new mutation rate to alleles of 36 or more repeats may be significantly higher than expected but may have been previously missed if many of these mutant alleles of less than 40 repeats were nonpenetrant. This view is supported by recent data from the Hayden laboratory that suggest that almost one-quarter of patients with HD do not have a family history of the disease—thus, new mutations may be more common than previously believed, even if reliable data were not available from many of these individuals or if some of their parents

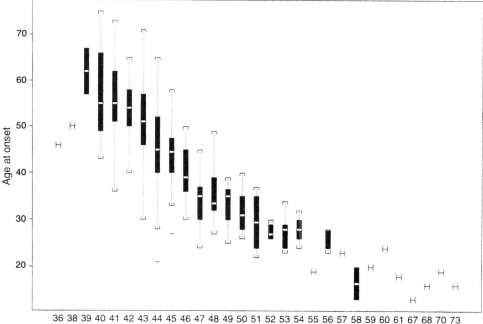

FIGURE 33.1 Relationship between CAG repeat number on HD chromosomes and age of onset of disease. For each repeat size, the median value is indicated as an open bar, the 95% confidence interval is indicated as the solid box, the range is indicated by brackets, and outlying points by solid lines. From Rubinsztein, D.C., Leggo, J., Chiano, M., Dodge, A., Norbury, G., Rosser, E., and Craufurd, D. (1997). Genotypes at the GluR6 kainate receptor locus are associated with variation in the age of onset of Huntington disease. (From *Proc. Nat. Acade. Sci. U.S.A.* **94**, 3872–6. With permission from the National Academy of Sciences, USA, Copyright (1997)).

died before disease may have presented (Almqvist et al., 2001). Chromosomes with 29–35 CAG repeats (at the upper limit of the normal range) are less stable when transmitted through the germline than shorter alleles and are thought to represent the pool of chromosomes from which the pathogenic "founder" mutations are derived. Thus, in HD and many other trinucleotide repeat diseases, there is a correlation between the frequency of "long normal" CAG repeats in a given population and the prevalence of the disease (Rubinsztein et al., 1994; Almqvist et al., 1995). Disease chromosomes with expanded CAG repeats show marked instability, particularly when passed through the male germline where expansions tend to occur more frequently than contractions (Ranen et al., 1995). Thus, CAG repeat numbers tend to increase in successive generations in families. This causes the phenomenon of anticipation, where the age of onset tends to decrease in successive generations.

## VI. DIAGNOSTIC AND PREDICTIVE TESTING

Prior to the identification of the CAG repeat mutation, definitive diagnosis of HD relied on postmortem pathological examination. However, now the mutation can be directly assayed. Three types of test can be offered for HD: diagnostic testing, predictive testing, and prenatal exclusion testing.

**Diagnostic testing**—The CAG repeats can be sized in individuals who have signs or symptoms of HD. In many centers it is suggested that patients have appropriate counseling and also give written informed consent. While the indications for testing are fairly straightforward in cases with typical presenting features and a family history, the diagnosis should also be considered in children with positive family histories and slightly atypical features (e.g., ataxia), or in adults with typical HD features without a family history (who may be new mutations).

**Predictive testing**—The CAG repeats can be sized in an asymptomatic individual who wishes to determine if he/she carries the HD mutation because of a family history of the disease. There are well-established international ethical guidelines for predictive testing (Craufurd, 1998). It is important that diagnostic testing is clearly distinguished from predictive testing, as they have different testing and counseling protocols.

In both diagnostic and predictive testing situations, it is desirable to have confirmation of the molecular diagnosis of HD in an affected relative. In a predictive testing situation, this is particularly important, since, without it, a negative result does not exclude the possibility that the individual has inherited another autosomal dominant neurodegenerative disorder.

We and others have sized the repeats in relation to the number of uninterrupted CAG repeats. There are more glutamines than CAG repeats in this part of the gene, as the polyglutamine tract is coded for in most chromosomes by (CAG)nCAACAG.

In diagnostic tests, symptomatic cases with greater than 35 repeats would have an HD mutation, which would be consistent with the diagnosis of HD. In symptomatic cases with less than 35 repeats, the individual is most unlikely to have HD. Alternative diagnoses should be considered. For instance, dentorubral-pallidoluysian atrophy (DRPLA) can mimic the features of HD (Becher et al., 1997), as can a number of HD-like disorders (see Section XV below).

In predictive tests, asymptomatic cases with >35 repeats would be considered to carry the mutation. In cases with 36–39 repeats the individual is at risk of developing HD but the mutation size is within the region of described nonpenetrance (Rubinsztein et al., 1996; McNeil et al., 1997). Thus, there is a finite probability of nonpenetrance during the normal life span of the individual. Unfortunately, the limited numbers of cases with repeats in this size range makes it difficult to precisely quantify risk.

No cases with confirmed HD have been described with 29–35 repeats. However, these alleles are at risk for expansion in subsequent generations into the disease size range. Thus, the possibility of prenatal diagnosis could be considered, if appropriate.

In rare cases, an individual has two alleles in the disease size range. This creates difficult counseling situations. The risks of passing the disease alleles to children is 100% and this may also uncover new relatives at risk, as the expansions would have come from both paternal and maternal sides of the proband.

**Prenatal exclusion testing**—Exclusion testing can be employed when an individual at risk of HD is expecting a child and does not wish to know his/her status but wants reassurance that the child will not get HD. This approach uses polymorphic markers closely linked to the HD gene. In a model scenario of a family where the father of the pregnant consultand had HD, one could test the fetus' DNA to see if it carried its maternal grandmaternal or maternal grandpaternal chromosome 4p16.3 region. If the unborn child carried the 4p16.3 region from one of its maternal grandmaternal chromosomes, then its HD risk would be close to 0%. If it inherited the 4p16.3 region from one of its maternal grandpaternal chromosomes, then the HD risk to the fetus would be close to 50% and a termination of pregnancy could be considered. An important feature of the design of such tests is that the analyses should not reveal which of the maternal grandpaternal chromosomes is associated with the HD mutation. A difficult situation can arise if the fetus is found to be at 50% risk and the parents do not terminate the pregnancy—if the at-risk parent subsequently develops HD, then the knowledge of the exclusion test result implies that the child will be at 100% risk. This scenario should be included in the genetic counseling and the issues surrounding termination should be carefully explored.

Using the direct test for the HD CAG expansion to test a pregnancy, where the at-risk parent is asymptomatic and does not wish to know his/her mutation status, is tantamount to performing an unauthorized predictive test on the at-risk parent, if the fetus is found to have a CAG repeat expansion.

Detailed recommendations for diagnostic testing can be found on http://cmgs.org/BPG/Guidelines/2nd_ed/huntington_disease.htm and guidelines have also been suggested by the American College of Human Genetics and the American Society of Human Genetics Huntington Disease Genetic Testing Working Group (1998).

## VII. GENE, NORMAL GENE, AND ABNORMAL GENE FUNCTION

### A. Evidence for "Gain of Function" Caused by Mutation

HD is an autosomal dominant condition—one mutated gene is sufficient to cause the disease, in spite of the presence of a normal gene inherited from the other parent. Genetic and transgenic data suggest that the mutation confers a novel deleterious function on the protein. In humans, hemizygous loss of one of the two huntingtin genes has been observed as a result of either a terminal deletion of one chromosome 4 (which includes the HD gene) in patients with Wolf-Hirschhorn syndrome (Harper, 1996), or a balanced translocation with a breakpoint between exons 40 and 41, physically disrupting the HD gene in one female patient (Ambrose et al., 1994). However, this hemizygous inactivation of huntingtin does not cause an HD phenotype. In addition, mice that have only one functioning HD gene do not show features of the disease (Duyao et al., 1995; Nasir et al., 1995; Zeitlin et al., 1995; White et al., 1997).

A gain-of-function is also supported by observations in spinobulbar muscular atrophy (SBMA), which is caused by a CAG repeat expansion in the androgen receptor. Males with SBMA develop progressive weakness and muscle atrophy due to loss of their motor nerve supply and mild androgen insensitivity (Beitel et al., 1998). This contrasts with the effects of point mutations elsewhere in the gene, which can result in severe androgen insensitivity but never cause a neuromuscular disease. Indeed, the neuromuscular features of SBMA are not even seen in a patient with a complete androgen receptor gene deletion.

This gain-of-function model was elegantly demonstrated in a mouse model, where 146 CAG repeat sequence was inserted into the hypoxanthine phosphoribosyltransferase (HPRT) gene, which is not involved in any CAG-repeat disorders. While previous work had shown that inactivation of the HPRT gene in mice does not cause deleterious effects, these mutant mice produced a polyglutamine-expanded form of the HPRT protein and developed a late-onset neurological phenotype that progressed to premature death (Ordway et al., 1997). Therefore, HD is not caused simply by hemizygous loss of huntingtin function and it is believed that a major effect of the CAG/polyglutamine mutation is to confer a novel deleterious function on the mutant protein.

### B. Function of Normal Huntingtin

Wild-type huntingtin is widely expressed and is found mainly in the cytoplasm, where, in neurons, it is associated with vesicle membranes and microtubules. Huntingtin appears to be associated with clathrin via the huntingtin interacting protein HIP-1 (Waelter et al., 2001; Metzler et al., 2001).

Wild-type huntingtin is necessary for development as homozygous knock-out mice show embryonic lethality (Nasir et al., 1995; Zeitlin et al., 1995; White et al., 1997). Furthermore, conditional inactivation of wild-type huntingtin in the brain and testis at postnatal day 5 led to neurological signs, neurodegeneration, and impaired spermatogenesis, suggesting that huntingtin is required in the postembryonic period. In 1995, Zeitlin et al. observed high levels of apoptosis in HD gene knock-out mice, leading them to suggest that huntingtin may be an antiapoptotic protein. Subsequently Goldberg et al. (1996) showed that huntingtin is a caspase substrate. The prediction of Zeitlin et al. (1995) has been supported by recent findings that overexpression of wild-type huntingtin protects cells against a variety of apoptotic stimuli (Rigamonti et al., 2000), probably by inhibiting procaspase-9 processing (Rigamonti et al., 2001). Wild-type huntingtin can protect against apoptosis in the testis of mice expressing full-length huntingtin transgenes with expanded CAG repeats (Leavitt et al., 2001) and overexpression of wild-type huntingtin significantly reduced the cellular toxicity of mutant HD exon 1 fragments in both neuronal and non-neuronal cell lines (Ho et al., 2001). This suggests that wild-type huntingtin can be protective in different cell types and that it can act against the toxicity caused by mutant huntingtin.

Wild-type huntingtin may also protect against cell death by binding to HIP-1. Free (unbound) HIP-1 binds to a protein called Hippi and HIP-1-Hippi heterodimers can form and recruit procaspase-8 into a complex, launching apoptosis through components of the "extrinsic" cell-death pathway (Gervais et al., 2002). Huntingtin with a polyglutamine expansion binds less strongly to HIP-1 compared to the wild-type form, and Gervais et al. (2002) have suggested that this may play a role in disease pathogenesis, since more free HIP-1 would be predicted to be present in the HD cases. While this mechanism may contribute to disease, it is unlikely to be either sufficient or necessary, as HD-like disease is seen in transgenic mice that have two endogenous wild-type HD genes as well as a mutant HD transgene, and hemizygous knock-out of the HD gene does not result in HD-like disease (see Sections VII.A and XII).

In addition to protecting against polyglutamine expansion toxicity, wild-type huntingtin may also play a role in HD by upregulating the transcription of brain-derived neurotrophic factor (BDNF); (Zuccato et al., 2001). This activity appears to be reduced in the disease state and Zuccato and colleagues (2001) have proposed that this may contribute to HD pathology by reducing neurotrophic support from cortical neurons to striatal neurons.

## VIII. HUNTINGTIN AGGREGATES

### A. Evidence for Protein Aggregation and the Role of Neuronal Intracellular Inclusions in the Pathogenesis of HD

A major step toward increased understanding of the molecular basis for the pathology of HD was made in 1996, when Bates and co-workers created a transgenic mouse which expressed exon 1 of the human HD gene containing different numbers of CAG repeats, under the control of the human huntingtin promoter (Mangiarini et al., 1996). Mice expressing 18 CAG repeats developed normally and remained healthy. By contrast, mice that expressed 113–156 CAG repeats developed progressive neurological symptoms together with some of the other clinical features seen in HD. These mice developed intraneuronal aggregates (inclusions) in nuclei and in neuronal processes (Davies et al., 1997). Similar aggregates have been identified in postmortem human HD brains in cortical and striatal neurons (DiFiglia et al., 1997) and in dystrophic neurites (Sapp et al., 1999), but not in the globus pallidus or cerebellum (Aronin et al., 1999). In HD brains these inclusions comprise truncated derivatives of the mutant proteins, which only appear to be recognized by antibodies to epitopes close to the expanded polyglutamines. The intranuclear inclusions are positioned variably in the nucleus, tend to be significantly larger than the nucleolus, and are not separated from the rest of the nucleoplasm by a membrane. The inclusions are spherical, ovoid, or elliptical in shape and are concentrated in neurons in areas of the brain which degenerate in HD. Electron microscopy revealed that the inclusions were heterogeneous in composition and contained a mixture of granules, straight and tortuous filaments, and many parallel and randomly oriented fibrils. Intraneuronal aggregates are a pathological common denominator for polyglutamine diseases, since similar structures are seen in brains from cases with all of these diseases.

A causal role for these inclusions in polyglutamine disease pathology has been suggested, as they appear before the signs of disease in a transgenic mouse expressing exon 1 of the HD gene with expanded repeats (Davies et al., 1997; 1998). In addition, the numbers of inclusions in the cortex of HD patients correlates with CAG repeat number (Becher et al., 1998). Inclusion formation in vitro in cultured cells also correlates with susceptibility to cell death (Paulson et al., 1997; Cooper et al., 1998; Igarashi et al., 1998; Martindale et al., 1998; Wyttenbach et al., 2000). In striatal cultures, neuritic aggregates block protein transport in neurites, and cause neuritic degeneration before nuclear DNA fragmentation occurs (Li et al., 2001). These authors suggest that the early neuropathology of HD may originate from axonal dysfunction and degeneration associated with huntingtin aggregates. Reduction of polyglutamine inclusion formation, by overexpression of heat shock proteins (HSPs) and bacterial and yeast chaperones (that are unlikely to directly affect cell-death pathways), is also associated with decreased cell death in vitro (Cummings et al., 1998; Chai et al., 1999a; Carmichael et al., 2000). While the effects of chaperones on aggregation and cell death strongly support a correlation between the appearance of aggregates and cell death/dysfunction, they may be compatible with another theoretical model where aggregation would not necessarily be causally related to cell death. The mutant monomeric proteins may exist in two sets of conformations: a properly folded nontoxic species and an aggregate-prone toxic form(s). If the chaperones promote the conversion of the toxic aggregate-prone monomers to the non-toxic form, then one would see a reduction of both aggregation and cell death in the presence of chaperones. This scenario may be difficult to distinguish from the model where aggregation is toxic, unless one can identify a means of distinguishing and quantifying different monomeric forms of these aggregation-prone proteins from cell models or in vivo experiments.

Saudou et al. (1998) queried the pathological role of inclusions in cell death using in vitro studies of primary striatal cultures, which suggested dissociation between inclusion formation and cell death. They tested the effect of inhibiting ubiquitination on inclusion formation and cell viability. The inclusions seen in patients' brains and in in vivo and in vitro models of polyglutamine diseases are ubiquitinated. This process is used by cells to tag misfolded proteins and target them for degradation. Saudou et al. (1998) showed that expression of a dominant-negative ubiquitin-conjugating enzyme mutant reduced the proportion of cells with aggregates but increased cell death caused by huntingtin constructs containing expanded repeats in the cells remaining on the dishes after 6 days. However, inhibition of ubiquitination also resulted in increased cell death in cells expressing "wild-type" huntingtin constructs with 17 repeats. It is not clear if the results were simply due to an additive effect of the polyglutamine insult and the defective ubiquitination, resulting in earlier death of cells expressing mutant huntingtin with smaller inclusion loads. This scenario would result in fewer adherent cells with visible inclusions after the 6 day experiment—dead cells do not remain attached to culture dishes for long.

Cummings and colleagues (1999) showed that a mutation in the E6-AP ubiquitin ligase reduced nuclear inclusion frequency but accelerated polyglutamine-induced pathology in SCA1 mice. While these data suggest that large visible inclusions may not be required for cell death, the authors consider other possibilities which are consistent with a pathological role for inclusions. For instance, ubiquitination of ataxin-1 may not be E6-AP dependent. The deletion of this enzyme may affect the turnover of other proteins (Cummings et al., 1999), which at abnormally high steady-state levels may enhance the cellular sensitivity to the SCA1 mutation (or aggregates).

Klement et al. (1998) suggested that aggregate formation may not be a prerequisite for pathology, since similar Purkinje cell pathology and ataxic phenotypes were observed in mice expressing the SCA1 gene with 77 repeats, with or without a deletion of the ataxin-1 self-association domain. The mice expressing transgenes without the self-association domain had no intranuclear inclusions in their Purkinje cells, in contrast to the mice expressing the entire mutant gene. This conclusion may be simplistic, since later reports suggest that the phenotype in the mice with the deleted self-association domain may be nonprogressive, in contrast to the mice with the full-length gene (Orr et al., 1998). Thus, inclusion formation may affect disease progression. These data need to be interpreted carefully, since no data were presented for mice with normal repeat lengths containing deletions of the self-association domain in SCA1, and Perutz (1999) has suggested that removal of the self-association domain from the normal SCA1 gene may itself cause abnormal protein folding and a SCA-like phenotype.

Further work needs to be done in order to clarify the controversy regarding the pathogenic role of aggregates. This may have a much wider relevance, since the phenomenon of ubiquitinated inclusion bodies is not confined to polyglutamine diseases. Indeed, it seems to be an emerging theme in many other late-onset neurodegenerative diseases like Alzheimer's disease, Parkinson's disease, amyotrophic lateral sclerosis, and prion diseases (Goedert, 1999).

Lansbury and colleagues have suggested that the aggregation process itself may be toxic, possibly with the most toxic species being oligomers/microaggregates, rather than aggregates that are easily visible using light microscopy (Conway et al., 2000). Microaggregates have been identified by electron microscopy in the YAC transgenic mouse model made by Michael Hayden's lab (Hodgson et al., 1999). They can be around 50 nm in diameter (Hodgson et al., 1999). Such microaggregates may be very difficult to detect or exclude in humans or animal models. A typical cortical neuron is about 25 μm in diameter and EM sections are usually 50 nm thick—500 sections per neuron. Thus, if there were one 50-nm microaggregate per cell (or 5 microaggregates in 20% of cells), one would only see such an aggregate in about 1/500 EM cell/sections. Thus, the failure to observe such microaggregates using EM does not exclude their existence even in the majority of neurons.

## B. The Role of Ubiquitin, Proteasomes, and Chaperones

Proteins with abnormally long polyglutamine tracts are probably misfolded. Cells have a complex array of chaperone proteins that assist in the folding of normal proteins, and recognition and handling of abnormally-folded proteins. Several of the HSPs are involved in modulation of protein folding pathways, promoting proper protein assembly in an ATP-dependent manner and preventing misfolding and aggregation under both normal and stress conditions (Hendrick and Hartl, 1993). HSP70 and HSP40 family members are sequestered to huntingtin exon 1 aggregates in cell models (Wyttenbach et al., 2000, 2001; Jana et al., 2000). It is possible that this redistribution into aggregates could sequester chaperones that would otherwise perform normal protective functions. Overexpression of the human HSP40 homologs HDJ-1 and HDJ-2 and/or HSP70 family members in vitro reduced aggregation of ataxin-1, ataxin-3, and the androgen receptor with expanded polyglutamine proteins in HeLa cells (Cummings et al., 1998; Chai et al., 1999b; Stenoien et al., 1999). However, overexpression of HDJ-2 did not alter the aggregation of exon 1 of huntingtin containing 72 repeats in two neuronal-precursor cell lines and actually increased aggregation in COS-7 cells (Wyttenbach et al., 2000). Thus, overexpression of this HSP may increase or decrease aggregates in different cell types or with different polyglutamine-containing proteins.

Another major intracellular pathway for reducing the levels of misfolded proteins is the ubiquitin-proteasome pathway. Intranuclear inclusions seen in patients' brains (Fig. 33.2) and in in vitro models of polyglutamine diseases are ubiquitinated and sequester components of the proteasome (Bailey et al., 1998; Cummings et al., 1998; Paulson et al., 1998, Wyttenbach et al., 2000). Proteasome inhibitors promote aggregation of proteins with expanded polyglutamines (Chai et al., 1999a; Cummings et al., 1999; Wyttenbach et al., 2000) consistent with the hypothesis that the proteasome degrades proteins containing expanded polyglutamine tracts.

A protein must be unfolded before entering the central 20S proteolytic subunit of the proteasome complex. Expanded polyglutamines may confer a conformational change,

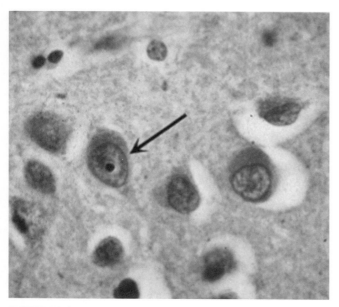

FIGURE 33.2 Intranuclear inclusion containing ubiquitin (indicated by arrow) in human HD brain, revealed by immunohistochemistry, using anti-ubiquitin antibody. From Ho, L.W., Carmichael, J., Swartz, J., Wyttenbach, A., Rankin, J., and Rubinsztein, D.C. (2001). The molecular biology of Huntington's disease. *Psychol. Med.* **31**, 3–14. (With permission from Cambridge University Press.)

resulting in restricted entry into the proteolytic chamber or incomplete degradation. If redistribution of proteasomes into inclusions depletes the neuron of functional proteolytic activity, the results would be deleterious. This possibility is supported by the work of Kopito and colleagues, who showed that cells containing polyglutamine (or other) inclusions have impaired proteasome activities, compared to cells without aggregates (Bence et al., 2001). A major role of the proteasome is to degrade short-lived proteins, like some transcription factors. The concentrations of some of these proteins needs to be carefully controlled, as they are critical regulators of cellular metabolism, hence the short half-lives. If the proteasome is impaired, the levels of such short-lived proteins rise abnormally and apoptosis often ensues.

A further link between proteasome malfunction and polyglutamine pathogenesis is suggested by the observation that arfaptin 2, a protein that appears to co-localize with huntingtin aggregates *in vivo* and *in vitro*, increases huntingtin aggregation by inhibiting proteasome function (Peters et al., 2002). It is interesting that arfaptin expression was increased in the striatum, cortex, and cerebellum in HD transgenic mice, compared to other regions (Peters et al., 2002).

## C. Proposed Mechanisms of Protein Aggregation

Two non-mutually exclusive mechanisms have been proposed to explain the aggregation of mutant huntingtin.

### 1. The Polar Zipper Model

One possibility is that the normal protein conformation is destabilized by the presence of the expanded polyglutamine tract, which, in turn, leads to abnormal protein-protein interactions and the formation of insoluble β-pleated sheets by linking β-strands together into barrels or sheets via hydrogen bonding, forming so-called polar zipper structures (Perutz et al., 1994; Stott et al., 1995). This model is supported by *in vitro* studies, which showed that purified recombinant proteins with glutamine expansions formed stable aggregates exhibiting polarization with Congo Red, characteristic of proteins that have β-pleated sheets (Scherzinger et al., 1997). Congo red staining has also been reported in HD brains (Huang et al., 1998).

### 2. The Role of Transglutaminases

Aggregation of mutant huntingtin products has been postulated to be mediated by transglutaminases (Kahlem et al., 1998), enzymes normally involved in cross-linking of glutamine residues in different proteins (Fesus and Thomazy, 1988). Huntingtin is a substrate of transglutaminase *in vitro* and the rate of the reaction increases with length of the polyglutamine tract (Kahlem et al., 1998). An expanded polyglutamine stretch could result in increased cross-linking between mutant huntingtin and itself or other proteins and precipitation with slow intraneuronal accumulation of huntingtin aggregates. The activity of transglutaminases has been seen to be increased in HD patients compared to normal individuals (Karpuj et al., 1999), and it has been proposed that the level of transglutaminase activity affects the age of onset in individuals with equal numbers of CAG repeats (Cariello et al., 1996). However, tissue transglutaminase is not required and does not facilitate huntingtin aggregate formation (Chun et al., 2001a), although this enzyme does appear to modify proteins that associate with truncated mutant huntingtin (Chun et al., 2001b). The data of Chun et al. (2001b) suggest that this modification is more likely to involve polyamination than cross-linking. It is tempting to speculate that this phenomenon may play a role in HD pathology, but further work is required.

## IX. CELL DEATH IN HD

### A. Is there Apoptosis in HD?

Apoptosis is programmed cell death—a conserved cellular mechanism initiated by diverse stimuli that leads to activation of aspartate-specific proteases (caspases), culminating in DNA fragmentation and cell death. A subset of neurons and glia in the neostriatum of postmortem HD brains show DNA strand breaks (as assayed with terminal transferase-mediated deoxyuridine triphosphate–biotin nick-end labeling (TUNEL); (Dragunow et al., 1995; Portera-Cailliau et al., 1995; Thomas et al., 1995). Isolated medium spiny neurons were stained, most intensely in the putamen, followed by the globus pallidus and caudate. Labeling was also increased in more advanced cases of the disease. There appears to be a positive correlation between the number of CAG repeats in HD and the degree of nuclear fragmentation in the HD striatum (Butterworth et al., 1998). However, TUNEL recognizes some forms of necrosis and is thus not selective for apoptosis (Nishiyama et al., 1996).

The presence of TUNEL positive neurons has, however, been disputed by Davies and colleagues (Turmaine et al., 2000), who found that the R6/2 HD transgenic mouse lines do develop late-onset neurodegeneration within the anterior cingulated cortex, dorsal striatum, and in the Purkinje neurons of the cerebellum. Dying neurons characteristically exhibited neuronal intranuclear inclusions, condensation of both the cytoplasm and nucleus, and ruffling of the plasma membrane while maintaining ultrastructural preservation of cellular organelles. These cells do not develop blebbing of the nucleus or cytoplasm, apoptotic bodies, or fragmentation of DNA. Neuronal death occurs over a period of weeks not hours. Degenerating cells of similar appearance were observed within these same regions in brains of patients who had died with HD. Thus, these workers suggested that the mechanism of neuronal cell death in both HD and a transgenic mouse model of HD is neither by apoptosis nor by necrosis (Turmaine et al., 2000). However, a failure to observe apoptotic cells may be a function of their rapid clearance. Therefore, it is not clear which cell death pathway(s) are operating in HD.

## B. The Role of Caspases

Caspases are a family of cysteine proteases which are activated in apoptosis. A number of studies have suggested a role for these proteases in polyglutamine diseases. During apoptosis, caspase-3 cleaves structural and nuclear proteins, as well as other caspases (Nicholson & Thornberry, 1997; Thornberry and Lazebnik, 1998). Caspase-3 also specifically cleaves huntingtin and long polyglutamine sequences make huntingtin more susceptible to this cleavage (Goldberg et al., 1996). In vitro studies suggest that N-terminal cleavage products of mutant huntingtin are more toxic and more prone to aggregate formation than the full-length protein. Toxic fragments from proteins with polyglutamine expansions may further activate caspases, which produce more toxic fragments in a positive feedback loop, ultimately resulting in cell death (Wellington et al., 1998).

In a transgenic HD mouse model, expression of a dominant negative caspase-1 mutant that inhibits the activity of caspase-1, delays onset of symptoms and extends survival. Direct intraventricular administration of a caspase-1 inhibitor also delayed disease progression and mortality in the mouse model of HD (Ona et al., 1999).

Caspase-8 appears to be a necessary mediator of death in primary rat neurons transiently transfected with a construct expressing an expanded polyglutamine repeat (Sanchez et al., 1999). This study showed that caspase-8 is recruited into inclusions and suggested that it is activated as part of the HD disease process.

Another pathway which may be relevant to HD pathogenesis is the activation of c-Jun amino-terminal kinases (JNK). This pathway is induced by a variety of oxidative stress stimuli and can induce apoptosis. Expression of mutant huntingtin with expanded polyglutamine repeats in rat hippocampal neuronal cells in vitro has been shown to stimulate JNKs activity and induce apoptotic cell death, whereas expression of normal huntingtin had no toxic effect. JNK activation preceded apoptosis and co-expression of a dominant negative mutant form of the stress signaling kinase (SEK1) almost completely blocked JNK activation and apoptosis (Liu, 1998).

## C. Cell Death versus Cell Dysfunction

Although there is cell death even in presymptomatic HD mutation carriers (Vonsattel and DiFiglia, 1998; Aylward et al., 1996, 2000), it is important to consider that cell dysfunction may significantly contribute to the early stages of disease. Guidetti et al., (2001) have demonstrated morphologic abnormalities that included a significant decrease in the number of dendritic spines and a thickening of proximal dendrites in striatal and cortical neurons in symptomatic HD transgenic mice before any detectable cell loss, a finding which is compatible with our observations in a cell model (Wyttenbach et al., 2001).

## D. Selective Cell Death and Protein-Protein Interactions

Different polyglutamine diseases exhibit distinct patterns of neurodegeneration, yet the encoded proteins are widely expressed. It is possible that selective expression of proteins which interact with huntingtin could account for the selective neuropathology in HD. One possibility is that the sequestration of such proteins may also contribute to the disease process (Cattaneo et al., 2001; Presinger et al., 1999) and the pathological specificity of the mutation may be partly determined by the differential susceptibility of different cell types to loss of specific proteins sequestered to aggregates/ mutant huntingtin.

It appears that full-length huntingtin needs to be cleaved to form an N-terminal fragment containing the glutamine repeats that is toxic (and which is also more prone to aggregation; Wellington *et al.*, 1998, 2000). Indeed, tissue-specific proteolysis occurs in HD brains (Mende-Mueller *et al.*, 2001). Li *et al.* (2000) compared the patterns of inclusion formation in HD transgenic mice expressing only exon 1 to knock-in mice expressing full-length mutant mouse huntingtin homologs. The former develop inclusions in most brain regions, while the knock-in mice develop inclusions that are confined to the striatum (Li *et al.*, 2000). While this study was not perfectly controlled, since the transgenes are expressed by different promoters (the mouse and human HD promoters differ; (Coles *et al.*, 1998), and inclusion formation is not a direct measure of cell dysfunction/death, it supports the hypothesis that tissue specificity of huntingtin proteolysis may be an important determinant of which cell populations are affected. The cleavage/toxic fragment model described above has recently been challenged by data suggesting that wild-type full-length huntingtin (with normal repeats) is more susceptible to cleavage compared to mutant full-length huntingtin (Dyer and McMurray, 2001). Further work will be required to resolve this important controversy.

## X. EARLY CHANGES IN GENE EXPRESSION

A number of studies have used cDNA expression and Affymetrix arrays to interrogate early changes in gene expression in HD transgenic mice (Iannicola *et al.*, 2000; Luthi-Carter *et al.*, 2000) and in stable inducible cell models (Wyttenbach *et al.*, 2001). While many different types of gene expression are perturbed, one of the classes of genes that are downregulated in mutant versus wild-type mice/cells are those that are controlled by cAMP response elements (CRE); (Luthi-Carter *et al.*, 2000; Wyttenbach *et al.*, 2001). A reduction of CRE-mediated transcription is also likely in human HD, since reduced levels of the CRE-responsive genes, somatostatin, corticotrophin-releasing hormone, proenkephalin, and substance P, are seen in HD versus control brains, even in early stages of the disease (De Souza *et al.*, 1987; Sapp *et al.*, 1995; Timmers *et al.*, 1996; Augood *et al.*, 1996). Such genes may be relevant to the increased susceptibility to cell death and decreased neurite outgrowth seen in HD and related cell models, as these phenotypes can be partly attenuated by stimulating CRE-mediated transcription by over-expressing transcriptional co-activators like CRE-binding protein (CREB) or $TAF_{II}130$, or by treating cells with cAMP or forskolin (which stimulates adenylyl cyclase); (McCampbell *et al.*, 2000; Shimohata *et al.*, 2000; Nucifora *et al.*, 2001; Wyttenbach *et al.*, 2001). Since CRE-mediated transcription appears to be impaired in three different polyQ disease models (HD, DRPLA, and SBMA), it will be important to test if this is one of the unifying features common to all polyQ diseases (McCampbell *et al.*, 2000; Shimohata *et al.*, 2000; Nucifora *et al.*, 2001; Wyttenbach *et al.*, 2001).

Decreased CRE-mediated transcription may also be important in relation to neurite outgrowth, which may impact on cell dysfunction that occurs in the early stages of polyQ diseases prior to cell death (Wyttenbach *et al.*, 2001; Guidetti *et al.*, 2001). This may be relevant *in vivo*, since CREB-mediated signaling is crucial for long-term potentiation (LTP), the synaptic analog for memory (Barth *et al.*, 2000). LTP is impaired in a number of HD mouse models and memory difficulties are a feature of HD (Usdin *et al.*, 1999; Murphy *et al.*, 2000).

There are a number of potential non-mutually exclusive mechanisms that may act upstream of CREB in polyQ diseases. Decreased adenylyl cyclase activity has been observed in HD transgenic mice (Luthi-Carter *et al.*, 2000). It is possible that some of these changes may be mediated by the increased levels of protein phosphatase 2A suggested by our expression assays (Wyttenbach *et al.*, 2001). This enzyme can deactivate phosphorylated CREB (Wadzinski *et al.*, 1993), one of the main transcription factors that bind to CRE elements. Another appealing model invokes the sequestration of co-activators like CREB-binding protein (CBP) and $TAF_{II}130$ by mutant polyQ stretches into inclusions, as these co-activators are important positive regulators of CREB-mediated transcription. CBP has been observed in aggregates in SBMA (McCampbell *et al.*, 2000), HD, and DRPLA (Steffan *et al.*, 2000; Nucifora *et al.*, 2001) and $TAF_{II}130$ in SCA3 and DRPLA (Shimohata *et al.*, 2000). It may be difficult to unravel the relative importance of these different mechanisms.

Transcription in HD may also be more widely affected, as the mutant protein binds to the acetyltransferase domains of histone acetylases like CBP and p300 (Steffan *et al.*, 2000). This is associated with impaired histone acetylation and this defect can be partially reversed in cell culture models by treating with histone deacetylase inhibitors. In a *Drosophila* HD model, histone deacetylase inhibitors slowed the progression of the disease, suggesting that such drugs may be beneficial in the human disease. The therapeutic potential of this approach will be strengthened if these drugs are found to be effective in HD mouse models.

Other genes whose expression is diminished in HD models include components of the neurotransmitter, calcium, and retinoid signaling pathways and these may be relevant to the disease phenotype (Luthi-Carter *et al.*, 2000).

## XI. EXCITOTOXICITY AND IMPAIRED ENERGY PRODUCTION

Exitotoxicity refers to death of neurons as a result of exposure to excitatory amino acids, like glutamate.

Excitotoxic pathways could be important modifiers of HD neuropathology, since animals injected intracranially with excitatory amino acids, such as kainate (Coyle and Schwarcz, 1976; McGeer et al., 1978) and quinolinic acid (Beal et al., 1986, 1989; Huang et al., 1995), have similar striatal pathology to that seen in HD. HD brains also show a loss of binding sites for excitatory amino acids (Young et al., 1988), suggesting that cells expressing these receptors are made vulnerable by the mutant HD protein. The role of excitotoxicity in HD is further supported by the finding that genotypes at the GluR6 kainate receptor locus may modify the age-at-onset of symptoms independently of the effect of the CAG repeat number (Rubinsztein et al., 1997; MacDonald et al., 1999). A contribution of excitotoxicity to HD pathology is supported by Zeron et al. (2001), who have shown that mutant full-length huntingtin enhances the excitotoxic cell death in HEK293 cells expressing NR1A2B glutamate receptors after treatment with 1 m$M$ glutamate.

Excitotoxicity in HD may be the result of a reduced threshold for glutamate toxicity that would occur in neurons with compromised energy metabolism, causing otherwise normal levels of this excitatory neurotransmitter to become toxic (Novelli et al., 1988; Beal, 1992). Impaired energy metabolism has been observed in the brains of HD patients using nuclear magnetic resonance spectroscopy (Jenkins et al., 1993), positron emission tomography (Grafton et al., 1992), and using biochemical methods (Gu et al., 1996; Tabrizi et al., 1999). The relevance of impaired energy metabolism to HD pathology is also suggested by the effects of the toxin 3-nitroprorionic acid (3-NP), which irreversibly inhibits succinate dehydrogenase, an enzyme involved in the tricarboxylic acid cycle and the electron transport chain during ATP synthesis. When 3-NP is administered chronically in low doses to animals it reproduces the slow progressive nature of human HD and the neuropathological and neurological outcomes closely mimic human HD (Brouillet et al., 1995; Borlongan et al., 1997). Humans surviving 3-NP toxicity develop choreiform movements and dystonia (Ludolph et al., 1991; He et al., 1995).

Oxidative stress has been implicated in late-onset neurodegeneration and there is *in vivo* support for oxidative damage to mitochondrial DNA in HD parietal cortex (Polidori et al., 1999), evidence for free radical production in the brains of HD patients and transgenic mice (Tabizi et al., 1999, 2000) and SOD-2 upregulation in an HD cell model (Li et al., 1999).

It is not clear if mitochondrial dysfunction and/or oxidative stress are primary changes or a consequence of the early neuropathological changes in HD. Guidetti et al. (2001) have argued that mitochondrial changes may be secondary events, since measurements of mitochondrial electron transport complexes I–IV did not reveal changes in the striatum and cerebral cortex in symptomatic HD transgenic mice without overt neuronal death. The neostriatum and cerebral cortex in human presymptomatic and pathological grade 1 HD cases also showed no change in the activity of mitochondrial complexes I–IV. Nevertheless, mitochondrial dysfunction, even if it is a secondary event, may impact on the pathological processes.

## XII. ANIMAL MODELS

A wide range of mouse models have been generated for HD, including animals where the huntingtin gene has been removed (knock-out), mice where a mutant HD gene or HD gene fragment has been added to the normal mouse genome that already has two wild-type mouse HD genes (transgenics), and mice where an endogenous mouse HD gene has been engineered to express an abnormally long polyglutamine tract (knock-in). These models have been summarized and compared by M. Chicuret, et al. on http://www.hdfoundation.org/PDF/hdmicetable.pdf. As described above, these models have provided important insights into processes associated with disease pathogenesis.

A powerful approach to identifying pathways involved in HD pathology and possible protective strategies is to perform suppressor screens (to identify genes that alleviate/modify the disease), for instance by using P element insertions in *Drosophila* (Kazemi-Esfarjani and Benzer, 2000). A number of groups have made *Drosophila* models of polyglutamine expansion diseases (Kazemi-Esfarjani and Benzer, 2000; Fernandez-Funez et al., 2001; Marsh et al., 2000; Warrick et al., 1998). These animals show many of the features of the human disease including cell death and aggregate formation. Suppressor screens have revealed several modifiers that highlight the role of protein folding and protein clearance in the development of SCA1 and HD (Kazemi-Esfarjani and Benzer, 2000; Fernandez-Funez et al., 2001). Other modifiers suggest novel mechanisms of polyglutamine pathogenesis, as some of the targets include genes that are involved in RNA processing, transcriptional regulation, and cellular detoxification (Fernandez-Funez et al., 2001).

The potential of such screens has stimulated the development of other models, including *Cenorhabditis elegans* (Faber et al., 1999). These screens have yielded and are likely to yield many interesting candidates and pathways for possible therapeutic interventions. However, these organisms may have important differences in the pathways that regulate and are perturbed by the HD mutation in man. Thus, ultimately such candidates from lower model organisms will need confirmation in mammalian models.

## XIII. GENOTYPE/PHENOTYPE/MODIFYING ALLELES

About 70% of the variance in the age at onset of HD can be accounted for by CAG repeat number. Family studies suggest that a component of this residual variance may be accounted for by additional genetic factors (Rosenblatt et al., 2001). One possible candidate modifying gene is the GluR6 kainate receptor (Rubinsztein et al., 1996). The effect that we reported of genotypes at this locus on age-at-onset of HD, after accounting for the CAG repeat length, has been replicated by MacDonald et al. (1999).

## XIV. TREATMENT

At present, there is no treatment that can arrest the course of HD. Recent trials in humans have used remacemide, a noncompetitive NMDA receptor antagonist, and lamotrigine, which blocks voltage-gated sodium channels inhibiting glutamate release—these compounds were selected on the assumption that excitotoxicity plays an important role in HD (Kremer et al., 1999; The Huntington Study Group, 2001). Coenzyme Q10, an antioxidant and cofactor involved in mitochondrial electron transfer has also been tested on its own and in combination with remacemide (Feigin et al., 1996, Kieburtz et al., 1996, Koroshetz et al., 1997, The Huntington Study Group, 2001). Although these and previous trials using baclofen (Shoulson et al., 1989) idebenone (Ranen et al., 1996) and vitamin E (Peyser et al., 1995) showed no clear effects, it is possible small beneficial effects may have been missed. For instance, the recent "large" trial of remacemide and coenzyme Q10 was designed with power to detect a 40% slowing of functional decline in a 30-month period on early HD patients (The Huntington Study Group, 2001). It is possible that such drugs may show clear beneficial effects if used for longer, or in larger studies, and may delay onset of disease if used in presymptomatic patients. Clearly, the logistical and cost implications of such large long-term trials in a comparatively rare disease will tend to restrict candidates to those which show major effects in symptomatic cases.

Another experimental approach that has been used to treat HD patients is striatal grafting. In rodents and nonhuman primates, striatal xenografts and allografts can survive, integrate into the host brain circuitry, and improve motor and cognitive functions in animals subjected to metabolic/excitotoxic lesions that mimic some of the features of HD (reviewed by Beal and Hantraye, 2001). Freeman and collaborators (2000) showed that grafted fetal striatal cells can survive and develop normally in the striatum of an HD patient. Recently Bachoud-Levi et al. (2000) reported promising pilot data from a small trial of five HD patients who received striatal grafting. Three of the five patients appeared to show a beneficial response to the procedure, while no overt functional benefit was seen in the other two patients. A number of centers are currently testing this procedure and it will be interesting to see data on a larger series of cases. It will be particularly important to investigate the long-term consequences of this procedure on dementia, as well as motor disturbances, since it is not clear to what extent striatal grafting could attenuate the cortical loss in HD.

The availability of a number of different mouse models of HD has provided powerful tools for preclinical testing of therapeutic strategies, since mice have uniform mutations (similar CAG lengths) and genetic backgrounds. Furthermore, compounds can be tested in the animals prior to onset of disease and data can be accumulated fairly rapidly, given the short life span of many of the HD mouse models. Promising results have been reported with minocycline, a tetracycline derivative that inhibits transcription of caspases and the inducible form of nitric oxide synthase, among other pathways, and creatine, which can buffer energy levels (Chen et al., 2000; Ferrante et al., 2000).

Other approaches that may be successful are those which inhibit polyglutamine aggregation. Huntingtin aggregation and cell death have been reduced in cell models with human single chain Fv intracellular antibodies (Lecerf et al., 2001), a polyglutamine antibody IC12 (Heiser et al., 2000), and peptides that interfere with polyglutamine aggregation (Nagai et al., 2000).

## XV. HD-LIKE DISORDERS

### A. Benign Hereditary Chorea

Benign hereditary chorea (or essential chorea) is another rare disease (or group of disorders) that is characterized by choreiform movement disorder. In contrast with HD, the onset of movement disorder is typically in early childhood and the severity peaks in the second decade after which it does not progress further. Patients have a normal life expectancy and there have been reports of disease improving with age. Recently, this disorder has been linked to a region of 14q in a large Dutch family (deVries et al., 2000) and the critical region was subsequently refined to a 6.93 cM region flanked by D14S1068 and D14S1064.

### B. Dominant Adult-Onset Basal Ganglia Disease

This recently described autosomal dominant disease presents variably with late-onset extrapyramidal features similar to those of HD or Parkinson's disease. It typically presents at 40–55 years with choreoathetosis, dystonia, spasticity, and rigidity and can show acute progression. However, it is not associated with significant cognitive decline or cerebellar involvement. It is caused by an adenine insertion in the ferritin light chain gene (19q13.3). The brain pathology is characterised by abnormal aggregates of ferritin and iron (Curtis et al., 2001).

## C. Neuroacanthocytosis

Neuroacanthocytosis is a rare degenerative autosomal recessive disorder characterized by the gradual onset of hyperkinetic movements and abnormal erythrocyte morphology. The neurological findings resemble those of HD, including dyskinetic choreiform movements and degeneration of the caudate nucleus. Other clinical features include psychiatric disturbances, epilepsy, self-mutilation, peripheral neuropathy and myopathy. The disease is associated with different mutations in the chorein protein gene that codes for 3096 residues (9q21–q22). Many of the mutations appear to be inactivating (Ueno *et al.*, 2001; Rampoldi *et al.*, 2001).

## D. McLeod Syndrome

McLeod syndrome is another disease associated with abnormal red cell morphology and is due to the loss of the XK antigen (Ho *et al.*, 1994). This X-linked disorder usually presents in males as a benign myopathy with areflexia. However, in some cases, the neurological symptoms are more severe and resemble those of acanthocytosis.

## E. Familial Prion Disease (Also Known as HD-Like-1)

A small proportion of families with HD-like signs do not have CAG repeat expansions in the HD gene. One such pedigree showed linkage to 20p12 and was found to have a 192-nucleotide insertion in the prion protein gene (PrP), which encodes an expanded PrP with 8 extra octapeptide repeats (Moore *et al.*, 2001).

## F. HD Like-2

Margolis *et al.* (2001) have described a family with HD-like symptoms with onset in the fourth decade that include involuntary movements and abnormalities of voluntary movement, psychiatric symptoms, weight loss, and dementia. The disease has a relentless course with death about 20 years after disease onset. Brain magnetic resonance imaging scans and an autopsy revealed marked striatal atrophy and moderate cortical atrophy, with striatal neurodegeneration in a dorsal to ventral gradient and occasional intranuclear inclusions. All tested affected individuals, and no tested unaffecteds, had a CAG/CTG trinucleotide repeat expansion of >40 in a variably spliced exon in junctophilin-3, a gene involved in the formation of junctional membrane structures. Depending on the splice acceptor site, this expansion is either in the 3′ UTR, in frame to code polyalanine, or in frame to code for polyleucine (Holmes *et al.*, 2001).

## G. Autosomal Recessive Huntington-Like Disorder (HD Like-3)

A disease that resembled juvenile HD was identified in a consanguineous family. Patients manifested at 3–4 years of age with pyramidal and extrapyramidal signs, including chorea, dytonia, ataxia, gait abnormailities, spasticity, seizures, mutism, and cognitive impairment. Brain MRI revealed progressive frontal cortical atrophy and bilateral caudate atrophy. Patients had normal repeat lengths in the HD. This autosomal recessive condition was mapped to a 7 cM region 4p15.3 (Kambouris *et al.*, 2000).

## Acknowledgments

I am grateful for Dr Kate Sugars for helpful suggestions and Mrs Denise Schofield for secretarial assistance. The work in my laboratory on HD is funded by Glaxo Wellcome, The Wellcome Trust, Medical Research Council, The Hereditary Disease Foundation, The Violet Richards Charity and The Isaac Newton Trust.

## References

Almqvist, E., Elterman, D., MacLeod, P., and Hayden, M. (2001). High incidence rate and absent family histories in one quarter of patients newly diagnosed with Huntington disease in British Columbia. *Clin. Genet.* **60**, 198–205.

Almqvist, E., Spence, N., Nichol, K., Andrew, S. E., Vesa, J., Peltonen, L., Anvret, M., Goto, J., Kanazawa, I., Goldberg, Y. P., and Hayden, M. R. (1995). Ancestral differences in the distribution of the delta 2642 glutamic acid polymorphism is associated with varying CAG repeat lengths on normal chromosomes: insights into the genetic evolution of Huntington disease. *Hum. Mol. Genet.* **4**, 207–214.

Ambrose, C. M., Duyao, M. P., Barnes, G., Bates, G. P., Lin, C. S., Srinidhi, J., Baxendale, S., Hummerich, H., Lehrach, H., Altherr, M., Wasmuth, J., Buckler, A., Church, D., Housman, D., Berks, M., Micklem, G., Durbin, R., Dodge, A., Read, A., Gusella, J. and MacDonald, M. E. (1994). Structure and expression of the Huntington's disease gene: evidence against simple inactivation due to an expanded CAG repeat. *Som. Cell Mol. Genet.* **20**, 27–38.

American College of Human Genetics and the American Society of Human Genetics Huntington Disease Genetic Testing Working Group (1998). ACMG/ASHG Statement Laboratory Guidelines for Huntington disease genetic testing. *Am. J. Hum. Genet.* **62**, 1243–1247.

Antonini, A., Leenders, K. L., Spiegel, R., Meier, D., Vontobel, P., Weigell-Weber, M., Sanchez-Pernaute, R., de Yebenez, J. G., Boesiger, P., Weindl, A., and Maguire, R. P. (1996). Striatal glucose metabolism and dopamine D2 receptor binding in asymptomatic gene carriers and patients with Huntington's disease. *Brain* **119**, 2085–2095.

Aronin, N., Kim, M., Laforet, G., and DiFiglia, M. (1999). Are there multiple pathways in the pathogenesis of Huntington's disease? *Philos. Trans. R. Soc. London, Ser.* **354**, 995–1003.

Augood, S. J., Faull, R. L., Love, D. R., and Emson, P. C. (1996). Reduction in enkephalin and substance P messenger RNA in the striatum of early grade Huntington's disease: a detailed cellular in situ hybridization study. *Neuroscience* **72**, 1023–1036.

Aylward, E. H., Codori, A. M., Barta, P. E., Pearlson, G. D., Harris, G. J., and Brandt, J. (1996). Basal ganglia volume and proximity to onset in presymptomatic Huntington disease. *Arch. Neurol.* **53**, 1293–1296.

Aylward, E. .H., Anderson, N. B., Bylsma, F. W., Wagster, M. V., Barta, P. E., Sherr, M., Feeney, J., Davis, A., Rosenblatt, A., Pearlson, G. D., and Ross, C.A. (1998). Frontal lobe volume in patients with Huntington's disease. *Neurology* **50**, 252–258.

Aylward, E. H., Codori, A. M., Rosenblatt, A., Sherr, M., Brandt, J., Stine, O. C., Barta, P. E., Pearlson, G. D., and Ross, C. A. (2000). Rate of caudate atrophy in presymptomatic and symptomatic stages of Huntington's disease. *Mov. Disord.* **15**, 552–560.

Bachoud-Levi, A. C., Remy, P., Nguyen, J. P., Brugieres, P., Lefaucheur, J. P., Bourdet, C., Baudic, S., Gaura, V., Maison, P., Haddad, B., Boisse, M. F., Grandmougin, T., Jeny, R., Bartolomeo, P., Dalla, Barba. G., Degos J. D., Lisovoski, F., Ergis, A. M., Pailhous, E., Cesaro, P., Hantraye, P., and Peschanski, M. (2000). Motor and cognitive improvements in patients with Huntington's disease after neural transplantation. *Lancet* **356**, 1975–1979.

Bailey, C. K., McCampbell, A., Madura, K. and Merry, D. E. (1998). Biochemical analysis of high molecular weight protein aggregates containing expanded polyglutamine repeat androgen receptor. *Am. J. Hum. Genet. Suppl.* **63**, A8.

Barth, A. L., McKenna, M., Glazewski, S., Hill, P., Impey, S., Storm, D., and Fox, K. (2000). Upregulation of cAMP response element-mediated expression during experience-dependent plasticity in adult neocortex. *J. Neurosci.* **20**, 4206–4216.

Beal, M. F., Kowall, N. W., Ellison, D. W., Mazurek, M. F., Swartz, K. J., and Martin, J. B. (1986). Replication of the neurochemical characteristics of Huntington's disease by quinolinic acid. *Nature* **321**, 168–171.

Beal, M. F., Ferrante, R. ., Swartz, K. J. and Kowall, N. W. (1989). Chronic quinolinic acid lesions in rats closely resemble Huntington's disease. *J. Neurosci.* **11**, 1649–1959.

Beal, M. F. (1992). Does impairment of energy metabolism result in excitotoxic neuronal death in neurodegenerative illnesses? *Ann. of Neurol.* **31**, 119–130.

Beal, M. F., and Hantraye, P. (2001). Novel therapies in the search for a cure for Huntington's disease. *Proc. Natl. Acad. Sci. U.S.A.* **98**, 3–4.

Becher, M. W., Rubinsztein, D. C., Leggo, J., Wagster, M. V., Stine, O. C., Ranen, N. G., Franz, M. L., Abbott, M. H., Sherr, M., MacMillan, J. C., Barron, L., Porteous, M., Harper, P. S., and Ross, C. A. (1997) Dentatorubral and pallidoluysian atrophy (DRPLA). Clinical and neuropathological findings in genetically confirmed North American and European pedigrees. *Mov. Disord.* **12**, 519–530.

Becher, M. W., Kotzuk, J. A., Sharp, A. H., Davies, S. W., Bates, G. P., Price, D. L., and Ross, C. A. (1998). Intranuclear neuronal inclusions in Huntington's disease and dentatorubral pallidoluysian atrophy: correlation between the density of inclusions and *IT15* CAG triplet repeat length. *Neurobio. Dis.* **4**, 387–397.

Beitel, L. K., Trifiro, M., and Pinsky, L. (1998). Spinobulbar muscular atrophy. In "Analysis of Triplet Repeat Disorders" (D.C. Rubinsztein and M.R. Hayden, eds.), pp. 85–103. Bios Scientific Publishers, Oxford.

Bence, N. F., Sampat, R. M., and Kopito, R. R. (2001). Impairment of the ubiquitin-proteasome system by protein aggregation. *Science* **292**, 1552–1555.

Borlongan, C. V., Koutouzis, T. K., and Sanberg, P. R. (1997). 3-Nitropropionic acid animal model and Huntington's disease. *Neurosci. Biobehav. Rev.* **21**, 289–293.

Brandt, J., Strauss, M. E., Larus, J., Jensen, B., Folstein, S. E., and Folstein, M. F. (1984). Clinical correlates of dementia and disability in Huntington's disease. *J. Clin. Neuropsychiatry* **6**, 401–412.

Brouillet, E., Hantraye, P., Ferrante, R. J., Dolan, R., Leroy-Willig, A., Kowall, N. W., and Beal, M. F. (1995). Chronic mitochondrial energy impairment produces selective striatal degeneration and abnormal choreiform movements in primates. *Proc. Nat. Acad. Sci. U.S.A.* **92**, 7105–7109.

Butterworth, N. J., Williams, L., Bullock, J. Y., Love, D. R., Faull, R. L., and Dragunow, M. (1998). Trinucleotide (CAG) repeat length is positively correlated with the degree of DNA fragmentation in Huntington's disease striatum. *Neuroscience* **87**, 49–53.

Cariello, L., de Cristofaro, T., Zanetti, L., Cuomo, T., Di Maio, L., Campanella, G., Rinaldi, S., Zanetti, P., Di Lauro, R., and Varrone, S. (1996). Transglutaminase activity is related to CAG repeat length in patients with Huntington's disease. *Hum. Genet.* **98**, 633–635.

Cattaneo, E., Rigamonti, D., Goffredo, D., Zuccato, C., Squitieri, F., and Sipione, S. (2001). Loss of normal huntingtin function: new developments in Huntington's disease research. *Trends Neurosci.* **24**, 182–188.

Carmichael, J., Chatellier, J., Woolfson, A., Milstein, C., Fersht, A.R., and Rubinsztein, D.C. (2000). Bacterial and yeast chaperones reduce both aggregate formation and cell death in mammalian cell models of Huntington's disease. *Proc. Natl. Acad. Sci. U.S.A.* **97**, 9701–9705.

Chai, Y., Koppenhafer, S.L., Shoesmith, S.J., Perez, M.K., and Paulson, H.L. (1999a). Evidence for proteasome involvement in polyglutamine disease: localization to nuclear inclusions in SCA3/MJD and suppression of polyglutamine aggregation *in vitro*. *Hum. Mol. Genet.* **8**, 673–682.

Chai, Y., Stacia, L. K., Bonini, N. M., and Paulson, H. L. (1999b). Analysis of the role of heat shock protein (Hsp) molecular chaperones in polyglutamine disease. *J. Neurosci.* **19**, 10,338–10,347.

Chen, M., Ona, V.O., Li, M., Ferrante, R.J., Fink, K. B., Zhu, S., Bian, J., Guo, L., Farrell, L. A., Hersch, S. M., Hobbs, W., Vonsattel, J. P., Cha, J. H., and Friedlander, R. M. (2000). Minocycline inhibits caspase-1 and caspase-3 expression and delays mortality in a transgenic mouse model of Huntington disease. *Nat. Med.* **6**, 797–801.

Chun, W., Lesort, M., Tucholski, J., Ross, C. A., and Johnson, G. V. (2001a). Tissue transglutaminase does not contribute to the formation of mutant huntingtin aggregates. *J. Cell Biol.* **153**, 25–34.

Chun, W., Lesort, M., Tucholski, J., Faber, P. W., MacDonald, M. E., Ross, C. A., and Johnson, G. V. (2001b). Tissue transglutaminase selectively modifies proteins associated with truncated mutant huntingtin in intact cells. *Neurobiol. Dis.* **8**, 391–404.

Coles, R., Caswell, R., and Rubinsztein, D. C. (1998). Functional analysis of the Huntington's disease (HD) gene promoter. *Hum. Mol. Genet.* **7**, 791–800.

Conway, K. A., Lee, S. J., Rochet, J. C., Ding, T. T., Williamson, R. E., and Lansbury, P. T., Jr. (2000). Acceleration of oligomerization, not fibrillization, is a shared property of both alpha-synuclein mutations linked to early-onset Parkinson's disease: implications for pathogenesis and therapy. *Proc. Natl. Acad. Sci. U.S.A.* **97**, 571–576.

Cooper, J. K., Schilling, G., Peters, M. F., Herring, W. J., Sharp, A. H., Kaminsky, Z., Masone, J., Khan, F. A., Delanoy, M., Borchelt, D. R., Dawson, V. L., Dawson, T. M., and Ross, C. A. (1998). Truncated N-terminal fragments of huntingtin with expanded glutamine repeats form nuclear and cytoplasmic aggregates in cell culture. *Hum. Mol. Genet.* **7**, 783–790.

Coyle, J. T., and Schwarcz, R. (1976). Lesion of striatal neurones with kainic acid provides a model for Huntington's chorea. *Nature* **263**, 244–246.

Cummings, C. J., Mancini, M. A., Antalffy, B., DeFranco, D. B., Orr, H. T., and Zoghbi, H. Y. (1998). Chaperone suppression of aggregation and altered subcellular proteasome localization imply protein misfolding in SCA1. *Nat. Genet.* **19**, 148–154.

Cummings, C. J., Reinstein, E., Sun, Y., Antalffy, B., Jiang, Y., Ciechanover, A., Orr, H. T., Beaudet, A. L., and Zoghbi, H. Y. (1999). Mutation of the E6-AP ubiquitin ligase reduces nuclear inclusion frequency while accelerating polyglutamine-induced pathology in SCA1 mice. *Neuron* **24**, 879–892.

Craufurd, D. (1998) Predictive testing for trinucleotide repeat disorders. In *Analysis of Triplet Repeat Disorders*. (D.C. Rubinsztein and M.R. Hayden, ed.), pp. 305–323. Bios Scientific Publishers, Oxford.

Curtis, A. R., Fey, C., Morris, C. M., Bindoff, L. A., Ince, P. G., Chinnery, P. F., Coulthard, A., Jackson, M. J., Jackson, A. ., McHale, D. P., Hay, D., Barker, W. A., Markham, A. F., Bates, D., Curtis, A., and Burn, J. (2001). Mutation in the gene encoding ferritin light polypeptide causes dominant adult-onset basal ganglia disease. *Nat. Genet.* **28**, 350–354.

Davies, S. W., Turmaine, M., Cozens, B.A., DiFiglia, M., Sharp, A. H., Ross, C. A., Scherzinger, E., Wanker, E. E., Mangiarini, L., and Bates, G.P. (1997). Formation of neuronal intranuclear inclusions underlies the neurological dysfunction in mice transgenic for the HD mutation. *Cell* **90**, 537–548.

Davies, S. W., Beardsall, K., Turmaine, M., DiFiglia, M., Aronin, N., and Bates, G.P. (1998). Are neuronal intranuclear inclusions the common neuropathology of triplet-repeat disorders with polyglutamine-repeat expansions? *Lancet* **351**, 131–133.

de la Monte, S. M., Vonsattel, J. P., and Richardson, E. P. Jr. (1998). Morphometric demonstration of atrophic changes in the cerebral cortex, white matter, and neostriatum in Huntington's disease. *J. Neuropathol. Exp. Neurol.* **47**, 516–525.

De Souza, E. B., Whitehouse, P. J., Folstein, S. E., Price, D. L., and Vale, W.W .(1987). Corticotropin-releasing hormone (CRH) is decreased in the basal ganglia in Huntington's disease. *Brain Res.* **437**, 355–359.

de Vries, B.B., Arts, W.F., Breedveld, G.J., Hoogeboom, J.J., Niermeijer, M.F., and Heutink, P. (2000). Benign hereditary chorea of early onset maps to chromosome 14q. *Am. J. Hum. Genet.* **66**, 136–142.

Dewhurst, K., Oliver, J. E., and McKnight, A. L. (1970). Sociopsychiatric consequences o Huntington's disease. *Br. J. Psychiatry* **116**, 255–258.

DiFiglia, M., Sapp, E., Chase, K. O., Davies, S. W., Bates, G. P., Vonsattel, J.-P., and Aronin, N.A. (1997). Aggregation of huntingtin in neuronal intranuclear inclusions and dystrophic neurites in brain. *Science* **277**, 1990–1993.

Di Maio, L., Squitieri, F., Napolitano, G., Campanella, G., Trofatter, J. A., and Conneally, P. M. (1993). Suicide risk in Huntington's disease. *J. Med. Genet.* **30**, 293–295.

Dragunow, M., Faull, R.L., Lawlor, P., Beilharz, E J., Singleton, K., Walker, E. B., and Mee, E. (1995). In situ evidence for DNA fragmentation in Huntington's disease striatum and Alzheimer's disease temporal lobes. *Neuroreport* **6**, 1053–1057.

Duyao, M. P., Auerbach, A. B., Ryan, A., Persichetti, F., Barnes, G. T., McNeil, S. M., Ge, P., Vonsattel, J.-P., Gusella, J. F., Joyner, A. L., and MacDonald, M. E. (1995). Inactivation of the mouse Huntington's disease gene homolog *Hd*h. *Science* **269**, 407–410.

Dyer, R. B. and McMurray, C. T. (2001) Mutant protein in Huntington disease is resistant to proteolysis in affected brain. *Nat. Genet.* **29**, 270–278.

Faber, P. W., Alter, J. R., MacDonald, M. E., and Hart, A. C. (1999). Polyglutamine-mediated dysfunction and apoptotic death of a Caenorhabditis elegans sensory neuron. *Proc. Natl. Acad. Sci. U.S.A* **96**, (1), 179–184.

Falush, D., Almqvist, E. W., Brinkmann, R. R., Iwasa ,Y. and Hayden, M. R. (2001). Measurement of mutational flow implies both a high new-mutation rate for Huntington disease and substantial underascertainment of late-onset cases. *Am. J. Hum. Genet.* **68**, 373–385.

Feigin, A., Kieburtz, K., Como, P., Hickey, C., Claude, K., Abwender, D., Zimmerman, C., Steinberg, K., and Shoulson, I. (1996). Assessment of coenzyme Q10 tolerability in Huntington's disease. *Mov. Dis.* **11**, 321–323.

Fernandez-Funez, P., Nino-Rosales, M. L., de Gouyon, B., She, W. C., Luchak, J. M., Martinez, P., Turiegano, E., Benito, J., Capovilla, M., Skinner, P. J., McCall, A., Canal, I., Orr, H. T., Zoghbi, H. Y., and Botas, J. (2001) Identification of genes that modify ataxin-1-induced neurodegeneration. *Nature* **408**, (6808), 101–106.

Ferrante, R. J., Andreassen, O. A., Jenkins, B. G., Dedeoglu, A., Kuemmerle, S., Kubilus, J. K., Kaddurah-Daouk, R., Hersch, S. M., and Beal, M. F. (2000). Neuroprotective effects of creatine in a transgenic mouse model of Huntington's disease. *J. Neurosci.* **20**, 4389–4397.

Fesus, L., and Thomazy, V. (1988). Searching for function of tissue transglutaminase: its possible involvement in biochemical pathway of programmed cell death. *Adv. Exp. Bio.* **231**, 119–134.

Folstein, S., Abbott, M. H., Chase, G. A., Jensen, B. A., and Folstein, M. F. (1983). The association of affective disorders with Huntington's disease in a case series and in families. *Psychol. Med.* **13**, 537–542.

Folstein, S. E., Chase, G., Wahl, W. E., McDonnell, A. M., and Folstein, M. F. (1987). Huntington's disease in Maryland: clinical aspects of racial variation. *Am. J. Hum. Genet.* **41**, 168–179.

Freeman, T. B., Cicchetti, F., Hauser, R. A., Deacon, T. W., Li, X. J., Hersch, S. M., Nauert, G. M., Sanberg, P. R., Kordower, J. H., Saporta, S., and Isacson, O. (2000). Transplanted fetal striatum in Huntington's disease: phenotypic development and lack of pathology. *Proc. Natl. Acad. Sci. U.S.A.* **97**, 13,877–13,882.

Gervais, F. G., Singaraja, R., Xanthoudakis, S., Gutekunst, C.-A., Leavitt, B. R., Metzler, M., Hackam, A., Tam, J., Vaillancourt, J. P., Houtzager, V., Rasper, D. M., Roy, S., Hayden, M. R., and Nicholson, D. W. (2002). Recruitment and activation of caspase-8 by the Huntingtin-interacting protein Hip-1 and a novel partner Hippi. *Nat. Cell Biol.* **4**, 95–105.

Goedert, M. (1999). Filamentous nerve cell inclusions in neurodegenerative diseases: tauopathies and alpha-synucleinopathies. *Philos. Trans. R. Soc. London, Ser. B:* **354**, 1101–1118.

Goldberg, Y. P., Nicholson, D. W., Rasper, D. M., Kalchman, M. A., Koide, H. B., Graham, R. K., Bromm, M., Kazemi-Esfarjani, P., Thornberry, N. A., Vaillancourt, J. P., and Hayden, M. R. (1996). Cleavage of huntingtin by apopain, a proapoptotic cysteine protease, is modulated by the polyglutamine tract. *Nat. Genet.* **13**, 442–449.

Grafton, S. T., Mazziotta, J. C., Pahl, J. J., St. George-Hyslop, P., Haines, J. L., Gusella, J., Hoffman, J. M., Baxter, L. R., and Phelps, M. E. (1992). Serial changes of cerebral glucose metabolism and caudate size in persons at risk for Huntington's disease. *Arch. Neurol.* **49**, 1161–1167.

Gu, M., Gash, M. T., Mann, V. M., Javoy-Agid, F., Cooper, M., and Schapira, A. H. V. (1996). Mitochondrial defect in Huntington's disease caudate nucleus. *Ann. Neurol.* **39**, 385–389.

Guidetti, P, Charles, V., Chen, E. Y., Reddy, P. H., Kordower, J. H., Whetsell, W. O. Jr., Schwarcz, R., and Tagle, D. A. (2001). Early degenerative changes in transgenic mice expressing mutant huntingtin involve dendritic abnormalities but no impairment of mitochondrial energy production. *Exp. Neurol.* **169**, 340–350.

Harper, P. S. (1996) *Huntington's Disease*, 2nd edition. W.B. Saunders, London.

He, F., Zhang, S., Qian, F., and Zhang, C. (1995). Delayed dystonia with striatal CT lucencies induced by a mycotoxin (3-nitropropionic acid). *Neurology* **45**, 2178–2183.

Heiser, V., Scherzinger, E., Boeddrich, A., Nordhoff, E., Lurz, R., Schugardt, N., Lehrach, H., and Wanker, E. E. (2000). Inhibition of huntingtin fibrillogenesis by specific antibodies and small molecules: implications for Huntington's disease therapy. *Proc. Natl. Acad. Sci. U.S.A.* **97**, 6739–6744.

Hendrick, J.P., and Hartl, F. U. (1993). Molecular chaperone functions of heat-shock proteins. *Ann. Rev. Biochem.* **62**, 349–384.

Ho, L. W., Brown, R., Maxwell, M., Wyttenbach, A., and Rubinsztein, D. C. (2001). Wild type huntingtin reduces the cellular toxicity of mutant huntingtin in mammalian cell models of Huntington's disease. *J. Med. Genet.* **38**, 450–452.

Ho, M., Chelly, J., Carter, N., Danek, A., Crocker, P., and Monaco, A. P. (1994). Isolation of the gene for McLeod syndrome that encodes a novel membrane transport protein. *Cell* **77**, 869–880.

Hodgson, J. G., Agopyan, N., Gutekunst, C. A., Leavitt, B. ., LePiane, F., Singaraja, R., Smith, D. J., Bissada, N., McCutcheon, K., Nasir, J., Jamot, L., Li, X. J., Stevens, M. E., Rosemond, E., Roder, J. C., Phillips, A. G., Rubin, E. M., Hersch, S. M., and Hayden, M. R. (1999). A YAC mouse model for Huntington's disease with full-length mutant huntingtin, cytoplasmic toxicity, and selective striatal neurodegeneration. *Neuron.* **23**, 181–192.

Holmes, S. E., O'Hearn, E., Rosenblatt, A., Callahan, C., Hwang, H.S., Ingersoll-Ashworth, R. G., Fleisher, A., Stevanin, G., Brice, A., Potter,

N. T., Ross, C. A., and Margolis, R. L. (2001) A repeat expansion in the gene encoding junctophilin-3 is associated with Huntington disease-like 2. *Nat. Genet.* **29**, 377–378.

Huang, Q., Zhou, D., Sapp, E., Aizawa, H., Ge, P., Bird, E. D., Vonsattel, J.-P., and DiFiglia, M. (1995). Quinolinic acid-induced increases in calbindin D28K immunoreactivity in rat striatal neurons *in vivo* and *in vitro* mimic the pattern seen in Huntington's disease. *Neuroscience* **65**, 397–407.

Huang, C. C., Faber, P. W., Persichetti, F., Mittal, V., Vonsattel, J. P., MacDonald, M. E., and Gusella, J. F. (1998). Amyloid formation by mutant huntingtin: threshold, progressivity and recruitment of normal polyglutamine proteins. *Som. Cell Mol. Genet.* **24**, 217–233.

Huntington's Disease Collaborative Research Group (1993). A novel gene containing a trinucleotide repeat that is expanded and unstable on Huntington's disease chromosomes. *Cell* **72**: 971–983.

Iannicola, C., Moreno, S., Oliverio, S., Nardacci, R., Ciofi-Luzzatto, A., and Piacentini, M. (2000). Early alterations in gene expression and cell morphology in a mouse model of Huntington's disease. *J. Neurochem.* **75**, 830–839.

Igarashi, S., Koide, R., Shimohata, T., Yamada, M., Hayashi, Y., Takano, H., Date, H., Oyake, M., Sato, T., Sato, A., Egawa, S., Ikeuchi, T., Tanaka, H., Nakano, R., Tanaka, K., Hozumi, I., Inuzuka, T., Takahashi, H., and Tsuji, S. (1998). Suppression of aggregate formation and apoptosis by transglutaminase inhibitors in cells expressing truncated DRPLA protein with an expanded polyglutamine stretch. *Nat. Genet.* **18**, 111–117.

Jana, N. R., Tanaka, M., Wang, Gh., and Nukina, N. (2000). Polyglutamine length-dependent interaction of Hsp40 and Hsp70 family chaperones with truncated N-terminal huntingtin: their role in suppression of aggregation and cellular toxicity. *Hum. Mol. Genet.* **9**, 2009–2018.

Jenkins, B. G., Koroshetz, W. J., Beal, M. F., and Rosen, B. R. (1993). Evidence for impairment of energy metabolism *in vivo* in Huntington's disease using localized $^1$H NMR spectroscopy. *Neurology* **43**, 2689–2695.

Jensen, P., Sorensen, S. A., Fenger, K., and Bolwig, T. G. (1993). A study of psychiatric morbidity in patients with Huntington disease, their relatives and controls. Admissions to psychiatric hospitals in Denmark from 1969 to 1991. *Br. J. Psychiatry* **163**, 790–797.

Kahlem, P., Green, H., and Dijan, P. (1998). Transglutaminase action imitates Huntington's disease: Selective polymerization of Huntington containing expanded polyglutamine. *Mol. Cell* **1**, 595–601.

Kambouris, M., Bohlega, S., Al-Tahan, A., and Meyer, B. F. (2000) Localisation of the gene for a novel autosomal recessive neurodegenerative Huntington-like disorder to 4p15.3. *Am. J. Hum. Genet.* **66**, 445–452.

Karpuj, M. V., Garren, H., Slunt, H., Price, D. L., Gusella, J., Becher, M. W., and Steinman, L. (1999). Transglutaminase aggregates huntingtin into nonamyloidogenic polymers, and its enzymatic activity increases in Huntington's disease brain nuclei. *Proc. Natl. Acad. Sci. U.S.A.* **96**, 7388–7393.

Kazemi-Esfarjani, P, and Benzer, S. (2000). Genetic suppression of polyglutamine toxicity in *Drosophila*. *Science* **287**, 1837–1840.

Kieburtz, K., Feigin, A., McDermott, M., Como, P., Abwender, D., Zimmerman, C., Hickey, C., Orme, C., Claude, K., Sotack, J., Greenamyre, J. T., Dunn, C., and Shoulson, I. (1996). A controlled trial of remacemide hydrochloride in Huntington's disease. *Mov. Dis.* **11**, 273–277.

Klement, I. A., Skinner, P. J., Kaytor, M. D., Yi, H., Hersch, S. M., Clark, H. B., Zoghbi, H. Y., and Orr, H. T. (1998). Ataxin-1 nuclear localization and aggregation: role in polyglutamine-induced disease in SCA1 transgenic mice. *Cell* **95**, 41–53.

Koroshetz, W. J., Jenkins, B. G., Rosen, B. R., and Beal, M. F. (1997). Energy metabolism defects in Huntington's disease and effects of coenzyme Q$_{10}$. *Ann. Neurol.* **41**, 160–165.

Kremer, B., Almqvist, E., Theilmann, J., Spence, N., Telenius, H., Goldberg, Y. P., and Hayden, M. R. (1995). Sex-dependent mechanisms for expansions and contractions of the CAG repeat on affected Huntington disease chromosomes. *Am. J. Hum. Genet.* **57**, 343–350.

Kremer, B., Clark, C. M., Almqvist, E. W., Raymond, L. A., Graf, P., Jacova, C., Mezei, M., Hardy, M. A., Snow, B., Martin, W., and Hayden, M. R. (1999). Influence of lamotrigine on progression of early Huntington disease: a randomized clinical trial. *Neurology*. **53**, 1000–1011.

Kuwert, T., Lange, H. W., Langen, K. J., Herzog, H., Aulich, A., and Feinendegen, L. E. (1990). Cortical and subcortical glucose consumption measured by PET in patients with Huntington's disease. *Brain*. **113**, 1405–1423.

Leavitt, B. R., Guttman, J. A., Hodgson, J. G., Kimel, G. H., Singaraja, R., Vogl, A. W., and Hayden, M. R. (2001). Wild-type huntingtin reduces the cellular toxicity of mutant huntingtin in vivo. *Am. J. Hum. Genet.* **68**, 313–324.

Lecerf, J. M., Shirley, T. L., Zhu, Q., Kazantsev, A., Amersdorfer, P., Housman, D. E., Messer, A., and Huston, J. S. (2001). Human single-chain Fv intrabodies counteract *in situ* huntingtin aggregation in cellular models of Huntington's disease. *Proc. Natl. Acad. Sci. U.S.A.* **98**, 4764–4769.

Li, S. H., Cheng, A. L., Li, H., and Li, X. J. (1999). Cellular defects and altered gene expression in PC12 cells stably expressing mutant huntingtin. *J. Neurosci.* **19**, 5159–5172.

Li, H., Li, S. H., Johnston, H., Shelbourne, P. F., and Li, X. J. (2000). Amino-terminal fragments of mutant huntingtin show selective accumulation in striatal neurons and synaptic toxicity. *Nat. Genet.* **25**, 385–389.

Li, H., Li, S. H., Yu, Z. X., Shelbourne, P., and Li, X. J. (2001). Huntingtin aggregate-associated axonal degeneration is an early pathological event in Huntington's disease mice. *J. Neurosci.* **21**, 8473–8481.

Lipe, H., Schultz, A., and Bird, T. D. (1993). Risk factors for suicide in Huntingtons disease: a retrospective case controlled study. *Am. J. Med. Genet.* **48**, 231–233.

Liu, Y. F. (1998). Expression of polyglutamine-expanded huntingtin activates the SEK1-JNK pathway and induces apoptosis in a hippocampal neuronal cell line. *J. Biol. Chem.* **273**, 28,873–28,877.

Ludolph, A. C., He, F., Spencer, P. S., Hammerstad, J., and Sabri, M. (1991). 3-Nitropropionic acid—exogenous animal toxin and possible human striatal toxin. *Can. J. Neurol. Sci.* **18**, 492–498.

Luthi-Carter, R., Strand, A., Peters, N. L., Solano, S. M., Hollingsworth, Z. R., Menon, A. S., Frey, A. S., Spektor, B. S., Penney, E. B., Schilling, G., Ross, C. A., Borchelt, D. R., Tapscott, S. J., Young, A. B., Cha, J. H., and Olson, J. M. (2000). Decreased expression of striatal signaling genes in a mouse model of Huntington's disease. *Hum. Mol. Genet.* **9**, 1259–1271.

McCampbell, A., Taylor, J. P., Taye, A. A., Robitschek, J., Li, M., Walcott, J., Merry, D., Chai, Y., Paulson, H., Sobue, G., and Fischbeck, K. H. (2000). CREB-binding protein sequestration by expanded polyglutamine. *Hum. Mol. Genet.* **9**, 2197–2202.

MacDonald, M. E., Vonsattel, J.-P., Shrinidhi, J., Couropmitree, N. N., Cupples, L.A., Bird, E. D., Gusella, J. F., and Myers, R. H. (1999). Evidence for the GluR6 gene associated with younger onset age of Huntington's disease. *Neurology* **53**, 1330–1332.

Mangiarini, L., Sathasivam, K., Seller, M., Cozens, B., Harper, A., Hetherington, C., Lawton, M., Trottier, Y., Lehrach, H., Davies, S. W., and Bates, G. P. (1996). Exon 1 of the HD gene with an expanded CAG repeat is sufficient to cause a progressive neurological phenotype in transgenic mice. *Cell* **87**, 493–506.

Margolis, R. L., O'Hearn, E., Rosenblatt, A., Willour, V., Holmes, S. E., Franz, M. L., Callahan, C., Hwang, H. S., Troncoso, J. C., and Ross, C.A. (2001). A disorder similar to Huntington's disease is associated with a novel CAG repeat expansion. *Ann. Neurol.* **50**, 373–380.

Marsh, J. L. Walker, H., Theisen, H., Zhu, Y. Z., Fielder, T., Purcel, L. J., and Thompson, L. M. (2000). Expanded polyglutamine peptides alone are intrinsically cytotoxic and cause neurodegeneration in Drosophila. *Hum. Mol. Genet.* **9**, (1), 13–25.

Martin, W. R., Clark, C., Ammann, W., Stoessl, A. J., Shtybel, W., and Hayden, M. R. (1992). Cortical glucose metabolism in Huntington's disease. *Neurology* **42**, 223–229.

Martindale, D., Hackam, A., Wieczorek, A., Ellerby, L., Wellington, C., McCutcheon, K., Singaraja, R., Kazemi-Esfarjani, P., Devon, R., Kim, S. U., Bredesen, D. E., Tufaro, F., and Hayden, M. R. (1998). Length of huntingtin and its polyglutamine tract influences localization and frequency of intracellular aggregates. *Nat. Genet.* **18**, 150–154.

McGeer, E. G., McGeer, P. L., and Singh, K. (1978). Kainate-induced degeneration of neostriatal neurons: dependency upon corticostriatal tract. *Brain Res.* **139**, 381–383.

McNeil, S. M., Novelletto, A., Srinidhi, J., Barnes, G., Kornbluth, I., Albuth, M. R., Wasmuth, J. J., Gusella, J. F., MacDonald, M. E., and Myers, R. H. (1997). Reduced penetrance of the Huntington's disease mutation. *Hum. Mol. Genet.* **6**, 775–779.

Mende-Mueller, L. M., Toneff, T., Hwang, S. R., Chesselet, M. F., and Hook, V. Y. (2001). Tissue-specific proteolysis of Huntingtin (htt) in human brain: evidence of enhanced levels of N- and C-terminal htt fragments in Huntington's disease striatum. *J. Neurosci.* **21**, 1830–1837.

Metzler, M., Legendre-Guillemin, V., Gan, L., Chopra, V., Kwok, A., McPherson, P. S., and Hayden, M. R. (2001). HIP1 functions in clathrin-mediated endocytosis through binding to clathrin and AP2. *J. Biol. Chem.* Aug 21 [epub ahead of print].

Moore, R. C., Xiang, F., Monaghan, J., Han, D., Zhang, Z., Edstrom, L., Anvret, M., and Prusiner, S. B. (2001). Huntington Disease Phenocopy Is a Familial Prion Disease. *Am. J. Hum. Genet.* Oct. 9 [epub ahead of print].

Murphy, K. S. J., Carter, R. J., Lione, L. A., Mangiarini, L., Mahal, A., Bates, G. P., Dunnett, S. B., and Morton, A. J. (2000). Abnormal synpatic plasticity and impaired spatial cognition in mice transgenic for exon 1 of the human Huntington's disease mutation. *J. Neurosci.* **20**, 5115–5123.

Nagai, Y., Tucker, T., Ren, H., Kenan, D. J., Henderson, B. S., Keene, J. D., Strittmatter, W. J., and Burke, J. R. (2000). Inhibition of polyglutamine protein aggregation and cell death by novel peptides identified by phage display screening. *J. Biol. Chem.* **275**, 10,437–10,442.

Nasir, J., Floresco, S. B., O'Kusky, J. R., Diewert, V. M., Richman, J. M., Zeisler, J., Borowski, A., Marth, J. D., Phillips, A. G., and Hayden, M. R. (1995). Targeted disruption of the Huntington's disease gene results in embryonic lethality and behavioural and morphological changes in heterozygotes. *Cell* **81**, 811–823.

Nicholson, D. W., and Thornberry, N. A. (1997). Caspases: Killer proteins. *Trends Biochem. Sci.* **22**, 299–306.

Nishiyama, K., Kwak, S., Takekoshi, S., Watanabe, K., and Kanazawa, I. (1996). In situ end-labelling detects necrosis of hippocampal pyramidal cells induced by kainic acid. *Neurosci. Lett.* **212**, 139–142.

Novelli, A., Reilly, J. A., Lysko, P. G., and Henneberry, R. C. (1988). Glutamate becomes neurotoxic via the *N*-methyl-*D*-aspartate receptor when intracellular energy levels are reduced. *Brain Res.* **451**, 205–212.

Nucifora, F. C., Jr., Sasaki, M., Peters, M. F., Huang, H., Cooper, J. K., Yamada, M., Takahashi, H., Tsuji, S., Troncoso, J., Dawson, V. L., Dawson, T. M., and Ross, C. A. (2001). Interference by huntingtin and atrophin-1 with cbp-mediated transcription leading to cellular toxicity. *Science* **291**, 2423–2428.

Ona, V. O., Li, M., Vonsattel, J.-P., Andrews, L. J., Khan, S. Q., Chung, W. M., Frey, A. S., Menon, A. S., Li, X. J., Stieg, P. E., Yuan, J., Penney, J. B., Young, A. B., Cha, J. H., and Friedlander, R. M. (1999). Inhibition of caspase-1 slows disease progression in a mouse model of Huntington's disease. *Nature* **399**, 263–267.

Ordway, J. M., Tallaksen-Greene, S., Gutekunst, C. A., Bernstein, E. M., Cearley, J. A., Wiener, H. W., Dure, L.S. 4th, Lindsey, R., Hersch, S. M., Jope, R. S., Albin, R. L., and Detloff, P. J. (1997). Ectopically expressed CAG repeats cause intranuclear inclusions and a progressive late onset neurological phenotype in the mouse. *Cell* **91**, 753–763.

Orr, H. T., Skinner, P. J., Klement, C. J., Cummings, C.J., and Zoghbi, H. Y. (1998). The role of ataxin-1 nuclear expression and aggregates in SCA1 pathogenesis. *Am. J. Hum. Genet. Suppl.* **63**, A8.

Paulson, H. L., Perez, M. K., Trottier, Y., Trojanowski, J. Q., Subramony, S. H., Das, S. S., Vig, P., Mandel, J.-L., Fischbeck, K. H., and Pittman, R. N. (1997). Intranuclear inclusions of expanded polyglutamine protein in spinocerebellar ataxia type 3. *Neuron* **19**, 333–344.

Paulson, H., Chai, Y., Gray-Board, G., and Bonini, N. (1998). Misfolding and aggregation in spinocerebellar ataxia type 3: a role for cellular chaperones in glutamine-repeat disease. *Am. J. Hum. Genet. Suppl.* **63**, A8.

Penney, J. B., Jr., Young, A. B., Shoulson, I., starosta-Rubenstein, S., Snodgrass, S. R., Sanchez-Ramos, J., Romos-Arroyo, M., Gomez, F., Penchaszadeh, G., Alvir, J., Esteves, J., DeQuiroz, I., Marsol, N., Moreno, H., Conneally, P.M., Bonilla, E., and Wexler, N. S. (1990). Huntington disease in Venezuela: 7 years of follow-up on symptomatic and asymptomatic individuals. *Mov. Dis.* **5**, 93–99.

Perutz, M. F., Johnson, T., Suzuki, M., and Finch, J. T. (1994). Glutamine repeats as polar zippers: Their possible role in inherited neurodegenerative diseases. *Proc. Nat. Acad. Sci. U.S.A.* **91**, 5355–5358.

Perutz, M.F. (1999) Glutamine repeats and neurodegenerative diseases: molecular aspects. *Trends Biochem. Sci.* **24**, 58–63.

Peters, P. J., Ning, K., Palacios, F., Boshans, R., Kazantzev, A., Thompson, L., Woodman, B., Bates, G. P., and D'Souza-Schorey, C. (2002) Arfaptin 2 regulates the aggregation of mutant huntingtin protein. *Nat. Cell. Biol.* Feb 18 [epub ahead of print].

Peyser, C. E., Folstein, M., Chase, G. A., Starkstein, S., Brandt, J., Cockrell, J. R., Bylsma, F., Coyle, J. T., McHugh, P. R., and Folstein, S. E. (1995). Trial of d-alpha-tocopherol in Huntington's disease. *Am. J. Psychiatry* **152**, 1771–1775.

Polidori, M. C., Mecocci, P., Browne, S. E., Senin, U., and Beal, M. F. (1999). Oxidative damage to mitochondrial DNA in Huntington's disease parietal cortex. *Neurosci. Lett.* **272**, 53–56.

Portera-Cailliau, C., Hedreen, J. C., Price, D. L., and Koliatsos, V. E. (1995). Evidence for apoptotic cell death in Huntington disease and excitotoxic animal models. *J. Neurosci.* **15**, 3775–3787.

Presinger, E., Jordan, B.M., Kazantsev, A., and Housman, D. (1999). Evidence for a recruitment and sequestration mechanism in Huntington's disease. *Philos. Trans. R. Soc. London, Ser. B:* **354**, 1029–1034.

Rampoldi, L., Dobson-Stone, C., Rubio, J. P., Danek, A., Chalmers, R. M., Wood, N. W., Verellen, C., Ferrer, X., Malandrini, A., Fabrizi, G. M., Brown, R., Vance, J., Pericak-Vance, M., Rudolf, G., Carre, S., Alonso, E., Manfredi, M., Nemeth, A. H., and Monaco, A. P. (2001). A conserved sorting-associated protein is mutant in chorea-acanthocytosis. *Nat. Genet.* **28**, 119–120.

Ranen, N. G., Stine, O. C., Abbott, M. H., Sherr, M., Codori, A.-M., Franz, M. L., Chao, N. I., Chung, A. S., Pleasant, N., Callahan, C., Kasch, L. M., Ghaffari, M., Chase, G. A., Kazazian, H. H., Brandt, J., Folstein, S. E., and Ross, C. A. (1995). Anticipation and instability of IT-15 (CAG)$_n$ repeats in parent-offspring pairs with Huntington's disease. *Am. J. Hum. Genet.* **57**, 593–602.

Ranen, N. G., Peyser, C. E., Coyle, J. T., Bylsma, F. W., Sherr, M., Day, L., Folstein, M. F., Brandt, J., Ross, C. A., and Folstein, S. E. (1996). A controlled trial of idebenone in Huntington's disease. *Mov. Disord.* **11**, 549–554.

Rigamonti, D., Bauer, J. H., De-Fraja, C., Conti, L., Sipione, S., Sciorati, C., Clementi, E., Hackam, A., Hayden, M. R., Li, Y., Cooper, J. K., Ross, C .A., Govoni, S., Vincenz, C., and Cattaneo, E. (2000). Wild-type huntingtin protects from apoptosis upstream of caspase-3. *J. Neurosci.* **20**, 3705–3713.

Rigamonti, D., Sipione, S., Goffredo, D., Zuccato, C., Fossale, E., and Cattaneo, E. (2001). Huntingtin's neuroprotective activity occurs via inhibition of procaspase-9 processing. *J. Biol. Chem.* **276**, 14,545–14,548.

Rodda, R. A. (1981). Cerebellar atrophy in Huntington's disease. *J. Neurol. Sci.* **50**, 47–57.

Rosenblatt, A., Brinkman, R. R., Liang, K. Y., Almqvist, E. W., Margolis, R. L., Huang, C. Y., Sherr, M., Franz, M. L., Abbott, M. H., Hayden, M. R., and Ross, C. A. (2001). Familial influence on age of onset among siblings with Huntington disease. *Am. J. Med. Genet.* **105**, 399–403.

Ross, C. A., and Hayden, M. R. (1998). Huntington's Disease. In *Analysis of Triplet Repeat Disorders* (D.C. Rubinsztein and M.R. Hayden, eds.), pp. 169–208. Bios Scientific Publishers, Oxford.

Rubinsztein, D. C., Amos, W., Leggo, J., Goodburn, S., Ramesar, R. S., Old, J., Bontrop, R., McMahon, R., Barton, D. E., and Ferguson-Smith, M.A. (1994). Mutational bias provides a model for the evolution of Huntington's disease and predicts a general increase in disease prevalence. *Nat. Genet.* **7**, 525–530.

Rubinsztein, D. C., Leggo, J., Coles, R., Almqvist, E., Biancalana, V., Cassiman, J.-J., Chotai, K., Connarty, M., Craufurd, D., Curtis, A., Curtis, D., Davidson, M. J., Differ, A.-M., Dode, C., Dodge, A., Frontali, M., Ranen, N. G., Stine, O. C., Sherr, M., Abbott, M. H., Franz, M. L., Graham, C. A., Harper, P. S., Hedreen, J. C., Jackson, A., Kaplan, J.-C., Losekoot, M., MacMillan, J. C., Morrison, P., Trottier, Y., Novelletto, A., Simpson, S. A., Theilmann, J., Whittaker, J. L., Folstein, S. E., Ross, C. A., and Hayden, M. R. (1996). Phenotypic characterisation of individuals with 30–40 CAG repeats in the Huntington disease (HD) gene reveals HD cases with 36 repeats and apparently normal elderly individuals with 36–39 repeats. *Am. J. Hum. Genet.* **59**, 16–22.

Rubinsztein, D. C., Leggo, J., Chiano, M., Dodge, A., Norbury, G., Rosser, E., and Craufurd, D. (1997). Genotypes at the GluR6 kainate receptor locus are associated with variation in the age of onset of Huntington's disease. *Proc. Nat. Acad. Sci. U.S.A.* **94**, 3872–3876.

Rubinsztein, D.C., and Hayden, M.R. (1998). Introduction. In *Analysis of Triplet Repeat Disorders* (ed. D.C. Rubinsztein and M.R. Hayden), pp. 1–12. Bios Scientific Publishers: Oxford.

Sanchez, I., Xu, C.J., Juo, P., Kakizaka, A., Blenis, J., and Yuan, J. (1999). Caspase-8 is required for cell death induced by expanded polyglutamine repeats. *Neuron* **22**, 623–633.

Sapp, E., Ge, P., Aizawa, H., Bird, E., Penney, J., Young, A.B., Vonsattel, J.P., and DiFiglia, M. (1995). Evidence for a preferential loss of enkephalin immunoreactivity in the external globus pallidus in low grade Huntington's disease using high resolution image analysis. *Neuroscience.* **64**, 397–404.

Sapp, E., Penney, J., Young, A., Aronin, N., Vonsattel, J.-P., and DiFiglia, M. (1999). Axonal transport of N-terminal huntingtin suggests early pathology of corticostriatal projections in Huntington disease. *J. Neuropathol. Exp. Neurol.* **58**, 165–173.

Saudou, F., Finkbeiner, S., Devys, D., and Greenberg, M. E. (1998). Huntingtin acts in the nucleus to induce apoptosis but death does not correlate with the formation of intranuclear inclusions. *Cell* **95**, 55–66.

Scherzinger, E., Lurz, R., Turmaine, M., Mangiarini, L., Hollenbach, B., Hasenbank, R., Bates, G. P., Davies, S. W., Lehrach, H., and Wanker, E. E. (1997). Huntingtin-encoded polyglutamine expansions form amyloid-like protein aggregates *in vitro* and *in vivo*. *Cell* **90**, 549–558.

Schilling, G., Becher, M. W., Sharp, A. H., Jinnah, H. A., Duan, K., Kotzuk, J.A., Slunt, H. H., Ratovitski, T., Cooper, J. K., Jenkins, N. A., Copeland, N. G., Price, D. L., Ross, C. A., and Borchelt, D. R. (1999). Intranuclear inclusions and neuritic aggregates in transgenic mice expressing a mutant N-terminal fragment of huntingtin. *Hum. Mol. Genet.* **8**, 397–407.

Sedvall, G., Karlsson, P., Lundin, A., Anvret, M., Suhara, T., Halldin, C., and Farde, L. (1994). Dopamine D1 receptor number—a sensitive PET marker for early brain degeneration in Huntington's disease. *Eur. Arch. Psychiatry Clin. Neurosci.* **243**, 249–255.

Shimohata, T., Nakajima, T., Yamada, M., Uchida, C., Onodera, O., Naruse, S., Kimura, T., Koide, R., Nozaki, K., Sano, Y., Ishiguro, H., Sakoe, K., Ooshima, T., Sat,o A., Ikeuchi, T., Oyake, M., Sato, T., Aoyagi, Y., Hozumi, I., Nagatsu, T., Takiyama, Y., Nishizawa, M., Goto, J., Kanazawa, I., Davidson, I., Tanese, N., Takahashi, H., and Tsuji, S. (2000). Expanded polyglutamine stretches interact with TAFII130, interfering with CREB-dependent transcription. *Nat. Genet.* **26**, 29–36.

Shoulson, I., Odoroff, C., Oakes, D., Behr, J., Goldblatt, D., Caine, E., Kennedy, J., Miller, C., Bamford, K., Rubin, A., *et al.* (1989). A controlled clinical trial of baclofen as protective therapy in early Huntington's disease. *Ann. Neurol.* **25**, 252–259.

Steffan, J. A., Kazantsev, A., Spasic-Boskovic, O., Greenwald, M., Zhu, Y.-Z., Gohler, H., Wanker, E., Bates, G. P., Housman, D. E., and Thompson, L. M. (2000). The Huntington's disease protein interacts with p53 and CREB-binding protein and represses transcription. *Proc. Natl. Acad. Sci. U.S.A.* **97**, 6763–6768.

Stenoien, D. L., Cummings, C.J., Adams, H. P., Mancini, M. G., Patel, K., DeMartino, G. N., Marcelli, M., Weigel, N. L., and Mancini, M. A. (1999). Polyglutamine-expanded androgen receptors form aggregates that sequester heat shock proteins, proteasome components and SRC-1, and are suppressed by the HDJ-2 chaperone. *Hum. Mol. Genet.* **8**, 731–741.

Stott, K., Blackburn, J. M., Butler, P. J., and Perutz, M. (1995). Incorporation of glutamine repeats makes protein oligomerize: implications for neurodegenerative diseases. *Proc. Nat. Acad. Sci. U.S.A.* **92**, 6509–6513.

Tabrizi, S. J., Cleeter, M. W., Xuereb, J., Taanman, J. W., Cooper, J. M., and Schapira, A. H. (1999). Biochemical abnormalities and excitotoxicity in Huntington's disease brain. *Ann. Neurol.* **45**, 25–32.

Tabrizi, .S J., Workman, J., Hart, P. E., Mangiarini, L., Mahal, A., Bates, G., Cooper, J. M., and Schapira, A. H. (2000). Mitochondrial dysfunction and free radical damage in the Huntington R6/2 transgenic mouse. *Ann. Neurol.* **47**, 80–86.

The Huntington Study Group. (2001). A randomized, placebo-controlled trial of coenzyme Q10 and remacemide in Huntington's disease. *Neurology* **57**, 397–404.

Thomas, L. B., Gates, D. J., Ritchfield, E. K., O'Brien, T. F., Schweitzer, J. B., and Steindler, D. A. (1995). DNA end labeling (TUNEL) in Huntington's disease and other neuropathological conditions. *Exp. Neurol.* **133**, 265–272.

Thornberry, N. A., and Lazebnik, Y. (1998). Caspases: Enemies within. *Science* **281**, 1312–1316.

Timmers, H. J., Swaab, D. F., van de Nes, J. A., and Kremer, H. P. (1996). Somatostatin 1–12 immunoreactivity is decreased in the hypothalamic lateral tuberal nucleus of Huntington's disease patients. *Brain Res.* **728**, 141–148.

Turjanski, N., Weeks, R., Dolan, R., Harding, A. E., and Brooks, D. J. (1995). Striatal D1 and D2 receptor binding in patients with Huntington's disease and other choreas. A PET study. *Brain* **118**, 689–696.

Turmaine, M., Raza, A., Mahal, A., Mangiarini, L., Bates, G. P., and Davies, S. W. (2000). Nonapoptotic neurodegeneration in a transgenic mouse model of Huntington's disease. *Proc. Natl. Acad Sci. U.S.A.* **97**, 8093–8097.

Ueno, S., Maruki, Y., Nakamura, M., Tomemori, Y., Kamae, K., Tanabe, H., Yamashita, Y., Matsuda, S., Kaneko, S., and Sano, A. (2001). The gene encoding a newly discovered protein, chorein, is mutated in chorea-acanthocytosis. *Nat Genet.* **28**, 121–122.

Usdin, M. T., Shelbourne, P. F., Myers, R. M., and Madison, D. V. (1999). Impaired synaptic plasticity in mice carrying the Huntington's disease mutation. *Hum. Mol. Genet.* **8**, 839–846.

Van Dijk, J. F., van der Velde, E. A., Roos, R. A. C., and Bruyn, G. W. (1986). Juvenile Huntington's disease. *Hum. Genet.* **73**, 235–239.

Vonsattel, J. P., Myers, R. H., Stevens, T. J., Ferrante, R. J., Bird, E. D., and Richardson, E. P. Jr. (1985). Neuropathological classification of Huntington's disease. *J. Neuropathol. Exp. Neurol.* **44**, 559–577.

Vonsattel, J. P., and DiFiglia, M. (1998). Huntington disease. *J. Neuropathol. Exp. Neurol.* **57**, 369–384.

Wadzinski, B.E., Wheat, W.H., Jaspers, S., Peruski, L.F. Jr., Lickteig, R.L., Johnson, G.L., and Klemm, D.J. (1993). Nuclear protein phosphatase 2A dephosphorylates protein kinase A-phosphorylated CREB and regulates CREB transcriptional stimulation. *Mol. Cell Biol.* **13**, 2822–2834.

Waelter, S., Scherzinger, E., Hasenbank, R., Nordhoff, E., Lurz, R., Goehler, H., Gauss, C., Sathasivam, K., Bates, G. P., Lehrach, H., and Wanker, E. E. (2001). The huntingtin interacting protein HIP1 is a clathrin and alpha-adaptin-binding protein involved in receptor-mediated endocytosis. *Hum. Mol. Genet.* **10**, 1807–1817.

Warrick, J. M., Paulson, H. L., Gray-Board, G. L., Bui, Q. T., Fischbeck, K. H., Pittman, R. N., Bonini, N. M. (1998). Expanded polyglutamine protein forms nuclear inclusions and causes neural degeneration in Drosophila. *Cell.* **93**, (6), 939–949.

Weeks, R. A., Piccini, P., Harding, A. E., and Brooks, D. J. (1996). Striatal D1 and D2 dopamine receptor loss in asymptomatic mutation carriers of Huntington's disease. *Ann. Neurol.* **40**, 49–54.

Weeks, R. A., Cunningham, V. J., Piccini, P., Waters, S., Harding, A. E., and Brooks, D. J. (1997). 11C-diprenorphine binding in Huntington's disease: a comparison of region of interest analysis with statistical parametric mapping. *J. Cereb. Blood Flow Metab.* **17**, 943–949.

Wellington, C. L., Singaraja, R., Ellerby, L., Savill, J., Roy, S., Leavitt, B., Cattaneo, E., Hackam, A., Sharp, A., Thornberry, N., Nicholson, D. W., Bredesen, D. E., and Hayden, M. R. (2000).Inhibiting caspase cleavage of huntingtin reduces toxicity and aggregate formation in neuronal and nonneuronal cells. *J. Biol. Chem.* **275**, 19,831–19,838.

Wellington, C. L., Ellerby, L. M., Hackam, A. S., Margolis, R. L., Trifiro, M. A., Singaraja, R., McCutcheon, K., Salvesen, G. S., Propp, S. S., Bromm, M., Rowland, K. J., Zhang, T., Rasper, D., Roy, S., Thornberry, N., Pinsky, L., Kakizuka, A., Ross, C. A., Nicholson, D. W., Bredesen, D. E., and Hayden, M. R. (1998). Caspase cleavage of gene products associated with triplet expansion disorders generates truncated fragments containing the polyglutamine tract. *J. Biol. Chem.* **273**, 9158–9167.

White, J. K., Auerbach, W., Duyao, M. P., Vonsattel, J.-P., Gusella, J. F., Joyner, A. L., and MacDonald, M. E. (1997). Huntingtin is required for neurogenesis and is not impaired by the Huntington's disease CAG expansion. *Nat. Genet.* **17**, 404–410.

Wyttenbach, A., Carmichael, J., Swartz, J., Furlong, R. A., Narain, Y., Rankin, J. and Rubinsztein, D. C. (2000). Effects of heat shock, Hsp40 and proteasome inhibition on protein aggregation in cellular models of Huntington's disease. *Proc. Nat. Acad. Sci. U.S.A.* **97**, 2898–2903.

Wyttenbach, A., Swartz, J., Kita, H., Thykjaer, T., Carmichael, J., Bradley, J., Brown, R., Maxwell, M., Schapira, A., Orntoft, T. F., Kato, K., and Rubinsztein, D.C. (2001). Polyglutamine expansions cause decreased CRE-mediated transcription and early gene expression changes prior to cell death in an inducible cell model of Huntington's disease. *Hum. Mol. Genet.* **10**, 1829–1845.

Young, A. B., Greenamyre, J. T., Hollingsworth, Z., Ablin, R., D'Amato, C., Shoulson, I., and Penney, J. B. (1988). NMDA receptor losses in putamen from patients with Huntington's disease. *Science* **241**, 981–983.

Zeitlin, S., Liu, J. P., Chapman, D. L., Papaioannou, V. E., and Efstratiadis, A. (1995). Increased apoptosis and early embryonic lethality in mice nullizygous for the Huntington's disease gene homolog. *Nat. Genet.* **11**, 155–163.

Zeron, M. M., Chen, N., Moshaver, A., Lee, A. T., Wellington, C. L., Hayden, M. R., and Raymond, L. A. (2001). Mutant huntingtin enhances excitotoxic cell death. *Mol. Cell Neurosci.* **17**, 41–53.

Zuccato, C., Ciammola, A., Rigamonti, D., Leavitt, B. R., Goffredo, D., Conti, L., MacDonald, M. E., Friedlander, R. M., Silani, V., Hayden, M. R., Timmusk, T., Sipione, S., and Cattaneo, E. (2001). Loss of huntingtin-mediated BDNF gene transcription in Huntington's disease. *Science* **293**, 493–498.

CHAPTER 34

# Paroxysmal Dyskinesias

KAILASH P. BHATIA
*University Department of Clinical Neurology*
*Institute of Neurology, Queen Square*
*University College London*
*London WC1N, United Kingdom*

I. Historical Aspects and Classification
   A. Paroxysmal Nonkinesigenic Dyskinesia (PNKD)/PDC
   B. PNKD Associated with Spasticity
   C. Paroxysmal Kinesigenic Dyskinesia PKD/(PKC)
   D. Paroxysmal Exercise-Induced Dyskinesia (PED)
   E. Genetics of PKD/PKC, the ICCA Syndrome, and PED (RE-PED-WC Syndrome)
   F. Paroxysmal Hypnogenic Dyskinesia (PHD)
II. Pathophysiology
   A. Epilepsy versus Basal Ganglia Dysfunction
   B. Similarities between Paroxysmal Dyskinesias and Some Neurological Channelopathies
III. Future Directions
IV. Concluding Summary
   References

Paroxysmal dyskinesias are a rare heterogenous group of conditions manifesting as abnormal involuntary movements that recur episodically and last only a brief duration (Fahn, 1994; Demerkirin and Jankovic, 1995). The abnormal movements may be choreic, dystonic, ballistic, or other or a mixture of these (Demirkirin and Jankovic, 1995). Between episodes the patient is generally normal. These conditions can be inherited or acquired.

## I. HISTORICAL ASPECTS AND CLASSIFICATION

In 1940, Mount and Reback described a 23-year-old man who had episodes of "choreo-dystonia" which could last several hours. More than twenty other family members were also affected with a clear autosomal dominant pattern of inheritance. Mount and Reback (1940) called these attacks *paroxysmal choreoathetosis* and thus gave the first clear descriptions of an episodic hyperkinetic condition. They and some others who described similar families (Forssman, 1961; Lance, 1963; Richards and Barnett, 1968) noted that these episodes of paroxysmal choreoathetosis seemed to be precipitated by drinking alcohol, coffee, or tea, and fatigue and smoking. Phenytoin and phenobarbitone were not found helpful. Richards and Barnett (1968), noting the torsion spasms and increased tone in the limbs, added "dystonia" to the description of Mount and Reback and this disorder thus came to be known as paroxysmal dystonic choreoathetosis (PDC).

In 1967 Kertesz described a new episodic disorder termed paroxysmal kinesigenic choreo-athetosis (PKC). Kertesz, noted that attacks in those affected were induced by sudden movement, i.e., *kinesigenic* and thus seemed different from PDC. As more cases and families were described (Lance, 1977) it became clear that this disorder responded well to anti-epileptics.

In 1977, Lance introduced a new form of paroxysmal dyskinesia describing a family who had attacks lasting between 5 and 30 minutes provoked by prolonged exercise and not by sudden movement. Lance referred to this paroxysmal exercise-induced dyskinesia (PED) as the "intermediate type" because the attack duration seemed longer then PKC but less than typical PDC. In the same paper, Lance gave the first clear classification of the paroxysmal dyskinesias based primarily upon the duration of attacks and divided these into three types: (1) PKC, in which there were brief attacks up to 5 minutes induced by sudden movement; (2) PDC, in which attacks were not induced by sudden movement and were of long duration up to 4 hours; and (3) PED, which was induced by exercise and the attack duration was between PKC and PDC. Over the years these terms have been used widely in the literature.

However, more recently, Demirkirin and Jankovic (1995) in a review pointed out that the attacks in these disorders were not necessarily choreic or dystonic and could be any form of dyskinesia. Hence they suggested classifying these disorders broadly into two main groups: paroxysmal kinesigenic dyskinesia (PKD) if the attacks were induced by sudden movement or paroxysmal non-kinesigenic dyskinesia (PKND) if they were not. These two broadly correspond to the terms paroxysmal kinesigenic choreoathetosis (PKC) and paroxysmal dystonic choreo-athetosis (PDC) of the earlier classification of Lance 1977. Apart from these two main forms, PED continued as a separate entity. Another paroxysmal disorder referred to as hypnogenic paroxysmal dyskinesia (HPD) in which dyskinetic episodes occurred only at night during sleep has also been recognized (Fahn, 1994; Demirkirin and Jankovic, 1995) and added to the main three. Each type can be further classified as either idiopathic or secondary (symptomatic) depending on the etiology. In the idiopathic form which is often familial, imaging and other investigations are unremarkable and there are no other signs to suggest a neurodegenerative or symptomatic cause. A whole host of different etiologies can cause the secondary paroxysmal dyskinesias. These include basal ganglia lesions (by stroke or other causes), demyelination, trauma, metabolic, drugs, and other causes (Demirkirin and Jankovic, 1995). This chapter, however, will only focus on the *idiopathic* four main varieties of paroxysmal dyskinesias mentioned above.

## A. Paroxysmal Nonkinesigenic Dyskinesia (PNKD)/PDC

### 1. Clinical Aspects (Phenotype) (Table 34.1)

PNKD is characterized by attacks of dyskinesia which are frequently precipitated by alcohol, caffeine, stress, or fatigue. The dyskinesia may be of any form but often tends to be more dystonic or choreic in nature. Patients with PNKD have longer (10 minutes to 6 hours) and less frequent attacks (1–3/day) as compared to PKD/PKC, followed by long attack-free intervals. More males than females are affected (1.4:1); (Fahn, 1994), and onset is usually in childhood with a tendency for the attacks to diminish with age (Jarman et al., 2000).

Since the initial description of Mount and Reback quite a few similar families have been described with an autosomal dominant inheritance with a fairly stereotyped

TABLE 34.1 Salient Clinical Features and Inheritance of the Familial Paroxysmal Movement Disorders

| Condition | Clinical features | Inheritance |
|---|---|---|
| PKD/PKC | Very brief movement-induced attacks, responsive to anticonvulsants | AD |
| ICCA syndrome | Benign infantile convulsions with episodes of dyskinesias like PKD, responsive to anticonvulsants | AD |
| PED | Attacks induced by exercise, lasting up to 1–2 hours, not responsive to anticonvulsants | AD |
| RE-PED-WC syndrome | Rolandic epilepsy, exercise-induced dyskinesias and writers cramp | AR |
| PNKD/PDC | Long duration attacks up to 6 hours, induced by caffeine, alcohol, fatigue; sleep produces benefit; usually no response to anticonvulsants | AD |
| PDC with spasticity | Single family in which patients with PDC had associated spasticity | AD |
| ANDFLE | Nocturnal episodes of dyskinesias, responsive to anticonvulsants | AD |

PKD/PKC = paroxysmal kinesigenic dyskinesia/chorea; PNKD/PKD = paroxysmal nonkinesigenic dyskinesia; PED = paroxysmal exercise-induced dyskinesia; ICCA = infantile convulsions and choreathetosis; RE-WC-PED = rolandic epilepsy, writers cramp, and paroxysmal exercise-induced dystonia syndrome.

clinical description between families (Mount and Reback, 1940; Richards and Barnett, 1968; Fink *et al.*, 1997; Jarman *et al.*, 2000). The clinical features of a large English family with 18 affected members were described recently (Jarman *et al.*, 2000). In all cases the onset of symptoms was very early in life in the second year in seven individuals and as early as age 6 months and 2 months in one individual each. Witnessed attacks consisted of generalized choreoathetosis in two individuals. In one case there was associated dysarthria in the attack. Attacks started in one limb or one side and progressed to become generalized. The attack duration varied from 10 minutes to 12 hours, however, the majority of attacks were between 30 and 180 minutes. All adults reported a decline in attack duration and frequency with age. One individual who was 85 had only 1 attack a year which was very mild. Coffee, alcohol, anger and excitement, hunger, and sleep deprivation were general precipitants as well as cold and exercise in two individuals. All affected reported remarkable response to sleep with 5–10 minutes of sleep being sufficient to abort an attack while some found drinking cold fluids or vigorous exercise while the attack was still mild could abort it. There was some diurnal fluctuation with a tendency to attacks in the afternoon or evening but not in the morning (Jarman *et al.*, 2000).

Generally PNKD cases have no detectable abnormalities between attacks, although there has been one report of a patient with PNKD who also had some interictal dystonia (Bressman *et al.*, 1988). There has also been a family with PNKD with additional myokimia (Byrne *et al.*, 1991).

With regard to investigations, EEGs and brain imaging in the idiopathic cases are normal. Pathological examination at autopsy in two cases revealed no abnormalities (Lance, 1977), nor were any neuropathological abnormalities seen in the basal ganglia in the mutant hamster which is an animal model of paroxysmal dystonia (Wahnschaffe *et al.*, 1990).

Regarding treatment most patients with PNKD generally do not benefit from anti-epileptics like carbamazepine as do patients with PKD, however, clonazepam can be helpful in some and clobazam was reportedly beneficial in one case (Jarman *et al.*, 2000). Some patients may respond to levodopa (Fink *et al.*, 1997). PNKD is thus more difficult to treat than PKD with drugs, however, over the years many patients learn to avoid precipitants and thus either avoid or even abort atttacks.

## 2. Genetics (Table 34.2)

Two groups separately linked families with PNKD to chromosome 2q (Fink *et al.*, 1996; Fouad *et al.*, 1996). Fink and co-workers performed a genome-wide search in a large American kindred of Polish descent with 28 affected members and mapped PNKD on chromosome 2q33–q35. Fouad *et al.* (1996) also showed tight linkage between PNKD and microsatellite markers on distal 2q (2q31–q36) in a 5-generation Italian family with 20 affected members. The smallest region of overlap of the candidate intervals identified by these 2 groups placed the PNKD locus in a 6–cM interval. In a 6-generation British family, Jarman and co-workers (1997) confirmed linkage to distal chromosome 2q and narrowed the candidate region to a 4-cM interval. Linkage to the same genetic location designated FPD1 (familial paroxysmal dyskinesia type 1) was also confirmed by Hofele *et al.* (1997) in a German family originally described by Przuntek and Monninger (1983) as classical of Mount and Reback type PNKD and other typical PNKD families one from North America of German descent (Raskind *et al.*, 1998) as well as a Japanese family (Matsuo *et al.*, 1999). Thus it has become clear that there is *genetic homogeneity* for classical familial PNKD/PDC.

As yet the gene has not been found but there are a number of ion channel genes in the area of linkage which may be suitable candidates (see Section II). Among them is a chloride/bicarbonate anion exchanger (SLC4A3), which has been considered a possible candidate gene particularly as a polymorphism within this gene and was thought to segregate with the disease (Raskind *et al.*, 1998; Matsuo *et al.*, 1999). Jarman *et al.* (1997) analyzed polymorphic tandem sequences within the gene and found no recombinations between PNKD and the intragenic polymorphisms. As yet it is not confirmed whether mutations in this gene cause the disorder, but it may have a putative role in the pathogenesis of PNKD. Similarly, Grunder *et al.* (2001) evaluated an acid-sensing ion channel (ASIC) 4 gene as a candidate gene but have not found this to be the cause of PDC. Further candidate genes are being studied.

### B. PNKD Associated with Spasticity

A large German family in whom affected members had choreo-dystonic attacks induced by alcohol, fatigue, and exercise and thus similar to PNKD was reported by Auburger *et al.* (1996). However, some affected, also had marked spastic paraparesis and other clinical features including perioral paresthesias, double vision, headache, and generalized myoclonic jerks and seizures were also present. This condition is therefore different from typical PNKD and can be considered as a PNKD *plus* syndrome. Not surprisingly this family was linked to a locus different to typical PNKD on chromosome 1p which has been designated CSE (choreoathetosis/spasticity, *episodic*). Linkage analysis in this family placed the disease locus in a 12–cM interval on chromosome 1p21 between flanking markers. The gene is yet to be determined but several potassium channel genes have been mapped to this region

TABLE 34.2 Mapped Loci/Genes for Familial Paroxysmal Dyskinesia Conditions

| Condition | Chromosome | Gene/ion channel |
|---|---|---|
| **PNKD** | | |
| Familial paroxysmal nonkinesiogenic dyskinesia (PNKD) | 2q33–35 | Not known |
| Paroxysmal choreoathetosis/spasticity | 1p | Not known |
| **PKD/ICCA** | | |
| Infantile convulsions and paroxysmal choreoathetosis (ICCA) | 16p12–q12 | Not known |
| Familial paroxysmal kinesigenic dyskinesias (PKD) | 16p11.2–q12.1 | Not known |
| **PED** | | |
| Autosomal recessive RE-WC-PED syndrome | 16p12–11.2 | Not known |
| **HPD** | | |
| Autosomal dominant nocturnal Frontal lobe epilepsy (ANDFLE) | 20q13 15q24 chr 1 | CHRNA4 ?CHRNA3 CHRNB2 |

RE-WC-PED = rolandic epilepsy, writers cramp, and paroxysmal exercise-induced dystonia syndrome; CHRN = neuronal nicotinic acetylcholine receptor; A4, A3, B2, = alpha 4, alpha 3 and beta 2 subunits, respectively.

and lie in a cluster. Further investigations are needed to determine the gene for this disorder. No other similar families have been reported so far.

## C. Paroxysmal Kinesigenic Dyskinesia PKD/(PKC)

### 1. Clinical Aspects (Phenotype)

The clinical aspects of this disorder were described in 10 cases by Kertesz (1967). Those affected had brief dyskinetic episodes precipitated by sudden movement and hence this condition was termed paroxysmal *kinesigenic* choreoathetosis. The kinesigenic form usually occurs from early childhood and in a recently reported large series of 26 idiopathic cases the mean age of onset was 13 years (range 1–39); (Houser *et al.*, 1999). If onset is much later in life it could suggest a symptomatic form of the disorder. In the same series there was a notable predominance of males (7:1) which has also been mentioned by earlier authors (Fahn, 1994). Most cases are idiopathic and apparently sporadic. Family history is present in about 23% of cases and usually follows an autosomal dominant pattern of inheritance (Houser *et al.*, 1999).

The attacks frequently manifest as dystonia or choreo-dystonia induced by a sudden change in position classically from a sitting to standing position or by a sudden change in velocity while walking or running. However, even startle, hyperventilation, and continuous exercise can trigger attacks. Rarely episodes occur at rest and even in sleep. A preceding "aura" like sensation in the limb which gets involved in an attack has been reported in 63% of cases with PKD (Houser *et al.*, 1999). Attacks commonly involve the hemi-body, in some almost always on the same side or alternating sides (Fahn, 1994; Demirkirin and Jankovic, 1995; Houser *et al.*, 1999). Rarely the episodes get generalized. Speech can be affected but consciousness is not lost. Typically PKD attacks are very brief and frequent and an attack will last from seconds to 1–2 minutes, occasionally up to 5 minutes. There can be dozens of attacks per day. Up to 26% of patients in the series by Houser *et al.* (1999) had other affected family members, with an autosomal dominant inheritance pattern in most.

Regarding treatment, PKD responds dramatically to different anti-epileptics, but there appears to be a particularly good response to carbamazepine even with relatively low doses of (Kato *et al.*, 1969; Wein *et al.*, 1996; Houser *et al.*, 1999).

Prognosis for the idiopathic form is good with the attack frequency tending to decrease with age and the condition may often abate in adult life.

### 2. The Association of PKC/PKD with Epilepsy

Until recently, epilepsy was not recognized to be an associated feature in both sporadic and familial cases with PKC. However, recently there have been a few reports describing families with "benign" infantile convulsions and later onset of episodes of paroxysmal choreoathetosis (called the *infantile convulsions and choreoathetosis*—ICCA syndrome); (Szeptowski *et al.*, 1997; Sadamatsu *et al.*, 1999; Hattori *et al.*, 2000; Caraballo *et al.*, 2001; Swoboda *et al.*, 2000, Thiriaux *et al.*, 2002). The first description was by Szeptowski *et al.* (1997) who found four French families said to have the ICCA syndrome as those affected had afebrile infantile seizures and paroxysmal involuntary movements. A report of a Chinese family by Lee *et al.* (1998) who had a similar condition soon followed. Although clinical details were somewhat sparse (in retrospect), the attacks of paroxysmal choreoathetosis in the ICCA syndrome resemble PKC in being very brief, frequent, and induced by sudden exertion and, interestingly also by ongoing exercise (Szeptowski *et al.*, 1997; Lee *et al.*, 1998). The families with the ICCA syndrome in both reports were linked to the pericentromic region of chromosome 16p12–q12 (Szeptowski *et al.*, 1997; Lee *et al.*, 1998). It does seem that for a while it was not appreciated that the paroxysmal dyskinesia in the ICCA syndrome had similarities to PKC. For example, soon after these reports, Sadamatsu *et al.* (1999) reported a large Japanese family with PKC and noted that some of them had infantile convulsions. However, although a negative search for linkage analysis to a variety of genes was performed (including the PNKD locus on chromosome 2q) they did not consider looking for linkage to chromosome 16 as they may have failed to recognize the similarity of the paroxysmal attacks in the ICCA syndrome to their family. Eventually but not surprisingly, families with typical PKD (without epilepsy) were also linked to the same region of chromosome 16 (Tomita *et al.*, 1999; Bennet *et al.*, 2000).

Other families with epilepsy and PKD have also been recently recognized. Hattori *et al.* (2000) described seven Japanese families and two sporadic cases with benign infantile convulsions and paroxysmal dyskinesias. Seventeen individuals developed afebrile complex partial infantile convulsions between age 3 and 12 months followed later by brief paroxysmal dyskinesia resembling PKC. Most families were autosomal dominant but in some only siblings were affected, suggesting a possible recessive inheritance and genetic heterogeneity. Another family with PKC and epilepsy has been described recently (Valente *et al.*, 2000). In this autosomal dominant family from India individuals with PKC did not have infantile convulsions but some had sporadic episodes of generalized tonic-clonic seizures in teenage years with spontaneous remission of their epilepsy in a few years although the PKC attacks continued (Valente *et al.*, 2000).

The genetics of PKC are discussed in more detail following the section on PED as there appears to be some genetic overlap between these disorders.

## D. Paroxysmal Exercise-Induced Dyskinesia (PED)

### 1. Clinical Aspects

Lance first described PED in a family in 1977. Affected members had attacks lasting between 5 and 30 minutes provoked by ongoing exertion like walking. The inheritance pattern was autosomal dominant. Plant *et al.* (1984) reported a mother and daughter with similar exercise-induced attacks and recently Munchau *et al.* (2000) reported an autosmal dominant family with PED who also had associated migraine. Sporadic examples of exercise-induced dystonia of the intermediate type of PDC are rare and only a few cases had been reported in the literature (Demerkirin and Jankovic, 1995; Bhatia *et al.*, 1997). PED is distinct from the kinesigenic form in that the attacks come on after 10 or 15 minutes of continuous exercise rather than at the initiation of movement. The attacks are usually dystonic and appear in the body part involved in the exercise, most commonly the legs after prolonged walking or running, and even focal dystonia of the jaw after chewing gum has been reported (Munchau *et al.*, 2000). Passive movements, vibration, and exposure to cold have also been reported to bring on attacks in affected individuals. The dystonic episodes usually cease in 10–15 minutes after stopping the exercise. There has been debate whether familial PED is a forme-fruste of PNKD (Lance, 1977; Kurlan *et al.*, 1987). However, in one family with classical PED linkage to the PNKD locus on chromosome 2q was excluded, although a genome-wide search to detect a novel locus could not be planned due to the small size of this family (Munchau *et al.*, 2000). If at all, there may be an overlap between PED and PKC as some of the affected in the ICCA families (see in the previous section) had attacks of dyskinesias precipitated not just by sudden movement but also by prolonged exercise. In this context the family described by Guerrini *et al.* (1999) characterized by rolandic epilepsy, episodes of exercise-induced dystonia, and writers cramp (so called RE-PED-WC syndrome) is relevant. Those affected (three members of the same generation) had not just PED attacks but also episodes induced by sudden movements. This autosomal recessive family has also been linked to chromosome 16p 12–11.2 (Guerrini *et al.*, 1999) in the same region as the families with the ICCA syndrome (Szepetowski *et al.*, 1998) and PKD (mentioned above) suggesting a possible overlap between these different disorders. It would be interesting to see whether the uncomplicated autosomal dominant families with PED as described by Lance (1977) and Plant *et al.* (1984) also link to the same region.

Regarding response to treatment, anticonvulsants may be useful in some but not all cases of PED (Bhatia *et al.*, 1997). Other drugs which may be helpful in some cases include levodopa, acetazolamide, and trihexiphenidyl which can be tried in turn. Generally PED is more difficult to treat then PKC. In drug refractory cases stereotactic surgery may be considered particularly if attacks are predominantly one sided. There has been report of unilateral pallidotomy being useful in a case with PED (Bhatia *et al.*, 1997).

## E. Genetics of PKD/PKC, the ICCA Syndrome, and PED (RE-PED-WC Syndrome)

These three disorders have to be considered together as they are all linked to the small arm of the pericentromic region of chromosome 16. Szepetowski and colleagues (1998) linked 4 French families with infantile convulsions and paroxysmal choreoathetosis called "the ICCA syndrome" to a 10–cM interval around the pericentromeric region of chromosome 16. Linkage to the same locus was further confirmed in a Chinese ICCA family (Lee *et al.*, 1998). Since the clinical characteristics of the paroxysmal dyskinetic episodes in the ICCA syndrome were very similar to those described for PKD, it was not surprising that eight Japanese families (Tomita *et al.*, 1999) as well as an African-American kindred (Bennett *et al.*, 2000) with typical PKC were mapped by linkage analysis to the same pericentromeric region of chromosome 16. The PKC region in the Japanese families spanned 12.4 cM and overlapped by 6.0 cM the ICCA region. As there was an increased prevalence of afebrile infantile convulsions in the Japanese families with PKC, Tomita *et al.* (1999) suggested the possibility that the same gene may be responsible for both PKC and ICCA. However, the PKC interval identified in the African-American family in which individuals have PKC alone (and no infantile seizures) overlaps by 3.4 cM with the ICCA region and by 9.8 cM with the PKC region identified in Japanese families. Thus it was unclear whether there were two (or more) genes giving rise to both ICCA and PKC in these families or a single gene in this interval. Furthermore, the autosomal recessive family with RE-PED-WC syndrome described by Guerrini *et al.* (1999; see Section I.D) was also linked to chromosome 16 within the ICCA region but outside the 3.4 cM overlap between ICCA and PKC. Thus, it appears that RE-PED-WC syndrome might also be allelic to ICCA but is probably not allelic to PKC. Also, as epilepsy is the most striking feature of both the ICCA and RE-PED-WC syndromes and some of the ICCA attacks were induced by exercise, perhaps suggests a common underlying gene(s) for these two conditions which may be different from that giving rise to isolated PKC.

Swoboda *et al.* (2000) pointed out that locus heterogeniety was likely reporting linkage data on 11 families with PKD with infantile convulsions also present in nine families. Ethnic background was diverse including two African-American, Taiwanese, Ashkenazi Jewish, Dutch, Mexican, and mixed European ancestry. Linkage was confirmed to chromosome 16 and overlapped the ICCA region which the authors narrowed down to 3.2 cM. However, one family with classic PKC and infantile convulsions failed to share a common haplotype thus suggesting heterogeneity. It is also interesting to note the paper by Valente et al. (2000) in this context. They recently identified a family with PKC from India and linked it to a second locus on the long arm of chromosome 16, distinct from the locus of the Japanese families with PKC and hence this is referred to as episodic kinesigenic dyskinesia 2 locus (*EKD2*). The localization of PKC in the African-American family (Bennett et al., 2000) overlaps with both these regions. The African-American PKC locus may thus be allelic with either the Japanese or Indian PKC locus or represent yet another gene altogether. These results further demonstrate *genetic heterogeneity* of PKC. In fact there are families with PKD not linked to chromosome 16 at all suggesting evidence of yet another locus (Spacey et al., 2002).

The gene(s) causing these disorders are yet unknown but there are a group of ion channel genes which lie in this pericentromeric region of chromosome 16, within the ICCA and PKC intervals which are good candidates and are being currently investigated.

Finally, it is also interesting to note that a recent paper described linkage of seven families with benign neonatal infantile convulsions (without paroxysmal dyskinesias) to the same region as ICCA on 16p12–q12 (Carballo et al., 2001) thus suggesting that there may be a family of genes causing different paroxysmal disorders on the pericentromc region of chromosome 16.

### F. Paroxysmal Hypnogenic Dyskinesia (PHD)

In PHD the attacks of paroxysmal dyskinesia occur at night in sleep (hence hypnogenic), and this disorder is often erroneously suspected to represent night terrors or some other sleep disorder (Scheffer et al., 1994). In a typical PHD attack the patient awakens with a cry followed by involuntary dystonic and ballistic limb movements which are very brief lasting seconds and mostly not over a minute. The movements often involve the legs. There is no loss of consciousness. Usually there are no detectable concurrent EEG abnormalities. Several attacks (sometimes even 20–25) can occur each night.

Lugaresi and Cirignotta (1981) gave one of the first clear descriptions of this condition in five patients who had attacks in sleep almost every night. Lee et al. (1985) and others also described similar familial cases with a clear autosomal dominant inheritance pattern. It has now become clear that in a large proportion of these cases, especially the familial variety, these nocturnal dyskinesias are due to mesial frontal lobe seizures which are difficult to pick up on surface EEG recordings (Scheffer et al., 1995). Describing the salient features in six families from Canada, Australia, and U.K., Scheffer et al. (1995) suggested the eponym autosomal dominant nocturnal frontal lobe epilepsy (ADNFLE) to describe this disorder. The gene responsible for ADNFLE has been discovered in a few families (see section on genetics below).

Anti-epileptics, particularly carbamazepine, are very effective in most cases (Scheffer et al., 1995).

### 1. Genetics

One ADNFLE locus was first mapped by Phillips et al. (1995) on chromosome 20q13.2 in an Australian family. The obvious candidate was the alpha 4 subunit of the neuronal acetylcholine receptor, a ligand gated channel gene. Two different mutations in two families in the alpha 4 subunit of the neuronal acetylcholine receptor (*CHRNA4*) on 20q13.2 were found. The mutations observed were either a missense mutation in an Australian family and a Japanese family and a three base pair insertion in the case of a Norwegian family (Hirose et al., 1999; Steinlein et al., 1995, 1997; Weiland et al., 1996). However, a family from U.K. with ADNFLE was not found to linked to *CHRNA4* on chromosome 20q but to a locus on chromosome 15q24 close to a *CHRNA3/CNRNA5/CHRNB4* nicotinic acetylcholine receptor gene cluster (Phillips et al., 1998). Also, seven other families with ADNFLE and seven sporadic cases were unlinked to these loci on chromosome 20q13.2 and 15q24 thereby suggesting the existence of at least a third ADNFLE locus and *genetic heterogeneity* of this disorder. Confirming this, recently mutations of beta subunit of the nicotinic acetylcholine receptor gene causing nocturnal frontal lobe epilepsy have been reported (De Fusco et al., 2000; Philips et al., 2001).

## II. PATHOPHYSIOLOGY

### A. Epilepsy versus Basal Ganglia Dysfunction

There has been great debate over the years about the pathophysiological mechanism of the paroxysmal dyskinesia particularly whether they represent a form of epilepsy or whether they are a basal ganglia disorder. Taking PKD/PKC as an example many authors regard this as a form of reflex epilepsy involving the thalamus or the basal ganglia (Lishman et al., 1962; Falconer et al., 1963; Kinast et al., 1980; Lombroso, 1995). Their main arguments are the paroxysmal nature of the attacks, the non-progressive and remitting character of the disorder, and the excellent response to anticonvulsants (Lishman et al., 1962). The arguments in favor of a subcortical focus and not a cortical one are the absence of seizure discharges on EEG in the majority of cases, the absence of evolution of the attacks into generalized focal convulsions, and the lack of an associated loss of consciousness or amnesia. In support of the epilepsy theory is the patient described by

Falconer *et al.* (1963). In this patient who had "seizures induced by movement" an excision of a cortical scar from the left supplementary motor cortex resulted in cessation of attacks.

Other authors believe that this disorder is due to basal ganglia disease in view of the electrophysiological studies in PKC by Franssen *et al.* (1983), who reported abnormalities of contingent negative variation (CNV) which normalized with phenytoin therapy in one patient with PKC. Similarly Houser *et al.* (1999) reported abnormalities of the premotor (*Bereitschafts*) potential in two cases. Also Kim *et al.* (1998) found abnormal decreased choline/creatine ratios in the basal ganglia in two patients with PKC on magnetic resonance spectroscopy. All this may suggest a possible functional disturbance of the prefrontal basal ganglia circuit.

### B. Similarities between Paroxysmal Dyskinesias and Some Neurological Channelopathies

The paroxysmal dyskinesias have many similarities to other episodic disorders of the nervous system like episodic ataxias and periodic paralysis (Bhatia *et al.*, 2000). Many of these paroxysmal neurological disorders are now known to be "channelopathies" due to mutations of genes regulating ion channels (Browne *et al.*, 1994; Ptacek *et al.*, 1994; Jurkat–Rott *et al.*, 1994; Ophoff *et al.*, 1996). In this regard it is interesting to note the many similarities between PKD and episodic ataxia type-1 (EA-1). Like PKD the episodes of ataxia in EA-1 are often provoked by kinesigenic stimuli and are brief lasting seconds to a few minutes and also can occur several times a day. (Brunt *et al.*, 1990). Both conditions have an early age of onset and tendency for both to abate in adulthood. EA-1 typically responds to acetazolamide as also to anticonvulsants (Griggs *et al.*, 1978; Brunt *et al.*, 1990). There are also reports of families with multiple episodic disorders, for example, paroxysmal dyskinesia in a family with episodic ataxia and association of episodic problems like migraine and epilepsy in families with paroxysmal ataxia or dyskinesias (Zubieri *et al.*, 1999; Valente *et al.*, 2000; Munchau *et al.*, 2000; Eunson *et al.*, 2000). Thus, the familial paroxysmal dyskinesias may also be due to defects in genes regulating ion channels. However, as mentioned earlier so far ADFNLE is the only condition where a mutation of a ligand gated ion channel gene has been identified and the genes for PKD and PNKD.

## III. FUTURE DIRECTIONS

The discovery of the gene will help better understand the pathophysiological mechanisms of these disorders. Even now given that there are some families linked to gene loci some speculations about pathophysiology have become possible. One example is PNKD linked to chromosome 2q. Fink *et al.* (1997) have tried to answer why alcohol and coffee precipitate attacks in their family. They have postulated a surge of dopamine release induced by alcohol, for example, followed by relative dopamine deficiency causes the dystonia. This notion is supported by the fact that some patients with PNKD respond to levodopa and there is sleep benefit (Fink *et al.*, 1997; Jarman *et al.*, 2000). A surge in CSF dopamine metabolites has been noted supporting the theory of Fink *et al.* in another family with PNKD also linked to 2q (Jarman *et al.*, 2000). Lombroso *et al.* (1999) have reported a PNKD patient in whom 18F-DOPA and an 11C raclopride PET scan revealed a marked reduction in the density of presynaptic dopa decarboxylase activity in the striatum, together with an increased density of postsynaptic dopamine D2 receptors. These findings may suggest a chronic upregulation of postsynaptic dopa receptors, either because of an increase in their numbers or changes in their affinity. All this suggests that there may be relative dopamine deficiency or some other mechanism involving the dopaminergic system (Fink *et al.*, 1997; Jarman *et al.*, 2000), but the exact pathophysiology is not yet clearly understood and will await the discovery of the genes.

The discovery of the genes will also lead to the possibility of developing animal models and to do functional studies at a cellular level particularly if the gene disorders are ion channel mutations as is expected. This has already become possible for ADNFLE which is known to be caused by mutations of the brain nicotinic acetyl choline receptor gene, a ligand gated channel. How these mutations cause epileptogenesis is not clearly understood but functional studies can be carried out expressing the mutant gene in cell lines and by doing single cell recordings using patch-clamp techniques to compare with the wild type. By this method, several different effects for different mutations of the alpha 4 or beta 2 subunits of the nicotinic acetyl choline receptor gene have been observed which include a decrease in maximal current amplitude (Bertrand *et al.*, 1998) or increase (Kuryatov *et al.*, 1997), decreases in acetylcholine affinity (Bertrand *et al.*, 1998), and reduction in calcium entry into cells (Kuryotov *et al.*, 1997; Steinlein *et al.*, 1997).

## IV. CONCLUDING SUMMARY

Given the advances in genetics and linkage for many of these disorders better decriptions of clinical phenotypes have become possible. Some disorders, for example, familial autosomal dominant PNKD, seem fairly distinct clinically and genetically, while there is possibly an overlap between PKD and PED. Although the genes for these have not yet been located the similarities between these paroxysmal dyskinesias and other intermittent neurological disorders like periodic paralysis and episodic ataxias is an indication that these are also likely to be caused by defective ion channel genes. Finding the genes will result in a better understanding of the pathophysiology, classification and treatment of these curious disorders.

## References

Auburger, G., Ratzlaff, T., Lunkes, A., Nelles, H. W., Leube, B., Binkofski, F., Kugel, H., *et al.* (1996). A gene for autososmal dominant paroxysmal choreoathetosis/spasticity (CSE) maps to the vicinity of a potassium channel gene cluster on chromososme 1p, probably within 2cM between D1S443 and D1S197. *Genomics* **31**, 90–94.

Bennett, L. B., Roach, E. S., and Bowcock, A. M. *et al.* (2000). A locus for paroxysmal kinesigenic dyskinesia maps to human chromosome 16. *Neurology* **54**, 125–130.

Bertrand, S., Weiland, S., Berkovic, S. F., Steinlein, O. K., and Bertrand, D. (1998). Properties of neuronal nicotinic acetylcholine receptor mutants from humans suffering from autosomal dominant nocturnal frontal lobe epilepsy. *Br. J. Pharmacol.* **125**, 751–760.

Bhatia, K. P., Griggs, R. C., and Ptacek L. J. (2000). Episodic movement disorders as channelopaties. *Mov. Disord.* **15**, 429–433.

Bhatia, K. P., Soland, V. L., Bhatt, M. H., Quinn, N. P., and Marsden, C. D. (1997). Paroxysmal exercise induced dystonia: eight new sporadic cases and a review of the literature. *Mov. Disord.* **12**, 1007–1012.

Bressman, S. B., Fahn, S., and Burke, R. E. (1988). Paroxysmal non-kinesigenic dystonia. *Adv. Neurol.* **50**, 403–413.

Browne, D. L., Gancher, S. T., Nutt, J. G., Brunt, E. R., Smith, E. A., Kramer, P. *et al.* (1994). Episodic ataxia myokimia syndrome is associated with point mutations in the human potassium channel gene, KCNA1. *Nat. Genet.* **8**, 136–140.

Brunt, E. R. P., and Van Weerden, T. W. (1990). Familial paroxysmal kinesigenic ataxia and continuous myokimia. *Brain* **113**, 1361–1382.

Byrne, E., White, O., and Cook, M. (1991). Familial dystonic choreoathetosis with myokimia; a sleep responsive disorder. *J. Neurol. Neurosurg. Psychiatry* **54**, 1090–1092.

Caraballo, R., Pavek, S., Lemainque, A., Gastaldi, M., Echenne, B., Motte, J., Genton, P., Cersosimo, R., Humbertclaude, V., Fejerman, N., Monaco, A. P., Lathrop, M. G., Rochette, J., and Szepetowski, P. (2001). Linkage of benign familial infantile convulsions to chromosome 16p12–q12 suggests allelism to the infantile convulsions and choreoathetosis syndrome. *Am. J. Hum. Genet.* **68**, 788–794.

De Fusco, M., Becchetti, A., Patrignani, A., Annesi, G., Gambardella, A., Quattrone, A., Ballabio, A., Wanke, E., and Casari, G. (2000). The nicotinic receptor beta 2 subunit is mutant in nocturnal frontal lobe epilepsy. *Nat. Genet.* **26**, 275–276.

Demirkirin, M., and Jankovic, J. (1995). Paroxysmal dyskinesias: clinical features and classification. *Ann. Neurol.* **38**, 571–579.

Eunson, L. H., Rea, R., Zuberi, S. M., Youroukos, S., Panayiotopoulos, C. P., Liguori, R., Avoni, P., McWilliam, R. C., Stephenson, J. B., Hanna, M. G., Kullmann, D. M., and Spauschus, A. (2000). Clinical, genetic, and expression studies of mutations in the potassium channel gene KCNA1 reveal new phenotypic variability. *Ann. Neurol.* **48**, 647–656.

Fahn, S. (1994). The paroxysmal dyskinesias. In *Movement Disorders 3* (C.D. Marsden and S. Fahn, eds) pp. 310–45. Butterworth-Heinmann Ltd, Oxford.

Falconer, M., Driver, M., and Serafetinides, E. (1963). Seizures induced by movement: report of case relieved by operation. *J. Neurol. Neurosurg. Psychiatry* **26**, 300–307.

Fink, J. K., Hedera, P., Mathay, J. G., and Albin, R. (1997). Paroxysmal dystonic choreoathetosis linked to chromosome 2q: clinical analysis and proposed pathophysiology. *Neurology* **49**, 177–183.

Fink, J. K., Rainier, S., Wilkowski, J., Jones, S. M., Kume, A., Hedera, P. *et al.* (1996). Paroxysmal dystonic choreoathetosis: tight linkage to chromosome 2q. *Am. J. Hum. Genet.* **59**, 140–145.

Forssman, H. (1961). Hereditary disorder characterized by attacks of muscular contractions, induced by alcohol amongst other factors. *Acta. Med. Scand.* **170**, 517–533.

Fouad, G. T., Servidei, S., Durcan, S., Bertini, E., and Ptacek, L. J. (1996). A gene for familial dyskinesia (FPD1) maps to chromosome 2q. *Am. J. Hum. Genet.* **5**, 135–139.

Franssen, H., Fortgens, C., Wattendorff, A. E., and Van Woerom, T. C. A. M. (1983). Paroxysmal kinesigenic choreoathetosis and abnormal contingent negative variation. A case report. *Arch. Neurol.* **40**, 381–385.

Griggs, R. C., Moxley, R. T. III, Lafralane, R. A., and McQuillen, J. (1978). Hereditary paroxysmal ataxia, response to acetazolamide. *Neurology* **28**, 1259–1264.

Grunder, S., Geisler, H. S., Ranier, S., Fink, J. K. (2001). Acid-sensing ionchannel (ASIC) 4 gene: Physical mapping, genomic organisation, and evaluation as a candidase for paroxysmal dystonia. *Eur. J. Hum. Genet.* **9**, 672–676.

Guerrini, R., Bonanni, P., Nardocci, N., Parmeggiani, L., Piccirilli, M., De Fusco, M. *et al.* (1999). Autosomal recessive rolandic epilepsy with paroxysmal exercise-induced dystonia and writers cramp: delineation of the syndrome and gene mapping to chromosome 16p12–11.2. *Ann. Neurol.* **45**, 344–352.

Hattori, H., Fujii, T., Nigami, H., Higuchi, Y., Tsuji, M., and Hamada, Y. (2000). Co-segregation of benign infantile convulsions and paroxysmal kinesigenic choreoathetosis. *Brain Dev.* **22**, 432–435.

Hirose, S., Iwata, H., Akiyoshi, H., Kobayashi, K., Ito, M., Wada, K., Kaneko, S., and Mitsudome, A. (1999). A novel mutation of CHRNA4 responsible for autosomal dominant nocturnal frontal lobe epilepsy. *Neurology* **53**, 1749–1753.

Hofele, K., Benecke, R., and Auburger, G. (1997). Gene locus FPD1 of the dystonic Mount-Reback type of autosomal-dominant paroxysmal choreoathetosis. *Neurology* **49**, 1252–1257.

Houser, M. K., Soland, V. L., Bhatia, K. P., Quinn, N. P., and Marsden, C. D. (1999). Paroxysmal kinesigenic choreoathetosis: a report of 26 cases. *J. Neurol.* **246**, 120–126.

Jarman, P. R., Bhatia, K. P., Davie, C., Heales, S. J., Turjanski, N., Taylor-Robinson, S. D., Marsden, C. D., and Wood, N. W. (2000). Paroxysmal dystonic choreoathetosis: clinical features and investigation of pathophysiology in a large family. *Mov. Disord.* **15**, 648–657.

Jarman, P. R., Davis, M. B., Hodgson, S. V., Marsden, C. D., and Wood, N. W. (1997). Paroxysmal dystonic choreoathetosis. Genetic linkage studies in a British family. *Brain* **120**, 2125–2130.

Jurkat-Rott, K., Lehmann-Horn, F., Elbaz, A., Heine, R., Gregg, R. G., Hogan, K., Powers, P. A. *et al.* (1994). A calcium-channel mutation causing hypokalemic periodic paralysis. *Hum. Mol. Genet.* **3**, 1415–1419.

Kato, M., and Araki, S. (1969). Paroxysmal kinesigenic choreoathetosis. Report of a case relieved by carbamazepine. *Arch. Neurol.* **20**, 508–513.

Kertesz, A. (1967). Paroxysmal kinesigenic choreoathetosis. An entity within the paroxysmal choreoathetosis syndrome. Description of 10 cases, including 1 autopsied. *Neurology* **17**, 680–690.

Kim, M. O., Im, J. H., Choi, C. G., and Lee, M. C. (1998). Proton MR spectroscopic findings in paroxysmal kinesigenic dyskinesia. *Mov. Disord.* **13**, 570–575.

Kinast, M., Erenberg, G., and Rothner, A. D. (1980). Paroxysmal choreo-athetosis, report of five cases and review of the literature. *Pediatrics* **65**, 74–77.

Kurlan, R., Behr, J., Medved, L., and Shoulson, I. (1987). Familial paroxysmal dystonic choreoathetosis: a family study. *Mov. Disord.* **2**, 187–192.

Kuryatov, A., Gerzanich, V., Nelson, M., Olale, F., and Lindstrom, J. (1997). Mutation causing autosomal dominant nocturnal frontal lobe epilepsy alters Ca2+ permeability, conductance, and gating of human alpha4beta2 nicotinic acetylcholine receptors. *J. Neurosci.* **17**, 9035–9047.

Lance, J. W. (1977). Familial paroxysmal dystonic choreoathetosis and its differentiation from related syndromes. *Ann. Neurol.* **2**, 285–293.

Lance, J. W., (1963). Sporadic and familial varieties of tonic seizures. *J. Neurol. Neurosurg. Psychiatry*, **26**, 51–59.

Lee, B. I., Lesser, R. P., Pippenger, C. E., Morris, H. H., Luders, H., Dinner, D. S. et al. (1985). Familial paroxysmal hypnogenic dystonia. *Neurology* **35**, 1357–1360.

Lee, W. L., Tay, A., Ong, H. T., Goh, L. M., Monaco, A. P., and Szepetowski, P. (1998). Association of infantile convulsions with paroxysmal dyskinesias (ICCA syndrome): confirmaton of linkage to human chromosome 16p12–q12 in a Chinese family. *Hum. Genet.* **103**, 608–612.

Lishman, W. A., Symonds, C. D., Whitty, C. W., and Wilson, R. G. (1962). Seizures induced by movement. *Brain* **85**, 93–108.

Lombroso, C. T., and Fischman, A. (1999). Paroxysmal non-kinesigenic dyskinesia: pathophysiological investigations. *Epileptic Disord.* **1**, 187–193.

Lombroso, C. T. (1995). Paroxysmal choreoathetosis: an epileptic or non-epileptic disorder? *Ital. J. Neurol. Sci.* **16(5)**, 271–277.

Lugaresi, E., and Cirignotta, F. (1981). Hypnogenic paroxysmal dystonia: epileptic seizure or a new syndrome? *Sleep* **4**, 129–138.

Matsuo, H., Kamakura, K., Saito, M., Okano, M., Nagase, T., Tadano, Y. et al. (1999). Familial paroxysmal dystonic choreoathetosis: clinical findings in a large Japanese family and genetic linkage to 2q. *Arch. Neurol.* **56**, 721–726.

Mount, L. A., and Reback, S. (1940). Familial paroxysmal choreo-athetosis. *Arch. Neurol. Psychiatry* **44**, 841–847.

Munchau, A., Valente, E. M., Shahidi, G. A., Eunson, L. H., Hanna, M. G., Quinn, N. P., Schapira, A. H. V., Wood, N. W., and Bhatia, K. P. (2000). A new family with paroxysmal exercise-induced dystonia and migraine: a clinical and genetic study. *J. Neurol. Neurosurg. Psychiatry* **68**, 609–614.

Ophoff, R. A., Terwindt, G. M., Vergouwe, M. N., et al. (1996). Familial hemiplegic migraine and episodic ataxia type-2 are caused by mutations in the $Ca^{2+}$ channel gene CACNLIA4. *Cell* **87**, 543–552.

Phillips, H. A., Scheffer, I. E., Berkovic, S. F., Holloway, V. E., Sutherland, G. R., and Mulley, J. C. (1995). Localization of a gene for autosomal dominant nocturnal frontal lobe epilepsy to chromosome 20q 13.2. *Nat. Genet.* **10**, 117–118.

Phillips, H. A., Scheffer, I. E., Crossland, K. M., Bhatia, K. P., Fish, D. R., Marsden, C. D. et al. (1998). Autosomal dominant nocturnal frontal-lobe epilepsy: genetic heterogeneity and evidence for a second locus at 15q24. *Am. J. Hum. Genet.* **63**, 1108–1116.

Phillips, H. A., Favre, I., Kirkpatrick, M., Zuberi, S. M., Goudie, D., Heron, S. E., Scheffer, I. E., Sutherland, G. R., Berkovic, S. F., Bertrand, D., and Mulley, J. C. (20001). CHRNB2 is the second acetylcholine receptor subunit associated with autosomal dominant nocturnal frontal-lobe epilepsy. *Am. J. Hum. Genet.* **68**, 225–231.

Plant, G. T., Williams, A. C., Earl, C. J., and Marsden, C. D. (1984). Familial paroxysmal dystonia induced by exercise. *J. Neurol. Neurosurg. Psychiatry* **47**, 275–279.

Przuntek, H., and Monninger, P. (1983). Therapeutic aspects of kinesio-genic paroxysmal choreoathetosis and familial paroxysmal choreo-athetosis of the Mount and Reback type. *J. Neurol.* **230**, 163–169.

Ptacek, L. J., Tawil, R., Griggs, R. C., Meola, G., McManis, P., Barohn, R. J., Mendell, J. R. et al. (1994). Sodium channel mutations in aceta-zolamide-responsive myotonia congenita, paramyotonia congenita, and hyperkalemic periodic paralysis. *Neurology* **44**, 1500–1503.

Raskind, W. H., Bolin, T., Wolff, J., Fink, J., Matsushita, M., Litt, M. et al. (1998). Further localization of a gene for paroxysmal dystonic choreo-athetosis to a 5-cM region on chromosome 2q34. *Hum. Genet.* **102**, 93–97.

Richards, R. N., and Barnett, H. J. (1968). Paroxysmal dystonic choreo-athetosis. A family study and review of the literature. *Neurology* **18**, 461–469.

Sadamatsu, M., Masui, A., Sakai, T., Kunugi, H., Nanko, S., and Kato, N. (1999). Familial paroxysmal kinesigenic choreoathetosis: an electro-physiologic and genotypic analysis. *Epilepsia* **40**, 942–949.

Scheffer, I. E., Bhatia, K. P., Lopes-Cendes, I. Fish, D. R., Marsden, C. D., Andermann, F. et al. (1994). Autosomal dominant frontal lobe epilepsy misdiagnosed as a sleep disorder. *Lancet* **43**, 515–517.

Scheffer, I. E., Bhatia, K. P., Lopes-Cendes, I., Fish, D. R., Marsden, C. D., Andermann, E. et al. (1995). Autosomal dominant nocturnal frontal lobe epilepsy. A distinctive clinical disorder. *Brain* **118**, 61–73.

Spacey, S. D., Valente, E. M., Wali, G. M., Warner, T. T., Jarman, P. R., Schapira, A. H. V., Dixon, P. H., Davis, M. B., Bhatia, K. P., and Wood, N. W. (2002). Genetic and clinical heterogeneity in paroxysmal kinesigenic dyskinesia: evidence for a third EKD locus. *Mov. Disord.* **17**, 717–725.

Steinlein, O. K., Mulley, J. C., Propping, P., Wallace, R. H., Phillips, H. A., Sutherland, G. R., Scheffer, I. E., and Berkovic, S. F. (1995). A missense mutation in the neuronal nicotinic acetylcholine receptor alpha 4 subunit is associated with autosomal dominant nocturnal frontal lobe epilepsy. *Nat. Genet.* **11**, 201–203.

Steinlein, O. K., Magnusson, A., Stoodt, J., Bertrand, S., Weiland, S., Berkovic, S. F., Nakken, K. O., Propping, P., and Bertrand, D. (1997). An insertion mutation of the CHRNA4 gene in a family with autosomal dominant nocturnal frontal lobe epilepsy. *Hum. Mol. Genet.* **6**, 943–947.

Swoboda, K. J., Soong, B. W., McKenna, C., Brunt, E. R., Litt, M., Bale. J, F. Jr., Ashizawa, T., Bennett, L. B., Bowcock, A. M., Roach, E. S., Gerson, D., Matsuura, T., Heydemann, P. T., Nespeca, M. P., Jankovic, J., Leppert, M., and Ptacek, L J. (2000). Paroxysmal kinesigenic dyskinesia and infantile convulsions. Clinical and linkage studies. *Neurology* **57**(Suppl. 4), S42–S48.

Szepetowski, P., Rochette, J., Berquin, P., Piussan, C., Lathrop, G. M., and Monaco, A. P. (1997). Familial infantile convulsions and paroxysmal choreoathetosis: a new neurological syndrome linked to the pericen-tromic region of human chromosome 16. *Am. J. Hum. Genet.* **61**, 889–898.

Thiriaux, A., de St. Martin, A., Vercueil, L., Battaglia, F., Armspach, J. P., Hirsch, E., Marescaux, C., and Namer, I. J. (2002). Co-occurrence of infantile epileptic seizures and childhood paroxysmal choreoathetosis in one family: Clinical, EEG, and SPECT characterization of episodic events. *Mov. Disord.* **17**, 98–104.

Tinuper, P., Cerullo, A., Cirignotta, F., Cortelli, P., Lugaresi, E., and Montagna, P. (1990). Nocturnal paroxysmal dystonia with short lasting attacks: three cases with evidence for an epileptic frontal lobe origin of seizures. *Epilepsia* **31**, 549–556.

Tomita, H., Nagamitsu, S., Wakui, K., Fukushima, Y., Yamada, K., Sadamatsu, M. et al. (1999). A gene for paroxysmal kinesigenic choreoathetosis mapped to chromosome 16p11.2–q12.1. *Am. J. Hum. Genet.* **65**, 1688–1697.

Valente, E. M., Spacey, S. D., Wali, G. M., Bhatia, K. P., Dixon, P. H., Wood, N. W., and Davis, M. B. (2000). A second paroxysmal kinesigenic choreoathetosis locus (EKD2) mapping on 16q13–q22.1 indicates a family of genes which give rise to paroxysmal disorders on human chromosome 16. *Brain* **123**, 2040–2045.

Wahnschaffe, U., Fredow, G., Heinz, P., and Losher, W. (1990). Neuro-pathological studies in a mutant hamster model of paroxysmal dystonia. *Mov. Disord.* **5**, 286–293.

Weiland, S., Witzemann, V., Villarroel, A., Propping, P., and Steinlein. O. (1996). An amino acid exchange in the second transmembrane segment of a neuronal nicotinic receptor causes partial epilepsy by altering its desensitization kinetics. *FEBS Lett.* **398**, 91–96.

Wein, T., Andermann, F., Silver, K., Dubeau, F., Andermann, E., Rourke-Frew, F., and Keene, D. (1996). Exquisite sensitivity of paroxysmal kinesigenic choreoathetosis to carbamazepine. *Neurology* **47**, 1104–1106.

Zuberi, S. M., Eunson, L. H., Spauschus, A., De Silva, R., Tolmie, J., Wood, N. W., McWilliam, R. C., Stephenson, J. P., Kullmann, D. M, Hanna, M. G. (1999). A novel mutation in the human voltage-gated potassium channel gene (Kv1.1) associates with episodic ataxia type 1 and sometimes with partial epilepsy. *Brain* **122**, 817–825.

# CHAPTER 35

# Primary Dystonias

ULRICH MÜLLER

*Institut für Humangenetik*
*Justus-Liebig-Universität*
*D35392 Giessen*
*Germany*

I. Autosomal Dominant Dystonias
   A. Dystonia 1 [Autosomal Dominant, Early-Onset Dystonia; Idiopathic Torsion Dystonia (ITD); Dystonia Musculorum Deformans 1: OMIM #128100]
   B. Dystonia 4 [OMIM [#128101]
   C. Dystonia 5 [Dopa-Responsive Dystonia (DRD); Hereditary Progressive Dystonia with Marked Diurnal Fluctuation (HPD); Segawa Syndrome; OMIM #128230]
   D. Dystonia 6 [Adult-Onset Idiopathic Torsion Dystonia of Mixed Type; OMIM #602629]
   E. Dystonia 7 [Focal, Adult-Onset Idiopathic Torsion Dystonia; Idiopathic Focal Dystonia (IFD); OMIM #602124]
   F. Dystonia 8 [Paroxysmal Dystonic Choreoathetosis (PDC); Paroxysmal Nonkinesigneic Dyskinesia (PNKD); Familial Paroxysmal Dyskinesia (FDP); Mount-Reback Syndrome; OMIM #118800]
   G. Dystonia 9 [Paroxysmal Choreoathetosis with Episodic Ataxia; and Spasticity (CSE); OMIM #601042]
   H. Dystonia 10 [Paroxysmal Familial Dystonia; Paroxysmal Kinesigneic Choreoathetosis (PKC); Periodic Dystonia; OMIM #128200]
   I. Dystonia 11 [Myoclonus-Dystonia; Alcohol-Responsive Dystonia; OMIM #159900]
   J. Dystonia 12 [Rapid-Onset Dystonia-Parkinsonism (RDP); OMIM #128235]
   K. Dystonia 13
II. Autosomal Recessive Dystonias
   A. Dystonia 2 [Autosomal Recessive Dystonia; Dystonia Musculorum Deformans 2; OMIM #224500]
III. X-Linked Recessive Dystonias
   A. Dystonia 3 [X-Linked Dystonia-Parkinsonism (XDP); "Lubag"; OMIM #314250]
   B. X-Linked Dystonia; Sensorineural Deafness, Mental Retardation, and Cortical Blindness; Dystonia Deafness Syndrome (DDS; DFN-1/MTS, OMIM #304700)
IV. Animal Models of Dystonia
References

Many neurogenetic disorders are clinically and genetically heterogeneous. Often clearcut genotype – phenotype correlations, however, cannot be established. Thus clinically different forms can be caused by mutations at the same locus and clinically identical variants can be due to mutations at different loci. In movement disorders this became quite obvious in the large group of spinocerebellar ataxias (SCA; Chapters 3 to 15). Failure of many attempts to arrive at a consistent clinical classification prompted Rosenberg in a visionary editorial to point out that "for the autosomal dominant phenotypes, the genotype is required to settle the issue of disease assignment and classification" (Rosenberg, 1990). Although this statement referred to SCAs at the time, it applies to many monogenic neurological disorders. Among movement disorders, the hereditary primary dystonias are another good example.

Clinically, dystonias can be classified based on affected body parts. Thus "focal dystonia" refers to involvement of a single region of the body, "segmental dystonia" denotes activity in two or more contiguous regions, and "multifocal dystonia" describes the affliction of two or more noncontiguous regions. Finally, generalized dystonia is defined by involvement of two or more body segments with the lower extremities always being affected.

Another clinical classification is based on age of disease onset distinguishing between childhood (0–12 years), adolescence (13–20 years), and adult onset. Childhood-onset dystonia tends to be most severe, frequently begins in the lower extremities, and often becomes generalized. Dystonia of adult-onset is frequently less severe, remains segmental or focal, and rarely begins in the lower limbs. The adolescence-onset form appears to lie between the two extremes, can manifest itself in either the lower or upper extremities, and tends to be less severe than childhood-onset dystonia (Fahn et al., 1987; Marsden, 1988; Marsden et al., 1976, 1986).

Associated signs, the mode of presentation of dystonic symptoms, and responsiveness to drugs and alcohol have been used for further classification of different forms of primary dystonias.

Molecular genetic findings of the last 10 years or so have now clearly demonstrated that clinical classifications are only of limited value. Thus genetically identical forms can manifest themselves in either childhood, adolescence, or adulthood. They can be generalized, focal, or even almost asymptomatic, and the site of onset can vary between individuals. Furthermore, associated signs such as parkinsonism are not necessarily obligatory manifestations in

TABLE 35.1 Classification of Primary Dystonias (AD = autosomal dominant, AR = autosomal recessive, XR = X-linked recessive)

| Designation | OMIM # | Mode of inheritance | Locus | Chromosomal location | Mutation in gene coding for |
|---|---|---|---|---|---|
| **Dystonia 1**—autosomal dominant, early onset dystonia; idiopathic torsion dystonia (ITD); dystonia musculorum deformans 1 | 128100 | AD | DYT1 | 9q34 | ATP-binding protein |
| **Dystonia 2**—autosomal recessive dystonia | 224500 | AR | DYT2 | — | — |
| **Dystonia 3**—X-linked dystonia parkinsonism (XDP); "lubag" | 314250 | XR | DYT3 | Xq13.1 | — |
| **Dystonia 4**—torsion dystonia 4 | 128101 | AD | DYT4 | — | — |
| **Dystonia 5**—dopa-responsive dystonia (DRD); hereditary progressive dystonia with marked diurnal fluctuation (HDP); Segawa syndrome | 128230 | AD | DYT5 | 14q22.1–q22.2 | GTP cyclohydrolase 1 |
| | 191290 | AR | — | 11p15.5 | Tyrosine hydroxylase |
| **Dystonia 6**—adult-onset idiopathic torsion dystonia of mixed type | 602629 | AD | DYT6 | 8p21–8q22 | |
| **Dystonia 7**—focal, adult-onset idiopathic torsion dystonia: idiopathic focal dystonia (IFD) | 602124 | AD | DYT7 | 18p | — |
| **Dystonia 8**—paroxysmal dystonic choreoathetosis (PDC); paroxysmal nonkinesigenic dyskinesia (PNKD); familial Paroxysmal dyskinesia (FDP); Mount-Reback syndrome | 118800 | AD | PNKCA | 2q34 | |
| **Dystonia 9**—paroxysmal choreoathetosis with episodic ataxia; kinesigenic choreoathetosis with episodic ataxia and spasticity (CSE) | 601042 | AD | CSE | 1p21–p13.3 | — |
| **Dystonia 10**—paroxysmal familial dystonia; paroxysmal kinesigenic choreoathetosis (PKC); periodic dystonia | 128200 | AD | DYT10, PKC | 16p11.2–q12.1 | — |
| **Dystonia 11**—myoclonus-dystonia; alcohol-responsive dystonia | 159900 | AD | DYT11 | 7q21 | ε-sarcoglycan |
| **Dystonia 12**—rapid-onset dystonia parkinsonism (RDP) | 128235 | AD | DYT12 | 19q13 | — |
| **Dystonia 13** | — | AD | DYT13 | 1p | — |
| **DFN-1/MTS**—deafness-dystonia syndrome/ Mohr-Tranebjaerg syndrome; deafness syndrome, progressive with blindness, dystonia, fractures, and mental deficiency | 304700 | XR | DDP | Xq22 | |

different members of the same families. Therefore, as in SCAs, the various forms of primary dystonias can only be distinguished beyond doubt by genetic criteria.

To date at least 13 monogenic primary dystonias have been delineated: 10 autosomal dominant, 1 autosomal recessive, and 2 X-linked recessive forms. They are designated as dystonias 1–13 (Müller *et al.*, 1998) and the corresponding disease loci can be referred to as *DYT1–DYT13* (Table 35.1). The disease gene is known in three forms of autosomal dominant dystonias, dystonia 1 (early-onset dystonia), dystonia 5 (dopa-responsive dystonia), and dystonia 11 (myoclonus-dystonia) and in the X-linked recessive DFN-1/MTS syndrome (X-chromosomal deafness dystonia syndrome).

In most primary dystonias there are no consistent neuropathological findings. Apart from dystonia 5 that is characterized by a dramatic therapeutic response to L-Dopa (Chapter 37), there are no causal therapies in most primary dystonias (Table 35.2).

## I. AUTOSOMAL DOMINANT DYSTONIAS

### A. Dystonia 1 [Autosomal Dominant, Early-Onset Dystonia; Idiopathic Torsion Dystonia (ITD); Dystonia Musculorum Deformans 1; OMIM #128100]

This form is the first inherited primary dystonia recognized (Schwalbe, 1908; Oppenheim, 1911) and is probably the one best studied. Based on the discovery of the disease gene in 1997 (Ozelius *et al.*, 1997) much work is now focusing on the molecular pathology of the disorder. In order to guarantee in depth treatment, dystonia 1 is discussed in a separate chapter (36).

### B. Dystonia 4 [OMIM#128101]

This designation has been reserved for autosomal dominant forms of dystonias in which known disease loci have been excluded. Since dystonia 4 potentially includes several different primary dystonias this designation is not useful. It should therefore be either abandoned or be applied to one of the newly recognized forms that are being discovered based on locus assignment by linkage analysis.

### C. Dystonia 5 [Dopa-Responsive Dystonia (DRD); Hereditary Progressive Dystonia with Marked Diurnal Fluctuation (HPD); Segawa Syndrome; OMIM #128230]

DRD can be caused by a mutation in the gene *GCH1*. Clinical manifestations of DRD vary widely and cover a broad spectrum of signs and symptoms ranging from generalized and focal dystonia via postural anomalies, parkinsonism, psychiatric symptoms, and subjective complaints to subtle signs only seen upon induction during clinical examination (Deonna, 1986; Nygaard *et al.*, 1988; Nygaard, 1993; Furukawa *et al.*, 1996; Bandmann *et al.*, 1996; Steinberger *et al.*, 1998, 1999). As a result of this highly variable phenotype DRD must be considered a differential diagnosis in most primary dystonias. This and the wealth of experimental data on the molecular pathology of DRD warrants separate treatment of this disorder (Chapter 37).

TABLE 35.2 Characteristic Manifestations in Various Forms of Dystonia

| Dystonia | Age of onset | Clinical features | Additional characteristics |
|---|---|---|---|
| 1 | Childhood | Generalized dystonia; onset lower body | |
| 3 | Adulthood | Generalized dystonia, parkinsonism | |
| 5 | Childhood | Generalized dystonia, diurnal fluctuation | Therapeutic response to L-Dopa |
| 6 | Adolescence, adulthood | Focal, segmental dystonia | |
| 7 | Adulthood | Focal dystonia | |
| 8 | Childhood, adolescence | Paroxysmal attacks of dystonia, chorea and ballism, nonkinesigenic | Precipitated by alcohol, caffeine, hunger, nicotine, fatigue, stress |
| 9 | Childhood, adolescence | Paroxysmal dyskinesia, kinesigenic | Precipitated by alcohol, fatigue, stress |
| 10 | Childhood | Frequent attacks of involuntary movement and posturing, kinesigenic seizures | Responsive to anticonvulsants |
| 11 | Childhood, adolescence | Focal, segmental dystonia, myoclonus | |
| 12 | Adolescence | Sudden onset of dystonia and parkinsonism, psychiatric involvement | |
| 13 | Adolescence, adulthood | Focal, segmental, sometimes generalized onset upper body | |
| DFN-1/MTS | Childhood | Dystonia, sensorineural deafness, spasticity, dysphagia, mental retardation, cortical blindness | |

Occasionally, the clinical phenotype of dopa-responsive dystonia is observed in patients with mutations of the *parkin* gene (Bonifati *et al.*, 2001). Most commonly, however, mutations in this gene cause an autosomal recessive, early-onset parkinsonism (see Chapter 28).

### D. Dystonia 6 [Adult-Onset Idiopathic Torsion Dystonia of Mixed Type; OMIM #602629]

This form was described in two not obviously related Mennonite families (Almasy *et al.*, 1997). The phenotype of dystonia 6 can be indistinguishable from that of dystonia 1 which is characterized by onset of symptoms in a limb and spreading to other limbs and the axial musculature, with laryngeal muscles being rarely affected. About half of the individuals from the Mennonite families, however, presented with cranial or cervical involvement, and even if symptoms started in an extremity, the head and neck became affected subsequently. Onset of dystonia 6 occurs during adolescence and early adulthood (average 18.9 years) and thus later than in dystonia 1 which typically has an onset in late childhood and early adolescence (average 13.6 years).

Linkage analysis in the two Mennonite families has mapped the disease gene to the pericentromeric region of chromosome 8 (8p21–8q22). The maximum combined two-point lod score of both families was 5.8 for marker D8S1797. Haplotypes across the candidate region were identical in affecteds from both families thus suggesting a founder effect and identical mutations. *DYT6* has not yet been identified in other families with autosomal dominant dystonia.

### E. Dystonia 7 [Focal, Adult-Onset Idiopathic Torsion Dystonia; Idiopathic Focal Dystonia (IFD); OMIM #602124]

Dystonias which remain focal usually have an onset during adulthood with a wide range from adolescence to late adult life. Frequently affected body parts include the neck (spasmodic torticollis), eyes (blepharospasm), and hands (writer's cramp). While the great majority of focal dystonias is sporadic, about 25% of cases have a positive family history (Waddy *et al.*, 1991).

One large family with adult-onset (range 28–70 years) autosomal dominant focal dystonia was studied in more detail. The family included seven members with definite dystonia, (six torticollis and one spasmodic dysphonia), and an additional six individuals with possible dystonia (five signs of torticollis such as muscular hypertrophy and minimal rotation of the head and one postural tremor of the hands). Linkage analysis in this family indicated a disease locus on the short arm of chromosome 18. The highest two-point lod score of 2.68 at $\theta = 0$ was found with marker D18S452. Haplotype analysis suggested location of the disease gene telomeric to locus D18S1153 (Leube *et al.*, 1996).

The same authors performed haplotype analyses with markers in the short arm of chromosome 18 in not obviously related patients with adult-onset idiopathic focal dystonia. Part of these patients originated from the same region in Northwestern Germany as the family with dystonia 7, some others came from Central Europe. The authors thought to have identified a common haplotype in some of these individuals and postulated a common founder of dystonia 7 in Europe (Leube *et al.*, 1997a,b). These findings were not confirmed in 85 individuals with focal dystonia from Northern Germany (Klein *et al.*, 1998).

In summary, delineation of dystonia 7 and assignment of *DYT7* to the short arm of chromosome 18 rests on linkage findings in a single family.

### F. Dystonia 8 [Paroxysmal Dystonic Choreoathetosis (PDC); Paroxysmal Nonkinesigenic Dyskinesia (PNKD); Familial Paroxysmal Dyskinesia (FDP); Mount-Reback Syndrome; OMIM #118800]

Paroxysmal (nonkinesigenic) dystonic choreoathetosis (PDC) is characterized by attacks of dystonia, chorea, and athetosis. Patients commonly present with a combination of dystonic posturing intermixed with chorea and ballism. Speech can be affected, but consciousness is always preserved. Attacks occur at rest and are not induced by sudden movements as, for example, in paroxysmal kinesigenic choreoathetosis/dyskinesea. There are, however, precipitating factors of PDC such as alcohol, caffeine, and to a lesser extent hunger, nicotine, fatigue, and emotional stress. Most patients are asymptomatic between attacks that last for 2 minutes to 4 hours. Age of onset is frequently during childhood or adolescence but adulthood-onset also occurs. Attacks tend to be more frequent at younger than at older age (Lance, 1977; Goodenough *et al.*, 1978; Bhatia, 1999).

PDC was first described by Mount and Reback in a large family with many affected members (Mount and Reback, 1940). Since then autosomal dominant inheritance of PDC has been demonstrated in many families. However, there are also sporadic cases of PDC. Linkage analyses in two large families have assigned the disease locus to the long arm of chromosome 2 (Fink *et al.*, 1996, 1997; Fouad *et al.*, 1996). One study mapped the *PDC* locus *(PNKCA)* to 2q33–q35 (Fink *et al.*, 1996) and the other study gives a more distal region in 2q (2q36) as the most likely critical interval of the disease gene. The slight discrepancies in locus assignment between the groups are likely due to analysis of different *STRP* (short tandem repeat polymorphism) loci that had been assigned to somewhat differing locations. The most recent linkage study in PDC also assigned *PNKCA* to 2q in a third family. The authors give a 5-cM interval in 2q34 as the most likely location of *PNKCA* (Raskind *et al.*, 1998). The authors of the first two studies speculated that one of the sodium channel genes in the region might be mutated in PDK. The investigators of the third study favored the anion exchanger SLC4A3 as a candidate. It appears, however, that there are no mutations in any one of these genes and the search for *PNKCA* is still on.

### G. Dystonia 9 [Paroxysmal Choreoathetosis with Episodic Ataxia; Kinesigenic Choreoathetosis with Episodic Ataxia and Spasticity (CSE); OMIM #601042]

Another syndrome of paroxysmal dyskinesia was reported by Auburger *et al.* (1996). The authors describe paroxysmal choreoathetosis, spasticity, and episodic ataxia in a large family. It appears that involuntary movements and dystonia in this family are similar to those in PDC. Episodes of involuntary movements, dystonic postures, dysarthria, paresthesias, and double vision lasted approximately 20 minutes and occurred between twice a day to twice a year. Age of onset was 2–15 years. Episodes could be induced by alcohol, fatigue, and emotional stress. Unlike PDC, physical exertion could also precipitate attacks and 5 of the 18 patients had spastic paraplegia both during and between episodes. Linkage analysis in this family assigned the disease locus *CSE* to a 12-cM interval on 1p (1p21–p13.3).

### H. Dystonia 10 [Paroxysmal Familial Dystonia; Paroxysmal Kinesigenic Choreoathetosis (PKC); Periodic Dystonia; OMIM #128200]

Kertesz (1967) and Walker (1981) delineated PKC as a separate entity quite different from PDC. Episodes are characterized by frequent, recurrent attacks of involuntary movement or posturing and are precipitated by sudden movements, thus being "kinesigenic". Other precipitating factors include startle, stress, anxiety, and excitement. Attacks are more frequent (up to 100 episodes per day) and of shorter duration than in PDC and are responsive to anticonvulsants. Although initially not considered part of the symptom spectrum of PKC, it is now recognized that abnormal EEG recordings and convulsions are common in PKC (Beaumanoir *et al.*, 1996; Sadamatsu *et al.*, 1999; Bhatia, 2001). Age of disease onset is during childhood with infantile convulsions occurring during the first two years of life (Swoboda *et al.*, 2000).

PKC can occur in a sporadic or familial fashion. Linkage analyses in large families with autosomal dominant transmission of the disorder have assigned a disease locus to the pericentromeric region of chromosome 16 (Szepetowski *et al.*, 1997; Tomita *et al.*, 1999; Swoboda *et al.*, 2000). Although Szepetowski *et al.* (1997) described the disorder that they mapped as a "new neurological syndrome characterized by infantile convulsions and paroxysmal choreoathetosis," they appear to have studied a variant or allelic form of PKC. This emphasizes the extremely variable expressivity observed in this disorder. Penetrance is reduced in PKC and was estimated at 0.74 for females and at 0.94 for males in 8 Japanese pedigrees (Tomita *et al.*, 1999). The sex difference of penetrance explains the more frequent occurrence of the disorder in males than in females (3:1 to 4:1; Kertesz, 1967; Lance, 1977; Goodenough *et al.*, 1978; Fahn, 1994; Marsden, 1996)

An autosomal recessive form of paroxysmal kinesigenic dystonia, i.e., rolandic epilepsy with paroxysmal exercise-induced dystonia and writer´s cramp (RE-PED-WC), has also been assigned to the pericentromeric region of chromosome 16 (16p12–p11.2); (Guerrini *et al.*, 1999). It is therefore possible that both autosomal recessive RE-PED-WC and autosomal dominant PKC are caused by mutations at the same locus on chromosome 16.

### I. Dystonia 11 [myoclonus-Dystonia; Alcohol-Responsive Dystonia; OMIM #159900]

Most cases of dystonia 11 (myoclonus-dystonia; M-D) are caused by mutations in the gene SGCE coding for ε-sarcoglycan (Zimprich *et al.*, 2001). Chapter 40 discusses the disorder in detail.

Findings in one family with M-D raise the possibility of genetic heterogeneity in this disorder. Thus linkage analysis in one large family with M-D resulted in a maximum multipoint lod score of 2.96 with markers of 11q thus tentatively assigning a disease locus to a 23-cM interval in 11q23. This region contains the gene coding for the D2 dopamine receptor (DRD2). Since this gene represented a good candidate for M-D, it was sequenced in patients and controls and a Val-Ile substitution was detected in a conserved region at the cytoplasmic end of transmembrane helix 4 of DRD2. To date this is the only amino acid change in DRD2 found in association with M-D (Klein *et al.*, 1999). It is presently not clear whether this exchange of two neutral amino acids has any functional consequences. In order to address this question, Klein *et al.* (2000) performed experiments with mutant and wild-type receptors expressed in a cell system. Their studies did not reveal differences in receptor-ligand binding and signaling between the wild-type and mutant receptor. Clearly, more studies are needed to settle the question whether DRD2 mutations can cause M-D or whether the amino acid change observed is a rare polymorphism possibly contributing to the development of M-D.

Interestingly, a study of polymorphisms in dopamine D1–5 receptor genes in patients with dystonia found one polymorphism (allele 2) in the D5 receptor gene that was overrepresented in patients with cervical dystonia as compared to controls (Placzek et al., 2001). If this finding is confirmed, it needs to be studied whether this polymorphism results in quantitative functional changes of the D5 receptor. Such hypothetical changes could be caused by variations in receptor density, ligand binding, and signal transduction.

### J. Dystonia 12 [Rapid-Onset Dystonia Parkinsonism (RDP); OMIM #128235]

RDP is characterized by the sudden onset of dystonia and parkinsonism over hours to days. Commonly observed dystonic symptoms include dysarthria, dysphagia, orofacial dystonia, dystonia of the upper extremities, and dystonic spasms. Common parkinsonian signs are bradykinesia and loss of postural reflexes. An additional hallmark of RDP is the frequent occurrence of psychiatric involvement, including depression, mood disturbances, and social anxiety and schizoid personality disorder. Disease onset is usually during adolescence but childhood- and adulthood-onset also occurs. The course of disease is relatively stable and the initial onset is followed by no or only moderate progression. Similar to other forms of dystonia, neuroimaging studies did not reveal any morphological changes in the basal ganglia (Kramer et al., 1999). Thus RDP is most likely due to a functional and not to a structural defect.

The disorder is inherited as an autosomal dominant trait with reduced penetrance and is quite rare. To date only three large families have been described (Kramer et al., 1999; Pittock et al., 2000). Linkage analysis in all three families assigned a disease locus to the long arm of chromosome 19 (19q13). The disease gene has not yet been identified.

Diagnosis relies on clinical evaluation of the patient and careful documentation of disease history. Differential diagnoses include other primary dystonias and juvenile-onset parkinsonism with mutations in the *parkin* gene.

### K. Dystonia 13

Dystonia 13 is a relatively benign form of dystonia of the head, neck, and upper limbs. Progression is mild and generalization appears to occur rarely. Age at onset is either during adolescence or adulthood. To date this form has been described in one large Italian family only. Penetrance was reduced in this family and expressivity was variable. The family included 11 definitely affected members. Linkage analysis resulted in a maximum two-point lod score of 3.44 between the disease locus and anonymous DNA marker D1S2667. This allowed assignment of *DYT13* to a 22-cM interval in the short arm of chromosome 1 (Valente et al., 2001).

## II. AUTOSOMAL RECESSIVE DYSTONIAS

### A. Dystonia 2 [Autosomal Recessive Dystonia; Dystonia Musculorum Deformans 2; OMIM #224500]

To date the existence of autosomal recessive forms of dystonia is controversial. Many cases of idiopathic torsion dystonia once thought to be inherited as autosomal recessives have been shown to be inherited as autosomal dominants with reduced penetrance. There are only a few pedigrees, mainly of consanguineous gypsies, still suggestive of autosomal recessive transmission of dystonia (Gimenez-Roldan et al., 1988). However, no homozygosity mapping has been performed in these families to map a disease locus. Hence, the occurrence of autosomal recessive inheritance remains uncertain.

#### 1. Tyrosine-Hydroxylase Deficiency

Mutations in the gene coding for tyrosine-hydroxylase have been reported in cases with dopa-responsive dystonia (Chapter 37). The investigations of van den Heuvel et al. (1998) and Swaans et al. (2000) suggest that deficiencies of this enzyme primarily present as autosomal recessive infantile parkinsonism.

## III. X-LINKED RECESSIVE DYSTONIAS

### A. Dystonia 3 [X-Linked Dystonia-Parkinsonism (XDP); "Lubag"; OMIM #314250]

X-linked dystonia-parkinsonism (XDP) is an adult-onset dystonia (age of onset 35.0 ± 8.0 years). XDP starts focally and generalizes after a median duration of 5 years (range 1–11) in almost all cases (Müller, 1999). A patient with generalized XDP is shown in Fig. 35.1. The first symptoms may either occur in the lower extremities (36%), the axial musculature (29%), the upper extremities (23%), or the head (12%). The site of onset does not affect the course of the disease, which is severely disabling in the majority of cases. "Parkinsonian symptoms" including bradykinesia, tremor, rigidity, and loss of postural reflexes were described in approximately one-third of a total of 42 cases studied. In those, parkinsonism was diagnosed as definitive (at least two parkinsonian symptoms including either bradykinesia or tremor) in 14%, probable (either tremor or bradykinesia alone) in 2%, and possible (rigidity and/or loss of postural reflexes) in 19% (Lee et al., 1991). More recent neurologic

FIGURE 35.1 Patient with generalized X-linked dystonia parkinsonism. Note oromandibular involvement.

examinations of additional patients suggest that concurrent parkinsonism is even more common in XDP with one or more parkinsonian symptoms being present in at least 50% of the cases (Kupke and Müller, unpublished). There is no correlation between occurrence of parkinsonism and site of onset, the tempo of progression, or the general severity of the disorder. The older the age of onset, however, the likelier concurrent parkinsonism seems to be. Duration of the illness may be as long as 40 years, but shorter courses are more common. Death usually occurs due to dysphagia and aspiration pneumonia.

XDP originated on the Philippine island of Panay from a single mutation (founder effect). Conservative estimates suggest 1 in 4000 males to be affected in Capiz Province, Panay (Kupke et al., 1990). XDP has also been reported in Filipino immigrants to the United States and Canada. Presently, the disorder has only been diagnosed in people of Filipino extraction.

XDP is inherited as an X-linked recessive trait. Penetrance is high, approaching 100% by the end of the fifth/beginning of the sixth decade. The disorder has primarily been described in males. In addition, three females with XDP have been reported (Kupke et al., 1990; Waters et al., 1993b). Linkage analysis assigned the XDP locus (DYT3) to the proximal long arm of the X chromosome (Xq12–q21). Analyses of allelic association and of haplotypes suggest a region within Xq13.1 as the most likely location of *DYT3* (Kupke et al., 1992; Graeber et al., 1992; Müller et al., 1994). This interval has been cloned in yeast artificial chromosmes (YACs) (Haberhausen et al., 1995), BACs/PACs and cosmids (Nemeth et al., 1999) and has been sequenced (accession number AL590763). Analysis of allelic association using STRPs from this region suggests that the critical interval covers approximately 300 kb and is flanked by *DXS6673E* proximally and by *DXS559* distally. Eight genes have been assigned to this region. They include *DXS6673E*, a gene of unknown function, *NonO* coding for p54$^{nrb}$ (a nuclear RNA binding protein), a gene coding for muscle-specific melusin (that interacts with integrin), *CCG1* coding for TATA binding protein-associated factor TAF$_{II}$$^{250}$, *ING2*, a putative tumor suppressor gene that is a member of the p33$^{ING1}$ gene family, *OGT* the gene for 0-linked N-acetylglucosamine transferase, *ACRC* coding for a putative nuclear protein containing an acidic amino acid repeat, and *CRK-L2* that codes for a G-protein coupled receptor (Müller et al., 2001; Nolte et al., 2001). Extensive sequence analysis of the expressed portions of all genes in XDP patients has not revealed a mutation. Thus XDP may be either caused by a mutation in a regulatory region, by a structural rearrangement, or by a mutation in a hither to unknown gene (Nolte et al., 2001).

Examination of patients by computer tomography (CT) and magnetic resonance imaging (MRI) did not demonstrate gross abnormalities. Neuropathological investigations of the brains of two XDP patients revealed neuronal loss and astrocytosis that were restricted to the caudate nucleus and the lateral putamen (Waters et al., 1993a). Consistently, neuroimaging in more advanced cases demonstrated "caudate and putaminal atrophy" (Lee et al., 2001). Positron emission tomography (PET) in three patients revealed selective reduction in normalized striatal glucose metabolism (Eidelberg et al., 1993). This suggests that abnormal striatal metabolism causes extra pyramidal signs in XDP.

As in most dystonias, there is no causative treatment. Of several medications tried in XDP patients including trihexyphenidyl, L-dopa, lorazepam, diazepam, and diphenhydramine, none affected the course of the disease. Even in patients with definitive concurrent parkinsonism (two treated), L-dopa did not result in noticeable improvement.

## B. X-linked Dystonia, Sensorineural Deafness, Mental Retardation, and Cortical Blindness; Dystonia Deafness Syndrome (DDS; DFN-1/MTS, OMIM #304700).

X-linked recessive dystonia also occurs in association with deafness and other symptoms. Putatively X-linked recessive dystonia deafness was first described by Scribanu and Kennedy (1976) in a small family with the proband and his maternal uncle being affected. A third male of the family (proband's nephew) only suffered from deafness. In both proband and his uncle deafness preceded dystonia. Dystonia started at age 7 and progressed rapidly in the patient. He was wheelchair-bound at age 9 and died at 11 years. Autopsy of the patient's brain revealed "neuronal loss and gliosis in both caudate nuclei, putamen, and globus pallidus." More recently, X-linked spasticity, dysphagia, mental retardation, and cortical blindness were observed in several pedigrees. In some patients, deafness and dystonia were the only symptoms. Other affected males displayed mental retardation in addition to deafness and dystonia but no blindness. Highly variable expressivity appears to be a characteristic of this syndrome which is referred to as DDS (dystonia deafness syndrome) when dystonia and deafness are the predominant findings in affecteds and as Mohr-Tranebjaerg syndrome (DFN-1/MTS) when mental retardation and cortical blindness occur. DDS/DFN-1/MTS appears to be a neurodegenerative disorder with corticospinal tract, brainstem, and the basal ganglia being affected (Mohr and Mageroy, 1960; Tranebjaerg et al., 1995; Jin et al., 1996).

The DFN-1/MTS locus was assigned to Xq21.3–q22 and a positional cloning approach identified the disease gene, DDP, in Xq22. Mutations in DDP were observed in both DDS and DFN-1/MTS (Ujike et al., 2001; Swerdlow and Wooten, 2001) thus proving that DDS is a milder manifestation of DFN-1/MTS and not a distinct syndrome. The gene DDP is composed of two exons separated by an intron of 2 kb. A transcript of 1.2 kb is detected in various tissues, including fetal and adult brain. In addition, alternative splice products of 600 and 400 bp are found in lower quantities in skeletal muscle, heart, and brain. DDP codes for a polypeptide of 97 amino acids with pronounced similarities to five yeast proteins (Tim8p, Tim9p, Tim 10p, Tim 12p, Tim 13p) of the mitochondrial intermembrane space (Koehler et al., 1999). These polypeptides interact to mediate the import of metabolite transporters from the cytoplasm into the inner mitochondrial membrane. The pronounced similarities of DDP with these polypeptides, especially with Tim8p, suggest that DDP plays an important role in human mitochondrial protein import as well. DFS-1/MTS is therefore most likely caused by disturbed mitochondrial function.

## IV. ANIMAL MODELS OF DYSTONIA

To date, no mouse models of dystonia have been reported by knock-out or knock-in of known disease genes, i.e., torsin A (DYT1), GCH1 (DYT5), SGCE (DYT11), and DDP (DFN-1/MTS). Therefore, animal studies into the pathophysiology of dystonia rely on the analysis of rodents carrying autosomal recessive single gene mutations that result in features more or less characteristic of various types of dystonias (Richter and Löscher, 2000). The animal models currently most useful for the investigation of dystonia are the $dt^{sz}$ hamster and the $dt$ rat.

$Dt^{sz}$ hamsters are characterized by attacks of twisting movements and abnormal postures that last for several hours. These attacks subside during sleep and can be induced by caffeine and stress. Originally, attacks were interpreted as seizures, hence the superscript "sz." Seizures have now been disproven by the documentation of normal EEG recordings during attacks. Their phenotype makes $dt^{sz}$ hamsters a good model of dystonia 8 (paroxysmal dystonic choreoathetosis). Standard neuropathological analyses of the brain of these animals did not reveal any morphological alterations. Neurochemical studies, however, detected changes in the striatum and in the ventral thalamic nuclei and point to disturbed GABAergic inhibition and enhanced dopaminergic activity. This is consistent with the recent observation of significant reduction of GABAergic interneurons in the striatum of mutant hamsters (Richter and Löscher, 2000). Furthermore, the density of dopamine D1 and D2 receptors is decreased in the dorsal striatum which might be due to receptor down-regulation in the presence of increased concentrations of dopamine (Nobrega et al., 1996). In agreement with these findings, dopamine receptor antagonists have a beneficial effect and agonists aggravate dystonia (Rehders et al., 2000). Studies in $dt^{sz}$ hamsters are greatly facilitated by their almost normal life span and fertility.

$Dt$ (dystonic) rats develop generalized dystonia due to abnormal cerebellar function. Specifically, cerebellar GABAergic inhibition appears to be altered. This in turn may result in disturbed function of the red and the thalamic nuclei as shown by the abnormal uptake of 2-deoxyglucose (2-DG). Significantly, neurolesions of the entopeduncular nucleus (corresponding to the internal segment of the globus pallidus in primates) in mutant rats aggravate dystonia (LeDoux and Lorden, 1998). Usefulness of $dt$ rats is limited due to their short life span. Death usually occurs within the first weeks of life shortly after disease onset at about 12 days. Furthermore, cerebellar atrophy is not a characteristic finding in human primary dystonia.

There are also mouse models of dystonia, i.e., the *dt*, the *wri* (wriggle), the Cacna1a null, and the mutScn8a mouse.

After manifestation of initial movement disturbances at 7–10 days of age, *dt/dt* homozygous mutant mice become progressively ataxic. Limbs are held in abnormal postures and there are twisting movements of the trunk while the animals attempt to walk. Neuropathological analysis of these animals demonstrated spinocerebellar lesions and degeneration of peripheral nerves (Duchen *et al.*, 1964). The *dt* gene codes for a neural isoform of bullous pemphigoid antigen 1 and is referred to as dystonin (Brown *et al.*, 1995). Given the nature of the neuropathological changes these animals might be a model of SCA rather than of dystonia.

The *wri* mouse has been less thoroughly studied but involuntary movements including tremor, dystonia, and myoclonic jerks are reminiscent of dystonia 11 (myoclonus dystonia); (Ikeda *et al.*, 1989). Before using the *wri* mouse to study aspects of dystonia 11 it needs to be excluded that the myoclonic jerks are in fact seizures. Given the recent isolation of the disease gene and the resulting opportunity to construct transgenic models of dystonia 11, the *wri* mouse will probably never become a widely used model of myoclonus dystonia. The usefulness of the *dt* and *wri* mouse for the study of abnormal movement is further limited by their dramatically reduced life spans.

Mutations in genes coding for ion channels can also result in features resembling human dystonia in mice. Thus homozygous inactivation of the gene encoding the P/Q-type voltage-dependent calcium channel Cacna1a results in ataxia, episodic dyskinesia, and cerebellar atrophy (Fletcher *et al.*, 2001). Homozygous mutations in the gene coding for the Scn8a sodium channel produce a wide spectrum of symptoms. Mice with null mutations suffer from motor neuron failure and loss of neuromuscular transmission and present with muscular atrophy, and lethal paralysis in addition to ataxia. Less severe mutations are associated with dystonia, ataxia, tremor, and muscle weakness (Meisler *et al.*, 2001). Of particular interest as a model of dystonia is the *med* mutation that results in truncation of the polypeptide as a result of abnormal splicing of the transcript. In these mice about 2% of the transcripts are correctly spliced thus guaranteeing residual activity of Scn8a. This residual activity protects homozygotes from paralysis. The animals do, however, exhibit severe muscle weakness and frequent episodes of sustained abnormal postures resembling dystonia. Little is known about the underlying pathophysiology and it might well be that the findings in these mice are of peripheral rather than of central origin. Thus, similar to the *dt* mouse, morphological and/or physiological changes as well as affection status suggest that mice with defects of the ion channels Cacna1a and Scn8a are better models for the study of ataxia than of dystonia.

# References

Almasy, L., Bressman, S. B., Raymond, D., Kramer, P. L., Greene, P. E., Heiman, G. A., Ford, B., Yount, J., de Leon, D., Chouinard, S., Saunders-Pullman, R., Brin, M. F., Kapoor, R. P., Jones, A. C., Shen, H., Fahn, S., Risch, N. J., and Nygaard, T. G. (1997). Idiopathic torsion dystonia linked to chromosome 8 in two Mennonite families. *Ann. Neurol.* **42**, 670–673.

Auburger, G., Ratzlaff, T., Lunkes, A., Nelles, H. W., Leube, B., Binkofski, F., Kugel, H., Heindel, W., Seitz, R., Benecke, R., Witte, O. W., and Voit, T. (1996). A gene for autosomal dominant paroxysmal choreo-athetosis/spasticity (CSE) maps to the vicinity of a potassium channel gene cluster on chromosome 1p, probably within 2 cM between D1S443 and D1S197. *Genomics* **31**, 90–94.

Bandmann, O., Nygaard, T. G., Surtees, R., Marsden, C. D., Wood, N. W., and Harding, A. E. (1996). Dopa-responsive dystonia in British patients: new mutations of the GTP-cyclohydrolase I gene and evidence for genetic heterogeneity. *Hum. Mol. Genet.* **5**, 403–406.

Beaumanoir, A., Mira, L., and van Lierde, A. (1996). Epilepsy or paroxysmal kinesigenic choreoathetosis? *Brain Dev.* **18**, 139–141.

Bhatia, K. P. (1999). The paroxysmal dyskinesias. *J. Neurol.* **246**, 149–155.

Bhatia, K. P. (2001). Familial (idiopathic) paroxysmal dyskinesias: an update. *Sem. Neurol.* **21**, 69–74.

Bonifati, V., De Michele, G., Lücking, C. B., Dürr, A., Fabrizio, E., Ambrosio, G., Vanacore, N., De Mari, M., Marconi, R., Capus, L., Breteler, M. M., Gasser, T., Oostra, B., Wood, N., Agid, Y., Filla, A., Meco, G., and Brice, A. (2001). The parkin gene and its phenotype. Italian PD Genetics Study Group, French PD Genetics Study Group and the European Consortium on genetic susceptibility in Parkinson's disease. *Neurol. Sci.* **22**, 51–52.

Brown, A., Bernier, G., Mathieu, M., Rossant, J., and Kothary, R. (1995). The mouse dystonia musculorum gene is a neural isoform of bullous pemphigoid antigen 1. *Nat. Genet.* **10**, 301–306.

Deonna, T. (1986). DOPA-sensitive progressive dystonia of childhood with fluctuations of symptoms—Segawa's syndrome and possible variants. Results of a collaborative study of the European Federation of Child Neurology Societies (EFCNS). *Neuropediatrics* **17**, 81–85.

Duchen, L. W., Strich, S. J., and Falconer, D. S. (1964). Clinical and pathological studies of an hereditary neuropathy in mice (dystonia musculorum) *Brain* **87**, 367–378.

Eidelberg, D., Takikawa, S., Wilhelmsen, K., Dhawan, V., Chaly, T., Robeson, W., Dahl, R., Margouleff, D., Greene, P., Hunt, A., Przedborski, S., and Fahn, S. (1993). Positron emission tomographic findings in Filipino X-linked dystonia-parkinsonism, *Ann. Neurol.* **34**, 185–191.

Fahn, S. (1994). The paroxysmal dyskinesias. *Movement disorders 3*. (C. D. Marsden, and S. Fahn, eds.), pp. 310–345. Butterworth-Heinemann, Oxford.

Fahn, S., Marsden, C. D., and Calne, D. B. (1987). Classification and investigation of dystonia. *Movement Disorders 2*. (C. D. Marsden, and S. Fahn, eds.), pp. 332–358. Butterworths, London.

Fink, J. K., Hedera, P., Mathay, J. G., and Albin, R. L. (1997). Paroxysmal dystonic choreoathetosis linked to chromosome 2q: clinical analysis and proposed pathophysiology. *Neurology* **49**, 177–183.

Fink, J. K., Rainier, S., Wilkowski, J., Jones, S. M., Kume, A., Hedera, P., Albin, R., Mathay, J., Girbach, L., Varvil, T., Otterud, B., and Leppert, M. (1996). Paroxysmal dystonic choreoathetosis: tight linkage to chromosome 2q. *Am. J. Hum. Genet.* **59**, 140–145.

Fletcher, C. F., Tottene, A., Lennon, V. A., Wilson, S. M., Dubel, S. J., Paylor, R., Hosford, D. A., Tessarollo, L., McEnery M. W., Pietrobon, D., Copeland, N. G., and Jenkins, N. A. (2001). Dystonia and cerebellar atrophy in Cacna1a null mice lacking P/Q calcium channel activity. *FASEB J.* **15**, 1288–1290.

Fouad, G. T., Servidei, S., Durcan, S., Bertini, E., and Ptácek, L. J. (1996).

A gene for familial paroxysmal dyskinesia (FTD1) maps to chromosome 2q. *Am. J. Hum. Genet.* **59**, 135–139.

Furukawa, Y., Shimadzu, M., Rajput, A. H., Shimizu, Y., Tagawa, T., Mori, H., Yokochi, M., Narabayashi, H., Hornykiewicz, O., Mizuno, Y., and Kish, S. J. (1996). GTP-cyclohydrolase I gene mutations in hereditary progressive and dopa-responsive dystonia. *Ann. Neurol.* **39**, 609–617.

Gimenez-Roldan, S., Delgado, G., Marin, M., Villanueva, J. A., and Mateo, D. (1988). Hereditary torsion dystonia in gypsies. *Adv. Neurol.* **50**, 73–81.

Goodenough, D. J., Fariello, R. G., Annis, B. L., and Chun, R. W. (1978). Familial and acquired paroxysmal dyskinesias: a proposed classification with delineation of clinical features. *Arch. Neurol.* **35**, 827–831.

Graeber, M. B., Kupke, K. G., and Müller, U. (1992). Delineation of the dystonia-parkinsonism syndrome (XDP) locus in Xq13. *Proc. Natl. Acad. Sci. U.S.A.* **89**, 8245–8248.

Guerrini, R., Bonanni, P., Nardocci, N., Parmeggiani, L., Piccirilli, M., De Fusco, M., Aridon, P., Ballabio, A., Carrozzo, R., and Casari, G. (1999). Autosomal recessive rolandic epilepsy with paroxysmal exercise-induced dystonia and writer's cramp: delineation of the syndrome and gene mapping to chromosome 16p12–11.2. *Ann. Neurol.* **45**, 344–352.

Haberhausen, G., Schmitt, I., Köhler, A., Peters, U., Rider, S., Chelly, J., Terwilliger, I. D., Monaco, A. P., and Müller, U. (1995). Assignment of the dystonia-parkinsonism locus, DYT3, to a small region within a 1.8 Mb YAC contig of Xq13.1. *Am. J. Hum. Genet.* **57**, 644–650.

Heuvel van den, L. W. P. J., Luiten, B., Smeitink, J. A. M., de Rijk-van Andel, J. F., Hyland, K., Steenbergen-Spanjers, G. C. H., Janssen R. J. T., and Wevers, R. A. (1998). A common point mutation in the tyrosine hydroxylase gene in autosomal recessive L-dopa-responsive dystonia in the Dutch population. *Hum. Genet.* **102**, 644–646.

Ikeda, M., Mikuni, M., Nishikawa, T., and Takahashi, K. (1989). A neurochemical study of a new mutant mouse presenting myoclonus-like involuntary movements: a possible model of spontaneous serotonergic hyperactivity. *Brain Res.* **495**, 337–348.

Jin, H., May, M., Tranebjaerg, L., Kendall, E., Fontán, G., Jackson, J., Subramony, S. H., Arena, F., Lubs, H., Smith, S., Stevenson, R., Schwartz, C., and Vetrie, D. (1996). A novel X-linked gene, DDP, shows mutations in families with deafness (DFN-1), dystonia, mental deficiency and blindness. *Nat. Genet* **14**, 177–180.

Kertesz, A. (1967). Paroxysmal kinesigenic choreoathetosis: an entity within the paroxysmal choreoathetosis syndrome. Description of 10 cases, including 1 autopsied. *Neurology* **17**, 680–690.

Klein, C., Brin, M. F., Kramer, P., Sena-Esteves, M., de Leon, D., Doheny, D., Bressman, S., Fahn, S., Breakefield, X. O., and Ozelius, L. J. (1999). Association of a missense change in the D2 dopamine receptor with myoclonus dystonia. *Proc. Natl. Acad. Sci. U.S.A.* **96**, 5173–5176.

Klein, C., Gurvich, N., Sena-Esteves, M., Bressman, S., Brin, M. F., Ebersole, B. J., Fink, S., Forsgren, L., Friedman, J., Grimes, D., Holmgren, G., Kyllerman, M., Lang, A. E., de Leon, D., Leung, J., Prioleau, C., Raymond, D., Sanner, G., Saunders-Pullman, R., Vieregge, P., Wahlström, J., Breakefield, X. O., Kramer, P. L., Ozelius, L. J., and Sealfon, S. C. (2000). Evaluation of the role of the D2 dopamine receptor in myoclonus dystonia. *Ann. Neurol.* **47**, 369–373.

Klein, C., Ozelius, L. J., Hagenah, J., Breakefield, X. O., Risch, N. J., and Vieregge, P. (1998). Search for a founder mutation in idiopathic focal dystonia from northern Germany. *Am. J. Hum. Genet.* **63**, 1777–1782.

Koehler, C. M., Leuenberger, D., Merchant, S., Renold, A., Junne, T., and Schatz, G. (1999). Human deafness dystonia syndrome is a mitochondrial disease. *Proc. Natl. Acad. Sci. U.S.A.* **96**, 2141–2146.

Kramer, P. L., Mineta, M., Klein, C., Schilling, K., de Leon, D., Farlow, M. R., Breakefield, X. O., Bressman, S. B., Dobyns, W. B., Ozelius, L. J., and Brashear, A. (1999). Rapid-onset dystonia-parkinsonism: linkage to chromosome 19q13. *Ann. Neurol.* **46**, 176–182.

Kupke, K., Lee, L. V., Viterbo, G. H., Arancillo, J., Donlon, T. A., and Müller, U. (1990). X-linked recessive torsion dystonia in the Philippines. *Am. J. Med. Genet.* **36**, 237–242.

Kupke, K. G., Graeber, M. B., and Müller, U. (1992). Dystonia-Parkinsonism syndrome (XDP) locus: Flanking markers in Xq12–q21.1. *Am. J. Hum. Genet.* **50**, 808–815.

Lance, J. W. (1977). Familial paroxysmal dystonic choreoathetosis and its differentiation from related syndromes. *Ann. Neurol.* **2**, 285–293.

LeDoux, M. S. and Lorden, J. F. (1998). Abnormal cerebellar output in the genetically dystonic rat. *Advances in Neurology.* (S. Fahn, C. D. Marsden, and M. R. DeLong, eds.), Vol. 78, pp. 63–78. Lippincott-Raven, New York.

Lee, L. V., Kupke, K. G., Caballar-Gonzaga, F., Hebron-Ortiz, M., and Müller, U. (1991). The phenotype of the X-linked dystonia-parkinsonism syndrome. An assessment of 42 cases in the Philippines. *Medicine* **70**, 179–187.

Lee, L. V., Munoz, E. L., Tan, K. T., and Reyes, M. T. (2001). Sex linked recessive dystonia parkinsonism of Panay, Philippines (XDP). *Mol. Pathol.* **54**, 362–368.

Leube, B., Rudnicki, D., Ratzlaff, T., Kessler, K. R., Benecke, R., and Auburger, G. (1996). Idiopathic torsion dystonia: assignment of a gene to chromosome 18p in a German family with adult onset, autosomal-dominant inheritance and purely focal distribution. *Hum. Mol. Genet.* **5**, 1673–1677.

Leube, B., Hendgen, T., Kessler, K. R., Knapp, M., Benecke, R., and Auburger, G. (1997a). Evidence for DYT7 being a common cause of cervical dystonia (torticollis) in Central Europe. *Am. J. Med. Genet.* **74**, 529–532.

Leube, B., Hendgen, T., Kessler, K. R., Knapp, M., Benecke, R., and Auburger, G. (1997b). Sporadic focal dystonia in northwest Germany: molecular basis on chromosome 18p. *Ann. Neurol.* **42**, 111–114.

Marsden, C. D. (1988). Investigation of dystonia. *Adv. Neurol.* **50**, 35–44.

Marsden, C. D. (1996). Paroxysmal choreoathetosis. *Adv. Neurol.* **70**, 467–470.

Marsden, C. D., Harrison, M. J. G., and Bundey, S. (1976). Natural history of idiopathic torsion dystonia. *Adv. Neurol.* **14**, 177–187.

Marsden, C. D., Lang, A. E., Quinn, N. P., McDonald, W. I., Abdallat, A., and Nimri, S. (1986). Familial dystonia and visual failure with striatal CT lucencies. *J. Neurol. Neurosurg. Psychiatry* **49**, 500–509.

Meisler, M. H., Kearney, J., Escayg, A., MacDonald, B. T., and Sprunger, L. K. (2001). Sodium channels and neurological disease: insights from Scn8a mutations in the mouse. *Neuroscientist* **7**, 136–145.

Mohr, J., and Mageroy, K. (1960). Sex-linked deafness of a possible new type. *Acta Genet. Stat. Med. (Basel)* **10**, 54–62.

Mount, L. A. and Reback, S. (1940). Familial paroxysmal choreoathetosis: preliminary report on a hitherto undescribed clinical syndrome. *Arch. Neurol. Psychiatry* **44**, 841–847.

Müller, U. (1999). Dystonia-parkinsonism syndrome, X-linked (XDP). In *Elsevier's Encyclopedia of Neuroscience.* (G. Adelman, and B. H. Smith, eds.), pp. 593–594. Elsevier Science, New York.

Müller, U., Haberhausen, G., Wagner, T., Fairweather, N., Chelly, J., and Monaco, A. P. (1994). DXS106 and DXS559 flank the X-linked dystonia-parkinsonism syndrome locus *(DYT3). Genomics* **23**, 114–117.

Müller, U., Steinberger, D., and Németh, A. H. (1998). Clinical and molecular genetics of primary dystonias. *Neurogenetics* **1**, 165–177.

Müller, U., Niemann, S., Ramser, J., Lehrach, H., Sudbrak, R., and Nolte, D. (2001). Expression map of the DYT3 critical region in Xq13.1. [Abstract] *Am. J. Hum. Genet.* **69**, 598.

Nemeth, A. H., Nolte, D., Dunne, E., Niemann, S., Kostrzewa, M., Peters, U., Fraser, E., Bochukova, E., Butler, R., Brown, J., Cox, R. D., Levy, E. R., Ropers, H. H., Monaco, A. P., and Müller, U. (1999). Refined linkage disequilibrium and physical mapping of the gene locus for X-linked dystonia-parkinsonism *(DYT3). Genomics* **60**, 320–329.

Nobrega, J. N., Richter, A., Tozman, N., Jiwa, D., and Löscher, W. (1996). Quantitative autoradiography reveals regionally selective changes in dopamine D1 and D2 receptor binding in the genetically dystonic hamster. *Neuroscience* **71**, 927–936.

Nolte, D., Ramser, J., Niemann, S., Lehrach, H., Sudbrak, R., and Müller, U. (2001). ACRC codes for a novel nuclear protein with unusual acidic repeat tract and maps to *DYT3* (dystonia parkinsonism) critical interval in Xq13.1. *Neurogenetics* **3**, 207–212.

Nygaard, T. G. (1993). Dopa-resonsive dystonia. Delineation of the clinical syndrome and clues to pathogenesis. *Adv. Neurol.* **60**, 577–585.

Nygaard, T. G., Marsden, C. D., and Duvoisin, R. C. (1988). Dopa-responsive dystonia. *Adv. Neurol.* **50**, 377–384.

Oppenheim, H. (1911). Über eine eigenartige Krampfkrankheit des kindlichen und jugendlichen Alters (Dysbasia lordotica progressiva, dystonia musculorum deformans). *Neurol. Centralbl.* **30**, 1090–1107.

Ozelius, L. J., Hewett, J. W., Page, C. E., Bressman, S. B., Kramer, P. L., Shalish, C., Leon D., Brin, M. F., Raymond, D., Corey, D. P., Fahn, S., Risch, N. J., Buckler, A., Gusella, J. F., and Breakefield, X. O. (1997). The early-onset torsion dystonia gene *(DYT1)* encodes an ATP-binding protein. *Nat. Genet.* **17**, 40–48.

Pittock, S. J., Joyce, C., O'Keane, V., Hugle, B., Hardiman, O., Brett, F., Green A. J., Barton, D. E., King, M. D., and Webb, D. W. (2000). Rapid-onset dystonia-parkinsonism. A clinical and genetic analysis of a new kindred. *Neurology* **55**, 991–995.

Placzek, M. R., Misbahuddin, A., Chaudhuri, K. R., Wood, N. W., Bhatia, K. P., and Warner, T. T. (2001). Cervical dystonia is associated with a polymorphism in the dopamine (D5) receptor gene. *J. Neurol. Neurosurg. Psychiatry* **71**, 262–264.

Raskind, W. H., Bolin, T., Wolff, J., Fink, J., Matsushita, M., Litt, M., Lipe, H., and Bird, T. D. (1998). Further localization of a gene for paroxysmal dystonic choreoathetosis to a 5-cM region on chromosome 2q34. *Hum. Genet.* **102**, 93–97.

Rehders, H. J., Löscher, W., and Richter, A. (2000). Evidence for striatal dopaminergic overactivity in paroxysmal dystonia indicated by microinjections in a genetic rodent model. *Neuroscience* **97**, 267–277.

Richter, A. and Löscher, W. (2000). Animal models of dystonia. *Funct. Neurol.* **15**, 259–267.

Rosenberg, R. N. (1990). Autosomal dominant cerebellar phenotypes: the genotype will settle the issue. *Neurology* **40**, 1329–1331.

Sadamatsu, M., Masui, A., Sakai, T., Kunugi, H., Nanko, S.-I., and Kato, N. (1999). Familial paroxysmal kinesigenic choreoathetosis: an electrophysiologic and genotypic analysis. *Epilepsia* **40**, 942–949.

Schwalbe, W. (1908). Eine eigentümliche tonische Krampfform mit hysterischen Symptomen. Inaugural-Dissertation. Schade, Berlin.

Scribanu, N., and Kennedy, C. (1976). Familial syndrome with dystonia, neural deafness, and possible intellectual impairment: clinical course and pathological findings. *Adv. Neurol.* **14**, 235–243.

Steinberger, D., Weber, Y., Korinthenberg, R., Deuschl, G., Benecke, R., Martinius, J., and Müller, U. (1998). High penetrance and pronounced variation in expressivity of *GCH1* mutations in five families with dopa-responsive dystonia. *Ann. Neurol.* **43**, 634–639.

Steinberger, D., Topka, H., Fischer, D., and Müller, U. (1999). *GCH1* mutation in a patient with adult-onset oromandibular dystonia. *Neurology* **52**, 877–879.

Swaans, R. J. M., Rondon, P., Renier, W. O., van den Heuvel, L. P. W. J., Steenbergen-Spanjers, G. C. H., and Wevers, R. A. (2000). Four novel mutations in the tyrosine hydroxylase gene in patients with infantile parkinsonism. *Ann. Hum. Genet.* **64**, 25–31.

Swerdlow, R. H., and Wooten, G. F. (2001). A novel deafness/dystonia peptide gene mutation that causes dystonia in female carriers of Mohr-Tranebjaerg syndrome. *Ann. Neurol.* **50**, 537–540.

Swoboda, K. J., Soong, B.-W., McKenna, C., Brunt, E. R. P., Litt, M., Bale, Jr., J. F., Ashizawa, T., Bennett, L. B., Bowcock, A. M., Roach, E. S., Gerson, D., Matsuura, T., Heydemann, P. T., Nespeca, M. P., Jankovic, J., Leppert, M., and Pták̆ek, L. J. (2000). Paroxysmal kinesigenic dyskinesia and infantile convulsions. Clinical and linkage studies. *Neurology* **55**, 224–230.

Szepetowski, P., Rochette, J., Berquin, P., Piussan, G., Lathrop, G. M., and Monaco, A. P. (1997). Familial infantile convulsions and paroxysmal choreoathetosis: a new neurological syndrome linked to the pericentromeric region of human chromosome 16. *Am. J. Hum. Genet.* **61**, 889–898.

Tomita, H., Nagamitsu, S., Wakui, K., Fukushima, Y., Yamada, K., Sadamatsu, M., Masui, A., Konishi, T., Matsuishi, T., Aihara, M., Shimizu, K., Hashimoto, K., Mineta, M., Matsushima, M., Tsujita, T., Saito, M., Tanaka, H., Tsuji, S., Takagi, T., Nakamura, Y., Nanko, S., Kato, N., Nakane, Y., and Niikawa, N. (1999). Paroxysmal kinesigenic choreoathetosis locus maps to chromosome 16p11.2–q12.1. *Am. J. Hum. Genet.* **65**, 1688–1697.

Tranebjaerg, L., Schwartz, C., Eriksen, H., Andreasson, S., Ponjavic, V., Dahl, A., Stevenson, R.E., May, M., Arena, F., Barker, D., Elverland, H.H., and Lubs, H. (1995). A new X-linked recessive deafness syndrome with blindness, dystonia, fractures, and mental deficiency is linked to Xq22. *J. Med. Genet.* **32**, 257–263.

Ujike, H., Tanabe, Y., Takehisa, Y., Hayabara, T., and Kuroda, S. (2001). A family with X-linked dystonia-deafness syndrome with a novel mutation of the DDP gene. *Arch. Neurol.* **58**, 1004–1007.

Valente, E. M., Bentivoglio, A. R., Cassetta, E., Dixon, P. H., Davis, M. B., Ferraris, A., Ialongo, T., Frontali, M., Wood, N. W., and Albanese, A. (2001). DYT13, a novel primary torsion dystonia locus, maps to chromosome 1p36.13–36.32 in an Italian family with cranial-cervical or upper limb onset. *Ann. Neurol.* **49**, 362–366.

Van den Heuvel, L. P. W. J., Luiten, B., Smeitink, J. A. M., de Rijk-van Andel, J. F., Hyland, K., Steenbergen-Spanjers, G. C. H., Janssen, R. J. T., and Wevers, R. A. (1998). A common point mutation in the tyrosine hydroxylase gene in autosomal recessive L-dopa-responsive dystonia in the Dutch population. *Hum. Genet.* **102**, 644–646.

Waddy, H. M., Fletcher, N. A., Harding, A. E., and Marsden, C. D. (1991). A genetic study of idiopathic focal dystonias. *Ann. Neurol.* **29**, 320–324.

Walker, E. S. (1981). Familial paroxysmal dystonic choreoathetosis: a neurologic disorder simulating psychiatric illness. *Johns Hopkins Med. J.* **148**, 108–113.

Waters, C. H., Faust, P. L., Powers, J., Vinters, H., Moskowitz, C., Nygaard, T., Hunt, A. L., and Fahn, S. (1993a). Neuropathology of Lubag (X-linked dystonia parkinsonism). *Mov. Disord.* **8**, 387–390.

Waters, C. H., Takahashi, H., Wilhelmsen, K. C., Shubin, R., Snow, B. J., Nygaard, T. G., Moskowitz, C. B., Fahn, S., and Calne, D. B. (1993b). Phenotypic expression of X-linked dystonia-parkinsonism (lubag) in two women. *Neurology* **43**, 1555–1558.

Zimprich, A., Grabowski, M., Asmus, F., Naumann, M., Berg, D., Bertram, M., Scheidtmann, K., Kern, P., Winkelmann, J., Müller-Myhsok, B., Riedel, L., Bauer, M., Müller, T., Castro, M., Meitinger, T., Strom, T. M., and Gasser, T. (2001). Mutations in the gene encoding epsilon-sarcoglycan cause myoclonus-dystonia syndrome. *Nat. Genet.* **29**, 66–69.

CHAPTER 36

# DYT1 Dystonia

**LAURIE J. OZELIUS**
*Department of Molecular Genetics*
*Albert Einstein College of Medicine*
*Bronx, New York 10461*

**SUSAN B. BRESSMAN**
*Department of Neurology*
*Beth Israel Medical Center*
*New York, New York 10003*

I. Summary
II. Phenotype
III. Gene
   A. Epidemiology
   B. Inheritance Pattern
   C. Genetic Linkage Studies and Linkage Disequilibrium
   D. Gene Identification
   E. Normal Gene and Protein Function
   F. Abnormal Gene and Protein Function
IV. Diagnostic and Ancillary Tests
   A. DNA Test
   B. Other Laboratory Tests
V. Neuroimaging
VI. Neuropathology
VII. Cellular and Animal Models of Disease
   A. *In Vitro*
   B. Invertebrates
   C. Mouse and Rat
VIII. Treatment
References

## I. SUMMARY

Dystonia is a movement disorder, characterized clinically by involuntary twisting and repetitive movements and abnormal postures. The DYT1 locus on human chromosome 9q34 is responsible for the most severe form of hereditary dystonia—early-onset, generalized torsin dystonia. The disorder is inherited as an autosomal dominant trait with reduced penetrance (30–40%). Onset almost always occurs before the age of 26 years typically beginning in an arm or leg with spread to other limbs, though it may remain localized as writer's cramp (Fig. 36.1). The incidence of this form of dystonia has been estimated at 1/160,000 in the general population with a higher frequency in the range of 1/6000–1/2000 in the Ashkenazi Jewish population due to a founder mutation. A 3-bp (GAG) deletion causing the loss of one of a pair of glutamic acids from the encoded protein torsinA, accounts for 50–70% of non-Jewish individual and over 90% of Ashkenazi Jewish individuals with this form of dystonia. One other mutation has been identified in this gene in a single family, an 18-bp in-frame deletion that removes six amino acids within the carboxy terminus. The DYT1 gene is a member of a larger gene family including three other mammalian members, TOR1B sharing 70% homology and TOR2A and TOR3A with about 50% identity. The proteins encoded by these genes show homology to the AAA+ superfamily of ATPases. Typically, AAA+ proteins form oligomeric complexes and may have roles in protein folding and degradation, cytoskeletal dynamics, membrane trafficking, and vesicle fusion, and response to stress. The function of torsinA remains elusive, however, descriptive studies show that it localizes to the lumen of the endoplasmic reticulum but can also associated with vesicles in neural cells and extends to the ends of processes. Both the message and protein are widely distributed throughout the brain primarily staining neurons and neuronal processes with the most intense expression localized in substantia nigra compacta dopamine neurons.

## II. PHENOTYPE

Oppenheim first used the term dystonia in 1911, when he described a childhood-onset syndrome that he called

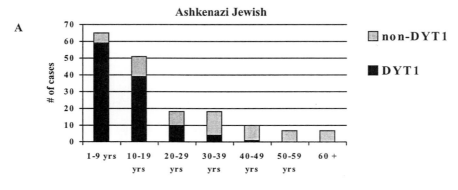

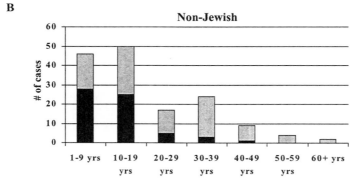

FIGURE 36.1 Role of the DYT1 gene in Ashkenazi Jewish and non-Jewish PTD. Graphs display the proportion of PTD cases by age (at onset) caused by the GAG deletion in the DYT1 gene in (A) the Ashkenazi Jewish population or (B) the non-Jewish population. The great majority of Ashkenazi Jewish cases involve the DYT1 gene because of the founder mutation in this population whereas only about half of the non-Jewish cases harbor the DYT1 mutation.

*dystonia musculorum deformans* (Oppenheim, 1911). Almost 100 years later, we now know that mutations in the DYT1 gene cause this disorder (Ozelius *et al.*, 1997a). Dystonia is a movement disorder characterized by sustained, involuntary muscle contractions generating twisting and repetitive movements or abnormal postures (Fahn *et al.*, 1987). Different muscle groups can be involved with variable extent and severity, ranging from intermittent contractions limited to single body region (focal dystonia) to generalized dystonia involving the axial and limb muscles.

DYT1 dystonia is categorized as one of the primary torsion dystonias (PTD) in which dystonia is the sole phenotypic manifestation (with the exception that tremor may be present) and there is no consistent associated pathological abnormality or evidence of neuronal degeneration. For PTD, the age at onset distribution is bimodal, with modes at age 9 (early-onset) and 45 (late-onset) divided by a nadir at age 27 (Bressman *et al.*, 1989). Further, there is a relationship between the age at onset of symptoms, body region first affected, and clinical progression of signs (Marsden and Harrison, 1974; Fahn *et al.*, 1987). When PTD begins in childhood or adolescence, it often starts in a leg or arm, and then progresses to involve multiple body regions; when PTD begins in adult years, symptoms first involve the neck, cranial, or arm muscles, and dystonia tends to remain localized (Greene *et al.*, 1995).

With the identification of the GAG deletion mutation in the DYT1 gene (see Section III.D) the phenotypic expression of this mutation could be studied directly and has been by investigators in the U.S., Europe, and Asia. The clinical expression of the DYT1 GAG deletion is generally similar across ethnic groups (Bressman *et al.*, 2000, Valente *et al.*, 1998, Slominski *et al.*, 1999, Leube *et al.*, 1999, Matsumoto *et al.*, 2001). In a U.S. study assessing 176 DYT1 mutation carriers, non-Jewish individuals had slightly more progressive disease with more leg onset compared to Ashkenazi cases (Bressman *et al.*, 2000). Other investigators have noted mild or localized phenotypes in select non-Jewish DYT1 families (Gasser *et al.*, 1998) but there have been no other large-scale systematic studies to further evaluate ethnic differences. One interesting and still unanswered question is whether the DYT1 phenotype or risk for developing dystonia (penetrance) differs between men and women. For the various non-DYT1 adult-onset focal and segmental PTD subtypes epidemiological studies have found gender-related differences (ESDE Collaborative Group, 1999). For DYT1, one study found that at least among non-Jews, men were somewhat more likely than women to have a DYT1 mutation and disease was more severe in the male gene-carrier group (Bressman *et al.*, 2000). However, other studies have not noted gender differences in either risk for having PTD due to DYT1 or disease expression.

The great majority of people with the DYT1 GAG deletion have early onset (before 26 years) with an average age of onset of 14 years. Later onset, up to age 44 years, however, has been described in affected relatives (Bressman *et al.*, 2000). Most cases begin in an arm or leg although onset in axial and cranial muscles may occur. Classically dystonia begins as an action dystonia that is evident with specific actions such as writing or walking. There is agonist and antagonist co-contractions leading to unusual posturing such as anterior displacement of the shoulder, elbow elevation, and ulnar deviation with writing or hip and knee flexion with ankle eversion on walking forward (Bressman and Fahn, 1997).

Disease severity can vary considerably among and within families. Usually but not always, dystonia progresses both in its temporal occurrence and extent of muscle involvement. More actions induce the movements and they may eventually occur at rest. Also the dystonia often spreads beyond the muscles first affected. About 65% of cases progress to a generalized or multifocal distribution (involving legs and at least one arm); (Bressman *et al.*, 1994, 2000). When viewed in terms of body regions ultimately involved, one or more limbs will almost always be affected (over 95% have an affected arm); the trunk and neck are affected in about 25–35% and the cranial muscles are less likely to be involved (< 15%). For the rest, contractions remain more localized as either segmental or focal dystonia. Writer's cramp and bibrachial dystonia are the most common forms of focal and segmental dystonia, respectively, and they may be the sole manifestation within a family (Gasser *et al.*, 1998).

In contrast, most patients with adult onset focal and segmental dystonia are not deletion carriers (Bressman *et al.*, 2000). Two studies looked specifically at adult-onset writer's cramp or musician's dystonia and found no involvement of the DYT1 gene in these patients (Kamm *et al.*, 2000; Friedman *et al.*, 2000). Finally, a single patient has been reported with a novel mutation in the DYT1 gene (see Section III.D) and the phenotype in this patient includes dystonia in both arms and neck as well as myoclonic jerks in the legs and head (Leung *et al.*, 2001; Doheny, in press).

## III. GENE

### A. Epidemiology

No studies have assessed the frequency of PTD due to DYT1 based on molecular diagnosis. However, using an associated haplotype of alleles surrounding DYT1 in Ashkenazi Jews (see Section III.C), the disease frequency in this population is estimated to be between 11–33 per 100,000 (Risch *et al.*, 1995). Based on previous (Zeman and Dyken, 1967; Zilber *et al.*, 1984) and unpublished (Risch, personal communication) studies, PTD due to DYT1 in the non-Ashkenazi Jewish and non-Jewish populations is estimated to be about 1/3 to 1/5 as common.

There is more known, however, about the prevalence of PTD, based on phenotypic features. Using medical records, a study from Rochester, Minnesota, calculated the prevalence to be 3.4 per 100,000 for generalized dystonia and 29.5 per 100,000 for focal dystonia (Nutt *et al.*, 1988). More recently, a European collaborative study (ESDE Collaborative Group, 2000) found a crude annual prevalence of 15.2 per 100,000 with most cases (11.7 per 100,000) being focal dystonia. However, this study was based on ascertainment from adult neurology and botulinum toxin clinics and is considered to significantly underestimate the true prevalence as family studies have shown that at least one-half of cases are either not medically diagnosed or misdiagnosed (Bressman *et al.*, 1989; Risch *et al.*, 1995).

### B. Inheritance Pattern

Initial reports of early-onset primary dystonia, over 90 years ago, described two important features of this disease that were not fully appreciated for decades: the disease is familial (Schwalbe, 1908), and it is more common in Jews of Eastern European ancestry (Zeman and Dyken, 1967). In 1970, Eldridge proposed that early-onset primary dystonia is inherited as an autosomal recessive trait in Jews and an autosomal dominant trait in non-Jews (Eldridge, 1970). For close to 20 years this view was widely accepted. Then studies from Israel (Zilber *et al.*, 1984) and the U.S. (Bressman *et al.*, 1989; Risch *et al.*, 1990), as well as re-analysis of Eldridge's original data (Pauls and Korczyn, 1990), were published indicating that the disorder is *not* autosomal recessive in Jews but rather autosomal dominant with reduced (30%) penetrance. The age-adjusted risk for first-degree relatives is about 15.5%, and for second-degree relatives it is 6.5%, with no significant sex differences. Parent, offspring, and sibling risks are not significantly different, consistent with dominant inheritance. Further, there is no evidence for sporadic cases or new mutations; that is, all cases are inherited (Bressman *et al.*, 1989; Risch *et al.*, 1990). The study of Bressman and colleagues found that the range of clinical features in the relatives of early-onset Ashkenazi probands is limited. Although the disease in relatives compared to probands is milder, more localized, and has a slightly later age-at-onset, most share a similar phenotype of early-limb-onset dystonia; symptom onset after age 40 and onset in cranial muscles are rare

(Bressman et al., 1989). This uniformity of symptoms among the Ashkenazi can be explained by the subsequent discovery of a founder mutation (see Section III.C) that underlies about 90% of early limb-onset cases of primary dystonia in this population (Bressman et al., 2000).

Early-onset dystonia in the non-Jewish Caucasian population is also inherited as an autosomal dominant trait with markedly reduced penetrance (Zeman and Dyken, 1967; Fletcher et al., 1990; Pauls and Korczyn, 1990; Kramer et al., 1994). A systematic analysis (Fletcher et al., 1990) of non-Jewish British probands with generalized, multifocal, and segmental dystonia (the majority had early-onset) concluded that approximately 85% of cases are inherited as an autosomal dominant trait with reduced penetrance of 40%; the remaining 15% are likely to be non-genetic phenocopies. Increased paternal age of singleton cases was found and about 14% of genetic cases are thought to be new mutations. As in the study of Ashkenazi families, there was variable and often milder expression in affected relatives although, unlike that population, a larger proportion (10–15%) of non-Jewish affected family members had late onset (>44 years). This heterogeneity is reflected by a greater genetic heterogeneity where only 50–70% of non-Jewish early-limb-onset PTD cases harbor the DYT1 GAG deletion mutation (Valente et al., 1998; Brassat et al., 2000; Kamm et al., 1999; Tuffery-Giraud et al., 2001; Major et al. 2001).

## C. Genetic Linkage Studies and Linkage Disequilibrium

The DYT1 gene was first linked to markers in a 30 cM (approximately 30 million base pairs) region on chromosome 9q32–q34 in a single large North American non-Jewish family of French-Canadian ancestry (Ozelius et al., 1989). Subsequent studies confirmed linkage to the same chromosomal region in early-onset Ashkenazi (Kramer et al., 1990; Ozelius et al., 1992), non-Jewish North American families (Kramer et al., 1994) and some, but not all, European non-Jewish families (Warner et al., 1993). Recombination events in these families placed the gene into a 5-cM interval on chromosome 9q34.

Strong linkage disequilibrium was observed in the Ashkenazi Jewish population between the disease gene and a particular haplotype of alleles at several polymorphic loci surrounding the gene including D9S62a, D9S62b, D9S63, ASS, ABL, and D9S64 (Ozelius et al., 1992; Bressman et al., 1994; Risch et al., 1995). This finding supports the idea that a single mutational event is responsible for most cases of early onset PTD in the Ashkenazi population. The presence of very strong linkage disequilibrium at a relatively large genetic distance (about 2 cM) also suggests that the mutation is recent. From this haplotype data, Risch and colleagues (1995) calculated that the mutation was introduced into the Ashkenazi population about 350 years ago and probably originated in Lithuania or Belorussia. They also argued that the high current prevalence of the disease in Ashkenazim (estimated to be about 1:3000–1:9000 with a gene frequency of about 1:2000–1:6000) is due to the tremendous growth of that population in the eighteenth century from a small reproducing founder population (Risch et al., 1995). A founder mutation and genetic drift (changes in gene frequency due to chance events such as migrations, population expansions), rather than a heterozygote advantage (i.e., nonpenetrant DYT1 carriers have some advantage that leads to carriers being more prevalent), is probably responsible for the high frequency of DYT1 dystonia in Ashkenazim.

Haplotype analysis of this founder mutation also allowed the definition of the most likely location of the DYT1 gene within a genomic contig generated using overlapping yeast artificial chromosomes (YACs) and cosmids spanning a 600-kb region of chromosome 9q34 (Ozelius et al., 1997b). Variations from the haplotype indicated "historic" recombination events that took place on either side of the DYT1 gene in preceeding generations. These historic recombination events placed the disease gene within a 150-kb interval (Ozelius et al., 1997a). A detailed genetic map of the 150-kb region was generated and screened for transcripts using the technique of exon amplification. Five transcripts were identified within the critical region and each was examined for mutations using single strand conformation polymorphism (SSCP) analysis.

## D. Gene identification

An in-frame trinucleotide GAG deletion in the coding region of a one of the transcripts was found in all affected individuals and obligate carrier members of chromosome 9-linked early-onset PTD families regardless of ethnic origin, but was never found on control chromosomes (Ozelius et al., 1997a). The GAG deletion results in the loss of one of a pair of glutamic acid residues near the carboxy terminus of a novel protein termed torsinA (Fig. 36.2; Ozelius et al., 1997a). Different haplotypes at three single nucleotide polymorphisms (SNPs) surrounding the GAG deletion indicated that this same mutation had arisen multiple independent times within these families. The finding of de novo GAG deletions in patients of Russian and Mennonite origin added definitive proof that this change causes early-onset PTD (Klein et al., 1998). Subsequently the GAG deletion has been found in non-Jewish individuals and families of diverse ethnic background (Ikeuchi et al., 1999; Valente et al., 1998; Slominski et al., 1999; Lebre et al., 1999; Major et al., 2001). Analyses of haplotypes confirmed that current deletions in the non-Jewish population originated from multiple independent mutation events. Further, although the GAG deletion in the great majority of Ashkenazim derives from the same founder mutation, a small number of Ashkenazi harbor GAG deletions arising from different mutation events (Bressman *et al.*, 2000; Lebre *et al.*, 1999). The increased frequency of this mutation is possibly due to genetic instability in an imperfect tandem 24-bp repeat in the region of the deletion (Klein *et al.*, 1998).

Several studies have been undertaken to determine if other mutations in the DYT1 gene can cause early-limb-onset dystonia (Ozelius *et al.*, 1999; Leung *et al.*, 2001; Tuffery-Giraud *et al.*, 2001). Extensive screening has revealed only a single family with another mutation in the DYT1 gene (Leung *et al.*, 2001). This mutation is similar to the GAG deletion in that it is also an in-frame deletion (18 bp) in the carboxy terminal region of the protein, which in the heterozygous state produces dystonia with reduced penetrance in other family members (Fig. 36.2). The fact that most cases screened were mutation negative implies that other loci exist that mimic the early onset phenotype, or are nongenetic phenocopies. In addition, the inability to identify additional mutations in the DYT1 gene suggests that the structural changes in the torsinA protein caused by these deletions may be functionally unique.

### E. Normal Gene and Protein Function

The DYT1 (also known as TOR1A) cDNA contains an open reading frame of 998 bp with two poly-A addition sites at the 3′ end. Northern analysis of adult and fetal human RNA revealed two ubiquitously expressed messages of about 1.8 and 2.2 kb (Ozelius *et al.*, 1997a). DYT1 is a member of a gene family consisting of three other highly homologous genes in the human genome. TOR1B is adjacent to DYT1 on chromosome 9q34 and is 70% identical at both the nucleotide and amino acid levels (Ozelius *et al.*, 1999). These two genes are in opposite orientations in the genome (tail to tail) and are assumed to

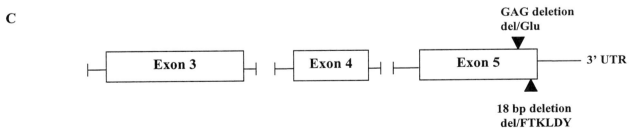

FIGURE 36.2 Mutations causing early-onset dystonia in the DYT1 gene. (A) Nucleotide sequence of the last 93 bp of exon 5 in the DYT1 gene. The position of the GAG and 18-bp deletions are indicated by the underline on the equivalent wild-type (WT) sequence. (B) Protein translation corresponding to the C terminal 40 amino acids of torsinA with the positions of the glutamic acid (E) deletion and the 6 amino acid (323FTKLDY328) deletion specified by asterisks. (C) A schematic showing the partial exon:intron structure of the DYT1 gene with the GAG and 18-bp deletions indicated. The 18-bp deletion removes a predicted casein kinase 2 phosphorylation site.

have arisen from a tandem duplication of an evolutionary precursor gene. Each of these genes has five exons spanning about 10 kb, with the same location of splice sites. They are separated from each other in the genome by about 2 kb. By Northern analysis TOR1B is also ubiquitously expressed. The other two members of the family, TOR 2A and TOR3A, share about 50% homology at the amino acid level with DYT1 (Ozelius et al., 1999; Dron et al., 2002; unpublished results). TOR2A resides on chromosome 9q34 about 10 cM centromeric to DYT1 and TOR1B. The cDNA has five exons, corresponding to a protein of 321 amino acids with homology to torsinA throughout its length. TOR3A is located on chromosome 1q24 and has been independently cloned by Dron et al. (2002) by virtue of transcriptional regulation in response to alpha-interferon and given the name ADIR1 (ATP-dependent interferon responsive gene). The message contains 6 exons, is about 2 kb in length, and encodes a protein of 397 amino acids. A splice variant has been described (ADIR2) in placenta that encodes a putative 336 amino acids protein. It differs from ADIR1 by an alternatively spliced sixth exon and different 3′ UTR. Both TOR2A and 3A are expressed ubiquitously on Northern blots (Dron et al. 2002, unpublished data). Genomic database searches have revealed torsin-like genes, all of unknown function, in mouse, rat, nematode, fruit fly, pig, cow, zebrafish, chicken, hamster, and Xenopus (Ozelius et al., 1999).

The DYT1 gene encodes a protein, torsinA, that is 332 amino acids long (~38 kDa), with potential sites for glycosylation and phosphorylation, as well as an amino terminal hydrophobic leader sequence consistent with membrane translocation/targeting (Ozelius et al., 1997a). Analysis of the primary amino acid sequence of the torsin family reveals domains with homology to functional regions (Ozelius et al., 1997a; Lupas et al., 1997) of the HSP100/Clp ATPase class of heat shock proteins (Schirmer et al., 1996) within the AAA+ (ATPases Association with a variety of cellular Activities) superfamily (Neuwald et al., 1999). These proteins are characterized by Mg++-dependent ATPase activity and typically form six-membered, homomeric ring structures. Although not highly conserved at the amino acid sequence level, they share a configurational similarity allowing them to be aligned at the three-dimensional level (Neuwald et al., 1999). Many of these proteins are chaperones that mediate conformational changes in target proteins. They are associated with a number of functions including protein folding and degradation, cytoskeletal dynamics, membrane trafficking and vesicle fusion and response to stress (Vale, 2000). However, the functions are so diverse that they provide little clue as to how a new member, like torsinA, might act.

Expression of torsinA has been examined in normal adult brain (Augood et al., 1998, 1999; Shashidharan et al., 2000a; Konakova et al., 2001). It is widely distributed, with intense expression in substantia nigra compacta dopamine neurons, cerebellar dentate nucleus, Purkinje cells, basis pontis, locus ceruleus, numerous thalamic nuclei, the pedunculopontine nucleus, the oculomotor nuclei, hippocampal formation, and frontal cortex. Both the mRNA and protein were localized to neurons and not to glia while the protein studies also showed torsinA in neuronal processes. Labeling was predominantly present in cytoplasm with some perinuclear staining (Shashidharan et al., 2000a; Konakova et al., 2001). A similar widespread pattern of expression was seen in both rat (Shashidharan et al., 2000a; Walker et al., 2001) and mouse (Konakova and Pulst, 2001) brains. The intense expression in nigral neurons suggests there may be dysfunction in dopamine transmission (Augood et al., 1999) while strong labeling of neuronal processes points to a potential role for torsin in synaptic functioning. Finally, two studies have further implicated torsinA in dopamine transmission with the finding of torsinA and alpha-synuclein immunoreactivity co-localized in Lewy bodies (Shashidharan et al., 2000b; Sharma et al., 2001).

## F. Abnormal Gene and Protein Function

The GAG deletion in the DYT1 gene results in the loss of one of a pair of glutamic acid residues in the C-terminal region of the protein (Ozelius et al., 1997a). Likewise, the 18-bp deletion identified in a single family where the proband had both dystonia and myoclonus, removes 6 amino acid (F323–Y328del) from the carboxy terminus (Leung et al., 2001). This latter mutation disrupts a predicted casein kinase 2 phosphorylation domain that could in turn, interfere with the normal function of torsinA. As stated in the previous section, torsinA is a member of the AAA/HSP/Clp ATPase protein superfamily. Although the exact function of torsinA remains elusive, a feature of AAA+ proteins may explain the dominant inheritance of DYT1 dystonia. AAA+ proteins are known to form oligomeric ring structures and the carboxy terminus of these proteins are important for this oligomerization (Whiteheart et al., 1994). In addition, the C-terminal region of these proteins is also crucial for the binding to interacting proteins as exemplified by mutations in this region of HS1U (Missiakas et al., 1996) and FtSH (Akiyama et al., 1994), two HSP/Clp/AAA+ family members, that interferes with binding to their partners and blocks proteolytic activity and protein translocation, respectively. Thus, dominantly inherited mutations in torsinA could cause disease in a dominant negative fashion by two mechanisms. First, by blocking ring formation, since even one defective member of the ring can inactivate the complex, resulting in a decrease of functional protein to less than 10% normal levels (Breakefield et al., 2001). Alternatively, if oligomeric rings formed but were composed of both wildtype and mutant protein, they might be unable to interact with partner proteins. For studies describing neuropathology of DYT1 brains, see Section VI.

## IV. DIAGNOSTIC AND ANCILLARY TESTS

### A. DNA Test

With the cloning of the DYT1 gene (Ozelius et al., 1997a) it is now possible to diagnose a leading cause of generalized primary dystonia. The DYT1 GAG deletion accounts for about 90% of early limb-onset cases in the Ashkenazi population but only about 50–70% of non-Jewish early limb-onset PTD, with the excess in Ashkenazim due to the founder mutation and genetic drift (see Section III.C); (Bressman et al., 2000). Because almost all dystonia with involvement of the DYT1 gene appears to be due to the same GAG deletion, screening is relatively easy and commercially available (Klein et al., 1999). The test should be considered the first diagnostic test to apply to all primary dystonia patients whether Ashkenazi or non-Jewish with onset by age 26 as these criteria result in 100% sensitivity with acceptable specificities ranging from 43% (in non-Jews) to 63% (in Ashkenazi Jews). If criteria are set more narrowly to include only those with early-limb-onset specificity improves (70–80%) sensitivity drops (94–96%); (Bressman et al., 2000). It may also be advisable to test individuals with later onset who have an early-onset blood relative as genetic studies have revealed that late-onset cases (most with writer's cramp) are found in the families of early-onset patients. Before diagnostic testing is done it is preferable to have genetic counseling performed so that the implications of both a positive and negative test can be explained. For example, even if the test is negative, a genetic etiology is not excluded and this needs to be discussed. If the test is positive, a diagnosis is secured but this diagnosis impacts on other at-risk family members. These members, even if asymptomatic, may wish carrier testing, and genetic counseling for all asymptomatic family members needs to be done first. The psychological and social implications of a disorder with autosomal dominant inheritance that has markedly reduced penetrance and very variable expression are complex.

### B. Other Laboratory Tests

To arrive at the clinical diagnosis of PTD, the examination should be normal except for dystonia and the history should not suggest another etiology. In this setting using the age-onset criteria stated above, DYT1 molecular testing is the first test to perform. Other tests (MRI, serum ceruloplasmin, levodopa trial) need only be performed if DYT1 testing is negative. Of course one's suspicion that dystonia is due to DYT1 is increased as the phenotype more narrowly conforms to predictive features (limb-onset, early-onset, multiple limb involvement).

## V. NEUROIMAGING

Routine CT and MRI scans are normal in DYT1 dystonia. Most glucose PET studies of PTD show lentiform hypermetabolism. Eidelberg specifically studied DYT1 GAG deletion mutation carriers, both affected individuals and their asymptomatic gene carrier relatives and as controls assessed noncarrier family members. Using [18F] fluorodeoxyglucose he identified two independent abnormal regional metabolic covariance patterns, movement free (MF) and movement related (MR); (Eidelberg et al., 1998). The MR pattern was identified in gene carriers affected with signs of dystonia. It showed increased metabolic activity in the midbrain, cerebellum, and thalamus. The MF pattern was characterized by increased metabolic activity in the lentiform nuclei, cerebellum, and supplementary motor areas and was seen in both affected and non-manifesting carriers. (Eidelberg et al., 1998). More recent PET studies using psychomotor testing show subtle abnormalities in sequence learning both in the motor performance and recruitment of brain networks (Carbon et al., 2002). These studies strongly support the presence of abnormal brain processing in gene carriers regardless of overt motor signs of dystonia, expanding the notion of penetrance and phenotype. It also raises the persistent enigma of why dystonia is only present in 30% of gene carriers as factors triggering or modifying expression remain unknown.

## VI. NEUROPATHOLOGY

No consistent neuropathological changes have been reported in the brains of patients with early-onset dystonia (Zeman, 1970; Hedreen et al., 1988), however, as stated above, there is some genetic heterogeneity, particularly among non-Jewish early-onset PTD cases. The discovery of the DYT1 gene and the fact that most cases are due to the same 3-bp deletion has allowed researchers to identify DYT1-positive brains for neuropathologic examination. Furukawa et al. (2000) found that nigral cellularity was normal and that subregional striatal dopamine and homovanillic acid levels, measured by high-performance liquid chromatography, were within the control range except for those in the rostral portions of the putamen and caudate nucleus which were slightly decreased compared to controls. Although these results suggest that the DYT1 mutation is not associated with significant damage to the nigrostriatal dopamine system, the study represented only a single patient. Another study, again examining a single DYT1-positive dystonia brain, observed no differences in immunoreactivity of torsinA compared to non-DYT1 brains. In addition, the authors found no evidence of intracellular aggregations, nor of specific co-localization with markers for endoplasmic reticulum (Walker et al., 2002) both of which have been reported in cell culture studies (see Section VII.A). A neuropathologic examination on autopsy brain specimens from two GAG-deleted dystonia patients also showed no signs of immunoreactive inclusions or neurodegeneration (Rostasy et al., submitted).

## VII. CELLULAR AND ANIMAL MODELS OF DISEASE

### A. In vitro

Cellular localization studies in culture indicate that most torsinA, either overexpressed or endogenous, resides in the lumen of the endoplasmic reticulum (ER) consistent with the deduced signal sequence and the observed high mannose content (Hewett et al., 2000; Kustedjo et al., 2000). However, it is also found associated with vesicles in neural cells and extends to the ends of processes (Hewett et al., 2000). This suggests it may be part of the tubular vesicular complex involved in anterograde membrane and vesicle movement (Ahmari et al., 2000). This is supported by the localization of torsin in neuronal processes and at synaptic endings in association with vesicles in human and nonhuman primate brain (Augood, personal communication). In contrast, overexpression of mutant torsinA containing the GAG deletion in cultured cells resulted in the formation of intracellular inclusions apparently derived from the ER and frequently associated with the nuclear membrane (Hewett et al., 2000; Kustedjo et al., 2000). It is speculated that these accumulations interfere with membrane trafficking in neurons (Breakefield et al., 2001). However, these results represent the effects of expressing torsinA at very high levels and to date, neuropathologic examination of very limited brain material from GAG-deleted dystonia patients shows no signs of inclusions or neurodegeneration (Walker et al., 2002; Rostasy et al., submitted).

Expression studies using an ADIR-EGFP (TOR3A) fusion protein expressed in HeLa cells was also shown to be associated with the ER. Suggesting that other torsin-family members many share similar functions (Dron et al., 2002).

### B. Invertebrates

As part of the torsin gene family, there is a single related gene in *Drosophila* and another in zebrafish, and three related genes in nematodes (Ozelius et al., 1999). All are predicted proteins of unknown function except the independently identified OOC-5, one of the three nematode genes (Basham and Rose, 1999). This is a dynamic ER protein, which is critical to rotation of the nuclear-centrosome complex with respect to cortical markers during early embryogenesis (Basham and Rose, 2001). A defect in rotation results in misorientation of the mitotic spindle and disruption of asymmetric cell division and cell fate determination. The involvement of OOC-5 in this pathway suggests a possible role for torsin-like proteins in the establishment of cell polarity (Breakefield et al., 2001; Basham and Rose, 2001). In addition, the fact that OOC-5 is an ER protein suggests that some essential ER-related function has been conserved throughout evolution in the torsin proteins.

### C. Mouse and Rat

Currently there are two genetic animal models related to early-onset dystonia but neither has neurophysiological features characteristic of early-onset dystonia. The dystonic (dt) rat exhibits an autosomal recessive, progressive motor syndrome with dystonic features derived primarily from cerebellar dysfunction (LeDoux et al., 1995). This rat results from a spontaneous mutation the nature of which is unknown, however, isolation and sequencing of the rat DYT1 cDNA from the brains of these animals revealed no defect in the coding region (Ziefer et al., 2002). A spontaneously occurring mouse mutant, termed dystonia musculorum (dMd), has a syndrome involving degeneration of sensory neurons with consequent ataxia; the protein responsible for this condition, dystonin, is involved in adherent junctions between cells, termed hemidesmosomes (Guo et al., 1995; Brown et al., 1995). As degeneration in the peripheral nervous system is not observed in dystonia, the relevance of this model is uncertain.

In order to understand the normal role of torsin and how the DYT1 mutation alters torsin function, Dauer and colleagues (DMRF/NINDS Workshop report, 2001) have begun to characterize a set of torsin mutant mice. Evaluation of these mice is in the preliminary stages, with primarily gross morphological and behavioral analysis to date. TorsinA knock-out mice appear grossly normal, but do not feed or vocalize normally, and die at P0. The precise reason for their early demise remains to be clarified (DMRF/NINDS Workshop report, 2001). Knock-down mice that express very low levels of torsinA display deficient habituation in open field studies (Dauer et al., 2000). Characterization of mice that contain the pathogenic GAG deletion in the native torsinA gene (knock-in mice) is at an early stage.

## VIII. TREATMENT

Treatment for most dystonia, including DYT1, is empirically based and aimed at controlling symptoms. For patients with childhood- and adolescent-onset dystonia, most of whom have segmental or generalized signs, oral medications are the mainstays of therapy, although there is increasing evidence that pallidal stimulation may provide dramatic improvement especially for those with disabling primary dystonia due to the DYT1 GAG deletion (Coubes et al., 2000). For those with adult-onset focal dystonias botulinum toxin injections are generally the treatment of choice.

Because of the dramatic response to levodopa in patients with dopa-responsive dystonia (DRD), most therapeutic recommendations advise trying levodopa to at least 300 mg levodopa combined with carbidopa (Bressman and Greene, 2000). However, DYT1 genetic testing is now readily available with fairly quick turn-around time. Thus, the rationale for starting levodopa in patients suspected clinically to be DYT1 is less compelling as DYT1 rarely substantively improves with low dose levodopa and may worsen. Nevertheless because DRD and DYT1 dystonia are similar clinically and there often is an urgency to treat, levodopa with carbidopa frequently is the first drug tried in childhood-onset DYT1 patients. The class of oral drug with the greatest clinical benefit is the anticholinergics with about 40–50% responding moderately (Burke et al., 1986, Greene et al., 1988). The dose is slowly titrated up to a maximum tolerated; central side effects such as memory impairment, confusion, and hallucinations usually limit dose, especially in older individuals. Peripheral side effects may be controlled with pyridostigmine.

If anticholinergics are not tolerated or not helpful the next drug tried is usually baclofen (Greene et al., 1988; Greene and Fahn, 1992). Although the drug appears to be less helpful than anticholinergics overall, fairly dramatic response may occur in children. Other drugs are then usually tried either in combination or alone and include mexiletine (Ohara et al., 1998; Lucetti et al., 2000), intravenous methylprednisolone (Kumar et al., 1997), clonazepam and other benzodiazepines, carbamazepine, and tetrabenazine (Greene et al., 1988). The latter may be used in primary dystonia in a "cocktail" that includes a dopamine blocker and an anticholinergic (Marsden et al., 1984). Tetrabenazine is a dopamine depletor, available in Europe and England but not the U.S., and like reserpine, another dopamine depletor, may be especially beneficial in tardive dystonia (Jankovic and Orman, 1988). In general, dopamine blocking agents are not recommended because of acute and tardive side effects, although risperidone has been reported to be useful in a short trial (Zuddas and Cianchetti, 1996). In some patients with generalized and segmental dystonia, botulinum toxin A to the most disabling or painful muscles may be given in conjunction with the above therapies.

Finally, if oral medications fail, intrathecal baclofen and surgery should be considered. Regarding the former, predicting response based on underlying cause remains to be determined (Ford et al., 1996; Walker et al., 2000) and only a minority of those responding to an intrathecal test dose appear to have clear-cut clinical benefit to chronic intrathecal administration via implanted pump (Walker et al., 2000).

Bilateral pallidotomy (Lozano et al., 1997; Lai et al., 1999) and pallidal stimulation (Kumar et al., 1999; Coube et al., 2000) are being assessed, particularly in patients with generalized dystonia with DYT1 unresponsive to other therapies. The most promising of these two appears to be globus pallidus (GPi) deep brain stimulation (DBS). Kumar et al. (1999) reported that GPi DBS in a patient with severe primary generalized dystonia resulted in immediate improvement of all aspects of dystonia. A second study also reported positive results of bilateral DBS of the GPi in 15 patients (mean age 14.2 years) with severe early-onset generalized dystonia (Coubes et al., 2000). Mean Burke-Fahn-Marsden Dystonia Rating Scale score was reduced from 69.5 to 11.1. Improvement was greatest for the 7 DYT1(+) individuals, who improved a mean of 90.3%. Thirteen patients could walk without aids, including several that were confined to bed before surgery. Patients reported rapid and complete disappearance of pain, and medications were reduced.

## References

Ahmari, S. E., Buchanan, J., and Smith, S. J. (2000). Assembly of presynaptic active zones from cytoplasmic transport packets. *Nat. Neurosci.* **3**, 445–451.

Akiyama, Y., Shirai, Y., and Ito, K. (1994). Involvement of FtsH in protein assembly into and through the membrane. *J. Biol. Chem.* **269**, 5225–5229.

Augood, S. J., Penney, J. B., Friberg, I., Breakefield, X. O., Young. A., Ozelius, L. J., and Standaert, D. G. (1998). Expression of the early-onset torsion dystonia gene (DYT1) in human brain. *Ann. Neurol.* **43**, 669–673.

Augood, S. J., Martin, D. M., Ozelius, L. J., Breakefield, X. O., Penney. J. B. J., and Standaert, D.G. (1999). Distribution of the mRNAs encoding torsinA and torsinB in the adult human brain. *Ann. Neurol.* **46**, 761–769.

Basham, S. E., and Rose, L. S. (1999). Mutations in ooc-5 and ooc-3 disrupt oocyte formation and the reestablishment of asymmetric PAR protein localization in two-cell *Caenorhabditis elegans* embryos. *Dev. Biol.* **215**, 253–263.

Basham, S. E., and Rose, L. S. (2001). The *Caenorhabditis elegans* polarity gene ooc-5 encodes a Torsin-related protein of the AAA ATPase superfamily. *Development* **128**, 4645–4656.

Brassat, D., Camuzat, A., Vidailhet, M., Feki, I., Jedynak, P., Klap, P., Agid, Y., Durr, A., and Brice, A. (2000). Frequency of the DYT1 mutation in primary torsion dystonia without family history. *Arch. Neurol.* **57**, 333–5.

Breakefield, X. O., Kamm, C., and Hanson, P. I. (2001). TorsinA: movement at many levels. *Neuron* **31**, 9–12.

Breakefield, X. O., Ozelius, L., Hanson, P., and Dauer, W. (2001). From gene to function in dystonia. DMRF/NINDS Workshop report.

Bressman, S. B., de Leon, D., Brin, M. F., Risch, N., Burke, R. E., Greene P, E., Shale, H., and Fahn, S. (1989). Idiopathic torsion dystonia among Ashkenazi Jews: Evidence for autosomal dominant inheritance. *Ann. Neurol.* **26**, 612–620.

Bressman, S. B., de Leon, D., Kramer, P. L., Ozelius, L. J., Brin. M. F., Greene, P. E., Fahn, S., Breakefield, X. O., and Risch, N.J. (1994). Dystonia in Ashkenazi Jews: clinical characterization of a founder mutation. *Ann. Neurol.* **35**, 771–771.

Bressman, S. B., and Fahn, S. (1997). Childhood Dystonia. In *Movement Disorders: Neurologic Principals and Practice*. (R.L. Watts, and W.C. Koller, eds.), pp. 419–428. McGraw-Hill, New York.

Bressman, S. B., Sabatti, C., Raymond, D., De Leon, D., Klein, C., Kramer, P. L., Brin, M. F., Fahn, S., Breakefield, X. O., Ozelius, L. J., and Risch. N. J. (2000). The DYT1 phenotype and guidelines for diagnostic testing. *Neurology* **54**, 1746–1752.

Bressman, S. B., and Greene, P. E. (2000). Dystonia. *Curr. Treat. Options Neurol.* **2**, 275–285.

Brown, A., Bernier, G., Mathieu, M., Rossant, J., and Kothary, R. (1995). The mouse dystonia musculorum gene is a neural isoform of bullous pemphigoid antigen 1. *Nat. Genet.* **10**, 301–306.

Burke, R. E., Fahn, S., and Marsden, C. D. (1986). Torsion dystonia: a double blind, prospective trial of high dosage trihexyphenidyl. *Neurology* **36**, 160–164.

Carbon, M., Ghilardi, M. F., Dhawan, V., Ghez, C. P., and Eidelberg, D. (2002). Brain networks subserving motor sequence learning in DYT1 gene carriers. *Neurology* Abstract P03.067.

Coubes, P., Roubertie, A., Vayssiere, N., Hemm, S., and Echenne, B. (2000). Treatment of DYT1—generalized dystonia by stimulation of the internal globus pallidus. *Lancet* **355**, 2220–2221.

Dauer, W. T., Kumar, P., Ozelius, L. J., Breakefield, X. O., Fahn, S., and Hen, R. (2000). Characterization of the motor phenotype of DYT1 mutant mice. *Soc. Neurosci. Abst.* **27** Program No. 572.8, 2000.

Doheny, D., Danisi, F., Smith, C., Morrison, C., Velickovic, M., de Leon, D., Bressman, S. B., Leung, J., Ozelius, L. J., Klein, C., Breakefield, X. O., Brin, M. F., Silverman, J. M. (in press). Clinical findings of a myoclonas-dystonia family with two distinct mutations. *Neurology*.

Dron, M., Meritet, J. F., Dandoy-Dron, F., Meyniel, J. P., Maury, C., and Tovey, M. G. (2002). Molecular cloning of ADIR, a novel interferon responsive gene encoding a protein related to the torsins. *Genomics* **79**, 315–325.

Eidelberg, D., Moeller, J. R., Antonini, A., Dhawan, V., Spetsieris, P., de Leon, D., Ghilardi, M. F., Ghez, C., Bressman, S., and Fahn, S. (1998). Functional brain networks in DYT1 dystonia. *Ann. Neurol.* **44**, 303–312.

Eldridge, R. (1970). The torsion dystonia: literature review: genetic and clinical studies. *Neurology* **20**, 1–78.

ESDE (Epidemiology Study of Dystonia in Europe Collaborative Group) (1999). Sex-related influences on the frequency and age of onset of primary dystonia. *Neurology* **53**, 1871–1873.

ESDE (Epidemiology Study of Dystonia in Europe Collaborative Group) (2000). A prevalence study of primary dystonia in eight European countries. *J. Neurol.* **247**, 787–792.

Fahn, S., Marsden, C. D., and Calne, D. B.(1987). Classification and investigation of dystonia. In "Movement Disorders 2" (C.D. Marsden, and S. Fahn, eds.), pp. 332–358. Butterworth, London.

Fletcher, N. A., Harding, A. E., and Marsden, C. D. (1990). A genetic study of idiopathic torsion dystonia in the United Kingdom. *Brain* **113**, 379–396.

Ford, B., Greene, P. E., Louis, E., Petzinger, G., Bressman, S. B., Goodman, R., Brin, M. F., and Fahn, S. (1996). Use of intrathecal baclofen in patients with dystonia. *Arch. Neurol.* **53**, 1241–1246.

Friedman, J. R., Klein, C., Leung, J., Woodward, H., Ozelius, L. J., Breakefield, X. O., and Charness, M. E..(2000). The GAG deletion of the DYT1 gene is infrequent in musicians with focal dystonia. *Neurology* **55**, 1417–1418.

Furukawa, Y., Hornykiewicz, O., Fahn, S., and Kish, S. J. (2000). Striatal dopamine in early-onset primary torsion dystonia with the DYT1 mutation. *Neurology* **54**, 1193–1195.

Gasser, T., Windgassen, K., Bereznai, B., Kabus, C., and Ludolph. A. C. (1998). Phenotypic expression of the DYT1 mutation: a family with writer's cramp of juvenile onset. *Ann. Neurol.* **44**, 126–128.

Greene, P. E., Shale, H., and Fahn, S. (1988). Analysis of open-label trials in torsion dystonia using high dosages of anticholinergics and other drugs. *Mov. Disord.* **3**, 46–60.

Greene, P. E., and Fahn, S. (1992). Baclofen in the treatment of idiopathic dystonia in children. *Mov. Disord.* **7**, 48–52.

Greene, P., Kang, U. J., and Fahn, S. (1995). Spread of symptoms in idiopathic torsion dystonia. *Mov. Disord.* **10**, 143–152.

Guo, L., Degenstein, L., Dowling, J., Yu, Q. C., Wollmann, R., Perman, B., and Fuchs, E. (1995). Gene targeting of BPAG1: abnormalities in mechanical strength and cell migration in stratified epithelia and neurologic degeneration. *Cell* **81**, 233–243.

Hedreen, J. C., Zweig, R. M., DeLong, M. R., Whitehouse, P. J., and Price, D. L. (1988). Primary dystonias: A review of the pathology and suggestions for new directions of study. *Adv. Neurol.* **50**, 123–132.

Hewett, J., Gonzalez-Agosti, C., Slater, D., Li, S., Ziefer, P., Bergeron, D., Jacoby, D. J., Ozelius, L. J., Ramesh, V., and Breakefield, X. O. (2000). Mutant torsinA, responsible for early onset torsion dystonia, forms membrane inclusions in cultured neural cells. *Hum. Mol. Genet.* **22**, 1403–1413.

Ikeuchi, T., Shimohata, T., Nakano, R., Koide, R., Takano, H., and Tsuji, S. (1999). A case of primary torsion dystonia in Japan with the 3-bp (GAG) deletion in the DYT1 gene with a unique clinical presentation. *Neurogenetics* **2**, 189–90.

Jankovic, J., and Orman, J. (1988). Tetrabenazine treatment in dystonia, chorea, tics, and other dyskinesias. *Neurology* **38**, 391–394.

Kamm, C., Castelon-Konkiewitz, E., Naumann, M., Heinen, F., Brack, M., Nebe, A. Ceballos-Baumann, A., and Gasser, T. (1999). GAG deletion in the DYT1 gene in early limb-onset idiopathic torsion dystonia in Germany. *Mov. Disord.* **14**, 681–683.

Kamm, C., Naumann, M., Mueller, J., Mai, N., Riedel, L., Wissel, J., and Gasser, T. (2000). The DYT1 GAG deletion is infrequent in sporadic and familial writer's cramp. *Mov. Disord.* **15**, 1238–1241.

Klein, C., Brin, M. F., de leon, D., Limborska, S. A., Ivanova-Smolenskaya, I. A., Bressman, S. B., Friedman, A., Markova, E. D., Risch, N. J., Breakefield, X. O., and Ozelius, L. J. (1998). De novo mutations (GAG deletion) in the DYT1 gene in two non-Jewish patients with early-onset dystonia. *Hum. Mol. Genet.* **7**, 1133–1136.

Klein, C., Friedman, J., Bressman, S., Vieregge, P., Brin, M. F., Pramstaller, P. P., de Leon, D., Hagenah, J., Sieberer, M., Fleet, C., Kiely, R. Xin, W. Breakefield, X. O., Ozelius, L. J., and Sims, K. B. (1999). Genetic testing for early-onset torsion dystonia (DYT1): Introduction of a simple screening method, experiences from testing of a large patient cohort, and ethical aspects. *Genet. Test.* **3**, 323–328.

Konakova, M., and Pulst, S. M. (2001). Immunocytochemical characterization of torsin proteins in mouse brain. *Brain Res.* **922**, 1–8.

Konakova, M., Huynh, D. P., Yong, W., and Pulst, S. M. (2001). Cellular distribution of torsin A and torsin B in normal human brain. *Arch. Neurol.* **58**, 921–927.

Kramer, L. P., de Leon, D., Ozelius, L., Risch, N. J., Bressman, S. B., Brin, M. F., Schuback, D. E., Burke, R. E., Fahn, S., and Breakefield, X. O. (1990). Dystonia gene in Ashkenazi Jewish population is located in chromosome 9q32-34. *Ann. Neurol.* **27**, 114–120.

Kramer, P. L., Heiman, G., Gasser, T., Ozelius, L., deLeon, D., Brin, M. F., Burke, R. E., Hewett, J., Hunt, A., Moskowitz, C., Nygaard, T. G., Wilhelmsen, K. Fahn, S.Breakefield, X. O., Risch, N. J., and Bressman, S. B. (1994). The DYT1 gene on 9q34 is responsible for most cases of early-onset idiopoathic torsion dystonia in non-Jews. *Am. J. Hum. Genet.* **55**, 468–475.

Kumar, R., Maraganore, D. M., Ahlskog, J. E., and Rodriguez, M. (1997). Treatment of putative immune-mediated idiopathic cervical dystonia with intravenous methylprednisolone. *Neurology* **48**, 732–735.

Kumar, R., Dagher, A., Hutchison, W. D., Lang, A. E., and Lozano, A. M. (1999). Globus pallidus deep brain stimulation for generalized dystonia:clinical and PET investigation. *Neurology* **11**, 871–874.

Kustedjo, K., Bracey, M. H., and Cravatt, B.F. (2000). Torsin A and its torsion dystonia-associated mutant forms are lumenal glycoproteins that exhibit distinct subcellular localizations. *J. Biol. Chem.* **275**, 27933–27939.

Lai, T., Lai, J. M., Grossman, R. G. (1999). Functional recovery after bilateral pallidotomy for the treatment of early-onset primary generalized dystonia. *Arch. Phys. Med. Rehab.* **80**, 1340–1342.

Lebre, A. S., Durr, A., Jedtnak, P., Ponsot, G., Vidailhaet, M., Agid, T., and Brice, A. (1999). DYT1 Mutation in French families with idiopathic torsion dystonia. *Brain* **122**, 41–45.

LeDoux, M. S., Lorden, J. F., and Meinzen-Derr, J. (1995). Selective elimination of cerebellar output in the genetically dystonic rat. *Brain Res.* **697**, 91–103.

Leube, B., Kessler, K. R., Ferbert, A., Ebke, M., Schwendemann, G., Erbguth, F., Benecke, R., and Auburger, G. (1999). Phenotypic variability of the DYT1 mutation in German dystonia patients. *Acta Neurol. Scand.* **99**, 248–251.

Leung, J., Klein, C., Friedman, J., Vieregge, P., Helfried, D., Kamm, C., DeLeon, D., Pramstaller, P., Penney, J., Eisengart, M., Jankovic, J., Gasser, T., Bressman, S., Corey, D., Kramer, P., Brin, M., Ozelius, L., and Breakefield, X.O. (2001). Novel mutation in the TOR1A (DYT1) gene in atypical, early onset dystonia and polymorphisms in dystonia and early onset parkinsonism. *Neurogenetics*, **3**, 133–143.

Lozano, A.M., Kumar, R., Gross, R. E., Giladi, N., Hutchinson, W. D., Dostrovsky, J. O., and Lang, A. E. (1997). Globus pallidus internus pallidotomy for generalized dystonia. *Mov. Disord.* **12**, 841.

Lucetti, C., Nuti, A., Gambacinni, G., Bernardini, S., Brotini, S., Manca, M. L., and Bonuccelli, U. (2000). Mexiletine in the treatment of torticollis and generalized dystonia. *Clin. Neuropharmacol.* **23**, 186–189.

Lupas, A., Flanagan, J. M., Tamura, T., and Baumeister, W. (1997). Self-compartmentalization proteases. *Trends Biochem. Sci.* **22**, 399–404.

Major, T., Svetel, M., Romac, S., and Kostic, V. S. (2001). DYT1 mutation in primary torsion dystonia in a Serbian population. *J. Neurol.* **248**, 940–943.

Marsden, C. D., and Harrison, M. S. G. (1974). Idiopathic torsion dystonia (dystonia musculorum deformans): a review of forty-two patients. *Brain* **97**, 793–810.

Marsden, C. D., Marion, M. H., and Quinn, N. (1984). The treatment of severe dystonia in children and adults. *J. Neurol. Neurosurg. Psychiatry* **36**, 160–164.

Matsumoto, S., Nishimura, M., Kaji, R., Sakamoto, T., Mezaki, T., Shimazu, H., Murase, N., and Shibasaki, H. (2001) DYT1 mutation in Japanese patients with primary torsion dystonia. *Neuroreport* **12**, 793–795.

Missiakas, D., Schwager, F., Betton, J-M., Georgopoulos, C., and Raina, J. (1996). Identification and characterization of HS1V HS1U (ClpQ ClpY) proteins involved in overall proteolysis of misfolded proteins in Eschericia coli. *EMBO J.* **15**, 6899–6909.

Neuwald, A. F., Aravind, L., Spouge, J. L ,and Koonin, E. V. (1999). AAA+: A class of chaperone-like ATPases associated with the assembly, operation, and dissassembly of protein complexes. *Genome Res.* **9**, 27–43.

Nutt, J. G., Muenter, M. D., Aronson, A., Kurland, L. T., and Melton, L. J. (1988). Epidemiology of focal and generalized dystonia in Rochester, Minnesota. *Mov. Disord.* **3**, 188–194.

Ohara, S., Hayashi, R., Momoi, H., Miki, J., and Yanagisawa, N. (1998). Mexilitine in the treatment of spasmodic torticollis. *Mov. Disord.* **13**, 934–940.

Oppenheim, H. (1911). Über eine eigenartige Krampfkrankheit des kindlichen und jugendlichen Alters (Dysbasia lordotica progressiva, dystonia musculorum deformans). *Neurol. Centralbl.* **30**, 1090–1107.

Ozelius, L., Kramer, P., Moskowitz, C. B., Kwiatkowski, D., Brin, M. F., Bressman, S. B., Schuback, D. E., Falk, C., Risch, N., de Leon, D., Burke, R. E., Haines, J., Gusella, J. F., Fahn, S., and Breakefield, X. O. (1989). Human gene for torsion dystonia located on chromosome 9q32-34. *Neuron.* **2**, 1427–1434.

Ozelius, L. J., Kramer, P. L., de Leon, D., Risch, N., Bressman, S. B., Schuback, D. E., Brin, M. F., Kwiatkowski, D. J., Burke, R. E., Gusella, J. F., Fahn, S., and Breakefield, X. O. (1992). Strong allelic association between the torsion dystonia gene (DYT1) and loci on chromosome 9q34 in Ashkenazi Jews. *Am. J. Hum. Genet.* **50**, 619–628.

Ozelius, L. J., Hewett, J. W., Page, C. E., Bressman, S. B., Kramer, P. L., Shalish, C., de Leon, D., Brin, M. F., Raymond, D., Corey, D. P., Fahn, S., Risch, M. J., Buckler, A. J., Gusella, J. F., and Breakefield, X.O. (1997a). The early onset torsion dystonia gene (DYT1) encodes an ATP-binding protein. *Nat. Genet.* **17**, 40–48.

Ozelius, L. J., Hewett, J., Kramer, P., Bressman, S. B., Shalish, C., de Leon. D., Rutter. M., Risch, N. J., Brin, M. F., Markova. E. D., Limborska, S. A., Ivanova-Smolenskaya, I. A., McCormack. M. F., Fahn, S., Buckler, A. J., Gusella, J. F., and Breakefield, X. O. (1997b). Fine localization of the dystonia gene (DYT1) on human chromosome 9q34:YAC map and linkage disequilibrium. *Genome Res.* **7**, 483–496.

Ozelius LJ, Page CE, Klein C, Hewett JW, Mineta M, Leung J, Shalish C, Bressman SB, de Leon D, Brin MF, Fahn S, Corey DP, Breakefield XO (1999). The TOR1A (DYT1) gene family and its role in early onset torsion dystonia. *Genomics* **62**, 377–384.

Pauls, D. L., and Korczyn, A. D. (1990). Complex segregation analysis of dystonia pedigrees suggests autosomal dominant inheritance. *Neurology* **40**, 1107–1110.

Risch, N. J., Bressman, S. B., de Leon, D., Brin, M. F , Burke, R. E., Greene, P. E., Shale, H., Claus, E. B., Cupples, L. A., and Fahn, S. (1990). Segregation analysis of idiopathic torsion dystonia in Ashkenazi Jews suggests autosomal dominant inheritance. *Am. J. Hum. Genet.* **46**, 533–538.

Risch, N., deLeon, D., Ozelius, L., Kramer, P., Almasy, L., Singer, B., Fahn, S., Breakefield, X., and Bressman, S. (1995). Genetic analysis of idiopathic torsion dystonia in Ashkenazi Jews and their recent descent from a small founder population. *Nat. Genet.* **9**, 152–159.

Rostasy, K., Augood, S., Hewett, J., Cheung-on Leung, J., Sasaki, H., Ozelius, L., Standaert, D., Breakefield, X. O., and Hedreen, J. Comparison of torsinA in the normal human brain with early onset generalized dystonia with GAG deletion, submitted.

Schwalbe, W. (1908). Eine eigentümliche tonische Krampfform mit hysterischen Symptomen. Thesis, Berlin.

Schirmer, E. C., Glover, J. R., Singer, M. A., and Lindquist, S. (1996). HSP100/Clp proteins: a common mechanism explains diverse functions. *TIBS* **21**, 289–296.

Sharma, N., Hewett, J., Ozelius, L. J., Ramesh, V., McLean, P. J., Breakefield, X. O., and Hyman, B. T. (2001). A close association of torsinA and alpha-synuclein in Lewy bodies: a fluorescence resonance energy transfer study. *Am. J. Pathol.* **159**, 339–344.

Shashidharan, P., Kramer, C., Walker, R., Olanor, C. W., and Brin, M. F. (2000a). Immunohistochemical localization and distribution of torsinA in normal human and rat brain. *Brain Res.* **853**, 197–206.

Shashidharan, P., Good, P. F., Hsu, A., Perl, D. P., Brin, M. F., and Olanow, C. W. (2000b). TorsinA accumulation in Lewy bodies in sporadic Parkinson's disease. *Brain Res.* **877**, 379–381.

Slominski, P. A., Markova, E. D., Shadrina, M. I., Illarioshkin, S. N., Miklina, N. I., Limborska, S. A., and Ivanova-Smolenskaya, I. A. (1999). A common 3-bp deletion in the DYT1 gene in Russian families with early-onset torsion dystonia. *Hum. Mutat.* **14**, 269.

Tuffery-Giraud, S., Cavalier, L., Roubertie, A., Guittard, C., Carles, S., Calvas, P., Echenne, B., Coubes, P., and Claustres, M. (2001). No evidence of allelic heterogeneity in the DYT1 gene of European patients with early onset torsion dystonia. *J. Med. Genet.* **38**, E35.

Vale, R. D. (2000) AAA proteins: Lords of the ring. *J. Cell. Biol.* **150**, F13–F19.

Valente, E. M., Warner, T. T., Jarman, P. R., Mathen, D., Fletcher, N. A., Marsden, C. D., Bhatia, K. P., and Wood, N. W. (1998). The role of primary torsion dystonia in Europe. *Brain* **121**, 2335–2339.

Walker, R. H., Danisi, F. O., Swope, D. M., Goodman, R. R., Germano, I. M., and Brin, M. F. (2000). Intrathecal baclofen for dystonia:benefits and complications during six years experience. *Mov. Disord.* **15**, 1242–1247.

Walker, R. H., Brin, M. F., Sandu, D., Gujjari, P., Hof, P. R., Warren Olanow, C., and Shashidharan, P. (2001) Distribution and immunohistochemical characterization of torsinA immunoreactivity in rat brain. *Brain Res.* **900**, 348–354.

Walker, R. H., Brin, M. F., Sandu, D., Good, P. F., and Shashidharan, P. (2002). TorsinA immunoreactivity in brains of patients with DYT1 and non-DYT1 dystonia. *Neurology* **58**, 120–124.

Warner, T., Fletcher, N. A., Davis, M. B., Ahmad, F., Conway, D,, Feve. A., Rondot, P., Kwiatkowski, D. J., Marsden, C. D., and Harding, A. E. (1993). Linkage analysis in British families with idiopathic torsion dystonia. *Brain* **116**, 739–744.

Whiteheart, S. W., Rossnagel, K., Buhrow, S. A., Brunner, M., Jaenicke, R., and Rothman, J.E. (1994). N-ethylmaleimids-sensitive fusion protein: A trimeric ATPase whose hydrolysis of ATP is required for membrane fusion. *J. Cell. Biol.* **125**, 945–954.

Zeman, W., and Dyken, P. (1967). Dystonia musculorum deformans; clinical, genetic and pathoanatomical studies. *Psychiatr. Neurol. Neurochir.* **70**, 77–121.

Zeman, W. (1970). Pathology of the torsion dystonias (dystonia musculorum deformans). *Neurology* **20**(No. 11 Part 2), 79–88.

Ziefer, P., Leung, J., Razzano, T., Shalish, C., LeDoux, M., Lorden, J., Ozelius, L. J., Breakefield, X. O., Standaert, D., and Augood, S. (2002). Molecular cloning and expression of rat torsinA in the normal and genetically dystonic (dt/dt) rat. *Mol. Brain Res.* **101**, 132–135.

Zilber, N., Korczyn, A. D., Kahana, E., Fried, K., and Alter, M. (1984). Inheritance of idiopathic torsion dystonia among Jews. *J. Med. Genet.* **21**, 13–20.

Zudas, A., and Cianchetti, C. (1996). Efficacy of risperidone in idiopathic segmental dystonia. *Lancet* **347**, 127–128.

CHAPTER 37

# Dopa-Responsive Dystonia

**HIROSHI ICHINOSE**
Institute for Comprehensive Medical Science
Fujita Health University
Toyoake, Japan

**TOSHIHARU NAGATSU**
Institute for Comprehensive Medical Science
Fujita Health University
Toyoake, Japan

**CHIHO SUMI-ICHINOSE**
Department of Pharmacology
School of Medicine
Fujita Health University
Toyoake, Japan

**TAKAHIDE NOMURA**
Department of Pharmacology
School of Medicine
Fujita Health University
Toyoake, Japan

I. Introduction
II. Phenotype and Treatment
III. Causative Gene
IV. Diagnosis
   A. Neuropathology
V. Genotype/Phenotype Correlation
VI. Mechanism of Dominant Inheritance
VII. The Mechanism of Neuronal Selectivity—A Study with an Animal Model of Biopterin Deficiency
   Acknowledgment
   References

The causative gene for dopa-responsive dystonia (DRD) was discovered in 1994 to be that for guanosine triphosphate cyclohydrolase I (GCH), an enzyme involved in tetrahydrobiopterin biosynthesis. DRD patients were heterozygous in terms of the mutations; i.e., patients possessed both the wild-type and mutated GCH genes. More than 70 mutations have been found in this gene in DRD patients. A defective GCH gene results in a decreased biopterin content and thus in a decreased dopamine production in the brain. Analysis of the molecular etiology of DRD should help us to understand the pathophysiology of basal ganglia disorders, including Parkinson's disease.

We suggest that the nigro-striatal dopaminergic neurons are highly susceptible to a deficiency of tetrahydrobiopterin and the resulting defect in dopamine production.

## I. INTRODUCTION

There are several types of dopa-responsive basal ganglia diseases. Parkinson's disease is the most well-known disease in that category. Sporadic Parkinson's disease develops its symptoms in senescence, and the symptoms, so called parkinsonism, are caused by selective degeneration of nigro-striatal dopaminergic neurons. DRD is a disorder characterized by childhood- or adolescent-onset of dystonia and by a dramatic response to low-dose L-dopa. DRD constitutes approximately 5–10% of primary dystonia of childhood and adolescence. DRD is also caused by dysfunction of nigro-striatal dopaminergic neurons, although its main symptom is dystonia, not parkinsonism. It has been noted that blocking of dopamine receptors by neuroleptics produces a dystonic reaction in childhood, whereas in adults it results in parkinsonism (Swett, 1995). Parkinsonian symptoms sometimes appears later in DRD patients in adolescence.

Segawa *et al.* (1971, 1976) first coined the term, hereditary progressive dystonia with marked diurnal fluctuation

(HPD), for a group of patients exhibiting a dystonia with marked diurnal fluctuation of symptoms and a marked response to L-dopa without any side effects. The term dopa-responsive dystonia was proposed by Nygaard et al. (1988) to describe dystonia responding to L-dopa. Because HPD and strictly defined DRD are thought to be identical from clinical and genetic points of view, we use the term DRD as the name for this disorder in the present review. DRD is a disorder inherited as an autosomal dominant trait with reduced penetrance.

## II. PHENOTYPE AND TREATMENT

Originally, clinical symptoms for DRD were described as follows (Segawa and Nomura, 1993; Nygaard, 1993):

1. Most cases have onset in childhood (in the first decade of life), whereas there are some cases with onset at adult ages.
2. Symptoms start with foot dystonia and gait disturbance, which usually progress to other body parts.
3. The symptoms are aggravated from afternoon toward the evening and alleviated in the morning after sleep; the diurnal fluctuation attenuates with age.
4. Patients show marked and sustained response to L-dopa without any unfavorable side effects.
5. Tremor and bradykinesia may appear in adult patients.

After the discovery of the causative gene, the clinical phenotypes of genetically confirmed DRD have been extended. They include focal dystonia, dystonia with relapsing and remitting course, dystonic symptoms with onset in the first year of life, adult-onset parkinsonian patients, and adult-onset oromandibular dystonia (Bandmann et al., 1998; Steinberger et al., 1998, 1999). Parkinsonism, such as postural instability, bradykinesia, and rigidity, are often present or develop early in the symptomatic course. A few cases have rest tremor, oculogyric crises (Nygaard, 1995), and spastic paraplegia (Tassin et al., 2000). Thus, the classical presentation of DRD is neither specific nor sufficient for a diagnosis of DRD. The hallmark feature of DRD is a marked and sustained response to small doses of L-dopa.

Recently, mutations in the *parkin* gene on chromosome 6 were found in families with autosomal recessive juvenile parkinsonism (ARJP); (Kitada et al., 1998). This is the most frequent form of monogenic parkinsonism so far identified. The clinical phenotype associated with parkin mutations is characterized by early-onset parkinsonism, good response to L-dopa, and slow course. Motor fluctuations and L-dopa induced dyskinesia are frequent in the patients with parkin mutation.

The phenotypical spectrum overlaps with DRD for early-onset cases, and most of parkin mutation patients had onset with dystonia (Tassin et al., 2000). Therefore, we can distinguish ARJP from DRD by the appearance of L-dopa-induced dyskinesias. This difference has been explained by the absence of dopaminergic cell loss in the substantia nigra in DRD. Because L-dopa uptake, decarboxylation, and storage mechanisms are thought to be intact in DRD patients, they would not develop L-dopa-induced dyskinesia. On the other hand, marked dopaminergic cell loss in the substantia nigra in a patient with 6q-linked ARJP was recently demonstrated (Mori et al., 1998). However, the presence of L-dopa-induced dyskinesias was reported in some DRD patients who carried a truncating mutation in the GCH gene (Tassin et al., 2000). In these cases, it is difficult to distinguish DRD from ARJP by clinical phenotypes.

The asymptomatic carrier status occurs more frequently in men than in women (Furukawa et al., 1998a). The penetrance in genetically proven patients with DRD was reported to be 90–100% in females and 40–55% in males, if minor symptoms and signs were considered (Steinberger et al., 1998; Furukawa et al., 1998a). It would be the reason of the female predominance in DRD, i.e., the female to male ratio in DRD patients has been reported to be approximately 4:1 (Segawa et al., 1986). Shimoji et al. (1999) reported that female mice expressed lower levels of GCH mRNA in monoaminergic neurons in the brain when compared to levels seen in male mice. Sex hormones and sexual differentiation of mesencephalic dopaminergic neurons might be involved in gender-related vulnerability to congenital BH4 deficiency.

Analysis of large DRD pedigrees revealed a highly variable expressivity in clinical symptoms in spite of the same mutation (Hahn et al., 2001). Whereas some patients are afflicted with generalized dystonia, others in the same pedigree have subtle signs such as abnormal writing test, even though they carry the same mutation (Steinberger et al., 1998). Epigenetic factors as well as genetic ones may affect the expressivity of GCH mutations.

Most of all symptoms of DRD respond to L-dopa with relatively low doses of 20 mg/kg per day of plain levodopa (with 4–5 mg/kg per day for levodopa/carbidopa); (Segawa, 2000). The response of motor syndromes sustain without any side effects. Administration of tetrahydrobiopterin (BH4) had been tried without a dramatic effect (LeWitt et al., 1986; Furukawa et al., 1996). The ineffectiveness of BH4 to DRD patients could be explained by low penetrance of BH4 into the brain and the decreased tyrosine hydroxylase (TH) protein in the striatum of the patients, as discussed later.

## III. CAUSATIVE GENE

In 1993, Nygaard et al. succeeded in mapping the causative gene for DRD to chromosome 14q by linkage analysis. Segawa, in collaboration with Nygaard and Tsuji, showed that the locus for the HPD was identical with that of DRD (Tanaka et al., 1995). In the same time period, we characterized the human and mouse GCH genes (Ichinose et al., 1995a), and mapped the human gene to 14q22.1–q22.2 within the DRD locus (Ichinose et al., 1994). GCH is the key enzyme for the biosynthesis of BH4, which is an essential cofactor for TH, the dopamine-synthesizing enzyme (Fig. 37.1).

We proved that the GCH gene is a causative gene for HPD, based both on the identification of a mutation in one allele of the gene in the patients, who also had the wild-type allele, and on a significant but partial decrease in the GCH activity expressed in their mononuclear blood cells after stimulation with phytohemagglutinin (PHA; Ichinose et al., 1994). First we found 6 independent mutations in 10 DRD families; 4 missense mutations due to single-base changes (L79P, R88W, D134V, G201E), and 2 frameshift mutations due to a 2-base insertion and a 13-base deletion (Ichinose et al., 1994, 1995b). All DRD patients were heterozygous in terms of these mutations, bearing both a mutated gene and a normal gene. None of these mutations were present on 108 chromosomes from 54 unrelated Japanese individuals. We overexpressed mutated GCH proteins in Escherichia coli, and found that R88W and G201E mutants had no substantial enzyme activity (Ichinose et al., 1994; Suzuki et al., 1999). In the remaining 4 of the 10 DRD families, we could not find any mutations in any of the exons or in the splicing junctions of the GCH gene.

The shortage of dopamine in the nigrostriatal dopaminergic neurons in the brain in DRD patients is thought to be caused by the partial decrease in BH4. Indeed, DRD patients show lower than normal levels of both biopterin and its metabolic by-product, neopterin, in their cerebrospinal fluid (CSF; Williams et al., 1979; LeWitt et al., 1986; Fink et al., 1988; Furukawa et al., 1993). In accordance with this concept, mutations in the GCH gene and a decrease in the dopamine and BH4 levels in the striatum to less than 20% of the normal values were confirmed in one autopsy case of DRD (Furukawa et al., 1999). Thus, dopamine deficiency in DRD patients was shown to be caused by mutated GCH, reduced GCH activity, a low BH4 content, and a low TH activity (Ichinose et al., 1994; Ozelius and Breakfield, 1994).

So far, more than 70 mutations had been reported to occur in the GCH gene from DRD patients (Ichinose et al., 1999; Blau et al., 2001a). Up-to-date information on mutations found in BH4-synthesizing enzymes is tabulated in a BIOMDB database (Blau et al., 2001b).

The location of mutations was widely distributed in all 6 exons of the GCH gene including splicing junctions (Fig. 37.2). We have not found any apparent difference in the occurrence of mutations in each exon. All kinds of mutations have been identified as a cause of DRD, including single nucleotide changes causing missense and nonsense mutations, nucleotide insertion and deletion leading to frameshift, and intron-exon boundary mutations causing exon skipping. There was no founder effect in DRD, because most of the mutations were different among families, whereas given family members, either affected or asymptomatic carriers, showed the identical mutation. Furukawa et al. (1998a) reported a high occurrence of sporadic (*de novo*) mutations in the GCH gene. These

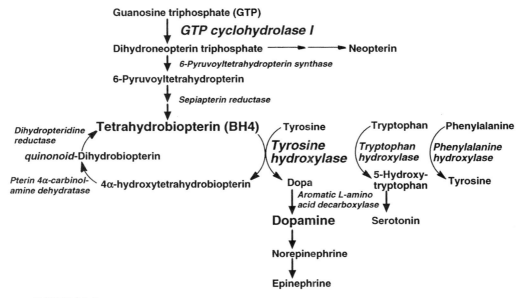

FIGURE 37.1 Biosynthesis of tetrahydrobiopterin (BH4) and its relationship with monoamine synthesis.

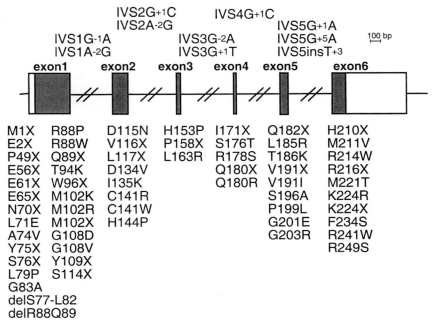

**FIGURE 37.2** Mutations in the GTP cyclohydrolase I gene found in DRD patients (Modified from Blau, N. et al. (2001a). *Mol. Genet. Metab.* **74**, 172–185; Blau, N. (2001b). BIOMDB: Database of mutations causing tetrahydrobiopterin deficiency. http://www.bh4.org/biomdb1.html.). A shaded box represents a protein-coding region, and an open box represents 5'- or 3'-untranslated region.

observations suggest that the frequency of mutation in the GCH gene may be higher than that in other genes.

Approximately 40% of DRD patients do not have an identified mutation in the coding region or splicing junctions of GCH (Ichinose *et al.*, 1995b; Bandmann *et al.*, 1996, 1998). In some of these patients, positive linkage to the GCH locus or dysfunction of the biopterin metabolism has been indicated (Bandmann *et al.*, 1996; Furukawa et al., 1999). There may be a mutation in noncoding regions of the GCH gene, such as 5'- or 3'-untranslated region, and intronic regulatory regions, in the patients. We reported that two alleles of the GCH gene were not equally expressed in a DRD patient (Inagaki *et al.*, 1999). The differential expression of each allele may suggest the presence of some novel mutations in an allele of the GCH gene.

In 1995, a mutation in the TH gene (Q381K) was reported in a patient with recessively inherited DRD (Lüdecke *et al.*, 1995). More than eight patients have been reported who had homozygous mutation in the TH gene. In other families with "TH deficiency," patients developed parkinsonism in infancy, severe motor retardation, spastic paraplegia, truncal hypotonia, and hypokinesia rather than dystonia (Lüdecke *et al.*, 1996; van den Heuvel *et al.*, 1998; Wevers *et al.*, 1999; Bräutigam *et al.*, 1998, 1999; Furukawa *et al.*, 2001). Thus, most of patients with TH deficiency do not show phenotypes similar to those seen in typical DRD. Patients with TH deficiency were treated with L-dopa, while one unresponsive case was also reported (DeLonlay *et al.*, 2000).

The recombinant TH Q381K mutant represented a kinetic variant form with a residual activity of about 15% of the wild-type TH (Knappskog *et al.*, 1995). Because the Q381K mutant possessed the residual activity, the patient had low but substantial ability to synthesize dopamine through dopa. Due to the partial and selective deficiency in dopamine, the clinical phenotypes of the Q381K homozygous patients were thought to be similar to typical phenotypes of DRD.

## IV. DIAGNOSIS

A therapeutic trial with L-dopa is a practical approach to the diagnosis of DRD. However, it is often difficult to diagnose patients only by clinical phenotypes including responsiveness to L-dopa, because phenotypes of DRD are highly variable. Tassin *et al.* (2000) reported the screening of 22 families with DRD for mutations. They found heterozygous mutations in the GCH gene in 12 families. Three out of the remaining 10 families had deletions in the parkin gene on chromosome 6, indicating how difficult it is to distinguish between DRD and parkin mutations in some cases (Tassin *et al.*, 2000). In these cases, molecular analysis would be of great help for giving a differential diagnosis, although approximately 40% of DRD patients have no mutations in the coding region (including the splicing junction) of the GCH gene.

The most informative biochemical examinations in patients with DRD are the measurement of pterins (neopterin and biopterin) and neurotransmitter metabolites (homovanillic acid, HVA—a major metabolite of dopamine, and 5-hydroxyindole acetic acid, 5HIAA—a major metabolite of serotonin) in cerebrospinal fluids (CSF); (Blau et al., 2001a). The levels of neopterin in the CSF reduced less than 20% of normal values in DRD patients (LeWitt et al., 1986; Furukawa et al., 1993). The levels of both HVA and 5HIAA in the CSF showed a severe to mild decrease in DRD patients. On the other hand, neopterin levels in the CSF of patients with juvenile parkinsonism were not severely decreased as those of DRD patients, while biopterin concentrations were decreased in both patients (Furukawa et al., 1993). Decreased CSF levels of HVA and 3-methoxy-4-hydroxyphenylethyleneglycol (MHPG), a major metabolite of norepinephrine, together with normal pterin and 5HIAA concentrations are the diagnostic hall marks of TH deficiency.

We reported reduced GCH activity in PHA-stimulated mononuclear blood cells in DRD patients (Ichinose et al., 1994, 1995c). Cultured skin fibroblasts are also a useful tool in the diagnosis of DRD (Bonafe et al., 2001). The phenylalanine loading test appears to be useful for the diagnosis of DRD, since the metabolism of phenylalanine in the liver is partly diminished in DRD patients because of the limited supply of BH4 (Hyland et al., 1997, 1999).

Neuroimaging methods such as CT scan and magnetic resonance imaging (MRI) are of limited help for the diagnosis of DRD. On the other hand, physiological imaging techniques such as positron emission tomography (PET) or single photon emission computed tomography (SPECT) provide the means to distinguish DRD from juvenile parkinsonism based on the nigral cell loss. PET scanning using the [$^{18}$F]-6-fluoro-dopa ([$^{18}$F]F-dopa) is a direct method of studying the nigrostriatal dopaminergic system in living subjects. $1R$-$2\beta$ Carbomethoxy-$3\beta$-(4-[$^{123}$I]iodophenyl) tropane ([$^{123}$I]$\beta$-CIT) is a ligand for the dopamine transporter using SPECT, and was found to be a useful marker for dopamine neurons. A pathological study suggests that there is no degeneration of the nigrostriatal dopaminergic system in DRD (Rajput et al., 1994). In good accordance with the study, DRD patients have a normal or very mild decrease of PET [$^{18}$F]F-dopa uptake, and a normal SPECT [$^{123}$I]-$\beta$-CIT binding (Snow et al., 1993; Jeon et al., 1998). In contrast, patients with juvenile parkinsonism had reduced striatal [$^{18}$F]F-dopa uptake (Snow et al., 1993). These results suggest that dopa uptake, decarboxylation, and storage systems are intact in DRD. This would explain the sustained response to low doses of L-dopa in DRD patients.

There was a hypothesis that DRD asymptomatic carriers have increased dopamine D2 receptors in the striatum that protect them from the clinical manifestations of dopaminergic deficiency. However, Kishore et al. (1998) reported that there was no significant difference in D2 binding between the symptomatic and asymptomatic DRD groups, while striatal D2-receptor binding was elevated by approximately 30% in both groups using [$^{11}$C]-raclopride PET.

### A. Neuropathology

There are two autopsy cases for DRD patients reported so far (Rajput et al., 1994; Furukawa et al., 1999). One case had a defective GCH mutation, but the other case had no mutation in the coding region (Furukawa et al., 1999). The neuropathological investigation demonstrated a normal population of cells with reduced melanin and no evidence of Lewy body formation in the substantia nigra. Biochemical studies revealed that striatal neopterin and biopterin levels were markedly reduced in both patients (less than 20% of control subjects in the biopterin content, and less than 45% in the neopterin content). In the putamen, both DRD patients had severely reduced TH protein levels but had normal concentrations of AADC protein, and dopamine and vesicular monoamine transporters (Rajput et al., 1994; Furukawa et al., 1999).

## V. GENOTYPE/PHENOTYPE CORRELATION

In DRD patients, only one allele of the GCH gene was affected. If both alleles of the GCH gene were defective, the patient develops hyperphenylalaninemia due to a defect in hydroxylation of phenylalanine in the liver, and the disorder was called "GCH deficiency" (Niederwieser et al., 1984). This disorder is inherited as an autosomal recessive trait. The phenotypes of GCH deficiency were hyperphenylalaninemia, severe retardation in development, severe muscular hypotonia of the trunk and hypertonia of the extremities, convulsions, and frequent episodes of hyperthermia without infections (Niederwieser et al., 1984).

In 1998, two patients were reported who were compound heterozygotes for GCH gene mutations (Furukawa et al., 1998b). The patients had maternally and paternally transmitted mutations in the GCH gene, and the mutations did not abolish the GCH activity completely as in the case of GCH deficiency. The patients showed decreased biopterin levels in plasma as well as in CSF, indicating that the residual GCH activities were lower than those in DRD. They presented with neurological features that were more severe than what is seen in DRD, such as developmental motor delay, truncal hypotonia, the complete lack of speech development, and oculogyric crises (Furukawa et al., 1998b). For the treatment, oral BH4 administration in addition to the levodopa therapy was required.

## VI. MECHANISM OF DOMINANT INHERITANCE

Although the molecular mechanism of DRD now seems to have become clear, there still remain several questions about this disorder. The first concern relates to the mechanism by which the symptoms appear in individuals in the heterozygous state, for DRD is a rare example of how partial reduction in a metabolic enzyme caused by disruption of one autosomal allele can lead to a disease phenotype (Ozelius and Breakefield, 1994). The second concern is the mechanism whereby only nigrostriatal dopaminergic neurons are affected in DRD.

In general, disorders caused by an inborn error of metabolism show recessive inheritance, because half of an enzyme activity is usually sufficient to maintain homeostasis *in vivo*. To the contrary, DRD is a dominant disorder with low penetrance, even though the causative gene for DRD is that for an enzyme, GCH. Because neopterin is a metabolite of dihydroneopterin triphosphate, the product of GCH, the neopterin content in the CSF is thought to reflect the GCH activity in the brain. The marked decrease (to approximately 20% of controls) in the neopterin content in the CSF from DRD patients suggests that GCH activity in the brain of DRD patients is also about 20% of that for normal individuals.

Genetic analysis of DRD families has proved the presence of asymptomatic carriers, who have the same mutations as the patients. Therefore, there must be some unknown biochemical difference between patients and asymptomatic carriers. Takahashi *et al.* (1994) reported that the CSF of an asymptomatic carrier showed a higher neopterin level than that of patients, although this was the only case examined.

The above observations support the idea that GCH activity in the brain is an important factor in the development of DRD symptoms. It seems to be essential that GCH activity in the brain must decrease to less than 20% of the normal level to cause the symptoms shown in DRD. Interestingly, it is known that there must be an 80% loss of striatal dopamine content before the appearance of symptoms in Parkinson's disease (Bernheimer *et al.*, 1973).

How can GCH activity in the patients be reduced to 20% of the normal value in spite of the presence of a normal allele? The answer usually given is the "dominant-negative effect," meaning that a mutant subunit interacts with a normal one in such a way as to inhibit (or interfere with) the activity of the normal one, since GCH forms a homodecameric structure (Yoneyama and Hatakeyama, 1998). Hirano *et al.* (1998) reported that co-expression of a truncated form of GCH with wild-type GCH in COS-7 cells reduced the enzyme activity. Hwu *et al.* (2000) also reported that a mutant, G201E, dramatically decreased the level of wild-type protein and GCH activity in co-transfection studies, suggesting that G201E exerts a dominant-negative effect on the wild-type protein, probably via an interaction between them. It might be possible that, through the dominant-negative mechanism, a single mutation could decrease GCH activity to less than 50% of normal.

However, we found that the amount of GCH protein in PHA-stimulated mononuclear blood cells was reduced not only in an DRD patient who had a missense mutation but also in one who had a frameshift mutation in the translational starting methionine (M1X; Suzuki *et al.*, 1999). Because the mutant subunit would not be able to interact with the wild-type subunit in the case of the frameshift mutation, a dominant-negative effect through formation of a chimeric protein does not seem to be the reason for the reduced enzyme activity in all cases of DRD patients. Furthermore, we cannot explain the presence of asymptomatic carriers by the dominant-negative mechanism, because their genotypes are the same. Asymptomatic carriers are thought to have higher enzyme activities in their brain than affected individuals.

We speculate that transcriptional regulation of the GCH gene is also an important factor in the development of the symptoms of DRD, although the defect in the GCH gene is essential. Lowered expression of the GCH gene would lead to further decrease in the enzyme activity. This suggestion would be reasonable because numerous different mutations, including nonsense mutations and frameshift mutations, have been identified in DRD patients with a similar clinical phenotype. We observed a decreased mRNA level of GCH in a DRD family (Inagaki *et al.*, 1999). We are now investigating the regulatory mechanisms of transcription or translation of GCH to clarify the molecular mechanisms in DRD.

## VII. THE MECHANISM OF NEURONAL SELECTIVITY—A STUDY WITH AN ANIMAL MODEL OF BIOPTERIN DEFICIENCY

In DRD patients, there are no symptoms other than dystonia/parkinsonism; and dysfunction of serotonergic neurons is not obvious. Biosynthesis of catecholamines and serotonin is mainly regulated by the activity of the hydroxylation reaction catalyzed by TH for catecholamine-synthesis and by tryptophan hydroxylase (TPH) for serotonin synthesis. BH4 is an essential cofactor for both TH and TPH. BH4 is synthesized from guanosine triphosphate through three enzymatic reactions catalyzed by GCH, pyruvoyltetrahydropterin synthase (PTS), and sepiapterin reductase (Fig. 37.1).

We established BH4-deficient mice by disrupting the *Pts* gene to investigate the effects of BH4 depletion on the animals and the involvement of BH4 in regulating neural systems (Sumi-Ichinose *et al.*, 2001). Homozygous $Pts^{-/-}$

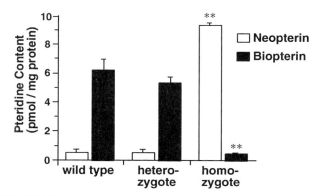

FIGURE 37.3  Pteridine contents in the brain of newborn mice with a disrupted *Pts* gene. Values are the means ± S.E., n = 6. Significant difference from the corresponding value for the wild-type is indicated by asterisks: **p < 0.01 (Student's *t* test).

mice were born at the almost expected Mendelian ratio without gross anomaly in major organs (wild-type; 27%, heterozygote 51%, homozygote 22%, n = 424); however, all homozygotes died within 48 hours after birth. Northern hybridization using total RNA extracted from the brain and the liver of newborn mice showed that homozygotes were null mutants, for *Pts* mRNA was not detected. As shown in Fig. 37.3, biochemical analysis revealed a dramatic decrease in the level of biopterin, and the great accumulation of neopterin. We assayed the levels of dopamine, norepinephrine, and serotonin. The monoamines were all decreased in homozygotes (15, 6, and 8% of wild-type, respectively), and there was no significant difference between heterozygotes and wild-type mice in the monoamine contents (Fig. 37.4). These results show that *in vivo* synthesis of catecholamines and serotonin was reduced by the lack of the hydroxylation cofactor, BH4, in homozygous mutant mice.

We assayed the *in vitro* activity of TH, TPH, NO synthase (NOS), aromatic L-amino acid decarboxylase (AADC), and GCH in the brain of newborn mice. The *in vitro* activities were assayed under the presence of sufficient amount of their cofactors. AADC decarboxylates L-dopa and 5-HTP to form dopamine and serotonin, respectively, and co-exists with TH and TPH in catecholaminergic and serotonergic neurons. Although both TH and TPH are BH4-dependent enzymes, only *in vitro* TH activity was severely impaired in the homozygotic mutant mice (7% of wild-type). TPH, AADC, and GCH activities were not altered significantly. The activity of NOS was decreased to 72% of the wild-type, but far less than that of TH. A reduction in TH protein in the brain of homozygous mutant newborn mice was shown by Western blotting, though we could not see a big difference in the mRNA level of the *Th* gene by Northern blot analysis.

We explored the localization of TH protein in the brain by using an immunohistochemical technique. Dopaminergic neural cell bodies located in the substantia nigra and in the ventral tegmental area in homozygotic mutant mice as well as in wild-type ones were stained with anti-TH antibody. To the contrary, in the striatum of wild-type mice, nerve fibers were strongly stained with anti-TH antibody, and many positive varicosities were easily seen at higher magnification. In the same region of the homozygotic mutant mice, nerve fibers were not stained; and instead, a few unusual TH-immunopositive cells were found in the striatum. In order to examine the presence of dopaminergic fibers, we stained the same sections with anti-AADC antibody. The anti-AADC antibody stained nerve fibers in the striatum in homozygotes similarly as in the wild type. These results indicate that dopaminergic neurons project to the striatum, and that the amount of TH protein was markedly decreased mainly in nerve terminals, but not in cell bodies, in homozygous *Pts*−/− mice.

Next, we examined the acute effect of administration of BH4 on dopamine and serotonin levels in the brain under the BH4-deficient state. We injected BH4 into newborn mice at a dose of 50 mg/kg body weight by the intraperitoneal route on postnatal day 0, and sacrificed them 60 min after the injection. Biopterin and monoamine contents in their brain were analyzed and compared with those of vehicle-injected mice (0.25% ascorbic acid used to protect BH4 from oxidization). The biopterin content in the brains of both BH4-injected wild-type and homozygous mutant mice was 6 to 7 times higher than that of the ascorbic-acid-injected wild-type mice. In spite of the great elevation in the amount of BH4, the dopamine content in the brain was not altered significantly in the wild-type mice (Fig. 37.5A), whereas the serotonin content was slightly increased (1.4 times; Fig. 37.5B). In the *Pts* knock-out mice, the serotonin level in the brain of BH4-injected animals was markedly elevated to more than 10 times the level for the ascorbic-acid-injected ones, reflecting a recovery up to 73% of the level for the wild-type injected with ascorbic acid (Fig. 37.5B). To the contrary, the

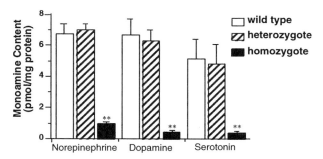

FIGURE 37.4  Monoamine contents in the brain of newborn mice with a disrupted *Pts* gene. Values are the means ± S.E., n = 6. Significant difference from the corresponding value for the wild-type is indicated by asterisks; **p < 0.01 (Student's *t* test).

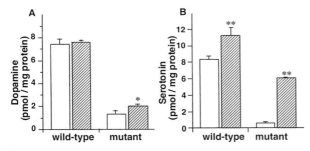

**FIGURE 37.5** Acute effect of intraperitoneal administration of BH4 on the brain of newborn mice with a disrupted *Pts* gene. Dopamine (A) and serotonin (B) contents in the brain of newborn mice at 60 min after an intraperitoneal injection of vehicle (ascorbic acid, open bar) or 50 mg/kg BH4 (hatched bar). Values are the means ± S.E., $n = 5$. Significant difference from vehicle-treated mice is indicated by asterisks, $*p < 0.05$, $**p < 0.01$ (Student's $t$ test).

increase in dopamine levels in the *Pts*-null mice was only 1.5-fold the original level (Fig. 37. 5A).

It has been reported that a reduction in the level of TH protein occurs mainly in the striatum in patients with DRD (Rajput *et al.*, 1994; Furukawa *et al.*, 1999). Autopsied brains of the DRD patients showed a marked decrease in the content of TH protein in the striatum, as seen in the $Pts^{-/-}$ mice. This suggests that the reduction in TH protein in the striatum of DRD patients resulted from partial deficiency in BH4 in the brain.

Our present results would explain the reason why only dysfunction of catecholaminergic neurons is apparent, and why serotonergic systems can function properly in DRD patients. At the nerve terminals, although the rate of *de novo* synthesis of BH4 is partly impaired in DRD patients, a small amount of BH4 is present there. The BH4 can be recycled after the hydroxylation reaction with TH and TPH through the action of dihydropteridine reductase (Fig. 37.1). In the case of serotonergic neurons, the number of the TPH molecules is not affected by depletion of BH4, and a sufficient amount of serotonin can be synthesized even under the decreased concentration of BH4. In catecholaminergic neurons, the ability to synthesize dopamine is not sufficient due to the decreased number of the TH molecules.

We think that diurnal fluctuation of symptoms in DRD may be derived from the decreased rate in the dopamine synthesis. The rate would not be high enough to supplement the consumption of dopamine during the day, thus aggravating symptoms toward the evening. BH4 and dopamine in the nigrostriatal dopaminergic neurons may be supplemented during sleep.

In spite of much research on DRD, we still cannot explain the difference among catecholaminergic neurons. We suggest high susceptibility of the nigrostriatal dopaminergic neurons to a deficiency of BH4 and dopamine. In other words, dystonia-parkinsonism would be the first symptom when BH4 is gradually depleted in the brain. Further research on the molecular mechanism of DRD should lead us to a better understanding of the pathophysiology of the basal ganglia.

## Acknowledgment

This work was supported by a Grant-in-Aid for Scientific Research on Priority Areas(C)—Advanced Brain Science Project—from the Ministry of Education, Culture, Sports, and Science and Technology of Japan.

## References

Bandmann, O., Nygaard, T. G., Surtees, R., Marsden, C. D., Wood, N. W., and Harding, A. E. (1996). Dopa-responsive dystonia in British patients: new mutations of the GTP-cyclohydrolase I gene and evidence for genetic heterogeneity. *Hum. Mol. Genet.* **5**, 403–406.

Bandmann, O., Valente, E. M., Holmans, P., Surtees, R. A., Walters, J. H., Wevers, R. A., Marsden, C. D., and Wood, N. W. (1998). Dopa-responsive dystonia: a clinical and molecular genetic study. *Ann. Neurol.* **44**, 649–656.

Bernheimer, H., Birkmayer, W., Hornykiewicz, O., Jellinger, K., and Seitelberger, F. (1973). Brain dopamine and the syndromes of Parkinson and Huntington. Clinical, morphological and neurochemical correlations. *J. Neurol. Sci.* **20**, 415–455.

Blau, N., Bonafé, L., and Thöny, B. (2001a). Tetrahydrobiopterin deficiencies without hyperphenylalaninemia: diagnosis and genetics of dopa-responsive dystonia and sepiapterin reductase deficiency. *Mol. Genet. Metab.* **74**, 172–185.

Blau, N., Thöny, B., and Dianzani, I. (2001b). BIOMDB: Database of mutations causing tetrahydrobiopterin deficiency. http://www.bh4.org/biomdb1.html

Bonafe, L., Thöny, B., Leimbacher, W., Kierat, L., and Blau, N. (2001). Diagnosis of dopa-responsive dystonia and other tetrahydrobiopterin disorders by the study of biopterin metabolism in fibroblasts. *Clin. Chem.* **47**, 477–485.

Brautigam, C., Wevers, R. A., Jansen, R. J., Smeitink, J. A., de Rijk-van Andel, J. F., Gabreels, F. J., and Hoffmann, G. F. (1998). Biochemical hallmarks of tyrosine hydroxylase deficiency. *Clin. Chem.* **44**, 1897–1904.

Brautigam, C., Steenbergen-Spanjers, G. C., Hoffmann, G. F., Dionisi-Vici, C., van den Heuvel, L. P., Smeitink, J. A., and Wevers, R. A. (1999). Biochemical and molecular genetic characteristics of the severe form of tyrosine hydroxylase deficiency. *Clin. Chem.* **45**, 2073–2078.

DeLonlay, P., Nassogne, M. C., van Gennip, A. H., van Cruchten, A. C., Billatte de Villemeur, T., Cretz, M., Stoll, C., Launay, J. M., Steenberger-Spante, G. C., van den Heuvel, L. P., Wevers, R. A., Saudubray, J. M., and Abeling, N. G. (2000). Tyrosine hydroxylase deficiency unresponsive to L-dopa treatment with unusual clinical and biochemical presentation. *J. Inherit. Metab. Dis.* **23**, 819–825.

Fink, J. K., Barton, N., Cohen, W., Lovenberg, W., Burns, R. S., and Hallett, M. (1988). Dystonia with marked diurnal variation associated with biopterin deficiency. *Neurology* **38**, 707–711.

Furukawa, Y., Nishi, K., Kondo, T., Mizuno, Y., and Narabayashi, H. (1993). CSF biopterin levels and clinical features of patients with juvenile parkinsonism. *Adv. Neurol.* **60**, 562–567.

Furukawa, Y., Shimadzu, M., Rajput, A. H., Shimizu, Y., Tagawa, T., Mori, H., Yokochi, M., Narabayashi, H., Hornykiewicz, O., Mizuno, Y., and Kish, S. J. (1996). GTP-cyclohydrolase I gene mutations in hereditary progressive and dopa-responsive dystonia. *Ann. Neurol.* **39**, 609–617.

Furukawa, Y., Lang, A. E., Trugman, J. M., Bird, T. D., Hunter, A., Sadeh, M., Tagawa, T., St George-Hyslop, P. H., Guttman, M., Morris, L. W.,

Hornykiewicz, O., Shimadzu, M., and Kish, S. J. (1998a). Gender-related penetrance and de novo GTP-cyclohydrolase I gene mutations in dopa-responsive dystonia. *Neurology* **50**, 1015–1020.

Furukawa, Y., Kish, S. J., Bebin, E. M., Jacobson, R. D., Fryburg, J. S., Wilson, W. G., Shimadzu, M., Hyland, K., and Trugman, J. M. (1998b). Dystonia with motor delay in compound heterozygotes for GTP-cyclohydrolase I gene mutations. *Ann. Neurol.* **44**, 10–16.

Furukawa, Y., Nygaard, T. G., Gutlich, M., Rajput, A. H., Pifl, C., DiStefano, L., Chang, L. J., Price, K., Shimadzu, M., Hornykiewicz, O., Haycock, J. W., and Kish, S. J. (1999). Striatal biopterin and tyrosine hydroxylase protein reduction in dopa- responsive dystonia. *Neurology* **53**, 1032–1041.

Furukawa, Y., Graf, W. D., Wong, H., Shimadzu, M., and Kish, S. J. (2001). Dopa-responsive dystonia simulating spastic paraplegia due to tyrosine hydroxylase (TH) gene mutations. *Neurology* **56**, 260–263.

Hahn, H., Trant, M. R., Brownstein, M. J., Harper, R. A., Milstien, S., and Butler, I. J. (2001). Neurologic and psychiatric manifestations in a family with a mutation in exon 2 of the guanosine triphosphate-cyclohydrolase gene. *Arch. Neurol.* **58**, 749–755.

van den Heuvel, L. P., Luiten, B., Smeitink, J. A., de Rijk-van Andel, J. F., Hyland, K., Steenbergen-Spanjers, G. C., Janssen, R. J., and Wevers, R. A. (1998). A common point mutation in the tyrosine hydroxylase gene in autosomal recessive L-DOPA-responsive dystonia in the Dutch population. *Hum. Genet.* **102**, 644–646.

Hirano, M., Yanagihara, T., and Ueno, S. (1998). Dominant negative effect of GTP cyclohydrolase I mutations in dopa-responsive hereditary progressive dystonia. *Ann. Neurol.* **44**, 365–371.

Hwu, W. L., Chiou, Y. W., Lai, S. Y., and Lee, Y. M. (2000). Dopa-responsive dystonia is induced by a dominant-negative mechanism. *Ann. Neurol.* **48**, 609–613.

Hyland, K., Fryburg, J. S., Wilson, W. G., Bebin, E. M., Arnold, L. A., Gunasekera, R. S., Jacobson, R. D., Rost-Ruffner, E., and Trugman, J. M. (1997). Oral phenylalanine loading in dopa-responsive dystonia: a possible diagnostic test. *Neurology* **48**, 1290–1297.

Hyland, K., Nygaard, T. G., Trugman, J. M., Swoboda, K. J., Arnold, L. A., and Sparagana, S. P. (1999). Oral phenylalanine loading profiles in symptomatic and asymptomatic gene carriers with dopa-responsive dystonia due to dominantly inherited GTP cyclohydrolase deficiency. *J. Inherit. Metab. Dis.* **22**, 213–215.

Ichinose, H., Ohye, T., Takahashi, E., Seki, N., Hori, T., Segawa, M., Nomura, Y., Endo, K., Tanaka, H., Tsuji, S., Fujita, K., and Nagatsu, T. (1994). Hereditary progressive dystonia with marked diurnal fluctuation caused by mutations in the GTP cyclohydrolase I gene. *Nat. Genet.* **8**, 236–242.

Ichinose, H., Ohye, T., Matsuda, Y., Hori, T., Blau, N., Burlina, A., Rouse, B., Matalon, R., Fujita, K., and Nagatsu, T. (1995a). Characterization of mouse and human GTP cyclohydrolase I genes. Mutations in patients with GTP cyclohydrolase I deficiency. *J. Biol. Chem.* **270**, 10,062–10,071.

Ichinose, H., Ohye, T., Segawa, M., Nomura, Y., Endo, K., Tanaka, H., Tsuji, S., Fujita, K., and Nagatsu, T. (1995b). GTP cyclohydrolase I gene in hereditary progressive dystonia with marked diurnal fluctuation. *Neurosci. Lett.* **196**, 5–8.

Ichinose, H., Ohye, T., Yokochi, M., Fujita, K., and Nagatsu, T. (1995c). GTP cyclohydrolase I activity in mononuclear blood cells in juvenile parkinsonism. *Neurosci. Lett.* **190**, 140–142.

Ichinose, H., Suzuki, T., Inagaki, H., Ohye, T., and Nagatsu, T. (1999). Molecular genetics of dopa-responsive dystonia. *Biol. Chem.* **380**, 1355–1364.

Inagaki, H., Ohye, T., Suzuki, T., Segawa, M., Nomura, Y., Nagatsu, T., and Ichinose, H. (1999). Decrease in GTP cyclohydrolase I gene expression caused by inactivation of one allele in hereditary progressive dystonia with marked diurnal fluctuation. *Biochem. Biophys. Res. Commun.* **260**, 747–751.

Jeon, B. S., Jeong, J. M., Park, S. S., Kim, J. M., Chang, Y. S., Song, H. C., Kim, K. M., Yoon, K. Y., Lee, M. C., and Lee, S. B. (1998). Dopamine transporter density measured by $[^{123}I]\beta$-CIT single-photon emission computed tomography is normal in dopa-responsive dystonia. *Ann. Neurol.* **43**, 792–800.

Kishore, A., Nygaard, T. G., de la Fuente-Fernandez, R., Naini, A. B., Schulzer, M., Mak, E., Ruth, T. J., Calne, D. B., Snow, B. J., and Stoessl, A. J. (1998). Striatal D2 receptors in symptomatic and asymptomatic carriers of dopa-responsive dystonia measured with $[^{11}C]$-raclopride and positron-emission tomography. *Neurology* **50**, 1028–1032.

Kitada, T., Asakawa, S., Hattori, N., Matsumine, H., Yamamura, Y., Minoshima, S., Yokochi, M., Mizuno, Y., and Shimizu, N. (1998). Mutations in the parkin gene cause autosomal recessive juvenile parkinsonism. *Nature* **392**, 605–608.

Knappskog, P. M., Flatmark, T., Mallet, J., Lüdecke, B., and Bartholomé, K. (1995). Recessively inherited L-DOPA-responsive dystonia caused by a point mutation (Q381K) in the tyrosine hydroxylase gene. *Hum. Mol. Genet.* **4**, 1209–1212.

LeWitt, P. A., Miller, L. P., Levine, R. A., Lovenberg, W., Newman, R. P., Papavasiliou, A., Rayes, A., Eldridge, R., and Burns, R. S. (1986). Tetrahydrobiopterin in dystonia: identification of abnormal metabolism and therapeutic trials. *Neurology* **36**, 760–764.

Lüdecke, B., Dworniczak, B., and Bartholomé, K. (1995). A point mutation in the tyrosine hydroxylase gene associated with Segawa's syndrome. *Hum. Genet.* **93**, 123–125.

Lüdecke, B., Knappskog, P. M., Clayton, P. T., Surtees, R. A., Clelland, J. D., Heales, S. J., Brand, M. P., Bartholome, K., and Flatmark, T. (1996). Recessively inherited L-DOPA-responsive parkinsonism in infancy caused by a point mutation (L205P) in the tyrosine hydroxylase gene. *Hum. Mol. Genet.* **5**, 1023–1028.

Mori, H., Kondo, T., Yokochi, M., Matsumine, H., Nakagawa-Hattori, Y., Miyake, T., Suda, K., and Mizuno, Y. (1998). Pathologic and biochemical studies of juvenile parkinsonism linked to chromosome 6q. *Neurology* **51**, 890–892.

Niederwieser, A., Blau, N., Wang, M., Joller, P., Atares, M., and Cardesa-Garcia, J. (1984). GTP cyclohydrolase I deficiency, a new enzyme defect causing hyperphenylalaninemia with neopterin, biopterin, dopamine, and serotonin deficiencies and muscular hypotonia. *Eur. J. Pediatr.* **141**, 208–214.

Nygaard, T. G., Marsden, C. D., and Duvoisin, R. C. (1988). Dopa-responsive dystonia. *Adv. Neurol.* **50**, 377–384.

Nygaard, T. G., Wilhelmsen, K. C., Risch, N. J., Brown, D. L., Trugman, J. M., Gilliam, T. C., Fahn, S. and Weeks, D. E. (1993). Linkage mapping of dopa-responsive dystonia (DRD) to chromosome 14q. *Nat. Genet.* **5**, 386–391.

Nygaard, T. G. (1993). Dopa-responsive dystonia: delineation of the clinical syndrome and clues to pathogenesis. *Adv. Neurol.* **60**, 577–585.

Nygaard, T. G. (1995). Dopa-responsive dystonia. *Curr. Opin. Neurol.* **8**, 310–313.

Ozelius, L. J., and Breakefield, X. O. (1994). Co-factor insufficiency in dystonia-parkinsonian syndrome. *Nat Genet.* **8**, 207–209.

Rajput, A. H., Gibb, W. R., Zhong, X. H., Shannak, K. S., Kish, S., Chang, L. G., and Hornykiewicz, O. (1994). Dopa-responsive dystonia: pathological and biochemical observations in a case. *Ann. Neurol.* **35**, 396–402.

Segawa, M., Ohmi, K., Itoh, S., Aoyama, M., and Hayakawa, H. (1971). Childhood basal ganglia disease with remarkable response to L-DOPA, hereditary basal ganglia disease with marked diurnal fluctuation (in Japanese). *Shinryo (Tokyo)* **24**, 667–672.

Segawa, M., Hosaka, A., Miyagawa, F., Nomura, Y., and Imai, H. (1976). In *Advances in Neurology*. (R. Eldrige and S. Fahn, eds.), Vol. 14, pp. 215–233. Raven Press, New York.

Segawa, M., Nomura, Y., and Kase, M. (1986). In *Handbook of Clinical Neurology*. (P. J. Vinken, G. W. Bruyn and H. L. Klawans, eds.). Elsevier Science Publishers, New York.

Segawa, M., and Nomura, Y. (1993). Hereditary progressive dystonia with marked diurnal fluctuation. In *Hereditary Progressive Dystonia With Marked Diurnal Fluctuation* (M. Segawa, ed.) pp. 3–19, Parthenon Publishing, New York.

Segawa, M. (2000) Hereditary progressive dystonia with marked diurnal fluctuation. *Brain Dev.,* **22** (Suppl. 1), S65–S80.

Shimoji, M., Hirayama, K., Hyland, K., and Kapatos, G. (1999). GTP cyclohydrolase I gene expression in the brains of male and female hph-1 mice. *J. Neurochem.* **72**, 757–764.

Snow, B. J., Nygaard, T. G., Takahashi, H., and Calne, D. B. (1993). Positron emission tomographic studies of dopa-responsive dystonia and early-onset idiopathic parkinsonism. *Ann. Neurol.* **34**, 733–738.

Steinberger, D., Weber, Y., Korinthenberg, R., Deuschl, G., Benecke, R., Martinius, J., and Muller, U. (1998). High penetrance and pronounced variation in expressivity of GCH1 mutations in five families with dopa-responsive dystonia. *Ann. Neurol.* **43**, 634–639.

Steinberger, D., Topka, H., Fischer, D., and Müller, U. (1999). GCH1 mutation in a patient with adult-onset oromandibular dystonia. *Neurology* **52**, 877–879.

Sumi-Ichinose, C., Urano, F., Kuroda, R., Ohye, T., Kojima, M., Tazawa, M., Shiraishi, H., Hagino, Y., Nagatsu, T., Nomura, T., and Ichinose, H. (2001). Catecholamines and serotonin are differently regulated by tetrahydrobiopterin: a study from 6-pyruvoyltetrahydropterin synthase knockout mice. *J. Biol. Chem.* **276**, 41,150–41,160.

Suzuki, T., Ohye, T., Inagaki, H., Nagatsu, T., and Ichinose, H. (1999). Characterization of wild-type and mutants of recombinant human GTP cyclohydrolase I: relationship to etiology of dopa-responsive dystonia. *J. Neurochem.* **73**, 2510–2516.

Swett, C., Jr. (1975). Drug-induced dystonia. *Am. J. Psychiat.* **132**, 532–534.

Takahashi, H., Levine, R. A., Galloway, M. P., Snow, B. J., Calne, D. B., and Nygaard, T. G. (1994). Biochemical and fluorodopa positron emission tomographic findings in an asymptomatic carrier of the gene for dopa-responsive dystonia. *Ann. Neurol.* **35**, 354–356.

Tanaka, H., Endo, K., Tsuji, S., Nygaard, T. G., Weeks, D. E., Nomura, Y., and Segawa, M. (1995). The gene for hereditary progressive dystonia with marked diurnal fluctuation maps to chromosome 14q. *Ann. Neurol.* **37**, 405–408.

Tassin, J., Durr, A., Bonnet, A. M., Gil, R., Vidailhet, M., Lucking, C. B., Goas, J. Y., Durif, F., Abada, M., Echenne, B., Motte, J., Lagueny, A., Lacomblez, L., Jedynak, P., Bartholome, B., Agid, Y., and Brice, A. (2000). Levodopa-responsive dystonia. GTP cyclohydrolase I or parkin mutations? *Brain* **123**, 1112–1121.

Wevers, R. A., de Rijk-van Andel, J. F., Brautigam, C., Geurtz, B., van den Heuvel, L. P., Steenbergen-Spanjers, G. C., Smeitink, J. A., Hoffmann, G. F., and Gabreels, F. J. (1999). A review of biochemical and molecular genetic aspects of tyrosine hydroxylase deficiency including a novel mutation (291delC). *J. Inherit. Metab. Dis.* **22**, 364–373.

Williams, A., Eldridge, R., Levine, R., Lobenberg, W., and Paulson, G. (1979). Low CSF hydroxylase cofactor (tetrahydrobiopterin) levels in inherited dystonia. *Lancet* **ii(8139)**, 410–411.

Yoneyama, T., and Hatakeyama, K. (1998). Decameric GTP cyclohydrolase I forms complexes with two pentameric GTP cyclohydrolase I feedback regulatory proteins in the presence of phenylalanine or of a combination of tetrahydrobiopterin and GTP. *J. Biol. Chem.* **273**, 20,102–20,108.

CHAPTER 38

# Hallervorden-Spatz Syndrome

SUSAN J. HAYFLICK

*Molecular and Medical Genetics, Pediatrics, and Neurology*
*Oregon Health & Science University*
*Portland, Oregon 97201*

I. Introduction
II. Phenotype
   A. Nosology
   B. Clinical Features
III. Gene
   A. Normal Gene Function
   B. Abnormal Gene Function
IV. Diagnostic and Ancillary Tests
   A. DNA
   B. Neuroimaging
V. Cellular and Animal Models of Disease
   A. *In Vitro*
   B. Invertebrates
   C. Mouse (KO, transgenic)
VI. Genotype/Phenotype Correlations/Modifying Alleles
VII. Treatment
   References

## I. INTRODUCTION

Hallervorden-Spatz syndrome (now called pantothenate kinase associated neurodegeneration) is an autosomal recessive neurodegenerative disorder affecting both children and adults. The phenotypic spectrum ranges from severe, early childhood dystonia with pigmentary retinopathy to adult-onset dysarthria, dystonia, and rigidity. Brain MRI changes are virtually pathognomonic, demonstrating bilateral areas of anteromedial hyperintensity surrounded by a region of hypointensity in the medial globus pallidus on T2-weighted images, the so-called "eye of the tiger" sign. These radiographic changes indicate deposition of iron in the basal ganglia, which is seen on pathologic study in association with axonal spheroids. The defective gene encodes pantothenate kinase, a key regulatory enzyme in the biosynthesis of coenzyme A. A *Drosophila* pantothenate kinase mutant shows neurologic incoordination and cell division cycle defects. A murine model, currently being developed, will facilitate the investigation of disease pathogenesis and new treatments.

## II. PHENOTYPE

### A. Nosology

As with many neurologic disorders that were delineated on clinical grounds and now have a defined genetic basis, diagnostic criteria must be revised to reflect this new knowledge. Hallervorden-Spatz syndrome (HSS) is a heterogeneous group of disorders that have historically been defined by clinical, radiographic, and pathologic data, with acknowledgement that this diagnosis is a "best fit" for many atypical patients. Interpretation of the medical literature on HSS is confounded by this lumping. Direct gene testing is now available, and only a subset of patients with HSS have mutations in the major gene, *PANK2*. In order to present clinically useful information, we have stratified patients into those with and those without mutations in *PANK2*. Nosologic recommendations reflecting this new knowledge follow.

The eponym for this disorder acknowledges its delineation by Julius Hallervorden and Hugo Spatz, German neuropathologists (Hallervorden and Spatz, 1922). Their shameless involvement in the active euthanasia of "mental defectives" has been brought to light, with appeals to rename the syndrome (Harper, 1996; Shevell, 1992, 1996; Shevell and Peiffer, 2001). Based upon these concerns and new information about the etiologies of several extrapyramidal disorders with high brain iron, we propose a new nosology and nomenclature for this group of disorders to aid diagnostic evaluation (Fig. 38.1).

Neurodegeneration with brain iron accumulation (NBIA) describes the group of progressive extrapyramidal disorders in which there is radiographic evidence of focal iron accumulation in brain, usually in the basal ganglia. NBIA is a clinical diagnosis, with patients stratified into two categories by age at onset (during or after the first decade) and rate of disease progression (loss of independent ambulation within or after 15 years of onset). The term, NBIA, already in use in the medical literature (Arawaka et al., 1998; Bruscoli et al., 1998; Galvin et al., 2000; Neumann et al., 2000; Swaiman, 2001; Zhou et al., 2001a), is sufficiently broad to include all disorders previously called HSS, along with other recently delineated disorders of high brain iron, including neuroferritinopathy caused by mutations in the ferritin light chain gene (Curtis et al., 2001) and aceruloplasminemia with mutations in the gene encoding ceruloplasmin (Gitlin, 1998). For patients with mutations in *PANK2*, we have proposed the designation pantothenate kinase associated neurodegeneration (PKAN); (Zhou et al., 2001b).

## B. Clinical Features[1]

PKAN is a progressive movement disorder. Features are mostly limited to the central nervous system; however, with the ability now to delineate a genetically homogeneous condition, it is clear that systemic manifestations occur but are uncommon. A diagnosis of PKAN is suspected on clinical grounds and strengthened by the finding of typical radiographic changes. Molecular testing can confirm the diagnosis.

Reliable prevalence figures for this rare disease are difficult to establish; however, a rate of 1-3 in 1,000,000 is proposed, based on observed cases in a population, assuming a small number of misdiagnoses and missed cases. This implies a general population carrier rate of approximately 1 in 275–500. The disorder has been documented in patients from all continents and most ethnic groups. There is no specific ethnic predilection; however, there may be a higher gene carrier frequency in Italy based on the observation of a relatively high number of cases. Alternatively, this may reflect better ascertainment and reporting. A founder effect has been observed to account for the most common mutation in the major gene. This mutation is associated with a conserved haplotype for markers near the gene, demonstrating linkage disequilibrium (Westaway et al., 2002). This haplotype primarily occurs in patients from Europe. Among patients in our Registry with *PANK2* mutations, one-third are the offspring of consanguineous unions.

Although the phenotype of patients with PKAN extends along a continuum, most patients clearly fall into one of two categories, distinguished by the age at onset of symptoms, rate of disease progression, occurrence of retinopathy, and predominant neurologic symptoms. Diagnostic criteria for HSS were established prior to the *PANK2* gene identification (Dooling et al., 1974, 1980; Swaiman, 1991, 2001). With modifications, they remain clinically useful. Refined diagnostic criteria are proposed after the two major phenotypes are discussed.

### 1. Early-Onset, Rapidly Progressive (Classical) PKAN

The clinical features of classical PKAN are remarkably homogeneous (Hayflick et al., 2002). Classical PKAN presents in early childhood, usually before age 6 years. Based on data in our registry, the mean age at onset in classical disease is 3.4 ± 3.0 years (range 0.5–12). The most common presenting symptom is impaired gait. Often the child is considered clumsy prior to the development of obvious problems. This symptom is due to a combination of lower extremity muscular rigidity, dystonia, and spasticity, as well as restricted visual fields in those children with retinopathy. Patients' limited ability to protect themselves during falls results in recurrent trauma, especially to the face and chin. Some children have developmental delay, which is primarily motor but occasionally global. Ocular symptoms may bring children with PKAN to medical attention (Battistella et al., 1998). In the index family used to identify *PANK2* (Zhou et al., 2001b), one of the seven

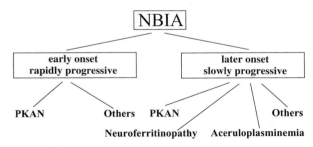

FIGURE 38.1 Proposed nosology for NBIA and PKAN. NBIA is a clinical entity that can be stratified by age at onset and rate of disease progression. NBIA is further stratified into distinct disorders based on genetic etiology.

[1]The information that follows is based on published reviews (Hayflick et al., 2002) as well as data from patients enrolled in the International Registry of PKAN and Related Disorders maintained by the author and her collaborators.

affected children had only retinopathy until the early teen years. She demonstrated a stepping gait and learned to avoid objects outside of her constricted visual fields. Toe-walking and upper extremity dystonia are less common presenting signs.

Though the overt clinical features of PKAN often are not evident until childhood, cell and tissue perturbations certainly begin earlier. In each of two families with affected children, a neurologically normal sibling demonstrated characteristic radiographic changes in the globus pallidus (Hayflick et al., 2001). This observation in presymptomatic patients indicates longstanding tissue injury. It is likely that the disease process begins prenatally.

The neurologic signs and symptoms of PKAN are primarily extrapyramidal and include dystonia, dysarthria and rigidity. In classical disease, dystonia is always present and usually is an early manifestation. Orobuccolingual and limb dystonia are frequent and can be extremely distressing to the patient and caregivers. Complications of these forms of dystonia may include recurrent trauma to the tongue, requiring full mouth dental extraction, or atraumatic long bone fracture resulting from the combination of extreme bone stress and osteopenia. Corticospinal tract involvement, including spasticity, hyperreflexia and extensor toe signs, is common in PKAN. Seizures are rare based on reports in the literature and the author's experience.

Intellectual impairment may or may not be a major feature of PKAN. Historically, it is associated with this diagnosis; however, the patient's movement disorder may limit their ability to perform intellectual testing. Newer methods of measuring cognitive function may help determine whether this is an essential component of the disorder.

Pigmentary retinal degeneration is a well-recognized feature of classical PKAN (Dooling et al., 1974; Luckenbach et al., 1983; Newell et al., 1979; Roth et al., 1971). Of 66 patients in our registry with classical disease and mutations in PANK2, 44 (68%) have clinical or electroretinographic evidence of retinopathy. The disease follows a typical clinical course, with nyctalopia leading to progressive loss of peripheral visual fields. Funduscopic changes initially include a flecked retina and later progress to bone spicule formation, conspicuous choroidal circulation, and "bull's-eye" annular maculopathy. Optic atrophy is less common in PKAN, occurring in about 3% of patients.

In classical PKAN, the electroretinogram shows characteristic changes associated with rod photoreceptor degeneration. ERG changes almost certainly predate funduscopic changes. The subset of classical patients with retinopathy develop these changes early, though they may be asymptomatic. Often retinopathy is not recognized until a full diagnostic evaluation is performed. As a corollary, it is uncommon for patients with a normal ophthalmologic examination at the time of diagnosis to develop retinopathy later.

The most commonly reported systemic manifestation is acanthocytosis, present in 8% of mutation-positive patients with classical disease. This feature is also seen in hypoprebetalipoproteinemia, acanthocytosis, retinitis pigmentosa, and pallidal degeneration (HARP) (Gitlin, 1998; Higgins et al., 1992; Orrell et al., 1995), in which we have demonstrated a mutation in PANK2 (Ching et al., 2002). Lymphocyte and histiocyte inclusions, reported in HSS, are probably not features of PKAN (Alberca et al., 1987; Swaiman et al., 1983; Zupanc et al., 1990). Since systemic signs are not sought routinely, their prevalence in mutation-positive patients is uncertain.

*a. Progression*

PKAN is a progressive disorder. The rate of progression seems to correlate with age at onset; patients with early symptoms have a more rapid pace of decline. As the disease advances, dystonia and spasticity compromise the patient's ability to ambulate, and most classical patients are wheelchair-bound by mid-teens, though for many this occurs much earlier. Classical PKAN progresses at a nonuniform rate. For reasons that are unclear, patients experience episodes of rapid deterioration, often lasting one to two months, interspersed with longer periods of relative stability. No cause for these periods of decline has been identified. Infection and other common etiologies of catabolic stress do not seem to precipitate episodic decline. Usually, lost skills are not regained.

Clinical description of early onset, rapidly progressive (classical) PKAN:

1. Onset in first decade
2. Extrapyramidal dysfunction
3. Corticospinal tract involvement
4. +/− Pigmentary retinopathy
5. Progression to loss of independent ambulation within 10–15 years of onset
6. Eye of the tiger on brain MRI

## 2. HARP syndrome

HARP is a rare syndrome with many clinical similarities to PKAN. Since first being described (Higgins et al., 1992), HARP syndrome was thought to fall within the spectrum of HSS. These disorders share many clinical and radiographic features, yet HARP is distinguished by a specific lipoprotein abnormality (Higgins et al., 1992; Orrell et al., 1995). HARP patients have decreased or absent prebetalipoproteins, a fraction that consists of very low density lipoproteins (VLDL). Recently, a PANK2 mutation was identified in the original HARP patient, confirming that this condition is part of the PKAN spectrum (Ching et al., 2002). What remains uncertain is how widespread this feature is and how pantothenate kinase 2 deficiency might lead to a decrease in prebetalipoproteins.

### 3. Late-Onset, Slowly Progressive (Atypical) PKAN

The clinical features of atypical PKAN are more varied than in classical PKAN (Hayflick *et al.*, 2002). By definition, onset is later, progression is slower, and presenting features are distinct. Genotype does correlate with phenotype to a limited degree, explaining the stratification of patients into two clinical categories. Patients predicted to have no functional protein have classical disease. Those predicted to have partial enzyme function generally fall into the category of atypical disease.

Based on registry data, the mean age at onset for atypical PKAN is 13.7 ± 5.9 years (range 1–28), significantly later than for classical disease. Presenting features differ, as well. Nine of twenty-three patients had difficulty with speech as either the sole presenting feature or part of their early disease. Six presented with palilalia, and three had dysarthria. No patients with classical PKAN presented with speech defects, though 16 developed these problems later.

Motor involvement is prominent in atypical disease and these patients are often described as clumsy in childhood and adolescence. Patients develop extrapyramidal deficits but they are generally less severe and more slowly progressive. Corticospinal tract involvement, including spasticity and hyperreflexia, is common and progressive, eventually limiting ambulation. Among the most striking manifestations of basal ganglia dysfunction in atypical PKAN are repetitive actions, freezing, and palilalia (Benke and Butterworth, 2001; Benke *et al.*, 2000), which can be profoundly debilitating. Freezing during ambulation, especially when turning corners, or encountering surface variations, is strikingly similar to changes seen in Parkinson's disease (Guimarães and Santos, 1999).

Psychiatric symptoms are more prominent in atypical disease than in classical disease (Morphy *et al.*, 1989; Szanto and Gallyas, 1966; Williamson *et al.*, 1982). They occur early and may delay diagnosis. Symptoms include personality changes with impulsivity and violent outbursts, depression, and emotional lability. For patients whose major early symptom is palilalia, psychiatric symptoms may be interpreted as the basis for their speech defect. As with classical disease, cognitive impairment may be part of the atypical PKAN phenotype, but additional investigations are needed.

Retinopathy is extremely rare in atypical disease (Coppeto and Lessell, 1990; Sethi *et al.*, 1988). Optic atrophy has not been associated with atypical disease. Strabismus and nystagmus have occurred in isolated cases but are not part of the PKAN phenotype in the vast majority of patients.

Clinical description of late onset, slowly progressive (atypical) PKAN:

1. Onset after first decade, commonly in second or third decade
2. Extrapyramidal dysfunction
3. Corticospinal tract involvement
4. Prominent speech disorder
5. Progression to loss of independent ambulation within 15–40 years of onset
6. Eye of the tiger on brain MRI

### 4. Diagnostic Criteria for PKAN

Diagnostic criteria were first proposed by Dooling *et al.* (1974) and later refined by Swaiman (1991). Dooling based her criteria on a review of 42 cases that formed a distinct clinicopathologic entity. When brain MRI became a valuable tool for diagnosing HSS, Swaiman revised the diagnostic criteria. Now that a genetically homogeneous disorder can be delineated, revisions to these criteria are needed. The proposed obligate and corroborative features (Table 38.1) replace the prior set and may be useful for identifying patients in whom molecular testing is indicated.

### 5. NBIA

NBIA includes patients who met earlier diagnostic criteria for HSS but who do not have a mutation in *PANK2* or in either of the other genes that can cause NBIA (*FTL* or *CP*). Despite the probable heterogeneity of this group, there are characteristics that distinguish them from patients with PKAN.

Historically, the diagnosis of NBIA has relied upon evidence for basal ganglia dysfunction and iron deposition. For this reason, the NBIA group shares many clinical features with PKAN. However, the two diagnostic groups can be distinguished by the changes seen on brain MRI. In NBIA, the globus pallidus is uniformly hypointense on T2-weighted images (Fig. 38.3). This change may indicate high iron; however, it is distinct from the eye of the tiger and is not seen in association with *PANK2* mutations. Iron

TABLE 38.1 Diagnostic Criteria for PKAN

| Obligate features | Corroborative features |
|---|---|
| Onset in the first three decades | Corticospinal tract involvement |
| Progression of signs and symptoms | Pigmentary retinopathy or optic atrophy |
| Evidence of extrapyramidal dysfunction | Family history suggestive of autosomal recessive inheritance |
| On T2-weighted brain MRI: bilateral anteromedial hyperintensity surrounded by a region of hypointensity in the medial globus pallidus | Acanthocytosis |

deposition in the red nucleus and dentate nucleus is common in NBIA but is not seen in PKAN. Radiographic and functional abnormalities in other brain regions, especially cerebellar atrophy, are common in the NBIA group and rare in PKAN. Finally, seizures are prominent in many NBIA patients but rarely occur in PKAN.

The differential diagnosis of NBIA includes several disorders with a similar radiographic appearance of the pallidum that probably does not represent iron. These include X-linked mental retardation with Dandy-Walker malformation (Pettigrew *et al.*, 1991) and α-fucosidosis (Provenzale *et al.*, 1995; Terepolsky *et al.*, 1996).

## III. GENE

### A. Normal Gene Function

PKAN is an autosomal recessive inborn error of pantothenate metabolism. It is caused by a defect in *PANK2*, which encodes pantothenate kinase 2 (Zhou *et al.*, 2001b). In humans, three additional genes are predicted to encode related proteins, based on homology studies, and are designated *PANK1* (Karim *et al.*, 2000; Zhou *et al.*, 2001b), *PANK3*, and *PANK4*, (Zhou *et al.*, 2001b). The murine homolog of *PANK1* encodes a pantothenate kinase (Rock *et al.*, 2000). The functions of PANK3 and PANK4 have not been determined. While *PANK2* maps to chromosome 20p13, *PANK1*, *PANK3*, and *PANK4* are located on chromosomes 10q23, 5q35, and 1p36, respectively. Only *PANK2* has been associated with a human disease.

The *PANK2* gene encodes a 1.85-kb transcript that is derived from seven exons spanning just over 35 kb of genomic DNA (Westaway *et al.*, 2002). Detailed sequence analysis reveals that *PANK2* is a member of a family of eukaryotic genes consisting of a group of six exons that encode homologous core proteins, preceded by a series of alternative initiating exons, some of which encode unique amino-terminal peptides. 5′ RACE and EST data provide evidence for at least five initiating exons for *PANK2*, but only one of these, exon 1C, has an open reading frame with potential initiation codons that splices in-frame to exon 2 (Zhou *et al.*, 2001b).

Pantothenate kinase is an essential regulatory enzyme in coenzyme A (CoA) biosynthesis, catalyzing the phosphorylation of pantothenate (vitamin $B_5$), *N*-pantothenoyl-cysteine and pantetheine (Fig. 38.2). Pantothenate kinase is

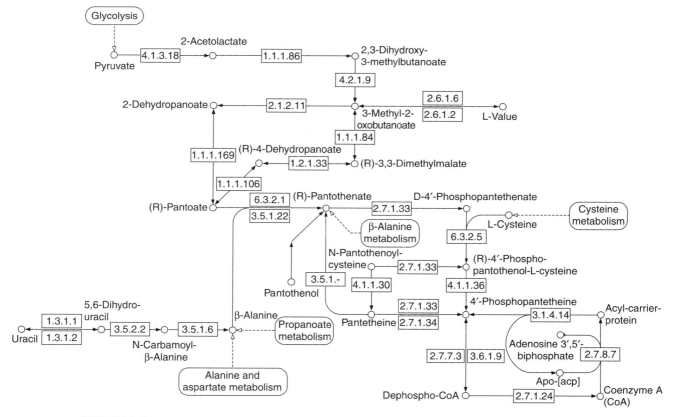

**FIGURE 38.2** The biosynthesis of coenzyme A from pantothenate. Pantothenate kinase is E.C.2.7.1.33. From the KEGG website http://www.biologie.uni-hamburg.de/b-online/kegg/kegg/Classes/dblinks_java/map/map00770.html.

regulated by acyl-CoA levels in prokaryotes and by acetyl-CoA levels in eukaryotes (Rock et al., 2000; Yun et al., 2000). The protein is a dimer in E. coli, and there is evidence that human PANK2 dimerizes, as well.

It is unclear why humans need four pantothenate kinase genes. Messenger RNA tissue distribution differs among the PANKs, which may show temporal-specific expression, as well (Zhou et al., 2001b). Another intriguing possibility is organellar specificity. The need for four PANK genes in vertebrates probably is key to understanding how abnormal gene function causes the PKAN phenotype.

### B. Abnormal Gene Function

PKAN is the first inborn error of coenzyme A metabolism. Mutations in PANK2 are predicted to result in deficient enzyme activity necessary to maintain health. This defect leads to product deficit and secondary metabolite accumulation as likely etiologies for the pathogenesis of disease. Since pantothenate is a water-soluble compound with no known toxicity even at megadose levels, its accumulation is unlikely to contribute to disease. Two other substrates for pantothenate kinase, N-pantothenoyl-cysteine and pantetheine are predicted to accumulate and may cause toxicity to cells.

N-pantothenoyl-cysteine and pantetheine are formed by the degradation of coenzyme A and acyl carrier protein (Fig. 38.2). In vertebrates, N-pantothenoyl-cysteine and pantetheine are recycled back into the CoA synthetic pathway by one of two enzymes, pantothenate kinase or pantetheinase. In humans, pantetheinase catalyzes the hydrolysis of pantetheine to pantothenate, leaving pantothenate kinase as an obligate enzyme in the processing of both of these compounds. Thus, a defect in pantothenate kinase is predicted to lead to their accumulation. High basal ganglia cystine and cysteine-mixed disulfides have been reported in a single patient with features of PKAN (Perry et al., 1985).

Just as cysteine can cause cytotoxicity, N-pantothenoyl-cysteine and pantetheine probably induce free radical damage in cells (Yang et al., 2000; Yoon et al., 2000). Cysteine chelates iron, and these compounds together produce reactive oxygen species leading to further toxicity. It is likely that iron is also chelated by the cysteine-containing compounds predicted to accumulate in PKAN. This prediction may explain the high levels of iron in tissue affected in this disorder. The primary enzyme deficiency results in substrate accumulation, which itself causes toxicity. The substrates are exposed to iron in this normally iron-rich part of the brain and chelate the metal, leading to further oxidative damage to the cell.

PANK2 mutations are predicted to result in CoA depletion and defective membrane biosynthesis in those tissues in which this is the major pantothenate kinase or in those tissues with the greatest CoA demand. Rod photoreceptors continually generate membranous discs; hence, the retinopathy frequently observed in classical PKAN may be secondary to this deficit. An animal model of PKAN will enable direct testing of these hypotheses of pathophysiology.

## IV. DIAGNOSTIC AND ANCILLARY TESTS

The diagnosis of PKAN is suspected from a combination of clinical and radiographic information. Brain MRI is a key diagnostic test. Molecular studies are necessary to confirm a suspected clinical diagnosis. Prenatal diagnosis using molecular testing is now available.

### A. DNA

Molecular testing currently focuses on identifying variations within the PANK2 coding sequence (Westaway et al., 2002). Mutation screening is done by polymerase chain amplification of patient genomic DNA using primers that hybridize to intronic sequence and enable analysis of 5′ and 3′ splice sites and exonic sequence. Clinical molecular diagnostic laboratories are listed at www.genetests.org.

The PANK2 gene comprises seven exons, and deleterious mutations have been found in all of them. Several splice site mutations have been identified, as well. Two specific sequence variations account for one-third of patient mutations (G411R and T418M). G411R occurs on a background of a conserved haplotype, indicating a founder effect. Of the remaining mutations, none occurs in more than 3% of patients, with the majority being found only in one family. PANK2 mutations are listed in Table 38.2, along with eight polymorphisms that have been identified.

To date, no biochemical perturbations specific to PKAN have been identified. There are no abnormalities in measures of systemic iron homeostasis (Dooling et al., 1974). The specific lipoprotein abnormality in HARP may illuminate part of the pathophysiology of PKAN. Based on knowledge of the enzymatic defect in PKAN, one can predict specific biochemical alterations (e.g., increased N-pantothenoyl-cysteine and pantetheine, and decreased CoA) and studies are underway to investigate these.

### B. Neuroimaging

#### 1. MRI

Magnetic resonance imaging has contributed to our understanding of PKAN. Disease diagnosis and delineation have benefited from this technology. Emerging neuroimaging technologies are likely to help define the pathophysiology of PKAN.

Brain MRI has become standard in the diagnostic evaluation of PKAN and NBIA. The technology first enabled premortem diagnosis (Drayer, 1989; Gallucci

TABLE 38.2 Mutations and Polymorphisms

| Nucleotide change | Exon/intron | Predicted result | Nucleotide change | Exon/intron | Predicted result |
|---|---|---|---|---|---|
| **Missense** | | | 606T→A | 2 | C202X |
| 71A→G | 1C | E24G | 664C→T | 3 | Q222X |
| 326G→t | 2 | G109V | 700A→T | 3 | K234X |
| 353T→C | 2 | F118S | 1021C→T | 4 | R341X |
| 370A→G | 2 | T124A | 1111C→T | 5 | R371X |
| 416G→C | 2 | R139P | | | |
| 460C→T | 2 | R154W | **Deletions/insertions/** | | |
| 502C→T | 2 | R168C | **duplications** | | |
| 503G→T | 2 | R168L | 206-228Δ23bp | 1C | Frameshift |
| 514C→G | 2 | L172V | 212-219Δ8bp | 1C | Frameshift |
| 519C→G | 2 | H173Q | 215∇A | 1C | Frameshift |
| 526C→T | 2 | R176C | 239∇A | 1C | Frameshift |
| 568A→G | 2 | R190G | 243ΔC | 1C | Frameshift |
| 635A→G | 2 | E212G | 285ΔG | 1C | Frameshift |
| 636G→T | 2 | E212D | 493-494Δ2bp | 2 | Frameshift |
| 650C→T | 2 | T217I | 597-603Δ7bp | 2 | Frameshift |
| 721T→C | 3 | S241P | 755-758Δ4bp | 3 | Frameshift |
| 734A→G | 3 | N245S | 794-822Δ29bp | 3 | Frameshift |
| 740G→A | 3 | R247Q | 846-847Δ2bp | 3 | Frameshift |
| 862G→A | 3 | A288T | 943-945Δ3bp | 4 | L315del |
| 881A→T | 3 | N294I | 993-996Δ4bp | 4 | Frameshift |
| 908T→C | 4 | L303P | 1112-1114Δ3bp | 5 | R371E372delinsQ |
| 953G→A | 4 | C318Y | 1171-1174dupl | 5 | Frameshift |
| 1009G→A | 4 | D337N | | | |
| 1082G→A | 4 | S361N | **Splice site** | | |
| 1145C→T | 5 | A382V | Ex2−1G→A | IVS1/Ex2 | Aberrant splicing |
| 1160T→C | 5 | I387T | Ex2+3A→G | Ex2/IVS3 | Aberrant splicing |
| 1169A→T | 5 | N390I | Ex5−1G→T | IVS4/Ex5 | Aberrant splicing |
| 1172T→C | 5 | I391T | Ex6−3C→G | IVS5/Ex6 | Aberrant splicing |
| 1196C→T | 5 | A399V | Ex7−2A→G | IVS6/Ex7 | Aberrant splicing |
| 1201A→G | 5 | N401D | | | |
| 1231G→A | 6 | G411R | **Polymorphisms** | | |
| 1253C→T | 6 | T418M | 2T→A | 1C | L1Q |
| 1264C→T | 6 | R422W | 47C→G | 1C | A16G |
| 1358T→C | 7 | L453P | Ex1+24C→A | Ex1/IVS1 | Unknown |
| 1379C→T | 7 | P460L | Ex1+34C→T | Ex1/IVS1 | Unknown |
| | | | Ex1+35C→T | Ex1/IVS1 | Unknown |
| **Nonsense** | | | 744A G | 3 | S248S |
| 118C→T | 1C | Q40X | Ex5−12insCCCCTT | IVS4/Ex5 | Unknown |
| 240C→G | 1C | Y80X | 1404C→T | 7 | Unknown |

From Westaway, S. K. et al. (2002), submitted.

et al., 1990; Mutoh et al., 1990; Rutledge et al., 1988; Schaffert et al., 1989; Scheer et al., 1988; Tabuchi et al., 1990); with 1.5 Tesla imaging units, brain iron was easily visualized. The radiographic changes that have been associated with NBIA are high iron in the basal ganglia, seen as hypointense lesions in the globus pallidus and substantia nigra pars reticulata on T2-weighted images (Drayer, 1987; Mutoh et al., 1988; Sethi et al., 1988; Tanfani et al., 1987) (Fig. 38.3). This finding was considered diagnostic and was used to expand the clinical features of the disease, with virtually all patients who had high basal ganglia iron being assigned this diagnosis.

Among the group of NBIA patients with hypointense lesions in the globus pallidus, a subset has the additional finding on T2-weighted images of bilateral anteromedial hyperintense lesions within the globus pallidus. This compound change, called the eye of the tiger sign by Sethi (1988), has been interpreted to represent tissue edema or vacuolization within the region where iron accumulates (Savoiardo et al., 1993) (Fig. 38.3). In presymptomatic patients, the hyperintense lesions predominate and as disease progresses the hypointensities appear and eventually dominate (Hayflick et al., 2001). There now is evidence that the central hyperintensity continues to diminish and may ultimately disappear as disease progresses (Gallucci et al., 1990; Hayflick et al., 2002). If one accepts the association of this hyperintense lesion with early tissue reactivity, then the evolving radiographic changes parallel

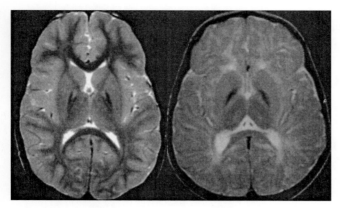

FIGURE 38.3 T2-weighted brain MRI demonstrating eye of the tiger with a central region of hyperintensity within the hypointense globus pallidus lesions (left) and hypointense globus pallidus lesions only (right).

the cellular and tissue disease process. Biochemical perturbations lead to early cell and tissue injury. Iron begins to accumulate causing further tissue destruction. Following extensive neuronal death, tissue reactivity subsides, and disease progression is marked by increasing iron deposition.

One of the most important associations to follow from the PKAN gene discovery is that of the eye of the tiger sign and *PANK2* mutations. Of 68 patients with eye of the tiger, all had mutations in *PANK2*. As a corollary, of the 16 patients who had been diagnosed with HSS but did not have eye of the tiger, none had an identified *PANK2* mutation. These data suggest that eye of the tiger correlates strongly with PKAN (Hayflick *et al.*, 2002).

## 2. FMRI and PET

These imaging modalities have only recently been explored in PKAN. In one study, PET was used to measure regional cerebral blood flow and dopaminergic function (Castelnau *et al.*, 2001). Though no pallidal abnormalities were demonstrated, significant hypoperfusion of the head of the right caudate nucleus, pons, and the cerebellar vermis was seen. Dopaminergic function of the basal ganglia was reported to be normal. In a second pair of studies, SPECT demonstrated no abnormalities in the nigrostriatal system, with normal dopamine transporter and D2 receptor scintigraphy (Hermann *et al.*, 2000a, b).

## 3. MR Spectroscopy

To date, no spectroscopic studies have been performed on patients with PKAN. Based on knowledge of the role of pantothenate kinase in CoA biosynthesis, it is possible to predict specific biochemical perturbations that would be likely to occur in patients with PKAN (e.g., low tissue CoA, high pantothenoyl cysteine and pantetheine). Many of these analytes are amenable to spectroscopic study, and plans for such studies are underway.

## 4. Neuropathology

Perhaps more than in any other area, the literature on neuropathology in NBIA is seriously limited by the heterogeneity of the entities being grouped under this diagnosis (Halliday, 1995). Data on the generally accepted spectrum of pathologic changes will be presented; however, it clearly does not represent a single disease entity and must therefore be viewed in this light. Despite this limitation, using clinical and radiographic data, one can select reports of patients likely to have had PKAN and can therefore begin to delineate the neuropathologic findings of this genetically homogeneous disease.

Before MRI became widely available as a diagnostic tool, NBIA was a postmortem diagnosis. On gross sectioning, the globus pallidus and the reticular zone of the substantia nigra show rust-brown pigmentation. Iron is the major component of this pigment (Hallervorden, 1924). Routine iron stains detect the metal mostly in microglia and macrophages, but scattered neurons also are reactive. Iron is also detected extracellularly, often concentrated around blood vessels.

In NBIA, iron accumulates in brain regions that are normally iron-rich. In normal human brain, iron is regionally distributed and highest in globus pallidus, substantia nigra pars reticulata, dentate nucleus, and red nucleus (Hill, 1988; Hill and Switzer, 1984). In HSS, the accumulation of iron is specific to the globus pallidus and substantia nigra; a global increase in brain iron is not seen (Spatz, 1922a, b). In HSS, the globus pallidus and substantia nigra contain approximately three times the normal amount of iron, yet iron content is normal in other regions of brain and in retina and optic nerve (Dooling *et al.*, 1974; Kornyey, 1964; Vakili *et al.*, 1977; Volkl and Ule, 1972). Systemic iron metabolism is normal, as well (Swaiman *et al.*, 1983; Szanto and Gallyas, 1966). In patients with NBIA, *in vivo* $^{59}$Fe-uptake studies suggest that accumulation of iron in the basal ganglia is secondary to increased iron uptake with normal turnover. (Swaiman *et al.*, 1983; Szanto and Gallyas, 1966; Vakili *et al.*, 1977).

In regions of massive iron accumulation, spheroid bodies, many positive for iron, are also seen (Koeppen and Dickson, 2001). Axonal spheroids are posited to represent swollen or bloated axons, possibly secondary to defects in axonal transport. They are seen in normal aging brain and in a number of other neurodegenerative disorders, including the neuroaxonal dystrophies. Other neuropathologic findings include demyelination, neuronal loss, and gliosis, which occur predominately within the globus pallidus and substantia nigra, where focal, symmetric destruction may be grossly evident. In addition to iron deposition and spheroid formation in the brains of people with NBIA, ceroid-lipofuscin and neuromelanin accumulate both intra- and extraneuronally.

Numerous papers on NBIA report the presence of Lewy bodies and neurofibrillary tangles with accumulations of tau and alpha-synuclein (Arawaka et al., 1998; Galvin et al., 2000; Neumann et al., 2000; Newell et al., 1999; Odawara et al., 1992; Saito et al., 2000; Sugiyama et al., 1993; Tuite et al., 1996; Wakabayashi et al., 1999, 2000). Yet, not a single case from this literature is likely to have PKAN, based on the clinical and radiographic information provided. This observation underscores the need to carefully re-examine our current knowledge of the neuropathology in NBIA in order to determine what can be applied to PKAN.

Nonspecific systemic cytologic abnormalities in NBIA include bone marrow macrophages containing ceroid-lipofuscin and circulating lymphocytes with vacuoles and cytosomic inclusion bodies, similar to those seen in ceroid-lipofuscin storage disorders (Swaiman et al., 1983). Acanthocytes have been reported in a subset of patients with NBIA (Higgins et al., 1992; Luckenbach et al., 1983; Roth et al., 1971; Swisher et al., 1972; Tripathi et al., 1992), many of whom probably had PKAN. Lipofuscin and acanthocytes both can result from lipid peroxidation, a process stimulated by iron (Defendini et al., 1973; Park et al., 1975). These cytologic abnormalities, while not prominent in the pathology of this disorder, may shed light on the underlying pathophysiology of HSS.

Since retinopathy is characteristic of PKAN, the pathology literature on this feature is likely to be specific to this disorder. Ophthalmoscopic examination in patients with HSS shows prominent bone spicule pigmentation and accumulation beneath the sensory retina of numerous yellowish-white globular masses (Luckenbach et al., 1983; Newell et al., 1979; Roth et al., 1971; Tripathi et al., 1992). On light microscopic study, there is total loss of the outer segment of the photoreceptor cells and near total loss of the inner segment. The outer nuclear and outer plexiform layers are thinned or absent. The retinal pigment epithelium comprises a population of enlarged epithelial cells that contain both individual pigment granules and large, round pigment aggregates, a change seen primarily in the equatorial and pre-equatorial regions. Similar pigment-laden cells are present in the outer retinal layers peripheral to the perimacular area. The pigment granule clusters are melanolipofuscin complexes and represent the pathologic correlate of the yellowish-white globular masses seen on funduscopic examination (Tripathi et al., 1992). There is migration of the retinal pigment epithelium into the inner retinal layers in perivascular regions (Luckenbach et al., 1983), which likely accounts for the ophthalmoscopic appearance of bone spicule pigmentation. No stainable iron is seen in any part of the eye; however, focal axonal degeneration and cytoid bodies (spheroids) are noted (Tripathi et al., 1992). Abnormal accumulation of lipofuscin is reported in conjunctival fibroblasts, retinal vessel pericytes and macrophages (Luckenbach et al., 1983). Glial proliferation occurs throughout the retina and around blood vessels.

## V. CELLULAR AND ANIMAL MODELS OF DISEASE

### A. In Vitro

No *in vitro* model of PKAN has been developed.

### B. Invertebrates

A *Drosophila melanogaster* hypomorphic mutant with deficient pantothenate kinase function was developed nearly a decade ago. This mutant has a cell division cycle defect leading to sterility, for which it has primarily been studied (Afshar et al., 2001; Castrillon et al., 1993). However, the fly is also described as uncoordinated, resulting in impaired ability to climb, fly, and mate (Afshar et al., 2001); (Afshar, personal communication). Further phenotypic studies are underway.

### C. Mouse (KO, Transgenic)

A murine model for PKAN is currently being developed. We expect that the knock-out mouse will indeed be a model for the human disease given the high degree of identity of *PANK2* gene homologs between humans and mouse along with evolutionary conservation of the entire pantothenate kinase gene family.

## VI. GENOTYPE/PHENOTYPE CORRELATIONS/ MODIFYING ALLELES

The PKAN phenotype correlates with the *PANK2* genotype and is based on the prediction of homodimerization of PANK2 protein and the hypothesis of interallelic complementation (Zhou et al., 2001b). Interallelic complementation results when mutations in domains that interact between protein subunits are able to restore partial function (Gravel et al., 1994).

The genotypes of patients with classical disease segregate into one of three categories: homozygous null, homozygous missense, and compound heterozygous missense/null. The functional result of these genotypes is production of either no protein or a single mutant monomer that cannot form a functional dimer without a complementing allele. In either case, the result is no functional protein. Patients predicted to have no functional protein usually have classical disease. As a corollary, we predict that the severe extreme of the phenotypic spectrum is represented by classical disease, and a connatal form of PKAN is unlikely to occur. In contrast, those patients predicted to have partial enzyme function usually have atypical disease.

Atypical patients generally are compound heterozygotes for missense mutations, enabling complementation between the products of their two alleles. These genotypes are predicted to lead to partial loss of function variants in which there is residual pantothenate kinase 2 enzymatic function. Patients with classical disease are much more likely than patients with atypical disease to have mutations that preclude interallelic complementation. A small number of classical patients are compound heterozygotes for missense mutations. These specific mutations may prevent complementation.

The incidence of retinopathy in classical versus atypical PKAN patients may be related to the extent to which PANK2 enzymatic function has been abated. The most abundant fatty acid in retinal photoreceptor membrane phospholipids is docosahexaenoic acid (DHA). Photoreceptor membranous discs have a fast turnover rate and are significantly impacted by low DHA levels. DHA deficiency is associated with progressive retinopathy in Zellweger syndrome and neonatal adrenoleukodystrophy (Harding et al., 1999). Because PANK2 is critical in the biosynthesis of CoA, a total loss of PANK2 enzymatic function may result in deficient CoA and thus deficient DHA in classical patients. Conversely, atypical patients may retain sufficient PANK2 enzymatic function to meet a threshold requirement for DHA in the membranous discs.

Genotype may determine response to therapies, as well. The prediction of residual enzyme activity in some patients with PKAN raises the possibility of treatment using high-dose pantothenate, the PANK2 enzyme substrate. Studies are underway to delineate biochemical markers of disease that might be tracked in clinical interventional trials. These markers will also help to delineate the pathophysiologic mechanism of PKAN as either substrate accumulation, product deficit, secondary metabolite accumulation, or a combination thereof. Therapeutic targets may then be more effectively sought. With an understanding of the metabolic perturbations in PANK2 deficiency, we may be able to devise neuroprotective interventions to stop or slow disease progression.

## VII. TREATMENT

Pharmacologic and surgical interventions have been aimed at palliation of symptoms. Many drugs have been tried and most offer little benefit. For almost all interventions that offer improvement of clinical symptoms, the period of benefit is limited, with symptoms returning to pretreatment levels usually within a year. To date, treatment of PKAN and NBIA has been disappointing.

Since the gene discovery in PKAN, several ideas for therapy have been proposed. These include pantothenate, phosphopantothenate, coenzyme A, and cysteamine. Phosphopantothenate is the product of the enzyme that is defective in PKAN. Though no commercial source for this compound has been identified, it is unlikely to be therapeutic based on the prediction that this phosphorylated form will not be transported across cell membranes and therefore cannot be used by the cell.

Pantothenate has a hypothetical role in the treatment of PKAN. The idea behind loading with pantothenate is to drive any residual enzyme activity by substrate overload. This is likely to benefit only those patients with low levels of enzyme function. However, since it is likely to be without toxicity, we recommend a trial of high-dose pantothenate in patients with PKAN. Since no toxicity has been reported in normal individuals even with doses of 5–10 g orally per day, we suggest starting at 250 mg per day and increasing to 2–5 grams per day. Formal clinical trials are planned.

Iron chelating agents have been tried in HSS without clear benefit (Albright et al., 1996; Dooling et al., 1974). Trials have been limited by the development of systemic iron deficiency before any clinical neurological benefits were evident. With the early experience of iron-chelating agents in aceruloplasminemia and the development of agents that are better able to reach the central nervous system, this mode of therapy may hold promise (Arthur et al., 1997; Miyajima et al., 1997).

The $\gamma$-aminobutyric acid (GABA) agonist baclofen provides the most consistent relief of disabling dystonia in both PKAN and NBIA. Oral dosing has been used for years. With the advent of intrathecal drug delivery systems, continuous direct infusion of baclofen in PKAN and NBIA has been tried with limited benefit (Albright et al., 1996).

The value of most other drugs is unpredictable. Therapies that generally do not help patients with PKAN may have a role in NBIA. These include levodopa/carbidopa, bromocriptine, and trihexyphenidyl. It is worth stating explicitly that L-dopa does not seem to benefit patients with PKAN. Seizures in NBIA are generally treatable with standard anticonvulsant therapies.

Stereotactic pallidotomy and thalamotomy have limited benefits in PKAN and NBIA (Justesen et al., 1999; Tsukamoto et al., 1992). Patients usually experience relief immediately following the procedure; however, most return to their baseline levels of dystonia within one to two years. Deep brain stimulation is currently being explored.

In these disorders where a significant level of cognitive function may be preserved, it is especially important to help patients maintain independence (Seibel et al., 1993). Regular review of communication needs and environmental adaptations is required.

## References

Afshar, K., Gonczy, P., DiNardo, S., and Wasserman, S. A. (2001). Fumble encodes a pantothenate kinase homolog required for proper mitosis and meiosis in Drosophila melanogaster. *Genetics* **157**, 1267–1276.

Alberca, R., Rafel, E., Chinchon, I., Vadillo, J., and Navarro, A. (1987). Late onset parkinsonian syndrome in Hallervorden-Spatz disease. *J. Neurol. Neurosurg. Psychiatry* **50**, 1665–1668.

Albright, A. L., Barry, M. J., Fasick, P., Barron, W., and Shultz, B. (1996). Continuous intrathecal baclofen infusion for symptomatic generalized dystonia. *Neurosurgery* **38**, 934–8; discussion 938–939.

Arawaka, S., Saito, Y., Murayama, S., and Mori, H. (1998). Lewy body in neurodegeneration with brain iron accumulation type 1 is immunoreactive for alpha-synuclein. *Neurology* **51**, 887–889.

Arthur, A. S., Fergus, A. H., Lanzino, G., Mathys, J., Kassell, N. F., and Lee, K. S. (1997). Systemic administration of the iron chelator deferiprone attenuates subarachnoid hemorrhage-induced cerebral vasospasm in the rabbit. *Neurosurgery* **41**, 1385–1391; discussion 1391–1392.

Battistella, P. A., Midena, E., Suppiej, A., and Carollo, C. (1998). Optic atrophy as the first symptom in Hallervorden-Spatz syndrome. *Childs Nerv. Syst.* **14**, 135–138.

Benke, T., and Butterworth, B. (2001). Palilalia and repetitive speech: two case studies. *Brain Lang.* **78**, 62–81.

Benke, T., Hohenstein, C., Poewe, W., and Butterworth, B. (2000). Repetitive speech phenomena in Parkinson's disease. *J. Neurol. Neurosurg. Psychiatry* **69**, 319–324.

Bruscoli, F., Corsi, A., Cavicchi, C., Carloni, R., Ferioli, I., Gemmani, A., Crociati, M., and Pompili, A. (1998). [Therapeutic objectives and strategies in NBIA 1 (Hallovorden-Spatz syndrome)]. *Minerva Anestesiol.* **64**, 529–534.

Castelnau, P., Zilbovicius, M., Ribeiro, M., Hertz-Pannier, L., Ogier, H., and Evrard, P. (2001). Striatal and pontocerebellar hypoperfusion in Hallervorden-Spatz syndrome. *Pediatr. Neurol.* **25**, 170–174.

Castrillon, D. H., Gonczy, P., Alexander, S., Rawson, R., Eberhart, C. G., Viswanathan, S., DiNardo, S., and Wasserman, S. A. (1993). Toward a molecular genetic analysis of spermatogenesis in Drosophila melanogaster: characterization of male-sterile mutants generated by single P element mutagenesis. *Genetics* **135**, 489–505.

Ching, K. H. L., Westaway, S. K., Levinson, B., Zhou, B., Higgins, J. J., Gitschier, J., and Hayflick, S. J. (2002). HARP syndrome is allelic with pantothenate kinase associated neurodegeneration, submitted.

Coppeto, J. R., and Lessell, S. (1990). A familial syndrome of dystonia, blepharospasm, and pigmentary retinopathy. *Neurology* **40**, 1359–1363.

Curtis, A. R., Fey, C., Morris, C. M., Bindoff, L. A., Ince, P. G., Chinnery, P. F., Coulthard, A., Jackson, M. J., Jackson, A. P., McHale, D. P., Hay, D., Barker, W. A., Markham, A. F., Bates, D., Curtis, A., and Burn, J. (2001). Mutation in the gene encoding ferritin light polypeptide causes dominant adult-onset basal ganglia disease. *Nat. Genet.* **28**, 350–354.

Defendini, R., Markesbery, W. R., Mastri, A. R., and Duffy, P. E. (1973). Hallervorden-Spatz disease and infantile neuroaxonal dystrophy. Ultrastructural observations, anatomical pathology and nosology. *J. Neurol. Sci.* **20**, 7–23.

Dooling, E. C., Richardson, E. P., Jr., and Davis, K. R. (1980). Computed tomography in Hallervorden-Spatz disease. *Neurology* **30**, 1128–1130.

Dooling, E. C., Schoene, W. C., and Richardson, E. P., Jr. (1974). Hallervorden-Spatz syndrome. *Arch. Neurol.* **30**, 70–83.

Drayer, B. (1987). Magnetic resonance imaging and brain iron: implications in the diagnosis and pathochemistry of movement disorders and dementia. *Barrow Neurol. Inst. Q.* 15–30.

Drayer, B. P. (1989). Magnetic resonance imaging and extrapyramidal movement disorders. *Eur. Neurol.* **29**, 9–12.

Gallucci, M., Cardona, F., Arachi, M., Splendiani, A., Bozzao, A., and Passariello, R. (1990). Follow-up MR studies in Hallervorden-Spatz disease. *J. Comput. Assist. Tomogr.* **14**, 118–120.

Galvin, J. E., Giasson, B., Hurtig, H. I., Lee, V. M., and Trojanowski, J. Q. (2000). Neurodegeneration with brain iron accumulation, type 1 is characterized by alpha-, beta-, and gamma-synuclein neuropathology. *Am. J. Pathol.* **157**, 361–368.

Gitlin, J. D. (1998). Aceruloplasminemia. *Pediatr. Res.* **44**, 271–276.

Gravel, R. A., Akerman, B. R., Lamhonwah, A. M., Loyer, M., Leon-del-Rio, A., and Italiano, I. (1994). Mutations participating in interallelic complementation in propionic acidemia. *Am. J. Hum. Genet.* **55**, 51–58.

Guimarães, J., and Santos, J. V. (1999). Generalized freezing in Hallervorden-Spatz syndrome: case report. *Eur. J. Neurol.* **6**, 509–513.

Hallervorden, J. (1924). Uber eine familiare Erkrankung im extrapyramidalen System. *Dtsch. Z. Nervenheilkd.* 204–210.

Hallervorden, J., and Spatz, H. (1922). Eigenartige Erkrankung im extrapyramidalen System mit besonderer Beteiligung des Globus pallidus und der Substantia nigra. *Z. Ges. Neurol. Psychiatr.* **79**, 254–302.

Halliday, W. (1995). The nosology of Hallervorden-Spatz disease. *J. Neurol. Sci.* **134** Suppl. 84–91.

Harding, C. O., Gillingham, M. B., van Calcar, S. C., Wolff, J. A., Verhoeve, J. N., and Mills, M. D. (1999). Docosahexaenoic acid and retinal function in children with long-chain 3-hydroxyacyl-CoA dehydrogenase deficiency. *J. Inherit. Metab. Dis.* **22**, 276–280.

Harper, P. S. (1996). Naming of syndromes and unethical activities—the case of hallervorden and spatz. *Lancet* **348**, 1224–1225.

Hayflick, S. J., Penzien, J. M., Michl, W., Sharif, U. M., Rosman, N. P., and Wheeler, P. G. (2001). Cranial MRI changes may precede symptoms in Hallervorden-Spatz syndrome. *Pediatr. Neurol.* **25**, 166–169.

Hayflick, S. J., Westaway, S. K., Levinson, B., Zhou, B., Johnson, M. A., Ching, K. H. L., and Gitschier, J. (2002). Genetic, clinical, and radiographic delineation of Hallervorden-Spatz syndrome. *N. Engl. J. Med.* (accepted).

Hermann, W., Barthel, H., Reuter, M., Georgi, P., Dietrich, J., and Wagner, A. (2000a). [Hallervorden-Spatz disease: findings in the nigrostriatal system]. *Nervenarzt* **71**, 660–665.

Hermann, W., Reuter, M., Barthel, H., Dietrich, J., Georgi, P., and Wagner, A. (2000b). Diagnosis of Hallervorden-Spatz disease using MRI, (123)I-beta-CIT-SPECT and (123)I-IBZM-SPECT. *Eur. Neurol.* **43**, 187–188.

Higgins, J. J., Patterson, M. C., Papadopoulos, N. M., Brady, R. O., Pentchev, P. G., and Barton, N. W. (1992). Hypoprebetalipoproteinemia, acanthocytosis, retinitis pigmentosa, and pallidal degeneration (HARP syndrome). *Neurology* **42**, 194–198.

Hill, J. M. (1988). "Brain Iron: Neurochemical and Behavioural Aspects." Taylor & Francis, London.

Hill, J. M., and Switzer, R. C., III. (1984). The regional distribution and cellular localization of iron in the rat brain. *Neuroscience* **11**, 595–603.

Justesen, C. R., Penn, R. D., Kroin, J. S., and Egel, R. T. (1999). Stereotactic pallidotomy in a child with Hallervorden-Spatz disease. Case report. *J. Neurosurg.* **90**, 551–554.

Karim, M. A., Valentine, V. A., and Jackowski, S. (2000). Human pantothenate kinase 1 (PANK1) gene: characterization of the cDNAs, structural organization and mapping of the locus to chromosome 10q23.2–23.31. *Am. J. Hum. Genet.* **67**, A984.

Koeppen, A. H., and Dickson, A. C. (2001). Iron in the Hallervorden-Spatz syndrome. *Pediatr. Neurol.* **25**, 148–155.

Kornyey, S. (1964). Die Stoffwechechselstorungen bei der Hallervorden-Spatzschen Krankheit. *Z. Neurol.*, 178–183.

Luckenbach, M. W., Green, W. R., Miller, N. R., Moser, H. W., Clark, A. W., and Tennekoon, G. (1983). Ocular clinicopathologic correlation of Hallervorden-Spatz syndrome with acanthocytosis and pigmentary retinopathy. *Am. J. Ophthalmol.* **95**, 369–382.

Miyajima, H., Takahashi, Y., Kamata, T., Shimizu, H., Sakai, N., and Gitlin, J. D. (1997). Use of desferrioxamine in the treatment of aceruloplasminemia. *Ann. Neurol.* **41**, 404–407.

Morphy, M. A., Feldman, J. A., and Kilburn, G. (1989). Hallervorden-Spatz disease in a psychiatric setting [see comments]. *J. Clin. Psychiatry* **50**, 66–68.

Mutoh, K., Okuno, T., Ito, M., and Mikawa, H. (1990). Somatosensory evoked potentials in Hallervorden-Spatz-neuroaxonal- dystrophy complex with dorsal column involvement. *Clin. Electroencephalogr.* **21**, 58–66.

Mutoh, K., Okuno, T., Ito, M., Nakano, S., Mikawa, H., Fujisawa, I., and Asato, R. (1988). MR imaging of a group I case of Hallervorden-Spatz disease. *J. Comput. Assist. Tomogr.* **12**, 851–853.

Neumann, M., Adler, S., Schluter, O., Kremmer, E., Benecke, R., and Kretzschmar, H. A. (2000). Alpha-synuclein accumulation in a case of neurodegeneration with brain iron accumulation type 1 (NBIA-1, formerly Hallervorden-Spatz syndrome) with widespread cortical and brainstem-type Lewy bodies. *Acta Neuropathol. (Berlin)* **100**, 568–574.

Newell, F. W., Johnson, R. O. D., and Huttenlocher, P. R. (1979). Pigmentary degeneration of the retina in the Hallervorden-Spatz syndrome. *Am. J. Ophthalmol.* **88**, 467–471.

Newell, K. L., Boyer, P., Gomez-Tortosa, E., Hobbs, W., Hedley-Whyte, E. T., Vonsattel, J. P., and Hyman, B. T. (1999). Alpha-synuclein immunoreactivity is present in axonal swellings in neuroaxonal dystrophy and acute traumatic brain injury. *J. Neuropathol. Exp. Neurol.* **58**, 1263–1268.

Odawara, T., Iseki, E., Yagishita, S., Amano, N., Kosaka, K., Hasegawa, K., Matsuda, Y., and Kowa, H. (1992). An autopsied case of juvenile parkinsonism and dementia, with a widespread occurrence of Lewy bodies and spheroids. *Clin. Neuropathol.* **11**, 131–134.

Orrell, R. W., Amrolia, P. J., Heald, A., Cleland, P. G., Owen, J. S., Morgan-Hughes, J. A., Harding, A. E., and Marsden, C. D. (1995). Acanthocytosis, retinitis pigmentosa, and pallidal degeneration: a report of three patients, including the second reported case with hypoprebetalipoproteinemia (HARP syndrome). *Neurology* **45**, 487–492.

Park, B. E., Netsky, M. G., and Betsill, W. L., Jr. (1975). Pathogenesis of pigment and spheroid formation in Hallervorden-Spatz syndrome and related disorders. *Neurology* **25**, 1172–1178.

Perry, T. L., Norman, M. G., Yong, V. W., Whiting, S., Crichton, J. U., Hansen, S., and Kish, S. J. (1985). Hallervorden-Spatz disease: cysteine accumulation and cysteine dioxygenase deficiency in the globus pallidus. *Ann. Neurol.* **18**, 482–489.

Pettigrew, A. L., Jackson, L. G., and Ledbetter, D. H. (1991). New X-linked mental retardation disorder with Dandy-Walker malformation, basal ganglia disease, and seizures. *Am. J. Med. Genet.* **38**, 200–207.

Provenzale, J. M., Barboriak, D. P., and Sims, K. (1995). Neuroradiologic findings in fucosidosis, a rare lysosomal storage disease. *AJNR* **16**, 809–813.

Rock, C. O., Calder, R. B., Karim, M. A., and Jackowski, S. (2000). Pantothenate kinase regulation of the intracellular concentration of coenzyme A. *J. Biol. Chem.* **275**, 1377–1383.

Roth, A. M., Hepler, R. S., Mukoyama, M., Cancilla, P. A., and Foos, R. Y. (1971). Pigmentary retinal dystrophy in Hallervorden-Spatz disease: clinicopathological report of a case. *Sur. Ophthalmol.* **16**, 24–35.

Rutledge, J. N., Hilal, S. K., Silver, A. J., Defendini, R., and Fahn, S. (1988). Magnetic resonance imaging of dystonic states. *Adv. Neurol.* **50**, 265–275.

Saito, Y., Kawai, M., Inoue, K., Sasaki, R., Arai, H., Nanba, E., Kuzuhara, S., Ihara, Y., Kanazawa, I., and Murayama, S. (2000). Widespread expression of alpha-synuclein and tau immunoreactivity in Hallervorden-Spatz syndrome with protracted clinical course. *J. Neurol. Sci.* **177**, 48–59.

Savoiardo, M., Halliday, W. C., Nardocci, N., Strada, L., D'Incerti, L., Angelini, L., Rumi, V., and Tesoro-Tess, J. D. (1993). Hallervorden-Spatz disease: MR and pathologic findings. *AJNR* **14**, 155–162.

Schaffert, D. A., Johnsen, S. D., Johnson, P. C., and Drayer, B. P. (1989). Magnetic resonance imaging in pathologically proven Hallervorden-Spatz disease. *Neurology* **39**, 440–442.

Scheer, P. J., Perz, A., Ebner, F., and Kratky-Dunitz, M. (1988). [Magnetic resonance tomography confirms the diagnosis of Hallervorden-Spatz disease]. *Padiatri. Padol.* **23**, 245–252.

Seibel, M. O., Date, E. S., Zeiner, H., and Schwartz, M. (1993). Rehabilitation of patients with Hallervorden-Spatz syndrome. *Arch. Phys. Med. Rehabil.* **74**, 328–329.

Sethi, K. D., Adams, R. J., Loring, D. W., and el Gammal, T. (1988). Hallervorden-Spatz syndrome: clinical and magnetic resonance imaging correlations. *Ann. Neurol.* **24**, 69–64.

Shevell, M. (1992). Racial hygiene, active euthanasia, and Julius Hallervorden [see comments]. *Neurology* **42**, 2214–2219.

Shevell, M. (1996). Naming of syndromes [letter; comment]. *Lancet* **348**, 1662.

Shevell, M. I., and Peiffer, J. (2001). Julius Hallervorden's wartime activities: implications for science under dictatorship. *Pediatr. Neurol.* **25**, 162–165.

Spatz, H. (1922a). Uber des Eisenmachweiss im Gehirn besonders in Zentren des extra-pyramidalmotorischen Systems. *Z. Gesamte Neurol. Psychiatry* **77**, 261.

Spatz, H. (1922b). Uber stoffwechseleigent Umlichkeiten in den Stammganglienen. *Z. Gesamte Neurol. Psychiatry* **78**, 641.

Sugiyama, H., Hainfellner, J. A., Schmid-Siegel, B., and Budka, H. (1993). Neuroaxonal dystrophy combined with diffuse Lewy body disease in a young adult. *Clin. Neuropathol.* **12**, 147–152.

Swaiman, K. F. (1991). Hallervorden-Spatz syndrome and brain iron metabolism. *Arch. Neurol.* **48**, 1285–1293.

Swaiman, K. F. (2001). Hallervorden-Spatz syndrome. *Pediatr. Neurol.* **25**, 102–108.

Swaiman, K. F., Smith, S. A., Trock, G. L., and Siddiqui, A. R. (1983). Sea-blue histiocytes, lymphocytic cytosomes, movement disorder and 59Fe- uptake in basal ganglia: Hallervorden-Spatz disease or ceroid storage disease with abnormal isotope scan? *Neurology* **33**, 301–305.

Swisher, C. N., Menkes, J. H., Cancilla, P. A., and Dodge, P. R. (1972). Coexistence of Hallervorden-Spatz disease with acanthocytosis. *Trans. Am. Neurol. Assoc.* **97**, 212.

Szanto, J., and Gallyas, F. (1966). A study of iron metabolism in neuropsychiatric patients. Hallervorden-Spatz disease. *Arch. Neurol.* **14**, 438–442.

Tabuchi, M., Mori, E., and Yamadori, A. (1990). [The role of magnetic resonance imaging in the diagnosis of Hallervorden-Spatz disease]. *Rinsho Shinkeigaku* **30**, 972–977.

Tanfani, G., Mascalchi, M., Dal Pozzo, G. C., Taverni, N., Saia, A., and Trevisan, C. (1987). MR imaging in a case of Hallervorden-Spatz disease. *J. Comput. Assist. Tomogr.* **11**, 1057–1058.

Terepolsky, D., Clarke, J. T. R., and Blaser, S. I. (1996). Evolution of the neuroimaging changes in fucosidosis type II. *J. Inherit. Metab. Dis.* **19**, 775–781.

Tripathi, R. C., Tripathi, B. J., Bauserman, S. C., and Park, J. K. (1992). Clinicopathologic correlation and pathogenesis of ocular and central nervous system manifestations in Hallervorden-Spatz syndrome. *Acta Neuropathol.* **83**, 113–119.

Tsukamoto, H., Inui, K., Taniike, M., Nishimoto, J., Midorikawa, M., Yoshimine, T., Kato, A., Ikeda, T., Hayakawa, T., and Okada, S. (1992). A case of Hallervorden-Spatz disease: progressive and intractable dystonia controlled by bilateral thalamotomy. *Brain Dev.* **14**, 269–272.

Tuite, P. J., Provias, J. P., and Lang, A. E. (1996). Atypical dopa responsive parkinsonism in a patient with megalencephaly, midbrain Lewy body disease, and some pathological features of Hallervorden-Spatz disease. *J. Neurol. Neurosurg. Psychiatry* **61**, 523–527.

Vakili, S., Drew, A. L., Von Schuching, S., Becker, D., and Zeman, W. (1977). Hallervorden-Spatz syndrome. *Arch. Neurol.* **34**, 729–738.

Volkl, A., and Ule, G. (1972). Trace elements in human brain: Iron concentrations of 13 brain areas as a function of age. *Neurology* **202**, 449–454.

Wakabayashi, K., Fukushima, T., Koide, R., Horikawa, Y., Hasegawa, M., Watanabe, Y., Noda, T., Eguchi, I., Morita, T., Yoshimoto, M., Iwatsubo, T., and Takahashi, H. (2000). Juvenile-onset generalized neuroaxonal dystrophy (Hallervorden-Spatz disease) with diffuse neurofibrillary and lewy body pathology. *Acta Neuropathol. (Berlin)* **99**, 331–336.

Wakabayashi, K., Yoshimoto, M., Fukushima, T., Koide, R., Horikawa, Y., Morita, T., and Takahashi, H. (1999). Widespread occurrence of alpha-synuclein/NACP-immunoreactive neuronal inclusions in juvenile and adult-onset Hallervorden-Spatz disease with Lewy bodies. *Neuropathol. Appl. Neurobiol.* **25**, 363–368.

Westaway, S. K., Levinson, B., Ching, K. H. L., Zhou, B., Johnson, M. A., Gitschier, J., and Hayflick, S. J. (2002). Molecular diagnosis of pantothenate kinase associated neurodegeneration, submitted.

Williamson, K., Sima, A. A., Curry, B., and Ludwin, S. K. (1982). Neuroaxonal dystrophy in young adults: a clinicopathological study of two unrelated cases. *Ann. Neurol.* **11**, 335–343.

Yang, E. Y., Campbell, A., and Bondy, S. C. (2000). Configuration of thiols dictates their ability to promote iron-induced reactive oxygen species generation. *Redox Rep.* **5**, 371–375.

Yoon, S. J., Koh, Y. H., Floyd, R. A., and Park, J. W. (2000). Copper, zinc superoxide dismutase enhances DNA damage and mutagenicity induced by cysteine/iron. *Mutat. Res.* **448**, 97–104.

Yun, M., Park, C. G., Kim, J. Y., Rock, C. O., Jackowski, S., and Park, H. W. (2000). Structural basis for the feedback regulation of Escherichia coli pantothenate kinase by coenzyme A. *J. Biol. Chem.* **275**, 28,093–28,099.

Zhou, B., Bae, S. K., Malone, A. C., Levinson, B. B., Kuo, Y., Cilio, M. R., Bertini, E., Hayflick, S. J., and Gitschier, J. M. (2001a). hGFRalpha-4: a new member of the GDNF receptor family and a candidate for NBIA. *Pediatr. Neurol.* **25**, 156–161.

Zhou, B., Westaway, S. K., Levinson, B., Johnson, M. A., Gitschier, J., and Hayflick, S. J. (2001b). A novel pantothenate kinase gene (*PANK2*) is defective in Hallervorden-Spatz syndrome. *Nat. Genet.* **28**, 345–349.

Zupanc, M. L., Chun, R. W., and Gilbert-Barness, E. F. (1990). Osmiophilic deposits in cytosomes in Hallervorden-Spatz syndrome. *Pediatr. Neurol.* **6**, 349–352.

# CHAPTER 39

# Genetics of Familial Idiopathic Basal Ganglia Calcification (FIBGC)

MARIA-JESUS SOBRIDO
Department of Neurology
Hospital Universitario de Salamanca
37000 Salamanca, Spain

DANIEL H. GESCHWIND
Department of Neurology
University of California
Los Angeles, California 90095

I. Clinical Phenotype
II. Genetics
  A. Linkage Studies and Candidate Gene Analysis
III. Diagnostic and Ancillary Tests
  A. DNA Tests
  B. Other Laboratory Tests; Differential Diagnosis
IV. Neuroimaging
V. Neuropathology
VI. Pathogenesis and Models of Disease
VII. Genotype-Phenotype Correlations
VIII. Treatment
  References

The syndrome of familial, nonarteriosclerotic, calcification of the basal ganglia not attributable to a metabolic, infectious, or toxic cause is a rare entity. About thirty families have been reported in the world literature that we are aware of, most of them with an autosomal dominant mode of inheritance (Foley, 1951; Bowman, 1954; Bruyn et al., 1964; Moskowitz et al., 1971; Boller et al., 1977; Ellie et al., 1989; Kobari et al., 1997). The clinical presentation is usually that of an adult-onset neurodegenerative condition with variable combination of neuropsychiatric and movement disorders. Characteristic calcium deposits in the basal ganglia and other brain structures can be visualized on neuroimaging and are the hallmark of the disease. Whether familial and sporadic cases of idiopathic calcification of the basal ganglia represent the same disease entity is currently unknown. Theodore Fahr's paper (1930) was not the first to describe the neurologic manifestations of basal ganglia calcification and did not describe the familial form of the disease. We thus support the view proposed by others that the eponym Fahr disease, frequently applied to designate any case of basal ganglia calcification of unknown etiology —whether familial or sporadic—is misleading and should be avoided. Although extra-striatopallidal calcifications often occur in the dentate gyrus and other brain areas, involvement of the basal ganglia is a constant feature of this condition. We therefore prefer the terms idiopathic basal ganglia calcification (IBGC), for the nonfamilial form and familial idiopathic basal ganglia calcification (FIBGC), for the familial form, over more complex names that have been used in the literature (Friede et al., 1961, Babbitt et al., 1969; Duckett et al., 1977; Manyam et al., 1992). While the first genetic locus for FIBGC was recently reported on chromosome 14q (*IBGC1*) (Geschwind et al., 1999), the causal gene or genes have yet to be identified. Exclusion of linkage to the *IBGC1* locus in at least three additional kindreds indicates that FIBGC is genetically heterogeneous (Sobrido, et al., 2000b; Brodaty et al., 2002). Future genetic research will likely help clarify the nosology of this group of disorders of the basal ganglia.

## I. CLINICAL PHENOTYPE

The first clinical manifestations of FIBGC typically present between the third and fifth decades after a normal physical and mental development in early years. Signs and symptoms are limited to the nervous system and may include a variable combination of neuropsychiatric, extrapyramidal, brainstem, and cerebellar signs (Foley, 1951; Moskowitz et al., 1971; Boller et al., 1977; Manyam et al., 1992; Aiello et al., 1981; Ellie et al., 1989). The range of movement disorders is broad, including parkinsonian features, choreoathetosis, muscle spasms, myoclonia, dyskinesia, and tremor. Some families have prominent dystonia (Caraceni et al., 1974; Larsen et al., 1985). Dysarthria, dysphagia, and cerebellar ataxia are also frequent and in some cases imbalance and other cerebellar disturbances predominate (Aiello et al., 1981). Seizures may be present, as well as headache (Ellie et al., 1989; Geschwind et al., 1999), urinary incontinence, and impotence (Manyam et al., 1992; Tokoro et al., 1993). Strength and sensation are generally preserved and there is no autonomic disorder. The association of basal ganglia calcification and motor neuron disease has been reported in a sporadic case (Eleopra et al., 1991). Cognitive and psychiatric symptoms are common and vary from mild concentration difficulties, changes in personality, mood, or behavior to psychosis, language impairment, and frontal-subcortical dysfunction, which can progress to frank dementia (Francis, 1979; Cummings et al., 1983; Trautner et al., 1988; Geschwind et al., 1999). The development of neuropsychiatric symptoms seems to be more likely in patients who become affected early in adulthood (Cummings et al., 1983). In a single multigenerational FIBGC kindred signs of frontal executive dysfunction were observed in the majority of affected subjects, suggesting that a simple neurobehavioral examination may be a sensitive diagnostic tool (Geschwind et al., 1999). The presence of subtle cognitive abnormalities needs to be explored more systematically in families that are currently under study. General physical examination is normal and there is no evidence of endocrine or systemic disorder. No specific blood or CSF abnormalities can be detected.

In summary, the onset age and spectrum of clinical manifestations is highly heterogeneous, both between and within families. Psychiatric manifestations, dystonia, and parkinsonism have been observed in individual affected members of one large family linked to chromosome 14q (Geschwind et al., 1999), again demonstrating the variable expressivity of this disorder, even when related to a single locus.

## II. GENETICS

Most of the published FIBGC kindreds exhibit an autosomal dominant inheritance. However, the transmitting parent may be clinically asymptomatic or may develop disease manifestations at a later age and/or with less severe expressivity. Thus, affected individuals may exist in whom the disorder has gone unrecognized. Several kindreds with recessive inheritance have also been reported (Fritzsche, 1935; Matthews, 1957; Bruyn et al., 1964; Smits et al., 1983). However, many of these reports predated the identification of parathyroid hormone or related metabolic abnormalities, which may be recessively inherited. Therefore the possibility that some of these cases actually represent variants of parathyroid diseases or other genetically distinct entities cannot be completely ruled out.

Although penetrance of CT-visible calcifications may vary between families, analysis of reported pedigrees indicates about 95% penetrance of basal ganglia calcification by age ≥50 years. Onset age is variable between and within families and the penetrance of overt clinical symptoms is lower than that of radiologic abnormalities. The observation of an earlier appearance of symptoms over successive generations in some FIBGC kindreds raises the possibility of anticipation, although this has not been widely reported (Geschwind et al., 1999).

### A. Linkage Studies and Candidate Gene Analysis

The first chromosomal locus for familial idiopathic basal ganglia calcification (*IBGC1*) was identified on chromosome 14 by linkage analysis of a large multigenerational pedigree, the FY1 family (Geschwind et al., 1999). The minimal affected haplotype spans a 17.1-cM region between D14s70 and D14s66. The responsible gene or genes are currently unknown. Mutations in the coding regions and promoters of several candidate genes, including *GTP cyclohydrolase I* (which causes dopa-responsive dystonia) *HSPA2*, and *BMP4* (bone morphogenetic protein 4) have been ruled out by sequencing (Sobrido et al., unpublished data; Geschwind et al., 1999). The observation of anticipation in some FIBGC families suggests that a triplet repeat expansion might be the mutational basis for this condition. Thirteen (CAG)n repeats with n >5 repeats were identified within the *IBGC1* region and analyzed in affected members of the FY1 family without evidence of expanded alleles (Sobrido et al., 2000a). The molecular pathogenesis of the disorder is, therefore, still unknown.

Linkage and haplotype analyses in three additional pedigrees performed in our lab (Sobrido et al., 2000b) as well as in several other kindreds by other laboratories ruled out *IBGC1* as the causative locus (Brodaty et al., 2002). This demonstrates that genetic heterogeneity exists within the autosomal dominant form of FIBGC providing evidence for at least one additional autosomal locus. The clinical presentation of the FY1 family overlaps that in other FIBGC kindreds, suggesting that indistinguishable clinical syndromes can be caused by mutations in different genes. It is likely that further linkage analysis in these and other families will eventually result in the identification of additional IBGC loci and, ultimately, of the responsible genes.

## III. DIAGNOSTIC AND ANCILLARY TESTS

### A. DNA Tests

Direct genetic analysis of affected FIBGC individuals is not available, since the responsible gene or genes are unknown. Linkage analysis may be pursued in informative families on a research basis.

## B. Other Laboratory Tests, Differential Diagnosis

Calcium deposits in the brain can be associated with a wide variety of disorders, both sporadic and familial (Table 39.1); (Lowenthal and Bruyn, 1968). Hypoparathyroidism (HP), idiopathic or postsurgical, is the most common cause of symmetric calcification of the basal ganglia (Bennett et al., 1959; Muenter and Whisnant, 1968; Illum and Dupont, 1985). Care should also be taken to rule out pseudohypoparathyroidism (PHP) and pseudo-pseudohypoparathyroidism (PPHP) since occasional variants of these disorders may present with few or no associated somatic abnormalities. In addition, these parathyroid defects are often familial and

TABLE 39.1 Differential Diagnosis of FIBGC

| Disorder | Main clinical features | Laboratory tests |
|---|---|---|
| Parathyroid disorders | | |
| HP | Onset in childhood or adolescence | ↓PTH, ↓Ca, ↑P |
| | Tetany, muscle weakness, mental retardation, cataracts, skin changes. | |
| PHP | Childhood onset, clinical manifestations similar to HP + Albright's osteodystrophy (short stature, obesity, short metacarpals or metatarsals, soft tissue calcification) | ↑PTH, ↓Ca, ↑P, reduced urinary phosphate and cAMP after exogenous PTH |
| PPHP | Albright's osteodystrophy | Normal serum Ca, P, and PTH. Normal PTH responsiveness. |
| Hyperparathyroidism | Lethargy, polyuria, polydipsia, constipation, vomitting | ↑PTH, ↑Ca, ↓P |
| Other Metabolic disorders | | |
| Wilson disease | Hepatic/neurological/ psychiatric | ↓sCu and ceruloplasmin, ↑uCu |
| Renal tubular acidosis | Bone defects, metabolic acidosis | Blood gases, electrolytes urine, acidification test |
| Mitochondrial disorders (KSS, MELAS, other) | Ophthalmoplegia, myopathy, lactic acidosis, sensorineural deafness, optic atrophy, retinopathy, diabetes mellitus, encephalopathy, seizures, ataxia, stroke-like episodes, etc. | Electrophysiological studies, muscle/nerve biopsy, DNA testing |
| Infectious diseases | | |
| Intrauterine/perinatal (toxoplasmosis, rubella, CMV, HSV) | Onset soon after birth, chorioretinitis, microcephaly, mental retardation | Serological tests |
| Postnatal/adult (AIDS, CMV, EBV, toxoplasmosis, neurocysticercosis, TB, brucellosis) | Associated symptoms/signs Usually asymmetric distribution of calcifications | Bacteriological/serological tests |
| Inherited Congenital or early-onset syndromes | | |
| Cockayne syndrome | Leukodystrophy, growth failure, characteristic facies, deafness, retinitis pigmentosa | DNA repair test |
| Aicardi-Goutieres syndrome | Leukodystrophy, microcephaly, CSF lymphocytosis | |
| Tuberous sclerosis | Abnormalities of skin, brain, kidney, heart | TSC1, TSC2, TSC3 mutations |
| Down syndrome | Characteristic phenotype | Chromosomal analysis |
| Other conditions | | |
| DRPLA | Ataxia, involuntary movements, seizures, psychosis, dementia | CAG repeat expansion |
| Traumatic/toxic insult to the brain (perinatal anoxia, carbon monoxide intoxication, ionizing radiation, etc.) | History of past exposure | |
| Systemic lupus erythematosus (SLE) | Additional systemic involvement | SLE autoantibodies |

*Note:* HP—hypoparathyroidism; PHP—pseudohypoparathyroidism; PPHP—pseuso-pseudohypoparathyroidism; PTH—parathyroid hormone; Ca—calcium; P—phosphorus; sCu—serum copper; uCu—urinary copper.

the co-existence of PHP and PPHP in the same family is not rare (Mann et al., 1962). Patients with hyperparathyroidism may also exhibit intracerebral calcifications (Margolis et al., 1980). Thus, in addition to routine hematologic and biochemical parameters, laboratory investigations of patients with basal ganglia calcification should include serum calcium, phosphorus, magnesium, alkaline phosphatase, calcitonin, and parathyroid hormone (PTH). Because the main concern is that of a parathyroid disorder, urinary excretion of phosphorus and cyclic-AMP should be determined at baseline and after PTH infusion (Ellsworth-Howard test). PTH responsiveness is within normal limits in individuals with FIBGC (Moskowitz et al., 1971; Boller et al., 1977; Ellie et al., 1989). Basal ganglia calcification can be occasionally encountered in Wilson's disease, although the pattern and extension of calcifications in Wilson's disease often differs from those in FIBGC (Harik and Post, 1981). However, because clinical and neuroimaging findings of both conditions can be somewhat overlapping and given the possibility of an effective therapy for Wilson's disease, it should always be included in the differential diagnosis.

Another important group of disorders to consider in the differential diagnosis of IBGC are mitochondrial cytopathies, since mineral deposits in the basal ganglia and other brain structures are frequently recognized in these conditions (Markesbery, 1979; Robertson et al., 1979). Mitochondrial disorders may present at any age and frequently exhibit multiple organ involvement, including the central and peripheral nervous systems. Neurophysiologic studies, muscle biopsy and molecular genetic testing of blood or muscle DNA may be necessary to investigate the possibility of a mitochondrial disease. Additional investigations on blood serum or CSF, as well as neurophysiologic tests may be necessary in order to rule out specific metabolic, immunological or infectious conditions (see Table 39.1). Electromyogram and nerve conduction studies may discover a latent tetany, myopathic changes, or polyneuropathy. These abnormalities, as well as alterations in somatosensory, auditory, or visual evoked responses should prompt the consideration of parathyroid dysfunction, mitochondrial disease, or other disorders associated with brain calcifications. Active viral, bacterial, and parasitic CNS infections should also be considered in the differential diagnosis, especially in apparently sporadic cases (Mousa et al., 1987; Morita et al., 1998). Calcification of the basal ganglia and cerebral white matter are frequently encountered in children with AIDS (Belman et al., 1986).

In summary, the diagnosis of IBGC can be made in those cases where bilateral calcification of the basal ganglia—with or without additional affected brain areas—are present in an individual with a movement and/or neuropsychiatric disorder, provided that metabolic, infectious, toxic, or traumatic causes have been ruled out. The diagnosis of FIBGC is appropriate if a positive family history of a similar disorder is documented. Congenital or early-onset findings, mental retardation, or systemic involvement should suggest the possibility of an alternative diagnosis. Finally, the sole presence of pallidal calcifications may be an incidental finding in about 0.5–1.5% of brain CT scans of otherwise asymptomatic, aged individuals (Koller et al., 1979; Sachs et al., 1979; Murphy, 1979; Harrington et al., 1981). These calcifications are generally small, confined to the globus pallidus, and do not have associated clinical findings.

## IV. NEUROIMAGING

The hallmark of FIBGC is the presence of characteristic calcium deposits in the basal ganglia and other brain areas, which can be visualized on brain CT scan (Aiello et al., 1981; Kazis, 1985). Neither the localization nor the extent of calcium deposits distinguishes the idiopathic from symptomatic (secondary) forms of basal ganglia calcification. The most commonly affected structure is the lenticular nucleus, especially the internal globus pallidus. Also frequent are calcified lesions of the caudate, putamen, thalamus, and dentate nuclei (Fig. 39.1). Calcifications may also be present in the cerebellar gyri, brainstem, and centrum semiovale white matter (Ellie et al., 1989; Manyam et al., 1992). On magnetic resonance imaging (MRI), calcium deposits typically appear as irregular areas of low or no signal on T2-weighted images and as a low- or high-intensity signal on T1-weighted images (Scotti et al., 1985; Avrahami et al., 1994; Ellie et al., 1989). Skull X-ray has been largely replaced by CT scan to detect calcified lesions affecting cerebral structures. Calcifications with symmetrical distribution can be evident in the region of the basal nuclei. Other, curved, calcifications may be located

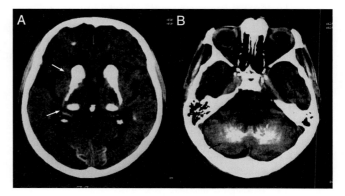

FIGURE 39.1 Brain CT scan of a patient with FIBGC. (A) Symmetrical calcification of the caudate nuclei and pulvinar regions of the thalami (arrows). (B) Bilateral calcification of the dentate nuclei in the same case.

more peripherally in the cerebral hemispheres (Aiello *et al.*, 1981). In addition, plain radiographs may be useful to rule out associated bone abnormalities suggestive of a metabolic disorder or other diagnoses.

There is no reliable correlation between extent of calcification, onset age, and severity of clinical manifestations. Asymptomatic individuals with calcifications on CT scan are frequently encountered within FIBGC kindreds, especially among the younger generations (Ellie *et al.*, 1989; Manyam *et al.*, 1992). Although most patients with calcifications eventually develop neurologic dysfunction, the type or severity of clinical symptoms cannot be predicted from the pattern of calcification. On the other hand, the observation of symptoms in FIBGC family members initially lacking visible calcium deposits but who later develop calcifications (Larsen *et al.*, 1985; Geschwind *et al.*, 1999), suggests that clinical manifestations are not entirely due to macroscopic calcification and further highlights the uncertain relationship between mineral deposits and neurological dysfunction.

## V. NEUROPATHOLOGY

Few autopsy cases of FIBGC have been published and most of the available pathologic analyses refer to nonfamilial cases of IBGC. Gross pathological examination generally shows a granular material and solid nodules accumulated in the striatum, internal capsules, and cerebellum. The calcification may also affect the cerebral white matter, with mineralized areas mostly at the gray-white junction. A mild lobar atrophy and ventricle enlargement are often encountered. Histological lesions consist of concentric calcium deposits within the walls of small and medium size arteries and, less frequently, veins (Norman and Urich, 1960; Löwenthal and Bruyn, 1968; Cervós-Navarro and Urich, 1995); (Fig. 39.2) Droplet calcifications can be observed along capillaries, which may eventually coalesce and obliterate the vessel. Mineralized aggregates are frequently present along the depths of sulci. Dense calcifications can also be observed within the deep white matter and nuclei, with surrounding vacuolization of the neuropil, neuronal loss, and reactive gliosis. On electron microscopy the mineral deposits appear as amorphous or crystal material surrounded by a basal membrane. Calcium granules may be seen within the cytoplasm of neuronal and glial cells (Guseo *et al.*, 1975; Cervós-Navarro and Urich, 1995).

## VI. PATHOGENESIS AND MODELS OF DISEASE

The primary pathophysiologic mechanisms of FIBGC have not been determined. The wide variety of insults that

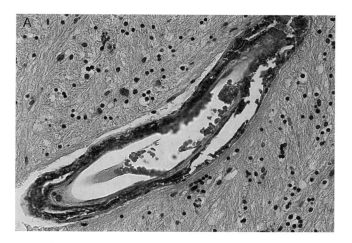

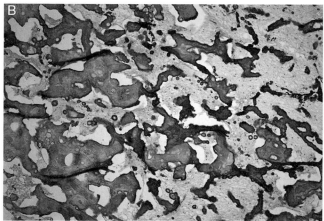

FIGURE 39.2 Histological sections of an autopsied case of FIBGC. (A) A medium size cerebral vessel with extensive medial calcification. (B) Dense mineral deposits replace the deep cerebellar white matter and nuclei. Hematoxylin-eosin stain, magnification = ×250.

may result in cerebral calcified lesions and lack of a good correlation between the degree of calcification and clinical dysfunction suggest that calcium deposits may be a secondary, relatively unspecific change rather than the primary pathogenic process. An undefined molecular mechanism might increase susceptibility to the brain mineralization observed in diverse disorders, with special vulnerability of the basal ganglia and other cerebral structures. A role for local inflammation and cytokines has been hypothesized for the basal ganglia calcification observed in cases of chronic CNS infection and AIDS (Morita *et al.*, 1998). Alternatively, a defect of mineral homeostasis and failure to maintain intracellular calcium levels might be the basic defect (McBurney and Neering, 1987; Beall *et al.*, 1989). The fact that parathyroid disorders are among the most frequent causes of basal ganglia calcification demonstrates that a disturbance in calcium metabolism may be an important etiological factor in many cases. Adachi and coworkers (1968) provided a histochemical study of a case of familial idiopathic basal ganglia calcification. The mineral deposits consisted mainly of calcium and iron, whereas adjacent areas contained high levels of a non-mineralized acid mucopolysaccharide. The appearances suggested an origin of the abnormal material in the glial cells with extension into the pericellular areas (Adachi *et al.*, 1968). In addition to calcium, iron, aluminum, magnesium, phosphorus, copper, zinc, and trace amounts of other elements were also disclosed in chemical examination of cerebral calcifications (Lowenthal and Bruyyn, 1968; Smeyers-Verbeke *et al.*, 1975; Duckett *et al.*, 1977). Whether the metal composition of the calcification actively contributes to its formation or if these elements have an indirect involvement, secondary to a pathological blood-brain barrier is unclear. Variation in the trace element content between different affected areas of the same brain suggests that the element deposition is not causally implicated in the pathogenesis of this disorder (Bouras et al., 1996). A pathogenic role has also been postulated for local inflammatory factors, disturbance of the blood-brain barrier, and mitochondrial function (Morgante *et al.*, 1986; Robertson *et al.*, 1979). Calcium accumulation can be observed in the mitochondria of cultured neurons under excitotoxic conditions (Dux *et al.*, 1996). Other investigators postulate that the primary defect is likely to be vascular (Norman and Urich, 1960). The fact that calcium deposits seem to be a common finding in the basal ganglia and other susceptible brain regions in response to hypoxia, inflammation, infection, and other processes suggests that mineral deposition is probably not the primary insult, but rather represents a final common phenomenon in the pathological process. Whether calcium deposition is the cause or the consequence of cell death is still unclear. Experimental calcification and neurodegeneration induced by excitotoxic lesions in the rat brain tissue may help understand the pathophysiologic mechanisms of basal ganglia calcification in humans (Korf and Postema, 1984; Stewart *et al.*, 1995; Mahy *et al.*, 1999). Additional investigation is needed to understand the pathway to cellular damage in both idiopathic and secondary forms of basal ganglia calcification. The identification of the genetic basis of FIBGC should provide important clues to the pathogenesis of this group of neurodegenerative disorders.

## VII. GENOTYPE-PHENOTYPE CORRELATIONS

Only one kindred with linkage to the *IBGC1* locus has been identified so far and the causal gene is still unknown. The clinical presentation of the IBGC1 family does not differ significantly from other FIBGC kindreds reported in the literature. Affected members presented variably with parkinsonism, dystonia, frontal-executive dysfunction, and psychosis (Geschwind *et al.*, 1999). Seizures, which have been reported in some families, have not been observed in this large family and, therefore may suggest different etiologies. The identification of families that are not linked to the *IBGC1* locus demonstrated that there are genetically different conditions underlying a similar phenotype (Sobrido *et al.* 2000b), and other loci are yet to be identified. More detailed studies using consistent measures between groups may reveal signficicant genotype-phenotype correlations in the future, but the distinctions between families presented thus far are quite subtle.

## VIII. TREATMENT

Treatment of FIBGC is symptomatic, since no curative therapy is currently available. Management of behavioral and neuropsychiatric manifestations is frequently the main concern and appropriate pharmacological treatment should be instituted to improve anxiety, depression, and other mood disturbances (Trautner *et al.*, 1988). Neuroleptic medication should be used cautiously, since patients may be very sensitive to the extrapyramidal side effects of these drugs (Francis, 1979). The response of parkinsonian features to levodopa therapy is generally poor (Klawans *et al.*, 1976; Berendes and Dörstelmann, 1978). Manyam and collaborators (2001) attributed a positive response to levodopa in an affected individual to the co-existence of basal ganglia calcification and idiopathic Parkinson disease. Symptomatic treatment may alleviate dystonia, chorea, and other involuntary movements. Anecdotally, a therapeutic trial with sodium etidronate in one patient resulted in partial improvement of speech and gait, but not of other neurologic symptoms and no reduction in the amount of calcification was observed (Loeb, 1998).

## References

Adachi, M., Wellman, K. F., and Volk, B. W. (1968). Histochemical studies on the pathogenesis of idiopathic non-arteriosclerotic cerebral calcification. *J. Neuropathol. Exp. Neurol.* **27**, 483–499.

Aiello, U., Bevilacqua, L., Boglium, G., Sanguineti, I., and Tagliabue, M. (1981). A case of Fahr's disease: clinical and CT study. *Ital. J. Neurol. Sci.* **4**, 395–398.

Avrahami, E., Cohn, D.-F., Feibel, M., and Tadmor, R. (1994). MRI demonstration and CT correlation of the brain in patients with idiopathic intracerebral calcification. *J. Neurol.* **241**, 381–384.

Babbitt, D. P., Tang, T., Dobbs, J., and Berk, R. (1969). Idiopathic familial cerebrovascular ferrocalcinosis (Fahr's disease) and review of differential diagnosis of intracranial calcification in children. *Am. J. Roentg. Rad. Ther. Nucl. Med.* **105**, 352–358.

Beall, S. S., Patten, B. M., Mallette, L., and Jankovic, J. (1989). Abnormal systemic metabolism of iron, porphyrin, and calcium in Fahr's disease. *Ann. Neurol.* **26**, 569–575.

Belman, A. L., Lantos, G., Horoupian, D., Novick, B. E., Ultmann, M. H., Dickson, D. W., and Rubinstein, A. (1986). AIDS: calcification of the basal ganglia in infants and children. *Neurology* **36**, 1192–1199.

Bennett, J. C., Maffly, A., and Steinback, H. L. (1959). The significance of bilateral basal ganglia calcification. *Radiology* **72**, 368–378.

Berendes, K., and Dörstelmann, D. (1978). Unsuccessful treatment with levodopa of a parkinsonian patient with calcification of the basal ganglia. *J. Neurol.* **218**, 51–54.

Boller, F., Boller, M., and Gilbert, J. (1977). Familial idiopathic cerebral calcifications. *J. Neurol. Neurosurg. Psychiatry* **40**, 280–285.

Bouras, C., Giannakopoulos, P., Good, P. F., Hsu, A., Hof, P. R., and Perl, D. P. (1996). A laser microprobe mass analysis of trace elements in brain mineralizations and capillaries in Fahr's disease. *Acta Neuropathol.* **92**, 351–357.

Bowman, M. J. (1954). Familial occurrence of "idiopathic" calcification of cerebral capillaries. *Am. J. Pathol.* **30**, 87–98.

Brodaty, H., Mitchell, P., Luscombe, G., Kwok, J.J., Badenhop, R.F., McKenzie, R., Schofield, P.R. (2002). Familial idiopathic basal ganglia calcification (Fahr's disease) without neurological, cognitive and psychiatric symptoms is not linked to the IBGC1 locus on chromosome 14q. *Hum. Genet.* **110**(1), 8–14.

Bruyn, G. W., Bots, G. T., and Staal, A. (1964). Familial bilateral vascular calcification in the central nervous system. *Psychiatr. Neurol. Neurochir.* **67**, 342–376.

Caraceni, T., Broggi, G., and Avanzini, G. (1974). Familial idiopathic basal ganglia calcifications exhibiting "dystonia musculorum deformans" features. Report on two cases. *Eur. Neurol.* **12**, 351–359.

Cervós-Navarro, J., and Urich, H. (1995). Disorders of mineral metabolism. In *Metabolic and Degenerative Diseases of the Central Nervous System. Pathology, Biochemistry and Genetics* (J. Cervós-Navarro, and H. Urich H, eds.), pp. 401–426. Academic Press, San Diego.

Cummings, J. L., Gosenfeld, L. F., Houlihan, J. P., and McCaffrey, T. (1983). Neuropsychiatric disturbances associated with idiopathic calcification of the basal ganglia. *Biol. Psychiatry* **18**, 591–601.

Duckett, S., Galle, P., Escourolle, R., Poirier, J., and Hauw, J. J. (1977). Presence of zinc, aluminium, magnesium in striopallidodentate (SPD) calcifications (Fahr's disease): electron probe study. *Acta Neuropathol.* **38**, 7–10.

Dux, E., Oschlies, U., Uto, A., Kusumoto, M., Siklos, L., Joo, F., and Hossmann, K. A. (1996). Serum prevents glutamate-induced mitochondrial calcium accumulation in primary neuronal cultures. *Acta Neuropathol.* **92**, 264–272.

Eleopra, R., Accurti, I., Neri, W., and Bazzocchi, O. (1991). Unusual case of Fahr syndrome with motoneuron disease. *Ital. J. Neurol. Sci.* **12**, 597–600.

Ellie, E., Julien, J., and Ferrer, X. (1989). Familial idiopathic striopallidodentate calcifications. *Neurology* **39**, 381–385.

Fahr, T. H. (1930). Idiopathische Verkalkung der Hirngefässe. *Zentrabl. Allg. Pathol. Pathol. Anat.* **50**, 129–133.

Foley, J. (1951). Calcification of the corpus striatum and dentate nuclei occurring in a family. *J. Neurol. Neurosurg. Psychiatry* **14**, 253.

Francis, A. F. (1979). Familial basal ganglia calcification and schizophreniform psychosis. *Br. J. Psychiatry* **135**, 360–362.

Friede, R. L., Magee, K. R., and Mack, E. W. (1961). Idiopathic non-arteriosclerotic calcification of cerebral vessels. Fahr's disease- a clinical and histological study. *Arch. Neurol.* **5**, 279–286.

Fritzsche, R. (1935). Eine familiar auftretende Form von Oligophrenie mit roentgenologisch nachweisbaren symmetrischen Kalkablagerungen in Gehirn, besonders in den Stammganglien. *Schweiz. Arch. Neurol. Neurochir. Psychiat.* **35**, 1–29.

Geschwind, D. H, Loginov, M., and Stern, J. (1999). Identification of a locus on chromosome 14q for idiopathic basal ganglia calcification (Fahr disease). *Am. J. Hum. Genet.* **65**, 764–772.

Guseo, A., Boldizsar, R., and Gellert, M. (1975). Elektronenoptische Untersuchungen bei striatodentaler Kalcifikation (Fahr). *Acta Neuropathol.* **31**, 305–313.

Harik, S. I,. and Post, M. J. (1981). Computed tomography in Wilson disease. *Neurology* **31**, 107–110.

Harrington, M. G., Macpherson, P., McIntosh, W. B., Allam, B. F,. and Bone, I. (1981). The significance of the incidental finding of basal ganglia calcification on computed tomography. *J. Neurol. Neurosurg. Psychiatry* **44**, 1168–1170.

Illum, F., and Dupont, E. (1985). Prevalence of CT-detected calcification in the basal ganglia in idiopathic hypoparathyroidism and pseudo-hypoparathyroidism. *Neuroradiology* **27**, 32–37.

Kazis, A. D. (1985). Contribution of CT scan to the diagnosis of Fahr's syndrome. *Acta Neurol. Scand.* **71**, 206–211.

Klawans, H. L., Lupton, M., and Simon, L. (1976). Calcification of the basal ganglia as a cause of levodopa-resistant parkinsonism. *Neurology* **26**, 221–225.

Kobari, M., Nogawa, S., Sugimoto, Y., and Fukuuchi, Y. (1997). Familial idiopathic brain calcifications with autosomal dominant inheritance. *Neurology* **48**, 645–649.

Koller, W. C., Cochran, J. W., and Klawans, H. L. (1979). Calcification of the basal ganglia: computerized tomography and clinical correlation. *Neurology* **29**, 328–333.

Korf, J., and Postema, F. (1984). Regional calcium accumulation and cation shifts in rat brain by kainite. *J. Neurochem.* **43**, 1052–1060.

Larsen, T. A., Dunn, H. G., Jan, J. E., and Calne, D. B. (1985). Dystonia and calcification of the basal ganglia. *Neurology* **35**, 533–537.

Loeb, J. A. (1998). Functional improvement in a patient with cerebral calcinosis using a biphosphonate. *Mov. Disord.* **13**, 345–349.

Lowenthal, A., and Bruyn, G. W. (1968). Calcification on the basal striopallidodentate system. In *Handbook of Clinical Neurology* (P .J. Vinken and G. W. Bruyn, eds.), pp. 703–729. North Holland, Amsterdam.

Mahy, N., Prats, A., Riveros, A., Andres, N., and Bernal, F. (1999). Basal ganglia calcification induced by excitotoxicity: an experimental model characterized by electron microscopy and X-ray microanalysis. *Acta Neuropathol.* **98**, 217–225.

Mann, J. B., Alterman, S. L., and Hills, A. G. (1962). Allbright's hereditary osteodystrophy comprising pseudo-hypoparathyroidism and pseudo-pseudohypoparathyroidism. *Ann. Intern. Med.* **56**, 315–342.

Manyam, B. V., Bhatt, M. H., Moore, W. D., Devleschoward, A. B., Anderson, D. R., and Calne, D. B. (1992). Bilateral striopallidodentate calcinosis: cerebrospinal fluid, imaging and electrophysiological studies. *Ann. Neurol.* **31**, 379–384.

Manyam, B. V., Walters, A. S., Keller, I. A., and Ghobrial, M. (2001). Parkinsonism associated with autosomal dominant bilateral striopallidodentate calcinosis. *Parkinsonism Relat. Disord.* **7**, 289.

Margolis, D., Hammerstad, J., Orwall, E., McClung, M., and Calhoun, D. (1980). Intracranial calcification in hyperparathyroidism associated with gait apraxia and parakinsonism. *Neurology* **30**, 1005–1007.

Markesbery, W. R. (1979). Lactic acidemia, mitochondrial myopathy, and basal ganglia calcification. *Neurology* **29**, 1057–1061.

Matthews, W. B. (1957). Familial calcification of the basal ganglia with response to parathormone. *J. Neurol. Neurosurg. Psychiatry* **20**, 172–177.

McBurney, R. N., and Neering, I. R. (1987). Neuronal calcium homeostasis. *Trends Neurol. Sci.* **10**, 164–169.

Morgante, L., Vita, G., Meduri, M., Di Rosa, A. E., Galatioto, S., Coraci, M.A., and Di Perri, R. (1986). Fahr's syndrome: local inflammatory factors in the pathogenesis of calcification. *J. Neurol.* **233**, 19–22.

Morita, M., Tsuge, I., Matsuoka, H., Ito, Y., Itosu, T., Yamamoto, M., and Morishima, T. (1998). Calcification in the basal ganglia with chronic active Epstein-Barr virus infection. *Neurology* **50**, 1485–1488.

Moskowitz, M.A., Winickoff, R.N., and Heinz, E.R. (1971). Familial calcification of the basal ganglions: a metabolic and genetic study. *N. Engl. J. Med.* **285**, 72–77.

Mousa, A. M., Muhtaseb, S. A., Reddy, R. R., Senthilselvan, A., Al-Mudallal, D. S., and Marafie, A. A. (1987). The high rate of prevalence of CT-detected basal ganglia calcification in neuropsychiatric (CNS) brucellosis. *Acta Neurol. Scand.* **76**, 448–456.

Muenter, M. D., and Whisnant, J.P. (1968). Basal ganglia calcification, hypoparathyroidism and extrapyramidal motor manifestations. *Neurology* **18**, 1075–1083.

Murphy, M.J. (1979). Clinical correlations of CT scan-detected calcifications of the basal ganglia. *Ann. Neurol.* **6**, 507–511.

Norman, R.M., and Urich, R. (1960). The influence of a vascular factor on the distribution of symmetrical cerebral calcification. *J. Neurol. Neurosurg. Psychiatry* **23**, 142–147.

Robertson, W. C., Jr, Viseskul, C., Lee, Y. E., and Lloyd, R. V. (1979). Basal ganglia calcification in Kearns-Sayre syndrome. *Arch. Neurol.* **36**, 711–713.

Sachs, C., Ericson, K., Erasmie, U., and Bergstrom, M. (1979). Incidence of basal ganglia calcifications on computed tomography. *J. Comput. Assist. Tomogr.* **3**, 339–344.

Scotti, G., Scialfa, E., Tampietti, D., and Landoni, L. (1985). MR imaging in Fahr disease. *J. Comput. Assist. Tomogr.* **9**, 790–792.

Smeyers-Verbeke, J., Michotte, Y., Pelsmaeckers, J., Lowenthal, A., Massart, D. L., Dekegel, D., and Karcher, D. (1975). The chemical composition of idiopathic nonarteriosclerotic cerebral calcification. *Neurology* **25**, 48–57.

Smits, M. G., Gabreëls, F. J. M., Thijsen, H. O. M., Lam, R. L., Notermans, S. L. H., ter Haar, B. G. A., and Prick, J. J. (1983). Progressive idiopathic strio-pallido-dentate calcinosis (Fahr's disease) with autosomal recessive inheritance. Report of three siblings. *Eur. Neurol.* **22**, 58–64.

Sobrido, M. J., Baquero, M., and Geschwind, D. H. (2000[a]). Genetics of Fahr's disease: refining the IBGC1 locus and candidate gene analysis. *Am. J. Hum. Genet.* **67**, 316.

Sobrido, M.J., Loginov, M., Baquero, M., Wszoleck, Z.K., Hutton, M., Boller, F., Gilbert, J., and Geschwind, D. H. (2000b). Genetic heterogeneity of idiopathic basal ganglia calcification (Fahr´s disease). *Ann. Neurol.* **48**, 466.

Stewart, G. R., Olney, J. W., Schmidt, R. E., and Wozniak, D. F. (1995). Mineralization of the globus pallidus following excitotoxic lesions of the basal forebrain. *Brain Res.* **695**, 81–87.

Tokoro, K., Chiba, Y., Ohtani, T., Abe, H., and Yagishita, S. (1993). Pineal ganglioglioma in a patient with familial basal ganglia calcification and elevated serum alpha-fetoprotein: case report. *Neurosurgery* **33**, 506–511.

Trautner, R. J., Cummings, J. L., Read, S. L., and Benson, D. F. (1988). Idiopathic basal ganglia calcification and organic mood disorder. *Am. J. Psychiatry* **145**, 350–353.

CHAPTER

# 40

# Myoclonus and Myoclonus-Dystonias

CHRISTINE KLEIN

*Department of Neurology*
*Medical University of Lübeck*
*23538 Lübeck, Germany*

I. Summary
II. Phenotype
   A. Definition, Terminology, and Classification
   B. Epidemiology
   C. Review of Selected Descriptions of Myoclonus and M-D
   D. Summary of the Clinical Features of M-D
   E. Additional Clinical Findings: Psychiatric Abnormalities in M-D
III. Gene(s)
   A. Summary of the Genetic Findings in M-D
   B. Normal Gene and Protein Function
   C. Abnormal Gene and Protein Function
   D. Evidence for Imprinting (SGCE Gene)
IV. Diagnostic and Ancillary Tests
   A. DNA Tests
   B. Other Laboratory Tests
V. Neuroimaging
VI. Neuropathology
VI. Cellular and Animal Models of Disease
VIII. Genotype/Phenotype Correlations
   A. M-D Families with Established Linkage to DYT11 or with Known SGCE Mutations
   B. M-D Family with a Mutation in the DYT1 (TOR1A) Gene
   C. M-D Family with a Mutation in the D2 Dopamine Receptor Gene
   D. M-D Families in whom Linkage to a Known M-D Gene Locus was Excluded
   E. M-D Families and Index Cases for whom Genetic Data are Currently Unavailable or Incomplete
   F. Sporadic and Atypical M-D Cases
   G. Psychiatric Disorders in M-D
   H. M-D Diagnostic Criteria Revisited
I. Differential Diagnosis
IX. Treatment
   Acknowledgments
   References

## I. SUMMARY

Myoclonus is characterized by rapid muscle jerks, while dystonia is defined as sustained twisting and repetitive movements, resulting in abnormal postures. In myoclonus-dystonia (M-D), a predominantly myoclonic syndrome is combined with dystonic features. Linkage of autosomal dominantly inherited M-D has been demonstrated to a locus on chromosome 7q, designated DYT11, followed by the recent identification of mutations in the ε-sarcoglycan gene (SGCE). The diagnostic criteria for M-D have been modified based on a review of 19 families with genetically proven M-D: (1) onset of myoclonus usually in the first or second decade of life; dystonic features are observed in more than half of the affected in addition to myoclonus and are rarely the only disease manifestation; (2) males and females equally affected; (3) a relatively benign course, often variable but compatible with an active life of normal span in most cases; (4) autosomal dominant mode of inheritance with variable severity, and incomplete penetrance which is dependent on the parental origin of the disease allele; affected individuals usually inherit the disease from their father; (5) absence of seizures, dementia, gross ataxia, and other neurological deficits; and (6) normal EEG, normal SSEP, normal results of neuroimaging studies. Two optional criteria include: (1) response of symptoms to alcohol; and (2) various personality disorders and psychiatric disturbances.

Two single families with M-D showed a mutation in the DYT1 gene (usually associated with early-onset torsion dystonia) and in the D2 dopamine receptor gene, respectively, suggesting some degree of genetic heterogeneity in M-D. However, SGCE clearly represents the major M-D gene. The function of its encoded protein, e-sarcoglycan, is largely unknown. It is a member of the sarcoglycan family, usually involved in muscular dystrophies, while M-D is a nondegenerative disorder of the central nervous system.

## II. PHENOTYPE

### A. Definition, Terminology, and Classification

#### 1. Definition of Myoclonus, Dystonia, and M-D

Myoclonus is a rapid, brief contraction ("fast lightning jerk") of one muscle or a group of muscles which may or may not result in observable movement at a joint. The movements can involve any muscle group but most commonly occur in proximal muscles of the limbs, and in the neck and trunk muscles (Daube, 1966). Dystonia, on the other hand, is characterized by sustained twisting and repetitive movements, resulting in abnormal postures (Fahn *et al.*, 1998). Different forms of dystonia classified based on the distribution of symptoms include focal, multifocal, segmental, hemidystonia, or generalized dystonia. Myoclonus and dystonia may occur together in several diseases (Quinn, 1996). Some patients with primary torsion dystonia display fast dystonic movements which may resemble myoclonus, and occasional myoclonic jerks may occur superimposed on dystonic movements (Obeso, 1983). Conversely, myoclonus may be associated with dystonic features and is then named myoclonus-dystonia (M-D); (see Section II.A.2). Myoclonus or dystonia, as well as myoclonus and dystonia in combination are considered primary if they are the only neurological feature, and the etiology is unknown or genetic.

Although several similar descriptions had appeared in the literature before 1967, Mahloudji and Pikielny were the first to set forth diagnostic criteria for M-D which they called "hereditary essential myoclonus":

1. Onset of myoclonus in the first or second decade
2. Males and females equally affected
3. A benign course, often variable but compatible with an active life of normal span
4. Dominant mode of inheritance with variable severity
5. Absence of seizures, dementia, gross ataxia, and other neurological deficits
6. Normal EEG

Three of these criteria (criteria 1, 4, and 6) have recently been modified by Gasser (1998) based on his review of the literature and on observations in several M-D families followed by himself:

1. Onset of myoclonus *usually* in the first or second decade of life; *mild dystonic features are observed in some affecteds in addition to myoclonus and may rarely be the only manifestation of the disorder*
4. Autosomal-dominant mode of inheritance with variable severity, *and incomplete penetrance*
6. Normal EEG, *normal SSEP*

These modified criteria by Mahloudji and Pikielny (1967); (Gasser, 1998) have been considered the current clinical diagnostic criteria for M-D. A patient with clinically typical M-D is shown in Fig. 40.1.

#### 2. Terminology

A large variety of different terms have been used in the past to describe M-D which has caused confusion with respect to the correct classification of patients with myoclonus, and particularly of patients with a combination of myoclonus and dystonia (Quinn *et al.*, 1988; Gasser, 1998). These terms include "paramyoclonus multiplex" (Biemond, 1963); "hereditary essential myoclonus" (Daube and Madison, 1966; Mahloudji and Pikielny, 1967; Fahn and Sjaastad, 1991), "familial essential myoclonus" (Korten *et al.*, 1974), "essential familial myoclonus" (Przuntek and Muhr, 1983), "dominantly inherited myoclonic dystonia with dramatic response to alcohol" (Quinn and Marsden, 1984), "hereditary myoclonic dystonia" (Quinn *et al.*, 1988; Quinn, 1996); "myoclonus-dystonia" (Lang, 1997); and "inherited myoclonus-dystonia syndrome" (Gasser, 1998). In their excellent review on the classification and terminology of the combination of muscle jerks and dystonia, Quinn *et al.* (1988) arrived at the conclusion that "a familial syndrome of isolated myoclonus is rare ... and ... is difficult to distinguish from hereditary myoclonic dystonia." Indeed, most cases of essential myoclonus (without dystonia, frequently with a later age of onset, and not responsive to alcohol) are not hereditary (Aigner and Mulder, 1960; Bressman and Fahn, 1986), suggesting a heterogeneity between hereditary and apparently nongenetic forms.

In a commentary on a review by Quinn (1996) on the differentiation between myoclonus and myoclonus associated with dystonia, Lang proposed to use a hyphenated combination of the two terms "myoclonus" and "dystonia." The order of presentation was used to emphasize the more constant of the two features, i.e., "dystonia-myoclonus" describes dystonia with some myoclonic jerks, while "myoclonus-dystonia" was suggested as the appropriate term for a predominantly myoclonic syndrome associated with dystonia (Quinn, 1996; Lang, 1997). Accordingly, the

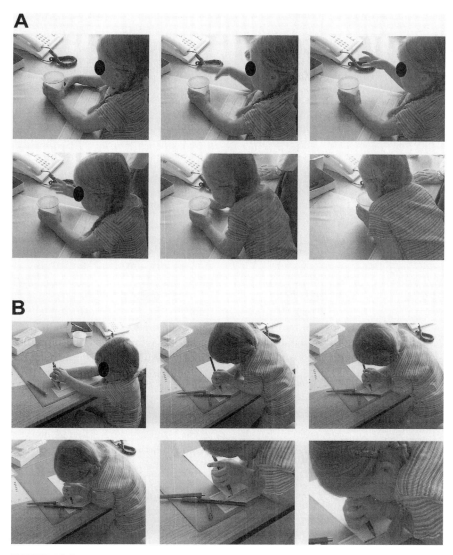

**FIGURE 40.1** Four-year-old girl with clinically typical M-D and a proven SGCE mutation. (A) Attempt to drink from a cup elicits myoclonus predominantly in the neck and arms. Due to the myoclonus and dystonia the patient is unable to lift the cup to the mouth. (B) Drawing an object provokes writer's cramp and myoclonus of the upper body half.

term myoclonus-dystonia shall be used in this review for all cases of myoclonus associated with dystonia. However, it has to be noted that in several M-D families, affected individuals may display myoclonus or dystonia as the sole symptom of the disease.

## 3. Classification

Recently, Fahn *et al.* (1988) have proposed a classification of dystonias that primarily relies on etiological and clinical features but begins to take genetics into account. It contains the following subgroups: primary (dystonia only), dystonia-plus (i.e., dystonia with parkinsonism, dystonia with myoclonic jerks), secondary (due to an environmental insult), and heredodegenerative (hereditary dystonia as the prominent clinical feature, secondary to neurodegeneration). According to this categorization, M-D falls into the dystonia-plus group. Owing to the advances in the molecular genetics of the dystonias in the past few years, a genetically based categorization has recently been introduced (Müller *et al.*, 1998; Klein *et al.*, 1999a). Currently, at least 13 different types of dystonia can be distinguished on a genetic basis, which are designated dystonia (DYT) 1–13 (Müller *et al.*, 1998; Klein *et al.*, 1999a; Valente *et al.*, 2001); M-D linked to chromosome 7q is synonymous with DYT11 dystonia (OMIM 159900).

Finally, based on the recent identification of the ε-sarcoglycan gene as the major M-D gene (Zimprich *et al.*, 2001), M-D may be grouped under the sarcoglycanopathies.

## B. Epidemiology

Little is known about the incidence, prevalence, and distribution of myoclonus and M-D in the general population. Based on a descriptive study with case ascertainment through a records-linkage system in Olmstedt County, Minnesota, the average annual incidence rate of pathologic and persistent myoclonus was calculated to be 1.3 cases per 100,000 person-years. The rate was higher in men and increased with age. The lifetime prevalence of myoclonus was estimated at 8.6 cases per 100,000 population and again increased with advancing age. However, essential myoclonus accounted for only 11% of the cases, with symptomatic myoclonus being by far the most common etiologic category. No case or family with M-D was ascertained through this study (Caviness *et al.*, 1999). However, familial M-D may be much more common than previously thought (Quinn, 1996; Caviness *et al.*, 1999) and underdiagnosed for a number of reasons:

1. The clinical picture is often mild and many patients may not seek medical advice.
2. The condition is not very well known, neither among pediatricians, nor among neurologists which may prevent establishment of the correct diagnosis.
3. The penetrance of M-D is reduced and, particularly in small families, the condition may appear sporadic rather than familial (Daube and Madison, 1966).
4. Because of the reduced penetrance, disease manifestation can skip several generations (for example, family MD2; (Zimprich *et al.*, 2001), again leading to pseudo-sporadic appearance if only a partial pedigree is available.
5. The frequency of *de novo* mutations in the e-sarcoglycan gene remains to be established. As shown for example in DYT1 early-onset torsion dystonia, *de novo* GAG deletions may erroneously suggest sporadic occurrence of the dystonia (Klein *et al.*, 1998).

## C. Review of Selected Descriptions of Myoclonus and M-D

### 1. Early History (From Friedreich, 1881 to Mahljoudi and Pikielny, 1967)

The history of myoclonus and M-D has been the subject of several previous reviews (for example, (Aigner and Mulder, 1960; Hallett, 1986; Quinn *et al.*, 1988, 1996; Gasser, 1998). In the present review on M-D, special emphasis shall be put on a selection of some of the seminal historical papers on the condition.

Friedreich coined the term "Paramyoklonus multiplex," later abbreviated as "myoclonus," to differentiate the "multiple symmetric muscle clonus" he had observed in a patient of his from all other types of "motor cramps" known to him at the time (Friedreich, 1881). He chose the prefix para to indicate the bilateral occurrence of the myoclonus and the suffix multiplex to refer to the involvement of multiple body sites. Friedreich's patient was a 50-year-old comb maker who was hospitalized because of acute pneumonia. Unrelated to the pneumonia, the patient reported a five-year history of a "peculiar affection of several muscles of the upper and lower extremities" which had first started two weeks after an episode of severe fright when he had suffered a near accident with a circular saw, however, without being hurt or injured in any way. Short, quick muscle contractions occurred at short intervals and exclusively involved the arms and thighs, leaving the truncal and facial muscles unaffected. Only some of the contractions resulted in a passive movement, while active movements of the affected limb completely abolished the involuntary muscle contractions. The condition was perfectly symmetric with the exact same muscles involved on both sides. The remainder of the neurological and psychiatric examination was normal with the exception of very brisk patellar reflexes bilaterally. Two months later the jerks disappeared completely after the application of electric therapy ("galvanizations"). However, Friedreich considered it to be a spontaneous remission of the condition which he thought arose in the spinal cord (Friedreich, 1881). Thus, Friedreich's well-described case provided the term myoclonus but is clearly different from classic M-D in many ways. It is unusual for M-D to be entirely symmetric and to exclusively involve the upper and lower limbs with all axial muscles spared, there is no description of dystonia in Friedreich's case, and the myoclonus resolved with action, rather than being elicited or worsened by active movements. Also, there was no family history, and a sudden complete remission is rare in M-D. In retrospect, it is difficult to establish what condition Friedreich's patient exactly suffered from; it may have been a form of "essential myoclonus" (Hallett, 1986).

Dawidenkow, a Russian neurologist in Moscow published, in German, the cases of a brother and sister with a condition he called "myoklonische Dystonie"—myoclonic dystonia (Dawidenkow, 1926). There were no other affected family members in four generations in his family, which led him to consider the mode of inheritance to be recessive. Both siblings developed severe axial dystonia at around 40 years of age. Slow tonic movements were interrupted by short, clonic jerks, for example, of the head, and improved when the patient actively moved his upper extremities. The patient drank moderate amounts of alcohol but no effect was reported on the involuntary movements. The sister had a very similar clinical picture, however, the disease completely resolved after one year. Dawidenkow concluded that his patients suffered from a "condition sui generis," i.e. myoclonic dystonia (Dawidenkow, 1926). Even though Dawidenkow introduced this term which has later been used to describe M-D for many years, there are several features of his patients' condition pointing away from a diagnosis of classic M-D, and the siblings most likely suffered from severe segmental dystonia with superimposed dystonic tremor or myoclonic jerks.

In 1960, Aigner and Mulder reviewed the subject of myoclonus and reinvestigated 94 of their own cases with various forms of myoclonus who were categorized as follows:

I. Myoclonus, seizures, and objective evidence of neurologic or mental deficits or both (n = 26)
II. Myoclonus and seizures without objective evidence of neurologic or mental deficits or both (n = 45)
III. Myoclonus alone, without evidence of seizures and neurologic or mental deficits or both (19 patients)
IV. Myoclonus and a neurologic or mental deficit but no seizures (4 patients); (Aigner and Mulder, 1960)

According to this classification, patients with M-D would fall into the third category. Average age of onset in this group (12 male, 7 female) was 20 years with a range of 4–86 years. The course of the condition was stationary and three patients even described improvement of the myoclonus. One patient reported a positive family history compatible with autosomal dominant inheritance with his father having been affected with myoclonus. Movement and nervousness increased the myoclonus in most patients.

Six years later, Daube and Peters described two families with autosomal dominant inheritance of "benign hereditary myoclonus." These authors were also the first to report alleviation of symptoms by alcohol. Case 3 of Fam. B drank moderate amounts of beer that provided some symptomatic relief (Daube and Madison, 1966). Although no dystonic signs were explicitly mentioned in these patients, "difficulty with handwriting" and "slowness of movements" (Daube and Madison, 1966) may be interpreted as dystonia, and it appears likely that these two families may have had M-D according to the current definition.

Based on their careful description of a family with what they called hereditary essential myoclonus and a review of the literature, Mahljoudji and Pikielny (1967) suggested dropping the term paramyoclonus multiplex and proposed the above-mentioned six diagnostic criteria for the condition. Again, dystonia was not regarded as part of the clinical syndrome. However, "a tendency to turn his head to the right" and "crooked walking and leaning to the right" are most likely descriptions of dystonic signs (Mahloudji and Pikielny, 1967).

## 2. Families with M-D who Fulfill the Current Diagnostic Criteria

Several review articles on M-D have carefully reconsidered families published with a clinical picture compatible with a diagnosis of M-D (Quinn et al., 1988; Gasser, 1998). Table 40.1 lists demographic and clinical features of nine large M-D families with five or more affected individuals

TABLE 40.1 Demographic and Clinical Features of Nine Large M-D Families (Five and More Affected Individuals), Published between 1963 and 1993, currently without Molecular Genetic Information

| Family no. | Ref. | No. of affected | Age of onset in years (range) | Myoclonus | Dystonia | Distribution | Alcohol response | Additional features |
|---|---|---|---|---|---|---|---|---|
| 1 | Lindemulder, 1933 | 5 | <20 | 5 | Hypertrophy of neck muscles | Face, neck, trunk, UL | ? | None reported |
| 2 | Biemond, 1963 | 11 | Juvenile | 11/11 | ? | Face, UL | ? | None reported |
| 3 | Daube and Peters, 1966 | 5 | (1–20) | 5/5 | None reported | Trunk, UL | + | Mild dysmetria |
| 4 | Daube and Peters, 1966 | 13 | (1–20) | 13/13 | None reported | Trunk, UL | + | Milkd dysmetria |
| 5 | Mahloudji and Pikielny, 1967 | 6 | 3.6±2.7 (1.5–7) | 6/6 | 2/6 | Face, trunk, UL, LL | ? | None reported |
| 6 | Korten et al., 1974 | 15 | (Congenital–55) | 15/15 (or "essential tremor" in 7/15) | None reported | Neck, trunk, UL | + | Essential tremor |
| 7 | Przuntek and Muhr, 1983 | 25 | (4–49) | 25/25 | 10/25 | Face, neck, trunk, UL, (LL) | + | Postural tremor |
| 8 | Quinn et al., 1988 | 5 | (3–?) | 5/5 | 3/5 | Neck, trunk, UL, (LL) | + | Melanoma |
| 9 | Fahn and Sjaastad, 1991 | 15 | 22.6±20.2 (6–60) | 15/15 | 3/15 | Face, neck, trunk, UL, LL | + | None reported |

*Note*: Abbreviations: UL = upper limbs; LL = lower limbs. For most families, age of onset is only available as range. Adapted from Gasser, T. (1998). *Adv. Neurol.* **78**, 325–334.

and no published molecular genetic information (Lindemulder, 1933; Biemond, 1963; Daube and Madison, 1966; Mahloudji and Pikielny, 1967; Korten et al., 1974; Przuntek and Muhr, 1983; Quinn et al., 1988; Fahn and Sjaastad, 1991). All of these families were included in a previous review by Gasser (1998). Confirming the notion that these families have typical M-D, linkage to the DYT11 locus or mutations in the SGCE gene have been found in all three families from his original list (Table 40.2 in Gasser, 1998), for whom molecular genetic results have become available (Feldmann and Wieser, 1964; Kurlan et al., 1988; Kyllerman et al., 1990; Gasser et al., 1996; Klein et al., 2000b; Zimprich et al., 2001). It may be anticipated that most, if not all, of the remaining families will also be shown to carry SGCE mutations.

Since the identification of the DYT11 locus in 1999 (Nygaard et al., 1999), linkage to this locus or mutations in the recently described SGCE gene were reported in 19 families (Table 40.2). The clinical features of these genetically proven M-D families are reviewed in detail in Section VIII.A. In addition, two M-D families have been described with a deletion in the DYT1 gene and a missense change in the D2 dopamine receptor gene, respectively (Table 40.2; Sections VIII.B and VIII.C). Finally, several recently published cases and families with M-D meet the current diagnostic criteria but molecular genetic information is either unavailable or incomplete. Details on these cases and families are listed in Table 40.3 and further outlined in Section VIII.E.

## D. Summary of the Clinical Features of M-D

Both the myoclonus and dystonia most frequently involve the neck and upper limbs, often the trunk and bulbar muscles and less commonly the lower limbs (Daube and Madison, 1966; Gasser, 1998) (Tables 40.2 and 40.3).

In the early-onset dystonias, especially in DYT1 dystonia, the abnormal movements usually start in a limb, later tend to involve more cranial muscle groups, and may become generalized as the disease progresses (Bressman et al., 2000). In M-D, the dystonia seems to follow this pattern at least in some cases. For example, in the largest family with M-D described to date (Kyllerman et al., 1990), in whom M-D was later linked to the DYT11 region (Klein et al., 2000b), leg dystonia or hemidystonia was present in the cases with the earliest ages of onset (one to four years), writer's cramp was observed in several teenagers and adults, and retro- and torticollis were found in adult cases only (Kyllerman et al., 1990). However, unlike in primary torsion dystonia, the dystonia did not tend to worsen or generalize in the course of the disease but rather gradually improved during motor development (Kyllerman et al., 1990). The myoclonus, however, frequently shows almost the opposite pattern: it usually starts in the upper body half and sometimes, although rarely, involves the legs in the course of the disease (Korten et al., 1974).

The involuntary movements are frequently precipitated or worsened by active movements of the affected body parts. Other factors eliciting or enhancing the hyperkinetic movements include certain postures, emotional stress (Korten et al., 1974; Kyllerman et al., 1990), anxiety, nervousness, fright (Daube and Madison, 1966; Fahn and Sjaastad, 1991), excitement (Mahloudji and Pikielny, 1967), fatigue (Kyllerman et al., 1990), sound, (Korten et al., 1974; Kurlan et al., 1988; Asmus et al., 2001; Trottenberg et al., 2001), startle (Korten et al., 1974), touch (Kurlan et al., 1988; Nygaard et al., 1999), and caffeine (Nygaard et al., 1999). In female patients, the involuntary movements were described to worsen premenstrually (Quinn and Marsden, 1984).

The myoclonus and dystonia disappear during sleep (Mahloudji and Pikielny, 1967). However, single violent jerks may occur at night which may be severe enough to awaken the patient (Kyllerman et al., 1990). This phenomenon was previously described by Friedreich (Friedreich, 1881). Response of the myoclonus and, to a lesser extent of the dystonia, to alcohol is a very frequent finding (Tables 40.1–40.3) but not invariably present (Klein et al., 2000b; Vidailhet et al., 2001). Some patients experience a rebound phenomenon after alcohol withdrawal (Trottenberg et al., 2001). By contrast, M-D is relatively unresponsive to medical treatment (reviewed in Section IX).

The course of the disease is often benign (Mahloudji and Pikielny, 1967; Asmus et al., 2001). An initial worsening of symptoms is almost always followed by stabilization of the condition in adulthood, with most patients remaining completely independent in the activities of daily living. This is illustrated by a 30-year follow-up of parts of Family MD2 (Feldmann and Wieser, 1964; Gasser et al., 1996) with a proven SGCE mutation (Zimprich et al., 2001). Symptoms in subjects who had been in their teens at the time of the initial examination have fluctuated slightly over the years but have not significantly changed (Gasser et al., 1996). The disease is compatible with an active life and normal life span (Nygaard et al., 1999) and even spontaneous remission of M-D has been reported in selected cases (Korten et al., 1974; Fahn and Sjaastad, 1991). In some patients, however, M-D may be gradually progressive (Kurlan et al., 1988; Quinn, 1996; Borges et al., 2000; Trottenberg et al., 2001) and may lead to considerable functional disability and result in early retirement (Borges et al., 2000). There is considerable inter- and intrafamilial phenotypic variation that is based on reduced penetrance of the SGCE gene (Section III.D) and variable expressivity.

Additional clinical features include alcohol abuse, personality disorders, and psychiatric disturbances (Section II.E). Unusual features have also been described in selected

TABLE 40.2  Demographic and Clinical Features of 21 Families with Genetically Proven M-D and Three or More Affected Family Members

| Family no. | Family designation | Ref. | No. of affected (sex) | Age of onset in years (range) | Myoclonus (distribution) | Dystonia (distribution) | No. of obligate carriers (sex) | Alcohol response | Additional features | Genetic features |
|---|---|---|---|---|---|---|---|---|---|---|
| 1 | None | Nygaard et al., 1999 (Saunders-Pullman et al., 2002) | 10 (7M, 3F) | 6.5±2.1 (4–10) | 9/10 affected (face, neck, trunk, UL, sometimes LL) | 5/10 affected (face, neck, UL, sometimes LL) | 10 (5M, 4F, 1U) | + | Psychiatric problems in 9/10 affected (major affective disorder, OCD anxiety disorder alcohol dependence) | Linkage to DYT11 |
| 2 | A (Fam. B) | Klein et al., 2000b (Klein et al., 2000a) | 4 (3M, 1F) | 6.4±3.8 (3–12) | 4/4 affected (predominantly neck and trunk) | 2/4 affected (probable dystonia predominantly neck and trunk) | 0 | + | None reported | Linkage to DYT11 |
| 3 | C | Klein et al., 2000b | 3 (1M, 2F) | 3.3±2.1 (1–5) | 3/3 affected (predominantly neck and trunk) | 1/3 affected 1 not examined for dystonia (predominantly neck and trunk) | 2 (1M, 1U) | + | None reported | Linkage to DYT11 |
| 4 | E | Klein et al., 2000b (Kurlan et al., 1988; Klein et al., 2000a) | 7 (5M, SF) | 9.5±4.8 (2–15) | 4/7 affected (head, trunk, UL, LL) | 7/7 affected (neck, trunk, UL, LL) | Unknown | + | None reported | Linkage to DYT11 |
| 5 | F | Klein et al., 2000b | 3 (3M) | 15.3±2.5 (13–18) | 3/3 affected (predominantly neck and trunk) | 2/3 affected (predominantly neck and trunk) | 3 (2M, 1U) | – | None reported | Linkage to DYT11 |
| 6 | G | Klein et al., 2000b | 3 (3F) | 18.3±16.2 (8–37) | 3/3 affected (predominantly neck and trunk) | 3/3 affected (predominantly neck and trunk) | Unknown | + | None reported | Linkage to DYT11 |
| 7 | H (Fam. S) | Klein et al., 2000b (Lundemo et al., 1985; Kyllerman et al., 1990; Wahlström et al., 1994; Klein et al., 2000a) | 21 (13M, 8F) | ~7 (0.5–17) | 20/21 affected 1 unknown (neck, UL, trunk) | 10/21 affected 1 unknown (neck, UL, LL) | 4 (2M, 2F) | + | Tremor in 3 affected, alcohol and drug abuse, mild mental retardation in several members, suicide attempt in 1 affected | Linkage to DYT11 |
| 8 | 8 | Vidailhet 2001 (Dürr et al., 2000) | 5 (3M, 2F) | 13.8±11.3 (4–23) | 5/5 affected (face, neck, trunk, UL) | 0/5 affected | Unknown | – | None reported | Linkage to DYT11 |
| 9 | 9 | Vidailhet et al., 2001 (Dürr et al., 2000) | 3 (3F) | 13.3±2.1 (11–15) | 2/3 affected (neck, trunk, back, UL, LL) | 3/3 affected (trunk, UL, LL) | Unknown | + | Postural tremor in all 3 affected | Linkage to DYT11 |
| 10 | 11 | Vidailhet et al., 2001 (Dürr et al., 2000) | 4 (2M, 2F) | 8.5±6.2 (1–15) | 4/4 affected (neck, UL) | 3/4 affected (neck, trunk, UL) | 1 (1M) | – | Deafness in one affected | Linkage to DYT11 |

(continues)

TABLE 40.2 (continued)

| Family no. | Family designation | Ref. | No. of affected (sex) | Age of onset in years (range) | Myoclonus (distribution) | Dystonia (distribution) | No. of obligate carriers (sex) | Alcohol response | Additional features | Genetic features |
|---|---|---|---|---|---|---|---|---|---|---|
| 11 | 12 | Vidailhet et al., 2001 (Dürr et al., 2000) | 3 (2M, 1F) | 14.8±20.3 (0.5–38) | 3/3 affected (face, neck, trunk, UL) | 2/3 affected (neck, UL) | 2 (2M) | + | None reported | Linkage to DYT11 |
| 12 | MD2 (Fam. II) | Zimprich et al., 2001 (Feldmann and Wieser, 1964); Gasser et al., 1996; Asmus et al., 2001) | 11 (9M, 2F) | 8.7±2.8 (4–12) | 11/11 affected | 4/1 affected (neck, UL) | 7 (4M, 3F) | + | Panic attacks, alcohol abuse | SGCE mutation (565delA) |
| 13 | MD6 (Fam. III) | Zimprich et al., 2001 (Asmus et al., 2001) | 3 (2M, 1F) | 3.5 | 2/3 affected | 3/3 affected (neck, UL) | 2 (2M) | + | OCD in 1 affected | SGCE mutation (R97X) |
| 14 | MD7 | Zimprich et al., 2001 | 7 (4M, 3F) | 4 | Present | Present (neck, UL) | 1 (1M) | Unknown | Anxiety disorder | SGCE mutation (R102X) |
| 15 | MD8 | Zimprich et al., 2001 | 4 (2M, 2F) | 7 | Present | Present (neck) | 1 (1M) | Unknown | None reported | SGCE mutation 488–97/del |
| 16 | MD9 | Zimprich et al., 2001 (Scheidtmann et al., 2000) | 4 (3M, 1F) | 4 (index case), age of onset for relatives unknown | Present | Present (neck) | 0 | + | Panic attacks, alcohol abuse | SGCE mutation (907 + 1G →A) |
| 17 | MD10 | Zimprich et al., 2001 | 8 (3M, 5F) | 8 | Present | Present (neck) | 1 (1M) | Unknown | None reported | SGCE mutation (R102X) |
| 18 | Fam. I | Asmus et al., 2001 | 5 (3M, 2F) | First or second decade | 5/5 affected | 3/5 affected | 0 | + | Alcohol abuse in 1 affected | Linkage to DYT11 |
| 19 | Fam. IV | Asmus et al., 2001 | 7 (3M, 4F) | Childhood | 6/7 affected (1 unknown) | 4/7 affected (1 unknown) | 0 | + | None reported | Linkage to DYT11 |
| 20 | None (F1) | Klein et al., 1999b (Doheny et al., 2000) | 8 (5M, 3F) | 8.1±4.7 (2–16) | 7/8 affected (neck, trunk, UL) | 5/8 affected (larynx, neck UL) | 2 (2M) | + | Depression, manic-depression, anxiety, panic attacks, OCD, alcohol/drug abuse | Mutation in the DRD2 gene (V151I) |
| 21 | | Leung et al., 2001 | 1 definite (F) 4 possible (3M, 1F) | 5 | Present (LL) | Present (LL) | Unknown (mother and grandfather possibly affected) | Unknown | None reported | Mutation in DYT1 F323-Y328del |

*Note:* Abbreviations: M = male; F = female; UL = upper limbs; LL = lower limbs.
Age at onset is given as mean ± standard deviation and the range in parentheses if it could be calculated from the data available. Distribution of myoclonus and dystonia is listed if available.

TABLE 40.3 Demographic and Clinical Features of Seven Recently Described Families and Eight Index Patients with Myoclonus or M-D and No or Incompleted Genetic Information

| Family case no. | Family case designation | Ref. | No. of affected (sex) | Age of onset in years (range) | Myoclonus (distribution) | Dystonia (distribution) | Alcohol oresponse | Additional features | Genetic features |
|---|---|---|---|---|---|---|---|---|---|
| 1 | 1 | Dürr et al., 2000 | 1 (1F) | 20 | Present | Present (predominant symptom) | Unknown | None reported | Mutations in DRD2 gene excluded |
| 2 | 2 | Dürr et al., 2000 | 1 (1F) | 28 | Present (predominant symptom) | Present | – | None reported | Mutations in DRD2 gene excluded |
| 3 | 3 | Dürr et al., 2000 | 1 (1F) | 18 | Present (predominant symptom) | Present | + | None reported | Mutations in DRD2 gene excluded |
| 4 | 4 | Dürr et al., 2000 | 1 (1F) | 41 | Present | Absent | – | None reported | Mutations in DRD2 gene excluded |
| 5 | 5 | Dürr et al., 2000 | 1 (1M) | 6 | Present | Present (predominant symptom) | + | None reported | Mutations in DRD2 gene excluded |
| 6 | 6 | Dürr et al., 2000 | 1 (1M) | 4 | Present | Present (predominant symptom) | + | None reported | Mutations in DRD2 gene excluded |
| 7 | 7 | Dürr et al., 2000 | 2 (2M) | 12 and 20 | Present | Absent | + | None reported | Mutations in DRD2 gene excluded |
| 8 | 10 | Dürr et al., 2000 | 2 (1M, 1F) | 1 and 4 | Present | Present (predominant symptom) | – | None reported | Mutations in DRD2 gene excluded |
| 9 | Fam. 1 | Borges et al., 2000 | 2 (2M) | 9 and 9 | 2/2 affected (neck, UL, LL) | 2/2 affected (neck, UL) | + | Alcohol-induced epilepsy | No genetic information available |
| 10 | Fam. 2 | Borges et al., 2000 | 1 (1M) | 15 | Present (neck, UL) | Present (UL) | + | None reported | No genetic information available |
| 11 | Fam. 3 | Borges et al., 2000 | 2 (2F) | Congenital and 8 | 2/2 affected (neck, UL, LL) | 2/2 affected (UL) | + | None reported | No genetic information available |
| 12 | Fam. 4 | Borges et al., 2000 | 1 (1M) | 10 | Present (neck, UL) | Present (UL) | + | Speech tremor | No genetic information available |
| 13 | F2 | Doheny et al., 2000 (Klein et al., 1999) | 8 (sex not reported) | 12.4±11.7 (range not reported) | 7/8 affected (upper body) | 7/8 affected (upper body) | + | Psychiatric disorder | No genetic information available |
| 14 | None | Trottenberg et al., 2001 | 5 definite, 2 possible (sex not reported) | 6 (index patient) | Index patient: present (face, neck, UL, minimal: trunk, LL) | Index patient: present (neck) | + | Mild action tremor of both hands | Linkage to 9q (DYT1) excluded |
| 15 | None | Grimes et al., 2001 | 9 definite (4M, 5F), 3 probable (1M, 2F) | 9.6±3.3 (5–15) | 12/12 affected (head, UL, generalized in 2) | 4/12 definite 4/12 probable (axial, UL, LL) | + | Mild learning difficulties in 3 affected, alcoholism in 2 | Linkage to 11q (DRD2) excluded |

genetically proven M-D families and include postural and other forms of tremor (Korten *et al.*, 1974; Kurlan *et al.*, 1988; Kyllerman *et al.*, 1990; Vidailhet *et al.*, 2001) and mild mental retardation (Kyllerman *et al.*, 1990). Although epilepsy is one of the exclusion criteria for M-D, epileptic seizures may occur in otherwise typical M-D patients due to alcohol withdrawal (Borges *et al.*, 2000).

### E. Additional Clinical Findings: Psychiatric Abnormalities in M-D

Upon careful review of the literature, many reports can be found of alcohol abuse and addiction to alcohol (Kyllerman *et al.*, 1990; Klein *et al.*, 1999b; Zimprich *et al.*, 2001), as well as of personality disorders and psychiatric abnormalities (Klein *et al.*, 1999b; Nygaard *et al.*, 1999; Doheny *et al.*, 2000; Saunders-Pullman *et al.*, 2002; Zimprich *et al.*, 2001). Psychiatric disorders include obsessive-compulsive disorder (OCD; (Klein *et al.*, 1999b; Saunders-Pullman *et al.*, 2002; Zimprich *et al.*, 2001), major affective and anxiety disorder (Klein *et al.*, 1999b; Saunders-Pullman *et al.*, 2002; Zimprich *et al.*, 2001), and panic attacks (Klein *et al.*, 1999b; Zimprich *et al.*, 2001). Surprisingly, the affective and psychiatric symptoms in M-D have been rather neglected. There are only two studies systematically evaluating psychiatric symptoms in large M-D families; one has been reported in abstract form only and one has very recently been published (Doheny *et al.*, 2000; Saunders-Pullman *et al.*, 2002). Both studies also aimed at determining whether the same genetic etiology underlies both neurologic and psychiatric signs (Section VIII.G).

## III. GENE(S)

### A. Summary of the Genetic Findings in M-D

#### 1. Early Linkage Studies

Early linkage studies in M-D failed to reveal any positive evidence of linkage but rather excluded large regions of the genome (Wahlström *et al.*, 1994; Gasser *et al.*, 1996). Although many large pedigrees suitable for linkage analysis had been published, it was not before 1999 that the first positive genetic findings were reported. (Table 40.4).

#### 2. Missense Change in the D2 Dopamine Receptor (DRD2) Gene in One M-D Family

Taking a candidate gene approach, the DRD2 gene appeared to be a particularly attractive candidate to be involved in the etiology of M-D. Indeed, strong evidence for linkage was found in one M-D family for markers on the long arm of chromosome 11, with a maximum two-point lod score of 2.96 at D11S897. The lod score of 2.96 closely approximated the maximum potential lod score for this particular family size and structure. Critical recombination events placed the putative M-D locus within a region of about 23 cM which contained the DRD2 gene (Grandy *et al.*, 1989a, b). Sequence analysis of the DRD2 coding region revealed a missense change (Val154Ile) in a conserved region which co-segregated with the disease and was not found in any of 250 control chromosomes (Klein *et al.*, 1999b).

Mutational screening of the DRD2 gene in 5 M-D patients from other families and linkage analysis using chromosome 11 markers surrounding this gene in another four clinically similar families were negative, suggesting locus heterogeneity for M-D (Klein *et al.*, 2000a). Similarly, mutations in the DRD2 coding region and/or linkage to DRD2 were later excluded in 12 M-D index cases, including eight familial cases (Dürr *et al.*, 2000; Vidailhet *et al.*, 2001) and one other large M-D family (Grimes *et al.*, 2001).

#### 3. Detection of a Major M-D Locus on Chromosome 7q (DYT11)

Confirming the idea of at least one other M-D gene, a second M-D locus was mapped to a 28 cM region on chromosome 7q21–q31 in a single family. The linked region was flanked by obligate recombination events at markers D7S2443 and D7S799 (Nygaard *et al.*, 1999). The

TABLE 40.4 Genes Involved in Inherited M-D

| Gene | Protein | Number of families | Comments |
| --- | --- | --- | --- |
| ε-Sarcoglycan (SGCE) (Zimprich *et al.*, 2001) | ε-Sarcoglycan | 19 (Proven SGCE mutation or linkage to DYT11) | Major M-D gene |
| DYT1 (Leung *et al.*, 2001) | TorsinA | 1 | Role remins to be established |
| D2 dopamine receptor gene (Klein *et al.*, 1999) | D2 dopamine receptor | 1 | Excluded in all other M-D families tested |

gene for the human metabotropic glutamate receptor type 3 was considered an appealing candidate and was located within the linked region. However, direct sequencing did not reveal any mutations in two affected and two unaffected family members (Nygaard et al., 1999). Subsequently, the linked region of 28 cM was narrowed by linkage studies in eight families with clinically typical M-D to a 14-cM interval between markers D7S2212 and D7S821 (Klein et al., 2000b). There was no shared haplotype within the linked region across families, excluding a founder effect. Candidate gene analysis of the two gamma subunits of guanine nucleotide-binding proteins (G-proteins GNG11 and GNGT1) did not reveal any mutations in index patients from the eight families (Klein et al., 2000b). However, in GNGT1, a 122G→A transition (R41K) was detected in index patients from two families. When other affected members of these families were typed, this change did not segregate with the disease. In addition, this same variation was seen in control individuals (Klein et al., 2000b). Both of these findings suggest that the R41K change is a polymorphism. These two genes were chosen as candidate genes based on the fact that G proteins are involved as modulators or transducers in various transmembrane signaling systems and by that could interact with the dopamine D2 receptor in second messenger pathways.

These linkage studies were followed by several others, all of which served to confirm DYT11 as the major M-D gene locus (Asmus et al., 2001; Vidailhet et al., 2001). Multipoint lod scores in four French M-D families ranged from 0.6 to 1.3 in individual families, and multipoint lod scores across families revealed a maximum cumulative value of 3.7. No common haplotype was observed among the four families, rendering a common founder mutation unlikely (Vidailhet et al., 2001). Linkage analyses in four German families with autosomal dominant M-D revealed a combined lod score of 5.99 across families and further narrowed the region of the putative M-D gene to a 7.2 area between the markers D7S652 and D7S2480 (Asmus et al., 2001). Mutations in the DRD2 gene were excluded in all four families by linkage analysis and direct sequencing of the coding region (Asmus et al., 2001).

### 4. Identification of the SGCE as the Major M-D Gene

Very recently, the gene causing M-D was identified to be the SGCE gene. For this, six families with established linkage to the DYT11 region on chromosome 7q showed recombination events that allowed further narrowing of the critical region of the putative M-D gene to approximately 3.2 Mb (Zimprich et al., 2001). This region contained 15 genes, 10 of which were completely sequenced in each of the index cases from the 6 families without detecting any mutations. However, loss-of-function mutations were found in all index patients in the SGCE gene. The mutations comprised two nonsense changes in exon three (289C→T; R97X and 304C→T;R102X), the latter of which was found in two families, a 1-bp deletion (565delA), a 10-bp deletion (488–97del), and a splice site mutation at the exon-intron junction of exon 6 (907+1G→A); (Zimprich et al., 2001). The mutations segregated in the families, and no sequence alterations were found in 72 control alleles in the coding region of the gene.

The SGCE gene clearly represents the major M-D gene, mutations in which probably account for most cases of clinically typical M-D. This finding has recently been confirmed by our group in several M-D families (Ozelius et al., submitted; Fig. 40.2). It remains to be investigated whether the original M-D family with a sequence change in the DRD2 receptor gene harbors an additional SGCE mutation.

### 5. A Novel Mutation in the DYT1 Gene Associated with M-D

It has recently been shown that a novel mutation in the DYT1 (TOR1A) gene may also be associated with an M-D phenotype (Leung et al., 2001). As part of a study on the phenotypic spectrum of DYT1 mutations, the DYT1 gene was screened in cases with early-onset dystonia and early-onset parkinsonism. Surprisingly, a novel 18-bp deletion (F323–Y328del) in DYT1 was found in a patient with early onset dystonia and additional myoclonic features.

Thus, although the SGCE gene has clearly been identified to represent the major M-D gene, there remains evidence for genetic heterogeneity in M-D with two M-D families harboring mutations in genes other than the SGCE gene (Klein et al., 1999b; Leung et al., 2001) and with clinically similar families not linked to any of the known loci (Asmus et al., 2001).

## B. Normal Gene and Protein Function

### 1. SGCE Gene and ε-Sarcoglycan Protein

The SGCE gene (accession number AF036364) is localized on chromosome 7q between the DNA markers

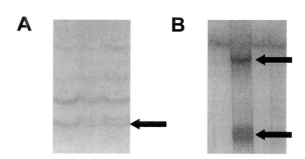

FIGURE 40.2 Silver stained polyacrylamide gel showing two band shifts (arrows) on single strand conformation polymorphism analysis of the SGCE gene (A: exon 6; B: exon 7). The observed band shifts correspond to SGCE mutations (data not shown; Ozelius et al., submitted).

D7S1513 and D7S1489. In addition, an anonymous CA repeat marker (69292) has been found in intron 3 (McNally et al., 1998). The SGCE is the fifth sarcoglycan gene to be identified (Ettinger et al., 1997; McNally et al., 1998), consists of twelve exons, and is highly homologous to the α-sarcoglycan gene. Both sequences predict a hydrophobic signal sequence followed by an extracellular domain with a conserved consensus site for asparagine-linked glycosylation, and four highly conserved cysteine residues. The overall gene structure is conserved which implies that both genes may have arisen through a gene duplication event. However, the introns in α-sarcoglycan are smaller than in SGCE which is estimated to span 70.9 kb. SGCE has a hydrophobic transmembrane domain and a short cytoplasmic domain. An alternative cytoplasmic domain, which was seen in one of four isolated cDNAs only, resulted from the inclusion of exon 10. Exon 10 is of particular note, as it is homologous to the Alu class of repetitive DNA sequences. A processed pseudogene, pseudo-ε-sarcoglycan, is located on human chromosome 2. The murine gene for ε-sarcoglycan maps to the syntenic region on mouse chromosome 6. SGCE is widely expressed in both embryonic and adult tissues (Ettinger et al., 1997), including smooth muscle (Straub et al., 1999), Schwann cell membranes of peripheral nerves (Imamura et al., 2000), and brain (McNally et al., 1998). Northern blot analysis showed broad expression of SGCE in various brain regions (putamen, temporal lobe, frontal lobe, occipital lobe, cerebral cortex, spinal cord, medulla, cerebellum; Zimprich et al., 2001).

The ε-sarcoglycan protein consists of 405 amino acids (Ettinger et al., 1997) and has a predicted molecular mass of 47 kDa (without exon 10 and post-translational modifications; McNally et al., 1998). The function of the ε-sarcoglycan protein is still largely elusive. It appears to be associated with β-, γ-, and δ-sarcoglycan (Durbeej and Campbell, 1999; Liu and Engvall, 1999). The other members of the sarcoglycan family, α-, β-, γ-, and δ-sarcoglycan, compose the heterotetrameric sarcoglycan complex which is an important component of the dystrophin-associated glycoprotein assembly in striated muscles. This complex and its associated components are thought to link the cytoskeleton with the extracellular basement membrane, thus stabilizing the muscle and protecting it from contraction-induced damage. SGCE appears to be functionally similar to α-sarcoglycan in skeletal muscle (Liu and Engvall, 1999); however, nothing is known about its function in the brain as yet.

## 2. DYT1 Gene and TorsinA

The DYT1 gene is discussed in detail in Chapter 36. Briefly, the DYT1 gene is localized on chromosome 9q and encodes a protein of 322 amino acids, designated torsinA, with a calculated molecular weight of 38 kDa. TorsinA is a novel member of the AAA+ chaperone family (Ozelius et al., 1999; Hewett et al., 2000), suggesting a role in manipulating the configuration of other proteins, for example, in recovery from cellular stress or movement/fusion of vesicles.

## 3. D2 Dopamine Receptor (DRD2) Gene and Protein

The DRD2 gene on chromosome 11q encodes the DRD2 protein, is highly expressed in the basal ganglia, and tightly involved in the control of movement (Missale et al., 1998). A number of observations made the DRD2 gene an attractive candidate to be mutated in M-D: dopamine receptor-blocking drugs can lead to dystonic symptoms (LeWitt, 1995), and patients with juvenile or young-onset parkinsonism frequently present with dystonia (Ishikawa and Tsuji, 1996; Leung et al., 2001). In addition, some forms of early-onset dystonia have been shown to involve dopaminergic transmission, including dopa-responsive dystonia with mutations in genes encoding enzymes in dopamine biosynthesis (Ichinose et al., 1994; Knappskog et al., 1995), and early-onset primary torsion dystonia, in which the transcript for the responsible gene, DYT1, is expressed at high levels in dopaminergic neurons of the substantia nigra (Augood et al., 1998). Finally, the DRD2 receptor is believed to have a role in affect and emotion and reward mechanisms, and has been implicated in the etiology of alcoholism and psychiatric disorders (for review see Missale et al., 1998).

## C. Abnormal Gene and Protein Function

### 1. SGCE and Sarcoglycanopathies

Previously, mutations in other members of the sarcoglycan family (α-, β-, γ-, and δ-sarcoglycan) have all been associated with limb girdle muscular dystrophies (sarcoglycanopathies); (Angelini et al., 1999). Mutations in any of the four genes encoding sarcoglycans cause a deficiency in all four sarcoglycans in the sarcolemma. M-D, on the other hand, is a nondegenerative disorder of the central nervous system. In addition, SGCE mutations may also play a role in a variety of neuropsychiatric disorders.

### 2. Mutated DYT1 Gene and TorsinA

A three base pair (GAG) deletion in exon 5 of the DYT1 gene results in the loss of one of a pair of glutamic acid residues in the C terminus of torsinA (Ozeliu et al., 1997). Most cases of early onset generalized dystonia world wide (about 70%) are caused by this mutation (Bressman et al., 2000). However, the recently described deletion of six amino acids in the carboxy terminus, 21 amino acids downstream from the glutamic acid deletion, resulted in an M-D phenotype (Leung *et al.*, 2001). This new mutation would remove a predicted casein kinase 2 phosphorylation domain in the protein. The deletion of a single or multiple amino acids in the carboxy terminus of torsinA would change the relative position and charge of the alpha helix in a conserved domain in this region, which could affect hexamer formation or binding to interacting proteins. In addition, the removal of a putative threonine phosphorylation site, TKLD, by the 18-bp deletion could also disturb the function of torsinA (Leung *et al.*, 2001). As the novel mutation has been described very recently, no functional studies have been published to date which investigate the effects of the six amino acid deletion on protein function.

### 3. Mutated DRD2 Gene and Protein

Although, at first sight, the valine-to-isoleucine change in DRD2 appeared to be minor, a number of facts point toward its potential significance. The mutation occurs in a region that is highly conserved in DRD2 across species; this change has never been reported as a polymorphism, although extensive screening for mutations of DRD2 has been performed in controls and patients with many diseases; the change in amino acid is located in a transmembrane domain which, in general, are of particular importance for agonist binding; and single base pair mutations causing amino acid substitutions in other G protein-coupled receptors have been shown to cause disease (Klein *et al.*, 1999b).

To further evaluate the role of DRD2 in M-D, a study was undertaken which investigated the biochemical properties of the mutant and wild-type receptors expressed in heterologous cells (Klein *et al.*, 2000a). In order to compare the expression and affinity of the mutant DRD2 to that of the wild-type receptor, expression constructs were used to transfect HEK293 cells, and binding of [$^3$H]spiperone was carried out over a range of concentrations. The affinity and the saturation binding capacity of the wild-type and mutant were similar. The binding of agonists (dopamine and quinpirole) to membranes of wild-type or mutant DRD2-transfected HEK293 cells was assessed by competition versus [$^3$H]spiperone, revealing similar Kis and slope factors. Finally, the functional properties of the wild-type and mutant DRD2 receptors were compared using a luciferase reporter construct assay. The ratio of the maximal effect of mutant receptor to that of the wild-type was not significantly different, and saturation binding of triple-transfected membranes from cells used for the functional assays showed similar expression levels for the wild-type and mutant receptors (Klein *et al.*, 2000a). Taken together, these binding and signal assays did not reveal any functional alteration of the mutant as compared to the wild-type receptor. However, this initial analysis does not exclude the possibility that differences may exist in other assays or expression systems between DRD2 wild-type and mutant. It should be noted that known functional differences between the DRD2 long and short isoforms, which differ by the presence or absence of a 29 amino acid segment in the third intracellular loop, would not be evident in the assays used. Therefore, further investigation of other possible functional effects of this mutation may be warranted.

### D. Evidence for Imprinting (SGCE Gene)

Reduced penetrance has been described in M-D pedigrees by many groups (Table 40.2). However, Zimprich *et al.* (2001) were the first to point out that the reduced penetrance was found particularly in the offspring of affected females, suggestive of possible maternal imprinting. Parental imprinting is defined as the epigenetic marking of certain subregions of the parental genome in mammals, i.e., imprinted genes are expressed from one allele only, according to their parent of origin (Reik and Walter, 2001). Imprinting plays an important role in embryogenesis and normal development and, if disregulated, may lead to disease. Well-described disorders of imprinting include Angelman and Prader-Willi syndrome (Shemer *et al.*, 2000).

Table 40.5 summarizes the distribution of parental alleles in eight informative families with genetically proven M-D and a total of 73 affected and 19 unaffected individuals. Clinical information on these families is listed in Table 40.2. Number of affected family members may vary for selected families between the two tables, as all available information on affected status and pedigree structure was used in the imprinting analysis which sometimes included deceased individuals. However, most of the latter were omitted from Table 40.2 as only incomplete clinical information was available.

Analysis of the distribution of parental alleles revealed that 64 out of 73 affected (87.7%) inherited the mutated allele from their father, while 17 out of the 19 unaffected family members (89.5%) inherited the mutated allele from their mother. These findings are highly significant ($p < 10^{-10}$ and $p < 0.001$, respectively) and strongly support the notion of differential expression of parental alleles which influences penetrance of the condition.

Interestingly, an imprinting mechanism has been described for the orthologous SGCE mouse gene which is predominantly expressed from its paternal allele in all adult mouse tissues, with the exception of weak expression from the maternal allele in mouse brain (Piras *et al.*, 2000). In addition, the existence of a new imprinted gene cluster on chromosome 7q21 has recently been proposed when a retrotransposon-derived gene, PEG10, was identified to be a novel imprinted gene which is localized near the SGCE gene (Ono *et al.*, 2001). Also, two other human genetic diseases are associated with genomic imprinting on the

TABLE 40.5  Evidence for Imprinting in Selected M-D Families (DYT11 Genetically Proven)

|  | No. of aff. F | No. of aff. M | No. of unaff. F | No. of unaff. M |
|---|---|---|---|---|
| Fam. 1 (Nygaard et al., 1999) | | | | |
| Mutation transmitted through F | 0 | 1 | 3 | 5 |
| Mutation transmitted through M | 3 | 6 | 0 | 0 |
| Fam. 7 (Klein et al., 2000; Kyllerman et al., 1990) | | | | |
| Mutation transmitted through F | 2 | 1 | 0 | 2 |
| Mutation transmitted through M | 6 | 14 | 2 | 0 |
| Fam. 8 (Vidailhet et al., 2001) | | | | |
| Mutation transmitted through F | 0 | 0 | 0 | 1 |
| Mutation transmitted through M | 3 | 2 | 0 | 0 |
| Fam. 12 (Zimprich et al., 2001) | | | | |
| Mutation transmitted through F | 0 | 0 | 2 | 3 |
| Mutation transmitted through M | 2 | 10 | 0 | 0 |
| Fam. 16 (Zimprich et al., 2001) | | | | |
| Mutation transmitted through F | 0 | 1 | 0 | 0 |
| Mutation transmitted through M | 1 | 3 | 0 | 0 |
| Fam. 17 (Zimprich et al., 2001) | | | | |
| Mutation transmitted through F | 0 | 0 | 0 | 1 |
| Mutation transmitted through M | 5 | 4 | 0 | 0 |
| Fam. 18 (Asmus et al., 1999) | | | | |
| Mutation transmitted through F | 1 | 2 | 0 | 0 |
| Mutation transmitted through M | 1 | 1 | 0 | 0 |
| Fam. 19 (Asmus et al., 2001) | | | | |
| Mutation transmitted through F | 1 | 0 | 0 | 0 |
| Mutation transmitted through M | 2 | 1 | 0 | 0 |

*Note*: Abbreviations: M = male; F = female.

7q21 region: choriocarcinoma (Ahmed et al., 2000) and Silver-Russel syndrome (Kotzot et al., 1995). Most imprinted genes show allelic differences in DNA methylation, however, the SGCE remains inactive despite its exposure to inhibitors of DNA methylation (El Kharroubi et al., 2001). These findings suggest that DNA methylation regulates the differential allelic expression of some but not all imprinted genes, and the exact mechanisms of imprinting of the SGCE gene remain to be established.

## IV. DIAGNOSTIC AND ANCILLARY TESTS

The diagnosis of M-D is established on clinical grounds and can now be confirmed by genetic analysis.

### A. DNA Tests

Genetic testing for M-D, i.e., a screen for mutations in the SGCE gene, is currently available on a research basis in selected laboratories specializing in the molecular analysis of movement disorders. In M-D patients who screened negative for SGCE mutations and who display predominant dystonia, the DYT1 gene should be tested for mutations. A test for the GAG deletion in DYT1 is commercially available. However, in M-D, the complete coding region of the DYT1 gene should be screened for mutations as the DYT1 mutation found in a family with M-D is different from the GAG deletion (Leung et al., 2001). Finally, although mutations have been excluded in all other M-D families tested to date, the DRD2 gene remains a candidate gene for M-D in families who are otherwise mutation-negative.

### B. Other Laboratory Tests

By definition, all laboratory tests are normal in primary M-D. However, routine laboratory testing should be performed to exclude secondary causes of the movement disorder. Liver function tests may be abnormal in patients who regularly consume alcohol.

## V. NEUROIMAGING

Likewise, neuroimaging studies, including CT and MRI are entirely unrevealing in M-D, with the possible exception of degenerative changes due to chronic alcohol intake. Currently, no functional imaging studies have been reported in M-D.

## VI. NEUROPATHOLOGY

The only neuropathologic examination of M-D reported to date has recently been published in abstract form (Brin et al., 2000). Postmortem examination was performed on a clinically unaffected mutation carrier from the M-D family with a missense change in DRD2 (Klein et al., 1999b) who died from a spontaneous frontal brain hemorrhage at age 76 (Brin et al., 2000). The cortical ribbon, hippocampal formation, and the deep gray matter structures were normal, as was the cerebellum. Microscopic examination showed mild to moderate neuronal loss in the substantia nigra with incontinentia pigmenti and Lewy bodies in occasional residual pars compacta neurons. No neocortical Lewy bodies or significant neurofibrillary pathology were identified (Brin et al., 2000). While the finding of Lewy bodies is interesting, it appears difficult to draw any conclusions from a single (unaffected) case and more postmortem studies will have to be performed in M-D.

## VII. CELLULAR AND ANIMAL MODELS OF DISEASE

Animal models of myoclonus (but not of M-D) have long been available but mostly study posthypoxic myoclonus and other secondary forms of myoclonus (Chadwick et al., 1978; Shohami et al. 1986; Kanthasamy et al., 1995, 2000). As the major M-D gene, SGCE, has only very recently been identified, no cellular or animal models are currently available for inherited M-D due to SGCE mutations. However, some clues to the function of SGCE can be garnered from $\alpha$, $\beta$, $\gamma$, and $\delta$-sarcoglycan-deficient mice which all develop severe muscular dystrophies (Duclos et al., 1998; Hack et al., 1998; Araishi et al., 1999; Coral-Vazquez et al., 1999). Despite the homology of SGCE and $\alpha$-sarcoglycan, endogenous SGCE is not able to rescue phenotypes associated with $\alpha$-sarcoglycan loss (Duclos et al., 1998). Both the *Drosophila melanogaster* and *Caenorhabditis elegans* genomes contain $\alpha/\varepsilon$-like sarcoglycans and thus may allow for functional studies in nonvertebrates.

## VIII. GENOTYPE/PHENOTYPE CORRELATIONS

### A. M-D Families with Established Linkage to DYT11 or with Known SGCE Mutations

#### 1. Original M-D Family in whom Linkage to the DYT11 Region was Established

Linkage of M-D to the DYT11 locus on chromosome 7q21–q31 was found in a large M-D family of European and Native American background from the Northwestern United States (Nygaard, 1999). The pedigree comprised 157 identified family members, including 10 affected individuals. The condition was transmitted in an autosomal dominant fashion. Five of the affected had myoclonus only, four had a combination of myoclonus and dystonia, and one had brachial dystonia only. However, four of the five individuals who were considered to have myoclonus only, showed occasional sustained muscle contractions and mild posturing, especially with writing, compatible with subtle dystonia. Myoclonus was sometimes present at rest but clearly exacerbated with action and showed overflow. Several affected family members reported other precipitating factors, including stress, sudden noise, caffeine, or touch. Alcohol relieved the symptoms in all individuals who tried it but it was unclear whether both myoclonus and dystonia responded. Therapy with clonazepam and valproate resulted in mild improvement, while levodopa did not have any beneficial effect (Nygaard et al., 1999). In addition, nine out of the ten affected family members described psychiatric problems that have been reviewed in Section II.E (Saunders-Pullman et al., 2002). The partial pedigree of the family reveals ten unaffected obligate carriers without any motor symptoms.

#### 2. Other Families with Confirmed Linkage to DYT11

The DYT11 locus was confirmed to be the main M-D locus by several independent groups (Klein et al., 2000b; Asmus et al., 2001; Vidailhet et al., 2001). Eight families with a total of 107 examined family members, including 42 affected individuals, all had a phenotype consistent with the phenotype of typical M-D families (Nygaard et al., 1999; Klein et al., 2000b). Mean age of onset was $8.1 \pm 5.9$ years with a range from 1–37 years. Myoclonus and dystonia, whether present individually or together, were most prominent in the neck or arm, followed by the trunk and face. A response to alcohol was noted in seven of the eight families (Klein *et al.*, 2000b). Detailed clinical findings including pedigrees of the following families have been described elsewhere: Families B and J (Klein *et al.*, 2000a), Family K (Kurlan *et al.*, 1988), and Family S (Kyllerman *et al.*, 1990). The mode of transmission was consistent with autosomal dominant inheritance with reduced penetrance in all eight families.

M-D was also shown to be linked to the DYT11 region in four French families (Vidailhet *et al.*, 2001). Dominant inheritance of the disorder could be established in three of the families. Age of onset was variable and spanned a range from as early as 6 months to 38 years. Some of the families showed slightly unusual clinical features: dystonia was absent in all members of Family 8, and there was no response of the myoclonus to alcohol. Interestingly, patients with a later age of onset of the myoclonus were less severely affected and vice versa, i.e., the younger the age at onset, the more severe the myoclonus. A similar tendency of symptoms being more severe in patients with an earlier age of onset has been described in various forms of dystonia. All affected members of Family 9, who showed otherwise typical M-D, had postural tremor in the upper limbs. Family 11 also displayed the typical clinical features of M-D, which was, however, unresponsive to alcohol (Vidailhet *et al.*, 2001).

### 3. Families with Proven SGCE Mutations

All six families with proven SGCE mutations had clinically typical, autosomal dominantly inherited M-D and were of German extraction. Pedigrees included four to eleven affected individuals, and average age of onset was in the first decade in all six families. Myoclonus and dystonia, namely torticollis (three families) or a combination of torticollis and writer's cramp (three families), were invariably present. In addition, four out of six families showed psychiatric disturbances such as OCD, anxiety disorder, panic attacks, or alcohol abuse (Zimprich *et al.*, 2001). Two of the families have previously been described in more detail elsewhere; (MD2 in Gasser *et al.*, 1996; MD9 in Scheidtmann *et al.*, 2000; and Asmus *et al.*, 2001). The largest family with eleven living affected members, MD2, was of Northern German extraction. Action-provoked bilateral, asymmetric, alcohol-responsive myoclonus was the prominent feature. Dystonic features, if present, were relatively mild (torticollis and writer's cramp) and evolved after the onset of the myoclonus. Several family members regularly consumed alcohol and panic attacks were also described (Zimprich *et al.*, 2001); however, except for one family member, all affected individuals remained employed and socially integrated (Gasser *et al.*, 1996). Seven obligate mutation carriers were clinically unaffected (Zimprich *et al.*, 2001). In the index patient of Family MD9, panic attacks, but not myoclonus led to admission to the neurology department. The 34-year-old patient reported a 30-year history of myoclonus, mainly at the proximal upper limbs which was followed by torticollis at the age of ten years. The myoclonus later spread to the neck and the distal parts of the upper limbs. The son of the patient was erroneously diagnosed as a case of "chorea"; an affected brother was addicted to alcohol. The affected father and grandfather were affected by history and not available for examination (Scheidtmann *et al.*, 2000).

### B. M-D Family with a Mutation in the DYT1 (TOR1A) Gene

Recently, a female patient of Polish-Lithuanian-Italian background with MD was shown to carry a novel 18-bp deletion in the DYT1 gene (Leung *et al.*, 2001). Age of onset was at five years, and the legs were the first body site to be involved. In addition, the patient's brother, mother, and maternal grandfather also showed the same mutation. All mutation carriers, including the index case, were heterozygous for the mutation. However, the index patient was the only definitely affected family member, while the other mutation carriers displayed possible myoclonus, possible dystonia, lip puckering, or tremulous voice with speech hesitation and were rated as possibly affected. This pattern is compatible with autosomal dominant inheritance of the disorder and reduced penetrance of the deletion (Leung *et al.*, 2001). It is of note that the condition started in the legs which is the typical site of onset in DYT1 dystonia due to the GAG deletion, especially when the movement disorder starts in early childhood (Bressman *et al.*, 2000). In typical M-D, however, involvement of the lower extremities is relatively rare, and the motor symptoms almost always start in the upper body half. It remains to be evaluated whether the 18-bp deletion represents a unique finding or whether this or other mutations in DYT1 are more frequently associated with cases of atypical M-D. It may be speculated that onset of dystonia and/or myoclonus in the legs have to be regarded as a "red flag," pointing away from M-D due to SGCE mutations.

### C. M-D Family with a Mutation in the D2 Dopamine Receptor Gene

The only family described to date with a mutation in the D2 dopamine receptor gene was of Welsh-Scottish-German origin with eight individuals showing alcohol-responsive M-D (five men, three women, mean age of onset 8.1 years +/− 4.7 years, range: 2–16 years) and two unaffected obligate carriers. The mode of transmission suggested autosomal dominant inheritance with reduced penetrance. Clinical features comprised predominant myoclonus in three, dystonia in one, and features of both in four individuals. Three of the affected individuals reported improved myoclonic symptoms after alcohol intake. However, there was no history of alcohol abuse. Affective traits, including depression, manic-depression, anxiety, panic attacks, and OCD were reported and subsequently formally evaluated by Doheny et al. (2000). No other additional or unusual features were described in this family, thus rendering the phenotype indistinguishable from that of M-D associated with SGCE mutations.

### D. M-D Families in whom Linkage to a Known M-D Gene Locus was Excluded

It was recently reported in abstract form that several but not all families tested for involvement of the DYT11 locus were linked to the chromosome 7q region (Asmus et al., 2000). However, no clinical details were available on the families for whom DYT11 linkage was excluded. Although it has been clearly demonstrated that SGCE represents the major M-D gene, this finding of M-D families without linkage to chromosome 7q further supports the notion of some degree of genetic heterogeneity in M-D.

### E. M-D Families and Index Cases for whom Genetic Data are Currently Unavailable or Incomplete

Table 40.3 lists demographic and clinical features of recently described families and index patients with myoclonus who meet the clinical criteria for M-D but no or incomplete genetic information is available as yet. Involvement of the SGCE gene may be assumed for the majority of these cases and families. Mutations in the DRD2 gene were excluded in 12 index patients with myoclonus or M-D, eight of whom were familial cases with established autosomal dominant inheritance in seven of the families. Subsequently, linkage to DYT11 was shown in four of the families, and no other genetic information is currently available on the remaining eight index cases. Among these, myoclonus was the only manifestation of the disease and dystonia was absent in one family and one index case, respectively. Interestingly, a response to alcohol was present in only five of the cases and families (Dürr et al., 2000). It appears likely that Patient 4 represents a case of essential myoclonus (absence of dystonia, late age of onset, no response to alcohol) rather than M-D. Molecular genetic analysis will allow for a definite classification of these eight families/cases.

Further demonstrating the worldwide occurrence of M-D, six Brazilian patients from four families were described to have myoclonic and dystonic movements, affecting predominantly the arms, which improved upon ingestion of alcohol (Borges et al., 2000). Age of onset was in the first or second decade, the involuntary movements were elicited by action, and all patients showed a predominance of myoclonus over dystonia. The sister of one of the index cases presented with postural tremor of the hands; head tremor was reported in two other relatives of index patients (Borges et al., 2000). Similarly, postural tremor was described as an additional feature in a patient with genetically proven M-D (see above; (Vidailhet et al., 2001). No genetic testing has been reported in any of the Brazilian families as yet.

Recently, a large family was described with autosomal dominant inheritance of M-D in eight definitely and four probably affected members with a mean age of onset of $9.6 \pm 3.3$ years (Grimes et al., 2001). All affected individuals had myoclonus, four had definite dystonia and four showed questionable dystonic signs only. Both the myoclonus and the dystonia predominantly affected the upper body half; the myoclonus was responsive to alcohol in half of the affected family members. Linkage to the DRD2 region on chromosome 11q was excluded, as was the previously reported missense change in exon 3 of the DRD2 gene (Grimes et al., 2001). The phenotype of this family would be compatible with an SGCE mutation; however, no DYT11 linkage data or mutational analysis has been reported as yet.

### F. Sporadic and Atypical M-D Cases

In addition to the pedigrees with obvious autosomal dominant inheritance of M-D, several sporadic cases with clinically typical M-D have appeared in the literature (for example, Dürr et al., 2000; see Section VIII.E). However, it is unknown whether they represent true sporadic cases or whether there is an underlying genetic defect and dominant inheritance is masked, for example, by the occurrence of a *de novo* mutation or by low penetrance of M-D within the family. Supporting the latter idea, we have recently detected an SGCE mutation in a 5-year-old patient with M-D and a negative family history. Interestingly, her unaffected father was also shown to carry the mutation, illustrating the fact that dominantly inherited M-D may appear "pseudo"-sporadic, especially in small families. The case of this girl is also an example of the observation that affected individuals tend to inherit the mutation from their fathers which is in keeping with the postulated imprinting mechanism of the SGCE gene. Therefore, it remains to be established whether true sporadic M-D phenocopies exist or not.

Two families of Saudi Arabian origin with consanguineous marriages displayed autosomal recessive inheritance of tremulous and myoclonic dystonia (Bohlega et al., 1995). Age of onset was in childhood in both families. Members of Family I exhibited a coarse, asymmetric, generalized tremor which involved the head, neck, and the extremities. In addition, all affected family members had focal or segmental dystonia affecting the upper body half.

One patient had spasmodic dysphonia. Patients of Family II showed tremulous jerky movements of the upper limbs and neck with rapid facial grimacing. Later in the course of the disease, the trunk and lower limbs also became affected. Symptomatic members of both families had normal intelligence and personalities. Extensive metabolic studies failed to detect any lysosomal, peroxisomal, mitochondrial, or other metabolic abnormalities. MRI scanning, however, revealed mild diffuse white matter changes (Bohlega et al., 1995). In summary, these families show some features of typical M-D, that is the combination of myoclonic jerks and dystonia, but differ from classic M-D in that the mode of inheritance was autosomal recessive, tremor was a predominant sign, and affected members showed abnormalities on MRI.

### G. Psychiatric Disorders in M-D

Currently, only two studies have formally addressed the occurrence and expression of psychiatric disorders in M-D (Doheny et al., 2000; Saunders-Pullman et al., 2002). In the first investigation, forty individuals from two M-D families underwent a neurological examination, videotaping, the Diagnostic Interviews for Genetic Studies, and psychiatric testing. FI was identical to the M-D family with a DRD2 gene mutation, previously described by Klein et al. (1999b) and comprised eight affected individuals. Psychiatric evaluation revealed alcohol/drug abuse in two, depressive disorder in all, personality disorder in one, anxiety/panic disorder in five, and OCD in five affected individuals. In addition, psychiatric disorder was diagnosed in two clinically unaffected individuals and in three family members who did not carry the mutation-associated haplotype of the DRD2 region (Doheny et al., 2000; Klein et al., 2000b)). FII also had eight affected members with an average age of onset of $12.4 \pm 11.7$ years and alcohol/drug abuse in six, depressive disorder in four, anxiety/panic disorder in one, and psychosis in one affected individual. Psychiatric abnormalities were also found in two family members without any motor symptoms. However, genetic status of this family was unavailable at the time of study (Doheny et al., 2000).

The second study directly correlated psychiatric symptoms with the known genetic status of the family members studied (manifesting carriers, nonmanifesting carriers, and noncarriers with established linkage to DYT11); (Saunders-Pullman et al., 2002). This family was originally described by Nygaard et al. and was the first family linked to the DYT11 region (Nygaard et al., 1999). The computerized version of the Composite International Diagnostic Interview was administered to all study subjects, and algorithms were used for the DSM-IV diagnosis of OCD, generalized anxiety disorder, major affective disorder, alcohol abuse, alcohol dependence, drug abuse, and drug dependence. A total of 55 participating individuals participated and included 16 manifesting carriers, 11 non-manifesting carriers, and 28 noncarriers. The rate of OCD was significantly higher in carriers (18.5%) compared to noncarriers (0%); ($p = 0.023$). It was also higher in the symptomatic gene carriers (25.0%) compared to the non-manifesting group (9.1%); ($p = 0.022$). The rate of generalized anxiety disorder was also elevated in symptomatic carriers compared to asymptomatic individuals (37.5% versus 12.8%); ($p = 0.061$), but was not increased in the gene carriers overall. Major affective disorder was increased in the symptomatic carriers (31.3%) versus the asymptomatic group (17.9%; not significant). Finally, alcohol dependence was increased in the symptomatic carriers (7:16); ($p = 0.027$), but not in the carrier group overall (7:27). There was also more drug use in the symptomatic carriers (31.3%) versus the asymptomatic group (12.8%), but this difference was not statistically significant (Saunders-Pullman et al., 2002). The results of these studies show that alcohol dependence/abuse is highly associated with motor symptoms of M-D. However, it currently remains unknown whether this is a feature of the disease or may be explained by self-"medication" to improve the motor symptoms of the disorder. Several psychiatric disorders, on the other hand, clearly appear to be associated with the genetically confirmed forms of M-D. These results need to be replicated in larger series, however, they are in keeping with the above-mentioned earlier observations in M-D families. Thus, unlike in other forms of early-onset dystonia, psychiatric abnormalities seem to be an integral part of the phenotypic spectrum of M-D.

### H. M-D Diagnostic Criteria Revisited

The original diagnostic criteria for M-D by Mahljoudji and Pikielny (1967) were recently modified by Gasser (1998). While Gasser (1998) had to choose 19 families from 14 publications for his summary of the spectrum of clinical M-D features based on the completeness of the clinical description and on the assumption that they might represent examples of a single disorder, the recent identification of M-D gene loci and genes now allows for a definition of the M-D phenotype on genetic grounds and thus ensures comparison of families with genetically identical disorders.

As the SGCE gene clearly represents the main M-D gene, and mutations in this gene appear to be by far the most common cause of M-D, the following is based on families with proven linkage to DYT11 or with SGCE mutations (Table 40.2; Families 1–19). Nineteen M-D families with a total of 115 affected family members had an average age of onset of about 9 years with a range of 0.5–38 years. Sixty-eight (59.1%) of the affected were male and 47 (40.9%) were female. Detailed information on the presence of myoclonus and dystonia in individual family members was available for 15 of these families with a total of 92 affected subjects; 84 of 92 (91.3%) had myoclonus and 52 of 92 (56.5%) showed dystonia. In 40 of the 92 (43.5%) affected individuals, myoclonus was the only movement disorder, while in 8 of 92 (8.7%) dystonia was the sole sign of M-D. In all families, both the myoclonus and the dystonia affected mainly, if not exclusively, the upper body half with predominant involvement of the neck and upper limbs. A response to alcohol was positive in 13 of the initial 19 families, negative in three and unknown in the remaining three families. Additional neurological features mainly included tremor and various psychiatric disturbances, although this was not investigated and/or reported systematically by all authors. In the light of this analysis of the genetically defined families, the diagnostic criteria of inherited M-D shall be further modified (Table 40.6).

Genetic analysis of the previously described (Table 40.1) and newly identified M-D families (Table 40.3) will serve to better define the phenotypic spectrum, will allow for more detailed phenotype-genotype correlations, and ultimately lead to further refinement of the diagnostic criteria of M-D.

TABLE 40.6 Diagnostic Criteria for M-D

**Cardinal clinical diagnostic criteria**
1. Onset of myoclonus usually in the first or second decade of life; dystonic features are observed in *more than half of the affected* in addition to myoclonus and may rarely be the only manifestation of the disorder.
2. Males and females *about* equally affected.
3. A *relatively* benign course, often variable but compatible with an active life of normal span *in most cases*.
4. Autosomal dominant mode of inheritance with variable severity, and incomplete penetrance *which is dependent on the parental origin of the disease allele; affected individuals usually inherit the disease from their father.*
5. Absence of seizures, dementia, gross ataxia and other neurological deficits.
6. Normal EEG, normal SSEP, *normal results of neuroimaging studies (CT or MRI).*

**Optional criteria**
1. *Response of symptoms (particularly of the myoclonus and to a lesser degree of the dystonia) to alcohol.*
2. *Various personality disorders and psychiatric disturbances.*

**Genetic testing**
1. *SGCE mutations confirm the diagnosis of M-D*

*Note*: Changes in the Gasser et al. modification are printed in italics. Adapted from Mahloudji, M., and Pikielny, R.T. (1967). *Brain* **90(3)**, 469–474; Gasser, T. (1998). *Adv. Neurol.* **78**, 325–334.

## I. Differential Diagnosis

The clinical picture with myoclonus, dystonia, or both, which often occur in a familial fashion, and the absence of any other neurological signs or symptoms is quite distinct. In most cases, the diagnosis of M-D can be established on clinical grounds with a relatively high degree of diagnostic certainty.

Most other conditions, in which myoclonus is a prominent feature, are characterized by a variety of other neurologic signs and symptoms not compatible with a diagnosis of M-D, such as epileptic seizures, ataxia etc. Genetically defined conditions with myoclonus as a major component include progressive myoclonus epilepsy, also known as Baltic myoclonus (EPM1), caused by mutations in the cystatin B gene (CSTB), myoclonus epilepsy of the Lafora type, associated with mutations in the EPM2A gene, and myoclonus epilepsy associated with ragged-red fibers (abbreviated as MERRF), due to mutations in mitochondrial genes. Finally, patients with the trinucleotide repeat disorder dentatorubro-pallidoluyisian atrophy (DRPLA) may also present with myoclonus.

## IX. TREATMENT

In general, medical treatment of M-D has proven to be rather ineffective. Uniformly, the best "treatment" results are achieved with alcohol (Tables 40.1–40.3). However, the risk of addiction to alcohol and the unacceptable side effects of this "medication" render it an unacceptable treatment option. Currently, the most promising drug appears to be gamma-hydroxy-butyric acid which has recently been shown to markedly improve symptoms of myoclonus, to a degree similar to alcohol (Priori et al., 2000). Selected patients have been reported to respond to various benzodiazepines (Kyllerman *et al.*, 1990; Fahn and Sjaastad, 1991; Nygaard *et al.*, 1999; Klein *et al.*, 2000b; Trottenberg *et al.*, 2001) or antiepileptic drugs such as valproate (Nygaard *et al.*, 1999). However, most patients with a good treatment response have been described on a case report basis only. Drugs reported to have a beneficial effect include baclofen (Korten *et al.*, 1974), and L-5-hydroxytryptophan (Przuntek and Muhr, 1983; Scheidtmann *et al.*, 2000). Finally, Korten *et al.* (1974) successfully applied very large doses of levodopa (5 g/day) which, however, cannot be recommended at such high doses and has otherwise been shown to be ineffective (Nygaard *et al.*, 1999).

The following drugs have been tested without any good effect on the involuntary movements: amphetamine, heroin, carbamazepine, propranolol, chlorazepate (Kyllerman *et al.*, 1990), and haloperidol (Przuntek and Muhr, 1983).

Stereotactic thalamotomy has improved myoclonus, but caused dysarthria in one patient and mild hemiparesis in another (Gasser et al., 1996). In two other M-D patients, myoclonus responded to thalamotomy but did not result in any significant gain in function (Suchowersky et al., 2000). Deep brain stimulation of the medial globus pallidus was shown to ameliorate both myoclonus and dystonia at an eight week follow-up (Liu et al., 2000). Finally, neurostimulation of the ventral intermediate thalamic nucleus has been shown to be a safe and efficacious treatment in a patient with medically intractable and progressing M-D (Trottenberg et al., 2001).

## Acknowledgments

I would like to thank Laurie Ozelius, Ph.D., Birgitt Müller, M. D., and Norman Kock, M. D. for critical reading of the manuscript and for their helpful comments and Thomas Piskol, M. S. for technical help with the figures. I gratefully acknowledge all patients and family members who participated in studies performed by our group for providing samples and inspiration. Parts of the work cited in this chapter were supported by research grants from the Deutsche Forschungsgemeinschaft and the Fritz Thyssen Stiftung.

## References

Ahmed, M. N., Kim, K. et al. (2000). Comparative genomic hybridization studies in hydatidiform moles and choriocarcinoma: amplification of 7q21–q31 and loss of 8p12–p21 in choriocarcinoma. *Cancer Genet. Cytogenet.* **116(1)**, 10–15.

Aigner, B. R, and Mulder, D. W. (1960). Myoclonus. *Arch. Neurol.* **2**, 600–615.

Angelini, C., Fanin, M. et al. (1999). The clinical spectrum of sarcoglycanopathies. *Neurology* **52(1)**, 176–179.

Araishi, K., Sasaoka, T. et al. (1999). Loss of the sarcoglycan complex and sarcospan leads to muscular dystrophy in beta-sarcoglycan-deficient mice. *Hum. Mol. Genet.* **8(9)**, 1589–1598.

Asmus, F., Zimprich, A. et al. (2001). Inherited myoclonus-dystonia syndrome: narrowing the 7q21–q31 locus in German families. *Ann. Neurol.* **49(1)**, 121–124.

Asmus, F., Zimprich, A. et al. (2000). Support for linkage to chromosome 7 in myoclonus-dystonia families. *Mov. Disord.* **15**, Suppl. 3, P774.

Augood, S. J., Penney, Jr., J. B. et al. (1998). Expression of the early-onset torsion dystonia gene (DYT1) in human brain. *Ann. Neurol.* **43(5)**, 669–673.

Biemond, A. (1963). Paramyoclonus multiplex (Friedreich). Clinical and genetical aspects. *Psychiatr. Neurol. Neurochir. (Amst)* **66**, 270–276.

Bohlega, S., Stigsby, B. et al. (1995). Familial tremulous and myoclonic dystonia with white matter changes in brain magnetic resonance imaging. *Mov. Disord.* **10(4)**, 513–517.

Borges, V., Ferraz, H. B. al. (2000). Alcohol-sensitive hereditary essential myoclonus with dystonia: A study of 6 Brazilian patients. *Neurol Sci* **21(6)**, 373–377.

Bressman, S., and Fahn S. (1986). Essential myoclonus. *Adv. Neurol.* **43**, 287–294.

Bressman, S. B., Sabatti, C. et al. (2000). The DYT1 phenotype and guidelines for diagnostic testing. *Neurology* **54(9)**, 1746–1752.

Brin, M., Doheny, D. et al. (2000). Neuropathology of myoclonic dystonia with the dopamine D2 receptor (DRD2) mutation. *Mov. Disord.* **15**, Suppl. 3, P800.

Caviness, J. N., Alving, L. I. et al. (1999). The incidence and prevalence of myoclonus in Olmsted County, Minnesota. *Mayo Clin. Proc.* **74(6)**, 565–569.

Chadwick, D., Hallett, M. et al. (1978). 5-hydroxytryptophan-induced myoclonus in guinea pigs. A physiological and pharmacological investigations. *J. Neurol. Sci.* **35(1)**, 157–165.

Coral-Vazquez, R., Cohn, R. D. et al. (1999). Disruption of the sarcoglycan-sarcospan complex in vascular smooth muscle: A novel mechanism for cardiomyopathy and muscular dystrophy. *Cell* **98(4)**, 465–474.

Daube, P., and Madison (1966). Hereditary essential myoclonus. *Arch. Neurol.* **15**, 587–594.

Dawidenkow, S. (1926). Auf hereditär-abiotrophischer Grundlage akut auftretende, regressierende und episodische Erkrankungen des Nervensystems und Bemerkungen über die familiäre subakute, myoklonische Dystonie. *Z. Ges. Neurol. Psychol.* **1926(104)**, 596–622.

Doheny, D., Brin, M., F. et al. (2000). Phenotypic features of myoclonic dystonia in two kindreds. *Mov. Disord.* **15** (Suppl. 3), 162.

Duclos, F., Straub, V. et al. (1998). Progressive muscular dystrophy in alpha-sarcoglycan-deficient mice. *J. Cell Biol.* **142(6)**, 1461–1471.

Durbeej, M., and Campbell, K. P. (1999). Biochemical characterization of the epithelial dystroglycan complex. *J. Biol. Chem.* **274(37)**, 26,609–26,616.

Dürr, A., Tassin, J. et al. (2000). D2 dopamine receptor gene in myoclonic dystonia and essential myoclonus. *Ann. Neurol.* **48(1)**, 127–128.

El Kharroubi, A., Piras, G. et al. (2001). DNA demethylation reactivates a subset of imprinted genes in uniparental mouse embryonic fibroblasts. *J. Biol. Chem.* **276(12)**, 8674–8680.

Ettinger, A. J., Feng, G. et al. (1997). Epsilon-sarcoglycan, a broadly expressed homologue of the gene mutated in limb-girdle muscular dystrophy 2d. *J. Biol. Chem.* **272(51)**, 32534–32538.

Fahn, S., Bressman, S. B. et al. (1998). Classification of dystonia. *Adv. Neurol.* **78**, 1–10.

Fahn, S., and Sjaastad, O. (1991). Hereditary essential myoclonus in a large Norwegian family. *Mov. Disord.* **6(3)**, 237–247.

Feldmann, H., and Wieser, S. (1964). Klinische Studie zur essentiellen Myoklonie. *Arch. Z. Ges. Neurol.* **205**, 555–570.

Friedreich, N. (1881). Neuropathologische Beobachtungen: Paramyoklonus multiplex. *Virchows Arch.* **86**, 421–434.

Gasser, T. (1998). Inherited myoclonus-dystonia syndrome. *Adv. Neurol.* **78**, 325–334.

Gasser, T., Bereznai, B. et al. (1996). Linkage studies in alcoholresponsive myoclonic dystonia. *Mov. Disord.* **11(4)**, 363–370.

Grandy, D. K., Litt, M. et al. (1989a). The human dopamine D2 receptor gene is located on chromosome 11 at q22–q23 and identifies a TaqI RFLP. *Am. J. Hum. Genet.* **45(5)**, 778–785.

Grandy, D. K., Marchionni, M. A. et al. (1989b). Cloning of the cDNA and gene for a human D2 dopamine receptor. *Proc. Natl. Acad. Sci. U.S.A.* **86(24)**, 9762–9766.

Grimes, D. A., Bulman, D. et al. (2001). Inherited myoclonus-dystonia: evidence supporting genetic heterogeneity. *Mov. Disord.* **16(1)**, 106–110.

Hack, A. A., Ly, C. T. et al. (1998). Gamma-sarcoglycan deficiency leads to muscle membrane defects and apoptosis independent of dystrophin. *J. Cell Biol.* **142(5)**, 1279–1287.

Hallett, M. (1986). Early history of myoclonus. *Adv. Neurol.* **43**, 7–10.

Hewett, J., Gonzalez-Agosti, C. et al. (2000). Mutant torsinA, responsible for early-onset torsion dystonia, forms membrane inclusions in cultured neural cells. *Hum. Mol. Genet.* **9(9)**, 1403–1413.

Ichinose, H., Ohye, T. et al. (1994). Hereditary progressive dystonia with marked diurnal fluctuation caused by mutations in the GTP cyclohydrolase I gene. *Nat. Genet.* **8(3)**, 236–242.

Imamura, M., Araishi, K. et al. (2000). A sarcoglycan-dystroglycan complex anchors dp116 and utrophin in the peripheral nervous system. *Hum. Mol. Genet.* **9(20)**, 3091–3100.

Ishikawa, A., and Tsuji, S. (1996). Clinical analysis of 17 patients in 12 Japanese families with autosomal-recessive type juvenile parkinsonism. *Neurology* **47(1)**, 160–166.

Kanthasamy, A. G., Matsumoto, R. R. et al. (1995). Animal models of myoclonus. *Clin. Neurosci.* **3(4)**, 236–245.

Kanthasamy, A. G., Nguyen, B. Q. et al. (2000). Animal model of post-hypoxic myoclonus: II. Neurochemical, pathologic, and pharmacologic characterization. *Mov. Disord.* **15** Suppl. 1, 31–38.

Klein, C., Breakefield, X. O. et al. (1999a). Genetics of dystonia. Inherited neurologic disorders. *Semin. Neurol.* **3**, 271–280.

Klein, C., Brin, M. E. et al. (1998). De novo mutations (GAG deletion) in the DYT1 gene in two non-Jewish patients with early-onset dystonia. *Hum Mol Genet* **7**(7): 1133–1136.

Klein, C., Brin, M. F. et al. (1999b). Association of a missense change in the D2 dopamine receptor with myoclonus dystonia. *Proc. Nat. Acad. Sci. U.S.A.* **96**(9), 5173–5176.

Klein, C., Gurvich, N. et al. (2000a). Evaluation of the role of the D2 dopamine receptor in myoclonus dystonia. *Ann. Neurol.* **47**(3), 369–373.

Klein, C., Schilling, K. et al. (2000b). A major locus for myoclonus-dystonia maps to chromosome 7q in eight families. *Am. J. Hum. Genet.* **67**(5), 1314–1319.

Knappskog, P. M., Flatmark, T. et al. (1995). Recessively inherited L-dopa-responsive dystonia caused by a point mutation (q381k) in the tyrosine hydroxylase gene. *Hum. Mol. Genet.* **4**(7), 1209–1212.

Korten, J. J., Notermans, S. L. H. et al. (1974). Familial essential myoclonus. *Brain* **97**, 131–138.

Kotzot, D., Schmitt, S. et al. (1995). Uniparental disomy 7 in Silver-Russell syndrome and primordial growth retardation. *Hum. Mol. Genet.* **4**(4), 583–587.

Kurlan, R., Behr, J. et al. (1988). Myoclonus and dystonia: a family study. *Adv. Neurol.* **50**, 385–389.

Kyllerman, M., Forsgren, L. et al. (1990). Alcohol-responsive myoclonic dystonia in a large family: Dominant inheritance and phenotypic variation. *Mov. Disord.* **5**(4), 270–279.

Lang, A. E. (1997). Essential myoclonus and myoclonic dystonia. *Mov. Disord.* **12**(1), 127.

Leung, J. C., Klein, C. et al. (2001). Novel mutation in the Tor1A (DYT1) gene in atypical early onset dystonia and polymorphisms in dystonia and early onset parkinsonism. *Neurogenetics* **3**(3), 133–143.

LeWitt, P. A. (1995). Dystonia caused by drugs. In *Handbook of Dystonia*, (J. King, C. Tsui, and D. B. Calne, eds.), pp. 227–240, Marcel Dekker, Inc., New York.

Lindemulder, F. (1933). Familial myoclonia occurring in three successive generations. *J. Nerv. Ment. Dis.* **77**, 489.

Liu, L. A. and Engvall E. (1999). Sarcoglycan isoforms in skeletal muscle. *J Biol Chem* **274**(53): 38171–38176.

Liu, X., Griffin, I. et al. (2000). Coherence of pallidal field potential with surface EMG in familial myoclonus dystonia. *Mov. Disord.* **15**, Suppl. 3, P473.

Mahloudji, M., and Pikielny, R. T. (1967). Hereditary essential myoclonus. *Brain* **90**(3), 669–674.

McNally, E. M., Ly, C. T. et al. (1998). Human epsilon-sarcoglycan is highly related to alpha-sarcoglycan (adhalin), the limb girdle muscular dystrophy 2d gene. *FEBS Lett.* **422**(1), 27–32.

Missale, C., Nash, S. R. et al. (1998). Dopamine receptors: from structure to function. *Physiol. Rev.* **78**(1), 189–225.

Müller, U., Steinberger, D. et al. (1998). Clinical and molecular genetics of primary dystonias. *Neurogenetics* **1**(3), 165–177.

Nygaard, T. G., Raymond, D. et al. (1999). Localization of a gene for myoclonus-dystonia to chromosome 7q21–q31. *Ann. Neurol.* **46**(5), 794–798.

Obeso, J. A. (1983). Myoclonic dystonia. *Neurology* **33**, 825–830.

Ono, R., Kobayashi, S. et al. (2001). A retrotransposon-derived gene, PEG10, is a novel imprinted gene located on human chromosome 7q21. *Genomics* **73**(2), 232–237.

Ozelius, L. J., Hewett, J. W. et al. (1997). The early-onset torsion dystonia gene (DYT1) encodes an ATP-binding protein. *Nat. Genet.* **17**(1), 40–48.

Ozelius, L. J., Page, C. E. et al. (1999). The Tor1a (DYT1) gene family and its role in early onset torsion dystonia. *Genomics* **62**(3), 377–384.

Piras, G., El Kharroubi, A. et al. (2000). Zac1 (lot1), a potential tumor suppressor gene, and the gene for epsilon-sarcoglycan are maternally imprinted genes: Identification by a subtractive screen of novel uniparental fibroblast lines. *Mol. Cell Biol.* **20**(9), 3308–3315.

Priori, A., Bertolasi, L. et al. (2000). Gamma-hydroxybutyric acid for alcohol-sensitive myoclonus with dystonia. *Neurology* **54**(8), 1706.

Przuntek, H., and Muhr, H. (1983). Essential familial myoclonus. *J. Neurol.* **230**(3), 153–162.

Quinn, N. P. (1996). Essential myoclonus and myoclonic dystonia. *Mov Disord* **11**(2): 119–124.

Quinn, N. P., and Marsden, C. D. (1984). Dominantly inherited myoclonic dystonia with dramatic response to alcohol. *Neurology* **34** (Suppl. 1), 236.

Quinn, N. P., Rothwell, J. C. et al. (1988). Hereditary myoclonic dystonia, hereditary torsion dystonia and hereditary essential myoclonus: an area of confusion. *Adv. Neurol.* **50**, 391–401.

Reik, W., and Walter, J. (2001). Evolution of imprinting mechanisms: the battle of the sexes begins in the zygote. *Nat. Genet.* **27**(3), 255–256.

Saunders-Pullman, R., Shriberg, J. et al. (2002). The spectrum of myoclonus dystonia: Possible association with OCD and alcohol dependence. *Neurology* **58**(2), 242–245.

Scheidtmann, K., Müller, F. et al. (2000). Familiäres Myoklonus-Dystonie-Syndrom assoziiert mit Panikattacken. *Nervenarzt* **71**(10), 839–842.

Shemer, R., Hershko, A. Y. et al. (2000). The imprinting box of the Prader-Willi/Angelman syndrome domain. *Nat. Genet.* **26**(4), 440–443.

Shohami, E., Evron, S. et al. (1986). A new animal model for action myoclonus. *Adv. Neurol.* **43**, 545–552.

Straub, V., Ettinger, A. J. et al. (1999). Epsilon-sarcoglycan replaces alpha-sarcoglycan in smooth muscle to form a unique dystrophin-glycoprotein complex. *J. Biol. Chem.* **274**(39), 27,989–27,996.

Suchowersky, O., Davis, J. L. et al. (2000). Thalamic surgery for essential myoclonus results in clinical but not functional improvement. *Mov. Disord.* **15**, Suppl. 3, P332.

Trottenberg, T., Meissner, W. et al. (2001). Neurostimulation of the ventral intermediate thalamic nucleus in inherited myoclonus-dystonia syndrome. *Mov. Disord.* **16**(4), 769–771.

Valente, E. M., Bentivoglio, A. R. et al. (2001). DYT13, a novel primary torsion dystonia locus, maps to chromosome 1p36.13–36.32 in an Italian family with cranial-cervical or upper limb onset. *Ann. Neurol.* **49**(3), 362–366.

Vidailhet, M., Tassin, J. et al. (2001). A major locus for several phenotypes of myoclonus-dystonia on chromosome 7q. *Neurology* **56**(9), 1213–1216.

Wahlström, J., Ozelius, L. et al. (1994). The gene for familial dystonia with myoclonic jerks responsive to alcohol is not located on the distal end of 9q. *Clin. Genet.* **45**(2), 88–92.

Zimprich, A., Grabowski, M. et al. (2001). Mutations in the gene encoding epsilon-sarcoglycan cause myoclonus-dystonia syndrome. *Nat. Genet.* **29**(1), 66–69.

CHAPTER 41

# Mitochondrial Mutations in Parkinson's Disease and Dystonias

DAVID K. SIMON

*Department of Neurology*
*Beth Israel Deaconess Medical Center and Harvard Medical School*
*Boston, Massachusetts 02115*

I. Mitochondrial Genetics
II. Parkinson's Disease
   A. Genetic Factors and PD
   B. Evidence that Mitochondrial Complex I Dysfunction Plays an Important Role in PD
   C. Connection between Mitochondrial Dysfunction and Alpha-Synuclein
   D. Origin of the Complex I Defect in PD; Evidence of a Role for mtDNA Mutations
   E. mtDNA Mutations in Familial Parkinsonism
   F. mtDNA Mutations in Sporadic PD
   G. Acquired mtDNA Mutations in PD
III. Dystonia
   A. Evidence for Mitochondrial Dysfunction in Dystonia
   B. Mitochondrial Toxins and Dystonia
   C. mtDNA Mutations and Rare Forms of Dystonia
   D. mtDNA Mutations in Idiopathic Dystonia
IV. Treatment Implications
V. Conclusions
   References

The role of mitochondrial DNA (mtDNA) mutations in human disease was first demonstrated in 1988 with the identification of a missense mutation in a mitochondrial complex I gene in patients with Leber's hereditary optic neuropathy (Wallace *et al.*, 1988) and of mtDNA deletions in patients with mitochondrial myopathies (Holt *et al.*, 1988) or Kearns-Sayre syndrome (Lestienne and Ponsot, 1988; Zeviani *et al.*, 1988). Since then, the rate of identification of new mtDNA mutations associated with various rare human diseases has rapidly increased (MITOMAP, 2001). It is increasingly being recognized that mitochondrial dysfunction and mtDNA mutations also may play a key role in much more common disorders such as Parkinson's disease (PD), dystonia, and even aging itself. In addition to PD and dystonia, mitochondrial dysfunction is hypothesized to play a key role in other disorders of the basal ganglia, including progressive supranuclear palsy (Albers *et al.*, 1999, 2000, 2001; Swerdlow *et al.*, 2000) and multiple system atrophy (MSA); (Blin *et al.*, 1994; Martinelli *et al.*, 1996; Swerdlow *et al.*, 1997), though this is controversial in the case of MSA (Benecke, *et al.*, 1993b; Gu *et al.*, 1997; Schapira *et al.*, 1992). Thus, the basal ganglia appear to be particularly susceptible to the effects of mitochondrial dysfunction.

The evidence is strongest in the case of PD (see Section II.B for details and references). Mitochondrial complex I is impaired in the brain in PD, and inhibitors of complex I induce parkinsonism in a variety of animal models. Indirect evidence from cell lines expressing mtDNA from PD patients indicates that mtDNA mutations account for the complex I dysfunction in PD, but extensive searches for mtDNA mutations in PD patients have failed to find mutation in the majority of patients. Accumulation of individually rare somatic mtDNA mutations, which might not be detectable by standard screening techniques, may account for the complex I dysfunction in PD. Dystonia is a clinical feature common to a variety of distinct neurological disorders, including PD. Several rare forms of dystonia are associated with mtDNA mutations. However, the role of mitochondrial dysfunction and mtDNA mutations in more common late-onset forms of dystonia remains unknown. This chapter will begin with a review of some basic principles of mitochondrial genetics, followed by a review of the evidence for a role for mtDNA mutations in PD, including a discussion of the possibility that the accumulation of somatic mtDNA mutations may account for the mitochondrial dysfunction in PD. Next, mtDNA mutations that have been linked to certain forms of dystonia will be presented, with a discussion of the possible role of mtDNA mutations in the more common late-onset forms of focal dystonia.

## I. MITOCHONDRIAL GENETICS

Understanding the potential role of mtDNA mutations in PD or dystonia requires awareness of several unique features of mitochondrial genetics. More detailed discussions of these issues are available in prior reviews (DiMauro and Moraes, 1993; Shoffner and Wallace, 1994; Simon and Johns, 1999). Mitochondria contain their own genome, encoding 37 genes. These include 13 genes encoding subunits of the mitochondrial electron transport chain, 22 transfer RNAs, and 2 ribosomal RNAs. MtDNA is inherited entirely from one's mother. Thus, classic mitochondrial genetic disorders such as mitochondrial encephalomyopathy, lactic acidosis, and stroke-like episodes (MELAS), myoclonic epilepsy with ragged red fibers (MERRF), leber's hereditary optic neuropathy (LHON), and neuropathy, ataxia, and retinitis pigmentosa (NARP) affect both men and women, but only women pass the mutation to their offspring. However, mtDNA mutations often are sporadically expressed, and thus a maternal inheritance pattern is not always apparent. One reason for this relates to heteroplasmy. Heteroplasmy refers to the co-existence within a cell or tissue of both normal and mutant mtDNA. When all copies of the mtDNA are identical at a given locus, this is termed homoplasmy. Unlike most nuclear genes, which are present in two copies per cell, several hundred mitochondria can be present in each cell, with several copies of the mitochondrial genome in each mitochondrion. The percentage of mtDNA molecules that are mutant is termed the "mutational burden." For most mtDNA mutations, there appears to be a threshold of expression, such that only mutations present at greater than a certain threshold mutational burden (usually around 50% or more) result in physiologically relevant mitochondrial dysfunction (Porteous et al., 1998; Rossignol et al., 1999; Vielhaber et al., 2000). The threshold varies between different tissues. This is complicated further by the fact that tissue mosaicism is common in individuals harboring an mtDNA mutation. Uneven distribution of mutant mtDNA during early cell divisions in the embryo can contribute to this. In addition, postmitotic tissues such as neurons and muscle tend to accumulate mutations to a greater extent than occurs in rapidly dividing cells such as blood cells.

Another reason that mtDNA mutations might not show maternal inheritance is that some mtDNA mutations are acquired. A propensity to develop acquired mtDNA mutations can be inherited in an autosomal dominant or autosomal recessive manner, due to the dependence of mtDNA integrity on products of nuclear-encoded genes. In addition, oxidative damage to mtDNA, which accumulates with age, can induce mtDNA mutations. The potential relevance of this to PD and other neurodegenerative disorders will be discussed further below.

## II. PARKINSON'S DISEASE

### A. Genetic Factors and PD

The importance of genetic factors in late-onset sporadic PD has been controversial. Although mutations in the alpha-synuclein gene play a role in very rare cases of familial PD (Kruger et al., 1998; Polymeropoulos et al., 1997), mutations in this gene are lacking in the vast majority of late-onset PD cases(Chan et al., 2000, 1998a, b; Higuchi et al., 1998; Lin et al., 1999; Lucotte et al., 1998; Munoz et al., 1997; Parsian et al., 1998; Scott et al., 1999; Vaughan et al., 1998a, b; Wang et al., 1998). Mutations in the parkin gene are relatively common in juvenile PD, but are rare in late-onset PD (Hattori et al., 1998a; Kitada et al., 1998; Lucking et al., 1998, 2000). The potential role of environmental factors in PD recently has received increased attention, in part due to a large study of twins showing similar concordance rates for late-onset PD in monozygotic and dizygotic twins (Tanner et al., 1999). The data from this well-designed study argue against a major role for high-penetrant *nuclear* genetic factors in late-onset PD and support a role for environmental factors. However, potentially important genetic susceptibility factors of low-penetrance would be unlikely to be detected by twin studies (Simon, et al., 2002), consistent with more recent results from linkage studies (DeStefano et al., 2001; Scott et al., 2001). Furthermore, the role of inherited *mitochondrial* genetic factors was not addressed as monozygotic and dizygotic twins each receive all of their mtDNA from their mother, and thus a role for mitochondrial genetic factors in PD would not be associated with altered concordance rates (Parker et al., 1999; Simon, 1999). The impact of *acquired* mtDNA mutations also could be missed by this type of study. These are critical issues in the case of PD, as there is evidence that mitochondrial dysfunction due to mtDNA mutations may play a key role in the pathogenesis of PD.

### B. Evidence that Mitochondrial Complex I Dysfunction Plays an Important Role in PD

Dysfunction of mitochondrial complex I is hypothesized to play a role in the pathogenesis of PD (Parker and

Swerdlow, 1998; Schapira et al., 1998; Simon and Beal, 2001). Mitochondrial electron transport chain activity decreases with age in the brain (Bowling et al., 1993; Chen et al., 1994). However, in PD patients, complex I activity is decreased in the substantia nigra to a greater extent than accounted for by normal aging (Janetzky et al., 1994; Mann et al., 1992; Schapira et al., 1990a, c; 1989). Most studies also find a decrease in complex I activity in platelets (Benecke 1993a; Haas et al., 1995; Krige et al., 1992; Parker et al., 1989; Yoshino et al., 1992), indicating a systemic defect, though there is some controversy regarding this point (Bravi et al., 1992; Mann et al., 1992). In the brain the decrease in complex I activity is specific for the substantia nigra (Mann et al., 1992; Schapira et al., 1990c). MSA patients have normal complex I activity in the substantia nigra, suggesting that the decrease in PD patients is not a nonspecific secondary consequence of neurodegeneration (Gu et al., 1997; Schapira et al., 1990c), (though others have reported evidence for mitochondrial dysfunction in MSA (Blin et al., 1994; Martinelli et al., 1996; Swerdlow et al., 1997)). The complex I defect is not secondary to levodopa therapy, as it does not correlate with levodopa dosage and is absent in MSA patients taking levodopa (Benecke et al., 1993b; Blin et al., 1994; Haas et al., 1995; Schapira et al., 1990c).

These data raise the question of whether or not complex I deficiency might be an important etiological factor in PD. Impaired complex I activity is associated with increased oxidative stress in *in vitro* and *in vivo* models (Andrews et al., 1996; Hasegawa et al., 1990; Pitkanen and Robinson, 1996; Poirier and Barbeau, 1985; Przedborski and Jackson-Lewis, 1998; Przedborski et al., 1992; Sinha et al., 1986; Zang and Misra, 1992). Similarly, in PD, biochemical markers of oxidative stress are increased in the substantia nigra (Schulz and Beal, 1994), providing a possible mechanism by which complex I deficiency might lead to neurodegeneration. However, complex I deficiency also can be a consequence rather than just a cause of oxidative stress (Jha et al., 2000).

A major clue that complex I impairment may represent an important step in the pathogenesis of PD and not just a secondary effect comes from the ability of 1-methyl-4-phenyl-1,2,3,6-tetrahydropyridine (MPTP) to induce clinical and pathological features similar to those of PD when systemically administered to man or nonhuman primates (Langston, 1996). MPTP is a potent inhibitor of mitochondrial complex I. This suggests that complex I deficiency can be a primary cause of neuronal degeneration. However, MPTP also inhibits alpha-ketoglutarate dehydrogenase (KGDH); (Mizuno et al., 1987). KGDH catalyzes the rate-limiting step in the Krebs cycle (Lai et al., 1977). Immunohistochemical staining of KGDH is decreased in postmortem PD substantia nigra (Mizuno et al., 1994), and a polymorphism in the E2 subunit of KGDH has been reported to occur at increased frequency in PD patients (though this has yet to be confirmed); (Kobayashi et al., 1998). Thus, the MPTP data alone do not conclusively implicate complex I dysfunction in PD. The identification of a second distinct complex I inhibitor that also reproduces features of PD in animals suggests that complex I dysfunction may, indeed, play a causal role in PD. Chronic systemic infusion in rats of low-dose rotenone, a specific and potent inhibitor of complex I, produces a progressive parkinsonian syndrome with loss of substantia nigra neurons (Betarbet et al., 2000). The ability of two distinct complex I inhibitors to induce a parkinsonian syndrome, together with the known complex I defect in the PD substantia nigra, emphasizes the importance of further investigation of complex I dysfunction as a potential pathogenic mechanism in PD.

### C. Connection between Mitochondrial Dysfunction and Alpha-Synuclein:

The parallel work on the role of alpha-synuclein in PD also now is revealing a strong connection to mitochondrial mechanisms. Inhibition of complex I (Betarbet et al., 2000; Kowall et al., 2000) or exposure to oxidative stress (Duda et al., 2000; Giasson et al., 2000; Hashimoto et al., 1999; Souza et al., 2000) promotes alpha-synuclein aggregation. Conversely, overexpression of mutant (Lee et al., 2001) or wild-type (Hsu et al., 2000) alpha-synuclein induces mitochondrial dysfunction and oxidative stress, and expression of mutant alpha-synuclein enhances susceptibility to oxidative stress (Ko et al., 2000). Oxidative ligation of dopamine to alpha-synuclein may induce the accumulation of alpha-synuclein protofibrils, which are hypothesized to be the toxic form of alpha-synuclein (Conway et al., 2001). These data suggest that mitochondrial dysfunction and the accompanying oxidative stress may be common factors in the pathogenesis of PD in both familial and sporadic cases.

### D. Origin of the Complex I Defect in PD; Evidence of a Role for mtDNA Mutations

Recognition of the potential pathogenic significance of complex I dysfunction in PD highlights the importance of understanding the origin of the complex I defect. Both environmental and genetic factors likely are important and interrelated, though this chapter focuses on the genetic factors. Genetic factors could include both nuclear and mitochondrial genes, as complex I is composed of over 40 subunits, 7 of which are encoded in the mitochondrial genome (the rest in the nuclear genome). A role for nuclear genetic factors is suggested by a report of a polymorphism in a nuclear-encoded subunit of complex I that is present at a higher frequency in Japanese patients with PD compared to controls (24% for PD versus 12% for controls for homozygous mutations) (Hattori et al., 1998b).

In addition, there is indirect evidence that mtDNA mutations may account for the complex I defect in most PD patients. Before discussing the evidence, this possibility immediately raises the issue of maternal inheritance. To the extent that mtDNA mutations influence the risk of PD, one would expect a maternal inheritance bias in cases of PD in which a parent also had PD. Though some prior studies failed to find such a bias (Maraganore et al., 1991; Zweig et al., 1992), a maternal inheritance bias in PD has been identified in more recent studies, consistent with an influence of mitochondrial genetic factors (Swerdlow et al., 2001; Wooten et al., 1997). This issue remains controversial. However, in any case, a maternal inheritance pattern may not be apparent as mtDNA mutations typically are sporadically expressed.

Evidence that mtDNA mutations may account for the complex I defect comes from studies of cytoplasmic hybrids ("cybrid") cell lines (Fig. 41.1). Cybrids provide a mechanism for studying the impact of differences in mtDNA between different subjects by expressing mtDNA from different subjects in cell lines that each contain identical nuclear DNA (King and Attardi, 1989; Miller et al., 1996). Cybrids can be prepared by exposure of tumor cells to low levels of ethidium bromide over several months, which preferentially eliminates the mtDNA leaving nuclear DNA relatively intact. The resulting "$\rho^0$" cells lack mtDNA, but can survive if provided supplemental uridine and pyruvate. These cells are then fused with platelets (which contain mitochondria and mtDNA, but no nuclear DNA). The successful fusions are selected by growing the post-fusion cells in media lacking uridine or pyruvate.

Swerdlow and colleagues (1996) reported that SH-SY5Y cybrid cell lines expressing mtDNA from PD patients have impaired complex I activity compared to cybrids expressing mtDNA from control subjects (Swerdlow et al., 1996). This defect remains present after numerous passages in culture, indicating that it is not the result of a toxin that is transferred during cybrid preparation. This work was subsequently confirmed by Gu and colleagues (1998) using a distinct $\rho^0$ cell line. The cybrid cell lines from the PD patients and controls have identical nuclei but differ only in the source of their mtDNA. Therefore, the ability to transfer the complex I defect of PD patients to cybrid cell lines expressing mtDNA from these patients indicates that the complex I defect originates from mtDNA mutations.

This conclusion recently has been challenged by another study of cybrids with different results. Aomi and colleagues (2001) transferred mtDNA from PD patients or controls to mtDNA-less ($\rho^0$) HeLa cells. They found wide variations in enzyme activities, but no significant difference in complex I activity in the PD cybrids versus controls. While this may seem to contradict the prior two studies that did demonstrate a complex I defect in PD cybrids, several factors potentially can account for this finding in HeLa-cell-derived cybrids. Most importantly, the biochemical and clinical expression of mtDNA mutations is highly dependent on tissue type and nuclear background (Cock et al., 1998; Hao, et al., 1999). Therefore, it is not surprising that a complex I defect would result from expression of mtDNA from PD patients in cybrids derived from human neuroblastoma or lung cancer cells but not in HeLa cells. The large body of data implicating a role for complex I dysfunction in PD, and the ability to transfer the complex I defect to at least some cybrid cell lines, indicate the importance of determining directly whether or not mtDNA mutations play a role in PD.

### E. MtDNA Mutations in Familial Parkinsonism

The data presented thus far indicate that a complex I defect is present in the substantia nigra of PD patients compared to age-matched controls. Pharmacological inhibition of complex I by MPTP or rotenone reproduces many features of PD in animal models. The complex I defect in PD can be transferred to cybrid cell lines expressing mtDNA from PD patients, indicating that the defect arises, at least in part, from mtDNA mutations. This work thus sets the stage for a search for the specific mtDNA mutations that account for the complex I defect in PD. Unfortunately, these mutations have been difficult to find.

Two approaches have been used in attempts to identify the relevant mtDNA mutations. Some studies have focused on exceptional families with maternally inherited parkinsonism, whereas others have attempted to identify mtDNA mutations associated with typical sporadic

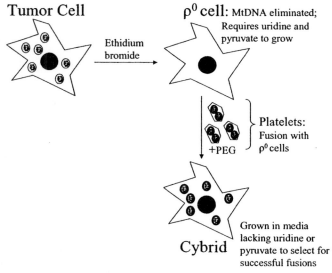

FIGURE 41.1 Cybrid preparation.

late-onset PD. A family with multiple maternally related members affected with PD over three generations was found to have a complex I defect (Swerdlow et al., 1998). Even asymptomatic young maternal relatives had impaired complex I activity. The complex I defect was transferable to cybrids. The coincidence of maternally inherited PD and complex I deficiency in this family supports the earlier work in cybrids from sporadic PD patients, and suggests that mtDNA mutations can cause complex I deficiency and PD. However, no specific mutations have yet been reported in this family.

Another multigenerational family has been identified with maternally inherited prominent parkinsonism (Fig. 41.2); (Simon et al., 1999). Unlike the previously discussed family, this family did not have typical PD, but also had other clinical features such as dementia, dystonia, and pyramidal dysfunction in some affected individuals. Interestingly, one family member developed external ophthalmoplegia and ptosis years before his parkinsonism developed. Another had psychosis and developed severe parkinsonism on neuroleptics. Tremors were levodopa-responsive in the one family member known to have tried levodopa. Postmortem examination of this patient's brain revealed severe loss of dopaminergic neurons in the substantia nigra as well as in the caudate and putamen, but no Lewy bodies. This family was found to harbor a heteroplasmic G to A missense mtDNA mutation at position 11778 (G11778A) in the mitochondrial gene encoding the ND4 subunit of complex I. The mutation converts a highly conserved arginine to a histidine. This mutation previously has been identified as the most common cause of LHON, a disorder that also is associated with complex I dysfunction but is generally restricted clinically to optic atrophy (Newman, 1993; Simon and Johns, 1999). It also has been reported in rare multiple sclerosis patients, particularly those with early and prominent optic nerve involvement (Bhatti and Newman, 1999; Harding et al., 1992; Kalman and Alder, 1998; Kalman et al., 1995; Kellar-Wood et al., 1994; Mayr-Wohlfart et al., 1996; Mojon et al., 1999; Ohlenbusch et al., 1998; Olsen et al., 1995; Vanopdenbosch et al., 2000). The identification of the G11778A mutation in this family demonstrates that an inherited mtDNA mutation can cause adult-onset parkinsonism. This conclusion is confirmed by a recent report of mtDNA deletions in two adult-onset parkinsonism patients (Siciliano et al., 2001). These data indicate the importance of searching for mtDNA mutations in patients with typical adult-onset PD.

In addition, multiple mitochondrial DNA deletions are present in rare families in which parkinsonism and other neurological deficits are inherited in an autosomal dominant

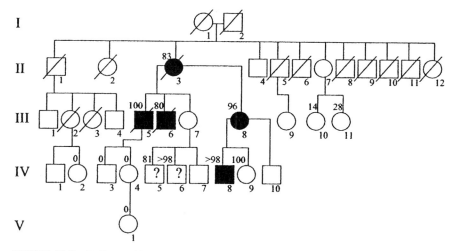

FIGURE 41.2 Pedigree of family harboring the G11778A complex I mutation. Darkened symbols represent clinically affected individuals. Question marks indicate two subjects with uncertain clinical status. The percentage of mutant DNA (in blood-derived DNA) is indicated to the top left of the symbols for each subject analyzed genetically. All affected individuals except IV–8 had prominent parkinsonism (patient III–6 by history alone). All had dysarthria. Subjects II–3 and III–8 had dementia and dystonia as well. Subject IV–8 had hyperreflexia and ataxia. Subject IV–6 had been diagnosed with schizophrenia and neuroleptic-induced parkinsonism. Subject IV–5 had congenital mental retardation without evidence of progression. Subject III–5 developed progressive external ophthalmoplegia and ptosis many years prior to developing parkinsonism. Neuropathology in this subject revealed marked loss of pigmented neurons in the substantia nigra along with changes in other brain regions, but no Lewy bodies. Ages of onset of neurological symptoms ranged from the teens (III–5) to 50 (II–3). Subject IV–9 was asymptomatic despite a homoplasmic mutation, but she was only 25 years old when last examined and thus could have been presymptomatic. Reprinted with permission (From Simon et al., (1999). *Neurology* **53**, 1787–1793. With permission.)

or autosomal recessive manner, indicating that a mutation in a nuclear-encoded gene predisposes to the development of multiple large deletions in mtDNA in these families (Casali et al., 2001; Chalmers et al., 1996; Checcarelli et al., 1994; Hara et al., 1994; Moslemi et al., 1999).

## F. mtDNA Mutations in Sporadic PD

The search for specific mtDNA mutations in PD began shortly after the identification of mitochondrial complex I deficiency in PD, even before the cybrid data suggested a role for mtDNA mutations. As early as 1990, a question arose as to whether or not a large mtDNA deletion known as the "common deletion" may play a role in PD. The common deletion is a 4977-bp deletion that is present in a tiny fraction of mtDNA molecules from controls (Corral-Debrinski et al., 1992; Cortopassi et al., 1992; Lee et al., 1994; Lezza et al., 1999; Simonetti et al., 1992). An increase in the common deletion has been reported in the striatum of PD patients compared to controls (Ikebe et al., 1990; Ozawa et al., 1990), but most of the PD cases in this study were substantially older than most controls in these studies. Other groups have found no difference in the proportion of mtDNA with the 4977 common deletion in PD compared to age-matched control striatum (Lestienne et al., 1990), substantia nigra (Kosel et al., 1997; Lestienne et al., 1990; Mann et al., 1992; Schapira et al., 1990b) or platelets (Sandy et al., 1993). A study of striatal mtDNA from one PD patient reported identification of numerous fragments representing variably sized deletions of mtDNA (Ozawa et al., 1997). Though this could reflect fragile mtDNA in PD patients, interpretation of these data is limited by the analysis of only a single PD patient and lack of comparison with striatal DNA from a normal control.

Other studies have compared the frequency of more common inherited polymorphisms in PD patients compared to controls, and several of these polymorphisms are reported to occur at an increased frequency in PD. A substitution of a G for an A at position 5460 of the mitochondrial gene encoding the ND2 subunit of complex I was reported to occur at a higher frequency in PD cases compared to controls (4 of 21 PD cases compared to 5 of 77 controls); (Kosel et al., 1996). However, this could not be replicated in two later studies (Bandmann et al., 1997; Simon et al., 2000). A mutation at position 4336 in the mitochondrial tRNA(Gln) gene has been reported in PD patients (Kosel et al., 1996; Shoffner et al., 1993), but again others failed to confirm an association of this mutation with PD (Bandmann et al., 1997; Simon et al., 2000). Either of two mutations in the mitochondrial tRNA(Thr) gene, one at position 15927 and one at 15928, eliminates an *HpaII* restriction site. Screening for the presence of either of these mutations by loss of this restriction site revealed loss of the site in 15 of 100 PD patients and in only 3 of 100 controls (Mayr-Wohlfart et al., 1997). A subsequent study was designed to distinguish between these two mutations by direct sequencing of any "positive" results from the restriction digest (Simon et al., 2000). In this study, the 15927 mutation was present in only 1 of 271 PD patients. The 15928 mutation was present in approximately 8% of PD patients and in a similar percentage of controls. Thus, no specific mtDNA point mutations have been consistently shown to be associated with PD.

Another strategy to identify mtDNA mutations in PD patients has been to sequence part or all of the mitochondrial genome in patients with PD (Brown et al., 1996; Kapsa et al., 1996; Kosel et al., 1998; Ozawa et al., 1991; Schapira et al., 1990b; Simon et al., 2000). These studies have revealed various point mutations, generally in tRNA or complex I genes, in a subset of PD patients, though the significance of these mutations remains uncertain. For example, in our study, we completely sequenced each of the mitochondrial complex I and tRNA genes in 28 PD patients, and then used restriction endonuclease assays to screen over 200 PD patients and over 200 controls for any potentially interesting mutations identified by sequencing (Simon et al., 2000). However, in this study, as in the others, the mutations identified were either found in only a single PD patient each, or were also found in controls. Interpretation of the results of these studies is complicated by the highly polymorphic nature of mtDNA. Polymorphisms, including missense mutations, are frequently found in controls, and insufficient controls have been sequenced to document all polymorphisms. This makes it difficult to distinguish between pathogenic mutations and clinically irrelevant polymorphisms (Chinnery et al., 1999; Florentz and Sissler, 2001; Simon et al., 2001b).

## G. Acquired mtDNA Mutations in PD

### 1. Theory of Acquired mtDNA Mutations in PD

The data discussed thus far reveal an apparent paradox. Two groups have independently demonstrated that the

TABLE 41.1 Reported Associations of Specific mtDNA Mutations and PD

| mtDNA mutations reported in PD | mtDNA mutations definitively linked to PD |
|---|---|
| "Common deletion" (4977 base pairs) | None |
| 4336 (tRNA-Gln) | |
| 5460 (complex I-ND2) | |
| 15927/15928 (tRNA-Thr) | |
| Multiple variable deletions | |
| Rare point mutations, each in 1 or 2 PD patients | |

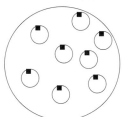 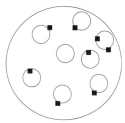

Homoplasmic mutation — The identical base pair change is present at the same position in every mtDNA molecule

Acquired mutations — Various base pair changes are present at different locations in different mtDNA molecules. Some mtDNA molecules may have multiple mutations. Others may have none.

**FIGURE 41.3** Homoplasmic versus low mutational burden acquired mutations.

complex I defect in PD patients can be transferred to cybrid cell lines expressing mtDNA from PD platelets, indicating that mtDNA mutations account for the defect (Gu et al., 1998; Swerdlow et al., 1996). Yet, extensive sequencing of mtDNA in PD patients has failed to reveal definitive evidence for mtDNA mutations that might account for this defect (Simon et al., 2000).

A potential explanation consistent with these observations is that *acquired* mtDNA mutations account for the complex I defect. Acquired randomly or semi-randomly positioned mutations may not reach a level detectable by standard methods at any single base pair site, though the cumulative burden of these mutations could be sufficient to cause mitochondrial dysfunction (Lin et al., 2002; Linnane et al., 1989; Simon et al., 2001a); (see Fig. 41.3). A likely source for acquired mtDNA mutations in PD is oxidative stress. Brain levels of OH$^8$dG, a marker of oxidative damage to DNA (Kasai, 1997), are increased 16-fold in mtDNA compared to nuclear DNA (Richter et al., 1988), increase with aging (Ames et al., 1993; Hayakawa et al., 1992; Mecocci et al., 1993), and increase further in neurodegenerative diseases (Alam et al., 1997b; Ferrante et al., 1997; Mecocci et al., 1994; Sanchez-Ramos et al., 1994), including PD (Alam et al., 1997b; Sanchez-Ramos et al., 1994; Shimura-Miura et al., 1999). OH$^8$dG can induce mutations during DNA replication as it can pair with adenine as well as cytosine, with almost equal efficiencies (Cheng et al., 1992; Kuchino et al., 1987; Maki and Sekiguchi, 1992; Shibutani et al., 1991). This results in G:C to T:A or T:A to G:C mutations. Mutations in the *E. coli* homolog of 8-oxo-dGTPase result in a dramatic increase in the frequency of spontaneous G:C to T:A (or T:A to G:C) mutations, and this mutator phenotype is partially suppressed (8-fold reduction) when an enzyme that specifically repairs oxidatively damaged DNA is expressed in the mutated

*E. coli* (Kakuma et al., 1995). Expression of the human homolog of this enzyme, 8-oxo-7, 8-dihydrodeoxyguanosine triphosphatase (8-oxo-dGTPase, or hMTH1), which is expressed specifically in the cytosol and mitochondria (Kang et al., 1995), is dramatically increased in the substantia nigra in PD, reflecting the increased oxidative damage to mtDNA (Shimura-Miura et al., 1999).

Though damage induced by oxidative stress is likely to be diffuse at a cellular level, a preferential effect on complex I is likely. Induction of oxidative stress by glutathione depletion in PC12 cells results in a selective inhibition of complex I activity (Jha et al., 2000). In this *in vitro* model, it appears that direct thiol oxidation of proteins accounts for the impaired complex I activity. However, in PD, studies of cybrid cell lines expressing mtDNA from PD patients indicate that mtDNA mutations account for the complex I defect. A model of the impact of randomly positioned mtDNA mutations induced by oxidative stress predicts a preferential impact of these mutations on complex I function (Cortopassi and Wang, 1995). In PD, markers of oxidative damage to both proteins (Alam et al., 1997a) and DNA (Alam et al., 1997b; Sanchez-Ramos et al., 1994) are increased in the substantia nigra, so both processes may contribute to complex I dysfunction. Inhibition of complex I leads to an increase in free radicals (Cleeter et al., 1992; Hasegawa et al., 1990; Pitkanen and Robinson, 1996; Takayanagi et al., 1980; Turrens and Boveris, 1980), and so a degenerative cycle can be envisioned in which initial complex I dysfunction leads to increased free radical damage to proteins and to mtDNA, leading to further mitochondrial dysfunction. Ultimately, this may lead to the accumulation of mtDNA mutations beyond a threshold at which cell death is triggered.

This hypothesized situation of multiple different individually rare mtDNA species induced by oxidative damage would be undetectable by standard sequencing methods, which detect mutations only when a specific mutation (at a particular base pair) is present in a large percentage of the mtDNA molecules. More sensitive techniques, such as two-dimensional denaturing gradient gel electrophoresis (van Orsouw et al., 1998), still may lack the sensitivity required to detect this situation. Therefore, the failure to identify these mutations in prior studies is not surprising, and future studies using techniques capable of detecting these types of mutations are required.

## 2. Do Somatic mtDNA Mutations Achieve High Mutational Burdens in the Brain?

The A3243G mutation, which accounts for about 80% of all cases of MELAS (Goto et al., 1990), is also present in a small percentage of mtDNA molecules in many tissues of control subjects. This percentage increases manyfold with age, but does not reach more than a tiny percentage (less than 1%) of the mtDNA molecules even in elderly subjects (Liu et al., 1997; Zhang et al., 1993, 1998). Interestingly, there is evidence that the A3243G mutation, when present at high mutational burdens, can induce the accumulation of additional somatic mutations, which may account for the late onset and progressive nature of MELAS (Kovalenko et al., 1996; Tanaka et al., 1996). A large 4977 base pair "common deletion" of mtDNA also is present in small percentage of mtDNA, and accumulates with age (Cortopassi et al., 1992; Meissner et al., 1997, 1999). In most tissues, including brain, the deletion is present in only a low percentage (generally 1% or less) of mtDNA molecules (Corral-Debrinski et al., 1992; Cortopassi et al., 1992; Lee et al., 1994; Lezza et al., 1999; Simonetti et al., 1992). This is well below the 50–55% level required to induce impaired energy metabolism in cybrids (Porteous 1998).

Though both the A3243G point mutation and the 4977 common deletion accumulate with age in normal subjects, neither is present in more than a tiny fraction of the mtDNA molecules even in elderly controls. In contrast, there is evidence from studies of human tumors that specific somatic point mutations can accumulate to reach high mutational burdens (Fliss et al., 2000; Polyak et al., 1998). The possibility that this might occur during normal aging as well was suggested by a study of mtDNA derived from fibroblasts. A heteroplasmic T to G substitution at nucleotide position 414 (T414G) in the noncoding mitochondrial D-loop was identified in 8 of 14 subjects over the age of 65 years, but in none of 13 younger subjects. The T414G mutation reached mutational burdens of up to 50%. These remarkable findings suggest that a single specific acquired mtDNA point mutation can achieve relatively high levels of mutational burden in a large fraction of elderly subjects. If a similar process occurs in the brain, then this might increase the chances that somatic mtDNA mutations might contribute to mitochondrial dysfunction in the brain. However, subsequent studies demonstrated that the T414G mutation is absent in mtDNA derived from brain tissue of elderly subjects with PD or AD, and controls (Chinnery et al., 2001; Simon et al., 2001a).

An interesting finding emerged from our study of the T414G mutation (Simon et al., 2001a). We confirmed absence of the T414G mutation in brain by cloning individual mtDNA molecules derived from the brain of elderly subjects, followed by sequencing of 70 of these clones. The sequenced region encompassed a portion of the D-loop including the 414 site. Consistent with the results from restriction endonuclease analyses, the T414G mutation was absent from every clone. However, a total of 11 other point mutations were found in 9 of these clones (2 of these clones each contained 2 point mutations). No particular point mutation was identified in more than a single clone (with the exception of known polymorphisms present in all clones from a given subject), suggesting that it is uncommon for any single specific somatic mtDNA mutation to reach high mutational burdens in brain. Because the study was conducted on DNA isolated from brain homogenate, the data do not rule out the possibility that specific mutations accumulate within individual neurons, as may be the case in muscle (Taylor et al., 2001). However, the data do suggest that multiple different, individually rare mutations may accumulate in the brain, so that the aggregate burden of these mutations could contribute to mitochondrial dysfunction in aging and age-related neurodegenerative diseases.

Studies involving sequencing of cloned mtDNA molecules are capable of detecting extremely low frequency mutations such as these, but must be interpreted with caution. One potential problem involves nuclear pseudogenes, which are evolutionary "fossils" of the mitochondrial genes that have been inserted into the nuclear genome (Perna and Kocher, 1996). Inadvertent amplification of nuclear pseudogenes can lead to the false appearance of mtDNA mutations. This type of artifact accounted for the initial misinterpretation of data as indicating the presence of heteroplasmic mtDNA mutations associated with Alzheimer's disease (Davis and Parker, 1998; Davis et al., 1997; Hirano et al., 1997). The potential for mutations introduced by PCR error also must be considered. A recent study accounting for each of these technical concerns has demonstrated the accumulation with age of point mutations in mtDNA in cortex, and an inverse correlation of the aggregate burden if these mutations with cytochrome oxidase activity (Lin et al., 2002). Further studies are needed to definitively determine the role of these individually rare, but potentially significant in combination, mtDNA mutations in aging and in neurodegenerative diseases such as PD.

## III. DYSTONIA

Recent advances in our understanding of the genetic causes of relatively rare forms of dystonia offer the promise for providing key insights into the pathogenesis of this disorder (Korf, 1998; Muller et al., 1998). Mutations in the nuclear gene encoding GTP cyclohydrolase (Ichinose et al., 1994) or tyrosine hydroxylase (Ludecke et al., 1995) are associated with dopa-responsive dystonia. A single codon (GAG) deletion in the DYT1 gene encoding torsinA is responsible for many cases of early-onset torsion dystonia (Ozelius et al., 1997), particularly in Ashkenazi Jews but in many non-Jewish families as well. In addition, linkage has been established to additional sites in several other families with dystonia, including some with late-onset cranial-cervical dystonia (Almasy et al., 1997; Leube et al., 1996; Valente et al., 2001). However, the cause of idiopathic adult-onset focal dystonia remains unknown in the vast majority of cases. This section focuses on the potential role of mtDNA mutations in late-onset sporadic dystonia.

## A. Evidence for Mitochondrial Dysfunction in Dystonia

Mitochondrial dysfunction is a feature of several rare forms of dystonia associated with mtDNA mutations, which are discussed below. In addition, mutations in a mitochondrial import protein (translocase of the inner mitochondrial membrane, or TIMM) that cause X-linked dystonia and deafness (also called Mohr-Tranebjaerg syndrome) in males (Jin et al., 1996, 1999; Koehler et al., 1999; Tranebjaerg et al., 2000) apparently also can cause isolated adult-onset focal dystonia in female "carriers" of the mutation (Swerdlow and Wooten, 2001). A mutation in this gene should be considered in any female with adult-onset dystonia and a brother with sensorineural deafness. These important observations suggest that a mutation in a nuclear-encoded gene important in mitochondrial function can cause adult-onset focal dystonia (Wallace and Murdock, 1999).

Mitochondrial dysfunction also may be a feature of more common forms of dystonia. Two reports indicate a defect in mitochondrial complex I in platelets of patients with dystonia. Schapira and colleagues (1997) reported a statistically significant 22% decrease in the activity of complex I, but not complex II/III or complex IV, in 20 patients with focal idiopathic dystonia compared to 22 controls. They did not find a significant decrease in patients with familial generalized dystonia, though the number of patients studied in this group was much lower (5 DYT1-linked and 4 non-DYT1-linked symptomatic cases). Benecke and colleagues (1992) also found a decrease specifically in complex I activity in patients with focal dystonia (mean decrease of 37%), and also reported a 67% mean decrease in complex I activity in 8 patients with segmental or generalized dystonia. In contrast, Reichmann and colleagues (1994) did not detect significant differences in platelet electron transport chain activities, including complex I, in 16 patients with dystonia (focal in 12, generalized in 4) compared to 8 age-matched controls. Many potential explanations account for the variability in these results, including the relatively low numbers of patients in some groups, differences in patient and control groups within and between studies, methods of mitochondria isolation, and technical aspects of the complex I assay. Unfortunately, the more relevant data on electron transport chain activities in the basal ganglia in idiopathic dystonia are not available.

## B. Mitochondrial Toxins and Dystonia

Though the extent of mitochondrial dysfunction in idiopathic dystonia is controversial, further data suggesting that mitochondrial dysfunction might play a causal role in dystonia comes from the ability of mitochondrial toxins to induce dystonia. The mitochondrial complex I inhibitor MPTP induces transient dystonia in baboons (Perlmutter et al., 1997). A delayed-onset progressive dystonia has been induced experimentally in monkeys following systemic injections of 3-nitroprionic acid (3-NP), a neurotoxin that irreversibly inhibits complex II of the mitochondrial electron transport chain (Palfi et al., 2000). Accidental ingestion by humans of sugar cane contaminated with 3-NP has been reported to induce an encephalopathy for several weeks followed by a progressive movement disorder including dystonia and chorea (Liu et al., 1992; Ludolph et al., 1991). A role for environmental factors in idiopathic dystonia is suggested by the incomplete concordance for dystonia in monozygotic twins, and the high degree of variability of symptoms among concordant twins (Wunderlich et al., 2001). However, there is no direct evidence for a role for 3-NP or other mitochondrial toxins in idiopathic dystonia.

## C. MtDNA Mutations and Rare Forms of Dystonia

The reports of mitochondrial dysfunction in dystonia patients, and the induction of dystonia by mitochondrial toxins, raise the question whether mtDNA mutations might play a role in dystonia. Though adult-onset dystonia usually is sporadic (rather than maternally inherited), mtDNA mutations often are of low penetrance, and can be highly dependent on environmental interactions, resulting in the appearance of sporadic expression. Dystonia, usually in combination with other features suggestive of a mitochondrial disorder, clearly can be a prominent manifestation of mtDNA mutations. A missense mutations at nucleotide position 14459 in a mtDNA-encoded complex I gene has been identified in several families with dystonia, LHON, or both (Jun et al., 1994; Shoffner et al., 1995). These patients have decreased complex I activity that is transferred to cell lines by transfer of their mtDNA, indicating that the complex I deficiency results from a mtDNA mutation(s) (Jun et al., 1996). A similar phenotype was reported in one family with a heteroplasmic mutation at nucleotide position 11696 along with a homoplasmic mutation at 14596 (De Vries et al., 1996). Parkinsonism and cervical dystonia developed in one LHON patient with the 3460 mutation (Nikoskelainen et al., 1995). Dystonia, along with other neurological deficits, also was reported in a subset of affected members of a family harboring the LHON-associated G11778A complex I mutation (Simon et al., 1999). Dystonia is commonly a prominent component of Leigh syndrome, which can be associated with various mtDNA point mutations or deletions (Hanna and Bhatia, 1997; Huang et al., 1995; Lera, et al., 1994; Macaya et al., 1993; Morris et al., 1996; Rahman et al., 1996; Santorelli et *al.*, 1993). In one study, dystonia was identified in 19 of 34 patients with Leigh syndrome (Macaya *et al.*, 1993). It is not clear if each of the mtDNA mutations linked to Leigh syndrome can be associated with dystonia, or if dystonia is a feature of only a subset of the genetic subgroups. Dystonia has been reported as the presenting feature of the 3243 "MELAS" mutation (Sudarsky *et al.*, 1999). A patient with classic clinical signs of Kearns-Sayre syndrome in addition to focal dystonia was found to harbor a large (5.9 kb) heteroplasmic mtDNA deletion (Marie *et al.*, 1999). Recently, we identified a patient with adult-onset focal dystonia and maternally inherited cataracts with a frame-shift mutation at position 3308 in the initiating methionine of the ND1 subunit of complex I (Simon *et al.*, 2001b). This represents the first-reported inherited frame-shift mtDNA mutation, though the pathogenicity of this mutation remains uncertain. A point mutation at this site previously was reported in a patient with ataxia, seizure, and marked generalized dystonia (Campos *et al.*, 1997), though the pathogenicity of the mutation has been questioned in this case as well (Rocha *et al.*, 1999). Recently, a missense mutation at nucleotide position 3796 in the gene encoding the NDI subunit of complex I was identified in 3 patients with adult-onset dystonia, though the association of this mutation with dystonia has yet to be confirmed (Simon *et al.*, 2002a) (Table 41.2).

### D. mtDNA Mutations in Idiopathic Dystonia

These rare cases reveal that mtDNA mutations can cause prominent dystonia, raising the question of whether or not mtDNA mutations contribute to the pathogenesis of more common idiopathic forms of dystonia. To address this question, Tabrizi and colleagues (1998) prepared cybrid cell lines expressing mtDNA from nine patients with focal dystonia. These cybrid cell lines did not manifest complex I deficiency. This has been interpreted as suggesting that mtDNA mutations do not account for the complex I defect in dystonia patients. However, while successful transfer of a complex I defect to the cybrids would be supportive of a role for mtDNA mutations in this defect, the failure to transfer complex I dysfunction cannot rule out a role for mtDNA mutations. This distinction results from important clinical and laboratory observations indicating that the biochemical expression of mtDNA mutations is strongly influenced by tissue type as well as nuclear background (Cock *et al.*, 1998; Hao *et al.*, 1999). The cybrid analyses of mtDNA from idiopathic dystonia patients were performed with only a single type of cell line, and thus the possible expression of complex I dysfunction in other nuclear backgrounds or in other tissue types was not addressed. Thus, the role of mtDNA mutations in the complex I defect of patients with idiopathic dystonia has not been definitively answered by the cybrid data.

There are no published reports of systematic screening of large numbers of patients with typical adult-onset idiopathic focal dystonia for mtDNA mutations. Shoffner *et al.* (1995) reported absence of known point mutations in eight dystonia patients, but only two of these had "pure" dystonia. Whether or not mtDNA mutations, together with other genetic and environmental factors, influence the risk of developing common forms of adult-onset dystonia is a question that needs to be addressed in future studies.

## IV. TREATMENT IMPLICATIONS

The ultimate goal of understanding the pathogenesis of PD and dystonia is to identify potential therapeutic interventions that improve the symptoms and, particularly in the case of PD, slow disease progression. Recognition of a role for mitochondrial dysfunction and mtDNA mutations in these disorders would suggest certain strategies. One mechanism by which mitochondrial dysfunction might lead to neuronal degeneration or dysfunction is by the increased production of free radicals (oxidative stress). Alpha-

TABLE 41.2 mtDNA Mutations Reported in Association with Dystonia, Usually Along with other Neurological Features (See Text for References)

| Syndrome/disease | Gene | Site of mutation | Types of mutations |
| --- | --- | --- | --- |
| MELAS | tRNA–Leu (UUR) | 3243 | A to G substitution |
| Dystonia + cataracts | ND1 (complex I) | 3308 | Frameshift |
| Adult-onset dystonia | ND1 (complex I) | 3796 | Missense |
| LHON | ND1 (complex I) | 3460 | Missense |
|  | ND4 (complex I) | 11778 | Missense |
| LHON + Dystonia | ND4 (complex I) | 11696 | Missense |
|  | ND6 (complex I) | 14459 | Missense |
|  | ND6 (complex I) | 14596 | Missense |
| Leigh's syndrome | tRNA-Lys | 8344 | A to G substitution |
|  | ATP-6 (complex V) | 8993 | Missense |
| Kearns-Sayre | Multiple |  | 5.9-kb deletion |

tocopherol (vitamin E), a free radical scavenger, showed no neuroprotective efficacy in PD in the DATATOP trial (1993). However, more potent free radical scavengers and ones with better penetration into the central nervous system are under development. Dietary supplementation with creatine, which enhances energy metabolism, has been shown in a randomized double-blinded trial to improve muscle strength in patients with mitochondrial myopathies (Tarnopolsky et al., 1997). Creatine also has shown neuroprotective efficacy in animal models of amyotrophic lateral sclerosis (Klivenyi et al., 1999) and Huntington's disease (Matthews et al., 1998), and against a variety of mitochondrial toxins (Tarnopolsky and Beal, 2001). Many additional strategies have been reported to show neuroprotective efficacy in the setting of mitochondrial dysfunction *in vitro*, in animal models, or in anecdotal reports in humans (Fadic and Johns, 1997).

In the case of PD, if oxidative stress-induced mtDNA mutations are shown to play a role, then DNA repair enzymes may provide additional therapeutic targets. An enzyme important in the repair of oxidatively damaged DNA (Hayakawa et al., 1995; Kakuma et al., 1995; Maki and Sekiguchi, 1992) and localized to the cytosol and mitochondria (Kang et al., 1995); (8-oxo-7,8-dihydrodeoxyguanosine triphosphatase, or "8-oxo-dGTPase"), appears to be upregulated in the substantia nigra of PD patients (Shimura-Miura et al., 1999), reflecting the increased oxidative damage to DNA in this region. Further upregulation of this enzyme, or improving mtDNA repair by other means, may prove to be protective. 8-oxoguanine DNA glycosylase is another enzyme that specifically repairs oxidatively damaged DNA, but normally is localized to the nucleus. Overexpression of recombinant human 8-oxoguanine DNA glycosylase specifically targeted to the mitochondria results in protection against oxidative stress-induced cell death *in vitro* (Dobson et al., 2000). This demonstrates the potential value of manipulating repair of oxidatively damaged DNA as a protective strategy in disorders associated with oxidative stress. Thus, though no treatments have yet been proven in humans to be neuroprotective against toxicities associated with mitochondrial dysfunction or mtDNA mutations, there are many promising avenues of research.

## V. CONCLUSIONS

The basal ganglia are a common site of pathology in relatively rare "classic" mitochondrial disorders such as Leigh syndrome. Mitochondrial dysfunction and mtDNA mutations also may play a role in more common basal ganglia disorders such as PD and dystonia. Complex I activity is impaired in the substantia nigra in PD and complex I inhibitors reproduce many features of PD in animal models, suggesting that the complex I defect may be important in the pathogenesis of PD. MtDNA mutations can be associated with parkinsonism in rare patients, but inherited pathogenic mtDNA mutations are absent in the majority of idiopathic PD patients. Further studies are needed to address the possibility that the accumulation of somatic mutations may play a role in PD. MtDNA mutations also can cause rare forms of dystonia, but the role of mitochondrial dysfunction and mtDNA mutations in more common forms of dystonia is uncertain.

## References

(1993). Effects of tocopherol and deprenyl on the progression of disability in early Parkinson's disease. The Parkinson Study Group. *N. Engl. J. Med.* **328(3)**, 176–183.

Alam, Z. I., Daniel, S. E., Lees, A. J., Marsden, D. C., Jenner, P., and Halliwell, B. (1997a). A generalised increase in protein carbonyls in the brain in Parkinson's but not incidental Lewy body disease. *J. Neurochem.* **69(3)**, 1326–1329.

Alam, Z. I., Jenner, A., Daniel, S. E., Lees, A. J., Cairns, N., Marsden, C. D., Jenner, P., and Halliwel, B. (1997b). Oxidative DNA damage in the parkinsonian brain: an apparent selective increase in 8-hydroxyguanine levels in substantia nigra. *J. Neurochem.* **69(3)**, 1196–1203.

Albers, D. S., Augood, S. J., Martin, D. M., Standaert, D. G., Vonsattel, J. P., and Beal, M. F. (1999). Evidence for oxidative stress in the subthalamic nucleus in progressive supranuclear palsy. *J. Neurochem.* **73(2)**, 881–884.

Albers, D. S., Augood, S. J., Park, L. C., Browne, S. E., Martin, D. M., Adamson, J., Hutton, M., Standaert, D. G., Vonsattel, J. P., Gibson, G. E., and Beal, M. F. (2000). Frontal lobe dysfunction in progressive supranuclear palsy: evidence for oxidative stress and mitochondrial impairment. *J. Neurochem.* **74(2)**, 878–881.

Albers, D. S., Swerdlow, R. H., Manfredi, G., Gajewski, C., Yang, L., Parker, Jr., W. D., and Beal, M. F. (2001). Further evidence for mitochondrial dysfunction in progressive supranuclear palsy. *Exp. Neurol.* **168(1)**, 196–198.

Almasy, L., Bressman, S. B., Raymond, D., Kramer, P. L., Greene, P. E., Heiman, G. A., Ford, B., Yount, J., de Leon, D., Chouinard, S., Saunders-Pullman, R., Brin, M. F., Kapoor R. P., Jones, A. C., Shen, H., Fahn, S., Risch, N. J., and Nygaard, T. G. (1997). Idiopathic torsion dystonia linked to chromosome 8 in two Mennonite families. *Ann. Neurol.* **42(4)**, 670–673.

Ames, B. N., Shigenaga M. K., and Hagen, T. M. (1993). Oxidants, antioxidants, and the degenerative diseases of aging. *Proc. Natl. Acad.Sci. U.S.A.* **90(17)**, 7915–7922.

Andrews, A. M., Ladenheim, B., Epstein, C. J., Cadet, J. L., and Murphy, D. L. (1996). Transgenic mice with high levels of superoxide dismutase activity are protected from the neurotoxic effects of 2'-NH2-MPTP on serotonergic and noradrenergic nerve terminals. *Mol. Pharmacol.* **50(6)**, 1511–1519.

Aomi, Y., Chen, C. S., Nakada, K., Ito, S., Isobe, K., Murakami, H., Kuno, S. Y., Tawata, M., Matsuoka, R., Mizusawa, H., and Hayashi, J. I. (2001). Cytoplasmic transfer of platelet mtDNA from elderly patients with Parkinson's disease to mtDNA-less HeLa cells restores complete mitochondrial respiratory function. *Biochem. Biophys. Res. Commun.* **280(1)**, 265–273.

Bandmann, O., Sweeney, M. G., Daniel, S. E., Marsden, C. D., and Wood, N. W. (1997). Mitochondrial DNA polymorphisms in pathologically proven Parkinson's disease. *J. Neurol.* **244(4)**, 262–265.

Benecke, R., Strumper, P., and Weiss, H. (1992). Electron transfer complex I defect in idiopathic dystonia. *Ann. Neurol.* **32(5)**, 683–686.

Benecke, R., Strumper P., and Weis, H. (1993a). Electron transfer complexes I and IV of platelets are abnormal in Parkinson's disease but normal in Parkinson-plus syndromes. *Brain* 116 (Pt 6):1451–1463.

Benecke, R., Strumper, P., and Weiss. H. (1993b). Electron transfer complexes I and IV of platelets are abnormal in Parkinson's disease but normal in Parkinson-plus syndromes. *Brain* 116 (Pt. 6), 1451–1463.

Betarbet, R., Sherer, T. B., MacKenzie, G., Garcia-Osuna M., Panov, A. V., and Greenamyre, J. T. (2000). Chronic systemic pesticide exposure reproduces features of Parkinson's disease. *Nat. Neurosci.* 3(12), 1301–1306.

Bhatti, M. T., and Newman, N. J. (1999). A multiple sclerosis-like illness in a man harboring the mtDNA 14484 mutation. *J. Neuroophthalmol.* 19(1), 28–33.

Blin, O., Desnuelle, C., Rascol, O., Borg, M., Peyro Saint Paul, H., Azulay, J. P., Bille F., Figarella, D., Coulom, F., Pellissier J. F. et al. (1994). Mitochondrial respiratory failure in skeletal muscle from patients with Parkinson's disease and multiple system atrophy. *J. Neurol. Sci.* 125(1), 95–101.

Bowling, A.C., Mutisya, E. M., Walker, L. C. et al. (1993). Age-dependent impairment of mitochondrial function in primate brain. *J. Neurochem.* 60, 1964–1967.

Bravi, D., Anderson, J. J., Dagani, F., Davis, T. L., Ferrari, R., Gillespie, M., and Chase, T. N. (1992). Effect of aging and dopaminomimetic therapy on mitochondrial respiratory function in Parkinson's disease. *Mov. Disord.* 7(3), 228–231.

Brown, M. D., Shoffner, J. M., Kim, Y. L., Jun, A. S., Graham, B. H., Cabell, M. F., Gurley, D. S., and Wallace, D. C. (1996). Mitochondrial DNA sequence analysis of four Alzheimer's and Parkinson's disease patients. *Am. J. Med. Genet.* 61(3), 283–289.

Campos, Y., Martin, M. A., Rubio, J. C., Gutierrez del Olmo M. C., Cabello, A., and Arenas, J. (1997). Bilateral striatal necrosis and MELAS associated with a new T3308C mutation in the mitochondrial ND1 gene. *Biochem. Biophys. Res. Commun.* 238(2), 323–325.

Casali, C., Bonifati, V., Santorelli, F. M., Casari, G., Fortini, D., Patrignani, A., Fabbrini, G., Carrozzo, R., D'Amati, G., Locuratolo, N., Vanacore, N., Damiano M., Pierallini, A., Pierelli, F., Amabile G. A., and Meco, G. (2001). Mitochondrial myopathy, parkinsonism, and multiple mtDNA deletions in a Sephardic Jewish family. *Neurology* 56(6), 802–805.

Chalmers, R. M., Brockington, M., Howard, R. S., Lecky, B. R., Morgan-Hughes, J. A., and Harding, A. E. (1996). Mitochondrial encephalopathy with multiple mitochondrial DNA deletions: a report of two families and two sporadic cases with unusual clinical and neuropathological features. *J. Neurol. Sci.* 143(1–2), 41–45.

Chan, D. K., Mellick, G., Cai, H., Wang, X. L., Ng, P. W., Pang, C. P., Woo, J., and Kay, R. (2000). The alpha-synuclein gene and Parkinson disease in a Chinese population. *Arch. Neurol.* 57(4), 501–503.

Chan, P., Jiang, X., Forno, L. S., Di Monte, D. A., Tanner, C. M., and Langston, J. W. (1998a). Absence of mutations in the coding region of the alpha-synuclein gene in pathologically proven Parkinson's disease. *Neurology* 50(4), 1136–1137.

Chan, P., Tanner, C. M., Jiang, X., and Langsto, J. W. (1998b). Failure to find the alpha-synuclein gene missense mutation (G209A) in 100 patients with younger onset Parkinson's disease. *Neurology* 50(2), 513–514.

Checcarelli, N., Prelle, A., Moggio, M., Comi, G., Bresolin, N., Papadimitriou, A., Fagiolari, G., Bordoni, A., and Scarlato, G. (1994). Multiple deletions of mitochondrial DNA in sporadic and atypical cases of encephalomyopathy. *J. Neurol. Sci.* 123(1–2), 74–79.

Chen, Y .I., Jenkins, B. G., Fink, S., and Rosen, B. R. (1994). Evidence for impairment of energy metabolism in Parkinson's disease using *in vivo* localized MR spectroscopy. *Proc. Soc. Mag. Res.* 1, 194.

Cheng, K. C., Cahill, D. S., Kasai, H., Nishimura, S., and Loeb, L.A. (1992). 8-Hydroxyguanine, an abundant form of oxidative DNA damage, causes G—T and A—C substitutions. *J. Biol. Chem.* 267(1), 166–172.

Chinnery, P. F., Howell, N., Andrews, R. M., and Turnbull, D. M. (1999). Mitochondrial DNA analysis: polymorphisms and pathogenicity. *J. Med. Genet.* 36(7), 505–510.

Chinnery, P. F., Taylor G. A., Howell, N., Brown, D. T., Parsons, T. J., and Turnbull, D. M. (2001). Point mutations of the mtDNA control region in normal and neurodegenerative human brains. *Am. J. Hum. Genet.* 68(2), 529–532.

Cleeter, M. W., Cooper, J. M., and Schapira, A. H. (1992). Irreversible inhibition of mitochondrial complex I by 1-methyl-4-phenylpyridinium: evidence for free radical involvement. *J. Neurochem.* 58(2), 786–789.

Cock, H. R., Tabrizi S. J., Cooper, J.M., and Schapira, A. H. (1998). The influence of nuclear background on the biochemical expression of 3460 Leber's hereditary optic neuropathy. *Ann. Neurol.* 44(2), 187–193.

Conway, K. A., Rochet, J. C., Bieganski R. M., and Lansbury, Jr., P. T. (2001). Kinetic Stabilization of the alpha-synuclein protofibril by a dopamine-alpha-synuclein adduct. *Science* 294(5545), 1346–1349.

Corral-Debrinski, M., Horton, T., Lott, M.T., Shoffner, J. M., Beal, M. F., and Wallace, D. C. (1992). Mitochondrial DNA deletions in human brain: regional variability and increase with advanced age. *Nat. Genet.* 2(4), 324–329.

Cortopassi, G., and Wang, E. (1995). Modelling the effects of age-related mtDNA mutation accumulation; complex I deficiency, superoxide and cell death. *Biochim. Biophys. Acta* 1271, 171–176.

Cortopassi, G. A., Shibata, D., Soong, N. W., and Arnheim, N. (1992). A pattern of accumulation of a somatic deletion of mitochondrial DNA in aging human tissues. *Proc. Natl. Acad. Sc. U.S.A.* 89(16), 7370–7374.

Davis, J. N., 2nd, and Parker, Jr., W. D. (1998). Evidence that two reports of mtDNA cytochrome c oxidase "mutations" in Alzheimer's disease are based on nDNA pseudogenes of recent evolutionary origin. *Biochem. Biophys. Res. Commun.* 244(3), 877–883.

Davis, R. E., Miller, S., Herrnstadt, C., Ghosh, S. S., Fahy, E., Shinobu, L. A., Galasko, D., Thal, L. J., Beal, M. F., Howell N., and Parker, Jr., W. D. (1997). Mutations in mitochondrial cytochrome c oxidase genes segregate with late-onset Alzheimer disease. *Proc. Natl. Acad. Sci. U.S.A.* 94(9), 4526–4531.

De Vries, D. D., Went L. N., Bruyn, G. W., Scholte, H. R., Hofstra R. M., Bolhuis P.A., and van Oost, B. A. (1996). Genetic and biochemical impairment of mitochondrial complex I activity in a family with Leber hereditary optic neuropathy and hereditary spastic dystonia. *Am. J. Hum. Genet.* 58(4), 703–711.

DeStefano, A. L., Golbe, L. I., Mark, M. H., Lazzarini, A. M., Maher N. E., Saint-Hilaire, M., Feldman, R. G., Guttman, M., Watts, R. L., Suchowersky O., Lafontaine, A. L., Labelle, N., Lew, M. F., Waters C. H., Growdon, J. H., Singer, C., Currie, L. J., Wooten, G F., Vieregge, P., Pramstaller, P. P., Klein, C., Hubble, J. P., Stacy, M., Montgomery, E., MacDonald, M. E., Gusella, J. F., and Myers, R. H. (2001). Genome-wide scan for Parkinson's disease: the GenePD Study. *Neurology* 57(6), 1124–1126.

DiMauro, S., and Moraes, C. T. (1993). Mitochondrial encephalomyopathies. *Arch. Neurol.* 50(11), 1197–1208.

Dobson, A. W., Xu, Y., Kelley, M. R., LeDoux, S. P., and Wilson, G. L. (2000). Enhanced mitochondrial DNA repair and cellular survival after oxidative stress by targeting the human 8-oxoguanine glycosylase repair enzyme to mitochondria. *J. Biol. Chem.* 275(48), 37518–37523.

Duda, J. E., Giasson, B. I., Chen, Q., Gur, T. L., Hurtig, H. I., Stern, M. B., Gollomp, S. M., Ischiropoulos H., Lee, V. M., and Trojanowsk, J. Q. (2000). Widespread nitration of pathological inclusions in neurodegenerative synucleinopathies. *Am. J. Pathol.* 157(5), 1439–1445.

Fadic, R., and Johns, D. R. (1997). Treatment of the mitochondrial encephalomyopathies. In *Mitochondria and Free Radicals in Neurodegenerative Diseases*. (M. F. Beal, ed.) Wiley-Liss, Inc.

Ferrante, R. J., Browne, S. E., Shinobu, L. A., Bowling, A. C., Baik, M. J., MacGarvey, U., Kowall, N. W., Brown, Jr., R. H., and Beal, M. F. (1997). Evidence of increased oxidative damage in both sporadic and familial amyotrophic lateral sclerosis. *J. Neurochem.* 69(5), 2064–2074.

Fliss, M. S., Usadel, H., Caballero, O. L., Wu, L., Buta, M. R., Eleff, S. M., Jen, J., and Sidransky, D. (2000). Facile detection of mitochondrial DNA mutations in tumors and bodily fluids. *Science* **287(5460)**, 2017–2019.

Florentz, C., and Sissler, M. (2001). Disease-related versus polymorphic mutations in human mitochondrial tRNAs. Where is the difference? *EMBO Rep.* **2(6)**, 481–486.

Giasson, B. I., Duda, J. E., Murray, I. V., Chen, Q., Souza, J. M., Hurtig H. I., Ischiropoulos, H., Trojanowski, J. Q., and Lee, V. M. (2000). Oxidative damage linked to neurodegeneration by selective alpha-synuclein nitration in synucleinopathy lesions. *Science* **290(5493)**, 985–989.

Goto, Y., Nonaka, I., and Horai, S. (1990). A mutation in the tRNA(Leu)(UUR) gene associated with the MELAS subgroup of mitochondrial encephalomyopathies. *Nature* **348(6302)**, 651–653.

Gu, M., Cooper, J. M., Taanman, J. W., and Schapira, A. H. (1998). Mitochondrial DNA transmission of the mitochondrial defect in Parkinson's disease. *Ann. Neurol.* **44(2)**, 177–186.

Gu, M., Gash M. T., Cooper J. M., Wenning, G. K., Daniel, S. E., Quinn N. P., Marsden, C. D., and Schapira, A. H. (1997). Mitochondrial respiratory chain function in multiple system atrophy. *Mov. Disord.* **12(3)**, 418–422.

Haas, R. H., Nasirian F., Nakano, K., Ward, D., Pay, M., Hill, R., and Shults, C. W. (1995). Low platelet mitochondrial complex I and complex II/III activity in early untreated Parkinson's disease. *Ann. Neurol.* **37(6)**, 714–722.

Hanna, M. G., and Bhatia, K. P. (1997). Movement disorders and mitochondrial dysfunction. *Curr. Opin. Neurol.* **10(4)**, 351–356.

Hao, H., Morrison, L. E., and Moraes, C. T. (1999). Suppression of a mitochondrial tRNA gene mutation phenotype associated with changes in the nuclear background. *Hum. Mol. Genet.* **8(6)**, 1117–1124.

Hara, K., Yamamoto, M., Anegawa, T., Sakuta, R., and Nakamura, M. (1994). [Mitochondrial encephalomyopathy associated with parkinsonism and a point mutation in the mitochondrial tRNA(Leu)(UUR)) gene]. *Rinsho Shinkeigaku* **34(4)**, 361–365.

Harding, A. E., Sweeney M. G., Miller D. H., Mumford, C. J., Kellar-Wood, H., Menard, D., McDonald, W. I., and Compston, D. A. (1992). Occurrence of a multiple sclerosis-like illness in women who have a Leber's hereditary optic neuropathy mitochondrial DNA mutation. *Brain* **115(Pt. 4)**, 979–989.

Hasegawa, E., Takeshige, K., Oishi, T., Murai, Y., and Minakami, S. (1990). 1-methyl-4-phenylpyridinium (MPP+) induces NADH-dependent superoxide formation and enhances NADH-dependent lipid peroxidation in bovine heart submitochondrial particles. *Biochem. Biophys. Res. Commun.* **170(3)**, 1049–1055.

Hashimoto, M., Hsu, L. J., Xia, Y., Takeda, A., Sisk, A., Sundsmo, M., and Masliah, E. (1999). Oxidative stress induces amyloid-like aggregate formation of NACP/alpha-synuclein *in vitro*. *Neuroreport* **10(4)**, 717–721.

Hattori, N., Kitada, T., Matsumine, H., Asakawa, S., Yamamura, Y., Yoshino, H., Kobayashi, T., Yokochi, M., Wang, M., Yoritaka, A., Kondo, T., Kuzuhara, S., Nakamura, S., Shimizu, N., and Mizuno, Y. (1998a). Molecular genetic analysis of a novel Parkin gene in Japanese families with autosomal recessive juvenile parkinsonism: evidence for variable homozygous deletions in the Parkin gene in affected individuals. *Ann. Neurol.* **44(6)**, 935–941.

Hattori, N., Yoshino, H., Tanaka, M., Suzuki, H., and Mizuno, Y. (1998b). Genotype in the 24-kDa subunit gene (NDUFV2) of mitochondrial, complex I and susceptibility to Parkinson's disease. *Genomics* **49(1)** 52–58.

Hayakawa, H., Taketomi, A., Sakumi K., Kuwano, M., and Sedkiguchi, M. (1995). Generation and elimination of 8-oxo-7,8-dihydro-2′-deoxy-guanosine 5′-triphosphate, a mutagenic substrate for DNA synthesis in human cells. *Biochemistry* **34**, 89–95.

Hayakawa, M., Hattori, K., Sugiyama, S., and Ozawa, T. (1992). Age-associated oxygen damage and mutations in mitochondrial DNA in human hearts. *Biochem. Biophys. Res. Commun.* **189(2)**, 979–985.

Higuchi, S., Arai, H., Matsushita, S., Matsui, T., Kimpara, T., Takeda, A., and Shirakura, K. (1998). Mutation in the alpha-synuclein gene and sporadic Parkinson's disease, Alzheimer's disease, and dementia with Lewy bodies. *Exp. Neurol.* **153(1)**, 164–166.

Hirano, M., Shtilbans, A., Mayeux R., Davidson M. M., DiMauro, S., Knowles, J. A., and Schon, E. A. (1997). Apparent mtDNA heteroplasmy in Alzheimer's disease patients and in normals due to PCR amplification of nucleus-embedded mtDNA pseudogenes. *Proc. Natl. Acad. Sci. U.S.A.* **94(26)**, 14894–14899.

Holt, I. J., Harding A. E., and Morgan-Hughes, J. A. (1988). Deletions of muscle mitochondrial DNA in patients with mitochondrial myopathies. *Nature* **331(6158)**, 717–719.

Hsu, L. J., Sagara, Y., Arroyo, A., Rockenstein, E., Sisk, A., Mallory, M., Wong, J., Takenouchi, T., Hashimoto, M., and Masliah, E. (2000). alpha-synuclein promotes mitochondrial deficit and oxidative stress. *Am J Pathol* **157** (2):401–410.

Huang, W. Y., Chi, C. S., Mak, S. C., Wu, H. M., and Yang, M. T. (1995). Leigh syndrome presenting with dystonia: report of one case. *Zhonghua Min Guo Xiao Er Ke Yi Xue Hui Za Zhi* **36(5)**, 378–381.

Ichinose, H., Ohye, T., Takahashi, E., Seki, N., Hori, T., Segawa, M., Nomura, Y., Endo K., Tanaka, H., Tsuji, S. et al. (1994). Hereditary progressive dystonia with marked diurnal fluctuation caused by mutations in the GTP cyclohydrolase I gene [see comments]. *Nat. Genet.* **8(3)**, 236–242.

Ikebe, S., Tanaka, M., Ohno K., Sato, W., Hattori K., Kondo, T., Mizuno Y., and Ozawa, T. (1990). Increase of deleted mitochondrial DNA in the striatum in Parkinson's disease and senescence. *Biochem. Biophys. Res. Commun.* **170(3)**, 1044–1048.

Janetzky, B., Hauck S., Youdim M. B., Riederer P., Jellinger K., Pantucek F., Zochling, R., Boissl, K. W., and Reichmann, H. (1994). Unaltered aconitase activity, but decreased complex I activity in substantia nigra pars compacta of patients with Parkinson's disease. *Neurosci. Lett.* **169(1–2)**, 126–128.

Jha, N., Jurm O., Lalli G., Liu Y., Pettus, E. H., Greenamyre, J. T., Liu, R. M., Forman, H. J., and Andersen, J. K. (2000). Glutathione depletion in PC12 results in selective inhibition of mitochondrial complex I activity. Implications for Parkinson's disease. *J. Biol. Chem.* **275(34)**, 26096–26101.

Jin, H., Kendall E., Freeman, T. C., Roberts R. G., and Vetrie, D. L. (1999). The human family of deafness/dystonia peptide (DDP) related mitochondrial import proteins. *Genomics* **61(3)**, 259–267.

Jin, H., May, M., Tranebjaerg L., Kendall, E., Fontan, G., Jackson, J., Subramony, S. H., Arena F., Lubs, H., Smith, S., Stevenson, R., Schwartz, C., and Vetrie, D. (1996). A novel X-linked gene, DDP, shows mutations in families with deafness (DFN-1), dystonia, mental deficiency and blindness. *Nat. Genet.* **14(2)**, 177–180.

Jun, A. S., Brown, M. D., and Wallace, D.C. (1994). A mitochondrial DNA mutation at nucleotide pair 14459 of the NADH dehydrogenase subunit 6 gene associated with maternally inherited Leber hereditary optic neuropathy and dystonia. *Proc. Natl. Acad. Sci. U.S.A.* **91(13)**, 6206–6210.

Jun, A. S., Trounce I. A., Brown, M. D., Shoffner, J. M., and Wallace, D. C. (1996). Use of transmitochondrial cybrids to assign a complex I defect to the mitochondrial DNA-encoded NADH dehydrogenase subunit 6 gene mutation at nucleotide pair 14459 that causes Leber hereditary optic neuropathy and dystonia. *Mol. Cell. Biol.* **16(3)**, 771–777.

Kakuma, T., Nishida, J., Tsuzuki, T., and Sekiguchi, M. (1995). Mouse MTH1 protein with 8-oxo-7,8-dihydro-2′-deoxyguanosine 5′-triphosphatase activity that prevents transversion mutation. cDNA cloning and tissue distribution. *J. Biol. Chem.* **270(43)**, 25942–25948.

Kalman, B., and Alder, H. (1998). Is the mitochondrial DNA involved in determining susceptibility to multiple sclerosis? *Acta Neurol. Scand.* **98(4)**, 232–237.

Kalman, B., Lublin F. D., and Alder, H. (1995). Mitochondrial DNA mutations in multiple sclerosis. *Mult. Scler.* **1(1)**, 32–36.

Kang, D., Nishida, J., Iyama, A., Nakabeppu, Y., Furuichi, M., Fujiwara, T., Sekiguchi, M., and Takeshige, K. (1995). Intracellular localization of 8-oxo-dGTPase in human cells, with special reference to the role of the enzyme in mitochondria. *J. Biol. Chem.* **270(24)**, 14659–14665.

Kapsa, R. M., Jean-Francois, M. J., Lertrit, P., Weng, S., Siregar, N., Ojaimi J., Donnan, G., Masters, C., and Byrne, E. (1996). Mitochondrial DNA polymorphism in substantia nigra. *J. Neurol. Sci.* **144(1–2)**, 204–211.

Kasai, H. (1997). Analysis of a form of oxidative DNA damage, 8-hydroxy-2'- deoxyguanosine, as a marker of cellular oxidative stress during carcinogenesis. *Mutat. Res.* **387(3)**, 147–163.

Kellar-Wood, H., Robertson, N., Govan, G. G., Compston, D. A., and Harding, A. E. (1994). Leber's hereditary optic neuropathy mitochondrial DNA mutations in multiple sclerosis. *Ann. Neurol.* **36(1)**, 109–112.

King, M. P., and Attardi, G. (1989). Human cells lacking mtDNA: repopulation with exogenous mitochondria by complementation. *Science* **246(4929)**, 500–503.

Kitada, T., Asakawa, S., Hattori, N., Matsumine, H., Yamamura, Y., Minoshima, S., Yokochi, M., Mizuno, Y., and Shimizu, N. (1998). Mutations in the parkin gene cause autosomal recessive juvenile parkinsonism. *Nature* **392(6676)**, 605–608.

Klivenyi, P., Ferrante, R. J., Matthews, R. T., Bogdanov, M. B., Klein A. M., Andreassen, O. A., Mueller, G., Wermer, M., Kaddurah-Daouk, R., and Beal, M. F. (1999). Neuroprotective effects of creatine in a transgenic animal model of amyotrophic lateral sclerosis. *Nat. Med.* **5(3)**, 347–350.

Ko, L., Mehta N. D., Farrer M., Easson, C., Hussey, J., Yen, S., Hardy J., and Yen, S. H. (2000). Sensitization of neuronal cells to oxidative stress with mutated human alpha-synuclein. *J. Neurochem.* **75(6)**, 2546–2554.

Kobayashi, T., Matsumine H., Matuda, S., and Mizuno, Y. (1998). Association between the gene encoding the E2 subunit of the alpha-ketoglutarate dehydrogenase complex and Parkinson's disease. *Ann. Neurol.* **43(1)**, 120–123.

Koehler, C. M., Leuenberger, D., Merchant, S., Renold, A., Junne, T., and Schatz, G. (1999). Human deafness dystonia syndrome is a mitochondrial disease. *Proc. Natl. Acad. Sc. U.S.A.* **96(5)**, 2141–2146.

Korf, B. R. (1998). The hereditary dystonias: an emerging story with a twist. *Ann. Neurol.* **44(1)**, 4–5.

Kosel, S., Egensperger R., Schnopp N. M., and Graeber, M. B. (1997). The 'common deletion' is not increased in parkinsonian substantia nigra as shown by competitive polymerase chain reaction. *Mov. Disord.* **12(5)**, 639–645.

Kosel, S., Grasbon-Frodl, E. M., Mautsch U., Egensperger R., von Eitzen, U., Frishman, D., Hofmann, S., Gerbitz K. D., Mehraein P., and Graeber, M. B. (1998). Novel mutations of mitochondrial complex I in pathologically proven Parkinson disease. *Neurogenetics* **1(3)**, 197–204.

Kosel, S., Lucking C. B., Egensperger, R., Mehraein, P., and Graeber, M. B. (1996). Mitochondrial NADH dehydrogenase and CYP2D6 genotypes in Lewy-body parkinsonism. *J. Neurosci. Res.* **44(2)**, 174–183.

Kovalenko, S. A., Tanaka, M., Yoneda, M., Iakovlev, A. F., and Ozawa, T. (1996). Accumulation of somatic nucleotide substitutions in mitochondrial DNA associated with the 3243 A-to-G tRNA(leu)(UUR) mutation in encephalomyopathy and cardiomyopathy. *Biochem. Biophys. Res. Commun.* **222(2)**, 201–207.

Kowall, N. W., Hantraye, P., Brouillet, E., Beal, M. F., McKee, A. C., and Ferrante, R. J. (2000). MPTP induces alpha-synuclein aggregation in the substantia nigra of baboons. *Neuroreport* **11(1)**, 211–213.

Krige, D., Carroll, M. T., Cooper, J. M., Marsden, C. D., and Schapira, A. H. (1992). Platelet mitochondrial function in Parkinson's disease. The Royal Kings and Queens Parkinson Disease Research Group. *Ann. Neurol.* **32(6)**, 782–788.

Kruger, R., Kuhn, W., Muller, T., Woitalla, D., Graeber, M., Kosel, S., Przuntek, H., Epplen, J.T., Schols L., and Riess, O. (1998). Ala30Pro mutation in the gene encoding alpha-synuclein in Parkinson's disease. *Nat. Genet.* **18(2)**, 106–108.

Kuchino, Y., Mori, F., Kasai, H., Inoue, H., Iwai, S., Miura, K., Ohtsuka, E., and Nishimura, S. (1987). Misreading of DNA templates containing 8-hydroxydeoxyguanosine at the modified base and at adjacent residues. *Nature* **327 (6117)**, 77–79.

Lai, J.C.K., Walsh, J. M., Dennis, S. C., and Clark, J.B. (1977). Synaptic and nonsynaptic mitochondria from rat brain: isolation and characterization. *J. Neurochem.* **28**, 625–631.

Langston, J. W. (1996). The etiology of Parkinson's disease with emphasis on the MPTP story. *Neurology* **47** (6 Suppl. 3), S153–S160.

Lee, H. C., Pang, C. Y., Hsu, H. S., and Wei, Y. H. (1994). Differential accumulations of 4,977 bp deletion in mitochondrial DNA of various tissues in human ageing. *Biochim. Biophys. Acta* **1226(1)**, 37–43.

Lee, M., Hyun, D., Halliwell, B., and Jenner, P. (2001). Effect of the overexpression of wild-type or mutant alpha-synuclein on cell susceptibility to insult. *J. Neurochem.* **76(4)**, 998–1009.

Lera, G., Bhatia, K., and Marsden, C. D. (1994). Dystonia as the major manifestation of Leigh's syndrome. *Mov. Disord.* **9(6)**, 642–649.

Lestienne, P., Nelson, J., Riederer, P., Jellinger, K., and Reichmann, H. (1990). Normal mitochondrial genome in brain from patients with Parkinson's disease and complex I defect. *J. Neurochem.* **55(5)**, 1810–1812.

Lestienne, P., and Ponsot, G. (1988). Kearns-Sayre syndrome with muscle mitochondrial DNA deletion. *Lancet* **1(8590)**, 885.

Leube, B., Rudnicki, D., Ratzlaff, T., Kessler, K. R., Benecke, R., and Auburger, G. (1996). Idiopathic torsion dystonia: assignment of a gene to chromosome 18p in a German family with adult onset, autosomal dominant inheritance and purely focal distribution. *Hum. Mol Genet.* **5(10)**, 1673–1677.

Lezza, A. M., Mecocci, P., Cormio, A., Beal, M. F., Cherubini, A., Cantatore P., Senin, U., and Gadaleta, M.N. (1999). Mitochondrial DNA 4977 bp deletion and OH8dG levels correlate in the brain of aged subjects but not Alzheimer's disease patients. *FASEB. J.* **13(9)**, 1083–1088.

Lin, J. J., Yueh, K. C., Chang, D. C., and Lin, S. Z. (1999). Absence of G209A and G88C mutations in the alpha-synuclein gene of Parkinson's disease in a Chinese population. *Eur. Neurol.* **42(4)**, 217–220.

Lin, M.T., Simon D. K., Ahn, C. H., Kim, L. M., and Beal M. F. (2002). High aggregate burden of somatic mtDNA point mutations in aging and Alzheimer's disease brain. *Hum. Mol. Genet.* **11(2)**, 133–145.

Linnane, A. W., Marzuki, S., Ozawa, T., and Tanaka, M. (1989). Mitochondrial DNA mutations as an important contributor to ageing and degenerative diseases. *Lancet* **1(8639)**, 642–645.

Liu, V. W., Zhang, C., Linnane, A. W., and Nagley, P. (1997). Quantitative allele-specific PCR: demonstration of age-associated accumulation in human tissues of the A→G mutation at nucleotide 3243 in mitochondrial DNA. *Hum. Mutat.* **9(3)**, 265–271.

Liu, X., Luo, X., and Hu, W. (1992). Studies on the epidemiology and etiology of moldy sugarcane poisoning in China. *Biomed. Environ. Sci* **5(2)**, 161–177.

Lucking, C. B., Abbas, N., Durr, A., Bonifati, V., Bonnet, A. M., de Broucker T., De Michele G., Wood, N. W., Agid, Y., and Brice, A. (1998). Homozygous deletions in parkin gene in European and North African families with autosomal recessive juvenile parkinsonism. The European Consortium on Genetic Susceptibility in Parkinson's Disease and the French Parkinson's Disease Genetics Study Group. *Lancet* **352(9137)**, 1355–1356.

Lucking, C. B., Durr, A., Bonifati, V., Vaughan, J., De Michele, G., Gasser, T., Harhangi, B. S., Meco, G., Denefle, P., Wood, N. W., Agid, Y., and Brice, A. (2000). Association between early-onset Parkinson's disease and mutations in the parkin gene. French Parkinson's Disease Genetics Study Group. *N. Engl. J. Med.* **342(21)**, 1560–1567.

Lucotte, G., Mercier, G., and Turpin, J. C. (1998). Lack of mutation G209A in the alpha-synuclein gene in French patients with familial and sporadic Parkinson's disease [letter]. *J. Neurol. Neurosurg. Psychiatry* **65(6)**, 948–949.

Ludecke, B., Dworniczak B., and Bartholome, K. (1995). A point mutation in the tyrosine hydroxylase gene associated with Segawa's syndrome. *Hum. Genet.* **95(1)**, 123–125.

Ludolph, A. C., He, F., Spencer, P. S., Hammerstad, J., and Sabri, M. (1991). 3-Nitropropionic acid-exogenous animal neurotoxin and possible human striatal toxin. *Can J Neurol Sci* **18(4)**, 492–498.

Macaya, A., Munell, F., Burke, R. E., and De Vivo, D. C. (1993). Disorders of movement in Leigh syndrome. *Neuropediatrics* **24(2)**, 60–67.

Maki, H., and Sekiguchi, M. (1992). MutT protein specifically hydrolyses a potent mutagenic substrate for DNA synthesis. *Nature* **355(6357)**, 273–5.

Mann, V. M., Cooper J. M., Krige, D., Daniel, S. E., Schapira, A. H., and Marsden, C. D. (1992). Brain, skeletal muscle and platelet homogenate mitochondrial function in Parkinson's disease. *Brain* **115(Pt 2**, 333–342.

Mann, V. M., Cooper, J. M., and Schapira, A. H. (1992). Quantitation of a mitochondrial DNA deletion in Parkinson's disease. *FEBS Lett.* **299(3)**, 218–222.

Maraganore, D. M., Harding A. E., and Marsden, C. D. (1991). A clinical and genetic study of familial Parkinson's disease. *Mov. Disord.* **6(3)**, 205–211.

Marie, S. K., Carvalho, A. A., Fonseca, L. F., Carvalho, M. S., Reed, U. C., and Scaff, M. (1999). Kearns-Sayre syndrome "plus". Classical clinical findings and dystonia. *Arq Neuropsiquiatr* **57(4)**, 1017–1023.

Martinelli, P., Giuliani, S., Lodi, R., Iotti, S., Zaniol, P., and Barbirol, B. (1996). Failure of brain and skeletal muscle energy metabolism in multiple system atrophy shown by in vivo phosphorous MR spectroscopy. *Adv. Neurol.* **69**, 271–277.

Matthews, R. T., Yang, L., Jenkins, B. G., Ferrante, R. J., Rosen, B. R., Kaddurah-Daouk, R., and Beal, M. F. (1998). Neuroprotective effects of creatine and cyclocreatine in animal models of Huntington's disease. *J. Neurosci.* **18(1)**, 156–163.

Mayr-Wohlfart, U., Paulus, C., Henneberg, A., and Rodel, G. (1996). Mitochondrial DNA mutations in multiple sclerosis patients with severe optic involvement. *Acta Neurol. Scand.* **94(3)**, 167–171.

Mayr-Wohlfart, U., Rodel, G., and Henneberg, A. (1997). Mitochondrial tRNA(Gln) and tRNA(Thr) gene variants in Parkinson's disease. *Eur. J.Med. Res.* **2(3)**, 111–113.

Mecocci, P., MacGarvey U., and Beal, M. F. (1994). Oxidative damage to mitochondrial DNA is increased in Alzheimer's disease. *Ann. Neurol.* **36(5)**, 747–751.

Mecocci, P., MacGarvey, U., Kaufman, A. E., Koontz, D., Shoffner J. M., Wallace, D. C., and Beal, M. F. (1993). Oxidative damage to mitochondrial DNA shows marked age-dependent increases in human brain. *Ann. Neurol.* **34(4)**, 609–616.

Meissner, C., von Wurmb, N., and Oehmichen, M. (1997). Detection of the age-dependent 4977 bp deletion of mitochondrial DNA. A pilot study. *Int. J. Lega. Med.* **110(5)**, 288–291.

Meissner, C., von Wurmb, N., Schimansky, B., and Oehmichen, M. (1999). Estimation of age at death based on quantitation of the 4977-bp deletion of human mitochondrial DNA in skeletal muscle. *Forensic Sci. Int.* **105(2)**, 115–124.

Miller, S. W., Trimmer, P.A., Parker, Jr., W. D., and Davis, R. E. (1996). Creation and characterization of mitochondrial DNA-depleted cell lines with "neuronal-like" properties. *J. Neurochem.* **67(5)**, 1897–1907.

MITOMAP. (2001). A Human Mitochondrial Genome Database. Center for Molecular Medicine, Emory University, Atlanta, GA. http://www.gen.emory.edu/mitomap.html.

Mizuno, Y., Matud S., Yoshino, H., Mori, H., Hattori, N., and Ikeb, S. (1994). An immunohistochemical study on alpha-ketoglutarate dehydrogenase complex in Parkinson's disease. *Ann. Neurol.* **35(2)**, 204–210.

Mizuno, Y., Saitoh, T., and Sone, N. (1987). Inhibition of mitochondrial alpha-ketoglutarate dehydrogenase by 1-methyl-4-phenylpyridinium ion. *Biochem. Biophys. Res. Commun.* **143**, 971–976.

Mojon, D. S., Herbert, J., Sadiq, S. A., Miller, J. R., Madonna, M., and Hirano, M. (1999). Leber's hereditary optic neuropathy mitochondrial DNA mutations at nucleotides 11778 and 3460 in multiple sclerosis. *Ophthalmologica* **213(3)**, 171–175.

Morris, A. A., Leonard, J. V., Brown, G. K., Bidouki, S. K., Bindoff, L. A., Woodward, C. E., Harding, A. E., Lake B. D., Harding, B. N., Farrell, M. A., Bell, J. E., Mirakhur M., and Turnbul, D. M. (1996). Deficiency of respiratory chain complex I is a common cause of Leigh disease. *Ann. Neurol.* **40(1)**, 25–30.

Moslemi, A. R., Melberg, A., Holme, E., and Oldfors, A. (1999). Autosomal dominant progressive external ophthalmoplegia: distribution of multiple mitochondrial DNA deletions. *Neurology* **53(1)**, 79–84.

Muller, U., Steinberger, D., and Nemeth, A. H. (1998). Clinical and molecular genetics of primary dystonias. *Neurogenetics* **1(3)**, 165–177.

Munoz, E., Oliva, R., Obach, V., Marti, M. J., Pastor, P., Ballesta, F., and Tolosa, E. (1997). Identification of Spanish familial Parkinson's disease and screening for the Ala53Thr mutation of the alpha-synuclein gene in early onset patients. *Neurosci. Lett.* **235(1–2)**, 57–60.

Newman, N. J. (1993). Leber's hereditary optic neuropathy. New genetic considerations. *Arch. Neurol.* **50(5)**, 540–548.

Nikoskelainen, E. K., Marttila, R. J., Huoponen, K., Juvonen, V., Lamminen, T., Sonninen, P., and Savontaus, M. L. (1995). Leber's "plus": neurological abnormalities in patients with Leber's hereditary optic neuropathy. *J. Neurol. Neurosurg. Psychiatry* **59(2)**, 160–164.

Ohlenbusch, A., Wilichowski, E., and Hanefeld, F. (1998). Characterization of the mitochondrial genome in childhood multiple sclerosis. I. Optic neuritis and LHON mutations. *Neuropediatrics* **29(4)**, 175–179.

Olsen, N. K., Hansen, A. W., Norby S., Edal, A, L., Jorgensen, J. R., and Rosenberg, T. (1995). Leber's hereditary optic neuropathy associated with a disorder indistinguishable from multiple sclerosis in a male harbouring the mitochondrial DNA 11778 mutation. *Acta Neurol. Scand.* **91(5)**, 326–329.

Ozawa, T., Hayakawa, M., Katsumata, K., Yoneda, M., Ikebe, S., and Mizuno, Y. (1997). Fragile mitochondrial DNA: the missing link in the apoptotic neuronal cell death in Parkinson's disease. *Biochem. Biophys. Res. Commun.* **235(1)**, 158–161.

Ozawa, T., Tanaka, M., Ikebe, S., Ohno K., Kondo, T., and Mizuno, Y. (1990). Quantitative determination of deleted mitochondrial DNA relative to normal DNA in parkinsonian striatum by a kinetic PCR analysis. *Biochem. Biophys. Res. Commun.* **172(2)**, 483–489.

Ozawa, T., Tanaka M., Ino, H., Ohno, K., Sano, T., Wada, Y., Yoneda, M., Tanno, Y., Miyatake, T., Tanaka, T. et al. (1991). Distinct clustering of point mutations in mitochondrial DNA among patients with mitochondrial encephalomyopathies and with Parkinson's disease. *Biochem. Biophys. Res. Commun.* **176(2)**, 938–946.

Ozelius, L. J., Hewett, J. W., Page, C. E., Bressman, S. B., Kramer, P. L., Shalish, C., de Leon, D., Brin, M. F., Raymond, D., Corey, D. P., Fan, S., Risch, N. J., Buckler, A. J., Gusella, J. F., and Breakefield, X. O. (1997). The early-onset torsion dystonia gene (DYT1) encodes an ATP-binding protein. *Nat. Genet.* **17(1)**, 40–48.

Palfi, S., Leventhal, L., Goetz C. G., Hantraye, T., Roitberg, B. Z., Sramek, J., Emborg, M., and Kordower, J. H. (2000). Delayed onset of progressive dystonia following subacute 3-nitropropionic acid treatment in Cebus apella monkeys. *Mov. Disord.* **15(3)**, 524–530.

Parker, W. D., Jr., Boyson, S. J., and Parks, J. K. (1989). Abnormalities of the electron transport chain in idiopathic Parkinson's disease. *Ann. Neurol.* **26(6)**, 719–723.

Parker, W. D., Jr., and Swerdlow, R. H. (1998). Mitochondrial dysfunction in idiopathic Parkinson disease. *Am. J. Hum. Genet.* **62(4)**, 758–762.

Parker, W. D., Jr., Swerdlow, R. H., Parks, J. K., Davis, J.N. 2nd, Trimmer, P., Bennett, J. P., and Wooten, G. F. (1999). Parkinson disease in twins [letter]. *JAMA* **282(14)**, 1328; discussion 1328–1329.

Parsian, A., Racette, B., Zhang, Z. H., Chakraverty S., Rundle M., Goate, A., and Perlmutter, J. S. (1998). Mutation, sequence analysis, and association studies of alpha-synuclein in Parkinson's disease. *Neurology* **51(6)**, 1757–1759.

Perlmutter, J. S., Tempel, L. W., Black, K. J., Parkinson, D., and Todd, R. D. (1997). MPTP induces dystonia and parkinsonism. Clues to the pathophysiology of dystonia. *Neurology* **49(5)**, 1432–1438.

Perna, N. T., and Kocher, T. D. (1996). Mitochondrial DNA: molecular fossils in the nucleus. *Curr. Biol.* **6(2)**, 128–129.

Pitkanen, S., and Robinson, B. H. (1996). Mitochondrial complex I deficiency leads to increased production of superoxide radicals and induction of superoxide dismutase. *J. Clin. Invest.* **98(2)**, 345–351.

Poirier, J., and Barbeau, A. (1985). A catalyst function for MPTP in superoxide formation. *Biochem. Biophys. Res. Commun.* **131(3)**, 1284–1289.

Polyak, K., Li, Y., Zhu, H., Lengauer, C., Willson, J. K., Markowitz, S. D., Trush, M. A., Kinzler K. W., and Vogelstein, B. (1998). Somatic mutations of the mitochondrial genome in human colorectal tumours. *Nat. Genet.* **20(3)**, 291–293.

Polymeropoulos, M. H., Lavedan, C., Leroy, E., Ide, S. E., Dehejia, A., Dutra, A., Pike, B., Root, H., Rubenstein, J., Boyer, R., Stenroos, E. S., Chandrasekharappa, S., Athanassiadou, A., Papapetropoulos, T., Johnson, W. G., Lazzarini A. M., Duvoisin, R. C., Di Iorio, G., Golbe L. I., and Nussbaum, R. L. (1997). Mutation in the alpha-synuclein gene identified in families with Parkinson's disease. *Science* **276(5321)**, 2045–2047.

Porteous, W. K., James, A. M., Sheard, P. W., Porteous, M., Packer, M. A., Hyslop, S. J., Melton, J. V., Pang, C. Y., Wei, Y. H., and Murphy, M. P. (1998). Bioenergetic consequences of accumulating the common 4977-bp mitochondrial DNA deletion. *Eur. J. Biochem.* **257(1)**, 192–201.

Przedborski, S., and Jackson-Lewis, V. (1998). Mechanisms of MPTP toxicity. *Mov. Disord.* **13** (Suppl. 1), 35–38.

Przedborski, S., Kostic, V., Jackson-Lewis, V., Naini, A. B., Simonetti S., Fahn, S., Carlson E., Epstein, C. J., and Cadet, J. L. (1992). Transgenic mice with increased Cu/Zn-superoxide dismutase activity are resistant to N-methyl-4-phenyl-1,2,3,6-tetrahydropyridine-induced neurotoxicity. *J. Neurosci.* **12(5)**, 1658–1667.

Rahman, S., Blok, R. B., Dahl, H. H., Danks, D. M., Kirby, D. M., Chow, C. W., Christodoulou, J., and Thorburn, D. R. (1996). Leigh syndrome: clinical features and biochemical and DNA abnormalities. *Ann. Neurol.* **39(3)**, 343–351.

Reichmann, H., Naumann, M., Hauck, S., and Janetzky, B. (1994). Respiratory chain and mitochondrial deoxyribonucleic acid in blood cells from patients with focal and generalized dystonia. *Mov. Disord.* **9(6)**, 597–600.

Richter, C., Park J. W., and Ames, B.N. (1988). Normal oxidative damage to mitochondrial and nuclear DNA is extensive. *Proc. Natl. Acad. Sci. U.S.A.* **85(17)**, 6465–6467.

Rocha, H., Flores, C., Campos, Y., Arenas, J., Vilarinho L., Santorelli F. M., and Torroni, A. (1999). About the "Pathological" role of the mtDNA T3308C mutationellipsis. *Am. J. Hum. Genet.* **65(5)**, 1457–1459.

Rossignol, Rodrigue, Malgat M., Mazat, J. P., and Letellier, T. (1999). Threshold Effect and Tissue Specificity. Implication for mitochondrial cytopathies. *J. Biol. Chem.* **274(47)**, 33,426–33,432.

Sanchez-Ramos, R.R., Overvik, E., and Ames, B.N. (1994). A marker of oxyradical-mediated DNA damage (8-hydroxy-2′-deoxyguanosine) is increased in nogr-striatum of Parkinson's disease. *Neurodegeneration* **3**, 197–204.

Sandy, M. S., Langston, J. W., Smith, M. T., and Di Monte, D. A. (1993). PCR analysis of platelet mtDNA: lack of specific changes in Parkinson's disease. *Mov. Disord.* **8(1)**, 74–82.

Santorelli, F. M., Shanske, A., Macaya, A., DeVivo, D. C., and DiMauro, S. (1993). The mutation at nt 8993 of mitochondrial DNA is a common cause of Leigh's syndrome. *Ann. Neurol.* **34(6)**, 827–834.

Schapira, A. H., Cooper, J. M., Dexter, D., Clark, J. B., Jenner P., and Marsden, C. D. (1990a). Mitochondrial complex I deficiency in Parkinson's disease. *J. Neurochem.* **54(3)**, 823–827.

Schapira, A. H., Cooper, J. M., Dexter, D., Jenner, P., Clark, J. B., and Marsden, C. D. (1989). Mitochondrial complex I deficiency in Parkinson's disease [letter] [see comments]. *Lancet* **1(8649)**, 1269.

Schapira, A. H., Gu, M., Taanman, J. W., Tabrizi, S. J., Seaton, T., Cleeter M., and Cooper, J. M. (1998). Mitochondria in the etiology and pathogenesis of Parkinson's disease. *Ann. Neurol.* **44** (3 Suppl. 1), S89–S98.

Schapira, A. H., Holt, I. J., Sweeney M., Harding, A. E., Jenner, P., and Marsden, C. D. (1990b). Mitochondrial DNA analysis in Parkinson's disease. *Mov. Disord.* **5(4)**, 294–297.

Schapira, A. H., Mann, V. M., Cooper, J. M., Dexter, D., Daniel, S. E., Jenner, P., Clark, J. B., and Marsden, C. D. (1990c). Anatomic and disease specificity of NADH CoQ1 reductase (complex I) deficiency in Parkinson's disease. *J. Neurochem.* **55(6)**, 2142–2145.

Schapira, A. H., Mann V. M., Cooper, J. M., Krige, D., Jenner, P. J., and Marsden, C. D. (1992). Mitochondrial function in Parkinson's disease. The Royal Kings and Queens Parkinson's Disease Research Group. *Ann. Neurol.* **32** (Suppl), S116–S124.

Schapira, A. H., Warner, T., Gash, M. T., Cleeter M.W., Marinho, C. F., and Cooper, J. M. (1997). Complex I function in familial and sporadic dystonia. *Ann. Neurol.* **41(4)**, 556–559.

Schulz, J. B., and Beal, M.F. (1994). Mitochondrial dysfunction in movement disorders. *Cur. Opin. Neurol.* **7**, 333–339.

Scott, W. K., Nance, M. A., Watts, R. L., Hubble, J. P., Koller, W. C., Lyons, K., Pahwa, R., Stern, M. B., Colcher, A, Hiner, B.C., Jankovic, J., Ondo, W. G., Allen, Jr., F. H., Goetz, C. G., Small, G. W., Masterman, D., Mastaglia, F., Laing, N. G., Stajich, J. M., Slotterbeck, B., Booze, M. W., Ribble, R. C., Rampersaud, E., West, S. G., Gibson, R. A., Middleton, L. T., Roses, A. D., Haines, J. L., Scott, B. L., Vance, J. M., and Pericak-Vance, M. A. (2001). Complete genomic screen in Parkinson disease: evidence for multiple genes. *JAMA* **286(18)**, 2239–2244.

Scott, W. K., Yamaoka, L. H., Stajich, J.M., Scott, B. L., Vance, J. M., Roses, A. D., Pericak-Vance M. A., Watts, R. L., Nance, M., Hubble, J., Koller, W., Stern, M. B., Colcher, A., Allen, Jr., F. H., Hiner, B. C., Jankovic, J., Ondo, W., Laing, N. G., Mastaglia, F., Goetz, G., Pappert, E., Small, G. W., Masterman, D., Haines, J. L., and Davies, T. L. (1999). The alpha-synuclein gene is not a major risk factor in familial Parkinson disease [letter]. *Neurogenetics* **2(3)**, 191–192.

Shibutani, S., Takeshita, M., and Grollman. A. P. (1991). Insertion of specific bases during DNA synthesis past the oxidation-damaged base 8-oxodG. *Nature* **349(6308)**, 431–434.

Shimura-Miura, H., Hattori, N., Kang, D., Miyako, K., Nakabeppu, Y., and Mizuno, Y. (1999). Increased 8-oxo-dGTPase in the mitochondria of substantia nigral neurons in Parkinson's disease. *Ann. Neurol.* **46(6)**, 920–924.

Shoffner, J. M., Brown, M. D., Stugard, C., Jun, A. S., Pollock, S., Haas, R. H., Kaufman, A., Koontz, D., Kim, Y., Graham, J. R. et al. (1995). Leber's hereditary optic neuropathy plus dystonia is caused by a mitochondrial DNA point mutation. *Ann. Neurol.* **38(2)**, 163–169.

Shoffner, J. M., Brown, M. D., Torroni, A., Lott, M. T., Cabell, M. F., Mirra, S. S., Beal, M. F., Yang, C. C., Gearing, M., Salvo, R. et al. (1993). Mitochondrial DNA variants observed in Alzheimer disease and Parkinson disease patients. *Genomics* **17(1)**, 171–184.

Shoffner, J. M., and Wallace, D. C. (1994). Oxidative phosphorylation diseases and mitochondrial DNA mutations: diagnosis and treatment. *Annu. Rev. Nutr.* **14**, 535–568.

Siciliano, G., Mancuso, M., Ceravolo, R., Lombardi, V., Iudice, A., and Bonuccelli, U. (2001). Mitochondrial DNA rearrangements in young onset parkinsonism: two case reports. *J. Neurol. Neurosurg. Psychiatry* **71(5)**, 685–687.

Simon, D. K. (1999). Parkinson disease in twins [letter]. *JAMA* **282(14)**, 1328; discussion 1328–1329.

Simon, D.K., and Beal, M. F. (2001). Pathogenesis: oxidative stress, mitochondrial dysfunction and excitotoxicity. In "Parkinson's Disease: Diagnosis and Clinical Management", S. A. Factor and W. J. Weiner, eds.), Chapter 27. Demos Vermande.

Simon, D. K., and Johns, D. R. (1999). Mitochondrial disorders: clinical and genetic features. *Annu. Rev. Med.* **50**, 111–127.

Simon, D. K., Lin M.T., Ahn, C.H., Liu, G.-J., Gibson, G. E., Beal, M. F., and Johns, D. R. (2001a). Low mutational burden of individual acquired mitochondrial DNA mutations in brain. *Genomics* **73**, 113–116.

Simon, D. K., Johns, D. R., Friedman, J. R., Breakfield, X. O., Brin, M. F., Charness, M. E., Tarsy, D., Tarnopolsky, M. A. (2002a). A novel heteroplasmic mitochondrial complex I gene mutation associated with adult-onset dystonia. *Neurology* **58 (suppl. 3)**, A324.

Simon, D. K., Lin, M. T., and Pascual-Leone, A. (2002b). "Nature vs. nurture" and incompletely penetrant mutations: lessons from twin studies of Parkinson's disease. *J. Neurol. Neurosurg. Psychol.* **72(6)**, 686–689.

Simon, D. K., Mayeux, R., Marder, K., Kowall, N. W., Beal, M. F., and Johns, D. R. (2000). Mitochondrial DNA mutations in complex I and tRNA genes in Parkinson's disease. *Neurology* **54**, 703–709.

Simon, D. K., Pulst, S. M., Sutton, J. P., Browne, S. E., Beal, M. F., and Johns, D. R. (1999). Familial multisystem degeneration with parkinsonism associated with the 11778 mitochondrial DNA mutation. *Neurology* **53(8)**, 1787–1793.

Simon, D. K., Tarnopolsky, M. A., Greenamyre, J. T., and Johns, D. R. (2001b). A frameshift mitochondrial complex I gene mutation in a patient with dystonia and cataracts: is the mutation pathogenic? *J. Med. Genet.* **38(1)**, 58–61.

Simonetti, S., Chen X., DiMauro, S., and Schon, E. A. (1992). Accumulation of deletions in human mitochondrial DNA during normal aging: analysis by quantitative PCR. *Biochim. Biophys. Acta* **1180(2)**, 113–122.

Sinha, B. K., Singh, Y., and Krishna, G. (1986). Formation of superoxide and hydroxyl radicals from 1-methyl-4-phenylpyridinium ion (MPP+): reductive activation by NADPH cytochrome P-450 reductase. *Biochem. Biophys. Res. Commun.* **135(2)**, 583–588.

Souza, J. M., Giasson, B. I., Chen, Q., Lee, V. M., and Ischiropoulos, H. (2000). Dityrosine cross-linking promotes formation of stable alpha-synuclein polymers. Implication of nitrative and oxidative stress in the pathogenesis of neurodegenerative synucleinopathies. *J. Biol. Chem.* **275(24)**, 18344–18349.

Sudarsky, L., Plotkin, G. M., Logigian, E. L., and Johns, D. R. (1999). Dystonia as a presenting feature of the 3243 mitochondrial DNA mutation. *Mov. Disord.* **14(3)**, 488–491.

Swerdlow, R. H., Golbe, L.I., Parks, J. K., Cassarino D. S., Binder, D. R., Grawey, A. E., Litvan, I., Bennett, Jr., J. P., Wooten, G. F., and Parker, W. D. (2000). Mitochondrial dysfunction in cybrid lines expressing mitochondrial genes from patients with progressive supranuclear palsy. *J. Neurochem.* **75(4)**, 1681–1684.

Swerdlow, R.H., Parker, W. D., Currie, L.J., Bennett, J. P., Harrison, M. B., Trugman, J. M., and Wooten, G. F. (2001). Gender ratio differences between Parkinson's disease patients and their affected relatives. *Parkinsonism Relat. Disord.* **7(2)**, 129–133.

Swerdlow, R. H., Parks, J. N., Davis, J. N. 2nd, Cassarino, D. S,. Trimmer, P. A., Currie, L. J., Dougherty, J., Bridges, W. S., Bennett, Jr., J. P., Wooten, G. F., and Parker, W. D. (1998). Matrilineal inheritance of complex I dysfunction in a multigenerational Parkinson's disease family. *Ann. Neurol.* **44(6)**, 873–881.

Swerdlow, R. H., Parks, J. K., Miller S. W., Tuttle, J. B., Trimmer, P. A., Sheehan, J. P., Bennett, Jr., J. P., Davis, R. E., and Parker, Jr., W. D. (1996). Origin and functional consequences of the complex I defect in Parkinson's disease. *Ann. Neurol.* **40(4)**, 663–671.

Swerdlow, R.H., Parks, J. K., Wooten, G. F., Miller, S. W., Davis, R. E., Parker, Jr, W. D. (1997). As in Parkinson's disease, a bioenergetic defect transfers with mitochondrial DNA of patients with multiple system atrophy. *Mov. Disord.* **12**(Suppl. 1)**(1)**, 3.

Swerdlow, R. H., and Wooten, G. F. (2001). A novel deafness/dystonia peptide gene mutation that causes dystonia in female carriers of Mohr-Tranebjaerg syndrome. *Ann. Neurol.* **50(4)**, 537–540.

Tabrizi, S. J., Cooper, J. M., and Schapira, A. H. (1998). Mitochondrial DNA in focal dystonia: a cybrid analysis. *Ann. Neurol.* **44(2)**, 258–261.

Takayanagi, R., Takeshige K., and Minakami, S. (1980). NADH- and NADPH-dependent lipid peroxidation in bovine heart submitochondrial particles. Dependence on the rate of electron flow in the respiratory chain and an antioxidant role of ubiquinol. *Biochem. J.* **192(3)**, 853–860.

Tanaka, M., Kovalenko, S. A., Gong, J. S., Borgeld, H. J. Katsumata, K., Hayakawa, M., Yoneda, M., and Ozawa, T. (1996). Accumulation of deletions and point mutations in mitochondrial genome in degenerative diseases. *Ann. N.Y. Acad. Sci.* **786**, 102–111.

Tanner, C. M., Ottman, R., Goldman, S. M., Ellenberg, J., Chan, P., Mayeux, R., and Langston, J. W. (1999). Parkinson disease in twins: an etiologic study. *JAMA* **281(4)**, 341–346.

Tarnopolsky, M. A., and Beal, M. F. (2001). Potential for creatine and other therapies targeting cellular energy dysfunction in neurological disorders. *Ann. Neurol.* **49(5)**, 561–574.

Tarnopolsky, M. A., Roy B. D., and MacDonald, J. R. (1997). A randomized, controlled trial of creatine monohydrate in patients with mitochondrial cytopathies. *Muscle Nerve* **20(12)**, 1502–1509.

Taylor, R. W., Taylor, G. A., Durham, S. E., and Turnbull, D. M. (2001). The determination of complete human mitochondrial DNA sequences in single cells: implications for the study of somatic mitochondrial DNA point mutations. *Nucleic Acids Res.* **29(15)**, E74–E74.

Tranebjaerg, L., Hamel, B. C., Gabreels, F. J., Renier, W. O., and Van Ghelue, M. (2000). A de novo missense mutation in a critical domain of the X-linked DDP gene causes the typical deafness-dystonia-optic atrophy syndrome. *Eur. J. Hum. Genet.* **8(6)**, 464–467.

Turrens, J. F., and Boveris, A. (1980). Generation of superoxide anion by the NADH dehydrogenase of bovine heart mitochondria. *Biochem. J.* **191(2)**, 421–427.

Valente, E. M., Bentivoglio, A. R., Cassetta, E., Dixon, P. H., Davis, M. B., Ferraris, A., Ialongo, T., Frontali, M., Wood, N. W., and Albanes, A. (2001). DYT13, a novel primary torsion dystonia locus, maps to chromosome 1p36.13–36.32 in an Italian family with cranial-cervical or upper limb onset. *Ann. Neurol.* **49(3)**, 362–366.

van Orsouw, N. J., Zhang, X., Wei, J. Y., Johns, D. R., and Vijg, J. (1998). Mutational scanning of mitochondrial DNA by two-dimensional electrophoresis. *Genomics* **52(1)**, 27–36.

Vanopdenbosch, L., Dubois, B., D'Hooghe, M. B., Meire, F., and Carton, H. (2000). Mitochondrial mutations of Leber's hereditary optic neuropathy: a risk factor for multiple sclerosis. *J. Neurol.* **247(7)**, 535–543.

Vaughan, J., Durr, A., Tassin, J., Bereznai, B., Gasser, T., Bonifati, V., De Michele G., Fabrizio, E., Volpe, G., Bandmann, O., Johnson, W. G., Golbe, L. I., Breteler M., Meco, G., Agid, Y., Brice, A., Marsden, C. D., and Wood, N. W. (1998a). The alpha-synuclein Ala53Thr mutation is not a common cause of familial Parkinson's disease: a study of 230 European cases. European Consortium on Genetic Susceptibility in Parkinson's Disease. *Ann. Neurol.* **44(2)**, 270–273.

Vaughan, J. R., Farrer, M. J., Wszolek, Z. K., Gasser, T., Durr, A., Agid, Y., Bonifati V., DeMichele G., Volpe, G., Lincoln, S., Breteler M., Meco, G., Brice, A., Marsden, C. D., Hardy, J., and Wood, N. W. (1998b). Sequencing of the alpha-synuclein gene in a large series of cases of familial Parkinson's disease fails to reveal any further mutations. The European Consortium on Genetic Susceptibility in Parkinson's Disease (GSPD). *Hum. Mol. Genet.* **7(4)**, 751–753.

Vielhaber, S., Kudin, A., Schroder, R., Elger, C. E., and Kunz, W. S. (2000). Muscle fibres: applications for the study of the metabolic consequences of enzyme deficiencies in skeletal muscle. *Biochem. Soc. Trans.* **28(2)**, 159–164.

Wallace, D. C., and Murdock, D. G. (1999). Mitochondria and dystonia: the movement disorder connection? *Proc. Natl. Acad. Sci. U.S.A.* **96(5)**, 1817–1819.

Wallace, D. C., Singh, G., Lott, M. T., Hodge, J. A., Schurr, T. G., Lezza, A. M., Elsas, L. J. D., and Nikoskelainen, E. K. (1988). Mitochondrial DNA mutation associated with Leber's hereditary optic neuropathy. *Science* **242(4884)**, 1427–1430.

Wang, W. W., Khajavi M., Patel, B. J., Beach, J., Jankovic J., and Ashizawa, T. (1998). The G209A mutation in the alpha-synuclein gene is not detected in familial cases of Parkinson disease in non-Greek and/or Italian populations. *Arch. Neurol.* **55(12)**, 1521–1523.

Wooten, G. F., Currie, L. J., Bennett, J. P., Harrison, M. B., Trugman, J. M., and Parker, Jr., W. D. (1997). Maternal inheritance in Parkinson's disease. *Ann. Neurol.* **41(2)**, 265–268.

Wunderlich, S., Reiners, K., Gasser, T., and Naumann, M. (2001). Cervical dystonia in monozygotic twins: case report and review of the literature. *Mov. Disord.* **16(4)**, 714–718.

Yoshino, H., Nakagawa-Hattori, Y., Kondo, T., and Mizuno, Y. (1992). Mitochondrial complex I and II activities of lymphocytes and platelets in Parkinson's disease. *J. Neural. Transm. Park. Dis. Dement. Sect.* **4(1)**, 27–34.

Zang, L. Y., and Misra, H. P. (1992). Superoxide radical production during the autoxidation of 1-methyl-4- phenyl-2,3-dihydropyridinium perchlorate. *J. Biol. Chem.* **267(25)**, 17547–17552.

Zeviani, M., Moraes, C. T., DiMauro, S., Nakase, H., Bonilla, E., Schon, E. A., and Rowland, L. P. (1988). Deletions of mitochondrial DNA in Kearns-Sayre syndrome. *Neurology* **38(9)**, 1339–1346.

Zhang, C., Linnane, A. W., and Nagley, P. (1993). Occurrence of a particular base substitution (3243 A to G) in mitochondrial DNA of tissues of ageing humans. *Biochem. Biophys. Res. Commun.* **195(2)**, 1104–1110.

Zhang, C., Liu, V. W., Addessi, C. L., Sheffield, D. A., Linnane, A. W., and Nagley, P. (1998). Differential occurrence of mutations in mitochondrial DNA of human skeletal muscle during aging. *Hum. Mutat.* **11(5)**, 36–71.

Zweig, R. M., Singh, A., Cardillo, J. E., and Langston, J. W. (1992). The familial occurrence of Parkinson's disease. Lack of evidence for maternal inheritance [published erratum appears in *Arch. Neurol.* (1993) Feb, **50(2)**, 153]. *Arch. Neurol.* **49(11)**, 1205–1207.

# CHAPTER 42

# Genetics of Gilles de la Tourette Syndrome

**DAVID L. PAULS**
*Department of Psychiatry
Massachusetts General Hospital
Harvard Medical School
Charlestown, Massachusetts 02129*

**STEFAN-M. PULST**
*Division of Neurology, Cedars-Sinai Medical Center
Departments of Medicine and Neurobiology
UCLA School of Medicine
Los Angeles, California 90048*

**MATTHEW W. STATE**
*Child Study Center
Yale University School of Medicine
New Haven, Connecticut 06520*

I. The CTS Phenotype
II. GTS is Heritable
   A. Family Studies
   B. Twin Studies
   C. A GTS Phenotypic Spectrum
III. Segregation Analyses of GTS Family Data
IV. The Search for Genes in GTS
   A. Association Studies
   B. Findings from Association Studies
   C. Linkage Analysis
  V. Cytogenetic and Molecular Cytogenetic Approaches
   A. Cytogenetic Findings
VI. Neuroimaging
VII. Treatment
VIII. Summary and Future Prospects
   Acknowledgments
   References

Gilles de la Tourette syndrome is a neuropsychiatric disorder with onset in childhood characterized by a waxing and waning clinical course. Research into the genetics of the disorder has led to several noticeable advances that have led to a better understanding of the patterns of transmission within families. These findings are summarized in the current article and future directions of research are discussed.

Gilles de la Tourette's syndrome (GTS) is a potentially debilitating neuropsychiatric disorder defined by the presence of both vocal and motor tics. Research into the genetics of GTS in the past two decades has led to several notable advances in the understanding of the syndrome: These include (1) a more accurate estimate of the prevalence of GTS and the realization that it is far more common than had previously been appreciated (Robertson and Stern, 1998); (2) a clearer elucidation of the range of clinical presentations that comprise the phenotype, including chronic tics (CT) and tic-related obsessive-compulsive (OC) symptoms (Comings and Comings, 1987; Pauls et al., 1991b, 1993, 1994; Eapen et al., 1997; Alsobrook and Pauls, 2002); (3) a more detailed documentation of the familial transmission of GTS and related conditions (Pauls and Leckman, 1986; Pauls et al., 1991a; Eapen et al., 1993; Hasstedt et al., 1995; Walkup et al., 1996; Hebebrand et al., 1997); and (4) the identification of several chromosomal regions that appear to contain susceptibility genes for GTS (Simonic et al., 1998, 2001; Barr et al., 1999; Tourette Syndrome Association International Consortium for Genetics, 1999, Mérette et al., 2000).

## I. THE GTS PHENOTYPE

The clinical hallmarks of GTS are chronic motor and vocal tics that typically begin in middle childhood. Motor tics are intermittent, stereotyped, and ultimately irrepressible fragments of normal motor movements. These may be simple in the sense of an eye blink or facial grimace. Alternatively, tics may take the form of more complex actions, such as bending down and touching the floor. In some cases, complex motor tics present as copropraxia (obscene gesturing), echopraxia (copying other's movements), or self-injurious behaviors.

Vocal tics are also intermittent and irrepressible. These too are classified as simple or complex. In the former case, common manifestations include throat clearing or coughing. The most dramatic examples of complex vocal tics include coprolalia, echolalia, or palilalia (the repetition of parts of words; Leckman and Cohen, 1999).

In addition to simple and complex tics, those with GTS also suffer from a range of associated somatosensory phenomena. Patients presenting with GTS commonly experience premonitory urges and the compelling need to have things "just right." Moreover, a very large number of patients seen in the clinic describe at least some of their tics as a voluntary acquiescence to internal tension or discomfort and may be able to suppress movements or noises for periods of time (Leckman et al., 1993, 1994). This seemingly purposeful component of the syndrome can be the source of significant difficulty for children and families particularly in the early phases of the disorder. Children may initially be viewed both at home and in school as engaging in willfully disruptive or challenging behavior, with the ability to suppress the tics cited as evidence. Most often, with appropriate education these misconceptions give way to an understanding of the compelling and largely irrepressible nature of the symptoms.

A large percentage of GTS patients seen in clinical settings suffer from co-morbid psychiatric disorders including obsessive-compulsive disorder (OCD), depression, and attention deficit hyperactivity disorder (ADHD). The nature of the relationship between GTS and these other Axis I disorders is the subject of ongoing investigation and is later discussed in more detail. From a clinical standpoint, it is very often the presence of these co-morbidities that both requires and directs psychiatric intervention.

DSM IV criteria require the presence of multiple motor tics and at least one vocal tic occurring daily over more than a year with associated functional impairment. There must not be a tic-free interval for three consecutive months. Just how relevant each of these criteria is in any given instance remains a question. For example, patients with chronic motor tics may demonstrate a constellation of complaints, a natural history of illness and a treatment response that is indistinguishable from GTS with the only observable difference being the absence of vocal tics. From the standpoint of genetic studies as well, twin, family, and cytogenetic data, as noted in Section II, support the notion that a range of clinical presentations represents a GTS phenotypic spectrum.

## II. GTS IS HERITABLE

The most important prelude to a search for genes in any disorder is a strong suggestion that genetic factors play an important role in its etiology. More than two decades of research has resulted in several complementary lines of evidence supporting the heritability of GTS and related conditions.

1. Multiple family studies have demonstrated the risks of having GTS are substantially greater among relatives of affected individuals than among the general population
2. Twin studies have shown that this familial aggregation is not simply the result of shared environment
3. Careful phenotypic analysis of GTS pedigrees indicate that some forms of OCD and/or CT are an alternative expression of genes that predispose to GTS
4. Segregation analysis has suggested particular modes of inheritance that underlie the syndrome

### A. Family Studies

Six family studies, completed over the past ten years, have provided overwhelming support for the hypothesis that GTS and related conditions are familial (Pauls et al., 1991b; Eapen et al., 1993; van de Wetering, 1993; Walkup et al., 1996; Hebebrand et al., 1997; Kano et al., 2001). Five of these studies focused on families from within the U.S. and Europe, and the sixth reported on families collected in Japan (Kano et al., 2001). All of the investigations relied on structured interviews and collected data directly from probands and available first-degree family members. In addition, best estimate diagnostic procedures were used in all six in order to achieve consensus diagnoses for all first-degree relatives.

The results from the five studies examining families from the U.S. and Europe were strikingly similar. The morbid risk for of GTS among relatives ranged from between 9.8–15% across all studies. These data are meaningful from a genetic standpoint only in the context of an understanding of the rate of GTS in the general population. For example, if the rate of disorder were found in general to be nearly 10%, then there would be no basis to conclude that genes play a major role in the etiology of GTS.

The question of how frequently GTS is found in the general population has historically been the subject of some debate. Early estimates were based on ascertainment through clinical samples and led to the conclusion that GTS was a rare condition, affecting approximately 5:1,000,000 people (Lucas et al., 1982). These figures have subsequently been challenged by a number of epidemiological studies. An investigation among Israeli teenagers undergoing physical examination for entrance into the army estimated the population prevalence at approximately 5:10,000 boys and 3:10,000 girls (Apter et al., 1993). More recent studies have placed the figure at close to 1:100 individuals (Robertson and Stern, 1998).

Nonetheless, even given this variability in estimates, a comparison of the figures for GTS in affected families versus the general population suggests that there is a 10 to 100-fold greater risk for a first-degree relative. Such findings provide strong support for a genetic contribution to the disorder.

It should be noted that in contrast to the studies in the U.S. and Europe, the rates of illness observed in Japanese families were significantly lower. The age-corrected rates of GTS among first-degree relatives were only 2.0%. If replicated, these data suggest that there may be differences in the etiology of GTS and related behaviors in the Japanese population when compared to populations of European descent. Importantly, it is not yet clear what the population prevalence of GTS is in Japan, and as noted above, the absolute figures for familial risk are truly informative only when they may be compared to the risks to the general population.

## B. Twin Studies

There are multiple reasons, apart from genetic influences, that illnesses may run in families. For instance, infectious diseases or those that are the result of an environmental exposure will show family aggregation. One way to begin to tease out the relative contributions of genes and the environment to the familiality of GTS is to assess whether or not twins who are identical in terms of their genetic material (monozygotic; MZ twins) are more likely to also share GTS, OCD, and/or CT than are twins who are on average only identical with respect to 50% of their genetic complement (dizygotic; DZ twins). The underlying rationale for these types of investigations is that both manner of twins are likely to share nongenetic risks as they have a common prenatal environment and are most often raised in a very similar fashion. Thus, differences observed between MZ and DZ twins in terms of their liability to a particular condition might be attributed to the degree to which they do or do not share genetic material.

Indeed when studies have examined how often both twins in a pair have a diagnosis of GTS, MZ twins have been found to be far more likely to be concordant for GTS than DZ twins (Hyde *et al.*, 1992; Price *et al.*, 1985; Walkup *et al.*, 1988). When strict diagnostic criteria have been used, MZ concordance rates have been found to be between 53 and 56% (Hyde, *et al.*, 1992; Price, *et al.*, 1985) versus less than 10% in DZ twins. When study methodology has included the diagnosis of CT, MZ concordance has been noted to be between 77 and 100% with DZ twins sharing tics approximately 20% of the time (Price *et al.*, 1985; Walkup, *et al.*, 1987). These investigations provide an additional line of evidence that familial aggregation of the GTS phenotype is not simply the result of shared environment.

## C. A GTS Phenotypic Spectrum

Individuals with GTS also suffer a very high rate of co-morbid psychiatric disorders. The finding of depression, ADHD and OCD is exceedingly common and far more likely than would be expected in the general population. In fact, these co-morbidities are so frequent that the question of whether they represent alternative manifestations of a GTS gene or genes has become an important area of inquiry. To date, the strongest evidence for this type of genetic relationship has been found with respect to OCD.

As many as 60% of individuals with GTS also have OCD. These rates compare to between 2 and 3% of the general population (Hebebrand *et al.*, 1997; Pauls and Leckman, 1986). Perhaps more importantly, when families of GTS patients are examined, there is a significant increase in the risk for OCD among relatives in addition to the increased risk with respect to GTS.

Certainly, this high risk for OCD among families who have members with GTS does not necessarily mean that the disorders are alternative expressions of a single genetic liability. The same finding could be explained by a shared pathophysiological mechanism apart from genes, or the clustering of environmental risk factors. However, if OCD is an alternate expression of a gene of major effect in GTS, one would expect that relatives of patients with GTS (without OC symptoms) would have higher rates of OCD (without tics) than would be found in unaffected families.

Multiple analyses have confirmed this hypothesis. Probands presenting with GTS alone have relatives with significantly elevated rates of OCD (without TS) compared with matched controls (Hebebrand, *et al.*, 1997; Pauls and Leckman, 1986; Pauls *et al.*, 1986b; Walkup, *et al.*, 1996).

A similar relationship has been suggested by the data regarding chronic tics. In the studies noted above, investigations in the U.S. and Europe found rates of "other tics" ranging from 15 and 20%. This compares with

The issue of whether ADHD and depression also are examples of variable expression of a GTS genotype has not yet been resolved. Evidence in support of this hypothesis has been presented by Comings and colleagues (Comings, 1995; Comings and Comings, 1984, 1990). However, other investigations have not replicated these results (Pauls *et al.*, 1986a). Though the frequency of ADHD in probands and families with TS has been found to be elevated, no increase was found in the rate of ADHD without tics. An additional analysis showed that depression was also likely to be found in GTS patients with OCD. However, the data did not support a conclusion that depression alone represented an alternative expression of a GTS genotype (Pauls *et al.*, 1994).

## III. SEGREGATION ANALYSES OF GTS FAMILY DATA

Segregation analyses are designed to test specific genetic hypotheses regarding the transmission of a disorder within families. Seven separate segregation analyses have been completed on GTS family data sets. All but one gave results that are consistent with the hypothesis that major genes are involved in the expression of GTS and related conditions. Three studies (Pauls and Leckman, 1986; Eapen et al., 1993; van de Wetering, 1993) reported a pattern consistent with autosomal dominant inheritance. Three others (Pauls et al., 1991a; Hasstedt et al., 1995; Walkup et al., 1996) found that the most likely mode of inheritance involved penetrance in heterozygote individuals that was intermediate between the two homozygote conditions. In addition, Walkup et al. (1996) observed evidence for a significant multifactorial (polygenic) background with the major locus accounting for more than half of the observed phenotypic variance. In contrast to all these studies, Seuchter et al. (2000) did not find evidence for Mendelian transmission of GTS and related conditions in 108 families ascertained as a result of having a proband with GTS. The recurrence risk for GTS and tics within in these families were similar to other reported rates (Hebebrand et al., 1997) and it is not immediately clear why the results of this analysis were so divergent from previous studies.

Nonetheless, the majority of studies do provide strong evidence for the effect of major genes in the expression of GTS. Taken together the results also suggest that the inheritance patterns of GTS and related conditions are more complex than were originally appreciated. For instance it appears increasingly likely that individual mutations in one of several different genes of major effect may be capable of leading to GTS phenotypes. In addition, it is probable that the interaction of genes and environment are an important source of clinical variability; and finally it may be that in some individuals and families, the genetic risk for GTS may involve subtle abnormalities in multiple genes acting together. All of these phenomena make the identification of specific genes contributing to the disorder somewhat more problematic than would be the case if only a single gene led to GTS and related conditions in a predictable fashion. Nonetheless, the past decade has been notable for significant advances in the hunt for the specific genes that contribute to these disorders. There efforts are reviewed in the following section.

## IV. THE SEARCH FOR GENES IN GTS

Three major strategies have been employed in the search for susceptibility genes for GTS and related conditions. These include (1) association studies, (2) linkage analysis, and (3) cytogenetic approaches. The rationale for each of these methodologies will be presented first with a subsequent review of the most pertinent findings over the past decade.

### A. Association Studies

Genetic association studies assess the risks in a given population of having GTS versus not having the disorder. These analyses include information regarding the genetic make-up of the individuals being compared and thus provide a means of investigating whether a particular genetic variable is associated with having the disorder in question. These types of studies are appealing because the analyses may be undertaken without the need to rely on specific assumptions about the mode of inheritance. Two types of studies have been pursued in this regard: case-control studies and family-based studies of association.

The advantage of the case-control design lies in the fact that cases may be readily obtained, efficiently genotyped, and compared with control samples. The disadvantage of this approach is the inherent difficulty involved in identifying control groups. In fact, association studies have a reputation for being prone to high rates of false-positive results, largely as a consequence of inappropriate matching of case and control individuals.

The challenge of identifying appropriate groups for comparison in genetic analysis is complicated by a phenomenon known as population stratification. If a control group that is thought to be genetically similar to the identified cases turns out to have a subpopulation of individuals with a distinctive genetic ancestry (i.e., their general range and frequency of polymorphisms) the result is that many irrelevant genetic differences between cases and controls appear erroneously to be associated with the disorder in question.

In the last several years, there have been a number of proposals in the literature that should help overcome this liability (see Grigorenko and Pauls, in press, for a more detailed discussion). One of the more promising has been suggested by Devlin and Roeder (1999) who describe a population-based association method using a "genomic control" (GC). The method requires the additional genotyping of markers that are unlikely to be involved in the etiology of the disorder (null loci). These data are then used to assess for the presence and extent of population stratification. If it is detected, a multiplier is derived that allows one to adjust the critical values for significance tests when assessing candidate loci that *are* hypothesized to be relevant to the disease in question. As a result, association analyses may be undertaken in the presence of population stratification without an increase risk for false positive findings.

In addition to the GC approach, two alternative methods have been developed that alleviate the need for stringently defined control samples: the haplotype relative risk (HRR) method (Falk and Rubinstein 1987; Terwilliger and Ott, 1992) and the transmission disequilibrium test (TDT) (Spielman et al.,1993, Spielman and Ewens, 1996). Both utilize samples of small families rather than unrelated cases and controls. To carry out these studies, DNA from affected probands and their parents is genotyped for the genes/ markers under investigation. The proband's two alleles define the "affected" group and the nontransmitted parental alleles constitute the control sample. Since both parents donate alleles equally to both groups they are, by definition, perfectly matched for ethnicity and race.

One particular type of association study known as a "candidate gene" analysis currently makes up the vast majority of the relevant literature. At present, limits on the availability of anonymous DNA markers makes high-resolution genome-wide association studies technically challenging. As a result, instead of evaluating anonymous DNA segments from across the genome, the majority of association studies evaluate specific genes or small genomic intervals to determine if an allele or alleles may be important in the manifestation of the disorder. The choice of these genes or genomic segments is typically based on available biological information about a condition (e.g., neurochemical levels in affected individuals, the mode of action of a pharmaceutical agent, or differences in brain structure and/or function). While this approach may be intellectually compelling, at the present time it is still fundamentally limited by gaps in the collective understanding of the biological underpinnings of GTS. On the other hand, once results of linkage and/or cytogenetic studies suggest regions of interest and those regions have been sufficiently narrowed, genes within these intervals will be excellent candidates for association methods.

## B. Findings from Association Studies

A number of candidate gene studies have been undertaken in GTS over the past decade and overall the results have been disappointing. The first candidate gene study was reported by Comings and colleagues. (Comings et al., 1991). The A1 allele at the dopamine DRD2 locus was suggested to be associated with the presence of GTS as well as with symptom severity. Subsequently, this hypothesis was examined using both linkage and association strategies. Linkage was excluded between GTS and the D2 dopamine receptor gene (DRD2; Gelernter et al., 1990; Devor et al., 1990; Brett et al., 1995). In addition, Gelernter et al. (1994) were not able to find an association between the A1 allele and symptom severity (Gelernter et al., 1994), and a family-based association study undertaken by Nöthen et al. (1994) was similarly unable to find any evidence for an association between DRD2 and GTS.

A number of other candidate genes have also resulted in negative findings. These include the dopamine receptors DRD1 (Gelernter et al., 1993; Brett et al., 1995), DRD3 (Hebebrand et al., 1993; Brett et al., 1995), and DRD5 (Brett et al., 1995; Barr et al., 1997), the dopamine transporter gene (Gelernter et al., 1995; Vandenbergh et al., 2000), the serotonin receptor gene (Gelernter et al., 1995), the serotonin transporter gene (Cavallini et al., 2000), the catechol-O-methyl transferase gene (Cavallini et al., 2000), the norepinephrine transporter gene (Stober et al., 1999), and the alpha1 subunit of the glycine receptor (Brett et al., 1997).

Results from one study (Grice et al., 1996) using a TDT approach were consistent with an apparent association between the dopamine D4 receptor (DRD4) locus and GTS. The DRD4*7R allele was found more frequently in the transmitted group of alleles than in the nontransmitted group ($\chi^2 =12.1$, $p < 0.0005$). Unfortunately, Hebebrand et al. (1997) were unable to replicate the association using a family-based paradigm and Barr et al. (1996) did not find any evidence for linkage in a sample of Canadian families using three different polymorphisms at the DRD4 locus.

## C. Linkage Analysis

Genetic linkage is one of the most powerful methods available for clarifying the role of genetic factors in the expression of human disorders. Linkage demonstrates the existence of an etiologically important locus even in the absence of a known biological abnormality. The objective of linkage studies is to demonstrate that a distinct polymorphic segment of DNA (i.e. a "DNA-marker") with a known chromosomal location co-segregates with a disease within a family. Genetic linkage is detectable if such a marker is sufficiently close to a susceptibility locus that the two alleles are transmitted together through the generations. The demonstration of genetic linkage provides *prima facie* evidence that a gene is present that is contributing to the expression of the disorder. Once located, the genetic marker points to the interval in which a specific gene may be identified. Typically, the initial result of a linkage study is a localization of a susceptibility locus to a particular region of a chromosome. Subsequent steps involve narrowing the chromosomal region from tens of millions of base pairs to tens of thousands, and eventually identifying the gene conferring the susceptibility.

Historically, large multigenerational families have been selected for genetic linkage studies. However, this approach is not without limitations (see Pauls, 1993 for a discussion). As an alternative to large families, many investigators have begun to use affected sib-pair families to complete initial genomic screens. Probably the most important advantage in the use of sib-pairs is that it is a model free procedure (i.e., it does not require any assumptions about the genetic mechanisms involved in the disease). The analytic approach relies on the comparison of the number of alleles at a given locus being shared by two affected sibs and does not require that the genetic model be specified.

### 1. Findings from Linkage Studies

Early segregation analyses suggested that GTS liability was transmitted as a single autosomal dominant gene. As a result, initial linkage studies focused on large multigenerational families. All linkage studies were done collaboratively by members of the Tourette Syndrome Association International Consortium for Genetics (TSAICG). Altogether, 31 multigenerational families were studied (Pauls et al., 1990; Pakstis et al., 1991; Heutink et al., 1995) and more than 800 genetic marker loci were screened. No strong positive evidence for linkage with GTS was observed. In fact, if the following assumptions were made: (1) that GTS was a homogeneous disorder; (2) that the genetic model parameters obtained from early segregation analyses were correct; and (3) that all CTs were related to GTS, then approximately 95% of the genome was excluded by these studies.

Given this lack of success, the TSAICG undertook an affected sib-pair study (TSAICG, 1999) of 76 families with 110 sib-pairs. Only families with at least one unaffected parent were included. Of the 76 families, 64 had only 2 affected children, 10 had 3 affected siblings, and the remaining two had 4 and 5 affected siblings, respectively.

A genome screen was completed using polymorphic markers at a 10-cM average density using a particular set of markers knows as the Marshfield Screening Set 8 (Yuan et al., 1997). Single-point analyses with dominance variance yielded maximum likelihood score (MLS) values of 2.38 and 2.09, for markers D4S1625 and D8S1106, respectively. Twelve additional markers also yielded MLS values between 1.0 and 2.0. Multipoint analyses revealed two regions with MLS values of 2.0 or greater; one on chromosome 4 with a peak MLS value of 2.3 at approximately 3 cM telomeric of D4S1625 and two adjacent regions on chromosome 8. MLS values reached 2.0 in two intervals that were bounded by markers D8S1106, D8S1145, and D8S136.

Another genome scan was completed by Simonic et al. (1998, 2001) on a sample of individuals with GTS from the South African Afrikaner population. Results of association analyses gave evidence for linkage or association for markers D2S139, GATA28F12, and D11S1377 on chromosomes 2p11, 8q22, and 11q23–24, respectively. Two-locus association analyses provided further evidence for linkage or association for markers on chromosome 2 and 8.

A third genome scan was completed by Barr and colleagues (1999) on data from seven multigenerational GTS families. In this study, two regions (19p13.3 and 5p13–q11.2) gave promising results. Of note, the markers on chromosome 19 which gave the most significant results in these large families were in a region that was also found to show some evidence for linkage in the TSAICG sib-pair analyses.

Finally, a linkage study using markers suggested by findings of Simonic et al., (1998) was completed in one large French Canadian kindred (Mérette et al., 2000). Parametric multipoint linkage analyses yielded a lod score of 3.24 for marker D11S1377 chromosome 11q23. Of note is that this marker showed significant linkage disequilibrium with GTS in a South African Afrikaner population (Simonic et al., 1998, 2001).

## V. CYTOGENETIC AND MOLECULAR CYTOGENETIC APPROACHES

An alternative to linkage and association studies involves the characterization of chromosomal abnormalities in rare individuals who present with both GTS and identifiable cytogenetic abnormalities. This research strategy presents something of a contrast to the two approaches described above in the sense that it does not rely at the outset on grouping a large number of individuals who share a phenotype. Instead, emphasis is placed initially on finding infrequent but potentially promising genetic abnormalities. While patients possessing these rearrangements are unlikely to account for a large percentage of the overall group of children with the diagnosis of GTS, the identification of genes that have been structurally or functionally disrupted by chromosomal abnormalities holds promise for providing insight into the basic biological mechanisms involved in more common pathways to these disorders.

The power of cytogenetic approaches for disease-gene identification in single-gene developmental disorders has been demonstrated for decades (Collins, 1992). Very recently, the identification of a gene involved in developmental dyspraxia has demonstrated the value of combining family genetic and cytogenetic methodologies in the study of *complex* phenotypes as well (Lai et al., 2001).

### A. Cytogenetic Findings

A small number of cytogenetic abnormalities have been reported in patients with GTS and related phenotypes over several decades A handful of examples of patients with aneuploidy (i.e., abnormal numbers of chromosomes) who also demonstrate vocal and motor tics has been documented (Barabas et al., 1986, Singh et al., 1982, Hebebrand et al., 1994). Less than a dozen other independent reports of chromosomal inversions, translocations, or deletions associated with GTS phenotypes are present in the literature. Within this small group, only genomic intervals on chromosomes 18, 7, and 8 are represented in more than one unrelated case.

In 1986, Comings et al. reported a translocation between chromosomes 7 and 18 in a family with GTS, CT, and OCD (see also Bhogosian-Sell et al., 1996). Shortly thereafter Donnai (1987) very briefly described a patient with an 18q21 deletion who had putative GTS phenotype. Following these reports, Heutink et al. (1990) examined chromosome 7 and 18 for linkage to GTS in 15 multigenerational Dutch families. No evidence was observed. Bhogosian-Sell et al. (1996)

mapped both the chromosome 7 and 18 to 1–2 million base pair intervals. Finally two cases of patients with OCD with or without CT have been identified with cytogenetic abnormalities in the 18q22 region. Molecular mapping has not yet identified a structurally disrupted gene in this interval (State, 2001).

Two cases have also been identified in which chromosome 7q has been identified as a possible GTS locus. A recent investigation by Petek *et al.* (2001) examined the region 7q22–31 that corresponded to the chromosome 7 breakpoint identified by Bhogosian-Sell *et al.* (1996). A 13-year-old male demonstrating TS, mild mental retardation, delayed language development, depression, strabismus, and minor physical anomalies presented with the following karyotype: 46,XY dup(7)(pter-q31.1::q31.1–q22.1::q31.1-qter). *IMMP2L*, a novel gene coding for the apparent human homolog of the yeast mitochondrial inner membrane peptidase subunit 2, was found to be disrupted by both the breakpoint in the duplicated fragment and the insertion site in 7q31. The authors did not present data on the analysis of *IMMP2L* in cytogenetically normal patients.

Two translocations involving chromosome 8 have also been reported in patients with GTS. One involved a t(3:8) (p21.3 q24.1) (Brett *et al.*, 1997). A second rearrangement involving chromosome 1 and 8q22.1 has been fine mapped. The investigators cloned and sequenced both translocation breakpoints from this family. The *CBFA2T1* gene was identified 11 kb distal to the 8q22.1 breakpoint. Sequence analysis of 37 unrelated GTS patients did not identify any relevant mutations (Matsumoto *et al.*, 2000).

## VI. NEUROIMAGING

Despite potential confounding factors such as the effects of co-morbid illnesses and adaptive responses several structural and functional imaging studies of patients with GTS have been conducted and revealed differences with studies obtained in normal controls. Overall these studies implicate the basal ganglia and their cortical and subcortical connections in the pathophysiology of GTS (reviewed in Peterson, 2001).

Further evidence for abnormalities in the cortico-striatal-thalamo-cortical pathways in GTS comes from volumetric studies. Fredericksen *et al.* (2002) identified larger right frontal white matter volumes in boys with GTS compared with boys with ADHD and GTS or ADHD alone. Studies of girls, however, failed to reveal clear differences in basal ganglia volumes, and only showed that girls had smaller lateral ventricular volumes than controls (Zimmerman *et al.*, 2000).

The involvement of the basal ganglia in GTS has also been confirmed by functional imaging studies, although the overall numbers for SPECT, PET, and functional MRI (fMRI) imaging studies are still relatively small. Reduced blood flow and metabolism are most frequently found in the ventral striatum. When asymmetry was noted, hypometabolism and reduced blood flow are usually seen in the left striatum. As Peterson (2001) points out, limited spatial resolution in some studies may have incorrectly assigned changes to the inferior insular cortex rather than the lateral putamen. Recent studies have also made use of combining "on-line" tasks such as prepulse inhibition (PPI) of the startle reflex and the power of functional MR imaging. Swerdlow *et al.* (2001) modified PPI for use in fMRI. Compared to control subjects children with GTS had decreased PPI, but a normal startle magnitude.

## VII. TREATMENT

The treatment of tics and co-morbid disorders is largely determined by their effect on quality of life as perceived by the patient rather than their absolute frequency or severity. Most patients experience only mild to moderate symptoms and treatment with medications should be reserved for those with disabling symptoms. Restructuring of the environment such as supportive counseling or small-group teaching can be a highly successful intervention.

Symptomatic therapy should be targeted to specific additional psychopathologies. Few double-blind controlled drug treatment studies have been published that specifically address treatment of tics or behavioral disorders in GTS and most of these did not evaluate long-term treatment (reviewed in Robertson and Stern, 2000).

The oldest and most prescribed drug, haloperidol, is not necessarily the first line of therapy for tics any more. Pimozide may be less sedating than haloperidol, but has potential cardiac toxicities. Risperidone has also shown promise. The centrally acting antiadrenergic agent, clonidine, has been used for the suppression of tics and may be particularly useful, when ADHD and GTS co-exist. A recent controlled trial found that combined treatment with clonidine and methylphenidate was superior to either drug alone for the treatment of attention deficits in GTS patients. Methylphenidate did not worsen tics in this study (The Tourette study group, 2002). Serotonin re-uptake inhibitors are primarily used for the treatment of obsessions and compulsions.

## VIII. SUMMARY AND FUTURE PROSPECTS

Significant progress has been made in the understanding of the genetics of GTS over the last decade. Several studies have reported results suggestive of genomic regions for GTS susceptibility loci. The replication of these initial linkage findings and the localization and characterization of genes responsible for the expression of GTS and related disorders is the next critical step in our understanding of the genetic/biological risk factors underlying the manifestation of this syndrome.

In addition, once susceptibility genes have been identified, future work will be possible that should help in the elucidation of nongenetic factors important for the manifestation or the amelioration of the symptoms of GTS and related conditions (Pauls, 1990). Examination of the gene products and their impact on the development of the GTS and related behaviors will allow much more incisive studies to illuminate the physiological/biochemical etiology of GTS. Furthermore, studies in which specific alleles at these genes can be investigated should allow investigators to document more carefully the interaction between genetic and nongenetic factors and its effect on the expression of GTS and related conditions.

While the past decade has been characterized by steady progress, there is building expectation that the pace of discovery will quicken dramatically in the coming years. The sequencing of the human genome is providing additional tools to genetic researchers that are having a profound effect on the speed and precision of genetic studies. The data infrastructure that is being put in place involves more than an elaboration of the raw sequence of the 3 billion bits of information that comprise the human genome. Very importantly, this data is rapidly being annotated with information regarding the precise locations of polymorphisms in the genome, the identification of countless new sites of genetic variation among individuals, and the location of specific genes within their overall genomic context. As genetics investigators in every medical and basic science discipline begin to take advantage of this information, the reservoir of knowledge regarding what genes are in the genome, how they vary between individuals, and how they function promises to grow exponentially.

In the context of the genetics of GTS, the combination of the approaches described above, including linkage, association, and cytogenetic approaches are all being made substantially more powerful by this knowledge explosion. The combination of these approaches over the next decade holds extraordinary promise for finally beginning to identify the specific genetic contributors to GTS and related conditions.

## Acknowledgments

This work was funded by grants from the National Institutes of Health (NS-16648, NS-40024 to DLP) and a Bovenizer Research Fellowship and Smart Family Foundation Grant (to M.W.S.).

## References

Alsobrook, J. P., II, and Pauls, D. L. (2002). A factor analysis of tic symptoms in Gilles de la Tourette's syndrome. *Am. J. Psychiatry* **159**, 291–296.

Alsobrook, J. P., II, and Pauls, D. L. (1997). The genetics of the Tourette syndrome. *Neurol. Clin.* **15**, 381–393.

Apter, A., Pauls, D. L., Bleich, A., Zohar, A. H., Kron, S., Ratzoni, G., Dycian, A., Kotler, M., Weizman, A., Gadot, N., et al. (1993). An epidemiologic study of Gilles de la Tourette's syndrome in Israel. *Arch. Gen. Psychiatry.* **50**, 734–738.

Barabas, G., Wardell, B., Sapiro, M., and Matthews, W. S. (1986). Coincident Down's and Tourette syndromes: three case reports. *J. Child. Neurol.* **1**, 358–360.

Barker, E., and Blakely, R. (1996). Identification of a single amino acid, phenylalanine 586, that is responsible for high affinity interactions of tricyclic antidepressants with the human serotonin transporter. *Mol. Pharmcol.* **50**, 957–965.

Baron, M., Shapiro, E., Shapiro, A. K. et al. (1981). Genetic analysis of Tourette syndrome suggesting major gene effect. *Am. J. Hum. Genet.* **33**, 767–775.

Barr, C. L., Wigg, K. G., Pakstis, A. J., Kurlan, R., Pauls, D. L., Kidd, K. K., Tsui, L.-C., and Sandor, P. (1999). Genome scan for linkage to Gilles de la Tourette Syndrome. *Am. J. Med. Genet. (Neuropsychiatr. Genet.)* **88**, 437–445.

Barr, C. L., Wigg, K. G., Zovko, E., Sandor, P., and Tsui, L.-C. (1997). Linkage study of the dopamine D5 receptor gene and Gilles de la Tourette syndrome. *Am. J. Med. Genet. (Neuropsychiatry Genet.)* **74**, 58–61.

Barr, C. L., Wigg, K. G., Zovko, E., Sandor, P., and Tsui, L.-C. (1996). No evidence for a major gene effect of the dopamine D4 receptor gene in the susceptibility to Gilles de la Tourette syndrome in five Canadian families. *Am. J. Med. Genet. (Neuropsychiatr. Genet.)* **67**, 301–305.

Bhogosian-Sell, L., Comings, D. E., and Overhauser, J. (1996). Tourette syndrome in a pedigree with a 7;18 translocation: identification of a YAC spanning the translocation breakpoint at 18q22.3. *Am. J. Hum. Genet.* **59**, 999–1005.

Brett, P., Curtis, D., Robertson, M., and Gurling, H. (1995). Exclusions of the 5-HT1A seratonin neuroreceptor and tryptophan oxygenase genes in a large British kindred multiply affected with Tourette's syndrome, chronic motor tics, and obsessive-compulsive behavior. *Am. J. Psychiatry.* **152**, 437–440.

Brett, P. M., Curtis, D., Robertson, M. M., Dahlitz, M., and Gurling, H. M. (1996). Linkage analysis and exclusion of regions of chromosomes 3 and 8 in Gilles de la Tourette syndrome following the identification of a balanced reciprocal translocation 46 XY, t(3:8)(p21.3 q24.1) in a case of Tourette syndrome. *Psychiatry. Genet.* **6**, 99–105.

Brett, P. M., Curtis, D., Robertson, M. M., and Gurling, H. M. (1997). Neuroreceptor subunit genes and the genetic susceptibility to Gilles de la Tourette syndrome. *Biol. Psychiatry* **42**, 941–947.

Buetow, K., Weber, J., Ludwigsen, S. et al. (1994). Integrated human genome-wide maps constructed using CEPH reference panel. *Nat. Genet.* **6**, 391–393.

Bull, L. N., van Eijk, M. J., Pawlikowska, L., DeYoung, J. A. et al. (1998). A gene encoding a P-type ATPase mutated in two forms of hereditary cholestasis. *Nat. Genet.* **18**, 219–224.

Carter, A.S., O'Donnell, D.A., Schultz, R.T., Scahill, L., Leckman, J.F., and Pauls, D.L. (2000). Social and emotional adjustment in children affected with Gilles de la Tourette's syndrome: Associations with ADHD and family functioning. *J. Child Psychol. Psychiatry Allied Discipl.* **41**, 215–223

Carter, A. S., Pauls, D. L., Leckman, J. F., and Cohen, D. J. (1994). A prospective longitudinal study of Gilles de la Tourette's syndrome. *J. Am. Acad. Child Adol. Psychiatry* **33**, 377–385.

Cavallini, M. C., Di Bella, D., Catalano, M., and Bellodi, L. (2000). An association study between 5-HTTLPR polymorphism, COMT polymorphism, and Tourette's syndrome. *Psychiatry Res.* **97**, 93–100.

Collins, F. S. (1992). Positional cloning: lets not call it reverse any more. *Nat. Genet.* **1**, 3–6.

Comings, D. E. (1995). Tourette's syndrome: a behavioral spectrum disorder. *Adv. Neurol.* **65**, 293–303.

Comings, D. E., Comings, B. G., Muhleman, D., Dietz, G., Shahbahrami, B., Tast, D., Knell, E., Kocsis, P., Baumgarten, R., Kovacs, B. W. *et al.* (1991). The dopamine D2 receptor locus as a modifying gene in neuropsychiatric disorders. *JAMA.* **266**, 1793–1800.

Comings, D. E., and Comings, B. G. (1987). A controlled study of Tourette syndrome. *Am. J. Hum. Genet.* **41**, 701–838.

Comings, D. E., and Comings, B. G. (1984). Tourette's syndrome and attention deficit disorder with hyperactivity: are they genetically related? *J. Am. Acad. Child Psychiatry* **23**, 138–146.

Comings, D. E., and Comings, B. G. (1990). A controlled family history study of Tourette's syndrome, III: Affective and other disorders. *J. Clin. Psychiatry* **51**, 288–291.

Comings, D. E., Comings, B. G., Devor, E. J. *et al.* (1984). Detection of a major gene for Gilles de la Tourette syndrome. *Am. J. Hum. Genet.* **36**, 586–600.

Comings, D., Comings, B., Dietz, G. *et al.* (1986). Evidence the Tourette syndrome gene is at 18q22.1. *Proc. VIIth Int. Congr. Hum. Genet.* Berlin, p. 620.

Comings, D. E., Comings, B. G., Muhleman, D., Dietz, G., Shahbahrami, B., Tast, D., Knell, E., Kocsis, P., Baumgarten, R., Kovacs, B. W. *et al.* (1991). The dopamine D2 receptor locus as a modifying gene in neuropsychiatric disorders. *JAMA.* **266**, 1793–1800.

de la Chapelle, A., and Wright, F. A. (1998). Linkage disequilibrium mapping in isolated populations: the example of Finland revisited. *PNAS (U.S.A.)* **95**, 12,416–12,423.

Devor, E. J., Grandy, D. K., Civelli, O., Litt, M., Burgess, A. K., Isenberg, K. E., van de Wetering, B. J., and Oostra, B. (1990). Genetic linkage is excluded for the D2-dopamine receptor lambda HD2G1 and flanking loci on chromosome 11q22–q23 in Tourette syndrome. *Hum. Hered.* **40**, 105–108.

Devor, E. J. (1984). Complex segregation analysis of Gilles de la Tourette syndrome: further evidence for a major locus mode of transmission. *Am. J. Hum. Genet.* **36**, 704–709.

Dib, C., Faure, S., Fizames, C., Samson, D., Drouot, N., Vignal, A., Millasseau, P., Marc, S., Hazan, J., Seboun, E., Lathrop, M., Gyapay, G., Morissette, J., and Weissenbach, J. (1996). A comprehensive genetic map of the human genome based on 5,264 microsatellites. *Nature* **380**, 152–154.

Donnai, D. (1987). Gene location in Tourette syndrome. *Lancet* **1**, 627.

Eapen, V., Pauls, D. L., and Robertson, M. M. (1993). Evidence for autosomal dominant transmission in Tourette's Syndrome—United Kingdom Cohort Study. *Br. J. Psychiatry* **162**, 593–596.

Eapen, V., Robertson, M. M., Alsobrook, J. P. II, and Pauls, D. L. (1997). Obsessive compulsive symptoms in Gilles de la Tourette's syndrome and obsessive compulsive disorder: differences by diagnosis and family history. *Am. J. Med. Genet. (Neuropsychiatr. Genet.)* **74**, 432–438.

Falk, C., and Rubinstein, P. (1987). Haplotype relative risks: an easy reliable way to construct a proper control sample for risk calculations. *Ann. Hum. Genet.* **51**, 227–233.

Fredericksen, K. A., Cutting, L. E., Kates, W. R., Mostofsky, S. H., Singer, H. S., Cooper, K. L., Lanham, D. D., Denckla, M. B., and Kaufman, W. E. (2002). Disproportionate increases of white matter in right frontal lobe in Tourette syndrome. *Neurology* **8**, 85–89.

Freimer, N. B., Reus, V. I., Escamilla, M. A., McInnes, L. A. *et al.* (1996). Genetic mapping using haplotype, association and linkage methods suggests a locus for severe bipolar disorder (BPI) at 18q22–q23. *Nat. Genet.* **12**, 436–441.

Gelernter, J., Kennedy, J., Grandy, D. *et al.* (1993). Exclusion of close linkage of Gilles de la Tourette syndrome to D1 dopamine receptor. *Am. J. Psychiatry* **150**, 449–453.

Gelernter, J., Pakstis, A. J., Pauls, D. L. *et al.* (1990). Gilles de la Tourette syndrome is not linked to D2 dopamine receptor. *Arch. Gen. Psychiatry* **47**, 1073–1077.

Gelernter, J., Pauls, D. L., Leckman, J. *et al.* (1994). D2 dopamine receptor (DRD2) alleles do not influence severity of Tourette's syndrome: results from four large kindreds. *Arch. Neurol.* **51**, 397–400.

Gelernter, J., Rao, P. A., Pauls, D. L., Hamblin, M. W., Sibley, D. R., and Kidd, K. K. (1995). Assignment of the 5HT7 receptor gene (HTR7) to chromosome 10q and exclusion of genetic linkage with Tourette syndrome. *Genomics* **26**, 207–209.

Grice, D. E., Leckman, J. F., Pauls, D. L., Kurlan, R., Kidd, K. K., Pakstis, A. J., Chang, F. M., Cohen, D. J., and Gelernter, J. (1996). Linkage disequilibrium between an allele at the dopamine D4 receptor locus with Tourette's syndrome by the transmission disequilibrium test. *Am. J. Hum. Genet.* **59**, 644–652.

Grigorenko, E. L., and Pauls, D. L. Analytical Methods Applied to Psychiatric Genetics. In Leboyer, M. and Bellivier, F. (eds.), *Psychiatry Genetics: Methods and Protocols*, Totowa, NJ. The Humana Press, in press.

Groenewald, J. Z., Liebenberg, J., Groenewald, I. M., and Warnich, L. (1998). Linkage disequilibrium analysis in a recently founded population: evaluation of the variegate porphyria founder in South African Afrikaners [letter]. *Am. J. Hum. Genet.* **62**, 1254–1258.

Gyapay, G., Morissette, J., Vigna, I. *et al.* (1994). Genethon human genetic linkage map. *Nat. Genet.* **7**, 246–339.

Hasstedt, S. J., Leppert, M., Filloux, F., van de Wetering, B. J. M., and McMahon W. (1995). Intermediate inheritance of Tourette syndrome, assuming assortative mating. *Am. J. Hum. Genet.* **57**, 682–689.

Hebebrand, J., Klug, B., Fimmers, R., Seuchter, S. A., Wettke-Schafer, R., Deget, F., Camps, A., Lisch, S., Hebebrand, K., von Gontard, A., Lehmkuhl, G., Poustka, F., Schmidt, M., Baur, M..P., and Remschmidt, H. (1997). Rates for tic disorders and obsessive compulsive symptomatology in families of children and adolescents with Gilles de la Tourette syndrome. *J. Psycholog. Res.* **31**, 519–530.

Hebebrand, J., Martin, M., Korner, J., Roitzheim, B., de Braganca, K., Werner, W., and Remschmidt, H. (1994). Partial trisomy 16p in an adolescent with autistic disorder and Tourette's syndrome. *Am. J. Med. Genet.* **54**, 268–270.

Hebebrand, J., Nöthen, M. M., Lehmkuhl, G., Poustka, F., Schmidt, M., Propping, P., and Remschmidt, H. (1993). Tourette's syndrome and homozygosity for the dopamine D3 receptor gene. German Tourette's Syndrome Collaborative Research Group. *Lancet* **341**, 1483–1484.

Heutink, P., van de Wetering, B. J. M., Breedveld, G. *et al.* (1990). No evidence for genetic linkage of Gilles de la Tourette syndrome on chromosome 7 and 18. *J. Med. Genet.* **27**, 433–436.

Heutink, P., van de Wetering, B. J. M., Pakstis, A. J., Kurlan, R., Sandor, P., Oostra, B. A., and Sandkuijl, L. A. (1995). Linkage studies on Gilles de la Tourette Syndrome: What is the strategy of choice? *Am. Hum. Genet.* **57**, 465–473.

Houwen, R. H. J., Baharloo, S., Blankenship, K., Raeymaekers, P., Juyn, J., Sandkuijl, L. A., and Friemer, N. B. (1994). Genome screening by searching for shared segments: mapping a gene for benign recurrent intrahepatic cholestasis. *Nat. Genet.* **8**, 380–386.

Hyde, T. M., Aaronson, B. A., Randolph, C. *et al.* (1992). Relationship of birth weight to the phenotypic of Gilles de la Tourette's syndrome in monzygotic twins. *Neurology* **42**, 652–658.

Kano, Y., Ohta, M., Nagai, Y., Pauls, D. L., and Leckman, J. F. (2001). A family study of Tourette syndrome in Japan. *Am. J. Med. Genet.* **105**, 414–421.

Kidd, K. K., and Pauls, D. L. (1982). Genetic hypotheses for Tourette syndrome, *Adv. Neurol.* **35**, 243–249.

Kidd, K. K., Prusoff, B. A., and Cohen, D. J. (1980). The familial pattern of Tourette syndrome. *Arch. Gen. Psychiatry* **37**, 1336–1339.

Kidd, K. K. (1993). Associations of disease with genetic markers: deja vu all over again [editorial]. *Am. J. Med. Genet. (Neuropsychiatr. Genet.)* **48**, 71–73.

Kidd, K. K. (1984). New genetic strategies for studying psychiatric disorders. In *Genetic Aspects of Human Behavior*. (T. Sakai and T. Tsuboi, eds.), pp 325–346. Igaku-Shoin LTD, Tokyo.

Kruglyak, L., and Lander, E. S. (1995). Complete multipoint sib-pair analysis of qualitative and quantitative traits. *Am. J. Hum. Genet.* **57**, 439–454.

Kurlan, R., Behr, J., Medved, L., Shoulson, I., Pauls, D. L., Kidd, J. R., and Kidd, K. K. (1986). Familial Tourette syndrome: Report of a large pedigree and potential for linkage analysis. *Neurology* **36**, 772–776.

Kurlan, R., Eapen, V., Stern, J., McDermott, M. P., and Robertson, M. M. (1994). Bilineal transmission in Tourette's syndrome families. *Neurology* **44**, 2336–2342.

Lai, C. S., Fisher, S. E., Hurst, J. A., Vargha-Khadem, F., and Monaco, A. P. (2001). A forkhead-domain gene is mutated in a severe speech and language disorder. *Nature* **413**, 519–523.

Lander, E. S., and Botstein, D. (1986). Mapping complex genetic traits in humans: new methods using a complete RFLP linkage map. *Cold Spring Harbor Symp. Quant. Biol.* **51**, 49–62.

LeBoyer, M., Bellivier, F., Nosten-Bertrand, M., Jouvent, R., Pauls, D. L., and Mallet, J. (1998). Psychiatric genetics: search for phenotypes. *Trends Neurosci.* **21**, 102–105.

Leckman, J. F., and Cohen, D. J. (1999). In "Tourettes Syndrome: Tics, Obsessions and Compulsions" (J. F. Leckman and D. J. Cohen, eds.), pp. 155–175. John Wiley and Sons, New York.

Leckman, J. F., Dolnansky, E. S., Hardin, M. *et al.* (1990). The perinatal factors in the expression of Tourette's syndrome. *J. Am. Acad. Child. Adol. Psychiatry* **29**, 220–226.

Leckman, J. F., Price, R. A., Walkup, J. T. *et al.* (1987). Birthweights of monozygotic twins discordant for Tourette's syndrome. *Arch. Gen. Psychiatry* **44**, 100.

Leckman, J. F., Sholomskas, D., Thompson, W. D. *et al.* (1982). Best estimate of lifetime psychiatric diagnoses: A methodological study. *Arch. Gen. Psychiatry* **39**, 879–883.

Leckman, J. F., Walker, D. E., and Cohen, D. J. (1993). Premonitory urges in Tourette's syndrome. *Am. J. Psychiatry* **150**, 98–102.

Leckman, J. F., Walker, D. E., Goodman, W. K., Pauls, D. L., and Cohen, D. J. (1994). "Just right" perceptions associated with compulsive behavior in Tourette's syndrome. *Am. J. Psychiatry* **151**, 675–80

Lucas, A. R., Beard, C. M., Raiput, A. H., and Kurland, L. T. (1982). Tourette syndrome in Rochester Minnesota, 1968–1979. *Adv. Neurol.* **35**, 267–269.

Matsumoto, N., David, D. E., Johnson, E. W., Konecki, D., Burmester, J. K., Ledbetter, D. H., and Weber, J. L. (2000). Breakpoint sequences of an 1:8 translocation in a family with Gilles de la Tourette syndrome. *Eur. J. Hum. Genet.* **8**, 875–883.

McDougle, C. J., Goodman, W. K., Leckman, J. F., Lee, N. C., Heninger, G. R., and Price, L. H. (1994). Haloperidol addition in fluvoxamine-refractory obsessive-compulsive disorder. A double-blind, placebo-controlled study in patients with and without tics. *Arch. Gen. Psychiatry* **51**, 302–308.

McMahon, W., Leppert, M., Filloux, F., van de Wetering, B. J. M., and Hasstedt, S. (1992). Tourette syndrome in 171 related family members. *Adv. Neurol.* **58**, 159–165.

McMahon, W., van de Wettering, B. J. M., Filloux, F., Bett, K., Coon, H., and Leppert, M. (1996). Bilineal transmission and phenotypic variation of Tourette's disorder in a large pedigree. *J. Am. Acad. Child Adol. Psychiatry* **35**, 672–680.

Mérette, C., Brassard, A., Potvin, A., Bouvier, H., Rousseau, F., Emond, C., Bissonnette, L., Roy, M. A., Maziade, M., Ott, J., and Caron, C. (2000). Significant linkage for Tourette syndrome in a large French Canadian family. *Am. J. Hum. Genet.* **67**, 1008–1013.

Nöthen, M. M., Hebebrand, J., Knapp, M., Hebebrand, K., Camps, A., von Gontard, A., Wettke-Schafer, R., Lisch, S., Cichon, S., Poustka, F. *et al.* (1994). Association analysis of the dopamine D2 receptor gene in Tourette's syndrome using the haplotype relative risk method. *Am. J. Med. Genet. (Neuropsychiatr. Genet.)* **54**, 249–252.

Pakstis, A. J., Heutink, P., Pauls, D. L., Kurlan, R., van de Wetering, B. J. M., Leckman, J. F., Sandkuijl, L. A., Kidd, J. R., Breedveld, G. J., Castiglione, C. M., Weber, J., Sparkes, R. S., Cohen, D. J., Kidd, K. K., and Oostra, B. A. (1991). Progress in the search for genetic linkage with Tourette syndrome: an exclusion map covering more than 50% of the autosomal genome. *Am. J. Hum. Genet.* **48**, 281–294.

Pauls, D. L., Alsobrook, J. P., II, Almasy, L., Leckman, J. F., and Cohen, D. J. (1991a). Genetic and epidemiological analyses of the Yale Tourette's Syndrome Family Study data. *Psychiatr. Genet.* **2**, 28.

Pauls, D. L., Alsobrook, J. P., II, Goodman, W., Rasmussen, S., and Leckman, J. F. (1995). A family study of obsessive compulsive disorder. *Am. J. Psychiatry* **152**, 76–84.

Pauls, D. L., Cohen, D. J., Heimbuch, R. *et al.* (1981). Familial pattern and transmission of Gilles de la Tourette syndrome and multiple tics. *Arch. Gen. Psychiatry* **38**, 1091–1093.

Pauls, D. L., Hurst, C. R., Kruger, S. D., Leckman, J. F., Kidd, K. K., and Cohen, D. J. (1986a). Gilles de la Tourette's syndrome and attention deficit disorder with hyperactivity. Evidence against a genetic relationship. *Arch. Gen. Psychiatry* **43**, 1177–1179.

Pauls, D. L., and Leckman, J. F. (1986). The inheritance of Gilles de la Tourette's syndrome and associated behaviors: Evidence for autosomal dominant transmission. *N. Eng. J. Med.* **315**, 993–997.

Pauls, D. L., Leckman, J. F., and Cohen, D. J. (1993). The familial relationship between Gilles de la Tourette's Syndrome, attention deficit disorder, learning disabilities, speech disorders and stuttering. *J. Am. Acad. Child. Adol. Psychiatry* **32**, 1044–1050.

Pauls, D. L., Leckman, J. F., and Cohen, D. J. (1994). Evidence against a genetic relationship between Gilles de la Tourette's syndrome and anxiety, depression, panic and phobic disorders. *Br. J. Psychiatry* **164**, 215–221.

Pauls, D. L., Leckman, J. F., Towbin, K. E., Zahner, G. E., and Cohen, D. J. (1986b). A possible genetic relationship exists between Tourette's syndrome and obsessive-compulsive disorder. *Psychopharmacol. Bull.* **22**, 730–733.

Pauls, D. L., Pakstis, A. J., Kurlan, R., Kidd, K. K., Leckman, J. F., Cohen, D. J., Kidd, J. R. *et al.* (1990). Segregation and linkage analyses of Tourette's Syndrome and related disorders. *J. Am. Acad. Child. Adol. Psychiatry* **29**, 195–203.

Pauls, D. L., Raymond, C. L., Leckman, J. F., and Stevenson, J. M. (1991b). A family study of Tourette's syndrome. *Am. J. Hum. Genet.* **48**, 154–163.

Pauls, D. L. (1993). Behavioural disorders: lessons in linkage. *Nat. Genet.* **3**, 4–5.

Pauls, D. L. (1990). Emerging genetic markers and their role in potential preventive intervention strategies. In *Conceptual Research Models for Preventing Mental Disorders*, (P. Muehrer, ed.). NIMH, Rockville, MD.

Petek, E., Windpassinger, C., Vincent, J. B., Cheung, J., Boright, A. P., Scherer, S. W., Kroisel, P. M., and Wagner, K. (2001). Disruption of a novel gene (IMMP2L) by a breakpoint in 7q31 associated with Tourette syndrome. *Am. J. Hum. Genet.* **68**, 848–858.

Peterson, B. S. (2001) Neuroimaging studies of Tourette syndrome: a decade of progress. *Adv. Neurol.* **85**, 179–196.

Price, R. A., Kidd, K. K., Cohen, D. J., Pauls, D. L., and Leckman, J. F. (1985). A twin study of Tourette syndrome. *Arch. Gen. Psychiatry* **42**, 815–820.

Price, R. A., Pauls, D. L., Kruger, S. D. *et al.* (1988). Family data support a dominant major gene for Tourette syndrome. *Psychiatr. Res.* **24**, 251–261.

Risch, N. (1990). Linkage strategies for genetically complex traits. I multilocus models. *Am. J. Hum. Genet.* **46**, 222–228.

Robertson, M. M., and Gourdie, A. (1990). Familial Tourette's syndrome in a large British pedigree: associated psychopathology, severity and potential for linkage analysis. *Br. J. Psychiatry* **156**, 515–521.

Robertson, M. M., and Stern, J. S. (1998). Tic disorders: new developments in Tourette syndrome and related disorders. *Curr. Opin. Neurol.* **11**, 373–380.

Robertson, M. M., and Stern, J. S. (2000). Gilles de la Tourette syndrome: symptomatic treatment based on evidence. *Eur. Child. Adol. Psychiatry* **9**, Suppl, 1, I60–75.

Santangelo, S. L., Pauls, D. L., Goldstein, J. M., Faraone, S. V., Tsuang, M. T., and Leckman, J. F. (1994). Tourette syndrome: What are the influences of gender and co-morbid OCD? *J. Am. Acad. Child Adol. Psychiatr.* **33**, 795–804.

Santangelo, S. L., Pauls, D. L., Lavori, P. L., Goldstein, J. M., Faraone, S. V., and Tsuang, M. T. (1996). Assessing risk for the Tourette spectrum of disorders among first-degree relatives of probands with Tourette syndrome. *Am. J. Med. Genet. (Neuropsychiatr. Genet.)* **67**, 107–116.

Schultz, R. T., Carter, A. S., Gladstone, M., Scahill, L., Leckman, J. F., Peterson, B. S., Zhang, H., Cohen, D. J., and Pauls, D. L. (1998). Visual-motor integration, visuosperceptual and fine motor functioning in children with Tourette syndrome. *Neuropsychology* **12**, 134–145.

Seuchter, S. A., Hebebrand, J., Klug, B., Knapp, M., Lehmkuhl, G., Poustka, F., Schmidt, M., Remschmidt, H., and Baur, M. P. (2000). Complex segregation analysis of families ascertained through Gilles de la Tourette syndrome. *Genet. Epidemiol.* **18**, 33–47.

Simonic, I., Gericke, G. S., Ott, J., and Weber, J. L. (1998). Identification of genetic markers associated with Gilles de la Tourette syndrome in an Afrikaner population. *Am. J. Hum. Genet.* **63**, 839–846.

Simonic, I., Nyholt, D. R., Gericke, G. S., Gordon, D., Matsumoto, N., Ledbetter, D. H., Ott, J., and Weber, J. L. (2001). Futher evidence for linkage of Gilles de la Tourette syndrome (GTS) susceptibility loci on chromosomes 2p11, 8q22 and 11q23-24 in South African Afrikaners. *Am. J. Med. Genet.* **105**, 163–167.

Singh, D. N., Howe, G. L., Jordan, H. W., and Hara, S. (1982). Tourette's syndrome in a black woman with associated triple X and 9p mosaicism. *J. Natl. Med. Assoc.* **74**, 675–682.

Spielman, R., and Ewens, W. (1996). The TDT and other family-based tests for linkage disequilibrium and association. *Am. J. Hum. Genet.* **59**, 983–989.

Spielman, R. S., McGinnis, R. E., and Ewens, W. J. (1993). Transmission test for linkage disequilibrium: the insulin gene region and insulin-dependent diabetes mellitus (IDDM). *Am. J. Hum. Genet.* **52**, 506–516.

State, M. W. (2001). Analysis of a chromosome 18 locus associated with Tourette syndrome phenotypes: Breakpoint characterization, transcript assessment and candidate gene analysis. Ph.D. thesis, Yale University Department of Genetics.

Swerdlow, N. R., Karban, B., Ploum, Y., Sharp, R., Geyer, M. A., and Eastvold, A. (2001). Tactile prepuff inhibition of startle in children with Tourette's syndrome: in search of an "fMRI-friendly" startle paradigm. *Biol. Psychiatry* **15**, 578–585.

te Meerman, G. J., van der Meulen, M. A., and Sandkuijl, L. A. (1995). Perspectives of identity by descent (IBD) mapping in founder populations. *Clin. Exp. Allergy* **25**, 97–102.

Terwilliger, J. D., and Ott, J. (1992). A haplotype-based "haplotype relative risk" approach to detecting allelic associations. *Hum. Hered.* **42**, 337–346.

The Tourette Syndrome Association International Consortium for Genetics. (1999). A complete genome screen in sib-pairs affected with the Gilles de la Tourette syndrome. *Am. J. Hum. Genet.* **65**, 1428–1436.

The Tourette study group (2002). Treatment of ADHD in children with tics: a randomized controlled trial. *Neurology* **58**, 527–536.

Vandenbergh, D. J., Thompson, M. D., Cook, E. H., Bendahhou, E., Nguyen, T., Krasowski, M. D., Zarrabian, D., Comings, D., Sellers, E. M., Tyndale, R. F., George, S. R., O'Dowd, B. F., and Uhl, G. R. (2000). Human dopamine transporter gene: coding region conservation among normal, Tourette's disorder, alcohol dependence and attention-deficit hyperactivity disorder populations. *Mol. Psychiatry* **5**, 283–292.

van de Wetering, B. J. M. (1993). A genetic study of Gilles de la Tourette syndrome in The Netherlands. Ph.D. thesis, Erasmus University, Rotterdam, The Netherlands.

van Schothorst, E. M., Jansen, J. C., Grooters, E., Prins, D. E. *et al.* (1998). Founder effect at PGL1 in hereditary head and neck paraganglioma families from the Netherlands. *Am. J. Hum. Genet.* **63**, 468–473.

van Tol, H. H. M., Caren, M. W., Guan, H.-C. *et al.* (1992). Multiple dopamine D4 receptor variants in the human population. *Nature* **358**, 149–152.

Walkup, J. T., LaBuda, M. C., Singer, H. S., Brown, J., Riddle, M. A., and Hurko, O. (1996). Family study and segregation analysis of Tourette syndrome: evidence for a mixed model of inheritance. *Am. J. Hum. Genet.* **59**, 684–693.

Walkup, J. T., Leckman, J. F., Price, R. A., Hardin, M., Ort, S. I., and Cohen, D. J. (1988). The relationship between obsessive-compulsive disorder and Tourette's syndrome: A twin study. *Psychopharmacol. Bull.* **24**, 375–379.

Yuan, B., Vaske, D., Weber, J. L., Beck, J., and Sheffield, V. C. (1997). Improved set of short-tandem-repeat polymorphisms for screening the human genome. *Am. J. Hum. Genet.* **60**, 459–460.

Zimmerman, A. M., Abrams, M. T., Giuliano, J. D., Denckla, M. B., and Singer, H. S. (2000). Subcortical volumes in firls with tourette syndrome: support for gender effect. *Neurology* **27**, 2224–2229.

CHAPTER 43

# The Genetics of Restless Legs Syndrome

ANDY PEIFFER

*Department of Pediatrics*
*Division of Medical Genetics*
*University of Utah*
*Salt Lake City, Utah 84112*

I. Introduction
II. Clinical Features
III. Prevalence and Progression
IV. Genetic Studies
    A. Families and Twin Studies
    B. Linkage Studies
V. Diagnosis
    A. Diagnostic and Ancillary Tests
VI. Neuroimaging and Neurophysiological Studies
VII. Treatment
VIII. Summary
    References

modes of inheritance have been proposed for RLS. Locus heterogeneity almost certainly exists for RLS and may explain the wide range of clinical severity. Varying age-of-onset and expressivity seen in large RLS families may also be caused by additional genetic influences or environmental factors. Large RLS families provide a means for identifying disease-causing genes using linkage studies. We have completed phenotypic assessments and collected DNA from seven large RLS families. Each family generated lod scores >3.0 in simulated linkage analyses assuming high disease penetrance and disease allele frequency. Genome-wide scans with microsatellite markers are being conducted in an effort to identify RLS loci. One recent linkage study in a large RLS family has reported significant LOD scores on chromosome 12 as well as two other putative disease loci.

## I. INTRODUCTION

Restless Legs Syndrome (RLS) is a common neurological disorder characterized by an unpleasant creeping sensation in the legs at rest and associated with an irresistible urge to move the legs. Most RLS patients also have repetitive jerking movements of the limbs during sleep. Neuroimaging studies have failed to identify structural abnormalities in RLS patients. However, studies have revealed differences in dopaminergic levels, brain iron metabolism, and cortical inhibition in patients suggesting pathophysiological bases for the symptomatology of RLS. Recent studies have placed the prevalence of RLS in the general population between 5 and 15%. Patients with primary (idiopathic) RLS often report a positive family history. Both autosomal dominant and autosomal recessive

## II. CLINICAL FEATURES

RLS is a neurological movement disorder characterized by paresthesia, usually of the legs, associated with an irresistible need to move that is worse when lying or trying to sleep (Ekbom, 1945). These symptoms cause sleep disruption and fatigue which are common reasons for which affected individuals consult doctors. Symptoms are often difficult to describe but can include terms such as a "unpleasant creeping, crawling" feelings most often in the legs (but also in the arms) that are relieved by moving or rubbing them (Walters, 1995; Michaud et al., 2000). The motor restlessness has been described by some clinicians as resembling neuroleptic-induced akathisia. These symptoms gradually build up when patients are at rest or sitting for long periods (i.e., driving long distances or flying) or during periods of stress and the associated urge to move soon becomes unbearable. Caffeine is also known to aggravate the symptoms of RLS (Lutz et al., 1978). RLS patients have described the severity of their symptoms as ranging from "mild" to "intolerable" (Ondo and Jankovic, 1996). Approximately 80–90% of affected individuals (or their spouses) also describe involuntary repetitive, periodic, jerking movements that occur most often at the onset of sleep (during non-REM sleep) and throughout the night. Such periodic limb movements of sleep (PLMS) have also been termed nocturnal myoclonus (Lugaresi et al., 1986), although PLMS may occur rhythmically at 20 to 30-second intervals for minutes to hours during sleep. Similar in nature to the stretching or swinging movements that often ease the symptoms during the day, the nocturnal movements are often extensor/flexion movements of the hips, knees, and ankles. RLS patients may awake frequently to stretch their legs or to get out of bed and walk in an effort to relieve symptoms. The sleep disruption and fatigue caused by PLMS in RLS patients can have an important impact on their ability to work, their social activities, and family life (Walters, 1995) and has led to the classification of RLS as a sleep disorder (American Sleep Disorders Association, 1997).

## III. PREVALENCE AND PROGRESSION

Recent studies have placed the prevalence of RLS in the general population between 5 and 15% (Lavigne and Montplaisir, 1994; Phillips et al., 2000). These figures are higher than those of earlier studies (Ekbom, 1950; Strang, 1967). Factors that may influence prevalence estimates include lack of self-reporting by patients, diagnostic proficiency of doctors, and the heterogeneous clinical presentation of RLS. It is also possible that some estimates include both primary (idiopathic) and secondary RLS patients (described in Section V). Men and women seem to be equally affected by primary RLS (Lazzarini et al., 1999; Phillips et al., 2000). Primary RLS has been reported in young children and it is now thought that children who suffer from RLS symptoms may be more prevalent than previously appreciated. Many children who develop RLS in adult life complain of severe growing pains or are diagnosed with attention deficit hyperactivity disorder (ADHD) as youngsters (Ekbom, 1975; Walters et al., 1994). One study has reported that a significant proportion of children with ADHD may have sleep problems and PLMS (Picchietti et al., 1999).

Up to 40% of patients with primary RLS will experience their first symptom before age 20 but the diagnosis is usually made when patients reach adult age because by then symptoms have become disruptive (Walters et al., 1996). Based on a study of 133 cases, Montplaisir et al. (1997) reported that RLS symptoms begin at a mean age of 27 years and before age 20 in 38.3% of patients. The prevalence of RLS seems to increase with age (Lavigne and Montplaisir, 1994) and the fact that older affected members in RLS families have more severe symptoms (Trenkwalder et al., 1996) supports this observation. Recently it has been shown that earlier-onset RLS progresses more slowly with age and that later-onset RLS progresses more rapidly (Allen and Earley, 2001). These authors postulate a relationship between these two types of RLS and serum iron status; the later-onset form is associated with lower ferritin values while the early onset form is not.

## IV. GENETIC STUDIES

### A. Family and Twin Studies

Many studies have shown that primary RLS has a strong hereditary component. Since its description, clinicians have recognized that members of the same family often have RLS (Bornstein, 1961; Ambrosetto et al., 1965; Clemençon, 1975; Jacobsen et al., 1986). Recent studies have reported that between 40 and 60% of patients with primary RLS have a positive family history (Ondo and Jankovic, 1996; Montplaisir et al., 1997; Winkelmann et al., 2000). In one study of primary RLS cases a positive family history of 92% was reported when all first-degree relatives were contacted and interviewed. (Walters et al., 1996). It is possible that in some cases, the high number of familial cases may be due to founder effects (Montplaisir et al., 1997), as has been reported for the Québec population or to local environmental factors. RLS patients with a positive family history are generally younger when their symptoms begin suggesting that the pathophysiological mechanism may be different than in people with primary RLS who develop symptoms later in life (Walters, 1995; Bassetti et al., 2001).

Genetic studies have also shown that primary RLS is inherited as an autosomal dominant trait with variable age of onset and high penetrance (Montplaisir et al., 1985; Walters et al., 1990). Transmission of RLS as an autosomal dominant trait has been confirmed in a more recent segregation analysis of RLS families (Winkelmann et al., 2001). Due to its high prevalence and variable expressivity, it is possible that RLS may also be inherited as an autosomal recessive trait or even in a pseudoautosomal manner upon homozygous-heterozygous matings. One recent study has reported lod scores suggestive of linkage in a family under autosomal recessive conditions (Desautels et al., 2001a); (see Section IV.B). Two family studies have reported earlier age of onset of RLS symptoms in successive generations suggesting the possibility of anticipation in RLS families (Trenkwalder et al., 1996; Lazzarini et al.,

1999). In a study of 12 sets of monozygotic twins, Ondo et al. (2000) report autosomal dominant inheritance, high concordance rates, and high penetrance for the RLS phenotype although age of onset varied.

Variable onset of age and phenotypic heterogeneity can make accurate assignment of affection status difficult thereby complicating genetic studies in RLS families. Affected individuals may have more nocturnal symptoms than daytime restlessness leading some authors (Walters et al., 1990) to conclude that all prospective study patients should undergo sleep studies. Obtaining a history of the sleeping patterns of the purported RLS patient from the spouse may also prove helpful in identifying nocturnal symptoms. Because RLS is so common, it is very likely that many disease-causing genes exist for this disorder. Therefore it is possible that segregation of several RLS genes might occur in the same family. Bilineality has been a challenge for genetic studies of other common neurological disorders such as Tourette's syndrome (McMahon et al., 1996). Nonetheless, for common disorders, the chances of identifying a disease gene locus using parametric linkage analysis are better when studying large, multigenerational families. For this reason, we are collecting large Utah families with RLS for use in our genetic studies. Collection of phenotypic data and DNA samples has been completed for some of these families all of whom have many affected members (Table 43.1). To overcome the problem of locus heterogeneity that may occur when studying many smaller families, we have included in our genetic study only families of sufficient size, each of which can yield significant linkage results. Age-of-onset of RLS symptoms for affected members within each family is similar.

The pedigree structure of one such family, Kindred 5332, is shown in Fig. 43.1. Neurologists collaborating with this project have interviewed family members from K5332 and reviewed questionnaires modeled after International RLS Study Group recommendations before assigning affection status. Family members less than age 20 years are considered of unknown affection status for purposes of this study since many RLS individuals do not experience symptoms until early adulthood. A simulated linkage analysis of K5332 assuming disease gene penetrance of 80% and a disease allele frequency of 5% has yielded a maximal optimal lod score of 6.45 at $\theta = 0.001$ for this family. Genome-wide scans are currently being conducted in our RLS families to search for disease-causing loci. Similar large, multigenerational Utah families of Northern European descent have been instrumental in allowing researchers to discover many genes including colon and breast cancer, neurofibromatosis, long QT syndrome, and epilepsy. It is also possible to link related RLS study families using a Utah genealogical database comprised of information from 2 million birth certificates, thereby creating extended kinships.

TABLE 43.1 Families Collected for the Utah RLS Genetic Project

| Kindred number | Number of family members participating | Number of affected family members |
|---|---|---|
| K4827 | 46 | 24 |
| K5006 | 34 | 20 |
| K5332 | 44 | 20 |
| K6245 | 22 | 10 |
| K6258 | 15 | 8 |
| K7150 | 33 | 23 |
| K7456 | 19 | 7 |

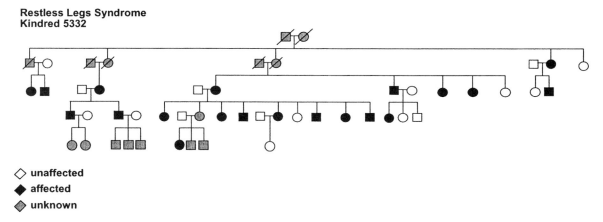

FIGURE 43.1 The pedigree structure of one family.

## B. Linkage Studies

Several linkage studies in RLS families have been conducted but few results suggestive of linkage have been obtained (Johnson et al., 1992; Dichgans et al., 1996). One report used the candidate gene approach to search for linkage to 22 different loci in a four-generation family comprised of 24 affected members. Dichgans et al. (1996) were able to exclude chromosomal regions carrying genes for tyrosine hydroxylase and other components of the dopaminergic system, several GABA receptors subunits and the gene for the α1 subunit of the glycine receptor. Similarly, in a genetic association analysis of 92 RLS patients, no differences in genotypic or allelic distributions, were found relative to controls in a study of 8 genes involved in dopaminergic transmission (DA-receptors D1 to D5, dopamine transporter, tyrosine hydroxylase, and dopamine β-hydroxylase) (Desautels et al., 2001b).

One successful linkage study conducted a genome-wide scan with 378 PCR-based markers in a large French-Canadian family with 14 affected individuals (Desautels et al., 2001a). These authors reported an lod score of 3.42 for the marker D12S1044 using two-point linkage analysis of this pedigree assuming autosomal recessive inheritance and a high disease allele frequency. Eight adjacent markers on chromosome 12q13-23 spanning a distance of 44 cM gave positive lod scores >1.0. Haplotype analysis of these markers shows a ~15 cM region of nonrecombination within which the disease gene in this family might lie. Desautels et al. (2001a) discuss several candidate genes found in this region as well as the need to replicate their findings in other large RLS families. Two other markers, D5S816 (at 5q31) and GGAT1A4 (at 10q22) also generated lod scores >1.0 in this family.

## V. DIAGNOSIS

The diagnosis of RLS is based primarily on patient history. To aid clinicians, the International Restless Legs Syndrome Study Group has established minimal criteria as a requirement for RLS diagnosis (Table 43.2); (Walters, 1995). These criteria are (1) desire to move the extremities, often associated with paresthesias/dysesthesias, (2) motor restlessness, (3) worsening of symptoms at rest with at least temporary relief by activity, and (4) worsening of the symptoms in the evening or at night. Other features important to the diagnosis include sleep disturbances, PLMS and similar involuntary movements while awake, a normal neurological exam in patients without secondary causes of RLS (see below), and a tendency for symptoms to be worse in middle to older age. Questionnaires used to collect information from persons with RLS symptoms have also been developed by this task force (Walters et al., 1996).

**TABLE 43.2  Clinical Criteria and Associated Features of RLS**

**Minimal criteria**
A compelling urge to move the limbs usually associated with paresthesias/dysesthesias
Motor restlessness as seen in activities such as floor pacing, tossing and turning in bed, and rubbing the legs
Symptoms that are worse or exclusively present at rest (i.e., lying, sitting) with variable and temporary relief by activity
Symptoms that are worse in the evening or at night

**Associated features**
Sleep disturbance and daytime fatigue
Normal neurologic examination (in patients with primary RLS)
Involuntary, repetitive, periodic, jerking movements, either in sleep or while awake and at rest

From Walters, A. S. (1995). *Mov. Disord.* **10**, 634–642. With permission.

These criteria have been validated in large studies using the Study Group guidelines and questionnaire to diagnose RLS patients (Montplaisir et al., 1997; Ulfberg et al., 2001) and should allow for easier diagnosis given that RLS and the resultant sleep disturbances are often underdiagnosed (Meissner et al., 1998).

RLS can either be primary (idiopathic) or symptomatic of an underlying disease (secondary). Secondary RLS is commonly associated with peripheral neuropathy (i.e., diabetics); (Rutkove et al., 1996), uremia (Callaghan, 1966; Wetter et al., 1998), or iron deficiency (Sun et al., 1998). RLS has also been reported in association with spinal cord and peripheral nerve lesions and may occur in patients with vertebral disc disease (Walters et al., 1996). RLS can also manifest during pregnancy; up to 19% of pregnant women report RLS symptoms that subside within a few weeks after giving birth (Goodman et al., 1988). Psychotropic medications such as antidepressants (i.e., tricyclics and selective serotonin reuptake inhibitors) and lithium are known to cause or aggravate RLS symptoms (Garvey and Tollefson, 1987; Bakshi, 1996; Terao et al., 1991).

### A. Diagnostic and Ancillary Tests

A physical exam should be done on all RLS patients to rule out secondary causes of RLS. The emphasis of the neurologic examination should be on spinal cord and peripheral nerve function. Peripheral pulses should be checked to rule out vascular disease. The association of RLS and decreased iron stores (in the absence of low hemoglobin) has led to the recommendation that newly diagnosed RLS patients should have their ferritin levels checked (NHLBI Working Group, 2000). Serum ferritin levels below 50 ng/mL are thought to be a secondary cause of RLS. Metabolic bloodwork should also rule out uremia and diabetes. Sleep studies are not routinely ordered in the work-up of RLS but some authors have recommended

EMG to exclude the possibility of peripheral neuropathy (Walters, 1995).

The differential diagnosis of RLS includes nocturnal leg cramping; this common condition consists of involuntary muscle contractions that are most often painful, limited to one leg and to one muscle group (Riley and Anthony, 1995). Vascular diseases such as deep vein thrombosis or arterial insufficiency often present as unilateral leg pain without sensory complaints. Peripheral neuropathy can cause leg symptoms that can confound clinicians because these patients may have sensory and RLS symptoms. Patients with long-standing diabetes often present with sensory symptoms such as numbness, tingling, or pain (van Alfen et al., 2001). Parkinson's disease patients with RLS have also been described (Lang, 1987). Restless legs symptoms have been reported in patients with spinocerebellar ataxias (SCAs) raising the possibility that SCA genes may play a role in the pathophysiology of RLS (Schols et al., 1998; Abele et al., 2001; van Alfen et al., 2001)

## VI. NEUROIMAGING AND NEUROPHYSIOLOGICAL STUDIES

The neurophysiological mechanisms of RLS are unknown. Neuroimaging studies have failed to identify structural lesions associated with RLS (Bucher et al., 1996). However, there is evidence that RLS may be the motor manifestation of a normal "periodicity" of the central nervous system that becomes disinhibited. The fact that dopaminergic drugs are used to treat RLS patients supports this theory. One report has found increased free dopamine levels and homovanillic acid in the CSF of an RLS patient (Montplaisir et al., 1985), although another study did not replicate these findings (Earley et al., 2001). Neuroimaging studies have shown that RLS patients have lower striatal D2 dopamine binding sites compared to controls (Turjanski et al., 1999). Interestingly, symptoms in patients with secondary RLS are often not relieved by treatment with dopaminergic agonists. A recent SPECT study, however, showed no difference in striatal dopamine transporter and dopamine D(2)-receptor binding between an RLS patient group and controls (Eisensehr et al., 2001). Ruottinen et al. (2000) reported reduced L-dopa uptake as measured by FDOPA PET in nigrostriatal structures of RLS patients suggesting dopaminergic hypofunction, although an earlier PET study found no differences in brain metabolic activity or presynaptic dopaminergic uptake between normal subjects and RLS patients (Trenkwalder et al., 1999).

Studies have begun to examine the role of iron metabolism in the pathophysiology of RLS following the observation that serum ferritin values were decreased in RLS patients (Ekbom, 1960; O'Keeffe et al., 1994). The increased incidence of RLS in pregnancy has been attributed to iron deficiency (Ekbom, 1960). Treatment with ferrous sulfate reduced the symptom severity in elderly patients with RLS (O'Keeffe et al., 1994). Subsequently Earley et al. (2000a) showed that CSF concentrations of ferritin and transferrin are altered in patients with primary RLS, compared to serum ferritin and transferrin levels, which were normal. The authors concluded that RLS patients may have altered iron transport across the blood-brain barrier with no manifestations of iron deficiency. Using a special MRI measurement, Allen et al. (2001) then showed that striatal brain iron concentrations were significantly decreased in RLS subjects. Several authors have recently advanced the "iron-dopamine model" of RLS that links decreased brain iron levels to decreased dopaminergic function as the cause of RLS (Earley et al., 2000b).

A growing body of literature focuses on transcranial magnetic stimulation (TMS) as another means of unraveling the pathophysiological basis of RLS. TMS of the brain allows for noninvasive evaluation motor cortex excitability (Rossini et al., 1991). TMS of the motor cortex during tonic contraction of a muscle produces an electromyographic response, a short latency motor-evoked potential (MEP) generated predominantly by trans-synaptic excitation of cortical motor neurons. A cortical silent period (CSP) follows, during which no EMG activity is noted. Several studies have shown that the CSP varies significantly in certain diseases affecting the basal ganglia (Priori et al., 1994; Prout and Eisen, 1994). One recent study has documented a shortened CSP during TMS in primary RLS patients (Entezari-Taher et al., 1999). This shortened CSP reflects a reduction of normal CNS motor inhibition above the level of the spinal cord. A shortened CSP has not been reported in patients with secondary RLS. Another TMS study revealed reduction of intracortical inhibition in RLS patients (Tergau et al., 1999). Interneuronal inhibitory circuits in the motor cortex depend on GABAergic systems. Dopaminergic pathways may alter the gain of GABAergic inhibitory circuitry at the motor cortex level (Ziemann et al., 1996). These studies suggest that genes encoding a component of the GABAergic inhibitory circuitry or the dopaminergic systems are candidate genes for RLS. Alternately, disease-causing genes may lie within the brain's sleep control centers or iron metabolic pathways.

## VII. TREATMENT

The severity of RLS symptoms varies from patient to patient and some may not require treatment. Several medications either taken alone or in combination have been shown effective in treating symptoms in patients with symptoms severe enough to disrupt their ability to live or sleep normally. Dopaminergic agents such as dopamine precursors or agonists are considered first-line drugs for most RLS patients (Brodeur *et al.*, 1988; Lin *et al.*, 1998; NHLBI Working Group, 2000). In cases where patients develop tolerance or require increasing doses, temporary withdrawal and reinstitution of therapy a few weeks later may improve efficacy. Alternately benzodiazepines, particularly clonazepam, can be used alone or in combination with dopaminergic agents. Commonly prescribed opioids such as codeine or hydrocodone also reduce the intensity of RLS symptoms (Hening *et al.*, 1986). Anticonvulsants such as carbamazepine or gabapentin can also be considered when dopaminergic agents fail to relieve RLS symptoms (Adler, 1997).

## VIII. SUMMARY

Recent genetic studies have shown that primary RLS is a common neurological disorder that has a strong hereditary component with up to 60% of patients reporting a positive family history. An autosomal dominant mode of inheritance within RLS families has been most commonly observed. However, autosomal recessive inheritance is also likely given the prevalence of RLS. The fact that RLS has variable expressivity and age-of-onset supports the likelihood that locus heterogeneity almost certainly exists for RLS. Therefore, one important approach to identifying RLS genes is to study families large enough to provide significant evidence for linkage. Studying families with similar clinical phenotypes further maximizes the chances that the same gene is causing the RLS phenotype in all affected family members. One recent linkage study in a large family has identified several putative RLS loci. Identifying RLS genes will lead to better clinical characterization of these patients as well as accurate diagnostic tests and more effective therapies for these patients.

## References

Abele, M., Burk, K., Laccone, F., Dichgans, J., and Klockgether, T. (2001). Restless legs syndrome in spinocerebellar ataxia types 1, 2, and 3. *J. Neurol.* **248**, 311–314.

Adler, C. H. (1997).Treatment of restless legs syndrome with gabapentin. *Clin. Neuropharmacol.* **20**, 148–151.

Allen, R. P., Barker, P. B., Wehrl, F., Song, H. K., and Earley, C. J. (2001). MRI measurement of brain iron in patients with restless legs syndrome. *Neurology* **56**, 263–265.

Allen, R.P., and Earley, C.J. (2001). Restless legs syndrome: a review of clinical and pathophysiologic features. *J. Clin. Neurophysiol.* **18**, 128–147.

Ambrosetto, C., Lugaresi, E., Coccagna, G., and Tassinari, C.A. (1965). Clinical and polygraphic remarks in the syndrome of restless legs. *Riv. Patol. Nerv. Ment.* **86**, 244–252.

American Sleep Disorders Association. (1997). The international classification of sleep disorders, revised: diagnostic and coding manual. Rochester, Minn.: American Sleep Disorders Association. 65–68.

Bakshi, R. (1996). Fluoxetine and restless legs syndrome. *J. Neurol. Sci.* **142**, 151–152.

Bassetti, C.L., Mauerhofer, D., Gugger, M., Mathis, J., and Hess, C.W. (2001). Restless legs syndrome: a clinical study of 55 patients. *Eur. Neurol.* **45**, 67–74.

Bornstein, B. (1961). Restless leg syndrome. *Psychiatr. Neurol.* **141**, 165–201.

Brodeur, C., Montplaisir, J., Godbout, R., and Marinier, R. (1988). Treatment of restless legs syndrome and periodic movements during sleep with L-dopa: a double-blind,controlled study. *Neurology* **38**, 1845–1848.

Bucher, S. F., Trenkwalder, C., and Oertel, W. H. (1996). Reflex studies and MRI in the restless legs syndrome. *Acta Neurol. Scand.* **94**, 145–150.

Callaghan, N. (1966). Restless legs syndrome in uremic neuropathy. *Neurology* **16**, 359–361.

Clemençon, J. C. (1975). Autosomal dominant inheritance of "restless legs" through three successive generations. *Arch. Genet. (Zur)* **48**, 27–46.

Desautels, A., Turecki, G., Montplaisir, J., Sequeira, A., Verner, A., and Rouleau, G. A. (2001a). Identification of a major susceptibility locus for restless legs syndrome on chromosome 12q. *Am. J. Hum. Genet.* **69**, 1266–1270.

Desautels, A., Turecki, G., Montplaisir, J., Ftouhi-Paquin, N., Michaud, M., Chouinard, V. A., and Rouleau, G. A. (2001b) Dopaminergic neurotransmission and restless legs syndrome: a genetic association analysis. *Neurology* **57**, 1304–1306.

Dichgans, M., Walther, E., Collado-Seidel, V., Trenkwalder, C., Muller-Myhsok, B., Oertel, W., and Gasser, T. (1996). Autosomal dominant restless legs syndrome: genetic model and evaluation of 22 candidate genes. *Mov. Disord.* **11**(Suppl. 1), 87.

Earley, C. J., Connor, J. R., Beard, J. L., Malecki, E. A., Epstein, D. K., and Allen, R. P. (2000a). Abnormalities in CSF concentrations of ferritin and transferrin in restless legs syndrome. *Neurology* **54**, 1698–1700.

Earley, C. J., Allen, R. P., Beard, J. L., and Connor, J. R. (2000b). Insight into the pathophysiology of restless legs syndrome. *J. Neurosci. Res.* **62**, 623–628.

Earley, C. J., Hyland, K., and Allen, R. P. (2001). CSF dopamine, serotonin, and biopterin metabolites in patients with restless legs syndrome. *Mov. Disord.* **16**, 144–149.

Eisensehr, I., Wetter, T. C., Linke, R., Noachtar, S., von Lindeiner, H., Gildehaus, F. J., Trenkwalder, C., and Tatsch, K. (2001). Normal IPT and IBZM SPECT in drug-naive and levodopa-treated idiopathic restless legs syndrome. *Neurology* **57**, 1307–1309.

Ekbom, K. A. (1945). Restless leg syndrome. *Acta Neurol. Scand.* **158** (Suppl.), 1–123.

Ekbom, K. A. (1950). Restless legs. A report of 70 new cases. *Acta Med. Scand.* **246** (Suppl.), 64–68.

Ekbom, K. A. (1960). Restless legs syndrome. *Neurology* **10**, 868–873.

Ekbom, K. A. (1975). Growing pains and restless legs. *Acta Paediatr. Scand.* **64**, 264–266.

Entezari-Taher, M., Singleton, J. R., Jones, C. R., Meekins, G., Petajan, J. H., and Smith, A. G. (1999). Changes in excitability of motor cortical circuitry in primary restless leg syndrome. *Neurology* **53**, 1201–1205.

Garvey, M. J., and Tollefson, G. D. (1987). Occurrence of myoclonus in patients treated with cyclic antidepressants. *Arch. Gen. Psychiatry.* **44**, 269–272.

Goodman, J. D., Brodie, C., and Ayida, G. A. (1988). Restless leg syndrome in pregnancy. *BMJ* **297**, 1101–1102.

Hening, W. A., Walters, A., Kavey, N., Gidro-Frank, S., Cote, L., and Fahn, S. (1986). Dyskinesias while awake and periodic movements in sleep in restless legs syndrome: treatment with opioids. *Neurology* **36**, 1363–1366.

Jacobsen, J. H., Rosenberg, R. S., Huttenlocher, P. R., and Spire, J. P. (1986). Familial nocturnal cramping. *Sleep* **9**, 54–60.

Johnson, W., Walters, A., Lehner, T., Coccagna, G., Ehrenberg, B., Lazzarini, A., Stenroos, E., Lugaresi, E., Hickey, K., Pichietti, D., Hening, W., Gilliam, C., Gold, B., Grin, M., Fazzini, E., McCormack,

M., Fahn, S, Ott, J., and Chokroverty, S. (1992). Affected only linkage analysis of autosomal dominant restless legs syndrome. *Sleep Res.* **21**, 214.

Lang, A. E. (1987). Restless legs syndrome and Parkinson's disease: insights into pathophysiology. *Clin. Neuropharmacol.* **10**, 476–478.

Lavigne G., and Montplaisir, J. (1994). Restless legs syndrome and sleep bruxism: prevalence and association among Canadians. *Sleep* **17**, 739–743.

Lazzarini, A., Walters, A. S., Hickey, K., Coccagna, G., Lugaresi, E., Ehrenberg, B. L., Picchietti, D. L., Brin, M. F., Stenroos, E. S., Verrico, T., and Johnson, W. G. (1999). Studies of penetrance and anticipation in five autosomal-dominant restless legs syndrome pedigrees. *Mov. Disord.* **14**, 111–116.

Lin, S. C., Kaplan, J., Burger, C. D., and Fredrickson, P. A. (1998). Effect of pramipexole in treatment of resistant restless legs syndrome. *Mayo Clin. Proc.* **73**, 497–500.

Lugaresi, E., Cirignotta, F., Coccagna, G., and Montagna, P. (1986). Nocturnal myoclonus and restless legs syndrome. *Adv. Neurol.* **43**, 295–307.

Lutz, E. G. (1978). Restless legs, anxiety and caffeinism. *J. Clin. Psychiatry* **39**, 693–698.

McMahon, W. M., van de Wetering, B. J., Filloux, F., Betit, K., Coon, H., and Leppert, M. (1996). Bilineal transmission and phenotypic variation of Tourette's disorder in a large pedigree. *J. Am. Acad. Child Adol. Psychiatry* **35**, 672–680.

Meissner, H. H., Riemer, A., Santiago, S. M., Stein, M., Goldman, M. D., and Williams, A. J. (1998). Failure of physician documentation of sleep complaints in hospitalized patients. *West J. Med.* **169**, 146–149.

Michaud, M., Chabli, A., Lavigne, G., and Montplaisir, J. (2000) Arm restlessness in patients with restless legs syndrome. *Mov. Disord.* **15**, 289–293.

Montplaisir, J., Godbout, R., Boghen, D., DeChamplain, J., Young, S. N., and Lapierre, G. (1985). Familial restless legs with periodic movements in sleep: Electrophysiological, biochemical, and pharmacological study. *Neurology* **35**, 130–134.

Montplaisir, J., Boucher, S., Poirier, G., Lavigne, G., Lapierre, O., and Lesperance, P. (1997). Clinical, polysomnographic, and genetic characteristics of restless legs syndrome: a study of 133 patients diagnosed with new standard criteria. *Mov. Disord.* **12**, 61–65.

NHLBI Working Group. (2000). Restless legs syndrome: detection and management in primary care. National Heart, Lung, and Blood Institute Working Group on Restless Legs Syndrome. *Am. Fam. Physician* **62**, 108–114.

O'Keeffe, S. T., Gavin, K., and Lavan, J. N. (1994). Iron status and restless legs syndrome in the elderly. *Age Ageing* **23**, 200–203.

Ondo, W. G., and Jankovic, J. (1996). Restless legs syndrome: clinical-etiologic correlates. *Neurology* **47**, 1435–1441.

Ondo, W. G., Vuong, K. D., and Wang, Q. (2000). Restless legs syndrome in monozygotic twins: clinical correlates. *Neurology* **55**, 1404–1406.

Phillips, B., Young, T., Finn, L. Asher, K., Hening, W. A., and Purvis, C. (2000). Epidemiology of restless legs symptoms in adults. *Arch. Intern. Med.* **160**, 2137–2141.

Picchietti D. L., Underwood, D. J., Farris, W. A., Walters, A. S., Shah, M. M., Dahl, R. E., Trubnick, L. J., Bertocci, M. A., Wagner, M., and Hening, W. A. (1999). Further studies on periodic limb movement disorder and restless legs syndrome in children with attention-deficit hyperactivity disorder. *Mov. Disord.* **14**, 1000–1007.

Priori, A., Beradelli, A., Inghilleri, M., Polidori, L., and Manfredi, M. (1994). Electromyographic silent period after transcranial magnetic stimulation in Huntington's Disease. *Mov. Disord.* **9**, 178–182.

Prout, A. J., and Eisen, A. A. (1994). The cortical silent period and ALS. *Muscle Nerve* **17**, 217–223.

Riley, J. D., and Anthony, S. J. (1995). Leg cramps: differential diagnosis and management. *Am. Fam. Physician* **52**, 1794–1798.

Rossini, P. M., Desiato, M. T., Lavaroni, F., and Caramia, M. D. (1991). Brain excitability and electroencephalographic activation: non-invasive evaluation in healthy humans via transcranial magnetic stimulation. *Brain Res.* **567**, 111–119.

Ruottinen, H. M., Partinen, M., Hublin, C., Bergman, J., Haaparanta, M., Solin, O., and Rinne, J. O. (2000). An FDOPA PET study in patients with periodic limb movement disorder and restless legs syndrome. *Neurology* **54**, 502–450.

Rutkove, S. B., Matheson, J. K., and Logigian, E. (1996). Restless legs syndrome in patients with polyneuropathy. *Muscle Nerve* **19**, 670–672.

Schols, L., Haan, J., Riess, O., Amoiridis, G., and Przuntek, H. (1998). Sleep disturbance in spinocerebellar ataxias: is the SCA3 mutation a cause of restless legs syndrome? *Neurology* **51**, 1603–1607.

Strang, R. R. (1967). The symptom of restless legs. *Med. J. Aust.* **1**, 1211–1213.

Sun, E. ., Chen, C. A., Ho, G., Earley C. J., and Allen, R. P. (1998). Iron and the restless legs syndrome. *Sleep* **21**, 371–377.

Terao, T., Terao, M., Yoshimura, R., and Abe, K. (1991). Restless legs syndrome induced by lithium. *Biol. Psychiatry* **30**, 1167–1170.

Tergau, F., Wischer, S., and Paulus, W. (1999). Motor system excitability in patients with restless legs syndrome. *Neurology* **52**, 1060–1063.

Trenkwalder, C., Seidel, V. C., Gasser, T., and Oertel, W. H. (1996). Clinical symptoms and possible anticipation in a large kindred of familial restless legs syndrome. *Mov. Disord.* **11**, 389–394.

Trenkwalder, C., Walters, A. S., Hening, W. A., Chokroverty, S., Antonini, A., Dhawan, V., and Eidelberg, D. (1999). Positron emission tomographic studies in restless legs syndrome. *Mov. Disord.* **14**, 141–145.

Turjanski, N., Lees, A. J., and Brooks, D. J. (1999). Striatal dopaminergic function in restless legs syndrome: 18F-dopa and 11C-raclopride PET studies. *Neurology* **52**, 932–937.

Ulfberg, J., Nystrom, B., Carter, N., and Edling, C. (2001). Prevalence of restless legs syndrome among men aged 18 to 64 years: an association with somatic disease and neuropsychiatric symptoms. *Mov. Disord.* **16**, 1159–1163.

van Alfen, N., Sinke, R. J., Zwarts, M. J., Gabreels-Festen, A., Praamstra, P., Kremer, B. P., and Horstink, M. W. (2001). Intermediate CAG repeat lengths (53,54) for MJD/SCA3 are associated with an abnormal phenotype. *Ann. Neurol.* **49**, 805–807.

Walters, A. S., Pichietti, D., Hening, W., and Lazzarini, A. (1990). Variable expressivity in familial restless legs syndrome. *Arch. Neurol.* **47**, 1219–1220.

Walters, A. S., Picchietti, D. L., Ehrenberg, B. L., and Wagner, M. L. (1994). Restless legs syndrome in childhood and adolescence. *Pediatr. Neurol.* **11**, 241–245.

Walters, A. S. (1995). Towards a better definition of restless legs syndrome. The International Restless Legs Syndrome Study Group. *Mov. Disord.* **10**, 634–642.

Walters, A. S., Hickey, K., Maltzman, J., Verrico, T., Joseph, D., Hening, W., Wilson, V., and Chokroverty, S. (1996). A questionnaire study of 138 patients with restless legs syndrome: the 'Night-Walkers' survey. *Neurology* **46**, 92–95.

Wetter, T. C., Stiasny, K., Kohnen, R., Oertel, W. H., and Trenkwalder, C. (1998). Polysomnographic sleep measures in patients with uremic and idiopathic restless legs syndrome. *Mov. Disord.* **13**, 820–824.

Winkelmann, J., Wetter, T. C., Collado-Seidel, V., Gasser, T., Dichgans, M., Yassouridis, A., and Trenkwalder, C. (2000). Clinical characteristics and frequency of the hereditary restless legs syndrome in a population of 300 patients. *Sleep* **23**, 597–602.

Winkelmann, J., Mueller-Myhsok, B., Wittchen, H. U., Hock, B., Prager, M., Pfister, H., Dichgans, M., Gasser, T., and Trenkwalder, C. (2001). Segregation analysis of families with restless legs syndrome provides evidence for a single gene and autosomal dominant inheritance. *Sleep* **24** (Suppl.), A102.

Ziemann, U., Bruns, D., and Paulus, W. (1996). Enhancement of human motor cortex inhibition by the dopamine receptor agonist pergolide: evidence from TMS. *Neurosci. Lett.* **208**, 187–190.

# CHAPTER 44

# Other Adult-Onset Movement Disorders with a Genetic Basis

JAMES P. SUTTON

*California Neuroscience Institute*
*Oxnard, California 93030*

I. Inborn Errors of Metabolism
   A. Adult-Onset Tay-Sachs Disease (OMIM 272800)
   B. Late-Onset Niemann-Pick Disease Type II (OMIM 257220)
II. Disorders of Heavy Metal Metabolism
   A. Neuroferritinopathy (OMIM 606159)
   B. Primary Ceruloplasmin Deficiency (OMIM 604290)
III. Movement Disorders Associated with Hematological Disease
   A. Adult-Onset Chediak-Higashi Syndrome (OMIM 214500)
   B. Choreoacanthocytosis (OMIM 200150)
   C. McLeod Syndrome (OMIM 314850)
   D. Acanthocytosis, Retinitis Pigmentosa, and Pallidal Degeneration (ARP)
IV. Other Rare Disorders
   A. Pallidopyramidal Degeneration
   B. Familial Parkinsonism and Dementia with Ballooned Neurons
   C. Fragile X Premutation Syndrome
V. Summary
   Acknowledgment
   References

A number of genetic movement disorders presenting in adolescence and adulthood are not discussed elsewhere in this text. This may be because they are extremely rare, poorly understood, or due to metabolic disorders not traditionally thought of as "movement disorders." A brief discussion of some of these conditions follows, broken down into inborn errors of metabolism, disorders of heavy metal metabolism, disorders associated with hematological disease, and other rare disorders.

## I. INBORN ERRORS OF METABOLISM

Labels such as "organic acidemia," "amino aciduria," and "lysosomal storage disease" were necessary when clinical diagnosis relied on detection of abnormal serum, urine, or tissue metabolites or enzyme activity. However, we can now diagnose inherited disease by genotype and understand the pathophysiology of a given patient's disorder at the molecular level. As a result, the distinction between "metabolic disorders" such as phenylketonuria and "movement disorders" such as dopa-responsive dystonia becomes blurred. The time has come to re-evaluate arbitrary and obsolete classifications of disease.

As an example, let us look at adult-onset Tay-Sachs disease (TSD) and autosomal recessive juvenile parkinsonism (ARJP); (Table 44.1). Each affects young adults and is slowly progressive. Each is a recessive disorder with reduced enzyme activity (hexosaminidase A versus alpha-synuclein ubiquitin ligase). Each results in disruption of a specific pathway of cellular catabolism (lysosomal versus ubiqitin-proteasomal). In one disorder, multiple neuronal populations are affected, whereas in the other, neuronal pathology is very focused. In one disorder undigested substrate accumulates and is visible with light microscopy, whereas in the other there are no obvious markers of substrate accumulation. By trying to understand the basis for the similarities and differences between these two disorders, we can learn much about each. The greatest benefit from studying the neurobiology of a rare disorder such as adult-onset TSD may be what we learn about more traditional movement disorders such as Parkinson's disease.

Although a detailed discussion of "traditional" metabolic disorders is beyond the scope of this text, it is worthwhile

TABLE 44.1  Adult-Onset Tay-Sachs Disease (AOTSD) versus Autosomal Recessive Juvenile Parkinsonisme (ARJP)

|  | AOTSD | ARJP |
| --- | --- | --- |
| Age of onset | Young adult | Young adult |
| Mode of inheritance | Autosomal recessive | Autosomal recessive |
| Conventional classification | Inborn error of metabolism | Neurodegenerative disorder |
| Gene locus | 15q23–q24 | 6q25.2–q27 |
| Gene name | HEXA | PARK1 |
| Normal gene product | Hexosaminidase type A | Parkin |
| Peptide function | Lysosomal enzyme | Ubiqitin-proteasomal enzyme |
| Peptide substrate | Glycolipid and glycoprotein | Alpha-synuclein |
| Type of mutation (allele 1) | Missense | Null |
| Type of mutation (allele 2) | Null or missense | Null |
| Intracellular storage product | Yes | No Lewy bodies |
| Neuronal population(s) affected | Nigral, retinal, thalamic, cerebellar, brainstem | Nigral |

to discuss two conditions that can present as adult-onset movement disorders, adult-onset GM2 gangliosidosis type 1 and Niemann-Pick type II. From a review of these disorders we can learn much about the molecular biology of adult-onset movement disorders.

## A. Adult-Onset Tay-Sachs Disease (OMIM 272800[1])

GM2 gangliosidosis type 1 or Tay-Sachs disease (TSD) is an autosomal recessive disorder caused by mutations of the HEXA gene on 15q23–q24, resulting in loss of hexosaminidase type A activity. The *infantile-onset* form is more common in the Ashkenazi Jewish population due to a high carrier frequency of two specific alleles. Homozygotes or compound heterozygotes with these mutations have less than 0.5% of normal hexosaminidase type A activity. Affected individuals present in early infancy with developmental regression, paralysis, dementia, and blindness, with death in the first few years of life. Funduscopy reveals retinal pallor due to lipid-laden ganglion cells. The contrast of this pallor with a relatively spared fovea centralis creates the illusion of a central "cherry-red" spot. This finding, along with an exaggerated startle reflex, strongly suggests the diagnosis. There is no effective treatment for the infantile form, which is invariably fatal by the second or third year of life.

There are several phenotypic variants. The *chronic* form is characterized by a slower rate of progression, with survival into adolescence or young adulthood (Rapin *et al.*, 1976). The *juvenile-onset* form begins in childhood or adolescence, and also has a more benign course. It may present with childhood-onset dystonia, dementia, spasticity, or an atypical juvenile-onset spinal muscular atrophy (Meek *et al.*, 1984; Nardocci *et al.*, 1992; Rondot *et al.*, 1997; Suzuki *et al.*, 1970). The *adult-onset* variant, although rare, can present with tremor, dystonia, and choreoathetosis, and is therefore important in the differential diagnosis of rare adult-onset movement disorders.

### 1. Phenotype

A detailed discussion of nonadult forms of TSD is beyond the scope of this text.

Adult-onset TSD typically presents with the onset in adolescence or young adulthood of various combinations of ataxia, amyotrophy, spasticity, sensorimotor neuropathy, and oculomotor disturbances. These combinations can mimic juvenile ALS, spinocerebellar ataxia, or the Kugelberg-Welander variant of spinal muscular atrophy (Harding *et al.*, 1987; Johnson *et al.*, 1982; Mitsumoto 1985; Parnes *et al.*, 1985). Psychosis is extremely common, occurring in 30–50% of adult-onset patients, and may lead to a misdiagnosis of schizophrenia (Lichtenberg *et al.*, 1988; MacQueen *et al.*, 1998; Navon 1986). Dementia, if present, is usually mild. Movement disorders are not uncommon, and may include tremor, dystonia, choreoathetosis, and myoclonus, although these features rarely predominate (Barnes *et al.*, 1991; Federico *et al.*, 1991; Oates *et al.*, 1986).

### 2. Genotype (HEXA, 15q23–q24)

*a. Infantile-Onset, Chronic, and Juvenile TSD*

The HEXA gene, located at 15q23–q24, codes for the alpha subunit of hexosaminidase A. In Ashkenazi Jews the most common mutation responsible for infantile-onset TSD is a 4-bp insertion in exon 11 of HEXA. The second most common mutation is a G-to-C substitution in the first nucleotide of intron 12 with defective mRNA splicing. Homozygotes of the first mutation are common, as are

---

[1] Online Mendelian Inheritance in Man, OMIM (TM). McKusick-Nathans Institute for Genetic Medicine, Johns Hopkins University (Baltimore, MD) and National Center for Biotechnology Information, National Library of Medicine (Bethesda, MD), 2000. World Wide Web URL: http://www.ncbi.nlm.nih.gov/omim

compound heterozygotes with one of each allele. All result in a null phenotype.

Many other mutations have been described; in general, missense mutations are more likely to be associated with less severe illness. The gly269ser mutation in particular is associated with both chronic TSD and adult-onset TSD, but not infantile-onset TSD (Paw et al., 1989; Paw et al., 1991). This suggests that this mutation codes for a protein with partial activity. In fact, clinical presentation correlates well with residual hexosaminidase A activity, with enzyme activity 0.1% of normal in infantile cases, 0.5% of normal in late-infantile cases, and 2–4% of normal in adult cases (Conzelmann et al., 1983).

*b. Adult-onset TSD*

In adult-onset TSD, one allele is always a missense mutation, and the other a missense or null mutation (Table 44.2). There appear to be partially protective as well as nonprotective missense mutations. The gly269ser mutation is protective; homozygotes have the most benign form of TSD (adult), and compound heterozygotes have either chronic or adult-onset TSD.

Nonprotective missense mutations include glu482lys and arg504cys, both associated with infantile-onset TSD. Glu482lys results in a change from a strong negative to a strong positive charge at position 482, blocking post-translational processing of the precursor beta-hexosaminidase alpha chain. This in turn results in a functionally null phenotype (Nakano et al., 1988; Proia and Neufeld, 1982). The arg504cys mutation is another example. If the second allele is arg504cys or IVS2, G-A, +1, the phenotype is infantile-onset, suggesting that the arg504cys mutation is functionally null. If the second allele is gly269ser, the phenotype is chronic or adult-onset.

### 3. Pathology

*a. Infantile-onset, chronic, and juvenile TSD*

Grossly, the brain may be small, normal, or enlarged, depending on the stage of illness at the time of death and whether neuronal loss or storage predominates. Ulegyria may be seen, as well as demyelination. The most striking changes, however, are microscopic. Neurons throughout the central and peripheral nervous system are ballooned, containing PAS-positive storage material. The storage material is lost on processing of routine paraffin sections, however, and is best seen on formalin-fixed frozen sections or cryostat preparations. The storage material stains strongly with Luxol fast blue and Sudan black. Microglial cells also contain PAS-positive storage material. Histochemistry of hexosaminidase activity does not reliably distinguish between Hex A and Hex B, and therefore cannot be used to diagnose TSD. Electron microscopy of the storage material demonstrates the characteristic and pathognomonic membranous cytoplasmic bodies (MCB) in both neurons and glia.

The pathology of TSD is generally limited to the central and peripheral nervous system; examination of liver, bone marrow, and lymphocytes is normal. Skin and conjunctival biopsy may reveal MCBs in Schwann cells, but these are not specific to TSD and can be seen in other storage disorders.

*b. Adult-onset TSD*

Typical MCBs of gangliosidosis are seen, although there are also less typical electron-dense heterogeneous conglomerates of neuronal inclusions. Cortical neurons are less involved than in early-onset cases. Storage neurons are seen in the thalamus, substantia nigra, brainstem nuclei, and cerebellum (Suzuki, 1991).

### 4. Pathophysiology

*a. Hexosaminidase Function*

There are three forms of hexosaminidase in man, hexosaminidase A (Hex A, consisting of three alpha- and three beta-subunits), hexosaminidase B (Hex B, consisting of six beta-subunits) and hexosaminidase S (consisting of six alpha-subunits); (Beutler et al., 1975). The active site on the alpha-subunit acts on sulfated substrates, whereas the

TABLE 44.2 Genotypes Associated with Adult-Onset TSD

| Allele 1 | Allele 2 | Phenotype | Ethnicity | N | Citation |
| --- | --- | --- | --- | --- | --- |
| tyr180his | Null | Adult | Ashkenazi | 2 | De Gasperi et al., 1996 |
| lys 197thr | | Adult | Non-Jewish | 1 | Akli et al., 1993 |
| gly269ser | Null | Adult | Ashkenazi | 8 | Navon and Proia 1989 |
| gly269ser | Null | Adult | Ashkenazi | 1 | Paw et al., 1989 |
| gly269ser | gly269ser | Adult | Non-Jewish | 3 | Navon et al., 1990 |
| gly269ser | | Adult | Non-Jewish | 3 | Navon et al., 1990 |
| gly269ser | arg504cys | Adult | Non-Jewish | 1 | Paw et al., 1991 |
| arg499cys | | Adult | Non-Jewish | 1 | Mules et al., 1992 |
| gly805ala | | Adult | Non-Jewish | 1 | Hechtman et al., 1992 |

beta-subunit primarily degrades neutral substrates. Only the alpha-subunit can degrade GM2 ganglioside. An activator protein is required for this reaction (O'Brien *et al.*, 1978). In TSD, a mutation in the HEXA gene causes a loss of alpha-unit activity and accumulation of GM2 ganglioside. In adult-onset TSD, missense mutations in the HEXA gene cause amino acid substitutions that severely depress the catalytic activity of the alpha-subunit (Navon and Proia, 1989) and may impair the association between alpha- and beta-subunits (Paw *et al.*, 1989).

### b. GM2 Ganglioside: Normal Function

The normal function of GM2 ganglioside is only now beginning to be understood. GM2 ganglioside is ordinarily synthesized in the Golgi body of pyramidal neurons, after which it undergoes exocytic trafficking to the somatic-dendritic plasmalemma (Walkley *et al.*, 2000). During normal development, an increase in GM2 expression precedes dendritic sprouting; after dendritic maturation is complete GM2 levels decrease significantly (Walkley *et al.*, 1995, 1998). This suggests that GM2 ganglioside acts as a regulatory signal for dendritogenesis in pyramidal neurons.

GM2 ganglioside may also play a role in neuronal survival. Binding of ciliary neurotrophic factor (CNTF) to the CNTF receptor (CTNF-R) on immortalized motor-neuron-like cells activates GM2 synthase (Usuki *et al.*, 1999, 2001). GM2 ganglioside in turn increases the binding affinity of the CNTF receptor five-fold. This could result in a feed-forward system where GM2 ganglioside synthesis facilitates neuronal survival during normal development.

### c. GM2 Ganglioside: Abnormal Function in TSD

GM2 accumulates in excess in TSD, and in association with cholesterol and phospholipid forms MCBs. The presence of numerous neurons filled with MCBs suggests that they may not actually be neurotoxic; otherwise one would not expect to see such large numbers of apparently viable storage neurons (compare, for example, with Lewy bodies in Parkinson's disease). It may be that the MCB is a way of sequestering excess GM2 ganglioside.

The excess accumulation of GM2 in TSD, however, is associated with a second, abnormal period of dendritogenesis in pyramidal neurons. Ectopic dendrites appear at axon hillocks, which gradually become covered with otherwise normal-appearing spines and synapses (Walkley *et al.*, 1990, 1998). Aberrant synaptogenesis may account for some of the clinical features of TSD, such as psychosis, dystonia, chorea, or myoclonus, although most of the phenotype appears to be due to progressive injury and death of pyramidal and other neurons. Whether cell death is due to accumulation of abnormal storage product, aberrant synaptogenesis, or other factors is unclear.

## 5. Diagnosis

### a. Clinical

The movement disorders specialist evaluating a patient with atypical dystonia, chorea, tremor, or myoclonus beginning in adolescence or adulthood should consider the possibility of adult-onset TSD, particularly if the patients is of Ashkenazi Jewish ancestry, has an affected sibling, or if there is a family history of consanguinity. The presence of dementia, ataxia, or ALS-like features should increase one's index of suspicion, and prompt serum hexosaminidase testing. Fifty percent of reported cases of adult-onset TSD are non-Jewish (Table 44.3). Although Ashkenazi Jewish ancestry should increase one's suspicion of the disorder, non-Jewish ethnicity should not exclude the diagnosis.

### b. Laboratory

The diagnosis of any forms of TSD, including the adult-onset form, is best made by measurement of serum hexosaminidase activity. In adult onset TSD, Hex A activity is 2–4% of normal, compared with more severe deficiency in the earlier, more severe forms. Leukocyte Hex A activity is also reduced, but this test is considerably more expensive than the serum assay, and does not appear to offer any advantages.

### c. Genetic Testing

Mutational analysis is not necessary to confirm the diagnosis of TSD in a symptomatic patient, as serum hexosaminidase A activity is an accurate assay of the molecular phenotype. However, once the enzymatic diagnosis is made, genotyping may provide additional useful information. Most clinical laboratories will screen for the six most common mutations, including three null alleles associated with infantile-onset TSD (+TATC1278, +1IVC12, +1IVS9), the gly269ser allele, and two pseudodeficiency alleles. Homozygosity or compound heterozygosity for the gly269ser allele would suggest a more benign prognosis, and could help with genetic counseling (see Section I.A.7).

### d. Imaging

Marked cerebellar atrophy can be seen on imaging studies (Mitsumoto *et al.*, 1985).

### e. Electrodiagnostic Testing

EMG can show complex repetitive discharges. NCV is consistent with an axonal sensorimotor polyneuropathy (Mitsumoto *et al.*, 1985).

### f. Biopsy

Suction rectal biopsy is a simple way to obtain ganglion cells for microscopic examination; the presence of PAS-positive neuronal inclusions on light microscopy or MCBs

TABLE 44.3 Arginine Substitutions in Non-Jewish Patients with TSD

| Allele 1 | Allele 2 | Phenotype | Ethnicity | N | Citation |
|---|---|---|---|---|---|
| arg178his | arg178his | Juvenile | Non-Jewish | 10 | dos Santos et al., 1991 |
| arg178his | Other | Infantile | Non-Jewish | 1 | dos Santos et al., 1991 |
| arg178his | arg178his | B1variant | Non-Jewishi | 1 | Tanaka et al., 1988 |
| arg178his | Other | B1variant | Non-Jewish | 3 | Tanaka et al., 1988 |
| arg178his | | B1variant | Non-Jewish | 1 | Ohno and Suzuki 1988 |
| arg178cys | | B1variant | Non-Jewish | 1 | Tanaka et al., 1990 |
| arg499his | Null | Juvenile | Non-Jewish | 2 | Paw et al., 1990 |
| arg499cys | | Adult | Non-Jewish | 1 | Mules et al., 1992 |
| arg499cys | Null | Infantile | Non-Jewish | 2 | Akli et al., 1993 |
| arg504his | arg504his | Juvenile | Non-Jewish | 2 | Paw et al., 1990 |
| arg504cys | gly269ser | Chronic | Non-Jewish | 1 | Paw et al., 1991 |
| arg504cys | gly269ser | Adult | Non-Jewish | 1 | Paw et al., 1991 |
| arg504cys | arg504cys | Infantile | Non-Jewish | 1 | Akli et al., 1991 |
| arg504cys | IVS2,G-A,+1 | Infantile | Non-Jewish | 1 | Akli et al., 1991 |

on electron microscopy is diagnostic of TSD. Skin biopsy for fibroblast culture may be assayed for both Hex A activity and quantity. Conjunctival biopsies may reveal Schwann cells containing MCBs. However, these tests do not appear to have any advantages over serum hexosaminidase testing, and no longer have clinical utility.

## 6. Treatment

There is no disease-modifying treatment for any form of TSD. Bone marrow transplant has been attempted without success (Platt and Butters, 1998). Although there are no specific therapies for the neurological complications of adult-onset TSD, there do not appear to be any contraindications to empiric trials targeted at specific symptoms, such as haloperidol or quetiapine for chorea; diazepam, oral baclofen, botulinum toxin, or intrathecal baclofen for dystonia; primidone, propanolol, or gabapentin for tremor; divalproex or clonazepam for myoclonus; and tizanidine, oral baclofen, intrathecal baclofen, botulinum toxin, or diazepam for spasticity. Neuropsychiatric symptoms should also respond to symptomatic therapy, such as anticholinesterase inhibitors for dementia; and quetiapine, olanzepine, or ziprasidone for psychosis. Sedative agents such as diazepam and oral baclofen should be used cautiously, given the possibility of clinical or subclinical cognitive impairment.

## 7. Genetic counseling

Patients and families should be offered professional genetic counseling regardless of whether genotype testing is anticipated. Counseling can help to inform, educate, and reassure concerned family members. Counseling should be tailored to the adult-onset genotype and phenotype and relevant prognostic information.

Asymptomatic adult siblings who request such information should be informed that they might be carriers, presymptomatic, or unaffected. They should be informed of various options to obtain more information, including Hex A testing and genotype testing.

Parents of an affected individual who are of childbearing age should be informed of the risk of having another child with adult-onset TSD, and should be offered the option of prenatal Hex A screening.

A spouse from a high-risk population may wish to be tested for carrier status, to aide in family planning. If the spouse is a noncarrier, children will be obligate carriers, carrying an adult-onset or infantile-onset allele. They will not be at risk of developing TSD, but should be informed of their carrier status when they come of age. If the partner is a carrier, children will have a 50% risk of developing TSD. Mutational analysis of either or both parents may be helpful. However, caution must be exercised if the couple already has children; Hex A and genotype testing of parents may provide unwanted information about a living child's risk of developing infantile- or adult-onset TSD.

A child of a person with adult-onset TSD should not be informed of any potential risk of developing TSD, and should not have Hex A or mutational analysis until he or she is of age and can actively participate in the process.

Prenatal testing should be offered only if the unaffected spouse is a carrier, the couple fully understands the implications of such testing, and if a positive test for TSD would alter the couple's decision to continue the pregnancy.

Genetic counseling for adult-onset TSD can be exceedingly complex, as patients may already have or may plan to have children. Mutational analysis, if informative, may greatly reduce the complexity of decision making in this situation. Genetic counseling in these situations is mandatory, regardless of whether genotype investigations are pursued.

## B. Late-Onset Niemann-Pick Disease Type II (OMIM 257220)

The designation of Niemann-Pick disease (NP) refers to an inherited disorder of lipid storage that affects the viscera and central nervous system, characterized by the presence of Niemann-Pick cells and visceral accumulation of sphingomyelin. Prior classification schemes were based upon the assumption that NP was a distinct genetic disorder with various phenotypes, distinguishable by age of onset, rate of progression, and presence or absence of neurological involvement. It has become clear, however, that NP is in fact two different diseases, distinguishable by distinct clinical features, biochemistry, and gene loci. It is therefore preferable to speak of Niemann-Pick disease type I (NP-I) and Niemann-Pick disease type II (NP-II), rather than types A, B, C, D, and E.

NP-I is characterized by early infantile onset, a retinal "cherry-red" spot, hepatosplenomegaly, retarded physical and mental growth, severe neurological disturbances, and absence of sphingomyelinase activity. Death usually occurs before age 3. NP-II includes cases previously classified as type C or type D, as well as cases previously described as juvenile dystonic lipidosis or sea-blue histiocytosis (Golde et al., 1975; Martin et al., 1984; Yan-Go et al., 1984). NP-II is characterized by a somewhat later onset, somewhat less severe involvement, normal sphingomyelinase activity, and the combination of ataxia and supranuclear vertical gaze palsy. There are three phenotypes seen in NP-II: these are (1) infantile-onset, rapidly progressive; (2) delayed-onset, slowly progressive; and (3) late-onset, slowly progressive (Fink et al., 1989; Vanier et al., 1991b)

### 1. Phenotype

The late-onset form of NP-II is characterized by ataxia and supranuclear downgaze palsy (Elleder et al., 1983; Lossos et al., 1997). Presentation is quite variable, however, and psychomotor retardation, dementia, or psychosis may predominate (Campo et al., 1998; Hulette et al., 1992; Shulman et al., 1995; Turpin et al., 1991). Dystonia may be the presenting feature (Elleder et al., 1983; Turpin et al., 1991). Unlike the infantile-onset form, splenomegaly, though present, is not pronounced, and may be missed on routine examination. Hepatomegaly is often absent.

### 2. Genotype (NPC1, 18q11-q12)

Sphingomyelinosis is the mouse equivalent of Niemann-Pick type II. The locus for murine sphingomyelinosis (spm) was shown to be on mouse chromosome 18 (Sakai et al., 1991b). Mouse chromosome 18 has extensive homology with human chromosomes 5 and 18, and the gene for Niemann-Pick type II was found to reside on human chromosome 18 (Carstea et al., 1993). The gene was named NPC1, as the phenotype at the time was known as Niemann-Pick type C (NPC). A second locus has been identified, NPC2, (Millat et al., 2001; Naureckiene et al., 2000), and determined to be identical to HE1 (human epididymis-1). This is a rare cause of NP-II, however, accounting for only 5% of cases, and does not appear to be associated with the late-onset form. Niemann-Pick type D was subsequently shown to be caused by a mutation in NPC1 (Greer et al., 1997, 1998, 1999b).

NPC1 spans more than 47 kb and contains 25 exons (Stephan et al., 2000). It has recently been sequenced and single nucleotide polymorphisms, pathological mutations, and haplotypes characterized (Bauer et al., 2002). The Niemann Pick type II phenotype is associated with missense mutations, small deletions, and intronic mutations, clustering in the carboxy terminal third of the NPC1 protein, suggesting this region is important in normal protein function (Greer et al., 1999a).

One case of adult-onset NP-II was reported to have a compound heterozygous mutation (Yamamoto et al., 1999). One allele was a missense mutation resulting in a val889met substitution. The other allele contained a splice site mutation in intron 54B that resulted in an in-frame deletion of 18 amino acids. Given the lack of data for adult-onset NP-II, no conclusions can be drawn about allelic variation and phenotype unlike with TSD.

### 3. Pathology

#### a. Hematological

Bone marrow of patients with NP-II contains foamy, vacuolated histiocytes, so-called "Niemann-Pick cells." On Wright-Giemsa stain, bone marrow histiocytes contain blue granules in a cytoplasm rich in free ribosomes, giving the appearance of so-called "sea-blue histiocytes." These blue granules are seen under electron microscopy to be cytosomes composed of lamellar fragments and a granular component. In addition, electron microscopy reveals smaller lamellar cytosomes and larger irregular cytosomes with amorphous and granular portions, often including fingerprint profiles (Golde et al., 1975; Martin et al., 1984; Yan-Go et al., 1984).

#### b. Neurological

Neurons contain abnormally increased stored cholesterol as well as neurofibrillary tangles (NFTs) with paired helical filaments similar to those seen in Alzheimer's disease. These NFTs are not seen in NP-I. Tau protein is present (Auer et al., 1995).

### 4. Pathophysiology

The primary biochemical findings in NP-II are sphingomyelin accumulation, normal sphingomyelinase activity, and a major block in cholesterol esterification (Vanier et al., 1988). The molecular basis for these abnormalities is being elucidated rapidly.

Cholesterol arrives in lysosomes via two pathways. Endogenous cholesterol is synthesized in the smooth endoplasmic reticulum, and then transported via the Golgi complex to late endosomes. The NPC1 peptide, one of a family of membrane-bound proteins containing sterol-sensing domains, travels a similar path. After tubulation and fission, rapid microtubular transport traffics these endosomes to lysosomes, whereupon the two fuse. The sterol-sensing domain of NPC1 peptide appears to be essential in this vesiculotubular transport process.

Exogenous cholesterol from low-density lipoprotein is internalized into lysosomes, where it undergoes esterification and subsequent translocation to other intracellular membrane sites.

In NPC1, there is a defect in the delivery of cholesterol-containing late endosomes from the Golgi complex to the lysosome, presumably due to abnormal vesiculotubular transport resulting from abnormal or absent NPC1 peptide. Unesterified cholesterol accumulates in the Golgi complex (Blanchette-Mackie et al., 1988). Internalization and lysosomal processing of exogenous lipoprotein cholesterol is normal (Pentchev et al., 1985), however, this cholesterol is not translocated from the lysosome to other membrane sites such as the plasma membrane and back to the endoplasmic reticulum (Cruz et al., 2000; Shamburek et al., 1997; Sokol et al., 1988). This suggests that intact NPC1 is not only required for vesiculotubular trafficking of endogenous cholesterol from the Golgi complex to the lysosome, but from the lysosome to other membrane sites.

The defect in lysosomal cholesterol esterification remains unexplained. One possibility it that loss of NPC1 function blocks vesiculotubular transport of a key enzyme from the endoplasmic reticulum to the lysosome.

Accumulation of sphingomyelin has been attributed to excess cholesterol inhibiting lysosomal sphingomyelinase activity.

In the brain NPC1 is present in astrocytic processes, closely associated with nerve terminals. A defect in NPC1-mediated vesicular trafficking in astrocytes may somehow result in nerve terminal degeneration (Patel et al., 1999).

## 5. Diagnosis

### a. Clinical

NP-II is an extremely rare cause of adolescent- or adult-onset movement disorders. However, there are cases presenting with prominent dystonia. The combination of dystonia and supranuclear downward gaze palsy should make one think of NP-II, especially in the presence of ataxia. Hepatosplenomegaly is supportive, although it may be absent. The differential diagnosis includes Wilson's disease, pantothenate kinase associated neurodegeneration (PKAN), juvenile HD, and the spinocerebellar ataxias; however, these are not characterized by prominent supranuclear vertical gaze palsy. The age of onset easily distinguishes this disorder from progressive supranuclear palsy. Pallidopyramidal degeneration is associated with supranuclear vertical gaze palsy, however, the disturbance in that disorder is of upgaze. In addition, it is characterized by spasticity and parkinsonism rather than dystonia and ataxia.

### b. Laboratory

There are no specific noninvasive laboratory tests that will allow confirmation of a suspected diagnosis of NP-II. Laboratory tests to exclude differential diagnoses such as Wilson's disease are appropriate.

### c. Genetic Testing

Mutation analysis is available for the most common pathological allele, ile1061thr, through the National Referral Laboratory for Lysosomal, Peroxisomal, and Related Genetic Disorders, Adelaide, Australia (http://www.health.adelaide.edu.au/NRL/nrl.htm). However, this mutation accounts for less than 15% of of all cases of NPC1 (Millat et al., 1999). Mutation scanning is also available from the same laboratory, at a cost of approximately $1000. Although less invasive than a skin biopsy, the utility of these investigations is unclear, given the low sensitivity of mutation analysis and high cost of mutation scanning.

### d. Imaging

Magnetic resonance imaging (MRI) can demonstrate cerebellar atrophy, but this is not specific for NP II (Lossos et al., 1997). MRI is more useful to exclude certain conditions such as PKAN.

### e. Electrodiagnostic Testing

There are no specific electrodiagnostic tests that are useful in confirming a diagnosis of NP-II.

### f. Biopsy

The most useful test in the diagnosis of NP-II is a skin biopsy, stained with filipin to demonstrate increased intra-vesicular cholesterol (Ledvinova and Elleder, 1993; Vanier et al., 1991a). Filipin is a polyene antibiotic derived from *Streptomyces filipinensis*. Its antibiotic properties derive from its ability to bind to cholesterol in bacterial cell membranes, effecting cytolysis. This cholesterol-binding property, along with natural fluorescence, makes it an excellent cytochemical assay of unesterified cholesterol in NP-II. The cytochemical finding is not specific for NPC, however, as hydrophobic amines (e.g., imipramine) can cause an acquired biochemical defect identical to NPC, with inhibition of cholesterol esterification and increased vesicular cholesterol.

Cultured fibroblasts from a skin biopsy may also be assayed to demonstrate the defect in cholesterol esterification.

Bone marrow biopsy, if performed, will show classic foamy Niemann-Pick cells as well as sea-blue histiocytes, but should not be necessary to make the diagnosis.

## 6. Treatment

There is no disease-modifying treatment for NP-II, nor are there any specific therapies for the neurological complications. However, empiric trials of diazepam, oral baclofen, botulinum toxin, or intrathecal baclofen may be considered for dystonia. As with adult-onset TSD, sedative agents such as diazepam and oral baclofen should be used cautiously, given the possibility of clinical or subclinical cognitive impairment.

## II. DISORDERS OF HEAVY METAL METABOLISM

The more common disorders of calcium, copper, and iron metabolism (Fahr's syndrome, Wilson's disease, PKAN) are discussed in Chapters 38, 31, and 39, respectively. Two rare disorders of heavy metal metabolism, which we shall discuss briefly, are neuroferritinopathy (adult-onset basal ganglia disease) and ceruloplasmin deficiency.

### A. Neuroferritinopathy (OMIM 606159)

Neuroferritinopathy (NFP, adult-onset basal ganglia disease) is a rare autosomal dominant disorder of iron metabolism. It has only been reported in one family and five other individuals from the Cumbria region of Northwest England (Curtis et al., 2001). All affected family members were found living within 25 miles of a common ancestor, identified from genealogical records going back to the 18th century.

### 1. Phenotype

Patients present between 40 to 55 years of age with variable degrees of choreoathetosis, dystonia, spasticity, or rigidity. Cognition and cerebellar function are spared. The extrapyramidal features are similar to those of Huntington's disease (HD) or atypical parkinsonism.

### 2. Genotype (FTL, 19q13.3–qter)

*a. Background*

Mammalian ferritin is a complex construct of heavy ferritin (H-ferritin, 21 kDa) and light ferritin (L-ferritin, 19 kDa). The heavy and light subunits have distinct loci; the gene for H-ferritin is at 11q13 and that for L-ferritin (FTL) is at 19q13.3–qter. Although it was previously thought that there were multiple ferritin light chain genes on at least three chromosomes, the others are pseudogenes caused by retroinsertion of cDNA into the genome. The pseudogenes lack introns, contain residual polyadenylic acid tails proving their mRNA origin, and are not expressed.

L-ferritin and H-ferritin mRNA both contain a non-translated iron-responsive element (IRE) at the 5′ end that controls peptide synthesis at the translational level. When intracellular iron stores are low, a 90-kDa iron-responsive protein (IRP) binds to the IRE, inhibiting synthesis of ferritin. When intracellular iron stores are high, cytosolic iron inactivates the IRP by catalyzing the formation of an intramolecular disulfide bond, resulting in synthesis of H- and L-ferritin.

*b. Previously Reported FTL Mutations*

The first human disease found to be associated with an FTL mutation was hyperferritinemia-cataract syndrome (HCS). HCS is caused by various mutations in the IRE of ferritin light chain mRNA. Eight point mutations and one deletion have been identified (Aguilar-Martinez et al., 1996; Beaumont et al., 1995; et al., Camaschella et al., 2000; Cazzola et al., 1997; Girelli et al., 1995, 1997, 2001; Martin et al., 1998; Mumford et al., 1998). The mutations are predicted to abolish or reduce binding of the IRP to L-ferritin IRE, resulting in constitutive L-ferritin synthesis. As the IRE is normally not translated, the mutations do not affect the structure or function of the L-ferritin that is synthesized, and there are no apparent abnormalities of iron metabolism. Aside from early-onset cataracts dues to lenticular L-ferritin crystallization, individuals are asymptomatic, coming to medical attention when hyperferritinemia is discovered on routine laboratory testing.

*c. FTL Mutations and NFP*

Based upon a possible role of iron in the pathophysiology of basal ganglia degenerations, the ferritin light chain gene was examined as a potential locus for neuroferritinopathy (Curtis et al., 2001). The disorder was mapped by linkage analysis to 19q13.3–q13.4, and an adenine insertion found at position 460–461. The authors found the same mutation in five apparently unrelated subjects with similar extrapyramidal symptoms. The mutation was not identified in over 300 controls from the same geographical region.

### 3. Pathology

Pathological data is scant, with only one neuropathological investigation being described in any detail (Curtis et al., 2001). On gross examination, there is a reddish discoloration of the basal ganglia with cystic cavitation of the lenticular nuclei. Iron- and ferritin-positive spherical inclusions are found in the forebrain, cerebellum, and globus pallidus. These inclusions, up to 50 mm in diameter, are predominantly extracellular, but also co-localize with microglia, oligodendrocytes, and neurons, especially in the pallidum. Axonal swellings are found in the pericystic areas and other white-matter tracts; these are immunoreactive for tau, ubiquitin, and neurofilaments.

## 4. Pathophysiology

### a. Normal Ferritin Structure and Function

Ferritin is the major iron-storage protein. It consists of a hollow shell 130 Å in diameter, composed of 24 H- and L-ferritin subunits, with a 60 Å central core of ferric-hydroxide-phosphate complexes. The core can increase and decrease in size according to physiological need. It usually contains less than 3000 iron atoms, however, the apoferritin shell can accommodate up to 4500 iron atoms. The E helix of L-ferritin appears to be essential for iron core formation.

There is considerable tissue variability in the composition of ferritin: for example, human spleen ferritin is rich in light subunits whereas heart ferritin is rich in heavy subunits. Its primary role is to sequester intracellular iron, providing a ready source of iron when needed and protecting the cell from iron toxicity. Neuronal ferritin is necessary for axonal transport of iron from the soma to the nerve terminal. In addition, ferritin has a particularly important role in neuronal health. Iron is neurotoxic in vitro and in vivo, presumably by increasing the formation of reactive oxygen species, and is a suspected co-factor in several neurodegenerative disorders including Parkinson's disease.

### b. Abnormalities of Ferritin in NFP

The mutation in the FTL gene predicts 22 amino acid substitutions at the C terminal of the FTL peptide, and extension of the peptide by 4 amino acids (Curtis et al., 2001). This in turn is predicted to disrupt the structure of the D helix, DE loop, and E helix. Mutant ferritin, composed of variable amounts of normal H-ferritin, normal L-ferritin, and abnormal L-ferritin, would be expected to be less able to form the normal iron core. Nonsequestered intracellular iron is a known endogenous neurotoxin. One can also speculate that the mutation disrupts axonal iron transport; intra-axonal release of iron from abnormal ferritin could contribute to the observed neuraxonal dystrophy.

## 5. Diagnosis

### a. Clinical

The presentation of a progressive adult-onset movement disorder with prominent chorea, dystonia, spasticity, or parkinsonism should always prompt a careful family history. If individuals in multiple generations are affected, NFP should be considered in the differential diagnosis, along with dentatorubralpallido-luysian atrophy (DRPLA), HD, and familial parkinsonism.

### b. Laboratory

Serum ferritin levels may support the diagnosis. In the original report, eight patients had examination of serum ferritin levels. All were abnormally low, with values ranging from 4 to 16 mcg/L. Iron, hemoglobin, and transferrin levels were normal.

### c. Genetic Testing

There is no commercially available genetic test for neuroferritinopathy at this time.

### d. Imaging

Magnetic resonance imaging demonstrates increased T2-weighted signal in the substantia nigra, globus pallidus, putamen, and central caudate, with decreased signal in the posterolateral caudate (Fig. 44.1). The areas of increased T2 signal correlate with areas of cavitating necrosis seen at autopsy.

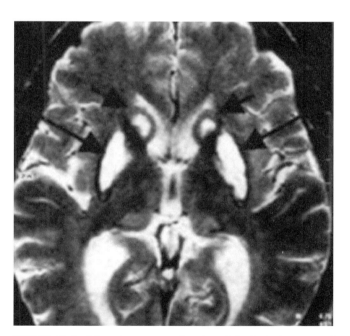

FIGURE 44.1 Magnetic resonance imaging of the brain in a case of neuroferritinopathy. T2-weighted MRI demonstrating increased signal in globus pallidus, putamen, and central caudate, with an area of decreased signal in the posterolateral caudate. The pattern of signal change, reminiscent of that seen in PKAN, correlates with cystic cavitation of the basal ganglia seen at autopsy. Clinical details of this specific case are not available. (From Curtis et al. (2002). Nat. Genet. **28**, 350–354. With permission.)

#### e. Electrodiagnostic Tests

These do not appear to be of diagnostic utility, as neuropathy and myopathy are not features of the disorder. EEG would not be expected to add to the work-up, as the diagnosis rests so heavily on the MRI findings.

#### f. Biopsy

As pathology is limited to the central nervous system, there is no role for tissue biopsy in confirming the diagnosis.

### 6. Treatment

Although there are no specific therapies for the neurological complications of NFP, there do not appear to be any contraindications to empiric trials targeted at specific symptoms. Parkinsonism can be treated with levodopa, dopamine agonists (pergolide, ropinerole, pramipexole), amantadine, deprenyl, or anticholinergic medications (trihexiphenidyl, benztropine). Chorea can be treated with high-potency dopamine antagonists such as haloperidol or atypical neuroleptics such as quetiapine. Dystonia can be treated with botulinum toxin or intrathecal baclofen. Dementia warrants a trial with an anticholinesterase inhibitor (donepezil, rivastigmine, galantamine).

This is a progressive and fatal disorder with no effective disease-altering therapy. Given that iron accumulation is likely to be pathogenic, iron chelation or therapeutic phlebotomy could be considered. However, serum iron levels are normal, and the likelihood of benefit appears low.

## B. Primary ceruloplasmin deficiency (OMIM 604290)

Even before the gene for Wilson's disease (WD) was cloned, a number of patients with neurologic degeneration and low or undetectable serum ceruloplasmin were suspected to have a distinct genetic disorder. Unlike in WD, these patients had normal copper metabolism but abnormal iron accumulation in liver and brain (Miyajima et al., 1987). One of these patients has since been shown to have a mutation in the ceruloplasmin (CP) gene (Harris et al., 1995). Since then additional mutations have been identified. All result in a primary deficiency of ceruloplasmin, unlike the secondary deficiency seen in WD, and may lead to either aceruloplasminemia or hypoceruloplasminemia.

### 1. Phenotype

The disorder presents between age 30 and 50 with a variable mixture of ataxia, hyperreflexia, dysarthria, rigidity, dementia, chorea, blepharospasm, retinal degeneration, diabetes mellitus, and hemosiderosis (Miyajima et al., 1987; Miyajima et al., 2001; Morita 1995).

### 2. Genotype (CP, 3q23–q25)

Consanguinity suggests an autosomal recessive mode of inheritance. Nonsense, frameshift, and splice site mutations have been described, predicting truncated or absent ceruloplasmin peptide (Harris et al., 1995; Okamoto et al., 1996; Takahashi et al., 1996; Yoshida et al., 1995). In addition, a missense mutation has been reported that results in peptide being retained in the endoplasmic reticulum (Hellman et al., 2002). Three cases of partial dominance due to a heterozygous nonsense mutation (TRP858TER) have been reported (Miyajima et al., 2001). The phenotype of this mutation is more benign than the homozygous state, consisting of ataxia, hyperreflexia, and dysarthria without extrapyramidal findings.

### 3. Pathology

Autopsy reveals severe destruction of the basal ganglia as well as the dentate nucleus and Purkinje cells of the cerebellum. There is considerable iron deposition in neurons and glia, particularly in the basal ganglia. (Morita et al., 1995; Miyajima et al., 2001)

### 4. Diagnosis

Ceruloplasmin levels may be absent or markedly reduced, although heterozygotes have levels as high as 36–41% of normal. Serum iron and copper are low, with elevated serum ferritin. Anemia may be present. MRI of the liver reveals abnormal iron deposition, with decreased signal on T1- and T2-weighted images (Morita et al., 1995). MRI of the brain reveals abnormal iron deposition in the caudate, putamen, thalamus, and dentate nucleus (Morita et al., 1995), although heterozygotes may only have cerebellar atrophy (Miyajima et al., 2001).

### 5. Treatment

Given the rarity of the disorder, there are no controlled trials of disease-modifying therapy. One patient treated with repeated administration of fresh frozen plasma containing ceruloplasmin had a reduction of liver iron levels and improvement in neurological symptoms (Yonekawa et al., 1999).

## III. MOVEMENT DISORDERS ASSOCIATED WITH HEMATOLOGICAL DISEASE

### A. Adult-Onset Chediak-Higashi syndrome (OMIM 214500)

This disorder was first reported by Beguez-Cesar, a Cuban pediatrician, in 1943 (Beguez-Cesar, 1943), and subsequently by Chediak and Higashi (Chediak, 1952; Higashi, 1954). Early descriptions focused on the hematological and dermatological manifestations, however, nystagmus was recognized early on as evidence of neurological involvement. The adult-onset phenotype was not described until much later, however. Only recently has the occurrence of extrapyramidal findings in this form received much attention. Specific allelic variants cause the adult-onset form of Chediak-Higashi syndrome (CHS), making this an interesting model of hereditary extrapyramidal disease.

## 1. Phenotype

### a. Childhood-Onset CHS

Typical Chediak-Higashi syndrome is a childhood disorder that presents with decreased pigmentation of hair and eyes (partial albinism); (Boissy and Nordlund, 1997; Oetting and King, 1999), and abnormal susceptibility to pyogenic infection. Ophthalmoscopic examination shows an abnormal optic disc, lack of pigment in the pigment epithelium, and relatively unaffected choroidal pigmentation (BenEzra et al., 1980). About 85–90% of patients develop an accelerated lymphoproliferative phase characterized by fever, jaundice, hepatosplenomegaly, lymphadenopathy, bleeding, pancytopenia, and generalized lymphohistiocytic infiltrates; the prognosis in this phase is grim, with a mean survival of about three additional years. In the past, death would often occur before age 7 due to infection or hemorrhage (Blume and Wolff, 1972), however new therapies such as bone marrow transplantation are improving the prognosis, however.

### b. Adult-onset CHS

Patients who do not develop the accelerated phase tend to have fewer or no infections and can survive to adulthood. Such patients present in the third or fourth decade with a progressive neurological disorder with variable signs, including ataxia, parkinsonism, dystonia, dementia, amyotrophy, oculogyric crisis, or axonal sensorimotor neuropathy (Hauser et al., 2000; Misra et al., 1991; Sheramata et al., 1971; Uchino et al., 1993; Uyama et al., 1991, 1994). The presentation can resemble spinocerebellar degeneration (Sheramata et al., 1971), parkinsonism-dementia (Uyama et al., 1991, 1994), or parkinsonism-dystonia (Hauser et al., 2000). Neutrophil function can be normal in this phenotype.

## 2. Genotype (CHS1, 1q42.1–q42.2)

### a. Gene Mapping

*Beige*, the murine analog of CHS, was found to be closely linked to TCRG (Holcombe et al., 1987) and NID (Jenkins et al., 1991) on mouse chromosome 13. In man, however, TCRG and NID are found on different chromosomes; TCRG is present on chromosome 7p14–p15 and NID at 1q43 (Bensmana et al., 1991; Olsen et al., 1989). Nonlinkage was found between CHS and TCRG; subsequently, homozygosity mapping of 4 inbred probands with childhood CHS showed linkage to 1q42–q44 (Fukai et al., 1996).

Further analysis narrowed the CHS locus to a 5-cM interval on chromosome 1q42.1–q42.2, allowing construction of a YAC contig covering the entire region (Barrat et al., 1996). The gene, named CHS1, was found to be identical to that of lysosomal trafficking regulator (LYST).

CHS1 consists of 55 exons, with 2 alternative 5′ untranslated regions whose sequences are encoded by a pattern of alternative promoters and alternative RNA splicing (Karim et al., 2002). There are 2 major mRNA isoforms of the CHS1 gene, 13.5 and 5.8 kb in size (Barbosa et al., 1997). They differ in sequence at the 3′ end of the coding domain. The smaller isoform results from incomplete splicing with reading through an intron.

### b. Childhood-Onset CHS

The first pathologic mutations in patients with CHS were identified using human cDNA homologous to the mouse beige gene (Nagle et al., 1996). In childhood-onset CHS, both homozygous and compound heterozygous mutations of CHS1 are found. All are nonsense or frameshift mutations predicted to result in either truncated or absent protein. (Barbosa et al., 1996, 1997; Dufourcq-Lagelouse et al., 1999; Karim et al., 1997; Nagle et al., 1996). In two patients with the accelerated phase of childhood-onset CHS, analysis of lymphoblastoid mRNA showed loss of the large mRNA transcript with normal amounts of the small mRNA transcript; it appears that the small mRNA isoform cannot complement CHS1 (Barbosa et al., 1997).

### c. Adult-Onset CHS

There is a significant correlation between allelic variation and age of onset in CHS. Patients with severe childhood-onset CHS tend to have functionally null mutant CHS1 alleles, whereas patients with adult-onset CHS are more likely to have missense mutant alleles; presumably the missense mutations encode CHS1 polypeptides with partial function (Karim et al., 2002). At least one nonsense mutation of CHS1 has been found to cause adult-onset CHS, however (Nagle et al., 1996). In addition, some cases of CHS may have a locus distinct from CHS1; in one investigation, homozygosity mapping of inbred patients with adult-onset CHS showed nonlinkage to markers in 1q42–q44 (Fukai et al., 1996).

## 3. Pathology

### a. Hematological Pathology

Peripheral leukocytes are characterized by giant lysosomal granules. Bone marrow reveals large eosinophilic, peroxidase-positive inclusion bodies in myeloblasts and promyelocytes (Windhorst et al., 1966).

### b. Other Tissue

Giant inclusion bodies are seen in all granule-containing cells, including granulocytes, lymphocytes, melanocytes, mast cells, neurons, renal cells, and conjunctival stromal fibroblasts (BenEzra et al., 1980; Fukai et al., 1996; Windhorst et al., 1966). In addition, abnormal giant melanosomes can be seen in retinal pigmentary cells (BenEzra et al., 1980).

### c. Neuropathology

Muscle biopsy shows neurogenic muscle atrophy due to peripheral neuropathy as well as widespread acid phosphatase-positive granules. With electron microscopy, these granules can be seen to be autophagic vacuoles containing glycogen particles and membranous structures (Uchino et al., 1993). If sensory impairment is present, sural nerve biopsy can reveal a loss of large myelinated nerve fibers and unmyelinated axons. The neuropathy appears to be axonal; electron microscopy and teased-fiber studies show normal myelination (Misra et al., 1991).

## 4. Pathophysiology

### a. Hematological Overview

CHS1 is caused by mutations in the gene coding for LYST. LYST is structurally similar to the yeast vacuolar sorting protein VPS15 (Nagle et al., 1996) and appears to be required for microtubule-dependent sorting of endosomal resident proteins into late multivesicular endosomes in leukocytes (Faigle et al., 1998). Defective LYST function results in a defect of the secretory lysosome pathway that governs endocytic protein regulation.

### b. Lymphocyte Pathophysiology

Cytotoxic T-lymphocyte-associated antigen-4 (CTLA4) is a membrane protein that plays a major role in the regulation of T-cell activation (Noel et al., 1996). CTLA4 is highly regulated by the endocytotic lysosome pathway. In CHS1, expression of CTLA4 on T cell membranes is abnormal (Barrat et al., 1999). In addition, peptide loading onto major histocompatibility complex class II molecules and antigen presentation are strongly delayed (Faigle et al., 1998). These findings may explain abnormal natural killer cell function in CHS (Roder et al., 1980, 1983). In addition, reduced T-cell membrane CTLA4 expression may play a role in the lymphoproliferative disease that characterizes the accelerated phase of CHS (Barrat et al., 1999).

### c. Neutrophil Pathophysiology

Neutrophils contain two principal types of granules, azurophil and specific (Gullberg et al., 1999). Azurophil granules appear early in neutrophil development and contain lysosomal enzymes, lysozyme, and myeloperoxidase. Cathepsin G, elastase, and defensins are major constituents of the azurophil granule. In CHS there is a deficiency of cathepsin G and elastase, whereas defensins are normal or only mildly decreased (Ganz et al., 1988). These abnormalities appear relevant to the immunocompromise found in patients with CHS.

### d. Neuronal Pathophysiology

Little is known about the role of the CHS1 peptide in the pathophysiology of neuronal involvement in CHS. The fundamental abnormality in CHS is a disturbance of microtubule structure and function (Oliver and Zurier, 1976). Disturbances of the microtubule associated protein TAU are important in degenerative disorders such as Alzheimer's disease (AD), progressive supranuclear palsy (PSP), and corticobasal degeneration (CBD). It is therefore tempting to speculate that patients with missense mutations in the CHS1 gene have an abnormality of neuronal microtubule function that leads to the slow degeneration of specific neuronal populations, similar to other neurodegenerative disorders with abnormal microtubule function.

## 5. Diagnosis

### a. Clinical

Adult-onset disorders of the basal ganglia associated with frequent pyogenic infections and partial albinism should raise the suspicion of CHS, especially with a history of affected siblings or consanguinity. However, immunological function may be normal, and the diagnosis may depend on further laboratory investigation.

### b. Laboratory

The pathognomonic cytoplasmic giant granules can be seen in peripheral blood leukocytes. Natural killer cell function is typically dramatically reduced or absent. This assay is of unclear utility, however, aside from academic interest, if giant granules are seen in peripheral leukocytes in a patient with partial albinism. Renal cells can be obtained from urinary sediment and examined, providing a non-invasive source of tissue for additional examination (Barrat et al., 1999).

### c. Genetic Testing

Genetic testing is not commercially available at this time.

### d. Imaging

MRI in one case of parkinsonism-dementia showed marked temporal dominant brain atrophy and diffuse spinal cord atrophy (Uyama et al., 1994).

### e. Electrodiagnostic Studies

In patients with sensory neuropathy nerve conduction studies and sural nerve biopsy are consistent with axonal neuropathy (Misra et al., 1991). Abnormalities of visual-evoked responses and electroretinography have been reported (BenEzra et al., 1980).

### f. Biopsy

Various tissues may be biopsied to look for giant inclusion bodies, including skin, conjunctiva, and bone marrow, however, such invasive procedures should not be necessary to make the diagnosis.

## 6. Treatment

### a. Treatment of Immune Compromise

Treatment of infection is based upon the site of infection and infectious agent. Prophylactic antibiotic therapy is indicated in certain situations. G-CSF treatment of one patient resulted in normalization of the white blood cell count, including the differential count, and a reduction in the number and severity of infections (Baldus et al., 1999).

Ascorbic acid has been shown to correct certain functional abnormalities of neutrophils (Boxer et al., 1976, 1979), and cholinergic agonists have shown to correct microtubule function and granule morphology *in vitro* (Oliver et al., 1976; Oliver, 1976; Oliver and Zurier, 1976). Neither of these therapies has been studied in humans with CHS, however.

### b. Treatment of Lymphoproliferative Phase

High-dose methylprednisolone may result in remission of up to 11 months, however, after relapse retreatment may not be successful (Aslan et al., 1996). High-dose intravenous gamma globulin has also been reported to be beneficial (Kinugawa and Ohtani 1985). Splenectomy can prolong survival for up to 29 months. Bone marrow transplantation performed prior to the accelerated phase can improve survival dramatically (Bujan et al., 1992; Filipovich et al., 1992; Haddad et al., 1995; Mottonen et al., 1995; Rappeport et al., 1983; Sullivan et al., 2000).

### c. Treatment of Neurological Symptoms

Although there are no specific therapies for the neurological complications of CHS1, parkinsonism may respond to levodopa (Hauser et al., 2000) and oral steroids may help the neuropathy (Fukuda et al., 2000). In addition, empiric therapy of parkinsonism with dopamine agonists (pergolide, ropinerole, pramipexole), amantadine, deprenyl, or anticholinergic medications (trihexiphenidyl, benztropine) is reasonable. Dystonia can be treated with botulinum toxin. Dementia can be treated with an anticholinesterase inhibitor (donepezil, rivastigmine, galantamine).

With improved management of childhood CHS and prolonged survival, we must be prepared to deal with the effects of the CHS1 null phenotype on neuronal function. If altered CHS1 peptide can cause adult-onset neurological disease, it may turn out that children with null CHS1 survive the immunodeficient and immunoproliferative phases only to succumb to progressive neurological impairment later in life. Children "cured" of CHS after bone marrow transplantation therefore warrant close observation for potential late-onset neurological complications.

## B. Choreoacanthocytosis (OMIM 200150)

There are three distinct clinical syndromes characterized by the presence of acanthocytes in the peripheral blood and a progressive movement disorder. These include autosomal recessive choreoacanthocytosis (CHAC), X-linked choreoacanthocytosis (McLeod syndrome), and the syndrome of hypo-prebetalipoproteinemia, acanthocytosis, retinitis pigmentosa, and pallidal degeneration (HARP syndrome). Abetalipoproteinemia (Bassen-Kornzweig syndrome, microsomal triglyceride transfer protein deficiency, MTP deficiency), characterized by acanthocytosis, ataxia, neuropathy, and retinitis pigmentosa (Stevenson and Hardie, 2001) is not associated with basal ganglia dysfunction.

In 1967 Critchley et al. reported on a family with an adult form of acanthocytosis associated with a progressive neurological disorder (Betts et al., 1970; Critchley et al., 1967, 1970). Five of ten siblings were affected by a variable combination of chorea, dystonia, dementia, psychosis, hypotonia and areflexia, beginning between ages 24 and 31, leading to death or severe disability within a few years of symptom onset. Mutilation of the tongue, cheek, and lips was common, apparently due to severe orobuccolingual dyskinesia with prominent tongue thrusting. Ataxia was not seen in these siblings. Peripheral acanthocytosis was demonstrated in both living affected siblings, as well as two of the five otherwise normal siblings.

Although the pedigree was interpreted as exhibiting an autosomal dominant mode of inheritance, it is possible that this family actually had the autosomal recessive disorder we now know as CHAC. First, neither parent was affected. Second, the only other affected family member was a niece with childhood-onset ataxia, areflexia, extensor plantar responses, dorsal column sensory loss, equinovarus foot deformity, and cardiomyopathy. She had only "occasional" acanthocytes, and no chorea, orobuccolingual dyskinesia, or dystonia. This phenotype was so dissimilar to other family members that one must consider the possibility that she had a distinct genetic disorder, e.g., Friederich's ataxia. Finally, although the mild acanthocytosis seen in the two neurologically normal siblings could have been due to reduced penetrance of an autosomal dominant disorder, it could also have been due to semidominance of acanthocytosis in an autosomal recessive disorder, a phenomenon that has in fact been demonstrated in other families with CHAC (Ueno et al., 2001).

Cederbaum (1971) reported the first clearly autosomal recessive pedigree: three children of a consanguineous relation with adult-onset dementia, chorea, and acanthocytosis. All subsequent pedigrees have been either autosomal recessive (Alonso et al., 1989; Bird et al., 1978; Rinne et al., 1994; Sorrentino et al., 1999; Vance et al., 1987; Villegas et al., 1987; Yamamoto et al., 1982) or X-linked (see Section III.C).

### 1. Phenotype

The clinical features of CHAC are those of a progressive neurological disorder, usually beginning between age 25 and 45, associated with acanthocytosis and normal serum lipoproteins. Neurological involvement is variable, and may include chorea, tics, ataxia, neurogenic amyotrophy, involuntary biting of the tongue, lips, and cheeks or a frontal subcortical dementia (Aminoff, 1972; Betts et al., 1970; Critchley et al., 1967, 1970; Kartsounis and Hardie, 1996; Spitz et al., 1985). Cases with prominent parkinsonism, dystonia or supranuclear ophthalmoplegia have been reported (Spitz et al., 1985). Muscle enzymes may be elevated, but unlike McLeod syndrome there is no myopathy.

### 2. Genotype (CHAC, 9q21)

Rubio et al. (1997) demonstrated linkage of CHAC to a 6-cM region of 9q21 in 11 families from diverse geographic areas, suggesting that CHAC was due to a single locus. The recently identified gene, named CHAC, (Rampoldi et al., 2001; Ueno et al., 2001) consists of an open reading frame of 9525 nucleotides organized into 73 exons. CHAC encodes two splice variants of a novel 3174 amino acid protein that has been named "chorein," Seventeen distinct mutations have been identified in individuals with CHAC. Rampoldi et al. (2001) found 16 different mutations. One of these, a compound heterozygote, was due to a 269T-A transversion in exon 4 (resulting in an ile90lys amino acid change) and a T insertion between nucleotides 6404 and 6405 in exon 48 (resulting in a frameshift mutation). Ueno et al. (2001) studied three Japanese families from a small geographic area with a founder effect. Affected homozygotes had a 260-bp deletion in a coding region of the cDNA, leading to a frameshift and production of truncated protein.

### 3. Pathology

#### a. Hematological

The hallmark of CHAC is the presence of acanthocytes on light microscopy. Transmission electron microscopy (TEM) of CHAC acanthocytes shows a spectrin network densely packed on the inner membrane surface, accumulation of spectrin in the thorn region, and regional variability in the compactness of the membrane structural protein meshwork (Hosokawa et al., 1992) (Terada et al., 1999).

#### b. Neurological

The neuropathology of CHAC is characterized by marked loss of small neurons in the caudate and putamen with associated gliosis (Alonso et al., 1989; Bird et al., 1978; Hardie et al., 1991; Iwata et al., 1984; Malandrini et al., 1993; Sato et al., 1984). Neuronal loss has also been demonstrated in the pallidum and substantia nigra (Hardie et al., 1991; Rinne et al., 1994).

Although the anterior horn of the spinal cord appears normal (Hardie et al., 1991), an axonal motor neuropathy has been demonstrated (Sato et al., 1984; Sobue et al., 1986). Muscle biopsy demonstrates neurogenic muscular atrophy (Alonso et al., 1989; Limos et al., 1982; Malandrini et al., 1993; et al., Ohnishi et al., 1981; Vita et al., 1989), although pseudomyopathic changes have also been found (Limos et al., 1982). Sural nerve biopsy reveals loss of large myelinated fibers and pathology of the axonal cytoskeleton (Hardie et al., 1991; Malandrini et al., 1993; Ohnishi et al., 1981; Sobue et al., 1986).

### 4. Pathophysiology

#### a. Hematological

Chorein is a evolutionarily conserved polypeptide that may be involved in protein sorting (Ueno et al., 2001). Although little is known about its role in normal erythrocyte structure and function, prior biochemical and ultrastructural investigations of the erythrocyte membrane in CHAC may provide clues (Araki et al., 1983; Asano and Oshima, 1990; Bosman and Kay, 1990; Gross et al., 1982; Hardie, 1998; Hardie et al., 1991; Hosokawa et al., 1992; Ishikawa et al., 2000; Limos et al., 1982; Ohnishi et al., 1981; Oshima et al., 1985; Sobue et al., 1986; Terada et al., 1999; Ueno et al., 1982; Vita et al., 1989).

Despite an early report of an abnormally high amount of a 100,000 molecular weight range RBC membrane protein (Copeland et al., 1982), other investigators have found overall protein composition to be normal (Hosokawa et al., 1992). One group found increases in membrane protein phosphorylation, specifically Ser/Thr phosphorylation, casein-kinase activity, and poly—(Glu, Tyr)—kinase activity (Olivieri et al., 1997).

Recent investigations have focused on the structure and function of the erythrocyte membrane glycoprotein Band 3. Band 3 is involved in the exchange of anions and glucose across the erythrocyte cell membrane, plays a central role in carbon dioxide metabolism, anchors the erythrocyte membrane skeleton, and is the precursor of erythrocyte senescent cell antigen (SCA); (Kay *et al.*, 1990c; Palumbo *et al.*, 1986) In CHAC, Band 3 appears to have normal molecular weight and concentration, and there are no obvious mutations of the Band 3 gene EPB3 (Olivieri *et al.*, 1997). Nevertheless, there is evidence that Band 3 is structurally abnormal in CHAC acanthocytes: the protein undergoes accelerated self-digestion (Asano *et al.*, 1985), is associated with increased fragmentation on immunoblotting (Bosman *et al.*, 1994; Kay *et al.*, 1990a, b), and has increased Tyr-phosphorylation in the cytoplasmic domain (Olivieri *et al.*, 1997), Furthermore, there are conformational abnormalities where Band 3 joins with cytoplasmic spectrin (Asano *et al.*, 1985). The cause of these structural changes is not known.

Functional abnormalities of Band 3 have also been found, with some investigators reporting decreased and others increased sulfate transport activity (Bosman *et al.*, 1994; Kay *et al.*, 1990a, b; Olivieri *et al.*, 1997). Of some interest in the observation of structural and functional abnormalities of Band 3 in patients with the neurological phenotype and normal acanthyocytes, suggesting a subclinical disorder of Band 3 that may be diagnostically useful in acanthocyte-negative patients (Bosman *et al.*, 1994).

Some investigators have found erythrocyte membrane lipids to be normal (Olivieri *et al.*, 1997; Villegas *et al.*, 1987). Others have found alterations in specific lipids, including an increased sphingomyelin/glycerophospholipid ratio, an increase in palmitic acid, and decrease in stearic acid covalently bound to membrane proteins, and decreased cholesterol and phospholipid levels (Clark *et al.*, 1989; Sakai *et al.*, 1991a).

Acanthocytic erythrocytes have impaired deformability with an increase in whole blood viscosity, and although this has been attributed to increased numbers of dehydrated cells containing high concentrations of hemoglobin (Clark *et al.* 1989), it is more likely due to a decrease in the fluidity of the interior of the erythrocytic membrane (Asano *et al.*, 1985; Villegas *et al.*, 1987).

### b. *Neurological*

Little is known about the pathophysiology of the movement disorder of CHAC. Hopefully, this will change now that the abnormal protein, chorein, has been identified. Two possibilities are a neurotoxic effect of mutated protein or an autoimmune process. The latter hypothesis is intriguing, as antibodies to both erythrocyte and brain Band 3 have been demonstrated in patients with CHAC. Similar immunoreactivity has been shown in clinically suspected individuals with and without acanthocytes (Bosman *et al.*, 1994). The primary antigen for this immunoreactivity is unknown, but it may reflect an immune response to structurally abnormal erythrocyte Band 3 protein. It is possible that antibrain Band 3 antibody is toxic to selected neuronal populations, by either an immune- or receptor-mediated mechanism. Band 3 is homologous to the GABA receptor. Anti-GABA receptor immunoreactivity could interfere with GABA inhibition, causing chorea and/or enhancing excitotoxic cell injury.

GM1 immunoreactivity has also been demonstrated, perhaps explaining the axonal sensorimotor neuropathy seen in this disorder (Hirayama *et al.*, 1997).

## 5. Diagnosis

### a. *Clinical*

The key to the diagnosis of CHAC is a history of a progressive neurological disorder beginning in young adulthood. The clinical presentation may mimic Huntington's disease (HD), however orobuccolingual dyskinesias are prominent, often severe enough to result in oral mutilation. A family history of affected siblings or consanguinity supports the diagnosis, as does the absence of the HD CAG repeat expansion.

### b. *Laboratory*

Laboratory screening in cases of suspected CHAC should include a peripheral blood smear. A fresh Wright stain should be examined, looking for peripheral acanthocytosis. Examination of a wet field preparation may increase the detection of acanthocytes. The degree of acanthocytosis should be quantified: greater than 3% acanthocytes is considered abnormal. If acanthocytes are present, the McLeod phenotype should be excluded by blood typing, and abetalipoproteinemia should be excluded by serum protein electrophoresis (SPEP). Other cause for acanthocytosis can be ruled out by physical examination and history.

The absence of acanthocytes should not exclude the diagnosis from consideration. Sorrentino *et al.* (1999) reported a patient with a behavioral disorder and orofaciolingual dyskinesia beginning at age 20, seizures from 28 to 33, and tongue biting and amyotrophy at 33. Blood smears at age 33 were normal; three years later a fresh Wright stain showed 51% acanthocytes. Whether this represents progression or temporal variability of acanthocytosis is unclear.

Investigations in possible acanthocyte-negative cases include studies of erythrocyte membrane structure and function and antibrain immunoreactivity against Band 3 protein. There is insufficient data on the specificity and sensitivity of these tests, however, and they are not commercially available.

GM1 ganglioside antibody testing is readily available, however, and should be obtained in all cases with clinical evidence of axonal neuropathy. If axonal neuropathy is absent or questionable, GM1 antibody testing may still be useful; if positive it supports the diagnosis of CHAC.

### c. Genetic Testing

Although there is no commercially available genetic test for CHAC at this time, it is reasonable to exclude HD and DRPLA.

### d. Imaging

Imaging studies may help to clarify the diagnosis. Computerized axial tomography (CT) demonstrates caudate atrophy, but this alone does not distinguish CHAC from HD (Hosokawa *et al.*, 1992; Kutcher *et al.*, 1999; Requena, *et al.*, 2000; Serra *et al.*, 1987). Early explorations of MRI of the brain also showed nonspecific atrophy of the caudate and putamen (Vance *et al.*, 1987). More recent efforts with high-field MRI reveal increased signal intensity in the caudate head and putamen on T2-weighted images, with scattered T2-bright/T1-dark spots in the striatum (Tanaka *et al.*, 1998); (Fig. 44.2). This pattern is quite distinct from that seen in HD, and may be pathognomonic for CHAC.

Positron emission tomography shows significantly reduced regional cerebral blood flow (rCBF) and oxygen metabolism in the caudate and putamen. Less marked but still significant findings include reduction of rCBF in bilateral frontal, left temporal and parietal, and bilateral thalamic areas, and reduced oxygen metabolism in bilateral frontal and left temporal areas (Hardie, 1998; Tanaka *et al.*, 1998).

### e. Electrodiagnostic testing

The disorder is typically associated with a sensorimotor axonal neuropathy (Spitz *et al.*, 1985; Vita *et al.*, 1989).

### f. Biopsies

Muscle biopsy should not be necessary, but if performed would be expected to be consistent with neurogenic atrophy.

## 6. Treatment

### a. Pharmacological

CHAC is a rare disorder, and there are no controlled clinical trials of specific treatments. As with other forms of acanthocytosis, the red blood cell abnormality is usually asymptomatic and specific treatment is not required. Chorea, if disabling, could presumably be treated with typical neuroleptics such as haloperidol or atypical neuroleptics such as clozapine (Wihl *et al.*, 2001). Other basal ganglia symptoms could also be treated symptomatically, e.g., botulinum toxin injections for disabling focal dystonia and levodopa or dopamine agonist therapy for parkinsonism.

### b. Surgical

There are limited reports of neurosurgical intervention in patients with chorea or trunk spasms. A patient with severe

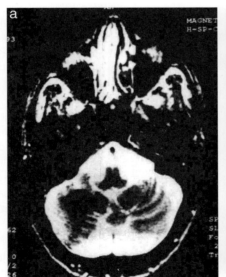 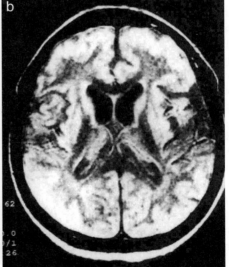

**FIGURE 44.2** Magnetic resonance imaging of the brain in a case of choreoacanthocytosis. (a) T2-weighted 1.5T MRI (t-rep 0.5 sec, t-echo 0.015 sec) demonstrating atrophy of the caudate head, diffusely increased signal in the putamen > caudate head, and scattered bright spots in the putamen. (b) T1-weighted 1.5T MRI (t-rep 2.5 sec, t-echo 0.09 sec) demonstrating low signal intensity in corresponding areas of the putamen, particulary in the areas seen on T2 as scattered bright spots. The patient presented at age 33 with chorea of the shoulder and neck spreading to the limbs and trunk. Other features were severe orobucco-lingual dyskinesias with vocalizations, areflexia in the lower extremities, elevated CPK, dementia, and seizures. MRI was obtained at age 40. (From Tanaka *et al.* (1998). *Mov. Disord.* **13**, 100–107. With permission.)

chorea underwent staged bilateral ablation of the posteroventral globus pallidus pars interna (GPi) with marked improvement (Fujimoto *et al.*, 1997). A second patient with severe chorea failed a trial of bilateral deep brain stimulation (DBS) of the GPi (Wihl *et al.*, 2001). A third patient with severe trunk spasms responded dramatically to DBS of the posteroventral oral nucleus of the thalamus, although dysarthria and hypotonia did not improve (Burbaud *et al.*, 2002). Because of the extremely limited data, the optimal target and treatment approach are not clear. However, there may be a role for ablation or DBS in carefully selected patients with disabling chorea.

## C. McLeod Syndrome (OMIM 314850)

McLeod syndrome is an X-linked disorder characterized by acanthocytosis, amyotrophy, and variable neurological involvement, associated with absence of the XK antigen (previously known as Kx or Kell blood group "precursor substance") on erythrocytes. The abnormality of XK antigen was first described in a blood donor named Hugh McLeod. Understanding of the phenotype has evolved since this initial observation, with gradual recognition of acanthocytosis (Wimer *et al.*, 1977), hemolytic anemia (Symmans *et al.*, 1979), myopathy (Marsh *et al.*, 1981), areflexia (Ueno *et al.*, 1982) and chorea (Faillace *et al.*, 1982) as important components.

Although myopathy and chorea were not attributed to the phenotype until the early 1980s, the first published report of McLeod syndrome with neurological involvement may have actually been the Levine pedigree (Levine *et al.*, 1968). Although an AD pattern of inheritance was inferred, no family member had more than 20% acanthocytes, and most had less than 3%. Neurological symptoms reported included "muscular atrophy of the Charcot-Marie-Tooth type, neck and shoulder girdle atrophy, choreiform movements of the Huntington's type, dementia, and nonfocal grand mal seizures." One patient was "described as a case of schizophrenia." Although a superficial inspection of the pedigree suggests 11 family members were affected in multiple generations, no details are provided regarding neurological findings in specific individuals.

Engel studied the same family and found 60–80% acanthocytes in two neurologically affected brothers and 40% acanthocytes in four neurologically normal females (Critchley *et al.*, 1967). These two males were reported to look like "Charcot-Marie-Tooth patients with tics" and to have had a relatively benign clinical course. Examination and EMG were consistent with lower motor neuron disease. Comparison with Levine's original paper reveals that the two descriptions of the same family are inconsistent. Unfortunately, there are no published reports that allow for further insight into the neurological findings of the original Levine pedigree. This begs the question of which family members truly had the neurological and hematological phenotype. It is intriguing that the description of Engel is consistent with an X-linked pattern of inheritance, erythrocyte mosaicism, and resultant semidominance for acanthocytosis, a pattern suggestive of McLeod syndrome (Symmans *et al.*, 1979). It is therefore conceivable that Levine's family, widely cited as an example of AD CHAC, may have actually been the first report of neurological involvement in McLeod syndrome.

### 1. Phenotype

The phenotype of McLeod syndrome is more benign and CNS involvement less common than in CHAC, although there is considerable overlap between the two disorders.

The most common neurological manifestations are peripheral, and include a type II fiber myopathy and areflexia due to an axonal peripheral neuropathy (Danek *et al.*, 1990, 2001a; Hanaoka *et al.*, 1999; Ho *et al.*, 1996; Jung *et al.*, 2001b; Swash *et al.*, 1983). Creatine phosphokinase and lactate dehydrogenase are typically elevated (Marsh *et al.*, 1981). The myopathy may be subclinical, manifested only by elevated muscle enzymes, but can occasionally be severe (Kawakami *et al.*, 1999; Supple *et al.*, 2001).

Neuropsychiatric disorders are common, including dementia, depression, bipolar disorder, and personality disorder (Danek *et al.*, 1994, 2001a; Hanaoka *et al.*, 1999; Ho *et al.*, 1996; Jung *et al.*, 2001a; Malandrini *et al.*, 1994). Seizures have been reported (Danek *et al.*, 2001a). Cardiomyopathy is common (Malandrini *et al.*, 1994)

Extrapyramidal involvement is common, particularly chorea and involuntary vocalizations or tics. Dystonia and parkinsonism have also been reported (Danek *et al.*, 1994, 2001a, b; Hanaoka *et al.*, 1999; *et al.*, Ho *et al.*, 1996; Malandrini *et al.*, 1994). In severe cases the phenotype can resemble HD or CHAC.

### 2. Genotype (XK, Xp21)

*a. Contiguous Gene Deletion Syndromes*

Some patients with chronic granulomatous disease (CGD) also have acanthocytosis and hemolysis due to lack of XK in red cells. This disorder, CGD II, was found to be due to a combined deletion of the McLeod and CGD loci on Xp21 (Frey *et al.*, 1988). Various combinations of the McLeod, CGD, and Duchenne muscular dystrophy (DMD) phenotypes were shown to be caused by contiguous gene deletions. Analysis of these deletions showed that the McLeod locus was within 500-kb of the CGD locus, in the direction of the DMD locus (Bertelson *et al.*, 1988).

Ho *et al.* (1992) were able to limit the McLeod locus to a 150- to 380-kb segment of Xp21 by constructing a long-range restriction map of the region, identifying CpG-rich islands, and using a new marker (DXS709). Creation of a cosmid contig of 360 kb encompassing this locus allowed detection of a 50-kb deletion containing two transcription units. One of these correlated closely with the McLeod phenotype and was given the designation "XK" (Ho *et al.*, 1994).

### b. Other Mutations

The XK gene is organized into three exons. Since identification of the locus, isolated mutations of the XK gene have been identified, including frameshift and nonsense mutations (Danek *et al.*, 2001a; Hanaoka *et al.*, 1999; Ho *et al.*, 1996; Jung *et al.*, 2001a; Supple *et al.*, 2001), point mutations in the donor and acceptor splice sites of intron 2 (Ho *et al.*, 1994), major deletions, and a missense mutation (Danek *et al.*, 2001a).

## 3. Pathology

### a. Muscle

Muscle biopsy demonstrates subtle myopathic changes, including central nucleation, fiber size variation, and sparce basophilic, necrotic fibers (Danek *et al.*, 2001a) Type II fiber atrophy is readily apparent (Jung *et al.*, 2001b), with negative immunological studies for dystrophin (Danek *et al.*, 1990).

### b. Nerve

Sural nerve biopsy demonstrates axonal and myelin loss, consistent with a sensory axonal neuropathy (Danek *et al.*, 2001a).

## 4. Pathophysiology

### a. Normal Function of XK

XK is a 37kDa integral membrane polypeptide that spans the erythrocyte membrane 10 times (Ho *et al.*, 1994). XK is a blood group antigen in its own right, expressing a single antigen determined by variant alleles at the XK locus. Previously known as Kx or Kell blood group "precursor substance," XK is not a precursor for Kell antigen, and the Kx label has been abandoned. Rather, it binds Kell antigen by a single disulfide bond between Kell Cys72 and XK Cys347, thereby attaching Kell to the erythrocyte surface. (Khamlichi *et al.*, 1995; Russo *et al.*, 1998). Thus, one function of XK may be to facilitate Kell-erythrocyte interactions. XK may also function as a chaperone or membrane transport protein (Ho *et al.*, 1994).

### b. Normal Function of Kell Glycoprotein

Kell is a 93-kDa type II glycoprotein coded by a single gene, KEL, which spans 19 exons. Mutations in the KEL gene result in single base changes that encode different amino acids resulting in various antigenic blood group phenotypes (Lee *et al.*, 1995). At least 23 antigens are associated with the Kell glycoprotein. The role of Kell in normal erythrocyte function, if any, however, is unclear, as patients with the Kell null phenotype (K0) have normal appearing erythrocytes.

Kell is an enzyme, homologous with zinc endopeptidases (Lee *et al.*, 1999). It preferentially cleaves big endothelin-3, an inactive precursor, to produce endothelin-3, a potent 21 amino acid peptide. Endothelins are potent vasoconstrictors, can act as mitogens, and are important in the migration of neural crest derived cells in normal development. Hereditary endothelin-3 deficiency may result in several congenital abnormalities, including idiopathic congenital central hypoventilation syndrome (Gozal and Harper, 1999), Hirschsprung's disease (Kusafuka and Puri, 1997), and bilateral congenital sensorineural hearing loss. However, the Kell null phenotype is not associated with any clinically significant abnormality, suggesting that Kell is not required for normal endothelin-3 synthesis.

Given that the Kell null phenotype is not associated with neurological or other clinically significant abnormality, it seems unlikely that Kell is involved in the pathophysiology of McLeod syndrome.

### c. XK and Acanthocytosis

Pathological mutations in the XK gene result in erythrocytes that lack the XK protein. These erythrocytes lack XK antigenicity, lack Kell antigenicity, are acanthocytic, and are prone to hemolysis. As Kell null erythrocytes are morphologically and functionally normal, it is unlikely that acanthocytosis in McLeod syndrome is due to lack of Kell. This suggests that XK may play a primary role in the maintenance of the normal erythrocyte membrane skeleton.

### d. XK and Myopathy

Jung *et al.* (2001b) studied the immunohistochemistry of Kell and XK. In normal muscle XK is localized to the sarcoplasmic reticulum of type muscle 2 fibers. In McLeod myopathy there is only a weak background signal without a specific staining pattern for XK. This suggests that the type 2 fiber atrophy seen in McLeod myopathy is related to lack of XK expression as opposeed to the axoral motor neuropathy. XK may play a role in the maintenance of normal myocyte structure and/or function.

### e. XK and Neuronal Pathophysiolog

There are high levels of XK expression in the brain, and it may be that XK has a role in maintenance of normal neuronal function. Similarities between the phenotype of CHAC and McLeod syndrome suggest a possible relationship between chorein, erythrocyte band 3 and XK.

## 5. Diagnosis

### a. Clinical

McLeod syndrome should be considered in any patient with an adult-onset myopathy of unknown etiology. The diagnosis becomes more important to consider if there is neuropathy, family history, neuropsychiatric features, seizures, or extrapyramidal findings such as prominent orobuccolingual dyskinesia, chorea, or tics. From the perspective of the movement disorder specialist, any patient with an adult-onset extrapyramidal disorder of unknown etiology should be examined closely for the presence of myopathy, especially if chorea is prominent. If clinical myopathy is present, further investigation for McLeod syndrome is warranted.

### b. Laboratory

Routine laboratory investigation may reveal acanthocytosis, normal serum lipoproteins, hemolytic anemia, or an elevated CPK. Blood group analysis will reveal the absence of the erythrocyte XK antigen, essentially diagnostic for the disorder.

### c. Genetic Testing

There is no commercially available DNA test for McLeod syndrome at this time; however, as the absence of erythrocyte XK antigen is diagnostic, there is no obvious need for such a test.

### d. Electrodiagnostic

EMG may reveal myopathic or neuropathic changes (Danek et al., 2001a; Swash et al., 1983). Nerve conduction studies are consistent with an axonal sensorimotor neuropathy, with normal conduction velocities and reduced amplitudes (Danek et al., 2001a).

### e. Imaging

Magnetic resonance imaging reveals atrophy of the caudate and putamen (Danek et al., 1994, 2001a; Jung et al., 2001a; Malandrini et al., 1994; Oechsner et al., 2001). A rim of increased T2-signal may be seen in the lateral putamen (Fig. 44.3); (Danek et al., 2001b), however, this same finding has been reported in a case of CHAC (Tanaka et al., 1998). SPECT imaging has shown reduced striatal dopamine D2-receptor binding (Danek et al., 1994). Positron emission tomography shows reduced striatal 2-fluoro-2-deoxy-glucose (FDG) uptake (Jung et al., 2001a; Oechsner et al., 2001).

Female carriers have variably reduced striatal FDG uptake without structural abnormalities, suggesting a semi-dominant effect at the neuronal level and raising the question of subclinical CNS dysfunction in female heterozygotes (Jung et al., 2001a)

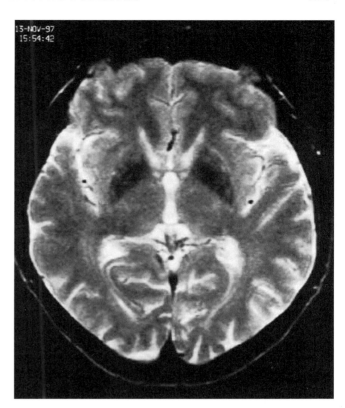

FIGURE 44.3 Magnetic resonance imaging of the brain in a case of McLeod syndrome with chorea. T2-weighted MRI, axial, demonstrating decreased signal of the globus pallidus and putamen bilaterally. There is a thin rim of increased signal at the lateral edge of the putamen, more pronounced on the right. Symptoms began at age 38 and the MRI was obtained at age 42. Findings at the time of MRI included elevated CPK, lower extremity areflexia, acanthocytosis, chorea, tics and lip biting. (From Danek et al. (2001b). *Mov. Disord.* **16**, 882–889. With permission.)

### f. Biopsy

Muscle biopsy is not required for the diagnosis, but if performed may reveal myopathy characterized by type II fiber atrophy (Jung et al., 2001b).

## 6. Treatment

There is no disease-modifying treatment for McLeod syndrome. As with other rare movement disorders, most treatment approaches are empiric and symptomatic.

Seizures have been reported to be controlled with carbamazepine, phenobarbital, lamotrigine, and phenytoin alone or in combination. Chorea was not controlled with tetrabenazine, buspirone, or verapamil, though tiapride and sulpiride were beneficial (Danek et al., 2001a). Haloperidol, quetiapine, and other neuroleptics can be tried for chorea or tics; diazepam, oral baclofen, botulinum toxin, or intrathecal baclofen can be tried for dystonia. Typical anti-parkinsonian therapy is a reasonable choice in a patient with features of parkinsonism, although caution should be used as these agents can provoke chorea. Amantadine is a good first choice, as it does not seem to cause chorea in Parkinson's disease. Sedative agents such as diazepam and oral baclofen should be used cautiously, given the possibility of clinical or subclinical cognitive impairment.

There are no reports of functional microsurgery for McLeod syndrome; however, it seems reasonable to consider this option in patients with refractory hyperkinesias (see CHAC treatment).

## D. Acanthocytosis, Retinitis Pigmentosa, and Pallidal Degeneration (ARP)

This is an extremely rare disorder; there are only eight reported cases (Higgins et al., 1992; Malandrini et al., 1996; Orrell et al., 1995). Sometimes referred to as HARP syndrome (hypoprebetalipoproteinemia, acanthocytosis, retinitis pigmentosa, and pallidal degeneration), the significance of the lipoprotein abnormality is unclear: two of the published cases had the lipid disorder but no retinal or neurological disease, and three had neurological and retinal disease with normal lipoproteins.

### 1. Phenotype

Clinically, the neurological disorder resembles PKAN, with the onset in childhood or adolescence of a progressive neurological disorder characterized by dystonia, dementia, dysarthria, and dysphagia. However, the presence of orobuccolingual dyskinesia/dystonia and peripheral acanthocytosis may distinguish this condition from PKAN. The retinopathy is characterized by diffuse mottling and macular granularity.

### 2. Genotype (PKAN2, 20p13–p12.3 )

In one family, the presence of the lipid abnormality in multiple generations suggested an AD or maternal mode of inheritance with variable penetrance. It is possible that the different features of the disorder have different loci, and that in some cases this is a contiguous gene syndrome due to multiple deletions. It is also quite possible, given the many phenotypic features identical to PKAN, that this simply represents a phenotypic variant of PKAN. Recently, a mutation was identified in the gene for PKAN2, suggesting that at least some cases of ARP are indeed simply allelic variants of PKAN (see Chapter 38 by Dr. Susan Hayflick).

### 3. Pathology

*a. Hematological*

On light and electron microscopy, acanthocytes and echinocytes are seen. It is not clear if the presence of echinocytes is specific for the disorder however.

*b. Neurological*

The neuropathology is identical to that seen in PKAN, evidence that these conditions may in fact be phenotypic variants of the same disorder.

### 4. Pathophysiology

The molecular mechanisms and pathophysiology of this disorder are unknown. Similarities to PKAN, CHAC, and McLeod syndrome suggest possible common mechanisms of cell death and injury, but the rarity of the disorder has precluded in-depth investigations.

### 5. Diagnosis

*a. Clinical*

The diagnosis should be suspected in cases of a childhood- or adolescent-onset disorder with prominent dystonia, dementia, dysarthria, and dysphagia, especially if orobuccolingual dyskinesia is prominent. The differential diagnosis includes PKAN, juvenile-onset HD, WD, CHAC, and McLeod syndrome.

*b. Laboratory*

Peripheral blood reveals acanthocytosis. High-field agarose-gel lipoprotein electrophoresis may show hypoprebetalipoproteinemia, although this is not required for diagnosis. Copper and ceruloplasmin should be requested to rule out WD. Normal ferritin levels will help to rule out NFP. Blood typing is useful to rule out the McLeod phenotype.

*c. Genetic Testing*

Genetic testing is not available, as the gene has not yet been identified. If the diagnostic features and/or family history are equivocal, genetic testing for HD will help to rule out the Westphal variant of this disorder. However, if MRI shows the characteristic "eye-of-the-tiger" sign, HD testing is not necessary.

*d. Neurodiagnostic Studies*

Electroretinography reveals absent scotopic and reduced photopic retinal functions, characteristic of tapetoretinal degeneration. EEG may show background slowing, however, this is not specific. EMG and nerve conduction studies are normal.

*e. Imaging*

CT scan of the brain may show calcification of the basal ganglia. T2-weighted MRI shows decreased signal intensity in the pallidal nuclei with central hyperintensity, identical to the so-called eye of the tiger sign seen in Hallevorden-Spatz syndrome (Fig. 44.4).

*f. Biopsy*

Normal liver, bone marrow, skin, and muscle biopsies have been reported. A liver biopsy might be helpful in ruling out WD, but in the presence of a characteristic MRI is not necessary. In general, tissue biopsy does not appear to be helpful in making the diagnosis.

## 6. Treatment

There is no known treatment for this disorder; treatment is therefore empiric and symptomatic. Botulinum toxin injection or intrathecal baclofen may be helpful for localized and generalized dystonia, respectively.

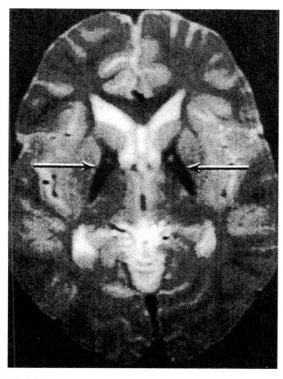

FIGURE 44.4 Magnetic resonance imaging of the brain in a case of acanthocytosis, retinitis pigmentosa, and pallidal degeneration. T2-weighted 1.5T MRI, axial. Arrows indicate the "eye of the tiger" sign. The patient presented with spasticity at age 3, followed by progressive dysarthria, dystonia, and orofacial movements. She was mute by age 8. MRI was obtained at age 10. (From Higgins et al. (1992). *Neurology* **42**, 194–198. With permission.)

## IV. OTHER RARE DISORDERS

### A. Pallidopyramidal Degeneration

Pallidopyramidal degeneration (PPD) is an extremely rare disorder; only 13 cases have been reported in 6 families (Davison, 1954; Horowitz and Greenberg, 1975; Hunt, 1917; Najim al-Din *et al.*, 1994; Tranchant *et al.*, 1991). Patients present with akinesia, rigidity, and spasticity in late childhood, adolescence, or early adulthood. Untreated, patients invariably progress to a bed-bound, immobile, locked-in state.

Ramsey Hunt first described PPD in 1917 (Hunt, 1917). Davison (1954) later described five affected cases in three families with a history of consanguinity, including Hunt's original case. In these families the illness began in the second or early third decade with parkinsonism and pyramidal tract signs. A recent report describes five affected siblings in a consanguineous Jordanian kindred (Najim al-Din *et al.*, 1994). The authors conclude that dementia and supranuclear upgaze palsy distinguish their family from prior reported cases of PPD, and propose the name "Kufor-Rakeb syndrome." However, profound motor impairment precluded IQ testing; their patients were essentially "locked-in" and dementia was only detected after initiation of levodopa therapy. If Hunt and Davison's patients had a similar phenotype, it is conceivable, even likely, that dementia was overlooked, as levodopa was not a treatment option. Similarly, a supranuclear vertical gaze palsy would have been difficult to detect in a patient with profound motor impairment. It is therefore likely that Kufor-Rakeb syndrome is the same as Hunt and Davison's PPD.

### 1. Phenotype

Patients present with akinesia, rigidity, spasticity, supranuclear upgaze paresis, and dementia in late childhood, adolescence, or early adulthood, with rapid progression of symptoms. Prolonged survival, up to age 65, distinguishes this disorder from WD. Exquisite levodopa responsiveness is common (Horowitz and Greenberg, 1975; Najim al-Din *et al.*, 1994; Tranchant *et al.*, 1991).

### 2. Genotype (Unknown, 1p36)

Autosomal recessive inheritance is likely given the prevalence of consanguinity. The locus was recently narrowed to a region of 9 cM between the markers D1S436 and D1S2843 on chromosome 1p36 (Hampshire *et al.*, 2001).

### 3. Pathology

The pathology is marked by pallor of the pallidal segments, thinning of the ansa lenticularis, neuronal loss in the substantia nigra, and early demyelination of the pyramids and crossed pyramidal tracts (Davison, 1954).

### 4. Pathophysiology

At this time, given the rarity of the disorder, and pending identification of the gene and gene product, there is little to inform a discussion of possible mechanisms of disease.

### 5. Diagnosis

*a. Clinical*

The diagnosis should be considered in patients presenting in childhood, adolescence, or young adulthood with a mixed picture of parkinsonism and pyramidal dysfunction. The added presence of dementia and a supranuclear palsy of upward gaze supports the diagnosis. It may not be possible to document dementia until after therapy with dopaminergic agents, however. A brief trial of levodopa (or a dopamine agonist) may be given for diagnostic purposes: dramatic and rapid improvement in parkinsonism strongly supports the diagnosis. The supranuclear vertical gaze palsy may suggest the possibility of NP-II, however, in NP-II the reported disturbance is of downgaze. Whether this reliably distinguishes between the disorders is unclear. Dystonia and ataxia are the motor hallmarks of NP-II, however, compared with parkinsonism and spasticity in PPD.

*b. Laboratory*

Normal serum copper and ceruloplasmin levels will help to rule out WD.

*c. Imaging*

Magnetic resonance imaging of the brain shows atrophy of the pyramidal tracts and globus pallidus and later generalized brain atrophy (Fig. 44.5).

*d. Biopsy*

There is no benefit in tissue biopsy to confirm the diagnosis. In cases with prominent supranuclear vertical gaze palsy, skin biopsy with filipin staining and and fibroblast culture to assay cholesterol esterification may rule out NP-II.

### 6. Treatment

Treatment with levodopa, even at low doses, can result in dramatic improvement in parkinsonism, typically within 48 hours As long-term levodopa therapy increases the risk of dyskinesia in PD (especially in younger patients), it is reasonable to try agonist monotherapy for PPD. An equi-

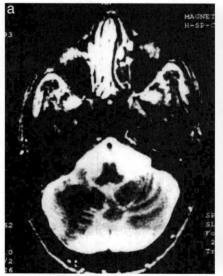
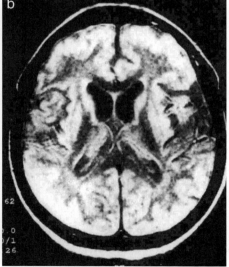

**FIGURE 44.5** Magnetic resonance imaging of the brain in a case of pallidopyramidal degeneration. (a) T2-weighted MRI, axial. Arrow demonstrates pyramidal greater than cerebellar atrophy (b) T1-weighted MRI, axial. Arrows demonstrate pallidal atrophy. The patient presented with bradykinesia at age 18, progressing to near-complete immobility over 1 to 2 years. At age 38, the patient was bedridden with severe akinesia and spasticity, anarthria, and dementia. Levodopa therapy improved parkinsonism but not spasticity. MRI was obtained at age 38. (From Najim Al-Din *et al.* (1994). *Acta Neurol. Scand.* **89**, 347–352. With permission.)

vocal response to agonist therapy, however, may require a brief test trial of levodopa, to pharmacologically characterize the patient as "levodopa responsive" or "levodopa nonresponsive."

Antispasticity management might include the use of pharmacological oral agents (clonidine, diazepam, baclofen, tizanidine), botulinum toxin injections, and intrathecal baclofen. Given the rarity of this disorder, there are no reports of these agents being used, let alone controlled trials.

### B. Familial Parkinsonism and Dementia with Ballooned Neurons

In 1993, Mizutani *et al.* reported on four patients with a new type of autosomal dominant parkinsonism and dementia (Mizutani *et al.*, 1993). As this condition resembles frontotemporal dementia with parkinsonism linked to chromosome 17, but appears to be a distinct disorder, it warrants a brief discussion.

The phenotype includes dopa-responsive parkinsonism, severe dementia, variable myoclonus, and autonomic disturbances. Leukocyte DNA analysis does not show any mutations in exons of the alpha-synuclein gene, however, further genetic analysis has not yet been reported.

The pathology is marked by the presence of ballooned neurons; in this respect, it is similar to the fronto-temporal dementia linked to chromosome 17. However, the paucity of tau pathology and presence of Lewy bodies mark it as a distinct disorder (Mizutani *et al.*, 1998, 1999).

Autopsy of two patients revealed symmetrical cerebral atrophy most prominent in the frontotemporal lobes, and marked depigmentation in the substantia nigra and locus ceruleus. Neuronal loss and gliosis were present in the cerebral cortex, amygdala, and substantia nigra. There were occasional Lewy bodies in the substantia nigra. In the cortex swollen or ballooned neurons were observed, with granulovacuolar change, argyrophilic intracytoplasmic inclusions, and neuropil threads.

Weakly tau-positive neurofibrillary tangles, consisting of 15-nm straight tubules, were numerous in upper cortical layers. Tau-negative astrocytic fibrillary tangles were frequent. The argyrophilic neuronal inclusions and weakly tau-positive neurofibrillary tangles were strongly immunoreactive to anti-alpha-synuclein antibody. Ballooned neurons and argyrophilic neuronal inclusions contained filamentous structures coated with fuzzy electron-dense deposits.

### C. Fragile X Premutation Syndrome

Fragile X syndrome or fragile X mental retardation syndrome is caused by a CGG repeat expansion that leads to methylation and transcriptional silencing of the FMR1 gene. Premutation alleles in the range of 50–200 repeats, however, are associated with elevated FMR1 mRNA levels. Hagerman *et al.* (2001) described a neurological syndrome in five elderly men with the fragile X premutation. The most obvious clinical feature was a slow intention tremor. Parkinsonian features included resting tremor, rigidity, and bradykinesia that was responsive to levodopa in one patient. Impairment of executive function was accompanied by generalized cerebral atrophy on MRI. In four individuals mRNA levels were found to be 2.4–4.4 times normal levels. The mechanism by which elevated FMR1 mRNA levels result in neurological impairment is unclear.

## V. SUMMARY

The rare and unusual movement disorders described in this chapter defy traditional classification. Despite their rarity, however, neurologists, neuroscientists, geneticists, and molecular biologists have made great progress toward identifying the clinical, genetic, and molecular features of each. New mechanisms of disease are coming to light, such as the defect of endosomal trafficking in Niemann-Pick type II and the second dendritogenesis seen in Tay-Sachs disease. This will lead to a reclassification of these disorders based upon molecular biology rather than clinical features and pathology. More importantly, it will lead to a better understanding of normal and abnormal neuronal function in common disorders such as Parkinson's disease.

### Acknowledgment

This chapter would not be possible without the courage of the patients and families who contributed to the research reviewed. They are the true heroes of the genetic revolution.

### References

Aguilar-Martinez, P., Biron, C., Masmejean, C., Jeanjean, P., and Schved, J. F. (1996). A novel mutation in the iron responsive element of ferritin L-subunit gene as a cause for hereditary hyperferritinemia-cataract syndrome. *Blood* **88**, 1895.

Akli, S., Chelly, J., Lacorte, J. M., Poenaru, L., and Kahn, A. (1991). Seven novel Tay-Sachs mutations detected by chemical mismatch cleavage of PCR-amplified cDNA fragments. *Genomics* **11**, 124–134.

Akli, S., Chomel, J. C., Lacorte, J. M., Bachner, L., Kahn, A., and Poenaru, L. (1993a). Ten novel mutations in the HEXA gene in non-Jewish Tay-Sachs patients. *Hum. Mol. Genet.* **2**, 61–67.

Alonso, M. E., Teixeira, F., Jimenez, G., and Escobar, A. (1989). Chorea-acanthocytosis: report of a family and neuropathological study of two cases. *Can. J. Neurol. Sci.* **16**, 426–431.

Aminoff, M. J. (1972). Acanthocytosis and neurological disease. *Brain* **95**, 749–760.

Araki, J., Tatsumi, Y., Sannomiya, Y., Yoshikawa, T., Hashimoto, K., Kawarabayashi, Y., Park, K., Kamitani, T., Im, T., and Tatsumi, N. (1983). [Chorea-acanthocytosis and deformability of erythrocytes]. *Rinsho Ketsueki* **24**, 1055–1059.

Asano, K., Osawa, Y., Yanagisawa, N., Takahashi, Y., and Oshima, M. (1985). Erythrocyte membrane abnormalities in patients with amyotrophic chorea with acanthocythosis. Part 2. Abnormal degradation of membrane proteins. *J.Neurol.Sci.* **68**, 161–173.

Asano, K., and Oshima, M. (1990). Protein extractabilities from the erythrocyte membranes in patients with chorea-acanthocytosis. *Jpn. J. Exp. Med.* **60**, 355–358.

Aslan, Y., Erduran, E., Gedik, Y., Mocan, H., and Yildiran, A. (1996). The role of high dose methylprednisolone and splenectomy in the accelerated phase of Chediak-Higashi syndrome. *Acta Haematol.* **96**, 105–107.

Auer, I. A., Schmidt, M. L., Lee, V. M., Curry, B., Suzuki, K., Shin, R. W., Pentchev, P. G., Carstea, E. D., and Trojanowski, J. Q. (1995). Paired helical filament tau (PHFtau) in Niemann-Pick type C disease is similar to PHFtau in Alzheimer's disease. *Acta Neuropathol.(Berlin)* **90**, 547–551.

Baldus, M., Zunftmeister, V., Geibel-Werle, G., Claus, B., Mewes, D., Uppenkamp, M., and Nebe, T. (1999). Chediak-Higashi-Steinbrinck syndrome (CHS) in a 27-year-old woman—effects of G-CSF treatment. *Ann. Hematol.* **78**, 321–327.

Barbosa, M. D., Barrat, F. J., Tchernev, V. T., Nguyen, Q. A., Mishra, V. S., Colman, S. D., Pastural, E., Dufourcq-Lagelouse, R., Fischer, A., Holcombe, R. F., Wallace, M. R., Brandt, S. J., de Saint, B. G., and Kingsmore, S. F. (1997). Identification of mutations in two major mRNA isoforms of the Chediak-Higashi syndrome gene in human and mouse. *Hum. Mol. Genet.* **6**, 1091–1098.

Barbosa, M. D., Nguyen, Q. A., Tchernev, V. T., Ashley, J. A., Detter, J. C., Blaydes, S. M., Brandt, S. J., Chotai, D., Hodgman, C., Solari, R. C., Lovett, M., and Kingsmore, S. F. (1996). Identification of the homologous beige and Chediak-Higashi syndrome genes. *Nature* **382**, 262–265.

Barnes, D., Misra, V. P., Young, E. P., Thomas, P. K., and Harding, A. E. (1991). An adult onset hexosaminidase A deficiency syndrome with sensory neuropathy and internuclear ophthalmoplegia. *J. Neurol Neurosurg. Psychiatry* **54**, 1112–1113.

Barrat, F. J., Auloge, L., Pastural, E., Lagelouse, R. D., Vilmer, E., Cant, A. J., Weissenbach, J., Le Paslier, D., Fischer, A., and de Saint, B. G. (1996). Genetic and physical mapping of the Chediak-Higashi syndrome on chromosome 1q42–43. *Am. J. Hum. Genet.* **59**, 625–632.

Barrat, F. J., Le Deist, F., Benkerrou, M., Bousso, P., Feldmann, J., Fischer, A., and de Saint, B. G. (1999). Defective CTLA-4 cycling pathway in Chediak-Higashi syndrome: a possible mechanism for deregulation of T lymphocyte activation. *Proc. Natl. Acad. Sci. U.S.A* **96**, 8645–8650.

Bauer, P., Knoblich, R., Bauer, C., Finckh, U., Hufen, A., Kropp, J., Braun, S., Kustermann-Kuhn, B., Schmidt, D., Harzer, K., and Rolfs, A. (2002). NPC1: Complete genomic sequence, mutation analysis, and characterization of haplotypes. *Hum. Mutat.* **19**, 30–38.

Beaumont, C., Leneuve, P., Devaux, I., Scoazec, J. Y., Berthier, M., Loiseau, M. N., Grandchamp, B., and Bonneau, D. (1995). Mutation in the iron responsive element of the L ferritin mRNA in a family with dominant hyperferritinaemia and cataract. *Nat. Genet.* **11**, 444–446.

Beguez-Cesar, A. B. (1943). Neutropenia cronica maligna familiar con granulaciones atipicas de los leucocitos. *Bol. Soc. Cubana Pediatr.* **15**, 900–922.

BenEzra, D., Mengistu, F., Cividalli, G., Weizman, Z., Merin, S., and Auerbach, E. (1980). Chediak-Higashi syndrome: ocular findings. *J. Pediatr. Ophthalmol. Strabismus* **17**, 68–74.

Bensmana, M., Mattei, M. G., and Lefranc, M. P. (1991). Localization of the human T-cell receptor gamma locus (TCRG) to 7p14—p15 by *in situ* hybridization. *Cytogenet. Cell Genet.* **56**, 31–32.

Bertelson, C. J., Pogo, A. O., Chaudhuri, A., Marsh, W. L., Redman, C. M., Banerjee, D., Symmans, W. A., Simon, T., Frey, D., and Kunkel, L. M. (1988). Localization of the McLeod locus (XK) within Xp21 by deletion analysis. *Am. J. Hum. Genet.* **42**, 703–711.

Betts, J. J., Nicholson, J. T., and Critchley, E. M. (1970). Acanthocytosis with normolipoproteinaemia: biophysical aspects. *Postgrad. Med. J.* **46**, 702–707.

Beutler, E., Kuhl, W., and Comings, D. (1975). Hexosaminidase isozyme in type O Gm2 gangliosidosis (Sandhoff-Jatzkewitz disease). *Am. J. Hum. Genet.* **27**, 628–638.

Bird, T. D., Cederbaum, S., Valey, R. W., and Stahl, W. L. (1978). Familial degeneration of the basal ganglia with acanthocytosis: a clinical, neuropathological, and neurochemical study. *Ann. Neurol.* **3**, 253–258.

Blanchette-Mackie, E. J., Dwyer, N. K., Amende, L. M., Kruth, H. S., Butler, J. D., Sokol, J., Comly, M. E., Vanier, M. T., August, J. T., and Brady, R. O. (1988). Type-C Niemann-Pick disease: low density lipoprotein uptake is associated with premature cholesterol accumulation in the Golgi complex and excessive cholesterol storage in lysosomes. *Proc. Natl. Acad. Sci. U.S.A* **85**, 8022–8026.

Blume, R. S. and Wolff, S. M. (1972). The Chediak-Higashi syndrome: studies in four patients and a review of the literature. *Medicine (Baltimore)* **51**, 247–280.

Boissy, R. E., and Nordlund, J. J. (1997). Molecular basis of congenital hypopigmentary disorders in humans: a review. *Pigment Cell Res.* **10**, 12–24.

Bosman, G. J., Bartholomeus, I. G., De Grip, W. J., and Horstink, M. W. (1994). Erythrocyte anion transporter and antibrain immunoreactivity in chorea-acanthocytosis. A contribution to etiology, genetics, and diagnosis. *Brain Res. Bull.* **33**, 523–528.

Bosman, G. J., and Kay, M. M. (1990). Alterations of band 3 transport protein by cellular aging and disease: erythrocyte band 3 and glucose transporter share a functional relationship. *Biochem. Cell Biol.* **68**, 1419–1427.

Boxer, L. A., Albertini, D. F., Baehner, R. L., and Oliver, J. M. (1979). Impaired microtubule assembly and polymorphonuclear leucocyte function in the Chediak-Higashi syndrome correctable by ascorbic acid. *Br. J. Haematol.* **43**, 207–213.

Boxer, L. A., Watanabe, A. M., Rister, M., Besch, H. R., Jr., Allen, J., and Baehner, R. L. (1976). Correction of leukocyte function in Chediak-Higashi syndrome by ascorbate. *N. Engl. J. Med.* **295**, 1041–1045.

Bujan, W., Ferster, A., Devalck, C., Vamos, E., Mascart, F., Denis, R., Azzi, N., Vergauwen, P., and Sariban, E. (1992). [Bone marrow graft in hereditary diseases]. *Rev. Med. Brux.* **13**, 207–211.

Burbaud, P., Rougier, A., Ferrer, X., Guehl, D., Cuny, E., Arne, P., Gross, C., and Bioulac, B. (2002). Improvement of severe trunk spasms by bilateral high-frequency stimulation of the motor thalamus in a patient with chorea-acanthocytosis. *Mov. Disord.* **17**, 204–207.

Camaschella, C., Zecchina, G., Lockitch, G., Roetto, A., Campanella, A., Arosio, P., and Levi, S. (2000). A new mutation (G51C) in the iron-responsive element (IRE) of L-ferritin associated with hyperferritinaemia-cataract syndrome decreases the binding affinity of the mutated IRE for iron-regulatory proteins. *Br. J. Haematol.* **108**, 480–482.

Campo, J. V., Stowe, R., Slomka, G., Byler, D., and Gracious, B. (1998). Psychosis as a presentation of physical disease in adolescence: a case of Niemann-Pick disease, type C. *Dev. Med. Child Neurol.* **40**, 126–129.

Carstea, E. D., Polymeropoulos, M. H., Parker, C. C., Detera-Wadleigh, S. D., O'Neill, R. R., Patterson, M. C., Goldin, E., Xiao, H., Straub, R. E., Vanier, M. T., and . (1993). Linkage of Niemann-Pick disease type C to human chromosome 18. *Proc. Natl. Acad. Sci. U.S.A* **90**, 2002–2004.

Cazzola, M., Bergamaschi, G., Tonon, L., Arbustini, E., Grasso, M., Vercesi, E., Barosi, G., Bianchi, P. E., Cairo, G., and Arosio, P. (1997). Hereditary hyperferritinemia-cataract syndrome: relationship between phenotypes and specific mutations in the iron-responsive element of ferritin light-chain mRNA. *Blood* **90**, 814–821.

Cederbaum, S. J., Heywood, D., Aigner, R., and Motulsky, A. G. (1971). Progressive chorea, dementia and acanthocytosis: a genocopy of Huntington's chorea. (Abstract). *Clin. Res.* **19**, 177.

Chediak, M. (1952). Nouvelle anomalie leucocytaire de caractere constitutionnel et familial. *Rev. Hemat.* **7**, 362–367.

Clark, M. R., Aminoff, M. J., Chiu, D. T., Kuypers, F. A., and Friend, D. S. (1989). Red cell deformability and lipid composition in two forms of acanthocytosis: enrichment of acanthocytic populations by density gradient centrifugation. *J. Lab Clin. Med.* **113**, 469–481.

Conzelmann, E., Kytzia, H. J., Navon, R., and Sandhoff, K. (1983). Ganglioside GM2 N-acetyl-beta-D-galactosaminidase activity in cultured fibroblasts of late-infantile and adult GM2 gangliosidosis patients and of healthy probands with low hexosaminidase level. *Am. J. Hum. Genet.* **35**, 900–913.

Copeland, B. R., Todd, S. A., and Furlong, C. E. (1982). High resolution two-dimensional gel electrophoresis of human erythrocyte membrane proteins. *Am. J. Hum. Genet.* **34**, 15–31.

Critchley, E. M., Clark, D. B., and Wikler, A. (1967). An adult form of acanthocytosis. *Trans. Am. Neurol. Assoc.* **92**, 132–137.

Critchley, E. M., Nicholson, J. T., Betts, J. J., and Weatherall, D. J. (1970). Acanthocytosis, normolipoproteinaemia and multiple tics. *Postgrad. Med. J.* **46**, 698–701.

Cruz, J. C., Sugii, S., Yu, C., and Chang, T. Y. (2000). Role of Niemann-Pick type C1 protein in intracellular trafficking of low density lipoprotein-derived cholesterol. *J. Biol. Chem.* **275**, 4013–4021.

Curtis, A. R., Fey, C., Morris, C. M., Bindoff, L. A., Ince, P. G., Chinnery, P. F., Coulthard, A., Jackson, M. J., Jackson, A. P., McHale, D. P., Hay, D., Barker, W. A., Markham, A. F., Bates, D., Curtis, A., and Burn, J. (2001). Mutation in the gene encoding ferritin light polypeptide causes dominant adult-onset basal ganglia disease. *Nat. Genet.* **28**, 350–354.

Danek, A., Rubio, J. P., Rampoldi, L., Ho, M., Dobson-Stone, C., Tison, F., Symmans, W. A., Oechsner, M., Kalckreuth, W., Watt, J. M., Corbett, A. J., Hamdalla, H. H., Marshall, A. G., Sutton, I., Dotti, M. T., Malandrini, A., Walker, R. H., Daniels, G., and Monaco, A. P. (2001a). McLeod neuroacanthocytosis: genotype and phenotype. *Ann. Neurol.* **50**, 755–764.

Danek, A., Tison, F., Rubio, J., Oechsner, M., Kalckreuth, W., and Monaco, A. P. (2001b). The chorea of McLeod syndrome. *Mov. Disord.* **16**, 882–889.

Danek, A., Uttner, I., Vogl, T., Tatsch, K., and Witt, T. N. (1994). Cerebral involvement in McLeod syndrome. *Neurology* **44**, 117–120.

Danek, A., Witt, T. N., Stockmann, H. B., Weiss, B. J., Schotland, D. L., and Fischbeck, K. H. (1990). Normal dystrophin in McLeod myopathy. *Ann. Neurol.* **28**, 720–722.

Davison, C. (1954). Pallido-pyramidal disease. *J. Neuropathol. Exp. Neurol.* **13**, 50–59.

De Gasperi, R., Gama Sosa, M. A., Battistini, S., Yeretsian, J., Raghavan, S., Zelnik, N., Leshinsky, E., and Kolodny, E. H. (1996). Late-onset GM2 gangliosidosis: Ashkenazi Jewish family with an exon 5 mutation (Tyr180→His) in the Hex A alpha-chain gene. *Neurology* **47**, 547–552.

dos Santos, M. R., Tanaka, A., Sa Miranda, M. C., Ribeiro, M. G., Maia, M., and Suzuki, K. (1991). GM2-gangliosidosis B1 variant: analysis of beta-hexosaminidase alpha gene mutations in 11 patients from a defined region in Portugal. *Am. J. Hum. Genet.* **49**, 886–890.

Dufourcq-Lagelouse, R., Lambert, N., Duval, M., Viot, G., Vilmer, E., Fischer, A., Prieur, M., and de Saint, B. G. (1999). Chediak-Higashi syndrome associated with maternal uniparental isodisomy of chromosome 1. *Eur. J. Hum. Genet.* **7**, 633–637.

Elleder, M., Jirasek, A., and Vlk, J. (1983). Adult neurovisceral lipidosis compatible with Niemann-Pick disease type C. *Virchows Arch. A: Pathol. Anat. Histopathol.* **401**, 35–43.

Faigle, W., Raposo, G., Tenza, D., Pinet, V., Vogt, A. B., Kropshofer, H., Fischer, A., Saint-Basile, G., and Amigorena, S. (1998). Deficient peptide loading and MHC class II endosomal sorting in a human genetic immunodeficiency disease: the Chediak-Higashi syndrome. *J. Cell Biol.* **141**, 1121–1134.

Faillace, R. T., Kingston, W. J., Nanda, N. C., and Griggs, R. C. (1982). Cardiomyopathy associated with the syndrome of amyotrophic chorea and acanthocytosis. *Ann. Intern. Med.* **96**, 616–617.

Federico, A., Palmeri, S., Malandrini, A., Fabrizi, G., Mondelli, M., and Guazzi, G. C. (1991). The clinical aspects of adult hexosaminidase deficiencies. *Dev. Neurosci.* **13**, 280–287.

Filipovich, A. H., Shapiro, R. S., Ramsay, N. K., Kim, T., Blazar, B., Kersey, J., and McGlave, P. (1992). Unrelated donor bone marrow transplantation for correction of lethal congenital immunodeficiencies. *Blood* **80**, 270–276.

Fink, J. K., Filling-Katz, M. R., Sokol, J., Cogan, D. G., Pikus, A., Sonies, B., Soong, B., Pentchev, P. G., Comly, M. E., and Brady, R. O. (1989). Clinical spectrum of Niemann-Pick disease type C. *Neurology* **39**, 1040–1049.

Frey, D., Machler, M., Seger, R., Schmid, W., and Orkin, S. H. (1988). Gene deletion in a patient with chronic granulomatous disease and McLeod syndrome: fine mapping of the Xk gene locus. *Blood* **71**, 252–255.

Fujimoto, Y., Isozaki, E., Yokochi, F., Yamakawa, K., Takahashi, H., and Hirai, S. (1997). [A case of chorea-acanthocytosis successfully treated with posteroventral pallidotomy]. *Rinsho Shinkeigaku* **37**, 891–894.

Fukai, K., Oh, J., Karim, M. A., Moore, K. J., Kandil, H. H., Ito, H., Burger, J., and Spritz, R. A. (1996). Homozygosity mapping of the gene for Chediak-Higashi syndrome to chromosome 1q42–q44 in a segment of conserved synteny that includes the mouse beige locus (bg). *Am. J. Hum. Genet.* **59**, 620–624.

Fukuda, M., Morimoto, T., Ishida, Y., Suzuki, Y., Murakami, Y., Kida, K., and Ohnishi, A. (2000). Improvement of peripheral neuropathy with oral prednisolone in Chediak-Higashi syndrome. *Eur. J. Pediatr.* **159**, 300–301.

Ganz, T., Metcalf, J. A., Gallin, J. I., Boxer, L. A., and Lehrer, R. I. (1988). Microbicidal/cytotoxic proteins of neutrophils are deficient in two disorders: Chediak-Higashi syndrome and "specific" granule deficiency. *J. Clin. Invest.* **82**, 552–556.

Girelli, D., Bozzini, C., Zecchina, G., Tinazzi, E., Bosio, S., Piperno, A., Ramenghi, U., Peters, J., Levi, S., Camaschella, C., and Corrocher, R. (2001). Clinical, biochemical and molecular findings in a series of families with hereditary hyperferritinaemia-cataract syndrome. *Br. J. Haematol.* **115**, 334–340.

Girelli, D., Corrocher, R., Bisceglia, L., Olivieri, O., De Franceschi, L., Zelante, L., and Gasparini, P. (1995). Molecular basis for the recently described hereditary hyperferritinemia-cataract syndrome: a mutation in the iron-responsive element of ferritin L-subunit gene (the "Verona mutation"). *Blood* **86**, 4050–4053.

Girelli, D., Corrocher, R., Bisceglia, L., Olivieri, O., Zelante, L., Panozzo, G., and Gasparini, P. (1997). Hereditary hyperferritinemia-cataract syndrome caused by a 29-base pair deletion in the iron responsive element of ferritin L-subunit gene. *Blood* **90**, 2084–2088.

Golde, D. W., Schneider, E. L., Bainton, D. F., Pentchev, P. G., Brady, R. O., Epstein, C. J., and Cline, M. J. (1975). Pathogenesis of one variant of sea-blue histiocytosis. *Lab Invest.* **33**, 371–378.

Gozal, D., and Harper, R. M. (1999). Novel insights into congenital hypoventilation syndrome. *Curr. Opin. Pulm. Med.* **5**, 335–338.

Greer, W. L., Dobson, M. J., Girouard, G. S., Byers, D. M., Riddell, D. C., and Neumann, P. E. (1999a). Mutations in NPC1 highlight a conserved NPC1-specific cysteine-rich domain. *Am. J. Hum. Genet.* **65**, 1252–1260.

Greer, W. L., Riddell, D. C., Byers, D. M., Welch, J. P., Girouard, G. S., Sparrow, S. M., Gillan, T. L., and Neumann, P. E. (1997). Linkage of Niemann-Pick disease type D to the same region of human chromosome 18 as Niemann-Pick disease type C. *Am. J. Hum. Genet.* **61**, 139–142.

Greer, W. L., Riddell, D. C., Gillan, T. L., Girouard, G. S., Sparrow, S. M., Byers, D. M., Dobson, M. J., and Neumann, P. E. (1998). The Nova Scotia (type D) form of Niemann-Pick disease is caused by a G3097→T transversion in NPC1. *Am. J. Hum. Genet.* **63**, 52–54.

Greer, W. L., Riddell, D. C., Murty, S., Gillan, T. L., Girouard, G. S., Sparrow, S. M., Tatlidil, C., Dobson, M. J., and Neumann, P. E. (1999b). Linkage disequilibrium mapping of the Nova Scotia variant of Niemann-Pick disease. *Clin. Genet.* **55**, 248–255.

Gross, K. B., Skrivanek, J. A., and Emeson, E. E. (1982). Ganglioside abnormality in amyotrophic chorea with acanthocytosis. *Lancet* **2**, 772.

Gullberg, U., Bengtsson, N., Bulow, E., Garwicz, D., Lindmark, A., and Olsson, I. (1999). Processing and targeting of granule proteins in human neutrophils. *J. Immunol. Methods* **232**, 201–210.

Haddad, E., Le Deist, F., Blanche, S., Benkerrou, M., Rohrlich, P., Vilmer, E., Griscelli, C., and Fischer, A. (1995). Treatment of Chediak-Higashi syndrome by allogenic bone marrow transplantation: report of 10 cases. *Blood* **85**, 3328–3333.

Hagerman, R. J., Leehey, M., Heinrichs, W., Tassone, F., Wilson, R., Hills, J., Grigsby, J., Gage, B., and Hagerman, P. J. (2001). Intention tremor, parkinsonism, and generalized brain atrophy in male carriers of fragile X. *Neurology* **57**, 127–130.

Hampshire, D. J., Roberts, E., Crow, Y., Bond, J., Mubaidin, A., Wriekat, A. L., Al Din, A., and Woods, C. G. (2001). Kufor-Rakeb syndrome, pallido-pyramidal degeneration with supranuclear upgaze paresis and dementia, maps to 1p36. *J. Med. Genet.* **38**, 680–682.

Hanaoka, N., Yoshida, K., Nakamura, A., Furihata, K., Seo, T., Tani, Y., Takahashi, J., Ikeda, S., and Hanyu, N. (1999). A novel frameshift mutation in the McLeod syndrome gene in a Japanese family. *J. Neurol. Sci.* **165**, 6–9.

Hardie, R. (1998). Cerebral hypoperfusion and hypometabolism in chorea-acanthocytosis. *Mov. Disord.* **13**, 853–854.

Hardie, R. J., Pullon, H. W., Harding, A. E., Owen, J. S., Pires, M., Daniels, G. L., Imai, Y., Misra, V. P., King, R. H., and Jacobs, J. M. (1991). Neuroacanthocytosis. A clinical, haematological and pathological study of 19 cases. *Brain* **114** (Pt. 1A), 13–49.

Harding, A. E., Young, E. P., and Schon, F. (1987). Adult onset supranuclear ophthalmoplegia, cerebellar ataxia, and neurogenic proximal muscle weakness in a brother and sister: another hexosaminidase A deficiency syndrome. *J. Neurol. Neurosurg. Psychiatry* **50**, 687–690.

Harris, Z. L., Takahashi, Y., Miyajima, H., Serizawa, M., MacGillivray, R. T., and Gitlin, J. D. (1995). Aceruloplasminemia: molecular characterization of this disorder of iron metabolism. *Proc. Natl. Acad. Sci. U.S.A* **92**, 2539–2543.

Hauser, R. A., Friedlander, J., Baker, M. J., Thomas, J., and Zuckerman, K. S. (2000). Adult Chediak-Higashi parkinsonian syndrome with dystonia. *Mov. Disord.* **15**, 705–708.

Hechtman, P., Boulay, B., De Braekeleer, M., Andermann, E., Melancon, S., Larochelle, J., Prevost, C., and Kaplan, F. (1992). The intron 7 donor splice site transition: a second Tay-Sachs disease mutation in French Canada. *Hum. Genet.* **90**, 402–406.

Hellman, N. E., Kono, S., Miyajima, H., and Gitlin, J. D. (2002). Biochemical analysis of a missense mutation in aceruloplasminemia. *J. Biol. Chem.* **277**, 1375–1380.

Higashi, O. (1954). Congenital gigantism of peroxidase granules: the first case ever reported of qualitative abnormality of peroxidase. *Tohoku J. Exp. Med.* **59**, 315–332.

Higgins, J. J., Patterson, M. C., Papadopoulos, N. M., Brady, R. O., Pentchev, P. G., and Barton, N. W. (1992). Hypoprebetalipoproteinemia, acanthocytosis, retinitis pigmentosa, and pallidal degeneration (HARP syndrome). *Neurology* **42**, 194–198.

Hirayama, M., Hamano, T., Shiratori, M., Mutoh, T., Kumano, T., Aita, T., and Kuriyama, M. (1997). Chorea-acanthocytosis with polyclonal antibodies to ganglioside GM1. *J. Neurol. Sci.* **151**, 23–24.

Ho, M., Chelly, J., Carter, N., Danek, A., Crocker, P., and Monaco, A. P. (1994). Isolation of the gene for McLeod syndrome that encodes a novel membrane transport protein. *Cell* **77**, 869–880.

Ho, M. F., Chalmers, R. M., Davis, M. B., Harding, A. E., and Monaco, A. P. (1996). A novel point mutation in the McLeod syndrome gene in neuroacanthocytosis. *Ann. Neurol.* **39**, 672–675.

Ho, M. F., Monaco, A. P., Blonden, L. A., van Ommen, G. J., Affara, N. A., Ferguson-Smith, M. A., and Lehrach, H. (1992). Fine mapping of the McLeod locus (XK) to a 150-380-kb region in Xp21. *Am. J. Hum. Genet.* **50**, 317–330.

Holcombe, R. F., Strauss, W., Owen, F. L., Boxer, L. A., Warren, R. W., Conley, M. E., Ferrara, J., Leavitt, R. Y., Fauci, A. S., and Taylor, B. A. (1987). Relationship of the genes for Chediak-Higashi syndrome (beige) and the T-cell receptor gamma chain in mouse and man. *Genomics* **1**, 287–291.

Horowitz, G. and Greenberg, J. (1975). Pallido-pyramidal syndrome treated with levodopa. *J. Neurol. Neurosurg. Psychiatry* **38**, 238–240.

Hosokawa, T., Omoto, K., Kanaseki, T., Sugi, Y., Wakamatsu, H., and Hamaguchi, K. (1992). [Studies on the erythrocyte membrane skeleton in a patient with chorea-acanthocytosis—theoretical speculation on the mechanism of neurological involvement]. *No To Shinkei* **44**, 739–744.

Hulette, C. M., Earl, N. L., Anthony, D. C., and Crain, B. J. (1992). Adult onset Niemann-Pick disease type C presenting with dementia and absent organomegaly. *Clin. Neuropathol.* **11**, 293–297.

Hunt, J. R. (1917). Progressive atrophy of the globus pallidus (primary atrophy of the pallidal system): a system disease of the paralysis agitans type, characterized by atrophy of the motor cells of the corpus striatum; a contribution to the functions of the corpus striatum. *Brain* **40**, 58–148.

Ishikawa, S., Tachibana, N., Tabata, K. I., Fujimori, N., Hayashi, R. I., Takahashi, J., Ikeda, S. I., and Hanyu, N. (2000). Muscle CT scan findings in McLeod syndrome and chorea-acanthocytosis. *Muscle Nerve* **23**, 1113–1116.

Iwata, M., Fuse, S., Sakuta, M., and Toyokura, Y. (1984). Neuropathological study of chorea-acanthocytosis. *Jpn. J. Med.* **23**, 118–122.

Jenkins, N. A., Justice, M. J., Gilbert, D. J., Chu, M. L., and Copeland, N. G. (1991). Nidogen/entactin (Nid) maps to the proximal end of mouse chromosome 13 linked to beige (bg) and identifies a new region of homology between mouse and human chromosomes. *Genomics* **9**, 401–403.

Johnson, W. G., Wigger, H. J., Karp, H. R., Glaubiger, L. M., and Rowland, L. P. (1982). Juvenile spinal muscular atrophy: a new hexosaminidase deficiency phenotype. *Ann. Neurol.* **11**, 11–16.

Jung, H. H., Hergersberg, M., Kneifel, S., Alkadhi, H., Schiess, R., Weigell-Weber, M., Daniels, G., Kollias, S., and Hess, K. (2001a). McLeod syndrome: a novel mutation, predominant psychiatric manifestations, and distinct striatal imaging findings. *Ann. Neurol.* **49**, 384–392.

Jung, H. H., Russo, D., Redman, C., and Brandner, S. (2001b). Kell and XK immunohistochemistry in McLeod myopathy. *Muscle Nerve* **24**, 1346–1351.

Karim, M. A., Nagle, D. L., Kandil, H. H., Burger, J., Moore, K. J., and Spritz, R. A. (1997). Mutations in the Chediak-Higashi syndrome gene (CHS1) indicate requirement for the complete 3801 amino acid CHS protein. *Hum. Mol. Genet.* **6**, 1087–1089.

Karim, M. A., Suzuki, K., Fukai, K., Oh, J., Nagle, D. L., Moore, K. J., Barbosa, E., Falik-Borenstein, T., Filipovich, A., Ishida, Y., Kivrikko, S., Klein, C., Kreuz, F., Levin, A., Miyajima, H., Regueiro, J., Russo, C., Uyama, E., Vierimaa, O., and Spritz, R. A. (2002). Apparent genotype-phenotype correlation in childhood, adolescent, and adult Chediak-Higashi syndrome. *Am. J. Med. Genet.* **108**, 16–22.

Kartsounis, L. D., and Hardie, R. J. (1996). The pattern of cognitive impairments in neuroacanthocytosis. A frontosubcortical dementia. *Arch. Neurol.* **53**, 77–80.

Kawakami, T., Takiyama, Y., Sakoe, K., Ogawa, T., Yoshioka, T., Nishizawa, M., Reid, M. E., Kobayashi, O., Nonaka, I., and Nakano, I. (1999). A case of McLeod syndrome with unusually severe myopathy. *J. Neurol. Sci.* **166**, 36–39.

Kay, M. M., Goodman, J., Goodman, S., and Lawrence, C. (1990a). Membrane protein band 3 alteration associated with neurologic disease and tissue-reactive antibodies. *Exp. Clin. Immunogenet.* **7**, 181–199.

Kay, M. M., Goodman, J., Lawrence, C., and Bosman, G. (1990b). Membrane channel protein abnormalities and autoantibodies in neurological disease. *Brain Res. Bull.* **24**, 105–111.

Kay, M. M., Marchalonis, J. J., Hughes, J., Watanabe, K., and Schluter, S. F. (1990c). Definition of a physiologic aging autoantigen by using synthetic peptides of membrane protein band 3: localization of the active antigenic sites. *Proc. Natl. Acad. Sci. U.S.A.* **87**, 5734–5738.

Khamlichi, S., Bailly, P., Blanchard, D., Goossens, D., Cartron, J. P., and Bertrand, O. (1995). Purification and partial characterization of the erythrocyte Kx protein deficient in McLeod patients. *Eur. J. Biochem.* **228**, 931–934.

Kinugawa, N. and Ohtani, T. (1985). Beneficial effects of high-dose intravenous gammaglobulin on the accelerated phase of Chediak-Higashi syndrome. *Helv. Paediatr. Acta* **40**, 169–172.

Kusafuka, T. and Puri, P. (1997). Mutations of the endothelin-B receptor and endothelin-3 genes in Hirschsprung's disease. *Pediatr. Surg Int.* **12**, 19–23.

Kutcher, J. S., Kahn, M. J., Andersson, H. C., and Foundas, A. L. (1999). Neuroacanthocytosis masquerading as Huntington's disease: CT/MRI findings. *J. Neuroimaging* **9**, 187–189.

Ledvinova, J., and Elleder, M. (1993). Filipin test for diagnosis of Niemann-Pick disease type C. *Sb. Lek.* **94**, 137–143.

Lee S., Wu, X., Reid, M., Zelinski, T., and Redman, C. (1995) Molecular basis of the Kell (K1) phenotype. *Blood* **85**, 912–916.

Lee, S., Lin, M., Mele, A., Cao, Y., Farmer, J., Russo, D., and Redman, C. (1999). Proteolytic processing of big endothelin-3 by the kell blood group protein. *Blood* **94**, 1440–1450.

Levine, I. M., Estes, J. W., and Looney, J. M. (1968). Hereditary neurological disease with acanthocytosis. A new syndrome. *Arch. Neurol.* **19**, 403–409.

Lichtenberg, P., Navon, R., Wertman, E., Dasberg, H., and Lerer, B. (1988). Post-partum psychosis in adult GM2 gangliosidosis. A case report. *Br. J. Psychiatry* **153**, 387–389.

Limos, L. C., Ohnishi, A., Sakai, T., Fujii, N., Goto, I., and Kuroiwa, Y. (1982). "Myopathic" changes in chorea-acanthocytosis. Clinical and histopathological studies. *J. Neurol. Sci.* **55**, 49–58.

Lossos, A., Schlesinger, I., Okon, E., Abramsky, O., Bargal, R., Vanier, M. T., and Zeigler, M. (1997). Adult-onset Niemann-Pick type C disease. Clinical, biochemical, and genetic study. *Arch. Neurol.* **54**, 1536–1541.

MacQueen, G. M., Rosebush, P. I., and Mazurek, M. F. (1998). Neuropsychiatric aspects of the adult variant of Tay-Sachs disease. *J. Neuropsychiatry Clin. Neurosci.* **10**, 10–19.

Malandrini, A., Cesaretti, S., Mulinari, M., Palmeri, S., Fabrizi, G. M., Villanova, M., Parrotta, E., Montagnani, M., Montagnani, M., Anichini, M., and Guazzi, G. C. (1996). Acanthocytosis, retinitis pigmentosa, pallidal degeneration. Report of two cases without serum lipid abnormalities. *J. Neurol. Sci.* **140**, 129–131.

Malandrini, A., Fabrizi, G. M., Palmeri, S., Ciacci, G., Salvadori, C., Berti, G., Bucalossi, A., Federico, A., and Guazzi, G. C. (1993). Choreoacanthocytosis like phenotype without acanthocytes: clinicopathological case report. A contribution to the knowledge of the functional pathology of the caudate nucleus. *Acta Neuropathol. (Berlin)* **86**, 651–658.

Malandrini, A., Fabrizi, G. M., Truschi, F., Di Pietro, G., Moschini, F., Bartalucci, P., Berti, G., Salvadori, C., Bucalossi, A., and Guazzi, G. (1994). Atypical McLeod syndrome manifested as X-linked choreaacanthocytosis, neuromyopathy and dilated cardiomyopathy: report of a family. *J. Neurol. Sci.* **124**, 89–94.

Marsh, W. L., Marsh, N. J., Moore, A., Symmans, W. A., Johnson, C. L., and Redman, C. M. (1981). Elevated serum creatine phosphokinase in subjects with McLeod syndrome. *Vox Sang.* **40**, 403–411.

Martin, J. J., Lowenthal, A., Ceuterick, C., and Vanier, M. T. (1984). Juvenile dystonic lipidosis (variant of Niemann-Pick disease type C). *J. Neurol. Sci.* **66**, 33–45.

Martin, M. E., Fargion, S., Brissot, P., Pellat, B., and Beaumont, C. (1998). A point mutation in the bulge of the iron-responsive element of the L ferritin gene in two families with the hereditary hyperferritinemia-cataract syndrome. *Blood* **91**, 319–323.

Meek, D., Wolfe, L. S., Andermann, E., and Andermann, F. (1984). Juvenile progressive dystonia: a new phenotype of GM2 gangliosidosis. *Ann. Neurol.* **15**, 348–352.

Millat, G., Chikh, K., Naureckiene, S., Sleat, D. E., Fensom, A. H., Higaki, K., Elleder, M., Lobel, P., and Vanier, M. T. (2001). Niemann-Pick disease type C: spectrum of HE1 mutations and genotype/phenotype correlations in the NPC2 group. *Am. J. Hum. Genet.* **69**, 1013–1021.

Millat, G., Marcais, C., Rafi, M. A., Yamamoto, T., Morris, J. A., Pentchev, P. G., Ohno, K., Wenger, D. A., and Vanier, M. T. (1999). Niemann-Pick C1 disease: the I1061T substitution is a frequent mutant allele in patients of Western European descent and correlates with a classic juvenile phenotype. *Am. J. Hum. Genet.* **65**, 1321–1329.

Misra, V. P., King, R. H., Harding, A. E., Muddle, J. R., and Thomas, P. K. (1991). Peripheral neuropathy in the Chediak-Higashi syndrome. *Acta Neuropathol. (Berlin)* **81**, 354–358.

Mitsumoto, H., Sliman, R. J., Schafer, I. A., Sternick, C. S., Kaufman, B., Wilbourn, A., and Horwitz, S. J. (1985). Motor neuron disease and adult hexosaminidase A deficiency in two families: evidence for multisystem degeneration. *Ann. Neurol.* **17**, 378–385.

Miyajima, H., Kono, S., Takahashi, Y., Sugimoto, M., Sakamoto, M., and Sakai, N. (2001). Cerebellar ataxia associated with heteroallelic ceruloplasmin gene mutation. *Neurology* **57**, 2205–2210.

Miyajima, H., Nishimura, Y., Mizoguchi, K., Sakamoto, M., Shimizu, T., and Honda, N. (1987). Familial apoceruloplasmin deficiency associated with blepharospasm and retinal degeneration. *Neurology* **37**, 761–767.

Mizutani, T. (1999). [Familial parkinsonism and dementia with ballooned neurons, argyrophilic neuronal inclusions, atypical neurofibrillary tangles, tau-negative astrocytic fibrillary tangles, and Lewy bodies]. *Rinsho Shinkeigaku* **39**, 1262–1263.

Mizutani, T., Inose, T., Nakajima, S., and Gambetti, P. (1993). Familial parkinsonism and dementia with "ballooned neurons." *Adv. Neurol.* **60**, 613–617.

Mizutani, T., Inose, T., Nakajima, S., Kakimi, S., Uchigata, M., Ikeda, K., Gambetti, P., and Takasu, T. (1998). Familial parkinsonism and dementia with ballooned neurons, argyrophilic neuronal inclusions, atypical neurofibrillary tangles, tau-negative astrocytic fibrillary tangles, and Lewy bodies. *Acta Neuropathol. (Berlin)* **95**, 15–27.

Morita, H., Ikeda, S., Yamamoto, K., Morita, S., Yoshida, K., Nomoto, S., Kato, M., and Yanagisawa, N. (1995). Hereditary ceruloplasmin deficiency with hemosiderosis: a clinicopathological study of a Japanese family. *Ann. Neurol.* **37**, 646–656.

Mottonen, M., Lanning, M., and Saarinen, U. M. (1995). Allogeneic bone marrow transplantation in Chediak-Higashi syndrome. *Pediatr. Hematol. Oncol.* **12**, 55–59.

Mules, E. H., Hayflick, S., Miller, C. S., Reynolds, L. W., and Thomas, G. H. (1992). Six novel deleterious and three neutral mutations in the gene encoding the alpha-subunit of hexosaminidase A in non-Jewish individuals. *Am. J. Hum. Genet.* **50**, 834–841.

Mumford, A. D., Vulliamy, T., Lindsay, J., and Watson, A. (1998). Hereditary hyperferritinemia-cataract syndrome: two novel mutations in the L-ferritin iron-responsive element. *Blood* **91**, 367–368.

Nagle, D. L., Karim, M. A., Woolf, E. A., Holmgren, L., Bork, P., Misumi, D. J., McGrail, S. H., Dussault, B. J., Jr., Perou, C. M., Boissy, R. E., Duyk, G. M., Spritz, R. A., and Moore, K. J. (1996). Identification and mutation analysis of the complete gene for Chediak-Higashi syndrome. *Nat. Genet.* **14**, 307–311.

Najim al-Din, A. S., Wriekat, A., Mubaidin, A., Dasouki, M., and Hiari, M. (1994). Pallido-pyramidal degeneration, supranuclear upgaze paresis and dementia: Kufor-Rakeb syndrome. *Acta Neurol. Scand.* **89**, 347–352.

Nakano, T., Muscillo, M., Ohno, K., Hoffman, A. J., and Suzuki, K. (1988). A point mutation in the coding sequence of the beta-hexosaminidase alpha gene results in defective processing of the enzyme protein in an unusual GM2-gangliosidosis variant. *J. Neurochem.* **51**, 984–987.

Nardocci, N., Bertagnolio, B., Rumi, V., and Angelini, L. (1992). Progressive dystonia symptomatic of juvenile GM2 gangliosidosis. *Mov. Disord.* **7**, 64–67.

Naureckiene, S., Sleat, D. E., Lackland, H., Fensom, A., Vanier, M. T., Wattiaux, R., Jadot, M., and Lobel, P. (2000). Identification of HE1 as the second gene of Niemann-Pick C disease. *Science* **290**, 2298–2301.

Navon, R., Argov, Z., and Frisch, A. (1986). Hexosaminidase A deficiency in adults. *Am. J. Med. Genet.* **24**, 179–196.

Navon, R., Kolodny, E. H., Mitsumoto, H., Thomas, G. H., and Proia, R. L. (1990). Ashkenazi-Jewish and non-Jewish adult GM2 gangliosidosis patients share a common genetic defect. *Am. J. Hum. Genet.* **46**, 817–821.

Navon, R., and Proia, R. L. (1989). The mutations in Ashkenazi Jews with adult GM2 gangliosidosis, the adult form of Tay-Sachs disease. *Science* **243**, 1471–1474.

Noel, P. J., Boise, L. H., and Thompson, C. B. (1996). Regulation of T cell activation by CD28 and CTLA4. *Adv. Exp. Med. Biol.* **406**, 209–217.

O'Brien, J. S., Tennant, L., Veath, M. L., Scott, C. R., and Bucknall, W. E. (1978). Characterization of unusual hexosaminidase A (HEX A) deficient human mutants. *Am. J. Hum. Genet.* **30**, 602–608.

Oates, C. E., Bosch, E. P., and Hart, M. N. (1986). Movement disorders associated with chronic GM2 gangliosidosis. Case report and review of the literature. *Eur. Neurol.* **25**, 154–159.

Oechsner, M., Buchert, R., Beyer, W., and Danek, A. (2001). Reduction of striatal glucose metabolism in McLeod choreoacanthocytosis. *J. Neurol. Neurosurg. Psychiatry* **70**, 517–520.

Oetting, W. S., and King, R. A. (1999). Molecular basis of albinism: mutations and polymorphisms of pigmentation genes associated with albinism. *Hum. Mutat.* **13**, 99–115.

Ohnishi, A., Sato, Y., Nagara, H., Sakai, T., Iwashita, H., Kuroiwa, Y., Nakamura, T., and Shida, K. (1981). Neurogenic muscular atrophy and low density of large myelinated fibers of sural nerve in chorea-acanthocytosis. *J. Neurol. Neurosurg. Psychiatry* **44**, 645–648.

Ohno, K., and Suzuki, K. (1988). Mutation in GM2-gangliosidosis B1 variant. *J. Neurochem.* **50**, 316–318.

Okamoto, N., Wada, S., Oga, T., Kawabata, Y., Baba, Y., Habu, D., Takeda, Z., and Wada, Y. (1996). Hereditary ceruloplasmin deficiency with hemosiderosis. *Hum. Genet.* **97**, 755–758.

Oliver, J. M. (1976). Impaired microtubule function correctable by cyclic GMP and cholinergic agonists in the Chediak-Higashi syndrome. *Am. J. Pathol.* **85**, 395–418.

Oliver, J. M., Krawiec, J. A., and Berlin, R. D. (1976). Carbamycholine prevents giant granule-formation in cultured fibroblasts from beige (Chediak-Higashi) mice. *J. Cell Biol.* **69**, 205–210.

Oliver, J. M., and Zurier, R. B. (1976). Correction of characteristic abnormalities of microtubule function and granule morphology in Chediak-Higashi syndrome with cholinergic agonists. *J. Clin. Invest.* **57**, 1239–1247.

Olivieri, O., De Franceschi, L., Bordin, L., Manfredi, M., Miraglia, D. G., Perrotta, S., De Vivo, M., Guarini, P., and Corrocher, R. (1997). Increased membrane protein phosphorylation and anion transport activity in chorea-acanthocytosis. *Haematologica* **82**, 648–653.

Olsen, D. R., Nagayoshi, T., Fazio, M., Mattei, M. G., Passage, E., Weil, D., Timpl, R., Chu, M. L., and Uitto, J. (1989). Human nidogen: cDNA cloning, cellular expression, and mapping of the gene to chromosome Iq43. *Am. J. Hum. Genet.* **44**, 876–885.

Orrell, R. W., Amrolia, P. J., Heald, A., Cleland, P. G., Owen, J. S., Morgan-Hughes, J. A., Harding, A. E., and Marsden, C. D. (1995). Acanthocytosis, retinitis pigmentosa, and pallidal degeneration: a report of three patients, including the second reported case with hypoprebetalipoproteinemia (HARP syndrome). *Neurology* **45**, 487–492.

Oshima, M., Osawa, Y., Asano, K., and Saito, T. (1985). Erythrocyte membrane abnormalities in patients with amyotrophic chorea with acanthocytosis. Part 1. Spin labeling studies and lipid analyses. *J. Neurol. Sci.* **68**, 147–160.

Palumbo, A. P., Isobe, M., Huebner, K., Shane, S., Rovera, G., Demuth, D., Curtis, P. J., Ballantine, M., Croce, C. M., and Showe, L. C. (1986). Chromosomal localization of a human band 3-like gene to region 7q35–7q36. *Am. J. Hum. Genet.* **39**, 307–316.

Parnes, S., Karpati, G., Carpenter, S., Kin, N. M., Wolfe, L. S., and Suranyi, L. (1985). Hexosaminidase-A deficiency presenting as atypical juvenile-onset spinal muscular atrophy. *Arch. Neurol.* **42**, 1176–1180.

Patel, S. C., Suresh, S., Kumar, U., Hu, C. Y., Cooney, A., Blanchette-Mackie, E. J., Neufeld, E. B., Patel, R. C., Brady, R. O., Patel, Y. C., Pentchev, P. G., and Ong, W. Y. (1999). Localization of Niemann-Pick C1 protein in astrocytes: implications for neuronal degeneration in Niemann-Pick type C disease. *Proc. Natl. Acad. Sci. U.S.A.* **96**, 1657–1662.

Paw, B. H., Kaback, M. M., and Neufeld, E. F. (1989). Molecular basis of adult-onset and chronic GM2 gangliosidoses in patients of Ashkenazi Jewish origin: substitution of serine for glycine at position 269 of the alpha-subunit of beta-hexosaminidase. *Proc. Natl. Acad. Sci. U.S.A.* **86**, 2413–2417.

Paw, B. H., Moskowitz, S. M., Uhrhammer, N., Wright, N., Kaback, M. M., and Neufeld, E. F. (1990). Juvenile GM2 gangliosidosis caused by substitution of histidine for arginine at position 499 or 504 of the alpha-subunit of beta-hexosaminidase. *J. Biol. Chem.* **265**, 9452–9457.

Paw, B. H., Wood, L. C., and Neufeld, E. F. (1991). A third mutation at the CpG dinucleotide of codon 504 and a silent mutation at codon 506 of the HEX A gene. *Am. J. Hum. Genet.* **48**, 1139–1146.

Pentchev, P. G., Comly, M. E., Kruth, H. S., Vanier, M. T., Wenger, D. A., Patel, S., and Brady, R. O. (1985). A defect in cholesterol esterification in Niemann-Pick disease (type C) patients. *Proc. Natl. Acad. Sci. U.S.A.* **82**, 8247–8251.

Platt, F. M., and Butters, T. D. (1998). New therapeutic prospects for the glycosphingolipid lysosomal storage diseases. *Biochem. Pharmacol.* **56**, 421–430.

Proia, R. L., and Neufeld, E. F. (1982). Synthesis of beta-hexosaminidase in cell-free translation and in intact fibroblasts: an insoluble precursor alpha chain in a rare form of Tay-Sachs disease. *Proc. Natl. Acad. Sci. U.S.A.* **79**, 6360–6364.

Rampoldi, L., Dobson-Stone, C., Rubio, J. P., Danek, A., Chalmers, R. M., Wood, N. W., Verellen, C., Ferrer, X., Malandrini, A., Fabrizi, G. M., Brown, R., Vance, J., Pericak-Vance, M., Rudolf, G., Carre, S., Alonso, E., Manfredi, M., Nemeth, A. H., and Monaco, A. P. (2001). A conserved sorting-associated protein is mutant in chorea-acanthocytosis. *Nat. Genet.* **28**, 119–120.

Rapin, I., Suzuki, K., Suzuki, K., and Valsamis, M. P. (1976). Adult (chronic) GM2 gangliosidosis. Atypical spinocerebellar degeneration in a Jewish sibship. *Arch. Neurol.* **33**, 120–130.

Rappeport, J. M., Smith, B. R., Parkman, R., and Rosen, F. S. (1983). Application of bone marrow transplantation in genetic diseases. *Clin. Haematol.* **12**, 755–773.

Requena, C., I, Arias, G. M., Lema, D. C., Sanchez, H. J., Barros, A. F., and Coton Vilas, J. C. (2000). [Autosomal recessive chorea-acanthocytosis linked to 9q21]. *Neurologia* **15**, 132–135.

Rinne, J. O., Daniel, S. E., Scaravilli, F., Harding, A. E., and Marsden, C. D. (1994). Nigral degeneration in neuroacanthocytosis. *Neurology* **44**, 1629–1632.

Roder, J. C., Haliotis, T., Klein, M., Korec, S., Jett, J. R., Ortaldo, J., Heberman, R. B., Katz, P., and Fauci, A. S. (1980). A new immuno-deficiency disorder in humans involving NK cells. *Nature* **284**, 553–555.

Roder, J. C., Todd, R. F., Rubin, P., Haliotis, T., Helfand, S. L., Werkmeister, J., Pross, H. F., Boxer, L. A., Schlossman, S. F., and Fauci, A. S. (1983). The Chediak-Higashi gene in humans. III. Studies on the mechanisms of NK impairment. *Clin. Exp. Immunol.* **51**, 359–368.

Rondot, P., Navon, R., Eymard, B., Fardeau, M., Turpin, J. C., Lefevre, M., Bathien, N., Wu, Y., and Baumann, N. (1997). [Juvenile GM2 gangliosidosis with progressive spinal muscular atrophy onset]. *Rev. Neurol. (Paris)* **153**, 120–123.

Rubio, J. P., Danek, A., Stone, C., Cha!mers, R., Wood, N., Verellen, C., Ferrer, X., Malandrini, A., Fabrizi, G. M., Manfredi, M., Vance, J., Pericak-Vance, M., Brown, R., Rudolf, G., Picard, F., Alonso, E., Brin, M., Nemeth, A. H., Farrall, M., and Monaco, A. P. (1997). Chorea-acanthocytosis: genetic linkage to chromosome 9q21. *Am. J. Hum. Genet.* **61**, 899–908.

Russo, D., Redman, C., and Lee, S. (1998). Association of XK and Kell blood group proteins. *J. Biol. Chem.* **273**, 13950–13956.

Sakai, T., Antoku, Y., Iwashita, H., Goto, I., Nagamatsu, K., and Shii, H. (1991a). Chorea-acanthocytosis: abnormal composition of covalently bound fatty acids of erythrocyte membrane proteins. *Ann. Neurol.* **29**, 664–669.

Sakai, Y., Miyawaki, S., Shimizu, A., Ohno, K., and Watanabe, T. (1991b). A molecular genetic linkage map of mouse chromosome 18, including spm, Grl-1, Fim-2/c-fms, and Mbp. *Biochem. Genet.* **29**, 103–113.

Sato, Y., Ohnishi, A., Tateishi, J., Onizuka, Y., Ishimoto, S., Iwashita, H., Kuroiwa, Y., and Kanazawa, I. (1984). [An autopsy case of chorea-acanthocytosis. Special reference to the histopathological and biochemical findings of basal ganglia]. *No To Shinkei* **36**, 105–111.

Serra, S., Xerra, A., Scribano, E., Meduri, M., and Di Perri, R. (1987). Computerized tomography in amyotrophic choreo-acanthocytosis. *Neuroradiology* **29**, 480–482.

Shamburek, R. D., Pentchev, P. G., Zech, L. A., Blanchette-Mackie, J., Carstea, E. D., VandenBroek, J. M., Cooper, P. S., Neufeld, E. B., Phair, R. D., Brewer, H. B., Jr., Brady, R. O., and Schwartz, C. C. (1997). Intracellular trafficking of the free cholesterol derived from LDL cholesteryl ester is defective *in vivo* in Niemann-Pick C disease: insights on normal metabolism of HDL and LDL gained from the NP-C mutation. *J. Lipid Res.* **38**, 2422–2435.

Sheramata, W., Kott, H. S., and Cyr, D. P. (1971). The Chediak-Higashi-Steinbrinck syndrome. Presentation of three cases with features resembling spinocerebellar degeneration. *Arch. Neurol.* **25**, 289–294.

Shulman, L. M., David, N. J., and Weiner, W. J. (1995). Psychosis as the initial manifestation of adult-onset Niemann-Pick disease type C. *Neurology* **45**, 1739–1743.

Sobue, G., Mukai, E., Fujii, K., Mitsuma, T., and Takahashi, A. (1986). Peripheral nerve involvement in familial chorea-acanthocytosis. *J. Neurol. Sci.* **76**, 347–356.

Sokol, J., Blanchette-Mackie, J., Kruth, H. S., Dwyer, N. K., Amende, L. M., Butler, J. D., Robinson, E., Patel, S., Brady, R. O., and Comly, M. E. (1988). Type C Niemann-Pick disease. Lysosomal accumulation and defective intracellular mobilization of low density lipoprotein cholesterol. *J. Biol. Chem.* **263**, 3411–3417.

Sorrentino, G., De Renzo, A., Miniello, S., Nori, O., and Bonavita, V. (1999). Late appearance of acanthocytes during the course of chorea-acanthocytosis. *J. Neurol. Sci.* **163**, 175–178.

Spitz, M. C., Jankovic, J., and Killian, J. M. (1985). Familial tic disorder, parkinsonism, motor neuron disease, and acanthocytosis: a new syndrome. *Neurology* **35**, 366–370.

Stephan, D. A., Chen, Y., Jiang, Y., Malechek, L., Gu, J. Z., Robbins, C. M., Bittner, M. L., Morris, J. A., Carstea, E., Meltzer, P. S., Adler, K., Garlick, R., Trent, J. M., and Ashlock, M. A. (2000). Positional cloning utilizing genomic DNA microarrays: the Niemann-Pick type C gene as a model system. *Mol. Genet. Metab* **70**, 10–18.

Stevenson, V. L., and Hardie, R. J. (2001). Acanthocytosis and neurological disorders. *J. Neurol.* **248**, 87–94.

Sullivan, K. M., Parkman, R., and Walters, M. C. (2000). Bone marrow transplantation for non-malignant disease. *Hematology. (Am. Soc. Hematol. Educ. Program)* 319–338.

Supple, S. G., Iland, H. J., Barnett, M. H., and Pollard, J. D. (2001). A spontaneous novel XK gene mutation in a patient with McLeod syndrome. *Br. J. Haematol.* **115**, 369–372.

Suzuki, K. (1991). Neuropathology of late onset gangliosidoses. A review. *Dev. Neurosci.* **13**, 205–210.

Suzuki, K., Rapin, I., Suzuki, Y., and Ishii, N. (1970). Juvenile GM2-gangliosidosis. Clinical variant of Tay-Sachs disease or a new disease. *Neurology* **20**, 190–204.

Swash, M., Schwartz, M. S., Carter, N. D., Heath, R., Leak, M., and Rogers, K. L. (1983). Benign X-linked myopathy with acanthocytes (McLeod syndrome). Its relationship to X-linked muscular dystrophy. *Brain* **106** (Pt. 3), 717–733.

Symmans, W. A., Shepherd, C. S., Marsh, W. L., Oyen, R., Shohet, S. B., and Linehan, B. J. (1979). Hereditary acanthocytosis associated with the McLeod phenotype of the Kell blood group system. *Br. J. Haematol.* **42**, 575–583.

Takahashi, Y., Miyajima, H., Shirabe, S., Nagataki, S., Suenaga, A., and Gitlin, J. D. (1996). Characterization of a nonsense mutation in the ceruloplasmin gene resulting in diabetes and neurodegenerative disease. *Hum. Mol. Genet.* **5**, 81–84.

Tanaka, A., Ohno, K., Sandhoff, K., Maire, I., Kolodny, E. H., Brown, A., and Suzuki, K. (1990). GM2-gangliosidosis B1 variant: analysis of beta-hexosaminidase alpha gene abnormalities in seven patients. *Am. J. Hum. Genet.* **46**, 329–339.

Tanaka, A., Ohno, K., and Suzuki, K. (1988). GM2-gangliosidosis B1 variant: a wide geographic and ethnic distribution of the specific beta-hexosaminidase alpha chain mutation originally identified in a Puerto Rican patient. *Biochem. Biophys. Res. Commun.* **156**, 1015–1019.

Tanaka, M., Hirai, S., Kondo, S., Sun, X., Nakagawa, T., Tanaka, S., Hayashi, K., and Okamoto, K. (1998). Cerebral hypoperfusion and hypometabolism with altered striatal signal intensity in chorea-acanthocytosis: a combined PET and MRI study. *Mov. Disord.* **13**, 100–107.

Terada, N., Fujii, Y., Ueda, H., Kato, Y., Baba, T., Hayashi, R., and Ohno, S. (1999). Ultrastructural changes of erythrocyte membrane skeletons in chorea-acanthocytosis and McLeod syndrome revealed by the quick-freezing and deep-etching method. *Acta Haematol.* **101**, 25–31.

Tranchant, C., Boulay, C., and Warter, J. M. (1991). [Pallido-pyramidal syndrome: an unrecognized entity]. *Rev. Neurol. (Paris)* **147**, 308–310.

Turpin, J. C., Masson, M., and Baumann, N. (1991). Clinical aspects of Niemann-Pick type C disease in the adult. *Dev. Neurosci.* **13**, 304–306.

Uchino, M., Uyama, E., Hirano, T., Nakamura, T., Fukushima, T., and Ando, M. (1993). A histochemical and electron microscopic study of skeletal muscle in an adult case of Chediak-Higashi syndrome. *Acta Neuropathol. (Berlin)* **86**, 521–524.

Ueno, E., Oguchi, K., and Yanagisawa, N. (1982). Morphological abnormalities of erythrocyte membrane in the hereditary neurological disease with chorea, areflexia and acanthocytosis. *J. Neurol. Sci.* **56**, 89–97.

Ueno, S., Maruki, Y., Nakamura, M., Tomemori, Y., Kamae, K., Tanabe, H., Yamashita, Y., Matsuda, S., Kaneko, S., and Sano, A. (2001). The gene encoding a newly discovered protein, chorein, is mutated in chorea-acanthocytosis. *Nat. Genet.* **28**, 121–122.

Usuki, S., Cashman, N. R., and Miyatake, T. (1999). GM2 promotes ciliary neurotrophic factor-dependent rescue of immortalized motor neuron-like cell (NSC-34). *Neurochem. Res.* **24**, 281–286.

Usuki, S., Ren, J., Utsunomiya, I., Cashman, N. R., Inokuchi, J., and Miyatake, T. (2001). GM2 ganglioside regulates the function of ciliary neurotrophic factor receptor in murine immortalized motor neuron-like cells (NSC-34). *Neurochem. Res.* **26**, 375–382.

Uyama, E., Hirano, T., Ito, K., Nakashima, H., Sugimoto, M., Naito, M., Uchino, M., and Ando, M. (1994). Adult Chediak-Higashi syndrome presenting as parkinsonism and dementia. *Acta Neurol. Scand.* **89**, 175–183.

Uyama, E., Hirano, T., Yoshida, A., Doi, O., Maruoka, S., and Araki, S. (1991). [An adult case of Chediak-Higashi syndrome with parkinsonism and marked atrophy of the central nervous system]. *Rinsho Shinkeigaku* **31**, 24–31.

Vance, J. M., Pericak-Vance, M. A., Bowman, M. H., Payne, C. S., Fredane, L., Siddique, T., Roses, A. D., and Massey, E. W. (1987). Chorea-acanthocytosis: a report of three new families and implications for genetic counselling. *Am. J. Med. Genet.* **28**, 403–410.

Vanier, M. T., Pentchev, P., Rodriguez-Lafrasse, C., and Rousson, R. (1991a). Niemann-Pick disease type C: an update. *J. Inherit. Metab. Dis.* **14**, 580–595.

Vanier, M. T., Rodriguez-Lafrasse, C., Rousson, R., Duthel, S., Harzer, K., Pentchev, P. G., Revol, A., and Louisot, P. (1991b). Type C Niemann-Pick disease: biochemical aspects and phenotypic heterogeneity. *Dev. Neurosci.* **13**, 307–314.

Vanier, M. T., Wenger, D. A., Comly, M. E., Rousson, R., Brady, R. O., and Pentchev, P. G. (1988). Niemann-Pick disease group C: clinical variability and diagnosis based on defective cholesterol esterification. A collaborative study on 70 patients. *Clin. Genet.* **33**, 331–348.

Villegas, A., Moscat, J., Vazquez, A., Calero, F., Alvarez-Sala, J. L., Artola, S., and Espinos, D. (1987). A new family with hereditary choreo-acanthocytosis. *Acta Haematol.* **77**, 215–219.

Vita, G., Serra, S., Dattola, R., Santoro, M., Toscano, A., Venuto, C., Carrozza, G., and Baradello, A. (1989). Peripheral neuropathy in amyotrophic chorea-acanthocytosis. *Ann. Neurol.* **26**, 583–587.

Walkley, S. U., Baker, H. J., and Rattazzi, M. C. (1990). Initiation and growth of ectopic neurites and meganeurites during postnatal cortical development in ganglioside storage disease. *Brain Res. Dev.* **51**, 167–178.

Walkley, S. U., Siegel, D. A., and Dobrenis, K. (1995). GM2 ganglioside and pyramidal neuron dendritogenesis. *Neurochem. Res.* **20**, 1287–1299.

Walkley, S. U., Siegel, D. A., Dobrenis, K., and Zervas, M. (1998). GM2 ganglioside as a regulator of pyramidal neuron dendritogenesis. *Ann. N.Y. Acad. Sci.* **845**, 188–199.

Walkley, S. U., Zervas, M., and Wiseman, S. (2000). Gangliosides as modulators of dendritogenesis in normal and storage disease-affected pyramidal neurons. *Cereb.Cortex* **10**, 1028–1037.

Wihl, G., Volkmann, J., Allert, N., Lehrke, R., Sturm, V., and Freund, H. J. (2001). Deep brain stimulation of the internal pallidum did not improve chorea in a patient with neuro-acanthocytosis. *Mov. Disord.* **16**, 572–575.

Wimer, B. M., Marsh, W. L., Taswell, H. F., and Galey, W. R. (1977). Haematological changes associated with the McLeod phenotype of the Kell blood group system. *Br. J. Haematol.* **36**, 219–224.

Windhorst, D. B., Zelickson, A. S., and Good, R. A. (1966). Chediak-Higashi syndrome: hereditary gigantism of cytoplasmic organelles. *Science* **151**, 81–83.

Yamamoto, T., Hirose, G., Shimazaki, K., Takado, S., Kosoegawa, H., and Saeki, M. (1982). Movement disorders of familial neuroacanthocytosis syndrome. *Arch. Neurol.* **39**, 298–301.

Yamamoto, T., Nanba, E., Ninomiya, H., Higaki, K., Taniguchi, M., Zhang, H., Akaboshi, S., Watanabe, Y., Takeshima, T., Inui, K., Okada, S., Tanaka, A., Sakuragawa, N., Millat, G., Vanier, M. T., Morris, J. A., Pentchev, P. G., and Ohno, K. (1999). NPC1 gene mutations in Japanese patients with Niemann-Pick disease type C. *Hum. Genet.* **105**, 10–16.

Yan-Go, F. L., Yanagihara, T., Pierre, R. V., and Goldstein, N. P. (1984). A progressive neurologic disorder with supranuclear vertical gaze paresis and distinctive bone marrow cells. *Mayo Clin. Proc.* **59**, 404–410.

Yonekawa, M., Okabe, T., Asamoto, Y., and Ohta, M. (1999). A case of hereditary ceruloplasmin deficiency with iron deposition in the brain associated with chorea, dementia, diabetes mellitus and retinal pigmentation: administration of fresh-frozen human plasma. *Eur. Neurol.* **42**, 157–162.

Yoshida, K., Furihata, K., Takeda, S., Nakamura, A., Yamamoto, K., Morita, H., Hiyamuta, S., Ikeda, S., Shimizu, N., and Yanagisawa, N. (1995). A mutation in the ceruloplasmin gene is associated with systemic hemosiderosis in humans. *Nat. Genet.* **9**, 267–272.

CHAPTER

# 45

# Ethical Issues in Genetic Testing for Movement Disorders

MARTHA A. NANCE
*Park Nicollet Clinic*
*St. Louis Park, Minnesota 55426*

THOMAS D. BIRD
*Department of Medicine and Neurology*
*University of Washington*
*Seattle, Washington 98108*

STEFAN-M. PULST
*Division of Neurology, Cedars-Sinai Medical Center*
*Departments of Medicine and Neurobiology,*
*UCLA School of Medicine*
*Los Angeles, California 90048*

I. Introduction
II. Understanding the Role of Molecular Diagnostics in the Management of Neurological Disorders
   A. Definition of a Genetic Test
   B. Diagnostic and Prenatal Genetic Testing
   C. Predictive Testing
   D. Carrier Testing and Risk Factor Testing
III. Ethical Principles
   A. Ethical Principles in Genetics
   B. Beneficence
   C. Privacy
   D. Autonomy
   E. Justice and Equality
IV. Conclusions
References

The ability to test for disease-causing alleles has facilitated the diagnosis of movement disorders, and has opened up the possibility of predictive, prenatal, and carrier testing for a number of adult-onset conditions. Testing for genetic risk factors for certain neurologic disorders will become increasingly available in the coming years. The goals and implications of genetic testing differ in important ways from those of other aspects of neurological practice. Ethical conflicts and dilemmas occur rather frequently in genetic medicine, in part because of a tension between the physician's duty to an individual patient and the fundamental involvement of the family in genetic matters, as well as the potential interest of third parties in an individual's genetic status. Incorporating genetic tests into neurologic practice requires physicians to understand both the underlying genetic principles and the ethical principles that guide their use.

## I. INTRODUCTION

The combined attention of the Decade of the Brain and the Human Genome Project in the 1990s led to an unprecedented explosion in our understanding of the genetic and molecular basis of many neurologic diseases, including—in particular—movement disorders. Once the province of specialized laboratories and specialized departments, molecular genetic methodologies are now widely used in both basic and clinical research, and university-based and commercial laboratories around the world perform DNA assays as a clinical diagnostic service. Molecular genetic testing for Huntington's disease (HD) has facilitated the diagnosis of this disorder and permitted at-risk individuals to undergo predictive testing, while the recognition that the genetic causes for adult-onset dominantly inherited ataxia number in the double digits has forced a genetic reclassification of this group of disorders. Multiple genes responsible for hereditary spastic paraplegia and dystonia have been localized, several of which, now identified, are clinically testable. The same research techniques that led to the identification of these "disease genes" are now being used to search for "risk factor genes"; genes for which certain variants or polymorphisms may confer an increased risk of developing a disease, without directly causing the disease. In the meantime, available gene tests are used not only for diagnostic purposes, but also for prenatal and predictive testing of clinically healthy individuals, a practice that we term "genetic medicine."

This chapter will focus on the ethical issues that attend the use of DNA tests for clinical purposes, emphasizing the ways in which genetic medicine differs from the usual and familiar practices of clinical neurology. While the ethical principles that govern genetic testing and genetic medicine are no different from the ethical principles that guide all of medical practice, ethical conflicts and dilemmas arise frequently and are of a nature that may be unfamiliar to neurologists and other clinicians.

After defining the term "genetic test," we will emphasize below that "genetic testing" is not a single entity. There are a variety of clinical situations in which gene tests are used, and the ethical devil is often buried in the details, the context in which the testing is performed. To illustrate our points, clinical vignettes will be presented, with modifications to protect the privacy of the individuals and families described. One author had the uncomfortable experience of meeting at a lay convention an individual whose difficult case had been described in a professional publication because of its unique privacy and ethical complexities. Recognizing the individual's story from its description in the journal, the author had to restrain the urge to say, "I've already read all about your case in Journal X!" This experience shows that even by illustrating for publication the ethical dilemmas we have faced in clinical practice, which we believe is important, as they validate the theoretical concerns, we threaten our patients' privacy.

## II. UNDERSTANDING THE ROLE OF MOLECULAR DIAGNOSTICS IN THE MANAGEMENT OF NEUROLOGICAL DISORDERS

### A. Definition of a Genetic Test

The United States Task Force for Genetic Testing defined a genetic test as "the analysis of human DNA, RNA, chromosomes, proteins, and certain metabolites in order to detect heritable disease related genotypes, mutations, phenotypes, or karyotypes for clinical purposes" (Task Force on Genetic Testing, 2000). While the discussion below will focus on DNA-based diagnostics, it is important to recognize that not all genetic tests are tests of genes. Measurement of cholesterol level in a child with a family history of familial hypercholesterolemia may confer a diagnosis of a genetic disorder and could be considered a genetic test. Some have argued that the Task Force definition is insufficiently broad—an MRI scan of the brain in a person at risk for tuberous sclerosis, for example, could be considered a genetic test (Zimmern, 1999). We recall a young woman whose "genetic test" for myotonic dystrophy took place in a medical school auditorium. During her examination as a normal control, intended to illustrate a contrast with affected family members, the neurologist unexpectedly demonstrated myotonia and weakness.

### B. Diagnostic and Prenatal Genetic Testing

#### 1. Limitations of Diagnostic Testing

Gene tests are used in five clinically distinct situations (Table 45.1), which can have widely varying implications for the immediate health, long-term prognosis, medical management, and reproductive risks of the individual tested. The technical and interpretative limitations of gene tests also impact on the ethical dilemmas that arise with their use. While physicians are accustomed to using tests of limited diagnostic sensitivity and specificity, the lack of published data about gene tests that incorporate this terminology could lead to the erroneous impression that they are 100% sensitive and 100% specific. In fact, there is a potential for both over- and under-interpretation of the significance of a gene test result for a particular patient. For example, a negative test result for the SPG4 gene does not mean that a patient cannot have a genetic form of spastic paraplegia, only that a SPG4 mutation was not detected. Previously unknown mutations in the SPG4 gene are still being identified, which may or may not be detected depending on the assay or assays used, and several other genetic forms of spastic paraplegia are known (Kobayashi *et al.*, 1996; Figlewicz *et al.*, 1999; Meijer *et al.*, 2002). A child with a very young onset age for HD, SCA2, or SCA7 may have a "negative" test result unless special diagnostic techniques are used, as the very large gene expansions associated with young-onset ages may not be detected using routine laboratory assays (Nance *et al.*, 1999; van de Warrenburg *et al.*, 2001; Mao *et al.*, in press). Because new genes are identified so quickly, because each gene and assay are liable to have unique idiosyncrasies that complicate the performance or interpretation of the assay, and

TABLE 45.1  Clinical Indications for Molecular Genetic Testing

| Type of test | Test candidate |
|---|---|
| Diagnostic testing | Symptomatic patient |
| Prenatal testing | Pregnant symptomatic or at risk individual or carrier couple |
| Predictive testing | Asymptomatic individual at risk for a specific disorder |
| Carrier testing | Parent, sibling, or other relative of individual with a autosomal or X-linked recessive disorder; screening of high-risk ethnic groups or population isolates |
| Risk factor assessment | Asymptomatic individual with (or without) a family history of a particular condition |

because clinical molecular diagnostic experience accrues in a piecemeal fashion across many laboratories, it is difficult for any physician to maintain a detailed working knowledge of the entire range of genetic tests, even in a small subspecialty such as the movement disorders. However, a modest understanding of the limitations of gene tests is important in the diagnostic setting, so that the physician can recognize test results that are incongruous with the patient before him, and know when he has exhausted the available diagnostic capabilities. This is analogous to the neurologist's need to understand enough details of brain MR imaging to decide which "white spots" require further evaluation in a particular patient, which are "just white spots," and at what point the limits of imaging technology have been reached.

Because the first ethical precept guiding physicians is to do no harm, there is an ongoing need for researchers who identify and study genes, laboratories that analyze genes for clinical purposes and render diagnostic interpretations, and physicians who translate these results into clinical care to (1) not use tests whose clinical significance is unknown, (2) convey clearly to patients the clinical limitations of a diagnostic test, and (3) re-evaluate and modify diagnostic interpretations as clinical experience accrues (analogous to post-marketing analysis of new therapeutic agents). In the excitement that attends the discovery of a new gene, researchers, clinicians, families, and lay disease organizations often push for a clinical test to be made available as soon as possible, sometimes before there is a good understanding of how the gene or its test would be relevant to a clinical population. Physicians must be aware of the "research-treatment gap" and help their patients to understand that the availability of a gene test does not imply that a treatment for the disease is near.

We have attempted to address the problem of haphazard post-marketing data collection by establishing two groups, the U.S. Huntington's Disease Genetic Testing Group and the Ataxia Molecular Diagnostic Testing Group, composed of clinicians and laboratories that use HD and ataxia gene tests, to monitor the use of the respective tests and to combine individual experiences into a larger whole (Potter *et al.*, 2000). Monitoring by the first group led to the formation of a working group that defined diagnostic boundaries for the HD gene test (American College of Medical Genetics/American Society of Human Genetics Huntington Disease Genetic Testing Working Group, 1998). A similar effort is currently underway for the ataxia genes. A similar analysis of genetic testing for torsion dystonia (DYT1) has been published (Klein *et al.*, 1999). An ongoing dialog between genetics practitioners, biotechnology companies, the insurance industry, and government exists, with the goal of helping all parties to ensure that the powerful tools of molecular diagnostics are neither suppressed nor overused (Finley Austin and Kreiner, 2002).

## 2. Responsibility to Other Family Members; Genetic Medicine

The neurologist may or may not perceive any responsibility to anyone else in the patient's family after diagnosing a genetic disorder; the unspoken contract in a neurologic encounter is made between the physician and the patient, and typically does not include other family members. Genetics specialists, in contrast, as an integral aspect of the practice of genetic medicine, acknowledge that the diagnosis of a genetic condition has immediate implications for others in the family, and ensure that counseling, educational materials, or appropriate referrals are available for the others. If the neurologist does not want to involve himself in the practice of genetic medicine, he should at least ensure, after making a genetic diagnosis, that referral to a genetics specialist is available to the family, much as he would refer a patient to an endocrinologist if hyperthyroidism was revealed to be the cause of a patient's tremor.

## 3. Prenatal Testing

Reproductive planning is often a major concern for patients with genetic disorders. While the neurologist typically does not play a major role in reproductive counseling or prenatal testing, he should refrain from letting his own cultural or religious beliefs interfere with the patient's reproductive freedom or access to prenatal testing or counseling.

## C. Predictive Testing

### 1. History and Definition of Predictive Testing

The use of genetic testing by an asymptomatic individual for the sole purpose of determining the presence or absence of a disease-causing allele is called predictive testing. The era of predictive medicine, which will soon permeate all areas of medicine, began in the mid-1980s after the localization of the HD gene, and hit its stride in neurology in the 1990s with the identification of the HD gene and several hereditary ataxia genes. Although the field has now expanded to include predictive testing for breast and colon cancer and other non-neurologic disorders, much of the early discussion and research on the ethical, legal, and social implications of predictive testing was built on the experience with HD. We will emphasize several points about predictive testing.

Predictive medicine differs from other medical transactions in several nontrivial ways, shown in Table 45.2. Predictive testing is a medical procedure performed on healthy individuals with no therapeutic goal other than whatever the patient's perceived psychosocial benefits may be. As it pertains to any of the movement disorders at

TABLE 45.2 Differences between Symptomatic, Preventive, and Predictive Medicine

|  | Symptomatic medicine | Preventive medicine | Predictive medicine |
| --- | --- | --- | --- |
| Target population | Sick inviduals | Healthy groups | Healthy individuals |
| Goal | To make well; to relieve discomfort | To prevent disease; early detection | Psychosocial benefits |
| Cost per individual | High | Low | Intermediate |
| Example | Treatment of stroke, brain tumor | MMR vaccine; Pap smear | Predictive test for HD |

this time, predictive testing does not lead to early diagnosis, improved or preventive medical treatment, or reduction in symptoms. Most physicians, neurologists included, have little experience in performing medical transactions with healthy individuals with no medical diagnosis, decision, or therapy at stake—there are no analogous encounters elsewhere in medicine. Because the physical procedure performed on the patient is a venipuncture, there is a tendency for both patients and physicians to equate a gene test with other blood tests, such as a cholesterol level or blood count. The minimally invasive nature of the physical procedure notwithstanding, we believe that the predictive test is more akin to a surgical biopsy than to a blood test. Once performed, a predictive genetic biopsy cannot be undone (just as an appendix cannot be replaced), nor can the result be influenced by medical treatment or acquired disease (unlike a cholesterol level or a white blood cell count). Finally, although a gene test is minimally invasive in a physical sense, it is a snip of the individual's very essence, and it can be invasive of family and personal relationships (Williams *et al.*, 2000; Sobel and Cowan, 2000). There are potential risks of predictive testing, primarily psychological and social ones. Fortunately, the risk of psychiatric hospitalization or death after predictive testing, at centers that provide genetic and psychological counseling as part of a testing protocol, is low (Almqvist *et al.*, 1999). Although there is reasonable ongoing concern about the possibility of discrimination against individuals who have undergone predictive tests, the number of patients who have experienced job-related or insurance discrimination following a gene test appears to be small (U.S. HD Genetic Testing Group, unpublished data).

Practitioners of predictive testing for HD have defined for the international medical and lay communities a safe and reasonable clinical approach to this new procedure (Anonymous, 1994; Huntington's Disease Society of America, 1994). HD predictive testing guidelines emphasize the process of genetic counseling, during which accurate information about the patient's disease risk, as well as the nature, limitations, uncertainties, and medical implications of the test, are provided. During the process of genetic counseling, a supportive rapport is also established with the patient, to help minimize the risk of psychosocial harm during or after the test. Autonomy—the free choice of a fully informed and consenting individual to complete the test—is also emphasized; family members, employers, insurers, and other third parties (including parents of minors) are not permitted to demand a test without the free consent of the individual being tested (Quaid, 1991).

## 2. Ethical Challenges in Predictive Testing

Challenges to autonomy arise frequently in the predictive testing setting. Often these are conflicts between a person who wants a test and a parent, twin, spouse, or other family member who does not want the test (or conversely, between a person who does not want to be tested and a parent, spouse, sibling, or other family member who wants the person to be tested). The testing of one sibling may put subtle (or overt) pressure on another sibling to undergo a test. A minority, perhaps 10–20%, of at-risk individuals throughout the world elects to undergo predictive testing for HD or the hereditary ataxias (Delatycki and Tassidker, 2001; Solis-Perez *et al.*, 2001), meaning that there is good reason to explore the motives of anyone who requests such a test. A recent case points out the potential for genetic tests to be used outside of the physician's office in ways that abrogate individual autonomy. The case concerned a man who was threatened with termination of employment by his employer when he refused to undergo testing for the gene mutation that causes hereditary neuropathy with a tendency to pressure palsies (HNPP) after filing a claim for work-related carpal tunnel syndrome. The case was settled out of court after representatives of the employee's union filed a lawsuit alleging that the employer coerced this employee and others into undergoing this test (Derse, 2001).

Like other physicians, geneticists strive to maintain their patients' right to privacy, which often presents a significant challenge, given that multiple family members with conflicting needs and requests may simultaneously be under the care of the same specialist. Some patients, because of particular concerns about privacy, have requested "anonymous" predictive testing. The challenges that attend anonymous testing have recently been discussed (Visintainer *et al.*, 2001). If we believe that documenting the patient's family history and experience with the disease, providing genetic information, and establishing a supportive rapport are important to ensure the safety and benefits of the predictive test, how should we manage a patient who will not provide family history information or discuss his life experiences, or who simply says, "I'm fine. I asked for the test. You don't have to know anything else. Do the test"?

## D. Carrier Testing and Risk Factor Testing

Two additional genetic testing indications remain to be discussed. One is carrier testing, relevant to (autosomal) recessively inherited diseases, for which a carrier of a single abnormal disease gene will not be affected with the disease, but may pass the gene to his children; and for X-linked disorders, in which a woman carrying one copy of an abnormal X-linked gene may not develop disease symptoms because of the presence of a second normal copy on her other X chromosome, but can have affected sons whose only X chromosome has the abnormal gene. In an individual family, the rationale for carrier testing is straightforward. However, problems can arise if third parties (employers, insurers), learning the results, interpret the presence of an abnormal gene to mean that the carrier has a disease. Carrier screening of ethnic groups with a high incidence of a particular genetic abnormality has also presented concerns, because of the risk of stigmatization of the group, or of discrimination against the group or against individuals within the group who have tested positive for the gene. A dark chapter in modern genetics occurred in the 1970s, when a well-intentioned sickle cell anemia screening program for African-Americans led to employment discrimination involving not only affected individuals but also carriers of a single abnormal sickle cell gene (Bowman, 1998a, b). Similar concerns pertain to genetic screening in the Ashkenazi Jewish population, where several identified genes occur in relatively high frequency (including the idiopathic dystonia gene, DYT1). In this example, in contrast to the previous one, religious community leaders have functioned proactively as leaders of the ethical debate, and the community as a whole has embraced the idea of genetic screening (Levin, 1999; Rothenberg and Rutkin, 1998).

The final indication for genetic testing is risk factor analysis. Many genes will be identified over the next 10 years which do not directly cause a disease in the classical Mendelian sense, but which increase the risk or influence the development of a disease in one way or another (Scott *et al.*, 2001; Li *et al.*, in press). Genetic epidemiologists have only recently begun to explore this very large area of inquiry. Perhaps the most well-known example of a neurogenetic risk factor is Apo E. The Apo E4 allele is a well-documented risk factor for Alzheimer's disease, but it is neither necessary nor sufficient to cause the disease. Some persons with the Apo E4 allele do not develop the disease, and some without the Apo E4 allele do. Thus, the predictive value of the test in asymptomatic persons is not high, and not well understood (Anonymous, 1995; Blacker, 2000). The incorporation of genetic risk factor analysis into neurological practice will require neurologists to convey to their patients concepts of molecular genetics, disease pathogenesis, and statistics in a succinct and clinically meaningful fashion.

## III. ETHICAL PRINCIPLES

### A. Ethical Principles in Genetics

Ethics is defined as a system of scholarly analysis of moral beliefs. We have already alluded to some of the universal ethical principles that guide medical practice in the United States in the early 21st century. The Committee on Assessing Genetic Risk suggested five principles that should govern ethical discussions of genetic testing: autonomy, confidentiality, justice or equity, privacy, and beneficence (Andrews *et al.*, 1994). Ethical dilemmas arise when the application of different equivalent ethical values or standards would result in opposing actions. In genetics, these ethical dilemmas may occur, for example, when the right of privacy of an individual conflicts with the right to beneficence of another individual. Dilemmas arise commonly in clinical practice, when real life decisions have to be made, and their recognition can be an important part of the genetic counseling process (Pembrey and Anionwu, 1996). Table 45.3 presents some clinical situations in which ethical conflicts or problems come to bear; the list is by no means exhaustive.

### B. Beneficence

#### 1. Beneficence in Diagnostic Testing

Beneficence is "doing good," which is the general goal of all medical encounters. A related idea, nonmaleficence, or the avoidance of harm, forms the first precept to physicians. Avoiding harm in the use of genetic tests primarily requires a properly informed laboratory, physician, patient, and public. While the overall educational task is a daunting one for geneticists, the assignment for the individual neurologist is not substantially different from familiarizing himself with a new drug, procedure, or diagnostic technology. There is a particularly great potential at this point in time for mistaken use of gene tests, mistaken interpretation of results, and misinformation to patients, when so many tests have recently become available but with relatively little cumulative worldwide experience.

We give as an example, a 72-year-old patient with dementia and subtle involuntary movements. The family history may include dementia in an elderly aunt; the patient's parents both died in their 60s of non-neurologic causes. The physician orders an Apo E test for Alzheimer's disease and an HD gene test. The Apo E report indicates that the patient has an Apo E3 and an Apo E4 allele,

TABLE 45.3  Ethical Conflicts in Clinical Neurogenetics

| Testing situation | Details | Principles |
|---|---|---|
| Diagnostic testing | Patient unexpectedly found to have two abnormal alleles for a dominantly inherited condition; how much information to give to give to which family members | Autonomy, justice, privacy, beneficence |
| Diagnostic testing | Interpretative uncertainty of test results | Beneficence |
| Predictive testing | One identical twin wants testing, the other does not | Autonomy, justice, privacy |
| Predictive testing | Testing an individual at 25% risk for a dominant disease (i.e., affected grandparent, but parent not yet showing symptoms) | Autonomy, justice, privacy |
| Predictive testing | Parental requests to test an at-risk child | Autonomy, beneficence |
| Prenatal testing | Testing the fetus of an at-risk person who does not want a predictive test | Autonomy, justice |
| Preimplantation testing | At-risk individual, who did not want predictive testing, is found to have two normal alleles (implying that preimplantation testing would not be necessary to ensure a fetus free of the disease) | Justice, beneficence |
| Carrier testing | (Mandatory) screening of high-risk populations | Justice, (automony) |
| Risk factor testing | Despite counseling, patient clearly misunderstands the meaning of the test | Beneficence |
| Risk factor testing | Pre-employment screening of genetic risk factors or genes | Autonomy, justice, privacy |

with the comment that this is strongly supportive of the diagnosis of Alzheimer's disease in an appropriate clinical context. The HD gene test indicates that the patient has one allele with 19 CAG repeats and one with 38 CAG repeats, and indicates that 38 repeats falls in the "indeterminate range." The physician concludes that the HD gene test is negative and the Alzheimer's test is positive and tells the family that the patient has Alzheimer's disease, which is usually not genetic. Although the test results are accurate, the conclusion is wrong! An HD gene with 38 repeats is abnormal, but "indeterminate" as to whether it will cause symptoms during a particular individual's lifetime; in a *symptomatic* person, the latter issue is moot, and the test result confirms the clinical diagnosis of HD. In contrast, Apo E4 is only a risk factor for Alzheimer's disease; its presence is not "diagnostic" of Alzheimer's disease, and does not exclude the possibility of another diagnosis.

We are accustomed to thinking of genes as normal or abnormal, mutated or not. For the trinucleotide (and now tetra- and penta-nucleotide) repeat disorders that form a growing group of neurogenetic disorders, however, the gene abnormality is a graded phenomenon, which may require more than a simple, "black and white," interpretation. Genes may contribute in a graded fashion to age of onset, nonpenetrance, or minimal disease expression may be present in some patients with abnormal alleles, and carriers of X-linked disease genes may show subtle evidence of disease, depending on who is looking and with what tools. The situation will grow more complex as more risk factor genes are identified. To help their patients and avoid harm, physicians who test patients for disease genes and risk factor genes will have to explain very carefully what the presence of a disease gene or risk factor does and does not imply, as well as alerting them to the possibility that third parties will misunderstand or misinterpret the same results.

## 2. Beneficial Effects of Predictive Testing

Knowledge gained through genetic testing has the potential to be positive for the individual, thus fulfilling the ethical principle of beneficence. Unfortunately, few studies have examined empirical outcomes following predictive genetic testing. Short-term follow-up of participants in the Canadian Predictive Testing Program for HD suggested that predictive testing showed beneficial effects in many individuals (Bloch *et al.*, 1992; Huggins *et al.*, 1992). Receipt of a decreased risk was associated with improved psychological well-being. Receipt of an increased risk did not result in the negative outcome that some had feared might result. However, a focus on physical symptoms and requests for frequent physical examinations were noted. Interestingly, about 10% of individuals receiving a decreased risk estimate had difficulties with facing a life that now needed planning beyond the likely onset age of HD. These studies were carried out prior to the identification of the HD gene. Later studies confirmed that the risk of catastrophic outcomes following predictive testing is low, suggesting that observed behavior may differ from preconceived notions about human behavior (Almqvist *et al.*, 1999). Importantly, Almqvist *et al.*(1999 were able to identify risk factors associated with adverse outcomes, including unemployment and prior psychiatric history. These studies and others (reviewed in Meiser and Dunn, 2000) form a basis for comparison with systematic studies of predictive testing as it is applied to other diseases, such as the ataxias, the dystonias, and the spastic paraplegias.

## C. Privacy

### 1. Infringing on Family Privacy for the Sake of the Patient

The unit of care in genetic medicine is really the family, and by its very nature and focus, genetics treads on the privacy of individuals. The tension between the primary contract between the physician and the patient and the perceived duty to others in the family underlies many of the ethical conflicts seen in clinical neurogenetics. This tension must be understood and acknowledged by the physician who cares for patients with genetic disorders. To begin with, even taking a family history, a critical step in establishing a patient's diagnosis, is invasive of the privacy of other family members, both living and deceased. We recall a patient with HD who indicated there was no family history of the disease. A few years later, a college research paper by the patient's nephew detailed how this family's progenitor, a leader of the community in which he was the first homesteader, was ridiculed and excluded from the town's history when HD led to intemperate behavior and strange movements. Even after several others over several generations developed the disease, most members of this family still do not acknowledge the existence of HD. The rest of us whose lives and families have not been subjected to public ridicule may simply prefer that others not know of our illegitimate children, past mental health problems, or current illnesses. But what better clue that a patient's diagnosis of Wilson's disease was incorrect than the history that three of her maternal uncles had a similar movement disorder, while her "unaffected" mother had died at age 35 in a car accident?

### 2. Confidentiality; Restricted Access to Genetic Information for Third Parties

A concept related to privacy is confidentiality, which requires the health professional to protect the information he has collected from being released to third parties without overwhelming reason. Although the challenges to the physician in ensuring confidentiality are no different for genetic information than for any other medical information, the perceived importance of confidentiality to the patient is much greater for genetic information, for good reason. We recall, as an example, the African-American patient who indicated that a dominantly inherited genetic disease had been introduced into his family a few generations ago by a prominent white community leader. Public release of this information could bring embarrassment to both families, as well as a major revision to the history of the community. As suggested above, the social stigma associated with the existence of a genetic disease in a family may extend over generations, and may include people who are not clinically affected with the disease, just because they are members of the family. Huntington's disease, along with epilepsy, mental retardation, and several other neurologic and neurosensory conditions, were included in the Racial Hygiene laws of 1930s Nazi Germany. Under these laws, individuals who were known to have specified conditions underwent involuntary sterilization (Harper, 1992). In the social environment of 21st century America, involuntary sterilization would not occur, but subtle and not-so-subtle forms of discrimination occur frequently in the workplace, at home, and in the social milieu.

Patients are particularly concerned about the possibility for adverse consequences if insurers become aware that they carry a gene responsible for a particular adult-onset condition. The concern is that health insurers could consider the condition to be pre-existing, and deny coverage of care for that condition. The Health Insurance Accountability and Portability Act of 1996 prohibits group health insurers from applying "preexisting conditions" exclusions to genetic conditions solely based on genetic tests in asymptomatic individuals. This protection, however, only applies to participation in group health insurance and does not cover individual health coverage. Over half of the states in the United States have enacted legislation to prevent insurers from using genetic information to assess risk in health insurance. However, these laws provide no protection against discrimination on the basis of family history of a genetic condition, nor, in general, do they relate to other types of insurance, such as life, disability, long-term care, and homeowner's insurance. On the other hand, preventing insurers from using the powerful genetic tools now available for determining health risk undermines the basis on which insurance is based. To date, the instances of genetic-based insurance discrimination appear to be few in number (Hall and Rich, 2000).

### 3. Using Patient Information for the Sake of the Family

Not only does the collection of genetic information impinge on the privacy of others, but so does a genetic diagnosis have immediate implications regarding the health risks of other family members. The physician who performs a gene test or makes a genetic diagnosis may acquire some degree of duty to these family members. The diagnosis of Friedreich's ataxia in a 10 year old implies that both parents are carriers of the FA gene and could have other affected children, and that siblings, aunts, and uncles, as well as grandparents and other more distant relatives could also be carriers. The correct diagnosis of HD in the woman described in Section III.C.1 suggests that her uncles also have HD, and that their children have a 50% risk of carrying the HD gene and one day developing HD as well. It can be difficult enough to explain to a single patient what his genetic diagnosis means to him; it is even more challenging to explain succinctly to the patient's brother, wife, and daughter who are in the room with him what the implications are to them. However, it is a lapse in justice and honesty not to inform, or at least provide an opportunity for informing, family members whose health or reproductive risks are directly affected by the patient's diagnosis.

Clinical situations can severely test the physician's desire, on the one hand, to be honest and to inform, with his duty, on the other hand, not to harm and to respect patient autonomy. One case involved a patient, being evaluated for possible symptoms of HD, who proved to have two abnormal HD alleles—confirming the diagnosis of HD, and also suggesting that all of his children had essentially a 100% risk of carrying an abnormal HD allele and therefore developing the disease. The specific question at hand, whether the patient had HD or not, could be answered without reference to the unusual details of the laboratory result. But while withholding information relevant to the children's disease risk would be dishonest, neither burdening the affected individual with the additional information that not only did he have HD, but that all his children would also one day develop the disease (unless they were not his biological children), nor reviewing the genetic details with the adult child who accompanied the parent to his appointment, would fit with the current views regarding justice, privacy, and autonomy in predictive genetic testing.

## D. Autonomy

### 1. Conflicts between Family Members

The principle of autonomy can be challenged when one family member's desire for genetic knowledge conflicts with another member's desire to remain ignorant of his genetic status. Autonomy is one concern in the case just described above. Other situations leading to ethical dilemmas or conflicts are shown in Table 45.3, including predictive or diagnostic testing of one identical twin and predictive testing of an individual at 25% risk for a dominantly inherited condition (Maat-Kievit et al., 1999; Benjamin and Lashwood, 2000). In these situations, a positive result for the requesting individual implies a positive result for a second individual, who has not undergone counseling or given consent for testing. A particularly difficult situation is one in which a woman requests a prenatal test for her fetus, while the at-risk partner does not want a predictive test. Genetically, this is simply a special case of testing a 25% risk individual, but socially there is a significant added twist, in that it is not the individual but one of two disagreeing parents requesting the test, and that the conflicting individuals likely live together and have shared responsibility for the outcome of the individual being tested.

### 2. Predictive and Carrier Testing of Children

Children are not considered to be autonomous in their decision-making ability, and they may not comprehend the implications of genetic testing. Recalling that a minority of adults chose to undergo predictive testing, and that there is no medical treatment to offer a minor who has had such a test, we believe that it is usually inadvisable for a parent to preempt the child's future right to choose whether or not to undergo predictive testing. Others have made a similar argument regarding parentally requested carrier testing of children for recessively inherited disorders (Davis, 1998). Exceptions can be made in some states for pregnant girls, who are regarded as emancipated minors for the purposes of genetic testing (Bird and Bennett, 2000). The balance between benefit and burden of childhood genetic testing is likely to change as preventive treatments for genetic disorders are developed, and as the interaction between genetic disorders and environmental factors are better understood. It may depend on the disorder and the reason for testing. One retrospective study of women who had undergone carrier testing for Duchenne muscular dystrophy as children found generally positive attitudes toward childhood testing (Jarvinen et al., 1999).

## E. Justice and Equality

Maintaining an appearance of justice and equality may also present a challenge, as alleles are not always distributed in a racially equal or politically correct fashion. Certain genetic conditions may be much more common in a particular ethnic group or population isolate. As long as individual practitioners care for individual patients in the same way, justice is not an issue. But if an employer, insurer, or researcher selects individuals of a particular religious or ethnic group to undergo a test or to be involved in a study, even if the goal is to improve the health of individuals or to prevent injury, injustice and infringement of autonomy may appear to be present. The principle of justice also requires that individual cases be managed in an equivalent fashion—the physician who agrees to perform a predictive genetic test on a mature 15 year-old because he knows the family well and thinks it would probably turn out okay, may be in trouble when the child's less mature 16 year-old sibling then requests a test.

Honesty is a derivative principle that sometimes conflicts with the physician's precepts not to harm and to preserve privacy. Paternalistic behavior can result when the physician, deviating from a course of complete honesty and full disclosure, decides what information or therapeutic choice is best for the patient, suppressing information that he thinks would be harmful. An example in Table 45.3 regards a couple requesting preimplantation testing for a dominantly inherited genetic disease. The husband is at risk for the disease, and the goal is to implant into the woman an embryo known to be free of the disease gene. The husband, after thinking carefully about it, states that he does not want to know his own gene status. The procedure is successful, and the couple bears an unaffected child. What they do not know, as they plan to proceed with another *in vitro* fertilization (IVF) procedure, is that all embryos formed in the first procedure were free of the disease gene (and that the laboratory, as a self-check process, tested both parents and knows that both are free of the disease gene). Is it ethically (or medically) correct for this couple to undergo the expense and risk of another IVF procedure when information is immediately available proving that they are not at risk for having a child with the disease? Should the IVF center simply provide the requested service, respecting the husband's desire not to know his true risk status (Braude *et al.*, 1998)?

## IV. CONCLUSIONS

The availability of genetic tests for diseases such as the hereditary ataxias, HD, and the dystonias, has greatly simplified the diagnostic process for these conditions. However, the ability to provide an accurate and specific diagnosis of a genetic disorder forces clinicians to recognize the implications of the diagnosis for other family members, thus moving them into the realm of genetic medicine. The goals of genetic medicine differ in important ways from those of general neurologic practice, and the ability to apply gene tests to healthy individuals is novel in medicine. As more gene tests become available for a wider range of disease-causing genes and disease-predisposing genetic variants, physicians will need to increase their understanding of genetics so that they can explain genetic concepts to their patients and make appropriate recommendations to them. Physicians must also be aware of the potential ethical conflicts presented by genetic medicine and genetic testing, so that they can maximize the benefits of these tests to their patients.

## References

Almqvist E. W., Bloch M., Brinkman R., Craufurd D., and Hayden M. R. (1999). A worldwide assessment of the frequency of suicide, suicide attempts, or psychiatric hospitalization after predictive testing for Huntington disease. *Am. J. Hum. Genet.* **64**, 1293–1304.

American College of Medical Genetics/American Society of Human Genetics Huntington Disease Genetic Testing Working Group (1998). Laboratory guidelines for Huntington disease genetic testing. *Am. J. Hum. Genet.* **62**, 1243–1247.

Andrews, L. B., Fullarton, J. E., Holzman, N. A., Motulsky, A. G., eds. (1994). Assessing genetic risks. Implications for health and social policy. Committee on assessing genetic risks. Institute of Medicine. Washington, D.C.: National Academy Press.

Anonymous (1994). Guidelines for the molecular genetics predictive test in Huntington's disease. International Huntington Association (IHA) and the World Federation of Neurology (WFN) Research Group on Huntington's Chorea. *Neurology* **44**, 1533–1536.

Anonymous (1995). Statement on use of apolipoprotein E testing for Alzheimer disease. American College of Medical Genetics/American Society of Human Genetics Working Group on ApoE and Alzheimer disease. *JAMA* **274**, 1627–1629.

Benjamin, C. M., and Lashwood, A. (2000). United Kingdom experience with presymptomatic testing of individuals at 25% risk for Huntington's disease. *Clin. Genet.* **58**, 41–49.

Bird T.D., and Bennett R.L. (2000). Genetic counseling and DNA testing. In *Neurogenetics*. (S. M. Pulst, ed.), pp. 433–442. Oxford University Press., New York.

Blacker, D. (2000). New insights into genetic aspects of Alzheimer's disease. Does genetic information make a difference in clinical practice? *Postgrad. Med.* **108**, 119–122, 125–126, 129.

Bloch M., Adam S., Wiggins S., Huggins M., and Hayden M. R. (1992). Predictive testing for Huntington disease in Canada: the experience of those receiving an increased risk. *Am. J. Med. Genet.* **42**, 499–507.

Bowman J. E. (1998a). Minority health issues and genetics. *Commun. Genet.* **1**, 142–144.

Bowman, J. E. (1998b). To screen or not to screen: when should screening be offered? *Commun. Genet.* **1**, 145–147.

Braude, P. R., De Wert, G. M., Evers-Kiebooms, G., Pettigrew, R. A., and Geraedts, J. P. (1998). Non-disclosure preimplantation genetic diagnosis for Huntington's disease: practical and ethical dilemmas. *Prenat. Diag.* **18**, 142–1426.

Davis, D. S.(1998). Discovery of children's carrier status for recessive genetic disease: some ethical issues. *Genet. Test.* **2**, 323–327.

Delatycki, M., and Tassidker, R. (2001). Adult onset neurological disorders. Predictive genetic testing. *Aust. Fam. Physician* **30**, 948–952.

Derse, A. R. (2001). The new genetic world and the law. *Wis. Lawyer* **74**, Number 4. Accessed through www.wisbar.org/wislawmag/2001/04.

Figlewicz, D. A., and Bird, T. D. (1999). "Pure" hereditary spastic paraplegias: the story becomes complicated. *Neurology* **53**, 5–7.

Finley Austin, M. J., and Kreiner, T. (2002). Integrating genomics technologies in health care: practice and policy challenges and opportunities. *Physiol. Genomics* **8**, 33–40.

Hall, M. A., and Rich, S. S. (2000). Laws restricting health insurers' use of genetic information: impact on genetic discrimination. *Am. J. Hum. Genet.* **66**, 293–307.

Harper, P. S. (1992). Huntington disease and the abuse of genetics. *Am. J. Hum. Genet.* **50**, 460–464.

Huggins, M., Bloch, M., Kanani, S., Quarrell, O. W., Theilmann, J., Hedrick, A., Dickens, B., Lynch, A., and Hayden, M. R. (1990). Ethical and legal dilemmas arising during predictive testing for adult-onset disease: the experience of Huntington disease. *Am. J. Hum. Genet.* **47**, 4–12.

Huggins, M., Bloch, M., Wiggins, S., Adam, S., Suchowersky, O., Trew, M., Klimek, M. L., Greenberg, C. R., Eleff, M., Thompson, L. P., Knight, J., MacLeod, P., Girard, K., Theilmann, J., Hedrick, A., and Hayden, M. R. (1992). Predictive testing for Huntington disease in Canada: adverse effects and unexpected results in those receiving a decreased risk. *Am. J. Med. Genet.* **42**, 508–515.

Huntington's Disease Society of America (1994). Guidelines for genetic testing for Huntington's disease. Huntington's Disease Society or America Inc., New York.

Jarvinen, O., Lehejoski, A. E., Lindlof, M., Uutela, A., and Kaariainen, H. (1999). Carrier testing of children for two X-linked diseases: a retrospective evaluation of experience and satisfaction of subjects and their mothers. *Genet. Test.* **3**, 347–355.

Klein, C., Friedman, J., Bressman, S., Vieregge, P., Brin, M. F., Pramstaller, P. P., De Leon, D., Hagenah, J., Sieverer, M., Fleet, C., Kiely, R., Xin, W., Breakfield, X. O., Ozelius, L. J., and Sims, K. B. (1999). Genetic testing for early-onset torsion dystonia (DYT1): introduction of a simple screening method, experiences from testing of a large patient cohort, and ethical aspects. *Genet. Test.* **3**, 323–328.

Kobayashi, H., Garcia, C. A., Alfonso, G., Marks, H. G., and Hoffman, E. P. (1996). Molecular genetics of familial spastic paraplegia: a multitude of responsible genes. *J. Neurol. Sci.* **137**, 131–138.

Levin, M. (1999). Screening Jews and genes: a consideration of the ethics of genetic screening within the Jewish community: challenges and responses. *Genet. Test.* **3**, 207–213.

Li, Y-J, Scott, W. K., Hedges, D. J., Zhang, F., Gaskell, P. C., Nance, M. A., Watts, R. L., Hubble, J. P., Koller, W. C., Pahwa, R., Stern, M. B., Hiner, B. C., Jankovic, J., Allen, F. A., Goetz, C. G., Mastaglia, F., Stajich, J. M., Gibson, R. A., Middleton, L. T., Saunders, A. M., Scott, B. L., Small, G. W., Reed, A. D., Schmechel, D. E., Welsh-Bohmer, K. A., Conneally, P. M., Roses, A. D., Gilbert, J. R., Vance, J. M., Haines, J. L., and Pericak-Vance, M. A. (2002). Onset in neurodegenerative diseases is genetically controlled. *Am. J. Hum. Genet.* **20**, 985–993.

Maat-Kievit, A., Vegter-van der Vlis, M., Zoeteweij, M., Losekoot, M., van Haeringen, A., and Roos, R. (1999). Predictive testing of 25 percent at-risk individuals for Huntington disease (1987–1997). *Am. J. Med. Genet.* **88**, 662–668.

Mao, R., Aylsworth, A. S., Potter, N., Wilson, W. G., Breningstall, G., Wick, M. J., Babovic-Vuksanovic, D., Nance, M. A., Patterson, M. C., Gomez, C. M., and Snow, K. (2002). Childhood-onset ataxia: testing for large CAG repeats in SCA 2 and SCA 7. *Am. J. Med. Genet.* **110**, 339–345.

Meijer, I. A., Hand, C. K., Cossette, P., Figlewicz, D. A., and Rouleau, G. A. (2002). Spectrum of SPG4 mutations ina large collection of North American families with hereditary spastic paraplegia. *Arch. Neurol.* **59**, 281–286.

Meiser, B., and Dunn, S. (2000). Psychological impact of genetic testing for Huntington's disease: an update of the literature. *J. Neurol. Neurosurg. Psychiatry* **69**, 57–578.

Nance, M. A., Mathias-Hagen, V., Breningstall, G., Wick, M. J., and McGlennan, R. C. (1999). Molecular diagnostic analysis of a very large trinucleotide repeat in a patient with juvenile Huntington disease. *Neurology* **52**, 392–394.

Pembrey, M. E., Anionwu, E. N. (1996). Ethical aspects of genetic screening and diagnosis. In *Principles and Practice of Medical Genetics*. (D. L. Rimoin, J. M. Connor, R. E. Pyeritz, eds.), New York: Churchill Livingstone, p. 641–653.

Potter, N. T., Nance, M. A., for the Ataxia Molecular Diagnostic Testing Group. (2000). Genetic testing for ataxia in North America. *Mol. Diagn.* **5**, 92–99.

Quaid, K. A. (1991). Predictive testing for HD: maximizing patient autonomy. *J. Clin. Ethics* **2**, 238–240.

Rosner, F. (1976). Tay-Sachs disease: to screen or not to screen? *J. Relig. Health* **15**, 271–281.

Rothenberg, K. H., and Rutkin, A. B. (1998). Toward a framework of mutualism: the Jewish community in genetics research. *Commun. Genet.* **1**, 148–153.

Scott, W. K., Nance, M. A., Watts, R. L., Hubble, J. P., Koller, W. C., Lyons, K., Pahwa, R., Stern, M. B., Colcher, A., Hiner, B. C., Jankovic, J., Ondo, W. G., Allen, F. H., Goetz, C.,G., Small, G. W., Masterman, D., Mastaglia, F., Laing, N. G., Stajich, J. M., Slotterbeck, B., Booze, M. W., Ribble, R. C., Rampersaud, E., West, S. G., Gibson, R. A., Middleton, L. T., Roses, A. D., Haines, J. L., Scott, B. L., Vance, J. M., and Pericak-Vance, M. A. (2001). Complete genomic screen in familial Parkinson Disease: evidence for multiple genes. *JAMA* **286**, 2239–2244.

Sobel, S. K., and Cowan, D. B. (2000). Impact of genetic testing for Huntington disease on the family system. *Am. J. Med. Genet.* **90**, 49–59.

Solis-Perez, M. P., Burguera, J. A., Palau, F., Livianos, L., Vila, M., and Rojo, L. (2001). Results of a program of presymtpomatic diagnosis of Huntington's disease: evaluation of a 6 year period. *Neurologia* **16**, 348–352.

Task Force on Genetic Testing. Promoting safe and effective genetic testing in the United States. http://www.nhgri.nih.gov/ELSI/TSGT_final/ (3/28/2000).

van de Warrenburg, B. P., Frenken, C. W., Ausems, M. G., Kleefstra, T., Sinke, R. J., Knoers, N. V., and Kremer, H. P. (2001). Striking anticipation in spinocerebellar ataxia type 7: the infantile phenotype. *J. Neurol.* **248**, 911–914.

Visintainer, C. L., Matthias-Hagen, V., and Nance, M. A. (2001). Anonymous predictive testing for Huntington's disease in the United States. *Genet. Test.* **5**, 213–218.

Williams, J. K., Schutte, D. L., Holkup, P. A., Evers, C., and Muilenburg, A. (2000). Psychosocial impact of predictive testing for Huntington disease on support persons. *Am. J. Med. Genet.* **90**, 353–359.

Zimmern, R. L. (1999). Genetic testing: a conceptual exploration. *J. Med. Ethics* **25**, 151–156.

# Index

## A

ABC7, 23
Abetalipoproteinemia
    microsomal triglyceride transfer protein deficiency, 237
    mouse model, 238
    treatment, 238
Acanthocytosis, retinitis pigmentosa, and pallidal generation, *see* HARP syndrome
ADCAs, *see* Autosomal dominant cerebral ataxias
Admixture, 11
Adrenoleukodystrophy (ALD), 6
    diagnosis, 244
    gene, 244
    phenotypes, 244
    treatment, 244–245
ALD, *see* Adrenoleukodystrophy
Allele
    definition, 8, 10
    frequency, 10
    haplotypes, 10
Allelic heterogeneity, 2
Alternative splicing, 8
Amplicon length, 12
Annealing, 12
Antibodies, 253, 257, 259; *see* autoantibodies
Anticipation
    ataxias, 25–27
    definition, 7
    dentatorubral-pallidoluysian atrophy, 144–145
    discovery, 25–26
    spinocerebellar ataxia 2, 48–49
    spinocerebellar ataxia 3, 62
    spinocerebellar ataxia 5, 77
    spinocerebellar ataxia 7, 88, 91
    spinocerebellar ataxia 10, 104
AOA, *see* Ataxia with oculomotor apraxia
ApoE4, Parkinson's disease allele association, 282–283
Apoptosis, Huntington's disease
    caspase mediation, 373
    cell-type specificity, 373–374
    evidence, 372–373
Aprataxin, mutation in ataxia with oculomotor apraxia, 22
ARJP, *see* Autosomal recessive juvenile parkinsonism
ARSACS/SACS, *see* Autosomal recessive spastic ataxia of Charlevoix-Saguenay
Ashkenazy Jews, 5
Association, definition, 10
Ataxia, *see also specific ataxias*
    animal models, 29
    anticipation, 25–27
    autosomal dominant ataxias, 19–21
    autosomal recessive ataxias, 19–20, 22–23
    classification, 19–21
    diagnostic evaluation
        antibody testing, 257
        cerebellar symptoms and signs, 254–255
        comparison of imaging and electrical data in inherited ataxias, 260
        differential diagnosis based on neural phenotype, 259, 262–263
        DNA testing, 27–28, 264, 266–267
        electrodiagnostics, 258–259
        genetic cause determination, 263–267
        neuroimaging, 28–29, 256–258
        neurological examination, 254–256
        pedigree construction, 264
        phenotypic templates for progressive ataxia categorization, 261–262
        sporadic ataxia, 28
        steps, 253
        testing after initial uninformative test, 267
    discovery of new types, 22–23
    distribution of genotypes, 264–265
    genotype/phenotype correlations, 29–30
    mitochondrial ataxias, *see* Mitochondrial disease
    modifying loci, 30
    prevalence and geographic distribution, 23–25
    prion diseases, *see* specific diseases
    progression and treatment, 30
    work-up of metabolic and mitochondrial ataxias, 247–248
Ataxia with oculomotor apraxia (AOA)
    loci, 22
    types, 22
Ataxia–telangiectasia
    animal models, 195, 200–201
    *ATM* gene
        DNA testing, 198
        locus, 197
        mutations, 197
        structure, 197
    ATM protein
        checkpoint control, 199–200
        cytoplasmic function, 200
        DNA repair, 198–199
        functional overview, 197
        kinase targets, 197–198
    clinical features
        cancer, 196
        diabetes, 197
        immune system dysfunction, 197
        infertility, 196
        neurological phenotype, 196
        onset, 196
        radiation sensitivity, 196
        telangiectasia, 196–197
    diagnosis, 198
    epidemiology, 195
    genotype/phenotype correlations, 201–202
    neuroimaging, 198
    reated syndromes, 202
    treatment, 202
Ataxia with hearing impairment and optical atrophy, 23
Ataxia with sideroblastic anemia, 23
Ataxia with vitamin E deficiency (AVED)
    animal models, 29, 184–185
    blood tests, 183
    clinical features

Ataxia with vitamin E deficiency (AVED) (*continued*)
    cardiac involvement, 181
    cerebellar syndrome, 180
    ocular signs, 181
    onset, 180
    overview, 179
    pyramidal syndrome, 180
    sensory disturbances, 180
    skeletal abnormalities, 181
    tend reflex changes, 180–181
  epidemiology, 180
  genotype/phenotype correlations, 185
  history of study, 179–180
  nerve biopsy, 184
  nerve conduction study, 183–184
  neuroimaging, 184
  neuropathology, 184
  somatosensory-evoked potentials, 184
  α-tocopherol transfer protein gene
    DNA testing, 183
    function, 181–182
    identification, 181
    mutations, 179, 182–183
  treatment, 185
Ataxin-1
  overexpression in cells, 38
  protein–protein interactions, 36–37
  RNA binding, 37
  spinocerebellar ataxia 1 pathogenesis, 38–41
  subcellular localization, 36–37, 39–40
  transgenic mouse, 38–41
Ataxin-2
  functions, 49–50
  homologous proteins, 49
  protein–protein interactions, 49
  RNA binding, 45
  spinocerebellar ataxia 2 pathogenesis, 50
  tissue distribution, 49–50
Ataxin-3
  function, 63
  misfolding and aggregation in spinocerebellar ataxia 3, 64
  neurotoxicity mechanisms, 64–65
  subcellular localization, 63–64
  tissue distribution, 63
Ataxin-7
  aggregation in spinocerebellar ataxia 7, 85, 87
  caspase cleavage, 90
  function, 86–87
  protein–protein interactions, 90
  subcellular localization, 87
  tissue distribution, 86–87
ATM, *see* Ataxia–telangiectasia
ATP7B, *see* Wilson disease
AT sequence, 11
ATTCT repeat, *see* Spinocerebellar ataxia 10
Atypical Parkinsonism, 3
A1Up, ataxin-1 interactions, 37
Autoantibodies
  cerebellar degeneration, 257, 259
Autosomal dominant cerebral ataxias (ADCAs), *see also specific spinocerebellar ataxias*
  clinical features, 21
  SCA4 locus linkage, 72
  type 2, *see* Spinocerebellar ataxia 7
  types, 21, 104
Autosomal dominant disease, penetrance, 4–5
Autosomal dominant progressive external ophthalmoplegia, features, 235
Autosomal recessive disease, heredity, 5
Autosomal recessive juvenile parkinsonism (ARJP), adult-onset Tay–Sachs disease comparison, 511–512
Autosomal recessive spastic ataxia of Charlevoix-Saguenay (ARSACS/SACS)
  clinical features, 189–191
  genotype/phenotype correlations, 192
  neuroimaging, 192
  neuropathology, 192
  neurophysiologic assessment, 191–192
  *SACS* gene
    DNA testing, 191
    mapping, 191
    mutations, 189
    structure, 191
    tissue distribution of expression, 191
AVED, *see* Ataxia with vitamin E deficiency

## B

Bell, Julia, 25
Beneficience, genetic testing
  diagnostic testing, 545–546
  predictive testing, 546–547
Benign hereditary chorea, features, 376
Biopterin deficiency, animal model, 424–426
Biotinidase deficiency, ataxia, 237
Botulinum toxin, essential tremor management, 360
Bovine spongiform encephalopathy (BSE), human transmission, 155
BSE, *see* Bovine spongiform encephalopathy

## C

*CACNA1A*, *see also* Spinocerebellar ataxia 6
  episodic ataxia type 2
    animal models, 210
    linkage analysis, 205
    molecular pathogenesis, 209–210
    mutations, 208
    phenotypes with mutations, 208
    protein function and topology, 207
Carbidopa, multiple system atrophy management, 226
Carrier, 5–6
Catechol-*O*-methyl transferase (COMT), Parkinson's disease allele association, 283
Cayman ataxia, 23
CBD, *see* Corticobasal degeneration
CDG, *see* Congenital disorders of glycsylation
Cerebellar atrophy, causes, 256–257
Cerebellar deficient folia mouse 29
Cerebellar hypoplasia, features, 247
Cerebrospinal fluid (CSF), prion disease diagnostics, 160
Cerebrotendinous xanthomatosis (CTX)
  animal models, 238
  clinical features, 238
  CYP27A1 deficiency, 238
  diagnosis, 238
  neuroimaging, 238
Ceruplasmin
  copper binding, 343–344
  deficiency
    clinical features, 520
    diagnosis, 520
    gene mutations, 350, 520
    pathology, 520
    treatment, 520
  functions, 343–344
  isoforms, 344
  plasma population, 344
CHAC, *see* Choreoacanthocytosis
Chain termination method, DNA sequencing, 13
Chediak–Higashi syndrome (CHS)
  clinical features, 521
  diagnosis, 522–523
  gene
    adult-onset disease, 521
    childhood-onset disease, 521
    mapping, 521
  pathology, 521–522
  pathophysiology, 522
  treatment, 523
  variants, 521
Choreoacanthocytosis (CHAC)
  clinical features, 524
  diagnosis, 525–526
  discovery, 523
  genotype, 524
  neuroimaging, 526
  pathology, 524
  pathophysiology, 524–525
  pedigree analysis, 523–524
  treatment, 526–527
Chimeric animal, 16
Chromosomal abnormalities, 11
CHS, *see* Chediak–Higashi syndrome
CJD, *see* Creutzfeldt–Jakob disease
Co-dominant, 3
Coenzyme Q
  coenzyme Q10 muscle deficiency and ataxia, 235
  Friedreich's ataxia management, 175
Complex genetic trait, 2
Compound heterozygosity, 3
Computed tomography, *see* Neuroimaging
COMT, *see* Catechol-*O*-methyl transferase
Congenital disorders of glycsylation (CDG)
  clinical features, 246
  diagnosis, 246
  *PMM* genes, 246
  types, 246
Consanguinity, 5
Copper
  absorption, 343
  ATP7B role in homeostasis, 345
  ceruloplasmin binding, 343–344
  chelation therapy, 349–350
  homeostasis, 343
  Wilson disease diagnosis, 346

Corticobasal degeneration (CBD)
  animal models, 335
  genetic testing
    clinical implications, 334
    ethical dilemmas, 334–335
    specificity, 333
  neuroimaging, 333
  *tau* mutations
    clinical features, 332
    frequency, 333
    neuropathology, 332–333
    types, 325–326, 332
  treatment, 335
Creatine kinase, 22
Cre-LoxP, 16
Cre-recombinase, 16
Creutzfeldt–Jakob disease (CJD)
  clinical features
    iatrogenic form, 154
    sporadic form, 153–154
    variant form, 155–156
  diagnosis, 159–160, 258
  familial *PRNP* mutations
    D178N, 157–158
    E200K, 156–157
    repeat expansion, 158
CSF, *see* Cerebrospinal fluid
CTX, *see* Cerebrotendinous xanthomatosis
CYP27A1 deficiency, *see* Cerebrotendinous xanthomatosis

## D

DDS, *see* Dystonia deafness syndrome
Deep brain stimulation
  dystonia 1 management, 415
  essential tremor management, 360
Deletion, 11, 15
Dentatorubral-pallidoluysian atrophy (DRPLA)
  animal models, 147–148
  anticipation, 144–145
  clinical features, 143–144
  diagnosis, 145–146
  electroencephalography, 145
  gene
    CAG repeat expansion, 144
    diagnostics, 145
    mapping, 144
    tissue distribution of expression, 144
  genotype/phenotype correlations, 144
  molecular pathology, 146–147
  neuroimaging, 145
  neuropathology, 146
  prevalence and geographic distribution, 23–25
  treatment, 148
Diabetes
  ataxia–telangiectasia patients, 197
  Friedreich's ataxia patients, 167
  neonatal diabetes, 247
Direct mutation detection, 14
Disequilibrium, linkage, 10–11
DNA polymorphism, 8
DNA sequencing
  chain termination method, 13
  mutation detection, 13–15
DNA testing, *see* Genetic testing; Polymerase chain reaction
Dominance
  gain of function mutations, 3
  loss of function mutations, 3
  mechanisms, 3
  overview, 2–3
Dominant adult-onset basal ganglia disease, features, 376
Dominant-negative mutation, 3
Dopamine, α-synuclein aggregation inhibition, 297
Dopamine D2 receptor, myoclonus dystonia
  genotype/phenotype correlation, 466–467
  mutations, 399–400, 460, 462–463
Dopa-responsive dystonia (DRD)
  animal model of biopterin deficiency, mechanism of neuronal selectivity, 424–426
  clinical features, 397, 419–420
  diagnosis, 422–423
  dominant inheritance mechanism, 424
  genotype/phenotype correlation, 423
  guanosine triphosphate cyclohydrolase I gene
    mapping, 420–421
    mutations, 398, 419, 421–422
    sporadic mutations, 421–422
  neuroimaging, 423
  neuropathology, 423
  treatment, 420
  tyrosine hydroxylase mutations, 422
DRD, *see* Dopa-responsive dystonia
*Drosophila* models
  dentatorubral-pallidoluysian atrophy, 147
  dystonia 1, 414
  Hallervoden–Spatz syndrome, 437
  spinocerebellar ataxia 1, 41
  spinocerebellar ataxia 2, 52
DRPLA, *see* Dentatorubral-pallidoluysian atrophy
Dynamic mutations, 1
Dysmorphologic feature, 11
Dystonia, *see also* specific dystonias
  animal models, 402–403
  classification
    age of onset, 396
    focal, 395
    generalized, 395
    multifocal, 395
    primary dystonia types, 396–397
    segmental, 395
  definition, 452
  mitochondrial mutations
    idiopathic dystonia, 482
    mitochondrial dysfunction evidence, 481
    mitochondrial toxin induction, 481
    rare dystonia forms, 481–482
    treatment implications, 482–483
Dystonia 1
  clinical features, 407–409
  diagnosis, 413
  epidemiology, 407, 409
  gene
    DNA testing, 413
    GAG deletion, 408, 410–412, 480
    homologous genes, 411–412
    identification, 410–411
    mutations, 407
    torsin A function and dysfunction, 412–413
  haplotype analysis, 410
  inheritance pattern, 409–410
  linkage disequilibrium, 410
  mitochondrial mutations, *see* Dystonia
  models
    cells, 414
    *Drosophila*, 414
    mouse and rat, 414
  neuroimaging, 413
  neuropathology, 413–414
  onset, 397, 409
  sex differences, 408–409
  treatment, 414–415
Dystonia 2
  herredity, 400
  tyrosine hydroxylase deficiency, 400
Dystonia 3, *see* X-linked dystonia-parkinsonism
Dystonia 4, classification, 397
Dystonia 5, *see* Dopa-responsive dystonia
Dystonia 6
  linkage analysis, 398
  onset, 398
Dystonia 7
  clinical features, 398
  linkage analysis, 398
Dystonia 8
  clinical features, 386–387
  genetics, 387, 398–399
  onset, 398
Dystonia 9, features, 399
Dystonia 10
  clinical features, 386–388, 399
  epilepsy association, 388, 390–391, 399
  genetics, 387, 389–390, 399
Dystonia 11, *see* Myoclonus dystonia
Dystonia 12, *see* Rapid-onset dystonia parkinsonism
Dystonia 13
  clinical features, 400
  linkage analysis, 400
Dystonia deafness syndrome (DDS)
  clinical features, 402
  genetics, 402

## E

EA-1, *see* Episodic ataxia type 1
EA-2, *see* Episodic ataxia type 2
EA-3, *see* Episodic ataxia type 3
EA-4, *see* Episodic ataxia type 4
Early-onset cerebellar ataxia with hypoalbuminemia, 22
EEG, *see* Electroencephalography
Electroencephalography (EEG)
  autosomal recessive spastic ataxia of Charlevoix-Saguenay, 191
  Creutzfeldt–Jakob disease, 258

dentatorubral-pallidoluysian atrophy, 145
prion disease, 160
Embryogenesis, 16
Embryonic lethality, 16
Embryonic stem cell, 15
Epilepsy, paroxysmal kinesigenic dyskinesia association, 388, 390–391
Episodic ataxia type 1 (EA-1)
   animal models, 210
   clinical features, 205–206
   diagnosis, 208–209
   *KCNA1* gene
      linkage analysis, 205
      molecular pathogenesis, 209
      mutation, 207
      protein function, 207
   treatment, 210–211
Episodic ataxia type 2 (EA-2)
   animal models, 210
   *CACNA1A* gene
      linkage analysis, 205
      molecular pathogenesis, 209–210
      mutations, 208
      protein function and topology, 207
   clinical features, 205–206
   diagnosis, 208–209
   treatment, 210–211
Episodic ataxia type 3 (EA-3)
   clinical features, 205–206
   diagnosis, 208–209
   gene discovery, 205
   treatment, 210–211
Episodic ataxia type 4 (EA-4)
   clinical features, 205–207
   diagnosis, 208–209
   gene discovery, 205
   treatment, 210–211
Essential tremor (ET)
   animal models, 357–358
   clinical features, 353–355
   course, 354
   diagnosis, 356
   epidemiology, 353
   etiology, 354
   genes, 353, 355–356
   genotype/phenotype correlations, 358
   heredity, 354–356
   neuroimaging, 356–357
   neuropathology, 357
   penetrance, 358
   treatment, 358–360
ET, *see* Essential tremor
Exon, 8
Expressivity, definition, 5

## F

Familial aggregation, 8
Familial hemiplegic migraine, 2
Familial idiopathic basal ganglia calcification (FIBGC)
   clinical features, 443–444
   differential diagnosis, 445–446
   gene
      candidate gene analysis, 444
      DNA testing, 444–445
      linkage analysis, 443–444
   genotype/phenotype correlations, 448
   inheritance, 444
   neuroimaging, 446–447
   neuropathology, 447
   pathogenesis, 447–448
   penetrance, 444
   treatment, 448
Familial isolated vitamin E deficiency, *see* Ataxia with vitamin E deficiency
Fatal familial insomnia (FFI)
   clinical features, 158
   diagnosis, 159–160
   histopathology, 158
   transgenic mouse models, 161
Ferritin
   neuroferritinopathy mutations, 518–519
   structure and function, 519
α-Fetoprotein, 22
FFI, *see* Fatal familial insomnia
FIBGC, *see* Familial idiopathic basal ganglia calcification
Flunarizine, essential tremor management, 360
Founder effect, 22
Fragile X syndrome, premutation alleles, 533
Frameshift mutation, 11
Frataxin
   Friedreich's ataxia pathogenesis, 165, 170
   functions, 170
   knockout mouse, 174
   point mutations, 169–170
FRDA, *see* Friedreich's ataxia
Friedreich's ataxia (FRDA)
   animal models, 29, 174
   biochemical investigations, 172
   carriers, 172
   clinical features
      diabetes mellitus, 167
      heart disease, 167, 173
      neurological signs and symptoms, 166
      onset, 166
      variant phenotypes, 168
   differential diagnosis, 168–169
   DNA testing, 27–28
   epidemiology, 165–166
   familial isolated vitamin E deficiency, *see* Ataxia with vitamin E deficiency
   *FRDA* gene
      DNA testing, 170–172
      GAA repeat expansion, 103, 169, 174
      point mutations, 169–170, 174
      protein, *see* Frataxin
      structure, 169
      tissue distribution of expression, 169
   genotype/phenotype correlations, 174
   neuroimaging, 172–173
   neurological examination, 166–167
   neurophysiological investigations, 167–168
   pathogenesis, 165, 170
   pathology, 173
   prognosis, 168
   treatment, 30, 174–175
Frontotemporal lobar dementia (FTLD)
   animal models, 335
   genetic testing
      clinical implications, 334
      ethical dilemmas, 334–335
      specificity, 333
   neuroimaging, 333
   *tau* mutations in FTDP-17
      clinical features, 331–332
      frequency, 333
      neuropathology, 332–333
      overview, 325–326
      translation defects, 330–331
   treatment, 335
Fruitfly models, 15; *see also* Drosophila models
FTLD, *see* Frontotemporal lobar dementia

## G

Gabapentin, essential tremor management, 360
Gain of function mutation, dominance, 3
Galactosemia, features, 247
GeneClinics, 16–17
Gene
   mutations, *see* Mutation
   structure, 8
   transcription, 8
Gene targeting, 16
Genetic testing, *see also* Polymerase chain reaction
   ataxia, 27–28, 264, 266–267
   ataxia–telangiectasia, 198
   ataxia with vitamin E deficiency, 183
   autonomy, 548
   autosomal recessive spastic ataxia of Charlevoix-Saguenay, 191
   beneficience
      diagnostic testing, 545–546
      predictive testing, 546–547
   carrier testing, 545
   children, 548
   definition, 540
   dystonia 1, 413
   familial idiopathic basal ganglia calcification, 444–445
   Friedreich's ataxia, 27–28, 170–172
   Hallervoden–Spatz syndrome, 434
   Huntington's disease, 541, 547–548
   indications, 542
   justice and equality, 548–549
   limitations of diagnostic testing, 542–543
   myoclonus dystonia, 464
   Parkinson's disease
      *parkin* mutations
         haplotype analysis, 309
         polymerase chain reaction, 308–309
         sequencing, 309
         sporadic patients, 309
      α-synuclein mutations, 298
   predictive testing
      definition, 543–544
      ethical challenges, 544–545
      historical perspective, 543–544
   prenatal testing, 543

privacy
  confidentiality, 547
  patient versus family, 547–548
responsibility to other family members, 543
risk factor testing, 545
spinocerebellar ataxia 7, 87–88
tauopathies
  clinical implications, 334
  ethical dilemmas, 334–335
  specificity, 333
Tay–Sachs disease counseling, 515
Wilson disease, 346
Genotype
  alleles, 10
  definition, 2
Germline mutation, 15
Gerstmann–Sträussler–Scheinker disease (GSS)
  clinical features, 158–159
  diagnosis, 159–160
  PRNP mutations
    F198S, 159
    P102L, 159
    Q217R, 159
Gilles de la Tourette syndrome (GTS)
  clinical features, 491–492
  cytogenetic findings, 496–497
  genes
    association studies, 494–495
    linkage analysis, 495–496
  heredity
    family studies, 492–493
    segregation analysis, 493–494
    twin studies, 493
  neuroimaging, 497
  prevalence, 491
  prospects for study, 497–498
  psychiatric disorder comorbidity, 493
  treatment, 497
Gliadin antibodies, 255, 257; see autoantibodies
Glial cytoplasmic inclusion (GSI), multiple system atrophy, 213, 223–224
γ-Globulin, 22
Glycosylation, see Congenital disorders of glycsylation
GM2 gangliosidosis, see Tay–Sachs disease
GSI, see Glial cytoplasmic inclusion
GSS, see Gerstmann–Strüssler–Scheinker disease
GTS, see Gilles de la Tourette syndrome
GT sequence, 11
Guanosine triphosphate cyclohydrolase I, see Dopa-responsive dystonia

# H

Hallervoden–Spatz syndrome (HSS)
  animal models, 437
  clinical features
    atypical PKAN, 432
    classical PKAN, 430–431
    HARP syndrome, 431–432
    neurodegeneration with brain iron accumulation, 432–433
    progression, 431
  diagnostic criteria, 432
  genotype/phenotype correlation, 437–438
  neurodegeneration with brain iron accumulation, 428, 432–433, 436–437
  neuroimaging
    functional imaging, 436
    magnetic resonance imaging, 429, 434–436
    magnetic resonance spectroscopy, 436
  neuropathology, 436–437
  nosology, 427–428
  PANK2
    abnormal functions, 434
    DNA testing, 434
    homologs, 433–434
    mutations, 427, 434–435
    protein function, 433–434
    structure, 433
  treatment, 438
Haplotype, definition, 10
Harding, Anita, 21
HARP syndrome
  clinical features, 530
  diagnosis, 530–531
  genotype, 530
  Hallervoden–Spatz syndrome, 431–432
  neuroimaging, 531
  pathology, 530
  pathophysiology, 530
  treatment, 531
Hartnup disease, features, 247
HD, see Huntington's disease
Hemizygosity, 6
Hereditary spastic ataxia, 22
Hereditary spastic paraplegia, 22
Heterozygote, 2
Hexosaminidase A deficiency, see Tay–Sachs disease
Homologous chromosome, 2
Homologous recombination, 16
HSS, see Hallervoden–Spatz syndrome
HUGO, see Human Genome Organization
Human Genome Organization (HUGO), 19
Huntingtin
  aggregation
    chaperones, 371
    degradation, 371–372
    mechanisms, 372
    neuronal inclusions and toxicity, 370–371
    transglutaminase role, 372
  function, 369–370
  mutation, see Huntington's disease
  proteolysis, 373–374
Huntington's disease (HD)
  animal models, 375
  apoptosis
    caspase mediation, 373
    cell-type specificity, 373–374
    evidence, 372–373
  cell dysfunction, 373
  clinical features, 365–366
  complementary DNA analysis of early changes in gene expression, 374
  diagnostic testing, 368–369
  excitotoxicity, 374–375
  gene
    CAG repeats, 367
    gain of function model, 369
    mapping, 2, 367
    protein, see Huntingtin
    sporadic mutations, 367–36
    structure, 367
  genetic testing ethics, 541, 547–548
  genotype/phenotype correlations, 375–376
  mitochondrial dysfunction, 375
  neuroimaging, 366–367
  neuropathology, 366
  predictive testing, 368
  treatment, 376
Huntington's disease-like disorders
  benign hereditary chorea, 376
  dominant adult-onset basal ganglia disease, 376
  HD like-1, 377
  HD like-2, 377
  HD like-3, 377
  McLeod syndrome, 377
  neuroacanthocytosis, 377
4-Hydroxybutyricaciduria
  animal models, 247
  clinical features, 246–247
  diagnosis, 247
L-2-Hydroxyglutaricacidemia, features, 245–246
Hypertrophic cardiomyopathy, 14

# I

Idiopathic focal dystonia, see Dystonia 7
Idiopathic torsion dystonia, see Dystonia 1
Imprinting
  definition, 7
  diseases, 8
Indirect mutation detection, 14, 15
Insertion, 11, 15
Insertional mutagenesis, 15

# J

Johannsen, 8

# K

KCNA1
  episodic ataxia type 1
    animal models, 210
    linkage analysis, 205
    molecular pathogenesis, 209
    mutation, 207
    protein function, 207
Kearns–Sayre syndrome (KSS), features, 232–233
KLHL1, see Spinocerebellar ataxia 8
Knock-in, 16
Knockout mouse, disease models, 16
KSS, see Kearns–Sayre syndrome
Kuru
  clinical features, 154–155
  diagnosis, 159–160

## L

LANP, see Leucine-rich acidic nuclear protein
Leaner mouse, 29
Leigh syndrome (LS), features, 234
Leucine-rich acidic nuclear protein (LANP), ataxin-1 interactions, 36–37, 41
Levodopa, multiple system atrophy management, 226, 300, 305, 306, 312, 316, 317, 319, 321
Linkage analysis
   disequilibrium, 10–11
   markers, 10
Locus, 2
Lod score, 10
Loss of function mutation, dominance, 3
LS, see Leigh syndrome
Lubag, see X-linked dystonia-parkinsonism

## M

Machado–Joseph disease, see Spinocerebellar ataxia 3
Magnetic resonance imaging, see Neuroimaging
Magnetic resonance spectroscopy, see Neuroimaging
Manifesting heterozygote, 6
MAO-B, see Monoamine oxidase type B
Marinesco–Sjogren syndrome, clinical features, 23
McLeod syndrome
   clinical features, 377, 527
   diagnosis, 529
   family studies, 527
   neuroimaging, 529
   pathology, 528
   pathophysiology, 528
   treatment, 529–530
   XK gene mutations, 527–528
MELAS, see Mitochondrial encephalopathy, lactic acidosis, and stroke-like episodes
Mendel, Gregor, 2
Mendelian trait, heredity, 2
MERFF, see Myoclonus-epilepsy with ragged red fibers
Mevalonic aciduria
   clinical features, 245
   diagnosis, 245
Missense mutation, 11, 15–16
Mitochondrial disease
   biotinidase deficiency, 237
   classification, 232
   coenzyme Q10 muscle deficiency, 235
   definition, 231–232
   dystonia mutations, see Dystonia
   genetics overview, 474
   genome, 7
   heredity, 7
   heteroplasmy, 474
   mitochondrial DNA mutations and ataxias
      genetics, 232
      hearing loss, ataxia, and myoclonus, 233–234
      Kearns–Sayre syndrome, 232–233
      mitochondrial encephalopathy, lactic acidosis, and stroke-like episodes, 233
      myoclonus-epilepsy with ragged red fibers, 233
      neurogenic weakness, ataxia, and retinitis pigmentosa, 233
   nuclear gene mutations and ataxias
      autosomal dominant progressive external ophthalmoplegia, 235
      defects in factors involved in mitochondrial biogenesis, 235
      Leigh syndrome, 234
      mitochondrial neurogastrointestinal encephalomyopathy, 235
   Parkinson's disease mutations, see Parkinson's disease
   pyruvate dehydrogenase deficiency, 236–237
   work-up of ataxias, 247–248
Mitochondrial encephalopathy, lactic acidosis, and stroke-like episodes (MELAS), features, 233
Mitochondrial neurogastrointestinal encephalomyopathy (MNGIE), features, 235
MJD1, see Ataxin-3; Spinocerebellar ataxia 3
MNGIE, see Mitochondrial neurogastrointestinal encephalomyopathy
Modified Harding phenotype, 21
Modifying allele, 2
Modifying gene
   ataxia loci, 30
   definition, 2
Modifying locus, 2
Molecular-genetics databases, Web sites, 16–17
Monoamine oxidase type B (MAO-B), Parkinson's disease allele association, 283
Mouse models, 15
MSA, see Multiple system atrophy
Multiple system atrophy (MSA)
   classification, 213–214
   clinical features
      autonomic dysfunction, 220
      cerebellar type, 214–215, 217
      parkinsonian type, 214
   diagnostic criteria
      consensus criteria, 219
      Quinn criteria, 217–219
   environmental risk factors, 225
   epidemiology, 224–225
   genetics, 225
   imaging
      magnetic resonance imaging, 220–221
      overview, 213
      photon emission tomography
         metabolic markers, 222
         striatal dopaminergic terminal markers, 222–223
         striatal postsynaptic dopaminergic receptor markers, 223
      single photon emission computed tomography
         brain imaging, 221–223
         cardiac imaging, 221
   management
      autonomic disorders, 226
      extrapyramidal features, 226
   neuropathology, 213, 223–224
Mutation
   deletions, 11
   direct detection, 14–15
   DNA sequencing, 13–15
   indirect molecular testing, 15
   insertions, 11
   missense mutations, 11
   nonsense mutations, 11
   point mutations, 11
   splice site mutations, 11
Myoclonus
   classification, 455
   definition, 452
   epidemiology, 45
Myoclonus dystonia
   animal models, 465
   classification, 453–454
   clinical features, 456, 460
   definition, 452
   diagnostic criteria, 451–452, 468–469
   differential diagnosis, 469
   epidemiology, 454
   family demographic and clinical features, 455–459
   genes
      DNA testing, 464
      dopamine D2 receptor, 399–400, 460, 462–463
      DYT11 gene, 460–462
      linkage analysis, 399, 460
      mutations, 399
      ε-sarcoglycan, 399, 451–452, 461–464
   genotype/phenotype correlations
      dopamine D2 receptor, 466–467
      DYT11, 465–467
      psychiatric disorders, 468
      ε-sarcoglycan, 466
      sporadic and atypical cases, 467–468
   historical cases, 454–456
   neuroimaging, 465
   neuropathology, 465
   terminology, 452–453
   treatment, 469–470
Myoclonus-epilepsy with ragged red fibers (MERFF), features, 233

## N

NARP, see Neurogenic weakness, ataxia, and retinitis pigmentosa
NBIA, see Neurodegeneration with brain iron accumulation
Negative test, 14
Nematode models, 15
Neuraminidase deficiency
   animal models, 243
   clinical forms, 242–243
   diagnosis, 243
Neuroacanthocytosis, features, 377
Neurodegeneration with brain iron accumulation (NBIA)
   Hallervoden–Spatz syndrome, 432–433

neuropathology, 436–437
nosology, 428
Neuroferritinopathy (NFP)
clinical features, 518
diagnosis, 519–520
FTL
ferritin structure and function, 519
mutations, 518
pathology, 518–519
Neurofilament, L- and M-subunit mutations in Parkinson's disease, 320
Neurogenic weakness, ataxia, and retinitis pigmentosa (NARP), features, 233
Neuroimaging
ataxia, 28–29, 256–258
ataxia–telangiectasia, 198
ataxia with vitamin E deficiency, 184
autosomal recessive spastic ataxia of Charlevoix-Saguenay, 192
cerebrotendinous xanthomatosis, 238
choreoacanthocytosis, 526
corticobasal degeneration, 333
dentatorubral-pallidoluysian atrophy, 145
dopa-responsive dystonia, 423
dystonia 1, 413
essential tremor, 356–357
familial idiopathic basal ganglia calcification, 446–447
Friedreich's ataxia, 172–173
frontotemporal lobar dementia, 333
Gilles de la Tourette syndrome, 497
Hallervoden–Spatz syndrome
functional imaging, 436
magnetic resonance imaging, 429, 434–436
magnetic resonance spectroscopy, 436
HARP syndrome, 531
Huntington's disease, 366–367
McLeod syndrome, 529
multiple system atrophy
magnetic resonance imaging, 220–221
overview, 213
photon emission tomography
metabolic markers, 222
striatal dopaminergic terminal markers, 222–223
striatal postsynaptic dopaminergic receptor markers, 223
single photon emission computed tomography
brain imaging, 221–223
cardiac imaging, 221
myoclonus dystonia, 465
pallidopyramidal degeneration, 532
Parkinson's disease, 275, 298, 311
progressive supranuclear palsy, 333
restless legs syndrome, 507
spinocerebellar ataxia 1, 37
spinocerebellar ataxia 2, 51
spinocerebellar ataxia 3, 61
spinocerebellar ataxia 5, 79
spinocerebellar ataxia 7, 88
spinocerebellar ataxia 8, 97–98
spinocerebellar ataxia 10, 109
spinocerebellar ataxia 11, 119
spinocerebellar ataxia 12, 129

spinocerebellar ataxia 13, 14, and 16, 136
spinocerebellar ataxia 17, 140
spinocerebellar ataxia 18, 20, 22
spinocerebellar ataxia 19, 20, 22
Wilson disease, 346–348
NFP, see Neuroferritinopathy
Niemann-Pick disease type C
animal models, 240
classification, 516
clinical features, 238–239, 516
diagnosis, 239–240, 517–518
*NPC1* gene and molecular pathology, 240, 516
pathology, 516
pathophysiology, 516–517
phenotypes, 239
treatment, 518
Nijmen breakage syndrome, features, 202
Non-allelic heterogeneity, 2
Non-coding DNA sequence, 8
Nonsense mutation, 11, 16
*NPC1*, see Niemann-Pick disease type C

## O

OMIM, *see* Online Mendelian Inheritance in Man
Online Mendelian Inheritance in Man (OMIM), 16, 21
Oocyte, 15

## P

Pallidopyramidal degeneration (PPD)
clinical features, 531
diagnosis, 532
genotype, 531
history of study, 531
incidence, 531
neuroimaging, 532
pathology, 531–532
pathophysiology, 532
treatment, 532–533
*PANK2*, *see* Hallervoden–Spatz syndrome
Pantothenate kinase associated neurodegeneration (PKAN), *see* Hallervoden–Spatz syndrome
Parkin
gene
cloning from animals, 307
structure, 306
Lewy body deposition, 296–297
Parkinson's disease
clinical features, 305–306
DNA testing
haplotype analysis, 309
polymerase chain reaction, 308–309
sequencing, 309
sporadic patients, 309
genotype/phenotype correlations, 312
immunoreactivity in brain, 308
mutations, 278–279, 305, 309–310
neuroimaging, 311
neuropathology, 311–312

polymorphisms, 309, 311
pseudoautosomal dominant inheritance, 309
treatment, 312
structure, 296, 306
subcellular localization, 307–308
α-synuclein interactions, 296–297
tissue distribution, 307
ubiquitin-protein ligase function and substrates, 308
Parkinson's disease (PD)
clinical features, 274
Contursi kindred, 288–289
definition, 299
diagnosis, 274–275
etiology, 273–274
familial aggregation studies
multiplex family studies, 278–280
populations not selected by familial patterns, 276–278
familial parkinsonism and dementia with ballooned neurons, 533
genes and phenotypes
genome-wide scans and candidate loci, 279–280, 319–3321
PARK1, *see* α-Synuclein
PARK2, *see* Parkin
PARK3, 279, 315–316
PARK4, 279, 316–317
PARK5, *see* Ubiquitin carboxy-terminal hydrolase-1
PARK6, 279, 317, 19
PARK7, 279, 319
PARK8, 319
PARK9, 319
single gene associations, 281–283
limitations in etiology and diagnosis studies, 275–276, 283
mitochondrial mutations
acquired mutations
burden in brain, 479–480
versus inherited mutations, 474
theory, 478–479
complex I dysfunction, 473–476
cybrid studies, 476
familial parkinsonism mutations, 476–478
inheritance, 280
sporadic parkinsonism mutations, 478
treatment implications, 482–483
α-synuclein dysfunction correlation, 475
neuroimaging, 275
treatment, 275, 288, 300, 312
twin studies, 280–281
Paroxysmal dyskinesias
classification, 385–386
history of study, 385–386
ICCA syndrome, 386, 388–390
paroxysmal exercise-induced dyskinesia
clinical features, 389
genetics, 389
paroxysmal hypnogenic dyskinesia
clinical features, 390
genetics, 390
paroxysmal kinesigenic dyskinesia, *see* Dystonia 10

paroxysmal nonkinesigenic dyskinesia, *see* Dystonia 8
paroxysmal nonkinesigenic dyskinesia associated with spasticity, 387–388
pathophysiology
    channelopathy similarities, 391
    epilepsy versus basal ganglia dysfunction, 390–391
    prospects for study, 391–392
PCR, *see* Polymerase chain reaction
PD, *see* Parkinson's disease
Pedigree analysis
    ataxia family construction, 264
    choreoacanthocytosis, 523–524
    multigeneration pedigree importance, 3–4
    restless legs syndrome, 505
    symbols, 4
Penetrance
    autosomal dominant disease, 4–5
    definition, 4–5
Penrose, Lionel, 25
PET scan, 213, 221–223, 226, 257, 497, 507
Phase of the marker allele, 15
Phenotype
    definition, 2
    genotype–phenotype correlations, 2
    variation in Mendelian disorders, 8
Photon emission tomography, *see* Neuroimaging
PKAN, *see* Pantothenate kinase associated neurodegeneration
Point mutation, 11
Polygenic trait, 8
Polyglutamine disease
    anticipation, 25–27
    phenotypic variation, 30
    types, 25, 35
Polymerase chain reaction (PCR)
    applications, 12
    ataxia assays, 27–28
    limitations, 12–13
    *parkin* mutation testing, 308–309
    principles, 12
    reverse transcription–polymerse chain reaction, 13
    spinocerebellar ataxia 1 diagnosis, 36–37
    spinocerebellar ataxia 2 diagnosis, 50
    spinocerebellar ataxia 8 diagnosis, 97
    spinocerebellar ataxia 10 diagnosis, 109
    spinocerebellar ataxia 12 diagnosis, 124, 128
    α-synuclein mutation testing, 298
Polymorphism
    definition, 10
    detection, 14
Polymorphic protein markers, 1
PPD, *see* Pallidopyramidal degeneration
*PPP2R2B*, *see* Spinocerebellar ataxia 12
P/Q-type calcium channel, *see* Spinocerebellar ataxia 6
Predisposing allele, 8
Prenatal diagnosis, 28
Pre-symptomatic disease, 28
Primidone, essential tremor management, 359
Prion disease, *see also PRNP*; specific diseases
    animal models, 160–161

clinical features, 153–159
diagnosis, 159–160
pathogenesis, 151–153
prevention, 161
treatment, 161
Privacy, genetic testing
    confidentiality, 547
    patient versus family, 547–548
*PRNP*
    locus, 151
    mutations in disease
        diagnostics, 159–160
        familial Creutzfeldt–Jakob disease, 156–158
        Gerstmann–Sträussler–Scheinker disease, 159
        overview, 153
        transgenic mouse models, 160
    protein
        pathogenesis, 152–153
        processing, 152
        structure, 152
    repeats, 151–152
    structure, 151
Progressive supranuclear palsy (PSP)
    animal models, 335
    genetic testing
        clinical implications, 334
        ethical dilemmas, 334–335
        specificity, 333
    neuroimaging, 333
    *tau* mutations
        clinical features, 332
        frequency, 333
        neuropathology, 332–333
        types, 325–326
    treatment, 335
Promoter, 15
Propranolol, essential tremor management, 359
Protein phosphatase 2A, *see* Spinocerebellar ataxia 12
PSP, *see* Progressive supranuclear palsy
Purkinje-cell-degeneration mouse, 29
Pyroglutamicaciduria
    clinical features, 245
    diagnosis, 245
    treatment, 245
Pyruvate dehydrogenase deficiency, ataxia, 236–237

## R

Random mutagenesis, 15
Rapid-onset dystonia parkinsonism (RDP)
    clinical features, 400
    diagnosis, 400
    genetics, 400
RDP, *see* Rapid-onset dystonia parkinsonism
Recessiveness, overview, 2–3
Recombination, 11
Refsum disease, features, 244
Restless legs syndrome (RLS)
    clinical features, 503–504
    diagnosis, 506–507

genetic studies
    family studies, 503–505
    heredity mode, 508
    linkage studies, 506
    twin studies, 505
neuroimaging, 507
neuropathology, 507
prospects for study, 508
treatment, 507–508
Restriction fragment lenth polymorphism, 2
RLS, *see* Restless legs syndrome

## S

*SACS*, *see* Autosomal recessive spastic ataxia of Charlevoix-Saguenay
ε-Sarcoglycan
    gene structure, 462
    imprinting of gene, 463–464
    myoclonus dystonia
        genotype/phenotype correlations, 465–468
        mutations, 399, 451–452, 461
    sarcoglycanopathies, 462
    tissue distribution, 462
SCA, *see* Spinocerebellar ataxia
SCA1, *see* Spinocerebellar ataxia 1
SCA2, *see* Spinocerebellar ataxia 2
SCA3, *see* Spinocerebellar ataxia 3
SCA4, *see* Spinocerebellar ataxia 4
SCA5, *see* Spinocerebellar ataxia 5
SCA6, *see* Spinocerebellar ataxia 6
SCA7, *see* Spinocerebellar ataxia 7
SCA8, *see* Spinocerebellar ataxia 8
SCA10, *see* Spinocerebellar ataxia 10
SCA11, *see* Spinocerebellar ataxia 11
SCA12, *see* Spinocerebellar ataxia 12
SCA13, *see* Spinocerebellar ataxia 13
SCA14, *see* Spinocerebellar ataxia 14
SCA16, *see* Spinocerebellar ataxia 16
SCA17, *see* Spinocerebellar ataxia 17
SCA18, *see* Spinocerebellar ataxia 18
SCA19, *see* Spinocerebellar ataxia 19
Selection. 11
Semi-dominant, 3
Short tandem repeat (STR), detection, 14
Sialidase deficiency, *see* Neuraminidase deficiency
Sialuria
    diagnosis, 244
    forms, 243
    neuropathology, 243–244
    *SLC17A5* gene, 244
Sickle cell anemia, recessiveness, 3
Single nucleotide polymorphism (SNP), detection, 14
Single photon emission computed tomography, *see* Neuroimaging
SNP, *see* Single nucleotide polymorphism
Southern blot, principles, 13
SPAR, 22
Spastic ataxia, types, 22
Spastic paraplegia type 4 (SPG4), 22
SPECT, 257–259, 260, 497, 507, 529
SPG4, *see* Spastic paraplegia type 4

Spinocerebellar ataxia (SCA)
   classification, 21
   diagnosis of type, 14
   loci discovery, 22
   prevalence and geographic distribution, 23–25
Spinocerebellar ataxia 1 (SCA1)
   ataxin-1 characteristics, 36–37
   cell culture studies, 38
   clinical features, 35–36
   *Drosophila* model, 41
   neuroimaging, 37
   neuropathology, 37–38
   *SCA1* gene
      CAG repeat expansion, 35–37
      mapping history, 1, 19, 36
      polymerase chain reaction diagnostics, 36–37
      protein, *see* Ataxin-1
   transgenic mouse models, 38–41
   treatment, 41
Spinocerebellar ataxia 2 (SCA2)
   anticipation, 48–49
   CACNA1A gene, 53
   clinical features
      ataxia, 46
      dementia, 47
      eye movements and retinal changes, 46
      movement disorders, 46
      neuropathy, 46–47
      onset, 48
      overview, 45–46
      progression, 47
   *Drosophila* model, 52
   epidemiology, 47
   geographic distribution, 45
   modifying alleles, 52, 53
   neuroimaging, 51
   neuropathology, 51
   neurophysiology, 51
   polymerase chain reaction diagnosis, 50
   *SCA2* gene
      alleles, 47–48
      genotype/phenotype correlations and modifying alleles, 52–53
      meiotic instability of repeat, 48–49
      protein, *see* Ataxin-2
      structure, 47, 49
      transcription, 49
      trinucleotide repeats, 47, 49, 52–53
   sporadic ataxia testing, 50
   transgenic mouse models, 51–52
   treatment, 53
Spinocerebellar ataxia 3 (SCA3)
   animal models, 64
   anticipation, 62
   clinical features
      age at onset, 58, 62–63
      autonomic deficits, 60
      bulbar signs, 59
      cerebellar signs, 59
      cognitive deficits, 59–60
      course, 59
      extrapyramidal signs, 59
      lower motor neuron signs, 59
      oculomotor signs, 59
      peripheral neuropathy, 59
      phenotypic variability, 60
      presenting symptoms, 58–59
      sleep disturbances, 60
      upper motor neuron signs, 59
   evoked potentials, 61
   haplotypes, 63
   heredity, 63
   history of study, 57
   *MJD1* gene
      CAG repeat expansion, 58, 62
      diagnostics, 60–61
      instability of repeat, 62
      protein, *see* Ataxin-3
   nerve conduction stsudies, 61
   neuroimaging, 61
   neuropathology, 61, 64–65
   prevalence, 58
   treatment, 65
Spinocerebellar ataxia 4 (SCA4)
   autosomal dominant cerebellar ataxia linkage, 72
   clinical feaures, 71
   diagnosis, 72
   gene locus, 71
   neuropathology, 72
Spinocerebellar ataxia 5 (SCA5)
   anticipation, 77
   clinical features, 75, 77–79
   gene
      mapping, 77–78, 80
      RAPID cloning, 78
      repeat expansion detection, 78
   neuroimaging, 79
   neuropathology, 79–80
   onset, 75, 77
   pedigree analysis of Lincoln family, 75–76
Spinocerebellar ataxia 6 (SCA6)
   animal models, 82
   *CACNA1A* gene
      CAG repeats, 81–82
      molecular pathogenesis, 82–83
      protein function in P/Q-type calcium channel, 81–82
      stability of repeat, 82
   clinical features, 81
   diagnosis, 82
   neuropathology, 83
   treatment, 82
Spinocerebellar ataxia 7 (SCA7)
   anticipation, 88, 91
   cell culture sudies, 89–90
   clinical features
      neurological signs, 86
      onset, 90–91
      overview, 85–86
      visual impairment, 86
   diagnosis, 87–88, 92
   gene
      CAG repeats, 85–86, 88, 91–92
      DNA tests, 87–88
      protein, *see* Ataxin-7
   genotype/phenotype correlations, 90–91
   modifying alleles, 91
   mouse models, 90
   neuroimaging, 88
   neuropathology, 87–89
   treatment, 91
Spinocerebellar ataxia 8 (SCA8)
   carriers, 99–100
   clinical features, 95–96, 100–101
   genotype/phenotype correlation, 98–101
   *KLHL1* gene
      allele penetration, 99
      antisense transcripts, 96
      CTG expansions, 95, 97–101
      polymerase chain reaction testing, 97
      protein function, 96–97
      RAPID cloning, 99
      tissue distribution of expression, 96
   models, 98
   neuroimaging, 97–98
   neuropathology, 98
   treatment, 101
Spinocerebellar ataxia 10 (SCA10)
   anticipation, 104
   clinical features, 103–104, 109, 111
   diagnosis, 108–109
   genotype/phenotype correlation, 109, 111
   neuroimaging, 109
   neurophysiology, 109
   population genetics, 111
   *SCA10* gene
      ATTCT repeats, 103, 105, 107, 111
      E46L discovery, 106
      instability of expanded repeat, 107–108
      mapping, 104–105
      molecular pathogenesis, 111–113
      polymerase chain reaction testing, 109
      structure, 111
   treatment, 113
Spinocerebellar ataxia 11 (SCA11)
   clinical features, 117–118
   diagnosis, 119
   gene mapping, 118
   neuroimaging, 119
   pedigree analysis, 117–118
Spinocerebellar ataxia 12 (SCA12)
   clinical features
      age of onset, 129
      case reports, 123
      overview, 121–123, 128
      psychiatric disorders, 129
   diagnosis, 128–129
   neuroimaging, 129
   neuropathology, 129
   pathogenesis models, 129
   pedigree analysis, 122, 128
   *PPP2R2B* gene
      CAG repeat expansion, 121
      conservation of promoter region and repeat in nonhuman primates, 126
      polymerase chain reaction detection, 124, 128
      promoter repeat localization, 124–126
      protein function, 127–128
      repeat expansion detection, 124
   treatment, 130
Spinocerebellar ataxia 13 (SCA13)

clinical features, 133–134
diagnosis, 136
gene mapping, 134, 136
genotype/phenotype correlations, 137
neuroimaging, 136
treatment, 137
Spinocerebellar ataxia 14 (SCA14)
clinical features, 133–134
diagnosis, 136
gene mapping, 136
genotype/phenotype correlations, 137
neuroimaging, 136
treatment, 137
Spinocerebellar ataxia 16 (SCA16)
clinical features, 133–134
diagnosis, 136
gene mapping, 136
genotype/phenotype correlations, 137
neuroimaging, 136
treatment, 137
Spinocerebellar ataxia 17 (SCA17)
clinical features, 139–140
diagnosis, 140
models, 140
neuroimaging, 140
neuropathology, 140
TATA-binding protein gene
CAG repeat expansion, 139–140
molecular pathogenesis, 140
treatment, 140
Spinocerebellar ataxia 18 (SCA18), 20, 22
Spinocerebellar ataxia 19 (SCA19), 20, 22
Splice site mutation, 11
Splicing, 8
Sporadic transmissible spongiform
encephalopathy
clinical features, 153–154
definition, 151
diagnosis, 159–160
*Stargazer* mouse, 29
Stochastic events, 2
Stop codon, 11
STR, *see* Short tandem repeat
Synphilin-1, mutations in Parkinson's disease, 320
α-Synuclein
aggregation
fibrillation and structural changes, 294–295
modulators
dopamine, 297
environmental factors, 297
mutations, 296
Parkin, 296–297
phosphorylation, 297–298
protein–protein interactions, 297
synuclein homolog inhibition, 296
neurotoxicity, 295–296
chaperone activity, 294
domains, 292–293
function, 293
glial cytoplasmic inclusions in multiple system atrophy, 224
homology with other synucleins, 292
Parkinson's disease

A30P mutation
clinical features, 291
discovery, 291
impact on Parkinson's disease research, 291–292
neuroimaging, 298
A53T mutation
age of onset, 289–290
Contursi kindred, 288–289
idiopathic Parkinson's disease similarities and differences, 289–290
motor features, 289
neuroimaging, 298
neuropathology, 290–291
neuropsychological features, 289
Parkinson's disease definition, 299–300
survival, 290
treatment, 300
tremor, 290
animal models, 298–299
ApoE4 allele association, 282–283
DNA testing, 298
Lewy body deposition, 278
mitochondrial dysfunction correlation, 475
mutations, overview, 278, 287
prospects for study, 300
protein–protein interactions, 293–294, 297
tissue distribution, 307

## T
TATA-binding protein (TBP)
CAG repeat expansion in gene, 139–140
spinocerebellar ataxia 17 molecular pathogenesis, 140
*Tau*
gene
exon 10 splicing defects, 329–330
missense mutations, 330
structure, 326
transcription regulation, 326–327
genetic disorders, *see also specific disorders*
animal models, 335
genetic testing
clinical implications, 334
ethical dilemmas, 334–335
specificity, 333
neuroimaging, 333
neuropathology, 332–333
overview, 325–326
treatment, 335
isoform expression scheme, 335–336
messenger RNA
splicing, 327, 329–330
translation, 327–328, 330–331
Parkinson's disease mutations, 320–321
phosphorylation, 328
Tay–Sachs disease
animal models, 242
autosomal recessive juvenile parkinsonism comparison, 511–512
clinical features, 240–241
diagnosis, 514–515
forms, 241

genetic counseling, 515
GM2 ganglioside function and dysfunction, 514
hexosaminidase A
deficiency, 241–242, 512–513
functions, 513–514
pathology, 513
treatment, 242, 515
variants, 512–513
TBP, *see* TATA-binding protein
TH, *see* Tyrosine hydroxylase
TMS, *see* Transcranial magnetic stimulation
α-Tocopherol transfer protein (α-TTP)
function, 179, 181–182
gene mutation, *see* Ataxia with vitamin E deficiency
knockout mouse, 184–185
Topiramate, essential tremor management, 360
Torsin A
dystonia 1
DNA testing, 413
GAG deletion, 408, 410–412, 480
gene identification, 410–411
homologous genes, 411–412
mutations, 407
protein function and dysfunction, 412–413
function, 412
myoclonus dystonia
genotype/phenotype correlation, 465–467
mutations, 461–462
Parkinson's disease mutations, 321
Tottering mouse, 29
Transgene, 15
Transcranial magnetic stimulation (TMS), restless legs syndrome studies, 507
Transgenic mouse
ataxia–telangiectasia, 200–201
dentatorubral-pallidoluysian atrophy, 147–148
disease models, overview, 15–16
prion disease models, 160–161
spinocerebellar ataxia 1, 38–41
spinocerebellar ataxia 2, 51–52
spinocerebellar ataxia 7, 90
Tay–Sachs disease, 242
Transglutaminase, huntingtin aggregation role, 372
Transmission disequilibrium test, 10
α-TTP, *see* α-Tocopherol transfer protein
Twin studies
complex inheritance pattern analysis, 8
essential tremor, 355
Gilles de la Tourette syndrome, 493
Parkinson's disease, 280–281
restless legs syndrome, 505
Tyrosine hydroxylase (TH)
biopterin deficiency animal model studies, 424–426
deficiency in dystonia, 400, 422

## U
Ubiquitin carboxy-terminal hydrolase-1
function, 317

Parkinson's disease mutations, 279, 317
tissue distribution, 307, 317
Unifactorial inheritance, 2

## V

Vertical transmission, 4
Vitamin E, *see also* Ataxia with vitamin E deficiency; α-Tocopherol transfer protein
function, 182
transport, 181–182

## W

WD, *see* Wilson disease
Western blot, principles, 13–14
Wildtype, 2
Wilson disease (WD)
aminoaciduria, 345–346
animal models, 341, 348
ATP7B
copper homeostasis role, 345
DNA testing, 346
gene structure, 344
mutations, 341, 344–345
subcellular distribution, 343
clinical features, 341–343
diagnosis, 346
history of study, 341
iron-binding globulin levels, 345
neuroimaging, 346–348
pathologic anatomy, 348
treatment, 348–350
Writer's cramp, DYT1 mutations 5

## X

XDP, *see* X-linked dystonia-parkinsonism
XK
function, 528
McLeod syndrome gene mutations, 527–528
X-linked disease
heredity, 6–7
X chromosome inactivation, 7
X-linked dystonia-parkinsonism (XDP)
clinical features, 400–401
geographic distribution, 401
linkage analysis, 401
neuropathology, 401
penetrance, 401
treatment, 401

## Z

Zebrafish models, 15
Zinc acetate, Wilson disease management, 349

ISBN 0-12-566652-7

90038

**Sidney Silverman Library
and Learning Resource Center
Bergen Community College
400 Paramus Road
Paramus, NJ 07652-1595**

www.bergen.edu

Return Postage Guaranteed